Biology

The cover and page ii of this book show two computer-generated representations of the DNA molecule—deoxyribonucleic acid, the genetic material found in all living organisms.

Cover
This image shows a model of 171 base pairs of DNA, smoothly bent into a superhelix fitting the dimensions of a nucleosome core particle in a chromosome. Blue spheres are carbon atoms, cyan are nitrogen, red are oxygen, and yellow are phosphorus; hydrogen atoms are not shown. The image is by Nelson Max, Lawrence Livermore National Laboratory.

Page ii
This image displays a side view of the DNA molecule and is the result of work performed by Robert Langridge, Ph.D., Computer Graphics Laboratory, University of California, San Francisco. Blue spheres are nitrogen, red are oxygen, green are carbon, and yellow are phosphorus; hydrogen atoms are not shown. (*Copyright: Regents of the University of California.*)

Leland G. Johnson

Augustana College

Biology

wcb **Wm. C. Brown Company Publishers** Dubuque, Iowa

Book Team

John Stout
Editor

Rebecca Strehlow
Associate Developmental Editor

Mary M. Monner
Production Editor

David A. Corona
Designer

James M. McNeil
Design Layout Assistant

Mary M. Heller
Visual Research Editor

Mavis M. Oeth
Permissions Editor

wcb group

Wm. C. Brown
Chairman of the Board

Mark C. Falb
Executive Vice President

wcb

*Wm. C. Brown Company Publishers,
College Division*

Lawrence E. Cremer
President

David Wm. Smith
Vice President, Marketing

David A. Corona
*Assistant Vice President, Production
Development and Design*

Marcia H. Stout
Marketing Manager

Janis M. Machala
Director of Marketing Research

William A. Moss
Production Editorial Manager

Marilyn A. Phelps
Manager of Design

*All text and visual credits begin on page
C-1 at the end of the book.*

To Rebecca,
who worked so very hard
to make it happen
and to Mike, Julie, and Frank,
who were very patient

Brief Contents

Expanded Contents

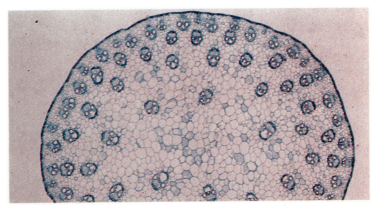

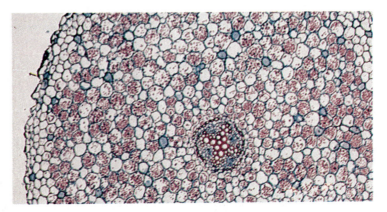

The search for truth is in one way hard and in another easy. For it is evident that no one can master it fully nor miss it wholly. But each adds a little to our knowledge of nature, and from all the facts assembled there arises a certain grandeur.

Aristotle

Preface

There are many levels of understanding of the living world. For most people a certain general understanding arises from casual and chance observation of living things around them. A more satisfying and deeper understanding of life comes from the study of biology. The goal of this book is to help students effectively plan and organize the observation, the analysis, and the search for understanding of the biological world and Aristotle's "certain grandeur."

This book is designed for biology majors, for majors in other science disciplines who take a biology course, for all categories of preprofessional students, and for many of the students in the nursing and paramedical fields. It also can be used in courses that majors and nonmajors take together.

The organismic aspects of biology are emphasized in this text. Whether an organism is unicellular, a loosely integrated multicellular form, or a complexly organized higher plant or animal, it is a living, functioning unit that must maintain the integrity of its internal environment and interact successfully with the external environment. The internal environment must be suitable for cellular life so that cells can carry on their individual functions and make their contributions to the life of the whole organism. To accomplish this, the organism integrates functions that control the internal environment and interacts successfully with its external environment in many ways.

Organisms also are evolutionary units of populations in that they are subjected to natural selection. Present-day organisms are the products of evolutionary processes, and some of them reproduce to become parents of future generations. Thus, organismic functions can also be thought of in this evolutionary context, that is, as adaptations whose suitability is tested by the forces of natural selection. This evolutionary theme is complementary to consideration of immediate solutions for organismic problems of maintenance and interaction with environment. It puts the life of the individual organism into perspective relative to the continuing life of the species.

At key points the book focuses directly on the human organism, including physiological function of organ systems, nervous and hormonal regulation, reproduction and development, and evolution. But human biology is always treated as part of biology in general because of the clear recognition that we humans are part of the world of life and that we have many problems and processes in common with other organisms.

The organization of this book is straightforward to allow for direct pursuit of the goals and principal themes. Chemistry and cell biology are introduced early as the foundation for analyzing the more complex organismic functions and processes. After the fundamental processes have been presented, the broader topics of development, evolution, and ecology are discussed. Surveys of organisms are presented at the end of the book for purposes of flexibility, since there is tremendous variability in the placement of the surveys in courses at various institutions. The survey material can be inserted anyplace that suits local course requirements. Within this framework, sections and individual chapters have become sufficiently autonomous to make the text functional in virtually any course organization.

Some new approaches have been taken to the old problems of organization and content. For example, a balance has been struck regarding the organization of various topics in the biology of plants and animals. Fundamental biological unity is recognized where it is scientifically and educationally sound to do so. Thus, plants and animals are considered together when discussing basic cell biology, genetics, reproduction, development, evolution, and ecology.

But plants and animals also differ in basic ways. For example, while water and mineral ion absorption by roots of vascular plants and the digestive and absorptive processes by which animals obtain nutrients are processes that might be broadly categorized as nutrient procurement, they both are worthy of careful, separate examination. Likewise, plant transport processes and animal circulation are fundamentally different sets of processes. Separate consideration also is given to plant hormonal regulation, which is primarily involved in growth control, and animal hormonal regulation, which is involved much more in short-term modulation of physiological processes.

A new organizational approach also has been taken to the discussions of chemical reaction mechanisms, photosynthesis, and cell respiration. Because the placement of these topics is a problem for many instructors, there is flexibility within that section of the text. Chapter 5 deals with energy relationships and chemical reaction mechanisms, and it includes information relevant to both chapter 6 (photosynthesis) and chapter 7 (cellular energy conversions). But chapters 6 and 7 are specifically designed to be interchangeable in sequence. Slight redundancy regarding phosphorylation mechanisms occurs in these chapters, but it is a small price to pay for the flexibility gained for users of the book. Most teachers find that the topic of phosphorylation mechanisms bears some repetition anyway.

A new approach has also been taken to one of the fastest expanding fields in modern biology: the study of resistance and immune mechanisms. Because expressions of immune responses involve developmental events that occur throughout the lives of organisms, these topics have been included in a chapter on lifelong developmental change (chapter 26). Chapter 26 also treats other developmental changes that occur throughout life, such as wound healing and regeneration, cancer and other abnormal growths, and aging.

The discussion of DNA structure, including the history of its discovery, is presented in chapter 2 as part of the general discussion of large, biologically important molecules. This strategy is logical from a biochemical viewpoint and avoids a prolonged digression into DNA structure that might interrupt the discussion of genetics, where only those aspects of DNA structure that are most directly relevant to genetic mechanisms are discussed.

Fast-developing fields of biology have been approached carefully so as to present solid introductions that should motivate students to read further on these topics, but yet not include some findings that are still so new or controversial that they may not stand unchallenged even until the book is in students' hands. This cautious approach has been taken with a number of topics, including the ecological significance of the various photosynthetic mechanisms, peptide hormones and their relationships to transmitters in animal brains, certain cellular secondary messengers, genetic engineering and ''split genes,'' and interferons.

Terminology in this text agrees with current international conventions for naming chemical compounds. Thus, for example, instructors will find discussion of fructose bisphosphate and ribulose bisphosphate instead of the older names fructose diphosphate and ribulose diphosphate. And glyceraldehyde phosphate (GAP) replaces phosphoglyceraldehyde (PGAL). This terminology has been adopted, with strong encouragement from reviewers, because it is universally used in current biological research literature and in virtually all modern advanced biology texts.

The International System of units (S.I. units) has been used for all physical measurements, although in the case of heat energy measurements references to both the more familiar calories, which have a strong reference framework in everyday life, and joules, the approved S.I. units for heat energy, are found.

Aids to the Reader

Text Introduction

The text opens with an introduction that sets the tone of the book and introduces the text's themes, goals, and principal ideas. This overview gives students an idea of the scope of the text and of the field of biology.

Part Introductions

Part introductions are brief overviews that provide continuing perspective for the reader and preview the part in terms of the part's organization and purpose. One biological topic can lead to others, and because biological topics often "build" on one another, chapter continuity is carefully emphasized in each part introduction.

Chapter Concepts

Each chapter of the book begins with a set of fundamental concepts that are stressed throughout each chapter. The reader should refer to these concepts before reading the chapter and return to them for review.

Chapter Introductions

Each chapter introduction previews the chapter's contents and reinforces the integration of the textual material.

Key Terms

Key terms appear in bold type throughout each chapter. The terms will help readers to recall the essential elements of each basic concept. They are listed in the student study guide and tested there as a further aid to retention.

Boxed Essays

Biology is an exciting field. In almost thirty places in this text, an intriguing story is told or a recent discovery is considered so that the reader can take a stimulating, closer look at the process or concept under study. The essays have *not* been used, as boxed essays are in some texts, to segregate "difficult or advanced material" to be assigned only to selected students or omitted entirely. Throughout preparation of this book, the goal was to provide clear and straightforward explanations of all concepts covered and explanations that are meaningful to all student readers in the book's intended audience. Boxed essays are set off from the chapter text and may well serve as starting points for further study.

Chapter Summaries

Summaries appear at the end of each chapter. Important terms, concepts, and interrelationships are reviewed in the sequence they appeared in the chapter, and a perspective on the chapter material is reinforced.

Chapter Questions

Questions at the end of each chapter review major concepts and suggest avenues for thought and further inquiry on the part of the reader.

Suggested Readings

The list of suggested readings at the end of each chapter is intended as a resource for further investigation of selected topics. The readings amplify concepts, processes, or discoveries covered in the chapter. The lists cite readable secondary sources and are intended to provide students with a realistic route to primary scientific literature.

Glossary

The glossary provides concise definitions of the more important scientific terms in the text. It is a valuable tool for refreshing and reinforcing key information.

Illustrations, Photographs, and Tables

The illustration and photograph program in this text is extensive and thorough. Organisms, processes, and interrelationships are clearly represented in the visual program both in artwork and in photos (many of them color). Tables appear throughout the text to organize information and to provide background data for many important discussions.

Additional Aids

The *student study guide* that accompanies this text has been carefully written to enhance the retention of information presented in the book. The study guide includes overviews, key concepts, key objectives, key terms, focus questions, study questions, suggestions for special activities, and a mastery test for each text chapter. Answers to the questions are also provided to encourage self-directed study.

The *instructor's resource manual* is closely keyed to text chapters. Possible course organizations for semester and quarter systems are suggested. Possible course organizations for different instructor approaches also are presented. Then, for each text chapter, the authors provide an overview, learning objectives, sample test items corresponding to learning objectives, a section on the use of student study guide materials, suggested classroom techniques, and a list of audiovisual resources.

The *student study guide* and *instructor's resource manual* were prepared by Joseph and Betty Allamong of Ball State University.

A complete *test item file* has been created to accompany the text. Thirty questions per chapter, written by experienced biology instructors and carefully reviewed, test major concepts. Multiple-choice, true-false, and matching questions are provided.

Two laboratory manuals also are available. The first contains thirty-five exercises and is designed for a two-term course. The second contains twenty exercises and is designed for a one-term course. In both manuals a balance is maintained between descriptive and physiological, classical and experimental exercises. The exercises are arranged in sequence to correspond to relevant sections of the text. Activities within each exercise may be modified to accommodate a range of instructional approaches and laboratory facilities. Each exercise includes a list of necessary materials and equipment, a discussion of the purpose of the exercise, instructions for preparation of laboratory reports, and a list of references and suggested readings. The manuals are illustrated with numerous drawings and photographs, including both scanning and transmission electron micrographs. The manuals approach biology as an experimental science and provide students with a thorough introduction to the practical aspects of the field. The laboratory manuals were prepared by Warren Dolphin of Iowa State University.

Over fifty *transparencies* are available for use in the classroom. They have been carefully chosen to illustrate key concepts clearly and informatively.

Acknowledgments

In any project as large as this one, an author becomes indebted to many people. It would not be possible to recognize all of them, but I do want to mention a few of the people who have given me considerable help.

Several reviewers made helpful suggestions that went well beyond the valuable and constructive criticism normally provided by reviewers. My special thanks to Joseph W. Vanable, Jr., of Purdue University for suggesting many useful explanations that significantly improved the text. Thanks also to Douglas Fratianne of Ohio State University for many excellent suggestions regarding both biology and scientific philosophy.

I am grateful to all of my departmental colleagues at Augustana College for their understanding, and especially want to acknowledge the continuing generous and patient assistance that I received from Lansing Prescott and Gilbert Blankespoor.

Several colleagues overseas also aided me in one way or another. Part of this book was written during two visits to Odense University in Denmark. Many helped to make me feel welcome there, but I want to acknowledge especially Raymond P. Cox of the Biochemistry Institute and Lars Kofoed, Bent Andersen, Axel Michelsen, Tariq Mustafa, Jens Olesen, and Lilian Brosbø and her coworkers at the Biology Institute. Also, Richard C. Newell of the Institute for Marine Environmental Research in Plymouth, England, generously shared his knowledge of physiology and the interactions of invertebrate animals with their environment, as well as his scientific writing skills, during our several collaborations.

As originally planned, this project was to have involved a small group of academic colleagues who would contribute original material to several sections of the book. I was to edit and rewrite this contributed material, thus creating a cohesive final product. The plans changed as the text grew and took shape. Some colleagues found that various demands on their time limited their participation, while others contributed more than I or they had intended. As it turned out, I wrote far more of the original text than I had planned to write, but that has resulted in a stronger continuity of style and depth within the text. In the final analysis, I take full responsibility for all errors of omission or commission that remain in the book.

It is a pleasure to acknowledge the following colleagues who contributed in many ways, both large and small, to the finished text:

Chapters 34 and 35, covering taxonomic principles, Monera, Protists, and Fungi, were written by Lansing M. Prescott, Augustana College.

The following people wrote and contributed basic material for various chapters:

Lansing M. Prescott
Augustana College
Chapters 3 and 13

Raymond P. Cox
Odense University, Denmark
Chapter 6

Maureen Diggins
Augustana College
Chapter 14

H. Jane Brockman
University of Florida
Chapter 30

Robert Leo Smith
West Virginia University
Chapters 31, 32, and 33

Material used in sections of various chapters was contributed by:

Lansing M. Prescott
Augustana College
Chapters 2, 4, and 5

Raymond P. Cox
Odense University, Denmark
Chapters 1, 2, and 5

Arlen Viste
Augustana College
Chapters 1, 2, and 5

William Klein
Northwestern University
Chapters 15, 16, and 17

Maureen Diggins
Augustana College
Chapters 15 and 16

Robert Chapman
Iowa State University
Chapters 19, 20, 21, and 22

Frederick Racle
Michigan State University
Chapters 27 and 28

Richard J. Seltin
Michigan State University
Chapter 29

Gilbert Blankespoor
Augustana College
Chapters 30, 31, 32 and 33

Suggestions concerning certain chapters or sections under development were made by:

Donald J. Farish
University of Rhode Island

F. M. Butterworth
Oakland University

Rollin Richmond
University of Indiana

John N. Farmer
University of Oklahoma

The following critical reviewers provided in-depth analyses of individual chapters and many valuable suggestions concerning the shape and content of the finished text:

Betty D. Allamong
Ball State University

Roger Anderson
Illinois State University

Robert Azen
Cypress College

Marilyn Bachmann
Iowa State University

Natalie Barish
California State University, Fullerton

William E. Barstow
University of Georgia

Gilbert W. Blankespoor
Augustana College

Richard K. Boohar
University of Nebraska

Eugene Bovee
University of Kansas

Roy L. Caldwell
University of California at Berkeley

Allen C. Cohen
Cypress College

Barbara Crandall-Stotler
Southern Illinois University

John Crowe
University of California at Davis

Charles D. Drewes
Iowa State University

Marvin Druger
Syracuse University

Mark Wm. Dubin
University of Colorado at Boulder

Donald J. Farish
University of Rhode Island

Milton Fingerman
Tulane University

Abraham S. Flexer, consultant

David Fox
University of Tennessee

Stuart Fox
Los Angeles City College

Robert G. Franke
University of Tennessee at Chattanooga

Douglas G. Fratianne
Ohio State University

Tim Gaskin
Cayahoga Community College

Merrill L. Gassman
University of Illinois, Chicago Circle

Gerald Gates
University of Redlands

Douglas Gelinas
University of Maine at Orono

Malcolm P. Gordon
University of California at Los Angeles

James L. Greco
Pennsylvania Electric Company/ Environmental Affairs

Phillip M. Groves
University of California at San Diego

Jack P. Hailman
University of Wisconsin at Madison

Jeffrey R. Hazel
Arizona State University

Dale L. Hoyt
Agnes Scott College

Robert J. Huskey
University of Virginia

Colin L. Izzard
State University of New York at Albany

Steven Jensen
Southwest Missouri State University

Russell Jones
University of California at Berkeley

G. W. Kalmus
East Carolina University

William R. Kodrich
Clarion State College

Joseph Kunkel
University of Massachusetts

Deforest Mellon, Jr.
University of Virginia

Thomas R. Mertens
Ball State University

Roger Milkman
University of Iowa

Robert Moore
Clarion State College

Brian Myres
Cypress College

Lansing M. Prescott
Augustana College

Rudolf A. Raff
Indiana University

William J. Rowland
Indiana University

Doug Smith
State University of New York at Stony Brook

Steven Strand
University of California at Los Angeles

Daryl Sweeney
University of Illinois at Urbana

Stan R. Szarek
Arizona State University

William W. Thompson
University of California at Riverside

Joseph Vanable
Purdue University

Donald Whitehead
Indiana University

Bert K. Whitten
Michigan Technological University

Varley E. Wiedeman
University of Louisville

John Williams
Clarion State College

Norman Williams
University of Iowa

C. Richard Wrathall
American University

John Zimmerman
University of Kansas

Many people at the Wm. C. Brown Company have contributed to this project. William C. Brown and Larry Cremer repeatedly expressed warm support and keen interest in my work. John Stout, biology editor during most of the project, provided direction, support, encouragement, and most of all, organizational skill throughout. Mary Heller searched diligently for photographs, and David Corona designed the book. Mary Monner copy edited the text, coordinated the illustration program, and generally held things together during production.

I am also grateful to people who helped with the preparation of manuscript at various stages. Kristin Von Seggern worked early in the project, and Paula Olson contributed considerably to early draft production. And throughout most of this project, a very special person, Christa Vollstedt, provided help and support of many kinds. Christa's untimely death at Christmas in 1980 was a saddening loss to all of us.

Having thanked many for their help, I now thank one for more than help. My wife Rebecca encouraged, aided, and supported me in every way, and she worked side by side with me on every phase of the project. She shared disappointment and satisfaction, criticism and praise. Without her, the book would not have materialized, and in every sense, this book is also hers.

Biology

Introduction

One day in the 1870s, a Spanish nobleman named Marcelino de Sautuola took his five-year-old daughter Maria with him on an exploring and excavating trip to a cave near Altamira in northern Spain. While Maria was playing in a chamber only a few meters from the entrance of the cave, she happened to look up at the ceiling and saw a magnificent series of animal pictures. Possibly, Maria was the first to notice them because this "picture gallery" is only two meters high in places. Maria's father reasoned that since only a few people could have entered the cave in modern times, the works of art found there were of ancient origin. However, controversy arose regarding the authenticity of these paintings; some scientists believed that someone could have forged the paintings in recent times. Only years later did experts finally accept the paintings discovered by Maria as authentic Stone Age art.

Even now, this cave in Altamira is recognized as one of the most important Stone Age "art galleries." But archaeologists have since discovered many more sites of Stone Age cave art in southwestern Europe and in other parts of the world. Many of these cave sites contain pictures of humans, animals, or hunting scenes in which humans are killing animals. The artists who created these works were probably members of advanced hunting-and-gathering societies, and their pictures may have been used in magic rituals performed to ensure hunting success. Whatever the reasons for the creation of these figures, they record impressions of human beings and other organisms that Stone Age artists observed in their environment. Such early observations of living organisms could be considered the beginning of biology, the study of life. People continued to observe and draw plants and animals, and in Neolithic cultures, they used these pictures to develop systems of communication.

Interest in biology later moved from the simple observation of nature and the performance of rituals that were intended to influence natural events to practical applications in the domestication and breeding of plants and animals. Thus, in one way or another, human interest in biology from prehistoric times was associated with meeting everyday, practical human needs, such as acquiring food and shelter.

Facing page Cave painting of the head and chest of a bull from the "picture gallery" in a cave at Lascaux, France. Note the use of perspective in representing the horns.

Painted buffalo hide showing a "winter count" of a tribe of Dakotah Indians. A "winter count" was a pictorial chronicle of events in the Indians' lives.

Record of early Egyptian agriculture, depicting the new interaction developing between humans and other organisms.

Aristotle, Greek philosopher and biologist.

William Harvey demonstrating experiments that he used to prove that human blood circulates and does not flow back and forth through vessels.

But a different kind of biological observation and study emerged in some ancient civilizations. The aim of this form of biology was to meet a less tangible yet very important human need: the need to satisfy human curiosity. Surviving records of ancient civilizations indicate that when societies advanced to the point at which some leisure time became available (at least to a privileged few), biological knowledge grew impressively.

Aristotle and Biological Observation

Some of the most impressive early studies in biology were the work of the Greek philosopher Aristotle (384–322 B.C.). Aristotle made many thorough observations of individual living organisms, and he interpreted these natural phenomena with remarkably keen insight. For example, he developed an interesting scheme for classifying organisms and proposed what some historians regard as a rudimentary theory of evolution. Aristotle also studied the biology and natural history of a large variety of marine organisms and was especially interested in their reproduction and development. He extended his studies of development to other kinds of organisms; for instance, he recorded a series of careful observations of chick embryo development.

Aristotle's studies were surprisingly detailed despite the fact that he made all of his observations with the naked eye. His achievements are even more amazing in light of the fact that, as Aristotle himself pointed out, there was no model of scientific study or observation for him to follow. Thus, in many cases, his observations represented the very first steps ever taken in the scientific investigation of the various organisms and phenomena he studied.

In the years following Aristotle's time, many curious-minded people added to the growing body of knowledge concerning living things. While the majority of these studies were simply biological observations, a tradition of experimental biology also developed. William Harvey's brilliant research in the seventeenth century on the circulation of blood is an outstanding example of this tradition. Harvey showed by careful observation and simple experimentation that blood did not flow back and forth through the same vessels, but rather passed through vessels in only one direction as it circulated throughout the body.

How has the study of biology progressed from the time of Aristotle, when only a few isolated individuals made biological observations, through the first attempts at simple experimentation, to the point where biological knowledge is increasing at the fastest rate in history? The story of the progress of science in general and biology in particular is a very long one; a reasonably complete history of biology alone would require at least one good-sized volume. But although a complete survey of the history of biology is not feasible here, a few key points in the history of the field will help to set the stage for the present-day approach to living organisms and the world in which they live.

A Change in World View: Darwin and Wallace

Have all the varieties of organisms living today always existed as they are, or are they products of change over the course of time? This deceptively simple question has intrigued biologists and other thoughtful people at least as far back as the time of Aristotle, but thinking on the subject took a decisive turn in the second half of the nineteenth century.

Early Evolutionary Theories

Long before the mid 1800s, some biologists theorized that living things are subject to gradual evolution or change. Some of these early scientists even made proposals regarding the mechanisms of evolution. Notable among these was the interesting but erroneous concept of the inheritance of acquired characteristics advocated by French biologist Jean Baptiste Lamarck (1744–1829). Lamarck suggested that gradual change took place because organisms were able to pass on to their offspring certain traits that they themselves had acquired during their lifetimes. For example, this theory might hold that giraffes have long necks because generation after generation of giraffes stretched their necks to reach for food, causing members of each subsequent generation to inherit longer necks.

But while a few scientists proposed various evolutionary theories, most people through the centuries continued to believe that living organisms had remained unchanged since a specific moment of creation. In fact, even nineteenth-century biologists found it difficult to accept the notion of evolutionary change, although they were aware of the accumulating fossil record and other evidence that indicated that there might have been some sort of evolutionary process. Part of this difficulty can be traced to their lack of understanding of the enormity of the geologic time scale and to the absence of a reasonable explanation of evolutionary mechanisms. And of course, for some biologists and many nonbiologists, there was an entirely different kind of barrier to the acceptance of evolutionary theory; they were comfortable with the idea that the natural world was constant and unchanging. This view lent stability to their understanding of nature, their ideas about various social relationships, and to some of their religious beliefs. The new evolutionary theory developed during the 1850s and 1860s thus caused a revolution not only in biology but also in the fabric of human thought.

Background for Discovery

The two naturalists who were instrumental in bringing about this revolution in thought were Charles Darwin (1809–82) and Alfred Russel Wallace (1823–1913), who proposed nearly identical explanations for the process of evolution at almost exactly the same time. Of course, Darwin and Wallace did not work in an intellectual vacuum; many other scientists contributed to their theories. Charles Lyell (1797–1875) expanded on the earlier work of James Hutton and wrote about the continuity of geological processes from past to present and suggested that the earth was many times older than had previously been thought. Thomas Malthus (1766–1834) researched the problem of potentially excessive population growth. Even Darwin's own grandfather, Erasmus Darwin (1731–1802), contributed to evolutionary theory: he proposed that all living things had descended from a primitive protoplasmic mass.

Darwin and Wallace, working independently, developed a set of postulates known as the theory of natural selection. This theory provides an explanation for the process of evolution. How did Darwin and Wallace come to the same conclusion regarding the evolution of life on earth at virtually the same time? They both were well read and had travelled widely and made careful biological observations around the world, but each developed his eventual statement of the principles of natural selection under quite different circumstances.

After his around-the-world voyage as a naturalist on the exploring ship H.M.S. *Beagle*, Darwin occupied himself for years systematically organizing the data that he collected during and after the voyage. He lived an isolated life in a country house in Downe, Kent, near London and spent his time writing up his observations and gradually putting all of the pieces of the theory of natural selection together in his mind. Darwin actually wrote a preliminary outline of the theory as early as 1842 but delayed publishing it so that he could add more supporting details.

In contrast to Darwin's long, systematic progression to a statement of natural selection, Wallace focused his thinking on the idea in a flash of inspiration that occurred one night in 1858 during a spell of tropical fever! He wrote that he was so excited by the idea that he lay impatiently waiting for the fever to lift so that he would have the strength to get up and write down an outline of his ideas.

Since Wallace knew of Darwin's interest in evolution, he sent Darwin an outline of his ideas. Darwin was so shocked to see ideas so similar to his own that he felt rather discouraged for a time. Because Wallace was in the South Seas, some of Darwin's friends suggested that he ignore Wallace's letter and proceed with publication of his own ideas. Instead, Darwin felt that he and Wallace should have their ideas presented at a joint public reading. Afterwards, he would proceed with publication of his own book on the subject.

(a)

(b)

(c)

(a) Charles Darwin. He thought out the principles of natural selection over a period of twenty or thirty years of quiet contemplation at his country house in Downe, Kent. (b) Darwin's study. He did most of his writing using the lapboard on the chair in the corner. (c) Saddleback tortoise from the Galápagos Islands. Darwin was very impressed that tortoises from each of the Galápagos Islands were clearly distinguishable from each other.

Alfred Russel Wallace. He thought out the principles of natural selection during a one-night bout of tropical fever in the South Seas.

Because Darwin was hurried by this series of events, his *Origin of Species,* published in 1859, was much shorter than he had originally intended, although it still comprised a very substantial book. This "abstract," as Darwin called it, proposed a theory that rocked the biological and intellectual worlds.

The Theory of Natural Selection

The **theory of natural selection,** which provides an explanation of the evolutionary process, is based on certain observations and a set of conclusions drawn from those observations. In developing this theory, Darwin and Wallace pointed out that organisms can reproduce at rates that could lead to enormous increases in the population sizes of most species. Yet, despite these vast reproductive potentials, population sizes tend to remain fairly constant.

Darwin and Wallace then noted that there is considerable variability among individuals within a species and that some variations appear to be advantageous in terms of survival. The two scientists proposed that individuals favored with these beneficial characteristics would be more likely to survive and succeed in reproducing themselves; thus, their characteristics would be perpetuated at a greater rate than the less advantageous characteristics of other individuals.

Gradually, the organisms possessing the favorable characteristics would make up a greater proportion of the population, and over a long period of time, this reproductive advantage could lead to a gradual modification of the entire population. However, if environmental conditions changed, new sets of characteristics might be favored and the whole process could move off in another direction. Darwin and Wallace suggested that these mechanisms have collectively guided the evolution of living things.

Reactions to Evolutionary Theory

The notion of evolution, or continuous change, went far beyond biology. It shook the foundations of social systems and religious beliefs and even raised questions about the stability and orderliness of Victorian English society. Thus, the publication of Darwin's work caused a great stir, and Darwin himself was personally attacked and vilified. No doubt Darwin must have sometimes wished that he had been able to follow an earlier plan to have his work published only after his death.

Theorists in various fields began to examine other questions in the light of natural selection. Unfortunately, many people have misapplied Darwin's principles throughout the years to provide convenient explanations or justifications for all sorts of social phenomena. For example, some politicians have applied the notion of "survival of the fittest" to justify exploitation that produces great wealth for a few people at the expense of the welfare and dignity of many others. Meanwhile, other social and political theorists have proposed that Darwin's and Wallace's ideas be interpreted in terms of survival of the "fittest social order." For these people, Darwinian theory supports the idea that class struggles, revolutionary activity, and other socially disruptive processes are not only justifiable but inevitable.

While Darwin and Wallace were not the first evolutionists, their statement of the principles of natural selection caused great changes in modern thinking. The time was right, the work was scientifically powerful, and Darwin's book received enough publicity to make it known to the general reading public. Darwin and Wallace's statement of the principles of natural selection was indeed a turning point in the history of human thought and a milestone in the history of biology.

Mendel: The Genetic Theory

In Darwin's day, most explanations of inheritance emphasized the mixing and blending of characteristics from both parents. This idea posed a problem for Darwin and his supporters, because the theory of natural selection requires that characteristics be passed on unaltered to off-spring for evolution to occur. In other words, evolution by natural selection depends on the transmission of single, specific variations from one generation to the next, while the blending of inherited traits would dilute and diminish such characteristics.

An Austrian monk named Gregor Mendel (1822–84) was already working on a solution to this problem at the time Darwin published his theory. However, Darwin was apparently un-aware of Mendel's research, as were most other biologists of that time.

Mendel worked in a monastery garden on a series of pea-breeding experiments that took him years to complete. He showed through his re-search that characteristics pass from one gener-ation to the next in discrete units. These units, now called *genes,* express themselves in different ways; for example, a *dominant* hereditary unit might mask the expression of a *recessive* unit in one generation. Still, as Mendel demonstrated, genes are not lost or in any way blended or changed. Their expression simply is unobservable as long as they are coupled with dominant genes for the same trait. Thus, Mendel's work proved that the traits favored by natural selection could be passed from generation to generation without change.

The existence of distinct hereditary units was critical to Darwin's theory of natural selec-tion, but the task of unifying the work of Darwin and Mendel fell to later biologists. Mendel turned to experiments on animal inheritance, working with bees for a time, but then was elected abbot of his monastery and apparently devoted much of the rest of his life to administrative duties.

Mendel apparently made repeated attempts to interest other biologists in his research on in-heritance because he periodically sent reprints of his work to other leading plant hybridizers. The journal containing Mendel's report did reach many important libraries, but his results did not receive the international attention they deserved. As is often the case with an innovative concept, the biologists of Mendel's time may not have been ready to accept and appreciate his conclusions.

Gregor Mendel.

Mendel's mathematical approach to inheritance was probably just too different from prevailing ideas to be readily understood and accepted. However, even before Mendel's time, certain bi-ologists were developing another powerful set of ideas that would eventually provide the frame of reference needed for Mendel's principles. This set of concepts would become known as the **cell the-ory.**

The Cell Theory

Are organisms homogeneously constructed, or do they have recognizable, fundamental structural units? If structural units exist, what are they, and how do they function? Nineteenth-century scientists answered these questions with the cell theory. But in contrast with the rather abrupt events that changed evolutionary theory and ge-netics, the development of the cell theory was a gradual process involving the work of many bi-ologists over many years. In fact, work leading to the cell theory actually began well before the time of Darwin and Mendel.

Early Observations of the Cell

In the seventeenth and eighteenth centuries, biologists began to use magnifying devices to study the structure of organisms. Some of these early microscopists made very important contributions to the development of the cell theory. For example, the Dutch merchant, wine taster, and amateur scientist Anton van Leeuwenhoek (1632–1723) reported his carefully and patiently made microscopic observations of living organisms in a series of colorful letters to the Royal Society in London. Leeuwenhoek observed and described many single-celled microorganisms but also reported that larger organisms were constructed of separate structural components.

Robert Hooke (1635–1703), a contemporary of Leeuwenhoek, made an important early contribution to cell theory in his treatise *Micrographia,* which was published when he was only twenty-nine years old. Among the many observations reported in *Micrographia* was Hooke's description of the microscopic structure of cork, which, he discovered, contained pores or spaces. Hooke named these compartments "cells" because they reminded him of small monastery rooms or cells. The term has been used ever since to describe the units that make up living organisms.

Cell Products and Cell Division

Through the years, other scientists continued to make observations regarding the globules or cells that make up organisms' bodies, but their ideas were not unified until two nineteenth-century biologists developed what is now called the cell theory. In 1838, M. J. Schleiden published his theory that cells were the component units that make up all plant bodies. One year later, Theodor Schwann published a similar proposal regarding animal bodies. However, Schwann's extensive observations led him to qualify earlier statements of the cell theory because he discovered that in some tissues there was some material other than

(a)

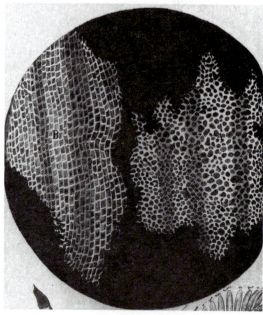

(b)

(a) The microscope that Robert Hooke used to make his observations of the cellular nature of cork. (b) The "cells" of cork as Robert Hooke saw them. The two parts of the picture represent thin slices of cork cut at different angles.

cells. Schwann correctly concluded that this material was produced by living cells, and he modified his statement of the cell theory to read, "All living things are composed of cells and cell products."

After Schwann and Schleiden developed the basic cell theory, other workers made observations on cell division and its role in the production of cells. It was then possible for Rudolf Virchow to hypothesize that "where a cell exists, there must have been a preexisting cell"; or that all cells are direct descendants of other cells.

The cell theory inspired cytology, the extensive research on internal cell structure that continued through the remainder of the nineteenth century and up to the present. With the description of the cell nucleus and chromosomes, it became possible for cytologists to make a connection between cell structure and genetic function. This connection added a new dimension to the understanding of Gregor Mendel's "units of inheritance."

Genes and Chromosomes

Where are the hereditary factors (genes) that Mendel described actually located within living organisms? A partial answer to this question was deduced by cytologists such as Boveri, Sutton, and E. B. Wilson, who developed the theory that genes are physically linked with **chromosomes,** the small bodies observed in cell nuclei stained with certain dyes.

But a more precise picture of the relationship between genes and chromosomes was formed later through the genetic analysis of Thomas Hunt Morgan and his colleagues. They proved that genes are distributed along chromosomes in linear fashion. A group of genes located on a single chromosome are part of a single physical unit called a **linkage group.**

Morgan and his coworkers completed a part of the work begun by Mendel because they proved beyond doubt that genes had specific locations in chromosomes. But questions still remained; for example, how does the genetic information stored in chromosomes actually control cell function and, ultimately, the functions of whole organisms?

Thomas Hunt Morgan. Morgan and his co-workers used experiments on *Drosophila* to clarify the physical relationships between genes and chromosomes. The "fly room" at Columbia was the center of the world of genetics research for a number of years.

Even as Morgan's successors were exploring the relationship between genetics and cytology, other workers were studying the chemical processes occurring in living cells. These researchers found that **enzymes,** protein molecules functioning as catalysts, played very important roles in most biochemical reactions. Then, during the 1940s, George W. Beadle and Edward L. Tatum performed a series of experiments on the mold *Neurospora* that showed that genes control the synthesis of enzymes. These experiments proved that genes are involved in the control of chemical processes within cells. Now the study of biology took another step forward, into the era of molecular biology.

Erwin Schrödinger. His provocative question, "What is life?" stimulated several of the early workers in the infant science of molecular biology.

The Development of Molecular Biology
Schrödinger's Question

In February of 1943, physicist Erwin Schrödinger delivered a series of lectures entitled *What Is Life?* at Trinity College, Dublin, Ireland. The following year, Schrödinger published a book based on these lectures that made his speculations on the relationship between living systems and the laws of physics and chemistry more widely available.

Schrödinger's question, "What Is Life?" has intrigued philosophers, theologians, and biologists through the centuries. Biologists often define life by listing the characteristics shared by all living things. However, although these characteristics convey an understanding of life, they do not really define it.

Organisms have specific, complex organizations and structural and functional features that adapt them to live successfully in their environments. Living systems use energy gained from their environments to maintain their many functions. The chemical reactions that convert this energy to a usable form are known collectively as an organism's metabolism.

Furthermore, living things are irritable; that is, they respond to changes in their environments, and some (but certainly not all) living organisms move. Living things grow and develop. But reproduction, the ability to make "copies" of oneself, is the most distinctive general characteristic of living things.

Thus, a definition of life might be based on characteristics of living systems, such as **complex organization, adaptation, metabolism, irritability, growth and development,** and **reproduction.** But Schrödinger dug even deeper in the attempt to determine the mechanisms controlling these characteristics.

Schrödinger was concerned about the spontaneous tendency of chemical systems to become increasingly less organized. This tendency toward disorganization, termed **entropy,** seemed to contradict the high degree of organization among many organisms. Schrödinger concluded that living things must continually obtain energy from their environments to counter the tendency toward entropy and maintain their highly ordered organization.

Schrödinger was also especially interested in the genetic information contained in every living thing. How can genetic data be so precisely coded and accurately transmitted by chemical systems? The organization of the genetic code must be maintained exactly because errors in transmission could be disastrous for the living organism. Surely, Schrödinger reasoned, the molecules involved in transmitting genetic information must be exceptionally stable and well protected from the tendency toward entropy. In *What Is Life?*, he attempted to show that the laws of chemistry and physics could explain the stability of genetic structure.

Many biologists feel that Schrödinger's work was invaluable in connecting the abstract concept of genes with the actual chemical processes involved in transmitting genetic data and regulating cell functions. Moreover, Schrödinger's book helped inspire others to push ahead with the development of molecular biology.

Watson and Crick: A Model of DNA

In the early 1950s, James Watson and Francis Crick proposed the double-helix model of the DNA molecule, the carrier of genetic information. This model not only described the structure of the DNA molecule, it also gave rise to testable predictions concerning the mechanisms of genetic (DNA) replication. Watson and Crick's model suggested approaches to questions about the transmission of genetic information during cell division.

The DNA structure that Watson and Crick proposed also provided direction for the study of genetic expression, which occurs through the now familiar molecular sequence: DNA-RNA-protein synthesis. Watson reports in his book *The Double Helix* that while he and Crick were working out the structure of the DNA molecule, he put a crudely lettered sign above his desk in Cambridge that said, "DNA$\rightarrow$ RNA$\rightarrow$ protein."

The Development of Molecular Biology

Although research has continued to clarify the actual connection between the transmission of genetic information and the chemical functions of cells, there are still many unknown facets of genetic regulation. For example, it is not yet known what determines the timing in which certain genes are expressed during development.

Molecular biologists have vigorously studied the mechanisms of genetic expression and the controls that regulate the timing and rate of flow of genetic information. In the 1960s, molecular biology became the dominant branch of the field. It seemed that the keys to understanding many of the secrets of life must be at hand.

Great progress has indeed been made in understanding a variety of cellular processes, but this progress has also raised many new areas of inquiry, and along with them, the danger that many fundamental questions will be lost in a tangle of details. Molecular biology proceeds by dissecting life processes more and more finely at the biochemical level. Biochemist Albert Szent-Györgyi warns that when the study of living things is separated into such minute components, "life may slip through your fingers someplace along the way." Thus, molecular biologists must be careful to keep the whole organism in mind even as they explore the most intricate subcellular processes.

Environmental Concerns

In the 1960s and 1970s, as molecular biology was reaching the peak of its dominance, biologists were confronted by another set of urgent practical problems. Not only scientists but others began to ask, "Are we in danger of making the world completely unfit for human habitation, or of creating an environment in which life will have lost most of its positive qualities?" In 1962, Rachel Carson issued a haunting warning of the "silent spring" to come; many biologists turned their attention to these long-ignored issues and to the equally large problem of alerting society, in general, to the problems' existence. The interrelated problems of population growth, famine, pollution, and the depletion of natural resources still threaten the stability of the modern world.

In some ways, an even more disturbing problem with modern industrial society is the callous attitude that humans sometimes display toward other living things. Very few people stop to think about or understand the role of humankind in the ecological system, or the relationship between humans and other organisms in the natural world. Modern civilization might learn a great deal from, for example, Native Americans, who have long appreciated the balance of nature and, in their culture and religion, express the need to "walk gently" through the world.

Some idealists have proposed a return to a "simple" life and a reversal of the trend toward urbanization and industrialization. Unfortunately, considering basic human nature and the probability that technological problems will require technological solutions, such a major reversal of human history seems a remote possibility—unless some worldwide catastrophe such as nuclear or biological warfare imposes that solution. Perhaps, a more constructive and realistic aim is to reduce humanity's impact on the environment to the point where natural corrective mechanisms can operate.

Because the world's ecological problems are far from solved, it is the responsibility of biologists as well as others to understand the problems and to contribute to the solutions. Environmental concerns will probably continue to play a central role in the study of biology for many years to come.

Is there danger that the world may become unfit for human habitation? The growing use of energy can stress and threaten the environment. This facility, Three Mile Island nuclear power plant near Harrisburg, Pennsylvania, had a malfunction in 1979 that threatened the area with serious nuclear contamination.

Humans have almost lost sight of the illuminating vision of oneness with other creatures in the natural world. How ironic that even the magnificent bald eagle, a national symbol of the United States of America, is endangered by environmental contamination, habitat destruction, and illegal hunting.

Unity and Diversity of Life

The principles developed during the recent history of biology have shaped the views of today's biologists. There is a unity among living things because they are all products of evolution and participants in the continuing evolutionary process.

Similarly, functional unity exists at the cellular level because a number of biochemical processes are very similar or identical in the cells of a great variety of organisms.

But along with unity, there is also diversity. Different types of organisms have different structural and functional adaptations that permit them to interact successfully with their own environments. In this book, a number of the specific adaptations of individual organisms and the interactions among organisms are examined. But this organismic diversity is presented within the framework of evolutionary and cellular unity.

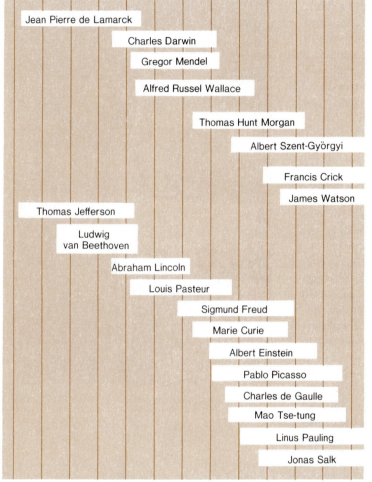

1740 1760 1780 1800 1820 1840 1860 1880 1900 1920 1940 1960 1980

Jean Pierre de Lamarck

Charles Darwin

Gregor Mendel

Alfred Russel Wallace

Thomas Hunt Morgan

Albert Szent-Györgyi

Francis Crick

James Watson

Thomas Jefferson

Ludwig van Beethoven

Abraham Lincoln

Louis Pasteur

Sigmund Freud

Marie Curie

Albert Einstein

Pablo Picasso

Charles de Gaulle

Mao Tse-tung

Linus Pauling

Jonas Salk

1740 1760 1780 1800 1820 1840 1860 1880 1900 1920 1940 1960 1980

The life spans of some biologists mentioned in this introduction (top) and some of their contemporaries (bottom).

Biologists' Methods

How, then, do modern biologists study life? Is there a single "scientific" method? Although people sometimes refer to the "scientific method" as if it were a single, special process, biologists actually use a variety of approaches and methods.

Some biologists devote themselves to making observations of nature and to recording and organizing the data that they gather. Usually, such biologists try to interfere as little as possible with the systems they study. Of course, the very act of observing might have an influence on the organism being studied, but observational biologists attempt to study nature as it is with a minimum of external interference.

Other biologists use means closer to what is commonly described as the scientific method. They gather and study facts about a biological process and think about possible explanations for it. From their study they develop **hypotheses,** or statements whose accuracy is testable. The test of a good hypothesis sometimes yields a definite positive or negative answer, but often more work is required to determine whether the hypothesis is valid.

A hypothesis is tested either by making more observations or, more commonly, by performing an experiment. When biologists have completed their tests, they draw conclusions from the results and often write papers to report their conclusions to other biologists. A good paper must discuss the hypothesis, the techniques used to test the hypothesis, the results of the test, and conclusions drawn from the results.

In addition to simply observing, or creating and testing hypotheses, other biologists use combinations of these methods, and still other biologists use methods that defy exact description or classification. But regardless of the methods a biologist uses, he or she has to meet several basic requirements to work successfully and to derive satisfaction from the work. First, biologists must have adequate background information on the processes or topics they are studying. Second, they must be able to use the "tools of the trade"—in modern biology, the physical sciences and mathematics in addition to basic training in biology. Third, biologists must recognize that what is thought and known today will be corrected and modified in the future, just as the viewpoints of earlier biologists have been corrected and modified. Present knowledge is only a set of ideas to work with, not unalterable truth. Sometimes emotional involvement in one's own scientific work makes it difficult to keep this perspective, but realizing that scientific knowledge itself is undergoing continuing evolution helps to promote objectivity.

Finally, and perhaps most of all, science requires an open and inquiring mind. Perhaps Albert Szent-Györgyi put it best in his succinct analysis of scientific work: "Discovery consists in seeing what everybody else has seen and thinking what nobody else has thought."

Albert Szent-Györgyi. "Discovery consists in seeing what everybody else has seen and thinking what nobody else has thought."

Suggested Readings

Books

Bandi, H. G. 1961. *The art of the stone age*. New York: Crown Publishers.

Bronowski, J. 1973. *The ascent of man*. Boston: Little, Brown & Co.

Gardner, E. J. 1972. *History of biology*. 3d ed. Minneapolis: Burgess Publishing.

Kormondy, E. J. 1966. *General biology: Molecules and cells*. Dubuque, Iowa: Wm. C. Brown Company Publishers.

Kuhn, T. S. 1970. *Structure of scientific revolutions*. 2d ed. Chicago: University of Chicago Press.

Sayre, A. 1975. *Rosalind Franklin and DNA*. New York: Norton.

Schrödinger, E. 1967. *What is life?* and *Mind and matter*. New York: Cambridge University Press.

Watson, J. D. 1968. *The double helix*. New York: Signet Books.

Articles

Eiseley, L. C. February 1956. Charles Darwin. *Scientific American* (offprint 108).

Eiseley, L. C. February 1959. Alfred Russel Wallace. *Scientific American*.

Gould, S. J. 1980. Wallace's fatal flaw. *Natural History* 89:26.

Keele, K. D. 1978. The life and work of William Harvey. *Endeavour* 2:104.

Marshack, A. 1975. Exploring the mind of ice age man. *National Geographic* 147:64.

Part 1

The Chemical and Cellular Basis of Life

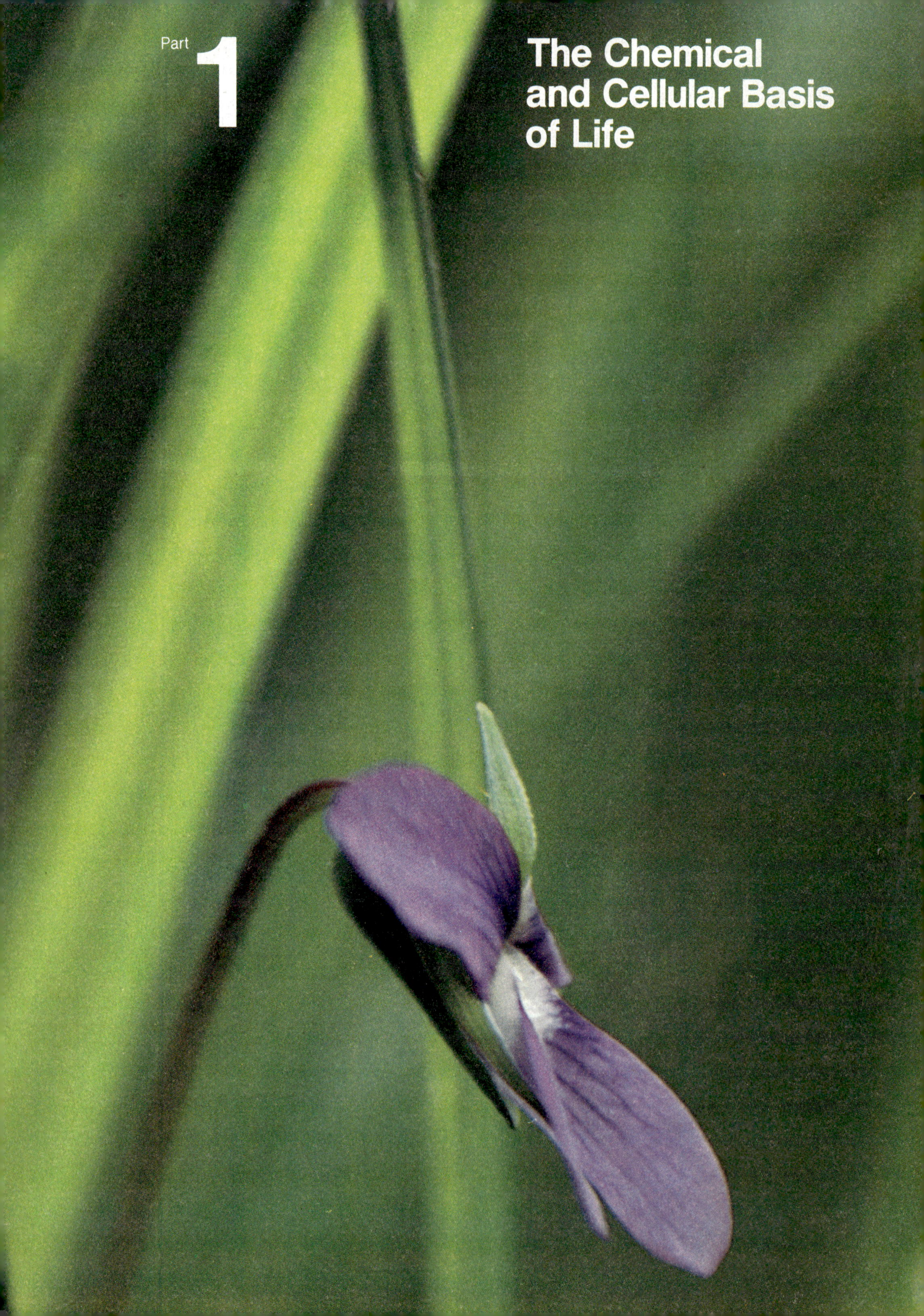

A fundamental principle of modern biology is that processes that occur in living things obey the laws of chemistry and physics, just as do processes in nonliving systems. Special properties of living things do not arise because of a unique set of chemical principles that apply only to living things, but because of the great complexity of the organization of living things and the chemical processes that occur in them. It is essential, therefore, that biologists understand certain principles of chemistry and that they be able to apply that understanding as they analyze the organization and functions of living things.

Another key to understanding living things is recognition of the role of cells. The cell is a fundamental organizational unit of life because all organisms are cells or aggregates of cells. Thus, the study of modern biology requires application of a sound understanding of cell function and structure. Many components of cells and many chemical processes that occur in cells are identical or very similar in cells from a variety of organisms. Recognition of these similarities at the cellular level provides an important unifying framework within which biologists approach the diversity of living things.

Another unifying concept of biology is that all living things require energy for maintenance, growth, and reproduction. Organisms must obtain energy from the outside world. Some do so by harnessing light energy for the production of organic material (photosynthesis). This organic material is then used in a variety of ways. Other organisms are not capable of using light energy in this way and thus must use photosynthesizing organisms as sources of energy and materials.

In chapters 1–7 some biologically important concepts of chemistry, some basic functional and structural characteristics of the cells that constitute living things, and some of the energy relationships and energy conversions upon which all of life depends are examined.

Atoms and Molecules

Chapter Concepts

1. Chemical processes in living things obey the laws of chemistry and physics.

2. Biologists must understand some characteristics of atoms and the ways in which atoms combine in molecules to understand and analyze the characteristics of whole organisms.

3. Water is the most common substance in living things and is the solvent in which most biochemical reactions occur. The properties of water determine many properties of life processes.

4. Most types of molecules in organisms are organic compounds, or compounds containing carbon. Many kinds of carbon compounds exist because of the diversity of carbon skeletons and functional groups and because of the formation of isomers.

A long-running argument among biologists of the past concerned the basic composition of living organisms. Many biologists believed that the substances composing living things are fundamentally different from those that make up inanimate objects. Those who accepted this idea thought there must be a special life chemistry, with its own set of laws differing from the laws of ordinary chemistry. They also proposed that there was a special physical property, a "vital force," that was unique to life. But such ideas have not withstood the tests of time and scientific investigation.

One of the most important principles of modern biology is that life processes *do* obey the laws of chemistry and physics; a special, different kind of chemistry is not necessary to study life. Today's biologists recognize that all matter consists of atoms, often combined in molecules, and that these atoms and molecules have certain fundamental properties that remain the same even when they are part of a living organism. Modern scientists also know that although there is no special "life chemistry," there are some special properties associated with life that arise from the complex organization of living things. Many of the molecules in living organisms are larger and more complex than those commonly found in nonliving systems. These relatively large and complex molecules are arranged into highly organized systems such as membranes and cells.

Because of the high degree of organization among their molecules, living things have properties that are not found in nonliving systems. Biologists believe that many of these properties can be explained on the basis of the chemical and physical properties of the component molecules. For instance, much progress has been made in describing many of the functional properties of living cells. However, it is not yet practical to try to explain some very complex phenomena, such as mating behavior, for example, in purely physical and chemical terms.

Thus, when modern biologists examine biological phenomena, they feel no need to postulate a special kind of chemistry or "vital force" unique to living things; but on the other hand, they cannot take the extreme viewpoint that life is nothing more than chemistry and physics, either. Perhaps the most practical point of view is that certain elementary concepts of chemistry are essential background for understanding of many principles in modern biology. A brief review of those concepts follows.

Elements and Compounds

All matter is composed of basic substances called **elements.** An element is a substance that cannot be broken down into simpler units by chemical reactions; it contains only one kind of atom. An **atom** is the smallest characteristic unit of an element.

There are ninety-two naturally occurring elements, as well as a number of artificial elements that can be made in laboratories. Each element is symbolized by one or two different letters that abbreviate the elements' English or Latin names. For example, the symbols for hydrogen (H) and oxygen (O) come from English names, while the symbols for sodium (Na, for *Natrium*) and potassium (K, for *Kalium*) are derived from Latin names.

A **compound** is a substance that can be split into two or more elements; or in other words, a compound is a combination of two or more kinds of atoms. Water is a compound because it can be split into its components, hydrogen and oxygen. The **formula** of a compound is comprised of information about the kinds and numbers of atoms that make up a **molecule**, the smallest characteristic unit of a compound. A formula contains the abbreviations for the atoms that make up the molecule, and subscripts that indicate the number of each kind of atom. For example, the formula for water, H_2O, indicates that a water molecule contains two hydrogen atoms and one oxygen atom. By the same token, the sugar glucose, $C_6H_{12}O_6$, contains six carbon atoms, twelve hydrogen atoms, and six oxygen atoms.

Atomic Structure

While atoms are the smallest units that have all the properties of elements, and thus can be considered to be the fundamental chemical units of matter, each atom is actually composed of still smaller particles. The numbers and kinds of subatomic particles in an atom determine its behavior in chemical reactions—including, of course, reactions that occur in living things.

Protons, Neutrons, and Electrons

Physicists have identified several kinds of subatomic particles, but only three are crucial to the properties of elements under consideration here: protons, neutrons, and electrons. **Protons** and **neutrons** are located in a central area in the atom called the **nucleus. Electrons**, the third type of particles, are located at various distances from the nucleus. Protons have a positive electric charge, and electrons have a negative charge; thus, protons attract electrons. Neutrons have no electric charge.

Protons and neutrons have nearly the same mass, and their mass is much greater than the mass of electrons (more than 1,800 times as great). Chemists customarily use the atomic mass unit, or **dalton**, to describe the mass of subatomic particles. Both the proton and the neutron have masses of approximately one dalton.

Each atom of an element has a characteristic number of protons in its nucleus, and this number is known as the element's **atomic number**; for example, because a carbon atom has six protons in its nucleus, carbon's atomic number is six. Normally, the number of electrons outside the nucleus of the atom is equal to the number of protons in the nucleus. Because the positive charges of the protons equal the negative charges of the electrons, the charges cancel one another and the atom is neutral. The atomic numbers of some biologically important elements are given in table 1.1.

The **mass number** of an element indicates how many protons *and* neutrons are in the nucleus of one of its atoms. For example, most carbon atoms contain six protons and six neutrons. Carbon's mass number is symbolized with a superscript numeral preceding the element's symbol: ^{12}C. Mass numbers are usually very close to but not exactly the same as the actual measured atomic weights of elements, which are also expressed in daltons.

Isotopes

All atoms of an element have the same number of protons, but different atoms of the same element can have different numbers of neutrons. Thus, an element can have different **isotopes**, or atoms with different mass numbers, because of

Table 1.1

Some Elements Important in Biology. H, C, O, N, P, and S are the major constituents of biological molecules, while Na, Mg, Cl, K, and Ca occur mainly as dissolved salts. Fe, Cu, and Zn are present in smaller quantities but play vital roles as components of certain proteins.

Element	Symbol	Atomic Number	Atomic Weight
Hydrogen	H	1	1.01
Carbon	C	6	12.01
Nitrogen	N	7	14.01
Oxygen	O	8	16.00
Phosphorus	P	15	30.97
Sulfur	S	16	32.06
Sodium	Na	11	23.00
Magnesium	Mg	12	24.31
Chlorine	Cl	17	35.45
Potassium	K	19	39.10
Calcium	Ca	20	40.08
Iron	Fe	26	55.85
Copper	Cu	29	63.54
Zinc	Zn	30	65.37

differences in the number of neutrons. A carbon isotope with six protons and eight neutrons is ^{14}Carbon, or simply ^{14}C (figure 1.1).

Some isotopes are stable, while others are unstable (or **radioactive**) and tend to decay. Radioactive isotopes decay by spontaneously emitting small particles. The medical uses of radioactive isotopes are well known, but their role in biological research is not as familiar. For example, radioactive isotopes are often used as "labels" in biochemical experiments. Because the different isotopes of an element react chemically in the same way, radioactive isotopes may be substituted for stable isotopes to make the atoms under observation easier to detect. Biologists use this technique to trace various atoms and molecules in living organisms; for example, ^{14}Carbon, which is radioactive, is often used in metabolism research.

Electrons and Orbitals

Although electrons have a much smaller mass than protons and neutrons, they are by no means less important in determining the chemical properties of atoms. For instance, the location of electrons in relation to the nucleus has a great influence on the way atoms and molecules react with each other.

Figure 1.1 Nuclei of carbon isotopes differ only in the number of neutrons present. Each has six protons, but ^{12}Carbon has six neutrons and ^{14}Carbon has eight neutrons. The shaded areas around the nuclei represent the areas in which six electrons are located in each case.

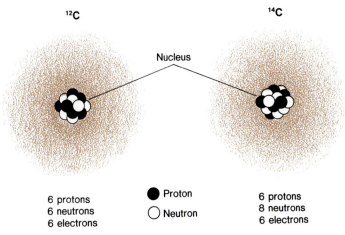

^{12}C

^{14}C

Nucleus

6 protons
6 neutrons
6 electrons

● Proton
○ Neutron

6 protons
8 neutrons
6 electrons

Electrons are continually in motion around the nucleus. Thus, it is impossible to make a precise statement about the location of an electron at any given moment. In describing the locations of electrons, chemists speak in terms of *electron clouds*. This does not mean that there literally is a cloud of electrons surrounding the nucleus; rather, it indicates that an electron may be located anywhere in the space of a certain area (the "cloud") at any given time. The location of electrons within these electron clouds, however, is not entirely random. It is possible to make a time-averaged description that depicts the space within which an electron is most likely to be located most of the time. The space an electron is located in 90 percent of the time is called the electron's **orbital** (figure 1.2).

Each orbital can contain only two electrons, and orbitals are grouped into **shells** of different energy levels. That is, the electrons located in these various shells have different energy levels. The shell closest to the nucleus has the lowest energy level, and each shell beyond has a progressively higher energy level as the distance between nucleus and shell increases.

The first shell, the one closest to the nucleus, contains only one orbital (a spherical one). Thus, the first shell can contain only two electrons. The second shell contains four orbitals, one spherical and the other three shaped like dumbbells. This second shell, then, can contain up to eight electrons. Although the third shell and those beyond can hold more than eight electrons, these shells are in a particularly stable condition when they do hold exactly eight electrons.

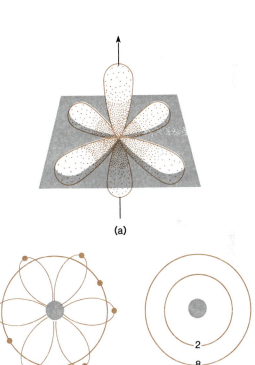

(a)

(b)   (c)

2
8

Figure 1.2 Electron orbitals. (a) The three dumbbell-shaped orbitals of the second shell are at right angles to one another. For clarity, the spherical orbital of the first shell is not included in this sketch. Densities of dots represent relative probabilities of electrons being located in various parts of the orbitals. (b) All four orbitals of the second shell on a flat plane. In this example, all orbitals are complete as each contains two electrons. The nucleus is in the center. (c) A simpler way of representing the orbital shells with the number of electrons contained in each. This sketch represents both the first and second orbital shells.

Box 1.1
In Pursuit of Atomic Structure

In 1894, a young New Zealander, who had recently graduated from the University of New Zealand in Wellington, arrived at Cambridge University in England to continue his study of physics. Ernest Rutherford, the son of a wheelwright, had come to what was then the center of the world of physics, the great Cavendish Laboratory at Cambridge. Rutherford there began his work with Professor J. J. Thomson, discoverer of the electron. Almost from the beginning, Rutherford was an acknowledged world leader in his field. He received the Nobel Prize in chemistry in 1908, but in retrospect it now seems that his greatest work actually came after he received the prize.

In 1911, Rutherford and his colleagues at Manchester University were actively exploring atomic structure. They were bombarding very thin sheets of metal foil with alpha particles and studying the deflection of those particles as they bounced off atoms during their passage through the foil.

It occurred to Rutherford that they should check whether any alpha particles were being bounced back toward their source, not a likely possibility in light of the basic ideas of atomic structure that

prevailed at the time. The atom was viewed as a diffuse, cloudlike, positively charged mass with electrons embedded in it. But Rutherford and his colleagues performed the experiment and found, to their surprise, that a very small percentage of the bombarding particles were bounced back as if they were striking much more dense objects than were the great majority of the particles that passed right on through the foil. The discovery of these rebounding alpha particles led Rutherford to conclude that most of the mass of each atom is concentrated in a central structure, the nucleus, from which alpha particles bounced back. He also concluded that electrons orbit around the nucleus, with by far the largest part of the atom's volume being the space in which the electrons are located. He calculated that the diameter of the nucleus was only a hundred-thousandth part of the atom's diameter.

But how are the electrons arranged in the space around the nucleus? The first part of the answer came from a young Danish physicist, Niels Bohr, who began working with Rutherford in 1911. Bohr approached the problem using the quantum theory that

In its normal state, an atom's electrons occupy the lowest-energy orbitals possible; that is, electrons tend to be found in the lowest energy shell in which orbital space is available. Each lower energy shell, therefore, is filled before the next shell, with its higher energy level, contains any electrons (figure 1.3). When an atom's outermost shell contains eight electrons, the atom is very stable and does not react readily with other atoms. This octet (eight) rule, as it is called, is very important in the associations of atoms to form molecules.

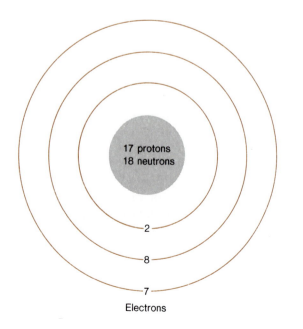

17 protons
18 neutrons

2

8

7

Electrons

had been developed by the German physicist Max Planck and refined by Albert Einstein. This theory stated that light energy of a given wavelength is absorbed or emitted by a body not in infinitely variable amounts, but as discrete units or bundles (quanta) of energy. Bohr extended the idea of quanta to the question of atomic structure and suggested that electrons occupy certain "allowed" orbits around the atomic nucleus. He proposed further that these different orbits represent different energy states and that a specific amount (quantum) of energy change occurs when an electron moves from one to another of these orbits. A quantum of energy must be absorbed to move an electron to a higher energy orbital. Conversely, a quantum of energy is radiated away from the atom when an electron moves from a higher to a lower energy orbital. Depending on the nature of the radiation released, it can be visualized as, for example, a piece of metal glows "red-hot."

The idea of atomic structure that emerged from the work of Rutherford and Bohr was a system that could be visualized as a miniature version of a solar system, a system in which electrons whirled around the nucleus in specific orbits. But with appropriate energy exchanges, the electrons could move from one orbit to another. The normal stable state of any atom, however, would be one in which the maximum energy was emitted and each electron occupied the lowest energy orbital available to it. Today there is a more refined view of orbital structure and of movements of the electrons, but Bohr's basic ideas about energy levels of orbitals have been retained.

Rutherford and Bohr, together with Einstein, Werner Heisenberg, and many others continued the quest for understanding of atomic structure, despite the disruptions of World War I and the problems that followed it. Today the subatomic particles of the 1920s are known to be made of still smaller particles. Students of atomic structure now deal with up quarks and down quarks, with leptons, and with carriers of force (for example, gluons) that hold the others together. This scientific revolution has led to the modern picture of the atom and, for better or worse, to the release of enormous quantities of energy by atomic fission.

Figure 1.3 Electron orbital shells of a chlorine atom. The chlorine atom illustrates the rule that if an atom has an incompletely filled orbital shell, it normally will be the outermost electron-containing shell that is not completely filled. The two inner shells are filled, having two and eight electrons respectively. The third shell contains only seven electrons. Electrons in the third shell are at a higher energy level than electrons in the two inner shells.

Chemical Bonding

Although atoms are the fundamental units, or building blocks, of all substances, free atoms are quite rare in nature—at least in the earth's crust and in living things. Instead, atoms usually bind together in molecules such as O_2 or H_2O, or in larger aggregates, which contain enormous numbers of atoms and molecules. Crystals of common table salt (NaCl) are good examples of such aggregates.

The forces that hold atoms together in molecules are **chemical bonds**. An important clue to understanding chemical bonds lies in the structure and characteristics of a very stable group of elements known as the **noble gases**. These relatively rare elements are exceptions to the generalization about free atoms because noble gases are **monatomic** (they do exist as single atoms).

Figure 1.4 Atomic structures of
three noble gases.

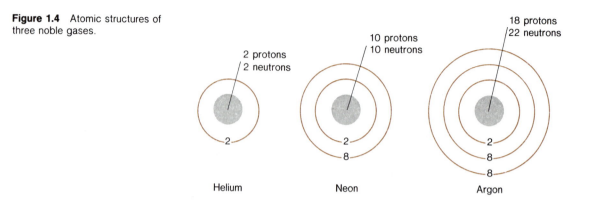

These six elements—helium (He), neon (Ne), argon (Ar), krypton (Kr), xenon (Xe), and radon (Rn)—have very little tendency to react chemically with other atoms. They all have a sufficient number of electrons to complete their outer shells, which accounts for their stability and low level of reactivity (figure 1.4).

By forming chemical bonds, other atoms tend to achieve or approximate the same stable electron configuration that the noble gases have. A stable molecule is formed when the uniting atoms can reshuffle their electrons so that each atom has a complete, or at least more stable, outer shell. Through such rearrangements of electrons, stable molecules are formed from relatively unstable atoms.

Atoms combining into molecules rearrange their electrons through different means. In **ionic bonding**, atoms attain a noble gas configuration by *transferring* electrons from one atom to another. In **covalent bonding**, pairs of electrons are *shared* between atoms to attain the electron configurations of noble gases.

Ionic Bonding

An illustration of ionic bonding can be obtained from the reaction between sodium (Na) and chlorine (Cl) in forming sodium chloride, or common table salt (NaCl). Sodium's atomic number is eleven; thus, a sodium atom has eleven protons in its nucleus and eleven electrons orbiting around the nucleus. The inner shell, containing two electrons, is complete, as is the second shell with eight electrons; but the third shell of a sodium atom contains only one electron. An atom with only one electron in its outer shell tends to be an electron donor—that is, it tends to get rid of its lone electron.

Chlorine has an atomic number of seventeen, which means that a chlorine atom has two electrons in its first shell, eight in its second shell, and seven in its outer shell. Because it needs only one electron to complete its outer shell, a chlorine atom tends to be an electron acceptor.

When a sodium atom and a chlorine atom come together, an electron is transferred from the sodium atom to the chlorine atom. This gives the sodium atom a stable configuration because its complete second shell becomes its outer shell. The chlorine atom has attained a stable electron configuration because its outer (third) shell is now complete with eight electrons. But this electron transfer changes the balance between the protons and electrons within each of the two atoms. The sodium atom is left with one more proton than it has electrons and the chlorine atom with one more electron that it has protons. The sodium atom is therefore left with a net charge of $+1$ (symbolized by Na^+), while the chlorine atom's net charge is -1 (symbolized Cl^-). Charged atoms are called **ions** and are attracted to one another by their opposite charges. Thus, an ionic compound such as sodium chloride (NaCl) is held together by the attraction between its positive and negative ions, that is, by an **ionic bond** (figure 1.5).

The ionic bonds of sodium chloride are quite strong when the salt exists as a dry solid, but when an ionic compound such as NaCl is dissolved in water, it separates into Na^+ and Cl^- ions. Ionic compounds are most commonly found in this **dissociated (ionized)** form in biological systems that consist mostly of materials dissolved or suspended in water.

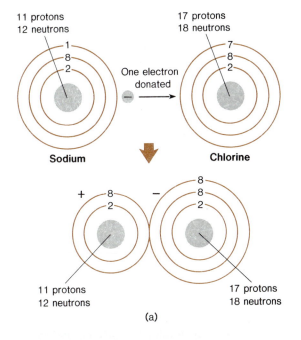

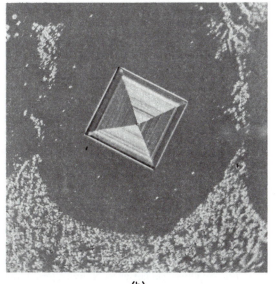

(b)

Figure 1.5 Ionic bonding. (a) Sodium donates an electron to chlorine. Thus, sodium has a stable configuration with eight electrons in its second shell, which is now its outer shell, and chlorine also has eight electrons in its outer shell. Achieving a noble gas configuration is also called satisfying the octet rule (octet for eight electrons in the outer shell). This electron transfer leaves both atoms with net charges; they are ions. Na^+ and Cl^- are strongly attracted to one another and form NaCl. (b) A sodium chloride crystal photographed by interference contrast microscopy. (Magnification $\times$ 240)

There are, of course, other biologically important ions in addition to Na^+ and Cl^-. Some, such as K^+, also are formed by the transfer of a single electron, but others (such as Ca^{2+} and Mg^{2+}) are formed when two electrons are transferred.

Covalent Bonding

A molecule such as fluorine (F_2) is formed by covalent bonding. Because each of the two fluorine atoms has nine electrons (with seven in its outer shell), there is no possible electron transfer that will lead to a noble gas configuration for both atoms. However, a stable electron configuration can be achieved if the two atoms *share* electrons. The following diagram, called a **Lewis structure,** illustrates the shared-electron, or covalent, form of bonding. The small circles in the diagram represent the electrons in the outer shell of each fluorine atom. Although the two atoms are actually identical, their electrons are represented as open and solid circles to more clearly demonstrate the bond.

Thus, a **covalent bond** consists of a pair of electrons that are shared between two atoms and that occupy two stable orbitals, one from each atom. If a single pair of electrons is shared between a pair of atoms, it is called a **single bond** (figure 1.6). But sometimes two atoms share two or three pairs of electrons to form **double or triple bonds** (figure 1.7). When two atoms are united by a single bond, they can rotate or move around one another. Atoms joined by multiple bonds, on the other hand, are fixed in relation to one another. This fact holds important consequences for the three-dimensional structures of proteins and other molecules.

Many molecules can only be represented by a single Lewis structure, but other molecules may have two or more reasonable Lewis structures. In structural chemistry, the quality of having more than one reasonable Lewis structure is called **resonance.** The structures of some molecules having several resonance structures are shown in figure 1.8. Figure 1.8b shows two resonance structures for benzene. This means that the electrons that form double bonds in the ring of carbon atoms in benzene are *delocalized;* that is, these electrons may be located at various positions within the ring.

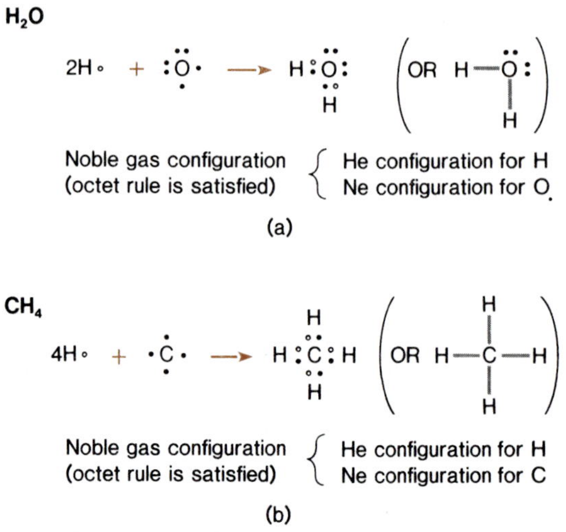

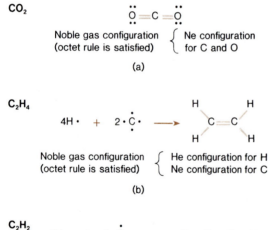

H₂O

$$2H\cdot \ + \ :\overset{..}{\underset{.}{O}}\cdot \ \longrightarrow \ H:\overset{..}{\underset{\underset{H}{..}}{O}}: \quad \left(OR \ H{-}\overset{..}{O}: \atop \qquad \ H \right)$$

Noble gas configuration   { He configuration for H
(octet rule is satisfied)   { Ne configuration for O.

(a)

CH₄

$$4H\cdot \ + \ \cdot\overset{..}{C}\cdot \ \longrightarrow \ H:\overset{\overset{H}{..}}{\underset{\underset{H}{..}}{C}}:H \quad \left(OR \ H{-}\overset{\overset{H}{|}}{\underset{\underset{H}{|}}{C}}{-}H \right)$$

Noble gas configuration   { He configuration for H
(octet rule is satisfied)   { Ne configuration for C

(b)

Figure 1.6 Lewis structures of covalently bonded compounds with single bonds. A dash (—) represents a shared pair of electrons in the alternative representation shown to the right in each case. (a) Water. (b) Methane.

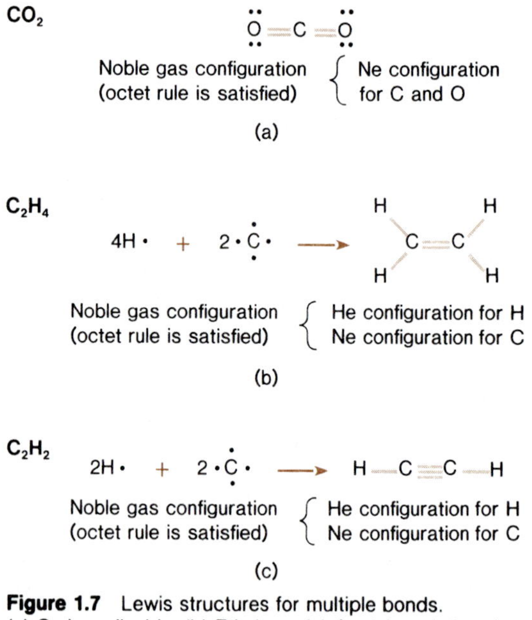

CO₂

$$\overset{..}{\underset{..}{O}}{=}C{=}\overset{..}{\underset{..}{O}}$$

Noble gas configuration   { Ne configuration
(octet rule is satisfied)   { for C and O

(a)

C₂H₄

$$4H\cdot \ + \ 2\cdot\overset{.}{C}\cdot \ \longrightarrow \ \overset{H \qquad\quad H}{\underset{H \qquad\quad H}{C{=}C}}$$

Noble gas configuration   { He configuration for H
(octet rule is satisfied)   { Ne configuration for C

(b)

C₂H₂

$$2H\cdot \ + \ 2\cdot\overset{.}{C}\cdot \ \longrightarrow \ H{-}C{\equiv}C{-}H$$

Noble gas configuration   { He configuration for H
(octet rule is satisfied)   { Ne configuration for C

(c)

Figure 1.7 Lewis structures for multiple bonds.
(a) Carbon dioxide. (b) Ethylene. (c) Acetylene (ethyne).

Various molecules containing delocalized electrons are essential to many biological processes. For example, ring structures in chlorophyll molecules contain delocalized electrons that can move within the molecule; if visible light energy strikes such molecules, electrons are moved from lower energy orbitals to higher ones. This absorption of light energy by chlorophyll molecules provides energy for the production of sugars in the process of photosynthesis.

Weaker Interactions

In addition to ionic and covalent bonds, there are also several types of weaker interactions among atoms and molecules. These weaker interactions are largely responsible for the three-dimensional shapes of many large biological molecules, such as proteins, and for some of their functional properties.

In some molecules that are formed by covalent bonds, the electrons tend to be attracted more towards one atom than the other. This unequal amount of attraction among the electrons is called **polarity.** Polarity results in molecules that have a small positive charge at one end and a small negative charge at the other (figure 1.9). Because of these slight charges at their ends or poles, polar molecules are attracted to each other. The weak electrical attractions between opposite charges on polar molecules are known as **dipole forces.** These forces are only effective at a short distance.

Hydrogen Bonding

A special case of attraction involving unusually strong dipole forces is **hydrogen bonding.** Hydrogen bonding occurs between a hydrogen atom that is already covalently bonded to nitrogen or oxygen and a *different* nitrogen or oxygen atom that has a lone pair of electrons (a pair of electrons not involved in bonding; see figure. 1.10). This type of bond can be formed between different molecules or between different parts of the same molecule. Both types of hydrogen bonding are important in biological processes. Hydrogen bonding between different molecules is crucial to the role of water as a solvent in living systems. Hydrogen bonding within molecules is especially important in the organization of protein and nucleic acid molecules, because hydrogen bonding within these large molecules is a key factor in determining their three-dimensional shapes.

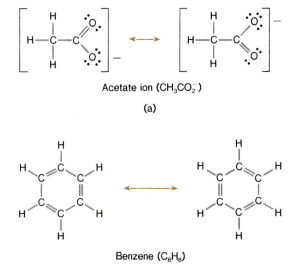

Acetate ion (CH₃CO₂⁻)

(a)

Benzene (C₆H₆)

(b)

Figure 1.8 Examples of resonance. Note the different positions of double bonds. (a) Acetate ion. (b) Benzene.

Polar molecules

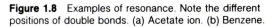

Nonpolar molecules

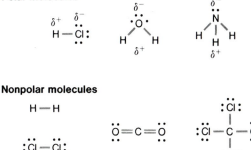

| Nonpolar bonds | Polar bonds, but nonpolar molecule; bond dipoles cancel by symmetry |

Figure 1.9 Examples of polar and nonpolar molecules. δ⁺ and δ⁻ symbolize small charge differences at different ends of polar molecules. Nonpolar molecules can be nonpolar either because they lack polar bonds or because they have polar bonds whose dipole forces are cancelled by symmetry of the molecules.

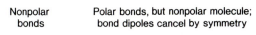

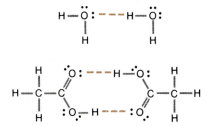

Figure 1.10 Hydrogen bonding (indicated by colored dashed lines).

London Dispersion Forces

London dispersion forces (also called **van der Waals forces**) are weak forces resulting from the interaction of temporary dipoles. They are only about one-fifth as strong as hydrogen bonds but, weak as they are, assume great importance in large molecules because there are so many of them per molecule. London dispersion forces also help hold aggregates of smaller molecules together. For instance, the phospholipids of cell membranes are held together largely by these forces.

In a biological context, bonding forces can be ranked from weakest to strongest as follows: London dispersion < hydrogen bonds < ionic < covalent. However, the strength of an ionic bond can change—in a dry, solid state, ionic bonds are roughly as strong as covalent bonds, but in solution, ionic bonds are just a bit stronger than hydrogen bonds.

Water in Living Systems
Some Properties of Water

Water is the most abundant material in living things (60 to 95 percent by weight). It is both the suspending medium and solvent in organisms, and the great majority of biological reactions occur in water. Understanding the properties of water as a solvent is thus essential to understanding virtually all processes that occur in living systems.

The structure of the water molecule is nonlinear (or bent) with a bond angle of 104.5° between the HOH (hydrogen-oxygen-hydrogen) atoms (figure 1.11). Water is a polar molecule, with the oxygen atom forming the negative end and the hydrogen atoms the positive end. Therefore, hydrogen bonding occurs between water molecules. The importance of hydrogen bonding in water is most strikingly seen by examining the structure of ice. In ice, each oxygen atom is surrounded tetrahedrally by four hydrogen atoms, two close (covalent bonds) and two distant (hydrogen bonds). As figure 1.11c shows, hydrogen bonding in ice produces a very open structure, resulting in a low density (or a large volume per molecule). The density of ice is, in fact, less than the density of water in its liquid state. Water reaches its maximum density at 4° C. This explains why ice floats on cold water and why lakes freeze at the top first, leaving water below.

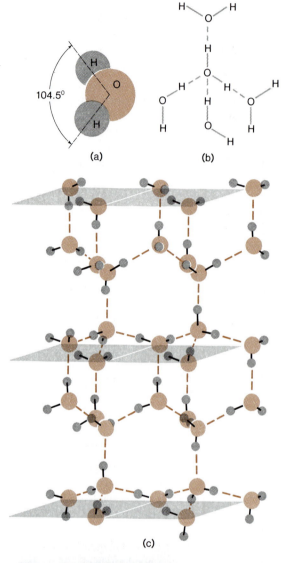

(a)   (b)

(c)

(d)

Figure 1.11 (a) Structure of a
water molecule. (b) Tetrahedral
hydrogen bonding of water
molecules in ice. Distances are in
nanometers. (c) The open structure
of ice. (d) An iceberg, a huge mass
of floating ice. Because of its
relatively open structure, ice has a
lower density than liquid water and
will float in water.

When a body of water does freeze on the surface, the ice acts as an insulator to prevent much of the water below it from freezing. If ice were denser than water, lakes and ponds would freeze solid, from the bottom up, and most aquatic organisms would die during the winter.

Several models have been proposed for liquid water, which has a less ordered structure than ice. The models differ in detail, but they generally show that some short-lived, icelike structures exist in the liquid. Such areas are probably small and temporary, rapidly forming and dispersing over and over again in the liquid.

Several other physical properties of liquid water are unusual. Because of the strong hydrogen bonds holding its molecules together, water has much higher melting and boiling points than would otherwise be expected for a molecule of its size. It also has a distinctively high **heat capacity** (the amount of heat energy that must be added or subtracted to raise or lower its temperature). The high heat capacity of water helps stabilize and protect organisms from rapid fluctuations in temperature and, on a larger scale, stabilizes the temperature of aquatic environments such as lakes and oceans.

Water also has a high **heat of vaporization.** Heat of vaporization is the amount of heat energy required to convert water from a liquid to a gaseous state. Water's high heat of vaporization makes cooling by evaporation an efficient process; for example, the evaporation of a single gram of water cools 540 grams of water by 1° C. This property gives organisms in hot environments an efficient cooling method, and processes such as perspiration can help to dissipate excess body heat.

Finally, the hydrogen bonding that takes place among water molecules also contributes to the large amount of **surface tension** in water. Surface tension is the *force per unit length* (or the amount of work per unit area) needed to expand a surface. High surface tension causes liquid surfaces to form shapes requiring a minimal amount of space; small spherical droplets of water or soap bubbles are good examples of this (figure 1.12a). Water's surface tension is directly exploited by insects such as the water strider, which can "skate" on a water surface, its small weight supported by the surface tension of the water (figure 1.12b).

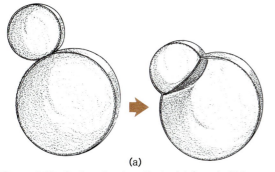

(a)

(b)

Figure 1.12 Surface tension effects. (a) Soap bubbles in contact with one another. Surface tension causes the bubbles to assume a shape with minimum surface area. (b) Water strider (Family Gerridae) supported by surface tension.

Water as a Solvent

The forces that attract polar molecules or portions of molecules to water are called **hydrophilic** (*water-loving*) interactions. Such interactions disperse molecules of a **solute** (or dissolved substance) among the molecules of the **solvent** (or dissolving substance—in this case, water). In a sense, solutes disrupt the structure of water because they interact with water molecules and become interspersed among them. Salt, ethanol, and ammonia are examples of water-soluble substances.

Hydrophobic (*water-hating*) interactions occur between water and nonpolar molecules or parts of molecules. Because the forces attracting water molecules to one another are stronger than those acting between water and nonpolar molecules, the nonpolar molecules are "squeezed out" from between the water molecules. In other words, nonpolar molecules do not effectively disrupt the structure of water. Examples of such substances include sulfur, fats and oils, starch, and nitrogen.

Some substances, such as ethanol (CH_3CH_2OH) have both hydrophilic (at the $-OH$ end) and hydrophobic interactions (at the CH_3CH_2- end of the molecule). In the case of ethanol, the nonpolar ($CH_3 CH_2-$) part of the molecule is small enough that its effect is overcome by the strong hydrophilic attraction between the polar OH group and the surrounding H_2O molecules. Thus, ethanol dissolves in water.

When an ionic compound, such as sodium chloride (Na^+Cl^-), is dissolved in water, its ions separate because the charged poles of the water molecules attract the ions more than the ions attract each other. Not only are the ions pulled apart in the process, but they become surrounded by layers of water molecules called **hydration spheres** (figure 1.13). In this particular type of hydrophilic interaction, the water molecules around the ions are oriented so that the negative (O) end of each water molecule is adjacent to the positive ion (Na^+). Similarly, the positive (H) end of each water molecule would be attracted to the negative ion (Cl^-). Hydration spheres effectively increase the size of ions and affect the movement of ions in biological systems, including the passage of ions through cell membranes. The hydration process also affects the structure of water because the water molecules in each sphere are oriented in one direction—with either their positive or negative ends attracted toward the ion. Because of their fixed orientation, the water molecules cannot, for example, take part so readily in ice formation (figure 1.11c). The higher the salt content of water, the more resistant it is to freezing; the freezing point of water is depressed by solutes.

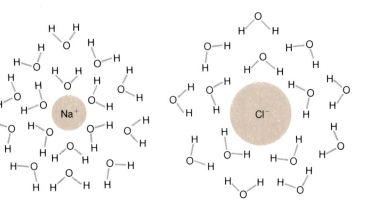

Figure 1.13 Hydration of ions, using Na$^+$ and Cl$^-$ ions as examples. Positive ions attract the negative end (O) of surrounding water molecules. Negative ions attract the positive end (H) of the water molecules. This process affects the structure of water because water molecules in the hydration spheres around ions are specifically oriented. They cannot, for example, take part so readily in ice formation (see figure 1.11c). The higher the salt content of water, the more resistant it is to freezing; the freezing point is depressed by solutes.

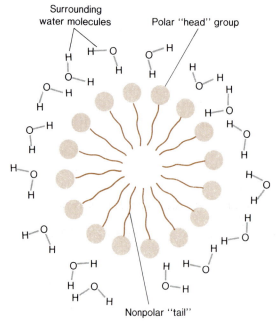

Figure 1.14 The formation of a micelle by molecules that have a polar "head" and a nonpolar "tail."

What happens when a molecule that has both a small, polar (hydrophilic) portion and a large, nonpolar (hydrophobic) portion is added to water? When molecules with polar "heads" and long, nonpolar "tails" are mixed with water, they form structures known as **micelles** (figure 1.14). In a micelle, the nonpolar tails turn toward each other and away from the H$_2$O molecules because the interactions among the tails are strong and because the interactions of the nonpolar tails and the water molecules are weaker than water-water interactions. However, hydrophilic attraction does occur between the polar head and adjacent H$_2$O molecules.

The action of a detergent provides a familiar, everyday example of micelle formation: a detergent disperses nonpolar grease by incorporating the grease molecules into the interiors of micelles. The phenomenon of micelle formation reveals much about the structure of biological membranes because biological membranes are largely made up of molecules that have both hydrophilic and hydrophobic ends.

Concentration of Solutions

The properties of a solution depend not only on the natures of the solvent and solute, but also on their concentration, or the amount of solute per unit amount of solvent or solution. Several units of measurement may be used to express the concentrations of solutions. Stating the concentration as a **percentage** (normally, weight of solute to volume of solvent) is one simple approach. Physiological saline solution, for example, is 0.9 percent (weight-to-volume) NaCl. Stated differently, every 100 ml of this solution contains 0.9 grams of NaCl dissolved in water.

Table 1.2
Ionic Composition of Seawater and Body Fluids, Expressed in Millimoles Per Liter (One Millimole = 10^{-3} Moles).

	Na^+	K^+	Ca^{2+}	Mg^{2+}	Cl^-	SO_4^{2-}
Seawater	417	9.1	9.4	50	483	30
Vertebrates						
Human	145	5.1	2.5	1.2	103	2.5
Frog (amphibian)	103	2.5	2.0	1.2	74	
Lophius (fish)	228	6.4	2.3	3.7	164	
Invertebrates						
Hydrophilus (insect)	119	13	1.1	20	40	0.14
Lobster (arthropod)	465	8.6	10.5	4.8	498	10
Venus (mollusc)	438	7.4	9.5	25	514	26

Condensed from Ariel G. Loewy and Philip Siekevitz, *Cell Structure and Function,* 2d ed. (New York: Holt, Rinehart & Winston, 1969), p. 87.

However, in describing chemical reactions, the **mole** is a more useful unit of measurement for the amounts of different substances. A mole is the mass of substance, expressed in grams, that is numerically equal to the **molecular (or formula) weight.** The formula weight of any molecule is equal to the sum of the atomic weights of the atoms in the molecule. Thus, the formula weight of NaCl is 58.5, and one mole of NaCl is 58.5 grams:

$$23.0(\text{atomic weight of Na}) +$$
$$35.5(\text{atomic weight of Cl}) = 58.5$$

The particular value of moles as units of measure is that they are proportional to the number of molecules of a substance that are present. One mole of any substance contains Avogadro's number (6.02×10^{23}) of molecules of the substance.

Molarity (M) is the number of moles of solute per liter of solution. Physiological saline solution contains 9 grams of NaCl per liter of solution, and thus, it is a 9/58.5, or 0.154, M NaCl solution. Moles and molarity can be used to describe accurately the small concentrations of many materials in living systems. For example, in table 1.2, the concentrations of ions in seawater and in the body fluids of several organisms are compared in terms of millimoles (1 millimole = 1/1,000 mole) per liter.

Acid-Base Phenomena

In biological systems, one of the most important ions is also the simplest: H^+, the hydrogen ion. The concentration of the hydrogen ion strongly affects the rates of many reactions, the shape and function of enzymes and other protein molecules, and even the integrity of the cell itself.

A hydrogen atom is the simplest of all atoms. It contains one proton in the nucleus and one electron in an orbital outside the nucleus. When the electron is lost (forming the H^+ ion), all that is left is, obviously, the proton in the nucleus. In fact, the H^+ ion is often simply called a proton.

The powerful effect of the H^+ ion on other molecules is due to its positive charge and its extremely small size. A proton (H^+) can often attach itself to a molecule by approaching a lone pair of electrons and forming a new covalent bond:

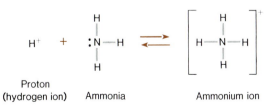

Proton
(hydrogen ion)   Ammonia   Ammonium ion

In the internal environment of organisms, the most abundant molecule is water, which carries two lone pairs of electrons.

A proton (H^+) is strongly attracted to a lone pair
of electrons on a water molecule.

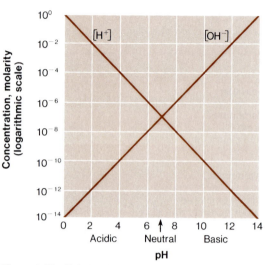

The proton affinity of H_2O is so strong that no
free H^+ ions exist in aqueous solution: the H^+
ion is always hydrated (attached to a water molecule). But, for convenience, the hydrogen ion is
represented simply as H^+.

H^+ Concentration and pH

Ions in solution conduct an electric current because they are mobile, charged particles. Tap
water, which usually contains mineral ions, conducts electricity quite well (as a person who is
careless with electricity while taking a bath soon
learns). The conductance of pure water is much
weaker but still demonstrates the presence of a
few ions. These ions result from the ionization of
a few water molecules:

$$H_2O \rightleftharpoons H^+ + OH^-$$

In pure water, this ionization occurs to such an
extent that

$$[H^+] = 10^{-7} M \text{ and } [OH^-] = 10^{-7} M$$

(Square brackets are conventionally used to de-
note concentrations.) Of course, the concentration of H^+ ions in pure water has to equal the
concentration of OH^- ions because neither can
be formed without the other.

An alternative way of designating the H^+
concentration in a solution is through the concept
of pH, or the negative logarithm of the hydrogen
ion concentration. A logarithm is the exponent
that indicates the power to which one number
must be raised to obtain another number (for
example, the logarithm of 100 to the base 10 is
2). Thus:

$$pH = \log_{10} \frac{1}{[H^+]}$$

$$(\text{or, } pH = -\log_{10} [H^+])$$

As an example, if $[H^+] = 10^{-3} M$ (.001 M) then
the pH = 3.

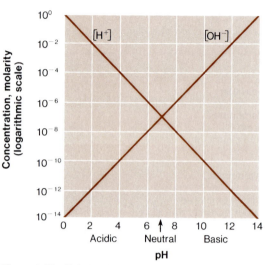

Figure 1.15   Relationships of pH, hydrogen ion
concentration [H^+], and hydroxide ion concentration
[OH^-].

A solution is *neutral* if its H^+ and OH^-
concentrations are equal. For water, $[H^+]$
$= [OH^-]$ when both exist in concentrations of
10^{-7} M, that is, when the pH is 7. A solution is
acidic if it has a higher H^+ concentration than
OH^- concentration, and *basic* if its H^+ concentration is lower than its OH^- concentration.
Therefore, acidic solutions have pH values below
7, and basic solutions have pH values above 7
(figure 1.15).

Acids and Bases

According to the *Brønsted definition* of acids and
bases, an **acid** is a *proton (hydrogen ion) donor*;
that is, it can contribute a proton in a reaction.
Some examples of acids are H_2CO_3 and NH_4^+:

$$H_2CO_3 \longrightarrow H^+ + HCO_3^-$$
Carbonic acid   Bicarbonate ion

$$NH_4^+ \longrightarrow H^+ + NH_3$$
Ammonium ion   Ammonia

An acid can be a neutral molecule or a positive or negative ion. The part of the acid that is
formed or left over when the acid gives up its
hydrogen ion (proton) is called its **conjugate base:**

$$\text{Acid} \rightleftharpoons \text{Proton} + \text{Conjugate base}$$

A **base** is a *proton acceptor*. Some examples of bases are HCO_3^- and NH_3:

$$HCO_3^- + H^+ \longrightarrow H_2CO_3$$
$$NH_3 + H^+ \longrightarrow NH_4$$

In an **acid-base reaction,** a hydrogen ion is transferred from an acid to a base. For example:

$$H_2CO_3 + NH_3 \longrightarrow NH_4^+ + HCO_3^-$$

Chemical Equilibrium

When a reaction is at **chemical equilibrium,** the concentrations of its components remain stable over time. But chemical equilibrium is a dynamic process, not a static condition: forward and reverse reactions occur at the same rate, so that while there is no net change in the concentrations of reactants and products, chemical reactions are constantly taking place. For example, in aqueous solutions, the reaction

$$H_2O \rightleftharpoons H^+ + OH^-$$

is at equilibrium because the rate of ionization of water molecules,

$$H_2O \longrightarrow H^+ + OH^-$$

is equal to the rate of reaction among the ions.

$$H^+ + OH^- \longrightarrow H_2O$$

Thus, there is no further *net* change in concentration, even though both reactions continually occur.

In the following example, acetic acid (an ingredient of vinegar) has just one ionizable hydrogen atom:

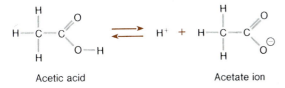

Acetic acid　　　　　　　　　　　　Acetate ion

Only the hydrogen of the polar O-H bond dissociates; the C-H bonds are nonpolar and unreactive. The effect of such an acid on the pH of a solution would depend on how many of the acetic acid molecules are ionized when the reaction is at equilibrium. Acetic acid, it turns out, is a **weak acid;** it ionizes only partially. In other words, chemical equilibrium is reached when considerable quantities of intact acetic acid molecules still remain in the solution.

Strong acids, on the other hand, are acids that are ionized much more completely at equilibrium and that therefore contribute many more protons than weak acids. Some examples of strong acids are HCl (hydrochloric acid), HNO_3 (nitric acid), and H_2SO_4 (sulfuric acid).

Strong and weak bases are analogous to strong and weak acids. In water, a **strong base** (such as NaOH) ionizes almost completely and produces a very high concentration of OH^- ions. A **weak base** (for example, NH_4OH) ionizes less completely, producing a much smaller OH^- concentration and a much less basic solution.

Buffers

Even small pH changes may have great effects on biological processes, particularly when these changes alter the shapes of enzyme molecules and change rates of reactions. Because many processes are sensitive to pH changes, it is important that the pH of body fluids remain stable. Cell interiors have an average pH of 7.0 to 7.3, and in humans, for example, blood and tissue fluids have a pH of 7.4 to 7.5. These values remain very stable despite the fact that many biochemical reactions either release or incorporate H^+ ions.

This pH stability is possible because organisms have built-in mechanisms to prevent pH change. The most important of these mechanisms are **buffers.** A buffer resists pH change by removing H^+ ions when the H^+ ion concentration rises and by releasing H^+ ions when the H^+ ion concentration falls. Buffers often consist of a weak acid along with its conjugate base. For example, the most important buffer system in human blood is the acid-conjugate base pair of carbonic acid and bicarbonate.

$$H_2CO_3 \rightleftharpoons H^+ + HCO_3^-$$
$$\text{(H}^+ \text{ donor)} \qquad \text{(H}^+ \text{ acceptor)}$$

In this example, if H^+ ions are added to the system, they combine with HCO_3^- to form H_2CO_3. This reaction removes extra H^+ ions and keeps the pH from changing. If H^+ ions are removed from the system (for example, if OH^- ions are added and H_2O is formed), more H_2CO_3 will ionize and replace the H^+ ions that were used. Again, pH stability is maintained.

Figure 1.16 Possible arrangements of bonds around the carbon atom. Single and double bonds are common in organic compounds found in living organisms, but triple bonds are not. In the methane molecule, the dashed line indicates that the carbon-hydrogen bond goes away from you, and the extended triangle indicates that the carbon-hydrogen bond comes toward you, out of the plane of the page.

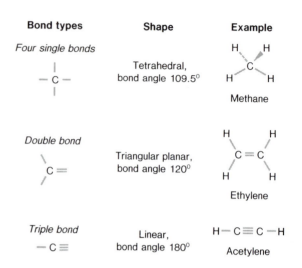

Bond types	Shape	Example
Four single bonds	Tetrahedral, bond angle 109.5°	Methane
Double bond	Triangular planar, bond angle 120°	Ethylene
Triple bond	Linear, bond angle 180°	Acetylene

Organic Molecules

Almost all of the chemical compounds in living organisms are **organic compounds,** or compounds that contain carbon. Technically, this definition can be expanded to say that organic compounds contain carbon and hydrogen since carbon dioxide (CO_2) is not considered an organic compound. Nevertheless, the most important characteristics of organic compounds depend on properties of the element carbon.

The carbon atom has an atomic number of six, with six protons and six electrons. Since such an atom has two electrons in the first shell and four electrons in the second shell, it can form covalent bonds with as many as four other atoms. A carbon atom can bond to a variety of elements, but it most commonly bonds to hydrogen, oxygen, nitrogen, and carbon.

The carbon-to-carbon bonding ability of carbon atoms makes possible carbon chains (sometimes called *carbon skeletons*) of various lengths and shapes. The carbon skeleton establishes the overall framework of an organic molecule. Carbon atoms can form single, double, or triple covalent bonds with one another. Various patterns of single, double, and triple carbon bonding also determine the characteristics of organic compounds. Possible types of carbon bonds are shown in figure 1.16.

Carbon-carbon bonds and carbon-hydrogen bonds are nonpolar, and thus are generally less reactive than other bonds that are found in organic molecules. Many organic compounds consist of just carbon and hydrogen, but many others contain **functional groups** as well. Functional groups are characteristic patterns of atoms other than those involved in carbon-carbon and carbon-hydrogen bonds. While carbon skeletons, which are made up mainly of carbon-carbon and carbon-hydrogen bonds, are relatively unreactive, functional groups are often the reactive sites within organic molecules. Some of the most important functional groups are shown in figure 1.17.

Literally millions of different organic compounds have either been isolated or synthesized. One reason there are so many organic compounds is that carbon forms chains of atoms better than any other element. An unusually long noncarbon chain might contain ten atoms. On the other hand, organic compounds with chains of more than fifty carbon-carbon bonded atoms are found in living systems, and chains containing carbon along with other elements such as nitrogen and oxygen are sometimes thousands of atoms long.

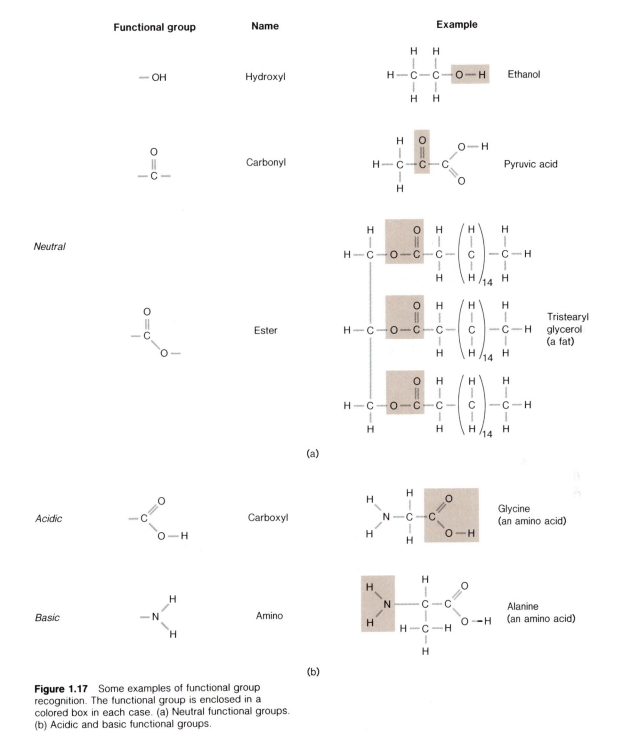

Figure 1.17 Some examples of functional group recognition. The functional group is enclosed in a colored box in each case. (a) Neutral functional groups. (b) Acidic and basic functional groups.

Figure 1.18 Types of isomers. (a) Structural isomers are molecules with the same molecular formula, but different pairs of atoms connected by bonds. (b) Geometric isomers are molecules that have different orientations of bonds in space but which are not mirror images. Each line indicates a C-C bond in retinal. Hydrogens are not shown. In the retina of the eye, *cis*-retinal is present (combined with a protein) as rhodopsin. A photon of light is absorbed by *cis*-retinal, converting it to *trans*-retinal. Thus, geometrical isomers are important in the initial step in vision. (c) Optical isomers are pairs of molecules that are nonsuperimpossible mirror images. Most optical isomers contain one or more asymmetric carbon atoms (indicated by an asterisk). An asymmetric carbon atom is a carbon bonded to four different groups in a molecule. L-alanine occurs commonly in organisms; D-alanine does not.

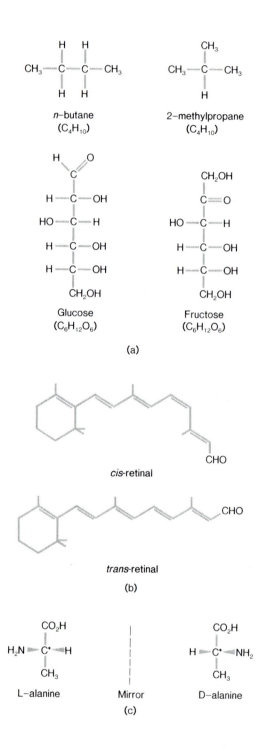

   Another factor contributing to the diversity of organic compounds is the presence of numerous **isomers.** Isomers are compounds having the same molecular formulas but different atom arrangements. *Structural isomers* are completely different chemical compounds; they share the same molecular formula but have different pairs of atoms connected by bonds (figure 1.18a). *Geometric isomers* differ only in the orientations of bonds between particular pairs of atoms (figure 1.18b). These differences are possible in the presence of certain rings or C=C double bonds. Although rotation around C-C single bonds occurs readily, rotation around C=C double bonds does not. *Optical isomers* occur as pairs of nonsuperimposible mirror images, similar to the relationship between a right hand and a left hand. In optical isomers, one or more asymmetric carbon atoms (carbon atoms attached to four different groups) may exist.

Summary

Because chemical processes in living organisms obey the laws of chemistry and physics, it is not necessary to propose a special life chemistry. Life's special properties arise instead from the complex and diverse organizations of molecules, cells, and organisms. The study of modern biology requires a background in the physical sciences, especially chemistry.

The basic substances composing all forms of matter are called elements. Atoms are the smallest units having all the characteristic properties of the elements, but atoms are made up of smaller units called subatomic particles. Three types of subatomic particles play crucial roles in determining the properties of different elements: protons (positively charged particles), neutrons, and electrons (negatively charged particles). Protons and neutrons are located in the nucleus of an atom and electrons circle the nucleus in orbitals in a series of shells.

An element's atomic number is equal to the number of protons in its nucleus. Normally, the number of electrons in shells around the nucleus is the same as the number of protons. Different isotopes of an element have different numbers of neutrons in their nuclei. Some isotopes are radioactive and spontaneously emit particles.

The numbers and locations of electrons in the orbital shells around different atoms determine the chemical properties of those atoms and their bonding characteristics. When an atom's outermost shell is complete, or carries the maximum number of electrons possible, the atom is stable (unreactive). The noble gases, with their complete outer shells, are examples of such atoms. Other atoms tend to achieve or approximate this stable condition in forming chemical bonds.

Ionic bonding involves the transfer of electrons between atoms. After such transfers, the atoms, now called ions, have a net electric charge. Ions with opposite charges combine to form ionically bonded compounds.

Covalent bonds form when atoms share pairs of electrons in their outer orbital shells. Single, double, and triple covalent bonds result from the sharing of one, two, or three pairs of electrons. Resonance, or the quality of having more than one possible structure, occurs because electrons can move from one bond to another in certain kinds of molecules.

Weaker bonds, such as dipole forces, also hold molecules together. The hydrogen bond is a relatively strong form of dipole force. Hydrogen bonding occurs both between molecules and within molecules, where it plays a role in establishing complex three-dimensional molecular structures.

Water is the most abundant material in living organisms and is the solvent in which most biochemical reactions occur. Water is made up of individual polar molecules, each with a negative charge at the oxygen end and positive charges at the hydrogen ends. This polarity explains water's solvent properties; since polar molecules are attracted to the charges in the water molecules, polar substances are soluble in water and nonpolar substances are not. Water's physical properties, such as high heat capacity, high heat of vaporization, and high surface tension also serve useful functions in living systems.

Biologists use several different kinds of units to express the concentrations of various substances in solutions. One of the most useful is molarity. One mole is the mass of substance, expressed in grams, that is numerically equal to the molecular weight of the substance. Molarity is the number of moles of solute per liter of solution.

Hydrogen ion concentration can be expressed through the concept of pH. Acidic solutions have high H^+ concentrations and low pH numbers, while basic solutions have high OH^- concentrations and high pH numbers. Changes in pH strongly affect biological processes. Therefore, buffer systems that resist pH changes in cells and in body fluids are very important for the maintenance of normal biochemical functions.

Almost all of the chemical compounds in living organisms are organic compounds, or compounds containing carbon. A carbon atom can form four covalent bonds and may bond to a variety of elements, especially hydrogen, oxygen, nitrogen, and carbon. Carbon-to-carbon bonding enables the formation of carbon chains, the "skeletons" of organic molecules.

Functional groups are the most reactive parts of organic molecules. Functional groups are characteristic patterns of atoms and bonds other than carbon-carbon and carbon-hydrogen bonds. Isomers are chemical compounds that have the same molecular formula, but different arrangements of atoms. Carbon-to-carbon bonding, functional groups, and isomers all contribute to the diversity of structures and functions among living systems.

Questions

1. Determine the number of neutrons in the nucleus of each of these isotopes. (Oxygen has an atomic number of 8, and phosphorus has an atomic number of 15.)
   a. ^{16}O; ^{18}O
   b. ^{31}P; ^{32}P; ^{33}P

2. How many electrons are there in the outer shell of each of the following atoms? (The atomic number of nitrogen (N) is 7, sodium (Na) 11, magnesium (Mg) 12, and calcium (Ca) 20.)
   a. N
   b. Na
   c. Na^+
   d. Mg
   e. Mg^{2+}
   f. Ca^{2+}

3. Explain what would happen to many aquatic organisms if ice had a greater density than liquid water.

4. The pH scale is a logarithmic scale. A change from pH 5 to pH 6 in a solution means that the H^+ ion concentration has increased by a factor of:
   a. 6
   b. 10
   c. 5
   d. 100

5. Draw Lewis structures for the formation of methane (CH_4) and ethane (C_2H_6) from carbon and hydrogen.

6. How many grams of potassium chloride (KCl) are in every liter of 0.1 molar potassium chloride solution? (The formula weight of KCl is 74.557.)

Suggested Readings

Books

Baker, J. J. W., and Allen, G. E. 1981. *Matter, energy, and life.* 4th ed. Reading, Mass.: Addison-Wesley.

Henderson, H. L. 1958. *The fitness of the environment.* Boston: Beacon Press.

Pauling, L., and Pauling, P. 1975. *Chemistry.* San Francisco: W.H. Freeman.

White, E. H. 1970. *Chemical background for the biological sciences.* 2d ed. Englewood Cliffs, N.J.: Prentice-Hall.

Articles

Frieden, E. July 1972. The chemical elements of life. *Scientific American.*

Milne, L. J., and Milne, M. April 1978. Insects on the water surface. *Scientific American* (offprint 1387).

Stillinger, F. H. 1980. Water revisited. *Science* 209:451.

Macromolecules, Lipids, and Membranes

Chapter Concepts

1. Many biologically important molecules are polymers, or chains of individual molecular units linked together.

2. Polysaccharides (polymers of sugars) function in nutrient storage in plants and animals and form major structural components in plants.

3. Proteins (polymers of amino acids) play structural and functional roles in all living things.

4. Nucleic acids (DNA and RNA) are polymers that code genetic information.

5. Lipids (compounds that are insoluble in water but soluble in hydrocarbon solvents) play important structural and storage compound roles. Certain lipids are essential components of cell membranes.

6. Cell membranes are orderly arrays of smaller molecular units and contain both lipid and protein constituents.

7. The properties of plasma membrane surfaces help to explain normal and abnormal cell functioning.

Most of the thousands of kinds of molecules found in living things are organic compounds, compounds based on the element carbon. The characteristics of the living cells of organisms depend on the properties and arrangements of various organic molecules. Although it would be impractical to attempt to list even a small percentage of the organic compounds that are known to exist in living things, it is important to become acquainted with some of the major groups of organic compounds.

Several prominent types of organic compounds in cells are very large molecules called **macromolecules.** Macromolecules are **polymers;** that is, they are large molecular chains of smaller organic molecules linked together. The smaller organic molecules that form the individual units of polymer chains are called **monomers.**

Monomers are joined together into polymers · by **condensation reactions** in which the components of water molecules are removed (figure 2.1). The reverse of a condensation reaction is **hydrolysis** (the addition of water molecules), which separates the monomers from one another. This pattern of condensation and hydrolysis is essential to many synthesis processes in living things, as well as to the breakdown of large molecules in processes such as digestion.

Certain organic compounds are organized into complex, orderly arrays of molecules that are not polymers, but rather aggregates of individual molecules. **Membranes** are especially good examples of this type of arrangement, in which small organic molecules (especially **lipids**) compose larger functional units. A membrane forms the boundary between each cell's interior and the external environment, and membranes also subdivide the complex internal environments of cells into separate compartments. The orderly molecular arrangements in membranes are thus a very fundamental organizational feature of cells.

To a large degree, the chemistry of biologically important organic molecules is the chemistry of several groups of organic polymers, polysaccharides, proteins, and nucleic acids, and of orderly lipid arrays in membranes. In this chapter, each of these types of organic molecules and their functions in living cells are examined.

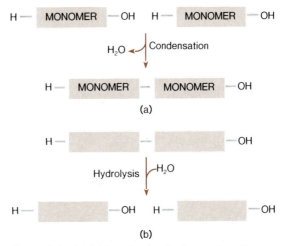

Figure 2.1   (a) Joining of molecules by condensation (H_2O is removed). Condensations can connect a series of monomers to produce a long-chain polymer.
(b) Hydrolysis (separation with addition of H_2O) is the reverse of condensation.

Polysaccharides

Polysaccharides ("many sugars") function as energy-storage and structural compounds in organisms. The monomer units that make up polysaccharides are molecules of single sugars (**monosaccharides**) such as glucose. Both polysaccharides and the sugar units from which they are constructed belong to the class of organic compounds called **carbohydrates** because they contain carbon and the components of water—hydrogen and oxygen. In carbohydrates, carbon, hydrogen, and oxygen occur in a 1:2:1 ratio; thus, a general formula for a carbohydrate is $(CH_2O)_n$. Although there are sugars that contain three, four, five, six, or more carbons, the common polysaccharides are polymers of six-carbon sugars. The polysaccharides described here are all polymers of glucose ($C_6H_{12}O_6$).

Actually, glucose is not the only sugar that has the molecular formula $C_6H_{12}O_6$; a number of six-carbon sugars (**hexoses**) share this molecular formula. But these hexoses are structural isomers (see chapter 1); that is, they are different chemical compounds with different structures even though they all contain the same numbers of carbon, hydrogen, and oxygen atoms. Glucose is the

monomer of several important types of polysaccharides. The several possible structures of glucose are shown in figure 2.2. Note especially the
two different ring structures (the α and β forms)
and the position of the hydroxyl (OH) group attached to the number 1 carbon in each structure.
It is important to understand this apparently
slight difference between these two forms of glucose in order to grasp some very significant differences among the structures of various glucose
polymers.

Linking Glucose Molecules

Polymers of glucose, like other biologically important polymers, are formed by condensation
reactions. A simple example of such a reaction
can be seen in the formation of **disaccharides**
(double sugars) from monosaccharides. For example, two glucose molecules may combine to
produce a molecule of maltose (malt sugar), a
disaccharide. The molecular formulas involved in
this condensation reaction are:

$$C_6H_{12}O_6 + C_6H_{12}O_6 \longrightarrow C_{12}H_{22}O_{11} + H_2O$$
(Glucose)   (Glucose)   (Maltose)

The structural formulas in figure 2.3a show
how the actual linkage of the glucose units in
maltose is formed between the number 1 carbon
of one glucose molecule and the number 4 carbon
of the other. Because these are α-glucose molecules (see figure 2.2), this is called an $\alpha(1\rightarrow4)$
linkage.

Sucrose is a disaccharide familiar to most
as common table sugar. Sucrose is a very common
sugar in nature because it is the major transport
sugar in plants. (When sugar is transported from
one part of a plant to another, it is usually in the
form of sucrose.) Sucrose consists of one molecule
of glucose and one molecule of fructose, another
common six-carbon sugar (figure 2.3b).

Starch and Glycogen

Starch and **glycogen** are two glucose polymers
that are used in carbohydrate storage in living
things. Plant cells store extra carbohydrates as
starch; for instance, when leaf cells are actively
producing sugar by photosynthesis, they store
some of it in the form of starch in their own cells.

Other sugars are transported to storage organs,
especially in roots and modified stems, where the
sugar is incorporated into starch and reserved in
specialized storage cells.

Starch is a polymer of glucose that actually
occurs in two forms, α-**amylose** and **amylopectin.**
The first form, α-amylose, consists of unbranched
chains of α-glucose units connected by $\alpha(1\rightarrow4)$
linkages. Because of the shape of these linkages,
α-amylose molecules are generally coiled. Amylopectin molecules are branched. The backbone
or stem of an amylopectin molecule is constructed
of $\alpha(1\rightarrow4)$ linkages, but the branch points are
$\alpha(1\rightarrow6)$ linkages. In other words, at a branch
point, the number 1 carbon of one glucose is
linked to the number 6 carbon of another (figure
2.4). In amylopectin, branch points occur at intervals of every twenty-four to thirty glucose
units.

Glycogen or "animal starch" is yet another
glucose polymer. Glycogen functions in carbohydrate storage in animals (for example, human
liver cells contain considerable quantities of glycogen). The glucose units in the backbone of a
glycogen molecule are linked by $\alpha(1\rightarrow4)$ linkages, but glycogen is a more compact and highly
branched molecule than amylopectin; a glycogen
molecule branches at about every eight to twelve
glucose units.

Cellulose

Cellulose is an important structural polysaccharide because it forms a major part of the rigid
cell walls surrounding plant cells. Because of its
structural role in all plant cells, cellulose is the
most abundant organic compound in the world.

Cellulose is a polymer of β-glucose (figures
2.2a and 2.5). The $\beta(1\rightarrow4)$ linkages in cellulose
are linear, so cellulose molecules tend to form in
straight chains. When these chains lie side by
side, they interact through extensive hydrogen
bonding. The bonds hold cellulose molecules
firmly together to form the strong cellulose fibers,
called **microfibrils,** found in cell walls.

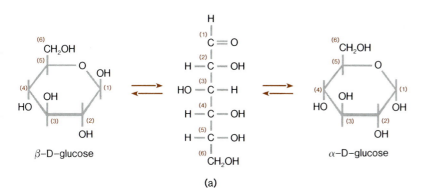

β-D-glucose

α-D-glucose

(a)

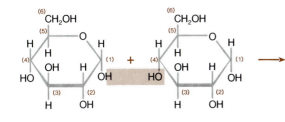

β-D-glucose

(b)

Figure 2.2 Representations of glucose molecular structure. (a) Conventional cyclic (ring) structures (Haworth projections) of β- and α-glucose with the straight-chain form between them. Note that in the ring structure, "corners" represent carbons and short, unlabeled lines show where hydrogens are attached. Colored numbers represent the conventional numbering scheme that permits identification of each carbon atom. The small D indicates that this is the D optical isomer of glucose, the more biologically important isomer. (b) A more accurate representation of the cyclic structure of β- D- glucose.

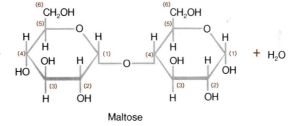

Maltose

(a)

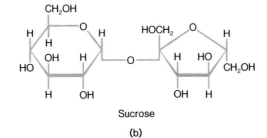

Sucrose

(b)

Figure 2.3 Disaccharides. (a) Maltose is formed from two molecules of glucose in a condensation reaction. Components of a water molecule (colored box) are removed as the bond is formed. Note that the α(1→4) bond is between the number 1 carbon of one glucose and the number 4 carbon of the other. (b) Sucrose, a disaccharide composed of one glucose molecule and one fructose molecule.

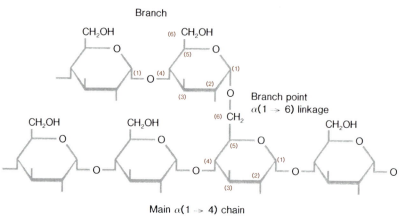

(a)

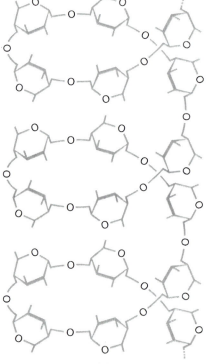

(b)

Figure 2.4 Starch. (a) A portion of an amylopectin molecule showing a branch point. Note that the branch point is an $\alpha(1{\rightarrow}6)$ linkage. The number 1 carbon of one glucose is linked to the number 6 carbon of another glucose. (b) Amylose is an unbranched chain like the main $\alpha(1{\rightarrow}4)$ chain of amylopectin shown. An amylose molecule actually coils to form a helix because of characteristics of the $\alpha(1{\rightarrow}4)$ linkages. Individual glucose units have been further simplified in this diagram. (c) Plant cells store carbohydrate as starch. This electron micrograph shows many starch grains inside chloroplasts in photosynthesizing cells in a corn *(Zea mays)* leaf. Plants also store considerable quantities of starch in specialized storage areas in roots or, in some cases, stems. (Magnification $\times$ 10,688)

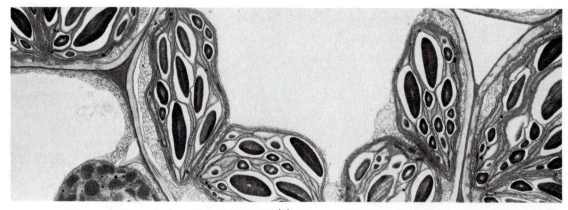

(c)

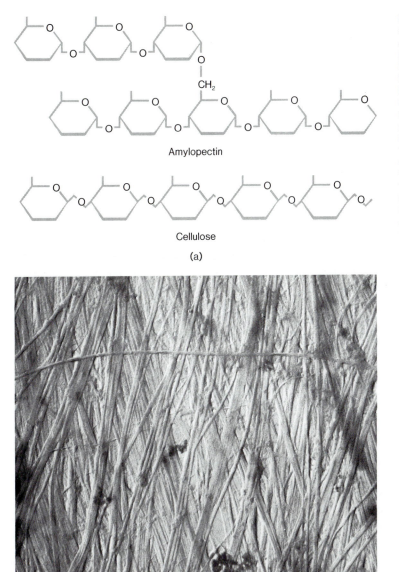

Amylopectin

Cellulose

(a)

Figure 2.5 Cellulose. (a) A highly diagrammatic comparison of the $\alpha(1{\rightarrow}4)$ linkages in starch (amylopectin) and the $\beta(1{\rightarrow}4)$ linkages in cellulose. Refer again to figure 2.2 to see the difference between α- and β-glucose. (b) A surface view of part of the cell wall of the alga *Valonia,* showing the arrangement of cellulose microfibrils. (Magnification $\times$ 27,200) (c) Animals that can derive nutritional value from cellulose. Some snails' digestive tracts secrete cellulases, enzymes that hydrolyze the $\beta(1{\rightarrow}4)$ linkages between glucose units in cellulose. Flagellated protozoans in the termite gut digest cellulose, thus permitting termites to thrive on a wood diet. Cows and other ruminant mammals (see chapter 10) have a permanent culture of cellulose-digesting microorganisms in the saclike rumen, a part of the stomach.

(b)

(c)

The cellulose microfibrils in each layer of a plant cell wall are parallel to one another, but the layers lie at different angles, adding to the strength of the wall. While a plant cell is growing, the cell wall is stretchable, expanding to allow cell growth, but eventually the wall becomes hard and very resistant to stretching. Thus, mature plant cells are enclosed in rigid boxes made of layers of cellulose fibers.

Although the variations between the kinds of linkages that join glucose molecules in various polysaccharides may seem insignificant, they actually make a great difference for animals that eat plants. Polysaccharides must be broken down into glucose molecules, or at least into disaccharides, before they can be absorbed in an animal's digestive system. The digestion of polysaccharides is accomplished by enzymes that catalyze hydrolysis of the bonds linking glucose molecules. These enzymes, however, are very specific in the types of bonds they will break. For example, enzymes that catalyze the hydrolysis of the bonds linking glucose units in starch will not catalyze the hydrolysis of cellulose.

Most animal digestive systems produce **amylases,** enzymes that split the $\alpha(1\rightarrow4)$ linkages in starch molecules. But **cellulases,** enzymes that break the $\beta(1\rightarrow4)$ linkages of cellulose molecules, are extremely rare in the animal kingdom. Because of the absence of cellulases, foods such as lettuce have practically no caloric value for humans and most other animals because they pass through the digestive tract with their cellulose virtually intact. But cellulose molecules would provide a great deal of energy if animals could use the glucose that cellulose contains.

Some animals do derive energy from the cellulose they eat. Certain snails and a very few other animals, for example, actually seem to produce cellulases and can digest cellulose. Other animals, such as termites, derive energy from cellulose through the microorganisms that live in their digestive tracts. These organisms produce cellulases, breaking down cellulose and making glucose available for themselves and their hosts.

Thus, when termites do their familiar damage to wood, which is mostly cellulose, they are dependent on the protozoans (single-celled organisms) in their digestive tracts to produce cellulase. Termites produce no cellulase of their own, and in fact, a young termite raised in sterile isolation from other termites will starve to death even when it is provided with all the cellulose it can eat. Apparently, termites actually must be "infected" with these gut-dwelling organisms early in life.

A ruminant mammal (such as a cow, sheep, or bison) has a special stomach chamber, the **rumen,** where a culture of microorganisms digests cellulose, making absorbable material available for the host. By this means, ruminant mammals convert the cellulose in plant material into edible meat. When people insist on eating animals that are fed grain—grain that could be used to nourish humans—they may aggravate the world's food-supply problem. Some biologists propose that humans adjust their tastes and learn to eat meat from basically grass-fed ruminants rather than grain-fed animals.

Microorganisms that produce cellulases may also eventually be used to help solve general energy supply problems. Scientists are trying to develop efficient processes for using inedible plant material to produce alcohol for fuel. But the structure of cellulose presents several obstacles that must be overcome before such processes will be practical. First, the hydrogen bonding that holds cellulose molecules together in fibers must be broken. This makes cellulose accessible and permits hydrolysis by cellulase, making glucose, the raw material for the fermentation reactions in alcohol production, available. There are a number of technical problems involved in conducting these reactions on a large scale, but finding the solutions seems worthwhile. The development of practical methods for alcohol production from cellulose would provide a large new source of liquid fuel and perhaps stop the use of edible grain in alcohol-production processes.

Proteins

It would be difficult to overemphasize the importance of proteins in living systems. Certain types of proteins form important parts of structural components in all cells and in spaces between cells. Other types of protein molecules act as enzymes; as the contracting elements in muscle cells and all other cells capable of movement; and as immunoglobulins, or antibodies, which are involved in resistance to infection. There are still other important structural and functional proteins that could be added to this list.

Proteins consist of precisely defined linear sequences of amino acids; that is, protein molecules are polymers made up of amino acids linked together by condensation reactions. An **amino acid** is an organic compound with an amino group (NH_2) and a carboxyl group (COOH) bonded to the same carbon atom. This particular carbon atom is designated as the alpha-carbon (α-carbon) and, therefore, amino acids in proteins are called α-**amino acids** (figure 2.6).

Although amino acids all have amino and carboxyl groups, different amino acids have a variety of side chains, (symbolized by R), which in turn have different chemical properties.

These various amino-acid side chains affect the shapes and properties of the protein molecules that contain them. Furthermore, individual proteins contain various amino acids in different sequences and proportions. As there are twenty types of amino acids, an enormous variety of protein molecules, all with different structures and properties, are possible. In this respect proteins contrast sharply with polysaccharides such as starch and cellulose, which are polymers constructed of only one type of monomer.

Amino acids are linked together in proteins by means of **peptide bonds,** which are formed by condensation reactions between carboxyl and amino groups. The formation and structure of peptide bonds are shown in figure 2.7. Because proteins are chains of amino acids connected by these bonds, proteins often are called **polypeptide chains** or simply **polypeptides.**

Most proteins are between about 40 or 50 and about 500 amino acids long, but there are some shorter and some longer polypeptides. Any one particular polypeptide always has exactly the same number of amino acids arranged in exactly the same sequence. This remarkable constancy in individual polypeptide structures proves to be an absolutely essential characteristic of protein structure and function.

Levels of Protein Structure

Proteins are not simple, straight chains of amino acids; their molecules are coiled, folded, and sometimes grouped together with other protein molecules in complex arrangements. This complexity has been cataloged into four structural levels called (quite logically) the primary, secondary, tertiary, and quaternary structures of proteins.

The **primary structure** of a protein is the sequence of the amino acids joined together by peptide bonds in the protein molecule. Determining the primary structure of a protein is almost like solving a chemical mystery story. To do so, it is necessary to remove just one amino acid (or a small group of amino acids) at a time from the end of the polypeptide chain for identification. Before this process was made much easier with modern analytical equipment, amino acid sequencing in proteins was an extremely difficult and painstaking job.

The first successful determination of the complete sequence of amino acids in a protein molecule was accomplished by Frederick Sanger and his colleagues at Cambridge University. For a period of about ten years, they selectively removed a number of small, overlapping peptide fragments of the **insulin** molecule (a protein hormone produced by the pancreas) and ascertained the sequence of amino acids in each fragment (figure 2.8). In 1953, Sanger and his associates were able to put together the pieces of the puzzle and describe the complete primary structure of this important hormone molecule; in 1958, Frederick Sanger received the Nobel Prize for his work. Although modern techniques for amino acid sequencing have made the work less laborious, the basic principles are still the same: proteins are sequenced by the hydrolysis of peptide bonds to produce peptide fragments for analysis.

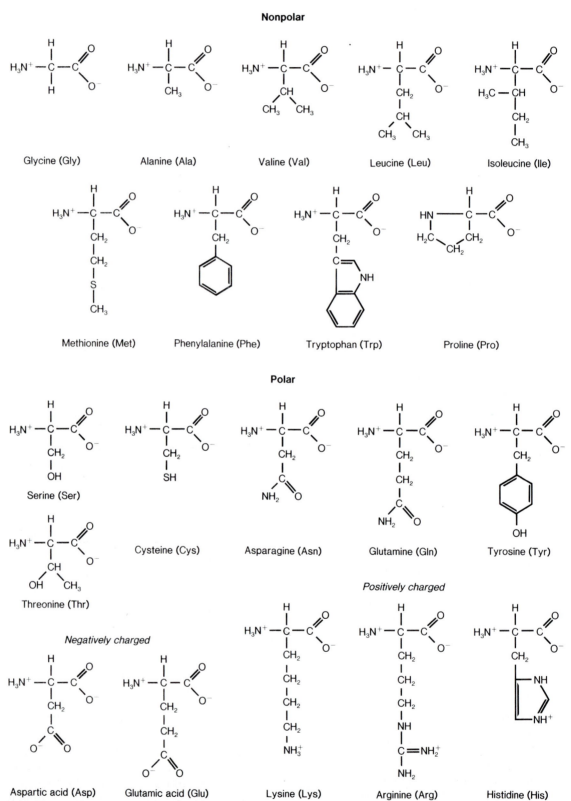

Figure 2.6 α-amino acids that occur in proteins. The amino and carboxyl groups are shown ionized as they are at intracellular pH. Three-letter abbreviations commonly used for the different amino acids are included. (Proline is, strictly speaking, an **imino acid** because its nitrogen is linked to two carbons (C—N—C).

Figure 2.7 Formation of the peptide bond from the amino and carboxyl groups (colored boxes) of two amino acids.

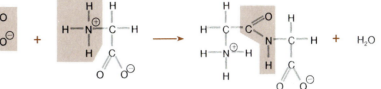

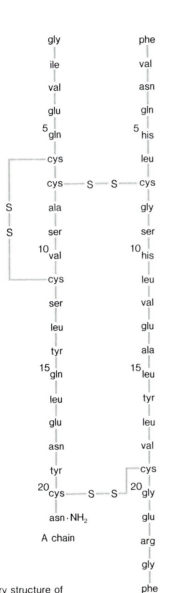

Figure 2.8 Primary structure of bovine (cow's) insulin. Two polypeptide chains are held together by hydrogen bonds (not shown) and by covalent disulfide bonds between cysteines. Note that there is also one disulfide bond between two cysteines of the A chain. Insulin actually is synthesized as part of a larger polypeptide, a portion of which is removed to make this active form (chapter 14).

The structure of insulin also illustrates another aspect of protein structure—that proteins often consist of more than one polypeptide chain. In insulin, two polypeptide chains are held together by two covalent bonds called **disulfide bonds** or **bridges** ($-S-S-$) and by numerous hydrogen bonds (not shown in figure 2.8). Disulfide bonds link units of the amino acid cysteine at two points in the insulin protein chains.

A polypeptide has a specific three-dimensional shape (conformation) that depends on several levels of organization beyond the primary structure. The **secondary structure** of a protein consists of specific three-dimensional folding arrangements within a polypeptide. These orderly arrangements are based on short-range interactions within the molecule and are repeated along a single axis. A common example of such a pattern is the alpha-helix (α-helix) found in many proteins (figure 2.9). The α-helix is a coil running in a single direction. A large number of weak bonds between amino acids in the polypeptide chain stabilize the α-helix.

The **tertiary structure** of a protein, which also is stabilized by a number of weak bonds between amino acids in different parts of the molecule, is formed when the polypeptide chain, with its orderly secondary structure, is bent and folded into an even more complex three-dimensional arrangement. One of the most powerful and useful techniques for determining the three-dimensional structure of a protein is **X-ray crystallography.** In this technique, an X-ray beam is passed through a crystal of the substance being studied. Part of the X-ray beam is scattered (diffracted) as it passes through the crystal; the way that it scatters depends on the molecule's structure. A photographic plate on the other side of the crystal records a pattern of spots representing the intensity of the emergent X rays (figure 2.10). This pattern reveals information about the locations of the various atoms in the crystal. This information can be used to determine the three-dimensional shapes of molecules.

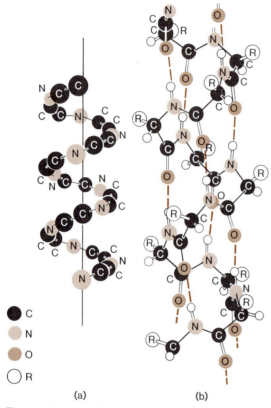

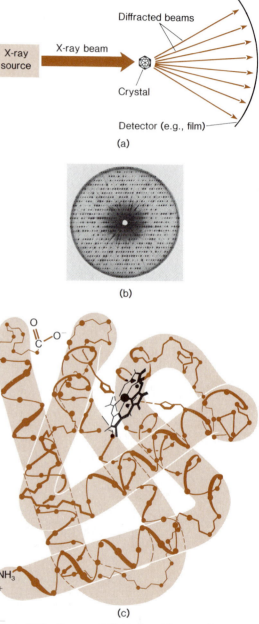

Figure 2.9 The alpha-helix (α-helix). (a) Carbon and nitrogen atoms are the "backbone" of the protein α-helix. The structure of the α-helix was discovered by Linus Pauling and Robert Corey. (b) More detailed model of the α-helix showing how it is stabilized by C-O---H-N hydrogen bonds (colored dashed lines). A few of the hydrogen atoms (small circles) are shown, including those that are involved in hydrogen bonding.

The form of a protein's amino acid chain in a crystal seems to be similar to its form in solution. Therefore, the information provided by X-ray crystallography is relevant to the study of protein structure in solution in living cells.

Forces Determining Protein Structure

Secondary and tertiary structures are formed by weak, noncovalent bonds. Three types of weak bonds are of particular significance—hydrogen bonds, London dispersion (van der Waals forces) and ionic bonds between charged amino-acid side chains.

The attractive forces associated with all of these types of bonds are relatively weak compared to the strengths of covalent bonds. Because their bond energies are so small, noncovalent

Figure 2.10 X-ray crystallography of the protein myoglobin, which reversibly stores oxygen in muscle tissue. The three-dimensional structure of myoglobin was the first protein structure to be determined by this technique. (a) Diagram of the X-ray diffraction technique. (b) X-ray diffraction picture of myoglobin. (c) Three-dimensional representation of myoglobin as it was determined by John Kendrew using X-ray diffraction techniques and complex mathematical analysis of the X-ray diffraction pictures that he obtained. Note the α-helix secondary structure and the folded tertiary structure. The heme group (which is involved in oxygen binding) is shown in black.

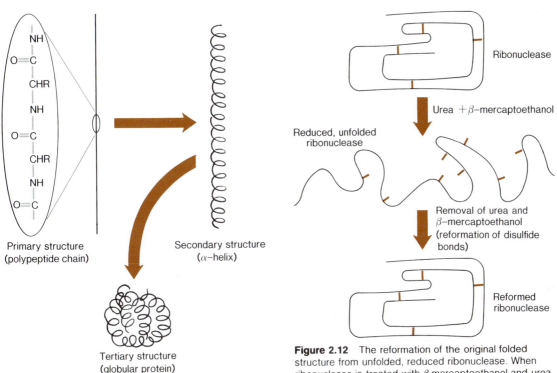

Figure 2.11 The relationship of primary, secondary, and tertiary protein structure. Expanded diagram of a small part of the polypeptide chain shows peptide bonds connecting several amino acids.

Figure 2.12 The reformation of the original folded structure from unfolded, reduced ribonuclease. When ribonuclease is treated with β-mercaptoethanol and urea, its disulfide (—S—S—) bonds (color) are broken and it is unfolded. If these reagents are carefully removed, the polypeptide chain will refold to form the native, catalytically active enzyme again. Thus, the sequence of amino acids in the unfolded chain determines the final structure of the refolded protein.

bonds constitute a significant factor in determining protein structure only when they are formed in relatively large numbers. Furthermore, the strength of these types of bonds decreases as their length increases. A protein molecule folds in such a way that the number of weak bonds is maximized and their bond lengths are optimized, or shortened. Thus, a protein chain, which may already contain stretches of α-helix, folds upon itself to form its tertiary structure, a more complex, often globular, shape (figure 2.11).

Weak bonds within protein molecules have another function beyond the establishment of secondary and tertiary structure; they are also involved in the interactions between proteins and other molecules. For example, the highly specific binding of enzymes with reactants depends on formation of many noncovalent bonds as does the very specific binding of antibodies with antigens. A protein will interact strongly with another molecule only if the surface shapes of the two molecules are complementary. That is, the two

molecules must fit together precisely—much like two puzzle pieces—before there will be a large enough number of weak bonds to make possible the formation of a stable compound. This need for a precise fit between molecules is the basis for the high degree of specificity that exists in the interactions between proteins and other molecules.

The studies of Christian Anfinsen on the enzyme ribonuclease revealed much about how a protein's complex structure comes into being. Ribonuclease, a digestive enzyme produced by the pancreas, hydrolyzes ribonucleic acid (RNA) in the digestive tract. It is a relatively small protein consisting of a single polypeptide chain made up of 124 amino acids and cross-linked by four disulfide bonds. Anfinsen discovered that when ribonuclease is treated with urea and β-mercaptoethanol (a reagent that breaks disulfide bonds), the peptide chain unfolds. When this happens, the protein is no longer able to act as an enzyme because its normal functioning depends upon its specific three-dimensional shape (figure 2.12).

Such a disruption of a protein's structure with an accompanying decrease in biological activity is called **denaturation.**

Anfinsen also discovered that if he slowly and carefully removed the urea and mercaptoethanol from the denatured ribonuclease, the disulfide bonds re-formed, and the polypeptide chains refolded into their proper secondary and tertiary structures—thus regaining their enzymatic ability. As a result of Anfinsen's experiments, it became clear that the amino acid sequence, which is the only aspect of the structure of ribonuclease left intact by this treatment, determines the structure of the fully folded protein. This has since been found to be true for other proteins as well.

The conclusion to be drawn from such experiments is that if the surrounding chemical conditions are appropriate, a protein molecule assumes its characteristic secondary and tertiary structure because of the attractive forces between various parts of the molecule. A polypeptide will fold so that it is in its most energetically stable condition, and this condition is reached when the maximum possible number of weak, noncovalent bonds are formed. This phenomenon is called the *principle of self-assembly* because the protein assumes its characteristic three-dimensional shape without any directing influence from outside the molecule.

Many proteins are constructed of two or more polypeptide chains, each with its own primary, secondary, and tertiary structures. These individual chains, called **subunits** or **protomers,** are associated with one another in specific ways. The spatial relationships among them are designated as the **quaternary structure,** the fourth level of protein structure. Proteins with quaternary structure, which, again, are really combinations of several individual polypeptides, are called **oligomeric** ("few unit") **proteins.** Interestingly, the subunits of an oligomeric protein, like the peptide chains in Anfinsen's experiments, assemble into an arrangement that permits the maximum number of weak bonds to form. In other words, under appropriate conditions, oligomeric proteins form spontaneously from their subunits, which are usually held together by noncovalent forces (although covalent bonds are also involved in some proteins). The formation of quaternary structure in proteins is thus another example of the principle of self-assembly in operation.

Fibrous Proteins

On the bases of structural organization and solubility, proteins can be divided into two major groups: **fibrous proteins** and **globular proteins.** Fibrous proteins are mainly structural proteins, while globular proteins generally have more dynamic functions involving interactions with other molecules (for example, enzymes are globular proteins). Fibrous proteins, unlike most globular proteins, are insoluble in dilute saline solutions. Characterized by parallel polypeptide chains lined up along an axis to form fibers or sheets, fibrous proteins are major components of the connective tissues of many living things.

Collagen is a good example of a fibrous protein. It is the most abundant protein in vertebrate animals (including humans), comprising as much as a quarter to a third of the total amount of protein in their bodies. Collagen is ideally suited for its connecting and supporting functions in the body because of its flexibility and great tensile strength (resistance to stretching).

Collagen fibers are found in all types of connective tissue, including ligaments, cartilage, bone, tendons, and the cornea of the eye. In these tissues, collagen is arranged in stable extracellular fibers to make the tissues resistant to tearing or other damage. In fact, it is possible to skin vertebrate animals primarily because of the large amount of collagen in the lower layer (dermis) of the skin; the collagen makes the skin so tough that it can be pulled away from underlying tissues without being torn.

How do stable collagen fibers develop in the spaces between cells? Once again, the structure of this protein is based on the principle of self-assembly. However, the self-assembly of collagen fibers takes place in spaces outside the cells. In the assembly process, three long polypeptide chains coil together to form a triple-stranded, helical, ropelike aggregate (figure 2.13). These triple-stranded protein aggregates, called **tropocollagen,** are rod-shaped and are about 280 nm (1 nanometer $= 10^{-9}$ meter) long. The three polypeptide strands are held together by hydrogen bonding (between the peptide NH and CO groups), by London dispersion forces and by covalent cross-links.

(a)

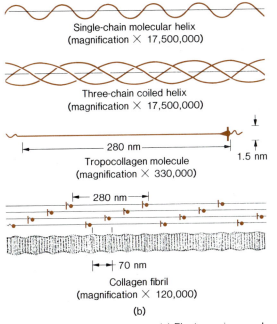

(Gly-X-Pro), (Gly-X-Hyp), (Gly-Pro-X)
Amino acid sequences in polypeptide chain

Single-chain molecular helix
(magnification $\times$ 17,500,000)

Three-chain coiled helix
(magnification $\times$ 17,500,000)

280 nm

Tropocollagen molecule
(magnification $\times$ 330,000)

1.5 nm

280 nm

70 nm

Collagen fibril
(magnification $\times$ 120,000)

(b)

Figure 2.13 Collagen structure. (a) Electron micrograph of skin collagen fibers that have been coated with metal. The repeat pattern (with a spacing of 70 nm) is created by the stacking of tropocollagen molecules in staggered, parallel arrays. (Magnification $\times$ 25,194) (b) The structure of collagen. Single polypeptide chains with often-repeated sequences (especially ones involving the amino acids glycine, proline, and hydroxyproline) are coiled to form a helix. Three of these helical polypeptide chains are then wound about one another to form a large, triple-stranded tropocollagen molecule. Tropocollagen molecules finally associate together in a staggered, parallel arrangement to produce a collagen fiber such as those seen in (a).

Tropocollagens serve as subunits in collagen assembly; collagen itself is composed of parallel tropocollagen units stacked end to end. Each tropocollagen unit overlaps with the next about a quarter of the way down its length, and the tropocollagen subunits are covalently cross-linked to one another. The staggered arrangement of tropocollagen subunits, together with the covalent cross-linking between them, makes collagen fibers very strong and stable.

In sum, collagen consists of polypeptide chains that first aggregate to form tropocollagen subunits. These subunits then assemble to form collagen, which means that collagen is really a composite of many individual polypeptide chains.

Globular Proteins

In contrast to fibrous proteins, the polypeptide chains in globular proteins are coiled into compact, spherical shapes. Globular proteins are much more flexible and fulfill more dynamic functions in organisms than fibrous proteins, such as collagen. Enzymes, antibodies, and transport proteins are all globular proteins.

The oxygen-transporting protein **hemoglobin** is a good example of a globular protein. Details of the three-dimensional structure of hemoglobin were determined through the X-ray analysis of Max Perutz and his colleagues at Cambridge. Hemoglobin is an oligomeric protein that has four subunits or polypeptide chains—two α- and two β-chains. The hemoglobin molecule is roughly spherical, with each α-chain in close contact with the two neighboring β-chains. Attached to each subunit, in a specific position, is a single **heme** group, containing an iron atom that is capable of binding with and separating from an oxygen molecule (figure 2.14).

Even though the amino acid sequences of hemoglobin polypeptide chains vary from species to species, the overall conformations of all hemoglobin molecules are very similar. However, the substitution of certain amino acids in the polypeptide chains of hemoglobin can have serious consequences. For example, a mutation that causes a polar amino acid to replace a nonpolar one can seriously disrupt the folding of a polypeptide chain and the structure of the entire hemoglobin molecule, thus interfering with its normal function in oxygen transport.

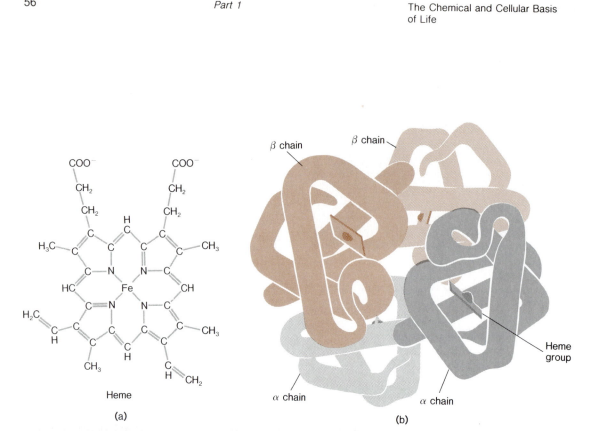

(c)

Figure 2.14 Hemoglobin structure.
(a) The heme group of hemoglobin.
This large, flat molecule has an iron
atom complexed with its central
nitrogens. The iron atom can
reversibly bind oxygen molecules. It
is this binding that makes it possible
for hemoglobin to function as an
oxygen-transporting molecule.
(b) The structure of oxygenated
hemoglobin. The heme groups are
pictured as rectangles with spherical
iron atoms at their centers. The
binding of oxygen to one heme
group changes the interactions of
the subunits with one another and
thus stimulates the binding of
oxygen by the other heme groups.
(c) Red blood cells of a person
suffering from sickle-cell anemia (left)
compared with normal red blood
cells (right). One amino acid
substitution in the hemoglobin
β-chains changes the properties of
sickle-cell hemoglobin and causes
the cell distortion that can result in
severe circulatory problems.

A striking example of the effect of amino
acid substitution in hemoglobin is **sickle cell
anemia.** In this human disease, a single amino
acid substitution in the β-chains (valine instead
of the glutamic acid normally present) has disastrous effects. When oxygen molecules leave the
hemoglobin of a person suffering from sickle cell
anemia, the defective hemoglobin molecules
change shape, actually altering the shape of
whole red blood cells; the normally spherical cells
elongate into "sickles" that have trouble passing
through small vessels (figure 2.14c). This difficulty can cause a painful sickle cell crisis, which
occurs when circulation to various body areas is
blocked and tissues are damaged. Thus, the dramatic consequence of the substitution of a single
amino acid in the β-chains of hemoglobin shows
clearly that higher levels of protein structure
arise from the primary structures, the specific
amino acid sequences, of polypeptides.

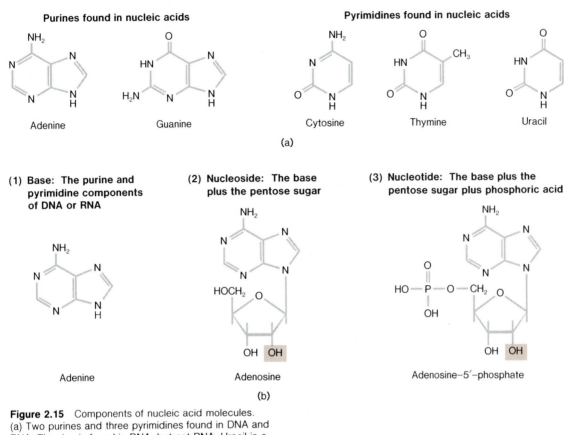

Figure 2.15 Components of nucleic acid molecules.
(a) Two purines and three pyrimidines found in DNA and
RNA. Thymine is found in DNA, but not RNA. Uracil is a
component only of RNA. (b) Nucleic acids are polymers
of nucleotides that are composed of purine or pyrimidine
bases, pentose sugars (ribose or 2-deoxyribose), and
phosphate. The diagram shows components of RNA. In
DNA, the OH in the colored box is replaced by H.

Nucleic Acids

Nucleic acids are macromolecules that carry genetic information. Structurally, nucleic acids are linear polymers of **nucleotides.** Each nucleotide is composed of three substances: a nitrogen-containing organic base (either a **purine** or a **pyrimidine**), a five-carbon sugar, and phosphoric acid (figure 2.15). The sequence in which these purine- or pyrimidine-containing nucleotides are arranged determines the genetic coding contained in nucleic acids.

There are two major types of nucleic acids, **deoxyribonucleic acid (DNA)** and **ribonucleic acid (RNA).** The first, DNA, contains the sugar 2-deoxyribose, and the bases adenine, guanine, cytosine, and thymine. The composition of RNA differs from that of DNA in two major ways: the pentose (five-carbon) sugar found in RNA is ribose rather than deoxyribose, and RNA contains uracil rather than thymine. As figure 2.15 shows, adenine and guanine are purines, while cytosine, thymine, and uracil are pyrimidines. Bases and sugars are linked together by phosphate groups to form polynucleotide chains in both DNA and RNA (figure 2.16).

DNA polynucleotide strand structure

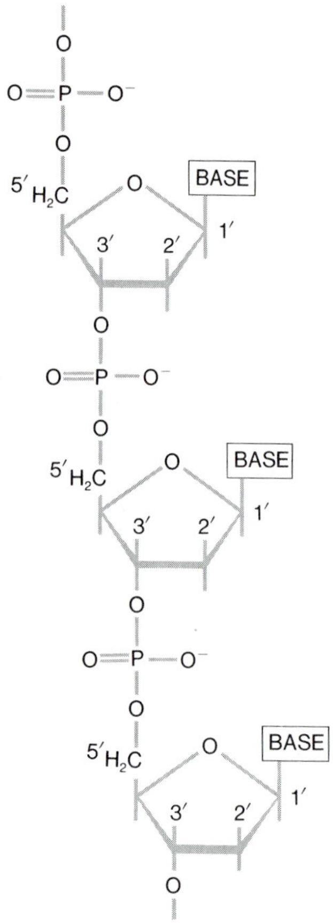

Figure 2.16 A short stretch of DNA. Each nucleoside in the polymer is linked to neighboring nucleosides by phosphate groups. These phosphates connect the 3′ carbon of one sugar with the 5′ carbon of the adjacent nucleoside sugar. The structure of an RNA strand is similar except that ribose replaces the deoxyribose.

The Search for Nucleic Acid Structure

The basic chemical composition of nucleic acids was elucidated in the 1920s through the efforts of P. A. Levene. However, at that time Levene and others believed that DNA was a very small molecule, probably only four nucleotides long, with its four bases occurring in equal proportions and arranged in a fixed, unchangeable sequence. This early view may partly explain why biologists were confident for years that nucleic acids were too simple in structure to convey complex genetic information. They concluded that genetic information must be coded in proteins rather than in nucleic acids.

But the 1940s witnessed the development of significant new analytical techniques in chemistry that led to great advances in the research on nucleic acid structure. One of the most important of the new developments was the invention of paper chromatography by Martin and Synge between 1941 and 1944. By 1948, chemist Erwin Chargaff had begun using paper chromatography to analyze the base composition of DNA from a number of species, and he soon found that the early work on DNA structure was in error. Chargaff found that the base composition of DNA varies from species to species, just as one would expect from a substance carrying genetic material. Furthermore, the chemist demonstrated that there were two consistencies in DNA composition, regardless of source: the total amount of purines always equals the total amount of pyrimidines, and both the adenine/thymine and guanine/cytosine molar ratios are 1.0 (table 2.1). In other words, adenine content equals thymine content ($[A] = [T]$) and guanine content equals cytosine content ($[G] = [C]$). These findings came to be known as *Chargaff's rules* and were the key to much later research.

The year 1951 marked another turning point in the research on DNA structure. In that year, Rosalind Franklin arrived at King's College, London, and joined Maurice Wilkins in efforts to prepare highly oriented DNA fibers and to study them by X-ray crystallography. This work culminated in the winter of 1952 to 1953 with the production of Franklin's X-ray diffraction photograph of DNA (figure 2.17).

It was also in 1951 that an American biologist, James Watson, went to Cambridge University and began to work with Francis Crick on DNA structure. After Watson and Crick had made a series of unsuccessful attempts to unravel the structure of DNA, Franklin's data finally provided them with the necessary clues. The cross-pattern of X-ray reflections in Franklin's photograph told Watson and Crick that DNA is helical. The black areas at the top and bottom of the photograph indicated that the purines and pyrimidines are regularly stacked next to each other, at a distance of 0.34 nm. Franklin had already concluded that the phosphate groups lay to the outside of the helical structure. It also became clear from her X-ray data and her determination of the density of DNA that the helix contained two strands, not three or more (as some chemists had proposed).

Table 2.1
DNA Base Composition.

Organism	Molar Ratio of Bases				$\dfrac{A + G}{C + T}$
	Adenine	Guanine	Cytosine	Thymine	
Escherichia coli	24.6	25.5	25.6	24.3	1.00
Saccharomyces cerevisiae (yeast)	31.3	18.7	17.1	32.9	1.00
Carrot	26.7	23.1	23.2	26.9	0.99
Rana pipiens (Frog)	26.3	23.5	23.8	26.4	0.99
Human liver	30.3	19.5	19.9	30.3	0.99

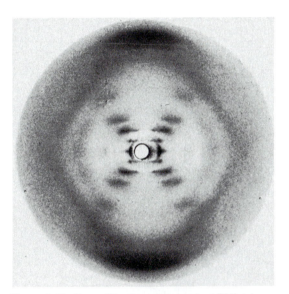

Figure 2.17 X-ray diffraction photograph of DNA taken by Rosalind Franklin. The crossing pattern of dark spots in the center of the picture indicates that DNA is helical. The dark regions at the top and bottom of the photograph show that the purine and pyrimidine bases are stacked on top of one another and are 0.34 nm apart.

Watson and Crick gathered information on DNA from diverse sources without actually doing any experiments themselves. They combined Chargaff's discoveries on base composition, Franklin's X-ray data, and various predictions of how genetic material might logically be expected to behave in constructing a model for DNA. Watson and Crick knew that the DNA double helix would have to be a smooth cylinder of constant diameter regardless of the base sequence. Chargaff's rule that [A] = [T] and [G] = [C] would hold true if each purine (adenine or guanine) was hydrogen bonded with its corresponding pyrimidine (thymine or cytosine). In fact, Watson and Crick found that only the A-T and G-C combinations would bond properly and not result in a helix of uneven diameter. Finally, when a complete model was constructed, it had to be compatible with the data obtained from Franklin's X-ray photographs.

The structure that Watson and Crick proposed is known as the double-helix model of DNA. Their model looks somewhat like a twisted ladder, with two polynucleotide strands running in opposite directions and winding about each other, and the A-T and G-C base pairs forming the "rungs" in the interior of the helix (figure 2.18). A purine (for example, adenine) on one strand is hydrogen bonded with the complementary pyrimidine (in this case, thymine) on the opposite strand.

The helix is 2.0 nm in diameter, and it makes a full spiraled turn every 3.4 nm, or every ten base pairs. Thus, the helix rotates about its longitudinal axis 36° with each base pair or rung; the spiral winds through a full circle (360°) in the length of helix occupied by ten base pairs.

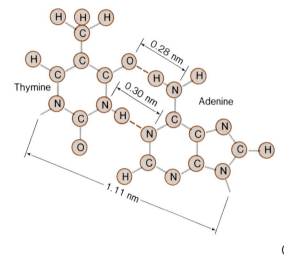

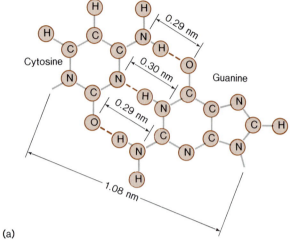

(a)

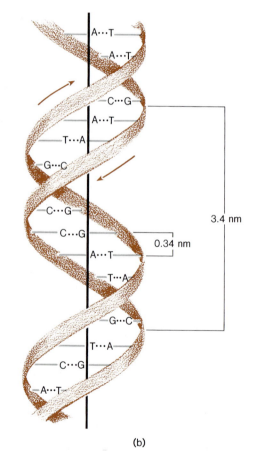

(b)

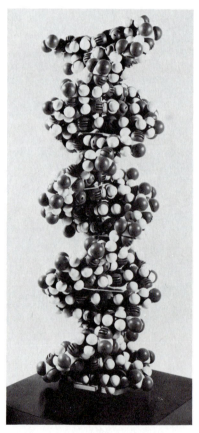

(c)

Figure 2.18   DNA structure. (a) Association of bases by
hydrogen bonding. (b) Diagrammatic structure of the
DNA double helix. The DNA double helix can be
visualized as a twisted ladder. (c) Space-filling model of
DNA.

Figure 2.19 Lipids. (a) Glycerol and a fat. Biochemists call fats triacylglycerols or triglycerides because a fat molecule is formed, by condensation, from glycerol and three molecules of fatty acid. (b) An example of a phospholipid, a phosphatidylcholine (lecithin). Such phospholipids are important constituents of biological membranes.

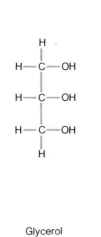

Glycerol

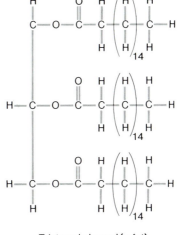

Tristearyl glycerol (a fat)

(a)

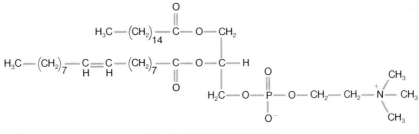

Phosphatidylcholine (lecithin)

(b)

The DNA helix is stabilized by hydrogen bonding between the A-T and G-C base pairs, by London dispersion forces between the stacked bases, and through general hydrophobic interactions that favor a structure with the bases to the inside and the sugar-phosphate backbone in contact with the surrounding water.

The overall structure of DNA with its hydrogen-bonded base pairs has enormous genetic implications because it provides a mechanism by which genetic material could be replicated. The two strands of the DNA molecule can separate and a new strand, complementary to each of the lone strands, can be made so that two new and identical DNA molecules exist where there was one. These implications, immediately recognized by Watson and Crick, were alluded to in their original paper: "It has not escaped our notice that the specific pairing we have postulated immediately suggests a possible copying mechanism for the genetic material." Thus, the work of five scientists—Watson, Crick, Franklin, Wilkins, and Chargaff—transformed the study of biology.

Lipids

Lipids are structurally much more diverse than polysaccharides, proteins, or nucleic acids. In fact, the term lipid is not much more than a convenient name for any compound in the cell that is insoluble in water but soluble in hydrophobic solvents. Some lipids are large molecules, but unlike polysaccharides, proteins, and nucleic acids, they are not polymers (chains of repeated smaller units).

Lipids resemble polysaccharides in fulfilling both structural and storage-compound roles. Fats, which are condensation products of glycerol and long-chain fatty acids, are used as storage compounds (figure 2.19). Because of variations in the number of carbon atoms, in the number of double bonds (C=C bonds), and in the positions of these bonds in the fatty acids, a wide range of possible structures exists for fats.

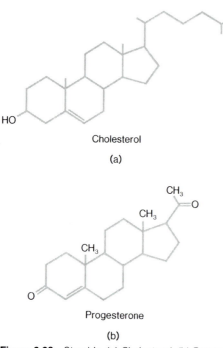

Cholesterol

(a)

Progesterone

(b)

Figure 2.20 Steroids. (a) Cholesterol. (b) Progesterone, a hormone. Progesterone and a number of other hormones are built around the basic cholesterol structure.

Since the components of fatty acid chains are generally hydrophobic, fat molecules are insoluble in water. However, if one of the fatty acids is replaced by a group that has an affinity for water, as is the case in phospholipids, the situation becomes more complex (figure 2.19b). The presence of both hydrophobic and hydrophilic tendencies in such compounds is important in the formation of biological membranes.

Biological membranes also contain lipids other than phospholipids; one of the most important is the **steroid** cholesterol (figure 2.20). Steroids contain four fused rings of carbon atoms in a characteristic arrangement. The basic steroid structure is also the foundation for many other biologically important molecules, such as the steroid hormones of vertebrate animals.

Membranes

Modern cell biology strongly emphasizes the importance of membranes in cell organization. The old-fashioned view of the cell as a sac of watery soup enclosed in a membrane has been replaced by the concept of an extremely complex array of membranes not only enclosing the cell, but arranged throughout the interior of the cell. In the most general terms, a biological **membrane** is a thin, hydrophobic layer separating two aqueous phases (areas with materials dissolved and suspended in water). Membranes therefore serve to separate the cell into a number of individual compartments since water-soluble molecules in one area cannot readily pass through a hydrophobic membrane into another area. But membranes are not just passive barriers; they can control the passage of molecules, and the membranes themselves are the site of some biochemical reactions.

Biological membranes are so thin that they cannot be seen with a light microscope. Biologists had to turn to physical and chemical analysis of membrane structure long ago. E. Overton proposed in 1895 that lipids were an essential membrane constituent, because he had noted that molecules could penetrate the **plasma membrane** (the membrane enclosing the cell) at rates that were closely correlated with their lipid solubility. In general, the more lipid-soluble a substance was, the faster it could enter a cell.

Overton's proposal was extended in 1925 by Gorter and Grendel. These scientists isolated the plasma membranes of erythrocytes (red blood cells) and extracted their lipids. They calculated that each erythrocyte membrane contained enough lipid to form a layer two molecules thick around the cell, and so they suggested that cell membranes are basically composed of a bimolecular lipid layer. But their model was a little too simple, since there was also evidence for the presence of proteins in membranes. In addition, in the 1930s Davson and Danielli carefully measured the surface tension of cell membranes and found that it was much lower than would be expected with a membrane composed of lipid only. They postulated that the lipid bilayer was covered on both surfaces with protein. In other words, Davson and Danielli were saying that membranes resemble protein-lipid "sandwiches" (figure 2.21).

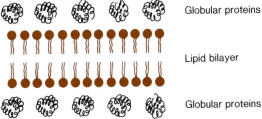

Globular proteins

Lipid bilayer

Globular proteins

Figure 2.21 The Davson-Danielli membrane model. The hydrophilic lipid heads are indicated by circles and the hydrophobic tails by lines. The hydrophobic tails are buried inside the membrane, away from water.

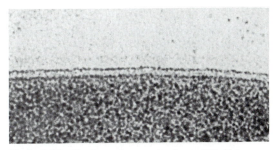

Figure 2.22 Electron micrograph of a portion of a plasma membrane showing the two outer electron-dense layers separated by a less dense central zone. (Magnification × 250,000)

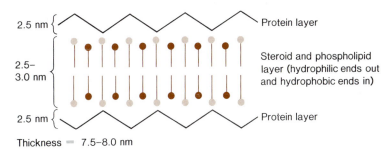

2.5 nm { — Protein layer

2.5–
3.0 nm { — Steroid and phospholipid layer (hydrophilic ends out and hydrophobic ends in)

2.5 nm { — Protein layer

Thickness = 7.5–8.0 nm

Figure 2.23 The unit membrane model of J. D. Robertson. This model differed from the Davson-Danielli membrane model because it proposed that proteins on the surface of membrane lipids are extended rather than globular. Robertson further proposed that these protein layers account for the two electron-dense lines in electron micrographs of membranes (figure 2.22).

Since that time, direct chemical analysis of isolated membranes has shown that both proteins and lipids are, in fact, present in membranes. For example, the erythrocyte plasma membrane is 60 percent protein and 40 percent lipid by weight. Over half of the lipid in an erythrocyte membrane is composed of phospholipids like phosphatidylcholine, and another one-quarter is made up of the steroid cholesterol (figure 2.20).

These membrane lipids are structurally asymmetric, with polar and nonpolar ends. The polar ends interact with water and are hydrophilic; the nonpolar, hydrophobic ends are insoluble in water and tend to associate with one another. In the Davson-Danielli model, the proteins and the hydrophilic ends of the lipids contact water. The hydrophobic tails are buried in the lipid bilayer, away from water (figure 2.21).

The Unit Membrane Model

The development of the electron microscope made it possible, at long last, to observe the plasma membrane directly. When membranes are treated with osmium tetroxide and viewed through an electron microscope, a membrane structure that looks somewhat like a sandwich,

about 7.5 to 8.0 nm thick, can be seen. This membrane consists of two electron-dense layers separated by a less dense central zone (figure 2.22).

J. D. Robertson used such observations to develop the **unit membrane theory** in the late 1950s. Robertson's theory is basically an elaboration of the Davson-Danielli model; it proposes that the two outer layers of a membrane consist of proteins in an extended (rather than globular) conformation. The central zone, theorized Robertson, is a lipid bilayer with its molecules oriented with their hydrophilic ends outward (figure 2.23). This three-layered structure was thought to be characteristic of all cell membranes.

Although the unit membrane theory has been a popular and fruitful concept, it is unsatisfactory (at least in its simplest form) for a number of reasons. For one, scientists now know that there is great variation in the chemical compositions of various membranes. The enzyme contents and functions of membranes are also variable. These differences are hard to reconcile with the simple, relatively uniform three-layer structure Robertson proposed for all membranes.

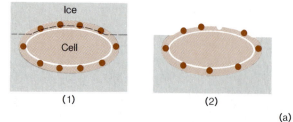

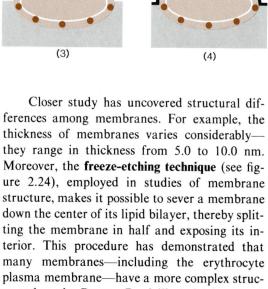

(b)

Figure 2.24 (a) The freeze-etch procedure. (1) The cells or tissue are frozen in liquid nitrogen at −196°C. (2) The specimen is then cut with a very cold microtome knife. The knife fractures it along lines of weakness, such as the center or face of membranes. (3) The specimen can then be etched to expose more intracellular surfaces and details. This is accomplished by subliming ice away at temperatures above −120°C. (4) A replica is made of the exposed structures by coating them with platinum and carbon. The replica may then be examined in the electron microscope. Particles within membranes are exposed and clearly visible in the replica. A and B correspond to the parts of the electron micrograph in part b. (b) This electron micrograph of a freeze-etched red blood cell plasma membrane shows the difference between the interior and the outer surface of the membrane. The interior of this membrane (surface A) has numerous globular particles. In contrast, the cell membrane's outer face (surface B) is relatively smooth. (Magnification × 27,170)

Closer study has uncovered structural differences among membranes. For example, the thickness of membranes varies considerably—they range in thickness from 5.0 to 10.0 nm. Moreover, the **freeze-etching technique** (see figure 2.24), employed in studies of membrane structure, makes it possible to sever a membrane down the center of its lipid bilayer, thereby splitting the membrane in half and exposing its interior. This procedure has demonstrated that many membranes—including the erythrocyte plasma membrane—have a more complex structure than the Davson-Danielli model predicted. Finally, the small, globular particles seen in these membranes (figure 2.24b) are now thought to be membrane proteins. This observation raises questions about the unit membrane theory because it indicates that membrane proteins actually lie within the membrane lipid bilayer and do not simply coat the surface of the membrane, as was earlier supposed.

The Fluid Mosaic Model

These discoveries have led to alternate models of membrane structure. One of the most widely accepted is the **fluid mosaic model** of S. J. Singer and G. L. Nicholson. Other current models are fairly similar to Singer and Nicholson's.

All evidence still indicates that membranes have a bimolecular lipid layer. The lipids are arranged with their hydrophilic ends at the surface and their hydrophobic ends in the interior of the membrane, just as earlier models suggested. But membrane proteins definitely do not spread in sheets over the surfaces of the lipid layers—they are globular.

Singer distinguishes between two types of membrane proteins. **Peripheral** or **extrinsic proteins** are relatively loosely connected to the membrane surface and can be easily removed. They are soluble in aqueous solutions and make up around 20 to 30 percent of the total membrane

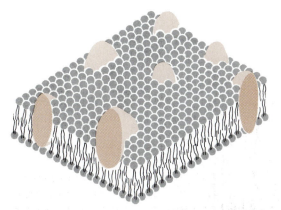

Figure 2.25 This diagram of the fluid mosaic model of membrane structure shows the integral proteins floating in a lipid bilayer. Some of these integral proteins extend all the way through the lipid layer. Extrinsic proteins have been omitted for reasons of clarity.

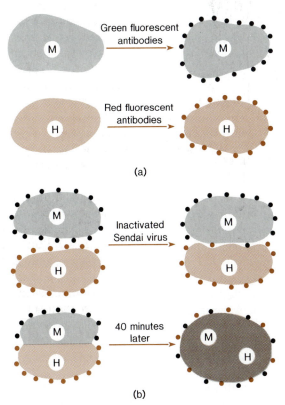

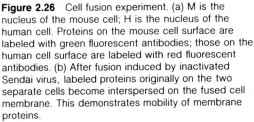

Figure 2.26 Cell fusion experiment. (a) M is the nucleus of the mouse cell; H is the nucleus of the human cell. Proteins on the mouse cell surface are labeled with green fluorescent antibodies; those on the human cell surface are labeled with red fluorescent antibodies. (b) After fusion induced by inactivated Sendai virus, labeled proteins originally on the two separate cells become interspersed on the fused cell membrane. This demonstrates mobility of membrane proteins.

protein. The other 70 to 80 percent of membrane proteins are **integral** or **intrinsic proteins.** These proteins are not easily extracted from membranes; in fact, detergents must be used to remove them.

Membrane proteins, like membrane lipids, are both hydrophobic and hydrophilic in character. Their hydrophobic regions are buried in the lipid bilayer, while the hydrophilic portions project from the membrane's surface. Some integral proteins even extend all the way through the lipid bilayer (figure 2.25). Integral proteins can move laterally around the membrane's surface, but they seldom if ever flip-flop through the lipid from one side of the membrane to the other. In other words, they can "float" laterally to new locations on the cell surface, but they cannot rotate through the lipid layer.

Lipid molecules can also move laterally to new positions within the membrane, but like membrane proteins, they do not seem to move readily from one side of the membrane to the other. If a lipid moleclule is on the interior side of the bilayer enclosing a cell, it will probably not move to the side of the bilayer facing the outside of the cell. This may be because the lipids on the two sides of membranes are chemically different. That is, the lipid composition of one half of the bilayer may differ from that of the other half.

The dynamic nature of the cell membrane has been strikingly demonstrated by experiments with **cell fusion.** To perform these experiments, membrane proteins of mouse cells are labeled with green fluorescent antibodies, while those of human cells are labeled red in the same way. This treatment makes it possible to trace the movements of membrane proteins and to keep track of which proteins were originally in the human cell membrane and which were in the mouse cell membrane. The two cell populations are next mixed in the presence of inactivated Sendai virus. Virus particles adhere to the cell surfaces and "glue" the two cells together; the result is a fusion of the cell membranes to form a single cell hybrid. At first the two types of color-labeled proteins lie in separate halves of the single fused cell. However, within about forty minutes the red-and-green-labeled proteins are distributed evenly over the outer cell surface. This experiment clearly demonstrates that membrane structure is dynamic: membrane proteins are not fixed in one position, they can move to new positions in the membrane of a hybrid cell (figure 2.26).

Some cell biologists have stated that membrane proteins are like "icebergs floating in a sea of lipids," but this analogy is somewhat inadequate. Although membrane proteins often do move within the lipid bilayer, in other cases they remain relatively fixed. Several different mechanisms are involved in controlling the positions of membrane proteins. Strong interactions among membrane protein molecules themselves may hold some proteins together in specific spatial relationships to one another. Other proteins may be held in place by the elements of a network of tubules and filaments that exist inside cells (chapter 3).

The functions of membrane proteins go beyond the simple maintenance of membrane structure. For example, plasma membrane proteins are responsible for the cell's active intake of nutrients. Other integral proteins of the plasma membrane serve as specific receptors for hormones and other molecules that transmit "messages" (see chapter 14); for example, insulin-receptor proteins can be isolated from the plasma membranes of cells that respond to insulin. Membrane proteins in some of the structures inside cells serve important functions in energy conversions during cell respiration and photosynthesis.

The Glycocalyx

Cell surfaces are being studied intensively, and knowledge about the outer surfaces of the plasma membranes surrounding cells is growing rapidly. The **glycocalyx**—the outer, carbohydrate-containing portion of a cell, sometimes called the **cell coat**—is a main focus of this research. The term *glycocalyx* is derived from the characteristics of some cell surface components: *glyco* ("carbohydrate") and *calyx* ("outer"). The terms **glycoprotein** and **glycolipid** are used to describe the plasma-membrane proteins and lipids covalently bonded to oligosaccharides (carbohydrates containing a few sugars) in the cell surface (figure 2.27).

The oligosaccharides of the glycocalyx are quite different from the carbohydrates discussed earlier in this chapter because these cell-surface oligosaccharides contain mixtures of sugars and sugar derivatives. Various functional groups such as amino groups and sulfates are attached to the sugars of the glycocalyx.

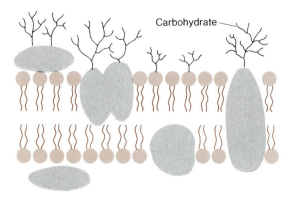

Figure 2.27 The glycocalyx. The oligosaccharide portions of glycoproteins and glycolipids are in black.

Glycoproteins are especially interesting because they appear to be involved in several kinds of cell recognition processes. For example, it is believed that during embryonic development, different kinds of cells develop specific types and patterns of glycoproteins on their surfaces. Normal development may depend on cells being able to "recognize" each other on the basis of these patterns and form the lasting connections that lead to normal tissue organization.

One human cell-surface glycoprotein that has been studied in detail is **glycophorin,** which plays a role in determining ABO blood type. Vincent Marchesi and his colleagues have determined the sequence of the 113 amino acids that make up the single polypeptide chain of glycophorin and have examined glycophorin's orientation in the membrane. The hydrophilic portion of glycophorin, projecting into the cell's environment, has sixteen branched oligosaccharide chains attached to it, with each chain containing from four to fifteen sugars. In people with blood type A, certain of these chains end with one set of sugar derivatives, and in people with blood type B, the chains end with other sugars. People with type O blood do not have the terminal sugars at all.

Cell recognition processes involving glycocalyx molecules form the basis for a number of other types of cell interactions, some of which are medically important. For example, tissue and organ transplants depend on proper matching of cell surface recognition factors. Viruses and bacteria selectively attach to specific cell-surface molecules when infecting cells. Certain parasites actually seem to incorporate host cell-membrane components into their own surface membranes so that they escape recognition by the host's normal defense mechanisms. And apparently cancer cell surfaces are very different from normal cell surfaces; certainly, some properties of cancer cells will be better understood when more is known about their surfaces.

Considering all of the important functions and relationships of the glycocalyx, it is small wonder that many biologists consider the study of cell-surface properties one of the most important areas of modern biological research.

Summary

Several major types of biologically important organic molecules are especially large and are called macromolecules. These macromolecules are polymers (chains of individual molecular units called monomers) and are formed by condensation reactions. Hydrolysis reactions perform the opposite role by separating the individual units of polymers.

Polysaccharides are polymers of sugar molecules. Monosaccharides (individual sugar molecules), disaccharides, and polysaccharides are all carbohydrates. The major polysaccharides are all polymers of glucose, which is a six-carbon sugar molecule.

The linkages that connect glucose molecules in starch and glycogen are different from those in cellulose. These differences give the various glucose polymers different properties and affect their digestion by the animals that consume them.

Proteins are polymers of amino acids. An amino acid has a carboxyl group and an amino group attached to the same carbon. Amino acids are linked together by peptide bonds into polypeptides.

The primary structure of a polypeptide is the sequence of amino acids in the protein. Secondary structure involves the orderly folding of the amino acid chain, in forms such as the α-helix. The folded polypeptide is then folded further into the tertiary structure of the protein; secondary and tertiary structures arise from weak interactions, such as hydrogen bonding, among the amino acids in the polypeptide. Many proteins are constructed of more than one polypeptide chain. The spatial relationships among these polypeptide subunits constitute the quaternary structure of proteins.

Fibrous proteins such as collagen tend to be structural proteins, while globular proteins function in dynamic roles such as enzymes, antibodies, and transport molecules.

Nucleic acids are polymers of nucleotides. A nucleotide is composed of three substances: a purine or pyrimidine, a pentose sugar (ribose or deoxyribose), and phosphoric acid. For a long time, nucleic acids were regarded as rather simple, uninteresting compounds. But a long search for nucleic acid structure reached a climax with the development of the Watson-Crick model of DNA structure. This model suggested that DNA carries and transmits genetic information, and thus, the model shaped much of modern molecular biology research.

Lipids are cell constituents that are insoluble in water. One of the most vital roles of lipids is in cell membranes. The current fluid mosaic model describes membranes as bimolecular lipid layers with protein molecules moving about in them.

The plasma-membrane surface with its glycoproteins and glycolipids is only partly understood. Many normal and abnormal cell interactions depend upon characteristics of the glycocalyx. Certainly, further investigation of the glycocalyx will do much to clarify the understanding of the dynamic structures and functions of biological membranes.

Questions

1. Although the general formula for carbohydrates is $(CH_2O)_n$, the molecular formula for sucrose and several other disaccharides is $C_{12}H_{22}O_{11}$. Explain this in terms of condensation reactions.

2. What biochemical evidence would convince a person concerned about world food shortages that eating grass-fed cattle does not directly deprive the hungry world of food in the same way that eating grain-fed animals does?

3. Draw structural formulas for the formation of a peptide bond between two amino acids.

4. Why is the principle of self-assembly especially significant in a protein such as collagen, which is exported by cells to its permanent extracellular location?

5. Explain the relationship of Chargaff's rules to the structure of DNA.

6. The hydrophilic portions of integral proteins stick out of membrane surfaces, while the hydrophobic portions of the proteins lie in the interior of the bimolecular lipid layer. Explain the existence of a "hydrophobic interior" in cell membranes.

Suggested Readings

Books

Dyson, R. D. 1978. *Cell biology: A molecular approach.* 2d ed. Boston, Mass.: Allyn & Bacon.

Lehninger, A. L. 1975. *Biochemistry.* 2d ed. New York: Worth Publishers.

Metzler, D. E. 1977. *Biochemistry: The chemical reactions of living cells.* New York: Academic Press.

Articles

Bauer, W. R.; Crick; F. H. C., and White, J. H. July 1980. Supercoiled DNA. *Scientific American.*

Capaldi, R. A. March 1974. A dynamic model of cell membranes. *Scientific American* (offprint 1292).

Koshland, D. E., Jr. October 1973. Protein shape and biological control. *Scientific American* (offprint 1280).

Lodish, H. F., and Rothman, J. E. January 1979. The assembly of cell membranes. *Scientific American* (offprint 1415).

Lucy, J. A. 1975. The plasma membrane. *Carolina Biology Readers* no. 81. Burlington, N.C.: Carolina Biological Supply Co.

Phillips, D. C., and North, A. C. T. 1973. Protein structure. *Carolina Biology Readers* no. 34. Burlington, N.C.: Carolina Biological Supply Co.

Raff, M. C. May 1976. Cell-surface immunology. *Scientific American* (offprint 1338).

Sharon, N. June 1977. Lectins. *Scientific American* (offprint 1360).

Cells

Chapter Concepts

1. The cell is the fundamental unit of life. All living things are cells or aggregates of cells and cell products.

2. The development of modern techniques of observation and analysis has greatly advanced the understanding of cell structure and function.

3. Eukaryotic cells contain a number of highly organized, membrane-enclosed compartments called organelles. Each type of organelle is associated with a specific set of functions in cells.

4. Although most plant and animal cell organelles are similar, there are some functionally significant differences between plant and animal cell structures.

5. Prokaryotic cells lack a true nucleus and membrane-bound organelles, but a number of biochemical processes occur in prokaryotic cells that are similar to those that take place in eukaryotic cells.

A biologist's study of cells is similar in some ways to the chemist's investigation of atomic structure and behavior. Just as matter is composed of atoms or groups of atoms, living things consist of cells, or aggregates of cells, and cell products. To learn how organisms grow, reproduce, and interact with their environments, it is necessary to understand cell structure and function.

Because cells are extremely diverse with respect to size, structure, and function, it is difficult to describe all cells with a single definition. In broad terms, a **cell** can be defined as a discrete body of self-reproducing, living material, bound by a selectively permeable plasma membrane.

The study of cells began with the development of the simple microscope in the seventeenth century, and the appearance of cells was first described in 1666 by Robert Hooke in his book *Micrographia*. However, Anton van Leeuwenhoek (1632–1723) of Delft, Holland, contributed the most to early cell research (figure 3.1). Leeuwenhoek made his living as a draper and haberdasher and held minor political posts in Delft— including that of official wine taster. Despite his many activities, Leeuwenhoek found time to construct a large number of simple microscopes with such high-quality lenses that some could magnify approximately 200 times (figure 3.2). Over a period of fifty years, this early microscopist observed literally hundreds of different cells and described them in amazing detail in letters to the Royal Society of London.

Despite these auspicious beginnings, it was not until more than 100 years after Leeuwenhoek's death that scientists generally recognized that cells are the fundamental units of all living things. At that point, the observations of the early microscopists were synthesized into one general statement known as the **cell theory.** In 1838 Matthias Jakob Schleiden published his theory that cells were the component units that make up all plant bodies; one year later, Theodor Schwann published a similar proposal concerning animal bodies, amending his statement to include cell products as well as cells. Finally, in 1858, Rudolf Virchow completed the cell theory when he hypothesized that "every cell comes from a preexisting cell." Thus, by the middle of the nineteenth century, biologists clearly recognized the cell as the fundamental unit of life.

Figure 3.1   Anton van Leeuwenhoek.

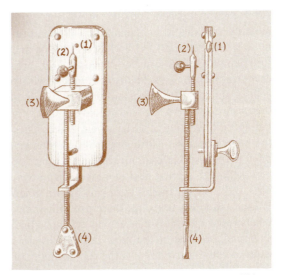

Figure 3.2   One of Leeuwenhoek's microscopes. This simple microscope consisted of (1) a plate with a single lens, (2) a mounting pin that held the object to be observed, (3) a focusing screw, and (4) a specimen-centering screw.

This view of the cell as the basic component of all living things stimulated biologists to undertake intensive research on the structures and functions of various kinds of cells. Unfortunately, the amount that biologists were able to learn about cells was limited for a long time by the research tools and methods available to them; the primary tool at their disposal was the light microscope, and even the best of these yielded only a limited amount of information about cells. Years passed with only limited advances made in understanding cell structure and function. Thus, when E. B. Wilson, a pioneer of cell biology in the United States, published a diagram in 1925 of the details of cell structure that were then known, his illustration was relatively simple (figure 3.3). "Wilson's cell" was a picture of a cell with a nucleus, cytoplasm, a few other internal structures, and a bounding membrane (which was inferred, since it could not actually be seen with the light microscope).

It is now known that the granular, relatively featureless cytoplasm depicted in Wilson's cell actually contains a complex array of membranous structures that are both structurally and functionally specialized for various vital processes in the cell. This "division of labor" within cells is similar to the division of labor in whole organisms, where various individual organs carry out specific functions for the whole body. Thus, the name **organelle** ("little organ") has been applied to these specialized intracellular structures.

Although a great deal more is known about cells and their organelles than was known when Wilson made his drawing, there are still many unanswered questions concerning cell structure and function. One of the most basic characteristics of cells—their size—is the root of many of the problems facing biologists who study cells.

Cell Size

Cells come in a variety of sizes and shapes, but almost all of them are small in terms of everyday frame of reference. A few gigantic cells, such as the yolk of a chicken's egg, are very familiar, but almost all other cells are too small to be seen with the unaided eye (figure 3.4).

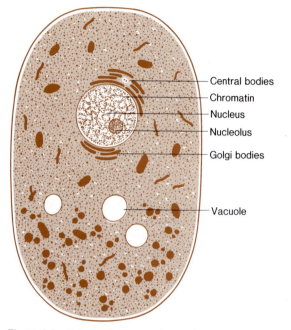

Figure 3.3 Cell structure as understood in 1925. This illustration is based on a diagram published that year by Edmund B. Wilson in his classic cell biology book *The Cell in Development and Heredity*. Central bodies are the pair of centrioles seen near the nucleus.

Central bodies
Chromatin
Nucleus
Nucleolus
Golgi bodies
Vacuole

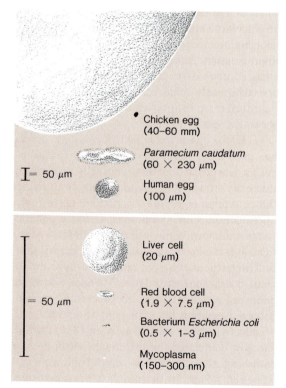

$\text{I}= 50\ \mu m$

Chicken egg
(40–60 mm)
Paramecium caudatum
(60 × 230 μm)
Human egg
(100 μm)

$= 50\ \mu m$

Liver cell
(20 μm)
Red blood cell
(1.9 × 7.5 μm)
Bacterium *Escherichia coli*
(0.5 × 1–3 μm)
Mycoplasma
(150–300 nm)

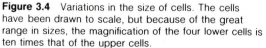

Figure 3.4 Variations in the size of cells. The cells have been drawn to scale, but because of the great range in sizes, the magnification of the four lower cells is ten times that of the upper cells.

The smallest free-living bacteria are around 0.2 to 0.3 μm in diameter, while bird eggs may reach 10 cm or more in diameter. Despite this enormous range in size, the great majority of cells fall into the size range between 0.5 μm and 40 or 50 μm in diameter. Most human cells, for example, are between 10 μm and 20 μm in diameter (table 3.1).

Even the smallest cell must be large enough to possess certain components necessary for life; a cell must have a functional plasma membrane, an adequate quantity of genetic material, the cellular machinery needed to synthesize nucleic acids and proteins, the enzymes necessary to derive energy from molecules obtained from the environment, and so forth. On the other hand, the factors limiting the sizes of very large cells are less clear. The ratio of the volume of a cell's nucleus to the volume of the cytoplasm seems to be one significant factor. The nucleus, the major regulatory center for the cell, apparently can exert control over only a certain amount of cytoplasm. Therefore, if a cell were to grow beyond a certain size, its nucleus might not be able to regulate it effectively. A cell division serves to readjust the ratio between the size of the nucleus and the amount of cytoplasm in a cell.

Another facet of cell volume also works to limit cell size. As the diameter of a cell becomes greater, its volume increases more rapidly than its surface area (figure 3.5). The demand for nutrients and the rate of waste production are proportional to the volume of the cell, but the cell takes up nutrients and eliminates wastes through its surface. If the cell becomes too large, its surface area may not be great enough to carry on an adequate exchange of nutrients and wastes. This is not an urgent problem in cells with relatively low metabolic activity, as they need less nutrients and generate fewer wastes; for example, certain bird eggs contain huge quantities of stored nutrient material (yolk) that is almost inert metabolically. Still, no cell can live without maintaining at least the minimum level of metabolism required to sustain essential functions.

Simple mechanical factors may also restrict cell size. The plasma membrane can directly support only a certain mass without rupturing, but it can be aided by the counterpressure of surrounding cells or special structures such as cell walls. Thus, when there is external support, cells may exceed the size restrictions imposed by the limited strength of their own plasma membranes.

Table 3.1
Units of Measurement.

Unit	Symbol	Value
Meter	m	39.37 inches
Centimeter	cm	0.01 meter
Millimeter	mm	0.1 cm (10^{-3} meter)
Micrometer (micron)	μm (μ)	0.001 mm (10^{-6} meter)
Nanometer	nm	0.001 μm (10^{-9} meter)
Angstrom	Å	0.1 nm (10^{-10} meter)

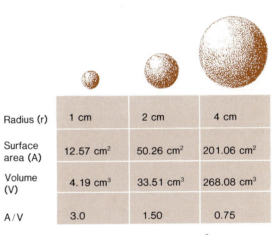

Radius (r)	1 cm	2 cm	4 cm
Surface area (A)	12.57 cm^2	50.26 cm^2	201.06 cm^2
Volume (V)	4.19 cm^3	33.51 cm^3	268.08 cm^3
A/V	3.0	1.50	0.75

Surface area of a sphere $= 4 \pi r^2$
Volume of sphere $= 4/3 \pi r^3$

Figure 3.5 Comparison of surface-to-volume ratios in spheres with different radii.

Methods Used in Studying Cells

Despite the very small size of most cells and the difficulties inherent in studying the details of such small, delicate objects, knowledge of cell structure and function has been expanding at a dizzying pace in recent years. This growth in knowledge is, to a great extent, the result of the development of new and more sophisticated techniques for studying cells. In fact, to appreciate modern cell biology, it is essential to have some understanding of the methods that cell biologists employ in their research.

Light Microscopy

One of biology's oldest and most familiar tools is the **transmitted light microscope.** However, although light microscopes are useful for certain kinds of studies, they are of very limited use in cell research. It is possible to build a light microscope that would magnify objects more than the microscopes currently in use do, but the greatest limitation of the light microscope is not magnification but resolution. **Resolution** is what makes it possible for an observer to discern separate, small objects that are very close together.

The resolving power of a light microscope is related to the wavelength of the light illuminating the object under observation. The shorter the wavelength, the higher the microscope's resolution. This is because shorter wavelengths of light are diffracted, or scattered, less than larger wavelengths as they pass through the material being observed. Therefore, shorter wavelengths of light distort the image less. A microscope equipped with a good lens, capable of a high degree of useful magnification, and using blue-green (short-wavelength) light will be able to resolve two points that are approximately 0.2 μm apart. Such a microscope allows the observer to study overall cell structure and some of the larger structures inside cells at magnifications up to about 1,000 to 1,500 times their actual size, but the resolving power of a light microscope is not great enough to permit detailed study of the many smaller structures in cells. Resolving power is limited by the wavelengths of *visible* light, since light microscopy must make use of wavelengths that the human eye can see.

Another problem with light microscopy is that cells and their constituent parts are relatively transparent and thus do not contrast much when viewed with a light microscope. For this reason, biologists often **stain** cells or components of cells with dyes, thereby increasing the contrast and making them easier to observe. Some stains specifically react with and color certain cell constituents. Thus, it is possible to pinpoint the locations of, for example, proteins, nucleic acids, carbohydrates, or lipids within cells. This is called **cytochemistry.**

Cytochemistry yields a great deal of information about the makeup and functioning of various parts of cells. Even the sites of specific enzyme-catalyzed reactions can be detected by cytochemistry. Although cytochemistry has been a boon to cell research done with light microscopes, even such specialized staining techniques do not solve the problem of resolution. To obtain greater effective magnification of cell components with an accompanying increase in resolving power, other methods of observation, the most effective of which is electron microscopy, have been developed.

Electron Microscopy

The **electron microscope** (figure 3.6) has a much greater resolving power than the light microscope—about 0.5 nm or less, as opposed to 0.2 μm. Hence, the electron microscope can be used to effectively magnify cells 200,000 times or more. This great increase in resolution is achieved through using an electron beam instead of a beam of light.

There are two basic types of electron microscopes, the **transmission electron microscope (TEM)** and the **scanning electron microscope (SEM).** In transmission microscopy, an electron beam passes through the object being studied, while in scanning electron microscopy, an electron beam strikes the surface of the specimen.

In a transmission electron microscope, the electron beam is focused on a very thin section or slice of the cell by means of electromagnets (figure 3.6b). After leaving the cell, the electron beam travels through two more magnetic lenses, which magnify the image and project it either on a fluorescent screen (resembling a television screen) or on photographic film.

In transmission electron microscopy, cells are not stained with dyes to increase contrast. Instead, during preparation, specimens are treated with heavy metals such as osmium or uranium, which attach to intracellular structures in the specimens. This heavy-metal treatment increases contrast among the cell components because electrons will not pass through the cell components where metal atoms are located. Thus, these components appear as dark (not fluorescent) objects on the screen.

The scanning electron microscope is used to study surfaces rather than thin sections of cells. SEM pictures, with their three-dimensional quality, reveal some remarkable details of the surfaces of cells and other objects. As in transmission electron microscopy, these surfaces are usually coated with a very thin layer of metal, often gold. In the SEM, electron beams scan over the surface of the specimen, and the striking electrons drive off electrons from the surface atoms. The pattern of these **secondary electrons** landing on a **detector** is amplified, and the image is displayed on a cathode ray tube much like that in a television set. Many previously unknown surface details have already been discovered using the scanning electron microscope, and SEM studies undoubtedly will yield a great deal more information about the surfaces of biological structures.

(a)

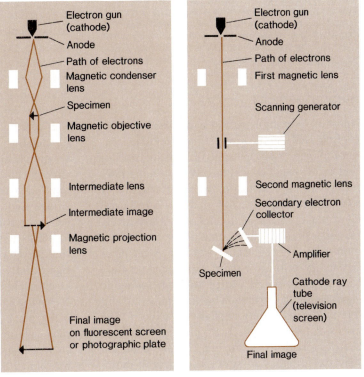

(b)

Figure 3.6 Electron microscopy. (a) A transmission electron microscope. (b) Highly diagrammatic and simplified comparison of principles of transmission (left) and scanning (right) electron microscopy. In transmission electron microscopy the focused electron beam passes through a very thin section of the specimen. A final image of the object is viewed on a flourescent screen or photographic plate at the bottom of the column. In scanning electron microscopy, the scanning generator makes the electron beam scan over the surface of the specimen. Secondary electrons come off atoms on the specimen's surface. The pattern of secondary electrons striking the secondary electron detector is amplified and displayed on a cathode ray tube.

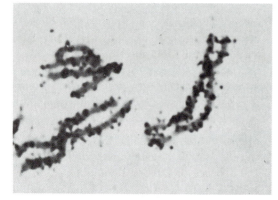

Figure 3.7 Autoradiograph of chromosomes during mitotic metaphase following exposure to the radioactive isotope. The cells were exposed to radioactively labelled thymidine and allowed to undergo cell division. They were then washed free of the labeled thymidine and allowed to begin a second mitotic cycle. After the chromosomes had condensed and lined up during metaphase, this autoradiograph was prepared. The labeling pattern shows that DNA synthesis has occurred in the chromosomes because radioactive thymidine was incorporated almost exclusively into the chromosomes.

Autoradiography

Another observation technique, **autoradiography,** is often used for studying the intracellular locations of cell constituents, or the specific sites of particular reactions within cells. In autoradiography, cells are incubated with molecules that will be specifically incorporated into certain cell constituents. These molecules are radioactively labeled—that is, they contain one or several atoms that are radioactive isotopes. After incubation, the cells are washed and placed on glass slides. Each slide is coated with **photographic emulsion** similar to that on photographic film and left in the dark long enough to "expose" the emulsion to any radioactivity present. Wherever there is a radioactive site in a cell, the emulsion reacts as if light had struck it, and each radioactive disintegration produces a particle track of silver grains in the emulsion. Thus, the locations of specific molecules and reactions within the cell can be determined through autoradiography.

An example may better demonstrate the usefulness of autoradiography. The pyrimidine thymine is found in DNA but not RNA, so incorporation of thymine into cells indicates specifically that DNA synthesis is occurring. The nucleotide containing thymine is called thymidine. Cells can be incubated with thymidine that contains the radioactive isotope tritium. These

cells will then use the radioactive thymidine (*tritiated thymidine*) to synthesize DNA. After the treated cells are washed and coated with photographic emulsion, the great majority of the silver grains are found over the cell's chromosomes. This demonstrates clearly, then, that DNA is synthesized in the chromosomes within the nucleus rather than in the cytoplasm (figure 3.7).

Cell Fractionation

All of the cell-research techniques described thus far—even such useful and informative techniques as electron microscopy and autoradiography—share one significant shortcoming: they all produce "stop-action" pictures of cell components. At some point, the cells are killed and prepared for observation. Thus, what is observed is the condition of cells and cell organelles at the moment the cells were killed. Since cells and organelles are structurally and functionally dynamic, ever-changing structures, there is obviously a need for techniques that allow observers to study cell and especially organelle functions over a period of time. This need has led to the development of **cell fractionation** techniques, or techniques that allow the isolation and study of various cell components. Cell fractionation permits biologists to analyze the chemical compositions and biochemical functions of isolated cellular constituents.

The first step in cell fractionation is to disrupt a cell and release cell components without significantly altering their structure and activity. One disruption process is called **homogenization**. In homogenization, a glass sleeve or tube fitted with a plastic or glass pestle is used to break open cells (figure 3.8). After homogenization, cell components such as nuclei and mitochondria may be isolated by **differential centrifugation**. In this separation technique, which makes use of centrifugal force, a container filled with a mixture of particles is spun in a centrifuge to drive the heavier particles to the bottom of the container. These heavier particles form a sediment or a **pellet**, while the lighter particles remain suspended in the surrounding solution, or **supernatant.** Nuclei, which are relatively large organelles, can readily be sedimented at forces of 1,000 times the force of gravity (g), while lighter particles,

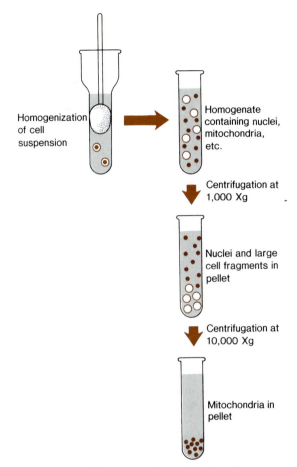

Homogenization of cell suspension

Homogenate containing nuclei, mitochondria, etc.

Centrifugation at 1,000 Xg

Nuclei and large cell fragments in pellet

Centrifugation at 10,000 Xg

Mitochondria in pellet

Figure 3.8 Cell homogenization and fractionation by differential centrifugation.

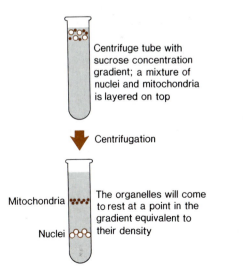

Centrifuge tube with sucrose concentration gradient; a mixture of nuclei and mitochondria is layered on top

Centrifugation

Mitochondria

Nuclei

The organelles will come to rest at a point in the gradient equivalent to their density

Figure 3.9 Cell fractionation by density gradient centrifugation. Sucrose in the tube increases linearly in concentration and therefore in density. The lowest concentration of sucrose is at the top and the highest at the bottom.

such as mitochondria, remain in the supernatant. Mitochondria can be collected by centrifuging the first supernatant again at 10,000 × g. If the resulting supernatant is centrifuged yet again, this time at forces around 100,000 × g, even the smallest organelles, membrane fragments, and macromolecules are sedimented. The remaining supernatant, which contains only dissolved molecules, is often called the **cytoplasmic matrix** (or **cytosol**). The cytoplasmic matrix probably represents the relatively featureless portion of the cytoplasm that fills the spaces between organized cell structures in electron microscope pictures of cells.

Another technique that is often used to isolate cell constituents is **density gradient centrifugation.** In this technique, a homogenate is placed on top of a sucrose solution that progressively increases in concentration and therefore in density from the top to the bottom of the container (figure 3.9). When this sucrose gradient is centrifuged at high speeds, each particle in the homogenate will move down in the tube and come to rest at the point in the gradient where its density equals that of the sucrose solution. The nuclei from a cell homogenate would thus be found at a lower point in the gradient than the mitochondria, which are less dense.

Once organelles have been isolated by differential centrifugation or by density gradient centrifugation, they can be studied and analyzed over a period of time to determine such properties as their chemical compositions and enzymatic activities. This kind of study can provide information concerning the functions of organelles within intact living cells. For example, if mitochondria are isolated from a cell and a biologist determines that all the enzymes involved in the Krebs cycle are located within them, it is very likely that the mitochondria are the site of Krebs cycle activity. Further experiments can be carried out with isolated mitochondria to substantiate this conclusion; in fact, this is exactly how Albert Lehninger discovered that mitochondria do contain the Krebs-cycle enzymes.

Cell fractionation makes possible dynamic functional studies of cell organelles—studies that complement the information obtained through techniques that give "stop-action" pictures of cells and organelles at single moments in time. The most complete picture of cell structure and function is obtained if several different techniques are employed because each technique contributes some valuable information. This is why biologists very often utilize more than one approach to a problem in cell biology.

All of the techniques discussed here, as well as others, have been used in developing the present knowledge of cells and their organelles. These techniques may also be combined with one another to examine various components of living cells.

Structural Components of Eukaryotic Cells

All cells can be placed in one of two classes. Bacteria and blue-green algae are **prokaryotic** (*pro* means *before; karyon* means *nucleus*) cells. Prokaryotic cells lack a distinct nucleus bound by a membrane and do not have organelles such as mitochondria. All other cells—the cells of plants, animals, protists, and fungi—are **eukaryotic** ("true nucleus") cells. Eukaryotic cells possess nuclei with membranes and are much more structurally complex. Because several of the different types of cell organelles in eukaryotic cells are surrounded by membranes or are entirely constructed of membranes, some readers may wish to refer to the general discussion of membrane structure in chapter 2 as they study these cell organelles.

The Plasma Membrane

One of the most important parts of a cell is the **plasma membrane,** which encloses the entire cell. The plasma membrane forms the cell's point of contact with its environment and determines what materials enter and leave the cell. In fact, the plasma membrane is vital to the life of the cell because it helps to ensure a stable internal cell environment amidst continually changing surroundings.

The plasma membrane resembles all cellular membrane structures (see chapter 2); it is an ordered but dynamic array of lipids and proteins.

The proteins of the membrane rest in a lipid bilayer composed principally of phospholipids and steroids.

The plasma membrane is a versatile organelle that fulfills a number of roles. It is, of course, the cell boundary in a mechanical sense and therefore holds all the cell constituents together. It also functions as a selectively permeable barrier that allows certain atoms and molecules to enter or leave the cell. Components of the plasma membrane also carry out active transport processes that move some substances into the cell and move others to the exterior. The movement of materials through the plasma membrane will be discussed in more detail in chapter 4.

Plasma membranes are involved in many cell interactions in multicellular organisms. A plasma membrane can maintain an electrical potential and, in certain specialized cells such as nerve and muscle cells, can conduct impulses. In this way, plasma membranes are involved in integrating and coordinating the functions of whole multicellular organisms.

The Cytoplasmic Matrix

When a eukaryotic cell is examined with an electron microscope, its organelles appear to lie in a featureless, homogeneous substance called the **cytoplasmic matrix,** or **cytosol** (sometimes also called the **hyaloplasm**). The cytoplasmic matrix, although superficially uninteresting in electron micrographs, is in reality a most important part of the cell. It is the "environment" of the organelles and the site of a number of processes.

Approximately 85 to 90 percent of the cytoplasmic matrix is water. A portion of this cellular water is **bound water,** held in hydration spheres around proteins and other macromolecules. Bound water is unable to act as a solvent for other intracellular molecules. The remainder of the matrix water is **free water** and can act as a solvent and participate in metabolic processes.

The supernatant remaining after a cell homogenate has been centrifuged at $100,000 \times g$ presumably contains the constituents of the cytoplasmic matrix. Analysis of this $100,000 \times g$ supernatant shows that it contains about a quarter of the cell's protein. Many of these protein molecules are enzymes that catalyze a number of metabolic reactions that take place in the matrix.

Protein molecules also contribute to some special properties of the cytoplasmic matrix because these proteins may not be in solution in the same way that ions and smaller molecules (such as NaCl or glucose) are. Some biologists think that the cytoplasmic matrix proteins are part of a special arrangement called a **colloid** or colloidal solution. Colloids have a more orderly arrangement than ordinary solutions and have certain other properties that may affect the structure and function of the cytoplasmic matrix. Colloids exist in two different forms, each with different physical properties. In one form, molecules of the colloidal system associate and remain associated to form a semisolid or solid system. This solid colloidal system is called a **gel.** The other colloidal form is a fluid system in which any two particles interact with each other only momentarily. This is called a **sol.** Protein solutions, like many other colloids, can exist in either the sol state or the gel state and often can be reversibly transformed from one to the other. An excellent example of converting a sol to a gel is making a gelatin dessert. The hot, dissolved gelatin is a protein sol. When it is cooled in the refrigerator, the colloid is transformed into a gel.

There is considerable evidence that the cytoplasmic matrix, which is composed of other molecules as well as proteins, behaves as a colloid. For example, a number of cells possess an outer gel layer (the **ectoplasm**) and an inner sol region (the **endoplasm**). The matrix may undergo reversible sol-gel transformations during amoeboid movement and other activities (figure 3.10).

The Nucleus and Nucleolus

Because it is visually prominent under the light microscope, the nucleus, the repository for most of the cell's genetic information, was discovered early in the study of cell structure. Nuclei are membrane-enclosed spherical bodies around 5 to 7 μm in diameter. Relatively dense, fibrous patches of **chromatin** are the DNA-containing regions of the nucleus. In nondividing cells, this chromatin exists in a dispersed condition, but during cell division, it condenses into **chromosomes**, which are familiar features of dividing cells (figure 3.11).

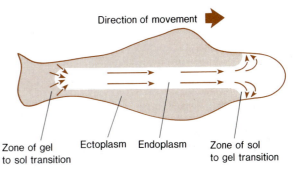

Direction of movement

Zone of gel to sol transition   Ectoplasm   Endoplasm   Zone of sol to gel transition

Figure 3.10   The role of sol-gel transformation in amoeboid movement of a cell. Ectoplasm portion of the cytoplasm is a gel, and endoplasm is a sol.

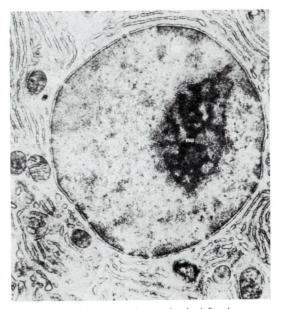

Figure 3.11   This nucleus has a clearly defined nucleolus (nu) and irregular patches of chromatin scattered throughout its matrix. The chromatin contains DNA and condenses to form chromosomes during mitosis.

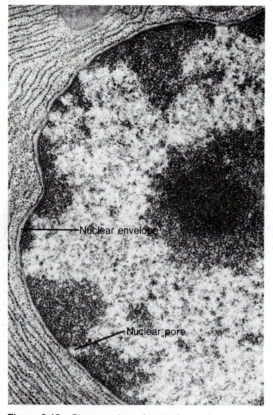

Nuclear envelope

Nuclear pore

Figure 3.12 Close-up view of part of a nuclear
envelope and its pores. The inner and outer membranes
are separated by a space. A number of pores penetrate
the envelope. Patches of chromatin are present in the
nuclear matrix. (Magnification × 19,845)

The nucleus is enclosed by a complex structure called the **nuclear envelope** (figure 3.12). This envelope consists of inner and outer membranes separated by a space 15 to 75 nm in width. The envelope is continuous with the *endoplasmic reticulum* at a number of points, and its outer membrane may be covered with *ribosomes* (both of which are discussed later in this chapter).

A large number of **nuclear pores** penetrate the surface of the nuclear envelope. Each pore is formed by a fusion of the outer and inner membranes and is around 70 nm in diameter. These nuclear pores are so numerous that they occupy from 10 to 25 percent of the surface of most nuclei. The pores appear to function as a route of transport between the nucleus and the surrounding cytoplasm. Electron microscope pictures sometimes show granular material in the pores; it is assumed that this material was in the process of moving through the pores when the cell was prepared for observation.

The **nucleolus** is a distinctive region of the nucleus that is present in nondividing cells but usually disperses during mitosis (cell division). After mitosis, the nucleolus, which is not membrane bound, re-forms around a certain part of a specific chromosome called the **nucleolar organizer.** The nuclei of many types of cells have more than one nucleolus, with two being a common number.

The nucleolus plays a major role in the formation of specialized bodies, the ribosomes, which are involved in protein synthesis in cells. DNA associated with the nucleolus directs the synthesis of **ribosomal RNA (rRNA),** an important constituent of ribosomes. It is not surprising that nucleoli in cells that are actively synthesizing proteins are larger than nucleoli in cells that are not doing so.

Ribosomes

Ribosomes are the sites of protein synthesis in cells. They may either be associated with an extensive network of membrane tubules and sacs called the **endoplasmic reticulum (ER),** or they may be free in the cytoplasmic matrix.

Each ribosome is a very complex entity that is about 20 nm in diameter and consists of two subunits. Each ribosomal subunit is composed of a large variety of proteins and ribosomal RNA molecules, with about 65 percent of each subunit being rRNA (figure 3.13). Under appropriate conditions, these subunits will actually assemble spontaneously to form ribosomes in a test tube. (This remarkable fact was demonstrated by M. Nomura, who has dissociated ribosomes into their components and then allowed them to reassociate to form complete, functional ribosomes.) The formation of ribosomes, then, is another clear example of the principle of self-assembly.

The Endoplasmic Reticulum

One principle that has arisen from modern studies of cells, especially examinations of cells with the transmission electron microscope, is that eukaryotic cells contain complex, dynamic membranous structures, the most extensive of which is the endoplasmic reticulum (ER). The endoplasmic reticulum, however, is not a single structure, but an irregular network of branching and fusing membranous tubules (which range between 40 nm and 70 nm in diameter) and a large number of flattened sacs called **cisternae** (figure 3.14). This network of tubules and cisternae permeates much of the cytoplasmic matrix.

Although the endoplasmic reticulum looks fixed and static in any cell prepared for electron microscopy, this appearance is misleading. The endoplasmic reticulum is continually changing in living cells as new ER is being produced to replace parts that detach or break down. The shape and complexity of the ER seems to vary with the functional activity of the cell (especially in cells that are actively synthesizing materials, because the ER is intimately involved in a cell's synthetic activities). In cells that are synthesizing a great deal of protein, a large part of the outside of the ER is coated with ribosomes and is called **rough** (or **granular**) **ER.** In other cells—for example, those producing lipids such as steroid hormones—the ER is virtually devoid of ribosomes.

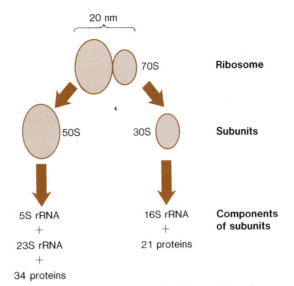

Figure 3.13 The components of a ribosome from the bacterium *Escherichia coli,* which lives in the human digestive tract. S = Svedberg unit, a quantity used to express the size of molecules and molecular aggregates. Svedberg values indicate how rapidly particles sediment in a centrifuge running at high speed. S values are approximately proportional to size; therefore, the greater the S value, the larger the particle, but S values are not strictly additive. For example, the *E. coli* ribosome is composed of a 50S and a 30S subunit, but the whole ribosome is a 70S, not an 80S, particle. The large subunit consists of two types of ribosomal RNA (rRNA) molecules and thirty-four protein molecules, while the smaller subunit has a different rRNA molecule and twenty-one protein molecules. Thus, bacterial ribosomes are very complex structures. Ribosomes of eukaryotic cells (cells with nuclei and internal membranes) have not been so thoroughly studied, but they are somewhat larger than bacterial ribosomes (80S as opposed to 70S), and their subunits also are complex aggregates of rRNA and protein molecules.

This is called **smooth** (or **agranular**) **ER** (figure 3.14).

In addition to being a site for synthesis reactions, the ER serves as a major transport route by which proteins, lipids, and other materials get from one part of the cell to another. This transport involves changes in the ER's shape and movement of parts of the ER through the cytoplasmic matrix. The transport functions of the endoplasmic reticulum are also related to the functions of another membranous organelle, the Golgi apparatus.

The Golgi Apparatus

The **Golgi apparatus** (also called **Golgi body**) is an organelle that was discovered near the turn of the century. Cell biologists such as Camillo Golgi, for whom it was named, described a distinctive area of the cytoplasm that reacted intensively with certain chemicals such as silver nitrate. But despite this early discovery, details of Golgi apparatus structure and function have been worked out only since electron microscopy and various autoradiographic and cytochemical techniques have become available.

The Golgi apparatus, a membranous organelle that is structurally and functionally related to the endoplasmic reticulum, is involved in the packaging and secretion of cell products. A Golgi body is composed of stacked saclike structures called cisternae. These cisternae are made of membranes that resemble smooth ER in that they lack ribosomes (figure 3.15). There are usually around four to eight cisternae or sacs in a stack, and at the edges of the cisternae, there is a complex network of tubules and vesicles (hollow membranous spheres).

The stack of cisternae in a Golgi apparatus has two ends or faces that are quite different from one another structurally. The sacs on the proximal or forming face are more closely associated with the ER. Apparently, the sacs move down the stack from one side to the other (away from the ER). When they reach the side opposite the ER, many spheres pinch off. These spheres are **secretory vesicles,** membrane-bound "packages" of material synthesized by the cell.

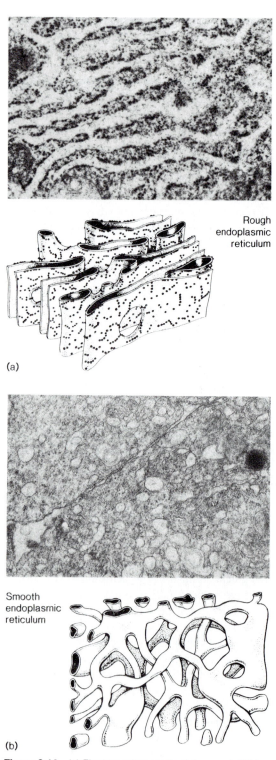

Rough
endoplasmic
reticulum

(a)

Smooth
endoplasmic
reticulum

(b)

Figure 3.14 (a) Electron micrograph of the rough ER in a cell. The cell's cytoplasm is packed with flattened cisternae that are coated with ribosomes. (b) An area of cytoplasm full of smooth ER tubules. These tubules lack ribosomes.

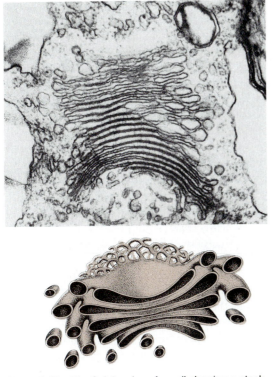

Figure 3.15 The Golgi region of a cell showing a stack of cisternae. The curved nature of the cisternae is obvious. The lower illustration represents the structure of the Golgi apparatus as a whole. (Magnification × 36,000)

The functioning of the Golgi apparatus in the packaging and secretion of cell products has been studied in detail with the cells of the pancreas that secrete digestive enzymes. If radioactively labeled amino acids are administered to the pancreas cells for a short time, the amino acids are incorporated into enzyme molecules. The positions of these labeled enzyme molecules can then be checked at certain time intervals using electron microscopic autoradiography. Within about five minutes after the labeled amino acids are administered to the cells, the most intensive radioactivity is found to be concentrated in the rough endoplasmic reticulum. A few minutes later the labeled compounds appear to have traveled to the vicinity of the Golgi apparatus. Finally, the radioactive proteins are packaged in secretory vesicles and are ready for release at the plasma membrane. These observations indicate that proteins are being synthesized on the ribosomes of the rough ER, transported to the Golgi, and packaged in some way by the Golgi apparatus (figure 3.16). Thus, membrane-enclosed materials are produced in the ER and transferred to a Golgi apparatus, where they are prepared for export from the cell.

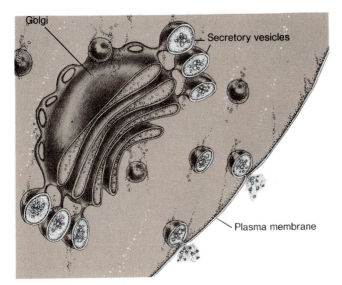

Golgi

Secretory vesicles

Plasma membrane

Figure 3.16 Golgi apparatus secretory function. Material synthesized in the endoplasmic reticulum is transferred to the Golgi apparatus, where it is "packaged" for export from the cell. Secretory vesicles move to the cell surface and fuse with the plasma membrane, and their contents are released to the outside.

Lysosomes

The Golgi apparatus and endoplasmic reticulum are responsible for much more than the construction of secretory vesicles of various types. They also produce another important cell organelle, the **lysosome.** Lysosomes are unique among cell organelles in that they were detected by microscopic observation only *after* their existence had been proposed as a result of cell fractionation and various types of biochemical research. Lysosomes serve as compartments that segregate various types of enzymes from the rest of the cytoplasm.

Christian de Duve found that when a homogenate of rat liver cells was fractionated, several types of enzymes that are active at acid pHs could be detected in a fraction of the sedimented material. But if the homogenate was treated harshly prior to centrifugation, these enzyme activities were found in the supernatant, not in the sediment. De Duve correctly concluded that there must be a membrane-bound organelle that normally encloses these enzymes, and he named this organelle a lysosome. When cell fractions known to contain high concentrations of these particular enzymes were examined with an electron microscope, de Duve's idea was supported: lysosomes were indeed found.

It now appears that almost all animal cells contain lysosomes. They are also found in some plant cells, in fungi, and in algae. Lysosomes are roughly spherical in shape and are enclosed in a single membrane (figure 3.17). They average about 500 nm in diameter, but may range from 50 nm to several μm in size.

Lysosomes contain the enzymes necessary to digest all types of macromolecules. Obviously, this is why protection for the rest of the cell, in the form of an enclosing membrane, is required. The lysosomal enzymes are manufactured by the ER and enclosed in membranes by the Golgi apparatus. Newly formed lysosomes, called **primary lysosomes,** sometimes fuse with phagocytic vesicles (figure 3.18), which are formed when the cell membrane enfolds around a small quantity of solid material and then pinches off into a separate body. Fusion with a lysosome exposes a vesicle's contents to the digestive enzymes contained in the lysosome. Such a lysosome, actually in the process of digestion, is called a **secondary lysosome.**

Lysosomes join with phagocytic vesicles not only to digest nutrients that will be absorbed by the cell, but also for defensive purposes. Invading bacteria, often picked up by phagocytic vesicles, are destroyed when a lysosome fuses with a vesicle and the lysosomal enzymes digest the bacteria. This is a common function of the leukocytes or white blood cells in mammals, which engulf and destroy bacteria and other foreign cells.

Cells can also selectively digest portions of their own cytoplasm in a type of secondary lysosome formed by an **autophagic vesicle.** These vesicles are evidently formed when the ER pinches off to enclose bits of cytoplasm. They subsequently fuse with primary lysosomes, in which the lysosomal enzymes digest the enclosed cytoplasm. Cells are dynamic, and various structures within them are constantly being replaced; autophagic vesicles probably play at least some role in the normal turnover or recycling of cell constituents. Autophagic vesicles also help prevent a cell from starving when it is not receiving adequate nutrients because a cell can selectively digest portions of itself to remain alive.

Mitochondria

Mitochondria (singular: **mitochondrion**) are frequently called the powerhouses of the cell since they are the site of major energy conversion reactions in cells. Most eukaryotic cells have many mitochondria—in fact, some cells may possess as many as a thousand or more. On the other hand, at least a few cells, such as some yeast cells, apparently contain a single giant tubular mitochondrion that forms a continuous network throughout the cytoplasm.

Most mitochondria are elongated structures measuring about 0.3 to 1.0 μm by 5 to 10 μm. The electron microscope reveals that mitochondria are complex membranous structures. Mitochondrial membranes isolate certain reactions from the rest of the cell and even compartmentalize certain functions within the mitochondrion itself. Each mitochondrion is enclosed by a double membrane, the **outer mitochondrial membrane,**

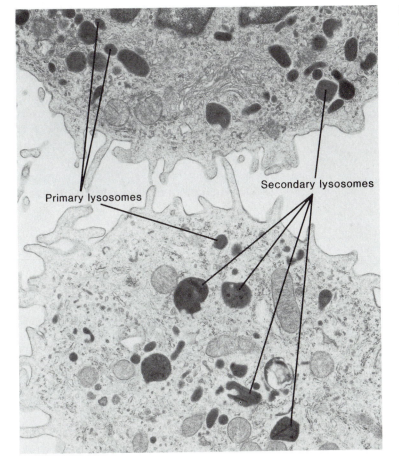

Primary lysosomes

Secondary lysosomes

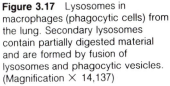

Figure 3.17 Lysosomes in macrophages (phagocytic cells) from the lung. Secondary lysosomes contain partially digested material and are formed by fusion of lysosomes and phagocytic vesicles. (Magnification × 14,137)

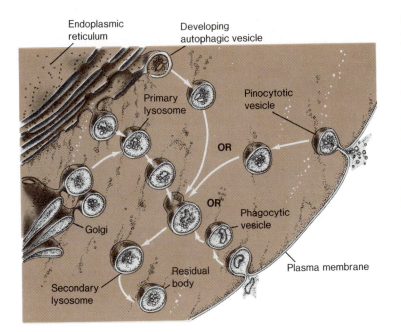

Endoplasmic reticulum

Developing autophagic vesicle

Primary lysosome

Pinocytotic vesicle

OR

OR

Phagocytic vesicle

Golgi

Residual body

Secondary lysosome

Plasma membrane

Figure 3.18 Lysosome formation and function. Secondary lysosomes form in several different ways. Phagocytic vesicles contain small amounts of solid material taken in from outside the cell, while pinocytotic vesicles contain fluid. Autophagic vesicles contain portions of the cell's cytoplasm enclosed in membrane pinched off from the ER. Primary lysosomes, which contain digestive enzymes, fuse with vesicles to form secondary lysosomes within which the original vesicle contents are digested. Following digestion, the material remaining in residual bodies is expelled from the cell.

Figure 3.19 Electron micrograph showing the structure of a typical mitochondrion. Note the outer and inner mitochondrial membranes, the cristae, the mitochondrial matrix, and dense mitochondrial granules, 30–50 nm in diameter, lying in the matrix. This mitochondrion is surrounded by rough ER. (Magnification × 40,000)

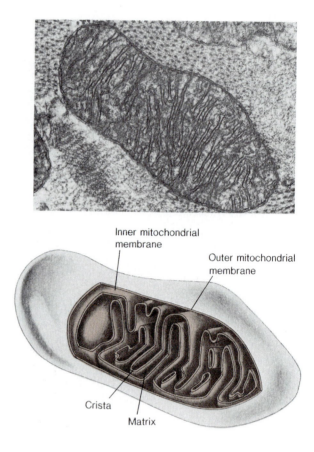

Inner mitochondrial membrane

Outer mitochondrial membrane

Crista

Matrix

which is separated from the **inner mitochondrial membrane** by a 6 to 8 nm space (figure 3.19). The inner membrane's surface area is greatly increased by foldings called **cristae.** The cristae vary in shape among various species and even among the cells in different tissues of a single organism. In mammals, cristae are often plate-like, while they are tubular in a number of protozoa and algae.

The inner membrane of a mitochondrion encloses a **mitochondrial matrix.** This matrix, which is fairly dense, contains ribosomes, DNA, and often large granules of calcium phosphate. It may seem logical to assume that since they contain DNA and ribosomes, mitochondria must contain genetic information independent of the nucleus and must be capable of synthesizing certain proteins using the ribosomes located in the matrix.

This indeed is the case. But mitochondrial ribosomes differ from ordinary cytoplasmic ribosomes; in fact, they resemble the ribosomes of bacteria in several ways, including their subunit composition (see figure 3.13). Similarly, mitochondrial DNA differs from that found in the nucleus. The DNA in a mitochondrion consists of a double-stranded molecule that forms a closed circle. These features have led many biologists to view mitochondria as self-reproducing units that live semi-independently within eukaryotic cells. This conclusion has important evolutionary implications, as will be discussed in dealing with the evolution of eukaryotic cells (chapter 35).

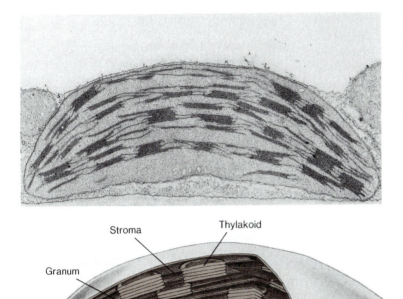

Figure 3.20 The structure of a chloroplast from a tobacco leaf *(Nicotiana tabacum)*. A complex network of grana, connected by membranous stroma lamellae, lies within the chloroplast matrix or stroma. The organization of this membranous network is diagrammed in the insert. (Magnification × 19,200)

Stroma

Thylakoid

Granum

Stroma lamella

Plastids

Plastids are plant cell organelles in which the synthesis and storage of food reserves takes place. There are two general classes of plastids—leucoplasts and chromoplasts. **Leucoplasts** are colorless plastids that are present in the cells of many plant tissues. These plastids manufacture and store starch, eventually developing into specialized starch-storing bodies. Leucoplasts also may serve as storage sites for fats, oils, and proteins. Many **chromoplasts** contain carotenoids and are yellow, orange, or red in color. These chromoplasts are the elements that give some fruits, flowers, seeds, and autumn leaves their hues. **Chloroplasts** are chromoplasts that contain large quantities of the green pigment chlorophyll, which is crucial to photosynthesis, the absorption of light energy to convert carbon dioxide and water to carbohydrates and oxygen.

In higher plants, chloroplasts are usually oval and around 2 to 4 μm by 5 to 10 μm. Like mitochondria, chloroplasts are encompassed by a double membrane (figure 3.20). Also like mitochondria, chloroplasts have a matrix, the **stroma,** that lies within the inner membrane and contains DNA and ribosomes. Chloroplasts and mitochondria both carry circular DNA molecules. The chloroplast matrix also contains starch granules.

Again like mitochondria, a chloroplast has a complex internal membrane system. The most prominent components of this system are the numerous **grana** (singular: **granum**). These grana are stacks of between five and thirty flattened, hollow sacs of vesicles called **thylakoids.** In an electron micrograph, a granum looks much like a stack of coins, about 300 to 600 nm in diameter.

Various reactions involved in photosynthesis are structurally isolated in the chloroplast, as are certain reactions occurring in the mitochondrion. The structure and function of chloroplasts will be further explored in the section on photosynthesis in chapter 6.

Cell Vacuoles

Most mature plant cells, and some animal and protist cells, contain **vacuoles** (figure 3.21). Vacuoles are much more prominent in plant cells because they occupy a greater proportion of the space in mature plant cells than the vacuoles in animal cells do. In fact, well over half of a plant cell's volume may be composed of a fluid-filled, central vacuole enclosed by a single membrane. This vacuolar fluid or **cell sap** contains relatively high concentrations of stored materials, such as sugars, organic acids, proteins, pigments, and salts.

In addition to acting as storage facilities, vacuoles also play a role in the growth of plant cells. A young plant cell, immediately after it is produced by cell division, generally has only small vacuoles scattered throughout its cytoplasm. As the cell matures, these small vacuoles fuse and increase in size because of water intake. The wall of a young cell is still elastic, and it stretches in response to the pressure caused by vacuole enlargement. Eventually, the whole cell enlarges as the vacuoles continue to grow.

The vacuoles of many plant cells contain water-soluble pigments called **anthocyanins.** These pigments are responsible for the blue, violet, and scarlet colors of parts of plants such as radishes, beets, plums, cherries, and a number of flowers. When anthocyanins are present in vacuoles along with carotenoids (yellow pigments) and xanthophylls (brown pigments), they give autumn leaves their bright and beautiful colors. These pigments accumulate in leaves during the growing season and then become visible when the chlorophyll that has masked them breaks down in the autumn.

Microfilaments and Microtubules

Many types of cells have minute protein filaments in their cytoplasmic matrices that are approximately 4 to 7 nm in diameter. These **microfilaments** may be scattered randomly in the matrix or may be organized in networks or parallel arrays (figure 3.22). Microfilaments are thought to play a major role in cell motion and shape changes. Examples of cellular movements that microfilaments are involved in include the movement of various granules, cytoplasmic streaming within cells, and the formation of the furrow that actually splits a cell in two during cell division. Experiments using the drug **cytochalasin B** provide evidence of the involvement of microfilaments in cell movement. Cytochalasin B disrupts microfilament structure and, at the same time, stops many cell movements.

Microfilaments have been isolated from a number of cells and analyzed chemically. Interestingly, they are very similar in composition to *actin,* one of the proteins responsible for contractions of muscle. Thus, it seems that the same kinds of proteins are involved in most, if not all, kinds of movement in living cells.

There is a second type of small organelle found in the matrices of many cells that is also involved in maintaining cell shape and in some cell movements. These organelles, the **microtubules,** are thin cylinders several times larger than microfilaments (about 25 nm in diameter). Microtubules are complex structures constructed of two types of slightly different spherical subunits, each made up of a protein called **tubulin.** Each subunit is approximately 4 to 5 nm in diameter. They unite to form the cylindrical shape of the microtubule, which is usually thirteen subunits in circumference (figure 3.22b).

Microtubules are located in long, thin cell structures that require support, such as the extended axons and dendrites of nerve cells and the axopodia (long, slender, rigid extensions) of some protozoa (figure 3.23). The drug **colchicine** has been used to demonstrate the supporting function of microtubules. Colchicine binds with microtubular subunits so that they can no longer combine to form microtubules: when cells are exposed to colchicine, they lose their microtubules and their characteristic shape. Colchicine-treated cells in a culture vessel wander as if they are incapable of directed movement without their normal form.

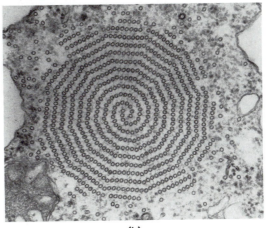

(a)

(b)

Figure 3.21 Development of the vacuole in a growing plant cell. A young plant cell (top) has several vacuoles scattered in its cytoplasm. These small vacuoles fuse into a single vacuole that expands due to water uptake. The vacuole occupies more than half of the volume of a mature plant cell (bottom).

Figure 3.22 Microfilaments and microtubules. These organelles penetrate much of the cytoplasmic matrix and are largely responsible for maintaining the shape of the cell. Thus, they collectively are called the cytoskeleton ("cell skeleton"). They also are involved in several types of movement of cells or organelles. (a) Microfilaments in a rat embryo cell. The microfilaments have been made visible by treatment with fluorescent antibodies that bind specifically with their protein and very little with other cell proteins. (Magnification × 632) (b) Microtubule structure. A microtubule is a thin cylinder composed of two types of spherical protein (tubulin) subunits, which are shown as colored and gray spheres. These subunits are stacked in a helix, and in most microtubules, the tubule wall is thirteen subunits in circumference.

25 nm

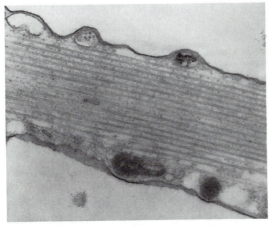

(a)

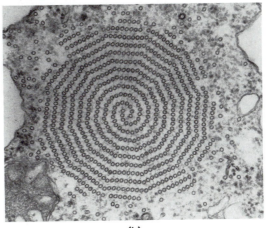

(b)

Figure 3.23 Electron micrographs of axopodia (long, slender cell extensions) from the protozoan *Echinosphaerium nucleofilum* showing microtubules. (a) Longitudinal section. (Magnification × 40,200) (b) Cross section through the axopodium. (Magnification × 44,400) The axopodium is filled with an array of microtubules organized in a spiral pattern.

Their microfilaments are still intact; but because of the disruption of their microtubules by colchicine, they no longer behave normally.

Microtubules also are found in cell structures that are clearly involved in cell or organelle movements. For example, the mitotic spindle (the complex structure to which chromosomes attach during cell division) is constructed of microtubules. When a dividing cell is treated with colchicine, the spindle is disrupted and chromosomal separation ceases. The movement of cilia and flagella also results from the action of microtubules; their movements are unaffected by the drug cytochalasin B but are inhibited by colchicine.

Cilia and Flagella

Cilia (singular: **cilium**) and **flagella** (singular: **flagellum**) are long, slender, hairlike organelles associated with cell movement. They are the means by which swimming cells—including many unicellular organisms—propel themselves; and in certain stationary cells, cilia and flagella move material over the cell's surface. In the epithelial cells lining human respiratory passages, for example, cilia carry mucus and dirt away from the lungs.

Cilia and flagella differ from one another in two ways. First, cilia may be only 10 to 20 μm long, while flagella are much longer (about 100 to 200 μm in length). Second, the two organelles move differently (figure 3.24). A flagellum moves in an undulating fashion—its "beat" can originate at either its base or its tip. A cilium's beat, on the other hand, is usually divided into two phases. In the effective stroke, the cilium pulls through the fluid around the cell somewhat like an oar, thereby propelling the cell through the fluid. Then the cilium bends along its length as it is pulled forward during the recovery stroke. Cells possessing cilia actually coordinate their beats so that some cilia are in the recovery phase while others are carrying out their effective strokes.

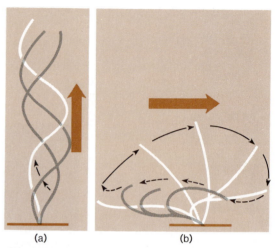

(a)   (b)

Figure 3.24 Flagella and cilia. (a) Flagellar movement often takes the form of waves that move either from the base of the flagellum to its tip or in the opposite direction. The motion of these waves propels the organism along. (b) The beat of a cilium may be divided into two phases. In the effective stroke (white), the cilium remains fairly stiff as it swings through the water. This is followed by a recovery stroke (gray) in which the cilium bends and returns to its initial position. The large arrows indicate the direction of water movement in these examples.

Despite their differences, cilia and flagella are very similar in ultrastructure (structure as revealed by the electron microscope). Both are membrane-bound cylinders (about 0.2 μm in diameter) that enclose a matrix. In this matrix is a complex structure, the **axoneme** or **axial filament,** which consists of nine pairs of microtubules arranged in a circle around two central tubules (figures 3.25 and 3.26). This construction is called a 9 + 2 pattern of microtubules. The microtubules in an axoneme are quite similar to those found in cytoplasm. Each microtubule pair or doublet also has pairs of arms projecting toward a neighboring doublet and spokes extending toward the central pair of microtubules. Recent evidence indicates that cilia and flagella actually move because the microtubule doublets slide along one another. The doublet arms and spokes seem to be involved in this sliding action, and ATP energy is required for these movements.

Each cilium or flagellum has a **basal body** lying in the cytoplasm at its base. A basal body is a short cylinder with a circular arrangement of nine microtubule triplets (a 9 + 0 pattern). This arrangement is also characteristic of another organelle, the centriole.

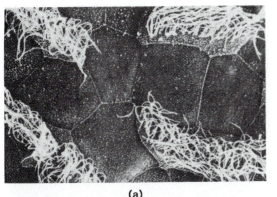

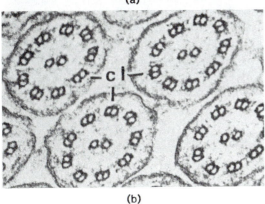

Figure 3.25 Cilia structure. (a) Scanning electron micrograph of ciliated cells on the gill of an amphibian (*Ambystoma mexicanum*) embryo. (Magnification × 366) (b) Electron micrograph showing cross sections through cilia (cl) of *Euplotes eurystomes*. Note the two central microtubules surrounded by nine microtubule doublets. (Magnification × 79,920)

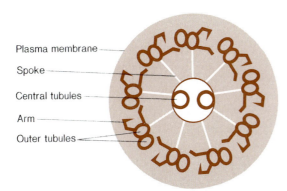

Plasma membrane

Spoke

Central tubules

Arm

Outer tubules

Figure 3.26 Ultrastructure of a cilium in cross section. Arms extend from each microtubule doublet toward a neighboring doublet, and spokes extend toward the central pair of microtubules.

Centrioles

Centrioles have essentially the same structure as basal bodies, but they usually occur in pairs. They are small cylinders (0.15 to 0.20 μm by 0.3 to 0.5 μm) that have nine tubule triplets embedded in a matrix (figure 3.27). The end of a centriole has a cartwheel-like appearance, but the "hub" of the wheel is not a microtubule. In fact, it is not clear what this central element actually is.

Centrioles give rise to the basal bodies, which direct cilia and flagella formation. Centrioles are also closely associated with the mitotic spindle in most dividing cells except those of higher plants. However, their actual role in cell division is not understood as yet.

The Plant Cell Wall

The most prominent structure of a plant cell, the **cell wall,** actually lies outside the plasma membrane. The cell wall is secreted by the cell and gives an individual plant cell its shape and rigidity. The structure and chemical nature of the cell wall varies depending on the type of cell and its function. For example, cell walls are thin in fruits and leaves, while cells in mature stems and nuts possess thick walls organized somewhat differently.

Cells from soft tissues have cell walls composed of two layers. Each cell produces a thin **primary wall**, and the primary walls of adjacent cells are cemented together by a layer called the **middle lamella.** In addition to primary walls and middle lamellae, the cells of hard, woody tissues form a **secondary cell wall** (figure 3.28) between the primary wall and the plasma membrane. The secondary cell wall may have several layers, depending on the type of tissue it is found in and the stage of tissue development. The secondary cell wall makes the boundary around each cell much thicker and harder than it would be if the wall consisted of a primary wall alone.

Figure 3.27 Electron micrograph of
a pair of centrioles. One is seen
from the end and the other from the
side. (Magnification $\times$ 142,000) The
accompanying sketch shows
microtubule triplets arranged in a
9 + 0 pattern in a centriole.
Centrioles are structurally similar and
functionally related to the basal
bodies of cilia and flagella that play
a role in assembling the
microtubules of those organelles.

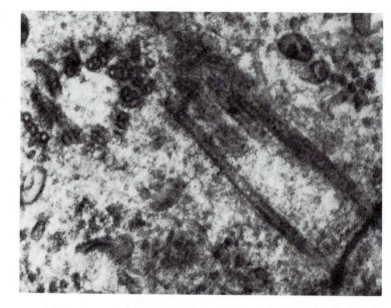

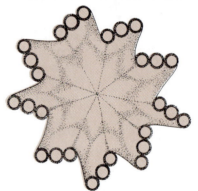

Figure 3.28 A diagrammatic
sketch showing the layers of a plant
cell wall.

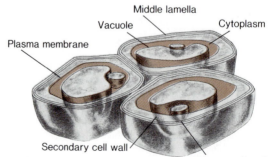

A young cell in a growth zone synthesizes a primary wall first. In the beginning, this wall consists of a loose network of fibrils that can expand and change shape as the cell enlarges. As the cell approaches maturity, it manufactures a secondary wall inside the primary wall, and the size and shape of the cell become much more rigid. In broad terms, the secondary cell wall consists of aggregates of cellulose (held together by other polysaccharides) and a substance called lignin. **Cellulose** is a large, linear polymer of glucose molecules (see chapter 2). The average length of a cellulose chain is about 3,000 glucose units. These long cellulose molecules associate with one another to form **microfibrils,** which are stabilized by hydrogen bonding between them. **Lignin** is a complex mixture of polymers that adds to the hardness of the cell wall. Some cell walls also contain other substances, such as pectin, which functions to cement cellulose microfibrils together, and wax.

The middle lamella is composed mainly of **calcium pectate** (the calcium salt of pectin) and small amounts of other polysaccharides, excluding cellulose. The physical consistency of a plant tissue depends heavily on the condition of the calcium pectate in the middle lamellae between its cells. For example, as fruits ripen, the pectin of the middle lamella is converted to a more soluble form and the cells become loosely associated. The effects of excessive pectin loss are present, for example, in overripe, mealy apples. Home canners are familiar with pectin because concentrated pectin extracts are added to fruit juices to make jelly. In woody tissues, the middle lamella also contains considerable amounts of lignin, which contributes even more hardness to the tissue.

Mature plant cell walls are quite rigid and may serve functions other than simple mechanical support. For example, the vascular tissue **xylem** provides support for the plant stem but also conducts water to various parts of the plant. It is composed of elongated tubes produced by cells that have died after producing very strong, thick secondary walls. Perforations at the ends of these dead cells transform chains of them into noncollapsible conducting tubes.

Cell-wall constituents from many plants are very useful in an economic sense. Wood used in construction is probably the first that comes to mind, but there are many other economically important cell-wall products. Cotton is almost pure cellulose. Paper is a cellulose product that is manufactured from wood treated with dilute acids or alkalies to remove other constituents. Lignin is extracted from wood and used to produce pigments, adhesives, vanilla flavoring, and many other items. And, as we noted earlier, pectins are important to the food industry as gelling agents used in making jams and jellies.

The Eukaryotic Cell: A Summary

Composite views of two eukaryotic cells, including many of the organelles just discussed, are pictured in figure 3.29. A comparison of this illustration with E. B. Wilson's sketch, done in 1925 (figure 3.3), indicates how far cell biology has progressed in recent years. Obviously, the cell is not a simple sac of granular fluid with a nucleus at its center, as it was once visualized. The development of many new research techniques has enabled modern biologists to construct a much more detailed and accurate picture of cell structure and function.

The most striking characteristics of the eukaryotic cell are the high degree of internal organization and the predominance of membranous structures. Membranes divide the cell into a myriad of individual compartments. Cell processes can take place simultaneously within various compartments, but still can be independent of one another because of their physical separation. This makes possible a greater variety and complexity of cell function than would otherwise be feasible.

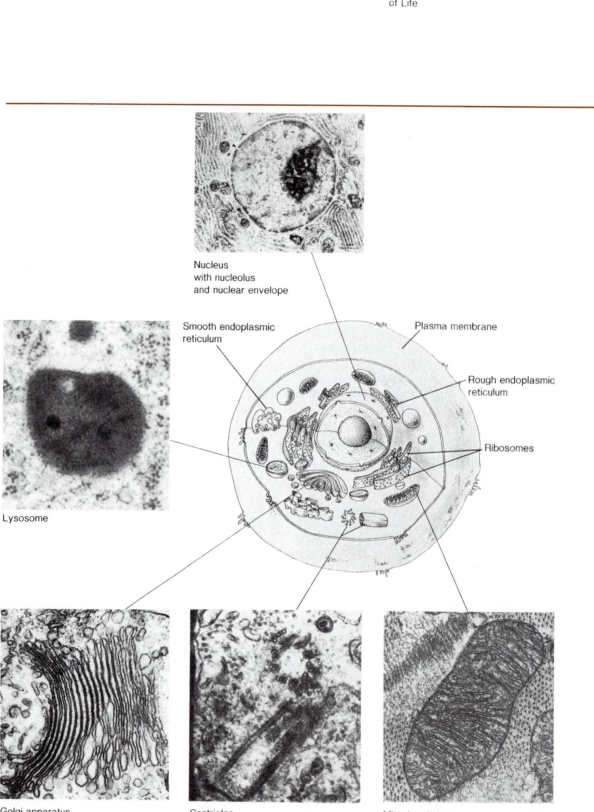

Nucleus
with nucleolus
and nuclear envelope

Smooth endoplasmic
reticulum

Plasma membrane

Rough endoplasmic
reticulum

Ribosomes

Lysosome

Golgi apparatus

Centrioles

Mitochondrion

(a)

Figure 3.29   Diagrammatic
representations of eukaryotic cells
based on electron micrographs.
(a) A generalized animal cell. (b) A
generalized plant cell. The vacuole
actually occupies a greater
percentage of the cell's volume in
the average mature plant cell than it
does in this sketch.

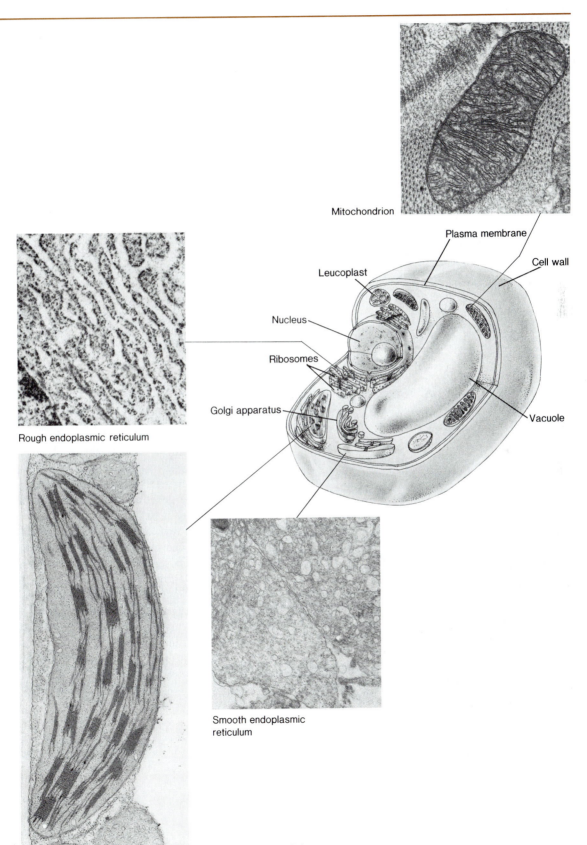

Mitochondrion

Plasma membrane

Cell wall

Leucoplast

Nucleus

Ribosomes

Golgi apparatus

Vacuole

Rough endoplasmic reticulum

Smooth endoplasmic
reticulum

Chloroplast

(b)

Figure 3.30 Electron micrograph of a prokaryotic cell undergoing cell division. Parts of the cell are identified in the accompanying sketch. This micrograph is of a bacterium, *Bacillus megaterium.* (Magnification × 63,960) Notice that there is a nuclear body (nucleoid) rather than a membrane-enclosed nucleus like that of a eukaryotic cell, and that the cell lacks membrane-bounded organelles. Mesosomes are inward-folded portions of the plasma membrane. Mesosomes may function to increase the surface area available for absorption of nutrients. They also may be involved in cell division processes.

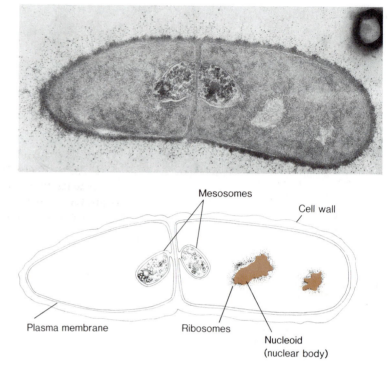

The Prokaryotic Cell

Thus far, the discussion of cell structure has focused on eukaryotic cells, cells that possess a membrane-enclosed nucleus and an extensive intracellular membrane system. However, there is another, fundamentally different type of cell called the **prokaryotic** cell. Prokaryotic cells lack a true membrane-enclosed nucleus. Bacteria and cyanabacteria (blue-green algae) are prokaryotes (figure 3.30). All other organisms—protists, fungi, plants, and animals—are eukaryotes. In other words, these organisms are eukaryotic cells or aggregates of eukaryotic cells.

Although the presence or absence of a true nucleus is the most obvious difference between these two cell types, there are other significant distinctions, as noted in table 3.2. As is clear from table 3.2, prokaryotic cells are structurally much simpler; they lack both the diverse collection of membrane-bound organelles and the high degree of functional compartmentalization that are features of eukaryotic cells.

Despite the many significant differences between these two basic cell forms, they are surprisingly similar at the biochemical level. Although prokaryotes carry circular DNA, as opposed to the linear DNA in the nuclei of eukaryotic cells, the genetic code is the same in both, and the way in which the genetic information in DNA is expressed is similar. The principles underlying the metabolic processes and a majority of the important metabolic pathways in both types of cells are identical. Thus, beneath the significant structural and functional differences that separate prokaryotes from eukaryotes, there is an even more fundamental molecular unity that is basic to life processes.

Table 3.2
Comparison of Prokaryotic and Eukaryotic Cells.

Property	Prokaryotes	Eukaryotes
Organization of Hereditary Material		
True nucleus	Absent	Present
DNA complexed with histones	No	Yes
Mitosis occurs*	No	Yes
Other Cell Components		
Mitochondria	Absent	Present
Chloroplasts	Absent	Present
Plasma membrane has sterols, such as cholesterol	Usually no	Usually yes
Flagella	Submicroscopic in size; composed of one fiber	Microscopic in size; membrane bound; usually twenty microtubes in 9 + 2 pattern
Ribosomes	70S†	80S (except in mitochondria and chloroplasts)
Lysosomes	Absent	Present
Microtubules	Absent or rare	Present

*It should be noted also that sexual reproduction with meiosis is absent in prokaryotes but present in eukaryotes.
†S = Svedburg unit, a measurement used to compare sizes of molecules and molecular aggregates (see figure 3.13).

Summary

Biologists study cells intensively because they are the fundamental units of life. All organisms are either cells or aggregates of cells and cell products.

Cells vary considerably in shape and size. Lower size limits are determined by certain minimal components a cell needs to be a functional unit. Upper size limits seem to be determined by factors related to the exchange and movement of materials in and out of the cell.

Modern cell biology has risen from the application of modern techniques to cell studies. Electron microscopy makes possible high-magnification studies of cell structure not possible with the light microscope. Cytochemical techniques and autoradiographs are used to determine the exact locations of specific substances inside cells. Cell fractionation techniques permit isolation and functional analysis of cell organelles.

Eukaryotic cells contain a number of types of structurally and functionally specialized, membrane-bound compartments called organelles. Organelles participate in an intracellular "division of labor"—different sets of functions are associated with different organelles.

The plasma membrane is a membranous enclosure around the cell. It holds cell constituents together and controls exchanges between the cell's interior and the external environment.

Organelles are suspended in the cytoplasmic matrix, a watery medium that contains proteins and other materials. The cytoplasmic matrix undergoes sol-gel transformations and shows other properties of colloids.

A nuclear envelope with many pores encloses the nucleus. Chromatin is the dispersed form of the cell's genetic material. Chromatin condenses into chromosomes during cell division. The nucleolus participates in ribosome synthesis and is thus deeply involved in cell protein synthesis also.

Ribosomes, the sites of protein synthesis in cells, are complex particles that are assembled from smaller subunits. Some protein synthesis occurs on ribosomes that are free in the cytoplasmic matrix, but most occurs on ribosomes bound to the complex cytoplasmic membrane network known as the endoplasmic reticulum (ER). Rough ER has ribosomes attached to it. Smooth ER carries no ribosomes.

The Golgi apparatus consists of a stack of saclike cisternae. It packages cell products for export from the cell or, as in the case of lysosomes, for isolation from other cell contents. Lysosomes, which contain hydrolytic enzymes that digest cell components, function in defense reactions, in the turnover of cell materials, and in the elimination of cell debris.

Important energy conversions occur in mitochondria. Mitochondria are complexly organized double-membrane organelles containing enzyme systems that function in cell respiration.

Plastids are a diverse group of plant organelles, many of which contain pigments. Some plastids have storage functions. Chloroplasts, the organelles involved in photosynthesis, have very complex membrane structures bearing chlorophylls, other pigments, and important enzyme systems.

Both plant and animal cells contain vacuoles, but plant-cell vacuoles are much more prominent organelles. The cell sap enclosed in plant vacuoles is important for plant-cell osmotic relationships.

Microfilaments and microtubules are involved in maintaining cell shape and in many types of cell movement. The two can be differentiated on the basis of different ultrastructures and different sensitivities to experimental inhibitors.

Cilia and flagella are both movement organelles. They beat differently but have a similar internal ultrastructural arrangement of microtubules. Their basal bodies are very similar to centrioles. Centrioles are involved in cell division in protist and animal cells, but not in higher plant cells.

Cell walls are multilayered enclosures around plant cells that vary in thickness and composition depending on the age and function of the cells.

Eukaryotic and prokaryotic cells differ in a number of structural and functional respects. The most obvious difference is that prokaryotic cells lack true nuclei and membrane-bound organelles; thus, they do not have the kind of functional compartmentalization that eukaryotic cells have. But many cell processes in the two types of cells are similar at the biochemical level. This demonstrates that though there are important structural and organizational differences between the two, some of their life processes share a fundamental molecular unity.

Questions

1. Lens systems could be built for light microscopes that provide more magnification than commonly used light microscopes have. Explain why a light microscope with much higher magnification can not be substituted for an electron microscope in the study of cell structure.

2. Lysosomes contain enzymes that catalyze the hydrolysis of many important types of molecules. Explain why other cell constituents are safe from digestion by these enzymes and how this illustrates the significance of compartmentalization in eukaryotic cells.

3. What distinguishes smooth endoplasmic reticulum (SER) from rough endoplasmic reticulum (RER)?

4. A number of plant and animal cell organelles are similar, but there are several significant differences between animal cells and the cells of higher plants. List and explain at least three of these differences.

5. One evolutionary hypothesis proposes that during the evolution of eukaryotic cells, prokaryotic organisms actually were incorporated into cells, where they became established as permanent cell constituents (organelles). This hypothesis suggests, for example, that mitochondria and chloroplasts may have originated in this way. How would evidence regarding DNA organization and ribosome structure support this hypothesis?

Suggested Readings

Books

Avers, C. J. 1976. *Cell biology*. New York: Van Nostrand.

DeRobertis, E. D. P., and DeRobertis, E. M. F., Jr. 1980. *Cell and molecular biology*. 7th ed. Philadelphia: Saunders.

Dyson, R. D. 1978. *Cell biology*. 2d ed. Boston: Allyn & Bacon.

Karp, G. 1979. *Cell biology*. New York: McGraw-Hill.

Wolfe, S. L. 1980. *Biology of the cell*. Belmont, Calif.: Wadsworth.

Articles

Allison, A. C. 1977. Lysosomes. *Carolina Biology Readers* no. 58. Burlington, N.C.: Carolina Biological Supply Co.

Cook, G. M. W. 1975. The Golgi apparatus. *Carolina Biology Readers* no. 77. Burlington, N.C.: Carolina Biological Supply Co.

Dustin, P. August 1980. Microtubules. *Scientific American* (offprint 1477).

Jordan, E. G. 1978. The nucleolus. *Carolina Biology Readers* no. 16. Burlington, N.C.: Carolina Biological Supply Co.

Lazarides, E., and Revel, J. P. May 1979. The molecular basis of cell movement. *Scientific American* (offprint 1427).

Porter, K. R., and Tucker, J. B. March 1981. The ground substance of the living cell. *Scientific American* (offprint 1494).

Satir, B. October 1975. The final steps in secretion. *Scientific American* (offprint 1328).

Homeostasis and Cell Exchanges

Chapter Concepts

1. Homeostasis is a dynamic steady state in which there is variation, but only within an acceptable range.
2. Homeostasis at the cellular level depends on the cell's selective exchange of materials with its environment.
3. Some materials penetrate the plasma membrane by simple diffusion. Water enters and leaves the cell by a special form of diffusion called osmosis. Other materials penetrate the cell membrane by carrier-mediated processes that are solute specific. No energy input is required in facilitated diffusion. In active transport, cells expand energy to move materials across membranes.
4. Endocytosis involves vesicle formation at the cell surface and permits materials that do not penetrate the plasma membrane by other means to enter the cell. Exocytosis involves vesicle fusion with the plasma membrane and is a mechanism for moving materials out of the cell.
5. In unicellular organisms, all necessary cell exchanges are made directly with the external environment.
6. In multicellular organisms, cell specialization results in a division of labor that contributes to the maintenance of a stable internal cellular environment.

All living cells are dynamic systems that carry on continuing interactions with their environments. To maintain its normal structure and function, a cell must obtain some materials from and release others into its environment. These materials include organic compounds, gases, certain ions, and wastes. The processes involved in exchanging such materials are basic to life at the cellular level.

All cells require energy, and they obtain it through reactions that degrade organic compounds, such as glucose, into simpler molecules, such as carbon dioxide and water. Thus, cells require a supply of reduced organic compounds to meet their energy requirements. Some cells meet their needs for these compounds through photosynthesis; they trap light energy for the synthesis of organic compounds (carbohydrates) from carbon dioxide and water. Other cells must obtain organic compounds from their environments.

Cells must also exchange two important gases, oxygen (O_2) and carbon dioxide (CO_2), with their environments. In addition to carbon dioxide, other metabolic wastes must be eliminated because many of them can accumulate to toxic levels if they are not released as they are produced.

In addition to organic compounds and gases, a number of cell functions require specific ions, so that a cell must maintain appropriate internal concentrations of these ions. Doing so requires the cell to work; because the ionic composition of the cell interior is usually not identical to the environment in either kind or quantity of ions, the cell must expend energy to regulate its internal ionic composition. For example, some marine algae maintain a much higher internal potassium ion (K^+) concentration than is found in the seawater around them. At the same time, their intracellular sodium ion (Na^+) concentration is much lower than that of the surrounding water. Such ionic differentials are critical for many specialized cell functions.

Living cells are usually able to meet all of these and other requirements for maintaining relatively stable internal conditions. But the environment inside a cell is certainly not absolutely constant. Fluctuations occur constantly in nutrient levels, waste accumulation, ionic concentrations, and many other parameters, but certain mechanisms normally prevent these fluctuations

from becoming excessive or dangerous. This condition, then, in which cells are relatively stable yet experience constant variation within certain ranges, might be termed *dynamic steady state*. Walter Cannon called this general state of controlled dynamic steady state **homeostasis.** Cannon actually applied this term to whole multicellular organisms that maintain stable inner environments by internal regulation and selective exchange with the environment. But the same principles apply to cells and their environments.

Exchanges between Living Cells and Their Environments

To maintain homeostasis, a cell must make appropriate exchanges with its environment that permit it to stabilize its internal environment even in the face of external changes. The plasma membrane of the cell is the chief region of contact between the cell and its environment. One of the most important functions of the plasma membrane is the regulation of the movement of substances into and out of the cell. There are several different means by which various materials pass through plasma membranes.

Diffusion

Diffusion is the tendency for materials to move from an area in which they are highly concentrated to an area of lower concentration. The phenomenon of *thermal agitation* plays an important role in diffusion.

All molecules (and ions) are in a state of constant motion because of heat energy. This movement, called thermal agitation, can be accelerated by applying more heat. Although increased heat makes the constituent molecules of a material move around more rapidly, the distance that molecules can move without encountering others of their kind depends on the nature of the material. In a solid, movement is limited; for example, the movement of water molecules within a piece of ice is severely restrained. There is more freedom of movement for molecules in a liquid, which is a less orderly arrangement of molecules, and even more freedom of movement in a gas, which is still less orderly.

The process of diffusion, and the part that thermal agitation plays in it, can be demonstrated by a container filled with water and a crystal of copper sulfate placed in one corner of the container. The crystal will dissolve, forming a halo of color as the blue cupric ions move slowly away from the crystal and into the surrounding water. The thermally agitated ions tend to spread from regions where they are highly concentrated to regions where they are less concentrated (figure 4.1). As long as there is a concentration gradient, the ions will move toward the low-concentration areas.

As the crystal of copper sulfate dissolves, each ion is buffeted about continually by thermally agitated water molecules and other cupric ions. Each ion appears to move about randomly—sometimes toward more concentrated areas and sometimes in other directions. The cupric ions will gradually move from the area of higher concentration to the area of lower concentration, because on the average, the ions can move farther in that direction before they collide with other cupric ions. Thus, the diffusion of the ions will eventually result in an equilibrium in which the cupric ions are evenly dispersed throughout the water.

The process of diffusion can also be described as the tendency to move from order to disorder. The cupric ions in the example were in a fairly orderly condition when the crystal first began to dissolve. After a time, when they had become dispersed in the water, the ions were in a much less orderly arrangement and could move about more without encountering other cupric ions. This is just one example of **entropy,** or the general tendency of systems to move from orderliness toward a state of increasing disorder. Entropy is one of the driving forces in chemical and physical processes, including those that occur in living things.

Some substances, such as oxygen and carbon dioxide, diffuse freely through plasma membranes. Biologists call this movement **simple (or passive) diffusion** because the material enters and leaves the cell without active involvement of the cell or its plasma membrane.

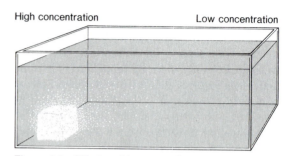

High concentration   Low concentration

Figure 4.1 Diffusion of ions away from a crystal of copper sulfate placed in water. A halo of blue color surrounds the crystal where cupric ions are most concentrated. On the average, ions can move further in the direction of decreasing concentration before colliding with other ions of their own kind, so long as a concentration gradient exists. Diffusion eventually produces an equilibrium in which ions are evenly distributed throughout the water.

Osmosis

Osmosis is a specialized case of diffusion involving the passage of a solvent such as water through a selectively permeable membrane. Osmosis can readily be demonstrated with an apparatus consisting of two compartments separated by a selectively permeable membrane—one that will allow water, but not a solute, to cross it. In figure 4.2, solution 1 is pure water, while solution 2 is a mixture of sucrose and water. The membrane will allow water to pass through, but not sucrose. In the context of osmosis, a solution is considered in terms of the concentration of solvent (water), rather than in terms of the concentration of solute. The concentration of the water in solution 1 is greater than in solution 2 because of the presence of solute molecules in the latter. Since any material will tend to diffuse from a region of higher concentration to one of lower concentration, the water will tend to flow from area 1 to area 2.

When water enters area 2 in figure 4.2, the volume of the solution there increases slightly and rises in the capillary tube. Eventually the solution reaches a height (h) at which an equilibrium is reached and, consequently, the solution stops rising. This equilibrium is reached when the weight of the column of liquid exerts exactly enough pressure to balance or counteract the tendency for net water movement from solution 1 into solution 2. The pressure required to halt net solvent movement across a true selectively permeable membrane is called **osmotic pressure**

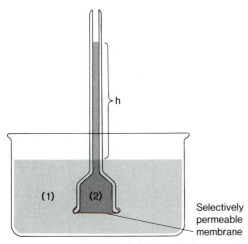

Figure 4.2 An osmometer. Solution (1) is pure water. Solution (2) is sucrose and water. The selectively permeable membrane is permeable to water but not to sucrose. The height (h) that the solution reaches is proportional to the osmotic pressure (π).

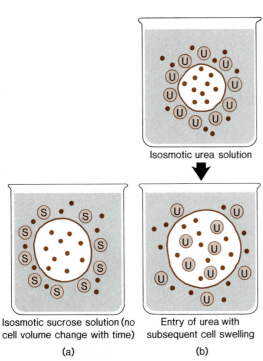

Figure 4.3 Isosmotic and isotonic solutions. In (a) the isosmotic sucrose solution does not cause a change in cell volume and is therefore isotonic as well. This is because sucrose molecules cannot penetrate the membrane. In contrast, the isosmotic urea solution in (b) is not isotonic because urea can enter the cell, which it does because of the urea concentration gradient between the outside and the inside of the cell. Water follows the urea osmotically, and the cell swells.

(π). Osmotic pressure increases linearly with solute concentration (because the higher the solute concentration, the lower the water concentration). A device like the one pictured in figure 4.2, which can measure the osmotic pressure of a solution, is an **osmometer.**

Osmosis is of great importance to cells and organisms because under some conditions cells face problems due to osmotic effects. When cells are exposed to solutions with the same osmotic pressure as their contents, there are different outcomes depending on membrane permeability to particular substances. If a solute does not pass through the membrane, osmosis does not occur. For example, if an erythrocyte (a red blood cell) is placed in a 0.31 M sucrose solution, the cell will not change size or shape. A 0.31 M sucrose solution is **isosmotic** to red blood cells. Not only do the cell's contents have the same osmotic pressure as the solution so that there is no net movement of water in or out of the cell, but the cell's membrane does not allow sucrose to freely enter the cell.

On the other hand, substances such as urea can move freely through the membrane of a red blood cell. When a red blood cell is placed in an isosmotic urea solution, urea molecules diffuse into the cell because of the urea concentration gradient between the outside and the inside of the cell. As the influx of urea changes the cell's internal osmotic pressure, water flows into the cell by osmosis and the cell swells, even though the original urea solution was isosmotic (figure 4.3).

Osmotic terms are used to describe the concentration of solute particles in solutions. But terms of **tonicity** (*isotonic, hypertonic, hypotonic*) describe the effects of solutions on cell volumes and are thus closely tied to the permeability characteristics of the plasma membrane. Tonicity is functionally important because cells must be in a stable water-balance relationship with their environment if they are to function normally.

Box 4.1
Hypertonic Medium, Shrunken Cells, and Sea Urchin Twins

Many years ago, German biologist Hans Driesch studied the effects of separating the two cells produced by the first division of a sea urchin egg following fertilization. Driesch separated these embryonic cells by shaking them vigorously in seawater in a covered container. Surprisingly, some cells survived this rough treatment and developed further. But instead of producing half of a sea urchin larva, as each cell normally would if the cells were left together, each cell divided, produced a multicellular embryo, and developed into a complete larva that was normal in appearance but one-half normal size.

In 1940, Ethel Browne Harvey developed a method for separating embryonic sea urchin cells that is far less likely to damage them than is the Driesch method. In Harvey's technique, just as the first cell division is being completed, the embryos are transferred from seawater to a hypertonic solution—water with twice the salt content of normal seawater. The cells lose water by osmosis and shrink so that the members of each cell pair pull apart and lose contact with each other. After five or ten minutes, the embryos are returned to normal seawater, where they gradually regain their normal water content and size. But the cells no longer lie in intimate contact with each other as they did before they were subjected to the hypertonic medium, and each cell develops independently, just as the cells did after they were shaken apart in Driesch's experiments. Again, each cell produces a normal but half-size larva (box figure 4.1A).

The Harvey technique makes it possible for biology students and researchers to separate embryonic sea urchin cells without seriously damaging them. This technique also allows researchers to study the development of the half-size "twin" embryos that are produced by the treatment. This useful separation technique is based on the osmotic effects of placing living cells in a hypertonic solution and then returning them to an isotonic medium.

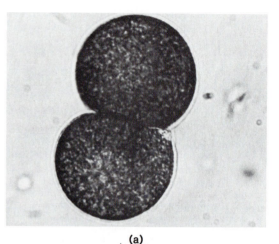

(a)

(b)

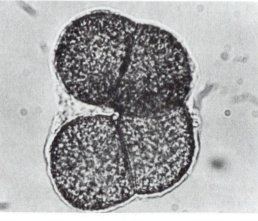

(c)

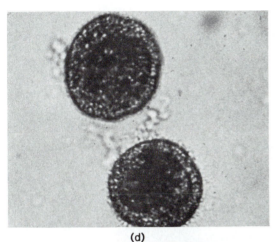

(d)

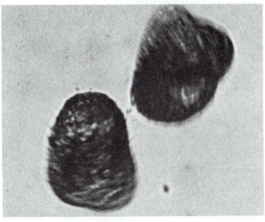

(e)

(f)

Box Figure 4.1A   Separation of embryonic sea urchin cells by Harvey's technique, which uses a hypertonic medium to shrink cells apart. (a) Two cells at the end of the first cell division in development of a sea urchin embryo. (A structure called the fertilization membrane, which normally surrounds sea urchin embryos, has been removed from this embryo.) (b) The two cells in hypertonic solution. Note the shrunken appearance of the cells and the way that they have pulled away from one another. (c) The cells after they have been returned to normal seawater and a second division has occurred. Cells regain most of their water and their total cell volume when returned to normal seawater, but they are not in close contact as they were before the treatment. Thus, each cell's development proceeds independently. (d) A stage (blastula) in development of the cells into two separate small embryos still attached to each other. (e) Swimming blastula stage. The physical force of swimming usually results in separation of the two embryos. (f) Metamorphosed Siamese urchin formed due to incomplete twinning, which sometimes occurs in such experiments.

A solution that does not cause volume changes in cells is called **isotonic.** In the red blood cell example, the 0.31 M sucrose solution is both isosmotic and isotonic, since sucrose molecules do not penetrate the red cell membrane and there is no volume change. But the isosmotic urea solution is not isotonic for red blood cells. As the urea and water enter the blood cell, the cell's volume increases.

A **hypertonic** solution is one in which the cell tends to lose volume because of water loss to its environment. When an erythrocyte is transferred from an isotonic solution to a hypertonic solution (for example, 0.4 M sucrose), water tends to leave the cell, and the cell shrinks (figure 4.4). In a **hypotonic** solution, water from the extracellular solution will flow into the cell. A red blood cell will eventually lyse (burst) and release its hemoglobin into the suspending medium if it remains in a hypotonic solution for any length of time.

Thus far, the examples used to illustrate osmosis have all involved animal cells. Plant cells also react to osmotic changes in their environments, but the situation is more complex in plant cells because of the presence of a cell wall around the plasma membrane. In a hypotonic solution, water enters a plant cell until the counterpressure of the plant cell wall equals the difference between the osmotic pressure of the cell and that of its environment (figure 4.5). In other words, when the restraining force of the cell wall equals the outward push of the water entering the cell by osmosis, the net water flow ceases. This pressure, exerted by plant cells against their cell walls, is called **turgor pressure.** Plant cells normally are rigid from turgor pressure since the fluid in their environments is usually hypotonic.

Plant cells respond to hypertonic solutions in the same way as animal cells. In a hypertonic environment water flows out of a plant cell, causing the cell to shrink in volume and to pull away from its cell wall. This is what happens when a plant goes without water for too long; the loss of turgor pressure in many of the cells causes wilting. When the environment around a shrunken cell becomes hypotonic once again, it expands to press against its cell walls and its turgor pressure is thus restored.

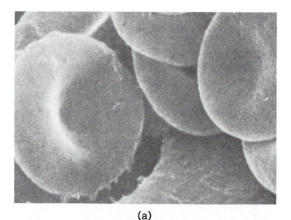

(a)

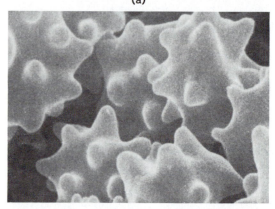

(b)

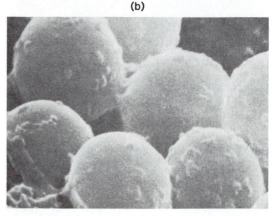

(c)

Figure 4.4   The effects of hypertonic and hypotonic solutions on erythrocytes (red blood cells). (a) Normal cells in an isotonic solution. (b) In hypertonic solutions, erythrocytes lose water and their plasma membranes become wrinkled (crenated). (c) In hypotonic solutions, erythrocytes gain water and become swollen. Eventually, they lyse and release their hemoglobin into the suspending medium. (The photo in (c) is a lower magnification than the photos in (a) and (b).)

Hypertonic solution (shrinkage)

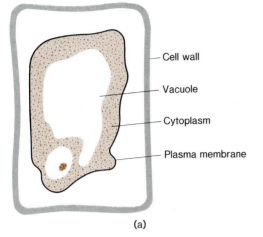

(a)

Hypotonic solution (turgor pressure)

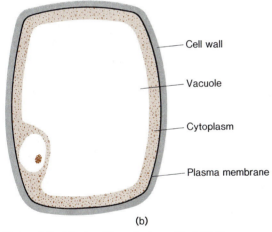

(b)

Figure 4.5 Effects of hypertonic and hypotonic solutions on plant cells. (a) In a hypertonic solution, the cell membrane pulls away from the cell wall because of a net water loss and shrinkage of the cell. (b) In a hypotonic solution, the cell is swollen, but it is prevented from bursting by the restraining force of the cell wall.

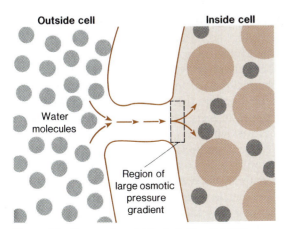

Figure 4.6 Proposed model for bulk flow of water through a membrane pore.

Bulk Flow of Water

Under certain conditions, water seems to move more quickly across membranes than can be accounted for by diffusion alone. Some biologists have suggested that water can enter cells by a different kind of process called **bulk flow.** In bulk flow, a volume of water moves as a unit in response to pressure of some sort. For instance, a very large osmotic-pressure gradient may develop between the water in a membrane pore and the solution into which the water is diffusing (figure 4.6). This large gradient would generate very rapid diffusion rates and draw the water molecules through the pore in bulk flow.

Many plant physiologists believe that bulk flow of solutions is a key factor in transporting material from one part of a plant to another. For example, this might account for the very rapid flow of nutrient solutions through plants (chapter 9).

Factors Affecting Permeability

Water, which is the solvent in biological systems, enters cells quite readily. The plasma membrane of a cell is also permeable to a number of types of solute molecules, which enter at varying rates. What determines the rates at which various solutes penetrate the plasma membrane?

A number of biologists have studied the permeability of the plasma membrane to various solutes. Large algae cells have been some of their favorite experimental subjects because the fluid in the vacuoles of these cells can be removed easily for chemical analysis. These analyses have shown that a number of properties of solute molecules affect the rate at which they enter cells. One of the more important factors is *lipid solubility.* The more soluble a molecule is in lipids, the more quickly it will penetrate cell membranes.

However, there are significant exceptions to this rule. Water and some small hydrophilic ("water-loving," or polar) molecules enter cells much more rapidly than might be expected, considering that there is a hydrophobic ("water-hating," or nonpolar) zone in the center of the plasma membrane. This discovery led some scientists to propose that plasma membranes possess small hydrophilic **pores** around 0.3 to 0.8 nm in diameter, through which small polar molecules can enter. Lipid solubility is still a major factor, however, in the case of larger molecules, which cannot pass through these small pores.

Ionization also affects permeability. In general, ionized substances penetrate cell membranes more slowly than un-ionized substances. In addition, anions (negatively charged ions) normally enter cells more rapidly than cations (positively charged ions) because the net charge on cell surfaces is positive and this tends to repel the like-charged cations. Ions in water are surrounded by layers of oriented water molecules called hydration spheres, and these spheres effectively increase the size of ions. The amount of charge that an ion carries also affects its penetration rate; cations such as Na^+ and K^+ or anions such as Cl^- enter cells more readily than Ca^{2+}, Mg^{2+}, or SO_2^{2-}.

If all other properties of two sets of solute molecules are relatively similar, molecular size is the key factor in accounting for the rates at which the solutes enter cells. The general rule is that penetration rate decreases as size increases. In fact, extremely large molecules such as polysaccharides and proteins cannot diffuse through membranes. They must enter cells by other means.

The Transport of Molecules across Membranes

Cells acquire some needed materials and eliminate certain wastes through simple diffusion. Although simple diffusion alone is sometimes adequate to meet the needs of cells, most often some other method is required. Even when a concentration gradient exists, many substances simply will not diffuse rapidly across the plasma membrane. Therefore, many types of lipid-insoluble molecules and various molecules of large size can traverse a plasma membrane only by means of membrane proteins called **carriers.** These carriers provide a hydrophilic route or passage through the membrane.

Carrier-mediated membrane transport is characterized by a number of features that distinguish it from simple diffusion. Carriers have a very high degree of specificity for the molecules or ions they transport. That is, a given carrier normally is involved in the transport of only one type of molecule or ion. Another characteristic of carrier-mediated transport involves penetration rate. The rate of simple diffusion increases linearly with solute concentration, but the response of carrier-mediated systems to solute concentration changes follows a different curve (figure 4.7). This is because carrier-mediated transport systems can be saturated. When the external solute concentration is very low, any increase in its level directly produces a proportional rise in its rate of entry. But with continuing solute concentration increases, the rate response reaches a plateau. Eventually, no rise in the velocity of entry occurs, even with further concentration increases.

Such behavior is indirect evidence that a carrier molecule is involved and that the carrier is binding the solute molecules at a specific site and then moving them across the membrane. The plateau in figure 4.7 is reached when the solute concentration is so great that each carrier binding site is saturated, or occupied by a new molecule as soon as the previous molecule has entered the cell.

There are several types of carrier-mediated transport systems that operate by somewhat different mechanisms. Two of the most widespread and important of these systems are facilitated diffusion and active transport.

Facilitated Diffusion

In **facilitated diffusion,** the carrier increases the rate at which a specific solute molecule can enter the cell by diffusion. In many cases, the carrier provides a route into the cell for a solute that could not diffuse through the plasma membrane by itself. In facilitated diffusion the solute is still traveling down a concentration gradient, so no energy input is required from the cell (figure 4.8). The rate of entry can be enhanced, however, if the transported molecule is chemically altered after it enters the cell, and this alteration may require an energy expenditure. For example, glucose, which enters some cells by facilitated diffusion, is phosphorylated after it enters a cell, forming glucose phosphate (figure 4.8b). This maintains the concentration gradient of glucose between the outside and the inside of the cell because glucose, as such, does not actually accumulate inside the cell.

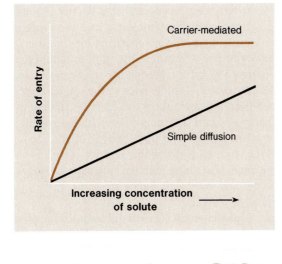

Figure 4.7 The concentration relationships of simple diffusion and carrier-mediated transport. The simple diffusion rate continues to increase linearly with increasing solute concentration. But the rate of entry by carrier-mediated processes does not increase beyond a certain level of solute concentration, which indicates that solute binding sites of carrier molecules can be saturated.

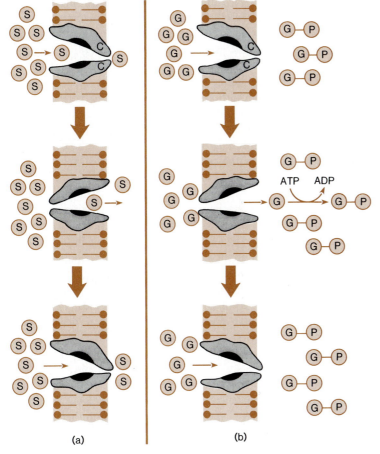

Figure 4.8 Facilitated diffusion. (a) The carrier molecule (C) simply speeds up the rate at which the solute (S) can cross a membrane in the direction of decreasing concentration. Facilitated diffusion occurs only when there is a concentration gradient across the membrane. (b) Sometimes the concentration gradient is maintained by altering the solute molecule after it has crossed the membrane. In this example, glucose (G) is phosphorylated to form glucose phosphate (G—P) immediately after entry. Phosphate groups are provided by adenosine triphosphate (ATP), which becomes adenosine diphosphate (ADP) after transfer of one phosphate to glucose.

Facilitated-diffusion carriers are present in the cells of many different organisms. When present, they may facilitate the entry of various sugars, amino acids, and nucleosides into cells.

Active Transport

The second carrier-mediated form of transport is called **active transport.** Active transport, like facilitated diffusion, involves carrier protein molecules that are specific for individual solutes. Some common solutes that are moved by active transport in various cells are Na^+ and K^+ ions, sugars, and amino acids.

Active transport differs from facilitated diffusion in that substances can be moved by active transport against a concentration gradient. Such movement requires the cell to expend energy, since the process is much like rolling a rock uphill. Normally, this energy is provided by the hydrolysis of molecules of **adenosine triphosphate (ATP),** the renewable "energy currency" of cells. ATP is replenished by cellular metabolic processes that degrade reduced organic compounds such as glucose to carbon dioxide and water. Thus, when a cell's metabolism is inhibited, active transport processes cease because of the lack of ATP and other energy sources.

Each individual active transport system is used to move a solute either into or out of cells, but not both. Some substances, such as sugars and amino acids, enter cells with the aid of active transport systems, but active transport may also be used to extrude other substances (sodium ions, for example) to the outside. In facilitated diffusion, solute movement will change direction if the concentration gradient across the membrane is reversed. But each particular active transport system operates irreversibly; that is, it moves a solute either into a cell or out of it.

How is metabolic energy actually used to move molecules across cell membranes? The solutes to be transported are relatively small compared to the 6.0 to 10.0 nm thick membrane and must, therefore, be moved a considerable distance by the carrier protein. Several models for the transport mechanism have been suggested.

It has been proposed that a carrier protein might bind the solute molecule and then rotate within the lipid bilayer so that its binding site is transferred to the other side of the membrane (figure 4.9). After releasing the solute molecule, the carrier would rotate back to the original side. However, it would take a considerable amount of energy to rotate a large protein with hydrophilic surfaces through the lipid bilayer of the membrane, with its hydrophobic interior. There is no evidence that such carrier-molecule rotations occur, and most cell biologists think that this model is unlikely.

Many biologists favor a different model, proposed by S. J. Singer and others, called the **fixed-pore mechanism.** Advocates of this model suggest that a membrane may possess hydrophilic pores, each of which is formed by two or more integral protein molecules. Singer has proposed that these integral proteins have specific binding sites for solute molecules. After a solute molecule has been bound, metabolic energy is used to induce a conformational change in the pore proteins (figure 4.9), and this change in pore shape transports the solute molecule to the other side of the membrane, where it is released. The cycle ends when the pore proteins return to their original conformation.

Although the issue is not settled, the fixed-pore mechanism seems more feasible than the rotation theory. Conformational changes in integral proteins would require a much smaller energy input than carrier rotation would. Experimental evidence also tends to support some sort of fixed-pore mechanism.

Actually, there are several means by which cellular energy expenditures can drive active transport processes. One excellent example is glucose transport by cells in the lining of the vertebrate small intestine, which is a physiologically important active transport process. The epithelial lining cells of the intestine take glucose out of the cavity of the intestine, even though their internal glucose concentration is greater than that in the cavity contents. In this case, the intake of glucose is coupled with the intake of sodium ions, and the carrier involved in glucose transport appears to have a sodium-ion binding site in addition to its glucose site (figure 4.10).

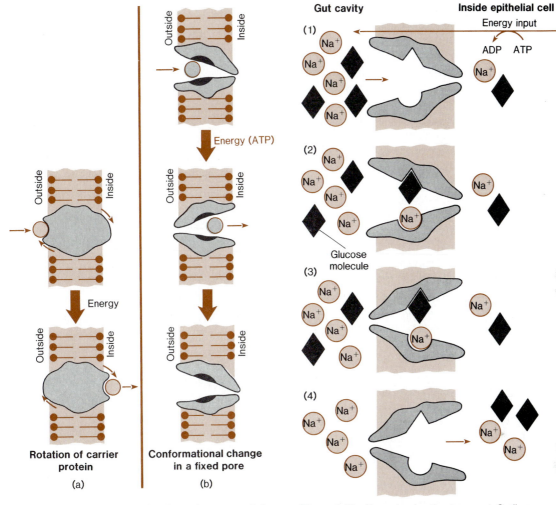

Figure 4.9 Two different models for active transport. It is very unlikely that carrier proteins rotate as proposed in the model in (a). Thus, the fixed pore model (b) is favored by most cell biologists. Active transport differs from facilitated diffusion in that active transport requires energy (ATP) input by the cell, and active transport can and does work against concentration gradients.

Figure 4.10 Example of active transport. Sodium-dependent sugar uptake by an intestinal epithelial (lining) cell. (1) ATP supplies energy for the sodium pump that actively expels sodium ions from the cell, thus maintaining a sodium ion concentration gradient between the outside and the inside of the cell. The tendency of Na^+ ions to move down this gradient makes possible glucose uptake by the cell in a coupled process. Glucose uptake remains possible as long as the Na^+ ion concentration gradient is maintained. (2) Sodium and glucose bind to their respective sites in the carrier complex, which is oriented toward the cell exterior. (3) The carrier complex changes conformation and directs the binding sites inward. (4) Sodium and glucose dissociate from the carrier, which then orients outward again.

For the carrier to function, a sodium ion must be transported at the same time that each glucose molecule is transported. The cell must first generate and maintain a sodium ion gradient, which it does by means of a sodium transport system or **sodium pump** located in the plasma membrane. This sodium pump uses ATP energy to expel Na^+ from the cell. It is entirely separate from the coupled glucose carrier, which moves material from the outside of the cell to the cell's interior. The sodium pump maintains a sodium concentration gradient so that Na^+ tends to move back into the cell if allowed to do so. This tendency toward Na^+ reentry is coupled with glucose acquisition, providing the driving force for glucose uptake.

This glucose active transport system is diagrammatically represented in figure 4.10. Sodium and glucose first bind to their respective sites on the carrier, which is oriented toward the outside. When both molecules have attached, the proteins change conformation so as to direct the binding sites inward. Because of its concentration gradient, the sodium ion will dissociate from its carrier and move into the cell. This presumably results in further conformational changes that cause the release of glucose as well. The carrier then returns to its outward-oriented conformation and is again ready to accept more sodium and glucose.

With a mechanism such as this, active glucose transport will continue as long as the Na^+ gradient is greater than the concentration gradient the glucose must move against. ATP energy is involved indirectly in glucose uptake; it helps to establish a sodium ion gradient through the plasma membrane sodium pump.

Intestinal epithelial cells also use the sodium-gradient mechanism for amino acid uptake. There are several different amino acid transport systems, each of which is specific for a different set of amino acids. The sodium-gradient mechanism is only one of the active transport processes by which cells take up several kinds of molecules.

Exocytosis and Endocytosis

There is another, very different, energy-dependent mechanism that cells employ to move substances across their plasma membranes. This mechanism involves the formation of membrane-bound vesicles that can be involved in either the secretion or the uptake of molecules. **Exocytosis** occurs when a vesicle approaches the inside of the plasma membrane and fuses with it to release its contents to the exterior (figure 4.11). Exocytosis is particularly evident in specialized secretory cells (such as those of the pancreas, which produce and release digestive enzymes).

Endocytosis occurs when extracellular material is enclosed in a vesicle and taken into the cell. There are two forms of endocytosis, **phagocytosis** and **pinocytosis.** In phagocytosis solid material is ingested, while in pinocytosis the cell engulfs a small quantity of fluid.

Phagocytosis is used by a wide variety of cells. For example, in many protozoa, phagocytosis serves as a basic feeding mechanism. On the other hand, the white blood cells (leukocytes) of vertebrates phagocytize invading pathogenic organisms such as bacteria and destroy them inside the vesicles that are formed.

Pinocytosis is apparently used by a number of cells to take in large molecules, such as proteins, that will not readily penetrate the plasma membrane. It also supplements active transport

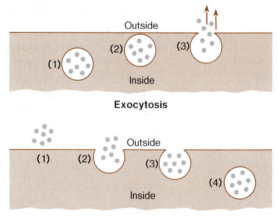

Figure 4.11 Exocytosis and endocytosis. These transport mechanisms involve membrane-bound vesicles.

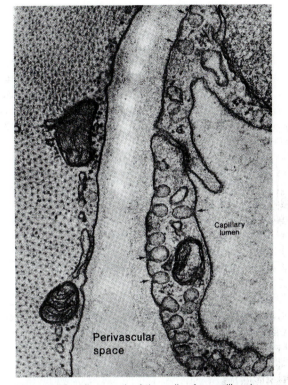

Figure 4.12 Micrograph of the cells of a capillary in heart muscle. Flask-shaped micropinocytotic vesicles (arrows) may be seen attached to the parts of the plasma membrane facing the capillary lumen (cavity inside the capillary) and the perivascular space (space outside the capillary). These vesicles seem to move from the capillary lumen to the perivascular space and transport larger molecules across the capillary wall in this way. (Magnification × 44,650)

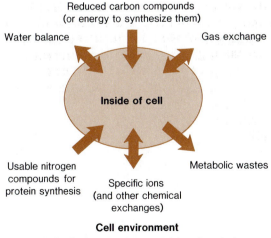

Cell environment

Figure 4.13 Exchanges that must take place between a living cell and its immediate environment.

as a means of taking in smaller molecules. Pinocytosis is normally induced by the presence of proteins, positively charged smaller molecules like amino acids, or sodium ions, all of which bind to the membrane surface in great quantities. The plasma membrane responds to the presence of these inducers by pinching off pinocytotic vesicles that are about 1 μm in diameter. Because it takes a piece of the plasma membrane to produce a pinocytotic vesicle, new membrane material must be synthesized to replace what was lost in the formation of the vesicle.

Pinocytosis may be used by the cell to take in proteins and other nutrients for the cell's own use, but it also can function in the transport of materials that are passed right through the cell. A good example of this is the pinocytosis carried out by the cells that line capillaries (figure 4.12). These cells form very small vesicles about 65 to 80 nm in diameter. Because these vesicles are smaller than the average pinocytotic vesicle, the process of pinocytosis in capillaries is referred to as **micropinocytosis.** After they are formed, the vesicles travel across the cell's cytoplasm and fuse with the plasma membrane on the opposite side to release their contents. Thus, the cell uses endocytosis (micropinocytosis) to take in material, moves the material through the cytoplasm to the opposite side of the cell, and then releases the material through the cell membrane (exocytosis). This is thought to be the major route for the movement of plasma proteins, hormones, antibodies, and other large molecules across capillary walls.

Cells, Organisms, and Environments

The wide variety of relationships among cells, organisms, and their environments is one of the most important concepts in the study of biology. It should be apparent by now that a cell can interact successfully with only a relatively narrow range of environments. A cell's environment must contain the materials that the cell needs in order to function, and materials leaving the cell must not build up to the point where they interfere with cell functions (figure 4.13).

There are many kinds of environments that various cells can survive in. In the case of unicellular organisms, cells equal organisms and must make all of their exchanges directly with the extraorganismic environment. But specialized, multicellular organisms have internal environments with which the majority of their cells make exchanges (figure 4.14). The internal environment of a multicellular organism provides a relatively stable, controlled habitat for cells.

Unicellular Organisms

In a unicellular organism, the cell equals the organism. Obviously, since the boundary of the cell is also the boundary of the organism, the whole organism is directly exposed to the cell's external environment. This means different things for different kinds of unicellular organisms.

Chlamydomonas, an aquatic green alga, is an **autotrophic** ("self-feeding") organism; that is, it produces reduced carbon compounds through photosynthesis (figure 4.15). The light energy needed to drive this process is absorbed directly from the environment. In *Chlamydomonas,* mineral nutrients are obtained by ion uptake from the surrounding water, and metabolic wastes are expelled by simple diffusion into the environment. Gas exchange is also made directly with the watery environment through the cell membrane and the cell wall.

An amoeba, on the other hand, is a **heterotrophic** ("other-feeding") organism (figure 4.15). It must take in reduced carbon compounds as well as all other nutrients from its environment. Amoebas feed by engulfing bits of debris or smaller organisms by phagocytosis. A particle of food material is enclosed in a vesicle inside the cell. Then the material is enzymatically digested and absorbed through the membrane of the vesicle. Of course, this limits an amoeba's food selection to particles that are considerably smaller than individual amoeba cells. In an amoeba, gas exchange occurs by simple diffusion and metabolic wastes can diffuse directly into the environment.

Whether autotrophic or heterotrophic, then, unicellular organisms must interact successfully with the external environment.

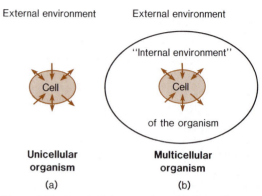

External environment   **External environment**

"Internal environment"

Cell   Cell

of the organism

Unicellular   **Multicellular**
organism   **organism**

(a)   (b)

Figure 4.14 In unicellular organisms (a), the cell surface is also the boundary between organism and environment (cell = organism). All exchanges are directly between the cell and the external environment. In multicellular organisms (b), many cells are exposed only to the "internal environment" of the organism, and they exchange materials with that environment.

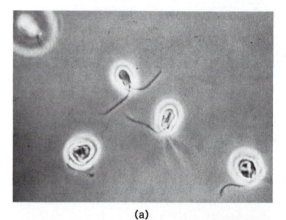

(a)

(b)

Figure 4.15 Examples of unicellular organisms in which cell = organism. In these cases all exchanges are made between cells and external environments.
(a) *Chlamydomonas,* an autotrophic organism that produces reduced carbon compounds photosynthetically. (b) An amoeba, a heterotrophic organism that ingests debris particles and smaller organisms and thereby obtains reduced carbon compounds.

Loosely Integrated Multicellular Organisms

The interactions of some cells with their environments fall somewhere between the direct environmental exchanges of unicellular organisms and the intricately regulated internal environments maintained by more highly organized plants and animals. For example, some small plants known as liverworts are composed of fairly simple, flat plates of cells called **thalli** (singular: **thallus**). In many cases the liverwort thallus is only a few cells in thickness. Though a liverwort's surface is quite permeable, and certain exchanges can be made directly between thallus cells and the external environment, there is still some degree of specialization among the cells for certain functions. Photosynthetic cells near the upper surface produce the reduced carbon compounds that supply the energy requirements of all the cells in the thallus. Extensions, called **rhizoids**, on the bottom of the thallus reach into the soil and anchor the thallus in place. Rhizoids also absorb mineral nutrients from the soil, which are then made available to all cells (figure 4.16).

(a)

(c)

Air chamber

Rhizoid

(b)

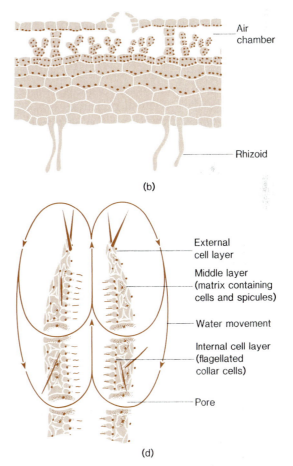

External cell layer

Middle layer (matrix containing cells and spicules)

Water movement

Internal cell layer (flagellated collar cells)

Pore

(d)

Figure 4.16 Examples of loosely integrated organisms in which some functions are carried out for the entire organism by specialized cells, while other functions are left entirely to individual cells. (a) The liverwort *Marchantia*. This picture shows a number of thalli, some of which bear stalked reproductive structures (see chapter 36). (b) Cross section of part of a *Marchantia* thallus. Chloroplasts (color) are present only in cells of the upper layers, especially in cells in the air spaces. These cells produce reduced organic compounds for all cells in the thallus. Rhizoids anchor the thallus and absorb mineral nutrients. (c) Simple sponges. (d) Section cut through a sponge showing the types of cells that make up the sponge's body. Note especially the flagellated cells that line the inner cavity of the sponge's body.

A sponge is an example of an animal in which the organism is a rather loose association of a large number of cells. Some of a sponge's environmental exchanges—for example, gas exchange and waste removal—occur between individual cells and the external environment (the surrounding water). But several of the cells that line the inner cavity of the sponge's body perform specialized functions. The beating flagellae that extend from lining cells keep a current of water flowing through the sponge's body. The cells then trap small particles from the water and enclose them in vesicles by phagocytosis. Some of these bits of food are transferred to wandering amoebalike cells and carried to other cells in the sponge's body. Thus, the cells in the sponge's body that are specialized for food gathering procure nutrients for all of the cells of the organism.

These two very different examples of loosely integrated organisms illustrate an intermediate degree of specialization in dealing with the environment. Some functions are carried out for the whole organism by certain cells, while other functions are handled as direct exchanges between individual cells and the external environment.

Complex Organisms with Regional Specializations

A distinct division of labor exists among the cells of more complex organisms. Many functions are localized in specific body areas; the cells in these areas are involved in *regional specialization.* In multicellular bodies, regional specialization is especially obvious in the form of discrete exchange areas, or localized boundaries across which exchanges of certain materials take place. These localized boundaries help to control the internal environment in which the cells live.

Some of the externally apparent regional specializations of a plant body are shown in figure 4.17. Roots take up water and minerals (in the form of ions) from the water around soil particles.

Leaves are flattened and spread as efficient absorbers of light energy for photosynthesis and are also specialized for the gas exchange that is vital to photosynthetic activity. These two major exchange boundaries—leaves and roots—efficiently solve the basic problem of nutrient procurement in terrestrial plants. Regional specialization ensures that necessary exchanges with the external environment take place both above and below ground.

Most of the rest of the surface of a plant's body is not significantly involved in exchanges with the external environment. In fact, most of the surface of a large terrestrial plant is almost impermeable to water and other substances, so that it forms a barrier between the plant's internal and external environments. The internal environment is regulated by exchanges with the external environment that occur mainly at the specialized exchange surfaces of roots and leaves. Most of the plant's cells live in this controlled internal environment, carrying on their own specialized activities and making special contributions to the maintenance of the whole organism. Their own cellular exchanges are not made with the external environment, but with the controlled internal environment of the plant.

Figure 4.18 shows some of the specialized exchange boundaries of a complex animal body. The animal's body surface, like that of the plant, is relatively impermeable in most places. Therefore, nutrient, gas, and waste exchanges between the inside and outside of the body all depend on specialized regional exchange boundaries.

Distance Problems and Transport Systems

Although regional specialization is an effective mechanism for regulating the internal environments of complex organisms, regional specialization may present a problem in itself in large organisms, where specialized body areas are quite distant from one another. This distance problem necessitates the efficient movement of materials from one part of the body to another. This need is met by transport systems in the bodies of large, complex organisms—vascular systems in plants and circulatory systems in animals.

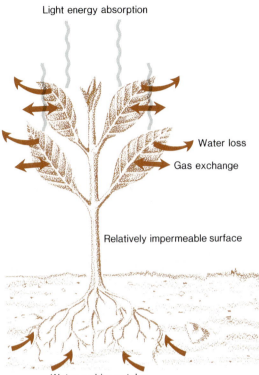

Light energy absorption

Water loss

Gas exchange

Relatively impermeable surface

Water and ion uptake

Figure 4.17 Generalized drawing of a complex plant body showing some of the exchanges that occur at specialized boundaries.

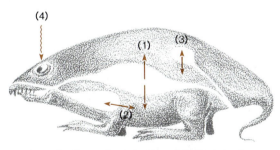

(4)

(1)

(3)

(2)

Figure 4.18 Generalized drawing of an "average" complex animal showing some of the exchanges that occur at specialized boundaries: (1) food absorption from the gut; (2) gas exchange in the lung; (3) metabolic waste excretion, ionic regulation, and maintenance of body water balance in the kidney; and (4) sensory reception of information about environmental conditions.

The Unity of the Body

Obviously, the many specialized functions of individual cells and groups of cells must be effectively coordinated. Regulating mechanisms control the body's specialized functions at every level, from simple individual enzyme actions right up to the functions of the whole body. The overall mechanisms of regulation are chemical control systems, or hormonal regulation, in both plants and animals.

In animals, the nervous system also plays a major regulatory role. In addition to playing its role in regulating the internal environment, an animal's nervous system gathers and processes information about the external environment. Sensory receptors act as the specialized boundaries through which this information is received and converted into a form that can be processed by the nervous system. The flow of information processed by sensory organs may then influence the organism's behavior.

Actually, sensory reception and behavior can be thought of as components of the *information exchange* an animal carries on with its environment. This exchange is crucial if the animal is to survive because most animals constantly face two problems that require information exchange: finding food and avoiding becoming food for other animals. In addition, many animals communicate with other members of their own species for a variety of purposes, including social behavior and reproduction. Thus, this information exchange is just as important to the maintenance of homeostasis in animals as are the specific exchanges of materials, internal transport systems, and regulatory mechanisms that coordinate specialized body parts.

Homeostasis: A Dynamic Steady State

The concept of homeostasis, or the maintenance of a relatively stable internal environment, as it relates to multicellular organisms has intrigued biologists for a long time. French physiologist Claude Bernard called this the *milieu intérieur* (internal environment) and said essentially that constancy of the internal environment is a prerequisite to life in varying external environments.

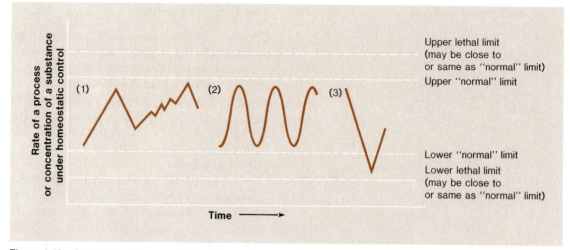

Figure 4.19   A generalized representation of some of the patterns of variation that can occur as part of homeostatic regulation: (1) short-term changes resulting from changes in activity or nutritional state; (2) cyclical regular changes with time of day or night (biological rhythm); (3) a change resulting from illness or injury.

But just as is the case with cell interiors, the internal environment of a complex body is not absolutely constant; it fluctuates. The rates of various metabolic processes, concentrations of various substances, and physical and chemical measures of the internal environment all change as the organism's nutritional state, activity levels, and environmental conditions change. Some may change in a systematic, cyclic way in response to the time of day (biological rhythms). However, each of these fluctuating levels is normally held within an acceptable range of values (figure 4.19). They are held within these ranges by controlled exchanges with the external environment and by internal regulation, which ensure the maintenance of organismic homeostasis, that dynamic steady state that assures that body cells live in a stable internal environment.

Homeostasis is important because it permits organisms to exploit more variable external environments. Through homeostasis, many organisms can maintain complex organization and a regulated internal environment even in the face of drastic environmental changes. Environmental changes may swirl about an organism, but its body cells live and function in a sheltered, stable internal environment. Environmental fluctuation is especially prevalent in terrestrial environments. Not surprisingly, therefore, complex multicellular organisms are the most successful occupants of terrestrial environments; the ability to maintain homeostasis in the face of environmental change is essential to life on land.

Summary

All living cells must make selective exchanges with their environments. They must obtain energy, exchange oxygen and carbon dioxide, release metabolic wastes, and regulate ionic composition in order to survive. Generally, cells meet these requirements and maintain relatively stable internal conditions. But although cells' interiors are reasonably stable, many factors inside cells fluctuate within tolerable ranges. Cells thus exist in a dynamic steady state, which is called homeostasis.

The plasma membrane is the cell's boundary with its environment. Properties of the plasma membrane determine characteristics of a cell's exchanges with its environment. In diffusion, particles tend to move from an area of greater concentration to an area of lower concentration. Some materials penetrate plasma membranes by simple diffusion, but the plasma membrane is selectively permeable—some materials do not diffuse through it readily. Osmosis is the diffusion of water through a membrane in response to solute concentration differences on the two sides of the membrane. The net diffusion of water through a membrane is toward the area with the greater solute particle concentration and the lower water concentration.

Osmotic terms describe the concentration of solute particles in solutions, but tonicity describes the actual effects that solutions have on cell volume. Cells in isotonic solutions do not show a net gain or loss of water or cell volume changes. Cells in hypertonic solutions lose water and shrink, and cells in hypotonic solutions gain water and swell.

Lipid solubility is an important determinant of membrane permeability, as is molecular size. Smaller molecules enter cells more readily than larger molecules. However, water and some small hydrophilic molecules enter cells even more rapidly than would be predicted. This suggests that there may be small hydrophilic pores in cell membranes. Bulk flow, or the flow of large numbers of molecules as a unit, has been proposed as a possible means of rapid water entry through membrane pores.

Carrier-mediated transport, or the movement of materials by carrier molecules, is specific for certain substances. In facilitated diffusion, the carrier increases the rate at which a material moves down a concentration gradient but does not require metabolic energy (ATP) input. In some cases, the diffusion gradient is maintained because the entering molecule is chemically altered after it is inside the cell. In active transport, materials can be moved across a membrane against a concentration gradient, which requires metabolic energy (ATP) input. A fixed-pore mechanism of active transport is favored over any form of carrier rotation.

Endocytosis and exocytosis are other mechanisms by which materials enter and leave cells. In phagocytosis, solid materials are incorporated into vacuoles. In pinocytosis, fluid is engulfed. Both of these forms of endocytosis are energy requiring.

Unicellular organisms exchange materials directly with the external environment, but many of the cells of multicellular organisms are specialized to contribute to the maintenance of the internal cellular environment. Loosely integrated organisms have specialized cells for nutrient acquisition, but the rest of the cells carry out other functions independently, often making exchanges directly with the external environment. In more complex plants and animals, a stable internal cellular environment is maintained through a high degree of regional specialization; internal body cells live in and make exchanges with a stable cellular environment.

Homeostatic maintenance of a stable internal cellular environment permits organisms to live in more varied and changeable environments.

This is especially important for terrestrial organisms, as their environment undergoes almost constant change.

Questions

1. Explain the error in the following statement: Homeostasis means the maintenance of absolutely constant internal conditions.
2. Explain why an isosmotic solution of sucrose is isotonic to red blood cells, while an isosmotic urea solution is not isotonic.
3. Why does the presence of a plateau in the curve for the rate of entry versus solute concentration signify that the solute enters cells by carrier-mediated transport rather than simple diffusion?
4. Explain why a metabolic poison such as cyanide that depresses cellular energy-conversion reactions also depresses active transport.
5. Why is a carrier-rotation model of active transport considered theoretically unlikely?
6. Explain the proposed roles of micro-pinocytosis and exocytosis in moving large molecules through the cells that line capillaries.
7. Name at least three factors that are regulated and stabilized in the internal cellular environment of a complex terrestrial animal.

Suggested Readings

Books

Christensen, H. N. 1975. *Biological transport.* 2d ed. Reading, Mass.: Benjamin/Cummings.

Giese, A. C. 1979. *Cell physiology.* 5th ed. Philadelphia: W. B. Saunders.

Hardy, R. M. 1976. *Homeostasis.* Studies in Biology no. 63. Baltimore: University Park Press.

Kennedy, D., ed. 1974. *Cellular and organismal biology.* San Francisco: W. H. Freeman.

Articles

Cram, W. J. 1980. Pinocytosis in plants. *The New Phytologist* 84:1.

Satir, B. October 1975. The final steps in secretion. *Scientific American* (offprint 1328).

Stein, W., and Lieb, W. 10 January 1974. How molecules pass through membranes. *New Scientist.*

Energy, Reactions, and Enzymes

Energy, the capacity to do work, is vital to all living things because all cells and organisms must work in order to live. Cells do chemical work when they synthesize larger molecules from smaller ones or when they assemble orderly aggregates of molecules into forms such as membranes. Organisms do mechanical work every time they move. And, as was discussed in chapter 4, living things expend energy to transport materials from one place to another.

The ultimate source of energy for life on earth is the sun—the radiant energy of this star bathes our planet. A portion of the sun's energy is trapped through **photosynthesis;** light energy is absorbed by chlorophylls and other pigments and converted into chemical energy. Plants use this chemical energy to combine carbon dioxide and other simple inorganic molecules into more complex compounds, such as carbohydrates, lipids, nucleic acids, and proteins. Then the plants, as well as the animals and other nonphotosynthetic organisms that consume the plants, can use these compounds as structural building blocks and as energy sources. Complex molecules can serve as energy sources because they can be degraded back to simpler molecules; this makes available energy to do work.

Biologists ask two fundamental questions about reactions in living things: (1) What factors determine whether a reaction will take place, and (2) what factors determine how fast that reaction will occur? The first of these questions is concerned to a large extent with energy change. **Thermodynamics** is the area of science that deals with energy exchanges and the relationships between various forms of energy. Biologists often use the term **bioenergetics** when referring to thermodynamic studies of biological processes. The second question, the one referring to the factors that determine how fast a reaction takes place, falls within the province of **chemical kinetics,** the study of reaction rates.

Thermodynamics

Any discussion of energy relationships and exchanges must be quantitative and must use specific energy units. Traditionally, the most commonly used energy unit is the **calorie.** A calorie is defined as the amount of heat energy required to raise the temperature of 1 gram of water 1° Celsius (or 1° Kelvin). Sometimes this unit is called a **gram calorie** or **small calorie** because another unit called a **large calorie** or **kilocalorie** (1,000 gram calories) also exists. Kilocalories are familiar in everyday life as the calories that people count when they are concerned about weight control.

Although the calorie has been the traditional energy unit, quantities of energy are also often expressed in terms of work done rather than in terms of heat. In such cases, the appropriate unit of energy is the **joule.** Modern international scientific usage is rapidly converting to the joule as the basic measure of energy. One calorie of heat energy is equal to 4.186 joules (table 5.1).

The science of thermodynamics analyzes energy changes in collections of matter called **systems.** A biologist might consider a whole organism to be a system undergoing energy changes and might study that organism from a thermodynamic viewpoint. More commonly, however, thermodynamic studies focus on individual chemical reactions or sets of chemical reactions. Whether the system in question is an entire organism or a single chemical reaction, all matter outside of the system is considered that system's **surroundings.**

An energy change is studied as a system goes from its **initial state** to its **final state.** Thermodynamics is *not* concerned with *how long* the process takes or what the route or mechanism of change is. It is concerned only with the actual energy difference between the initial and final states.

Table 5.1
Units of Energy Measurement.

---

1 gram calorie = 4.186 joules
1 gram calorie = 4.186 × 10^7 ergs of mechanical work
    1 joule = 10^7 ergs

The First Law of Thermodynamics

The science of thermodynamics is based on two fundamental laws. The **first law of thermodynamics** states that energy can be neither created nor destroyed. Thus, during a chemical or physical process, the total energy of a system and its surroundings remains constant, although the energy may change in form or the system itself may gain or lose energy. If the total energy of the system decreases, the energy of its surroundings increases. Conversely, if the total energy of the system increases, the energy of its surroundings decreases.

Chemical reactions are, in fact, normally accompanied by either the absorption of energy from the system's surroundings or the release of energy from the system to the surroundings. Very often these energy changes are expressed in terms of heat. When a chemical reaction releases energy as heat, the reaction is **exothermic.** A reaction that absorbs heat from the surroundings is **endothermic.** Such heat changes can be measured fairly easily and are usually expressed in terms of kilocalories or kilojoules per mole of reacting material.

The Second Law of Thermodynamics

Because the first law of thermodynamics cannot explain all observed phenomena, a second law is needed. The **second law of thermodynamics** is illustrated by the example shown in figure 5.1. Two cylinders are connected by a tube with a valve, and one of the two cylinders is filled with gas. If the valve is opened, gas will move into the empty cylinder until the amount of gas (pressure) in each cylinder is approximately equal; as long as one cylinder contains more gas, the net movement of gas tends to be toward the cylinder with less gas in it. Theoretically, this gas movement could be used to do work. For example, if a miniature windmill were placed in the tube between the cylinders, the movement of the gas would rotate the windmill and make possible some kind of work.

However, when the gas pressure in the two cylinders is equal, the system no longer has the potential to do work as that potential depended upon the *difference* between the two cylinders.

The system has not lost any total energy because the moving gas molecules in the two chambers still have the same total energy that they originally had, but their energy is now equally distributed between the two cylinders. This outcome is consistent with the first law of thermodynamics because no energy has been created or destroyed—it has just been redistributed. But what makes the gas molecules redistribute themselves? The movement of gas molecules between the two cylinders can be accounted for by the second law of thermodynamics.

According to the second law, there is a state of matter called **entropy,** which is basically a measure of randomness or disorder. All chemical and physical processes proceed in such a way that the amount of entropy (randomness) in the universe increases. When a process has made its maximum possible contribution to the randomness of the universe, a stable condition (equilibrium) is reached. In the example, equilibrium is reached when the gas pressure is equal in the two cylinders. This tendency to maximize entropy is one of the driving forces for physical and chemical processes.

Free Energy

To relate the concepts of thermodynamics to reactions that occur in living things, another factor, called **free energy,** must be introduced. Free energy can be identified in the following equation:

$$\Delta G = \Delta H - T\Delta S$$

In the equation, G is the free energy and ΔG is the change in free energy in a system such as a chemical reaction. (The Greek letter Δ is a general symbol for change and is read as "change in.") Free energy change is the total energy change in the system available to do useful work as the system proceeds to equilibrium.

The equation factor ΔH stands for the change in **enthalpy.** In biological reactions, which generally take place at constant pressure, enthalpy represents the heat content of the system and is essentially equal to the total energy content of the system. The change in enthalpy (ΔH) represents the heat of reaction and can be negative if the reaction is exothermic (gives off heat energy) or positive if the reaction is endothermic (absorbs heat energy). Change in enthalpy is measured in calories/mole or joules/mole.

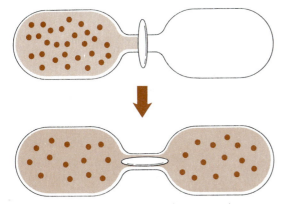

Figure 5.1 The expansion of gas into an empty cylinder when the valve is opened increases the entropy (randomness) of the system.

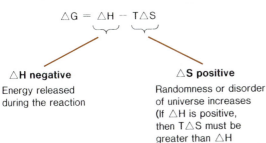

$$\triangle G = \triangle H - T\triangle S$$

△H negative

Energy released during the reaction

△S positive

Randomness or disorder of universe increases (If △H is positive, then T△S must be greater than △H for △G to be negative)

Figure 5.2 Thermodynamic conditions that favor reaction spontaneity. These factors tend to make △ G negative.

S_o

S_s

Figure 5.3 The relationship between an organism and its surroundings. Growing organisms are systems with increasing order (decreasing entropy). Organisms increase the entropy of their surroundings. S_o symbolizes the entropy of the organism. S_s symbolizes the entropy of the surroundings.

Finally, ΔS is the change in entropy, measured in calories or joules/mole/degree. T is the temperature in degrees Kelvin, rather than in degrees Celsius ($°K = 273.15 + °C$).

A process such as a chemical reaction will occur spontaneously if the free energy of the system decreases during the process; that is, the process will be spontaneous if ΔG is negative. Change in free energy is negative if either the decrease in the enthalpy or the increase in the entropy of the system is sufficiently large. In other words, the magnitude of ΔG is determined by the sum of the enthalpy and entropy components. This relationship is expressed in the free energy equation and is illustrated in figure 5.2.

In figure 5.2, if ΔH is positive (the reaction is endothermic), it does not mean that ΔG cannot be negative (the reaction cannot be spontaneous). ΔG can still be negative as long as ΔS, the increase in entropy, is large enough.

An everyday example may help illuminate these abstract concepts. As ice melts to liquid, it absorbs a great deal of heat energy; that is, the process of ice melting has a positive ΔH. But as the ice melts, there is also a great increase in entropy because the molecules in liquid water are in a much less orderly state than in ice (water molecules are much freer to move about randomly in liquid water than in ice, see p. 29). The increase in entropy (ΔS) in the melting process is large enough to give the process a negative ΔG; thus, the melting process occurs spontaneously. The same conditions are true when liquid water evaporates to become vapor.

One additional general point must be made about entropy. The second law of thermodynamics states only that the entropy of the universe (system plus surroundings) must increase during any process. This does *not* mean that the entropy of a specific system such as a living organism cannot actually decrease—of course, this does indeed happen. A growing organism, for example, is a system that gains increasing order.

However, this by no means implies that living systems are exceptions to the second law of thermodynamics. When a system gains order (decreasing its entropy), the entropy of the surroundings must increase. Animals, for instance, take in complex food molecules, degrade them, and release simpler waste molecules. Thus, they increase the entropy of their surroundings, and they use the free energy they obtain to maintain their own order (figure 5.3).

Photosynthetic organisms also must constantly obtain and use free energy to minimize their entropy (maintain their organization). But in contrast with animals, they trap and use light energy to accomplish this.

Equilibrium

Free energy is also related to the direction of a chemical reaction and the nature of the final equilibrium reached. Consider the following reaction:

$$A + B \rightleftharpoons C + D$$
$$\text{(Reactants)} \qquad \text{(Products)}$$

In this reaction, reactants A and B are being converted to products C and D, but the reverse reaction is also occurring: C and D are being converted to A and B. Such a reaction reaches equilibrium when the rate of the forward reaction (the formation of products) equals the rate of the reverse reaction (the formation of reactants) and the net concentrations of all four components change no further. This stable situation is represented by an **equilibrium constant,** symbolized as K_{eq}. The equilibrium constant for the reaction is calculated by multiplying the product concentrations and the reactant concentrations and dividing the former by the latter:

$$K_{eq} = \frac{[C]\,[D]}{[A]\,[B]}$$

In a reaction that reaches equilibrium with a high concentration of products compared to reactants, [C] [D] will be larger than [A] [B], and K_{eq} for the reaction will be greater than 1.0. Conversely, in a reaction that reaches equilibrium with a low concentration of products compared to reactants, K_{eq} will be less than 1.0. The equilibrium constant thus represents in quantitative terms the final state of a reaction.

What then determines the final state of a reaction (and its equilibrium constant)? The equilibrium constant is directly related to the free energy change that occurs during the reaction. Table 5.2 shows this relationship. When a reaction has a negative free-energy change ($-\Delta G$), it is an **exergonic reaction** and will spontaneously proceed toward completion (its K_{eq} is greater than 1.0). If a reaction has a positive free-energy change ($+\Delta G$), it is an **endergonic reaction** and will not spontaneously move far toward completion (its equilibrium constant is less than 1.0).

Table 5.2

The Relationship between the Equilibrium Constant (K'_{eq}) and Standard Free-Energy Changes ($G^{o'}$) of Reactions*

K'_{eq}	$G^{o'}$ (kcal/mole)	
0.001	+4.09	
0.10	+1.36	Endergonic ($K'_{eq} < 1.0$)
1.00	0.00	
10.00	−1.36	
1000	−4.09	
1×10^5	−6.80	Exergonic ($K'_{eq} > 1.0$)
1×10^6	−8.18	

*The symbol $G^{o'}$ represents the standard free-energy change for a reaction. It is the amount of free-energy loss or gain per mole of reaction under certain specified conditions. Similarly, K'_{eq} symbolizes the equilibrium constant under the standard conditions used for such measurements.

Free energy must be supplied to the system to make an endergonic reaction proceed. Thus, exergonic reactions are energy-yielding reactions, while endergonic reactions are energy-absorbing reactions.

Reaction Rates

The concepts of thermodynamics can help determine whether or not a particular reaction will occur and what the concentrations of the individual components will be when equilibrium has been reached, but thermodynamic studies do not indicate how fast a reaction will proceed. Even if a reaction is thermodynamically spontaneous, it is not certain that it will take place at a measurable rate. For example, combining hydrogen and oxygen to form water is a highly exergonic reaction, yet when these two gases are carefully mixed together, nothing happens. However, once the reaction starts (if, for instance, a match is lit), it proceeds so rapidly that an explosion results! The concepts of thermodynamics lend little help in understanding why this is so.

However, the principles of **chemical kinetics,** the study of reaction rates, do help determine whether a particular reaction will occur at a measurable rate. In chemical kinetics, there are two ways of viewing the course of a chemical reaction. These two views are described as **collision theory** and **transition state theory.**

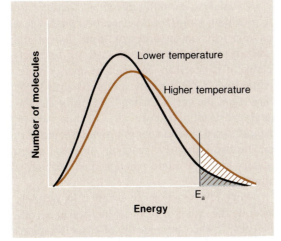

Figure 5.4 The Boltzmann distribution: the distribution of energies in populations of molecules at two different temperatures. The proportion of the molecules with an energy greater than the activation energy (E_a) is crosshatched. Because more molecules have energy contents above the reaction energy, the reaction proceeds more rapidly at the higher temperature.

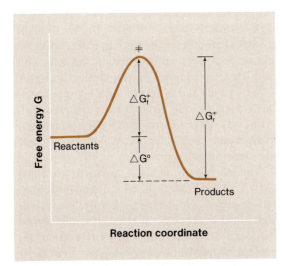

Figure 5.5 A reaction coordinate. This curve shows the course of a single set of reactants on their way to becoming products. ‡ denotes the transition state or activated complex. The transition state has a higher energy than reactants or products. $\triangle G_f^{\ddagger}$ denotes free energy of activation for forward reaction. $\triangle G_r^{\ddagger}$ denotes free energy of activation for reverse reaction. $\triangle G^{\circ}$ denotes free energy change for net reaction. $\triangle G^{\circ} = \triangle G_f^{\ddagger} - \triangle G_r^{\ddagger}$.

According to collision theory, chemical reactions take place by means of effective collisions between molecules. The reaction rate will, therefore, never be faster than the rate at which the reactants collide with each other. Moreover, according to the theory, the reactants must collide in an orientation favorable for reaction. For example, in the reaction of bromine atoms with hydrogen bromide, the Br presumably must collide with the Br end of HBr for the reaction to occur.

$$Br + HBr \longrightarrow Br_2 + H$$

Along with colliding in the right position, reactants must also collide with at least a certain critical amount of energy in order to react. This critical energy level is called the **activation energy** ($\mathbf{E_a}$) of the reaction. Only a certain percentage of molecules have this critical energy level because not all molecules at the same temperature have the same energy.

Molecules at various energy levels can be represented by a **Boltzmann distribution.** The Boltzmann distribution in figure 5.4 shows how applying heat affects a chemical reaction. Since more molecules at the higher temperature have energy contents above the necessary activation energy, the reaction will normally proceed at an increased rate as the temperature increases. This is because increasing the temperature increases the energy of the individual molecules.

Another way of viewing the rate at which chemical reactions occur is provided by the **transition state theory.** An example is the free energy (G) of a set of molecules as they are transformed from reactants to products through breaking and forming bonds. Figure 5.5 shows the progress of a single set of reactant molecules along the way toward becoming products. The **transition state** or **activated complex** (symbolized by ‡) represents the substances at the highest free-energy level reached along the way. The transition state complex is, structurally speaking, intermediate between reactants and products. If the transition state complex could actually be seen, it would look a little like the reactants and a little like the products. This activated complex is inherently unstable because the bonds in the reacting molecules are strained. The complex is always in the act of decomposing, either to products or back to reactants. In most reactions, the activated complex is at equilibrium with the reactants.

The governing idea in transition state theory is that the reaction rate is directly proportional to the concentration of the activated complex, symbolized by [$\neq$]; the higher the concentration of the activated complex, the faster the reaction proceeds. If the **free energy of activation** (ΔG^+_f) is large, the concentration of the activated complex is small, and the reaction is slow. This is exactly the situation in the example of the hydrogen and oxygen mixture mentioned before. Reactants such as hydrogen and oxygen might be able to react with a large negative ΔG^0 but do not react rapidly because very few molecules have sufficient energy to reach the transition state (or to exceed activation energy, ΔG^+_f). However, a lighted match can provide enough energy to trigger the reaction. The match raises the temperature enough so that more molecules reach or exceed the activation energy and a significant reaction begins. Because the reaction is highly exergonic, reacting molecules generate enough heat to raise the energy of still more molecules and the rate of the reaction increases explosively.

On the other hand, a relatively smaller free energy of activation (ΔG^+_f) leads to a faster reaction because more activated complexes form. This means that a reaction rate can be increased by lowering the energy of activation (ΔG^+_f).

Many important reactions in living things have relatively high activation energies. How then can they proceed at rapid rates as they do in many cases? Obviously, it is not possible to apply large quantities of heat energy to the reactants. Conditions inside living things are mild and stable compared to the conditions used to promote reactions in the chemical laboratory. Instead, in living organisms, reaction rates are increased by lowering activation energies through the action of catalysts.

Catalysts

Catalysts are substances that effectively lower activation energies. A catalyst affects the rate of a reaction without being changed by it; the catalyst remains in its original form at the end of the reaction. This does not mean, however, that the catalyst is totally unreactive.

A catalyst normally affects a reaction rate by providing an *alternative reaction pathway,* a reaction mechanism different from that of the uncatalyzed reaction. For the catalyst to be effective, it is necessary for the free energy of activation ($\triangle G^+_f$) for this catalyzed path to be *lower* than the corresponding $\triangle G^+_f$ for the uncatalyzed reaction (figure 5.6). In other words, in a catalyzed reaction, more activated complexes are formed, and the reaction proceeds more quickly than the same reaction does without a catalyst.

Very few reactions in living cells actually occur by the same pathway as they would outside the cells. In the vast majority of cases, reactions in cells are catalyzed by specific catalysts called **enzymes** that are produced by living cells.

Enzymes

The study of enzymes—and to a great extent the study of the molecular nature of the cell—began in 1897 with the work of Eduard Buchner. Buchner was attempting to preserve an extract of yeast that he had prepared by grinding the cells with sand. He tried adding a large quantity of sucrose to the extract and observed with some surprise that carbon dioxide and alcohol were produced. In this way, Buchner accidentally discovered that fermentation could take place in an extract, isolated from intact living cells.

Prior to this time, most scientists, including Louis Pasteur himself, had been convinced that such processes as fermentation were inseparable from intact living cells. Buchner's finding opened the door to the analytical chemical study of individual metabolic processes isolated from other cellular activities. These studies eventually led to the discovery of the enzymes that catalyze chemical reactions in living things.

The field of enzymology came of age in August 1926, when James B. Sumner announced that he had crystallized an enzyme, urease, and that this enzyme was a protein molecule. Sumner's enzyme preparation, which was pure protein, would decompose urea to CO_2 and NH_3 at a rapid rate (figure 5.7). This first enzyme purification was the culmination of nine years of work under somewhat primitive conditions. Sumner did not even have an ice chest at first—he cooled his solutions by placing them on the window sill at night.

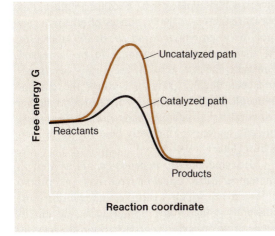

Figure 5.6 Catalysts provide a reaction pathway with a lower free energy of activation. Thus, the catalyst makes possible a large increase in the fraction of molecules having sufficient energy for the reaction to take place. As a result, the reaction moves toward equilibrium much more quickly.

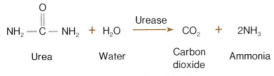

Figure 5.7 The reaction catalyzed by the enzyme urease. James Sumner's crystallization of urease in 1926 was a milestone in the study of enzymes. Urease catalyzes only the hydrolysis of urea to carbon dioxide and ammonia. It has a great specificity for both reaction and substrate.

Although other chemists had previously contributed some evidence to the notion that enzymes might be proteins, Sumner's data were challenged by a number of scientists, among them the renowned biochemist Richard Willstätter, who had attempted to isolate enzymes without success. Sumner responded by producing more evidence that urease is a protein, and eventually gained corroborative support from J. H. Northrop, who crystallized the digestive enzymes pepsin, trypsin, and chymotrypsin in the 1930s. Even so, it was not until 1946 that Sumner and Northrop received the Nobel Prize in recognition of their work.

It is difficult to comprehend how radical Sumner's original claim was at the time. A. H. T. Theorell told the following story about an encounter between Sumner and The Svedberg, the inventor of the ultracentrifuge:

Sometime in the late 1920s The Svedberg had a ring on the door to his office in the Uppsala Institute. He opened and there stood a man, unknown to him, saying: "My name is James B. Sumner; I have crystallized an enzyme." Svedberg immediately drew the conclusion that the man was mentally ill, so he said: "Yes, yes, one moment,"—shut the door and locked it from the inside!

Often, progress in science is accompanied by struggle and controversy; Sumner's case is a classic example. But Sumner's ideas eventually won out, and the principle that most reactions in living things are catalyzed by protein molecules called enzymes is one of the central concepts of modern biology.

The Structure of Enzymes

Enzymes are globular protein molecules. They have specific regions known as **catalytic sites** (or **active sites**) that bind reactants (**substrates**) during enzyme-catalyzed reactions. An enzyme's specificity depends on the shape of the catalytic site and the shape of the substrate molecule because enzyme and substrate must fit together in a very specific way, much as puzzle pieces fit together.

Many functional enzymes contain a nonprotein portion in addition to protein. These nonprotein components, which must be present for the enzymes to catalyze reactions, are called **cofactors.** The cofactors required by many enzymes are metal ions, such as potassium, magnesium, or zinc ions, but other reactions require specific molecules as cofactors. One of the most common types of cofactors, called a **coenzyme,** is a relatively small organic molecule that can readily dissociate from the enzyme protein and that participates directly in the enzyme's catalytic action. In addition, coenzymes actually transfer electrons, atoms, or molecules from one enzyme to another in some reactions. Coenzyme-mediated transfers are vital to a number of reactions in the energy exchanges of photosynthesis and cell respiration.

Coenzyme research offers a good illustration of how basic research can sometimes help resolve questions in seemingly unrelated areas—in this instance, in the field of nutrition. **Vitamins** are relatively small organic molecules that are required in trace amounts in the diets of animals, including humans, for health and physical fitness. The study of nutrition is a very old discipline;

even the ancient Greeks emphasized the relationship between food and health. By the eighteenth century, specific dietary remedies were being recommended for particular health problems—for example, lime juice was used to cure scurvy. But even though research on dietary requirements continued, it was not until the 1930s that biochemists discovered that many vitamins serve as components in the synthesis of coenzymes. For example, the B vitamin niacin is used in the synthesis of an important electron-carrying coenzyme called nicotinamide adenine dinucleotide (NAD). The diverse effects of niacin deficiency in animals are due to inadequate NAD synthesis in all body cells. Several other vitamins are also precursors to various coenzymes.

General Properties of Enzymes

All enzymes, regardless of the reactions that they catalyze, have certain characteristics in common. Enzymes, like other catalysts, do not make possible reactions that are thermodynamically impossible, nor do they alter the equilibrium constants of reactions. Enzymes simply increase the rate at which equilibrium is reached; that is, they change the kinetics of reactions.

Enzymes characteristically have a high degree of specificity, but the nature and degree of this specificity varies considerably from enzyme to enzyme. Some enzymes are specific for a particular type of reaction, but relatively nonspecific for various substrates. For example, D-amino acid oxidase, which oxidizes D-amino acids to α-keto acids and ammonia, is specific for the reaction that it catalyzes (figure 5.8). However, D-amino acid oxidase will act on a variety of substrates of the same general type (different D-amino acids). Other enzymes may be absolutely specific for both the reaction and the substrate. Urease (figure 5.7) catalyzes only the hydrolysis of urea to carbon dioxide and ammonia, and therefore belongs in this category. Enzyme specificity is based on the structure of the enzyme's catalytic site, where weak interactions such as hydrogen bonds are involved in binding enzyme and substrate to one another.

A number of environmental factors can affect an enzyme's activity, either by altering its conformation or by directly affecting the catalytic mechanism. Alterations in temperature and pH changes are examples of such factors. Up to a point, an increase in temperature results in an increase in enzyme activity because a higher temperature increases the frequency of effective contacts between enzyme and substrate molecules. As the temperature continues to be increased, however, enzyme activity eventually levels out and then declines rapidly because the enzyme molecule itself is gradually denatured at higher temperatures. Its shape changes, and its catalytic site can no longer efficiently bind substrate molecules (figure 5.9).

Enzyme activity depends on pH as well as on temperature. Each enzyme has an optimal pH—a pH at which it catalyzes its reaction most efficiently. Since some amino acid side chains have amino or carboxyl groups, their charges depend on the pH levels of their environments. When their charges change, there are changes in the weak interactions that maintain the three-dimensional conformation of the enzyme molecule. This can result in a change in the shape of the enzyme's catalytic site, reducing its activity (figure 5.9).

It seems logical that an enzyme's activity would rise with increasing substrate concentration, and this is precisely what happens. The true measure of enzyme activity is the **initial velocity**, or the rate at which the reaction proceeds just after its initiation, before much substrate has yet been converted to product.

Initial velocities are used to quantitatively describe enzyme-catalyzed reactions. The initial velocity of a particular enzyme-catalyzed reaction is determined for each of a series of reaction mixtures. Each reaction mixture contains the same small amount of enzyme and a different concentration of substrate; all other conditions affecting the rate are kept constant. The measured individual velocities then are used to prepare a velocity versus substrate concentration plot. At low substrate levels, a linear increase in initial velocity occurs with increasing substrate concentration (figure 5.10). But as the substrate concentration is increased, the initial velocity is increased more slowly until it finally reaches a **maximum velocity.**

These results may be explained most simply as follows. When the substrate concentration is low, the frequency of collision between substrate and enzyme is low. Increasing the substrate concentration increases the frequency of collision so that empty catalytic sites on enzymes are filled with substrate molecules for a larger proportion of the time. Thus, each enzyme molecule can

Figure 5.8 An enzyme that is specific for a kind of reaction. D-amino acid oxidase will catalyze the oxidation of a variety of D-amino acids. (These are the opposite optical isomers to the L-isomers found in proteins.) The amino acid may be D-tyrosine, D-proline, D-methionine, D-alanine, D-histidine, D-isoleucine, or any one of many others. This enzyme thus has great specificity for the reaction, but not the substrate.

$$R-\underset{\underset{NH_2}{|}}{CH}-COOH \; + \; H_2O \; \xrightarrow{\text{D-amino acid oxidase}} \; R-\underset{\underset{O}{\|}}{C}-COOH \; + \; NH_3$$

Amino acid   Water   Keto acid   Ammonia

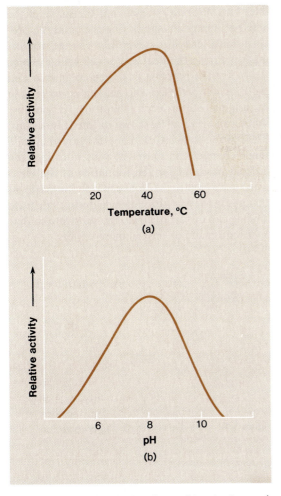

Figure 5.9 Examples of the effects of temperature and pH on enzyme activity. (a) Enzyme activities generally increase with temperature up to a certain point and then decline because the enzyme becomes unstable. The temperature of maximum activity may depend on how long the enzyme is kept at that temperature because loss of activity is usually relatively slow. Thus, it is not always possible to speak of an optimal temperature.
(b) Profiles of activity against pH are often bell-shaped because as the pH changes from the optimal value, groups on the substrate or on the enzyme change their ionization from that needed for catalysis.

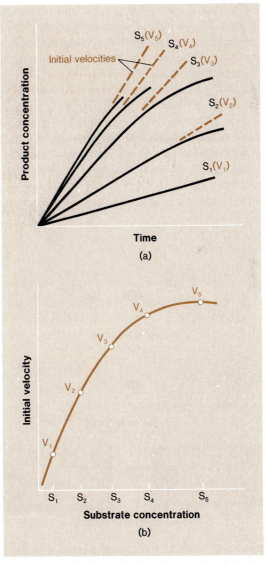

Figure 5.10 The effect of substrate concentration on the rate of an enzyme-catalyzed reaction.
(a) Measurements of the initial velocities of reaction (V_1, V_2, etc.) at different concentrations of substrate (S_1, S_2, etc.). (b) Initial velocities plotted against substrate concentration. The result is a hyperbolic curve.

participate in more catalytic events per unit time. When the concentration of substrate is so great that the enzyme's catalytic sites are virtually continuously filled with substrate, the enzyme's activity will increase no further; its maximum velocity has been reached. At this point, the enzyme molecules' catalytic sites are **saturated**. A similar kind of site saturation occurs with carrier-mediated transport processes in cells (chapter 4).

The Mechanism of Enzyme Action

Living cells use enzymes to lower the activation energies of specific reactions so that they may proceed more rapidly toward equilibrium at normal cellular temperatures. An enzyme can reduce the activation energy of a reaction in a variety of ways. Often, the reaction between two substances is accelerated simply by increasing their concentrations. This raises the frequency of collision between the reactants and therefore increases the overall reaction rate. When an enzyme binds the substrates at its catalytic site, it *concentrates* them (by literally holding them close to one another) and thereby enhances the rate of reaction.

Enzymes can do much more, however, than simply hold substrate molecules close together. For two molecules to react with one another, their atoms must be *oriented* in such a fashion that the proper bonds can be formed. Without the correct orientation, the substrates cannot react even if they are brought close together. Enzyme catalytic sites can bind substrate molecules in precisely the proper arrangement for the substrate molecules to react with one another (figure 5.11).

The conformation of an enzyme is often flexible. This flexibility means that the shape of the active site may change during the catalytic process; some enzymes reversibly change conformation during the course of a catalytic cycle so that after the reaction has been completed and the products released, the active site returns to its original state. Such alterations in enzyme shape can serve at least two different functions. First, when substrates bind, enzyme molecule shape changes can move the substrates into the optimal positions for the formation of the transition state (figure 5.12). Second, conformational changes in the enzyme molecule during substrate binding can distort or strain the substrate so that bonds are made or broken more readily.

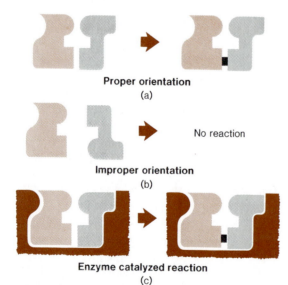

Proper orientation
(a)

Improper orientation
(b)

No reaction

Enzyme catalyzed reaction
(c)

Figure 5.11 Catalysis by substrate orientation. (a) When two substrates are properly oriented with respect to one another, the reaction can take place. (b) There will be no reaction with improperly oriented substrates. (c) An enzyme molecule may function by bringing the substrates together in the proper orientation for reaction.

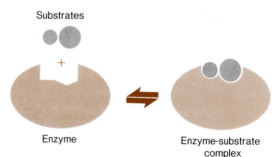

Substrates

Enzyme

Enzyme-substrate
complex

Figure 5.12 The role of enzyme shape change in catalysis. The binding of the substrate changes the shape of the active site into a conformation that allows the reaction to occur.

Thus, there are a variety of means by which an enzyme may lower the activation energy of a reaction, thereby increasing its rate. Many enzymes employ several different means at one time, each of them making its own contribution to the overall rate increase produced by the enzyme.

Feedback Inhibition and Allosteric Control

Many enzyme-catalyzed reactions are parts of **metabolic pathways**, or series of reactions in cells. Very often in these pathways, the product of one enzyme-catalyzed reaction is a substrate for the enzyme that catalyzes the next reaction in the series. In such a case, each reaction effectively removes the product of the reaction that preceded it in the sequence; and with the product constantly being removed, the reaction does not come to equilibrium, but continues to convert substrate to product. However, it is not economical for cells to have enzyme-catalyzed reactions occurring at maximum rates until every last available molecule of substrate has reacted. There must be some way to regulate the catalytic activity of enzymes so that the output of a reaction series can be correlated with the chemical requirements of the cell.

Temperature and pH changes, among other things, affect the catalytic efficiency of enzymes. But cells cannot produce (or tolerate) significant temperature or pH changes for purposes of enzyme regulation. Instead, mechanisms such as feedback inhibition and allosteric control regulate enzyme activity.

In some cases, rate regulation takes place at several points in a metabolic pathway, but in other cases, the entire reaction series is regulated by a single enzyme in the pathway. These key enzymes are called **regulatory enzymes.** Their activity is affected by the presence of **activators** and **inhibitors.**

Quite often, the first enzyme in a metabolic pathway is a regulatory enzyme that is specifically inhibited by the final product of the sequence. This phenomenon is called **feedback inhibition** or **end product inhibition.** A metabolic pathway that is regulated in this way functions normally until its final product accumulates beyond a crucial level. At that point, the end product significantly inhibits the activity of the first

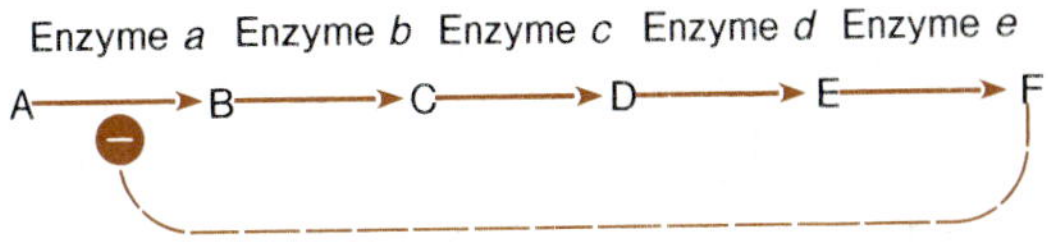

Figure 5.13 Diagrammatic representation of feedback inhibition of a simple metabolic pathway. The product F inhibits its own formation by reducing the rate of the first reaction in the pathway, catalyzed by enzyme a.

enzyme in the pathway, which in turn slows the whole reaction series, and thus slows the synthesis of the end product (figure 5.13). Feedback inhibition ensures that nutrients and energy are not wasted in the synthesis of unnecessarily high quantities of any particular cell constituent.

A thoroughly studied case of feedback inhibition is found in the control of pyrimidine synthesis in the bacterium *Escherichia coli,* which inhabits the human gut. This metabolic pathway is important because pyrimidines are precursors of nucleic acids. One of the products of this pyrimidine-synthesizing pathway, cytidine triphosphate (CTP), inhibits the first enzyme in the pathway, aspartate transcarbamoylase (ATCase), and thus controls the rate of its own synthesis (figure 5.14).

Aspartate transcarbamoylase is similar to most regulatory enzymes in that it is an **allosteric enzyme.** Allosteric enzymes characteristically have sites separate from their catalytic sites for the binding of regulatory molecules. The inhibitor CTP binds to such a regulatory site and alters the protein's conformation in such a way that the shape of the catalytic site is changed (figure 5.15). This results in a decrease in catalytic efficiency.

The enzyme ATCase is a very large oligomeric protein, or a combination of several individual polypeptides. It has two catalytic subunits and three regulatory subunits. Conformational changes in the regulatory subunits, which are induced when CTP binds to them, produce conformational changes in the catalytic subunits and thus influence the rate of the catalytic reaction. Although this case is an example of an allosteric change that inhibits an enzyme's action, in some other enzymes, allosteric effectors can stimulate enzyme activity. Interactions and changes such as these in the quaternary structures of proteins are of great importance in the control of enzyme function.

Box 5.1
The Mechanisms of Lysozyme Activity

Lysozyme, an enzyme that catalyzes the hydrolysis of a complex polysaccharide found in bacterial cell walls, was discovered by Alexander Fleming in 1922. Fleming added a few drops of nasal mucus to a bacterial culture plate to see what would happen. He found to his surprise that the bacteria nearest the mucus were lysed (destroyed). Fleming therefore named the substance responsible *lysozyme*. It has since been learned that lysozyme is an enzyme that destroys bacteria by severing a specific bond in a polysaccharide found in bacterial cell walls. The bacteria osmotically rupture after their walls have been destroyed.

Egg whites are a particularly good source of lysozyme and have been used as a source of the enzyme for various studies.

Box Figure 5.1A Lysozyme structure and activity. (a) Three-dimensional structure of lysozyme determined from X-ray diffraction studies. Only the backbone of the molecule is shown. The numbers represent positions of some of the 129 amino acids in the polypeptide chain. The substrate is shown in gray. Letters represent hexose rings in the substrate. The amino acids at the enzyme's active site that are involved in the reaction (glutamic acid, 35, and aspartic acid, 52) are in darker color. Dashed lines indicate weak bonds holding the enzyme and substrate molecules together. (b) Steps in the postulated mechanism of action of lysozyme. Letters correspond to hexose rings shown in A.

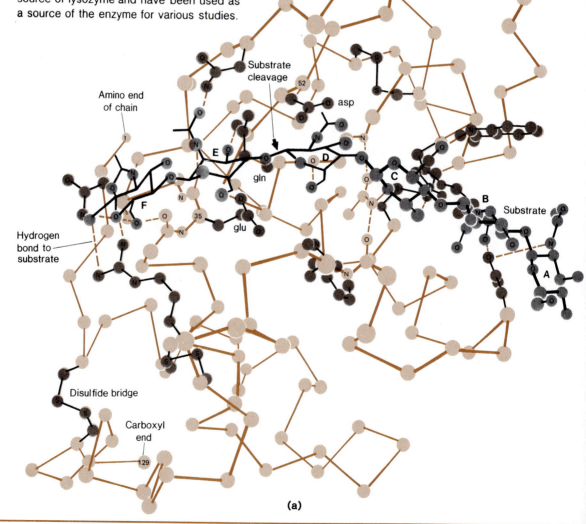

(a)

D. C. Phillips and his colleagues at Oxford have determined the structure of lysozyme by X-ray diffraction analysis (box figure 5.1A). The enzyme is a relatively small protein (molecular weight 14,600), and its active site is a long cleft that extends across one side of the molecule. The substrate is held in the cleft by hydrogen bonding and London dispersion (van der Waals) forces in such a way that the hexose ring next to the susceptible bond is distorted or forced out of shape. The carboxyl group of glutamic acid, 35 (the thirty-fifth amino acid in the polypeptide chain), then donates a proton to the oxygen atom of the bond between the two sugars (step 1 in figure 5.1A(b)). This causes the bond to break and leaves a positive charge on the adjacent hexose ring (now called a **carbonium ion** because it contains a positively charged carbon atom). This carbonium ion is stabilized by a negative charge on the carboxyl group of a molecule of aspartic acid, 52 (the fifty-second amino acid of the polypeptide chain). The polysaccharide has now been cleaved, and the part attached to the C_4 atom of the broken bond can diffuse away. A hydroxyl ion (step 2 in part b) then attaches to the carbonium ion, regenerating the normal hexose ring. The part of the carbohydrate attached to the C_1 of the original bond can now also diffuse away. The proton corresponding to the hydroxyl ion attaches to glutamic acid, 35 (step 3). The original state of the enzyme has thus been regained.

This mechanism illustrates some important aspects of enzyme activity. The combined strength of the individually weak noncovalent forces involved in binding the substrate is sufficient to distort the hexose ring and to form the transition-state carbonium ion. The precise orientation of the substrate on the enzyme enables the positive carbonium ion to be stabilized further by the negatively charged aspartic acid. Thus, lysozyme activity illustrates the roles of enzyme conformation changes and the formation of a transition state in an enzyme-catalyzed reaction.

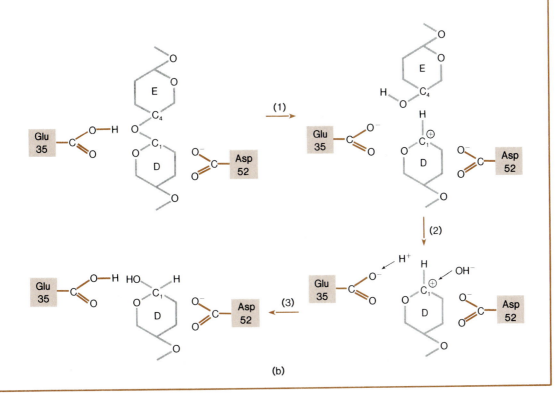

(b)

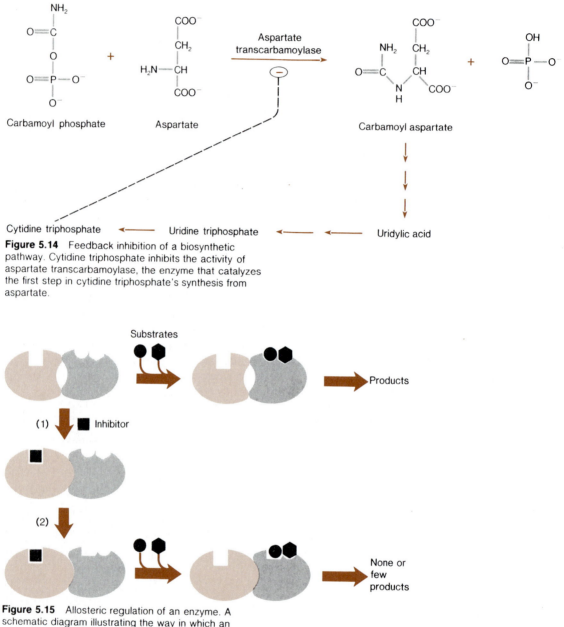

Figure 5.14 Feedback inhibition of a biosynthetic pathway. Cytidine triphosphate inhibits the activity of aspartate transcarbamoylase, the enzyme that catalyzes the first step in cytidine triphosphate's synthesis from aspartate.

Figure 5.15 Allosteric regulation of an enzyme. A schematic diagram illustrating the way in which an oligomeric allosteric enzyme, such as aspartate transcarbamoylase (ATCase), is regulated. (1) The inhibitor binds to a regulatory subunit (in color) and induces a conformational change in the regulatory subunit. (2) This in turn alters the catalytic subunit. The modified catalytic subunit is not as catalytically active.

Oxidation-Reduction Reactions

Many of the enzyme-catalyzed reactions in living cells involve oxidation and reduction. **Oxidation** is the loss of electrons; **reduction** is the gain of electrons. These two processes are the key to many chemical energy changes in living things.

A specific example is cytochrome c, a protein found in most organisms that functions in cellular energy conversions. Cytochrome c contains an iron atom in a heme group similar to that found in the hemoglobin molecule. This iron atom may be in either the reduced (Fe^{2+}) or the oxidized (Fe^{3+}) state. The difference in charge exists because Fe^{2+} has one more electron than Fe^{3+}.

The reduction and oxidation of cytochrome c's iron atom is symbolized thus:

$$Fe^{2+} \text{ cytochrome } c \rightleftharpoons$$
$$Fe^{3+} \text{ cytochrome } c + \text{Electron}$$

Since the electron does not appear free in solution, this represents only a **half reaction.** An oxidation-reduction reaction does not occur in isolation; it must take place in conjunction with another oxidation-reduction reaction.

Another pair of substances that differ by the gain or loss of an electron are the oxidized and reduced forms of plastocyanin, a protein that contains copper. These substances undergo the following half reaction:

$$Cu^{2+} \text{ plastocyanin} + \text{Electron} \rightleftharpoons$$
$$Cu^{+} \text{ plastocyanin}$$

The cytochrome c and plastocyanin half reactions can combine to form a complete oxidation-reduction reaction:

$$Fe^{2+} \text{ cytochrome } c + Cu^{2+} \text{ plastocyanin} \rightleftharpoons$$
$$Fe^{3+} \text{ cytochrome } c + Cu^{+} \text{ plastocyanin}$$

The different forms of a compound that appear on opposite sides of an oxidation-reduction equation (for example, Fe^{2+} cytochrome c and Fe^{3+} cytochrome c) are known as a **couple.** One member of a couple is a reduced form that can donate an electron; this form is called a **reductant** or **reducing agent.** In the example, Fe^{2+} cytochrome c is a reductant because it can donate an electron. The other member of a couple is the oxidized form, which can accept an electron; this form is called an **oxidant** or **oxidizing agent.** In the example, Fe^{3+} cytochrome c is an oxidant.

Thus, a couple includes an oxidant and a reductant.

Oxidation-reduction reactions are essential to the energy transformations in photosynthesis and cellular respiration. It is thus important to understand how various couples participate in oxidation-reduction reactions. An oxidation-reduction reaction contains two couples, and each side of an oxidation-reduction equation contains a reductant and an oxidant. In principle, the transfer of an electron from one compound to another in such a reaction can go either way. But the final equilibrium position depends on the relative reducing and oxidizing powers of the two component couples in any given oxidation-reduction reaction.

The reducing and oxidizing power of the two members of a couple is measured and stated quantitatively as the **redox potential** (reduction-oxidation potential). The redox potentials of the couples involved in any oxidation-reduction reaction determine the equilibrium that reaction will reach. Redox potentials are expressed in volts because they can be considered in terms of potential charges generated in an electrical cell.

Redox potentials and the equilibria of oxidation-reduction reactions can be put in a biological context by considering the oxygen/water couple:

$$O_2 + 4 \text{ Electrons} + 4H^+ \rightleftharpoons 2H_2O$$

This couple has a strong oxidant (O_2) and a weak reductant (H_2O). Its redox potential is $+0.816$ volts at a pH of 7.0. This relatively large positive redox potential is a result of the strong oxidizing ability of one component and the relatively weaker reducing ability of its partner. It signifies that oxygen will readily accept electrons from any electron donor and be reduced to water. This is the basis for oxygen's role as an electron acceptor in cell respiration.

The redox potentials of some biologically important couples are given in table 5.3. *The couples with more negative redox potentials will donate electrons to couples with more positive potentials.* For example, reduced cytochrome c will donate an electron to oxidized plastocyanin

Table 5.3

Some Biologically Important Redox Potentials. These values have been determined for a pH of 7 and a temperature of 25 to 30°C.*

Electrode Equation	Redox Potential, V
$2H^+ + 2e^- \rightleftharpoons H_2$	-0.421
Ferredoxin (oxidized) $+ e^- \rightleftharpoons$ Ferredoxin (reduced)	-0.42
α-Ketoglutarate $+ CO_2 + 2H^+ + 2e^- \rightleftharpoons$ Isocitrate	-0.38
$NAD^+ + 2H^+ + 2e^- \rightleftharpoons NADH + H^+$	-0.320
$NADP^+ + 2H^+ + 2e^- \rightleftharpoons NADPH + H^+$	-0.324
Pyruvate $+ 2H^+ + 2e^- \rightleftharpoons$ Lactate	-0.185
Oxaloacetate $+ 2H^+ + 2e^- \rightleftharpoons$ Malate	-0.166
Fumarate $+ 2H^+ + 2e^- \rightleftharpoons$ Succinate	-0.031
Ubiquinone $+ 2H^+ + 2e^- \rightleftharpoons$ Ubiquinol	$+0.10$
Cytochrome c (Fe^{3+}) $+ e^- \rightleftharpoons$ Cytochrome c (Fe^{2+})	$+0.254$
Plastocyanin (Cu^{2+}) $+ e^- \rightleftharpoons$ Plastocyanin (Cu^+)	$+0.37$
$\frac{1}{2}O_2 + 2H^+ + 2e^- \rightleftharpoons H_2O$	$+0.816$

Adapted from Albert L. Lehninger, *Biochemistry*, 2d ed. (New York: Worth Publishers, 1975), p. 479.

*Redox potentials all are relative values. They are comparisons with the redox potential of the H_2/H^+ couple, which is arbitrarily assigned a value of 0.0 when it is measured at a pH of 0. The redox potential for the H_2/H^+ couple is affected by changing pH, which is a change in H^+ ion concentration. Thus, the redox potential for the H_2/H^+ couple at a pH of 7 is -0.421, not 0.0.

or to molecular oxygen. As table 5.3 shows, the redox potential of the oxygen/water couple is relatively strong. The strong positive redox potential of this couple is extremely important in cell respiration, in which oxygen accepts electrons removed from other molecules and is reduced to water. In fact, the strong redox potential of the oxygen/water couple helps explain the basis for oxygen's essential role in the chemistry of living things.

Although this discussion has so far considered reduction and oxidation in terms of the transfer of a single electron, oxidation-reduction reactions are often more complicated. Electrons may be transferred in pairs and are sometimes accompanied by protons. For example, ubiquinone, a small, lipid-soluble molecule involved in biological electron transfer, has oxidized and reduced forms that differ by two protons and two electrons (figure 5.16):

Ubiquinone + 2 Electrons + $2H^+ \rightleftharpoons$ Ubiquinol

Another example of a more complex oxidation-reduction reaction involves **nicotinamide adenine dinucleotide (NAD)** and the related **nicotinamide adenine dinucleotide phosphate (NADP).** These two molecules both occur in forms differing by one proton and two electrons (figure 5.17):

$$NAD^+ + 2 \text{ Electrons} + H^+ \longrightarrow NADH$$
$$NADP^+ + 2 \text{ Electrons} + H^+ \longrightarrow NADPH$$

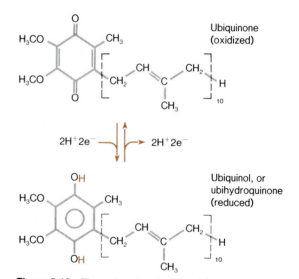

Figure 5.16 The reduced and oxidized forms of ubiquinone. This is the formula for the most common form in mammalian mitochondria. Other forms with different side-chain lengths are found in various other organisms.

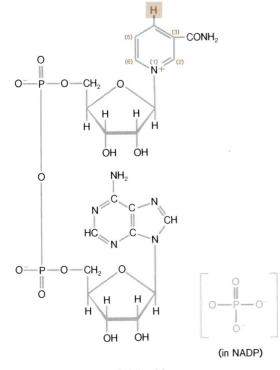

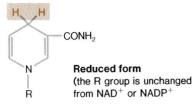

Figure 5.17 The structure of nicotinamide adenine dinucleotide (NAD). The upper figure shows the oxidized form (NAD⁺) and the lower figure the difference when it is reduced to NADH. The closely related nicotinamide adenine dinucleotide phosphate (NADP) has an extra phosphate group attached, shown in gray.

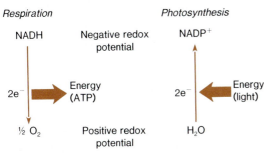

Figure 5.18 The relationship between electron flow and energy. The long arrows represent series of oxidation-reduction reactions known as electron transport systems.

NAD and NADP act as intermediates in cellular reactions involving electron transfers; they are sometimes regarded as the "reductant currency" of the cell. This means that the reduced forms of NAD and NADP carry electrons that can be donated in a variety of reactions. It has been noted several times that living things obtain usable energy during reactions in which reduced carbon compounds are degraded. These carbon compounds are degraded by oxidation, or the removal of electrons. Many of the electrons removed from reduced carbon compounds in various enzyme-catalyzed reactions are transferred to NAD to produce NADH.

When a reduced electron carrier with a relatively large negative redox potential, such as NADH, transfers electrons to an oxidized substance with a positive redox potential, such as oxygen, a great deal of free energy may be released. The actual energy relationships of the reduction of oxygen by NADH is an example:

$$NADH + H^+ + \tfrac{1}{2}O_2 \longrightarrow H_2O + NAD^+$$

The standard free-energy change ($\Delta G^{0\prime}$) for this reaction is quite large (-52.6 kcal/mole). Living cells take advantage of this large free-energy change. However, electrons are passed from reduced NAD to oxygen not in a single reaction but by a series of intermediate redox reactions involving several different electron carriers. Part of the energy made available during these electron transfers can be used to produce adenosine triphosphate (ATP) from adenosine diphosphate (ADP) in a process called **oxidative phosphorylation** (figure 5.18). ATP is called the cell's "energy currency" because it serves as a direct energy source for many processes in living things.

On the other hand, if energy is released when electrons are transferred from a molecule with a more negative redox potential to one with a more positive potential, then, obviously, energy is required to drive electrons in the opposite direction. In photosynthesis, sunlight provides the energy to do precisely this.

Thus, oxidation-reduction reactions are essential to the processes by which living things obtain the supply of energy they continually require. These reactions play an especially important role in photosynthesis and cell respiration (chapters 6 and 7).

Box 5.2

The Bombardier Beetle's Chemical Blaster

Many animals produce toxic substances that are used for capturing prey and for self-defense. Toxins produced by snakes, spiders, and bees are familiar examples. And when some millipedes are irritated, they release hydrogen cyanide (HCN), a deadly metabolic poison.

However, some of the most dramatic defense mechanisms are the chemical sprays that animals such as certain scorpions and some insects direct at their attackers. One of the most spectacular of these defense sprays is the hot, irritating mist produced by the bombardier beetle—a mist that is emitted in explosive blasts of fine spray at a temperature of 100°C (box figure 5.2A).

The bombardier beetle in box figure 5.2A, *Stenaptinus insignis,* is from Kenya. It has a pair of double-chambered glands that open at the tip of its abdomen. The inner chamber of each gland contains a mixture of hydroquinones and hydrogen peroxide (H_2O_2). The outer chamber contains enzymes that catalyze the breakdown of hydrogen peroxide to release oxygen (O_2) as a gas and the oxidation of hydroquinones to quinones with the release of hydrogen gas (H_2).

A bombardier beetle points its abdomen at a target and "fires" its chemical spray by squeezing the hydrogen peroxide and hydroquinones from the inner chambers to the outer chambers of its glands. This instantly sets off an explosive set of reactions as oxygen and hydrogen are released. The gases cause a dramatic pressure increase as they combine in a highly exothermic reaction ($H_2 + \frac{1}{2}O_2 \longrightarrow H_2O$). The energy released from the reactions in each chamber heats the water and quinones to 100°C. Finally, the gas pressure buildup shoots the hot mixture out as a fine mist. This explosion causes an audible bang.

The bombardier beetle's spray is doubly effective in repelling attackers. Quinones are irritating to other animals, and the tremendous heat of the spray also is likely to discourage a would-be predator. The beetle can turn its abdomen in any direction to aim accurately (as it is doing in box figure 5.2A, where the "attacker" is a forceps grasping one of its legs), and it can spray repeatedly in quick succession. Thus, the beetle's chemical defense system functions as a formidable repellent to would-be attackers.

Box figure 5.2A   The bombardier beetle *Stenaptinus insignis.* This animal is about 20 mm long. It is shown here spraying the forceps with which an experimenter is grasping one of its legs.

ATP: The Cell's Energy Currency

Many essential processes in cells, such as chemical synthesis, transport of materials, and movement, do not proceed spontaneously. They require energy input; that is, they are endergonic (have a positive ΔG). Living things obtain the energy used to carry on these normal functions from oxidation-reduction reactions that proceed with large negative changes in free energy ($-\Delta G$). What mechanism makes free energy derived from such reactions available for endergonic processes in cells? Usable free energy is provided in the form of **adenosine triphosphate (ATP)**, the cellular "energy currency."

Adenosine triphosphate (ATP) is the triphosphate of the nucleoside adenosine (figure 5.19). The terminal phosphate bond of an ATP molecule can be hydrolyzed at a pH of 7.0, and the free-energy change when the bond is hydrolyzed is large and negative. The products of the hydrolysis of ATP are inorganic phosphate (P_i) and adenosine diphosphate (ADP).

$$ATP + H_2O \rightleftharpoons ADP + P_i$$
$$\Delta G^{0\prime} = -7.3 \text{ kcal/mole } (-30.5 \text{ kjoule/mole})$$

In the cell, the free-energy change from ATP hydrolysis is certainly much higher, probably around -12 kcal/mole (-50 kjoule/mole), because the conditions inside cells are very different from those employed for standard free-energy calculations.

Bonds such as those in ATP that have negative $\Delta G^{0\prime}$ values of at least -7.3 kcal/mole sometimes are loosely called "high-energy bonds" and are designated by a $\sim$ symbol. This name is misleading because these bonds do *not* have a great deal of energy packed into them; there is simply a large difference in free-energy content between ATP and its cleavage products. This difference makes the equilibrium for ATP hydrolysis or phosphate transfer from ATP to a suitable acceptor molecule lie far toward completion.

However, ATP is surprisingly stable under cellular conditions, and the rate of spontaneous breakdown is low. The hydrolysis of ATP generally occurs only in a closely coupled relationship with an energy-requiring (endergonic) process in the cell. Thus, ATP is an excellent energy currency because it is not needlessly wasted through uncontrolled hydrolysis, but is instead used when and where it is needed.

ATP is a versatile energy currency that can be used in many different kinds of energy-requiring processes. The free energy inherent in ATP is expended in endergonic synthesis reactions. Hydrolysis of ATP provides the energy required for muscle contraction and other forms of movement; ATP is also a major source of energy for transporting substances across cell membranes. ATP is therefore the common energy currency for chemical, mechanical, and transport work in living cells.

Figure 5.19 Structure of adenosine triphosphate (ATP).

Adenosine triphosphate

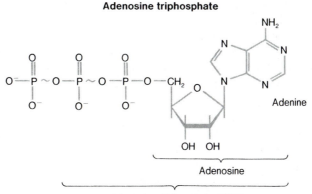

Adenosine diphosphate (ADP)

Summary

Energy is the capacity to do work. All physical and chemical processes occur as the result of the application or transfer of energy, and living things require a continuing supply of energy to function normally. Thermodynamics is the area of science that deals with energy exchanges. Bioenergetics is the term applied to thermodynamic studies of biological energy exchanges.

The ultimate source of energy for life is the radiant energy of the sun. Photosynthetic organisms use sunlight to provide energy for the synthesis of organic compounds. These compounds in turn provide a source of chemical energy for all forms of work in organisms.

Thermodynamics involves the measurement of energy changes in systems and their surroundings. The first law of thermodynamics states that energy can be neither created nor destroyed. Although the energy content of a particular system may change as a result of energy exchanges with the surroundings, the total energy content of the universe remains constant. The second law of thermodynamics states that all chemical and physical processes proceed in such a way that the entropy of the universe increases. A living organism may decrease its entropy, but it does so by increasing the entropy of its surroundings.

Free-energy change determines the direction and equilibrium of a chemical reaction. An exergonic reaction $(-\Delta G)$ can proceed spontaneously. An endergonic $(+\Delta G)$ reaction requires energy input to proceed.

Chemical kinetics is the study of reaction rates. Collision theory explains reaction rates in terms of collisions between reactants. The rate of collision, the orientation of the reactants, and the activation energy all affect the number of effective collisions and thus affect the reaction rate. Transition state theory deals with the course of an individual reaction. The transition state or activated complex is an intermediate state between reactants and products. It represents the state of highest free energy in the route from reactants to products. Large free energies of activation lead to slow reactions. Catalysts provide alternate reaction routes with lower free energies of activation.

The discovery of enzymes was a turning point in the study of the chemistry of life processes. Enzymes are globular protein molecules that act as catalysts in living cells. They have catalytic sites that complement the shapes of their substrates. Cofactors such as coenzymes are essential to the functions of many enzymes. Environmental factors such as temperature and pH also affect the catalytic efficiency of enzymes.

Enzymes lower activation energies by binding (and thus concentrating and orienting) substrate molecules. Some enzyme conformations change after binding substrates to bring the substrates into optimal orientation or to distort the substrates so that bond making or breaking is easier.

Enzyme activities are often under the control of activators or inhibitors that act on regulatory enzymes. Most regulatory enzymes are allosteric; that is, they have separate binding sites for regulatory molecules. When a regulatory site is occupied, the conformation of an enzyme changes, thus altering the catalytic site and decreasing the enzyme's catalytic efficiency.

Oxidation-reduction reactions involve the transfer of electrons. Oxidation is the loss of electrons; reduction is the gain of electrons. An oxidation-reduction couple is composed of an oxidant and reductant. The stronger the oxidizing ability of the oxidant and the weaker the reducing ability of the reductant, the more positive the redox potential of the couple. Reductants in couples with more negative redox potentials tend to donate electrons to oxidants in couples with more positive redox potentials. Some oxidation-reduction reactions in biological systems involve the transfer of a single electron, but in many biological oxidation-reduction reactions, pairs of electrons are transferred.

When reduced electron carriers that have negative redox potentials (such as NADH) transfer electrons to oxidized substances that have positive potentials (such as oxygen), free energy is released. Living cells use free energy released from oxidation-reduction reactions for ATP production.

ATP has a relatively large standard free-energy of hydrolysis and provides energy for many kinds of cell work. For this reason, ATP is sometimes called cellular "energy currency."

Questions

1. Which characteristics of chemical reactions are described in terms of thermodynamics (bioenergetics), and which are described in terms of chemical kinetics?

2. Growing organisms are systems that clearly have decreasing entropy. Is life therefore an exception to the second law of thermodynamics? Why or why not?

3. Describe and explain the effect that a change in the Boltzmann distribution, which is caused by increasing temperature, has on a reaction rate.

4. What is the significance of feedback inhibition in terms of energy conservation within cells?

5. Explain how allosteric enzymes are regulated.

6. Referring to redox potentials, explain why electrons are transferred from reduced ubiquinone to oxidized cytochrome *c*.

Suggested Readings

Books

Baker, J. J. W., and Allen, G. E. 1981. *Matter, energy, and life*. 4th ed. Reading, Mass.: Addison-Wesley.

Christensen, H. N. 1967. *Enzyme kinetics*. Philadelphia: W. B. Saunders.

Christensen, H. N., and Cellarius, R. A. 1972. *Introduction to bioenergetics: Thermodynamics for the biologist*. Philadelphia: W. B. Saunders.

Lehninger, A. L. 1971. *Bioenergetics*. 2d ed. New York: Benjamin.

Stryer, L. 1981. *Biochemistry*. 2d ed. San Francisco: W. H. Freeman.

Wynn, C. H. 1976. *The structure and function of enzymes*. Studies in Biology no. 42. Baltimore: University Park Press.

Articles

Chappell, J. B. 1977. ATP. *Carolina Biology Readers* no. 50. Burlington, N.C.: Carolina Biological Supply Co.

Hollaway, M. R. 1976. The mechanism of enzyme action. *Carolina Biology Readers* no. 45. Burlington, N. C.: Carolina Biological Supply Co.

6

Photosynthesis

Chapter Concepts

1. In photosynthesis, light energy is used to produce reduced carbon compounds that provide chemical energy for all living things.
2. The leaves of terrestrial plants serve as absorbers and converters of light energy.
3. Chloroplasts are the sites of photosynthetic activity in plant cells. Pigments involved in the light reactions are associated with the internal membrane structures of the chloroplast. Enzymes involved in carbon dioxide fixation are in the aqueous stroma matrix of the chloroplast.
4. Specific photochemical reactions of pigments result in production of reduced NADP and made energy available for ATP production.
5. Calvin cycle reactions use reduced NADP and ATP carbon dioxide into carbohydrate molecules.
6. The C_4 and CAM pathways provide alternative carbon dioxide incorporation mechanisms. Under certain environmental conditions, plants possessing these mechanisms have an advantage over plants not possessing them.

One of the most fundamental characteristics of living things is their need for a continual input of energy. This energy is required not only for growth, but for most other activities as well. The study of energy relationships is thus a central topic in all parts of biology.

"All flesh is grass." This statement captures an important biological truth. The energy that all living organisms use comes directly or indirectly from the light of the sun. Sunlight provides a continuous input of energy for the biosphere (the whole world of living organisms). In the process known as **photosynthesis,** green plants convert light energy to chemical energy in the form of reduced organic compounds. These compounds serve as the energy source for the rest of the biological world; for parts of the plant not capable of photosynthesis, for organisms that eat plants, and for other organisms that feed on those that eat plants.

Broadly speaking, the reverse of photosynthesis is **respiration,** or the breakdown of organic compounds to provide energy. Together, photosynthesis and respiration convert light energy to usable chemical energy in cells (figure 6.1).

Photosynthesis forms organic compounds from carbon dioxide and water. If the product is considered to be a carbohydrate such as glucose, photosynthesis can be represented by this summary statement:

$$6CO_2 \;+\; 6H_2O \xrightarrow{\text{Light}} C_6H_{12}O_6 \;+\; 6O_2$$

(Carbon dioxide) (Water)   (Glucose)   (Oxygen)

Photosynthesis can thus be regarded as the reduction of CO_2. Respiration is roughly the opposite process; it involves the oxidation of organic compounds. This oxidation process produces usable energy, often in the form of the cellular "energy currency," ATP. Respiration can be represented by this summary statement:

$$C_6H_{12}O_6 + 6O_2 \rightleftarrows 6CO_2 + 6H_2O$$

Usable energy (ATP)

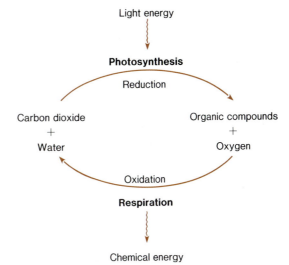

Figure 6.1 The relationships between photosynthesis and respiration.

There are a number of general similarities between the processes of photosynthesis and respiration. For example, each can be divided into two phases. One phase is intimately associated with biological membranes—in photosynthesis, the chloroplast membranes, and in respiration, the membranes of the mitochondria. The membrane-associated phase of each process involves electron transport between special electron-carrier molecules. The other phase involves a number of small, stepwise changes in organic molecules. In each process, the reactions that bring about these changes occur in aqueous solution and are catalyzed by specific enzymes. And in both photosynthesis and respiration, the reactions occurring in solution and in the membrane are linked by a "reductant currency," either NAD or NADP.

Figure 6.2 is a simplified overview of the linked processes of photosynthesis and respiration.

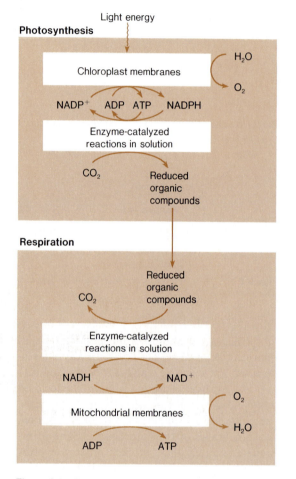

Figure 6.2 Simplified overview of the linked processes of photosynthesis and respiration.

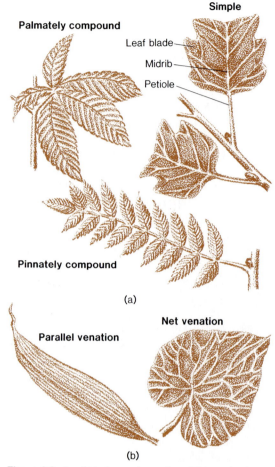

Figure 6.3 Leaf blades and venation. (a) In a simple leaf, the blade is in one piece. Leaflets are attached to the petiole in different ways in different types of compound leaves. (b) Parallel and net venation compared.

The Leaf

In most vascular plants, plants with vascular (transport) tissue, photosynthesis occurs mainly in a specialized organ, the **leaf.** The leaf's structural organization is well-suited for gathering the raw materials of photosynthesis and for exporting the products. The usual shape of a leaf is a flattened blade, which may be a single structure (**simple leaves**) or an arrangement of several leaflets (**compound leaves**) (figure 6.3a).

Because the leaf is flat and thin, it provides a large area for light absorption and a small diffusion pathway to all internal parts of the leaf for gases exchanged with surrounding air. Each leaf cell is close to the vascular tissue that transports mineral salts and water to the leaf and distributes the products of photosynthesis (mainly in the form of sucrose) to other parts of the plant.

The arrangement of vascular tissue elements (**vascular bundles** or **veins**) is called the **venation** of the leaf. There are a variety of leaf venation patterns depending on the type of plant, but all ensure a distribution of vascular tissue to all parts of the leaf (figure 6.3b). In addition to the conducting tissues involved in the transport of materials, vascular bundles contain fibers that provide structural support for the leaf. The vascular tissue connects the leaf with the stem through the **petiole.**

Photosynthesis requires CO_2, which is present only in very small amounts in the air (about 0.03%), and this presents the plant with a major problem. CO_2 enters leaf cells after first being dissolved in the water that covers their surfaces, and efficient transfer of CO_2 into the leaf cells

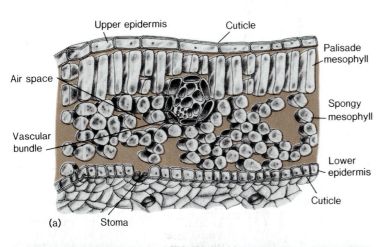

Upper epidermis   Cuticle

Palisade mesophyll

Air space

Spongy mesophyll

Vascular bundle

Lower epidermis

Cuticle

(a)   Stoma

Figure 6.4   (a) Leaf structure. Two guard cells surround each stoma. (b) Scanning electron micrograph showing palisade and spongy mesophyll. (Magnification $\times$ 1,150)

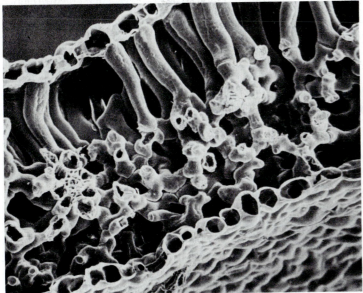

(b)

requires a large area of contact between air and water on leaf cell surfaces. But when a large amount of wet surface area is exposed to the air, water is inevitably lost by evaporation. Some of this water loss is actually necessary for the movement of water and mineral salts through the plant, but excessive loss will cause **wilting.** Wilted leaves become limp because water loss leads to a loss of turgor pressure (see p. 106), and the rate of metabolism in a wilted leaf is much lower than normal. Plants can recover from wilting if the condition does not continue for too long, but after a few days, cells in wilted plants begin to die, and the plant itself eventually dies as a result. Water loss from land plants is reduced because most of the leaf surface is covered by a waxy **cuticle** with a low permeability to water. Unfortunately, this protective cuticle is also poorly permeable to gases. Because of this, the exchange of gases in a land plant takes place mainly via a few specialized controllable pores called **stomata** (singular: **stoma**). In many species, stomata are located mostly or even exclusively on the lower surface of the leaf.

The diagram in figure 6.4 shows a cross-section of a typical leaf; the upper and lower epidermis are both protected by a waxy cuticle on the outer side. Ordinary epidermal cells do not contain chloroplasts, the organelles where photosynthesis takes place. However, the cells surrounding the stomata, called **guard cells,** do contain chloroplasts.

The photosynthetic cells in the body of the leaf are divided into two areas; toward the top of the leaf are regularly arranged **palisade** layers of **mesophyll** cells and underneath is a more loosely packed **spongy mesophyll.** Cells in the palisade mesophyll layers have more chloroplasts than cells in the spongy mesophyll. The somewhat loosely packed arrangement of the spongy mesophyll cells increases the efficiency of CO_2 uptake into the photosynthesizing cells by increasing the amount of cell surface in contact with the air spaces. The air spaces are in turn connected with the external atmosphere via the stomatal pores.

Stomatal Functioning

The activity of the stomata is very important in the life of plants. The size of each stomatal opening affects the rates at which CO_2, O_2, and water vapor move in and out of leaves. The actual physical opening and closing of the stomata is understood better than the mechanisms that control the process.

The opening and closing of the stomata depend on a combination of structural and functional features of the guard cells. Stomata open when guard cells take up water osmotically and swell outward against their walls. Although it seems logical that swelling in the guard cells would instead tend to *close* the stomata, this is prevented by several special features of each guard cell's walls. One of these features is the arrangement of many of the cellulose microfibrils in the walls. These microfibrils form rings around the circumference of each guard cell, with the rings radiating outward from the center of the stoma (figure 6.5). The microfibrils prevent the guard cell from increasing its girth.

A second structural feature of the guard cells that helps the stomata to open is that guard cells are very tightly connected to one another at each end of a stoma. As water osmotically enters guard cells and they become more turgid, the cells apply increased pressure against their cell walls. Although the radially arranged microfibrils prevent guard cell walls from stretching much in diameter, the walls can lengthen slightly. But since the guard cells are firmly bound together at their ends, the lengthening cell walls cause the cells to tend to curve outward, thereby opening the stoma. Stomata thus open when guard cells take up water, and the control of stomatal opening is dependent on the regulation of the osmotic pressure of guard cells.

When stomata open, the rate of gas exchange between the leaf interior and the atmosphere increases. Because of the obvious importance of stomatal regulation, biologists are interested in the mechanisms controlling the opening and closing of stomata, but the process is only partly understood.

It has long been known that stomata open when light shines on leaves. However, the cellular mechanisms that actually cause stomatal opening are not entirely clear. The key question is, what actually causes the osmotic pressure change inside the guard cells that results in stomatal opening? Unlike ordinary epidermal cells, guard cells have chloroplasts. Thus, when leaves are illuminated, guard cells begin to carry on photosynthesis. It was thought for some time that the sugar produced by photosynthesis in illuminated guard cells was a major factor in the osmotic change. But stomatal opening occurs too rapidly for this to be the case.

More recently, biologists have focused their attention on other mechanisms. Stomatal opening is closely correlated with decreasing CO_2 concentration in the guard cells themselves and in the air spaces inside the leaf. In fact, decreasing CO_2 content in the air can cause stomatal opening even in the dark. (Of course, in an illuminated leaf, the CO_2 content falls as CO_2 is used in photosynthesis.) Decreasing CO_2 somehow causes guard cells to take up potassium ions (K^+). Energy-requiring K^+ uptake occurs in the guard cells adjacent to opening stomata, and most biologists now believe that K^+ uptake is involved in turgor pressure changes within the guard cells. However, the immediate cause of K^+ uptake is presently unknown. The mechanisms controlling stomatal closure will also require further investigation.

There is some evidence that one additional factor may be involved in stomatal regulation under certain conditions. Cells in wilting leaves produce increased quantities of a hormone called **abscisic acid (ABA),** which causes stomata to close. But abscisic acid's role in stomatal regulation seems to be limited to water-stress situations, and ABA probably is not involved in routine stomatal opening and closing.

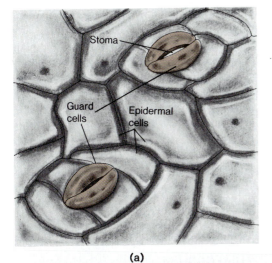

(a)

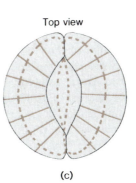

(c)

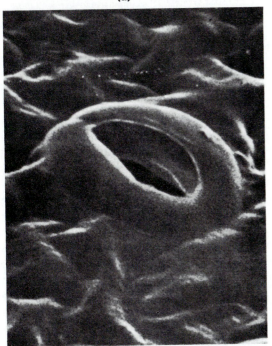

(b)

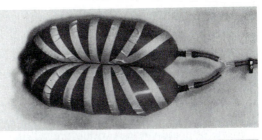

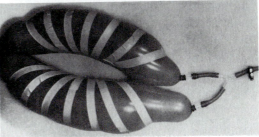

(d)

Figure 6.5 Stomata. (a) Closed and open stomata on the bottom of a leaf. (b) Scanning electron micrograph of an open stoma showing its two guard cells. (Magnification × 2,380) (c) Guard cell movement during stomatal opening. Dashed lines indicate position of guard cells when the stoma is nearly closed. Radial arrangement of cellulose microfibrils in guard cell walls is shown in color. Guard cells are fastened together tightly at their ends. K^+ uptake and osmotic entry of water cause guard cell swelling, but cells cannot increase in girth because of the cellulose microfibrils running around their walls. They do increase in length, but curve outward because of the connections at their ends. (d) A two-balloon model of the guard cells of a stoma. Masking tape represents the radial cellulose microfibrils. Balloons were glued together at their ends with rubber cement before inflation.

The essentially impermeable cuticle and controllable stomata seem adequate for most land plants, but many **xerophytes,** plants that live in particularly dry environments, have extra anatomical adaptations such as sunken stomata and rolled leaves (figure 6.6). These adaptations decrease the rate of water loss because the air in contact with the leaf cells contains more water vapor than the surrounding atmosphere, and water loss by diffusion is reduced. In addition, such plants often have specialized physiological adaptations in the photosynthetic process itself that also aid in water conservation. Some will be mentioned later.

Because of the role of stomata in plant/water relationships and photosynthetic productivity, the study of stomatal regulation is a very important aspect of modern research in plant biology. There may well be practical agricultural applications for what is learned in the future about stomatal regulation.

Chloroplast Structure

The site of photosynthesis within the cell is the **chloroplast.** Chloroplasts are plastids, distinctive organelles found in green plants (see chapter 3). Four concentric "layers" can be distinguished in a chloroplast (figure 6.7).

The whole chloroplast is surrounded by the **chloroplast envelope,** a double membrane that separates the chloroplast from the rest of the cell. The raw materials for photosynthesis enter the chloroplast across this membrane. Similarly, photosynthetic products cross the membrane when they are exported to the rest of the cell.

The second layer is the area within the chloroplast envelope, called the **stroma matrix.** This area contains a number of soluble enzymes responsible for the catalysis of various reactions involved in the incorporation of CO_2 into organic compounds.

Inside the chloroplast is a complicated array of parallel membranes forming the third layer. These membranes are arranged in structures that are visible by light microscopy as darker grains and hence are called **grana** (singular: **granum**). The more detailed pictures of an electron microscope show that these grana are composed of stacks of flattened bags or discs called **thylakoids.**

The various grana stacks are connected by membranous structures called **stroma lamellae.** The pigments that give the chloroplast and the plant its characteristic green color, as well as numerous proteins that play a role in photosynthesis, are contained in the membranes of the thylakoids and the stroma lamellae.

The space inside the thylakoids, the **intrathylakoid space,** is the fourth layer of the chloroplast. The movement of ions across the thylakoid membrane between the stroma matrix and the intrathylakoid space is very important in photosynthesis.

Photosynthesis

Because of the work of a number of scientists over the years, some basic characteristics of photosynthesis were well understood by the beginning of this century. It was known, for instance, that green plants use water from the soil and carbon dioxide from the air to synthesize carbohydrates. It was also known that oxygen is a byproduct of this process and that this oxygen is released into the atmosphere. Further, it was well established that photosynthetic activity is centered in the chloroplasts of green plant cells. Beyond these basics, however, practically nothing was known about the details of the actual chemical processes in photosynthesis. For example, there was a vigorous debate between biologists who believed that the oxygen released during photosynthesis comes from water molecules and those who believed that it comes from carbon dioxide.

Modern analytical research on photosynthesis began early in this century with the work of F. F. Blackman. On the basis of his research on photosynthesis under various light and temperature conditions, Blackman proposed in 1905 that there are two basically different kinds of reactions in photosynthesis: a **light reaction** that requires light and a **dark reaction** that does not require light. Blackman based his proposal on the results of temperature and light changes on photosynthetic rates. He found that increasing the light intensity increased the rate of photosynthesis up to a certain maximum rate; this level of intensity is the **light saturation level.** But Blackman also discovered that the maximum rate

Figure 6.6 Cross section of a leaf of the grass *Ammophila arenaria* showing sunken stomata and leaf rolling, which increases the diffusion distance for H_2O from the leaf cells to the atmosphere.

Figure 6.7 (a) Chloroplast structure showing the four concentric areas. (b) Electron micrographs of chloroplasts from broad bean. (1) The whole chloroplast. This chloroplast contains stored carbohydrate in the form of two starch granules. (Magnification × 10,500) (2) Detail to show stacking of thylakoids to form grana.

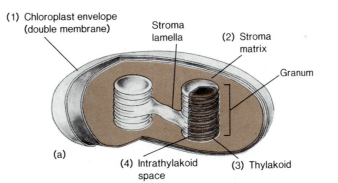

(1) Chloroplast envelope (double membrane)

Stroma lamella

(2) Stroma matrix

Granum

(a)

(4) Intrathylakoid space

(3) Thylakoid

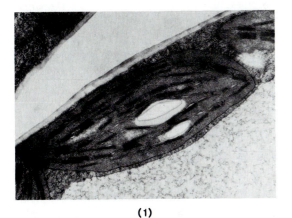

(1)

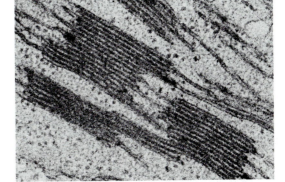

(2)

(b)

of photosynthesis changed when the experiments were done at different temperatures. Because light-activated chemical (photochemical) processes are essentially temperature independent, Blackman reasoned that the temperature dependence of the maximum photosynthetic rate was evidence for a second type of reaction, a nonphotochemical reaction, which he called the "dark reaction" (figure 6.8).

Blackman's work was an important step toward understanding photosynthesis because it has led to the analysis of the light and dark reaction sets—that is, the photochemical and nonphotochemical aspects of photosynthesis. Details of these processes have been pieced together over the years. Microbiologist C. B. van Niel made an important contribution in the 1930s in relation to the origin of the oxygen released during photosynthesis. Actually, van Niel was not working on green plants at all, but on a group of photosynthesizing bacteria called purple sulfur bacteria. Van Niel found that these bacteria released sulfur and not oxygen during photosynthesis. He summarized the process of photosynthesis in these organisms as follows:

$$CO_2 + 2H_2S \xrightarrow{\text{Light}} (CH_2O) + H_2O + 2S$$

Van Niel concluded that light energy was used to decompose hydrogen sulfide (H_2S) and that the hydrogen atoms were used in the reduction of CO_2 to produce carbohydrates (here symbolized (CH_2O)). He went on to propose that this and other photochemical reactions in photosynthesis could be generally represented as follows:

$$CO_2 + 2H_2A \xrightarrow{\text{Light}} (CH_2O) + H_2O + 2A$$

where H_2A is any oxidizable substance that can supply electrons and protons. In green plants, this substance is water.

$$CO_2 + 2H_2O \xrightarrow{\text{Light}} (CH_2O) + H_2O + O_2$$

This strongly suggested to van Niel that oxygen released by green plants during photosynthesis comes from water and not from CO_2. Van Niel's hypothesis was a stimulus for further research and was indeed proven correct in later experiments. Other scientists were able to determine the fate of oxygen from water and carbon dioxide in photosynthesis by using a radioactive isotope of oxygen, [18]oxygen.

A landmark in the study of the light reactions in photosynthesis was the discovery in 1937 by Robert Hill that under certain circumstances, chloroplasts isolated from disrupted leaf cells would release oxygen when they were illuminated. This demonstrated that chloroplasts contain all of the elements necessary to split water and release oxygen, even in the absence of the rest of the cell. Hill used oxidation-reduction indicators (electron acceptors) in his experiments and found that the electron acceptors were reduced when the illuminated chloroplasts were releasing oxygen. Further, he found that oxygen was not released if he did not supply an electron acceptor. Hill therefore concluded that there must be a naturally occurring electron acceptor in chloroplasts that acts in the same way as the added electron acceptor in his experiments.

It should be noted that Hill's extracted chloroplasts generated oxygen without incorporating CO_2 into carbohydrates. Probably the chloroplasts did not incorporate CO_2 because they had been physically damaged by the extraction procedure. But this lack of CO_2 incorporation confirmed the existence of the separate light and dark reactions that Blackman had proposed more than thirty years earlier. Hill's experiments

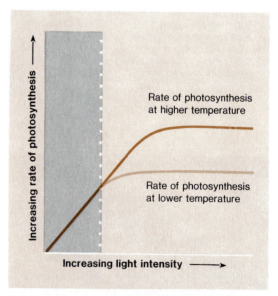

Figure 6.8 Results of Blackman's experiments. At lower light intensities (gray shaded area of figure), light intensity limits photosynthetic rate. Above saturating light intensity (gray broken line), the maximum photosynthetic rate depends on experimental temperature. On the basis of these results, Blackman proposed that there are separate "light" and "dark" reactions in photosynthesis.

showed that light reactions could be separated from dark reactions and could even occur in their absence.

Using improved chloroplast extraction techniques, Daniel Arnon, Severo Ochoa, and others analyzed these reactions further. They discovered that the key electron acceptor in intact chloroplasts is nicotinamide adenine dinucleotide phosphate (NADP). Thus, the light reactions in photosynthesis produce reduced NADP (NADPH). It was also discovered that ADP is phosphorylated to produce ATP during the light reactions.

This led Arnon to draw an important conclusion and to test an additional hypothesis. He concluded that the role of the light reactions was to supply NADPH as a "reduction currency" and ATP as an "energy currency," both of which the chloroplast could spend on CO_2 reduction in the manufacture of carbohydrates. Arnon then proposed that if these materials were present, chloroplasts should be able to use CO_2 to produce carbohydrates even in the absence of light. He tested his hypothesis by supplying chloroplasts in the dark with NADPH, ATP, and CO_2, and found that they did indeed synthesize sugars in the dark.

It is now known that the CO_2-incorporating reactions (dark reactions) occur in the stroma matrix of a chloroplast. It is also known that the light reactions occur within the thylakoid membranes and supply reduced NADP and ATP for the reduction and incorporation of CO_2 into carbohydrate molecules (figure 6.9). Actually, the term "light reactions" is somewhat imprecise because only the earliest steps in the light reactions actually require light.

Light Interactions with Molecules

Radiation can be classified in terms of the **electromagnetic spectrum,** which includes radiation ranging from X rays and gamma rays with very short wavelengths to radio waves with very long wavelengths. One of the fundamental properties of electromagnetic radiation is that it comes in discrete packets called **quanta.** The energy of one quantum is inversely proportional to the wavelength of the radiation; that is, a quantum of long-wavelength radiation has a lower energy level than a quantum of shorter-wavelength radiation.

Visible light (figure 6.10) is the portion of the electromagnetic spectrum to which human

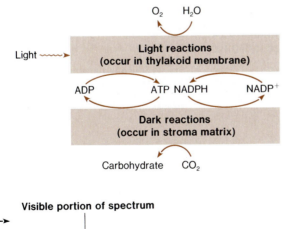
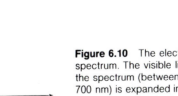

Figure 6.9 The light and dark reactions of photosynthesis. The light reactions supply reduced NADP (NADPH) and ATP for the reduction and incorporation of CO_2 into carbohydrate molecules in the dark reactions.

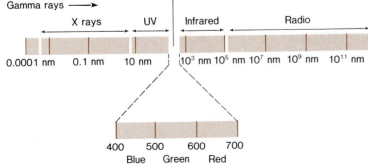

Figure 6.10 The electromagnetic spectrum. The visible light portion of the spectrum (between 400 nm and 700 nm) is expanded in the lower part of the diagram. Color is the interpretation of what is seen when radiation of a given wavelength strikes the eyes. Thus, it is only a matter of convenience to speak of blue light or red light when describing light of a particular range of wavelengths.

eyes are sensitive; interestingly, this corresponds to a large part of the sun's radiation that penetrates through the atmosphere and reaches the surface of the earth. Visible light includes wavelengths of light that have quantum energy levels that permit interaction with molecules in human eyes.

The energy of a quantum or packet of light affects the way in which it can interact with molecules. Quanta of high energy, such as those of short-wavelength ultraviolet radiation, can break chemical bonds—for example, bonds in DNA molecules. This is why such radiation is harmful to living cells. Quanta of longer wavelength, such as those of infrared radiation, have much less energy and are only capable of increasing the vibrational or rotational energy of a molecule. Between these two extremes are quanta with sufficient energy to set off specific photochemical reactions. In such a reaction, an electron is promoted to a higher energy orbital (figure 6.11). In molecules with delocalized orbitals, in which an electron can be shared between several atoms, the energy required to promote the electron is often low enough that the absorption of a single quantum of visible light can cause the transition. Light-energy absorbing compounds in which this occurs are called **pigments.**

Both the photosynthetic apparatus and human eyes have to operate with the radiation available to them on the surface of the earth, and in each case, the less energetic infrared radiation is not strong enough to cause the specific photochemical responses upon which these functions depend. Thus, photosynthetically effective radiation and visible light fall within the same range of wavelengths.

Leaf Pigments

Light interacts with pigments, and the pigments necessary for photosynthesis are found in the thylakoid membrane. The green color of plants is due mainly to a pigment called **chlorophyll** (figure 6.12). Chlorophyll's absorbance spectrum peaks in the blue and red regions (figure 6.12b), which means that light of these wavelengths is absorbed, while the unabsorbed or reflected light appears green. Vascular plants have two types of chlorophyll designated a and b. These two types of chlorophyll differ slightly in their chemical structure and hence in their spectra.

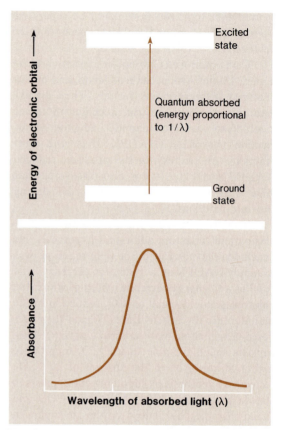

Figure 6.11 The relationship between the electronic energy levels and the absorbance spectrum of a molecule. Light of a particular wavelength is absorbed if the energy of its quanta corresponds to an energy difference between different electronic configurations of the molecule. The ground state is the usual configuration of the molecule. In the excited state, an electron has been raised to a higher energy level (a different orbital). The wavelength that causes this change is the absorbed wavelength. Light of longer wavelengths has too little energy to be absorbed; light of shorter wavelengths has too much energy to be absorbed.

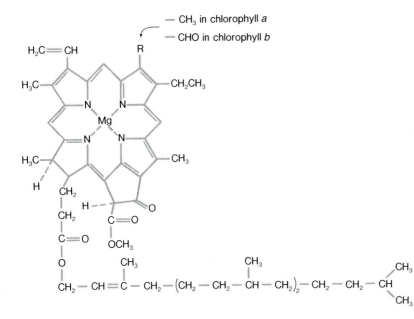

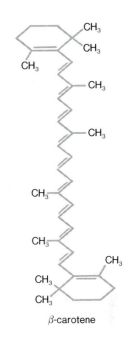

β-carotene

(a)

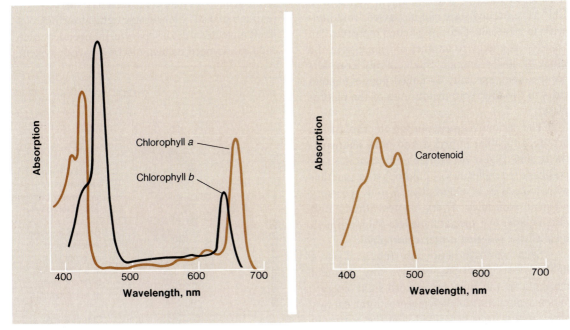

(b)

Figure 6.12   Pigments. (a) Structural formulas of chlorophylls *a* and *b* and β-carotene. (b) Absorption spectra of chlorophylls *a* and *b* and β-carotene (a carotenoid).

It is apparent from figure 6.12b that there is a large gap in the visible spectrum where the chlorophylls do not absorb light efficiently. Other pigments, known as **accessory pigments**, absorb some of the light energy in the wavelength range between the peaks of chlorophyll absorption and make some of that energy available for use in photosynthesis. In green plants, for example, **carotenoids** such as β-carotene absorb some of the light that is not absorbed by the chlorophylls. (β-carotene gives carrots their orange color but is also present in all green plant tissue.) The carotenoids, like the chlorophylls, are hydrophobic molecules located in the thylakoid membranes.

The Photochemical Reactions

When a molecule absorbs a quantum of light, an electron is raised to a higher energy level. This higher level is known as the **excited state** (see figure 6.11) and is designated by an asterisk; thus, excited chlorophyll (Chl) is symbolized Chl*. The excited state can be passed from molecule to molecule, since an excited molecule can transfer its energy to another molecule with the same or lower energy. Such energy transfers occur among specially arranged pigment molecules in the thylakoid membranes of the chloroplast.

The various pigments in the chloroplast are arranged together in arrays known as **photosynthetic units** (figure 6.13). Each unit contains several hundred molecules of chlorophyll *a* together with molecules of chlorophyll *b* and accessory pigment molecules. Every photosynthetic unit also includes one special molecule of chlorophyll *a* called the **reaction center chlorophyll.**

Each photosynthetic unit functions as a light-harvesting system because any pigment molecule in the unit can absorb light energy. When a pigment molecule absorbs light energy, an excited state is generated, and the energy in this excited molecule can be transferred to adjacent pigment molecules. Chlorophyll *b* and accessory pigments such as β-carotene have higher energy levels than the main body of chlorophyll *a* molecules, and the chlorophyll *a* molecules, in turn, have higher energy levels than the reaction center chlorophyll. Because energy transfer occurs more readily to a lower-energy molecule, the excited state is funneled down to the reaction

center chlorophyll, which has the lowest energy level in the unit. Thus, light energy absorbed by any pigment molecule in the unit eventually causes the excitation of the reaction center chlorophyll. This molecule acts as a "trap" for light energy absorbed by any molecule in the unit.

Some biologists say that the molecules in a photosynthetic unit essentially function as an antenna system. Another good analogy is that a photosynthetic unit functions much like a large upright funnel leading to a small container set out in the rain. Raindrops land in various places in the funnel, but they all run down through the funnel into the container. Similarly, light energy absorbed anywhere in the photosynthetic unit eventually "funnels down" to the reaction center chlorophyll.

The light-gathering process that takes place in photosynthetic units seems to add efficiency to the light reaction mechanism. Each reaction center chlorophyll is associated with a system of electron carriers that are involved in further light reaction steps. If every single light-absorbing pigment molecule were a separate, independent unit, each would need its own electron carrier system.

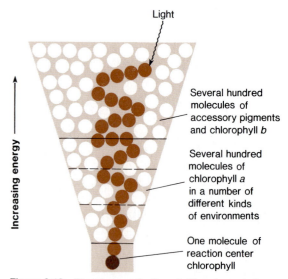

Figure 6.13 The photosynthetic unit. The pigments in the thylakoid membrane are arranged in units, each comprised of several hundred molecules. Light absorbed by any one of these molecules can eventually cause the formation of the excited state of one special chlorophyll molecule, the reaction center chlorophyll. The chain of molecules shown in color is an example of the way that the excited state resulting from absorption of one quantum of light energy can pass from molecule to molecule through the unit to the reaction center molecule.

Many of these electron carrier systems would remain idle most of the time because they would function only when their own individual pigment molecules absorbed light energy. Thus, the arrangement of several hundred pigment molecules within the light-harvesting photosynthetic unit eliminates the need for multiple sets of electron carriers to be used in subsequent reaction steps.

The excited state of a molecule has different chemical properties than the same molecule has in its normal ground state. For example, excited Chl* is a much better reducing agent than chlorophyll in the ground state; that is, Chl* loses an electron more readily than Chl does. It is this Chl* that, through a series of intermediate steps, donates electrons for the reduction of $NADP^+$. After donating an electron, the oxidized chlorophyll is reduced again to the ground state chlorophyll by a weaker reductant (figure 6.14). Thus, energy from light is used to transfer an electron "uphill" from a weak reducing agent to a stronger one (Chl*) that can then donate an electron for the reduction of $NADP^+$.

The overall result of the light-induced electron transfer reactions in chloroplasts is the reduction of $NADP^+$ and the oxidation of water. In other words, water is the weak reductant that serves as an electron donor for the photochemical reactions. However, the photosynthetic system actually requires more energy than is provided by a single quantum of the red light absorbed by chlorophyll to transfer an electron all the way from water to $NADP^+$. The reaction therefore occurs in two stages and involves two different photochemical events that together provide *more than enough* energy for the transfer. (The "extra" energy will be accounted for later.) There are two different types of reaction center chlorophylls, each with its own photosynthetic unit. One **photosystem** reduces $NADP^+$. Its oxidized reaction-center chlorophyll is then reduced again by the Chl* of the other photosystem. That chlorophyll, in turn, is reduced by water. A byproduct of this reaction is the release of oxygen gas.

$$2H_2O \longrightarrow 4H^+ + 4e^- + O_2$$

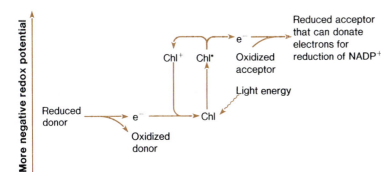

Figure 6.14 How the absorption of light by chlorophyll can be coupled to the transfer of an electron from a weak reducing agent to form a stronger one, an energy-requiring process. Absorption of radiation forms the excited state chlorophyll (Chl*) with an electron in a more energetic orbital. It can donate an electron (e^-) for the reduction of $NADP^+$ with the production of oxidized chlorophyll (Chl^+). Only a weak reducing agent (electron donor) is needed to convert Chl^+ back to the original "ground state" chlorophyll (Chl).

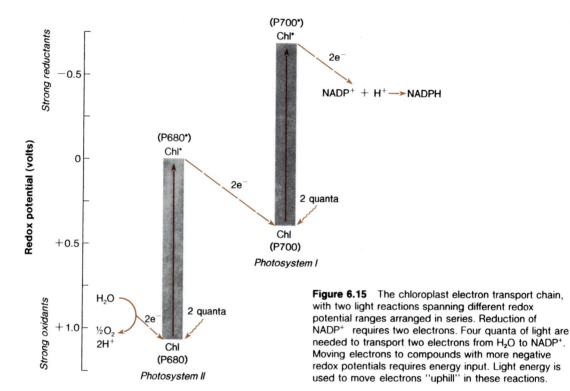

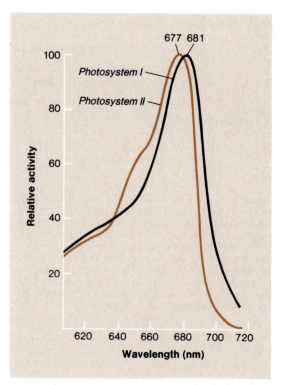

Figure 6.15 The chloroplast electron transport chain, with two light reactions spanning different redox potential ranges arranged in series. Reduction of $NADP^+$ requires two electrons. Four quanta of light are needed to transport two electrons from H_2O to $NADP^+$. Moving electrons to compounds with more negative redox potentials requires energy input. Light energy is used to move electrons "uphill" in these reactions.

The photosystem that directly reduces $NADP^+$ is called **Photosystem I.** Its reaction center chlorophyll has an absorbance maximum around 700 nm and so is called **P700** (P stands for pigment). **Photosystem II** is the system directly involved in oxygen release; its reaction center chlorophyll absorbs at slightly shorter wavelengths (680 nm) and is called **P680** (figure 6.15).

Because some of the chlorophyll molecules in the photosynthetic unit of Photosystem I absorb at longer wavelengths than those associated with Photosystem II, far-red light is able to activate P700 more efficiently than P680. In other words, Photosystems I and II have different **action spectra** (figure 6.16). An action spectrum is obtained by measuring the magnitude of an action (such as photosynthetic activity) obtained by using different wavelengths of light.

Electron Transport

As was just discussed, to transfer electrons "uphill" from water to $NADP^+$, two light reactions are coupled together. These reactions produce an oxidant that can oxidize water and a reductant that can reduce $NADP^+$. However,

Figure 6.16 Action spectra of Photosystems I and II. Photosystem I functions more efficiently at slightly longer wavelengths than Photosystem II.

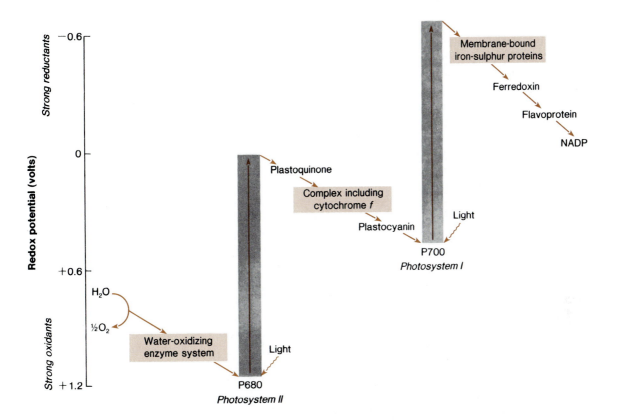

water and NADP$^+$ do not react directly with P680$^+$ and P700*, respectively; neither does P680* transfer its electrons directly to P700$^+$. Instead of single-step oxidation-reduction reactions, series of electron carriers transfer electrons through stepwise sequences of oxidation-reduction reactions.

Such a series of oxidation-reduction reactions can be represented schematically for the hypothetical electron carriers A, B, C, and D:

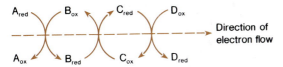

In this example, A, B, C, and D are progressively weaker reducing agents and better oxidizing agents (see p. 135). Thus, the reactions proceed in the directions shown by the arrows and the overall result is the transfer of the electron from A to D.

In spite of an intensive research effort, there are still a number of gaps in the understanding of the arrangement of the various electron transport components involved in photosynthesis. Some of the known elements are shown in figure 6.17.

Figure 6.17 The photosynthetic electron transport chain from water to NADP$^+$. The positions of the components on the redox potential scale are only approximate. Because biologists represent these electron transports in this standard way, this arrangement is sometimes called the "Z-scheme." This is only a way of illustrating a relationship and should not be taken as representing the real positions of these elements relative to one another. Electron transfer from plastoquinone to plastocyanin involves cytochrome *f* and possibly other components as well. Cytochromes are proteins that contain an iron atom in a heme coenzyme group. Cytochromes are important in mitochondria and also in chloroplasts, as will be seen in the next chapter.

The electron transport chain between the two photosystems contains **plastoquinone** (figure 6.18). This relatively small molecule can accept two electrons and two protons and be reduced to a quinol. Plastoquinone has a long hydrophobic side chain that makes it insoluble in water but soluble in the lipid phase of the membrane. Later it will become apparent that the involvement of both protons and electrons in the reduction and oxidation of plastoquinone is an important factor in the processes leading to the formation of ATP. A plastoquinone molecule in a special environment, possibly attached to a protein, is probably the primary acceptor of Photosystem II; that is, it is most likely the compound that is first reduced by P680*. The electron transport chain between the two photosystems also includes **plastocyanin,** a soluble protein that donates electrons to P700.

The primary electron acceptor of Photosystem I is evidently a protein-bound iron atom, and the electron from this acceptor is eventually transferred to **ferredoxin.** Ferredoxin is a soluble iron-sulfur protein found in the stroma matrix. The two iron atoms in ferredoxin are attached to the protein by cysteine amino acid residues. The iron atoms also are connected to each other by two sulfur atoms:

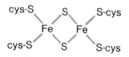

Ferredoxin is an important branch point for the electron transport chain. Although most reduced ferredoxin passes its electrons on to $NADP^+$ via a specific **flavoprotein enzyme,** ferredoxin can also be a carrier in a **cyclic pathway** of electron flow around Photosystem I, which will be discussed later.

Finally, the mechanism of the water oxidation process, the reaction in which oxygen is released and electrons donated to $P680^+$, is another unsolved problem in photosynthesis. This reaction was the original source of the oxygen in the earth's atmosphere, and it continues to replace oxygen used up in respiration. The photosynthetic production of reduced NADP must be balanced by the oxidation of some other component.

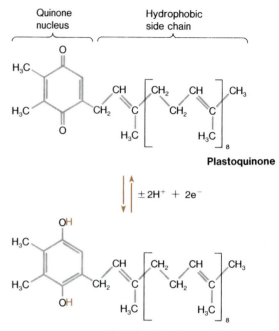

Figure 6.18   Plastoquinone and its reduced form (plastoquinol or plastohydroquinone). The structure of the most common form of plastoquinone is shown; there are other types with minor differences in the side chains.

Since green plants can oxidize water, they can use water, a relatively abundant material, as an electron donor. However, little is known about the chemical nature of the components involved in oxygen evolution. The only clue is that manganese is involved in some way.

Photophosphorylation

Photosynthetic light reactions supply the NADPH and ATP that are used in CO_2 incorporation during the dark reactions. The reduction of $NADP^+$ by Photosystem I in the chloroplasts has already been discussed. Illuminated chloroplasts also produce ATP. This light-related ATP formation is called **photophosphorylation** and is coupled with the electron flow that occurs in the photosystems.

Coupling of electron flow and the phosphorylation of ADP to form ATP occurs in both chloroplasts and mitochondria and is a crucial part of the energy exchange process in living things.

However, the nature of the coupling mechanism remains one of the central problems of modern biology. The major question is: What is the link between the transfer of electrons and the combination of ADP and inorganic phosphate to form ATP? Two major hypotheses have been proposed.

One proposal is that free energy released during electron transport makes possible the formation of a "high-energy" compound. Then, during the return of this intermediate compound to its original state, enough free energy becomes available to allow ATP production.

There are actually two related but different possible explanations for this basic idea. The **chemical hypothesis** proposes that a special "high-energy" chemical compound is actually formed during electron transport and degraded during phosphorylation, with adequate free energy changes in each case. The **conformational hypothesis** suggests instead that an endergonic change in shape takes place to form a "high-energy" conformation of a protein molecule during electron transport. The exergonic return of the molecule to its original state supplies energy for phosphorylation.

There has been a long and intensive search for such "high-energy" intermediates, and so far, no one has found unequivocal evidence for such a compound or for conformation changes in chloroplasts and mitochondria. In all fairness, however, the instability of such intermediates would make it very difficult to detect or isolate them if they do exist.

The second major theory concerning the link between electron transport and ATP production was proposed by British biochemist Peter Mitchell, who was awarded the Nobel Prize in 1978. In Mitchell's **chemiosmotic** scheme, the intermediate is a *transmembrane electrochemical gradient of H^+ ions*. The gradient exists in conjuction with membrane-bound transport system. When such a gradient is established, the differences in proton concentration (the pH) between the two sides of the membrane, combined with the electrical forces across the membrane, tend to cause the positively charged H^+ ions to move in a particular direction. But how is such a transmembrane proton gradient established, and how is it associated with ATP production? The simplest proposed explanation for this phenomenon suggests that electron flow through an electron carrier system generates a proton gradient as a result of the way the different components are arranged in the membrane. Then, in a separate reaction, protons returning through the membrane are thought to cause the formation of ATP via a special enzyme system spanning the membrane (figure 6.19).

There is no doubt that electron flow in chloroplasts can be linked to proton transport. Illuminated suspensions of thylakoid membranes take up protons into the intrathylakoid space (shown, for example, by a pH increase in an unbuffered medium). It is also known that an artificially generated pH gradient is capable of causing ATP synthesis in the dark. This was shown by André Jagendorf in a famous experiment. In Jagendorf's experiment, chloroplasts were allowed to come to equilibrium with a buffer at pH 4, so that the proton concentration inside the chloroplasts was high. The external pH was then increased, and ADP and phosphate were added to the external solution. ATP was produced until pH equilibrium was reached again (figure 6.20).

Figure 6.21 summarizes the difference between the chemiosmotic theory and the other theories. Both types of schemes are based on the existence of electrochemical gradients, but they differ on the proposed role of the gradients in phosphorylation. The chemical and conformational hypotheses hold that the "high-energy" intermediate formed during electron transport can either provide energy reversibly for production of an electrochemical gradient or directly for phosphorylation. The electrochemical gradient thus acts as an energy reserve that can be used to produce more of the "high-energy" intermediate. In contrast, the chemiosmotic hypothesis holds that the electrochemical gradient *is* the "high-energy" intermediate and that it provides energy directly for phosphorylation.

Figure 6.19 (a) General model for the linkage of light-induced electron transport and ATP synthesis in chloroplasts by proton movements according to the chemiosmotic hypothesis. (b) How reduction and oxidation of a quinone (Q) at opposite sides of a membrane could cause the transport of hydrogen ions across the membrane according to the chemiosmotic scheme. An electron carrier on one side of the membrane reduces the quinone. This requires two electrons, and two protons that are taken up from the aqueous phase. The reduced quinone (QH_2) then diffuses across the membrane. On the other side it donates its electrons to a second electron carrier. Two protons are released into the aqueous phase on that side. Finally, the oxidized quinone (Q) diffuses back across the membrane to complete the cycle. In the chloroplast, the aqueous phase to the left would be the stroma matrix, the membrane would be the thylakoid membrane, and the aqueous phase to the right would be intrathylakoid space.

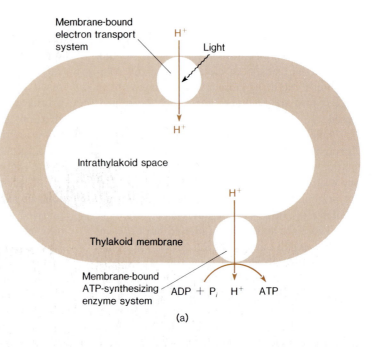

(a)

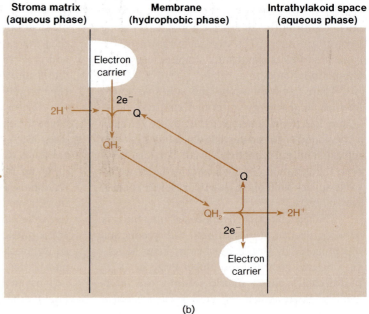

(b)

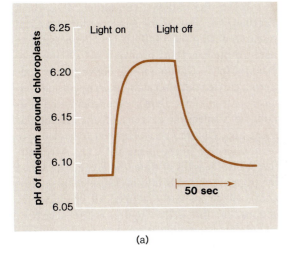

(a)

Figure 6.20 Relationships between electron transport, proton gradients, and ATP formation in chloroplasts. (a) Illuminated chloroplasts take up protons, and this can change the pH of the medium in which they are suspended. In the dark the protons leak out again, and the pH returns to the original value. (b) Jagendorf's "acid-bath" experiment, showing that an artificially generated pH gradient can cause ATP formation. The entire experiment was conducted in the dark.

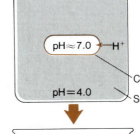

Spinach chloroplasts are incubated at pH 4 in the dark for a prolonged period.

As a result of inward diffusion of protons, the internal pH of the chloroplast approaches that of the medium.

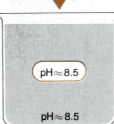

ADP and P$_i$ are added together with sufficient alkali to bring the external pH to 8.5. This creates a momentary pH gradient across the membrane, and ATP is formed from ADP and P$_i$ as the protons flow out, down their concentration gradient.

ATP production ends when pH equilibrium is reached.

(b)

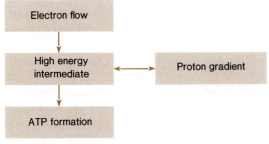

"Chemical" or "conformational" schemes

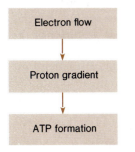

"Chemiosmotic" scheme

Figure 6.21 Relationship of intermediates between electron flow and ATP formation according to the chemical (or conformational) and the chemiosmotic models.

Types of Phosphorylation

Whatever the coupling mechanism, one thing is clear: electron transport and phosphorylation are coupled in chloroplasts. When photochemical events cause electron transfers in the photosystems of chloroplasts, ATP is produced. Biologists recognize two types or categories of photophosphorylation, known respectively as **noncyclic** and **cyclic photophosphorylation.**

Noncyclic photophosphorylation occurs as electrons flow from water to $NADP^+$ through Photosystems II and I. It is noncyclic in the sense that it is linked to the one-way flow of electrons from water to NADP (figure 6.22a).

The free energy needed for ATP production is available because, as noted earlier, the quanta of red light used in photosynthesis contain more energy than is needed to transfer electrons from water to form NADPH. The excess can be used to form the "high-energy" intermediate and thus to phosphorylate ADP.

Cyclic photophosphorylation involves a somewhat different process. Under some conditions, Photosystem I "short circuits." Although photochemical events occur, there is no net production of oxidized or reduced compounds. Electrons return to their original sources, the chlorophyll (P700) molecules, probably via ferredoxin and plastoquinone (figure 6.22b). This cyclic pathway of electron flow involves a large enough free-energy change to form ATP. Although the phosphorylation itself is not cyclic, this process is called cyclic photophosphorylation because electrons return to their original source. In itself, cyclic photophosphorylation *does not* result in water oxidation and the release of oxygen or NADPH production. It *does*, however, provide a mechanism for the generation of ATP that is separate from the linear electron-transport pathway through Photosystems I and II.

Carbon Dioxide Incorporation

In the dark phase of photosynthesis, the NADPH and ATP produced in the light reactions are used to reduce CO_2 during its incorporation into carbohydrate molecules. Carbon dioxide incorporation (often called **CO_2 fixation**) occurs in the stroma matrix of the chloroplast. It is a several-step process that is catalyzed by a series of soluble enzymes, with a specific enzyme catalyzing each reaction.

When the radioactive isotope of carbon, ^{14}carbon (^{14}C), was made available for biological research during the 1940s, it became possible to analyze the reactions involved in CO_2 incorporation. Melvin Calvin and his colleagues in California carried out a series of experiments on CO_2 incorporation. In Calvin's experiments, an illuminated suspension of cells of the unicellular green alga *Chlorella* was exposed to radioactive carbon dioxide in the form of $^{14}CO_2$. After different brief lengths of time, the algae were killed by dropping the cells into boiling alcohol (methanol), which instantly stopped all chemical reactions in the cells (figure 6.23). Carbon-containing compounds were then extracted from the cells and separated by paper chromatography.

Calvin and his coworkers were able to trace the reaction pathway through which CO_2 is incorporated by measuring the amount of radioactive carbon in various compounds after different periods of exposure to $^{14}CO_2$. If the cells were exposed to $^{14}CO_2$ for a very short time, only a few compounds contained radioactive carbon; after a slightly longer time, more compounds contained radioactive carbon; and so forth. Calvin reasoned that the compound containing ^{14}C after the shortest possible exposure would be the first stable compound actually formed during CO_2 incorporation and that further steps in the pathway could be determined by using progressively longer exposure periods.

The first radioactive compounds detected, after a very short exposure to $^{14}CO_2$, contained three carbon atoms; in fact, the whole pathway is often called the **C_3 pathway.** (It also is known as the **Calvin cycle.**) It seems reasonable to believe that CO_2 is bound to a two-carbon compound to make the three-carbon product, but this is not the case. The immediate precursor of the C_3 compound is a C_5 compound. Thus, the "carbon arithmetic" of the CO_2 fixation reaction must proceed as follows:

$$C_5 \quad + \quad C_1 \quad \longrightarrow \quad 2C_3$$

(Five-carbon molecule)   (CO$_2$)   (Two three-carbon molecules)

During the incorporation process, a six-carbon intermediate molecule is apparently formed, but it exists for such a short time before it splits into two three-carbon molecules that it has not been possible to isolate or identify it.

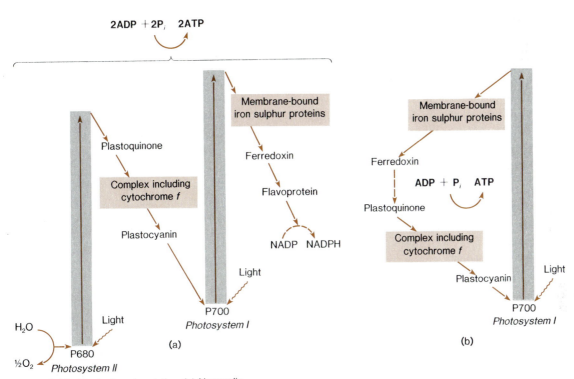

Figure 6.22 Photophosphorylation. (a) Noncyclic photophosphorylation. (b) Cyclic photophosphorylation. Possible pathway of cyclic electron flow around Photosystem I if ferredoxin donates its electrons to plastoquinone instead of to $NADP^+$. The components involved in electron transport between ferredoxin and plastoquinone remain to be established. Note that electrons return to the P700 chlorophyll, making this a cycle.

Figure 6.23 The apparatus used by Melvin Calvin for his classic experiments. The algae were contained in the flat flask, which was illuminated by the two lamps. $^{14}CO_2$ was added with a syringe, and the algae were killed after various periods by opening the stopcock and allowing them to fall into the beaker containing methanol.

For the CO_2 incorporation process to continue, an adequate supply of the original C_5 molecule must be maintained. Thus, only some of the molecules produced by the incorporation reactions become available to the cell for other uses. The majority of the molecules produced are used to regenerate the supply of C_5 molecules. This continuing return to an original point means that the process is cyclic (hence the name Calvin *cycle*).

The CO_2 fixation reaction involves the five-carbon precursor molecule **ribulose 1,5-bisphosphate (RuBP)**[1] and is catalyzed by the enzyme **ribulose bisphosphate carboxylase.** The product is two molecules of **3-phosphoglycerate (PGA),** a three-carbon compound.

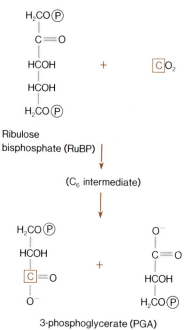

Ribulose
bisphosphate (RuBP)

(C_6 intermediate)

3-phosphoglycerate (PGA)

This is one of the most important of all biochemical reactions. The great majority of all carbon atoms in organic compounds in living organisms are derived from CO_2 through this reaction.

The 3-phosphoglycerate is phosphorylated by ATP to **1,3-diphosphoglycerate,** which is then reduced by NADPH to **glyceraldehyde-3-phosphate (GAP).** These two steps use some of the reductant and energy currency generated by the light stages of photosynthesis.

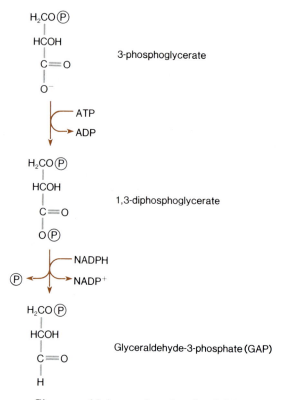

3-phosphoglycerate

1,3-diphosphoglycerate

Glyceraldehyde-3-phosphate (GAP)

Since combining each molecule of CO_2 with ribulose bisphosphate gives rise to two molecules of 3-phosphoglycerate, this sequence of reactions requires two ATP and two NADPH units per molecule of CO_2 incorporated into carbohydrate.

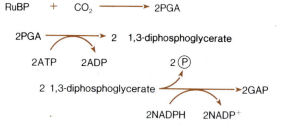

ATP, however, also is required for the regeneration of the RuBP supply. Thus, RuBP regeneration must be discussed before the "ATP arithmetic" of CO_2 incorporation can be summarized.

---

[1]Ribulose bisphosphate was formerly called ribulose diphosphate, but it has been renamed in accordance with a new standard biochemical terminology. The general distinction between di- and bis- is drawn from the positions of two similar groups on complex molecules. For example, adenosine diphosphate is adenosine-Ⓟ-Ⓟ (the two phosphates are attached to one another), while ribulose bisphosphate is Ⓟ-ribulose-Ⓟ (the two phosphates are separate). The older term, ribulose diphosphate, may still be encountered in some texts and in older journal articles.

RuBP Regeneration

To start the reaction pathway just discussed, there must logically be one RuBP molecule to react with each CO_2 molecule. The products of the CO_2 incorporation reactions, three-carbon molecules, enter a series of Calvin cycle rearrangements to regenerate RuBP molecules. In fact, a full five-sixths of the molecules produced in CO_2 incorporation must be used to keep the supply of RuBP constant. In simple terms, five three-carbon molecules are transformed to three five-carbon molecules. The five-carbon molecules that are produced by these rearrangements are **ribulose (mono) phosphate** molecules (figure 6.24). A final step is needed to produce ribulose bisphosphate:

Ribulose phosphate + ATP $\longrightarrow$ RuBP + ADP

Thus, the regeneration of each molecule of RuBP also requires one ATP.

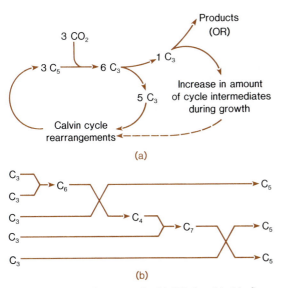

(a)

(b)

Figure 6.24 The Calvin cycle. (a) CO_2 is added to five-carbon precursor molecules (RuBP), producing three-carbon molecules. Five out of every six product molecules must be converted back to the precursor molecule. By a series of rearrangements, five C_3 molecules are used to produce three C_5 molecules. The other one-sixth of the product molecules can either be used for synthesis of other compounds or to increase the amount of the precursor, and hence the rate of CO_2 fixation during growth. (b) A skeleton outline of the Calvin cycle rearrangement reactions, showing how five C_3 molecules are used in reactions involving a number of intermediate compounds to produce three C_5 molecules. Each arrow intersection represents a specific, enzyme-catalyzed reaction. Each of the C_5 molecules produced is ribulose phosphate, and one ATP is needed to convert each of them to ribulose bisphosphate (RuBP).

ATP and NADPH Sums

Since the Calvin cycle requires two ATP and two NADPH for each molecule of CO_2 incorporated into carbohydrate, and since the production of ribulose bisphosphate requires one ATP, a total of three ATP and two NADPH are used for each CO_2 molecule incorporated into carbohydrate.

There is another useful way of looking at ATP and NADPH "arithmetic." Glucose is a product of photosynthesis but is not produced in the Calvin cycle itself. Glucose and other compounds can be produced, however, using glyceraldehyde-3-phosphate (GAP). But the withdrawal of GAP must be balanced with CO_2 incorporation if the pool of Calvin cycle compounds is to be kept constant so that the process continues to operate at a steady rate. How much does it cost in terms of ATP and NADPH to incorporate the three carbons of three CO_2 molecules, thus making one GAP molecule available for withdrawal?

If the incorporation of one CO_2 molecule requires three ATP and two NADPH, the incorporation of three CO_2 molecules will require three times as much energy. That is, nine ATP and six NADPH coming from the light reactions must be used in dark reactions to produce a net gain of one GAP. This GAP molecule can then be withdrawn from the Calvin cycle pool of compounds and used by the cell for other purposes.

Uses of GAP

As was previously mentioned, at least five-sixths of the GAP produced by CO_2 fixation reactions must be used in Calvin cycle rearrangements to produce ribulose bisphosphate. What happens to the other one-sixth?

While the plant is growing, the capacity for photosynthesis must continually be increased, and this is possible if more than five-sixths of the GAP is converted back to RuBP. In such a case, the *rate* of CO_2 fixation will also increase for each cycle because the size of the pool of Calvin cycle compounds will increase. Any surplus glyceraldehyde-3-phosphate, once requirements for increased photosynthetic capacity have been met, can be used to form other compounds. Within the chloroplast, it may be converted to **glucose phosphate**. Glucose phosphate is then converted to **starch**, a glucose polymer that acts as a store of carbohydrate inside the chloroplasts. Plants accumulate starch during the day, when light

Box 6.1
Photorespiration and Photosynthetic Efficiency

Photorespiration is a process that occurs in photosynthesizing plant cells. It is the opposite of photosynthesis in that it results in the production of CO_2 and the consumption of O_2 in light. The discovery of this apparent cause of inefficiency in the photosynthetic mechanism has recently aroused much interest. If ways of reducing or eliminating it could be found, the result might well be a dramatic improvement in agricultural productivity.

Photorespiration is intimately connected with photosynthetic mechanisms. It should not be confused with CO_2 production and O_2 consumption caused by the ordinary processes of cell respiration. The latter processes involve the mitochondria and also occur in the dark and in nongreen tissues, while photorespiration occurs only in photosynthesizing cells (box figure 6.1A).

The explanation of the consumption of O_2 in the light relates to the built-in inefficiency of the enzymatic functioning of ribulose bisphosphate carboxylase. CO_2 is a small molecule that does not contain charged groups or other distinctive features, and the active site of the enzyme cannot completely distinguish it from O_2. Under normal conditions (when the ratio of O_2 to CO_2 in the air is 21 percent : 0.03 percent, or 700:1) the ribulose bisphosphate carboxylase enzyme also acts as an oxygenase, sometimes incorporating oxygen in place of CO_2. When this happens, instead of two molecules of phosphoglycerate, the reaction produces one molecule of phosphoglycerate and one of phosphoglycollate.

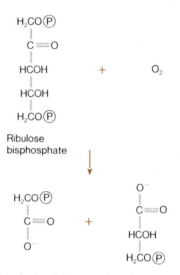

Phosphoglycollate   Phosphoglycerate

Phosphoglycollate loses its phosphate group to become glycollate. Some algae excrete glycollate into their environments as a waste product, but the cells of higher plants convert some of this material back to intermediates in the Calvin cycle. This conversion takes place in a complex series of reactions involving mitochondria, chloroplasts, and other compartments of the cell, especially the glyoxisomes, a specialized type of small organelle found in plant cells. At one stage during these reactions, two two-carbon molecules are converted into a three-carbon molecule, and one CO_2 molecule is released. This accounts for the CO_2 production that occurs during photorespiration.

The photorespiration processes lag a little behind those involved in photosynthesis. If a photosynthesizing leaf is suddenly put into the dark, a transient burst of CO_2 production takes place. Biologists measure this burst of CO_2 production by a previously photosynthesizing leaf to determine how much photorespiration occurs in various photosynthesizing leaves.

Photorespiration seems to be an inevitable result of the atmospheric O_2/CO_2 ratio. It is possible, though, to reduce photorespiration and increase the efficiency of photosynthesis by altering this ratio in favor of CO_2. This is what happens in photosynthesizing C_4 plants because they concentrate CO_2 in their bundle sheath cells, which are the sites of Calvin cycle reactions. In fact, photorespiration has not been detected in C_4 plants.

Knowledge of the effect of altered CO_2 ratios has even been used commercially to a limited extent (for example, by enriching greenhouse atmospheres with CO_2).

There are other factors limiting the efficiency of plant growth under natural conditions. For photosynthesis, the limiting factor changes with the conditions in the plant's environment. The photosynthetic process may be limited by low light intensity, especially in the morning and evening, on cloudy days, and in shade. But increasing the light intensity beyond a certain level has no effect on photosynthetic rate.

Under such light-saturated conditions, however, the rate of photosynthesis can be increased by increasing the concentration of CO_2. This suggests that CO_2 concentration becomes the limiting factor in photosynthesis when light is saturating the light-reaction mechanisms (box figure 6.1B).

Still, the success of a plant in a particular environment depends on a great number of factors apart from the efficiency of the photosynthetic process. Thus, even a dramatic increase in the efficiency of CO_2 fixation might have only a marginal effect on the competitive ability of the plant under natural conditions.

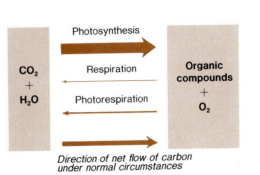

Direction of net flow of carbon under normal circumstances

Box Figure 6.1A The relation between photosynthesis, photorespiration, and "dark" respiration (mainly mitochondrial). The thickness of the arrows gives a measure of the magnitude of the various processes.

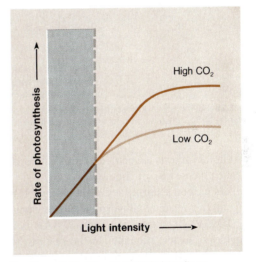

Box Figure 6.1B The effect of light intensity on photosynthesis at different CO_2 concentrations. At lower light intensities (gray area of figure), light intensity limits photosynthetic rate. Above saturating light intensity (gray broken line), CO_2 concentration becomes a limiting factor.

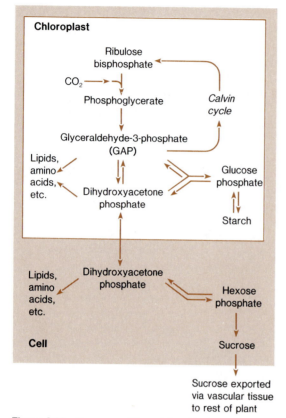

Figure 6.25 Summary of the relationship between photosynthetic CO_2 fixation and other cell metabolism.

their own proteins. Although chloroplasts can survive for some time outside the plant cell, they are not capable of growth and division under such conditions. Because certain of their proteins must be synthesized in the cytoplasm, chloroplasts are dependent on the rest of the cell.

An interesting example of the chloroplast's ability to survive outside the plant cell is provided by the marine mollusc *Elysia*. An *Elysia* looks green because it contains chloroplasts in the cells lining its gut. These chloroplasts are derived from the seaweeds on which the organism feeds, and they survive for several months within the molluscan cells, photosynthesizing and providing reduced carbon compounds for the host. *Elysia* is a nudibranch, a type of mollusc with no shell. Thus, light can readily penetrate its body to reach these chloroplasts.

The C_4 Pathway

The mechanism of CO_2 incorporation established through Calvin's classic experiments on the green alga *Chlorella* was subsequently found in many other plants. More recently, however, it was discovered that in some plants, particularly those growing in hot climates, the first detectable products of CO_2 incorporation are C_4 acids. Examples of such plants are corn (*Zea mays*) and sugar cane.

The "C_4 plants" share a number of distinctive characteristics. Most have a specialized leaf anatomy in which the vascular bundles are surrounded by distinctive **bundle sheath** cells that contain chloroplasts (figure 6.26). In cross section, this sheath looks somewhat like a wreath; the German word for wreath, *Kranz*, is applied to this type of structure. Thus, C_4 plants are said to display ***Kranz* anatomy.**

The C_4 plants have an extra set of reactions, in addition to the Calvin cycle, by which CO_2 is incorporated into organic compounds. Although this mechanism also involves a cyclic series of reactions, the extra cycle does not replace the Calvin cycle. Instead, it seems to *concentrate* the CO_2 available to the ribulose bisphosphate carboxylase of the Calvin cycle (figure 6.27).

The first fixation of CO_2 occurs in the **mesophyll** cells in the body of the leaf, cells that lack the normal Calvin cycle enzymes. CO_2 is converted to bicarbonate ions (HCO^-_3) through a reaction involving water:

$$CO_2 + H_2O \rightleftharpoons H_2CO_3 \rightleftharpoons H^+ + HCO^-_3$$

allows photosynthesis to occur, and break it down to release glucose during the night.

Fixed carbon seems to leave the chloroplast mainly in the form of **dihydroxyacetone phosphate** (DHAP) (figure 6.25). Thus, DHAP is a chloroplast's "export product" to the rest of the cell. DHAP is an isomer of GAP and can be formed from it. In the cytoplasm outside the chloroplast, DHAP can be used to make the six-carbon sugars glucose and fructose, which are then joined to form **sucrose**. Sucrose is the major form in which fixed carbon is transported from the leaf to other parts of the plant.

Other Chloroplast Metabolism

Any glyceraldehyde-3-phosphate surplus over that needed for the regeneration of ribulose bisphosphate can be converted to a variety of other types of compounds, including lipids and amino acids, within the chloroplast. In addition to being able to synthesize many small molecules, chloroplasts have their own DNA, RNA, and ribosomes and therefore can synthesize certain of

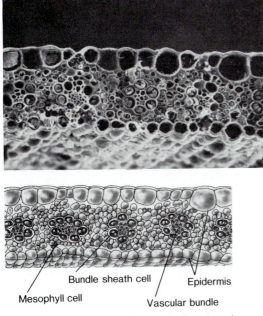

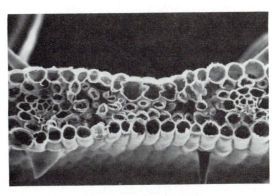

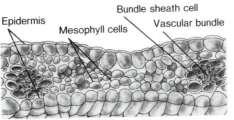

Figure 6.26 Comparison between leaf anatomy in plants using C_3 and C_4 pathways of CO_2 fixation. Scanning electron micrographs of cryofractured leaves from a C_4 plant bluestem, *Andropogon* (left) and a C_3 plant brome grass, *Bromus* (right). (Magnification $\times$ 277 and 267, respectively) Accompanying sketches highlight some major structural differences. Vascular bundles are separated by only a few (usually five or less) mesophyll cells in C_4 plants' leaves. C_4 bundle sheath cells contain large numbers of chloroplasts. Mesophyll cells form a concentric ring around the bundle sheath and contribute to the ''wreath'' appearance. When a bundle sheath is present in a C_3 plant, its cells contain no chloroplasts, and vascular bundles usually are separated by a number of mesophyll cells.

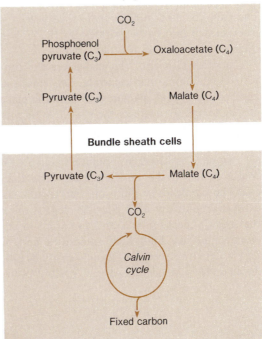

Figure 6.27 Summary diagram of C_4 metabolism in some C_4 plants. Other C_4 plants have a related scheme in which the amino acids aspartate and alanine take the place of malate and pyruvate.

The bicarbonate ions are then used to convert a C_3 acid, **phosphoenol pyruvate (PEP),** to the C_4 acid **oxaloacetate,** in a reaction catalyzed by an enzyme called **phosphoenol pyruvate carboxylase (PEP carboxylase).**

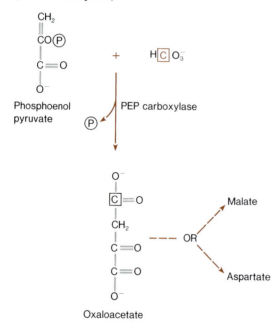

In many C_4 plants the oxaloacetate is reduced to **malate,** another C_4 acid. Malate can then be transferred to the bundle sheath cells and broken down into CO_2 and **pyruvate (C_3).** To complete the cycle, the pyruvate is then returned to the mesophyll cells and converted back to PEP in a reaction that requires two molecules of ATP. The CO_2 liberated in the bundle sheath cells is incorporated into carbohydrates by the Calvin cycle enzymes, just as it is in C_3 plants.

In other C_4 plants, the oxaloacetate is converted to the amino acid **aspartate.** Aspartate and another amino acid, **alanine,** then take the place of malate and pyruvate as the shuttle system in the transfer and release of CO_2 into the bundle sheath cells.

These C_4 mechanisms produce a higher CO_2 concentration in the bundle sheath cells than could be obtained by diffusion, but require an extra expenditure of ATP. Many biologists think that the C_4 mechanism is an adaptation to environments with high light intensities and/or high water stress (extra ATP could easily be produced under such conditions as a result of cyclic

photophosphorylation). Such environments place an extra premium on efficient CO_2 fixation because this will reduce the plant's water loss through the open stomata.

It is likely that the advantages of higher CO_2 fixation efficiency are finely balanced with the extra energy requirements (and perhaps other handicaps); otherwise, C_4 plants should be more successful than C_3 plants whenever they occupy the same geographical range. This is not the inevitable result, however, when C_4 and C_3 plants live together.

Crassulacean Acid Metabolism

Many succulent plants growing in extremely arid environments, including some in the family Crassulaceae, are characterized by **crassulacean acid metabolism (CAM).** CAM is similar to C_4 metabolism in that CO_2 is first fixed into C_4 acids, but the two differ in that the characteristic *Kranz* anatomy of the leaf is lacking in CAM plants, and separation of initial CO_2 incorporation from Calvin cycle activity is temporal rather than spatial in CAM plants (in other words, the two mechanisms operate at different times in CAM plants rather than in different locations in the plant, as they do in C_4 plants). This temporal separation is advantageous for plants living in very arid climates such as deserts. During the cool desert night, when the danger of water loss decreases, plants with CAM open their stomata and fix large amounts of CO_2 into oxaloacetate. The oxaloacetate is then converted into malate and aspartate, which are stored in high concentrations in the cell vacuoles of the mesophyll cells. Succulent plants have fleshy leaves because their leaves contain large quantities of these mesophyll cells.

During the day, the C_4 acids in CAM plants are broken down, providing a source of CO_2 for fixation by the Calvin cycle enzymes. To prevent water loss, the stomata are firmly closed during the day, and any daytime exchange of CO_2 and water is restricted to movement through the cuticle, which is very slow (figure 6.28).

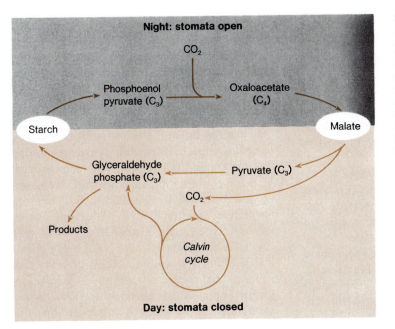

Figure 6.28 Summary of crassulacean acid metabolism. Note especially that the initial CO_2 incorporation occurs at night, while Calvin cycle activity occurs during the day. Malate accumulates during the night and is broken down in the daytime to make CO_2 available for Calvin cycle activity. Starch accumulates during the daytime and is broken down at night to produce phosphoenol pyruvate (PEP).

Plant Products

Photosynthesis is the source of virtually all reduced carbon compounds in the world because reduced carbon compounds produced in photosynthesis make up the bodies of photosynthesizing organisms and are consumed by nonphotosynthesizing organisms. Thus, it is not surprising that from a biochemical point of view, plants and animals are quite similar in many respects; they both contain a number of the same types of molecules.

There are, however, a number of interesting compounds that are found only in plants. One of these is **cellulose,** a polymer of glucose and the most abundant of all organic compounds. Cellulose and related compounds form the major part of the rigid cell walls that surround plant cells. Many biologists believe that it should eventually be possible to develop economically feasible techniques for cellulose hydrolysis to make these enormous quantities of glucose available. This glucose supply could then be used as a raw material for alcohol fuel production.

In addition to the compounds common to the majority of plants, various types of plants also synthesize a wide range of **secondary plant products,** which are more or less specific to particular groups of plants. In some cases these secondary products have obvious functions—for example, pigments in flowers attract pollinating insects, while other secondary products may help repel predation by herbivores, especially insects. But in many cases, the functions of secondary plant products are uncertain.

Secondary plant products often are of use to human societies. Drugs such as **quinine** are used in the treatment of illnesses, and products such as natural rubber serve a variety of uses. The desert shrub *Euphorbia* contains a sap that can be distilled to produce hydrocarbons similar to those in petroleum. Melvin Calvin and others propose that *Euphorbia* might be cultivated commercially in "petroleum plantations," using land that is too arid for other commercial crops, to reduce dependence on present petroleum supplies.

To meet human society's future requirements for energy and raw materials, it may well be that more efficient harvesting of the sun's energy will become possible not through construction of bigger and better solar energy collection devices, but by learning more about, and taking advantage of, the remarkable energy conversions of photosynthesis.

Summary

A fundamental characteristic of living organisms is their need for a continual input of energy, and the energy that all living organisms use comes directly or indirectly from sunlight. Reduced organic compounds produced in photosynthesis provide chemical energy for all organisms. These reduced organic compounds are oxidized in respiration, resulting in the formation of ATP.

Leaves are photosynthesizing organs. Most leaves are flat and thin with a large surface area exposed to light. Photosynthesizing mesophyll cells in the leaf interior are surrounded by air spaces that connect with the outside atmosphere through openings called stomata. Guard cells surrounding the stomata undergo turgor pressure changes that cause the guard cells to change in shape and to regulate the opening and closing of the stomata. Stomatal control is important in regulating the CO_2 supply for photosynthesis and in preventing excessive water loss from leaves.

Chloroplasts are distinctive membrane-bound organelles containing complex internal membrane structures, the thylakoids and stroma lamellae. The stroma matrix is the fluid area surrounding these membrane structures. Pigment molecules involved in light reactions are associated with the thylakoids. Enzymes involved in CO_2 fixation are in the stroma matrix.

Parts of the visible light spectrum have the appropriate quantum energy levels to cause specific photochemical reactions in pigment molecules. Electrons are moved to higher-energy orbitals in these molecules to produce molecules in an "excited" state.

Pigments in chloroplasts are arranged in photosynthetic units, each containing hundreds of pigment molecules and one reaction center molecule. The units function as light-harvesting systems. Because the reaction center molecule has the lowest energy level of all molecules in the unit, a photochemical event anywhere in the unit causes energy transfers leading to an excited state in the reaction center molecule. And since the molecules function as a unit, there is no need for multiple sets of electron carriers to be used in subsequent reaction steps.

Excited chlorophyll molecules are strong reducing agents that readily donate electrons. Photosystem I reduces $NADP^+$. Photosystem II is directly involved in oxygen evolution. Photosystem II gains electrons from water and transfers them to Photosystem I to replace those given up to $NADP^+$ by the chlorophyll in Photosystem I. These electron transfers are several-step processes involving a series of oxidation-reduction reactions.

Photophosphorylation, or ATP production, occurs as a result of free energy changes during electron transport. The details of the phosphorylation mechanism are unknown, but several hypotheses attempt to explain the link between electron transport and ATP production. Two of these hypotheses suggest that electron transport results in the formation of a high-energy chemical intermediate, while the third suggests that a transmembrane electrochemical gradient is formed as protons are moved across the thylakoid membrane during electron transport. The hypotheses go on to say, respectively, that energy from either a chemical intermediate or the electrochemical gradient is then used to phosphorylate ADP.

Noncyclic photophosphorylation occurs as electrons flow from water to $NADP^+$ through Photosystems II and I. In cyclic photophosphorylation, photochemical events in Photosystem I lead to electron transport and ATP production, but the electrons return to the chlorophyll molecule that begins the cycle. Thus, neither water oxidation nor NADPH production occur in cyclic photophosphorylation.

In the Calvin (C_3) cycle, CO_2 is combined with a C_5 precursor, ribulose bisphosphate. The first stable products of this combination are two molecules of phosphoglycerate, a three-carbon compound. ATP and NADPH are used in subsequent phosphorylation and reduction steps to produce glyceraldehyde-3-phosphate (GAP). Five-sixths of the GAP produced is used in regeneration reactions to resupply the C_5 precursor. A final step in this regeneration process also requires ATP. The other one-sixth of the product molecules can be withdrawn from the Calvin cycle. Some is used for glucose and starch synthesis in the chloroplast. Some may be exported from the chloroplast to the rest of the cell. The cell, in turn, may export carbohydrates to other cells in the form of sucrose. The products of photosynthesis are also used to produce a wide range of other compounds in chloroplasts and cells.

In plants using the C_4 photosynthetic pathway, mesophyll cells incorporate CO_2 into C_4 compounds. These compounds then enter the bundle sheath cells, where they release CO_2 that is incorporated in carbohydrates in the Calvin cycle. The CO_2-concentrating mechanism in C_4 plants permits them to photosynthesize at lower CO_2 concentrations; thus, stomatal opening and the resulting water loss are decreased. Because of this mechanism, C_4 plants can live more successfully than C_3 plants in hot, dry environments.

In a plant with CAM metabolism, initial CO_2 incorporation occurs while the stomata are open at night. The compounds formed are then stored until daytime, when they are broken down to supply CO_2 for fixation by Calvin cycle enzymes. This mechanism permits CAM plants to keep their stomata closed in the daytime when water loss would be the greatest.

In addition to the reduced carbon compounds found in all living organisms, plants produce many useful secondary plant products. Common plant substances such as cellulose may gain in economic importance as methods are developed to use plant products in new ways for fuel production. Other secondary plant products have the potential to replace increasingly scarce materials now obtained from other sources.

Questions

1. Explain in terms of quantum energy levels why visible light is visible. How can visible light be involved in the photochemical reactions of photosynthesis, while radiation of longer and shorter wavelengths cannot?

2. Can Photosystem I continue to reduce $NADP^+$ if Photosystem II is not operating?

3. Explain the basic difference between the chemical or conformational hypotheses and the chemiosmotic hypothesis of phosphorylation.

4. Why does cyclic photophosphorylation not result in water oxidation or the reduction of $NADP^+$?

5. Explain how three ATP and two NADPH molecules are used for the incorporation of one CO_2 molecule into carbohydrate through the Calvin cycle.

6. How do C_4 and CAM metabolism differ from one another in terms of the separation of CO_2 concentration and CO_2 fixation?

Suggested Readings

Books

Bonner, J., and Varner, J. E., eds. 1976. *Plant biochemistry*. 3d ed. New York: Academic Press.

Hall, D. O., and Rao, K. K. 1977. *Photosynthesis*. 2d ed. Studies in Biology no. 37. Baltimore: University Park Press.

Lehninger, A. L. 1975. *Biochemistry*. 2d ed. New York: Worth Publishers.

Articles

Björkman, O., and Berry, J. October 1973. High-efficiency photosynthesis. *Scientific American* (offprint 1281).

Black, C. C., Jr. 1973. Photosynthetic carbon fixation in relation to net CO_2 uptake. *Annual Review of Plant Physiology* 24:253.

Calvin, M. 1979. Petroleum plantations for fuel and materials. *BioScience* 29:533.

Goldsworthy, A. 1976. Photorespiration. *Carolina Biology Readers* no. 80. Burlington, N.C.: Carolina Biological Supply Co.

Govindjee, and Govindjee, R. December 1974. Primary events of photosynthesis. *Scientific American* (offprint 1310).

Heath, O. V. S. 1975. Stomata. *Carolina Biology Readers* no. 37. Burlington, N.C.: Carolina Biological Supply Co.

Miller, K. R. October 1979. The photosynthetic membrane. *Scientific American*.

Whittingham, C. P. 1977. Photosynthesis. 2d ed. *Carolina Biology Readers* no. 9. Burlington, N.C.: Carolina Biological Supply Co.

Cellular Energy Conversions and Organismic Metabolism

Chapter Concepts

1. Reduced organic compounds, originally produced by photosynthetic activity, are broken down in a stepwise series of oxidation reactions to provide energy for ATP production.

2. The Embden-Meyerhof pathway uses a small portion of the potential energy of glucose molecules to produce ATP and yields two molecules of pyruvic acid for each molecule of glucose processed.

3. In fermentation, which proceeds without oxygen, reduced electron carriers donate electrons to organic molecules.

4. Aerobic respiration involves the Krebs cycle reactions and the mitochondrial electron transport system. In the Krebs cycle, decarboxylations shorten carbon chains by producing CO_2, and oxidations supply electrons to the mitochondrial electron transport system.

5. Oxidative phosphorylations in the electron transport system greatly increase the ATP energy yield over that obtained in the Embden-Meyerhof pathway.

6. Fats and proteins can be oxidized in the same metabolic pathways as carbohydrates, but they must undergo preparatory reactions before entering the metabolic pool.

7. Compounds can also be withdrawn from the metabolic pool for synthesis.

8. Organismic metabolism is the sum total of all cellular processes in the body.

9. Metabolic rates are strongly affected by temperature. Different organisms have different temperature-metabolism relationships.

In photosynthesis, light energy is used to synthesize reduced carbon compounds from simple, inorganic precursors: CO_2 and water. The breakdown and oxidation of these reduced carbon compounds yields energy for the many energy-requiring processes that take place in the cells of living organisms (figure 7.1).

The gradual breakdown of reduced carbon compounds is a key characteristic of these energy-yielding processes. They release the chemical energy of carbohydrates and other organic molecules in "packets" of manageable size through a series of oxidation-reduction reactions. These controlled energy conversions result in the formation of the cell's "energy currency," adenosine triphosphate (ATP), which then serves as a direct energy source for several kinds of biological processes.

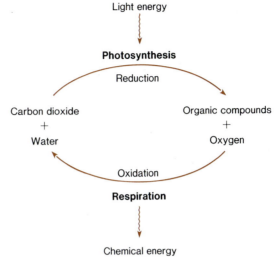

Figure 7.1 The relationships between photosynthesis and respiration.

An Overview of Glucose Oxidation

It is best to begin a consideration of cellular energy conversions with the utilization of glucose. The first series of reactions in the oxidative breakdown of glucose is called the **Embden-Meyerhof (E-M) pathway** after two German biochemists who studied these processes in the 1920s and 1930s. The term **glycolysis** ("glucose breakdown") is sometimes used as a synonym for the Embden-Meyerhof pathway, but biochemists usually apply the term to the E-M pathway plus a form of fermentation, which together lead from glucose to lactate. The E-M pathway is a several-step process that occurs in the cell's cytoplasmic matrix. It can proceed either in the presence or absence of oxygen and, by itself, makes available only a small portion of the potential energy that could be extracted from glucose. The reactions of the E-M pathway are thought to be older in an evolutionary sense than the oxygen-requiring processes of respiration, because the E-M reactions could have occurred in the primitive environment of earth before free oxygen became available in the atmosphere.

The formation of two three-carbon molecules of **pyruvic acid** is the final step of the E-M pathway (figure 7.2), so the process does not decrease the number of carbon atoms in organic molecules (glucose is a six-carbon molecule; $C_6 \rightarrow C_3 + C_3$). Pyruvic acid formation is an important step in utilizing carbohydrates. Different cells further metabolize pyruvic acid in different ways.

Under **anaerobic** conditions (conditions lacking oxygen), pyruvic acid is further processed through one of several types of **fermentation.** Two kinds of fermentation are identified in figure 7.2. In **alcoholic fermentation,** which occurs in many microorganisms (for example, brewer's yeast), pyruvic acid is converted to ethanol (ethyl alcohol). **Lactic fermentation,** which occurs in some microorganisms and some animal cells (for instance, in muscle cells during heavy exercise, when the oxygen supply is insufficient), produces lactic acid.

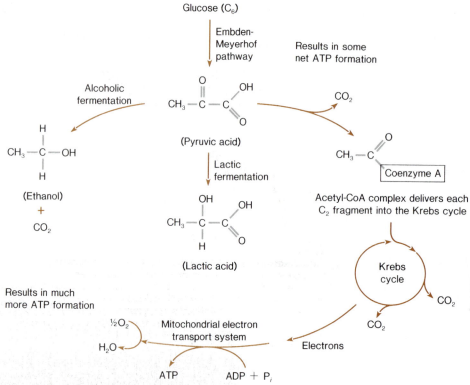

Figure 7.2 Preview of carbohydrate breakdown
pathways. Alcoholic and lactic fermentation occur in
certain cells under anaerobic conditions. The Krebs
cycle is part of aerobic respiration because electrons
from Krebs cycle oxidation reactions must go through
the electron transport system to oxygen.

Under **aerobic** conditions (in the presence of
oxygen), oxygen-requiring processes collectively
known as **respiration** use much more of the potentially available chemical energy to form ATP.
Pyruvic acid enters another series of reactions
that occur inside the mitochondria. This reaction
series is the **Krebs cycle,** named for Sir Hans
Krebs, who clarified the cyclical nature of the
reactions involved. In the Krebs cycle, organic
molecules are further degraded by the removal
of one-carbon fragments as CO_2. This process is
called **decarboxylation.**

Oxidations of organic molecules in the Krebs
cycle provide electrons that are transferred to
electron carriers in the mitochondrial membranes. There, a series of oxidation-reduction reactions occurs as electrons are passed from
carrier to carrier in an **electron transport system.**
During these transfers, free-energy changes make
possible the production of considerable amounts
of ATP.

These electron carriers finally donate electrons to oxygen, which combines with hydrogen
ions (protons) to form water. The reduction of
oxygen, which serves as the *final electron acceptor,* and the subsequent formation of **metabolic
water** is a fundamental characteristic of aerobic
respiration. The formation of metabolic water is
summarized in the following equation:

$$O_2 + 4e^- + 4H^+ \longrightarrow 2H_2O$$

Thus, the final products of the oxidative
pathways are CO_2 and water, the CO_2 produced
in decarboxylation reactions and the water produced by the reduction of oxygen. Energy made
available during the oxidation of organic compounds and subsequent electron transport events
is used to produce ATP, the cellular "energy currency."

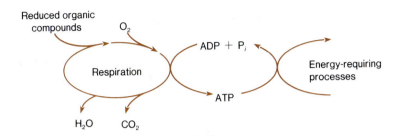

Figure 7.3 ATP as energy currency. Oxidation of reduced organic compounds provides energy to combine ADP and phosphate to replenish the supply of ATP. Hydrolysis of ATP to ADP and phosphate provides energy to drive energy-requiring processes in cells.

The Central Role of ATP

It is difficult to overemphasize the central role of ATP in cellular energy economy. The energy from the **exergonic** (energy-yielding) reactions of cellular respiration generally cannot directly provide the energy needed for various **endergonic** (energy-requiring) processes in living cells. Rather, ATP is the directly usable energy currency that can drive a variety of endergonic reactions.

It is commonplace in the field of biology to refer to the "high-energy bonds" of ATP and to designate ATP as a "high-energy compound." The application of the term "high-energy" to ATP really means, however, that the hydrolysis of ATP molecules, when coupled with other reactions, makes available a considerable amount of energy to do work such as synthesis, transport, and movement (figure 7.3).

The Embden-Meyerhof Pathway

The progressive breakdown of glucose in the Embden-Meyerhof pathway begins with preparatory reactions that "activate" the molecule and make it possible for the succeeding reactions to take place. Since these preparatory reactions are phosphorylations (that is, phosphate groups are attached to carbohydrate molecules), they use ATP. The ATP molecules donate their terminal phosphate groups in the reactions. Thus, the Embden-Meyerhof pathway reactions actually begin with an expenditure of ATP. But these priming reactions provide activation energy at the beginning to make all the subsequent reactions possible.

In figure 7.4, glucose molecules are shown in their normal ring form. In analyzing E-M pathway reactions, however, it is helpful to think of carbohydrate molecules as linear chains of carbon atoms, omitting the hydrogen and oxygen atoms of the molecules. For example, the hexokinase-catalyzed reaction of ATP and glucose, which results in the formation of **glucose 6-phosphate** (glucose with a phosphate group attached to the number 6 carbon) can be symbolized simply as:

$$\text{C-C-C-C-C-C} + \text{ATP} \xrightarrow{\text{Hexokinase}}$$
(Glucose)

$$\text{C-C-C-C-C-C-}\textcircled{P} + \text{ADP}$$
(Glucose 6-phosphate)

Two additional preparatory reactions follow the phosphorylation of glucose. A molecular reorganization yields **fructose 6-phosphate,** and then the transfer of another phosphate group from ATP produces **fructose 1,6-bisphosphate.** Returning again to the simplified notion of carbon chains, the latter reaction can be symbolized this way:

$$\text{C-C-C-C-C-C-}\textcircled{P} + \text{ATP} \xrightarrow{\text{Phosphofructokinase}}$$
(Fructose 6-phosphate)

$$\textcircled{P}\text{-C-C-C-C-C-C-}\textcircled{P} + \text{ADP}$$
(Fructose 1,6-bisphosphate)

To this point, then, two ATP molecules have been expended in preparatory reactions.

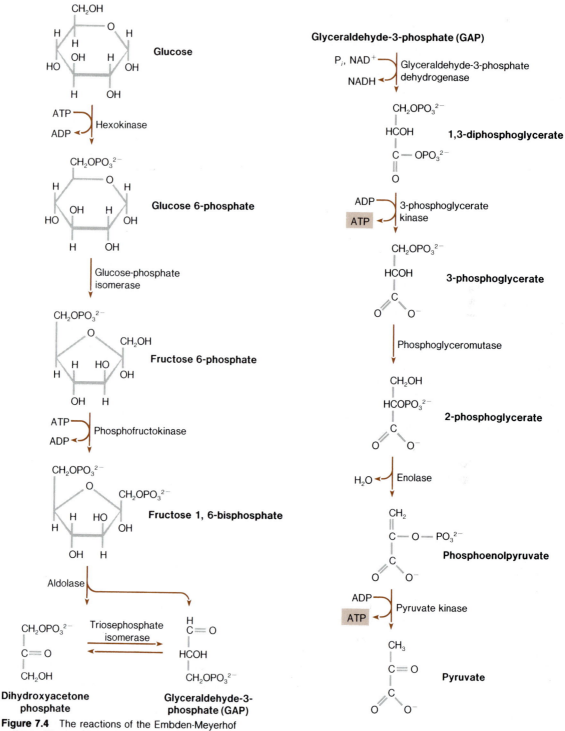

Figure 7.4   The reactions of the Embden-Meyerhof
pathway. The positions of the carbons in the glucose
molecule are represented symbolically as the
"corners" of the ring. Note that acids such as pyruvic
acid actually are dissociated at cellular pH's. Thus, the
pyruvate ion, which is produced by pyruvic acid
dissociation, is pictured rather than the pyruvic acid
molecule. (The "-ate" ending is used generally to
indicate ionic forms of dissociated acids.)

Subsequent reactions in the Embden-Mey-erhof pathway are energy yielding. Fructose 1,6-bisphosphate is split into two three-carbon molecules, **dihydroxyacetone phosphate** and **glyceraldehyde-3-phosphate (GAP)**. Again, this reaction can be symbolized using carbon chains:

$$\text{(P)-C-C-C-C-C-C-(P)} \xrightarrow{\text{Aldolase}}$$
(Fructose 1,6-bisphosphate)

$$\text{(P)-C-C-C} \quad + \quad \text{C-C-C-(P)}$$
(Dihydroxyacetone phosphate and GAP)

These three-carbon molecules are readily interconvertible in a reversible reaction catalyzed by the enzyme **triose phosphate isomerase.** Since one product, GAP, reacts in the next step of the pathway and is thus used up, the dihydroxyacetone phosphate is converted to GAP. Thus, two GAP molecules must be accounted for in determining energy relationships of subsequent reactions.

Further processing of C_3 molecules begins with a dehydrogenation reaction. Dehydrogenation is an oxidation in which the transfer of electrons involves the simultaneous removal of hydrogen ions. In this particular reaction, two electrons, with an accompanying hydrogen ion, are transferred to the electron carrier nicotinamide adenine dinucleotide (NAD). At the same time, a phosphate group is attached to the molecule, so the products of this reaction are a C_3 molecule with two phosphate groups, **1,3-diphosphoglycerate,** and a reduced molecule of NAD (NADH). The value of the reduced NAD molecule in energy retrieval depends on whether the reactions are taking place under aerobic or anaerobic conditions.

In the next reaction in the sequence, the free-energy change is large enough to allow a removed phosphate group to be transferred to ADP by phosphorylation, thus yielding a molecule of ATP. This is the first of two such transfers that are essential to energy retrieval in glycolysis. Because this phosphorylation is associated directly with a reaction involving a substrate in the pathway, it is called a **substrate-level phosphorylation.**

Two molecular reorganizations precede the second phosphorylation, which is coupled to the reaction catalyzed by pyruvate kinase. The reaction yields **pyruvate** (the ionized form of pyruvic acid that occurs in solution), which is a key compound in cellular metabolism because it stands at the junction of several biochemical pathways.

Fermentation

Under anaerobic conditions, pyruvate is converted into one of several products. Whether the end product of anaerobic metabolism is lactate or ethanol depends, of course, on the enzymatic capabilities of the cells in question. Figure 7.5 pictures the reaction steps involved in the anaerobic conversion of pyruvate into ethanol as it occurs in yeast and some other cells. The figure also shows lactate formation, which occurs, for example, in muscle cells that must rely on anaerobic metabolism processes during vigorous exercise.

In both of the processes shown in figure 7.5, there is a reaction in which reduced NAD (NADH) donates electrons. This is critical in keeping the Embden-Meyerhof pathway reactions going, because oxidized NAD (NAD^+) must be available or the whole process is stopped at the step converting glyceraldehyde-3-phosphate to 1,3-diphosphoglycerate. The quantities of NAD^+ available in cells are very small, so the constant recycling is necessary (see figure 7.6). However, this recycling does eliminate the possibility of using the energy potentially available in a reduced NAD molecule.

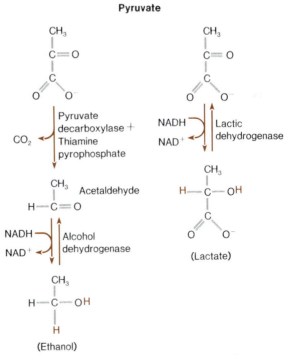

Figure 7.5 Reactions in ethanol or lactate formation under anaerobic conditions. The hydrogens shown in color are the ones attached when electrons are donated by reduced NAD. The pathway from glucose to lactate, which involves both the Embden-Meyerhof pathway and lactic fermentation reactions, is often called **glycolysis** (literally "glucose breakdown").

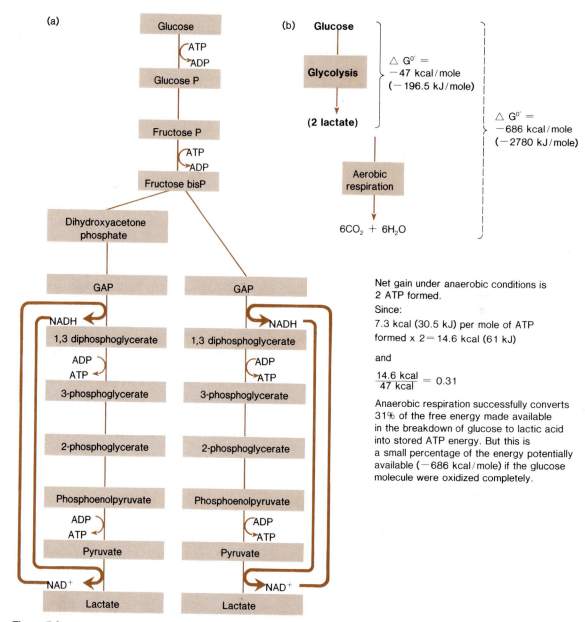

Figure 7.6 (a) Summary of anaerobic metabolism of glucose using lactate as end product. Parallel reaction series are shown to account for the two C_3 chunks produced when fructose diphosphate is split in two. (b) Energy sums on anaerobic processes and a comparison with the possibilities in aerobic respiration.

Net gain under anaerobic conditions is 2 ATP formed.
Since:
7.3 kcal (30.5 kJ) per mole of ATP formed x 2 = 14.6 kcal (61 kJ)

and

$$\frac{14.6 \text{ kcal}}{47 \text{ kcal}} = 0.31$$

Anaerobic respiration successfully converts 31% of the free energy made available in the breakdown of glucose to lactic acid into stored ATP energy. But this is a small percentage of the energy potentially available (-686 kcal/mole) if the glucose molecule were oxidized completely.

Two conclusions about energy conversion in anaerobic metabolism emerge from examining figure 7.6. First, only a portion of the free energy that could be made available if glucose were completely broken down to CO_2 and water actually is released in the conversions from glucose to lactate or ethanol (both of which are still reduced carbon compounds). Second, only a small portion of the energy released in the conversions from glucose to lactate or ethanol is captured in the form of ATP energy. For every molecule of glucose that undergoes these conversions, four molecules of ATP are formed, but two molecules of ATP must be expended at the beginning of the process. That means that the net gain of energy available for other uses is only two molecules of ATP per molecule of glucose. Clearly, there is more potential energy in glucose molecules that might be retrieved by more efficient energy-conversion mechanisms. Aerobic respiration is just such a mechanism. In aerobic respiration, pyruvate is oxidized further, and the additional NADH produced donates electrons to an electron transport chain, which in turn transfers them to oxygen, and additional ATP is formed in the process.

gain or lose an electron. One example of such a carrier is the **cytochromes,** which contain an iron atom in a heme group. The word *cytochrome* literally means "cell color"; the name was applied because of the pronounced pink color of the reduced forms of cytochromes. The electron-transporting activity of the cytochromes depends on the alternate reduction and oxidation of the iron atoms:

$$Fe^{2+} \underset{\text{Reduction}}{\overset{\text{Oxidation}}{\rightleftharpoons}} Fe^{3+} + e^-$$

(Ferrous ion)　　　　(Ferric ion) (Electron)

Another particularly important element in the electron transport system is **ubiquinone** (formerly called coenzyme Q). Ubiquinone is not a protein but a small molecule with a quinone nucleus and a hydrophobic side chain that makes it soluble in the lipids of the membrane but insoluble in water (figure 7.7). Ubiquinone is closely related to plastoquinone, which is found in chloroplasts, and its oxidation and reduction involve both protons and electrons. The substance illustrates an important fact about the electron transport system: electrons are actually passed through the system not singly but in pairs.

Oxygen, Electron Transport, and ATP Production

In aerobic respiration (which occurs inside mitochondria), pyruvate, the product of the Embden-Meyerhof pathway, is not converted to lactate or ethanol, as it would be under anaerobic conditions. Instead, it is degraded further under aerobic conditions with a great increase in the ATP yield. The primary factor making this possible is a mechanism that uses the energy potentially available in reduced coenzymes such as NADH to phosphorylate ADP molecules, thus producing ATP.

The reduced coenzymes do not react directly with oxygen. Instead, electrons are passed down a series of electron carriers known as the **electron transport system** or **respiratory chain.** In chapter 6, a similar arrangement of electron carrier molecules in the chloroplast was discussed. The types of electron carriers found in mitochondria and chloroplasts are similar; many are membrane-bound proteins containing metal ions that can

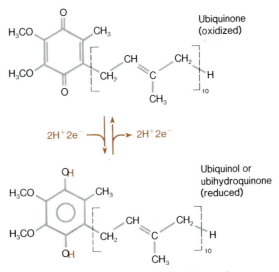

Figure 7.7 The oxidized and reduced forms of ubiquinone. This is the formula for the most common form in mammalian mitochondria. Other forms with different side-chain lengths are found in various other organisms.

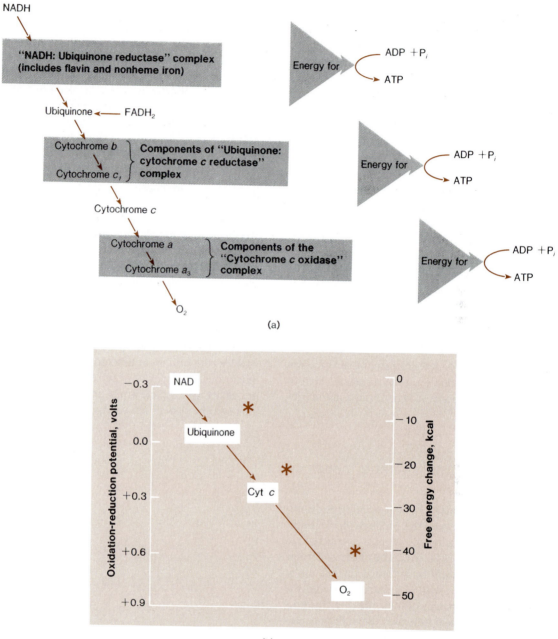

(a)

(b)

Figure 7.8 Electron transport system (respiratory chain). (a) The mitochondrial electron transport system through which pairs of electrons are passed to oxygen. Note the presence of the three complexes joined by ubiquinone and cytochrome c. The complexes contain other components apart from those shown. Electrons from reduced FAD enter the chain at the level of ubiquinone. (b) Energetic relationships of the electron transport system. The free energy changes in the electron transport system are adequate to permit ATP formation (phosphorylation), which is indicated by asterisks.

The actual physical arrangement of the various elements in the mitochondrial electron transport chain is not known, but these elements are associated with the internal membrane systems of mitochondria. In the transport system between NADH and oxygen, there seem to be three complexes, each containing several components, and these complexes are linked by ubiquinone and cytochrome c (figure 7.8a). The chain is also branched; that is, electrons from several sources enter through different components. For example, enzymes containing the important coenzyme **flavin adenine dinucleotide (FAD)** donate electrons to ubiquinone.

At the end of this electron transport system, oxygen serves as the final electron acceptor. Electrons are transferred to oxygen from the cytochrome oxidase complex, thus reducing the oxygen atoms. Hydrogen ions join the oxygen atoms along with the electrons, forming what is called *metabolic water*.

Though a good deal of the energy potentially available in the reduced NAD molecule is dissipated as heat, at least part of it is captured in the form of ATP energy that is available to do cell work. For every pair of electrons passed through the electron transport system from NADH to oxygen, enough free energy becomes available to cause the production of three molecules of ATP from ADP and phosphate (figure 7.8b). This form of phosphorylation, associated with the oxidation-reduction reactions in the electron transport system, is called **oxidative phosphorylation**. This energy conversion is the basic function of the electron transport system in aerobic respiration.

The **Krebs cycle** completes the oxidation of carbohydrates and supplies electrons to be passed through the electron transport system. Enzymes catalyzing Krebs cycle reactions are located inside the mitochondria, in the nonmembranous matrix portion.

The Krebs Cycle

The cyclical nature of the reactions by which further oxidations beyond pyruvate occur was established after a series of experimental analyses by Albert Szent-Györgyi, Hans Krebs, and others.

Szent-Györgyi found that several four-carbon acids (succinic, fumaric, malic, and oxaloacetic acids) were actively metabolized by suspensions of minced muscle that he studied. He also discovered an interesting additional effect of these acids on muscle metabolism: they caused a general increase in carbohydrate oxidation when they were added to the muscle suspensions. Adding C_4 acids somehow caused the muscle suspensions to more rapidly oxidize glucose and pyruvate.

Hans Krebs pursued this research on carbohydrate metabolism by testing other organic acids, and he found that several of them had the same effect. Citric acid as well as several other C_6 acids and α-ketoglutaric acid (C_5) also stimulated carbohydrate oxidation. Krebs correctly deduced that these effects were obtained because the reactions occurred as a cyclical process. Therefore, adding the acids that are intermediates in the cycle, as Szent-Györgyi and Krebs did in their experiments, increased the total available pool of reacting molecules and accelerated the use of pyruvate. More detailed analysis of Krebs cycle reactions has revealed the roles of these organic acids in the oxidation of pyruvate.

Entering the Krebs Cycle

Entry into the Krebs cycle requires preparation of a two-carbon fragment from the pyruvate that enters the mitochondrion. This preparation actually requires several reaction steps, which are summarized as one reaction in figure 7.9. Several

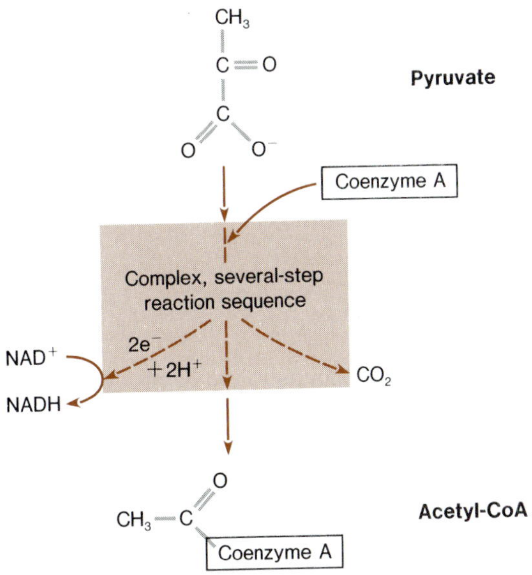

Figure 7.9 Summary of the reaction sequence converting pyruvate into acetyl CoA, which enters the Krebs cycle. Several enzymes are involved in these reactions.

facts about this preparatory reaction complex are noteworthy. First, a decarboxylation takes place; that is, the carbon chain is shortened by one carbon with the production of a molecule of CO_2. Second, a molecule of NAD is reduced as the remaining two-carbon fragment (**acetyl group**) is attached to its carrier, **Coenzyme A.** This complex, made up of the two-carbon fragment and Coenzyme A and called **acetyl CoA** for short, is now in an "active" form, ready to act as a substrate in a Krebs cycle reaction.

In the Krebs Cycle: Carbon Arithmetic

Although biologists refer to the "turns" of the Krebs cycle as if it were a wheel, obviously this analogy should be used only in the loosest sense. What is meant by a turn of the Krebs cycle is illustrated in skeletal form in figure 7.10. An acetyl group (C_2) from acetyl CoA is fed into the Krebs cycle and combined with a C_4 molecule to produce a C_6 molecule. Subsequent reactions of the Krebs cycle involve two decarboxylations that produce two molecules of CO_2 and reduce the length of the carbon chains by two carbons. This brings the cycle back to the starting point, where a four-carbon molecule is available to combine with another acetyl group and to start the whole process over again.

For every molecule of glucose, two molecules of pyruvate are produced; and for every molecule of pyruvate being processed, one molecule of CO_2 is produced during the formation of acetyl CoA and two molecules of CO_2 are produced during the ensuing turn of the Krebs cycle. This accounts for all three carbons present in each molecule of pyruvate and for all six carbons in the original glucose molecule ($3CO_2 \times 2 = 6CO_2$). Thus, the flow of carbon has been traced throughout the respiratory pathway.

In the Krebs Cycle: Energy Conversions

At four separate points in the Krebs cycle, pairs of electrons are transferred to coenzymes and, eventually, to the electron transport system (respiratory chain). In three cases, the coenzyme serving as initial electron acceptor is NAD^+, and in the other case (the step from succinate to fumarate, catalyzed by the enzyme succinic dehydrogenase), the coenzyme FAD is reduced as the

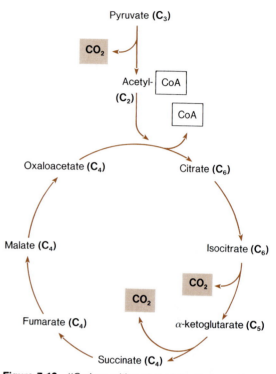

Figure 7.10 "Carbon arithmetic" of the Krebs cycle. Three CO_2 molecules are produced for each pyruvate molecule. Regeneration of oxaloacetate at the end of the reaction series prepares for another "turn" of the cycle.

initial electron acceptor and passes electrons to ubiquinone (figure 7.11). Figure 7.8b shows that for every pair of electrons passed through the respiratory chain from NADH to oxygen, three molecules of ATP are formed. However, the electron transport resulting from succinate oxidation permits just two phosphorylations because the electrons enter the electron transport system at a "lower" point. Thus, in terms of ATP formation, reduced FAD from the succinic dehydrogenase system is not as energy rich as reduced NAD.

In each turn of the Krebs cycle, one substrate-level phosphorylation also occurs. Guanosine diphosphate (GDP) is phosphorylated to produce guanosine triphosphate (GTP), which can then transfer its terminal phosphate to ADP to make ATP. But this substrate-level phosphorylation accounts for only a small part of the energy conversion in the Krebs cycle because the bulk of the ATP synthesis is associated with the electron transport system.

Enzymes

 I = Citrate synthase

 II = Aconitase

 III = Isocitrate dehydrogenase

 IV = α-ketoglutarate dehydrogenase

 V = Succinate dehydrogenase

 VI = Fumarase

 VII = Malate dehydrogenase

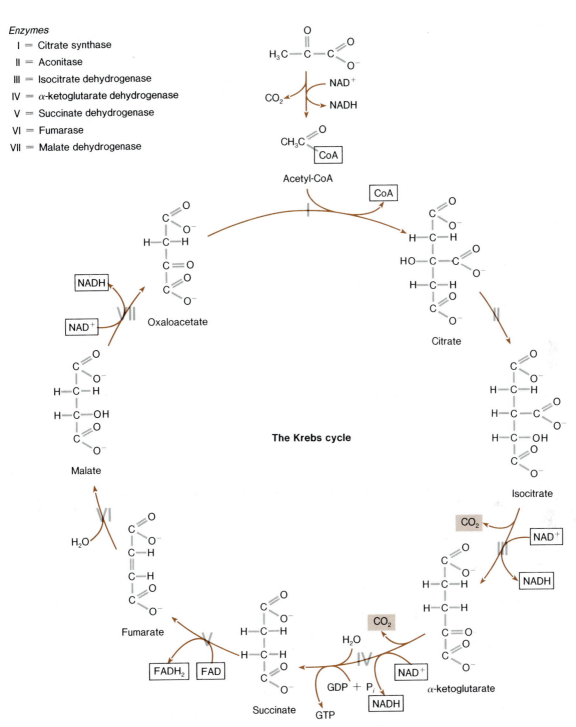

Figure 7.11 The Krebs cycle, including molecular structures of compounds as well as decarboxylation sites and coenzyme reduction sites. Enzymes are identified by Roman numerals. The Krebs cycle often is called the **citric acid cycle,** and sometimes it is called the **TCA cycle** (for tricarboxylic acid).

Figure 7.12 summarizes ATP formation and the Krebs cycle. When the substrate-level phosphorylation, the phosphorylations occurring as a result of electron transport initiated by reduced coenzymes from the Krebs cycle reactions, and the NAD reduction in the preparatory reactions are taken into account, a total of fifteen molecules of ATP are formed for every molecule of pyruvate entering the reaction sequence inside the mitochondrion.

Overall Energy Sums

It is now possible to summarize the energy return from the complete oxidation of a glucose molecule under aerobic conditions, and to make some comparisons between the efficiencies of aerobic and anaerobic respiration. Aerobic conditions change the energetics story of the Embden-Meyerhof pathway itself somewhat because the NADH produced during the E-M reactions can transfer electrons to the electron transport systems inside mitochondria, thereby salvaging

more usable energy than is possible under anaerobic conditions, where the electrons are accepted during the production of organic end products (lactate or ethanol). But to do so, electrons must be transferred into the interiors of the mitochondria, where they become available to the electron transport systems that are located in the internal membranes of the mitochondria. There are several different ways to accomplish this transfer.

One common way, known as the **glycerol phosphate shuttle**, picks up electrons from NAD, carries them inside the mitochondria and donates them not to NAD, but to ubiquinone. In this case, the electrons do not start at the "top" of the electron transport system and the possibility for one phosphorylation is bypassed (see figure 7.8). If mitochondria use this particular mechanism, a reduced NAD in the cytoplasm ends up being worth only two ATP molecules.

The E-M pathway itself, therefore, produces a greater net gain of ATP molecules per glucose molecule when it is associated with aerobic respiration than it does under anaerobic conditions.

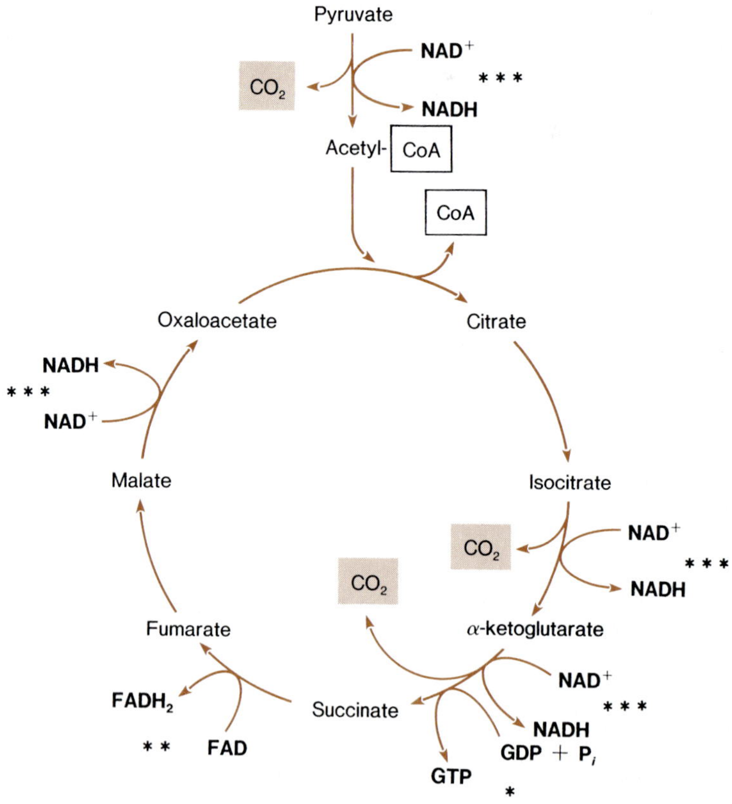

Figure 7.12 ATP arithmetic of the Krebs cycle. Each star is worth one ATP formed. Each reduced NAD molecule permits three phosphorylations in the electron transport system. A reduced FAD permits two phosphorylations in the electron transport system. A substrate-level phosphorylation forms GTP, which can transfer phosphate to ADP to produce ATP. Thus, one turn of the Krebs cycle produces twelve ATPs, or, starting from pyruvate, fifteen ATPs are formed.

Under aerobic conditions, the E-M pathway produced a net gain of six ATP molecules per glucose molecule processed (figure 7.13): four direct (substrate-level) phosphorylations plus four ATPs formed because of the two NAD reductions (one NAD reduction for each three-carbon carbohydrate molecule), *minus* the two ATP molecules used initially for activation. If the ATP output from the E-M pathway is added to the ATP yield from preparing and feeding pyruvate products into the Krebs cycle, a total of thirty-six molecules of ATP is formed for each molecule of glucose oxidized completely to CO_2 and water under aerobic conditions (table 7.1). Since the anaerobic process yield a net gain of only two molecules of ATP per molecule of glucose, the aerobic process yields a net gain of eighteen times as much usable energy as did strictly anaerobic processes.

The Efficiency of Respiration

How efficient is aerobic respiration as a system for converting energy from one form to another? It has been established that thirty-six molecules of ATP are formed for every molecule of glucose metabolized in aerobic respiration. Of course, the energy relationships of individual reacting molecules are too small to be measured. Therefore, measurements in terms of *moles* of the substances involved are used, keeping in mind that the behavior of moles (gram molecular weights) are proportional to the behavior of individual molecules. The standard free energy ($\Delta G^{0'}$) of oxidation of a mole of glucose to CO_2 and water is -686 kcal (-2870 kJ). The formation of a mole of ATP requires an energy input of 7.3 kcal (30.5 kJ) per mole. Thus, an efficiency figure for respiration can be calculated as follows:

$$\frac{36 \times 7.3}{686} \times 100 = 38\%$$

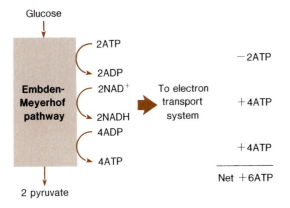

Figure 7.13 ATP production from the Embden-Meyerhof pathway when it is associated with aerobic respiration.

Table 7.1
ATP Formation during Aerobic Oxidation of Glucose.

Embden-Meyerhof Pathway	ATP Calculations	Sums
ATP expended to phosphorylate sugar molecules	−2	
Substrate-level phosphorylations	4	
ATP formed as a result of electron transport from two NADs reduced during E-M pathway reactions (two ATP each)	4	
Net gain from E-M pathway under aerobic conditions		6
Two Pyruvate to two Acetyl CoA		
Two NAD reduced (three ATP each)	6	
		6
Krebs Cycle (two turns)		
Six NAD reduced (three ATP each)	18	
Two FAD reduced (two ATP each)	4	
Two substrate-level phosphorylations (one ATP each)	2	
Total from Krebs Cycle (two turns)		24
Total ATP Formation per Molecule of Glucose		36

However, most biologists think, for several reasons, that 38 percent represents a minimum estimate. First, the number of ATP molecules formed per molecule of glucose may actually be thirty-eight rather than thirty-six in some situations (when different mechanisms are employed to transfer electrons from reduced NAD in the cytoplasm to the electron transport system inside the mitochondrion). Secondly, because of the effects of concentration and the efficiently coupled utilization of the ATP formed in respiration, these ATP units may really be worth more to living cells than 7.3 kcal per mole in terms of providing usable energy for cell work. Thus, respiration is probably 38 percent efficient at the very least.

The remainder of the free energy made available in glucose oxidation, that energy not converted via ATP production, is released as heat.

Mitochondrial Organization and Phosphorylation

While the Embden-Meyerhof pathway reactions occur in the cytoplasmic matrix, the Krebs cycle reactions, electron transport, and oxidative phosphorylation all take place in the mitochondria. To fully understand these processes, it is necessary to know something about the structure of a mitochondrion.

Mitochondrial Structure

The mitochondrial membrane system consists of two parts: the outer membrane is smooth, while the inner membrane is folded (figure 7.14). These folds or **cristae** may form flat plates, or they may be tubular. The **mitochondrial matrix**, the material enclosed inside the inner mitochondrial membrane, is gel-like and made up of about 50 percent protein. Part of this protein is a network of structural fibers and part is enzyme protein. Most of the Krebs cycle enzymes are located in the mitochondrial matrix.

Electron carriers of the respiratory chain are associated with the inner membrane. Apparently, they are arranged in clusters with the carriers organized in the proper sequences within each cluster, so that the transfer of electrons from one carrier to the next is very efficient.

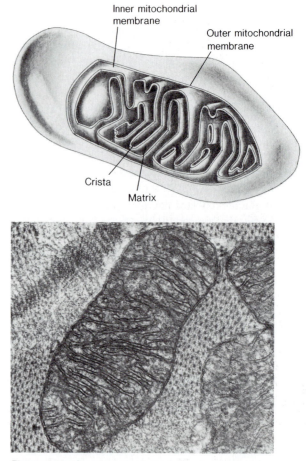

Figure 7.14 Structure of a typical mitochondrion. Note inner and outer mitochondrial membranes, cristae, and matrix.

More detailed examination of the inner mitochondrial membrane and its cristae has revealed a number of knoblike particles attached to the inner surface (the matrix side) of the membrane (figure 7.15a). These particles, called F_1 **factors** are associated with oxidative phosphorylation. When it was first discovered that the F_1 factors played a role in ATP formation, it was proposed that they might contain both the electron carriers and the enzymes involved in the actual phosphorylation of ADP. But the experimental isolation by Ephraim Racker and his colleagues of different parts of mitochondria has shown that this is not the case.

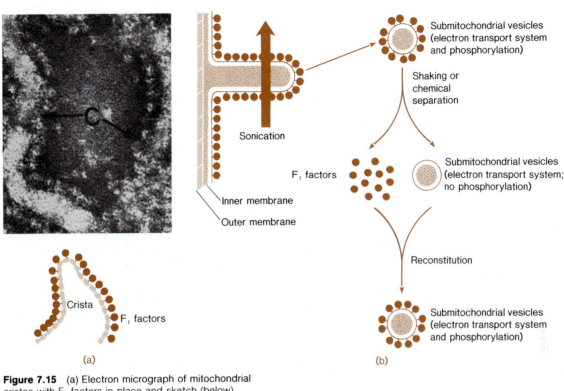

Figure 7.15 (a) Electron micrograph of mitochondrial cristae with F_1 factors in place and sketch (below) showing relationship of factors to crista. (b) Sonication (treatment with high frequency sound) is thought to produce vesicles of inner membrane by shearing cristae. Intact vesicles (top) carry on electron transport and phosphorylation. Vesicles stripped of F_1 factors by shaking or chemical treatment (middle) carry on electron transport but not phosphorylation. Reconstituted vesicles (bottom) again can do both.

Racker found that mitochondria can be treated with chemicals or with high-frequency sound so that bits of cristae material break off and form small spheres or **vesicles** (figure 7.15b). These vesicles have F_1 factor particles on their surfaces and can carry on both electron transport and oxidative phosphorylation. However, if these spheres are shaken or treated with the protein-hydrolyzing enzyme trypsin or with urea (so that the F_1 factors come off their surfaces), electron transport still occurs, but oxidative phosphorylation does not. Then if the F_1 factors are returned to the surfaces of the spheres, phosphorylation occurs once again. These experiments therefore demonstrate that electron transport assemblies must be in the inner mitochondrial membrane itself and not in the F_1 factor particles. But the F_1 factor particles are necessary for the **coupling** of electron transport and oxidative phosphorylation.

The Mechanism of Oxidative Phosphorylation

The way in which electron transport is coupled with the production of ATP remains one of the major unsolved problems of cell biology. The problem is basically the same in mitochondria as in chloroplasts and the same competing hypotheses are relevant to the question of mitochondrial phosphorylation mechanisms (see chapter 6).

One hypothesis suggests that during certain steps in the electron transport system, enough energy becomes available either to form a specific chemical intermediate compound or to alter the conformational state of a macromolecule. These two related hypotheses are known respectively as the **chemical** and **conformational** hypotheses. They propose that a "high-energy" chemical intermediate returns to its initial state with a great enough negative free energy to permit the phosphorylation of ADP.

Box 7.1
Respiratory Control and Metabolic Poisons

Do cells keep on using glucose as long as there is any available to them? It only takes a moment's thought to realize that the answer must be no. The oxidation of nutrients must be correlated with a cell's energy requirements; therefore, cellular energy conversions must be regulated.

Several reaction steps in the oxidation process are under allosteric feedback regulation (see p. 131). For other reactions, an adequate supply of oxidized coenzymes is critical. In fact, there is a whole network of regulatory interactions that controls the rates at which the Embden-Meyerhof pathway and the Krebs cycle proceed.

One interesting way in which aerobic respiration is regulated is called *respiratory control*. This is a slightly misleading name, but it is in common use among biologists. Respiratory control, which depends on the coupling of reactions, is a form of regulation that depends on the ATP to ADP ratio. Normally, electron flow through the electron transport system is tightly coupled to phosphorylation. That is, electron movement from one electron carrier to another will not occur unless ATP is being formed simultaneously. When there is a high ATP to ADP ratio, electron transport is slowed because there simply is not enough ADP available for the coupled reactions of electron transport and phosphorylation to proceed. On the other hand, if a great deal of ATP-requiring work is going on, a

relatively large supply of ADP will be available for phosphorylation and, consequently, electron flow through the electron transport system will speed up.

However, when the electron transport system is "backed up" because ADP is not available for phosphorylation, reduced NAD and FAD cannot unload electrons. As a result, the entire respiratory process is slowed because all reactions requiring coenzymes as electron acceptors are inhibited. Through this mechanism, then, the rate of ATP formation can be modulated by the rate that ATP is being used for work in the cell (box figure 7.1A).

One of the metabolic poisons, **2,4-dinitrophenol (DNP),** uncouples phosphorylation from electron transport. This uncoupling allows the metabolism to run wild because there is no "braking" effect or control over the rate of electron transport. Of course, ATP production ceases when oxidative phosphorylation is uncoupled from electron transport.

Even when electron transport is coupled with phosphorylation, quite a bit of energy is released in the form of heat. When these processes are uncoupled, the remainder of the free energy released in the electron transport system also is dissipated as heat. It is thus not surprising that elevated body temperature is a major symptom of DNP poisoning.

Another metabolic poison, **cyanide,** affects the electron transport system by interfering with the cytochrome oxidase complex so that electron transfer to oxygen is inhibited. The cell's inability to reduce oxygen when poisoned by cyanide has widespread effects. Krebs cycle reactions are shut down because the Krebs cycle is an aerobic process—it depends on coenzymes (NAD and FAD) that must become oxidized by donating electrons to oxygen through the electron transport system. Oxidative phosphorylation decreases with cyanide poisoning, and no ATP-requiring cell work can be done because the supply of ATP is quickly diminished.

Uncoupling and electron-transport blocking agents are important research tools used to study the mechanics of electron transport systems. Many of the facts known about these important oxidation-reduction reaction series have been gleaned through experiments employing metabolic poisons.

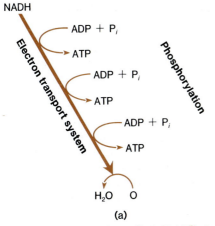

(a)

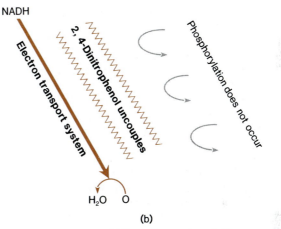

(b)

Box Figure 7.1A   "Respiratory Control." (a) Electron transport and phosphorylation are coupled. ADP availability controls rate of electron flow through the electron transport system. This is a control relationship in which the cell's energy requirements adjust the rate of electron transport. (b) Uncoupling agents such as 2,4-dinitrophenol remove feedback control by uncoupling electron transport and phosphorylation. With its "brakes" off, the electron transport system runs at full speed.

An alternative to these two hypotheses is the **chemiosmotic hypothesis.** This hypothesis has been especially vigorously supported by Peter Mitchell (the 1978 Nobel Chemistry Prize winner) and, in fact, is frequently called the *Mitchell hypothesis* by some biologists. The chemiosmotic hypothesis proposes that a *transmembrane electrochemical gradient* functions as a "high-energy" intermediate and is critically important to the phosphorylation process. This hypothesis does not rule out the possibility that chemical or conformational effects might be involved in the final stages of phosphorylation, but it emphasizes the importance of the electrochemical gradient (figure 7.16).

Mitchell proposes that the ionic gradients are produced by the precise arrangement of the electron transport chain across the mitochondrial membrane. Figure 7.17 shows a hypothetical example of how the reduction of a mobile proton carrier, such as ubiquinone, on one side of the membrane and its oxidation on the other could lead to the transport of a pair of hydrogen ions across the inner mitochondrial membrane. A continuing series of such reactions would produce a hydrogen ion gradient across the membrane.

It is further suggested that there are special channels through which the H^+ ions can move in response to the forces acting on them (which are both electrical and concentration differences). In some as yet undetermined way, the movement of protons back through the membrane in this way is coupled to the phosphorylation of ADP to form ATP (figure 7.18). This could be the point where a chemical intermediate might be involved.

The evidence for the chemiosmotic hypothesis in mitochondria is not as strong as it is in the case of chloroplasts. Experiments have shown, though, that protons indeed are "pumped" across mitochondrial membranes during electron transport. There is thus some experimental support for the chemiosmotic hypothesis, but it is by no means proven, nor are other hypotheses of ATP production proven incorrect. Although many aspects of cellular oxidations and energy conversions are now understood, there are large gaps in knowledge about the actual production of ATP, which is, after all, the key outcome of all of the metabolic energy conversions.

The Oxidation of Other Organic Compounds

Other compounds can be oxidized in the same metabolic pathways as carbohydrates. In each case, there must be preparatory steps to produce molecules that can enter the pathways.

The Oxidation of Fats

The metabolic utilization of a **triglyceride,** a common fat, can serve as an example of how lipids are prepared for oxidation. A **lipase** enzymatically catalyzes the hydrolysis of the triglyceride molecule to produce one molecule of **glycerol** and three **fatty acid** molecules (figure 7.19a). In most animal triglycerides, these fatty acids have even-numbered chains of carbon atoms (that is, they are likely to be made up of chains of C_{16} or C_{18}, or some other even number, rather than C_{15} or C_{17}, or some other odd number). This fact is important in the carbon arithmetic of fatty acid oxidation.

The glycerol derived from triglyceride hydrolysis can enter the Embden-Meyerhof pathway after it is used to produce dihydroxyacetone phosphate (DHAP) (figure 7.19b). Fatty acids are metabolized through a process that cleaves their carbon chains into two-carbon fragments in a repeating series of reactions (figure 7.19c). These two-carbon fragments end up as acetyl CoA, which then may enter the Krebs cycle. Fatty acid oxidation also yields reduced NAD and FAD, which then donate electrons to the electron transport system.

Lipids are essential storage materials, and their usefulness for that purpose rests on the relatively large amount of energy that can be derived through their oxidation. Because they are more highly reduced compounds than carbohydrates, lipids have the potential for greater energy conversion by oxidation. In practical terms, this means that the oxidation of lipids yields more energy per unit of weight than the oxidation of carbohydrates. Representative average values for energy yields from the oxidation of lipids and carbohydrates are 9 kcal (37.7 kJ) per gram for fats and 4 kcal (16.7 kJ) per gram for carbohydrates.

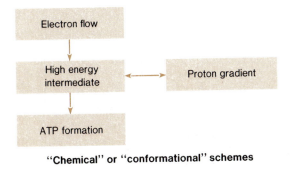

"Chemical" or "conformational" schemes

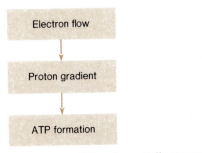

"Chemiosmotic" scheme

Figure 7.16 Relationship of intermediates between
electron flow and ATP formation according to the
chemical (or conformational) and the chemiosmotic
models. Note especially that the proton gradient is not
thought to be directly involved in ATP formation in the
chemical or conformational schemes, while it is of central
importance in the chemiosmotic scheme.

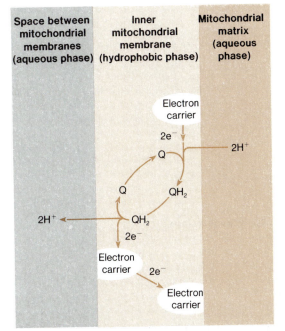

Figure 7.17 How reduction and oxidation of a quinone
(Q) on opposite sides of a membrane could lead to
transport of hydrogen ions across the membrane
according to the chemiosmotic scheme.

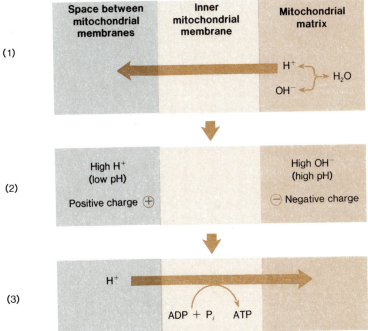

Figure 7.18 Summary of the
mechanism of oxidative
phosphorylation according to
chemiosmotic ideas. (1) Electron
flow results in H^+ moving across
the membrane, leaving an excess of
OH^- behind. (2) This results in a
gradient of pH and electrical
potential. (3) Protons moving back
down this gradient can be coupled
to the formation of ATP.

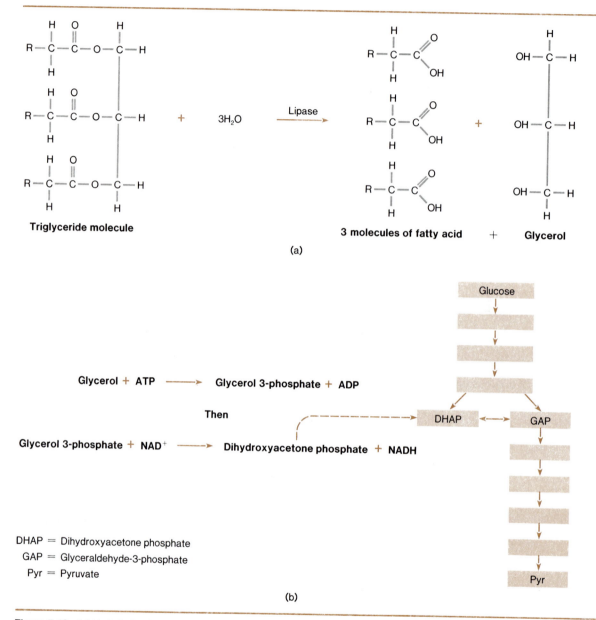

DHAP = Dihydroxyacetone phosphate
GAP = Glyceraldehyde-3-phosphate
 Pyr = Pyruvate

(b)

Figure 7.19 (a) Hydrolysis of a triglyceride (fat) into glycerol and three fatty acid molecules. The ''R'' in each case represents an even-numbered chain of carbons. (b) The reactions by which glycerol enters oxidative pathways. (c) ''β-oxidation'' of a fatty acid (palmitic acid is the example). The name β-oxidation comes from another system for identifying carbon atoms. The carbons are numbered to show how they are removed two at a time. β-oxidation results in the formation of acetyl CoA, NADH, and FADH₂, all of which can be used in the energy conversions described earlier.

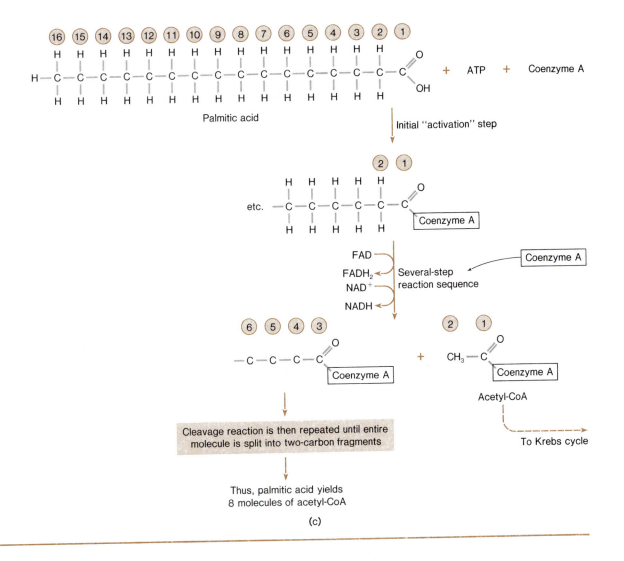

Palmitic acid

Initial "activation" step

etc.

Coenzyme A

FAD
FADH₂
NAD⁺
NADH

Several-step
reaction sequence

Coenzyme A

Coenzyme A

Acetyl-CoA

To Krebs cycle

Cleavage reaction is then repeated until entire
molecule is split into two-carbon fragments

Thus, palmitic acid yields
8 molecules of acetyl-CoA

(c)

Thus, lipids can be considered as relatively dense energy-storage compounds. It is not surprising that except for some short-term glycogen storage, most animals depend more on fats than on carbohydrates for storage. Plants, on the other hand, use carbohydrates (especially starch) as their main storage compounds. Interestingly, some animals such as clams, which live a relatively sedentary life, do store considerable amounts of carbohydrates. Obviously, such animals do not have to carry the extra weight and can afford to store extra fuel in a bulkier form.

The Metabolism of Proteins

There are a number of situations in which amino acids from protein molecules are used for energy-yielding oxidations, and there are several methods for feeding these amino acids into the various metabolic pathways. The amino acids come from digested dietary proteins as well as from body proteins that become available because of continuing protein replacement in cells.

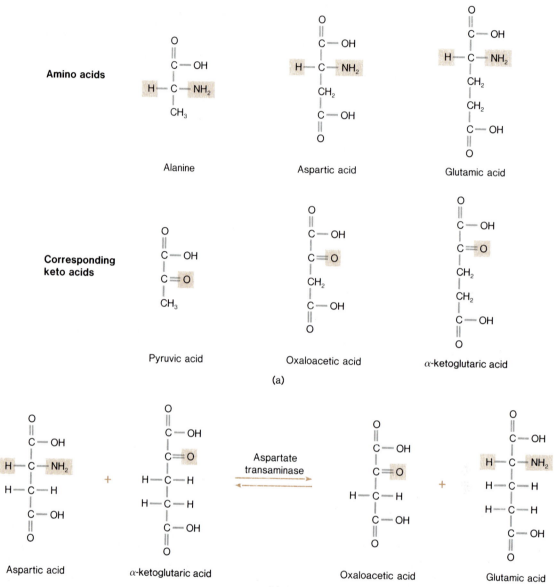

Figure 7.20 (a) Three amino acids with corresponding keto acids. Products of deamination of some amino acids can enter the Krebs cycle directly. (b) A transamination reaction in which an amino group is exchanged for an oxygen atom.

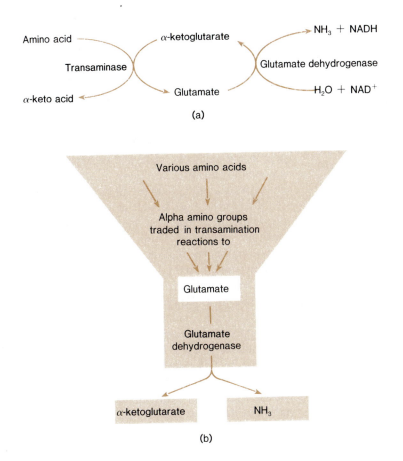

Figure 7.21 Amino acid metabolism. (a) Amino acids are prepared for oxidation as amino groups are transferred to glutamate through transamination reactions. The key oxidative deamination reaction is the deamination of glutamate. (b) The "glutamate funnel." (c) Urea, a nitrogenous waste carrier molecule.

$$\text{Urea} \qquad H_2N - \overset{\overset{\displaystyle O}{\|}}{C} - NH_2$$

(c)

An amino acid is prepared for oxidation in the Krebs cycle or in other oxidative reactions by the removal of its amino group in a **deamination** reaction. In this type of reaction, amino groups are replaced by oxygen atoms; that is, amino acids are converted to keto acids (figure 7.20a).

One type of deamination involves trading amino groups; this is known as a **transamination reaction.** In transaminations, amino groups and oxygen atoms are traded between amino acids and keto acids. These trades, which are catalyzed by enzymes called **transaminases,** can be made between a number of combinations of molecules. One example is shown in figure 7.20b.

Another type of deamination reaction is **oxidative deamination.** The most common and most important oxidative deamination reaction is the deamination of **glutamate** to produce **α-ketoglutarate:**

$$\text{Glutamate} + NAD^+ + H_2O \underset{\overset{\text{dehydrogenase}}{\longrightarrow}}{\overset{\text{Glutamate}}{\rightleftharpoons}}$$

$$\alpha\text{-ketoglutarate} + NH_3 + NADH$$

Actually, the amino groups of most of the amino acids are not directly removed to prepare the molecules for oxidation, but amino groups are passed to α-ketoglutarate to form glutamate. Glutamate then is oxidatively deaminated. Because of this, amino acids are sometimes said to be "poured through the glutamate funnel" on their way to the oxidative pathways (figure 7.21).

The ammonia produced during glutamate deamination is incorporated into nitrogenous waste carrier molecules such as **urea.** In the human body, glutamate deamination and urea production occur mainly in liver cell mitochondria.

The Metabolic Pool and Biosynthesis

The oxidative pathways described thus far are essentially open systems, with a two-way flow of materials in and out of them. Various compounds can enter each pathway at different points so that carbohydrates, fats, and amino acids can all be oxidized. Moreover, some of the intermediate products from the system can be withdrawn and used in synthesis reactions. The idea that various compounds can enter the metabolic reactions for oxidation and that various substrate molecules can be withdrawn for use in synthesis is called the **metabolic pool** concept (figure 7.22). Two examples may help to clarify this idea.

At times when more than adequate supplies of reduced organic compounds are present in the body, fat (triglyceride) synthesis begins. Glycerol is produced for fat synthesis from dihydroxyacetone phosphate. Excess acetyl CoA makes two-carbon fragments available for incorporation into the even-number carbon chains of fatty acids. The components of fat molecules, glycerol and fatty acids, are thus produced from material withdrawn from the oxidative pathways.

A second example of the metabolic pool concept is related to the need for amino acids to be used in protein synthesis. In animals, a large proportion of these amino acids are normally provided by the digestion of dietary proteins. But in plants, and in part in animals, intermediate products from the oxidative pathways are used for amino acid synthesis. Carbon skeletons are available in the metabolic pool, and these molecules are used in **amination** reactions, in which the addition of an amino group produces amino acids.

Plant cells must be able to synthesize all of the kinds of amino acid molecules that they need for protein synthesis because they have no external source of amino acids. But in the cells of many animals and microorganisms, the ability to synthesize amino acids is limited; that is, these organisms lack the enzyme systems needed to synthesize certain amino acids. Thus, it is essential that those particular amino acids be obtained from outside sources. For example, in animals, there are certain **essential amino acids** that the animal cannot synthesize and that are therefore absolute dietary requirements; **nonessential amino acids** *can* be synthesized.

Figure 7.22 The metabolic pool. The metabolic pathways function as an open system. Various compounds enter at different points for oxidation, and intermediate products can be used for synthesis. Reactions involved in synthesis usually are not the reverse of reactions preparing compounds for entry into oxidative pathways because different enzyme systems are involved.

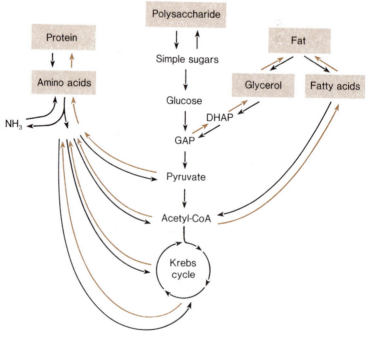
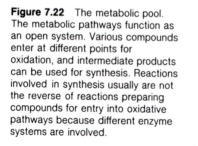

Table 7.2
Human Dietary Requirements Among the Twenty
Amino Acids That Occur in Proteins.

Nonessential	Essential
Alanine	Arginine
Asparagine	Histidine
Aspartate	Isoleucine
Cysteine	Leucine
Glutamate	Lysine
Glutamine	Methionine
Glycine	Phenylalanine
Proline	Threonine
Serine	Tryptophan
Tyrosine	Valine

Humans (and, incidentally, white rats) can synthesize ten of the amino acids found in protein molecules but must obtain the other ten through diet. For humans, the second group of ten is made up of the essential amino acids (see table 7.2). It is possible to eat large quantities of proteins and still suffer from deficiency symptoms if those proteins do not contain adequate quantities of the essential amino acids.

Temperature, Size, and Metabolism

The sum of all of the cellular processes discussed thus far constitutes the **metabolism** of the organism. Various chemical regulatory mechanisms control the cellular metabolic processes, but other factors affect the metabolisms of whole organisms.

Body Temperature

As was previously stated, only about 40 percent of the energy potentially available in reduced organic nutrients is used for ATP production. Most of the remaining 60 percent is emitted as heat, which, in the great majority of organisms, including all plants and most animals, is quickly lost to the environment. These organisms are in thermal equilibrium with their environments. When the environmental temperature rises, their body temperatures rise, and when the environmental temperature falls, their body temperatures fall. Organisms with environmentally influenced, variable body temperatures are called **heterothermic** or **poikilothermic** (both of which mean "different temperature").

Some animals, however, specifically birds and mammals, conserve body heat and regulate their metabolisms so that they maintain stable, relatively constant body core temperatures. These animals are called **homeothermic** ("same temperature").

Temperature and Metabolic Rates

Enzymes have optimal temperatures at which they catalyze reactions most efficiently. At a temperature lower than optimum, the rate of an enzyme-catalyzed reaction is low, but as the temperature increases toward the optimum, the reaction rate increases. Temperature changes affect reaction rates in all body cells. Thus, the effects of environmental temperature change are reflected in an organism's metabolism.

Metabolism is the sum total of all cellular activities in the body, and the **metabolic rate** is a collective measure of these processes. Metabolic rate is sometimes measured as heat released from the body, but in organisms using aerobic metabolism, it is usually easier to measure oxygen consumption or CO_2 release or both as an indication of metabolic rate.

Metabolic rates change with an organism's activity level. When an organism is very active, its energy requirements and its metabolic rate might be many times greater than when it is resting. In humans, for example, oxygen consumption during heavy work can be 20 times the resting rate. A butterfly in flight consumes 100 times as much oxygen as when it is resting. Plainly, metabolic rates depend on measurement conditions. In laboratory studies, the resting metabolism of an organism carrying on only life maintenance functions is called **standard metabolism.** Standard metabolism is comparable to **basal metabolic rate** in humans. Human basal metabolic rates are usually determined in mid-morning, with the person relaxed but awake and having eaten nothing since the preceding evening. The discussion of temperature relationships and metabolism in this text will be limited to temperature effects on standard metabolism.

Heterotherms

Because the rates at which chemical reactions proceed in living organisms are temperature dependent, environmental temperature changes and subsequent body temperature changes have a significant effect on the metabolic rates of heterothermic organisms. Suppose that the metabolic rate of a hypothetical heterothermic animal is measured at a series of different temperatures. Suppose further that this organism can tolerate an unusually broad range of temperatures, say from 5°C to 45°C. Figure 7.23 shows the type of curve that is obtained when the organism's metabolic rates are plotted against a series of measurement temperatures. Such a curve is called a rate-of-reaction-to-temperature curve or simply a **rate:temperature (R:T) curve.**

The metabolic rates of real organisms seldom follow a rate:temperature curve like that in figure 7.23 because organismic metabolism is the composite of hundreds or even thousands of individual processes having many different sets of temperature characteristics. Further, the range of **thermal tolerance** usually is more restricted than it is in the hypothetical organism. Beyond the lower and upper extremes of the acceptable temperature range, some key process or processes will slow down below a critical rate, depressing the activity and the metabolism of the whole organism. **Cold torpor** and **heat coma** are names given to the inactive states near the lower and upper extremes of the range of thermal tolerance. **Cold death** or **heat death** will occur when temperatures are extreme enough to cause extensive, irreversible damage. Figure 7.24 shows another hypothetical but much more realistic curve that represents these metabolic levels for a heterothermic organism.

However, a single rate: temperature curve does not thoroughly account for temperature effects on the metabolism of all heterotherms.

Enzyme systems and various other physiological processes with very different temperature characteristics have evolved in various organisms. Tropical organisms may have rate: temperature curves that extend over an entirely different range than those of arctic organisms. For example, in water at 4°C, an arctic fish thrives and carries out life activities normally. A tropical fish placed in water at 4°C would be paralyzed or worse by

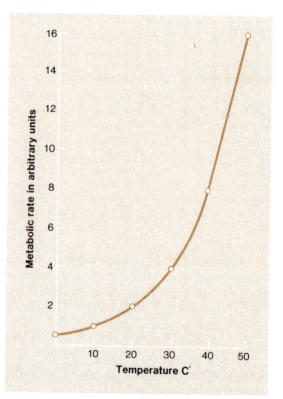

Figure 7.23 Rate of reaction to temperature (R:T) curve for metabolic rates at various temperatures in a hypothetical case where metabolism is quite temperature dependent and the organism can tolerate a very large range of temperatures. Note that in this case the metabolic rate doubles each time the temperature is increased by ten degrees. This is a commonly observed degree of temperature dependence in studies of heterothermic organisms' responses to temperature change.

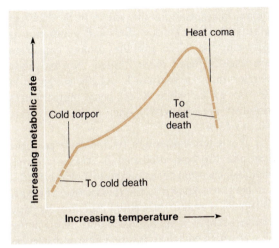

Figure 7.24 A more realistic, hypothetical
rate:temperature curve that shows the limits imposed by
cold torpor and heat coma at the ends of the curve.

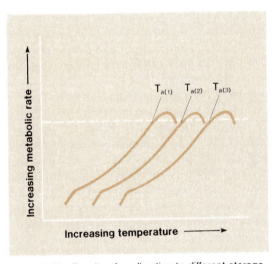

Figure 7.25 Results of acclimation to different storage
(acclimation) temperatures: $T_{a(1)}$, $T_{a(2)}$, and $T_{a(3)}$. The
broken line intersects each R:T curve at the acclimation
temperature of that group of animals. This demonstrates
that acclimatory compensation has occurred, making
rates measured at the acclimation temperature the same
in each case, even though measurement temperatures
are different.

the cold. On the other hand, a tropical fish lives
normally in 20°C water. An arctic fish placed in
20°C water would suffer heat death.

Such genetic differences in temperature op-
tima have evolved over long periods of time. How
do organisms that are subjected to significant
short-term temperature changes, such as the seasonal changes that affect organisms in freshwater
lakes and ponds, survive?

Acclimatization

Many organisms that live in environments with
changeable temperatures make physiological adjustments that permit them to continue to function normally even though their environmental
temperature goes through a definite annual temperature cycle. This physiological adjustment
made under natural conditions is called **acclimatization.**

Often, responses to changing thermal environments are studied experimentally in laboratories where all conditions except temperature
are kept constant. Metabolic adjustment under
laboratory conditions to changes in one factor
such as temperature (rather than a full range of
environmental influences) is known as **acclimation.** Scientists distinguish acclimation from acclimatization because the latter adjustment is
made under natural conditions when the organism is subject to the full range of changing environmental factors.

In an acclimation experiment, a group of
organisms is divided into several samples, each
of which is held at a different temperature for a
period of time (the acclimation period) while
other factors are kept the same for all groups.
Then, metabolic rates, or the rates of various
other physiological responses, are measured for
organisms from the various samples. The result
of one such experiment is shown in figure 7.25.
The rate: temperature curves for metabolism of
organisms from the different groups are located
in different parts of the temperature scale depending on the temperatures at which the organisms had been held. As the figure shows, the
metabolic rate:temperature relationships of many
organisms adjust so that they function efficiently
at the environmental temperatures that they are
experiencing.

Behavioral Thermoregulation

What forms of adjustment are available to heterothermic organisms who face hourly temperature changes on a daily basis? Some heterothermic animals adjust their body temperatures by behaving in specific ways. Such animals are capable of **behavioral thermoregulation.**

On cool mornings, snakes and lizards that are made very sluggish by low temperatures bask in the sunshine until their body temperatures rise to the point where they can carry out their normal activities. Then, later in the day they seek shelter in cool burrows if their body temperatures rise above a certain critical point (figure 7.26).

Another form of behavioral thermoregulation is the warm-up period that some insects go through on a cool morning. Butterflies spread their wings and vibrate them slowly until the heat generated by the metabolism of their muscle cells raises their temperature. When their body temperature is high enough to permit the intense muscular activity needed for flight, a butterfly can take off. Such forms of behavioral thermoregulation help heterothermic animals cope with changing environmental temperatures.

Homeotherms

A different strategy for dealing with environmental temperature fluctuations is to maintain a constant body core temperature. Animals that do this are called **homeotherms.** Temperatures in areas near the body's surface may vary, but the temperature deep in a homeotherm's body, the **core temperature,** remains relatively constant. Thus, major internal body organs function in an environment where the temperature changes very little.

Homeotherms (that is, birds and mammals) have relatively higher metabolic rates than heterothermic organisms, and many of them have insulating layers of fat, feathers, or fur to help conserve body heat. Furthermore, they have nervous and hormonal mechanisms that regulate their metabolic rates in response to body temperature changes. The temperature-maintenance strategies of homeothermic animals use "a furnace rather than a refrigerator"; the body temperature is set and kept near the high end of the organism's normal environmental temperature range. The reason for this seems to be that regulation by heat conservation and generation is very efficient, while animals' cooling capacities are usually much more limited. Normal body temperatures of mammals usually range from 36° to 39°C, while birds' body temperatures generally are somewhat higher, in the range from 40° to 43°C.

For a homeotherm, a balance between heat loss and the heat produced by normal resting metabolism occurs in a range of environmental temperatures known as the **thermoneutral range.** Within that range, the animal regulates its body temperature by making small adjustments in the rate at which body heat is lost. These adjustments are made by altering the amount of blood flow near the body surface. When the temperature of the environment is in the lower part of the thermoneutral temperature range, blood vessels near the body's surface are constricted. When the temperature is in the upper part of the range, the vessels are dilated so that there is greater blood flow near the body surface and greater heat loss to the environment. The thermoneutral range may be rather small for some homeotherms. For example, in humans it extends only from 26° or 27° to 31°C (figure 7.27). When the temperature falls below 26°C, a nude, resting person will begin to lose heat faster than he or she is producing it. The nervous system responds to falling body temperature by initiating **shivering,** which is, in reality, a series of involuntary muscle contractions. Shivering produces heat that helps to compensate for heat loss. When a person is quite chilled, shivering can be very vigorous, even to the point that his or her teeth "chatter" when the jaw muscles become involved. Humans artificially extend the lower end of the thermoneutral zone by wearing clothing and increasing their insulation against heat loss. Above 31°C, the body initiates **sweating,** which dissipates a great deal of body heat by evaporative cooling. Both shivering and sweating require energy expenditures and raise the metabolic rate.

Some mammals have larger thermoneutral ranges than humans do because they have much more effective body insulation. For example, some arctic mammals do not show metabolic rate increases until their environmental temperature falls well below 0°C (even as low as −20°C in the case of the arctic fox).

(a)

(b)

Figure 7.26 Behavioral thermoregulation. (a) A lizard basks in the sun during cool morning hours. (b) During the heat of the day, the same lizard enters a cool underground burrow when its body temperature rises above a certain critical point.

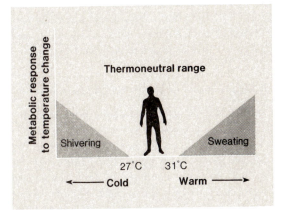

Figure 7.27 Human responses to temperature change. Within the thermoneutral range, small adjustments in heat loss from superficial blood vessels are adequate to regulate temperature. Beyond lower or upper critical temperatures, other physiological mechanisms come into play.

Figure 7.28　Metabolic rates of some mammals as oxygen consumed per unit body weight, *plotted on a logarithmic scale.* Metabolic rates clearly are inversely related to size.

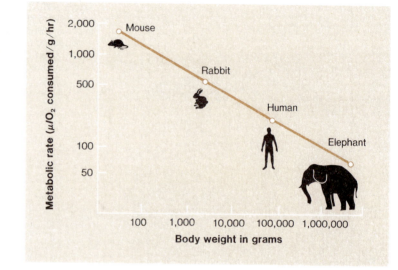

Size and Metabolism

Although homeothermic animals can, to an extent, escape the tyranny of temperature, they must expend energy to maintain their high body temperatures, and this energy cost can be enormous, especially for a small mammal. For example, a shrew weighing around four grams must consume close to the equivalent of its body weight in food every day to meet its energy requirements. Such an animal can literally starve to death on any given day if it falls short of its required food intake, so the shrew's life is a continuous, almost desperate, hunt for food. The problem is not so severe for larger mammals because they have more favorable surface-to-volume ratios and do not suffer as much heat loss. This is fortunate because if large mammals such as elephants had to consume the equivalent of their weight in food every day the earth would be unable to support many organisms for long. Large mammals can afford to go for considerable periods of time with little or no food.

The different heat-production requirements of large and small mammals are reflected in marked differences in their respective metabolic rates (figure 7.28). These metabolic-rate differences are expressed in terms of oxygen consumed per unit weight. As figure 7.28 shows, larger mammals have strikingly lower metabolic rates than small mammals.

Size-related metabolic differences have other physiological consequences. For example, the same amounts of various drugs and other chemicals have drastically different effects on animals of different sizes. In veterinary medicine, the normal drug dose for a rabbit cannot be multiplied by a simple weight factor to determine the proper dose for a horse. Because of metabolic rate differences, such a simple calculation would very likely lead to a serious overdose for the horse.

Hibernation and Other Avoidance Strategies

Considering the great metabolic cost of maintaining a high, constant body temperature, it is not surprising that some homeotherms avoid the struggle to maintain body temperature part of the time. Bats, for example, let their regulatory processes "slip" on a daily basis. They fly about actively and feed at night, then return in the day to roost in cool places such as caves, letting their daytime body temperatures fall to within a few degrees of the environmental temperature (figure 7.29). This normal, daily period of cold torpor seems to be a mechanism for energy conservation.

Some mammals let their body temperature and metabolism fall for much longer periods of time. These long periods of **hibernation** usually occur in winter, and the organism's temperature may drop to just a couple degrees above the environmental temperature. Hibernating mammals can live for long periods on the metabolic reserves they accumulated before entering hibernation.

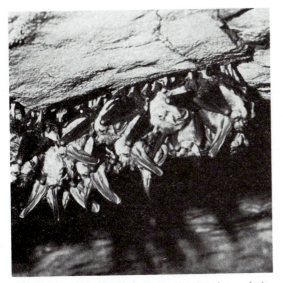

Figure 7.29 Roosting bats. During daytime hours, bats allow their body temperatures to fall near the temperature of their environment.

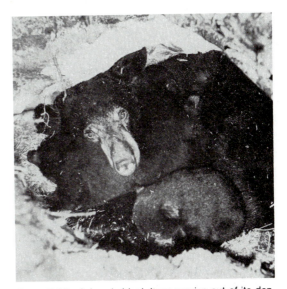

Figure 7.30 A female black bear peering out of its den in northern Minnesota in mid-April. The cub with her was born in January. While a bear's body temperature may drop a few degrees during its winter sleep, its is not hibernating.

During hibernation, an animal's heart rate may fall to only a few beats per minute, and its oxygen consumption is only a fraction of the normal rate (sometimes as little as 1 percent). However, a hibernating animal has not entirely stopped regulating its temperature. In fact, sharp decreases in environmental temperatures that could threaten to freeze body tissues cause quick metabolic increases. Sometimes, a large temperature drop will even arouse hibernating animals.

Some mammals, such as bears, accumulate huge metabolic reserves in heavy fat deposits and simply sleep the winter away. While a bear's body temperature may drop several degrees and it may be difficult to awaken, a bear is not in an entirely altered metabolic state, as hibernating animals are. Bears and other winter sleepers do not become torpid; woodchucks, ground squirrels, and other true hibernators do (figure 7.30).

In both winter sleepers and hibernators, metabolic reserves are conserved so that adequate quantities of reduced organic compounds are available for vital cellular energy conversions. In these organisms, as in all other living things, a continuing supply of the energy necessary for biological work is one of the fundamental requirements of life.

Summary

Reduced organic compounds such as glucose, originally produced by photosynthetic activity, are oxidized in a stepwise set of reactions. These cellular oxidation-reduction reactions result in ATP production.

The first series of reactions in the oxidative breakdown of glucose is the Embden-Meyerhof pathway. The E-M pathway reactions occur in the cytoplasmic matrix. Phosphorylations at the beginning of the Embden-Meyerhof pathway require two ATP molecules per molecule of glucose. After some molecular reorganizations, a six-carbon molecule is split into two three-carbon molecules. These molecules are processed further to produce pyruvate.

In the steps leading to pyruvate, an NAD reduction and two substrate-level phosphorylations occur for each three-carbon molecule. Two pyruvate molecules are produced from each molecule of glucose entering the reaction series. Pyruvate stands at an important branchpoint in carbohydrate metabolism.

Under anaerobic conditions, pyruvate may undergo either alcoholic or lactic fermentation. In the fermentation reactions, reduced NAD donates electrons to organic molecules, thus renewing the supply of oxidized NAD for Embden-Meyerhof reactions.

The net gain in ATP production gained through the E-M reactions and fermentation is only two ATP molecules per molecule of glucose oxidized. Although four ATPs are produced by substrate-level phosphorylations, two ATPs are spent in activating reactions.

Much more ATP can be produced if pyruvate is oxidized in aerobic respiration. Respiration releases much more of the potential energy of carbohydrates. In aerobic respiration, coenzymes, reduced during Krebs cycle reactions, donate electrons to an electron transport system, which passes them on to oxygen. During the transfers in the electron transport system, free energy changes make possible phosphorylations of ADP to produce ATP.

In addition to oxidation-reduction reactions, Krebs cycle decarboxylations produce carbon dioxide and shorten the carbon chains of carbohydrate molecules. Overall, six CO_2 molecules are produced for each glucose molecule entering the oxidative pathways.

As a result of all the phosphorylations occurring in conjunction with the electron transport system and the one substrate-level phosphorylation in the Krebs cycle, a total of fifteen ATP molecules are produced per molecule of pyruvate. Embden-Meyerhof reactions, under aerobic conditions, result in a net gain of six ATP molecules. Thus, overall, a total of thirty-six ATP molecules are gained from the aerobic oxidation of one glucose molecule.

Krebs cycle reactions occur in the mitochondrial matrix. Electron transport and oxidative phosphorylation are associated with the inner mitochondrial membrane. The mechanism of oxidative phosphorylation is not well understood. The chemical and conformational hypotheses propose that either a chemical compound or an altered molecule formed during electron transport returns to its original state, providing energy for ADP phosphorylation. The chemiosmotic hypothesis proposes that electron transport produces a proton gradient across the inner mitochondrial membrane and that the movement of protons back through the membrane results in ATP production.

Other organic compounds can be prepared for oxidation in the same reactions that oxidize carbohydrates. Glycerol and fatty acids from fats enter oxidative pathways as dihydroxyacetone phosphate and acetyl CoA, respectively. Fats are more highly reduced compounds than carbohydrates and thus yield more calories per gram oxidized.

Proteins are hydrolyzed into amino acids, which are then deaminated prior to entry into the oxidative pathways. Excess amino groups produced during amino acid deamination are removed from the body in nitrogenous waste carrier molecules such as urea.

Various molecules can be withdrawn from the metabolic pool for synthesis. Embden-Meyerhof and Krebs cycle intermediates are used as carbon skeletons in fat and amino acid syntheses.

Organismic metabolism is the sum total of all cellular processes and can be measured either in terms of heat produced or gas exchanged.

Most organisms are in a temperature equilibrium with their environments and thus have variable body temperatures. The metabolic rates of these heterothermic organisms are strongly affected by environmental temperature. Acclimatization is an adjustment of the body temperature range in which an organism can function. This is a critical process for many organisms experiencing drastic seasonal temperature changes. Shorter-term adjustments are often made through behavioral thermoregulation.

Homeothermic organisms conserve body heat and regulate their metabolic rates to produce stable body-core temperatures. Homeotherms can thus remain continuously active despite short-term environmental temperature variations. But small homeotherms that lose a considerable amount of heat because of unfavorable surface-to-volume ratios must maintain such high metabolic rates that they have to consume large quantities of food to supply themselves with fuel for cellular oxidations.

Many mammals temporarily avoid the problem of meeting the energy cost of a continuously maintained, high metabolic rate. Some let their body temperatures fall during daily periods of inactivity, and others let their temperatures fall to lower levels for long periods during hibernation. These avoidance techniques conserve body nutrient reserves during periods of adverse environmental conditions.

Questions

1. Why is CO_2 a product of alcoholic
fermentation and not of lactic fermentation?
2. Why do the Embden-Meyerhof pathway
reactions yield more ATP when they are
associated with aerobic respiration than
when associated with fermentation?
3. How many molecules of metabolic water are
produced as a result of coenzyme reductions
and electron transport activity during the
complete processing of one molecule of
pyruvate in aerobic respiration?
4. Explain how six molecules of CO_2 are
produced for each molecule of glucose
oxidized through the Embden-Meyerhof
pathway and the Krebs cycle.
5. Most fatty acids in animal fats have even
numbers of carbon atoms, for example, C_{16}
or C_{18}. How does this relate to the ways in
which fatty acids are oxidized and
synthesized?
6. Explain the metabolic pool concept.
7. Give at least one advantage and one
disadvantage of being homeothermic.
8. Which would most probably starve to death
first if totally deprived of food, a moose or a
mouse? Why?

Suggested Readings

Books

Baker, J. J. W., and Allen, G. E. 1981. *Matter,
energy, and life.* 4th ed. Reading, Mass.:
Addison-Wesley.
Becker, W. M. 1977. *Energy and the living cell: An
introduction to bioenergetics.* New York: J. B.
Lippincott.
Larner, J. 1971. *Intermediary metabolism and its
regulation.* Englewood Cliffs, N.J.: Prentice-
Hall.
Lehninger, A. L. 1975. *Biochemistry.* 2d ed. New
York: Worth Publishers.

Articles

Chappell, J. B., and Rees, S. C. 1972. Mitochondria.
Carolina Biology Readers no. 19. Burlington,
N.C.: Carolina Biological Supply.
Dickerson, E. March 1980. Cytochrome *c* and the
evolution of energy metabolism. *Scientific
American* (offprint 1464).
Heinrich, B. June 1981. The regulation of
temperature in a honeybee swarm. *Scientific
American* (offprint 1499).
Hinkle, P. C. and McCarty, R. E. March 1978. How
cells make ATP. *Scientific American* (offprint
1383).
Krebs, H. A. 1970. The history of the tricarboxylic
acid cycle. *Perspectives in Biology and
Medicine* 14:154.
Nicholls, P. 1975. Cytochromes and biological
oxidation. *Carolina Biology Readers* no. 66.
Burlington, N.C.: Carolina Biological Supply.

Part 2

Function
and Structure
of Living
Organisms

To maintain stable internal conditions, living things must selectively exchange materials with their environments. In single-celled organisms, the individual cell's plasma membrane is the boundary between organism and outside world. But in multicellular organisms, specific, localized body surfaces function as specialized exchange boundaries between organisms and environment. And materials are moved among various body areas, including these exchange boundaries, by specialized transport mechanisms.

Land plants obtain water and mineral nutrients from the soil via their roots and transport them through vascular systems to aboveground body parts, especially to the leaves, where photosynthesis occurs. Products of photosynthesis also are transported to other parts of plants.

Animals obtain reduced organic compounds and other materials from their environments, and they excrete waste materials into the environment. These exchanges occur at specialized exchange boundaries that are connected with other body areas by circulatory systems.

In chapters 8–13 the specialized exchanges, transport mechanisms, and internal processing of materials involved in maintenance of the stable internal environments that are essential to the lives of plants and animals are examined.

Roots, Water, and Mineral Nutrients

Chapter Concepts

1. A vascular plant's body is specialized to obtain and transport materials required for autotrophic existence.

2. Plant bodies consist mainly of molecules containing atoms of carbon and oxygen (from carbon dioxide) and hydrogen (from water).

3. Open stomata in leaves permit carbon dioxide diffusion into the leaves. At the same time there is transpiration (evaporative water loss from the leaves).

4. A stream of water continuously moves up through the vascular plant, replacing water lost by transpiration and carrying mineral ions from the roots to the rest of the plant.

5. Soil water moves freely through spaces in the outer portions of the root, but must cross a plasma membrane and move through cytoplasm to reach the vascular elements in the root's center.

6. Mineral ions may enter roots by diffusion under certain circumstances, but most ions are absorbed by active transport that requires the expenditure of ATP.

7. Despite natural replacement processes, such as the nitrogen cycle, fertilizers must be used to replace depleted mineral nutrients in soil used for agriculture.

Plants are **autotrophic** organisms: they can produce organic compounds from inorganic substances obtained from the environment. Animals are **heterotrophic** organisms: they must obtain ready-made organic compounds from their environments. This distinction between plants and animals is fundamental to an understanding of basic plant biology.

Plants must obtain from the environment all substances that are required for their autotrophic existence. This chapter on plant structure and function examines the ways in which plants obtain these necessary raw materials from their environment.

Organization of Vascular Plants

Vascular plants have distinctive regions of specialization. **Roots** absorb water and mineral ions from the water that surrounds soil particles. **Leaves** absorb light energy and are specialized for the gas exchange required for photosynthesis. **Stems** connect roots and leaves and support the leaves in an elevated position, where the leaves are well exposed to light.

Vascular tissues run through the roots, stems, and leaves, and transport materials from one part of the plant to another. Many cells in plant bodies are far from either roots or leaves (the exchange boundaries between the plant and the external environment). Thus, movement of water and minerals through the vascular tissues to cells that need these materials is vital to the life of the plant.

The two types of vascular tissue are xylem and phloem. Water and minerals from the soil are transported through **xylem** in only one direction, from the roots upward through the plant body. **Phloem**, on the other hand, can move dissolved material from place to place in the plant. For example, sugar produced in photosynthesizing leaf cells can be transported down through the phloem to the roots where it is stored, or it can be transported back up from the roots through the plant. (See figure 8.1.)

The conducting elements of xylem are hollow structures made up of cell walls that remain after the degeneration of cytoplasm and membranes during development. Phloem transporting tissue consists of living, cytoplasm-containing cells.

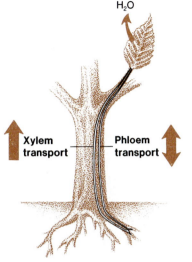

H_2O

Xylem transport   Phloem transport

Water and mineral ions in the soil

Figure 8.1 Generalized sketch of a vascular plant. Various specialized regions of plants are connected by the vascular tissues xylem and phloem.

The Construction of Plants— Carbon, Hydrogen, and Oxygen

Plants are chemically versatile; they use the products of photosynthesis—carbohydrates—to synthesize the entire array of organic compounds required for their structure and function. Chemical analysis of a typical vascular plant reveals that the greatest percentage of the plant's dry weight (total weight minus water) consists of carbohydrates, such as sugars and other relatively small molecules, as well as starch, and large quantities of cellulose. Carbon, hydrogen, and oxygen also are found in all of the other organic compounds in plants. The carbon and oxygen in all of these organic compounds are derived from atmospheric carbon dioxide (CO_2), while the hydrogen comes from water taken up from the soil. The only other nutrients that plants need are inorganic ions obtained from their soil environment.

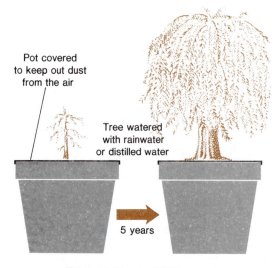

Pot covered
to keep out dust
from the air

Tree watered
with rainwater
or distilled water

5 years

Tree weight gain: 74 kilograms

Soil weight loss: 0.057 kilograms

Figure 8.2 Van Helmont's experiment. Van Helmont
demonstrated that weight gained by a growing plant
was not equaled by weight loss in the soil in which it
was grown. Although he concluded mistakenly that the
weight gained by the plant came entirely from the water
added during its growth, he dispelled the long-
established notion that plants derived all their substance
from the soil.

The discovery that a large proportion of
every plant, even the most massive tree, comes
from a minor component of the surrounding air
(carbon dioxide makes up only 0.03 percent of
the atmosphere) was a long time coming. Biologists had assumed that plants were constructed
of materials taken up from the soil. This seemed
a logical assumption, but had they weighed the
total amount of plant material produced each
year on a given area of land, they might have
wondered how year after year the soil could lose
an amount of weight equal to the plant material
grown.

In 1648 J. B. van Helmont, a Flemish physician, reported the results of a simple experiment
that challenged the idea that plants were constructed entirely of materials taken up from the
soil. Van Helmont had planted a willow in a covered pot to which he added only rainwater or
distilled water when rainwater was not available.
After five years, van Helmont found that the
plant had increased in weight by over 74 kg, while

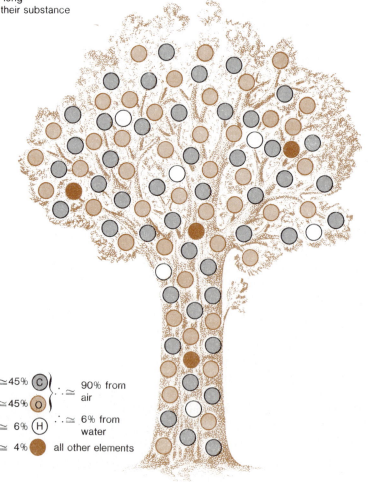

Figure 8.3 The proportions of a
plant body's dry weight derived from
the CO_2 and water incorporated
through photosynthesis.

$\simeq$45% C $\Big\}$ $\therefore \simeq$ 90% from air

$\simeq$45% O

$\simeq$ 6% H $\therefore \simeq$ 6% from water

$\simeq$ 4% all other elements

the soil was only about 0.057 kg lighter than it had been at the beginning of the experiment (figure 8.2). He concluded, mistakenly, that the plant's weight gain was almost entirely due to the incorporation of water, since he did not know anything about the role of carbon dioxide in photosynthesis.

Van Helmont's mistaken conclusion does not reduce the importance of his work because he showed clearly that plant growth was not entirely achieved by absorbing and incorporating soil material. Over the years other researchers clarified the role of carbon dioxide in photosynthesis. Today, the idea that even huge, heavy trees are produced mostly from water and a gas, carbon dioxide, is fully accepted (figure 8.3).

It is important to remember, however, that the soil does provide certain nutrients that are vital to plant life, even though they are required in only small quantities.

Obtaining Carbon Dioxide

Because leaves usually are thin organs, the distance that carbon dioxide must diffuse between the outside air and any individual leaf cell is not great. Leaves' high surface-to-volume ratios, however, also can result in serious water loss from terrestrial plants. Leaf surfaces are protected against water loss by a covering, the **cuticle,** which is heavily impregnated with a paraffin-containing substance called **cutin.** Cutin is quite impermeable to water, thus reducing general leaf-surface water loss, but is also poorly permeable to CO_2. Therefore, CO_2 must enter leaves through special regulated openings, known as **stomata** (see chapter 6 for more details). The stomata provide CO_2 diffusion routes that connect the outside air with air spaces inside leaves (figure 8.4). Photosynthetically active cells adjoin the air spaces inside leaves and absorb CO_2 from them. Before CO_2 is absorbed by leaf cells,

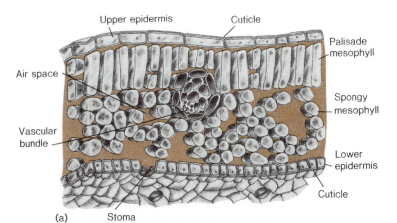

(a)

Figure 8.4 (a) Leaf structure. The cuticle contains cutin, a paraffinlike substance that is impermeable to water and poorly permeable to CO_2. Carbon dioxide enters the leaves mainly through stomata. (b) A stoma. This scanning electron micrograph shows the two guard cells that control the size of the stomatal opening. In transpiration, water evaporates from leaves through stomata. (Magnification $\times$ 2,660)

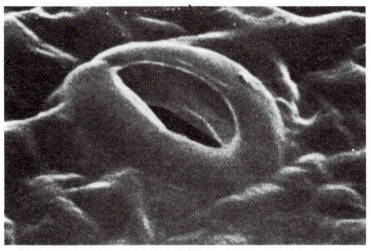

(b)

however, it must go into solution in the film of water that covers the surface of each cell. Therefore, cell surface water films are necessary if CO_2 is to be available to photosynthesizing cells.

Water evaporates from the water films that cover the surfaces of internal leaf cells. This results in a water-saturated or nearly saturated environment in the adjoining air spaces. In all but the most humid environments, therefore, there is a substantial difference in the water vapor content of the leaf air spaces and the outside air, and water diffuses out of the leaves whenever stomata are open. This evaporation of water from leaves to the outside air is called **transpiration.**

Water Loss and Replacement

Although plants continuously lose water by transpiration when the stomata are open, the water films on the surfaces of internal leaf cells are not depleted. A **transpiration stream** of water continuously moves up through the xylem from the roots. The transpiration stream replaces water lost by transpiration or consumed in photosynthesis, and also supplies the water that is retained in plant tissue (figure 8.5).

Most of the water absorbed by the roots is used to replace the water being lost by transpiration; only a very small percentage of the absorbed water is used in photosynthesis. For example, it is estimated that corn (*Zea mays*) plants growing in a moderately dry climate will transpire 98 percent of all the water that their roots absorb. Much of the remaining 2 percent of the absorbed water is retained in cell and tissue fluids. As little as 0.2 percent actually is consumed in photosynthesis.

The total amount of water lost by a plant over a long period of time is surprisingly large. A single *Zea mays* plant loses somewhere between 135 and 200 l of water through transpiration during a growing season. If this water loss is multiplied by the number of corn plants in a heavily planted cornfield, it is easy to understand why agricultural demands on soil water are so great. Another important consideration in agriculture is the amount of water required by different crops to produce a given quantity of food.

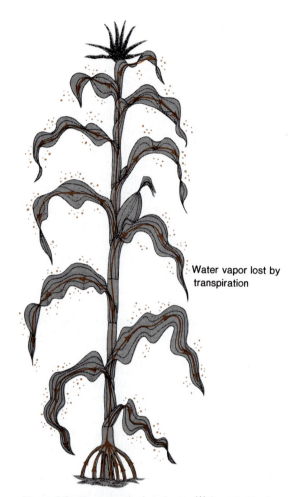

Water vapor lost by transpiration

Figure 8.5 The transpiration stream. Water moves up from the roots to replace water lost to the atmosphere by transpiration. This flow continues as long as stomata are open. A single corn *(Zea mays)* plant such as this one loses between 135 and 200 l of water by transpiration during a growing season.

These amounts vary considerably for different crops: millet requires about 225 kg of water for every kilogram of food produced, wheat requires about 500 kg of water per kilogram of food, and potatoes require about 800 kg of water per kilogram of food. Most of this water goes into the atmosphere by transpiration.

This considerable water loss is not such a negative factor for plants as it first might seem. The transpiration stream of water rising through a plant plays an important additional role because it carries minerals absorbed from the soil upward through the plant. The urgent functional necessity of the seemingly wasteful water loss by transpiration is explored further in chapter 9.

Soil and Soil Water

A soil's water-holding capacity and amount of aeration help to determine the soil's suitability for plant growth.

Soil water usually forms a thin film on the surface of soil particles and fills some of the smallest spaces among soil particles. Water is held in these spaces by **capillarity,** a tendency of water to enter and adhere to the walls of small spaces. Following a rain, all of the spaces among soil particles are filled with water, but most of this rainwater percolates (drains) down through the soil, leaving air in many of the spaces among soil particles.

The water-holding capacity of a soil depends on the kinds of particles in the soil. In sandy soil with large particles (20 to 2,000 micrometers in diameter), less total particle surface area is available than around the more numerous smaller particles in soils that contain more silt (particles 2 to 20 μm in diameter) or clay (particles less than 2 μm in diameter). Also, the larger particles of sandy soils are more loosely packed and have fewer small, water-holding spaces (capillary spaces) between particles. A dried-out, sandy soil provides only a barren and hostile environment for plant growth because of its low water-holding capacity.

Water-holding capacity is not the only factor to be considered in judging the physical suitability of a soil for plant growth. Root cells must have oxygen for respiration, and in clay, for example, the very tight packing of the small particles can reduce air spaces and gas diffusion to the point that aeration is not adequate to provide the oxygen needed for root respiration.

For optimal plant growth, soil should have a mixture of particles of different sizes so that there are enough very small capillary spaces among particles to hold adequate quantities of water, as well as enough larger air spaces to provide adequate oxygen supply to roots.

Roots and Root Systems

One of the most striking features of root function is that roots are not static; they grow continuously. Growing root tips constantly move through the soil and squeeze their way into new capillary spaces among soil particles. They continually push into new areas of the soil, where moisture may be more plentiful.

Root growth begins in all young plants with the growth of a single **primary root,** but subsequently, roots grow differently in various plants. In some plants, the primary root grows straight down and remains the dominant root of the plant, with much smaller **secondary roots** growing out from it. This arrangement is called a **taproot system** (figure 8.6a). In other plants, a number of slender roots develop, and no single root dominates. These slender roots and their lateral branches make up a **fibrous root system** (figure 8.6b). Yet another type of root system develops when roots grow out of the stem or other plant parts (leaves, for example) later in development. Such roots are called **adventitious roots** (figure 8.6c).

Root growth carries the roots far out to the sides of the plant in some cases. For example, corn roots spread as much as a meter in all directions from the stem. A dense tangle of intertwined root networks exists under the neatly rowed orderliness of a cornfield. The roots of other plants grow far down into the soil. Alfalfa (*Medicago sativa*) roots reach a depth of 6 m, and there are reports that the roots of some desert plants penetrate to several times that depth.

Because of their extensive growth, roots have a very large surface area exposed to the soil, but this is only the beginning. The total root surface area is increased dramatically by outgrowths of surface cells. These outgrowths, called **root hairs,** are found just behind the growing tips of the roots.

Some astonishing estimates have been made of the actual length and total surface area of root systems in some plants. In his study of rye roots in the 1930s, Dittmer estimated that a single four-month-old rye plant had roots totaling more than 600 km in length. There were literally millions of branched and rebranched roots. Further, he estimated that the plant had about 14×10^9 root hairs and that the total surface area of roots and root hairs was almost 640 m². Root hairs add tremendously to the total absorbing surface area of roots and to absorption functions of roots.

Root Tips and Root Hairs

Near the tip of each growing root is an area of active cell division, the **meristematic region,** that continuously produces new cells. Throughout a

(a)

(b)

(c)

Figure 8.6 Root systems. (a) The taproot system of a dandelion. (b) The fibrous root system of a grass. (c) Prop roots of a screw pine.

root's growing season, some of these newly produced cells are added to a protective cover over the tip of the root, called the **root cap.** Root cap cells are ground off by the wear and tear of being pushed through spaces among the rough soil particles, and they must be replaced constantly. Other cells produced by the cell division in the meristematic region are added to the length of the root itself.

One externally visible feature of growing root tips is the presence of root hairs a few millimeters back from the tip. New root hairs form as cell division and subsequent elongation of the newly formed cells lengthens the root and pushes its tip further through the soil. The area where root hairs are found, the **region of maturation** or the **root-hair zone,** is only a short section (1 to 6 cm long) of the length of the root, because even

as new root hairs are being added near the tip, older ones further from the tip are withering and falling off (figure 8.7). *Individual root hairs actually function for only a few days.* The root-hair cycle is continuous: new root hairs are produced, they mature and function briefly, and then they wither and fall off. New root hairs develop on the portion of the growing root that is pushing into new areas of the soil. As a result, they encounter new sources of soil water in capillary spaces among soil particles.

Internal Root Structure

At any given time, all stages of the root-hair cycle can be seen on the surface of the root: developing root hairs; mature, functional root hairs; and old, withering root hairs all are evident. Similarly,

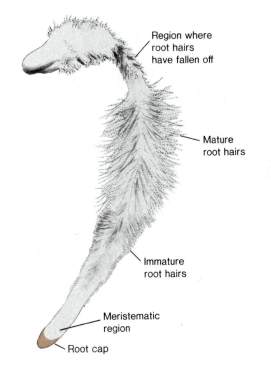

Region where root hairs have fallen off

Mature root hairs

Immature root hairs

Meristematic region

Root cap

Figure 8.7 Root hairs on the primary root of a developing radish seedling.

there is evidence of progressive internal maturational change (**differentiation**) of new cells being produced by cell divisions in the root-tip meristem (**apical meristematic region**). These cells differentiate into the functional internal tissues of the root.

A section cut lengthwise through a root tip illustrates the differences among cells in regions that are various distances from the meristematic zone (figure 8.8). Cells in and near the meristematic zone are small and rather homogeneous in appearance, while those further from the meristematic region are longer. These longer cells are in a zone of cell growth that is called the **region of elongation.** Although the meristematic region and the region of elongation make up only a tiny part (usually 5 or 6 mm together) of a root's length, cell division in the meristematic region and cell lengthening in the region of elongation account entirely for the lengthening of the growing root tip that pushes the root through the soil.

Further away from the meristematic region, structural specializations of the cells make up various functional tissues. This progressive differentiation, illustrated in figure 8.8, produces the structural organization that permits the root to function as an absorbing and transporting organ. Figure 8.9 is a cross section of a root cut through the region of maturation to show this structural organization.

Epidermis

The outer layer of the root, the **epidermis,** consists of only a single layer of cells. The majority of the epidermal cells are thin-walled and rectangular, but in the region of maturation, many epidermal cells are specialized as **root hair cells.** They have extensions, the root hairs, which project out as far as 5 to 8 mm among the soil particles.

The Cortex

Inside the epidermis are large, thin-walled, polyhedral cells (called **parenchyma cells**) that make up the **cortex** of the root. These irregularly shaped cells are loosely packed so that a network of intercellular air spaces exists among the cells. These spaces, and especially the free surface areas of the cortical cells, play an important role in water and mineral uptake. Though the parenchyma cells do not pack together neatly, they do have areas of definite cell-to-cell contact, and

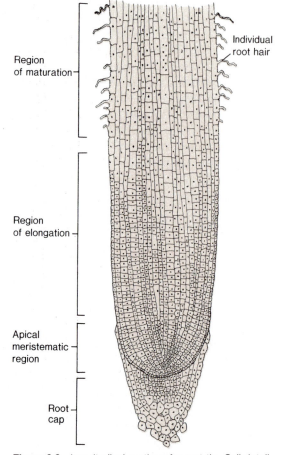

Region of maturation

Individual root hair

Region of elongation

Apical meristematic region

Root cap

Figure 8.8 Longitudinal section of a root tip. Cell detail shows progressive cell specialization in cells further from apical meristem. Root hairs develop as extensions of epidermal cells.

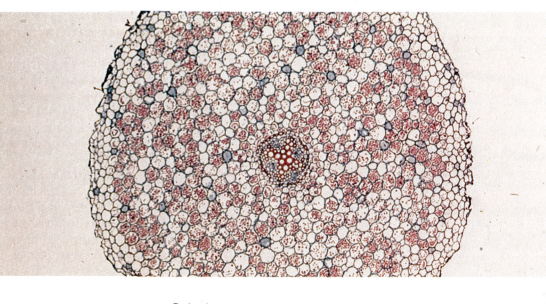

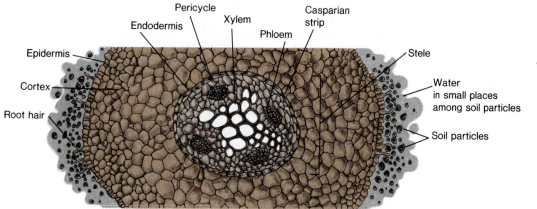

Figure 8.9 Cross sections of a root cut through the region of maturation.

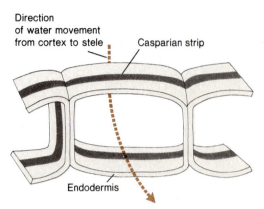

Direction
of water movement
from cortex to stele

Casparian strip

Endodermis

Figure 8.10 The Casparian strip. Fused endodermal cell walls impregnated with fatty material make an impermeable boundary between free spaces in the cortex and in the stele. Water and solutes must pass through the cytoplasm of endodermal cells to enter the stele.

these contact areas also are functionally very important in the transport of materials absorbed from the soil. Cortical cells also function in food storage, since they can store large quantities of starch.

The innermost layer of the cortex, the **endodermis,** forms a distinct boundary between the loosely organized parenchyma cells of the cortex and the **stele,** the central core of the root. Endodermal cells fit snugly together, and their walls are fused with one another. Since their top, bottom, and side walls are impregnated with waxy material, water and solutes cannot pass through spaces between the endodermal cells (figure 8.10). This cell wall barrier to water and solute movement is known as the **Casparian strip.** Because of the organization of the Casparian strip, water or minerals moving from the cortex to the stele must cross a cell membrane and pass through the cytoplasm of the endodermal cells. Thus, water and, especially, mineral movements are subject to controls that cells exert over movement of materials through membranes and cytoplasm. There are no uncontrolled, easy-access routes from the cortex to the stele.

The Stele

Just inside the endodermis is a layer of cells known as the **pericycle.** Pericyclic cells can become meristematic and initiate the development of secondary or branch roots (figure 8.11).

The central portion of the stele is occupied mainly by the two vascular tissues, xylem and

phloem (figure 8.9). (The cross section shown in figure 8.9 is only a thin slice of the root, which does not portray the true nature of the xylem and phloem elements seen there since they are parts of continuous tubular transport systems that extend all the way up through the root and stem into the leaves.) Xylem transports water and minerals upwards from the roots, while phloem moves dissolved materials from place to place in the plant.

A single layer of tissue, called the **cambium,** lies between the xylem and phloem. Cambial tissue is especially important in perennial plants because it is meristematic. Cell divisions in the cambium add more layers of xylem and phloem and thereby increase the thickness of the root.

Water Absorption

Biologists believe that the water deficiency that results when plants are unable to replace water lost through transpiration is the most common cause of plant death. So water movement into the roots is vitally important both for maintenance of transport up through the plant and for meeting the demands of transpirational water loss. Water must move from the spaces around soil particles into the root, up through the xylem, and eventually out into the air around the leaves.

Water Movement in the Root

There are two physically segregated areas in the cortex: the free space (figure 8.12b) and the cytoplasmic network (figure 8.12c). These two areas are separated from one another by cell membranes.

The **free space** is the space among the loosely packed cells of the cortex. Soil water readily enters this free space and forms films over the surfaces of cortical cells. Water also moves freely into microspaces among the cellulose fibrils that make up the walls of the cortical cells. All of this free space is open to water movement. There is no barrier between water in spaces among soil particles outside the root and water in this free space. This free movement of water ends, however, at the Casparian strip, which remains an impermeable barrier between the free space of the cortex and similar free space within the stele. Water must enter the stele via a cytoplasmic network.

Figure 8.11 Branch roots originate from areas of the pericycle that become meristematic. Because branch roots originate deep inside the root, they are connected to vascular tissue of the stele from the beginning. They push physically and erode by enzyme action to force their way through other tissue to the surface and out into the soil.

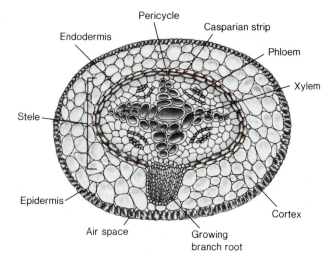

Figure 8.12 Relationships of free space and cytoplasmic network in a root. (a) All elements present in a small part of a root cross section. (b) The free space in the cortex and in the stele. Note that the Casparian strip segregates the two free spaces from one another. (c) The continuous cytoplasmic network of the root. Plasmodesmata connect cytoplasm of cells throughout both cortex and stele. Endodermis with its Casparian strip does not interrupt cytoplasmic network.

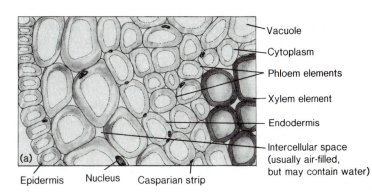

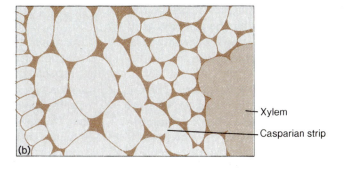

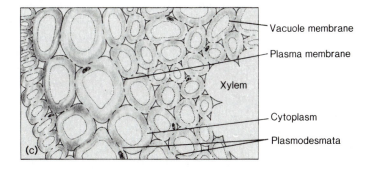

A network of cytoplasmic interconnections called **plasmodesmata** allow individual cortical cells to communicate with one another. Small holes in cell walls permit cell contact with actual cytoplasmic fusion, and a continuous **cytoplasmic network** exists in which the cells of the cortex are connected with one another. This cytoplasmic network functions as a single unit.

Some water moves through the free space of the cortex and enters the cytoplasmic network through the membranes of cells within the cortex. Once it is in the cytoplasmic network, the water moves through the cytoplasm of endodermal cells and into the stele (figure 8.13a).

Alternatively, water can also reach the stele by entering the cytoplasm of surface cells, especially root hair cells, by osmosis. These cells are in intimate contact with soil particles. Once water has entered these surface cells, it can move through the cytoplasmic network of the cortex and the cytoplasm of endodermal cells and into the stele (figure 8.13b).

To reiterate, the boundary that water crosses to enter the plant is neither the root epidermis nor the cell walls of cortical cells because water flows freely through them. Cell membranes form the actual boundary between the outside and the inside of the plant. Soil water, then, must cross a cell membrane to enter the cytoplasmic network of the cortex if it is to reach the stele and eventually move up through the plant. Water can enter either through membranes of cells at the root surface or of cells within the cortex; that is, it enters the cytoplasmic network directly from the water films on soil particles or from the free space of the cortex.

Water tends to flow inward from the soil because water is continually being drawn from the root and up through the plant by transpiration. Another force involved in the movement of water is the osmotic pressure difference between soil water and the cytoplasm of root cells. The cytoplasm normally contains more dissolved and suspended material than soil water, and water tends to enter the root cells osmotically.

The total mineral content of soil influences water entry into plants, and the osmotic pressure of salty soil water actually can prevent water from entering roots. The critical salinity level is different for different plants, but salinity higher than the critical level can cause roots to lose water to salty soil. The ancient Romans applied this principle when they salted the fields of Carthage so that the fields could not be used for agriculture.

A modern practical problem with soil salinity is the soil salting that can occur in irrigated fields. Certain combinations of soil properties, mineral content of the irrigation water, and evaporation and runoff conditions can result in gradual buildup of salt deposits in the soil. These salt deposits threaten future crop production either because of osmotic changes that interfere with water entry into roots or toxic oversupply of some particular mineral element or elements.

Soil chemistry, physical structure of the soil, and mineral content of available water can limit the possibility of bringing additional arid or semi-arid land under irrigation, and some land cannot be irrigated for reasons other than the lack of an economically feasible water supply. Increasing usable agricultural land by irrigation is not simply a matter of getting just any water supply to just any piece of dry land.

Mineral Absorption

Mineral ions available for absorption by roots are in solution in the water in the capillary spaces around soil particles. To enter plants, ions must pass through the plasma membranes of cells in the epidermis or cortex. Ions are not simply swept along in the streams of water moving into and through plants. Sometimes when particular ions are highly concentrated in soil water, as they can be after fertilizer application, the ions enter the cytoplasmic network by simple diffusion. Simple diffusion, however, is less important than two types of carrier-mediated processes: facilitated diffusion and active transport (chapter 4). In these carrier-mediated processes, specific carrier molecules in root cell membranes are involved in the passage of ions through the membranes.

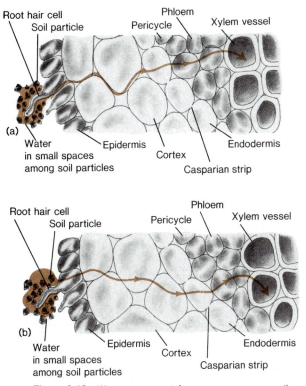

Figure 8.13 Water movement from spaces among soil particles to xylem. (a) One route is through the free space of the cortex with entry into internal cortex cells. Water movement through free space stops at the endodermis. (b) Another route is movement into surface cells and through the cytoplasmic network of the cortex.

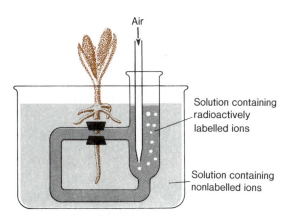

Figure 8.14 Experiment designed to determine specific level of root at which most active mineral absorption takes place. In different experiments, different parts of roots of other plants are exposed to the radioactively labeled ions. Such experiments demonstrate that most mineral ions enter in the first few centimeters above the root tip, especially in the region of maturation where root hairs are the most numerous.

Facilitated diffusion involves movement of material down a concentration gradient. An ion must be more concentrated in soil water than in root cell cytoplasm if it is to enter by facilitated diffusion.

The most important means by which ions are moved from soil water or the cortex free space into the cytoplasmic network of root cells is by active transport across root cell membranes. Active transport requires that root cells possess specific carrier molecules in their plasma membranes. The process is selective: some ions are actively transported into the root's cytoplasmic network while others are not. Active transport also requires energy (ATP) input. This can be demonstrated experimentally. If a respiratory inhibitor such as cyanide is applied, thereby cutting off the supply of ATP, active transport is slowed or stopped. Considerable amounts of energy are expended if active transport works against a concentration gradient, and it often has to do so in ion absorption by roots because the ions being absorbed frequently are less concentrated in soil water then in root cell cytoplasm.

Experiments using radioactively labelled ions have helped to determine the levels of the root where the most active absorption takes place. Labelled ions are applied selectively to restricted areas of a root while nonlabelled ions are applied to the rest of the root (figure 8.14). These experiments verify that the area of most extensive entry is in the first few centimeters of the root, especially in the region of maturation where the root hairs are most numerous. Root hairs, however, are not absolutely required for ion uptake. In fact, plants being grown in certain types of liquid culture commonly do not develop root hairs and yet take up ions. Also water and ions enter roots even during the winter when new roots with active root hairs are not being produced.

The older tissue above the area where the most active absorption takes place has **suberized** (waxy) cell walls, and, therefore, is less permeable; there is much less mineral absorption in these suberized areas.

Once ions have entered the cytoplasmic network, they move through the cytoplasmic network of the cortex, through the cytoplasm of endodermal cells, and into the cytoplasm of cells in the stele. Ion movement through the cytoplasm, like water movement through the cytoplasm, is not blocked by the Casparian strip. Because plasmodesmata connect all of the cells, the cytoplasm is continuous and the Casparian strip can be bypassed. Ions diffusing through the cytoplasmic network travel down a concentration gradient. The gradient apparently is maintained because there is active secretion of ions from stele cells into the xylem. Also, ion diffusion toward the stele probably is aided by **cyclosis** (cytoplasmic streaming within cells) in the cytoplasmic network because experimental inhibition of cyclosis slows the movement of ions across the root toward the stele.

Delivery of Ions to the Stele

The release of ions into the xylem vessels is not thoroughly understood, but an active, energy-requiring process (a process that is essentially the opposite of the process by which the ions first entered the cytoplasmic network) is probably necessary. Ions may be released directly into xylem vessels, or they may be actively transported out of cells into the free space of the stele, where the transpiration stream simply carries them along into the xylem. This release maintains the concentration gradient inside the cytoplasm, and ions continue moving by diffusion through the cytoplasmic network into the neighborhood of the xylem.

Some ions move inward across the root and are transported up through the xylem directly, but others are incorporated into organic compounds before they are transported through the xylem. For example, nitrogen, although commonly absorbed from the soil as nitrate ions, commonly is transported up through the xylem only after it has been incorporated into a molecule of glutamic acid, an amino acid. The conversion of nitrate nitrogen into the amino acid transport form is a several-step process. Nitrate (NO_3^-) is reduced to ammonia in a reaction catalyzed by the enzyme nitrate reductase. Once ammonia (NH_3) is produced, it is used by the glutamate dehydrogenase enzyme system (chapter 7) in the synthesis of glutamic acid.

Symbiosis and Mineral Nutrition

Nitrogen in a usable form is one of plants' most important nutritional requirements. Because nitrogen is a plentiful substance (about 78 percent of the air is nitrogen) this requirement would not seem to present any particular problem for plants. But plants cannot absorb molecular nitrogen (N_2) from the atmosphere. They require nitrogen in the form of nitrate (NO_3^-) or ammonium (NH_4^+) ions. Plants, therefore, are dependent on the activity of soil microorganisms that incorporate atmospheric nitrogen by the process of **nitrogen fixation.** After nitrogen fixation has occurred, usable nitrogen compounds (NO_3^- and NH_4^+) are available for both the microorganisms and the plants.

Many of these nitrogen-fixing microorganisms are free-living in the soil, but interesting symbiotic relationships (functional interdependencies) between plants and nitrogen-fixing microorganisms are important in the mineral nutrition of some vascular plants. For example, plant tissue and nitrogen-fixing bacteria are found together in swellings called **nodules** that develop on the roots of legumes such as beans and peas (figure 8.15a). The nitrogen-fixing bacteria convert molecular nitrogen (N_2) into forms required by the plant. At the same time, the plant provides the microorganisms with a supply of reduced organic compounds (produced photosynthetically) that meets the energy requirements of the microorganisms.

Another nutritionally important symbiotic relationship is the association of nonpathogenic fungi with roots in **mycorrhizae** (singular: **mycorrhiza**; meaning literally "fungus-root"). The fungi envelop large parts of root systems, including virtually all of the absorbing surfaces (figure 8.15b). All materials absorbed into the roots must pass first through these fungal sheaths. The fungus receives organic nutrients and growth factors from the host plant. A mycorrhizal infection is beneficial to a plant in situations where nutrients are deficient or the plant faces strong competition from other organisms, because the fungus increases the surface area available for mineral and water uptake. The mycorrhizal relationship seems particularly important for certain trees, and some trees do not grow or grow very poorly if the fungus is not present. In forestry, some trees are deliberately infected with mycorrhizal fungi when they are introduced into new habitats (figure 8.15c).

(a)

Figure 8.15 Symbiosis and mineral nutrition. (a) Nodules on soybean roots. Nodules are made up of plant tissue and nitrogen-fixing bacteria. These nodules are red because cells in the nodule produce a red pigment, leghemoglobin, during the time that active nitrogen fixation is occurring. Leghemoglobin resembles the hemoglobin found in vertebrate red blood cells. (b) Part of a beech *(Fagus sylvatica)* root showing that much of the root system, including the main absorption areas, is entirely enclosed in mycorrhizal sheath. All smaller branches are covered with fungal tissue. (c) Root systems of loblolly pine seedlings. Seedlings on the left are untreated. Seedlings on the right were inoculated with a soil fungus and show extensive mycorrhizal development.

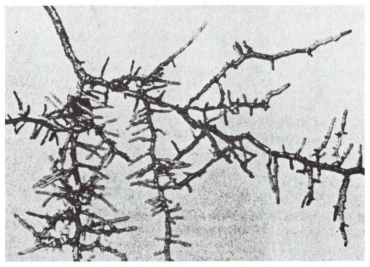

(b)

(c)

Foliar Fertilization

Nearly all plants are able, to some extent, to absorb mineral nutrients applied to leaf surfaces. Ions may enter through stomata, directly through the cuticle, or both. Foliar fertilization (leaf fertilization) has been used to supplement soil fertilization both in agriculture and forestry. Of course, in agriculture, a number of other substances, such as herbicides, growth regulators, and a variety of pesticides, also are applied as foliar sprays and absorbed.

Mineral Requirements of Plants

Carbon, hydrogen, and oxygen incorporated into plants by photosynthesis account for a very large part of a plant's weight. But other elements obtained from the soil, though present in much smaller proportions, also are vital for plant life.

One method of determining the mineral nutrient requirements of plants might be to burn the plant and chemically analyze the ash. During the ashing of the plant, the carbohydrates are lost as carbon dioxide, and most of the nitrogen is given off as ammonia or other gases. Analysis of the ash then permits quite accurate determination of the plant's mineral contents.

Unfortunately, the presence of a particular element in a plant's ashes does not necessarily establish that it plays a role in the plant's normal physiology or growth. Some elements not usually found in plants are incorporated because they substitute readily for more common elements in reactions occurring in plants. For example, the radioactive element ^{90}strontium, which is found in the fallout produced after atmospheric nuclear testing, substitutes for calcium in many reactions. Accumulation of ^{90}strontium in plants can become an important human health problem if milk cows eat the plants. The cows then produce milk that contains ^{90}strontium.

Other elements are taken up by plants even though they do not seem to participate in reactions in plants. Even an element such as plutonium, which does not occur naturally and is a product of nuclear reactors, can be taken up if it becomes available to plants.

Because of these uptake patterns, direct determination of the elements present in a plant is not a good index of the necessity of those elements for the plant's normal functioning. A better method of determining the mineral nutrient requirements of plants is **hydroponics**—growing plants in water cultures containing the same ions that occur in soil. Plants' responses to addition or deletion of various elements can be used to determine mineral requirements of the plant (figure 8.16).

Macronutrients

During the nineteenth century, hydroponics showed that seven elements are essential nutrients for plants: nitrogen, phosphorus, potassium, calcium, magnesium, sulfur, and iron. The first six of these elements are required in the greatest quantities by most vascular plants and are called **macronutrients**. Iron is required in far smaller quantities than any of the other six, and thus usually is placed in a group of elements called **micronutrients**, which need to be supplied to plants in much smaller amounts.

Micronutrients

Identification of the micronutrients has been a harder job, and only six more elements (chlorine, manganese, boron, zinc, copper, and molybdenum) have been added to the list of a plant's required nutrients since the 1920s (figure 8.17). In addition to these micronutrients are several other elements, for example, sodium and cobalt, that seem to be required by some plants and not by others.

There are two reasons for the uncertainty that existed for some time about the identification of the micronutrients and for the current indecision about other elements that may or may not be essential for some or many plants. First, the amounts needed of some elements are so small that a clear-cut experimental determination is very difficult. Second, small, contaminating quantities of micronutrients probably were present in the salts used to provide macronutrients in most of the early water culture experiments, and these minute quantities of contaminants may have been adequate to meet the plants' requirements. Determination of a micronutrient requirement requires very careful purification of salts being used to prepare solutions,

Figure 8.16 A hydroponics (liquid culture) experiment. Buckwheat plant on the left was grown in a nutrient solution without potassium. Culture on the right had a complete nutrient solution.

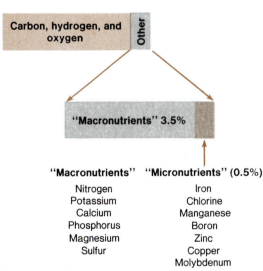

Figure 8.17 The proportional weights of various elements in a plant body. Some plants do accumulate significant amounts of certain other elements, such as silicon (as much as 1 percent of the dry weight of some grasses) and aluminum. The role of these other elements is not clear.

as well as fastidious washing of all culture vessels and the plant materials used to initiate the cultures. Even when these precautions have been taken, minute quantities of the elements in question still might be present in the seeds or cuttings used to begin the experiment. Such nutrients could be mobilized and transported to other parts of the plant and be adequate to prevent deficiency symptoms. Clearly, the study of plant micronutrients is difficult and demanding work.

Mineral Deficiencies in Plants

Mineral deficiencies cause a variety of symptoms in plants. While a thorough discussion of mineral deficiency effects on plants is beyond the scope of this book, the physiological roles of macronutrients (table 8.1), some problems of macronutrient deficiency that have direct bearings on agricultural production, and a few selected examples of the effects of some micronutrient deficiencies are discussed here.

Because of the importance of nitrogen as a constituent of proteins, nucleic acids, chlorophyll, and several coenzymes, it is not surprising that nitrogen deficiency causes a variety of problems. Nitrogen-deficient plants are generally stunted and weak, and they have small, distorted leaves that are subject to yellowing (chlorosis).

Phosphorus is important because of its occurrence in nucleic acids and in the phospholipids of cell membranes. The importance of ATP and ADP also is obvious, and a number of metabolic processes require that sugars and other substances be phosphorylated. Phosphorus deficiency is characterized by brown, dying patches of leaf tissue, or a darker-than-normal, blue green coloration of the leaves.

Potassium's role in activation of a number of key enzymes makes it absolutely vital to normal plant functioning. Potassium also accumulates in large quantities in the vacuoles of plant cells, where it seems to be important for osmotic regulation. In addition, potassium may have a role in the opening and closing of stomata (chapter 6). Potassium-deficient plants show generally reduced growth. Their leaves develop yellow spots, and then dying patches appear around the edges of leaves.

Table 8.1

Mineral Nutrients Required by Plants, the Ions That Plants Absorb, and the Established or Probable Roles of Thirteen Nutrients. Other Elements May Be Required in Very Low Concentrations.

Element	Form Absorbed from Soil	Some Functions in Plants
Macronutrients		
Nitrogen	NO_3^- or NH_4^+	Components of amino acids, proteins, coenzymes, chlorophyll, nucleotides, nucleic acids
Potassium	K^+	Activator of many enzymes; cellular osmotic regulation; regulation of stomata; possibly others
Calcium	Ca^{2+}	Calcium salts in cell walls, also in cell membranes; possible role in pH regulation in cell sap; enzyme activation
Phosphorus	$H_2PO_4^-$, HPO_4^{2-}	Component of phospholipids of membranes; nucleotides, nucleic acids; phosphorylated sugars; coenzymes; ADP and ATP
Magnesium	Mg^{2+}	Component of chlorophyll; enzyme activator; protein synthesis
Sulfur	SO_4^{2-}	Component of some amino acids, proteins, coenzymes; synthesis of hormones, cytochromes
Micronutrients		
Iron	Fe^{2+}, Fe^{3+}	Component of cytochromes and ferredoxin; cofactor of peroxidase and some other enzymes; chlorophyll synthesis; nitrogen metabolism
Chlorine	Cl^-	Uncertain, but may be involved in photophosphorylation
Manganese	Mn^{2+}	Enzyme activator; electron transfers
Boron	BO_3^{3-} or $B_4O_7^{2-}$	Uncertain, but may be involved in regulation of enzyme function; possible role in sugar transport; synthesis of pyrimidine bases
Zinc	Zn^{2+}	Enzyme activator; maintenance of ribosome structure
Copper	Cu^{2+}	Enzyme activator; may function as intermediate electron acceptor
Molybdenum	MoO_4^{2-}	Nitrogen fixation (where present); nitrate reduction

(a)   (b)   (c)

Figure 8.18 Symptoms of mineral nutrient deficiencies.
(a) Calcium deficiency symptoms in tomato fruits.
"Blossom end rot" forms on the side away from the
stem. (b) Boron deficiency symptoms include early death
of stem tips. Leaves become crinkled, and stems and
leaf stalks crack. (c) Iron deficiency symptoms in tomato
leaves. Chlorosis (yellowing) occurs between veins.
There is little sign of dying tissue.

Calcium, a structural component of cell walls, also is the activator of several important enzymes. In calcium-deficient plants, stem and root tip meristematic regions die early and leave both the stem and roots stunted. Leaf tips hook back and the margins of the leaves become chlorotic. Other common calcium-deficiency symptoms are seen in fruits. One of these is "blossom end rot," a familiar, dark, mushy degeneration of tomatoes (figure 8.18a).

There are many micronutrient deficiency syndromes. They frequently have very characteristic symptoms, which are recognized readily by plant nutrition experts (figure 8.18). Many gardeners have seen these symptoms without knowing what they were.

Nutrient Mobilization

The mobilities of various elements within the plant body are very different. Nitrogen, phosphorus, and potassium can be mobilized in the plant; they can be removed from older tissues and transported preferentially to newer, actively growing areas. Therefore, symptoms caused by deficiencies of any of these three elements are likely to show up first in older tissues. Younger tissues, such as new leaves, appear to be unaffected because they are being supplied at the expense of older tissues. This mobilization and transport to growing regions is under hormonal control.

When the element in short supply (sulfur, for example) cannot be mobilized and transported from one part of the plant to another, the deficiency symptoms are seen first in young, growing tissue, such as the newest leaves, instead of in older tissues that may contain adequate quantities of the nutrient.

Mineral Nutrition and Modern Agriculture

Mineral salts that are ionized in soil water are derived originally from the weathering of rocks. This physical and chemical breakdown of rocks accounts for the origin of most of the important mineral nutrients, but one nutritionally important element—nitrogen—is not available in rocks, and nitrogen nutrition is of central importance to plant productivity.

Box 8.1
The Nitrogen Cycle

Nitrogen is an abundant element. Nitrogen
gas (N_2) makes up about 78 percent of the
atmosphere by volume, and yet, soil
nitrogen deficiency commonly is a factor
that limits plant growth and productivity.
The reason for this is that vascular plants
cannot incorporate N_2 into organic
compounds. Plants absorb nitrogen from the
soil mainly as nitrate (NO_3^-) or ammonium
(NH_4^+) ions. Thus, these usable ions must
be available in adequate quantities in the
soil.

In the environment nitrogen
continuously is being converted from one
form to another through biological and
physical processes. All of these processes
constitute the **nitrogen cycle** (box figure
8.1A). How do the processes of the nitrogen
cycle make available the nitrogen that
plants can absorb?

Small quantities of NH_4^+ and NO_3^- are
washed out of the atmosphere and carried
to the soil by rain. The atmospheric NH_4^+
comes from gasoline engine exhaust,
industrial combustion, and similar processes,
as well as from natural processes, such as
forest fires and volcanic eruptions.

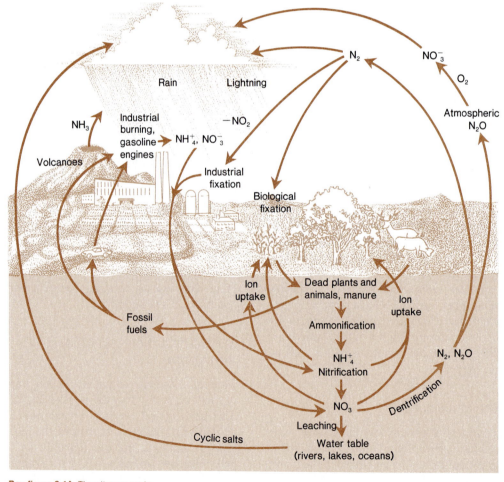

Box figure 8.1A The nitrogen cycle.

Atmospheric NO_3^- is produced through oxidation of N_2 in the atmosphere by oxygen (O_2) or ozone (O_3). Energy for such oxidations comes from lightning or ultraviolet radiation.

But the bulk of the nitrogen compounds absorbed by living plants is not provided by these physical processes. Biological **nitrogen fixation** by microorganisms (bacteria and cyanobacteria, or blue-green algae) provides most of the absorbable nitrogen compounds. Some of these microorganisms are free-living; others live symbiotically with plant cells in root nodules of various plants. In nitrogen fixation, ATP and reduced ferredoxin (see page 158) are used in the conversion of N_2 to NH_4^+.

Once NO_3^- or NH_4^+ is absorbed by plants, the plants—or the animals that eat the plants—can synthesize amino acids, proteins, and other nitrogenous compounds.

Some of these organic nitrogen compounds quickly return to the soil as fallen leaves or animal manure. Others return only when plants or animals die.

Decay processes then break down organic compounds in several stages. In the last of these stages, which is called **ammonification,** soil microorganisms break down organic nitrogen compounds to release NH_4^+. Often, other bacteria further oxidize NH_4^+ to produce NO_3^- in a process called **nitrification.** Plants absorb some of the NO_3^- produced, but NO_3^- is also removed from soil by still other microorganisms that use it as an electron acceptor during respiration, thus converting it to N_2 or N_2O, both of which are lost to the atmosphere. This latter process is called **denitrification.**

The formation of nitrogen compounds for plant nutrition is part of a complex network of processes in the nitrogen cycle. Physical processes in the environment and human activities, including addition of fertilizers, play important roles, but biological processes, especially microbial actions in the soil, are the heart of the nitrogen cycle.

The basic source of nitrogen is atmospheric nitrogen gas (N_2). But, as explained earlier, plants cannot absorb and use N_2. Plants are dependent on the activity of microorganisms that incorporate atmospheric nitrogen through the process of nitrogen fixation, which yields usable nitrogen compounds (NO_3^- and NH_4^+). Nitrogen, once it is incorporated into living material, moves from organism to organism in a **nitrogen cycle.** Nitrogen compounds previously part of living organisms are returned to the soil by decay processes following death.

Other minerals also are constantly recycled through decay processes that return them to the soil after the death of plants and animals. Despite these natural replacement processes, intensive modern agriculture depletes soil mineral elements so drastically that they must be replaced through other means to maintain productivity of the land.

Nitrogen, phosphorus, and potassium are the three macronutrients that most frequently must be supplemented by fertilization of agricultural soils. Fertilizers commonly are labeled with a formula that indicates the percentage of each of these three. A formula of 20-10-10 indicates that the fertilizer contains 20 percent nitrogen, 10 percent phosphoric acid, and 10 percent soluble potassium (as K_2O). Often, a fourth element, calcium, also is included in fertilizers, and occasionally other nutrients may be added as well.

Fertilization, however, is not the simple solution it appears to be for replenishing these elements. Heavy nitrogen fertilization is absolutely essential for the maintenance of high levels of agricultural production, both in countries with advanced agricultural technologies and in less developed countries where the "green revolution" has been introduced. But the energy input required to produce and deliver nitrogen fertilizers is becoming a critical problem. Production of reduced nitrogen fertilizers consumes large amounts of fossil fuel, and transport from production sites to fields requires an additional heavy energy input. The constraints of soil chemistry and plant mineral nutrition coupled with increasing energy costs and decreasing supplies of fuel reserves further complicate the problem of meeting the food requirements of a growing world population.

Summary

Plants are autotrophic organisms that must supply themselves with inorganic molecules to be used in the synthesis of organic compounds. Vascular plants have specialized regions, such as leaves, roots, and stems, that obtain specific materials from the environment.

Water and mineral ions are obtained from the soil by root systems and are moved up through vessels to reach the rest of the plant. Carbon dioxide is incorporated into organic compounds during photosynthesis in leaves. Organic compounds and other materials are moved from one part of a plant to another through the phloem.

The bulk of plant bodies consists of carbohydrates. The carbon and oxygen in these molecules comes from carbon dioxide, and the hydrogen comes from water. Plant growth is not due mainly to incorporation of large quantities of soil constituents, but of elements from water and a minor component of the atmosphere, carbon dioxide.

Carbon dioxide is obtained by diffusion through stomata. But stomatal opening results in evaporative loss of water from leaves, called transpiration. Replacement of this loss necessitates movement of large quantities of water up through the xylem from the roots. Most of the water absorbed by the roots replaces water lost by transpiration. Only a small proportion of the water is retained in the cells and tissues, and an even smaller proportion is used in photosynthesis.

The transpiration stream carries minerals absorbed by the roots up through the plant and, therefore, is more than just a mechanism for replacement of lost water.

Soil water usually is located in the small spaces among soil particles. The water-holding capacity of soil depends on the sizes of particles in the soil. Sandy soil with primarily large particles has less water-holding capacity than soils with smaller particles.

An important part of root function is the continuous growth of root tips that brings the tips into contact with new soil areas. This growth results from cell divisions in the meristematic region and subsequent elongation of cells in the region of elongation.

Root hairs, which increase the absorbing area of roots, function for only a few days and then wither. New root hairs then develop and function in the region of maturation not far from the growing root tip that pushes into new soil areas. This is called the root-hair cycle.

Root structural organization is correlated with root functioning in absorption and transport. The epidermis with root hairs exposes a relatively large surface area to soil water. The bulk of the cortex is composed of loosely packed cells and much free space among them. The innermost layer of the cortex, the endodermis, has waxy endodermal cell walls that fit together snugly and form a barrier, the Casparian strip, between the cortex and the stele. The stele contains xylem and phloem and lies in the center of the root.

Water readily moves into the free space of the cortex, but cannot penetrate the endodermis until it crosses cell membranes to enter the cytoplasmic network of the roots. The only route from the cortex into the stele is through the cytoplasm of endodermal cells. Water tends to flow into the plant because water is being drawn up from the root as long as stomata are open and transpiration is occurring.

High soil salinity can produce a high enough osmotic pressure to overcome the forces causing water movement into the plant. Sometimes salt accumulation due to excessive fertilizer application or irrigation problems causes this to happen in agricultural soil.

Mineral ions can enter roots by simple diffusion or facilitated diffusion, but the bulk of ion uptake is accomplished by active transport. Active transport requires ATP energy input and can work against concentration gradients. The process is selective: some ions are actively transported into the root's cytoplasmic network while others are not.

Ions diffuse through the cytoplasmic network of the cortex and endodermal cells into the stele. Ions probably are released into the xylem or are actively transported into the free space of the stele where the transpiration stream carries them into and up through the xylem.

Several kinds of symbiotic relationships contribute to nutrient uptake by roots. Nitrogen-fixing microorganisms live symbiotically in root nodules of some plants and provide a supply of usable nitrogen compounds to the plants.

Fungus tissue enclosing roots in mycorrhizae increases the efficiency of ion absorption. The fungus benefits from this symbiosis because it receives organic nutrients from the plant, and some plants are very dependent on the fungal mineral absorption.

Mineral nutrient requirements have been studied by chemical analysis of plants and, especially, by hydroponics experiments. The macronutrients have been known for some time and their deficiency symptoms are well recognized. But micronutrient requirements are difficult to determine because only very small quantities are required in many cases, and it is difficult to exclude trace amounts from culture experiments.

Nitrogen deficiency is a common limiting factor in plant growth and productivity. Nitrogen gas is plentiful but not usable by plants. Thus, biological nitrogen fixation as well as decay processes that produce usable nitrogen compounds are essential for soil fertility. Intensive modern agriculture requires supplementation of the natural usable nitrogen supply, but economic and energy problems are affecting our ability to provide nitrogen fertilizers and are threatening world food supplies.

Questions

1. How much of the 74 kg of dry weight gained by van Helmont's growing willow plant was made up of carbon and oxygen?
2. Explain the positive value of transpirational water loss.
3. Why would millet be a more desirable food calorie producer for a semiarid area than potatoes?
4. Explain the relationship between soil particle size and soil water-holding capacity. Also explain the significance of soil particle size and aeration.
5. What is the relationship between the root-hair cycle and absorption as the root tip pushes into new areas of the soil?
6. Explain the significance of the Casparian strip in the movement of water and mineral ions within the root.
7. Why do plants often suffer from nitrogen deficiency when nitrogen is such a plentiful element in the environment?

Suggested Readings

Books

Raven, P. H.; Evert, R. F.; and Curtis, H. 1981. *Biology of plants*. 3d ed. New York: Worth Publishers.

Salisbury, F. B., and Ross, C. W. 1978. *Plant physiology*. 2d ed. Belmont, Calif.: Wadsworth.

Sutcliffe, J. F., and Baker, D. A. 1974. *Plants and mineral salts*. Studies in Biology no. 48. Baltimore: University Park Press.

Articles

Brill, W. J. March 1977. Biological nitrogen fixation. *Scientific American* (offprint 922).

Epstein, E. May 1973. Roots. *Scientific American* (offprint 1271).

Harley, J. L. 1971. Mycorrhizae. *Carolina Biology Readers* no. 12. Burlington, N.C.: Carolina Biological Supply Co.

Heslop-Harrison, Y. February 1978. Carnivorous plants. *Scientific American* (offprint 1382).

Jarnick, J.; Noller, C. H.; and Rhykerd, C. I. September 1976. The cycles of plant and animal nutrition. *Scientific American*.

Stems and
Plant Transport

Chapter Concepts

1. Stems are important to vascular plants for transport of materials and for support of aboveground parts.

2. Primary growth of stems is produced by cell divisions in the apical meristem. Secondary growth is produced by cell divisions in the cambium, a cylinder of meristematic tissue within the stem.

3. Xylem transports water and mineral nutrients up from the soil. Transporting elements of the xylem consist of cell walls left after the cytoplasmic degeneration that occurs during xylem development.

4. Evaporation of water during transpiration pulls water molecules out of plant cells. Cohesion of water molecules causes water to be drawn up through the xylem in response to the pulling force of transpiration.

5. Phloem translocates organic molecules, mineral nutrients, and hormones from one part of a plant to another. The conducting elements in phloem are living, but nonnucleated cells.

6. Material is moved through phloem as a result of mass flow of water. Osmotic water entry at one end of the phloem produces pressure that drives water out of the other end. Water flows through the phloem toward the end where water is being lost, and solutes are swept along in the flow.

A basic problem for terrestrial plants is that the water and mineral nutrients that plants require are obtained by roots located underground, where light energy is not available for photosynthesis. Photosynthesis occurs in leaves, which usually are located well above the ground, where light is available. Integrated functioning of the plant body requires that these widely separated areas of the plant body be connected. This connection is provided by vascular tissues. Water and minerals move up from the roots to the leaves and other aboveground parts of the plant through the **xylem.**

Carbohydrates and other complex molecules produced during photosynthesis must be transported from the photosynthesizing cells to other parts of the plant for metabolic use or for storage in specialized food storage areas. When needed, stored food reserves are mobilized, and carbohydrates are transported from storage areas to other parts of the plant. This mobilization involves the breakdown of insoluble starch molecules to produce soluble sugar molecules. Sugar and various other soluble organic molecules are moved from place to place in the plant body through the **phloem,** another type of vascular tissue. Phloem transport is called **translocation.** In addition to carbohydrates, the phloem translocates other organic molecules, such as amino acids and hormones, as well as inorganic ions (mineral nutrients). (See figure 9.1.)

The focus in this chapter is on the structure and function of the vascular tissues in the stem, but it is important to bear in mind that xylem and phloem transport are responsible for movement throughout the entire plant. Xylem and phloem extend from the root, through the stem, and on through the vascular bundles (veins) of the leaves.

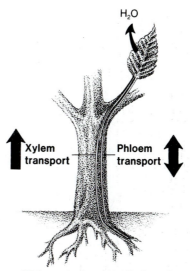

Figure 9.1 Generalized sketch of a vascular plant. Root absorption and storage areas are connected to photosynthesizing and transpiring surfaces of the leaf by vascular tissue. Xylem transports water and minerals up from the roots. Phloem translocates materials up or down, from one part of the plant to another.

Stem Structure

Typically, **annual** and **biennial** plants have **herbaceous** (not hardened with woody material) **stems.** In annuals, the plant's entire life, from germination to the production of seed, is lived out in one growing season. Underground parts of biennials, however, survive over the winter and give rise to aboveground parts in the second season. In both annuals and biennials, the basic stem

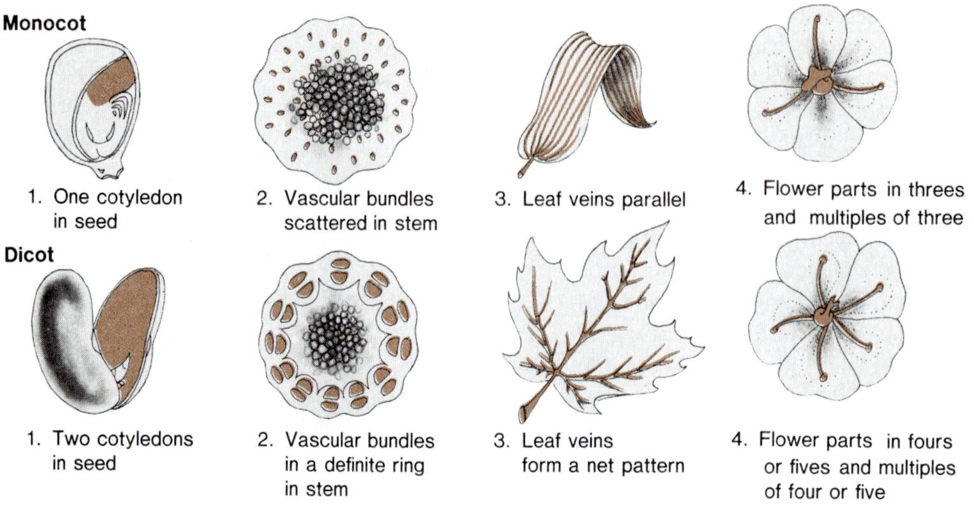

Monocot

1. One cotyledon
   in seed

2. Vascular bundles
   scattered in stem

3. Leaf veins parallel

4. Flower parts in threes
   and multiples of three

Dicot

1. Two cotyledons
   in seed

2. Vascular bundles
   in a definite ring
   in stem

3. Leaf veins
   form a net pattern

4. Flower parts in fours
   or fives and multiples
   of four or five

Figure 9.2 Comparison of monocots and dicots, the two major groups of flowering plants. In addition to differences pictured here, dicots possess cambium, a meristematic tissue not found in monocots. Therefore, dicots are capable of secondary growth, while monocots do not exhibit secondary growth.

structures are laid down by cell divisions occurring in a meristematic region at the tip of the growing shoot, called the **apical meristem.** These tissues, whose cells are produced by cell divisions in the apical meristem, are classified as **primary tissues.**

In contrast, **woody stems** characteristically are found in **perennial** plants (plants that grow for several to many years). Woody stems have **secondary tissues** that are produced during each growing season. Secondary tissues develop as more cells are added after the original stem structure has been produced by primary growth. Secondary growth of stems results from cell division in meristematic tissue within the stem. This meristematic tissue is the **cambium.**

There are two major groups of flowering plants, the **dicotyledonous** plants and the **monocotyledonous** plants. Plant biologists often call these two groups of plants **dicots** and **monocots** for short. Their names derive from differences in seed structure. Monocots have only one **cotyledon** ("seed leaf"), and dicots have two cotyledons (figure 9.2). Dicots and monocots have fundamentally different stem structures.

Herbaceous Dicot Stems

The herbaceous dicot stem illustrates a basic pattern of stem structure against which other types of stems can be contrasted.

One obvious feature of the herbaceous dicot stem shown in figure 9.3 is a discontinuous ring of vascular tissue (**bundles** of xylem and phloem) between the **cortex** and the large central **pith,** which is made up of thin-walled **parenchyma** cells. A thin layer of cambium (meristematic) tissue (the **vascular cambium**) lies between the xylem and phloem in each vascular bundle.

The supporting cells in the outer portion of the cortex, along with the xylem and various fibers, make up the **mechanical tissue** of the stem. Mechanical tissue helps to keep the stem upright and to support the weight of the entire aboveground structure of the plant. The other major supporting force in stems is turgor pressure of cells in other, nonmechanical tissues.

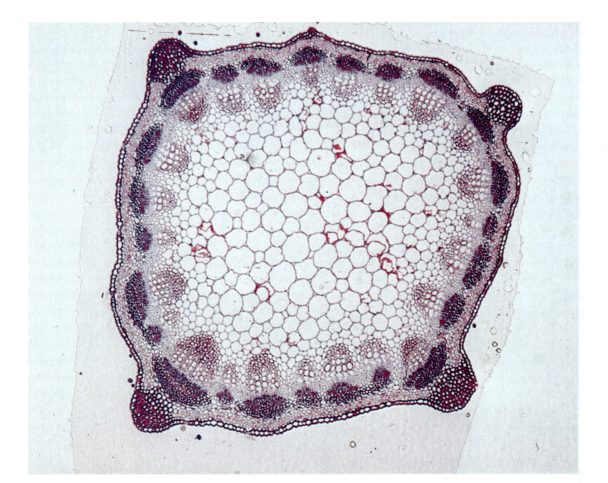

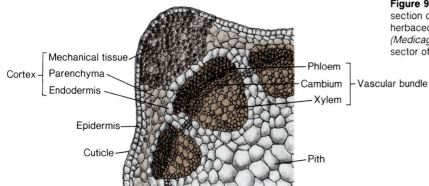

Figure 9.3 Photograph of a cross section of a representative herbaceous dicot stem—alfalfa *(Medicago)*—and a sketch of a sector of the stem.

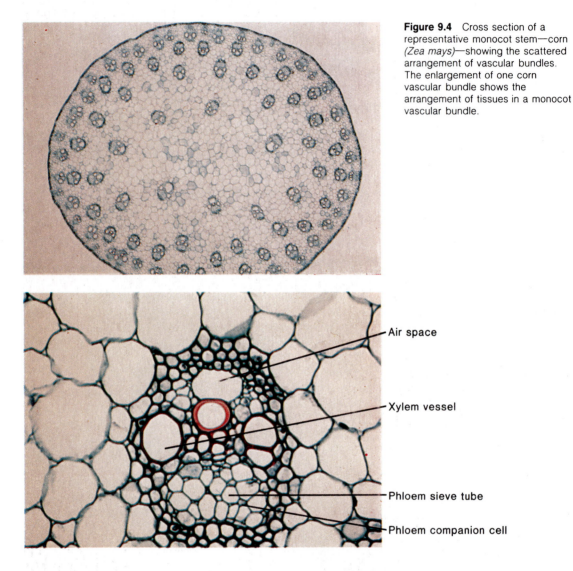

Figure 9.4 Cross section of a representative monocot stem—corn (*Zea mays*)—showing the scattered arrangement of vascular bundles. The enlargement of one corn vascular bundle shows the arrangement of tissues in a monocot vascular bundle.

Air space

Xylem vessel

Phloem sieve tube

Phloem companion cell

Monocot Stems

The basic pattern of a typical monocot stem is shown by a cross section of a corn (*Zea mays*) stem (figure 9.4). The monocot stem's vascular tissues are not arranged in a ring but are scattered throughout the stem as discrete bundles. This means that there is no clear distinction between cortex and pith in the monocot stem. All of the space around the vascular bundles is filled with parenchyma cells. Monocot stems do not have vascular cambium; all of the stem is produced by primary growth, that is, by the activity of the apical meristem of the shoot.

Support in monocot stems is provided by thick-walled cells that make up the mechanical tissue of the vascular bundles, by supporting cells in the outer layers of the stem, and by turgor pressure of cells in other tissues. An intercellular air space in each vascular bundle permits some gas movement up and down the inside of the stem.

Woody Stems

The stems of woody perennial plants persist from season to season and differ considerably from the herbaceous stems discussed thus far. Woody stems grow longer because they develop buds whose apical meristems become active at the beginning of each growing season and continue to add new tissue throughout the season. Buds are located at the tip of the stem (**terminal buds**) and just above the points where leaves are attached (**lateral** or **axillary buds**) (figure 9.5). Terminal buds add to the length of the original stem, while certain of the lateral buds are responsible for the growth of stem branches.

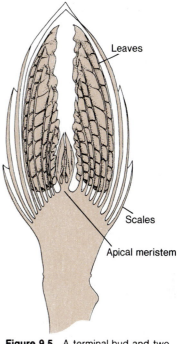

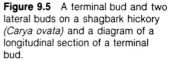

Figure 9.5 A terminal bud and two lateral buds on a shagbark hickory *(Carya ovata)* and a diagram of a longitudinal section of a terminal bud.

Each terminal bud contains the beginnings of a new section of stem. Leaves and their leaf stalks (**petioles**) encircle the central region of the tip, the apical meristem. Cell division in the apical meristem and elongation of the cells that it produces lengthens the stem, and differentiation of these cells produces new primary tissue in the growing stem. As the stem lengthens, the leaves, which are closely packed within the bud, space out at regular intervals along the stem. The points of leaf attachment are known as **nodes,** and the spaces between nodes are called **internodes.** Clearly then, all growth in length is due to growth occurring in apical regions. There is no factual basis, for example, for the popular misconception that a tree gets taller because the growth at the base of its trunk pushes it upward.

Buds form at the end of one growing season and become active in the next season. Thus, they must be adapted for overwintering. Buds are enclosed within several layers of highly modified leaves, the **bud scales,** which provide a protective coat around them. These scales may be waxy, or hairy, or modified in some other way to protect the delicate embryonic leaves and apical meristem in the bud. In the spring, buds swell as the internal elements begin to grow again following their winter dormancy. This swelling ruptures the coating layer of bud scales, and they eventually drop off. The area where these scales were attached to the stem is marked by the **terminal bud scale scar.** Each terminal bud scale scar identifies the beginning point of a season's growth, and these circular scars can be used as annual time markers on the surface of a young stem. Small

Figure 9.6 Generalized sketch of a woody stem. Terminal bud scale scars mark the beginning of each season's growth in length.

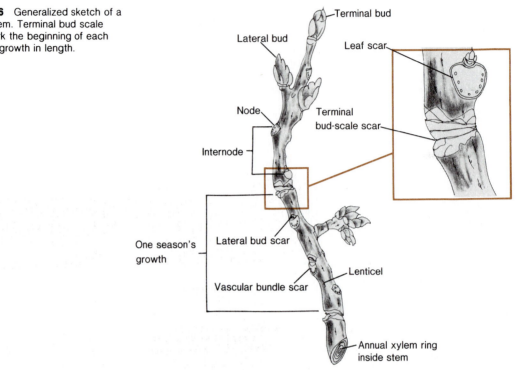

structures called **lenticels** also are found on woody stems. Lenticels are small patches of loosely organized tissue that function as pores through which gas exchange takes place. Lenticels provide necessary routes of oxygen and carbon dioxide diffusion between stem cells and the atmosphere. (See figure 9.6.)

Woody Stem Growth and Internal Organization

During its first growing season, a woody stem's basic organization is quite similar to that of an herbaceous dicot stem. But woody stems may persist for many growing seasons and have considerable secondary growth produced by proliferation of a cambial ring. This ring develops before the end of the first growing season as the areas of vascular cambium connect to produce a continuous cylinder in the stem. From then on, concentric rings of new tissue are produced each year, with secondary xylem being laid down inside the cambium ring and secondary phloem being produced outside it (figure 9.7). During

each growing season, xylem elements produced in the spring usually are larger than those produced later in the season, so the xylem formed in one growing season can be distinguished from the adjacent growth of another growing season. These accumulating thick-walled xylem cells become impregnated with **lignin** and make up what is commonly called the **wood** of a tree or shrub. The yearly deposits of xylem constitute the **annual growth rings** in the wood and form an accurate record of the age of the plant.

Only the more recently formed part of the xylem, the **sapwood,** actually contains conducting xylem elements that still function in water transport. In the older, inner part, the **heartwood,** the xylem becomes plugged with deposits of resins, gums, and other secondary plant products.

As secondary xylem is produced on the inside of the vascular cambium, the diameter of the wood stem increases. Thus, the circumference of the circle of cambium tissue increases because of the accumulating annual rings of xylem inside it.

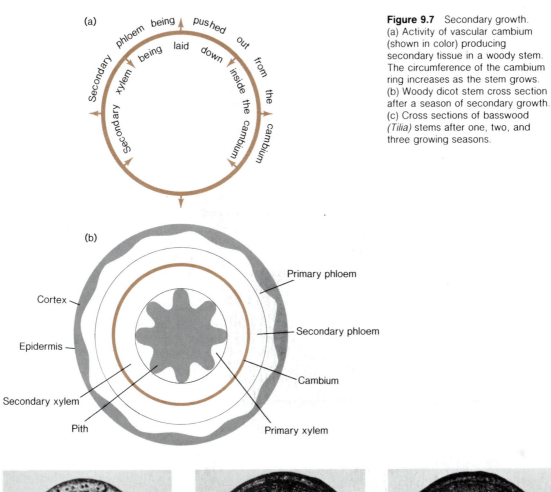

(a)

Secondary phloem being pushed out from the cambium

Secondary xylem being laid down inside the cambium

(b)

Cortex

Epidermis

Secondary xylem

Pith

Primary phloem

Secondary phloem

Cambium

Primary xylem

Figure 9.7 Secondary growth. (a) Activity of vascular cambium (shown in color) producing secondary tissue in a woody stem. The circumference of the cambium ring increases as the stem grows. (b) Woody dicot stem cross section after a season of secondary growth. (c) Cross sections of basswood *(Tilia)* stems after one, two, and three growing seasons.

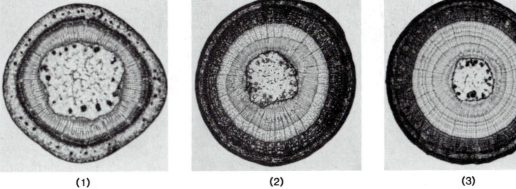

(1)   (2)   (3)

(c)

Figure 9.8 Bark formation. Bark is a secondary covering of the stem and replaces the epidermis, which was a primary tissue. Bark consists of **cork cells** produced by a ring of cork cambium, which becomes active in the outer part of the cortex. As cork cells are pushed out from the cork cambium, they become impregnated with a waxy material that makes the bark quite impermeable to water. Surface cork cells die and fall off, but the cork cambium continues to produce replacement cells.

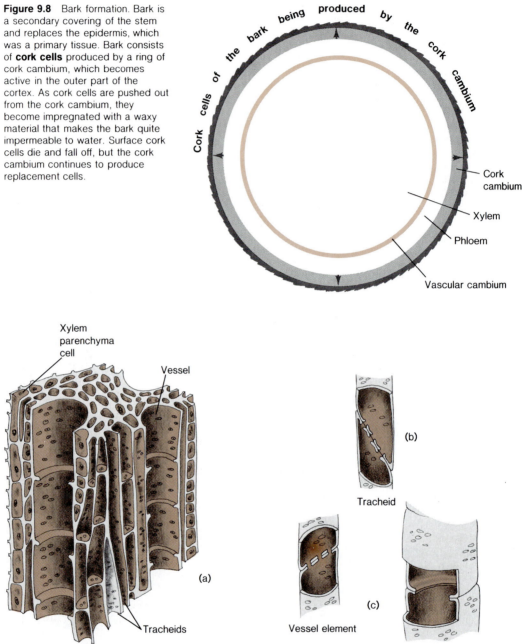

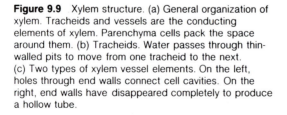

Figure 9.9 Xylem structure. (a) General organization of xylem. Tracheids and vessels are the conducting elements of xylem. Parenchyma cells pack the space around them. (b) Tracheids. Water passes through thin-walled pits to move from one tracheid to the next. (c) Two types of xylem vessel elements. On the left, holes through end walls connect cell cavities. On the right, end walls have disappeared completely to produce a hollow tube.

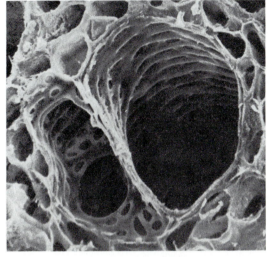

Figure 9.10 Xylem transporting elements as "cellulose pipes." Scanning electron micrograph of xylem vessels from a cucumber root. Note the sculptured appearance of the inside of these vessels. (Magnification × 1,050)

Over the growing seasons, phloem tissue does not accumulate as xylem tissue does. On the average, in woody stems, only one phloem cell is produced for every six xylem cells formed. Thus, the sheer bulk of xylem tissue is much greater than phloem tissue. Phloem, which consists of relatively thin-walled and delicate transport cells, is pushed outward as the diameter of the woody xylem area increases, and eventually the outermost parts of the phloem are crushed against the bark that covers older stems (figure 9.8).

Phloem tissue is active in transport only during the year that it is formed. However, since old phloem is crushed, there are no obvious annual rings in the phloem as there are in the xylem. Only the most recently formed phloem, that portion nearest the vascular cambium, is active transporting tissue.

Xylem

The xylem of flowering plants contains two basic types of conducting tubes, **tracheids** and **vessels** (figure 9.9a). During the development of a flowering plant, xylem tracheids develop earlier than xylem vessels. Botanists believe that tracheids also are older in an evolutionary sense and that vessels represent a more recent evolutionary development. Interestingly, conifers (evergreens) and their relatives and the vast majority of the non-seed-producing vascular plants such as ferns have only tracheids in their xylem.

Tracheids are hollow, elongate, spindle-shaped cells that have drawn out, flattened ends (figure 9.9b). There are a number of thin spots (known as **pits**) in the walls of tracheids. They are particularly numerous in the end walls of tracheids where the cells overlap one another. In pits, tracheids are separated only by the very thin walls of adjacent cells, which allow water to flow through the pits from one tracheid to another. Thus, tracheid cells are linked in water-conducting chains, but water must pass through pits and follow a somewhat meandering route through tracheids.

Vessels develop from end-to-end chains of cells. In some plants, these ends develop perforations as development proceeds, but in more highly specialized xylem arrangements, the end walls of the cells disappear entirely. While the end walls thin out and eventually open up, the side walls thicken. After the cytoplasm disappears, hollow vessels that may be several meters in length remain (figure 9.9c). Individual vessel cells are called **vessel elements** (or **vessel members**) since they no longer are individual cells but are nonliving parts of a structure derived from many cells. Literally, vessels are "cellulose pipes" running through the plant. Water moving up vessels flows through straight, open channels (figure 9.10).

Root Pressure and Guttation

Over the years, biologists have investigated the forces that cause movement of **xylem sap** (water and minerals) up through the xylem. One force results from osmotic pressure differences between the contents of root cells and the soil water. Usually, root cells contain a higher concentration of dissolved and suspended material than does soil water, and water tends to move into root cells in response to differences in osmotic pressure. This osmotic water movement force, called **root pressure,** is generated at the bottom of the xylem, and tends to push xylem sap upward. In fact, cut ends of stems will "bleed" xylem sap for some time because of the push applied by root pressure. Although stem **bleeding** may appear to be a very gentle fluid oozing, a considerable driving force is involved, as can be demonstrated in a very simple experiment. If a piece of glass tubing is sealed onto the cut stem of a tomato plant, root pressure will raise bleeding xylem sap to as high as a meter above the cut. In some kinds of vines, xylem sap will rise to several times that height above a cut stem.

Root pressure also is responsible for an interesting phenomenon that occurs when transpiration is not taking place and the soil is well watered. Drops of water are forced out of vein endings along the edges of leaves as a consequence of root pressure. This bleeding at leaf vein endings is called **guttation** (figure 9.11).

Calculations have shown, however, that root pressure alone simply is not strong enough to move water very far up a tall plant, even though it can be an important factor in small plants.

Figure 9.11 Young oat plants with drops of guttation water at their leaf tips.

Transpiration Pull/Cohesion of Water Hypothesis

If root pressure does not generate enough force to raise water to the tops of large plants, how then is it done?

Water rises through vessels and tracheids as a result of a pull applied to water at the top of the xylem and the very strong tendency of water molecules to stick to one another. This attraction among molecules of the same kind is called **cohesion.** Liquid water is not simply water molecules randomly floating around one another. There is considerable attraction between the hydrogen atoms of one water molecule and the oxygen atom of an adjacent molecule. This attraction, called hydrogen bonding (see p. 28), is not limited to a one-to-one relationship but is distributed to give the effect of a network of mutually attracted water molecules. That is how water molecules "stick together."

The cohesive force among water molecules is so great that it is legitimate to think of the column of water inside a xylem vessel as behaving almost like a chain of water molecules that can be pulled up through the xylem. In fact, it is easier to pull apart the molecules in fine wires made of some common metals than it is to disrupt a column of water in a small-diameter, airtight tube.

If cohesive forces among water molecules are great enough to hold a column of water together even when it is being pulled with enough force to raise it to the top of a large tree, what then supplies the actual pulling force? Stephen Hales, an English botanist, first suggested near the beginning of the eighteenth century that transpiration pulls water up through plants. Then, in 1894, Dixon and Joly coupled the idea of **transpiration pull** with knowledge that had become available by then about the **cohesion of water.**

The **transpiration pull/cohesion of water hypothesis** says that evaporation of water molecules from the surface water films of leaf cells pulls water from within those leaf cells to replenish the surface water films. This pull is transmitted in a chain reaction as water is drawn through leaf cells from the xylem in the leaf veins. Because the xylem contains continuous tubular structures that run from leaf veins down through the stem to the roots, this pull (tension) is transmitted down through the entire plant, so that whole columns of water literally are pulled up from the top.

After considering the idea that water was pulled up through the xylem under tension, D. T. MacDougal thought of a simple, but intriguing experiment. He reasoned that the tension on the water, which is being drawn upward, should produce inward pull on all xylem vessels, and he proposed that this would result in a decrease in the total diameter of a tree trunk. He designed a sensitive device, which he called a dendrograph, to measure small changes in trunk diameter; he found that the trunk diameter does indeed decrease in the daytime when the transpiration rate increases.

Given that the cohesion of water molecules could hold together a tall column of water under tension, researchers have tried to demonstrate that transpiration applies a strong enough pull to lift water columns to the height required in tall plants. Transpiration pull can be measured experimentally using either a living shoot or a physical apparatus (figure 9.12). To quantify transpiration pull, a pan filled with mercury is placed under a tube that holds a twig with leaves or a tube with a porous clay bulb. Mercury rises in the tube as the tube is evacuated by the withdrawal of water from the top, and the lifting force of transpiration pull can be expressed in terms of the height to which the mercury column is lifted. In the case of simple suction (figure 9.12b(1)), the mercury column rises to a height of only 76 cm (one atmosphere of pressure). But when a transpiration pull is added, either in the form of a twig with leaves or a porous clay bulb (figures 9.12b(2 and 3)), the mercury column rises to a height of 100 cm or more. If the data from mercury columns is translated into the heights of water columns, which are the concerns in xylem transport, it is plain that atmospheric pressure alone (or atmospheric pressure combined with root pressure, for that matter) could not raise water to the treetops. There must definitely be a pulling force at the top of the water columns in the xylem, and it is likely that the force is provided by transpiration pull.

Unfortunately, it is extremely difficult to measure the pulling force inside xylem vessels of living plants because the process of making the measurements disrupts the system so much that the validity of the measured values must be doubted. Most biologists agree that adequate pulling force can be generated by transpiration and that cohesive forces among water molecules are adequate to hold a column of liquid together under tension *if* the columns of liquid in the xylem are continuous. Many xylem tubes, however, seem to have at least some space occupied by water-saturated air. Critics of the transpiration pull/cohesion of water hypothesis say that when air interrupts the liquid columns, serious doubts are raised about the cohesion part of the hypothesis since water molecules would no longer be in contact with each other in a continuous chain.

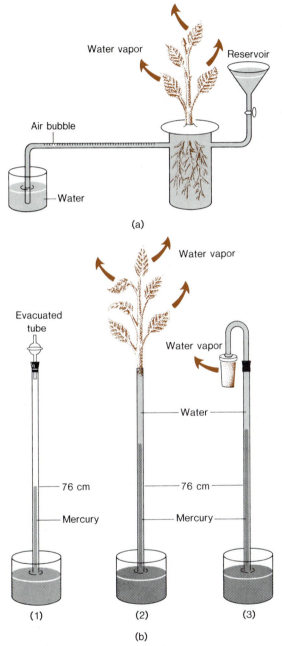

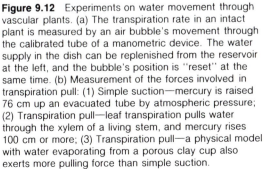

Figure 9.12 Experiments on water movement through vascular plants. (a) The transpiration rate in an intact plant is measured by an air bubble's movement through the calibrated tube of a manometric device. The water supply in the dish can be replenished from the reservoir at the left, and the bubble's position is "reset" at the same time. (b) Measurement of the forces involved in transpiration pull: (1) Simple suction—mercury is raised 76 cm up an evacuated tube by atmospheric pressure; (2) Transpiration pull—leaf transpiration pulls water through the xylem of a living stem, and mercury rises 100 cm or more; (3) Transpiration pull—a physical model with water evaporating from a porous clay cup also exerts more pulling force than simple suction.

Proponents of the hypothesis maintain, however, that at any one time, continuous water columns exist in an adequate proportion of the xylem vessels to supply the needs of the plant. Also, when an air bubble blocks flow through a given xylem vessel, water can move through pits in lateral walls of xylem elements and pass from that vessel to another nearby vessel and thus continue its upward flow.

The number of fluid-filled xylem vessels varies with conditions and with the time of year; for example, water content of stems usually is higher in spring and summer than in fall. Loggers find, in fact, that some kinds of logs float at some times of the year, but will not float at other times when a high percentage of xylem vessels are filled with water.

Even though the xylem vessels often are likened to miniature pipes and the forces raising fluid through the xylem can be described in terms of physical forces such as tension and cohesion, it is important to remember that this is water flow through a living organism. Plants are not passive water tubes. Water movement through the root into the bottom of the xylem must be through the cytoplasmic network in living cells of the root cortex and endodermis (chapter 8). Living cells at the root end of the plant clearly have potential for control. And plants also control the rate of transpiration from the leaves by opening and closing stomata.

Thus, a century after it was proposed, the transpiration pull/cohesion of water hypothesis still seems a good explanatory model for fluid movement through the xylem. Yet there remain intriguing questions about the hypothesis, and further research on xylem transport is needed.

Phloem

During early spring growth, organic solutes, especially sucrose, are translocated from plant storage areas and from developed leaves to young, actively growing areas of the plant that are not yet synthesizing adequate quantities of carbohydrates to meet their own needs. Later in the season, organic molecules produced in mature leaves are available for storage because a mature, actively photosynthesizing plant may produce twenty or more times the amount of carbohydrates needed to meet its own immediate requirements. Solutes are then translocated to storage tissues, which develop in modified stems or roots, or to seeds, which also become very significant storage depots. How are these organic compounds moved from their **sources** (the production sites in the leaves) to **sinks** (the actively growing areas or storage tissues of seeds, roots, or stems)?

The phloem was identified as the translocating tissue long ago when the results of **girdling** (removal of a ring of bark) were analyzed. Girdling removes the phloem but leaves the xylem intact, and it eventually kills a tree. If a tree is girdled below the level of the majority of leaves, the bark, with the underlying phloem, swells just above the cut, and sugar accumulates in the swollen tissue. This finding led biologists to the conclusion that the phloem is the carbohydrate translocating tissue.

Radioactive tracer studies have confirmed this. When ^{14}carbon-labelled CO_2 is supplied to mature leaves, radioactively labelled sugar produced in the leaves is soon found moving down the stem and into the roots, and this labelled sugar is found mainly in the phloem, not in the xylem.

Other radioactive isotope studies have confirmed the phloem's role in transport of other organic molecules, such as amino acids, and in mineral ion transport. Mineral nutrients can be mobilized in one part of the plant and moved to another part. For example, some mineral nutrients are translocated out of leaves before they fall in the autumn. In some cases, hormones also are transported through the phloem from their production sites to target areas where they exert their regulatory influences.

Phloem Cells

Material moved through the phloem is translocated through the **sieve tubes. Sieve tube elements**, the cells that make sieve tubes, are highly specialized cells that are arranged in linear arrays running vertically through the length of the phloem. In flowering plants, the sieve tube elements are lined up end-to-end with flat areas of contact where the end walls of cells abut one another (figure 9.13). As noted earlier, sieve tubes have a short functional life; usually they transport materials only during the year in which they are formed.

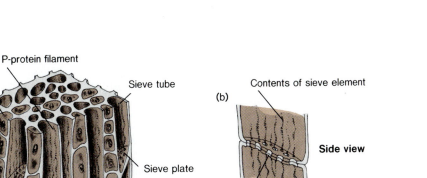

Figure 9.13 Phloem structure. (a) Arrangement of cells in phloem. Phloem fibers are supporting cells not involved in translocation. (b) Sketches of a sieve plate.

There are prominent holes in the end walls of sieve tube elements, and adjacent cells contact each other directly through those holes. Because of its sievelike appearance, this contact area is called the **sieve plate,** and the name sieve tube also is derived from this structure. Cytoplasmic connections run through the pores of each sieve plate, connecting the cytoplasm of the sieve elements adjacent to the plate. Thus, there is cytoplasmic continuity from cell to cell in the sieve tubes.

Individual sieve tube elements are structurally different from the majority of plant cells, because during their differentiation, each sieve tube element loses both its vacuole and its nucleus. Fully differentiated, functional phloem tissue, then, contains end-to-end rows of nonnucleated, nonvacuolated, but still living, sieve elements. This special structural arrangement is very different from that of xylem vessels and tracheids, which consist of empty cell walls.

Sieve elements contain some special filamentous structures that seem to run lengthwise through the cells. When the sieve tubes are physically damaged, these filaments disperse into what appear to be beaded chains. These filaments are known as **P-protein filaments** (P for phloem), and some researchers think that they are involved in phloem transport. Most biologists, however, favor a different hypothesis for phloem transport.

Because sieve tube elements lose their nuclei during their differentiation, they are irreversibly specialized for translocation. They are no longer capable of carrying out nucleus-directed protein synthesis or other functions controlled by nuclear genetic information. Continuing genetic control of sieve tube activity, then, may come from another source. Each sieve tube element has one or more **companion cells** adjacent to it. Unlike sieve tube elements, companion cells possess all the normal cellular components. They have nuclei and vacuoles as well as other organelles normally found in plant cells. It seems likely that the companion cells exert some control over sieve tube functioning. And almost certainly, companion cells can transfer material into sieve tubes through pores that connect them.

In addition to sieve tubes and companion cells, phloem tissue, like xylem tissue, contains fiber cells that serve a supporting function and thin-walled parenchyma cells.

Studying Phloem Function

Phloem function is difficult to study because the cells involved are delicate and easily damaged. The slightest disturbances of sieve tubes lead to several types of physiologically preprogrammed injury responses that shut down phloem translocation and conserve phloem sap. The dispersion of the P-protein filaments into beaded chains following injury has already been mentioned, but there is another visible injury response. A slimy plug made of a substance called **callose** (a glucose polymer) develops in each plate pore when the phloem is injured. The size of the callose plug formed in experiments varies dramatically, depending on the way in which the phloem cells are killed. The callose plug apparently develops in injured phloem in nature just as it does in experiments.

Because injury responses in phloem occur quickly, researchers studying phloem function had to find a way to study phloem without injuring it. One of the most clever methods of doing this was first applied by Mittler in 1957. Mittler's method takes advantage of the feeding habits of **aphids**, small insects that insert long pointed feeding devices, known as **stylets,** into the phloem and feed on phloem sap. The aphid stylet penetrates the tissue lying over the phloem and terminates in an individual phloem sieve tube (figure 9.14). An aphid stylet can enter a sieve tube without causing the injury response. Thus, an aphid can remain "plugged in" to a sieve tube and feed on phloem sap at leisure for hours. In experimental studies, aphids are anesthetized and cut away from their stylets. This leaves experimenters with small open channels through which phloem sap can be withdrawn from sieve tubes without concern about injury responses.

(a)

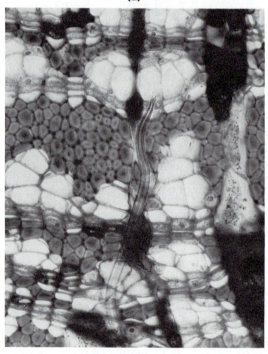

(b)

Figure 9.14 Aphids on phloem. (a) An aphid feeding on a tree branch. A droplet of "honeydew" is being exuded from the animal. (b) This section of the tree branch shows that the aphid stylet has penetrated to a phloem sieve tube cell. Biologists have studied phloem function by cutting aphids away from their stylets. This leaves the stylets as small open channels into the phloem.

Characteristics of Phloem Transport

Any explanation of phloem transport must account for the movement of fairly large amounts of organic material through long distances (many meters in some trees) in relatively short periods of time.

As long ago as 1922, on the basis of calculated results of translocation, plant physiologists showed that the rates and amounts of organic material moved in plants required explanations other than simple diffusion. For example, Dixon and Ball measured the total amount of carbohydrate in potato tubers and the total cross-sectional area of the phloem in the stem through which the stored carbohydrate had moved. They assumed that the phloem sap was a 2 percent sugar solution and then calculated a minimum flow rate required to deliver the amount of carbohydrate through the phloem to the tubers in the length of time that the tubers were growing. They concluded that diffusion alone was several thousand times too slow to account for movement of the required amounts of carbohydrate. Actually, their estimate of the sugar content of the phloem sap was too low because phloem sap contains up to 30 percent sucrose in some plants, but even that correction does not alter the basic conclusion that diffusion is an inadequate explanation of phloem transport.

Ziegler and Vieweg heated small sections of phloem slightly and applied sensitive thermocouples (temperature-measuring devices) further down the stem to detect the arrival of the warmed sap. In more recent experiments, the rate of movement of ^{14}carbon-labelled sugar has been determined by analysis of sap withdrawn through aphid stylets at two levels of the stem. Both of these techniques for measuring rates of movement, as well as other studies, have yielded values for phloem transport rates that are many times faster than simple diffusion. Materials appear to be moved through phloem at point to point rates of 60–100 cm/hr and possibly up to 300 cm/hr.

Over the years, various researchers have proposed that **cyclosis,** the streaming movement of cytoplasm, occurs in sieve elements and augments diffusion in phloem transport, but even the highest estimates of cyclosis movement rates are inadequate to explain the observed rates of transport through the phloem.

Another property of phloem transport also must be accounted for. The direction of flow apparently is one way in each sieve tube (though not everyone agrees on that point). But within a single bundle of phloem sieve tubes, some tubes may be translocating material in one direction while nearby sieve tubes are carrying material in the opposite direction. Also, an individual sieve tube can reverse its direction of transport at different times.

One factor that clouds interpretation of direction studies is that there is some lateral movement of materials between adjacent phloem sieve tubes. There is even a little lateral movement between xylem and phloem, with materials moving through the cambium layer. Some of the confusing observations that suggest that different materials simultaneously move in opposite directions through the same sieve tubes might be explained by such lateral transfer between various conducting elements. Despite these problems in interpretation, it is generally agreed that transport within each sieve tube involves one-way flow.

The Mass Flow Hypothesis

The most widely accepted hypothesis explaining phloem transport is called the **mass flow hypothesis,** first proposed by Münch in 1927. Despite disagreements and difficulties regarding interpretation, calculations indicate that mass flow through sieve tubes could account for the flow speed and the quantity of material transported in the phloem.

Figure 9.15 Phloem transport.
(a) Model system in which mass flow
occurs. (b) Proposed mass flow of
water and solute through the
phloem. Colored arrows represent
active transport of sucrose; gray
arrows represent flow of water. At a
source (for example, an actively
photosynthesizing leaf), sucrose is
actively transported into phloem
sieve tubes. At a sink (for example,
in a root), sucrose is actively
transported out of phloem sieve
tubes. Thus, the osmotic pressure is
lower in sieve tubes at the sink end
of the phloem than at the source
end. So much water enters the
source end of the phloem by
osmosis that water is forced out the
sink end. This results in flow of
water through the phloem that
sweeps sucrose along with it. (In this
example, water forced out of the
phloem in the roots can return up
the plant through the xylem.)

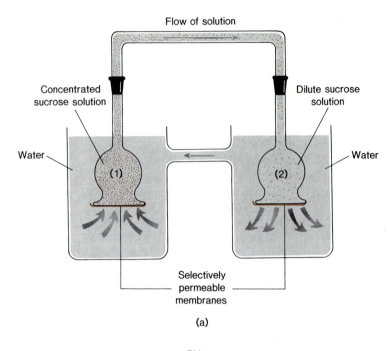

(a)

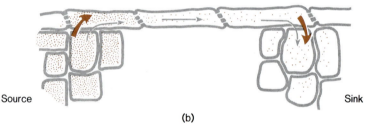

(b)

The hypothesis depends on the existence of a system bounded by differentially permeable membranes in which there are marked differences in solute concentrations at two separated points in a system (figure 9.15a is an experimental model of such a system). When this system is submerged in distilled water, water tends to enter osmotically through both differentially permeable membranes, but this tendency is much stronger at membrane (1) because the osmotic gradient across membrane (1) is so much greater. Thus, water moves into (1) much more rapidly than into (2), and soon pressure differences in the system result in a flow of water from (1) to (2). This flow occurs because pressure actually drives water out through the membrane of (2), even though water is being forced to move against an osmotic gradient. As water flows through the system it sweeps along solute, so that this mass flow of water tends to drive the model system toward equilibrium in a relatively short time.

These same principles can be applied to the mass flow hypothesis of phloem transport. The hypothesis states that in phloem the activities of living cells can maintain the concentration differences between different parts of the system, so that instead of moving toward equilibrium as the model system in figure 9.15a does, the differential is maintained and mass flow continues. During vigorous photosynthetic activity, for example, the leaf can serve as a "source" of sucrose that is loaded into the phloem sieve tubes by active transport across sieve tube element membranes (figure 9.15b). Then at another point in the phloem, active transport processes remove sucrose from the sieve tubes and move it toward a storage "sink" in the starch-storing plastids of cells, which are in storage areas such as root cells.

The phloem sieve tubes, then, can be seen as fluid-filled conduits, bounded by differentially permeable membranes, running through the plant from solute source to sink. A relatively more concentrated sucrose solution is maintained by addition of sucrose at one end of the phloem transport system, and sucrose is actively removed at the other end. Thus, concentration differentials are maintained by energy-requiring active transport, and the mass flow of phloem sap continues. The plant works to maintain a situation in which a spontaneous physical response sets up a flow through the system. Such a continuing flow would sweep along solutes at rates that biologists calculate would be adequate to agree with experimentally determined phloem flow rates.

The mass flow hypothesis also can account for reversal of the flow through any given chain of sieve tube elements. All that is required is a reversal of the source-sink relationship; that is, sucrose is loaded into the opposite end of the phloem and actively removed from the end that previously was the source.

Intriguing, unanswered questions remain about the role of the companion cells in all of this. In the mature leaf, which is a source area, the ratio of the relative diameters of companion cells and sieve tubes is about 1:1. But in midstem, the ratio is something like 1:100. The significance of this is not entirely clear, but the nucleated companion cells, which supposedly are regulators of phloem function, are much more prominently developed in the area where the sieve tubes are being loaded with sucrose. Possibly, companion cells help move sucrose from production sites in photosynthesizing leaf mesophyll cells to the sieve tubes, where it is loaded for transport.

If active, energy-requiring processes maintain the concentration differentials at the source and sink ends of phloem-conducting elements, and these concentration differentials account for the maintenance of mass flow through the phloem sieve tubes, then do the sieve tubes serve simply as living, but passive pipes through the plant? Maybe not, because if metabolic poisons are applied to a ring around a tree or if a ring is heated with steam, flow through the phloem stops. This implies that living, metabolically active phloem tissue is needed for phloem transport to continue. But the injury shutdown response of phloem could account for these effects of metabolic poisons as well. Clearly, there still is much to be learned about phloem transport and its control.

Summary

Water and mineral nutrients absorbed from the soil by roots are transported up through the xylem to aboveground parts of the plant. Carbohydrates and other organic molecules produced in leaves are transported to other plant parts through the phloem. Phloem also translocates inorganic ions, amino acids, and hormones from one part of the plant to another.

Stems are important transporting and supporting structures. Annual and biennial plants characteristically have herbaceous stems, while woody stems are normally found in perennial plants.

Dicots and monocots are two major groups of flowering plants that have fundamental differences in stem structure, as well as other structural differences.

In herbaceous dicot stems, vascular bundles are arranged in a circle that lies between the cortex and the large central pith. Vascular bundles contain xylem and phloem separated by cambium. Supporting cells in the xylem, xylem conducting cells themselves, and supporting cells in the outer portion of the stem constitute the mechanical (supporting) tissue of the stem.

Monocot stems have scattered vascular bundles and lack a clear distinction between cortex and pith.

Woody stems persist from year to year. Primary growth, initiated by buds, adds to stem length each growing season. Secondary growth, due to proliferation of cambium, increases the thickness of stems by adding new xylem and phloem tissue.

Each season's xylem growth accounts for one growth ring. Only the more recently produced xylem, the sapwood, is functional transporting tissue. Old phloem tissue is pushed outward and crushed against the bark that covers the stem surface. Only the phloem produced during the current growing season is functional transporting tissue.

Xylem contains two basic types of conducting tubes, the tracheids and the vessels. Both consist of empty cell walls when mature. Tracheids are elongate, spindle-shaped cells that are lined up with their ends overlapping. Water moves from one tracheid to the next through thin-walled pits. Xylem vessels are end-to-end rows of cells that are open at their ends to produce "cellulose pipes" running through plants.

The tendency of water to enter roots osmotically results in root pressure that can cause fluid to rise a short distance through the xylem. But root pressure alone cannot raise fluid to the heights required in large plants.

The best current explanation of the force that causes fluid to rise through the xylem is the transpiration pull/cohesion of water hypothesis. This hypothesis states that water evaporating from leaf cell surfaces pulls water out of leaf cells. The cohesion of water molecules causes this pulling force (tension) to be transmitted to the xylem where it pulls fluid up from the roots.

Phloem translocates material from sources to sinks. The material moves through sieve tubes that consist of living but nonnucleated cells. Sieve tube elements have cytoplasmic connections that run through holes in the sieve plates located at the ends of cells. Nucleated companion cells apparently exert some control over sieve tube functioning.

Phloem function is difficult to study because the slightest disturbances of sieve tubes lead to several types of injury responses that shut down phloem translocation. However, some useful techniques, such as the use of aphid stylets, allow phloem to be studied without injury.

Phloem translocation is too rapid to result from diffusion and cyclosis (cytoplasmic streaming). Currently, characteristics of phloem transport seem to be best explained by the mass flow hypothesis. This hypothesis proposes that solute is actively transported into phloem sieve tubes at a source and actively transported out of them at a sink. Water tends to enter sieve tubes by osmosis at the source end with such pressure that water is forced out of the sink end of the tubes, and the flow is maintained. This flowing water is believed to carry solutes with it.

Although the mass flow hypothesis is considered a likely explanation of phloem transport, further research is needed on many details of phloem function.

Questions

1. How could you determine the length that a twig on a tree grew during the growing season two years ago?
2. Why are no annual growth rings seen in the phloem of woody stems?
3. Explain the cause of root pressure.
4. Explain the relationship between the stomatal opening and the movement of xylem sap up through the xylem tracheids and vessels.
5. Explain the adaptive significance for plants of the phloem injury response.
6. What effect would applying a metabolic poison to phloem cells in the sink end of the phloem have on phloem transport? Why?

Suggested Readings

Books

Bidwell, R. G. S. 1979. *Plant physiology.* 2d ed. New York: Macmillan.

Esau, K. 1977. *Anatomy of seed plants.* 3d ed. New York: Wiley.

Milburn, J. A. 1979. *Water flow in plants.* New York: Longman.

Salisbury, F. B., and Ross, C. W. 1978. *Plant physiology.* 2d ed. Belmont, Calif.: Wadsworth.

Articles

Cohen, I. B. May 1976. Stephen Hales. *Scientific American.*

Rutter, A. J. 1972. Transpiration. *Carolina Biology Readers* no. 24. Burlington, N.C.: Carolina Biological Supply Co.

Wooding, F. B. P. 1978. Phloem. *Carolina Biology Readers* no. 15. Burlington, N.C.: Carolina Biological Supply Co.

Zimmerman, M. H. March 1963. How sap moves in trees. *Scientific American* (offprint 154).

Animal Nutrition and Digestion

Chapter Concepts

1. Animals are heterotrophic organisms that must obtain organic compounds from their environment.

2. Besides organic nutrients needed for energy and synthesis, animals require vitamins, minerals, and water.

3. Animal digestive tracts are specialized to transport, process, and absorb the nutrients that must be supplied to all body cells.

4. The kind and degree of specialization in any animal's digestive tract is related to the animal's metabolic requirements and the nature of its food supply.

5. Human digestion takes place in a highly specialized, complete digestive tract. Enzymes hydrolyze large organic nutrient molecules to produce smaller absorbable molecules.

6. Despite efficient protective mechanisms, the digestive tract can be damaged by its own digestive secretions and also is subject to occasional infection.

All organisms require energy for various functions that maintain **homeostasis,** that vital stable condition of the internal body environment. Organisms obtain this energy by oxidizing reduced organic compounds. Organisms also need a supply of raw material for growth and for replacement of worn-out or injured structures. All organisms can be separated into two categories—autotrophic or heterotrophic—on the basis of the methods by which they obtain the energy and raw materials that they require.

Autotrophic ("self-feeding") **organisms** use simple inorganic substances to synthesize reduced organic compounds that meet their energy and material requirements. Plants are the most familiar autotrophs; they use light energy, in photosynthesis, to produce reduced organic compounds. Plants are chemically versatile because they can produce the full range of their required organic compounds using only CO_2, H_2O, and various mineral ions taken up from the soil as starting materials.

The focus of this chapter is on **heterotrophic** ("other-feeding") **organisms,** such as animals, who are unable to synthesize reduced organic compounds from inorganic precursors. Heterotrophs must meet their energy and raw material requirements by obtaining already-formed reduced organic compounds. Some animals (**herbivores**) accomplish this by eating only plants. Other animals (**carnivores**) eat only other animals that have eaten plants or other autotrophic organisms. Still other animals (**omnivores**), such as humans, eat mixed (plant and animal) diets. Thus, animals ultimately are dependent on synthesis of organic compounds by autotrophic organisms.

Animals oxidize some ingested organic nutrients and thereby obtain energy (in the form of ATP) for the work that must be done by cells if organisms are to maintain themselves. Other organic nutrients are used as raw materials for synthesis of components of the animals' cells, tissues, and organs.

Usually, the organic nutrients that animals ingest are not in forms that can readily be absorbed into their bodies. Nutrients must be converted into absorbable forms so they can be transported across the plasma membranes of cells lining the digestive tract. **Digestion** is the sum total of physical and chemical processes that convert the food that animals eat into these absorbable forms. Thus, animals (and other heterotrophs) require **digestive systems** or, at least, cells capable of digestion to make these conversions of food material.

Nutrients

The three major categories of nutrients are: (1) organic molecules used in energy conversions and as raw material in synthesis; (2) minerals, typically in the form of inorganic salts or ions; and (3) vitamins (see table 10.1).

The first category is subdivided into four groups, three of which are familiar to almost everyone. They are **carbohydrates, proteins,** and **fats.** The fourth group of organic molecules contains the **nucleic acids,** compounds that are not commonly mentioned in discussions of diet and nutrition, but that nevertheless also are familiar because of their genetic roles.

Carbohydrates

Strictly speaking, carbohydrates such as sugars and starch are not essential nutrients for most animals, since carbohydrates can be synthesized in animal cells from proteins or fats. But carbohydrates are important nutrients for many organisms because they provide the bulk of the organisms' chemical energy supply. Carbohydrates are reduced carbon compounds that can be oxidized directly and conveniently in cellular energy conversion pathways in virtually all cells. This carbohydrate utilization centers around the reaction series of the Embden-Meyerhof pathway and the Krebs cycle (chapter 7) and uses the simple sugar (monosaccharide) **glucose** as a starting point. Thus, one carbohydrate, glucose, is a centrally important nutrient because it is used in cellular energy conversions in the cells of all organisms, whether the glucose is obtained in the diet or is synthesized from other organic compounds.

Glucose and other sugars are included in the diets of many animals, but many of the carbohydrates that animals ingest are in the form of **polysaccharides,** such as **starch, cellulose,** and **glycogen** (chapter 2). Polysaccharides are polymers of glucose; that is, they are chains of glucose molecules fastened together. Glucose and other monosaccharides are readily absorbed from the digestive tract, but polysaccharides are not. Polysaccharides must be enzymatically broken down (**hydrolyzed**) into sugar units that can be absorbed (figure 10.1).

Not all of the carbohydrates ingested by animals are immediately metabolized for energy. Especially in organisms that feed only periodically, carbohydrates must be accumulated and released slowly during times when the organisms are not feeding. The vast majority of animals store glucose as glycogen, which is hydrolyzed as needed to keep glucose levels in the body stable. Maintenance of a steady supply of glucose to body cells and tissues is an important homeostatic mechanism. In vertebrate animals, glycogen is stored in the liver, and the liver releases glucose into the blood. But most animals store

only a limited amount of carbohydrate and use it to meet short-term energy requirements. They convert the bulk of excess carbohydrates to fat, which is used for longer-term storage.

Some carbohydrates are not metabolized for energy but are used in assembling structural molecules such as the glycoproteins (carbohydrate and protein) and glycolipids (carbohydrate and lipid) of cells' plasma membranes. Glycoproteins also are important structural materials in the extracellular portions of tissues such as cartilage. In addition, some animals synthesize chitin, a polymer of modified glucose molecules, which is a major component of the hard, impermeable external skeletons (exoskeletons) of insects and other arthropods.

Proteins

Proteins are important structural elements of all cells, and enzymes, the vital organic catalysts in living organisms, also are proteins. Thus, all organisms must obtain the raw materials for protein synthesis. Proteins are polymers of **amino acids,** and animals obtain amino acids by eating and digesting the proteins of other organisms.

Table 10.1
Functions and Sources of Nutrients.

Nutrient Category	Physiological Roles	Some Major Food Sources
Carbohydrates	Primary source of energy; structural components of membranes and intercellular material; structural components of nucleic acids and ATP	Starch from grains and certain vegetables; glycogen from meats and seafoods; sugars from cane sugar, beet sugar, molasses, honey, and various fruits
Fats	Source of energy; energy reserve; carry fat-soluble vitamins; primary components of cell membrane	Meat, eggs, fats and oils, milk, cheese, nuts
Proteins	Source of essential amino acids for synthesis of enzymes, structural proteins, antibodies, and contractile proteins	Meat, fish, poultry, eggs, dairy and cereal products, nuts, legumes
Minerals	Structure of bones and teeth; internal pH regulation; components of hormones, cofactors of enzymes; transmission of nerve impulses; muscle contraction	Dairy products, vegetables, eggs, poultry, fish, meat, table salt, fruits, nuts, grains
Vitamins	See table 10.5	See table 10.5

Proteins are enzymatically hydrolyzed to yield amino acids that are then absorbed through digestive tract linings (figure 10.2).

Most animals can use carbohydrate starting material to synthesize some of the amino acids needed for protein synthesis, but they cannot synthesize certain other amino acids. The amino acids that animals cannot synthesize are known as **essential amino acids** (see chapter 7); therefore, essential amino acids must be present in the diet. Different animals have different numbers and different sets of essential amino acids. Ten amino acids are essential in the human diet.

The specific list of essential amino acids for any given species is believed to reflect the evolutionary history of that species. For example, lysine and tryptophan are two of the essential amino acids for humans. Because most plant proteins contain very little lysine or tryptophan, humans obtain these amino acids mainly by eating meat. This suggests that animal protein has been a part of human diets for a long period of time and that our very early ancestors were omnivores who included at least some meat in their diets.

A related consequence of the low lysine and tryptophan content of plant proteins is that humans on strictly vegetarian diets have difficulty obtaining adequate quantities of these essential amino acids in their diets. People who choose to become vegetarians in countries with adequate food supplies can solve the problem by including eggs or milk in their diets or by very carefully selecting the few plant products that do contain larger quantities of lysine and tryptophan.

No such options are open to many people in underdeveloped countries, where neither adequate supplies of animal proteins nor a wide selection of plant products are available. Many such people around the world suffer from **kwashiorkor,** a protein deficiency disease that occurs in people who eat a mainly starchy diet (even though their diets may contain enough calories). Growing children with kwashiorkor become anemic and listless, and they have very low resistance to even the mildest diseases and infections. Even if they survive the childhood diseases, many of them become mentally retarded and unhealthy adults. Quality dietary protein supply is one of the world's greatest health problems.

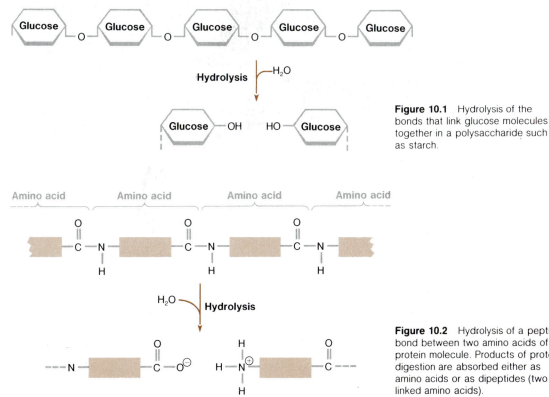

Figure 10.1 Hydrolysis of the bonds that link glucose molecules together in a polysaccharide such as starch.

Figure 10.2 Hydrolysis of a peptide bond between two amino acids of a protein molecule. Products of protein digestion are absorbed either as amino acids or as dipeptides (two linked amino acids).

Fats

Fats are very energy rich because fat molecules are highly reduced compounds; they yield more energy per unit weight when they are oxidized than do carbohydrates or proteins (table 10.2). Not surprisingly, animals' energy reserves generally are stored as fat. Fat's energy-to-weight ratio is more favorable than that of other possible organic storage compounds; fats simply are easier to carry around.

Fats are important structural components, especially in cell membranes. Fat also functions as an insulating material that helps to stabilize body temperatures of many organisms.

A molecule of the kind of fat commonly eaten by animals contains glycerol and three fatty acids. While some fat is absorbed intact as small droplets that pass through the digestive tract linings, much of the ingested fat is enzymatically hydrolyzed to yield glycerol and fatty acids that are then absorbed (figure 10.3).

As is the case with some of the amino acids required for protein synthesis, certain animals cannot synthesize some fatty acids required for fat synthesis. Those fatty acids must be included preformed in the diet, and their absence can produce deficiency symptoms. For example, human diets must include linoleic acid, a fatty acid that humans cannot synthesize.

Nucleic Acids

Despite their biological significance as genetic and regulatory molecules, the nucleic acids DNA and RNA are nonessential components of animal diets. This is because the cells of most, if not all, animal species can synthesize nucleic acids from other organic molecules. Nevertheless, nucleic acids are always present in animal diets because heterotrophs consume the cells of other organisms, and all cells contain nucleic acids. The energy contribution of nucleic acids is minimal, however, because they are present in such low concentrations in foods.

Minerals

One way to analyze the mineral nutrient requirements of an organism might be to determine the organism's mineral content (table 10.3). This can be misleading, however, because the mere presence of an element in an organism's body does not necessarily indicate that the element is required. Indeed, a chemical analysis of a typical human from an industrialized society probably would reveal trace amounts of mercury and lead. These elements definitely are not essential; in fact, they are poisons. However, they are not easily excreted and tend to accumulate in stable tissues such as hair and fat.

Table 10.2
Energy Yielded by Oxidation of Organic Nutrients.

Nutrients	Average Energy Yield in Kilocalories (and Kilojoules) Per Gram
Carbohydrates	4.1 (17.2 kJ)
Proteins	4.4 (18.4 kJ)
Fats	9.3 (38.9 kJ)

Table 10.3
Weight of Components of a 70 kg Adult Human Male.

Components	Number of Grams
Water	41,400
Fat	12,600
Protein	12,600
Carbohydrate	300
Na	63
K	150
Ca	1,160
Mg	21
Cl	85
P	670
S	112
Fe	3
I	0.014

Excluding an element from the diet experimentally and then monitoring the consequences is one way to distinguish between the mere presence of an element in the body and a nutritional requirement for that element. This technique demonstrates that some elements are required in relatively large quantities, others (the **trace elements**) are required in much smaller quantities, and yet others may or may not be required. The functional roles of some minerals, such as calcium, potassium, and sodium, are quite obvious, but the roles of certain trace elements are poorly understood (table 10.4).

Phosphorus and sulfur enter heterotrophs' bodies as components of organic molecules, but calcium, potassium, sodium, chlorine, and the trace elements are absorbed as ions.

Vitamins

Vitamins are a set of relatively simple organic molecules that are required in small quantities for a variety of biological processes. Because vitamins are compounds that cannot be synthesized in the body, they must be present in small quantities in an organism's diet. The lists of organic compounds classified as vitamins are different for different species because of differing capabilities for synthesis. For example, rats can synthesize

ascorbic acid, but humans cannot. Ascorbic acid is a vitamin (**vitamin C**) for humans and must be included in the diet, but it is not classified as a vitamin for rats. The fat soluble vitamins (page 261) required by vertebrate animals can be synthesized by most invertebrate animals and are not classified as vitamins for invertebrates.

Vitamins differ from essential amino acids and fatty acids in that vitamins are not oxidized for energy or used as building blocks of larger structural molecules. But vitamins are vital for a wide variety of biological processes. For example, B vitamins are essential as coenzymes of enzymes involved in cell metabolism, vitamin A is required for photochemical reactions in vision, and vitamin C must be present for maintenance of connective tissues and promotion of normal healing. Vitamin molecules can function over and over in these roles and, thus, are not depleted as quickly as other nutrients are. Maintenance of normal vitamin levels requires only a small, steady supply from the diet, and serious, noticeable vitamin deficiency symptoms develop only after extended periods of dietary vitamin deprivation.

Surprisingly, the need for vitamins has been recognized, at least in a vague way, for centuries. The ancient Egyptians determined that certain eye problems could be cured by eating liver,

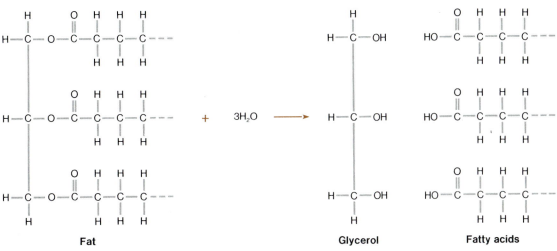

Fat   **Glycerol**   **Fatty acids**

Figure 10.3 Hydrolysis of fats yields glycerol and fatty acids. Actually, most digested fats are absorbed as fatty acids and monoglycerides, glycerol with one fatty acid still covalently bonded to it. Fatty acids are drawn in their unionized form to show how water addition occurs during hydrolysis. Normally, they would be ionized
$$(-C\overset{O}{\underset{O^-}{\parallel}}\ \ + H^+).$$

Table 10.4

Physiological Roles and Sources of Major Mineral Nutrients and Some Trace Minerals.

Minerals	Physiological Roles	Food Sources
Major Minerals		
Sodium (Na)	Major extracellular fluid cation; osmotic regulation; absorption of glucose into cells; transmission of electrochemical impulses in muscles and nerves	Table salt (NaCl), milk, meat, eggs, baking soda, baking powder, carrots, celery, beets, spinach
Potassium (K)	Major intracellular fluid cation; regulation of nerve and muscle function; glycogen formation; protein synthesis	Whole grains, fruits (especially bananas), meat, legumes, vegetables
Calcium (Ca)	Inorganic salts in bones and teeth; blood clotting; muscle contraction; nerve impulse transmission; cell membrane permeability; enzyme activation (ATPase)	Milk, cheese, leafy green vegetables, egg yolk, whole grains, legumes, nuts
Magnesium (Mg)	Constituent of bones and teeth; activator and enzyme cofactor in carbohydrate and protein metabolism	Whole grains, nuts, meat, legumes, milk
Phosphorus (P)	Bone formation; phosphorylation of glucose, glycerol, fatty acids; aids in absorption, transport, and metabolism; buffer system; energy metabolism (enzymes, ATP)	Cheese, milk, meat, egg yolk, whole grains, legumes, nuts
Chlorine (Cl)	Major extracellular fluid anion; buffering; water balance; hydrochloric acid in stomach	Table salt
Sulfur (S)	Component of some amino acids; activates enzymes	Meat, eggs, cheese, milk, nuts, legumes
Trace Minerals		
Iron (Fe)	Component of the heme group found in hemoglobin, myoglobin, and cytochromes	Liver, meats, egg yolk, whole grains, enriched bread and cereal, legumes, dark green vegetables, nuts
Copper (Cu)	Associated with iron in hemoglobin synthesis and the transport and absorption of iron; present in cytochromes, red blood cells; involved in bone formation, maintenance of nervous tissue	Liver, meat, seafood, whole grains, legumes, nuts
Iodine (I)	Component of thyroid hormone, which regulates cellular respiration	Iodized salt, seafood
Manganese (Mn)	Ions necessary in urea formation, protein metabolism; Embden-Meyerhof pathway and Krebs cycle	Legumes, cereals, soybeans, nuts, tea, coffee
Cobalt (Co)	Constituent of vitamin B_{12}, essential for red blood cell formation	Vitamin B_{12} in meat
Zinc (Zn)	Essential enzyme constituent	Widely distributed in foods, liver, seafoods
Molybdenum (Mo)	Constituent of specific enzymes involved in nitrogen metabolism	Organ meats, milk, whole grains, leafy vegetables, legumes

which is now known to contain a great deal of vitamin A. Over 200 years ago, an English physician noted that sailors who ate citrus fruits did not develop **scurvy,** a disease whose symptoms include anemia, slow healing, painful joints, gum bleeding, and eventual loss of teeth (figure 10.4). After this discovery, scurvy, formerly the scourge of sailors of the day, was prevented because the British Admiralty insisted that all ships carry limes on board to be included in the sailors' rations. This regulation accounts for the British (and especially British sailors) being called "Limeys." Vitamin C, the nutrient that had been missing from the sailors' diets, was not chemically identified as ascorbic acid until 1933.

In 1886, a group of Dutch scientists was sent to investigate the crippling disease **beriberi** in the East Indies. Originally, they assumed that beriberi was caused by a microorganism. But then one of them, Christian Eijkman, noted that chickens developed beriberilike symptoms when they were fed the same kind of polished rice eaten by people suffering from beriberi. Polished rice is simply rice from which the hull has been removed to retard spoilage. When the chickens were fed rice that still had the hulls intact, they quickly recovered; the same treatment proved effective for human victims of beriberi. Eijkman mistakenly concluded that rice kernels contained a nerve toxin that caused beriberi symptoms and that the hulls contained a factor that neutralized the toxin. Years later, the antiberiberi factor was identified as **thiamine (vitamin B₁).** Because most of the thiamine of rice is contained in the hull, and because much of that remaining in the rice kernels is lost in cooking (thiamine is water soluble), polished rice, like other milled grains, is vitamin deficient. This also explains why white flour typically must be "vitamin enriched." All the vitamins originally present in the grain are lost during the milling process, and vitamins must be added to the flour, which, without them, would be essentially starch with some protein.

Over the years, the list of vitamins required for humans has grown and is today deemed virtually complete, although the functions of a few of the vitamins (such as vitamin E) are still not well understood.

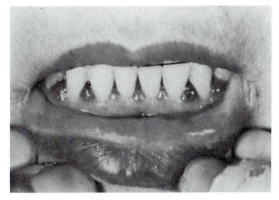

Figure 10.4   Scurvy. Gum bleeding is one of the most characteristic symptoms of severe vitamin C deficiency.

The vitamins are divided into two major groups: those that are soluble in organic solvents (**fat-soluble vitamins**) and those that dissolve in water (**water-soluble vitamins**). Fat-soluble vitamins include A, D, E, and K. Water-soluble vitamins include the vitamins of the B complex and Vitamin C. Information about vitamins required by humans is summarized in table 10.5. Two common vitamin deficiency syndromes are shown in figure 10.5.

Minimum daily requirements (now usually referred to as "**recommended daily allowance**" or **RDA**) for most of the vitamins have been calculated for humans (table 10.6). In a sense, however, these values are really only educated guesses. An individual ingesting the recommended daily allowance for vitamin C presumably will never develop scurvy, but might there be milder, less dramatic consequences of ingesting no more than the minimum amount? Some scientists believe that the recommended daily allowances are barely adequate to prevent deficiency symptoms and that they are well below the optimal level. For example, Linus Pauling, winner of two Nobel Prizes, believes that humans actually require vitamin C in much larger doses than the recommended daily allowance. Pauling says that failure to meet this larger "requirement" does not result in any life-threatening disease, but he believes that large doses of vitamin C increase resistance to upper respiratory diseases such as the common cold.

Table 10.5
Functions, Deficiency Symptoms, and Sources of Vitamins.

Vitamin	Physiological Role	Deficiency Symptoms	Some Major Food Sources
Water-Soluble Thiamine (B_1)	Coenzyme in carbohydrate metabolism	Beriberi, neuritis, loss of appetite, heart failure, indigestion, fatigue, edema, mental disturbance	Whole grains, organ meats, yeast, nuts, pork
Riboflavin (B_2)	Coenzyme in protein and carbohydrate metabolism, as part of FAD	Vascularization of the cornea; inflammation and fissuring of the skin; sores on the lip, swollen tongue	Milk, cheese, eggs, liver, yeast, leafy vegetables, wheat germ
Niacin (B_3)	Coenzyme in energy metabolism; part of NAD and NADP	Pellagra, skin eruptions, neuritis, fatigue, digestive disturbances	Whole grains, yeast, liver and other meats
Pyridoxine (B_6)	Coenzyme in many phases of amino acid metabolism	Anemia, convulsions, neuritis, dermatitis, impairment of antibody synthesis	Whole grains, kidney, liver, fish, yeast
Pantothenic acid	Forms part of Coenzyme A	Neuromotor and cardiovascular disorders, impairment of antibody synthesis	Present in most foods, especially meat, whole grains, eggs
Biotin	Coenzyme in decarboxylation and deamination	Dermatitis, muscle pains (deficiency is rare)	Egg yolk, liver, yeast
Folic acid	Coenzyme in formation of nucleotides and heme	Anemia, impairment of antibody synthesis	Leafy vegetables, liver
Cobalamin (B_{12})	Coenzyme in formation of nucleic acids and proteins	Pernicious anemia	Liver and other organ meats
Ascorbic acid (C)	Vital to collagen and intercellular cement for bone, teeth, cartilage, and blood vessels; maintains resistance to infection; frees iron to make hemoglobin	Scurvy, anemia, slow wound healing	Citrus fruits, tomatoes, green leafy vegetables
Fat-Soluble A (retinol)	Formation of visual pigments; maintenance of normal epithelial structure	Night blindness; dry, flaky skin and mucous membranes	Green and yellow vegetables, dairy products, egg yolk, fruits, butter, fish-liver oil
D (calciferol)	Increases absorption of calcium and phosphorus and their deposition in bones	Rickets	Fish oils, liver, egg yolk, milk and other dairy products, action of sunlight on lipids in the skin
E (tocopherol)	Antioxidant; protects red blood cells, vitamin A, and unsaturated fatty acids from oxidation	Fragility in red blood cells, male sterility (in rats)	Widely distributed, especially in meat, egg yolk, green leafy vegetables, seed oils
K (menadione)	Needed in synthesis of prothrombin, which is necessary for blood clotting	Slow blood clotting and hemorrhage	Green leafy vegetables, synthesis by intestinal bacteria

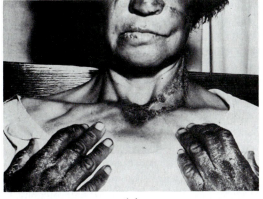

(a)

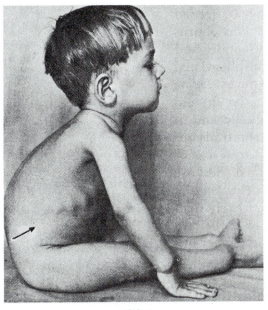

(b)

Figure 10.5 Common vitamin deficiency syndromes. (a) **Pellegra,** due to niacin deficiency. An obvious symptom of pellegra is dermatitis, a severe skin irritation, that develops in areas exposed to sunlight. (b) **Rickets,** due to vitamin D deficiency. Because a vitamin D deficiency reduces the body's ability to absorb and use calcium and phosphorus, bones become soft and deformed. This child has deformed ribs (arrow), ankles, and wrists, as well as bowing of the legs, which is a characteristic symptom of rickets.

Table 10.6
United States Recommended Daily Vitamin Allowances (U.S. RDA).

Vitamins	Unit of Measurement*	Children under Four Years of Age	Adults and Children Four or More Years of Age	Pregnant or Lactating Women
Water-Soluble				
Thiamine (B$_1$)	mg	0.7	1.5	1.7
Riboflavin (B$_2$)	mg	0.8	1.7	2.0
Niacin (B$_3$)	mg	9.0	20.0	20.0
Pyridoxine (B$_6$)	mg	0.7	2.0	2.5
Pantothenic acid	mg	5.0	10.0	10.0
Biotin	mg	0.15	0.3	0.3
Folic acid	mg	0.2	0.4	0.8
Cobalamin (B$_{12}$)	μg	3.0	6.0	8.0
Ascorbic acid (C)	mg	40.0	60.0	60.0
Fat-Soluble				
A (retinol)	IU	2,500	5,000	8,000
D (calciferol)	IU	400	400	400
E (tocopherol)	IU	10	30	30

Source: U.S. Department of Health, Education, and Welfare, Food and Drug Administration, Revised January 1976.

*Unit of measurement: IU = International Units; μg = micrograms; mg = milligrams

Serious problems can develop from taking large doses of vitamins (**megavitamin therapy**), however. Although excess amounts of the water-soluble vitamins are excreted in the urine, excess amounts of some of the fat-soluble vitamins accumulate in fat tissues in the body. Death by massive vitamin overdose is rare, but vitamin overdoses can cause toxicity responses. Furthermore, there is little or no clear-cut evidence that megavitamin therapies actually produce the benefits claimed by their proponents.

Digestion in Animals

Each unicellular heterotroph obtains nutrients directly from its environment, and its nutritional needs must be supplied by its own activities. But in even the simplest animals, which are multicellular heterotrophs, some cells that are specialized for digestion surround and line a digestive cavity or **gut.** Thus, one group of cells in one part of the body is specialized to obtain nutrients for all cells in the body. This specialization provides other body cells with a nutrient supply so that they can carry on other specialized functions.

More complex digestive tracts with various specializations for obtaining, digesting, and absorbing nutrients have paralleled the evolution of larger multicellular animal bodies, and there is considerable regional specialization in different parts of such digestive systems. But the actual processes of nutrient absorption, which are the key final steps of nutrient procurement, depend on properties and activities of the individual cells that line the digestive cavity. Thus, examination of nutrient absorption of unicellular heterotrophs can provide good background information for study of animal digestive tracts.

Unicellular heterotrophs obtain nutrients from their environments in several ways. Some live in environments that are so rich in nutrient molecules that the nutrients are more concentrated outside the cells than inside. In such cases,

nutrients enter by simple **diffusion** or by **facilitated diffusion** (see chapter 4), which involves carrier molecules but does not require direct energy expenditure by the cells. Other unicellular heterotrophs exist in environments in which nutrient molecules are relatively scarce. Nutrients do not diffuse into cells against concentration gradients, and therefore, they are moved across plasma membranes by **active transport.** That is, cells expend energy (in the form of ATP molecules) and work to bring in nutrient molecules.

Larger quantities of nutrient material are brought into some unicellular heterotrophs by specialized, energy-requiring activities that enclose the material in membrane-bounded spheres (figure 10.6). Liquid droplets can be taken in by **pinocytosis,** and nutrients are then absorbed into the cell through the membrane of the **pinocytotic vesicle.** Still other unicellular heterotrophs engulf solid food material by **phagocytosis.** After a **food vesicle** has been formed by phagocytosis, the vesicle fuses with a **lysosome** and digestive enzymes are introduced. Enzymes supplied by the lysosome digest the material in the food vesicle, and nutrient molecules are absorbed through the vesicle membrane (p. 84).

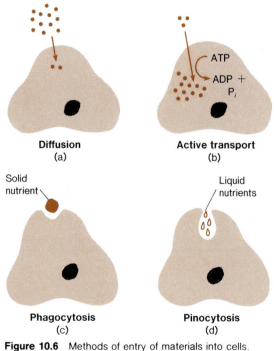

Figure 10.6   Methods of entry of materials into cells. (a) Diffusion. (b) Active transport. (c) Phagocytosis. (d) Pinocytosis.

Box 10.1
"Predatory" Fungi

One other kingdom of organisms, in addition to animals, consists entirely of heterotrophic organisms. This is the kingdom Fungi, and it includes organisms such as molds, rusts, smuts, and mushrooms.

What is the source of nutrient supply for the familiar bracket fungus growing on the side of a dead tree (box figure 10.1A)? Obviously, the answer must be the dead tree, since the fungus lacks chlorophyll and cannot photosynthesize. The fungus secretes enzymes that hydrolyze large organic molecules of the tree. This **extracellular** (outside the cell) **digestion** produces a supply of absorbable nutrients that can be utilized by the fungus. Fungi that live on dead organisms and derive their nutrients by participating in decay processes are called **saprophytes.**

Other fungi, including many that grow on living trees, are **parasites.** Parasitic fungi grow on or inside many different living plants or animals. The invasive and digestive activities of these parasitic fungi can cause significant damage to the host organism.

Some of the most unusual of fungal nutritional adaptations are seen in a group of soil fungi that capture living animals. These "predatory" fungi produce a sort of hangman's noose (box figure 10.1B) that tightens around the body of a soil roundworm and traps it. Once a worm has been secured, fungal cells grow into the worm and secrete hydrolytic enzymes that begin to digest the worm's body. Capturing a worm in this way provides the fungus with a concentrated source of organic nutrients.

Box figure 10.1A A bracket fungus.

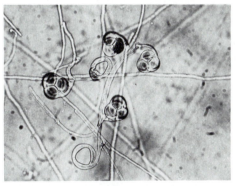

(a)

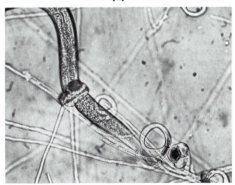

(b)

Box figure 10.1B (a) A fungus that traps roundworms (nematodes). The "nooses" consist of three fungal cells. (Magnification × 275) (b) When a roundworm enters the loop, the cells swell and the worm is held firmly in place. Fungal cells invade the body and secrete enzymes that digest it. (Magnification × 275)

This pattern of **intracellular** ("inside the cell") **digestion** is also seen in several types of cells in animal bodies. Some animals depend heavily on the intracellular digestion of gut-lining cells for a major part of their digestive activity. Phagocytic cells that function in body defense mechanisms (for example, some white blood cells of vertebrate animals) also perform intracellular digestion on invading microorganisms that they have engulfed.

For organisms obtaining nutrients by phagocytosis, one might assume, logically, that food items or prey should be smaller than the attacking organism since the food cannot be taken in bit by bit. However, some protozoans can engulf prey larger than themselves. Members of the protozoan genus *Didinium* prey upon members of the genus *Paramecium*. In feeding, the smaller *Didinium* first positions the *Paramecium* horizontally to the axis of the *Didinium,* and then, through processes not yet fully understood, folds the *Paramecium* longitudinally. As the collapsed *Paramecium* is engulfed, the *Didinium* elongates to accommodate its massive meal (figure 10.7).

Animal Digestive Tracts

In the most primitive animals, the gut is a blind (closed) sac with only one opening that must serve as both entrance and exit (**mouth** and **anus**). Undigested residues must be ejected through the same opening through which the food originally entered. Such digestive tracts are said to be **incomplete.**

Larger, more complex multicellular animals have gut tubes that are tunnellike. Food enters at one end through a mouth and flows through the gut where digestion and absorption occur. Undigested residues pass out through a posterior exit, the anus. Such flow-through systems are called **complete digestive systems.** The flow-through arrangement permits greater digestive efficiency because it involves regional specialization of parts of the gut that carry out such functions as physical breakdown of food, temporary storage, chemical digestion, and absorption. These regional specializations are not seen in incomplete digestive systems.

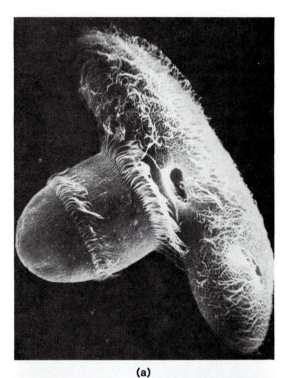

(a)

(b)

Figure 10.7   (a) A *Didinium* attacks a *Paramecium* twice its size. (Magnification × 520) (b) The *Paramecium* is folded in half, and the *Didinium* increases in length and engulfs it. (Magnification × 480)

Incomplete Guts

The coelenterates (the animal phylum that includes *Hydra,* jellyfish, sea anemones, and corals) are among the simplest of multicellular animals, and their digestive systems mirror this simplicity. Coelenterates all possess incomplete digestive tracts—digestive cavities that are blind sacs with only one opening, the mouth. Body projections called **tentacles** surround the mouth opening. Tentacles have specialized cells that contain stinging devices, the **nematocysts,** that paralyze and entangle prey. After a prey organism is captured, the tentacles push the prey through the mouth opening into the digestive cavity, which is called the **gastrovascular cavity** because it fulfills both digestive ("gastro") and transport ("vascular") functions (figure 10.8).

Some specialized cells in the cavity lining secrete digestive enzymes into the gastrovascular cavity, where digestion of a food particle or prey organism begins. Such digestion, outside cells and in a digestive cavity, is called **extracellular digestion.** But the bulk of digestion in coelenterates is **intracellular digestion.** Phagocytic cells in the lining of the gastrovascular cavity engulf pieces of food material and continue digestion inside food vacuoles. These phagocytic cells, in turn, relay the products of digestion to other cells of the organism.

Similar digestive patterns are seen in the flatworms (the animal phylum that includes the free-living planarians and the parasitic flukes and tapeworms). **Planarians,** for example, possess an incomplete gut that is much more structurally

Hydra

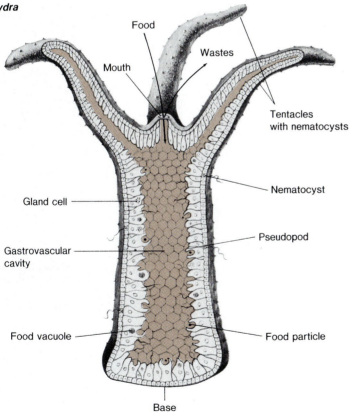

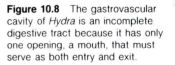

Figure 10.8 The gastrovascular cavity of *Hydra* is an incomplete digestive tract because it has only one opening, a mouth, that must serve as both entry and exit.

Food

Mouth

Wastes

Tentacles with nematocysts

Nematocyst

Gland cell

Pseudopod

Gastrovascular cavity

Food vacuole

Food particle

Base

elaborate than that of the coelenterates because
it is branched and lobed. But, functionally, the
planarian gut is quite similar to that of coelenterates. Extracellular digestion is limited; the
bulk of digestion is intracellular and is accomplished inside phagocytic cells in the gut lining.
There is no significant regional specialization of
the gut, and the problem of having to eject undigested residues through the mouth is the same
as it is for the coelenterates (figure 10.9).

One group of parasitic flatworms, the **flukes,**
have guts much like those of planarians, and they
obtain nutrients by sucking in tissue fluids and
blood of their hosts. **Tapeworms** are highly modified (degenerate) intestinal parasites of vertebrate animals. They live in a nutrient broth of
partially digested contents of their hosts' guts.
Tapeworms have no digestive systems at all and
absorb nutrients directly through their body surfaces by diffusion and active transport.

Complete Guts

The development of the anus and complete digestive tracts was an evolutionary breakthrough. In
an incomplete digestive tract, there is no effective
means of segregating recently ingested food from
food ingested much earlier. A complete digestive
tract, however, permits one-way flow because undigested food does not have to pass back out the
same way that it came in (figure 10.10).

Complete digestive tracts also can have progressive digestive processing in specialized regions along the gut. These specialized regions
amount to separated digestive microenviron-
ments with different conditions (for example,
different pH's). Thus, food can be digested
efficiently in a series of distinctly different steps.

There are many modifications of the basic
tubular plan of a complete digestive tract, and
the structure is closely correlated with different
animals' food-gathering mechanisms and their
diets.

Figure 10.9 A planarian's
gastrovascular cavity is more
complex than that of *Hydra* because
it branches extensively, but the two
are functionally similar because they
are both incomplete digestive tracts
with only one opening.

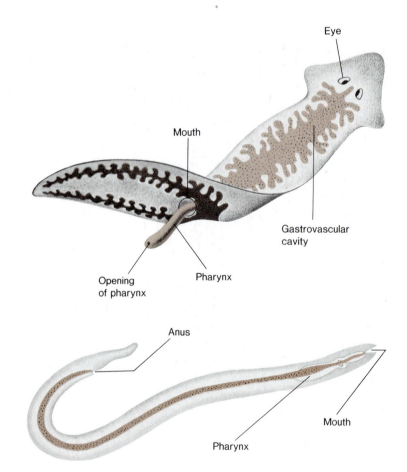

Figure 10.10 Simple complete
digestive tract of the parasitic
roundworm *Ascaris*.

Continuous Versus Discontinuous Feeding

One variable related to the structure of the gut is whether the organism in question is a **continuous** or **discontinuous feeder.** Typically, discontinuous feeders have more gut specialization than continuous feeders because discontinuous feeders take in large meals that must be either ground up, stored, or both.

Many continuous feeders are slow moving or even completely sessile (they remain permanently in one place). Aquatic **filter feeders** such as clams and barnacles remain in one place and continuously "strain" small food particles out of the water (figure 10.11).

Discontinuous feeders tend to be active, sometimes highly mobile, organisms. For example, many carnivores pursue and capture relatively large prey. When they are successful, they must eat large meals so that they need not spend all their time in constant or even frequent pursuit of prey. Thus, carnivores must have digestive tracts that permit storage and gradual digestion of large, relatively infrequent meals.

Herbivores spend more time eating than carnivores do, but they nevertheless are also discontinuous feeders. They have to move from area to area when food is exhausted and, at least in natural settings, must limit grazing time to avoid excessive exposure to predators. Their digestive tracts, like those of the carnivores, must permit relatively rapid food gathering and gradual digestion.

(a)

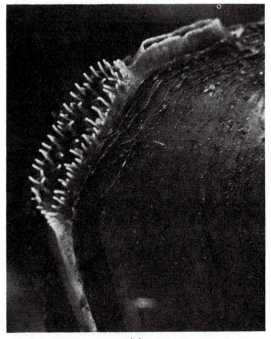

(b)

Figure 10.11 Two types of filter feeders. (a) "Gooseneck" barnacles *(Lepas)*. Barnacles feed by sweeping feathery appendages through the water and straining out food particles. These appendages are shown partially extended here. (b) A freshwater clam with its siphons extended. Clams move a stream of water over their gills, which filter out food particles. This water enters through one siphon and leaves through the other.

Figure 10.12 Crops are expanded regions in the anterior part of the digestive tract that function as food storage organs. Animals that have crops include: (a) earthworms, (b) grasshoppers, and (c) birds.

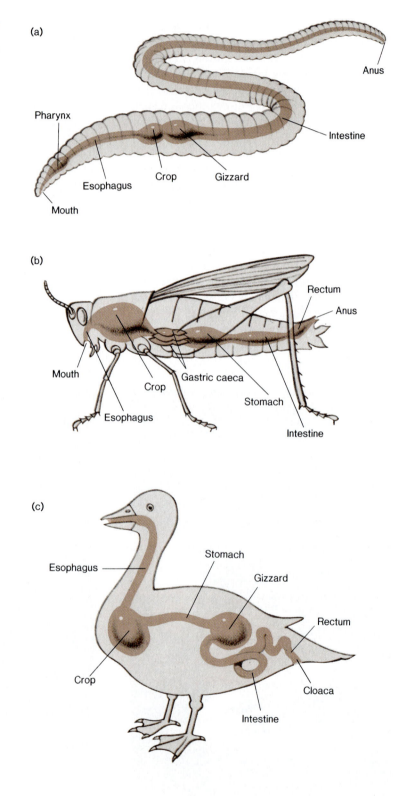

While **stomachs** provide storage for large meals eaten by many discontinuous feeders, there are also other specialized storage mechanisms. Chipmunks, hamsters, and other rodents pack food into massive cheek pouches and eat later in the safety of their burrows. Another storage mechanism found in many types of animals is the **crop.** A crop is an enlargement of the digestive tract anterior to the stomach. Such diverse animals as earthworms, insects, and birds have crops (figure 10.12).

Method of Physical Breakdown of Food

Another factor that influences digestive tract structure and function is the size of food items. If solid food items are included in the diet, a mechanism for physical breakdown (**maceration**) of food is often present. Digestion is more efficient if the food is fragmented, thereby exposing more surface area to the action of enzymes.

Teeth are the most commonly thought of macerating device. But not all animals' teeth are used for grinding up food. Many animals use their teeth for killing or for ripping chunks out of their prey. These chunks are then swallowed whole (figure 10.13).

Another macerating device is the **gizzard,** an organ found in the digestive tracts of many kinds of animals. Gizzards have very muscular walls between which gut contents are rubbed back and forth vigorously, thus grinding the food. Hard particles swallowed with food aid this grinding action. The small stones that an earthworm swallows as it eats its way through the soil help to grind up particles of food in the earthworm's gizzard. Birds eat small stones regularly to supply similar grinding material for their gizzards.

(a)   Horse

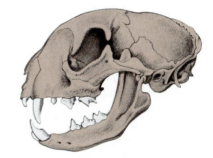

(b)   Lion

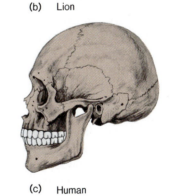

(c)   Human

Figure 10.13 Comparison of teeth of several mammals. (a) Horse, a grazing herbivorous mammal that clips off plants with its incisors and grinds them with its flat molars. (b) Lion, a carnivorous mammal that uses its teeth for killing and for tearing off chunks of flesh that are swallowed whole. (c) Human, an omnivorous mammal.

Many species of animals do not face the problem of food maceration at all because they are **fluid feeders.** Fluid feeders may be highly mobile discontinuous feeders, such as vampire bats and mosquitoes, or relatively inactive continuous feeders, such as aphids and scale insects (figure 10.14). Fluid feeders also include animals that, in effect, create their own fluids for food. Spiders, for instance, produce a series of powerful **proteolytic** (protein-hydrolyzing) **enzymes** that they inject into their victims. When the internal organs of its prey are liquified, a spider drains its victim dry, leaving nothing but an empty exoskeleton in the case of insect prey.

Type of Food   The type of food to be digested is another factor that influences digestive tract structure and function.

Carnivores' food (flesh) is relatively easier to digest than the food of herbivores or omnivores, and as a consequence, carnivores tend to have relatively shorter and less complicated guts.

Herbivores tend to have longer, more complex digestive tracts. Herbivores digest the contents of plant cells that are broken open during chewing and during further processing in the herbivores' digestive tracts. In addition, herbivores have mechanisms that allow them to derive nutrition from the cellulose in plant cell walls. Microorganisms are housed in special saclike regions (such as the **rumens** of cows, sheep, deer, and goats) of the herbivores' digestive tract. Symbiotic bacteria and protozoa living there produce **cellulase,** an enzyme that hydrolyzes cellulose (figure 10.15). Thus, herbivores can derive energy from cellulose even though their own digestive tracts do not produce cellulase. This adaptation is valuable because cellulose is abundantly available to animals.

Omnivores are halfway between carnivores and herbivores in a dietary sense, and not surprisingly, their gut lengths generally are intermediate between the two extremes. Omnivores do not have mechanisms for cellulose utilization so they can digest only the intracellular (cell contents) portion of the plant material in their diets.

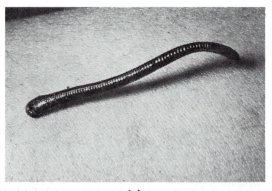

(a)

(b)

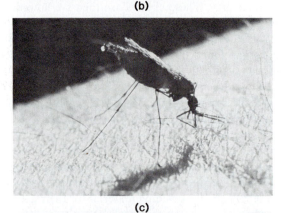

(c)

(d)

Figure 10.14   Fluid feeders. Leeches (a), vampire bats (b), and mosquitoes (c) are active, discontinuous feeders that feed on blood. Aphids (d) feed on plant sap for long periods.

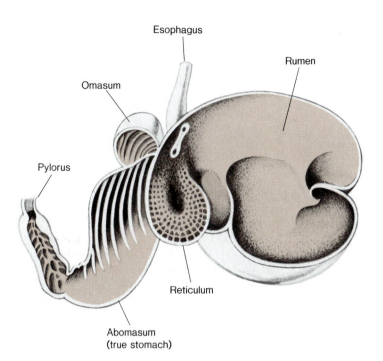

Figure 10.15 Stomach compartments of a ruminant mammal. Bacteria and protozoa, living in a dilute salt solution in the rumen and reticulum, hydrolyze cellulose in plant material. Food is regurgitated repeatedly for "cud" chewing and is swallowed again. Eventually, food is swallowed into the omasum, which churns it up before it passes into the abomasum. The abomasum is a typical vertebrate glandular stomach.

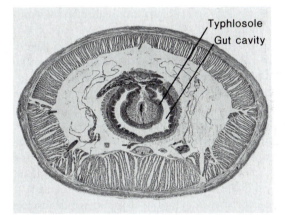

Figure 10.16 Cross section of an earthworm showing the typhlosole, which is a longitudinal fold that greatly increases the gut's absorptive surface area.

Surface Available for Absorption Absorption is the payoff of all activities of the digestive tract. Thus, it is not surprising that many digestive tract specializations that provide additional exposed surface area for absorption have evolved. Perhaps the simplest way to have a large surface area for absorption is to have a long absorptive section in the gut. This keeps food in contact with absorptive surfaces for a longer time. Long, coiled guts have evolved in many different animal groups.

Surface areas are increased in the relatively straight, uncoiled guts of some animals by other means. For example, most earthworms have a longitudinal fold called a **typhlosole** that gives their gut nearly twice as much surface area as a simple round tubular gut of similar length (figure 10.16). Sharks have a complex folded membrane called a **spiral valve** that forces food to follow a route resembling a spiral staircase as it passes

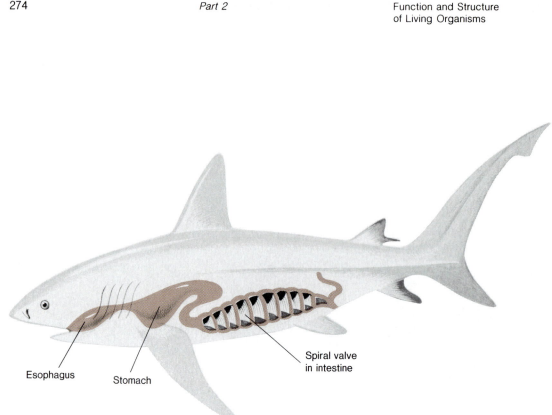

Esophagus   Stomach   Spiral valve
                                in intestine

Figure 10.17 The spiral valve greatly increases the
internal surface area of the shark intestine.

through the gut. As a result, the food is exposed
to a much larger surface area (figure 10.17).

The problem of sufficient absorptive surface
is most pronounced in large mammals. They
must supply nutrients to maintain large body
masses and the relatively high metabolic rates
required for their body temperatures. The human
gut is one example of how the surface area problem in mammals has been solved. The human
small intestine is not a simple, smoothly lined
tube; it has ridges and furrows that give it an
almost corrugated appearance. On the surface of

these ridges and furrows are small fingerlike projections called **villi.** Cells on the surfaces of villi
have minute projections called **microvilli** (figure
10.18). These structural features increase the
actual absorptive surface of the human small intestine to about 120 times the area of a simple
smooth tube of similar length. A smooth tubular
small intestine would have to be five to six
hundred meters long to have a surface area comparable to the human small intestine. Such a
huge, long gut would be hard to carry around,
and if the movement rate were the same as it is
in a normal human gut, it would require weeks
for food to traverse the small intestine.

(a)

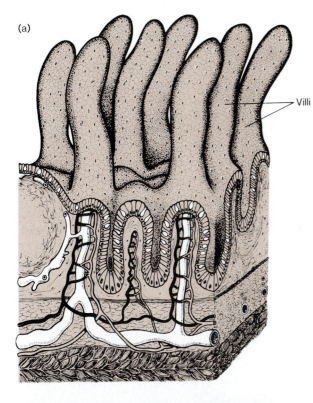

Villi

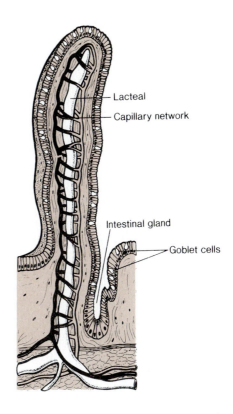

Lacteal

Capillary network

Intestinal gland

Goblet cells

Mv

(b)

Figure 10.18 Structures that increase surface area for absorption. (a) Sketch of several villi in the human small intestine (left) and details of structure of one villus (right). **Lacteals** are vessels of the lymph system. **Goblet cells** secrete mucus. (b) Scanning electron micrograph of microvilli (Mv), which are tiny projections on cells of villi. (Magnification × 8,412)

Digestion in Humans

The human digestive tract (figure 10.19) is specialized to fulfill the following basic functions of digestion: food acquisition, food storage and transport, physical breakdown, chemical digestion, absorption of nutrients, and concentration and evacuation of wastes.

The Mouth

Food enters the human digestive tract through the mouth. While in the mouth, it is manipulated by the muscular tongue and chewed by a set of teeth that are specialized for several different actions. The sharp **incisors** in the front of the mouth are used for biting; that is, they cut off chunks

of food from larger pieces. Pointed **canine** teeth are used for tearing food, and the flattened **premolars (bicuspids)** and **molars** located in the posterior part of the mouth are specialized for grinding and crushing (figure 10.20a).

The tongue has sensory functions in addition to its food-manipulating role. It has touch and pressure receptors similar to those in the skin, and **taste buds** are scattered over its surface. Taste buds are chemical receptors that provide sensory information about the chemical composition of food (figure 10.20b).

Three pairs of **salivary glands** secrete **saliva** into the mouth where it mixes with the food. Saliva is a watery liquid that moistens food so that it can move on more smoothly. Saliva also contains an enzyme, **salivary amylase,** that begins

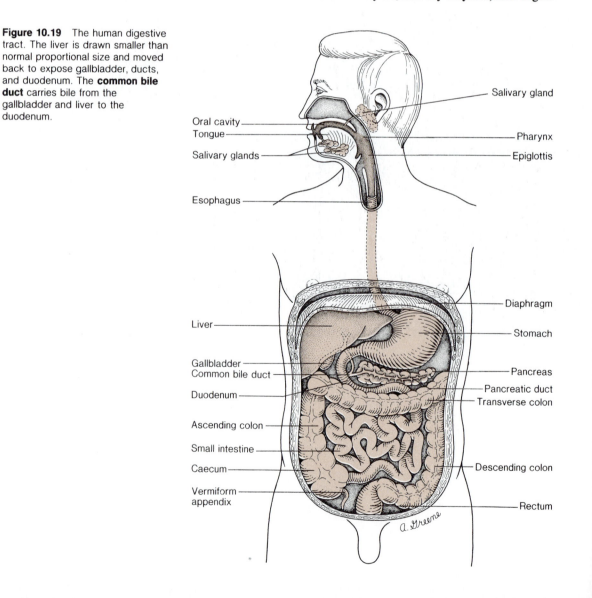

Figure 10.19   The human digestive tract. The liver is drawn smaller than normal proportional size and moved back to expose gallbladder, ducts, and duodenum. The **common bile duct** carries bile from the gallbladder and liver to the duodenum.

Oral cavity
Tongue
Salivary glands
Esophagus

Liver

Gallbladder
Common bile duct
Duodenum

Ascending colon

Small intestine

Caecum

Vermiform
appendix

Salivary gland

Pharynx

Epiglottis

Diaphragm

Stomach

Pancreas

Pancreatic duct
Transverse colon

Descending colon

Rectum

a. Greene

the process of starch digestion by hydrolyzing some of the bonds between the glucose units making up starches.

When food has been chewed and mixed with saliva, the tongue initiates the process of **swallowing** by pushing the food back through the **pharynx** toward the **esophagus.**

The Pharynx and Esophagus

The digestive and respiratory passages cross in the pharynx (figure 10.21). Thus, swallowing poses a potential problem because food might enter the **trachea** and block the path of air to the lungs. Normally, however, swallowing involves a set of reflexes that close off the opening into the **larynx,** which lies at the top of the trachea. A flap of tissue, the **epiglottis,** covers the opening as muscles move the food mass through the pharynx into the esophagus.

Only the upper portion of the esophagus is under voluntary control. Food is moved through the remainder of the esophagus by involuntary muscle contractions. That is, food is transported by muscular movements that under ordinary circumstances cannot be consciously controlled. Involuntary muscle contractions in the esophagus, and throughout the remainder of the digestive tract, occur in rhythmic waves. This rhythmic involuntary muscle contraction that moves material along the digestive tract is called **peristalsis** (figure 10.22).

(a)

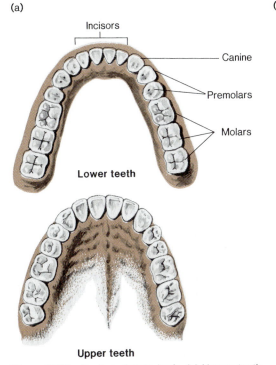

Lower teeth

Upper teeth

(b)

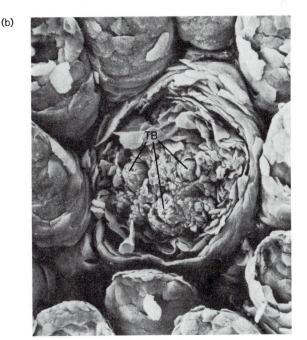

Figure 10.20 Teeth and taste buds. (a) Human teeth. (b) Scanning electron micrograph showing three taste buds (TB) exposed after some tongue surface cells have been removed. (Magnification × 480)

Figure 10.21 Swallowing.
Respiratory and digestive passages
cross in the pharynx (a). During
swallowing (b), the epiglottis covers
the opening into the trachea and
prevents food from entering it.

(a)

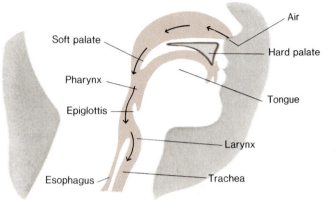

Soft palate

Pharynx

Epiglottis

Esophagus

Air

Hard palate

Tongue

Larynx

Trachea

(b)

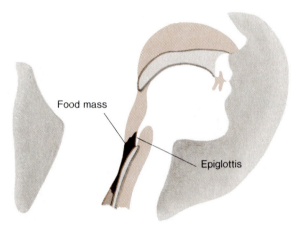

Food mass

Epiglottis

Figure 10.22 Peristalsis in the
esophagus. Rhythmic waves of
muscle contraction move material
along the digestive tract. The three
sketches show how a peristaltic
wave of contraction moves through
a single section of gut over time (left
to right).

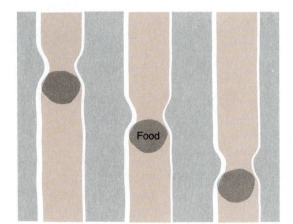

Food

The Stomach

The stomach is a storage organ that can hold up to two liters of material, or even more in some cases. Because of the stomach's storage capacity, humans can eat relatively large meals periodically and spend the rest of their time at other activities.

But the stomach is much more than a storage organ. Its muscular walls contract vigorously and mix food with secretions from glandular (secreting) cells in the stomach wall. This continues the physical breakdown begun by chewing. Because of the mixing, the contents of the stomach come to have a thick, soupy consistency. This thoroughly mixed suspension of food and **gastric secretions** (secretions of the stomach) is called **chyme.**

Some stomach lining cells secrete **mucus** that lubricates food as it is being mixed. Mucus also forms a protective coating over the stomach lining and prevents the stomach from being damaged by the other secretions of glandular cells in the stomach. The inner lining of the stomach, and of the rest of the digestive tract as well, is called the **mucosa,** taking its name from this characteristic mucous coating.

Other secretory cells in the stomach lining secrete **hydrochloric acid (HCl)** into the **lumen** (internal cavity) of the stomach. They secrete so much HCl that the fluid in the lumen of the stomach routinely has a pH of 2.0 or less. This highly acid environment kills most living cells in the food, as well as bacteria and other microorganisms taken in with the food. The acid also erodes and loosens up food material and exposes more surface area, thus facilitating further digestion.

The low pH level of the stomach effectively ends the activity of salivary amylase, which functions optimally at the near-neutral pH of saliva, but low pH provides a favorable environment for other enzymes. Certain secretory cells in the stomach lining secrete protein-hydrolyzing enzymes, collectively known as **pepsin,** that function well at such low pH levels.

Pepsin begins the process of protein digestion by hydrolyzing some of the peptide bonds that bind amino acids together in protein molecules. Pepsin is an **endopeptidase.** That is, it hydrolyzes bonds within a protein molecule, breaking up the molecule into shorter chains of amino acids (peptides). Pepsin does not attack the bonds that bind the terminal amino acids of these chains. Other enzymes, known as **exopeptidases,** which act in the small intestine, hydrolyze those particular bonds. Thus, pepsin breaks up protein molecules into peptides, but further digestion by exopeptidases in the small intestine is required to yield the individual amino acids that can be absorbed through the intestinal wall.

Pepsin is a general endopeptidase; it hydrolyzes peptide bonds in many different kinds of protein molecules. But that does not mean that pepsin is a nonspecific enzyme. Its specificity is for the particular bonds within protein molecules that it attacks. Pepsin hydrolyzes only peptide bonds that are adjacent to certain amino acids, notably tyrosine and phenylalanine, both of which contain six-carbon rings.

Pepsin is especially important for digestion of collagen (see chapter 2), which is a major extracellular protein in many animal tissues and constitutes a large percentage of the total protein in the diet of carnivores and omnivores.

But if pepsin is a general endopeptidase that hydrolyzes bonds in many kinds of protein molecules, what is to prevent it from attacking the cells that produce it or from digesting the lining of the stomach itself? One protection against self-digestion is that pepsin is produced and secreted as a **zymogen,** which is an enzymatically inactive form of an enzyme molecule. The zymogen molecule in this case is called **pepsinogen.** Pepsinogen does no damage inside the cells that synthesize it and is converted into the active enzyme pepsin only after it is in the stomach lumen (cavity). There, a forty-two amino acid segment of the polypeptide chain is removed from pepsinogen to produce pepsin, which is thus a considerably smaller molecule than pepsinogen. This conversion requires a low pH like that of the stomach contents. The stomach lining is protected from hydrolysis by the active pepsin in the stomach cavity because, as mentioned earlier, the lining is coated with a layer of mucus. Normally, these safeguards effectively protect the stomach from self-digestion by pepsin.

Chyme, the soupy mixture of food and digestive secretions, leaves the stomach through the **pyloric valve,** a narrow opening controlled by a ring of powerful muscles called the **pyloric sphincter.** The sphincter relaxes as a wave of contraction moves through the stomach wall toward it. This relaxation allows a small quantity of chyme to squirt through the valve into the **duodenum,** the first part of the small intestine. When chyme enters the duodenum, it sets off a neural reflex that causes the muscles of the sphincter to contract vigorously and temporarily close the pyloric valve. Then the sphincter relaxes again and allows more chyme to squirt through. Because of this alternating relaxation and contraction, the stomach empties very gradually, and the efficiency of intestinal digestion is increased because the rate at which food enters the small intestine is controlled.

The Small Intestine

When chyme enters the duodenum, proteins and carbohydrates have been only partly digested and are not yet in absorbable forms. Fat digestion has not begun at all. Considerably more digestive activity is required before these nutrients can be absorbed through the intestinal wall.

The duodenum is the short (about 20 cm) first section of the human small intestine. Ducts carrying **bile** and **pancreatic juice,** the complex secretions of the liver and pancreas, respectively, empty into the duodenum about midway along its length. These fluids contain a number of substances needed for digestion, including sodium bicarbonate, which neutralizes acid in chyme. This neutralization is essential because the enzymes that act in the small intestine function most effectively in the range of pH 7.8 to 8.0.

Bile also contains **bile salts,** which are fat-emulsifying agents. Bile salts break fat globules into smaller droplets and thus expose more surface to the action of fat-digesting enzymes secreted by the pancreas. Bile also contains **bile pigments,** which are breakdown products of hemoglobin from red blood cells that have been removed from the circulation. Bile pigments that pass on through the digestive tract are at least partly responsible for the dark color of **feces,** the digestive wastes. Liver problems sometimes prevent this normal removal of bile pigments via the digestive tract. When this happens and the bile pigments get into the circulation, they can cause a yellow discoloration of the skin called **jaundice.**

Enzymes in pancreatic juice, together with enzymes produced by glands in the intestinal wall, hydrolyze each of the major classes of nutrients to absorbable forms (table 10.7).

Lipase hydrolyzes fat molecules to fatty acids, glycerol, and some monoglycerides (glycerol with one fatty acid still bonded to it).

Pancreatic amylase completes the process, begun by salivary amylase, of splitting starch into maltose, a disaccharide made up of two glucose units. Then maltose and other disaccharides, such as sucrose, are hydrolyzed by various specific **disaccharidases** (table 10.7) to yield monosaccharides (simple sugars) that are absorbed through the intestinal lining.

Protein digestion, started by pepsin in the stomach, continues in the small intestine, although pepsin itself is inactivated by the slightly basic pH of the small intestine. (Pepsin functions actively at the very acidic pH levels found in the stomach contents.) Two pancreatic endopeptidases, **trypsin** and **chymotrypsin,** hydrolyze peptide bonds in proteins and peptides, thus producing smaller peptide fragments. These smaller peptides are digested by **exopeptidases** secreted by intestinal glands. **Carboxypeptidases** are exopeptidases that hydrolyze the peptide bond nearest the free carboxyl end of a peptide, and **aminopeptidases** are exopeptidases that attack the opposite ends (amino ends) of peptides (figure 10.23). These enzymes thus remove individual amino acids from peptides. Their action completes the process of protein hydrolysis to amino acids, the absorbable end products of protein digestion.

Trypsin and chymotrypsin, like pepsin, are general endopeptidases that hydrolyze peptide bonds in many different kinds of proteins. Thus, they present a threat of self-digestion in the pancreas and the small intestine similar to that posed by pepsin in the stomach. These enzymes, like pepsin, are secreted as zymogens, enzymatically inactive molecules. These zymogens, **trypsinogen** and **chymotrypsinogen,** are converted to active enzymes in the lumen of the small intestine. Trypsinogen is converted to trypsin by an enzyme called **enterokinase** produced by intestinal cells.

Table 10.7
Digestive Enzymes in Humans.

Digestive Organs and Enzymes	Source of Enzyme	Action
Mouth Amylase	Salivary glands	Hydrolyses starch to dextrins (small polysaccharides) and the disaccharide maltose
Stomach Pepsin (pepsinogen)	Stomach mucosa	An endopeptidase; splits proteins into smaller peptides; hydrolyzes peptide bonds adjacent to tyrosine and phenylalanine*
Small Intestine Endopeptidases Trypsin (trypsinogen)	Pancreas	Hydrolyzes peptide bonds adjacent to lysine and arginine*
Chymotrypsin (chymotrypsinogen)	Pancreas	Hydrolyzes peptide bonds adjacent to tyrosine, phenylalanine, leucine, and methionine*
Enterokinase	Intestinal glands	Converts trypsinogen to trypsin
Exopeptidases Various carboxypeptidases	Intestinal glands and pancreas	Hydrolyze peptide bond nearest free carboxyl ends of amino acid chains*
Various aminopeptidases	Intestinal glands and pancreas	Hydrolyze peptide bond nearest free amino ends of amino acid chains*
Lipase	Pancreas	Hydrolyzes fat to glycerides, fatty acids, and glycerol
Amylase	Pancreas	Hydrolyzes starch to the disaccharide maltose
Disaccharidases, including Sucrase	Intestinal glands	Hydrolyzes sucrose to glucose and fructose
Maltase	Intestinal glands	Hydrolyzes maltose to two glucose units
Lactase	Intestinal glands	Hydrolyzes lactose to glucose and galactase
Ribonuclease	Intestinal glands and pancreas	Hydrolyzes RNA to nucleotides
Deoxyribonuclease	Intestinal glands and pancreas	Hydrolyzes DNA to nucleotides

*See figure 10.23

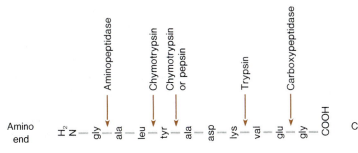

Figure 10.23 Different protein-hydrolyzing enzymes break peptide bonds of a polypeptide chain in different places. This sketch of a small peptide illustrates some of these points. For example, trypsin hydrolyzes bonds adjacent to the amino acids lysine (lys) or arginine.

Trypsin converts chymotrypsinogen to chymotrypsin. Both of these conversions involve removal of some amino acids from larger molecules to produce the smaller molecules that are enzymatically active.

Intestinal cells also produce mucus that lubricates intestinal contents and protects the intestinal lining from the protein-hydrolyzing action of trypsin and chymotrypsin.

Nutrient Absorption

Only a few substances such as ethanol are absorbed through the stomach wall. The bulk of nutrient absorption occurs in the small intestine.

Simple sugars (monosaccharides) and amino acids are absorbed into intestinal lining cells by active transport. This sugar and amino acid transport requires expenditure of chemical energy in the form of ATP and is coupled with sodium transport that pumps sodium out of the intestinal lining cells into the gut cavity. (See chapter 5 for more details of active transport mechanisms.) Sugars and amino acids are passed to blood vessels in the intestine and are transported away toward the liver.

Products of fat digestion are processed and transported in a different way. Fat molecules are reassembled in intestinal lining cells, and protein coats form around the fats to produce little droplets called **chylomicrons.** These droplets enter vessels of the lymph system, which carry them into the chest, where they pass into veins near the heart. Sometimes after a fatty meal, so many chylomicrons enter the circulation that they give the blood an almost milky appearance.

The Large Intestine

About 1.5 l of water enter the digestive tract daily as a result of eating and drinking. An additional 8.5 l enter the digestive tract each day carrying the various substances secreted by the digestive glands. About 95 percent of this water is reabsorbed into cells of the large intestine (**colon**). This water reabsorption is essential. Failure to reabsorb water can result in **diarrhea,** which can lead to serious dehydration and also ion loss, especially in children.

In addition to water, the large intestine absorbs some sodium and other ions from the material passing through it. At the same time, colon cells excrete still other ions into the wastes leaving the body. Thus, the colon functions both in water conservation and ion regulation.

Vitamin K, which is produced by intestinal bacteria, is also absorbed in the colon.

Digestive wastes (feces) eventually leave the body through the rectum and anus. Feces are about 75 percent water and 25 percent solid matter. Almost one-third of this solid matter is made up of intestinal bacteria. The remainder is undigested plant material, fats, waste products (such as bile pigments), inorganic material, and proteins, especially mucus and dead intestinal lining cells.

Control of Digestive Tract Secretions

The secretory activity of the gut linings and the accessory glands is regulated so that, under normal circumstances, the secretions are released only when food is present.

Salivary gland secretion is controlled by reflex activity of the nervous system. Saliva flows in response to the taste, smell, or sometimes even the thought of food. In fact, saliva sometimes flows in response just to thinking about saliva flowing.

Several mechanisms regulate secretion in the stomach (gastric secretion). Neural reflex responses to the sight, smell, and taste of food initiate secretion by glandular cells in the stomach, and then, when food reaches the stomach, both nervous and hormonal mechanisms regulate continuing gastric secretion. Food in the stomach stimulates sensory receptors whose activation causes additional reflex nervous stimulation of gastric secretion.

But a major factor in gastric secretory regulation is a hormone, **gastrin,** which is produced in the stomach itself. When proteins contact the stomach mucosa, gastrin-producing cells are stimulated to release gastrin into the bloodstream. As soon as gastrin circulates through blood vessels and reaches the acid- and enzyme-secreting cells of the stomach lining, these cells respond by secreting large quantities of HCl and pepsin. As the stomach empties, both the neural reflexes and gastrin release subside, and less HCl and pepsin are secreted.

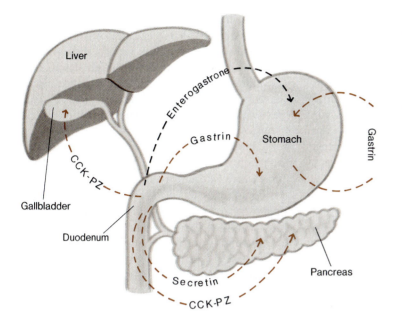

Figure 10.24 Hormonal control of digestive secretions. Colored arrows indicate stimulatory relationships. The black arrow indicates that enterogastrone inhibits stomach secretion and mobility. CCK-PZ stands for cholecystokinin-pancreozymin.

The duodenum also may produce some gastrin when proteins and peptides reach it, but this intestinal gastrin secretion is of minor importance in the overall regulation of gastric secretion. Some duodenal cells produce the hormone **secretin** that stimulates the pancreas to release pancreatic juice, especially the bicarbonate component. Other cells release a hormone called **CCK-PZ** (formerly called **cholecystokinin**) that stimulates the release of bile. Bile, produced by the liver, is stored in the **gallbladder,** a saclike structure embedded in the surface of the liver. Thus, CCK-PZ stimulates the gallbladder to empty its contents through the **common bile duct** into the duodenum. This same hormone also stimulates the pancreas, especially the enzyme secretion of the pancreas, and, therefore, it was renamed CCK-PZ, which stands for **cholecystokinin-pancreozymin,** in recognition of this dual role.

A hormone called **enterogastrone** is released into the circulation by duodenal cells in response to the presence of fatty food in the intestine. Enterogastrone inhibits stomach gland secretion and slows muscular movements of the stomach. This slows stomach emptying, thus allowing more time for processing meals that contain large amounts of fat, which is digested more slowly than other nutrients. (See figure 10.24.)

Cell Replacement in the Gut Lining

All along the digestive tract, cells in the lining epithelium (the part of the mucosa adjacent to the lumen) continuously are being replaced. This replacement occurs at a surprising rate in all parts of the digestive tract, but it is especially rapid in the small intestine. Mitotic activity produces new cells that move over the surface of intestinal villi and replace the millions of cells that shed from the tips of the villi into the gut lumen each day. Cells on the surface of villi in the small intestine, on the average, live less than forty-eight hours, a very short cell life span relative to other cell types. For example, red blood cells usually are considered relatively short-lived cells, but they live about 120 days. Cell replacement in the gut may be a mechanism that repairs normal wear and tear caused by abrasion and the actions of digestive enzymes on intestinal lining cells despite the protective mucous coating that covers them. But it might also be that only young cells are efficient enough to carry out the vital absorptive processes in the intestine. Two-day-old cells may have "aged" so much that they can no longer function adequately.

Unfortunately, the vital importance of this continuing cell replacement in the digestive tract sometimes is demonstrated in cancer patients receiving chemotherapy. These treatments suppress cell division throughout the body, both in the cancer that is the target of the therapy and in other areas where continuing cell division is necessary for normal cell replacement. One of the side effects of cancer chemotherapy can be intestinal problems associated with failure of the gut to produce new lining cells.

Liver and Pancreas

The liver is an extremely important and versatile chemical-processing center. Blood leaving the intestinal tract reaches the liver by way of the **hepatic portal system.** This blood, carrying recently absorbed nutrients, comes into contact with liver cells, which can make a variety of exchanges with it (figure 10.25).

For example, the liver plays an important role in **blood glucose** regulation. Many cells in the body are sensitive to changes in the level of the glucose being supplied to them by the circulatory system; brain cells are especially dependent on a stable glucose supply, and they can be seriously damaged by unusual fluctuations in blood glucose level.

Blood glucose level is under complex control by several hormones (chapter 14), and the liver is one of the sites where this control is expressed. When blood sugar levels are high, the liver absorbs glucose and synthesizes glycogen for storage. If the liver has stored adequate quantities of glycogen to meet the body's needs for a few hours and excess nutrients continue to be available, liver metabolism is switched to alternate pathways, and the liver stores some fat and also exports fat via the blood to fat tissue elsewhere in the body. When blood sugar levels fall, the liver breaks down glycogen and releases glucose into the blood. The liver continues to respond to the stimulus of lowered blood glucose even when the glycogen reserve has been used up. Stored fats are then used for glucose synthesis. In starvation, blood glucose eventually is maintained by synthesizing glucose from deaminated amino acids (chapter 7). This depletes body proteins and seriously damages the body if continued too long.

The liver also plays an important role in processing the body's nitrogenous wastes. During times of adequate nutrient supply, there may be excesses of particular amino acids. Liver cells deaminate these amino acids to produce usable carbohydrates. The amino groups removed during deamination are incorporated into **urea** molecules (chapter 13).

In addition to these and other metabolic functions, the liver collects and breaks down many toxic substances that enter the body.

The pancreas, like the liver, is a functionally complex organ that not only makes digestive contributions but also is involved in another way in meeting nutritional needs of the cells of the human body. Specialized cell populations in the pancreas, known as the **islets of Langerhans,** produce the hormones **insulin** and **glucagon.** These pancreatic hormones are involved in the control of blood glucose level and several related aspects of metabolism. The pancreatic hormones are examined in more detail in chapter 14.

Problems with the Digestive Tract

The gut tube, for all its twists and expansions, is really just a tunnel through the body. In the strict sense, the contents of the gut are outside the body and remain so unless they pass across cell membranes of gut-lining cells. Indeed, the contents of the gut must not come into direct contact with the blood or other interior tissues of the body, because the chyme contains bacteria and other potentially hazardous components. The nature of this risk is graphically illustrated by two rather common digestive ailments—appendicitis and ulcers.

The **vermiform appendix,** or just appendix for short (figure 10.26), is a short blind sac off the beginning of the colon, just past its junction with the small intestine. The human appendix does contain lymphoid tissue and might, therefore, be involved in resistance to infection, but its function is poorly understood. The appendix all too often becomes infected and develops an inflammation called **appendicitis.** Once appendicitis is diagnosed, standard medical procedure calls for surgical removal of the infected appendix. If untreated, the inflamed appendix can rupture,

(a)

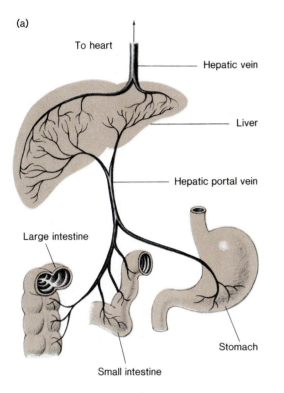

To heart

Hepatic vein

Liver

Hepatic portal vein

Large intestine

Stomach

Small intestine

(b)

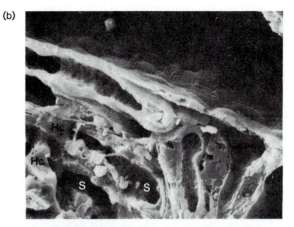

Figure 10.25 Liver. (a) Blood leaving the digestive tract flows through the hepatic portal vein to the liver, where it enters spaces (called **sinusoids**) among the liver cells. (b) Scanning electron micrograph showing the sinusoid (S) in the liver. Important chemical exchanges occur between blood in the sinusoids and liver, or hepatic, cells (Hc). Small channels collect bile secreted by liver cells and carry it toward the gallbladder. (Magnification × 592)

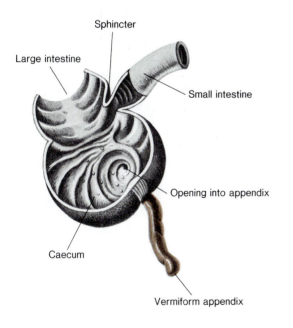

Sphincter

Large intestine

Small intestine

Opening into appendix

Caecum

Vermiform appendix

Figure 10.26 The vermiform appendix is a short blind sac off the caecum at the beginning of the colon.

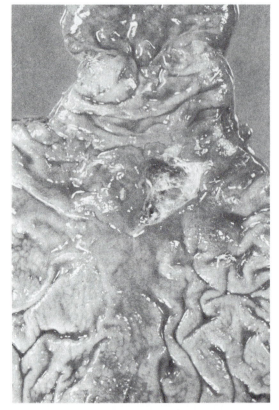

Figure 10.27 Part of a stomach lining showing an ulcer
(eroded area).

allowing gut contents to come in contact with the
peritoneum, the tissue that covers all the digestive
organs and lines the body cavity. The material
that escapes from a ruptured appendix causes
peritonitis, a life-threatening infection.

Ulcers are breaches, of varying depth and
position, in the wall of the gut. Most are located
in the duodenum and are called **duodenal ulcers**
(figure 10.27). Somewhat more rarely, **gastric
ulcers** may form in the stomach itself. Ulcers are
caused by the combined effect of acid and enzymes eroding the protective lining of the gut.
However, what is not well understood is why in
some individuals the normal mechanisms that
protect the linings of the digestive tract function
adequately for a lifetime, and in others they do
not. It is widely believed that worry, stress, or
frustration can contribute to ulcer development
because gastric juice secretion is at least partly
controlled by the nervous system. In fact, one of
the more drastic treatments for serious ulcers is
to cut the vagus nerve, the nerve that stimulates
stomach secretion.

Most ulcers are really rather mild, albeit
painful, abrasions in the gut lining. Healing can
be rapid because the cells lining the gut tube are
quickly replaced, but the causative features
(often extreme stress) must be removed and the
patient must switch to a bland diet that does not
stimulate excess acid secretion. The real dangers
come when ulcers begin to bleed or when they
perforate (produce a hole directly through the
gut wall). Peritonitis can develop following per-
foration unless prompt medical care is received.
Such serious cases require removal of the source
of stomach acid. This can be accomplished by
cutting the vagus nerve or by removing part of
the stomach. An interesting new technique accomplishes the same thing without surgery. The
patient swallows a balloon and then alcohol at a
temperature cold enough to freeze and kill stomach lining cells is poured into the balloon. This
destroys a large percentage of the acid-secreting
cells and has the same effect as surgical treatments. By the time these cells are replaced, the
ulcer has had a chance to heal.

Appendicitis and ulcers are not the only
digestive ailments. Problems also can occur in the
digestive glands. For example, the liver and its
bile-storing structure, the gallbladder, are subject
to several types of problems. One common condition is **gallstones,** which result from the precipitation of bile salts in the gallbladder or in the
bile duct. Blockage of the bile duct may cause
bile to back up and ultimately leads to jaundice.
Surgical removal of the gallbladder is the usual
solution.

Jaundice also may be a symptom of various
types of liver failure, a much more serious condition, given the overall importance of this organ.
Inflammation of the liver is known as **hepatitis,**
and it may result from a virus infection (viral
hepatitis), poison (toxic hepatitis), or protein deficiency (deficiency hepatitis). The liver does have
considerable regenerative ability, but when liver
cells are destroyed, they sometimes are replaced
by scar tissue. This scarring is called **cirrhosis** of
the liver. All of these liver problems weaken the
liver's metabolic capacities and greatly reduce its
ability to deal effectively with toxins entering the
body. Thus, persons suffering from liver disease,
whatever the original cause, are strongly urged
not to drink alcohol because alcohol normally is

metabolized almost entirely by liver cells. A damaged liver may well succumb to the onslaught of alcohol in the blood, with death as the inevitable result.

In addition to being susceptible to problems such as appendicitis, ulcers, gallstones, and liver disease, the digestive tract is simply an ideal environment for many disease organisms, from viruses to tapeworms.

Despite all of these potential problems, most human digestive tracts function normally and, except for occasional minor upsets, seldom cause much concern. A normally functioning digestive tract is essential if the body is to receive an uninterrupted supply of vital nutrients.

Summary

Heterotrophic organisms must obtain organic compounds from their environment to meet their energy and raw material requirements. The organisms or parts of organisms that animals eat to provide these organic compounds are processed in digestive systems. Large organic molecules are broken down into absorbable forms in animal digestive tracts.

Carbohydrates are important nutrients mainly because they provide chemical energy sources. Much ingested carbohydrate consists of polysaccharides that must be hydrolyzed to absorbable sugars.

Animals must ingest proteins to obtain amino acids for their own protein synthesis. Animals' cells can synthesize some amino acids from other compounds, but other amino acids are essential dietary constituents. Essential amino acids are vital components of human diets.

Fats are highly reduced organic compounds that are important for energy and synthesis of structural components. Certain animals cannot synthesize some fatty acids, and these fatty acids must be included preformed in the diet.

Some minerals are required in large quantities, but only trace amounts of others are needed. A few minerals enter the body in organic compounds; others are absorbed as ions.

Vitamins are relatively simple organic compounds that must be present in the diet to maintain health. Vitamins are used up slowly, and thus, deficiencies develop slowly, but vitamin deficiencies cause complex sets of serious problems.

All unicellular heterotrophs absorb nutrients directly from their environment. In multicellular heterotrophs only the cells that line digestive tracts are specialized for nutrient procurement.

Incomplete digestive tracts have only one opening, a mouth. Complete digestive tracts have both a mouth and an anus. In complete digestive tracts, food can move through and be processed in assembly-line fashion in differently specialized regions.

All animals secrete some enzymes into digestive cavities where the enzymes hydrolyze food molecules. Such digestion, outside cells and in a digestive cavity, is called extracellular digestion. Intracellular digestion is processing of nutrients inside vesicles in individual gut lining cells.

Modifications of the basic tubular plan of a complete digestive tract depend on the need for food storage in discontinuous feeders, the need for food maceration, the type of food, and the relative need for large absorption surfaces.

In the human digestive tract, food is chewed and manipulated in the mouth. Then it is swallowed through the esophagus to the stomach.

The stomach stores food and mixes it with mucus and gastric juice to produce chyme. Protein digestion begins in the stomach.

Chyme passes through the pyloric valve to the duodenum. There, bile, pancreatic juice, and intestinal gland secretions are added to the chyme and begin to act upon it. Enzymes hydrolyzing all of the organic nutrients act in the small intestine.

Most nutrient absorption takes place in the small intestine, but some water and mineral ions are absorbed in the colon.

Digestive wastes leaving the colon contain water, bacteria, undigested material, fats, waste products, inorganic material, proteins, and dead intestinal lining cells.

Neural and hormonal regulation of digestive tract secretion ensures that digestive substances such as enzymes are released only when food is present.

Cells in the digestive tract lining have very short life spans and must continuously be replaced.

The liver is involved in chemical processing of absorbed food and in maintenance of stable concentrations of nutrients such as glucose in the blood. The liver also processes nitrogenous wastes and breaks down toxins. The pancreas produces hormones that are involved in the control of carbohydrate metabolism in the body.

Problems with the digestive tract include possible infection by a wide range of organisms, as well as erosion of the tract's protective lining by its own secretions, which results in ulcers. These problems are serious; a normally functioning digestive tract is essential if the body is to receive an uninterrupted supply of vital nutrients.

Questions

1. Why can a protein deficiency disease such as kwashiorkor develop even when a person is regularly eating considerable amounts of plant protein?

2. Why are overdoses of vitamin C much less likely to cause serious problems than overdoses of vitamin A?

3. Explain the difference between extracellular and intracellular digestion and contrast the relative importance of each in planarian and human digestion.

4. Which would you expect to be longer: the digestive tract of a two kilogram cat or that of a two kilogram rabbit? Why?

5. List several structural features that add to the absorptive area in the human small intestine.

6. What normal protective mechanisms guard against self-digestion of digestive glands and the linings of the digestive tract?

7. What are the steps in the digestion and absorption of starches, proteins, and fats in the human digestive tract?

Suggested Readings

Books

Eckert, R., and Randall, D. 1978. *Animal physiology*. San Francisco: W. H. Freeman.

Guyton, A. C. 1979. *Physiology of the human body*. 5th ed. Philadelphia: W. B. Saunders.

Schmidt-Nielsen, K. 1979. *Animal physiology*. 2d ed. Cambridge, Mass.: Cambridge University Press.

Articles

Harpstead, D. D. August 1971. High-lysine corn. *Scientific American* (offprint 1229).

Loomis, W. F. December 1970. Rickets. *Scientific American* (offprint 1207).

McMinn, R. M. H. 1974. The human gut. *Carolina Biology Readers* no. 56. Burlington, N.C.: Carolina Biological Supply Co.

Moog, F. November 1981. The lining of the small intestine. *Scientific American*.

Scrimshaw, N. S., and Young, V. R. September 1976. The requirements of human nutrition. *Scientific American*.

Sernka, T. J. 1979. Claude Bernard and the nature of gastric acid. *Perspectives in Biology and Medicine* 22: 523.

Sherlock, S. 1978. The human liver. *Carolina Biology Readers* no. 83. Burlington, N.C.: Carolina Biological Supply Co.

Gas Exchange and Transport in Animals

Chapter Concepts

1. Gas exchange is an urgent, continuing requirement for animal life.
2. Size, shape, internal transport capabilities, and metabolic requirements of an animal, along with the type of gas-carrying medium in which the animal lives, dictate the method of gas exchange used.
3. Small animals can exchange gases directly through body surfaces, but larger animals have specialized gas exchange mechanisms.
4. Aquatic animals generally exchange gases through gills, while terrestrial animals have tracheal systems or lungs.
5. Special gas transport mechanisms in the blood make internal transport of adequate quantities of oxygen and carbon dioxide possible in large animals such as humans.
6. Because lungs are warm, moist, blind sacs, they are particularly susceptible to infection and to degenerative diseases caused by inhaled irritants.

A person who stops breathing is assumed to be either dead or very close to death because continuing gas exchange is a basic requirement of human life. Humans and many other animals are utterly dependent on continuing gas exchange. This dependence is much more immediately urgent than the need for food or water. Humans can survive for many days without eating, or for several days without drinking, but they can live for only a very few minutes without breathing.

This requirement for continuing gas exchange arises from the use of aerobic respiration. Oxygen is continuously being used in animal cells as an electron acceptor in metabolic oxidation-reduction reactions that produce usable energy (ATP) for a variety of energy-requiring processes. These cellular metabolic processes must continue if life is to continue.

Oxygen is not essential for all organisms; in fact, oxygen is toxic for some microorganisms, primarily some bacteria (members of the Kingdom Monera). Monerans first evolved more than 3,000 million years ago when there apparently was no free oxygen in the atmosphere. Oxygen present in the earth's crust was combined with other elements in compounds. Thus, the metabolism of those early organisms could not depend on using free oxygen, and many of their present-day descendants still live without oxygen.

With the evolution of photosynthesis, increasing amounts of molecular oxygen (O_2) were released into the atmosphere, and free atmospheric oxygen was available during the subsequent evolution of microorganisms, as well as during the evolution of eukaryotic cells and multicellular organisms. Metabolic processes in these organisms could use oxygen in the efficient oxidation of reduced nutrients.

Today many microorganisms and all plants and animals need oxygen. Because these organisms generally have very limited gas storage capacities, continuing gas exchange is a necessity of life for all of them.

Another requirement for gas exchange results from the decarboxylation of carbon compounds, which yields carbon dioxide as a metabolic waste product. This CO_2 must be released to the environment as rapidly as it is produced because excessive CO_2 accumulation in the cellular environment inside the body is potentially harmful.

In this chapter, the mechanisms by which animals accomplish this vital gas exchange are discussed.

Factors Affecting Methods of Gas Exchange

A combination of factors determine what mechanisms are adequate to allow an animal to obtain and deliver enough oxygen to the cells of its body:

1. Availability of oxygen in the organism's immediate environment. This depends largely on whether the gas-carrying medium is air or water.
2. Size and shape of the organism.
3. Presence or absence of an internal gas transport mechanism.
4. Metabolic rate.

Oxygen Availability: Air versus Water

Air has three distinct advantages over water as an oxygen source. First, air contains roughly 21 percent oxygen (210 parts per 1,000) while water, at the most, contains only 6 to 8 parts per 1,000 oxygen. Natural aquatic systems may contain considerably less than that under many circumstances (tables 11.1 and 11.2). Second, oxygen diffuses approximately 300,000 times faster in air than it does in water. Thus, an organism in air seldom faces the problem of depleting the oxygen supply immediately around its body, while aquatic organisms often experience oxygen depletion of the water immediately around them. Third, air is a much less dense medium than water, and much less energy is required to move air over gas exchange surfaces than to move water. Mammals use only 1 or 2 percent of their total energy output for breathing, but fish use up to 25 percent of theirs to propel the much more dense (and much less oxygen-rich) water over their gills.

Table 11.1
Composition of Dry Atmospheric Air at Sea Level.

Gas	Percentage	Partial Pressure (mm Hg)*
Nitrogen	78.09	593.5
Oxygen	20.95	159.2
Carbon dioxide	0.03	0.2
Others (mostly argon)†	0.93	7.1
Total	100.00	760.0

*Partial pressure is that part of atmospheric pressure exerted by a gas. For example, oxygen's partial pressure is 0.2095 × 760 = 159.2. Actually, all atmospheric air contains water vapor in varying amounts, and the partial pressures of other gases are reduced as the partial pressure of water vapor increases.
†The less common noble gases—helium, neon, krypton, and xenon—together constitute only 0.002 percent of the atmosphere.

Table 11.2
Amount of Oxygen Dissolved in Fresh Water and Seawater in Equilibrium with Atmospheric Air.

Temperature (C°)	Fresh Water (ml O_2 Per Liter Water)	Seawater (ml O_2 Per Liter Water)
0	10.29*	7.97
10	8.02	6.35
15	7.22	5.79
20	6.57	5.31
30	5.57	4.46

*These values are very low compared to the oxygen content of air, which is the same at all of these temperatures and is equal to 209.5 ml O_2 per liter of air.

However, air has one significant disadvantage as an oxygen-carrying medium. Oxygen must go into solution before it can diffuse into cells, and therefore, gas exchange surfaces must be kept moist. This produces a continuing water loss problem for organisms that exchange gases with air because water evaporates into the air over the moist gas exchange surfaces. Animals that exchange gases with air, then, must have some kind of protection against desiccation. The majority of terrestrial animals have internal gas exchange areas that are protected by dry, relatively impermeable skins. This adaptation reduces water loss due to evaporation.

Body Size and Shape

The sizes and shapes of animal bodies are other factors that influence the characteristics of an animal's gas exchange mechanism.

Oxygen diffuses very slowly through the watery medium in which it is dissolved once it enters an animal's body. For this reason, diffusion alone does not move oxygen very far through animal bodies. Only animals with small bodies in which all cells are relatively near gas exchange surfaces can depend on diffusion as the sole means for gas movement to body cells.

Several adaptations reduce the distance that oxygen must diffuse from the body surface to reach individual body cells. Some animals have very flattened bodies in which all body cells are relatively close to the body surface. The appropriately named flatworms, such as planarians, are excellent examples of this arrangement. Other animals, such as coelenterates like *Hydra*, are hollow, and some of their cells can exchange gases with water in an internal cavity. (See figure 11.1).

But diffusion distances severely limit the size of all animals that depend solely on diffusion of gases between body surfaces and individual body cells. All large animals must have arrangements that bring oxygen to internal body cells (and carry carbon dioxide away). Many of these animals have **circulatory systems** with special gas transport mechanisms that carry oxygen from specialized exchange areas to cells throughout the body.

Gas Transport Systems

The presence or absence of an internal gas transport mechanism is another factor that must be considered in determining how an organism obtains and delivers adequate oxygen to its body cells, and removes CO_2 from them.

The evolution of circulatory systems that can transport gases generally has paralleled the evolution of specialized gas exchange organs located in specific areas of the body. Exchange organs, such as **gills** and **lungs**, are richly supplied with blood vessels (highly vascularized) in most cases. Thus, a great deal of blood can be brought near the surfaces where gas exchange with water or air takes place. The circulatory system then carries the blood into the immediate environment of body cells that exchange gases with it (figure 11.2).

        Gas Exchange and Transport
in Animals

(a)

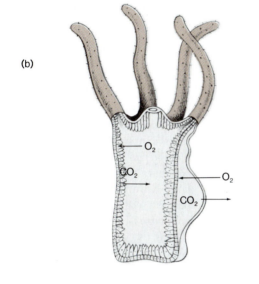

(b)

Figure 11.1 Body shape
adaptations that reduce oxygen
diffusion distances from outer body
surfaces to body cells. (a) The cells
of planarians and other flatworms
are spread out in their flattened
bodies so that no cells are far from
an outer body surface. (b) *Hydra*
has a hollow internal cavity. Some
cells exchange gases with water
inside this cavity.

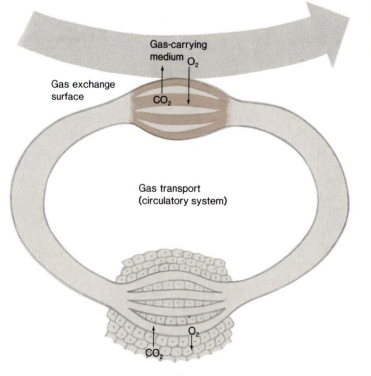

Cells

Figure 11.2 A circulatory system
brings blood to a highly vascular
gas exchange surface; then blood is
carried to body tissues where cells
exchange gases with blood.

Because of the low diffusion rate of oxygen in water and the fact that animal bloods are watery solutions, circulatory systems that carried only dissolved oxygen would not be very efficient. Most animals that depend on circulatory gas transport have specialized **respiratory (oxygen-carrying) pigments** in their blood. These carriers reversibly bind oxygen molecules and greatly increase the oxygen-carrying capacity of the blood in which they occur. In some animals, blood containing a respiratory pigment can carry up to 50 or even 100 times as much oxygen as the blood could carry in dissolved form alone (table 11.3).

Insects do not depend on their circulatory systems for gas transport. Instead, their bodies are penetrated by a network of fine tubules called the **tracheal system**. The tracheal system carries air deep into all parts of the body, and cells exchange gases directly with the air inside tracheal tubules. But few insects are larger than several centimeters in length. Larger animals usually have either gills or lungs and circulatory systems that transport gases efficiently.

Metabolic Rates

To a great extent, the demands that an animal puts on its gas exchange mechanism depend on its **metabolic rate**. The metabolic rate is a determination of the sum of all cellular processes occurring in the body. It can be measured as oxygen consumed per unit time, at least for organisms using aerobic respiration, and this measure gives a good indication of the demands that must be met by gas exchange systems. Animals that consume more oxygen require more efficient gas exchange mechanisms than organisms that consume less oxygen.

There is a wide range of variations in metabolic rates. Animals of different species and of different sizes have different metabolic rates. In addition, each individual animal can display a substantial range of metabolic rates depending on several environmental factors as well as changing activity levels. For example, humans consume twenty or more times as much oxygen during vigorous exercise as they do at rest (figure 11.3).

Gas exchange mechanisms must be efficient enough to meet the greatly increased demands that occur during periods of vigorous activity.

To simplify comparison, metabolic rates usually are measured when animals are relaxed and resting, but not sleeping. Such resting metabolism, the energy spent on ordinary life maintenance functions, is called **standard metabolism,** and physiologists usually use standard metabolic rates when making comparisons among animals.

Most animals are **heterothermic**; their body temperatures rise and fall with environmental temperatures (see chapter 7). Because metabolic processes are temperature dependent, heterothermic animals' standard metabolic rates increase as environmental temperatures increase.

For example, low environmental temperatures can so depress a heterothermic animal's metabolism that it is incapable of moving. This cold paralysis is relieved only when the animal's body is warmed. Trout fishers who catch grasshoppers for their day's bait usually try to fill their bait boxes in the morning chill before the sun warms the mountain meadows around their favorite trout streams. Later in the day, chasing sun-warmed grasshoppers can become hard and discouraging work. When a heterothermic animal is warm, its increased metabolic rate puts increased demands on its gas exchange mechanisms (figure 11.4). Gas exchange mechanisms, therefore, must be flexible enough to meet these increased demands.

Heat affects aquatic organisms more drastically than it affects terrestrial animals because it alters oxygen solubility in water. The percentage of oxygen in air remains the same when air temperature increases, but oxygen solubility in water decreases as water temperature increases (see table 11.2). Thus, at the very time that heterothermic aquatic animals must obtain more oxygen to meet increased metabolic demands, there is less oxygen available in the water around them.

Table 11.3
Effect of Respiratory Pigments on Oxygen-Carrying Capacity of Blood.

Pigment	Color	Site	Animal	Oxygen (ml O$_2$/100 ml Blood)
Hemoglobin	Red (iron pigment)	Erythrocytes (red blood cells)	Mammals	15–30
			Birds	20–25
			Amphibians	3–10
			Fishes	4–20
		Blood plasma	Annelids (for example, earthworms)	1–10
			Molluscs (some clams)	1–6
Hemocyanin	Blue (copper pigment)	Blood plasma	Molluscs	
			Gastropods (snails)	1–3
			Cephalopods (for example octopus)	3–5
			Crustaceans (for example, lobster)	1–4
No pigment	—	—	Insects and other terrestrial arthropods*	> 1

*Circulatory systems of insects and other terrestrial arthropods do not transport oxygen because their cells exchange gases with air inside tracheal systems.

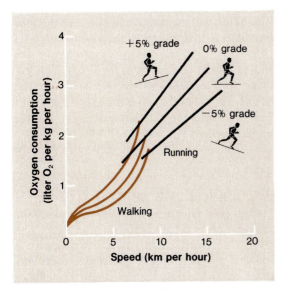

Figure 11.3　The energy expenditure of a person walking or running on three different grades.

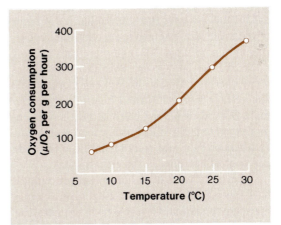

Figure 11.4　Rates of oxygen consumption of the Colorado potato beetle *(Leptinotarsa decemlineata)* at various temperatures between 7°C and 30°C. Before the experiment the animals had been maintained at 8°C. Note that oxygen consumption at 20°C is more than twice as much as it is at 10°C. At high temperatures, increased metabolic rates of heterothermic animals place great demands on their gas exchange mechanisms.

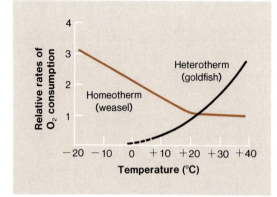

Figure 11.5 Comparison of the effects of environmental temperature on a homeothermic animal and a heterothermic animal. Note that these are relative rates of oxygen consumption that can be used to compare an individual animal's metabolic rates at different temperatures. On an absolute scale, the metabolic rate of a heterothermic animal does not exceed that of a homeothermic animal in the higher parts of the temperature range shown here.

Homeothermic animals (birds and mammals) maintain stable, relatively high, body core temperatures (see chapter 7 and figure 11.5). This is possible because of several differences between heterothermic and homeothermic animals. First, homeothermic animals have higher standard metabolic rates than heterothermic animals of the same size. Metabolic reactions are occurring at a more rapid rate, and more heat is being released from body cells. This also means that homeothermic animals must take in more oxygen and release more CO_2 than heterothermic animals. Second, insulation helps homeothermic animals to maintain a stable body core temperature. Feathers or hair, and body fat insulate the body and control heat loss to the environment.

Beyond these basics, homeothermic animals have specialized responses to temperature change that help to maintain a stable body temperature. Some of these are alterations in blood flow near the body surface, shivering, and sweating. All of these mechanisms require energy, and their metabolic costs include a need for increased oxygen and carbon dioxide exchange. Clearly, homeothermic animals require very efficient gas exchange and transport mechanisms because of their special metabolic characteristics.

Gas Exchange Organs

The four basic categories of animal gas exchange specializations are **integumentary (skin) exchange, gills, tracheal systems,** and **lungs** (figure 11.6).

Integumentary Exchange

Integumentary exchange (exchange through the skin) is the principal method of gas exchange only for quite small animals and for a small number of relatively larger animals that live in moist environments. For example, earthworms have capillary networks just under their skin surfaces, and they exchange gases with the air in spaces among soil particles. This method of gas exchange puts earthworms in a rather precarious position because they must maintain moist body surfaces if oxygen is to go into solution before it diffuses into surface cells. Earthworms risk serious desiccation if they enter dry soil areas or remain exposed to dry air above-ground for any length of time.

On the other hand, heavy rains threaten earthworms with suffocation because soil air spaces that they depend on for gas exchange become filled with water. To survive, earthworms must emerge from the ground and exchange gases with the air until the rainwater percolates down (drains) out of the upper layers of soil. This explains why earthworms are scattered on the ground and sidewalks after a heavy rain. Until they can reenter the soil, they are very vulnerable to predators.

Gills

Most aquatic animals carry out gas exchange through **gills.** Gills are body surface outgrowths (evaginations) that are exposed to the water. The gas exchange surfaces of gills are very thin (one cell thick in many cases), and thus the gas diffusion distance across them is small. The gills of many aquatic animals are very richly supplied with blood vessels.

The simplest gills are small individual projections of the skin, such as the tiny gills of starfish (figure 11.7). Individual starfish gills are very delicate, but they are protected from injury by larger spines and small, pincerlike structures, called pedicellariae, that surround them.

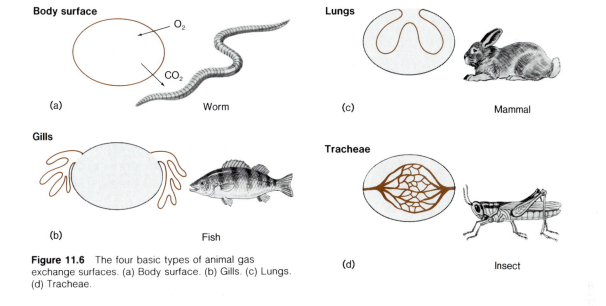

Body surface

(a)   Worm

Gills

(b)   Fish

Lungs

(c)   Mammal

Tracheae

(d)   Insect

Figure 11.6 The four basic types of animal gas
exchange surfaces. (a) Body surface. (b) Gills. (c) Lungs.
(d) Tracheae.

Figure 11.7 Starfish gills are small, delicate structures,
but they are protected by spines and the small,
pincerlike **pedicellariae.**

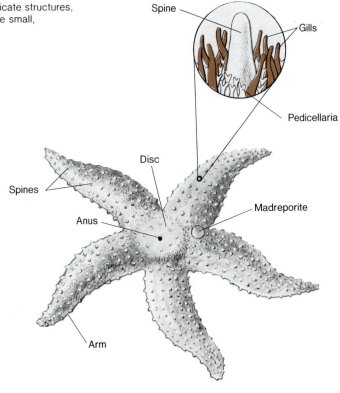

Aboral surface

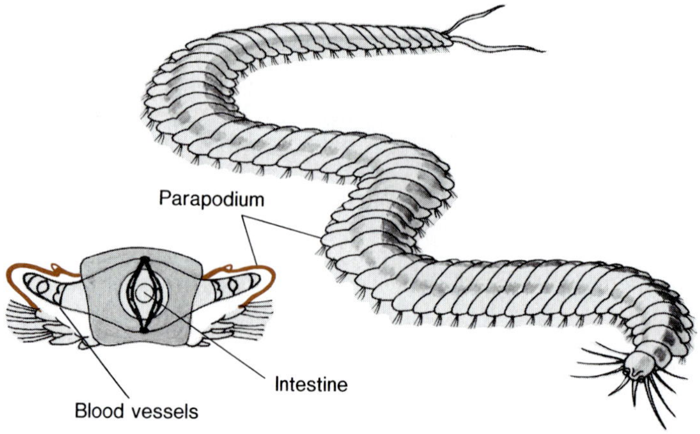

Figure 11.8 Parapodia are
projections of the body wall of
marine annelid worms. Parapodia are
highly vascular and function as
effective gas exchange devices.

Parapodium

Intestine

Blood vessels

Marine annelid worms have prominent lateral body projections called **parapodia** (figure 11.8). Parapodia are richly supplied with blood vessels, and they function as gills. While parapodia provide considerable surface for gas exchange, they are soft body surface areas that are very vulnerable to injury or damage by predators. Many animals have gills that are more compact and that are protected in some way with hard covering devices.

For compact gills to have adequate surface area for exchange, they must be finely divided into highly branched structures. For example, crayfish gills are complex, delicate, feathery structures attached at the bases of the legs. They are covered by a hard flap that protects them within a semienclosed **gill chamber**. Paddlelike structures beat back and forth, moving a stream of water in over the legs and out through the front of the gill chamber (figure 11.9a).

This one-way flow is typical of the general pattern of **gill ventilation** (water movement over gills) in animals. Water is moved over protected gills in a one-way stream rather than being moved in and out of the gill area through a single passageway, as air is moved in and out of lungs. A great deal more energy is required to move water back and forth rather than in one-way streams over gills, and such ebb and flow types of gill ventilations are extremely rare.

Clams and other molluscs have their gills located in protected **mantle cavities** inside their shells. Cilia move a stream of water over a clam's gills. The water comes into the mantle cavity through an **incurrent siphon**, passes over the gills, and passes out through an **excurrent siphon** (figure 11.9b). Clam gills also are food-filtering devices. Small food particles trapped in the gills pass on into the digestive tract.

The gills of bony fish also are compact and protected. Fish gills consist of many layers of highly vascular gill filaments with numerous delicate plates that protrude all along them and greatly increase the surface area for gas exchange. A hard, bony cover, the **operculum**, encloses and protects the gills. Fish gills are ventilated by the action of muscles in the mouth and pharynx. Water is taken in through the mouth opening. Then a valve flap closes the mouth, and muscles contract, forcing water through the gills and out a slit at the back of the operculum (figure 11.10).

Gas exchange in fish gills is very efficient because of a **countercurrent** flow arrangement through the gill. Blood flowing in vessels and capillaries inside of gills moves in the opposite direction of water flow over the gill surface. This increases the efficiency of gas exchange because as blood flowing through the gill gains more oxygen, it encounters water with still higher oxygen content, and a diffusion gradient is maintained.

(a)

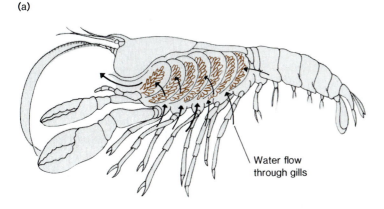

Water flow
through gills

(b)

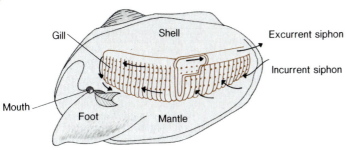

Gill

Shell

Excurrent siphon

Incurrent siphon

Mouth

Foot

Mantle

Figure 11.9 (a) The delicate, feathery gills of crayfish are protected inside a gill chamber. The cover of the gill chamber has been removed in this diagram to show the gills. Water is moved through the chamber to ventilate the gills. (b) A clam with its shell removed to show gills located inside the mantle cavity. Clams' gills function in gas exchange and filter feeding. Water enters an incurrent siphon, passes through the gills, and leaves through an excurrent siphon.

(a)

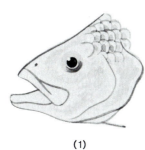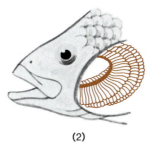

(1)

(2)

(b)

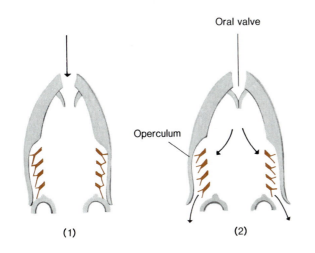

Oral valve

Operculum

(1)

(2)

Figure 11.10 Fish gills. (a) The operculum (1) covers and protects several layers of delicate gills. The operculum has been removed from (2) to expose the gills. (b) Water is taken into the pharynx through the mouth (1). Then the oral valve closes the mouth from the inside and muscles contract, forcing water over the gills and out through the operculum (2).

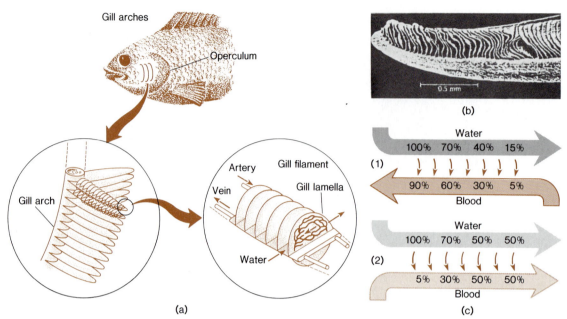

Figure 11.11 (a) Several gill arches are under the operculum. Each arch has two rows of **gill filaments,** and each filament has many thin, platelike **lamellae.** Gases are exchanged between blood inside the lamellae and the water that flows between them. Note that the blood inside the lamellae flows in a direction opposite to the direction of water flow. This countercurrent flow increases the efficiency of gas exchange in fish gills. (b) Scanning electron micrograph of the tip of a single gill filament from a trout. Note the regularly arranged lamellae. Water flows through the filaments in a direction perpendicular to the plane of this page (See (a)). (c) Diagram (1) shows how countercurrent flow permits oxygen uptake throughout the time that blood and water are in close proximity in the gill lamellae. This allows oxygen content of the blood to reach the highest possible level. Water may leave the gill having lost 90 percent of its initial oxygen content. Compare this with a hypothetical arrangement (2) in which blood and water move through gills in the same direction. No further exchange of oxygen would take place after equal concentrations were achieved.

Thus, the water continues to lose oxygen and the blood continues to gain oxygen throughout the time that they are in close proximity because, at any given point, the amount of oxygen in the water is always higher than in the nearby blood vessel. If both streams moved in the same direction, they would reach an equilibrium beyond which no further exchange of oxygen could take place. A great deal of oxygen would still be in the water as the water left the gill (figure 11.11)

Tracheal Systems

Tracheal systems are one of two major types of gas exchange systems in terrestrial animals. Tracheal systems are networks of air tubes that penetrate the tissues of insects and other terrestrial arthropods (centipedes, millipedes, and some spiders). These tubes, called **tracheae** (singular, **trachea**), open to the outside air through holes in the body surface called **spiracles.** Oxygen diffuses into the tracheal system, and carbon dioxide diffuses out through the spiracles (figure 11.12).

Animals with tracheal systems do not depend on gas transport by the circulatory system because the finest end branches of the system penetrate all parts of the body, and as a result, no cell is more than one or two cells away from a tube. Thus, exchange occurs directly between body cells and air in the tracheal system.

For small animals, diffusion through the tracheal tubes is adequate to assure oxygen supply for cells throughout the body, but larger insects, such as bees and grasshoppers, pump air in and out of the system with muscle contractions. Some of these larger insects have enlarged collapsible bags (air sacs) on some of their tracheal tubes, and muscular pressure on these bags aids the pumping process. Even with such modifications, the efficiency of tracheal systems is limited. Most biologists believe that dependence on this method of gas exchange has been a key factor in limiting insect body size.

(a)

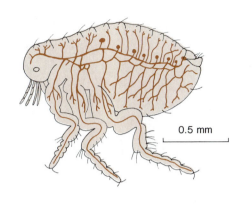

0.5 mm

(b)

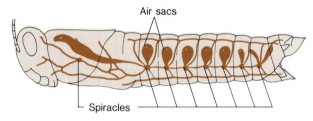

Air sacs

Spiracles

Figure 11.12 Tracheal systems. (a) The tracheal system of a small insect. Tracheae penetrate all parts of the body, and body cells exchange gases with the air inside the tracheal system. The tracheae open to the outside air through spiracles. (b) Larger insects such as grasshoppers pump air through their tracheal systems. This pumping occurs when contracting muscles put pressure on collapsible bags (air sacs) that are part of the tracheal system. This sketch shows larger respiratory structures of the left side of the body. (c) Photomicrograph of part of an insect tracheal system. Walls of tracheae contain **chitin,** a polymer synthesized from glucose. Heavy deposits of chitin make structures rigid. Note the ringed appearance of the tracheae. The rings are chitinous thickenings of the walls that help to prevent collapse of the tubes.

(c)

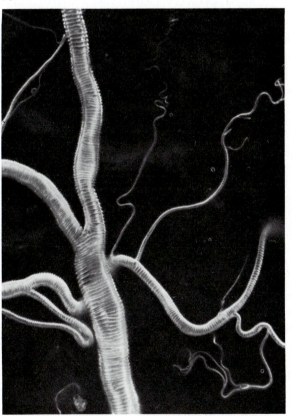

Figure 11.13 Gas exchange adaptations of aquatic insects. (a) A mosquito larva's tracheae open through a spiracle at the end of a tube that is extended above the water surface. (b) Diving beetles such as this *Dytiscus* trap air bubbles under their wings, where spiracle openings are located.

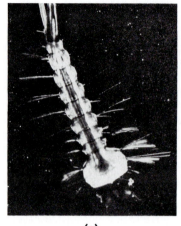

(a)   (b)

Although the majority of animals with tracheal systems are terrestrial, some aquatic insects have special adaptations of the tracheal system that permit gas exchange in an aquatic environment. For example, some insect larvae hang head-down just under the water surface. Each larva has a posterior **siphon tube** extended into the air above the water surface. The siphon tube has a single spiracle that opens into the tracheal tubes inside it. If the larva is disturbed, it can close the spiracle and dive, but it must soon return to the surface to extend its "snorkle" into the air for gas exchange (figure 11.13a).

Other aquatic insects have tracheal gills. Tracheal gills are expansions of the body surface that contain networks of tracheal system tubes (rather than the blood vessels found in most other animals' gills). Gases are exchanged with the surrounding water through the gill surfaces, and the oxygen moves in gaseous form into the tracheal system. Because of the low oxygen content of water, this mechanism is adequate only for very small insects.

Perhaps the most unusual and interesting adaptation of insects for gas exchange in water is seen in some diving beetles. They trap bubbles of air over their spiracle openings in the space under their wings and carry the bubbles as they dive. The beetles can utilize the oxygen in the bubbles and any additional oxygen that diffuses into the bubbles from the surrounding water (figure 11.13b).

Lungs

Lungs are **invaginations** (ingrowths) of the body surface. Thus, lungs are just the opposite of gills, which are outgrowths of the body surface (see figure 11.6). Lungs provide a highly vascular, moist gas exchange surface in a sheltered internal environment.

Lungs are superior to gills as gas exchange organs for terrestrial animals for several reasons. First, lungs are connected to the outside air through relatively narrow tubes. This helps to reduce water loss because some of the water evaporated from lung linings condenses in these tubes and can thus be conserved in the body. Water loss to air from large exposed, moist surfaces would be a huge problem for a terrestrial animal with gills. Second, gas exchange surfaces are delicate and highly vascular and thus very susceptible to injury by contact with dry surfaces in the terrestrial environment. It is advantageous to have them located in a protected internal position.

A few kinds of invertebrate animals, such as some snails (known as pulmonate snails), have lungs, but lungs are primarily a vertebrate adaptation.

Evolution of Vertebrate Lungs

A discussion of the evolution of vertebrate lungs must include information about some organs very similar to lungs called **swim bladders.** The swim bladder is an air sac located dorsal to the digestive tract in the bodies of many modern fish.

Box 11.1
Fish That Can Drown

A number of modern fish can breathe air, but lungfish probably are the best known. One genus of true lungfish is found on each of three continents: Africa, South America, and Australia (box figure 11.1A).

When biologists began to study the physiology of these rather curious creatures, they encountered some peculiar problems. They found that many African lungfish died in the nets that were used to catch them and that the lungfish that were caught alive often died on the way to the laboratory. This seemed strange because, in their natural environment, lungfish appear to be quite hardy animals. Because they breathe air at the surface, they can survive in very stagnant water with an extremely low oxygen content. Sometimes, they even wriggle across the ground to a new pond if the pond that they are in is drying up. Thus, it was hard to understand why lungfish almost invariably died, no matter how gently they were handled, while they were being carried to the laboratory in buckets of water.

Eventually it became clear that the lungfish died because in the buckets they could not bend their bodies to reach the surface of the water. They are obligate air breathers; that is, they must breathe air to live. If prevented from breathing air, African lungfish die for lack of oxygen—they drown!

South American lungfish also are obligate air breathers. They must be able to surface and breathe air because they cannot survive using only their gills for gas exchange. Australian lungfish, on the other hand, can survive without breathing air. They seem to use air breathing only as an emergency means of supplementing gas exchange through their gills.

Another interesting aspect of the biology of lungfish is their response to periods of extreme drought. When the water around a lungfish dries up completely, the lungfish crawls into the bottom of the pond and secretes material that causes the mud around it to form a sort of cocoon. This cocoon hardens as the mud dries, and the metabolic rate of the lungfish inside the cocoon slows to a fraction of its normal level. The lungfish can survive long periods of hot dry weather in this inactive state, which is called **estivation.** When the rains come and the lake or pond refills, the lungfish emerges from its cocoon and resumes its normal activity.

Some biologists who study lungfish collect them in their cocoons and carry them to laboratories in other parts of the world. The lungfish can be kept inside their cocoons on laboratory shelves until the biologists are ready to study them. Then it is a simple matter to put the cocoons in water and study the active lungfish that emerge. These fascinating animals have been stored in this way for years and still successfully revived. A great deal must yet be learned about the "suspended animation" and other features of lungfish biology.

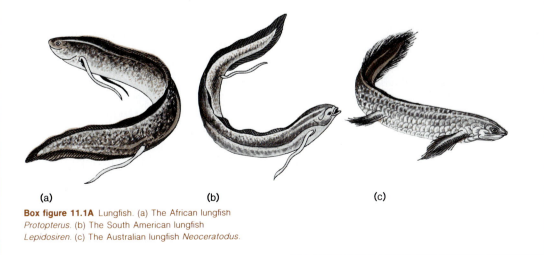

(a)   (b)   (c)

Box figure 11.1A Lungfish. (a) The African lungfish *Protopterus.* (b) The South American lungfish *Lepidosiren.* (c) The Australian lungfish *Neoceratodus.*

Figure 11.14 A possible interpretation of the evolution of swim bladders and lungs from simple ventral sacs attached to the pharynx of primitive bony fish. Swim bladders of modern fish are single structures located dorsal to the digestive tract, while lungs are paired structures attached to the ventral side of the digestive tract.

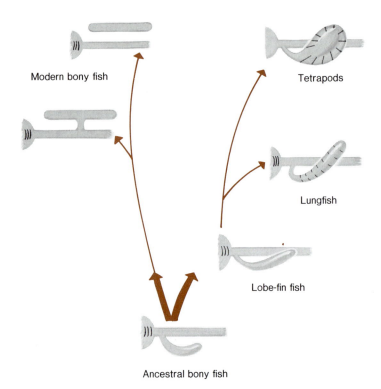

Biologists who study vertebrate evolution think that both lungs and swim bladders evolved from a lunglike structure that was present in primitive fish that were ancestors of both present-day fish and the **tetrapods** ("four-legged" vertebrates—amphibians, reptiles, birds, and mammals). These ancestral fish probably had a sac growing out from the ventral side of their pharynx. This structure was a supplementary gas exchange organ used when the fish could not obtain adequate oxygen through their gills because of stagnant or oxygen-depleted water. The fish could go to the surface, gulp air into this sac, and exchange gases through its wall.

Further evolution of this blind sac off the pharynx proceeded in two different directions (figure 11.14). The first was toward the arrangement seen in the majority of modern bony fish. These fish have swim bladders, single sacs lying *dorsal to the digestive tract*. In some kinds of fish (for example, sturgeons, salmon, and carp), it is connected to the digestive tract by a tube. In other kinds of fish (for example, perch and cod), it begins its development as an outgrowth from the digestive tract of the embryo but later loses its connection to the gut. Thus, in the adult body, it is a separate, hollow sphere with no connecting passage to the gut. In most of these modern fish,

the swim bladder functions in buoyancy adjustment, not in gas exchange. Fish vary the amount of gas in their swim bladders and thus change their buoyancy. Such buoyancy adjustments permit fish to maintain themselves at different depths in the water without having to expend energy to avoid floating up or sinking (figure 11.15).

The second evolutionary direction led to lungs, which are *ventral to the digestive tract*. Ventral lungs are found in a few present-day fish and in the tetrapods. The evolution of structurally complex lungs has paralleled the evolution of the large body sizes and high metabolic rates found in homeothermic vertebrates (birds and mammals).

Care must be taken not to imply that any present-day animal species might be an ancestor of any other present-day species. They are contemporaries, not ancestors and descendants. But by examining the lungs of several living vertebrates, a series of steps that may have occurred during lung evolution can be postulated. A few living fish have lungs that are smoothly lined sacs. Such lungs probably are similar to the lungs of ancestral fish. Frogs' lungs are subdivided into compartments by partitions (figure 11.16a). This partitioning increases the internal surface area

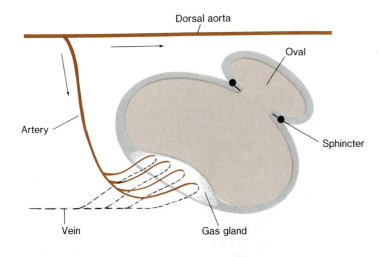

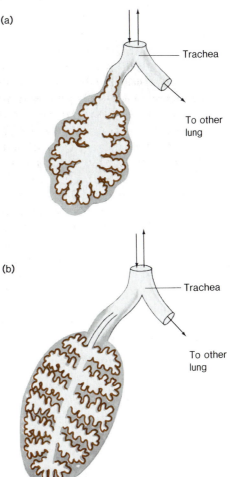

Dorsal aorta

Oval

Artery

Sphincter

Vein

Gas gland

Figure 11.15 Swim bladder function. Gas content of the swim bladder can be changed to change a fish's buoyancy. Gas volume is increased by gas secretion from the gas gland. Gas is removed from the swim bladder by absorption through the wall of the oval. A sphincter constricts the opening into the oval when gas is being added or when the volume is being kept stable.

(a)

Trachea

To other lung

(b)

Trachea

To other lung

Figure 11.16 Lung types. (a) Frog. (b) Turtle.

available for gas exchange. The lungs of turtles and other reptiles are subdivided by branched and rebranched partitions (figure 11.16b). These lungs have significantly more gas exchange surface area than do frog-type lungs of similar size. The lungs of birds and mammals are even more elaborately subdivided into small passageways and spaces. If the lungs of the ancestors of contemporary birds and mammals could be examined, a trend toward greater complexity and increased surface area probably would be seen that would be very similar to the changing complexity and surface area observed when comparing present-day fish lungs, frog lungs, turtle lungs, and bird or mammal lungs.

Lung Ventilation

Even though diffusion of oxygen in air is very rapid, only a few of the animals that have lungs (for example, pulmonate snails) depend on diffusion alone for movement of oxygen into their lungs. Vertebrate animals **breathe.** That is, they **ventilate** their lungs by moving air in and out of them. If vertebrate lungs were not ventilated regularly, oxygen depletion and carbon dioxide buildup would occur quickly in the air inside the lung cavities. Even with ventilation by ebb and flow breathing, the oxygen level is lower and the

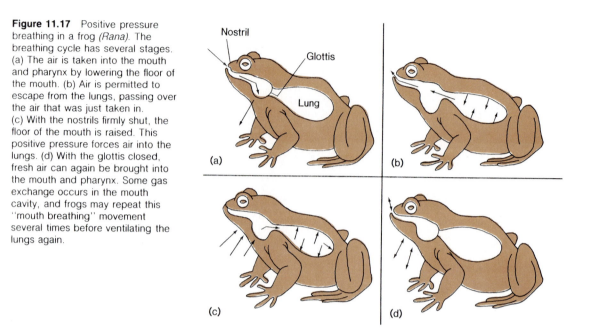

Figure 11.17 Positive pressure breathing in a frog *(Rana)*. The breathing cycle has several stages. (a) The air is taken into the mouth and pharynx by lowering the floor of the mouth. (b) Air is permitted to escape from the lungs, passing over the air that was just taken in. (c) With the nostrils firmly shut, the floor of the mouth is raised. This positive pressure forces air into the lungs. (d) With the glottis closed, fresh air can again be brought into the mouth and pharynx. Some gas exchange occurs in the mouth cavity, and frogs may repeat this "mouth breathing" movement several times before ventilating the lungs again.

carbon dioxide level higher inside vertebrate lungs than in the outside air. This is because there is never complete air exchange during breathing; the lungs are not completely emptied and refilled during each breathing cycle.

Flow-through lungs with air coming in one opening, passing over gas exchange surfaces, and leaving the body through a different opening would provide more efficient gas exchange because there would be continuing complete replacement of the air over the gas exchange surfaces. It is estimated that human lungs, for example, could be about one-tenth their present size if humans had a hole in their back and a flow-through lung arrangement. But flow-through lungs would result in much greater water loss because there would not be the opportunity for the water recovery that our present arrangement, with reverse flow through the same passageway, makes possible. Thus, vertebrate lung structures and ventilation mechanisms represent an evolutionary compromise between two urgent requirements—the need for gas exchange efficiency and the need to conserve body water.

Two different mechanisms for lung ventilation are found among vertebrate animals. Frogs and some reptiles use a **positive pressure** pumping mechanism (figure 11.17). They push air into their lungs. The majority of reptiles and all birds and mammals use a **negative pressure** system that draws air into the lungs by suction. Drawing air in is called **inhalation.**

Reptiles have jointed ribs that can be raised or lowered slightly, thus increasing the space and lowering the pressure around their lungs. When the pressure in the lungs is lowered, atmospheric pressure forces air in. This pressure change is very small because the pressure inside the lungs drops only by one or two mm of mercury, relative to an outside atmospheric pressure, which is 760 mm of mercury at sea level. But this small pressure change is enough to cause air to enter the lungs. Inhalation stretches the walls of the lungs, which are very elastic. When the rib muscles relax and the ribs lower, the lung tissue squeezes down on the air in the lung cavity and forces air out. Breathing out is called **exhalation.**

The entire rib cage of mammals can be lifted at the points of articulation of ribs and vertebral column, and the mammalian **thoracic** (chest) cavity has a muscular floor, the **diaphragm,** that can be flattened. These two movements supplement each other and decrease pressure around the lungs, thereby causing inhalation. Exhalation follows the relaxation of the rib cage and diaphragm muscles (figure 11.18).

Like other vertebrates, birds move air in and out in an ebb and flow breathing pattern, but birds have a special lung ventilation mechanism that permits one-way flow of air over the surfaces where gas exchange actually occurs. This mechanism makes birds' lungs more efficient than mammals' lungs (figure 11.19). Thus, birds'

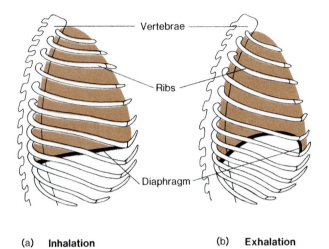

Figure 11.18   Ventilation of human lungs, an example of breathing in a mammal. (a) In inhalation, muscle contractions lift the ribs up and out, and lower the diaphragm. These movements increase the size of the thoracic cavity and decrease the pressure around the lungs. This negative pressure causes more air to enter the lungs. (b) Exhalation follows the relaxation of the rib cage and diaphragm muscles.

(a)   **Inhalation**   (b)   **Exhalation**

Mammal   **Bird**

0.5 mm   0.5 mm

(a)   (b)

Figure 11.19   (a) The gas exchange surfaces in mammals' lungs are in saclike **alveoli.** Ventilation is by an ebb and flow mechanism (arrows), and air replacement inside alveoli can never be complete. (b) The finest passages in birds' lungs are tubes that are open at both ends. Ventilation is by one-way flow (arrow), and there is continual complete replacement of air in the tubes.

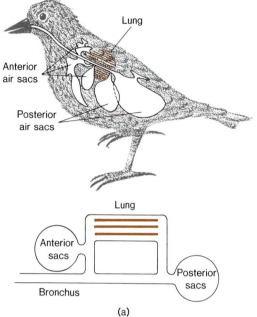

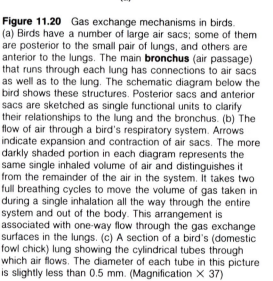

(a)

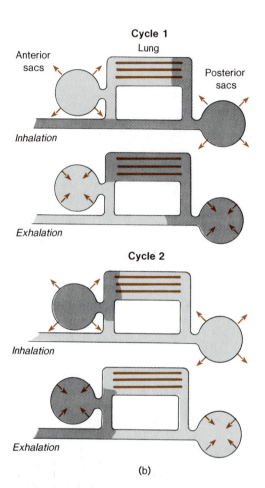

(b)

Figure 11.20 Gas exchange mechanisms in birds.
(a) Birds have a number of large air sacs; some of them
are posterior to the small pair of lungs, and others are
anterior to the lungs. The main **bronchus** (air passage)
that runs through each lung has connections to air sacs
as well as to the lung. The schematic diagram below the
bird shows these structures. Posterior sacs and anterior
sacs are sketched as single functional units to clarify
their relationships to the lung and the bronchus. (b) The
flow of air through a bird's respiratory system. Arrows
indicate expansion and contraction of air sacs. The more
darkly shaded portion in each diagram represents the
same single inhaled volume of air and distinguishes it
from the remainder of the air in the system. It takes two
full breathing cycles to move the volume of gas taken in
during a single inhalation all the way through the entire
system and out of the body. This arrangement is
associated with one-way flow through the gas exchange
surfaces in the lungs. (c) A section of a bird's (domestic
fowl chick) lung showing the cylindrical tubes through
which air flows. The diameter of each tube in this picture
is slightly less than 0.5 mm. (Magnification × 37)

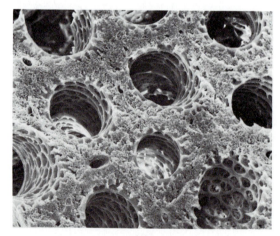

(c)

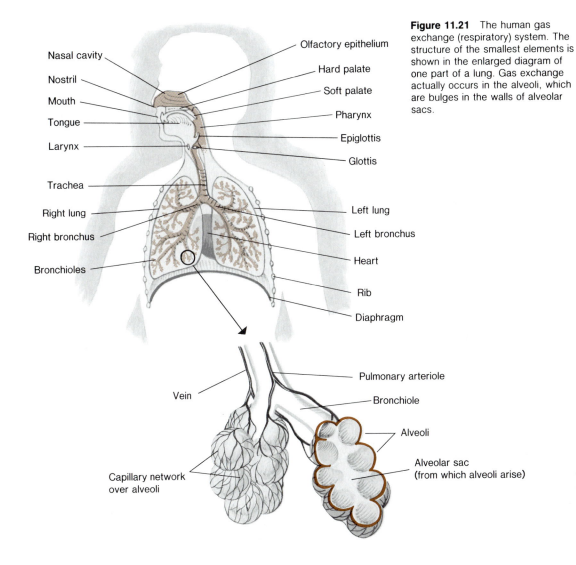

Figure 11.21 The human gas exchange (respiratory) system. The structure of the smallest elements is shown in the enlarged diagram of one part of a lung. Gas exchange actually occurs in the alveoli, which are bulges in the walls of alveolar sacs.

lungs are smaller than the lungs of mammals of comparable body size. Birds' lungs have tunnel-like passages in which gas exchange occurs, and one-way air flow through them is made possible by the arrangement and functioning of a system of **air sacs.** These air sacs are collapsible, and they open and close as a result of muscle contractions around them. Inhaled air bypasses the lungs and enters the posterior air sacs. It then passes through the lungs into the anterior air sacs. Finally, it is exhaled from the anterior air sacs. This process requires two complete breathing cycles (see figure 11.20).

The Human Gas Exchange System

The human gas exchange system (**respiratory system**) is similar to the gas exchange systems of other mammals both in terms of basic structure and functional aspects of lung ventilation (figure 11.21). Human lungs are paired sacs attached to the ventral side of the pharynx, and they are ventilated by an ebb and flow mechanism that uses negative pressure inhalation.

Humans generally breathe through their noses. During inhalation, air enters through **external nares (nostrils)** and passes through the **nasal cavities.** The linings of the nasal cavities near the nostrils have hairs, and the linings deeper in the cavities have cilia on their surfaces.

Hair and cilia, as well as mucus on the lining surfaces, trap dust particles and other foreign material. This filtering action prevents much of the particulate material in air from passing further into the respiratory passages. Air is also warmed and moistened in the nasal passages. Blood vessels in the linings of the nasal cavities lose heat to the air, and water evaporates off moist surfaces into the entering air. By the time that air enters the lungs, it has been cleaned and is more than 99 percent saturated with water vapor. Conversely, air leaving the lungs cools as it passes out through the nasal passages. It loses heat to the linings, and it loses water by condensation as it cools.

Specialized sensory cells that detect odors (various chemicals in the entering air) are located in the olfactory epithelium lying in the top of each of the nasal cavities. Many other mammals have better senses of smell than humans, but our sense of smell does provide us with useful information about food and, occasionally, with a warning about a potentially harmful substance in the air.

The nasal passages are separated from the mouth by the **hard** and **soft palates.** Air in the nasal passages is isolated from food in the mouth, and it is possible to chew and breathe at the same time. But air passes from the nasal passages into the **pharynx,** where it crosses the path of food. This intersection of food and air passages necessitates the swallowing mechanism, which temporarily closes the way to the lungs while food is passing through the pharynx (see page 277). It may seem inefficient for food and air to pass through the same space, and certainly there is danger of choking if food accidentally enters the passage leading toward the lungs, but this arrangement does have the advantage of permitting mouth breathing. Mouth breathing is critically important if the nostrils or nasal cavities become plugged. In addition, it permits greater air intake during heavy exercise when greater gas exchange is required.

Air passes on from the pharynx through the **glottis,** an opening into the **larynx.** The larynx, or "voice box," lies at the top of the **trachea.** During swallowing, the larynx is raised and a flap, the **epiglottis,** covers the glottis, temporarily closing it. The larynx is a rigid, cartilage-supported structure. It contains a pair of elastic ridges, the **vocal cords,** that vibrate when currents of air pass between them and produce sounds.

Their vibration pattern depends upon their tension, and humans can produce different sound tones by varying the tension of the vocal cords.

Air passes on through the larynx into the trachea ("windpipe"). The trachea is a tube that is always kept open because a series of C-shaped cartilages in its wall prevent it from collapsing. The tracheal lining is covered with cilia-bearing cells and mucus-secreting cells. The cilia beat in waves and move mucus and any particles that may have been trapped in it up toward the larynx (figure 11.22). From the larynx, mucus and foreign material can be cleared into the pharynx. This mechanism helps to protect the lungs from any particles that escape filtration in the nose. The tracheal lining cells also trap many potentially harmful microorganisms before they reach the lungs.

The lower end of the trachea divides into two **bronchi** (singular: **bronchus**) that carry air to the right and left lungs. Within the lungs, the bronchi branch and rebranch into a network of smaller passages called **bronchioles.** The bronchi and larger bronchioles have cartilage rings like those of the trachea, but the smaller bronchioles do not have such supports. The bronchi and bronchioles also have ciliated linings that function much like the lining of the trachea.

The smallest bronchioles terminate in **alveolar sacs** that have tiny bulges called **alveoli** (singular: **alveolus**) on their surfaces. Alveoli are the chambers within which gas exchange between air and blood actually occurs. Although each alveolus is very small (human alveoli are about 0.25 mm in diameter), there are hundreds of millions of alveoli in a pair of human lungs, thus making the total surface area of the alveolar linings enormous. Total surface areas are related to body size and physical condition, but alveolar surface areas in human individuals range from 50 to 100 m² with the average probably being about 70 to 80 m², an area about forty times as large as the skin surface area of a human adult.

Alveoli are well adapted for their role in gas exchange because they are highly vascular. More of their lining surface area is occupied by capillaries than by spaces between capillaries. Alveoli also are very thin-walled. The alveolar linings consist of a single layer of very flattened cells, as do the capillary walls. Thus, the distance that gases must diffuse between air and blood is very short, about 0.5 μm (figure 11.23).

(a)   (b)

Figure 11.22   The tracheal lining. (a) Scanning electron
micrograph of the tracheal lining showing cilia and
mucus-secreting goblet cells (GC). (Magnification × 567)
(b) More detailed view of part of the tracheal lining
showing cilia (Ci) and goblet cells (GC). Goblet cells also
have microvilli (Mv) on their surfaces. (Magnification ×
2,075)

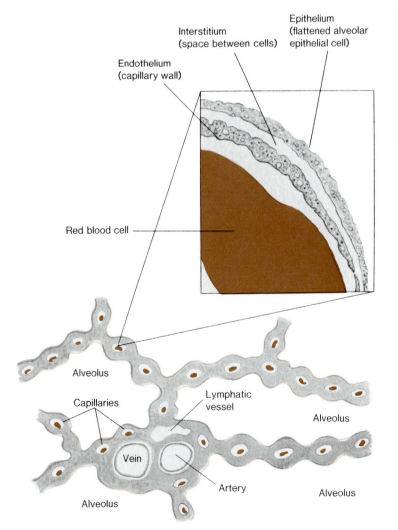

Epithelium
(flattened alveolar
epithelial cell)

Interstitium
(space between cells)

Endothelium
(capillary wall)

Red blood cell

Alveolus

Capillaries

Lymphatic
vessel

Alveolus

Vein

Artery

Alveolus   Alveolus

Figure 11.23   The exchange
surfaces in alveoli. The inset shows
the layers that lie between the air
inside the alveolus and the blood.
The epithelium lining the alveolus
consists of a single layer of flattened
cells, and the capillary wall is very
thin because it also consists of a
single layer of flattened cells. Note
the red blood cells (erythrocytes)
inside the capillaries.

Human Breathing

Humans breathe by the same mechanisms used by all mammals. The volume of the thoracic cavity is increased by muscular contractions that lower the diaphragm and raise the ribs (see figure 11.18). These movements create a negative pressure in the thoracic cavity around the lungs. Atmospheric pressure then forces air into the lungs. The walls of all of the air spaces in the lungs are elastic, and this inhaled air expands and stretches them. When rib and diaphragm muscles relax, air is exhaled as a result of increased pressure in the thoracic cavity and the recoil of the stretched elastic walls of the lungs.

The lungs have a very smooth, thin, membranous outer covering, and a similar smooth membrane lines the thoracic cavity. These lining membranes are called **pleura.** During breathing movements, the lungs move within the thoracic cavity. The pleura secrete a small amount of lubricating fluid that allows the pleura covering the lungs and those lining the thoracic cavity to slide past one another very smoothly.

The total capacity of the lungs and respiratory passages of an adult human is about 6 l of air, but a person has to inhale very forcefully to reach that capacity. If such a forceful inhalation is followed by a forceful exhalation, 1¼ l of air still remain in the lungs. It is not possible to force out this **residual volume** of air. Thus, humans have total breathing capacity of about 4¾ l, but this **vital capacity** is only reached by maximum, forceful inhalation and exhalation. A single, normal breath moves only a fraction of the vital capacity. The **tidal volume** moved by an ordinary breath is about 500 ml. However, only part of this air reaches the alveoli because the nasal passages, pharynx, trachea, bronchi, and bronchioles hold about 150 ml of air. Because very little gas exchange occurs through the walls of these passages, they represent dead space. The **alveolar tidal volume** of 350 ml, therefore, is the effective volume of gas movement in each breath.

A resting adult breathes about fourteen times per minute. With an alveolar tidal volume of 350 ml per breath, there is a functional gas movement in and out of the lungs of about 5 l per minute. During vigorous exercise, the breathing rate may rise to 100 breaths per minute, and the volume of air moved during each breath may be near the vital capacity. This means that humans have the remarkable capacity to increase the volume of air moved from 5 l to 475 l of air per minute, nearly a hundredfold increase.

Breathing control mechanisms are very responsive to gas content of blood, especially to the amount of CO_2 in the blood. For this reason, the mechanisms by which gases are transported in blood are examined before the control of breathing rates is considered.

Gas Transport in Blood

In alveolar gas exchange, oxygen diffuses from the air inside the alveoli into the blood in capillaries, and carbon dioxide diffuses from blood to the air. Thus, blood leaving the alveoli contains more oxygen and less CO_2 than blood arriving at the alveoli (table 11.4). Biologists describe the concentrations of gases in blood in terms of gas **tension** or **partial pressure** (partial pressure is explained in table 11.1). In this text partial pressures expressed as millimeters of mercury (mm Hg) are used in discussing gas concentrations in blood.

Oxygen

Most of the oxygen entering the blood combines with hemoglobin to form **oxyhemoglobin.** Each molecule of hemoglobin contains four polypeptide chains, and each of these is folded around an iron-containing **heme** unit (see chapter 2 for details). Oxygen molecules bind to the iron atoms of the hemes; each molecule of hemoglobin can carry four molecules of oxygen. Since there are about 280 million hemoglobin molecules in each red blood cell (**erythrocyte**), each cell is capable of carrying more than one thousand million molecules of oxygen.

The oxygen-binding characteristics of hemoglobin are rather complex, and some of these complexities are functionally important because they affect the "loading" and "unloading" of oxygen carried by hemoglobin. Oxygen-binding characteristics of hemoglobin can be studied by examining oxyhemoglobin **dissociation curves,** which show what percentage of the oxygen-binding sites of hemoglobin are carrying oxygen at each of a series of oxygen partial pressures (P_{O_2}) in the environment around the hemoglobin. Such dissociation curves are not linear, but S-shaped.

Table 11.4
Partial Pressures* of Oxygen and Carbon Dioxide.

	Body Tissues	Blood Entering Lungs	Alveolar Air	Blood Leaving Lungs
Partial Pressure of Oxygen (P_{O_2})	40	40	100	100
Partial Pressure of Carbon Dioxide (P_{CO_2})	45	45	40	40

*These are all approximate average values under normal conditions. All the values given for blood actually are gas tensions. Gas tension is the amount of gas in a solution that is in equilibrium with an atmosphere having the stated partial pressure. Dissolved gas as such exerts no measurable pressure. But in practice, it is easier to discuss all values using the term partial pressure.

At partial pressures such as those encountered in the lungs, hemoglobin becomes practically saturated with oxygen. As blood moves through the circulatory system, it loses very little oxygen, even when it enters environments where partial pressures are around 60 mm Hg. But when the blood flows through capillaries in body tissues where partial pressures between 20 and 40 mm Hg are encountered, hemoglobin quickly gives up much of its oxygen. In fact, oxygen release is very sensitive to small P_{O_2} changes within that range. Oxygen release, therefore, is responsive to P_{O_2} decreases caused by increased oxygen consumption by cells, such as occurs during heavy exercise.

Oxyhemoglobin dissociation curves, then, reveal several important properties of the oxygen-transporting mechanism. Hemoglobin is quickly and nearly completely loaded with oxygen in the lungs. In the body tissues, oxygen unloading is very responsive to P_{O_2} changes that reflect oxygen requirements of the cells.

Another functionally important aspect of oxyhemoglobin dissociation is the effect of pH on the process. Lowering the pH shifts the oxyhemoglobin dissociation curve to the right (figure 11.24). This change is called the **Bohr effect.** It causes oxygen to be unloaded more readily in tissues that require more oxygen because metabolic wastes, especially CO_2, lower the pH of blood in actively metabolizing tissues.

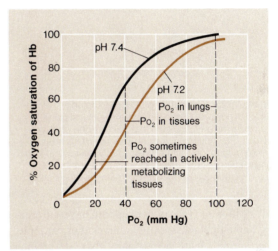

Figure 11.24 Oxyhemoglobin dissociation curves for human hemoglobin measured at two different pH's. The lower pH encountered in actively metabolizing tissues causes oxyhemoglobin to dissociate more readily. This "Bohr effect" promotes "unloading" of oxygen in tissues.

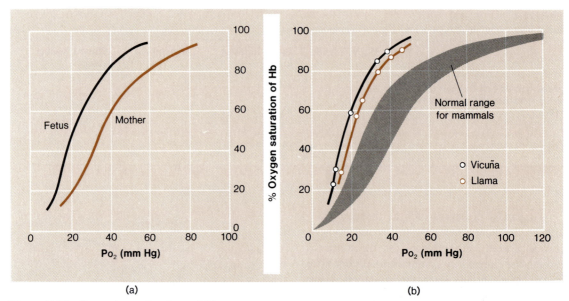

(a)

(b)

Figure 11.25 Comparisons of oxyhemoglobin dissociation curves for different hemoglobins. (a) Fetal mammals have hemoglobin that has a higher affinity for oxygen than adult hemoglobin. This permits a fetus's hemoglobin to load with oxygen at Po_2 values that cause dissociation of oxyhemoglobin in its mother's blood. Thus, oxygen tends to leave the maternal blood and move into the fetal blood. (b) The hemoglobins of llamas and vicuñas have higher affinities for oxygen than the hemoglobins of most other mammals. Thus, their hemoglobins can efficiently take up oxygen even at the low Po_2 levels of the high-altitude atmosphere where they live.

Different types of hemoglobin have different oxyhemoglobin dissociation curves. For example, the special **fetal hemoglobin** that developing mammals produce before birth is functionally different from normal adult hemoglobin. For the developing fetus to obtain oxygen, there must be a transfer of oxygen from the mother's blood to the fetus's blood in the placenta. This transfer is possible because fetal blood has a higher affinity for oxygen than adult blood at the Po_2 levels found in the placenta. Fetal hemoglobin loads with oxygen at the same time that the mother's hemoglobin is unloading oxygen. The difference in dissociation curves is essential for the transfer of oxygen during fetal development (figure 11.25a). After birth, fetal hemoglobin gradually is replaced with adult-type hemoglobin.

Another example of differences in oxyhemoglobin dissociation curves is seen in animals that live at high altitudes. Because atmospheric pressure is lower, the partial pressure of oxygen is lower at high altitudes than it is at lower altitudes. Some animals that normally live under these conditions of relative oxygen deficiency have oxyhemoglobin dissociation curves that are quite different from those found in animals living at lower altitudes. For example, llamas and vicuñas live high in the Andes mountains of South America, often at altitudes above 5,000 m. Their hemoglobins take up oxygen at the low pressures encountered there much more efficiently then the hemoglobins of other mammals. This is a permanent, genetically transmitted adaptation because llamas raised in zoos at much lower altitudes have hemoglobin that shows the same sort of dissociation curve as that of the hemoglobin of llamas living in their native, high altitude habitat.

Other mammals, including humans, can adjust to the low Po_2 values of high altitudes, but the adjustment does not involve producing new, llamalike hemoglobin. Instead, many more red blood cells are produced by the bone marrow, thus increasing the amount of hemoglobin available for carrying oxygen. The blood of people who make this adjustment literally becomes thicker because there are more erythrocytes per unit volume of blood.

Muscle cells contain an oxygen-binding pigment called **myoglobin.** Myoglobin resembles a single unit of a hemoglobin molecule. That is, myoglobin has one folded polypeptide chain and one heme unit. Myoglobin seems to fulfill a very special reserve oxygen-holding role in muscle cells. Its oxygen dissociation curve is well to the

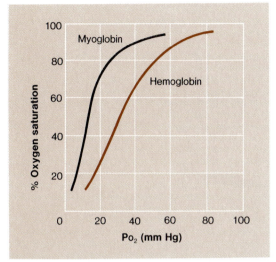

Figure 11.26 Comparison of the dissociation curves of myoglobin and hemoglobin. Because of their different affinities for oxygen, hemoglobin will give up oxygen to myoglobin in muscles. Myoglobin functions as a backup oxygen source because it dissociates only when the P_{O_2} in the muscle cells falls very low, as it does during heavy exercise.

left of that of hemoglobin (figure 11.26). Functionally, this means that myoglobin loads with oxygen at the expense of hemoglobin. It also means that myoglobin gives up its oxygen only when the P_{O_2} falls very low in the cells, as it does during heavy exercise. Thus, myoglobin provides an excellent reserve oxygen source for muscle cells when they are actively contracting and metabolizing rapidly.

Other gases besides oxygen bind to hemoglobin. **Carbon monoxide (CO)** is a very dangerous substance because it binds to hemoglobin even more readily than oxygen. Thus, carbon monoxide can "outcompete" oxygen for sites on hemoglobin molecules even when the CO is present at much lower partial pressures than oxygen. Carbon monoxide can be present in cigarette smoke and smoke from other incomplete combustions, but the most dangerous source of this colorless, odorless gas is exhaust from automobile engines. Carbon monoxide may become bound to so much of the hemoglobin that not enough hemoglobin is available for adequate oxygen transport. Symptoms of oxygen deprivation (**asphyxiation**), such as drowsiness and impaired vision and hearing, develop quickly. Recovery from carbon monoxide poisoning is possible because

oxygen gradually replaces the CO when the victim breathes fresh air. But carbon monoxide poisoning often leads to death because the victims soon become too weak to move away from the carbon monoxide source.

Carbon Dioxide

In addition to transporting oxygen to the body tissues, blood also transports carbon dioxide from body tissues to the lungs. There are several mechanisms of carbon dioxide transport. A small amount of CO_2 (about 7 percent of the total) dissolves in blood plasma. Somewhat more CO_2 (about 10 to 15 percent of the CO_2 transported) is bound to hemoglobin to produce **carbaminohemoglobin.** The union between CO_2 and hemoglobin is rather loose and does not involve the heme groups that bind oxygen. Carbon dioxide binds to amino groups of amino acids in the polypeptide chains of hemoglobin molecules.

But dissolved carbon dioxide and carbaminohemoglobin together account for less than one-fourth of the CO_2 transported by the blood. Most of the CO_2 carried by the blood is transported as **bicarbonate ions** (HCO_3^-). Bicarbonate ions are produced because carbon dioxide tends to combine with water to produce **carbonic acid:**

$$H_2O + CO_2 \rightleftharpoons H_2CO_3$$

The reaction of carbon dioxide and water occurs spontaneously (but relatively slowly) in the blood plasma. However, inside red blood cells, this reaction is catalyzed by an enzyme, **carbonic anhydrase,** and it occurs much more rapidly.

At the pH levels found in blood, carbonic acid tends to be dissociated to hydrogen ions (H^+) and bicarbonate ions:

$$H_2CO_3 \rightleftharpoons H^+ + HCO_3$$

The following equation summarizes what happens when carbon dioxide enters the blood flowing through body tissues:

$$H_2O + CO_2 \xrightleftharpoons[\text{}]{\text{Carbonic anhydrase}} H_2CO_3 \rightleftharpoons H^+ + HCO_3^-$$

The dissociation of carbonic acid slightly lowers the pH of the blood and has an important effect on delivery of oxygen to the tissues because of the Bohr effect mentioned earlier.

Figure 11.27 Summary diagram
showing circulation and gas
exchange in lungs and tissues.

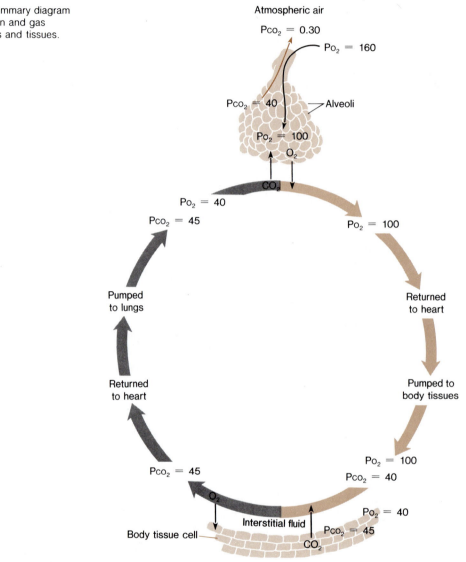

However, body cells cannot withstand significant changes in blood pH, and therefore, serious blood pH fluctuations are prevented by buffers. A large part of the buffering capacity of the blood is provided by hemoglobin and other blood proteins.

Hemoglobin (Hb) in blood actually is in the form of an almost completely ionized potassium salt:

$$KHb \rightleftharpoons K^+ + Hb^-$$

Carbonic acid reacts with potassium hemoglobin to produce potassium bicarbonate ($KHCO_3$), which is almost completely ionized as K^+ and HCO_3^-, and acid hemoglobin (HHb):

$$H^+ + HCO_3^- + K^+ + Hb^- \rightleftharpoons$$
$$K^+ + HCO_3^- + HHb$$

This buffering system amounts to exchanging one acid for another, that is, carbonic acid for acid hemoglobin. How does such an exchange prevent large pH changes? Acid hemoglobin is a much weaker acid than carbonic acid, and therefore, within the normal pH range of blood, very little acid hemoglobin is dissociated. Thus, this buffer system, which involves formation of acid hemoglobin at the expense of carbonic acid, reduces the number of hydrogen ions in solution in the blood.

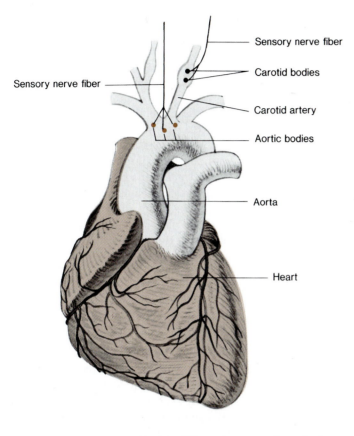

Sensory nerve fiber

Sensory nerve fiber

Carotid bodies

Carotid artery

Aortic bodies

Aorta

Heart

Figure 11.28 Location of the chemoreceptors that detect changes in CO_2, H^+, and O_2 concentrations in the blood. The aortic bodies are in the aorta, the major artery carrying blood from the heart to all body tissues. The carotid bodies are in the carotid arteries, which supply the brain and other parts of the head. Sensory nerves carry information to brain reflex centers that control breathing functions.

In body tissues, carbon dioxide diffuses into the blood, but in the lungs, the opposite occurs. Carbon dioxide pressure in alveolar air is lower than in the blood. Therefore, CO_2 diffuses out of the blood. This loss of CO_2 favors reversal of all of the chemical reactions that occur in the tissues when CO_2 is entering the blood and results in rapid release of CO_2 from the blood. (See figure 11.27.)

Control of Breathing Rate

The breathing rate and the volume of air inhaled and exhaled are controlled by a **respiratory center** in the brain (more specifically, in the medulla). Several other automatic reflex centers in the brain also interact with the respiratory center. Although breathing rate and depth of breathing can be consciously controlled, breathing itself is automatically under the control of these reflex centers when attention turns to other things and during sleep or unconsciousness.

The respiratory center controls the action of muscles in the ribs and diaphragm that are responsible for breathing movements. Increases in the breathing rate and the air volumes being breathed are made automatically by the respiratory center in response to changes in the gas content of the blood.

Blood gas content is monitored by **chemoreceptors** called the **aortic bodies** and **carotid bodies,** specialized structures located in the walls of major arteries. Some chemoreceptors detect changes in CO_2, H^+, and O_2 concentrations. Information from these chemoreceptors goes to the respiratory center and influences the breathing rate. The respiratory center itself also is sensitive and responsive to changes in the chemical content of the blood reaching it (figure 11.28).

The main chemical stimulus that changes the breathing rate is variation in blood P_{CO_2} and H^+ ion concentration (see page 315). Even small increases in blood P_{CO_2} and H^+ concentration cause rapid increases in breathing. Oxygen receptors, however, are sensitive only to large changes in blood P_{O_2}; they become stimulated only when the P_{O_2} of arterial blood has fallen below 70 mm Hg. Since changes in blood P_{CO_2} and P_{O_2} normally are proportional to one another, breathing generally is regulated by CO_2 receptors because they are much more sensitive than the O_2 receptors.

Respiratory System Problems

Any disorder that interferes with continuous gas exchange is serious and can quickly become life threatening.

Because their warm, moist, blind sac interiors provide very favorable environments for growth of microorganisms, the lungs are quite susceptible to infection, and infections of the lungs can produce a condition known as **pneumonia.** In pneumonia, alveolar sacs fill with fluid and dead white blood cells. This reduces the amount of surface available for efficient gas exchange because gases must diffuse through this fluid to reach the alveolar surfaces. Thus, pneumonia can result in seriously reduced blood O_2 levels. Bacterial pneumonia is not so great a threat now as it once was because bacterial infections of the lungs can be treated with antibiotics, but viral pneumonia still causes serious problems. Another once-dreaded bacterial disease of the lungs is **tuberculosis.** But this degenerative disease also responds to antibiotic therapy and is a serious health problem only for people who lack adequate medical attention.

Unfortunately, reduction of the prevalence of bacterial infections of the lungs has not relieved concern about disorders of the respiratory tract. In recent years, there has been a marked increase in the incidence of other lung disorders, such as **emphysema** and **lung cancer.** These problems seem to have increased as a result of certain features of life in modern, technological societies. Another lung disorder, infant **respiratory distress syndrome (RDS)**, has been treated more successfully in recent years, but much research remains to be done.

Emphysema

Emphysema is a progressive disease in which alveolar walls degenerate and lose their elasticity. Some alveolar walls break down, and larger chambers form (figure 11.29). Gas diffusion through the damaged membranes is reduced. Even mild exercise can cause emphysema sufferers to gasp for breath because it is very difficult for them to increase their oxygen and carbon dioxide exchange enough to meet the demands created by exercise.

Because of the loss of lung elasticity, exhalation does not occur by the normal passive mechanisms, and people with advanced stages of emphysema must make an effort to exhale. The extra work required for breathing combined with the reduced efficiency of gas exchange essentially incapacitates many emphysema victims.

Although the causes of emphysema are not clear, many scientists believe that prolonged exposure to respiratory irritants in polluted air, or tobacco smoke, or a combination of the two play a major role in the disease's development. Cigarette smoke is especially suspect because it contains irritating particulate matter and substances that inhibit the activity of the cilia that normally move debris up out of the respiratory passages.

Lung Cancer

Lung cancer is the most common form of cancer in human males, and its incidence is increasing rapidly in human females as well. As with other cancers (see chapter 26), the cause or causes of lung cancer are not completely established, but irritations such as those mentioned in connection with emphysema almost certainly play a role. For example, the incidence of lung cancer in cigarette smokers is several times higher than the incidence in nonsmokers.

Lung cancer involves uncontrolled growth of cells that replace normal functional lung cells and reduce the gas exchange surface area of the lungs. Lung cancers can grow to the point where they actually block bronchioles and prevent air movement in and out of entire areas of lungs. Spreading (**metastasis**) of cancer cells through the circulatory system can establish abnormal growths in other parts of the body.

Infant Respiratory Distress Syndrome

Infant respiratory distress syndrome (RDS) is sometimes called **hyaline membrane disease** because of the glassy, translucent layer that forms inside the alveoli of infants suffering from the condition. RDS develops most often in premature newborns because their lungs are not mature enough to function normally.

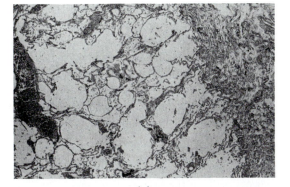

(a)

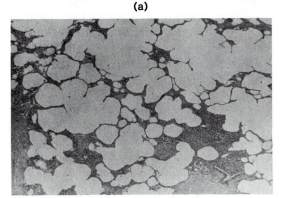

(b)

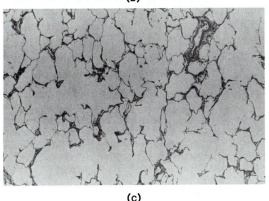

(c)

Figure 11.29 Development of emphysema. (a) Normal alveoli. (Magnification × 8) (b) Breakdown of some alveolar walls to produce enlarged chambers. Lungs lose both gas exchange surface area and elasticity as emphysema develops. (Magnification × 8) (c) Advanced stage of emphysema in which functional alveolar tissue breaks down completely, leaving only connective tissue, thus eliminating this part of the lung as a functional gas exchange area. (Magnification × 8)

The first few breaths that a newborn infant (**neonate**) takes are very labored. There is some fluid in the unexpanded lungs and respiratory passages, and the surface tension of this fluid is a significant force that must be overcome as the alveoli are initially expanded. Normally, substances called **surfactants** are present in the alveoli and reduce this fluid surface tension so that once the alveoli are expanded, they tend to stay expanded rather than collapsing under surface tension after each breath.

The respiratory distress syndrome develops when these surfactants are not present in adequate quantities to prevent alveolar collapse due to fluid surface tension. This makes each breath of the newborn nearly as difficult as the first one. The infant quickly becomes exhausted from the hard work of breathing and also suffers the effects of inefficient gas exchange. If an infant survives the initial problems associated with alveolar collapse, a new set of problems develops in a few hours. Fluid with a high protein content seeps into the alveoli and forms the hyaline membrane inside the alveoli. This aggravates the infant's problems because it interferes with oxygen diffusion and further depresses blood O_2.

The outlook for babies who develop infant respiratory distress syndrome was very bleak in the past, but some modern treatments offer hope for many of them. Smaller, weaker infants are given a treatment called continuous positive airway pressure (CPAP). A mask over the baby's face supplies oxygen-supplemented air under pressure. This continuously applied extra air pressure in the respiratory tract prevents alveolar collapse. Larger, stronger infants can be treated with positive end expiratory pressure (PEEP), which is a less drastic manipulation of the breathing cycle. In the PEEP procedure, a tube is passed down through the air passages to the top of the lungs. A small puff of oxygen-supplemented air is given through the tube just at the end of the infant's exhalation. This prevents alveolar collapse until the next inhalation is started. These treatments aid breathing until the lungs have had time to mature enough to function without assistance. Even with these treatments, infants who recover from RDS still have respiratory problems during their early childhood. They are more susceptible to various respiratory difficulties than other children.

Summary

Gas exchange is an urgent requirement for animal life. All animals must continuously obtain oxygen and release carbon dioxide.

The characteristics of an animal's gas exchange mechanism are determined by the medium with which the animal exchanges gases, the size and shape of the animal, the animal's internal gas-transporting capability, and the animal's metabolic rate.

The four basic categories of animal gas exchange mechanisms are integumentary exchange, gills, tracheal systems, and lungs.

Some relatively small animals that live in moist environments conduct all of their gas exchange through their body surfaces, but most larger animals have more localized, special gas exchange areas.

Most aquatic animals exchange gases through gills that are delicate, finely branched, highly vascular extensions of the body surface. Most gilled animals have a gill ventilation mechanism that produces a one-way flow of water over the gills. In many cases blood circulating in vessels inside the gills moves in the opposite direction of water flowing over the gill surfaces. This countercurrent flow mechanism increases gas exchange efficiency.

Tracheal systems are networks of air tubes that penetrate the body tissues of insects and other terrestrial arthropods. Gas diffusion through the tubes meets the needs of smaller insects, but many larger insects have pumping mechanisms that force air through the tubes.

Lungs are highly vascular, thin-walled invaginations of the body surface. Water loss by evaporation from the moist respiratory exchange surfaces of lungs is minimized because lungs are ventilated through narrow passages to the outside air. Lungs are found primarily in terrestrial vertebrate animals.

Because vertebrates' lungs are blind sacs connected to the outside air by narrow passages, breathing mechanisms are necessary for lung ventilation. Some vertebrates push air into their lungs using a positive pressure pumping mechanism. However, most vertebrates inhale by muscular movements that produce negative pressures around the lungs so that atmospheric pressure forces air into the lung. When the breathing muscles relax, air is exhaled because of the increased pressure in the lung cavity and the recoil of the elastic walls of the lungs.

Birds have a series of air sacs that move air in a one-way flow over the gas exchange surfaces of their lungs.

The human respiratory system is similar to the gas exchange systems of other mammals. Air is cleaned, warmed, and moistened during inhalation. Gas exchange takes place between air in alveoli and blood in capillaries in the alveolar walls. Only a small portion of the total gas volume in the lungs and respiratory passages is replaced in each normal breath. Breathing volume and breathing rate increase dramatically during heavy exercise, when demands for gas exchange are increased.

Most of the oxygen transported in blood is bound to hemoglobin as oxyhemoglobin. Hemoglobin loads with oxygen quickly and completely at the oxygen partial pressures of the lungs. Most of the oxygen remains bound until the blood reaches capillaries in body tissues. Small P_{O_2} changes within the ranges found in tissues cause large changes in oxyhemoglobin dissociation. Thus, the process is very responsive to the oxygen requirements of the cells. The lower pH resulting from higher CO_2 production in actively metabolizing tissues also increases oxyhemoglobin dissociation.

Carbon dioxide is carried in the blood mainly as bicarbonate ions. Bicarbonate ions are produced because CO_2 and water combine to produce carbonic acid. Carbonic acid molecules dissociate to hydrogen ions and bicarbonate ions. The buffering action of hemoglobin and other protein molecules prevents the significant pH changes that would otherwise be caused as a result of CO_2 entering the blood from body tissues.

A respiratory center in the brain along with several automatic brain reflex centers adjust breathing rates to meet gas exchange requirements. Chemoreceptors that respond to CO_2, hydrogen ion, and oxygen contents of the blood supply information to these brain reflex centers.

The warm, moist, blind sac interiors of the lungs are susceptible to infection. Pneumonia resulting from infection decreases gas exchange efficiency. Emphysema is a degenerative disease that also reduces gas exchange efficiency. Emphysema is caused, or at least aggravated, by chemical irritants and particulate matter in inhaled air. Lung cancer also is associated with chronic lung irritation.

Often, premature infants suffer from infant respiratory distress syndrome, which develops when the infants' lungs have not matured enough before birth to function normally. The syndrome is characterized by very labored breathing due to alveolar collapse after each exhalation.

Questions

1. Compare and contrast water and air as gas-carrying media with which animals exchange gases.
2. Why do mammals use only 1 to 2 percent of their total energy output for breathing while fish use up to 25 percent of their energy output for gill ventilation?
3. Why is a thin oil film spread over the surfaces of lakes and ponds to control mosquitoes?
4. Considering the size of the alveolar tidal volume in normal breathing, why must a person inhale and exhale quite forcefully to breathe through a tube, such as a snorkel? Why are there limits on the length and diameter of tubes through which a person can breathe?

5. Explain the factors that make the unloading of oxygen in tissues very responsive to the needs of cells in local areas of the body.
6. Explain why athletes participating in endurance events perform very poorly at high altitudes unless they have previously trained at comparable altitudes for an extended period of time. Explain why their performances improve after high-altitude training.
7. How does the fact that acid hemoglobin is a weaker acid than carbonic acid make possible hemoglobin's role in blood buffering?

Suggested Readings

Books

Krogh, A. 1968. *The comparative physiology of respiratory mechanisms.* New York: Dover.
Schmidt-Nielsen, K. 1979. *Animal physiology.* 2d ed. Cambridge, Mass.: Cambridge University Press.

Articles

Hughes, C. M. 1973. The vertebrate lung. *Carolina Biology Readers* no. 59. Burlington, N.C.: Carolina Biological Supply Co.
Johansen, K. October 1968. Air-breathing fishes. *Scientific American* (offprint 1125).
Perutz, M. F. December 1978. Hemoglobin structure and respiratory transport. *Scientific American* (offprint 1413).
Rahn, H.; Ar, A.; and Paganelli, C. V. February 1979. How bird eggs breathe. *Scientific American* (offprint 1420).
Schlesinger, R. B. 1982. Defense mechanisms of the respiratory system. *Bioscience* 32: 45.
Schmidt-Nielsen, K. December 1971. How birds breathe. *Scientific American* (offprint 1238).
Schmidt-Nielsen, K. May 1981. Countercurrent systems in animals. *Scientific American* (offprint 1497).
Wigglesworth, V. B. 1972. Insect respiration. *Carolina Biology Readers* no. 48. Burlington, N.C.: Carolina Biological Supply Co.

Circulation and Blood

Chapter Concepts

1. Circulatory systems move blood through animals' bodies, transporting materials between body exchange areas and internal body cells.

2. Blood, which is composed of plasma and formed elements, functions in the transport of a variety of molecules and in body defense reactions.

3. In open circulatory systems, blood leaves vessels and circulates among body organs; in closed circulatory systems, blood is continuously enclosed inside vessels.

4. The human heart is an efficient pumping device that is laterally partitioned so that blood on one side, being pumped to the pulmonary circulation of the lungs, is separated from blood being pumped to the systemic circulation.

5. Blood pressure forces blood through the circulatory system and provides the driving force for fluid movement from the capillaries into the body tissues.

6. The blood-clotting mechanism protects the body from excessive blood loss but can cause serious problems when it functions improperly.

7. The human circulatory system normally functions as an efficient transport mechanism, but it is subject to problems that affect the blood itself or the condition of the blood vessels.

All cells of living organisms must exchange materials with their immediate environments. Cells obtain nutrients and oxygen, and they release carbon dioxide and other metabolic wastes. Unicellular organisms can make these exchanges directly with the external environment, but in multicellular organisms, many body cells are far from external body surfaces. Multicellular organisms have an internal body environment with which individual body cells exchange materials.

Nutrients and oxygen are supplied to the internal body environment and wastes are removed from it at specialized exchange surfaces of the body (for example, gases are exchanged in gills or lungs). Organisms with relatively large bodies require **transport mechanisms** to carry materials between specialized exchange areas and the immediate environments of cells throughout the body.

The transport mechanisms that carry materials throughout plant bodies are discussed in chapter 9. In this chapter the transport mechanisms that move materials around animal bodies are explored.

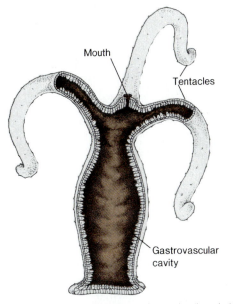

Figure 12.1 The gastrovascular cavity (in color) of *Hydra* functions in transport as well as digestion. Nutrients are distributed through the cavity. Cells lining the cavity can exchange gases with the water in the cavity and release wastes into it.

Transport Systems in Invertebrates

Although this chapter concentrates on vertebrate animals' circulatory systems, some important principles of animal transport can be understood by first examining transport systems of several invertebrate animals.

Gastrovascular Cavities

Coelenterates such as the fresh water *Hydra* do not have a specialized transport system, but they do have a fluid-filled internal cavity, the **gastrovascular cavity,** that functions to a certain extent in transport as well as in digestion. Most of *Hydra's* gastrovascular cavity is located inside the main body stalk, but the cavity has extensions out into the tentacles (figure 12.1). Digestion in the gastrovascular cavity supplies nutrients for all body cells. Cells also exchange materials with the water in the gastrovascular cavity. For example, cells lining the cavity can obtain oxygen from water in the cavity and release CO_2 and other cell wastes into it. Fluid in the cavity does not circulate in any regular way, but body movements put pressure on the fluid and move it

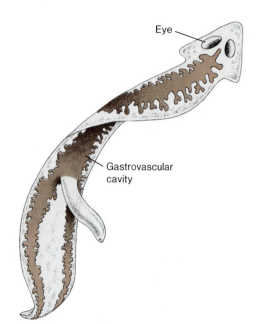

Eye

Gastrovascular
cavity

Figure 12.2 The planarian gastrovascular cavity (color)
functions much as *Hydra's* does. The planarian cavity
has branches that penetrate much of the body.

around. No cell in a *Hydra* is far from the water
outside the body or from the water inside the
gastrovascular cavity.

Planarian worms have gastrovascular cavities that are more complex than that of *Hydra*
because branches penetrate all parts of the body
(figure 12.2). Although the body of a planarian
is somewhat thicker than that of a *Hydra,* the
branched gastrovascular cavity extends close to
all body cells, and diffusion distances for nutrients and other materials are not great. Body
movements cause fluid movements that help to
distribute materials to various parts of the gastrovascular cavity.

Circulatory Systems

Circulatory systems are transport mechanisms by
which fluid is moved through the body in a regular fashion by the pumping action of a muscular
heart. A circulatory system with a heart provides
far more efficient transportation than the sluggish
fluid movement in the gastrovascular cavities of
Hydra and planaria because the heart keeps fluid
moving even when the animal is not. The circulating fluid, called **blood,** transports materials to
and from body cells. Besides nutrients, gases, and
wastes, blood carries chemical regulator substances (hormones) around the body.

The two basic types of circulatory system
arrangements are **open circulatory systems** and
closed circulatory systems.

Open Circulatory Systems

In open systems, the heart pumps blood out into
the body cavity, or at least, through large parts
of the body cavity, where the blood bathes organs
and tissues of the body. In closed circulatory systems, blood circulates only within the confines of
a system of tubular **blood vessels.**

Insect circulatory systems are open. An insect's heart pumps blood through vessels that
open into the body cavity, or **hemocoel** ("blood
cavity"). Body movements also help to propel
blood through the hemocoel. Blood that has circulated through the hemocoel enters the posterior
part of the heart through openings called **ostia,**
which are one-way valves that prevent backward
flow when the heart beats (figure 12.3).

Blood circulates more slowly in open systems, and for this reason they are less efficient
than closed systems. Since some insects are very
active and have very high metabolic rates, especially when they are flying, it might seem that
open circulatory systems would not be adequate
for them. But insects do not depend on circulatory
systems for oxygen and carbon dioxide transport.
They use tracheal systems (chapter 11) for gas
exchange. Insect blood, in fact, is unusual among
animal bloods because it is colorless; it does not
contain any **respiratory pigments,** the colored
substances that function in oxygen transport in
the blood of many other animals.

Closed Circulatory Systems

Annelids, such as the common earthworm, have
closed circulatory systems. An earthworm has
five pairs of pulsating hearts that keep blood
moving through its circulatory system. Peristaltic
waves of contraction pass through these hearts,
forcing blood out to a ventral blood vessel.
Branches from the ventral blood vessel carry
blood into the various body tissues. In closed circulatory systems, blood is not pumped out into
the spaces around body cells as it is in open circulatory systems. Exchanges between body cells
and the blood are made through the walls of the
smallest blood vessels, called **capillaries.** A capillary wall is only one cell thick. Since capillaries
form diffuse networks in all body tissues, no body

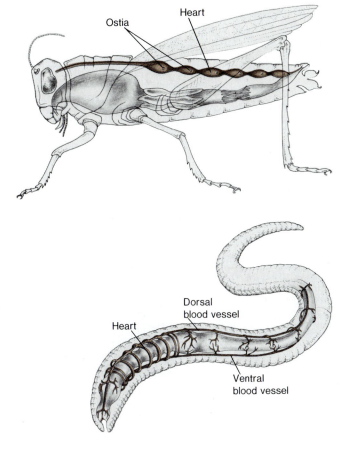

Ostia

Heart

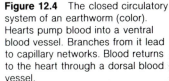

Figure 12.3 The dorsal heart of a grasshopper (color) pumps blood through an open circulatory system. Blood moves forward from the heart, through the hemocoel, and back into the heart through the ostia.

Dorsal blood vessel

Heart

Ventral blood vessel

Figure 12.4 The closed circulatory system of an earthworm (color). Hearts pump blood into a ventral blood vessel. Branches from it lead to capillary networks. Blood returns to the heart through a dorsal blood vessel.

cell is far from circulating blood. After blood passes through the capillary networks in an earthworm's tissues, it collects in vessels leading to a major dorsal blood vessel that returns blood to the heart (figure 12.4).

Earthworm blood is red because it contains the respiratory pigment hemoglobin. Earthworm hemoglobin functions in oxygen transport just as the hemoglobin in vertebrate blood does, but it is found in the blood plasma instead of inside blood cells.

Vertebrate Circulatory Systems

All vertebrate animals have closed circulatory systems. The vertebrate circulatory system has a strong, muscular heart with **valves** that prevent backward flow when the heart beats. Blood travels away from the heart through **arteries** to smaller vessels called **arterioles.** Arterioles lead to capillary beds that are sites of exchange between blood and the various body cells. From

capillaries, blood collects in **venules** that flow together into **veins,** the major vessels carrying blood toward the heart. (See figure 12.5.)

Vertebrate animals have specialized gas exchange organs, either gills or lungs, that contain capillary beds where gas exchange between the blood and the gas-carrying medium occurs. The major differences among vertebrate circulatory systems involve the relationship between circulation through those gas exchange organs and circulation to the rest of the body.

The Single Circuit Plan of Fish

In fish, the circulatory plan involves a single "loop" through the body. Blood leaves the heart and goes forward through a large vessel, the **ventral aorta.** The ventral aorta carries blood to the gill area where smaller vessels distribute it to the gill capillary networks. In the gills, blood is **oxygenated** (it picks up oxygen) and gives off CO_2 at the same time. Blood leaves the gills through vessels that flow into a large **dorsal aorta.** The

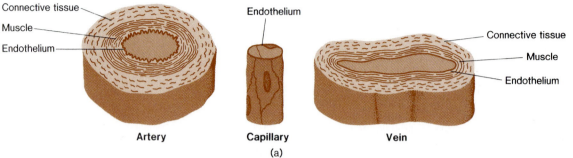

Artery   Capillary   Vein

(a)

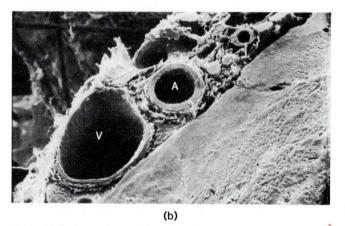

(b)

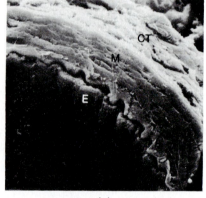

(c)

Figure 12.5 Comparison of the walls of arteries,
capillaries, and veins. (a) Diagrammatic comparison in
which capillary is drawn much larger than its normal
proportional size. Artery walls are more rigid and do not
expand or change shape as readily as vein walls.
Capillary walls consist of a simple **endothelium** that is
only one cell thick and corresponds to the lining surface
of the innermost layer of arteries and veins. (b) Scanning
electron micrograph of an artery (A) and its companion
vein (V), illustrating the difference in wall thickness and
lumen size. (Magnification × 56) (c) Scanning electron
micrograph showing layers in the wall of an artery. The
inner layer with an endothelium (E) on its surface, the
muscular layer (M), and the connective tissue layer (CT)
are shown. (Magnification × 1,316)

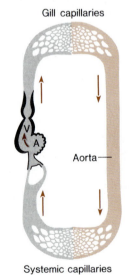

Gill capillaries

Aorta

Systemic capillaries

Figure 12.6 The single-circuit
circulatory arrangement of fish.
Blood passes from the heart,
through the gill circulation, and
directly on to the systemic
circulation. Oxygenated blood is
indicated by color.

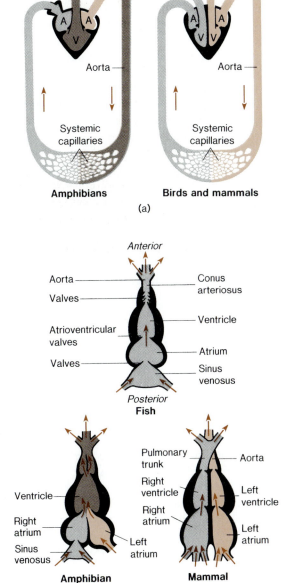

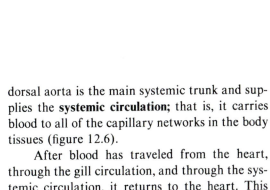

dorsal aorta is the main systemic trunk and supplies the **systemic circulation;** that is, it carries blood to all of the capillary networks in the body tissues (figure 12.6).

After blood has traveled from the heart, through the gill circulation, and through the systemic circulation, it returns to the heart. This single-circuit circulatory arrangement in which blood is oxygenated and then flows on directly to the systemic circulation has a significant drawback because of its relatively low systemic blood pressure.

Blood pressure is the force that drives blood through the circulatory system. Because blood pressure decreases as blood passes through the gill capillary network, the blood pressure in the fish circulatory system beyond the gills is reduced. A fish's systemic circulation is a relatively low pressure system compared with the systemic circulation of other vertebrates. While this low pressure systemic circulation is adequate for fish, it could not deliver oxygen and nutrients and remove metabolic wastes rapidly and efficiently enough to meet the needs of other vertebrates.

Separate Pulmonary and Systemic Circulations

Other vertebrates have a two-circuit (double) circulatory arrangement. Blood passes through the circulation of the lungs (**pulmonary circulation**), where it is oxygenated, and it returns to the heart. Then the heart pumps the blood out through the systemic circulation. In amphibians, this separation of pulmonary and systemic circulations is not complete, but in birds and mammals it is (figure 12.7a).

Figure 12.7 Separation of pulmonary and systemic circulations in vertebrates. (a) Ventral diagrammatic views of the circulatory arrangements of amphibians (left) and birds and mammals (right). (b) Diagrams of vertebrate hearts showing chambers stretched out for clarity. Fish heart is shown for purposes of comparison. In amphibians, oxygenated blood (color) returns from the lungs to the left atrium. Deoxygenated blood returns from the systemic circulation through the sinus venosus to the right atrium. Blood mixes in the ventricle so that partially oxygenated blood goes to both the lungs and the systemic circulation. In birds and mammals, complete separation is maintained in the heart because there are separate right and left ventricles. There is no conus arteriosus or sinus venosus in the heart of an adult bird or mammal.

Because of the separation of pulmonary and systemic circulations, the blood pressure in the systemic circulation is relatively high. Thus, circulatory supply to body tissues is more efficient than it is in fish.

The main pumping portion of a vertebrate heart is the **ventricle,** a muscular chamber that forces blood out through the blood vessels of the body. Before a vertebrate ventricle contracts, blood drains into the relaxed ventricle from a thin-walled chamber called the **atrium** (plural: **atria**). Then the atrium contracts, forcing any remaining blood into the ventricle. The ventricle then contracts, forcing blood out of the heart into the blood vessels.

Although fish have additional heart areas that are not present in adult bird and mammal hearts (that is, a posterior **sinus venosus** and an anterior **conus arteriosus,** figure 12.7b), the foregoing simplified description of a vertebrate heartbeat also adequately describes the flow of a single stream of blood through a fish heart.

The hearts of other vertebrates are partially (amphibians and some reptiles) or completely (some reptiles, all birds and mammals) partitioned into right and left halves. Bird and mammal hearts have right and left atria and right and left ventricles. The two atria beat simultaneously and then the two ventricles beat, pumping blood into the vessels leading to pulmonary and systemic circuits. In such **four-chambered hearts,** the complete partitioning into right and left halves permits the separation of two streams of blood passing through the heart that is required in the double circulatory pattern of birds and mammals.

Human Circulation

As is the case in all vertebrate animals, human circulation is a closed system. Blood flows away from the heart through arteries and arterioles to capillary networks, where exchanges between the blood and body cells take place. Blood leaves the capillaries and returns to the heart through venules and veins. Like other mammals, humans have a double circulatory arrangement with separate pulmonary and systemic circuits. Humans also have a typically mammalian four-chambered heart with right and left atria and right and left ventricles (figure 12.8).

The Human Circulatory Pattern

In humans, blood returns from the systemic circulation to the right atrium of the heart by way of two large veins, the **anterior (superior) vena cava** and the **posterior (inferior) vena cava.** The anterior vena cava returns blood from the head, arms, and upper trunk while the posterior vena cava returns blood from the remainder of the systemic circulation. All of this blood entering the right atrium is oxygen-poor (deoxygenated) blood. From the right atrium, blood passes into the right ventricle, which then pumps the blood out through the **pulmonary trunk.** The pulmonary trunk branches into two pulmonary arteries that carry blood to the lungs. After passing through lung capillaries around the alveoli (see p. 310), blood returns to the left atrium of the heart through **pulmonary veins.** Blood returning from the pulmonary circulation is oxygenated. This oxygenated blood passes into the left ventricle, which then pumps it out through the **aorta,** the large arterial trunk that supplies the entire systemic circulation.

The aorta sends branches to all parts of the body. The first branches off the aorta are **coronary arteries.** Because individual heart cells cannot exchange material with the blood being pumped through the chambers of the heart, coronary arteries must provide **coronary circulation** to the heart itself (figure 12.9). Normal blood flow through coronary arteries is critical for normal functioning of the heart. Blockage of any of the coronary arteries quickly results in damaged heart muscle and impaired heart function.

Anterior to the heart, the aorta arches around to the left and then passes posteriorly through the body. Branches off this **arch of the aorta** supply anterior parts of the body. As the aorta descends through the trunk, branches run to the digestive organs, the kidneys, the body wall, the legs, and all other posterior parts of the body.

Blood returns from many of these structures directly through veins that flow into the posterior vena cava. The digestive tract, however, has a special circulatory arrangement. Blood leaving the capillary networks of the major digestive organs—the stomach and the small and large intestines—flows into veins leading to a large **hepatic portal vein.** The hepatic portal vein does not empty into the posterior vena cava. Instead,

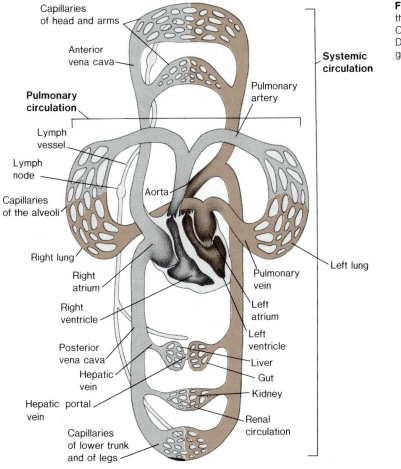

Capillaries
of head and arms

Anterior
vena cava

**Pulmonary
circulation**

Lymph
vessel

Lymph
node

Capillaries
of the alveoli

Right lung

Right
atrium

Right
ventricle

Posterior
vena cava

Hepatic
vein

Hepatic portal
vein

Capillaries
of lower trunk
and of legs

**Systemic
circulation**

Pulmonary
artery

Aorta

Left lung

Pulmonary
vein

Left
atrium

Left
ventricle

Liver

Gut

Kidney

Renal
circulation

Figure 12.8 Schematic diagram of
the human circulatory system.
Oxygenated blood is shown in color.
Deoxygenated blood is shown in
gray.

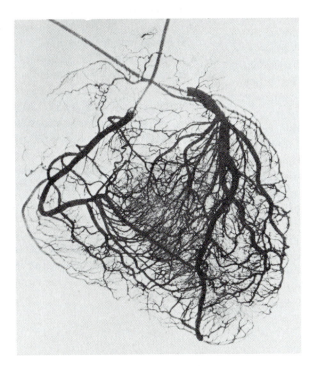

Figure 12.9 X-ray photograph of
the coronary arteries taken after
injection of the arteries with barium
sulphate.

it carries blood to a network of capillarylike spaces (**sinusoids**) inside the liver. In the sinusoids, materials are exchanged between the blood and the cells of the liver that are located among the sinusoids. This arrangement is efficient because it facilitates processing by the liver of nutrients just absorbed from the digestive tract. After passing through liver sinusoids, blood flows into **hepatic veins** that carry it to the posterior vena cava. This special circulatory arrangement involving the veins of the digestive tract, the hepatic portal vein, and the liver is called the **hepatic portal system. A portal system** is a portion of a circulatory system that begins in a capillary bed and leads not to vessels that carry blood directly toward the heart, but rather to a second capillary bed. Only after it has traversed this second capillary bed does blood from a portal system proceed through veins toward the heart.

The Human Heart

The human heart is a remarkably efficient and reliable pumping device, and its regular, continual beating is essential to life. Cessation of heartbeat is one of the most obvious indications of death. The heart can never rest or stop beating for even a short time because while some organs can survive brief pauses in circulatory supply, others cannot. The brain, for example, is critically sensitive to even short failures of circulation. If circulation to the brain stops for as little as five seconds, consciousness is lost. A four-minute break in circulation to the brain causes death of significant numbers of brain cells, and a nine- or ten-minute circulatory failure causes massive, irreversible brain damage.

During each minute of life, a normal, resting person's heart beats about seventy times and pumps out a total of about 5 l of blood, a volume approximately equal to the total blood volume in the body. All of this work is done by a relatively small organ—the average human heart weighs only 300 g, less than 0.5 percent of total body weight. Even the heart of a highly trained athlete weighs only about 1 percent of total body weight. This small organ sustains its heavy work load from the time that it begins to beat early in embryonic development until the time of death, and it never rests for even a whole second.

The Heartbeat

Each heartbeat involves a cycle of contraction (**systole**) and relaxation (**diastole**) of the atria, and systole and diastole of the ventricles. Although the human heart has two completely separated streams of blood flowing through its right and left sides, the two sides beat in unison. That is, right and left atria beat at the same time, and then slightly later, right and left ventricles beat in unison.

There is a brief pause between heartbeats when atria and ventricles both are relaxed. Blood from the atria begins to fill the ventricles during this pause. The relaxed ventricles are 70 to 75 percent filled during this period. Then atrial systole forces blood still contained in the atria into the ventricles and completes their filling.

The atria relax as the ventricles begin to contract. The beginning of ventricular systole (contraction) increases pressure in the ventricles and would force blood back into the atria except for **valves** that shut to prevent backward flow. These valves are the **tricuspid valve,** located in the canal between the right atrium and right ventricle, and the **bicuspid (mitral) valve,** located in the canal between the left atrium and left ventricle.

As ventricular systole continues, the pressure in the ventricles becomes greater than the blood pressure inside the pulmonary trunk and the aorta. This overcomes the resistance of arterial blood pressure, and blood is forced out into the major arteries. The elastic walls of the arteries expand as this volume of blood is pushed into them. When ventricular diastole (relaxation) begins, the artery walls recoil. Blood would be forced back into the ventricles except that here again valves prevent backward flow. Sets of cuplike **semilunar valves** at the bases of the pulmonary trunk and the aorta slam shut and prevent blood from returning to the ventricles. (See figure 12.10.)

The complete **cardiac cycle** described in the previous paragraphs lasts only about 0.8 second; then there is a period of about 0.4 second between the end of ventricular systole and the start of the next atrial systole at the beginning of the next cardiac cycle.

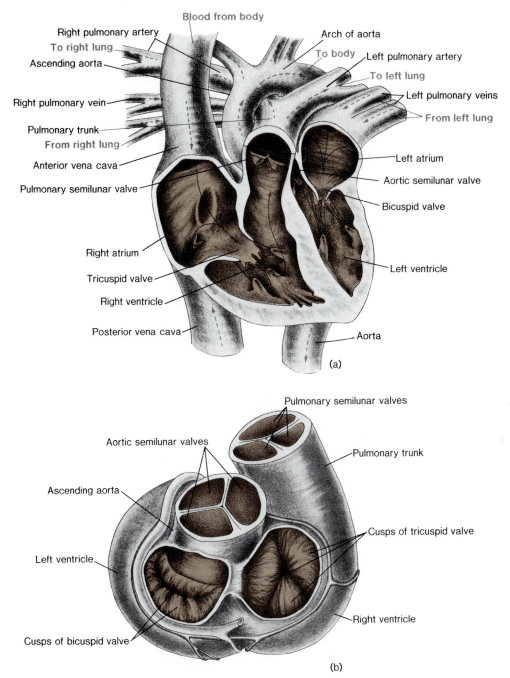

Right pulmonary artery
To right lung
Ascending aorta
Right pulmonary vein
Pulmonary trunk
From right lung
Anterior vena cava
Pulmonary semilunar valve

Right atrium
Tricuspid valve
Right ventricle
Posterior vena cava

Blood from body
Arch of aorta
To body
Left pulmonary artery
To left lung
Left pulmonary veins
From left lung
Left atrium
Aortic semilunar valve
Bicuspid valve

Left ventricle

Aorta

(a)

Pulmonary semilunar valves
Aortic semilunar valves
Ascending aorta
Pulmonary trunk
Cusps of tricuspid valve
Left ventricle
Right ventricle
Cusps of bicuspid valve

(b)

Figure 12.10 Circulation through the human heart.
(a) The paths of blood flow. This diagram shows the
veins entering the heart, the arterial trunks leaving the
heart, and the locations of valves in the heart. This is a
ventral view with several parts of the wall of the heart
removed. (b) Valves. This is a view from above the
heart with the valves exposed.

During the course of each heartbeat, the heart emits several characteristic **heart sounds** that can be heard through a stethoscope placed against the chest. The first heart sound is described as a "lubb" and is relatively longer and lower pitched than the second heart sound. The second heart sound, described as a "dup," is shorter, higher pitched, and louder. The first heart sound is the sound of the tricuspid and bicuspid valves closing during ventricular systole. The second sharper sound is the sound of the semilunar valves slamming shut. Normal heart sounds indicate normal valve functioning, while defective valves produce altered heart sounds. When valve closing is incomplete, there is a heart **murmur.** This hissing sound is blood leaking back through a defective valve into an emptied heart chamber. Sometimes, defective valves are birth defects. In other cases, valve damage occurs as a result of disease.

Control of the Heartbeat

Each heartbeat is initiated in a specialized mass of tissue called the **sinoatrial (S-A) node** located in the wall of the right atrium. Sometimes the S-A node is called the heart's "pacemaker," and it has an interesting developmental history. During development, a human embryo's heart has a sinus venosus very much like the sinus venosus that is a separate region in the hearts of some other vertebrates (see figure 12.7). In the human embryo's heart, as in the hearts of those animals, impulses for heartbeats are initiated in the sinus venosus and spread through the rest of the heart. But the sinus venosus is not retained as a separate chamber in the fully developed human heart. Instead, it is incorporated into the wall of the right atrium as the sinoatrial node. Thus, even though the S-A node becomes a part of the atrium, it retains the beat-generating function of the embryonic sinus venosus from which it was derived.

It can be demonstrated experimentally that individual heartbeats clearly are initiated inside the heart and do not depend on specific stimuli from the nervous system. If a frog's heart is carefully detached, removed from the body, and placed in an appropriate medium, it continues to beat for some time. Similar experiments can be done successfully with turtle hearts or the hearts of a number of other vertebrates. Even small mammals' hearts can be removed and kept beating, although the culture conditions required to maintain mammalian hearts are more complex and exacting. More evidence for the independent generation of heartbeats within the heart is the fact that the hearts of vertebrate embryos begin to beat early in development, before they have received their normal nerve connections. The heart, however, is not entirely independent of control by the nervous system. The nervous system influences several aspects of heart activity, including the heart's beating rate.

Tissue of the S-A node sends out an electrical impulse (excitation) to heart muscle cells. This impulse spreads along the membranes of muscle cells in the atria and causes them to contract. Excitation spreading through the atria eventually reaches a second mass of nodal tissue located in the wall between atria and ventricles. This nodal tissue, the **atrioventricular (A-V) node,** is the only electrical connection between atria and ventricles. The impulse is delayed slightly in the A-V node. This delay allows time for completion of atrial systole before ventricular contraction begins. Then the impulse is transmitted by conducting fibers so rapidly that muscle in all parts of the ventricle is stimulated almost simultaneously. The ventricular beat, then, is very efficient because muscle cells in all parts of the ventricles contract at the same time. (See figure 12.11.)

Cardiac Output

A resting person's heart pumps about 300 l of blood per hour, but the human heart can pump up to five times that much during heavy exercise or during times of stress or fright. Obviously, **cardiac output,** the total amount of blood pumped per unit time, can be adjusted to meet a wide range of requirements. Cardiac output depends on two factors, heart rate and the **stroke volume** (the volume pumped during each heartbeat), both of which can be adjusted to change cardiac output.

During exercise or stressful emergency situations, the nervous system stimulates release of the hormone **epinephrine** (also called **adrenalin)** from the **adrenal glands.** Epinephrine causes several kinds of circulatory adjustments, including increases in heart rate as well as contraction strength.

Heart rate also is influenced by two sets of nerve fibers that connect to the S-A node. One set of fibers is controlled by a **cardioacceleratory**

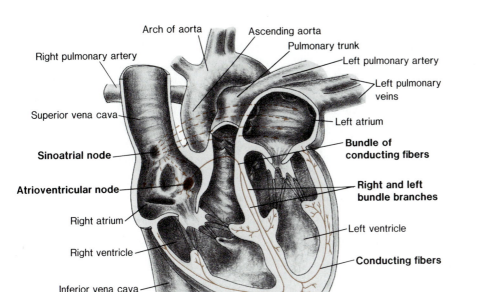

Arch of aorta   Ascending aorta
Pulmonary trunk
Right pulmonary artery
Left pulmonary artery
Left pulmonary veins
Superior vena cava
Left atrium
Sinoatrial node
Bundle of conducting fibers
Atrioventricular node
Right and left bundle branches
Right atrium
Left ventricle
Right ventricle
Conducting fibers
Inferior vena cava

Figure 12.11 Beat generation and the conducting system of the heart. Beat is generated at the sinoatrial node. The impulse spreads over the membranes of muscle cells in the atrium. When the impulse reaches the atrioventricular node, it is delayed briefly. Then it spreads quickly through the conducting system to all parts of the ventricle.

center in the brain. These fibers are part of the sympathetic division of the nervous system (page 446) and cause an increase in heart rate. Other fibers reaching the S-A node are controlled by a **cardioinhibitory center** in the brain and cause a slowing of heart rate. These fibers are part of the parasympathetic division of the nervous system (page 446). Through this double, antagonistic nerve supply, the heart rate can be adjusted by the nervous system to meet changing situations.

The stroke volume of the heart can be increased by intensifying the force of ventricular contraction. Epinephrine, which has already been noted as a factor in increasing heart rate, also increases the strength of ventricular contractions. Another factor that increases stroke volume is a response of the heart muscle cells themselves. Heart muscle contracts with greater force if it is stretched at the moment when contraction begins. This response of heart muscle to stretching is called **Starling's law of the heart,** after the English physiologist who discovered it. During exercise, venous return to the heart increases and thus the ventricles are filled with more blood. This greater ventricular filling stretches the heart muscle and causes more forceful contractions.

Blood Pressure and Human Circulation

Pressure generated by the heartbeat forces blood into the aorta and supplies the driving force that causes blood to move through the entire circulatory system. Blood pressure is highest during each ventricular systole, when blood is being forced into the aorta. At this time there is a resultant temporary expansion of the aortic wall, which passes like a wave along the wall of the aorta and the walls of each of its larger branches. This wave of expansion is the **pulse** that can be detected in certain arteries near the body surface. Because there is one arterial pulse per ventricular systole, the arterial pulse rate can be used to determine the heart rate. Arterial wall expansion is only temporary because arterial walls are elastic and quickly return to their original diameter. This elasticity of arterial walls is very important functionally because it helps to maintain blood pressure between heartbeats.

Figure 12.12 Blood pressure in different parts of the human circulatory system. The pulse pressure decreases as blood flows into smaller arteries and disappears as blood enters capillaries, where it flows smoothly and steadily. Pressure in veins very near the heart may actually fall below zero.

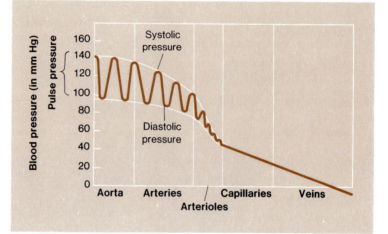

Thus, two different pressure levels alternate in large arteries. **Systolic pressure** results from blood being forced into the arteries during ventricular systole, and **diastolic pressure** is the blood pressure in the arteries during ventricular diastole. Human blood pressure is measured using a device called a sphygmomanometer. A sphygmomanometer has a pressure cuff that permits measurement of the amount of pressure required to stop the flow of blood through an artery. Blood pressures normally are measured on the brachial artery, an artery of the upper arm, and are stated in millimeters of mercury (Hg). A blood pressure reading includes two numbers, for example 120/80, and these numbers represent systolic and diastolic pressures, respectively.

Blood flows from the region of highest pressure, the ventricle, to the region of lowest pressure in the veins leading to the atria, and measurements of blood pressure at various points along the circulatory system show a progressive decrease in blood pressure (figure 12.12). Pressure is highest in the first part of the aorta, where it is about 140/120. As blood moves through the aorta and then spreads through smaller arteries and arterioles, pressure falls. At the same time, the **pulse pressure,** the difference between systolic and diastolic pressure, decreases. Pulse pressure finally disappears in the capillaries, where there is no pulse, but only a smooth, steady flow of blood.

Blood pressure in the veins is very low (5 to 10 mm Hg), and pressure finally falls to 2 mm Hg or less in the vena cava. This very low blood pressure in the veins is not adequate to move blood back to the heart, especially from lower parts of the body. Venous return to the heart depends in part on mechanisms that apply force from outside the circulatory system to keep blood moving. Walls of veins are much thinner and less elastic than the walls of arteries and, therefore, are quite easily collapsed. When body muscles that are near veins contract, they put pressure on the veins' collapsible walls and on the blood contained in the veins. Because veins have **valves** that prevent backward flow, pressure from muscle contractions effectively helps to move blood through veins toward the heart (figure 12.13). During periods of inactivity, blood moves only very sluggishly through veins in some parts of the body and can accumulate in the veins. For example, blood accumulates in leg veins during long periods of standing.

Capillary Circulation and Blood Pressure

Blood pressure changes when heart rate and stroke volume change. But another mechanism that can be altered to change blood pressure is the **peripheral resistance,** the resistance that blood vessels offer to the blood flowing through them. The main sites of changes in peripheral resistance are the arterioles that supply capillary beds. Arterioles have circular **sphincter muscles**

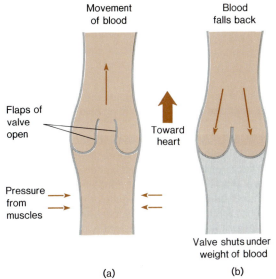

Figure 12.13 The role of valves in movement of blood through veins. Blood pressure is low in veins. Muscle contractions near veins put pressure on the blood and force it through veins (a). Blood moves only toward the heart because valves prevent backward flow when muscle pressure decreases (b).

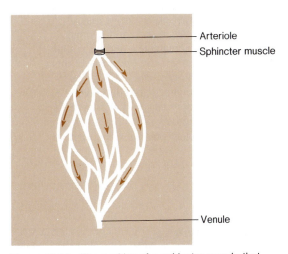

Figure 12.14 The position of a sphincter muscle that controls blood flow into a capillary bed. Nervous system control over sphincters throughout the body makes it possible to regulate the blood supply to different body regions in response to changing circulatory requirements.

in their walls that can contract and close down the arterioles, thus restricting flow through the capillaries. These arteriole sphincters are controlled by the nervous system. When more sphincters contract, general peripheral resistance increases and blood pressure rises. When more sphincters open (dilate) to allow blood into more capillary beds, the opposite occurs. (See figure 12.14.)

Actually, at any given moment, many of these sphincters must be contracted because if blood entered all of the body's capillary beds at once, there would be a disastrous drop in blood pressure. Furthermore, the total volume of the body's capillary beds is so great that they could contain all of the blood in the circulatory system if all of the arteriole sphinctors opened at once.

Control over arterioles is not simply a matter of their opening or closing randomly, here and there all over the body, to bring about pressure changes. The nervous system's control over arterioles is selective; it functions to meet specific circulatory requirements in specific areas of the body, and those requirements change with time. For example, in resting muscles, blood flow is maintained through only certain more or less permanently open capillary beds, while there is little or no blood flow through many other capillary beds. But when exercise begins, many more sphincters open, and circulation begins through those arterioles and the capillary beds that they supply. At the same time, capillary circulation is decreased selectively in other parts of the body, notably in the digestive tract. Conversely, following a large meal, many arterioles open to increase the flow through capillary beds in the digestive tract, thereby increasing the blood supply for transport of absorbed products of digestion.

Dilation and contraction of arterioles also function in body temperature regulation. Dilation of many arterioles near the skin surface allows much more blood flow through superficial capillaries and results in greater heat loss from the body. People who are in warm places or are exercising heavily have reddened skins because of the increased amount of blood circulating just under the skin. On the other hand, heat loss is decreased by closing down more superficial capillary beds.

Box 12.1
On Keeping a Cool Head or a Cold Flipper

Because blood is a watery material, it has a high heat capacity. Therefore, blood is a factor in stabilizing body temperature, and blood circulation is a means of distributing heat to various parts of the body. This heat-transporting role of blood can have both positive and negative aspects. On the positive side, it is essential for regulating heat loss from a mammal's body surface when constriction or dilation of superficial arterioles is involved. But the heat-carrying function of blood can also have a negative aspect. For example, actively contracting muscles generate a great deal of heat that warms blood circulating through them. This warmed blood is then carried to other parts of the body, where it can cause problems because some body organs cannot tolerate temperature elevation. Mammals' brains, in particular, are damaged when brain temperatures rise more than a few degrees above normal body temperature.

What happens when a mammal runs in a warm environment, as hoofed grazing mammals (ungulates) must do when pursued by predators? This problem was studied experimentally in the small African antelopes known as Thomson's gazelles, which can experience up to a forty-fold increase in metabolic rate when they run. C. R. Taylor and C. P. Lyman found that when a Thomson's gazelle ran for five minutes at 40 km per hour, heat buildup in its body raised the temperature of arterial blood in the body core from the normal 39° C to 44° C. But during the same period, the gazelle's brain temperature did not even reach 41° C. How could this differential between brain temperature and body core temperature be maintained when the temperature of the blood that was being pumped into arteries leading to the brain was 44° C?

Most of the blood supplied to the gazelle's brain flows through the external carotid arteries, which divide into hundreds of small vessels at the base of the skull. These many small arteries lie in a sinus (an expanded, blood-containing cavity) through which flows venous blood returning from the nasal area. The nasal passages are cooled by evaporation during breathing and thus the blood returning from them is relatively cool. Because the cooler venous blood flowing through the sinus bathes the surfaces of the small arterial branches of the carotid artery, there is heat transfer from the warmer arterial blood to the cooler venous blood. The small arterial branches then join to reform an artery, and this artery, which supplies the brain, contains blood that has been cooled several degrees by loss of heat to cool venous blood. (See box figure 12.1A.)

Selective cooling of the brain's blood supply is an adaptation that protects Thomson's gazelles and many other animals from one of the most serious hazards of body overheating.

At the other end of the temperature scale is the danger of excessive loss of body heat, and this is an urgent, continuing problem for many mammals and birds that live in cold environments. For many of them, appendages are major sites of heat loss because they generally are not as well insulated as body trunks and also have far less favorable surface to volume ratios, from a heat conservation point of view, than

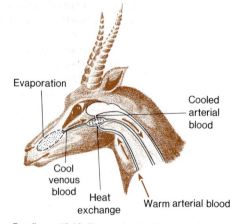

Box figure 12.1A The mechanism that cools blood being supplied to a gazelle's brain. Arterial blood passes through small arteries that are bathed by cooled venous blood that drains from the nasal region where evaporative cooling occurs. Following this heat exchange, blood going to the brain is as much as three degrees cooler than arterial blood coming from the heart.

body trunks. Thus, sending warm blood into feet or flippers can produce a tremendous heat-loss problem for birds and mammals in the cold. How can this dangerous heat loss be minimized?

Several solutions have evolved. Arctic wolves, for instance, have a mechanism that limits blood flow to their feet when they are cold. Only enough blood enters their legs to keep the temperature of their exposed footpads just above the freezing point.

Other animals have mechanisms that reduce heat loss without such severe restriction of blood flow to limbs. A countercurrent heat exchange mechanism is part of the limb circulation in animals ranging from ducks that swim with feet in icy water to dolphins whose flippers are not nearly so well insulated with fat as their body trunks. In all of these animals, arteries carrying blood toward the extremities flow alongside veins that return blood from them. Because of this close proximity of vessels, blood in the arteries loses heat to blood returning in the veins. Heat is conserved because arterial blood has given up much of its heat before it reaches the exposed extremities, and venous blood is warmed as it returns from the limb toward the body core. (See box figure 12.1B).

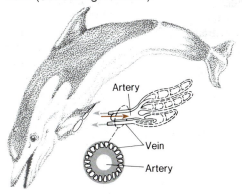

Box figure 12.1B In the flipper of a dolphin, each artery is surrounded by several veins. Venous blood is warmed by heat transfer from the arterial blood. This heat exchange is very efficient because this is a countercurrent mechanism; that is, fluid streams flow by each other in opposite directions, and heat is transferred all along the area of contact.

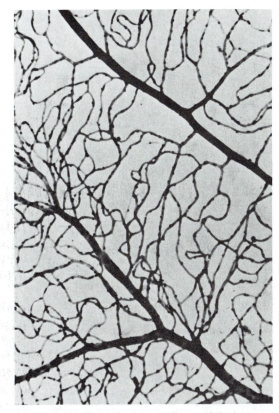

Figure 12.15 Circulation in a small area of the human retina. This specimen has been injected with a material that makes the blood vessels more clearly visible. At the top is an arteriole with its branches, the terminal arterioles, which lead to capillaries. Capillaries form a dense network and are uniform in diameter. The capillaries flow together into the branches of a venule (bottom). (Magnification $\times$ 67)

Capillary Function

Capillaries are distributed throughout every tissue of the body so that no cell is far from one or several capillaries. These capillaries form very complex networks, and the total number of capillaries in various tissues is so great as to defy imagination, much less accurate counting (figure 12.15). For example, in resting guinea pig muscle, a cross-sectional area of 1 mm^2 contains about 100 open capillaries. During exercise, more arterioles open so that more capillaries have blood flowing through them. An area of 1 mm^2 in a cross section of muscle prepared during maximal exercise may contain more than 3,000 open capillaries. The lead in an ordinary pencil has a cross-sectional area of about 3 mm^2. Thus, a piece of muscle of that same size prepared during maximal exercise would contain close to 10,000 open capillaries!

Individual capillaries are very small, barely large enough for red blood cells to squeeze through in single file, but they are so numerous that their total cross-sectional area is larger than that of any other part of the circulatory system. Table 12.1, which compares cross-sectional areas and velocities of blood flow in different parts of the circulatory system, shows that blood flows slowest while it is in the capillaries. This slow flow rate allows adequate time for exchange of materials between blood and tissue cells.

Capillary exchange also is facilitated by the very large total surface area of the numerous capillaries. This large surface area is exposed to the fluid present in the spaces around cells, the **interstitial fluid.** Much of the exchange between blood and interstitial fluid occurs by diffusion through capillary walls, which are very thin and consist of a single layer of flattened endothelial cells. Lipid-soluble substances pass freely through the membranes of cells in the capillary walls, but water-soluble material diffuses mainly through pores that are located in junctions between cells in the capillary wall. There also is some transfer of materials by **pinocytotic vesicles** (see page 113). These vesicles form on one surface of an endothelial cell in a capillary wall, move across the cytoplasm, and then empty their contents on the other side of the capillary wall. (See figure 12.16).

Another important factor in capillary exchange is the movement of fluid back and forth between blood and interstitial fluid through capillary walls. Water and dissolved materials are forced out through the capillary wall at the end of the capillary nearer the arteriole, and they flow into the capillary at the end nearer the venule.

Two forces cause these movements of fluid through the capillary wall. The first, **hydrostatic pressure,** is the outward pressure exerted by a liquid against the walls of a container that holds it. In this case, the container is the capillary, and hydrostatic pressure is the blood pressure inside the capillary. The hydrostatic pressure at the arteriole end of capillaries is about 35 mm Hg. Pressure decreases as blood flows through the capillary so that near the venule end, the hydrostatic pressure has fallen to 15 mm Hg. This differential in hydrostatic pressure between the ends of the capillary is very important for fluid movement.

Table 12.1
Velocities of Blood Flow.

Vessels	Approximate Cross-Sectional Area	Velocity*
Aorta	2.5 cm²	40 cm/second
Arteries	20 cm²	40-10 cm/second
Arterioles	40 cm²	10-0.1 cm/second
Capillaries	2,500 cm²	0.1 cm/second
Venules	250 cm²	0.3 cm/second
Veins	80 cm²	0.3-5 cm/second
Venae Cavae	8 cm²	5-20 cm/second

*Velocity values with two numbers represent changes in velocity as blood passes through that part of the circulatory system. For example, 40-10 indicates a decrease in velocity from 40 cm/second to 10 cm/second.

The second force involved in fluid movement through the capillary wall is **osmotic pressure.** The total concentration of ions and small molecules is very similar in blood and interstitial fluid, but the fluid portion of the blood, the blood **plasma,** has a relatively high concentration of protein molecules. These protein molecules do not pass freely through the capillary wall. The same kinds of protein molecules are found in interstitial fluid but in a much lower concentration. This difference in protein concentration results in an osmotic pressure differential; the osmotic pressure of blood plasma is about 25 mm Hg higher than that of interstitial fluid.

Hydrostatic pressure and osmotic pressure act to produce fluid movement through capillary walls. Hydrostatic pressure tends to force fluid out through the pores in the capillary wall. This is a *filtering* process because water and small dissolved molecules are forced out while protein molecules and blood cells are kept in. Osmotic pressure, on the other hand, is an opposing force that tends to cause water to move from the interstitial fluid to the blood because of the osmotic pressure differential. The direction of fluid movement through the capillary wall thus depends upon the relative strength of these two opposing forces. At the arteriole end of a capillary, hydrostatic pressure (the force tending to drive water out) is 35 mm Hg, and osmotic pressure (the force tending to cause water to move in) is overcome by hydrostatic pressure. Thus, water flows outward, carrying dissolved materials with it.

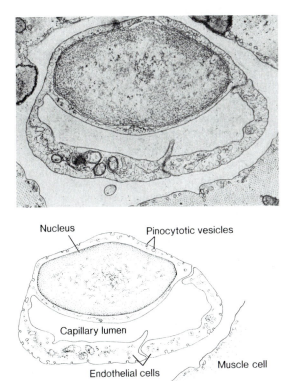

Nucleus   Pinocytotic vesicles

Capillary lumen

Endothelial cells   Muscle cell

(a)

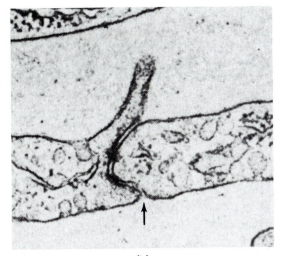

(b)

Figure 12.16 Electron micrographs of a cross section of a capillary in heart muscle. (a) The wall of a capillary consists of a single layer of flattened endothelial cells. This picture shows parts of two endothelial cells: the nucleus and part of the cytoplasm of one cell, and a small part of the cytoplasm of another cell, in the lower right part of the capillary. The pinocytotic vesicles have transported fluid across the cell from the capillary and are releasing their contents through the endothelial cells' plasma membranes into the interstitial fluid. (b) Enlarged view of part of this capillary wall showing the place where two endothelial cells join (arrow). Water and solutes flow through pores in these junction regions. Note the number of pinocytotic vesicles.

Midway along the capillary, where hydrostatic pressure is lower, the two forces essentially cancel one another, and there is no net movement of water. At the venule end, where hydrostatic pressure has fallen to 15 mm Hg, osmotic pressure is greater than hydrostatic pressure, and water moves inward (figure 12.17). Almost the same amount of fluid that left the capillary returns at the venule end. Thus, there is only a small net loss of fluid from the blood as it passes through the capillaries. The small quantity of fluid left in the body tissues is discussed in the next section on the lymphatic system.

Regular fluid movement through the capillary wall is important because water carries nutrients and other dissolved materials from the blood to the interstitial spaces, and water returning to the capillary carries metabolic wastes. These essential movements of fluid depend upon maintenance of normal blood pressure and osmotic pressure relationships, and significant changes in either can seriously alter fluid movement.

At arteriole end

Hydrostatic pressure > Osmotic pressure

25 35

25 25

25 15

Osmotic pressure > Hydrostatic pressure

At venule end

Figure 12.17 Forces causing flow of water and solutes through a capillary wall. Near the arteriole end, hydrostatic pressure (smaller white arrow) is greater than the osmotic pressure difference (colored arrow) so that there is a net movement of water (larger white arrow) and solvents out of the capillary. Midway along the capillary, the forces cancel each other, and there is no net movement. Near the venule end of the capillary, the osmotic pressure difference is greater than hydrostatic pressure, and water moves into the capillary.

The Lymphatic System

When fluid movement between capillaries and interstitial spaces was described, it was noted that most, but not quite all, of the fluid forced out of capillaries returns to them. What happens to the extra fluid left in the interstitial spaces? Vertebrate animals have a special system of vessels, the **lymphatic system,** that drains fluid from interstitial spaces in the tissues and returns it to the blood, thus maintaining a balance between blood volume and interstitial fluid volume in the body.

There are **lymph capillaries** in all parts of the body. Lymph capillaries have blind ends; that is, they are closed at one end. Their walls are permeable to water and small molecules, and they also are very permeable to protein molecules. The lymphatic system returns excess interstitial fluid to the blood, and it also transports proteins from the interstitial spaces to the blood. This mechanism is important because a few protein molecules leak through the capillary walls, and they must be returned to the blood if the normal and functionally necessary osmotic differential between blood and interstitial fluid is to be maintained.

Once interstitial fluid enters the lymphatic system, it is called **lymph.** Lymph moves through lymph capillaries and **lymph vessels** as a result of pressure applied by muscle contractions near the vessels and a system of valves that prevent backward flow (figure 12.18). Some vertebrate animals have pulsating lymph hearts that move lymph along through their lymphatic systems, but human lymph movement is completely passive and depends on the pressure of muscles contracting adjacent to the vessels. Lymph moves through smaller vessels that unite to form larger vessels. Finally, two major **lymph ducts** empty into large veins near the heart (figure 12.19).

In addition to returning fluids from the interstitial spaces to the circulatory system, the lymphatic system has several other important functions. Lymph capillaries of the digestive tract, known as **lacteals,** are distributed in the intestinal villi, where they are involved in transport of absorbed digested material. Specifically, the lacteals carry products of fat digestion and absorption from the intestine to lymph vessels and eventually to the circulatory system.

The lymphatic system also has a role in filtering of the lymph. **Lymph nodes** located along the lymph vessels contain a network of connective tissue fibers with **phagocytic cells** scattered among them. These phagocytic cells engulf dead cells, cell debris, and foreign objects such as bacteria as the lymph filters through the lymph nodes. In addition to this general phagocytic activity, cells that are present in the lymph nodes as well as in the blood, the **lymphocytes,** are involved in specific defensive responses to infection. Development and function of these specific defense responses are discussed in detail in chapter 26. Because the lymph nodes are very active in defense responses, they often become swollen and tender during infections. The so-called "swollen glands" that often accompany a sore throat are not glands, but active lymph nodes.

Problems with the lymphatic system can interfere with fluid balance in the body. Anything that interferes with the flow of lymph through a lymph vessel or one of the lymph nodes along that vessel causes fluid accumulation in the part of the body drained by that vessel. Not only is fluid drainage impaired; inadequate protein removal alters the osmotic balance between blood and interstitial fluid so that even more fluid accumulates in the tissue. Such fluid accumulation is called **edema,** and it often is visible externally as a swelling in that part of the body. An extreme example of this is a condition caused by a parasitic worm in tropical areas. The worms block lymphatic drainage so that portions of the body become severely swollen, sometimes to the point of becoming grotesquely misshapen. (see page 1043). This condition is known as **elephantiasis.**

Human Blood

Human blood is a red liquid made up of **formed elements** (cells and cell fragments) and an extracellular fluid, the blood plasma. The formed elements are **erythrocytes (red blood cells), leukocytes (white blood cells),** and **platelets,** which are small fragments of cells. These formed elements make up about 45 percent of the blood volume, and the plasma makes up the remaining 55 percent.

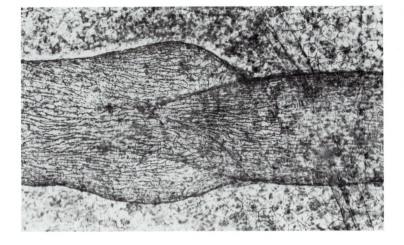

Figure 12.18 Photomicrograph of a valve in a lymph vessel. Lymph is forced through lymph vessels as contracting muscles exert pressure on the vessels' walls. Valves maintain one-way flow through lymph vessels by preventing backward flow. (Magnification × 45)

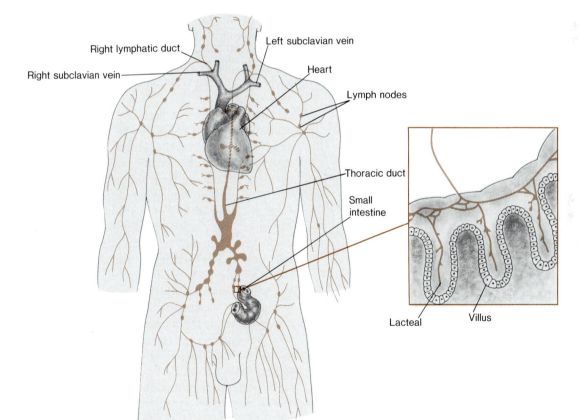

Right lymphatic duct

Left subclavian vein

Right subclavian vein

Heart

Lymph nodes

Thoracic duct

Small intestine

Lacteal

Villus

Figure 12.19 The lymphatic system. Lymph vessels flow into two main channels. The thoracic duct drains lymph vessels from all parts of the body except the upper right portion. Vessels from that area drain through the right lymphatic duct. These two large lymphatic ducts drain into the subclavian veins, thus returning fluid from the tissues to the circulatory system. Nodes filter debris and bacteria out of lymph. The inset is an enlarged sketch of part of the intestinal lining showing the location of lacteals in the intestinal villi. Products of fat digestion pass through the lacteals to lymph vessels and eventually to the circulatory system.

Red Blood Cells

Erythrocytes (red blood cells) are the most numerous of the formed elements. Although the actual red cell count varies from person to person and changes in response to a number of physiological factors, usually there are from four to six million red blood cells per cubic millimeter of human blood. Generally, men have a somewhat higher red cell count than women. A healthy man has about 5.4 million erythrocytes per cubic millimeter, while a healthy woman has about 4.8 million per cubic millimeter, but these values vary with time and activity. For example, erythrocyte counts increase during exercise or following meals. At these times additional blood cells are mobilized from storage sites, especially the **spleen,** an organ near the stomach that normally stores considerable quantities of blood cells. People living at high altitudes consistently have higher red cell counts than people living at lower altitudes. This adjustment helps to ensure delivery of an adequate oxygen supply to body tissues despite problems caused by the lower atmospheric oxygen pressures found at high altitudes.

A human erythrocyte is a thin, biconcave disc with a very thin center and a thicker rim. A mature circulating erythrocyte does not have a nucleus because the nucleus degenerates during the erythrocyte's development (figure 12.20). Approximately one-third of an erythrocyte's volume is taken up by about 280 million hemoglobin molecules.

The hemoglobin inside erythrocytes reversibly binds with oxygen to form **oxyhemoglobin** (see page 312). When most of its hemoglobin is in the form of oxyhemoglobin, blood takes on a bright red color. Such red, **oxygenated blood** is found in the pulmonary veins, which return blood from the lungs, and in the systemic arteries. Blood with relatively little oxyhemoglobin has a darker, more purple color. This is called **deoxygenated blood.** Characteristically, blood in systemic veins is deoxygenated because it has passed through capillary beds where it gave up oxygen to body tissues.

Hemoglobin is packed into erythrocytes instead of simply being carried in the blood plasma for a good reason. If all of the hemoglobin contained inside erythrocytes were to be in the plasma, the blood would be much more viscous (thick and syrupy), and blood pressure would have to be much higher to force blood through the circulatory system. Blood vessels would have a hard time withstanding the pressures involved. Because hemoglobin is packed inside erythrocytes, human blood can contain enough hemoglobin to carry adequate quantities of oxygen while still remaining a fluid that flows freely in response to moderate pressure.

Erthrocytes normally are quite elastic; even though they are bent as they pass through small blood vessels, they spring back into shape. As erythrocytes age, however, they become more fragile, and such bending can damage or even rupture them. Damaged erythrocytes are withdrawn from circulation as the blood passes through the liver or spleen. Large phagocytic cells called **macrophages,** which are especially common in the liver, phagocytize and destroy damaged or ruptured erythrocytes, and the hemoglobin from the erythrocytes is broken down by liver cells (figure 12.21). Iron, recovered from the heme portions of hemoglobin molecules, is exported from the liver and used again for hemoglobin synthesis by developing erythrocytes. The remainder of the heme parts of hemoglobin molecules are broken down by the liver cells and excreted as the greenish-brown bile pigments, which are part of the bile that passes through the common bile duct to the intestine (page 276).

How old is an "aging" erythrocyte? The average life span of human erythrocytes is about 120 days. Thus, erythrocytes are removed from circulation after they have functioned for about four months. New erythrocytes must be produced as fast as old ones are destroyed, and they develop in the **bone marrow,** which is located in cavities inside bones. About two million new erythrocytes are produced in the human body per second, a rate that equals the rate of red cell destruction in a normal healthy person. This tremendous cell production is necessary if each of the 25×10^{12} erythrocytes present in a human circulatory system at any given time are to be replaced after about 120 days.

Because of the critical importance of erythrocytes for oxygen transport, their numbers must be precisely controlled. The most important factor affecting the rate of erythrocyte production is a mechanism that is stimulated by an oxygen deficiency in tissues. This oxygen deficiency may develop, for example, when erythrocytes are not being produced as rapidly as they are being destroyed. Kidney cells, and to a lesser extent, liver cells respond to oxygen deficiency by releasing a factor into the blood that causes a hormone called

Red blood cell

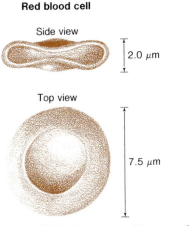

Side view

2.0 μm

Top view

7.5 μm

Figure 12.20 Erythrocytes. Human erythrocytes are flattened, biconcave discs. This side view is sectional and shows how thin the center of the erythrocyte is. Erythrocyte shape apparently is an adaptation for more efficient gas diffusion.

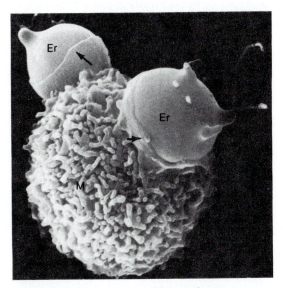

Figure 12.21 A macrophage (M) ingesting two erythrocytes (Er). Cytoplasmic extension (arrows) of the macrophage are in the process of surrounding the erythrocytes. Macrophages continually remove aging and misshapen erythrocytes from circulation. (Magnification × 5,180)

erythropoietin to be released from the carrier molecule that normally binds it. Erythropoietin stimulates erythrocyte production in the bone marrow and thus increases the number of erythrocytes being added to the circulation.

Sometimes the mechanisms that normally balance erythrocyte production and destruction fail. For example, nutritional deficiencies or diseases cause slower than normal erythrocyte production. This results in a condition known as **anemia** in which there are fewer than the normal number of erythrocytes (or, in some cases, reduced hemoglobin content in the erythrocytes). People suffering from anemia often are fatigued or chilled because their tissues lack adequate oxygen for cellular energy conversions or heat production. Anemia is a symptom of a more fundamental physiological problem, and proper treatment requires identification and correction of the underlying cause.

Human Blood Groups   The surfaces of human erythrocytes contain certain genetically determined sets of molecules, and people can be divided into groups or **blood types** depending on which sets of molecules are present. Because these sets of molecules are genetically determined, a person's blood type is inherited. Two major classifications of human blood types are based on **ABO** grouping and the **Rh** system.

ABO grouping is based on differences in the type of glycoprotein molecule (see page 66) present on erythrocyte surfaces. Type A people have A type of glycoprotein molecules on their erythrocyte surfaces; type B people have B type glycoproteins; type AB people have both of these glycoproteins; type O people have neither of them.

The A and B glycoproteins function as **antigens**; that is, they combine specifically with **antibody** molecules. **Anti-A** antibody and **anti-B** antibody are protein molecules that occur in human plasma, and they combine specifically with the A and B antigens, respectively. Antigens and the specific antibodies with which they combine are said to be **complementary**. When A and B antigens combine with anti-A and anti-B antibodies, erythrocytes clump together. This happens because antibodies can combine with antigens on the surfaces of two different red cells, thus causing the cells to be bound by antibody molecules. Whole clumps of red cells can be stuck together in this way. This cell clumping is called **agglutination.**

Type A people have the anti-B antibody in their plasma; type B people have anti-A antibody; type AB people have neither of these antibodies; type O people have both antibodies (table 12.2). ABO blood types must be determined before medical personnel can initiate any blood transfusions. Donor and recipient must be compatible, or a severe agglutination reaction can occur. For example, if type A blood is given to a type O person, the anti-A antibody in the recipient's blood combines with the A antigen on the surface of the donor cells and causes them to agglutinate. The clumps of cells formed can block blood vessels and cause serious circulatory problems. The donor's erythrocyte antigens must be compatible with the recipient's plasma; that is, the recipient's plasma must not contain an antibody against an antigen on donor cell surfaces (figure 12.22). Whenever possible, recipients should be given blood of the same ABO type as their own, but there are possible alternatives. AB people can receive blood of any of the ABO types because their plasma has neither of the antibodies. In an emergency, type O blood can be given to a person having any ABO blood type because type O erythrocytes have neither of the antigens on their surfaces.

The second major classification of human blood types is based on the **Rh** system. (Rh stands for **rhesus;** this system was first identified in the blood of the rhesus monkey.) Rh classification depends on the presence or absence of the **Rh antigen** on red blood cells. **Rh-positive (Rh$^+$)** people have the Rh antigen on their erythrocytes while **Rh-negative (Rh$^-$)** people do not. About 85 percent of all people in the United States are Rh-positive.

The Rh system is different from the pattern in the ABO groupings because Rh-negative people do not normally have anti-Rh antibodies in their plasma. They must be exposed to the Rh antigen for the antibodies to develop. For example, anti-Rh antibodies develop in an Rh-negative person over a period of several months following an accidental transfusion of Rh-positive blood. Once this has happened, a second transfusion of Rh-positive blood can result in a severe agglutination reaction. Because such reactions occur only after two transfusions of a wrong blood type, they are very rare.

A problem more likely to develop from Rh incompatibility is **erythroblastosis fetalis.** It can occur during a pregnancy in which an Rh-nega-

tive mother carries an Rh-positive fetus. This is a very likely problem for Rh-negative women because the odds are high that their husbands are Rh-positive (more than 85 percent of all people are Rh-positive), and the gene for development of the Rh antigen is dominant. Thus, these women's children would likely be Rh-positive. Usually, no problem develops during a first pregnancy of this sort because even if some fetal blood cells do reach the mother's bloodstream late in pregnancy or during delivery, antibody production is too slow to cause problems for the infant. But a subsequent pregnancy can be a very different matter because a second exposure to Rh antigen results in much more rapid antibody production. Anti-Rh antibodies can then pass through the placenta and enter the fetal circulation, where they bind to fetal erythrocytes. This reaction causes the fetus to begin to destroy large quantities of its own erythrocytes. At the time of birth, the infant is anemic and quite ill. It has a jaundiced (yellowish) appearance because pigments produced during destruction of its own erythrocytes are circulating in its blood. These babies require a rapid series of transfusions to replace a large percentage of the blood containing the anti-Rh antibodies.

A preventative for erythroblastosis fetalis is now available because an Rh-negative mother can be treated with a substance called RhoGAM after delivering her first Rh-positive child. RhoGAM binds Rh antigen so that it does not stimulate production of significant quantities of anti-Rh antibodies. Thus, there is far less danger of erythroblastosis fetalis during the woman's second pregnancy.

White Blood Cells

Leukocytes, or white blood cells, differ from erythrocytes in that mature leukocytes contain nuclei and do not contain hemoglobin. Leukocytes are not involved in transport functions of the blood; instead, they participate in body defense reactions. In carrying out these defense functions, leukocytes are capable of leaving the circulatory system; they travel by amoeboid movement and slip out of capillaries by squeezing between the endothelial cells in capillary walls. Thus, leukocytes can act both inside blood vessels and in the spaces around tissue cells.

Table 12.2
Antigens and Antibodies in Human ABO Blood Groups.

Blood Type	Antigen Present on Erythrocyte Membranes	Antibody in Plasma	Incidence of Type in the United States	
			Among Whites	*Among Blacks*
A	A	Anti-B	41%	27%
B	B	Anti-A	10%	20%
AB	A and B	Neither	4%	7%
O	Neither	Anti-A and Anti-B	45%	46%

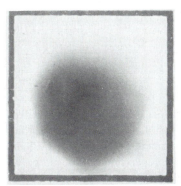
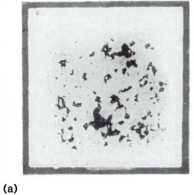

(a)

Figure 12.22 Erythrocyte agglutination reactions. (a) Comparison of unreacted blood sample (left) and blood sample with agglutinated erythrocytes (right). Agglutination occurs when cells bearing particular cell surface antigens are mixed with plasma containing specific antibodies against their antigens. (b) A diagrammatic representation of ABO blood typing reactions. Plasma antibody types used in the tests are indicated at the top of the columns.

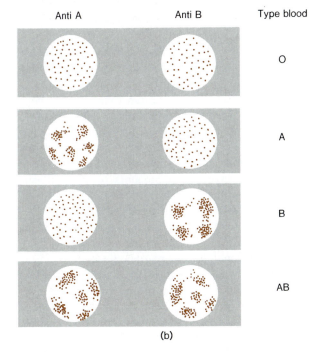

(b)

Leukocytes are far less numerous than erythrocytes. Normally, from 7,000 to 8,000 leukocytes per cubic millimeter are in the blood. Red cells thus outnumber white cells by about 700 to 1.

Leukocytes can be grouped into two categories: granular and agranular (table 12.3). Granular leukocytes (or granulocytes), so named because they contain distinctive cytoplasmic granules, are also called **polymorphonuclear leukocytes** because their nuclei have lobes and are highly variable in shape (polymorph = "many form" or "many shape"). The various types of granular leukocytes are shown in figure 12.23.

Agranular leukocytes include the **monocytes** and the **lymphocytes.** These cells do not contain cytoplasmic granules such as those of the granular leukocytes, and each monocyte or lymphocyte has a single, relatively large nucleus. Monocytes regularly move in and out of the circulatory system through capillary walls. Many of them remain in body tissues where they enlarge to become the important phagocytic cells, the **macrophages.** Lymphocytes are involved in very specific defensive responses to foreign antigens, such as those of infectious disease organisms. Some lymphocytes mount direct cellular attacks on foreign cells while other lymphocytes differentiate to become **plasma cells,** the cells that manufacture specific antibodies against foreign antigens.

The defense responses and the development and functioning of various leukocytes, especially the lymphocytes, are covered in more detail in chapter 26.

Platelets

Platelets are small disc-shaped cell fragments that lack nuclei. They average from 2 to 4 μm in diameter, and there are between 200,000 and 400,000 platelets per cubic millimeter of blood. Platelets are involved in the formation of clots and in this way function in preventing fluid loss from the circulatory system. Platelets are produced in the bone marrow, where they are pinched off as small fragments of large cells called **megakaryocytes.** They must be produced in large quantities because they have a life span of only about one week.

Table 12.3
Leukocytes.

Cell Type	Percent of Leukocytes in Blood
Granular Leukocytes (Polymorphonuclear Cells)	
Neutrophils	55 to 65
Eosinophils	2 to 3
Basophils	about 0.5
*Agranular Leukocytes**	
Lymphocytes	25 to 33
Monocytes	4 to 7

*Agranular leukocytes also are called mononuclear cells because their nuclei are not subdivided into lobes, as are the nuclei of the granular leukocytes.

Blood Plasma

Although blood plasma is 90 to 92 percent water, it is much more than just a liquid medium that suspends the formed elements of the blood. Blood plasma contains a complex mixture of inorganic and organic molecules, including a number of inorganic ions, various organic nutrients, nitrogenous wastes, special products such as hormones, plasma proteins, and dissolved gases.

Ions and salts make up about 0.9 percent of the plasma. By far the most numerous ions are sodium (Na^+) and chloride (Cl^-) ions. Other important ions in the plasma are potassium (K^+), calcium (Ca^{2+}), magnesium (Mg^{2+}), bicarbonate (HCO_3^-), phosphate ($H_2PO_4^-$ and HPO_4^{2-}), and sulfate (SO_4^{2-}). These ions contribute to the normal osmotic pressure of the blood and are involved in pH regulation. Some of these ions are required in different amounts for normal cell functioning, and they must be delivered to all body tissues by way of the blood.

Some of the important organic nutrients in the blood are glucose, amino acids, and various lipids, including fats, phospholipids, and cholesterol. Nutrients are transported in the blood following absorption in the digestive tract. They also are transported by the blood from nutrient storage sites, such as the liver, to various body tissues.

Most of the nitrogenous waste in human blood is urea. Urea is formed in the liver using amino groups removed from amino acid molecules. Amino acids can be prepared by deamination for oxidation in cellular energy conversion reactions such as the Krebs cycle.

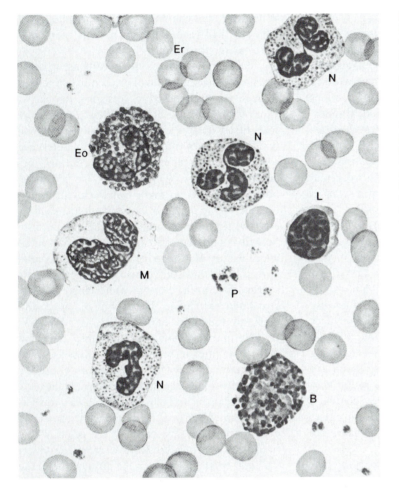

Figure 12.23 A representation of human blood cells. Shown are erythrocytes (Er), three kinds of granular leukocytes (eosinophil—Eo, neutrophils—N, and a basophil—B), a monocyte (M), a lymphocyte (L), and platelets (P). The names of the granular leukocytes are based on the different ways that the cells stain when stained blood smears are prepared. Neutrophils are the most numerous of all leukocytes. Neutrophils are important phagocytic cells that engulf and destroy bacteria, viruses, and scraps of damaged tissue both inside the circulatory system and in spaces around tissue cells.

Plasma proteins make up 7 to 9 percent of the blood plasma. The three main types of plasma proteins are the **albumins,** the **globulins,** and **fibrinogen.** About 60 percent of the total plasma proteins are albumins, which are produced by the liver. The albumins are particularly important in maintaining the blood's osmotic pressure, which, as discussed previously, helps to balance the fluid movement between capillaries and interstitial spaces. Some globulins function in transport because various molecules are transported attached to globulin molecules. But probably the most familiar of the globulins are the **gamma globulins.** Gamma globulins include specific antibody molecules, or **immunoglobulins.** Immunoglobulin molecules are produced by plasma cells in response to the presence of foreign antigens. (see page 726). Fibrinogen, which is synthesized in the liver and is the largest of the plasma protein molecules, makes up only about 4 percent of the total plasma proteins, but it plays a critical role in the blood-clotting process.

Blood Clotting

Blood must be a free-flowing liquid if it is to circulate easily through blood vessels under moderate blood pressure. But this liquidity also can cause serious problems. Any injury that breaks blood vessel walls can quickly lead to serious blood loss. This life-threatening danger of excessive blood loss is countered by the complex **clotting (coagulation)** mechanism. Clots are plugs that form temporary barriers to blood loss until vessel walls have healed.

The key reaction in blood clotting is the conversion of the plasma protein fibrinogen to a gel-like form called **fibrin.** Fibrin fibers form a sort of mesh that traps erythrocytes and becomes a solid barrier to blood loss (figure 12.24). But blood clotting is not just a single reaction that converts fibrinogen to fibrin. Blood clotting involves a chain of reactions, each of which depends on completion of the reaction that precedes it in the sequence. Furthermore, a number of substances either promote or inhibit clotting. Promoting substances are called **procoagulants,** and inhibiting substances are called **anticoagulants.**

Over thirty known substances are involved in blood clotting, and the effects and interactions of a number of these substances are poorly understood. This complexity seems to have evolved because of the requirement for a delicate balance between quick and efficient clot formation as a response to a significant blood vessel break and the prevention of the accidental formation of clots that interfere with normal circulation.

Although much research remains to be done on the details of human blood clotting, some of the major steps in clot formation are quite well understood. When there is damage to the wall of a blood vessel, the normally very smooth wall of the blood vessel is interrupted, and blood flowing in the area encounters much rougher surfaces than those normally found inside intact blood vessels. These rough surfaces have a particularly marked effect on the blood platelets, which are very fragile. Platelets stick to these rough surfaces and break open, releasing their contents; platelets accumulate to form a fragile plug that can close a very small break in a blood vessel wall.

Platelet rupturing also is an important step in fibrin clot formation because ruptured platelets release platelet factors, including **thromboplastins,** that are involved in the clot-forming reactions. Thromboplastins affect one of the globulin group of plasma proteins, **prothrombin,** in a very specific way. In the presence of thromboplastin and calcium ions (Ca^{2+}), prothrombin is converted to an active form called thrombin. This activation of prothrombin to form thrombin is essential for clot formation because thrombin catalyzes the conversion of fibrinogen to fibrin, and fibrin forms the fibrous mesh of a blood clot. (See figure 12.25.) The fibrin mesh traps blood cells and forms a plug that prevents further blood loss. Within a few minutes after a clot forms, it begins to contract, and within thirty to sixty minutes, it squeezes out most of the fluid that it contains. The fluid expressed from a clot is called **serum**, and it does not contain fibrinogen or other clotting factors. Thus, the term blood serum refers to blood plasma minus the elements involved in the clotting process. A contracted clot forms a tight barrier to blood loss that remains in place until normal healing of the blood vessel occurs.

When healing is complete, an enzyme called **plasmin,** which also is in the blood plasma, eventually dissolves the fibrin clot.

Clots must form readily in response to damage if excessive blood loss is to be prevented, but a grave danger is posed if clots form too easily. Clots formed inside blood vessels can interfere with blood flow or even entirely shut off blood flow to an area of the body. Because platelets are so fragile, some platelet breakage routinely occurs inside intact blood vessels. This breakage, however, does not normally lead to clot formation because of the action of anticoagulants, such as **heparin,** that are present in the blood. Because heparin is a quick-acting anticoagulant, it is medically useful. It can be extracted from donated human blood and used, for example, during heart surgery to prevent clotting while blood is being circulated through a heart-lung machine and back into the body.

Despite protective mechanisms, abnormal clots sometimes form inside blood vessels. Such clots are most likely to form in areas where the inner surface of a blood vessel is roughened. A clot on the inner wall of a vessel is called a **thrombus** (figure 12.26). A thrombus can impede or even block blood flow through the vessel where it forms, or it can break loose from the vessel wall and begin to flow with the blood through the circulatory system. Such a free-moving clot in the circulatory system is called an **embolus** (plural: **emboli**). Emboli generally flow along in the blood until they reach a small narrow point where they lodge in a vessel and stop flow through it. Emboli originating in systemic veins or in the right side of the heart usually lodge in the vessels of the lungs. Occasionally, a very large clot develops in a leg vein and breaks loose to become a large embolus. When such a large embolus reaches the pulmonary circulation, it may entirely block one (or both) of the pulmonary arteries with disastrous consequences. This is called a **massive pulmonary embolism.**

Not all problems with blood clotting involve extra, undesirable coagulation. Sometimes, clotting is so inefficient that excessive bleeding is likely. A vitamin K deficiency results in poor clotting because vitamin K is required for normal prothrombin synthesis by the liver, and lowered plasma prothrombin levels lead to poor clotting.

Another potential cause of excessive bleeding is a platelet deficiency. When the platelet count falls below 50,000 per mm^3, as compared

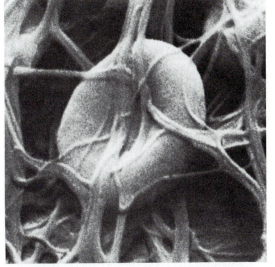

Figure 12.24 Scanning electron micrograph showing an erythrocyte caught in the fibrin threads of a clot. Fibrin threads form a mesh that catches many blood cells. Then the clot contracts, squeezing out serum and forming a solid plug that prevents further blood loss. (Magnification × 7,600)

(1) Damage to vessel wall and subsequent platelet rupture release thromboplastin

(2) Prothrombin $\xrightarrow[\text{Ca}^{2+}]{\text{Thromboplastin}}$ Thrombin

(3) Fibrinogen $\xrightarrow{\text{Thrombin}}$ Fibrin

(4) Fibrin mesh traps erythrocytes

(5) Clot contracts and becomes an effective plug that prevents further blood loss

Figure 12.25 Major steps in blood clotting.

Figure 12.26 Scanning electron micrograph of a thrombus, a blood clot formed inside an undamaged blood vessel. Blood cells are trapped in a mesh of fibrin threads. Erythrocytes (Er) often become misshapen (crenated) when they are in an altered environment such as the inside of this thrombus. Some crenated erythrocytes (CE) are labelled. (Magnification × 625)

with the normal range of 200,000 to 400,000 per mm^3, there is a greatly increased tendency for excessive bleeding. Most cases of platelet deficiency arise from unknown causes, although it can be the result of exposure to ionizing radiation or reactions to certain drugs.

Another condition that results in excessive bleeding is **hemophilia,** a hereditary problem that occurs almost exclusively in males. Victims of hemophilia bleed excessively because they have a deficiency of one of the plasma factors required for clotting. Even the slightest injury can lead to massive blood loss (**hemorrhage**). Very often, hemophiliacs die at a young age following repeated episodes of severe bleeding. Hemophilia influenced European history during the nineteenth century because young male members of several of the royal families of Europe inherited the problem.

Problems of the Circulatory System

Although some problems and abnormalities of the circulatory system, including blood clotting, have already been discussed, many more remain. Some of these problems affect the blood, and others affect blood vessels and the general circulation.

Infectious Mononucleosis

Mononucleosis is a virus-induced disease in which an abnormally high leukocyte count develops. The total leukocyte count usually is elevated to about 15,000 per mm^3. Most of the extra leukocytes are oversized lymphocytes or abnormal monocytes. There is no specific treatment for mononucleosis, but most people recover in a few weeks if they rest and receive treatment for any secondary problems that might result from their reduced defenses against other invading organisms.

Leukemia

Leukemia is a form of cancer in which there is a continuing excessive production of abnormal white blood cells. Production of these extra white cells so dominates the function of the bone marrow that the normal production of erythrocytes and platelets is depressed. Thus, leukemia victims often become anemic, and they are susceptible to hemorrhage because of platelet deficiency and the resultant faulty blood clotting. A common cause of death in leukemia is brain damage resulting from cerebral hemorrhaging. Infection is another frequent cause of death because there are insufficient normal leukocytes for body defense mechanisms.

Atherosclerosis

Atherosclerosis is an arterial disease that contributes to nearly one-half of all deaths in the United States. This common condition involves the formation of soft masses of fatty material in blood vessel linings. These fatty masses, called **plaques,** contain large quantities of cholesterol. They often form in arteries, and where they form, the arterial lining becomes much rougher than normal. This inner surface roughening tends to promote thrombus formation and leads to problems with embolisms. As a plaque develops, it decreases the diameter of the blood vessel and impedes blood flow. The formation of calcium deposits in the plaque and degenerative changes in the arterial wall lead to definite hardening of the artery. Hardened (sclerotic) arteries lose elasticity and are especially susceptible to rupture. The rupture of hardened cerebral arteries, which often results in brain damage, is called a **stroke.** (See figure 12.27.)

Plaque also can break loose and circulate until it blocks a small blood vessel or its rough surfaces cause formation of a clot that blocks a vessel. This is especially devastating if the blocked vessel is one of the coronary arteries of the heart. The portion of the heart muscle denied a blood supply is said to be **infarcted,** and the whole process is called **myocardial** (heart muscle) **infarction.** There often are simultaneous disturbances of the impulse conducting system so that rapid and uncoordinated heart beating occurs. These are symptoms of a heart attack. If the victim survives the initial crisis of a heart attack, there is a long recovery period during which dead muscle tissue in the infarcted area is replaced by fibrous tissue. People who recover from a myocardial infarction can return to normal restful activities, but they must be careful not to engage in activities that overtax their hearts. Their hearts have a reduced ability to meet extra circulatory requirements because part of the heart muscle is permanently lost.

Hypertension

The blood vessel and heart problems previously discussed all are related to or aggravated by another common circulatory problem, high blood pressure (**hypertension).**

Hypertension is a very common disease affecting the heart and blood vessels. Hypertension is diagnosed when the diastolic blood pressure continually exceeds 95 mm Hg. It is estimated that as many as one in every ten American adults suffer from hypertension, and fewer than one-half of these people are aware of it. Hypertension increases the possibility that a blood vessel will rupture with subsequent damage to the tissue supplied by the vessel. High blood pressure is a major contributing cause of strokes.

About 85 percent of all hypertension cases cannot be attributed to any organic cause. The remaining 15 percent are caused by hardening of the arteries, kidney disease, or hormonal imbalances. In these latter cases, hypertension can be treated by treating the condition that causes it. But the majority of cases must be treated directly with drugs that reduce blood pressure.

Circulatory System Health

The circulatory system is prone to many types of problems. Nothing can be done about inherited circulatory system tendencies, but positive steps can be taken to improve and maintain circulatory system health.

Dietary moderation is a good general rule. Obesity increases circulatory demands and the heart's work load. Also, people whose diets contain large quantities of fat are more likely to develop atherosclerosis than people with low-fat diets. High salt consumption can be another problem because it is strongly correlated with various circulatory problems, especially hypertension.

Not smoking or stopping smoking also can preserve circulatory system health. Cigarette smoking is very highly correlated with an increased risk of heart and blood vessel disease.

Finally, sensible exercise programs can contribute substantially to maintenance of circulatory system health.

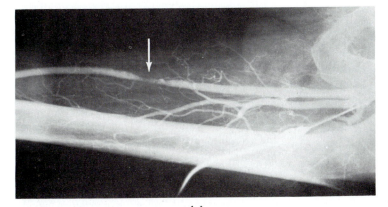

(a)

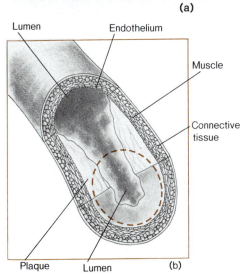

Lumen   Endothelium

Muscle

Connective
tissue

Plaque   Lumen   (b)

Figure 12.27 Atherosclerosis.
(a) An X-ray picture taken after a substance that appears opaque on X-ray photographs was injected into the bloodstream. An area where plaque formation has decreased the diameter of the femoral artery in the thigh is indicated by the arrow. (b) A diagram showing plaque accumulation in an artery. The circular broken line indicates the approximate size of the normal lumen. Note that the lining in the plaque area is much rougher than the normal endothelium. (c) Changes in arteries affected by atherosclerosis. (1) A normal artery. (2) Advanced atherosclerosis has led to wall hardening in this artery. (3) This sclerotic artery is completely occluded by a clot.

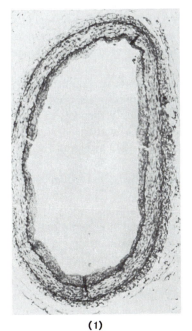

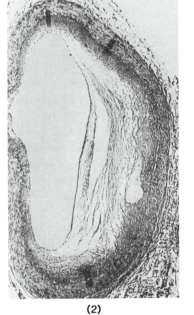

(1)   (2)   (3)

(c)

Summary

Transport mechanisms are necessary in the bodies of larger animals because materials must move between body surfaces and individual cells faster than they can move by diffusion. A circulatory system is a transport system in which blood is moved through the body in a regular fashion. Complex circulatory systems have hearts that pump blood through a network of blood vessels.

In open circulatory systems, blood leaves blood vessels and moves through parts of the body cavity, where it bathes organs and tissues. In closed circulatory systems, blood does not leave the confines of blood vessels; exchanges between the blood and body cells take place through vessel walls.

Circulatory systems of vertebrate animals are closed and have a heart, arteries, capillaries, and veins. In fish, blood passes through gill circulation, where it is oxygenated, and then moves on to the systemic circulation supplying the rest of the body. Vertebrate animals that have lungs for gas exchange have separate pulmonary and systemic circulatory loops. This arrangement permits more efficient, higher pressure, systemic circulation. Separation of pulmonary and systemic circulatory loops is associated with a partitioning of the heart into right and left chambers.

Human circulation resembles that of other vertebrates that have four-chambered hearts and separate pulmonary and systemic circuits.

The human heart has a regular cycle of contraction and relaxation in which atria contract and then ventricles contract. This cardiac cycle requires about 0.8 second. The heartbeat is initiated in the sinoatrial node of the right atrium, and the beat-generating impulse spreads rapidly through a system of conducting fibers to all of the muscle cells in the atria. The impulse is delayed briefly at the atrioventricular node and then spreads rapidly through the ventricles.

Cardiac output depends on heart rate, which is influenced by the nervous system and by hormones, and on the stroke volume of each heartbeat.

Blood pressure provides the driving force that propels blood through the circulatory system. Systolic pressure caused by ventricular contraction sends a pulse through the major arteries as arterial walls expand temporarily. Diastolic pressure is the pressure that can be measured during ventricular diastole. Pressure falls as blood moves through capillary beds and is so low in veins that movement through veins toward the heart depends on body muscle contraction and a system of valves that prevent backward flow.

Sphincter muscles in arterioles regulate flow through capillary beds and collectively are responsible for changing peripheral resistance in the circulatory system and thus altering blood pressure.

Individual capillaries are very small and very numerous. No body cell is far from one or several capillaries. Much exchange between body cells and the blood is by diffusion through capillary walls, but solute-carrying fluid also regularly moves out from the capillaries into tissue spaces and returns to capillaries near their venule ends. This fluid movement greatly increases the efficiency of exchanges.

Some fluid from the blood plasma does not return to the capillaries and remains in the tissues. This excess fluid is drained by the lymphatic system, through which it returns to the circulation. Lymph vessels also transport fats absorbed in the intestine. The lymphatic system is especially important for its role in body defense reactions to infection.

Human blood consists of formed elements (the blood cells and platelets) and the plasma.

Erythrocytes (red blood cells) are the most numerous of the formed elements, and they function in gas transport. Erythrocytes live for only about 120 days. Massive numbers of replacement erythrocytes are continually produced in the bone marrow.

Human blood groups are determined on the basis of sets of antigens present on erythrocyte surfaces and specific plasma antibodies to those antigens. Two major classifications of human blood types are based on ABO grouping and the Rh system.

Leukocytes (white blood cells) are much less numerous than erythrocytes. They participate mainly in body defense reactions.

Platelets are small cell fragments that are critically important for blood clotting.

Blood plasma is a watery material containing ions and salts, organic nutrients, nitrogenous wastes, hormones, plasma proteins, and dissolved gases.

Because blood is a free-flowing liquid, any break in vessel walls can lead to life-threatening blood loss. Blood clots form in response to injury of vessel walls, and these clots plug holes and prevent blood loss until healing occurs. A clot forms when a series of clotting reactions leads to formation of a mesh of sticky fibrin threads. This mesh traps blood cells and then contracts to become a solid barrier to further blood loss.

Abnormal clot formation inside unbroken vessels can block blood vessels and cause severe tissue damage. Clotting deficiencies, such as hemophilia, allow excessive blood loss in response to even minor injuries.

The circulatory system is subject to a number of diseases and problems. Some conditions, such as mononucleosis and leukemia, involve blood cells. In leukemia, excessive production of abnormal leukocytes interferes with production and functioning of normal blood cells.

Other kinds of circulatory problems involve the blood vessels. Atherosclerosis is a condition in which fatty deposits develop in blood vessel linings. Atherosclerosis can cause or contribute to a variety of circulatory problems, including those associated with hypertension. Hypertension overloads the heart and strains arterial walls. It can lead to heart disease or fatal blood vessel ruptures that can cause stroke.

Genetic factors are involved in the development of some heart and circulatory problems, but proper health habits can contribute to maintenance of a healthy and efficient circulatory system.

Questions

1. Explain the advantages of a circulatory system with separate pulmonary and systemic loops over a system with a single loop, such as that of fish.

2. When patients are immobilized in hospital beds, arrangements often are made to elevate their feet and legs higher than their trunks at least part of the time. What normal circulatory mechanism does this procedure replace?

3. Sometimes, problems develop with the heart's impulse conducting system. How do you think the efficiency of the heart's pumping would be affected if the atrioventricular node did not delay the impulse as it normally does?

4. Infections that affect lymph nodes sometimes produce a swollen puffiness in some parts of the body that cannot be explained by enlargement of the lymph nodes themselves. How would you explain the condition?

5. Explain why a person with any ABO blood type can be given type O blood in an emergency.

6. Explain how atherosclerotic plaques contribute to thrombus formation and blood vessel blockage.

7. Can you explain why such a great percentage of embolisms are pulmonary embolisms?

Suggested Readings

Books

Hill, R. W. 1976. *Comparative physiology of animals: An environmental approach.* New York: Harper and Row.

Ramsay, J. A. 1968. *Physiological approach to the lower animals.* 2d ed. Cambridge, Mass.: Cambridge University Press.

Tortora, G. J., and Anagnostakos, N. P. 1981. *Principles of anatomy and physiology.* 3d ed. New York: Harper and Row.

Articles

Baker, M. A. May 1979. A brain-cooling system in mammals. *Scientific American* (offprint 1428).

Bardell, D. 1978. William Harvey, 1578–1657, discoverer of the circulation of blood: In commemoration of the 400th anniversary of his birth. *BioScience* 28:257.

Doolittle, R. F. December 1981. Fibrinogen and fibrin. *Scientific American.*

Lassen, N. A.; Ingvar, D. H.; and Skinhøj, E. October 1978. Brain function and blood flow. *Scientific American* (offprint 1410).

Neil, E. 1975. The mammalian circulation. *Carolina Biology Readers* no. 82. Burlington, N.C.: Carolina Biological Supply Co.

Perutz, M. F. December 1978. Hemoglobin structure and respiratory transport. *Scientific American* (offprint 1413).

Schmidt-Nielsen, K. May 1981. Countercurrent systems in animals. *Scientific American* (offprint 1497).

Zucker, M. B. June 1980. The functioning of blood platelets. *Scientific American* (offprint 1472).

Excretion and Body Fluid Homeostasis

Chapter Concepts

1. Body fluid regulation is necessary to protect cells against harmful changes in their surroundings and to provide an environment with which cells can selectively exchange materials.

2. Excretory structures remove nitrogenous wastes and other by-products of cellular metabolic activity. They also function in maintenance of water balance and regulation of ionic composition and pH of body fluids.

3. The nature of the problems involved in maintenance of body fluid homeostasis is, to a large extent, determined by the type of environment that the organism inhabits.

4. The human kidney, with its tubular nephrons, functions in nitrogenous waste excretion and in ionic and pH regulation.

5. Nephrons are adapted to produce hyperosmotic urine, thereby accomplishing the water conservation that is vital in terrestrial animals such as humans.

6. Even brief impairment of kidney function interferes with body fluid homeostasis, and kidney failure causes serious illness that can quickly lead to death.

All cells of living organisms must selectively exchange materials with their immediate environments. Materials that cells require must be obtained from the cellular environment, and cells' waste products must be released into the environment around them. These wastes must then be removed from the cell environment so that they do not reach harmfully high concentrations.

When considering cell environments, it is best to go back to the environment in which living cells first arose. Most biologists agree that early phases of the evolution of living organisms took place in ancient seas, which very likely were quite stable environments, just as today's oceans are. In these seas, temperatures remained relatively constant over vast regions of water for long periods of time. Because concentrations of the many salts and ions in seawater also remained quite constant, ionic and osmotic conditions did not change abruptly. Huge quantities of water provided room for waste products to diffuse away from cells. Seawater also has a stable pH.

Living cells arose in the seas under a set of stable environmental conditions to which they were well adapted. As time passed, multicellular organisms evolved, and they also adapted to life in the stable ocean environment. Present-day marine invertebrate animals still have internal ionic compositions that are quite similar to the composition of seawater (table 13.1). Furthermore, their body fluids are isosmotic with seawater, and thus they do not suffer water gain or loss due to osmotic effects.

During the course of further evolution, organisms moved into new environments, such as the brackish water found in areas where fresh water mixes with seawater as rivers pour into the ocean. Eventually, freshwater environments also were occupied. Finally, organisms moved onto dry land, where they encountered a new set of problems associated with terrestrial existence.

Throughout all of these evolutionary changes, however, the requirement for a stable relationship between cells and their immediate environment has remained as urgent as ever. For example, unicellular organisms and the cells of simpler multicellular organisms living in fresh water have to be adapted to function in a hypoosmotic environment because fresh water contains much less dissolved and suspended material than living cells.

Table 13.1

Ionic Composition of Seawater and Some Invertebrate Animals' Body Fluids.

Ion	Concentration (mM)*		
	Seawater	*Lobster*	*Venus (a Clam)*
Na^+	417	465	438
K^+	9.1	8.6	7.4
Ca^{2+}	9.4	10.5	9.5
Mg^{2+}	50	4.8	25
Cl^-	483	498	514
SO_4^{2-}	30	10	26

Data from A. G. Loewy and P. Siekevitz, *Cell Structure and Function*, 2d ed. (New York: Holt, Rinehart & Winston, 1970).

*All data are expressed in millimoles (millimole = 10^{-3} mole) per liter.

Table 13.2

Composition of Seawater and Human Body Fluids.

Ion	Concentration (mM)			
	Seawater	*Serum*	*Interstitial Fluid*	*Intracellular Fluid*
Na^+	417	138	141	10
K^+	9.1	4	4.1	150
Ca^{2+}	9.4	4	4.1	—
Mg^{2+}	50	3	3	40
Cl^-	483	102	115	15
PO_4^{3-}	—	2	2	100
SO_4^{2-}	30	1	1.1	20
HCO_3^-	2.3	26	29	10

Data from James A. Wilson, *Principles of Animal Physiology*, 2d ed. (New York: Macmillan, 1979); and E. E. Selkurt, *Physiology*, 4th ed. (Boston, Mass.: Little, Brown & Co., 1976).

Multicellular organisms with specialized body parts took another evolutionary route. These organisms maintain a stable internal cellular environment, even in the face of highly changeable external conditions, through coordinated activities of the various specialized parts of the body. Such internal stability is a great advantage for organisms living in changeable environments because individual cells in their bodies are not exposed to changes in such factors as osmotic or ionic characteristics of their immediate environment.

The body fluid compositions of all organisms have not remained the same throughout evolutionary history, nor do body fluids so closely resemble seawater that they should be called "internal oceans" (table 13.2). But if body cells are to carry out their specialized functions normally, the internal body fluids of specialized multicellular animals still must have the *same degree of stability* that characterized the ancient oceans in which the first living cells arose.

More than one hundred years ago, the great French physiologist Claude Bernard described the stability of this *milieu interieur* ("internal environment") and emphasized its importance for organisms that encounter changeable environments outside their bodies. But it remained for Walter Cannon and others to expand the idea of a stable internal environment into the broader concept of **homeostasis.** They realized that conditions in the internal environments fluctuate within tolerable ranges, making homeostasis a dynamic steady state.

Homeostasis depends on precisely regulated division of labor among body parts, and the importance and the complexity of this division of labor and its regulation are particularly evident in the maintenance of body fluid composition. Body fluid homeostasis is the principal topic of this chapter.

Body Fluid Compartments

Approximately 60 percent of human body weight is water. This water is found in body fluids in several compartments—fluid-containing areas that are separated from each other by plasma membranes or by layers of cells.

About two-thirds of the fluid (around 40 percent of body weight) is **intracellular fluid**—the fluid contained inside body cells. Intracellular fluid is enclosed by plasma membranes of cells, and all exchanges between intracellular fluid and the fluid outside cells must be made through the plasma membranes. Selective activities of plasma membranes thus play a role in determining the makeup of the intracellular fluid, which is quite different in composition from other body fluids (table 13.2).

About one-third of the body fluid (around 20 percent of body weight) is extracellular fluid—fluid that is outside body cells. Extracellular fluid can be subdivided into two categories. About 80 percent of extracellular fluid is located in the spaces that surround body cells and is known as **interstitial fluid.** The remainder of the extracellular fluid is blood **plasma,** the fluid portion of blood. Often, biologists also refer to the

term **serum,** which is simply plasma minus the proteins responsible for blood clotting. (See figure 13.1.)

Body fluid homeostasis is dynamic, and it involves exchanges among the body fluid compartments. It depends on the specialized activities of several body organs and systems and on exchanges made between the body's internal environment and the external environment at specialized exchange boundaries (figure 13.2).

Some of the organs and systems involved in maintaining body fluid homeostasis include the gut, the lungs, the liver, the skin, and the kidneys. The gut digests food into simpler constituents that are absorbed as nutrient supplies for the body. The lungs exchange oxygen and carbon dioxide. The liver reversibly stores nutrients and processes metabolic waste products. Water, some ions, and some waste molecules are lost through the skin. The kidneys eliminate nongaseous body wastes, conserve body water and nutrient molecules, and are involved in maintenance of ion balance and in pH regulation in the body.

All exchanges with the external environment involve only one of the body fluid components, the blood plasma. The precise regulation of exchanges between the blood plasma and the external environment and between the plasma and the other fluid compartments keeps the total volume of body water and the composition of body fluids relatively stable while, at the same time, required materials are continuously being taken in and transported to body cells, and wastes are being removed.

Although functions of the gut, the liver, the lungs, and the skin are important in maintaining homeostasis, this chapter explores the contributions of kidneys and other specialized excretory organs of animals.

Excretion of Nitrogenous Wastes

Excretion is the term used to identify the processes that remove wastes, excess materials, and toxic substances from the body's cells and extracellular fluids. The two major kinds of wastes produced as a result of cellular metabolic activities are carbon dioxide and **nitrogenous** (nitrogen-containing) **wastes.** Carbon dioxide is a byproduct of cellular oxidation of sugars and various other organic molecules. It is released into the atmosphere through specialized gas exchange organs or the skin.

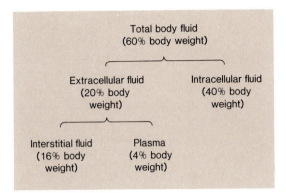

Figure 13.1   Body fluid compartments.

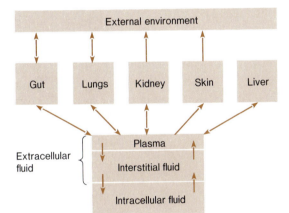

Figure 13.2 Relationships of body fluid compartments to one another and to the organs involved in body fluid regulation in a vertebrate animal. Blood plasma is the only fluid directly modified by external exchanges. Materials are exchanged with the external environment at specialized exchange boundaries (gut, lungs, kidneys, skin). Activities of the liver continuously modify plasma composition although the liver is not an external exchange boundary of the body.

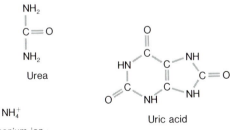

Figure 13.3 Chemical formulas for the three primary nitrogenous wastes of animals. Note that at pH's found inside animals' bodies, ammonia gains a proton to become an ammonium ion.

Nitrogenous waste compounds are products of cellular protein metabolism. Amino acids, derived from protein in food, can be used by cells for synthesis of new body protein or other nitrogen-containing molecules. The amino acids not used for synthesis are oxidized to generate energy or converted to fats or carbohydrates, which can be stored. Nitrogen-containing **amino groups** ($-NH_2$) must be removed from amino acids before oxidation, or conversion to fats or carbohydrates can occur. This amino group removal (**deamination**) during amino acid breakdown is the source of the nitrogenous wastes. In the human body, for example, deamination reactions occur mainly in the liver. Details of deamination are discussed in chapter 7, but the following key deamination reaction illustrates how nitrogenous wastes arise:

$$\text{Glutamate} + NAD^+ + H_2O \xrightleftharpoons{\text{Glutamate dehydrogenase}}$$

$$\alpha\text{-ketoglutarate} + NH_3 + NADH$$

Glutamate is an amino acid; glutamate dehydrogenase is the enzyme that catalyzes the deamination of glutamate; α-ketoglutarate is an organic acid that can be oxidized in the Krebs cycle; and **ammonia** (NH_3) is the nitrogenous waste molecule produced by this reaction.

Ammonia is a product of deamination; it also is the simplest kind of nitrogenous waste compound. Because ammonia is quite toxic and is very water soluble, it can be used as a nitrogenous excretory product only if there is a good deal of water available in which it can diffuse away. This limits direct ammonia excretion to organisms that live in water, such as protists, aquatic invertebrates, bony fish, and aquatic amphibians.

Other organisms must use different nitrogen excretion strategies. These animals incorporate waste ammonia into organic molecules, mainly either **urea** or **uric acid**, which are then excreted (figure 13.3). Synthesis of these organic waste molecules, however, requires that these animals expend more metabolic energy on nitrogen excretion than ammonia-excreting organisms.

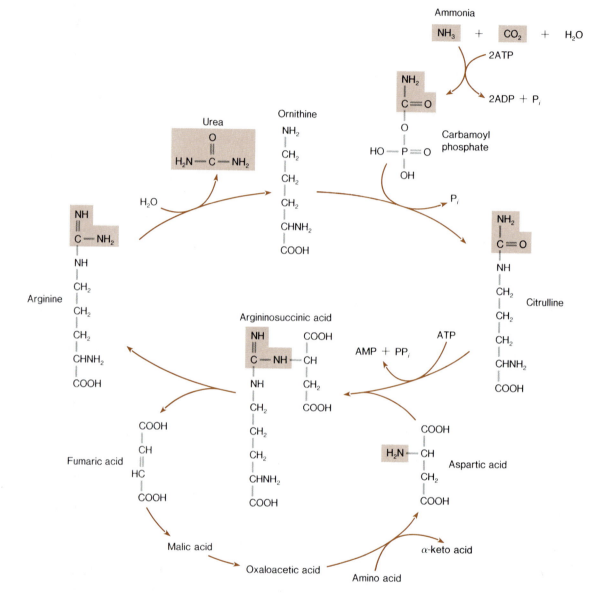

Figure 13.4 The urea cycle is a series of reactions by which urea is formed. In vertebrate animals, urea is produced in the liver. The parts of the urea molecule being assembled in this reaction are enclosed in colored boxes. After fumaric acid has been formed from argininosuccinic acid, it can be converted back to oxaloacetate by the activity of Krebs cycle enzymes. The oxaloacetate is then ready to accept an amino group from an amino acid. The ammonia used in carbamoyl phosphate synthesis can arise from the deamination of glutamic acid, as well as possibly from other sources. Note that this is an energy-requiring process because ATP must be used at several points.

Excretion of urea has both advantages and disadvantages. Urea is much less toxic than ammonia and thus can be excreted in a moderately concentrated solution. Body water is conserved, which is an important advantage for terrestrial animals with limited access to water. For this reason, a number of terrestrial organisms, such as mammals and adult amphibians, excrete urea as their main nitrogenous waste product.

Urea is synthesized in the livers of mammals and amphibians by a set of reactions known as the **urea cycle.** The urea cycle was discovered by Sir Hans Krebs, the same biochemist who played a major role in the discovery of the reaction series now called the Krebs cycle. As can be seen in figure 13.4, two amino groups are incorporated

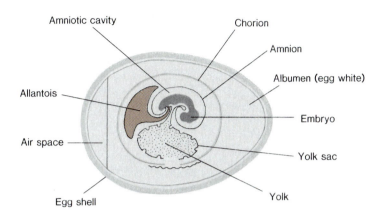

Amniotic cavity

Chorion

Amnion

Albumen (egg white)

Allantois

Embryo

Air space

Yolk sac

Egg shell

Yolk

Figure 13.5 A cleidoic egg. All nutrients and water required during development must be enclosed inside the eggshell because only oxygen and carbon dioxide diffuse freely through the shell. Various membranes partition the space inside the shell. The yolk sac encloses the yolk and functions in absorption and transport of nutrients. The amnion encloses an amniotic cavity filled with fluid that protects the developing embryo. The chorion encloses the whole developing system. Nitrogenous wastes produced by the embryo are stored as uric acid in the saclike allantois (shown in color). At hatching, the allantois is detached from the body and left inside the broken shell.

into each urea molecule. The urea cycle prevents accumulation of potentially toxic ammonia, but the process does use ATP. Since several ATPs are used in the synthesis of each urea molecule, this is a much more metabolically "expensive" way of excreting excess nitrogen than is the use of ammonia. Presumably, for an organism with a limited supply of water, the lower toxicity of urea and water balance factors make the use of urea as a nitrogenous waste molecule worth the "cost."

Uric acid (see figure 13.3) is the third major nitrogenous excretory product. Uric acid is not very toxic and also is poorly soluble in water. Poor solubility is an advantage in terms of water conservation because uric acid can be concentrated more readily than urea can. In fact, uric acid leaves the bodies of animals that excrete it as a damp mass of semicrystalline material. Animals that excrete uric acid recover most of the water from this semisolid form of urine before it leaves their bodies, and thus, uric acid excretion is advantageous for animals living in very dry environments. Reptiles, birds, and insects excrete mainly uric acid. This form of excretion is adaptive even for birds that have access to plentiful water supplies because extra water in the form of liquid urine would mean extra weight to carry during flight.

One disadvantage of uric acid is that it is synthesized by a long, complex series of reactions that require expenditure of even more ATP than does urea synthesis. Here again, there is a trade-off between the advantage of water conservation

and the disadvantage of energy expenditure for synthesis of an excretory waste molecule.

However, one other aspect of uric acid excretion adds to its adaptive value. Patterns of nitrogen excretion are related not only to the availability of water in the adult environment but also to the embryonic environments of animals. For example, reptile and bird embryos develop inside completely enclosed eggs. All nutrients and water required for metabolism and growth of the embryo must be inside the egg before embryonic development begins. Such **cleidoic** ("boxlike") **eggs** must also have a mechanism for nitrogenous waste storage because there is no way for nitrogenous waste to leave the confines of the shell until the embryo hatches. Production of insoluble, relatively nontoxic uric acid waste is advantageous for such embryos. Excreted ammonia would accumulate and reach toxic levels inside the egg, while urea would require storage of considerable quantities of waste liquid. Uric acid is stored in a highly concentrated form inside a saclike structure called the **allantois** that is attached to the body of the embryo during development (figure 13.5). As it hatches, the embryo breaks the connection with the allantois and leaves it and the uric acid that it contains behind inside the broken eggshell.

Table 13.3
Major Nitrogen Excretory Products of Selected Groups of Organisms.

Animal	Nitrogenous Excretory Product	Adult Habitat*	Embryonic Environment
Protozoa (protists)	Ammonia	FW/SW/T	
Aquatic invertebrates	Ammonia	FW/SW	
Earthworms	Ammonia, urea	T	
Planarians	Urea	FW/T	
Insects	Uric acid	T	
Teleost (bony) fish	Ammonia, some urea	FW/SW	Aquatic
Elasmobranchs (sharks, rays, and skates)	Urea	FW/SW	Aquatic
Amphibians, larval	Ammonia	FW	Aquatic
Amphibians, adult	Urea	T	Aquatic
Turtles	Urea, uric acid	T	Cleidoic egg
Lizards	Uric acid	T	Cleidoic egg
Snakes	Uric acid	T	Cleidoic egg
Birds	Uric acid	T	Cleidoic egg
Mammals	Urea	T	Aquatic (inside female body)

Adapted from K. Schmidt-Nielsen, *Animal Physiology*, 3d ed. © 1970, p. 477. Reprinted with permission of Cambridge University Press.

*FW = fresh water; SW = seawater; T = terrestrial.

A developing mammal releases its nitrogenous wastes through the placenta into its mother's blood, which transports the wastes to the mother's kidneys, where the wastes are excreted. For this reason, developing mammals produce urea just as adult mammals do. (See table 13.3.)

Water and Ion Balance

Because of the importance of water in life processes, maintenance of **water balance** is vital for all living organisms. An organism is in water balance when its body fluid compartments contain appropriate quantities of water, and the organism's water loss to its environment equals its water gain from the environment.

In addition to specific functional roles, ions in body fluids contribute to osmotic properties of the organism. Thus, regulation of the ionic composition of body fluids is interrelated with maintenance of water balance. What then are some of the osmotic and ionic problems faced by animals living in various environments, and what are some solutions to these problems?

Marine Animals

Marine invertebrate animals have body fluids that are isosmotic and isotonic with the seawater in which they live; that is, these animals do not tend to gain or lose water by osmosis. (Tonicity terms—hypertonic, hypotonic, and isotonic—are used specifically to describe situations in which water movement is caused in living things by osmotic pressure differences. See chapter 4 for more details.) Most marine invertebrate animals, however, must carry on some ionic regulation, because while the ionic composition of their body fluids is very similar to seawater, *it is not identical to seawater*. In other words, marine invertebrates' body fluids are isosmotic with seawater because they contain the same total quantities of ions, but concentrations of individual ions are quite different from those found in seawater. For example, magnesium ions (Mg^{2+}) are much less concentrated inside many invertebrates than in the surrounding seawater. Lobsters' body fluids contain only about one-tenth as much Mg^{2+} as seawater (see table 13.1). Such ionic differences are vitally important to the lives of these organisms because body cells function normally only when they are in an appropriate ionic environment. Thus, the specific ionic content of body fluids must be maintained.

Marine invertebrates have problems with ion homeostasis because of the tendency of ions to diffuse from regions of higher concentration to regions of lower concentration and because their body surfaces are somewhat permeable to various ions. Invertebrate animals maintain the important ionic differences between their body fluids and the surrounding seawater by using **active transport** processes to move ions across body membranes. In this way, they are able to maintain a constant body fluid composition that is different from the seawater around them. Because active transport requires energy (ATP) expenditure, even marine invertebrates, which are isosmotic with their seawater environment, must expend energy to maintain body fluid homeostasis.

Animals in Brackish Water

Marine animals whose body fluids are isotonic with normal seawater may at times be exposed to the diluted seawater in the brackish water regions along ocean shorelines where fresh water flows in from rivers. Many marine animals have no special mechanisms to deal with the hypotonic conditions encountered in brackish water. They gain weight rapidly. The bodies of some marine worms, for example, swell and take on a bloated appearance in brackish water.

The weight gain is due to water entering their bodies. Animals that respond in this way are **osmoconformers**; they conform to their osmotic environments. If they are exposed to hypotonic environmental conditions, water tends to enter their bodies by osmosis until their body fluids have been diluted to the same concentration as the surrounding brackish water. Osmoconformers' body fluids remain isosmotic with the external environment, changing as the osmotic conditions of the environment change.

Some osmoconformers can avoid water-weight gain by shielding their bodies when the surrounding environment is too dilute. For example, the common mussel *Mytilus edulis* is a strict osmoconformer, but it simply shuts its shell when the seawater around it has been diluted (figure 13.6a). But many other osmoconformers have no such evasive techniques, and for them, exposure to brackish water means water gain, body fluid dilution, and in some cases, death.

(a)

(b)

Figure 13.6 Osmoconformers and osmoregulators. (a) The edible blue mussel *Mytilus* is an osmoconformer, but it evades osmotic problems by closing its shell tightly when its environment becomes hypoosmotic. Many other osmoconformers, however, cannot evade the problem. For example, some marine worms swell until they look like tightly stuffed sausages when they are in diluted seawater. (b) The fiddler crab *Uca* is an osmoregulator. *Uca* and other osmoregulators maintain a stable osmotic concentration of body fluids even when environmental osmotic conditions change.

Some marine animals are **osmoregulators**; they are not at the mercy of their osmotic environments in the same way that osmoconformers are. Osmoregulators maintain the integrity of their body fluids even when they encounter changing external conditions. Many shoreline animals have at least some osmoregulatory ability, and some, such as the fiddler crab *Uca pugnax*, are very good osmoregulators (figure 13.6b).

Osmoregulators have excretory organs that produce large quantities of urine. They are able to "pump" water out of their bodies. This prevents their body fluids from becoming diluted as a result of osmotic water gain. Osmoregulators in hypotonic environments also must actively obtain ions from their environments. They need to replace ions lost with their urine and ions that diffuse out of their bodies. Both urine production

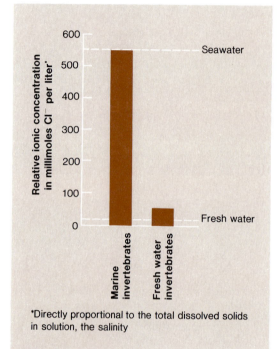

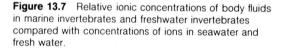

*Directly proportional to the total dissolved solids in solution, the salinity

Figure 13.7 Relative ionic concentrations of body fluids in marine invertebrates and freshwater invertebrates compared with concentrations of ions in seawater and fresh water.

and ion uptake require energy. Thus, osmoregulators have to expend considerable amounts of metabolic energy (ATP) to maintain the integrity of their body fluids.

Freshwater Animals

Animals living in fresh water face severe continuing problems because of their hypotonic environment. They tend to gain water osmotically, and body salts leach out because ions diffuse out through body surfaces.

It seems likely that brackish shoreline waters provided an evolutionary stepping stone between the oceans and freshwater environments. Osmoregulatory capabilities that evolved in brackish waters developed further over a long period of time until animals that could maintain body fluid homeostasis in freshwater environments evolved.

The ionic concentrations of freshwater invertebrates' body fluids are only about one-tenth as great as those of marine invertebrates (figure 13.7). Some dilution of the body fluids in freshwater animals seems to have been an evolutionary

Box 13.1
Cryptobiosis:

A fascinating array of adaptations for dealing with environmental water deficiencies has evolved in the animal kingdom. While some animals migrate to new habitats where water is available, others handle the problem in a different way.

For example, frogs and toads that live in desert areas burrow down into the bottom mud as their ponds dry up. There they enter a resting state called **estivation.** Animals in this "summer sleep" have very low metabolic rates and can remain inactive for long periods. When rain falls and refills their ponds, they quickly become active and resume their normal lives. Sometimes this includes a flurry of reproductive activity followed by very rapid development of a new generation so that the young individuals are mature enough to enter estivation when the ponds dry up again.

African lungfish are famous among biologists for their ability to survive years of estivation inside their special cocoons and then to emerge and become active very quickly when water is available (see box 11.1).

But possibly the most remarkable adaptations to periods of water deficiency are those of the microscopic tardigrades (box figure 13.1A). These animals, which are between 0.1 and 1 mm long, are found in almost every environment—marine, fresh water, and terrestrial. They have four pairs of stumpy legs and move with a lumbering, bearlike motion. Hence, they have been called "water bears."

When a tardigrade's surroundings begin to dry up, instead of moving on or attempting to conserve water, it simply contracts into a barrellike shape and becomes inactive. Its body dries until it contains barely detectable quantities of water, as little as 3 percent of its predrying water content. A tardigrade can remain in this dehydrated condition for years and yet be ready to rehydrate and become active quickly when water is once again available.

The Hidden Life
of the Water Bear

A museum specimen of moss that had been stored for 120 years yielded living tardigrades when it was placed in water. Clearly, tardigrades do not age at the normal rate while they are dehydrated (box figure 13.1B).

This dormant state in which the animal has extremely low metabolic activity is called **cryptobiosis** (meaning "hidden life"). Not only do cryptobiotic tardigrades somehow escape normal aging, they also are very resistant to harmful conditions that quickly kill normal, active tardigrades, or any other animals for that matter. Cryptobiotic tardigrades can survive for several minutes at 151°C or for long periods at temperatures almost −273°C. It requires an X-ray exposure of 570,000 roentgens to kill half of a cryptobiotic tardigrade population. For comparison, 500 roentgens would have an equivalent effect on a human population!

The ability to become cryptobiotic has great adaptive value because it allows tardigrades to survive drought conditions that would destroy populations of many other organisms. Furthermore, tardigrades are not harmed by heat or solar radiation to which they are exposed while dry. In the dried condition they are very light and can be blown about by the wind. Thus, cryptobiosis also serves as a dispersal mechanism that permits tardigrades to enter new habitats.

There still are many features of this adaptation to adverse conditions that are not well understood. How is it that tardigrades, some small worms, and a few other small animals can shut down their metabolism and survive with practically no body water? What protects their cells and tissues from complete disruption during drying and rehydration? Why are other animals not able to dry out and recover in the same way? Further research will be required to answer these and other questions about cryptobiosis.

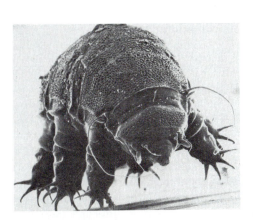

Box figure 13.1A Scanning electron micrograph of a tardigrade, *Echiniscus arctomys*. This is the normal appearance of the animal when it is active. Tardigrades live in very moist environments and their bodies contain about 85 percent water. (Magnification × 60)

Box figure 13.1B Scanning electron micrograph of a tardigrade in cryptobiosis. The animal has assumed a barrellike shape and has dehydrated as its environment dried up. A cryptobiotic tardigrade's body contains only about 3 percent water. (Magnification × 60)

trend. This has reduced the severity of osmotic problems of life in fresh water, but freshwater animals must, nevertheless, be good osmoregulators.

Terrestrial Animals

Terrestrial animals face the constant threat of dehydration. Relatively little water is available in terrestrial environments to make up for terrestrial animals' body water losses from nitrogenous waste excretion and evaporation. Thus, water conservation is a key requirement for fluid homeostasis in terrestrial animals.

Terrestrial animals have adapted in several ways to conserve body water. Their body surfaces usually are fairly impermeable to water. This reduces general evaporation, although there is always at least some water loss from the thin moist surface areas used for gas exchange. Many terrestrial animals conserve body water by producing urine that is a very concentrated urea solution, or by excreting uric acid, which reduces the water loss associated with excretion further still. Behavioral patterns also reduce water loss. Many terrestrial animals avoid situations in which water loss is greatest. For example, many desert animals are active during cool nighttime hours, and they rest in cool underground burrows during the hottest daytime hours.

Despite these adaptations, dehydration remains a serious threat to terrestrial animals, and water availability is a major determining factor in the distribution of various animals in terrestrial environments.

Excretory Organs

Most animals have specialized excretory structures that are involved in body fluid homeostasis. Some simple multicellular animals do not have specialized excretory structures because cell wastes simply diffuse away into the water that surrounds these animals. In animals such as sponges, which do not have specialized excretory structures, individual body cells have organelles that rid the cells of excess water that enters osmotically. These excretory organelles are called **contractile vacuoles.**

Similar contractile vacuoles are present in various protists, such as the protozoan *Paramecium*. Because the *Paramecium* contractile vacuole has been studied extensively, it is used here as a model of the regulatory role of contractile vacuoles in the individual cells of various organisms.

Contractile Vacuoles

Contractile vacuoles are energy-requiring pumping devices that expel excess water from individual cells that are exposed to hypotonic environments.

Functioning of the *Paramecium* contractile vacuole is shown in figure 13.8. Fluid collects in the canals that surround a vacuole. Then the vacuole swells as the canals empty into it. Finally, the vacuole contracts, expelling its contents through a pore in the cell's outer surface. When the vacuole contracts, it seems to collapse and disappear, but actually this is an active, forceful contraction that requires use of metabolic energy (ATP).

Analysis of fluid collected from contractile vacuoles shows that it has only about one-third as much dissolved material as the cytoplasmic fluid of the cell. Thus, contractile vacuoles eliminate very dilute fluid from the cell. This indicates that contractile vacuoles help to eliminate the excess water that enters *Paramecium* cells from their environment. Furthermore, contractile vacuole activity changes when osmotic conditions are altered; the vacuoles contract more frequently when the culture fluid becomes more dilute. For example, placing *Paramecium* cells in distilled water causes a marked increase in their vacuolar contraction rate.

Flame-Cell Systems

Most animal excretory systems consist of tubules or collections of tubules. Among the simplest of these tubular excretory systems are **flame-cell systems,** such as those found in planarian worms (figure 13.9).

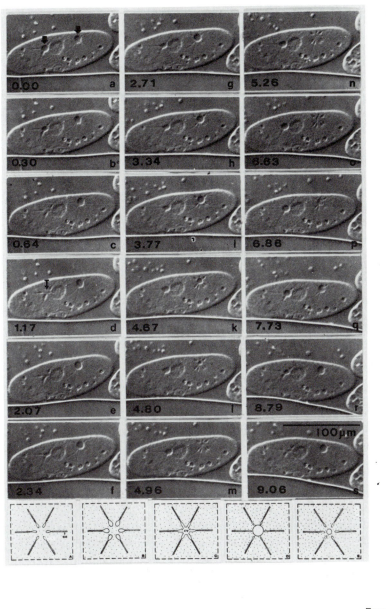

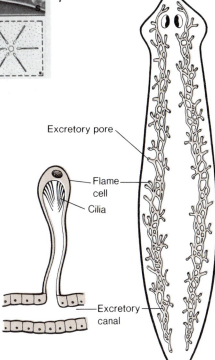

Figure 13.8 The function of contractile vacuoles in *Paramecium caudatum*. Each *Paramecium* cell has two vacuoles (thick arrows in (a)). Fluid accumulates in collecting canals and then enters the vacuole. When the vacuole is full, it contracts, forcing fluid out through a pore in the cell surface. The numbers in the lower left corners indicate the time (in seconds) counted from 0.00 seconds. The essential phases of the functional cycle of the contractile vacuole are represented schematically in the sketches labelled A–E at the bottom. (Compare A with c, B with d, C with g, D with p, E with r. All sketches illustrate the status of the left vacuole.) cc = collecting canal; cv = contractile vacuole. (Magnification × 220)

Figure 13.9 The flame-cell excretory system of a planarian worm.

The planarian excretory system is composed of a network of tubular excretory canals that open to the outside of the body through excretory pores. Located along the excretory canals are the bulblike **flame cells** from which the system takes its name. These bulbs are called flame cells because each one contains a cluster of beating cilia that looks like a flickering flame when viewed under a microscope.

Functioning of flame-cell systems is not completely understood, but apparently, fluid filters into the flame cells from the surrounding interstitial fluid. The beating of flame cell cilia propels this fluid through the excretory canals and out of the body. It is likely that flame-cell systems, like contractile vacuoles, function mainly in eliminating excess water that has entered the body osmotically.

The flame-cell system also is involved in ion regulation, but not in nitrogenous waste excretion. Nitrogenous wastes diffuse out through the body surface into the surrounding water.

Earthworm Nephridia

Earthworms have more complex excretory systems than planarian worms. Earthworm excretory structures are tubules called **nephridia** (singular: **nephridium**). The earthworm body is divided into segments by a series of partitions (septa), and each body segment has a pair of nephridia (figure 13.10).

Each nephridium begins with a ciliated funnel, the **nephrostome**, that opens from the body cavity of a segment into a coiled tubule. Fluid from the body cavity flows through the nephrostome into the nephridium. As it is moved along through the tubule by beating cilia, ions are reabsorbed and carried away by a network of capillaries that surround the tubule. In addition, waste materials, delivered by the capillary network, are added to the fluid. The close relationship between earthworm nephridia and their associated capillary networks makes possible these complex exchanges between blood and the fluid in the tubules.

The tubule leads to an enlarged bladder that empties to the outside of the body via an opening called the **nephridiopore. Urine,** the excretory fluid that leaves the body, is a dilute fluid containing metabolic wastes and is quite different in composition from the fluid that originally entered the nephrostome.

In this way, the earthworm functions much like a freshwater animal, even though it is a soil-dwelling terrestrial organism. Since the earthworm's environment is saturated with soil water, which is hypotonic to the earthworm's body fluids, the worm experiences osmotic water gain. The removal of excess fluid through the nephridia is important for maintenance of water balance.

Earthworm nephridia are functionally complex excretory structures that are involved in waste excretion, as well as in osmotic and ionic regulation. This same set of functions is encountered in excretory organs of more complex animals.

Malpighian Tubules

Insects have an excretory system that is quite different from those considered so far. Insects' excretory ducts do not open to the outside of the body. Instead, the system consists of a cluster of long, thin **Malpighian tubules** that are attached to the gut. The Malpighian tubules and the gut together function as the insect excretory system (figure 13.11).

Excretion begins with active transport of potassium ions (K^+) from the blood around the tubules across the tubule walls into the interiors of the tubules. (Because insects have an open circulatory system, blood bathes the surface of body structures such as Malpighian tubules.) This active transport of K^+ increases the ionic concentration inside the Malpighian tubules and establishes an osmotic differential that causes water to flow into the tubules osmotically. This water carries nitrogenous waste (uric acid), and also sugar and amino acids that are present in the blood. Fluid moves through the Malpighian tubules and into the gut. In the gut some of the water and some ions, sugar, and amino acids are recovered, but all of the uric acid remains and eventually passes out of the body.

← Anterior

Posterior →

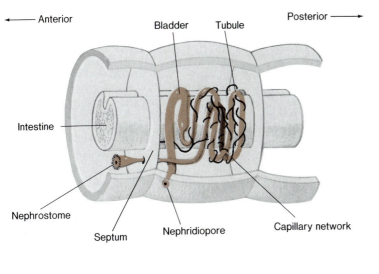

Bladder   Tubule

Intestine

Nephrostome

Septum   Nephridiopore

Capillary network

Figure 13.10 The earthworm nephridium. The nephridium opens by a ciliated nephrostome into the cavity of one segment, and the next segment contains the nephridiopore. The main tubular portion of the nephridium is coiled and surrounded by a capillary network. Urine can be stored in a bladder before being released to the outside. Most segments contain two nephridia. Note the close relationship between blood vessels and the tubules.

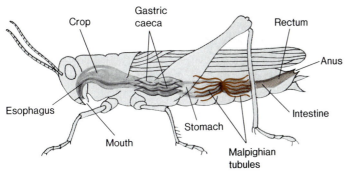

Crop   Gastric
        caeca

Rectum

Anus

Esophagus

Intestine

Mouth   Stomach

Malpighian
tubules

Figure 13.11 Malpighian tubules in the grasshopper. The Malpighian tubules are attached to the junction of the midgut (stomach) and the hindgut (intestine). Both Malpighian tubules and posterior portions of the digestive tract are involved in excretory functions. The Malpighian tubules are named for the Italian microscopist Marcello Malpighi (1628–1694), who discovered them.

The amount of water recovery in the gut depends on water availability. Insects living in water or insects eating large quantities of moist food produce quite watery urine. But insects in dry environments reabsorb most of the water from their urine and excrete a dry, semisolid mass of precipitated uric acid. For example, mealworms, which are the larva of a flour beetle (*Tenebrio molitor*), live in dry flour and have no access to water or moist food throughout their entire lives. Yet they manage their water balance by efficiently reabsorbing water and producing a very dry excretory product. The excreted material is so dry that it actually absorbs water from air, if the air has a relative humidity of 90 percent or greater.

Fluid Homeostasis in Fish

In the previous section, the excretory systems of invertebrate animals and the ways that these systems function in maintaining body fluid homeostasis were explored. Vertebrate excretory systems are the focus of this section. Because vertebrates inhabit marine, freshwater, and terrestrial environments, there is considerable diversity in the osmotic and ionic balance problems that they face, and in their nitrogen excretion patterns.

All vertebrates have specialized excretory structures called **kidneys.** Kidneys are complex organs containing many tubular units, the **nephrons.** In all cases, kidneys play a central role in body fluid homeostasis, but the specific functional adaptations of various vertebrates' kidneys are related to the animals' environmental adjustments. This diversity of adaptations to environments is clearly illustrated among the fishes.

Figure 13.12 Relative ionic concentrations of body fluids in vertebrate animals compared with concentrations of ions in seawater and fresh water (units as in figure 13.7). The upper portion of the bar for sharks and rays represents the contribution of urea to the total osmotic concentration of their body fluids.

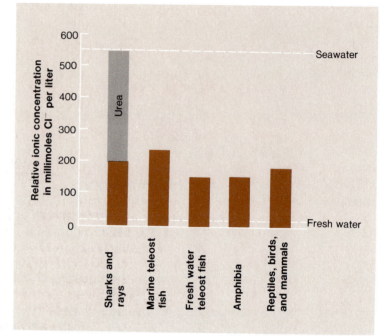

Figure 13.13 Water and salt balances in teleost fish. This figure illustrates the water and ion balance problems faced by marine and freshwater teleosts and also summarizes the ways in which teleosts meet these problems. The hollow arrows represent water and ion gains or losses resulting from the nature of the fish's environment. The solid arrows depict the active functions of the fish that counteract environmental pressures. Marine fish excrete Mg^{2+} and SO_4^{2-} ions via their kidneys, but the primary solution to the problem of excess body salts, caused by drinking seawater, is the active excretion of Na^+ and Cl^- ions through their gills.

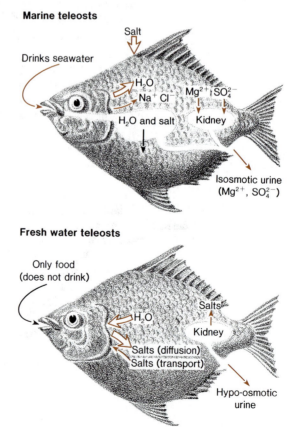

Both marine and freshwater bony fish (teleosts) have body fluids that are about one-third the osmotic concentration of seawater (figure 13.12). Because marine bony fish live in a hypertonic environment and freshwater teleosts live in a hypotonic environment, however, they face entirely different sets of ionic and osmotic regulation problems.

Marine teleosts continuously lose water to their environment by osmosis, especially through their gills. They replace lost water by drinking large quantities of seawater. On the average, marine teleosts swallow an amount of water estimated to be equal to about 1 percent of their body weight every hour, and some drink considerably more than that. This is approximately equivalent to an average-size human male drinking about 700 ml of water every hour around the clock.

Because of this drinking of seawater, a great deal of salt enters marine teleosts' bodies. Salt also enters by diffusion through marine teleosts' gills and skin, both of which are relatively permeable to ions. Therefore, marine teleosts also face the problem of continuous, unavoidable salt gain. It would be ideal if their kidneys could produce very concentrated urine, thus ridding their bodies of a great deal of salt without much water loss, but this is not the case. Because marine teleosts produce urine that is isosmotic with their body fluids, urinary salt excretion cannot solve the problem of excess body salt content. Marine teleosts' gills, however, actively secrete sodium (Na^+) and chloride (Cl^-) ions into the surrounding seawater, and this active ion "pumping" is the basis of marine teleosts' ability to drink seawater to replace the water lost by osmosis (figure 13.13).

Freshwater teleosts face an opposite set of problems from marine teleosts. Freshwater teleosts are threatened with excess water gain and diffusion of salts out of their bodies.

Freshwater teleosts are adapted to meet the problem of excess water gain. They never drink water and eliminate excess water through production of large quantities of urine that is hypoosmotic to body fluids. They can discharge a quantity of urine equal to one-third of their body weight each day.

Freshwater teleosts lose a significant amount of salt in their urine, even though their urine is quite dilute. This urinary salt loss, coupled with salt loss by diffusion through body surfaces, represents a serious ion balance problem. Some salt is replaced by salt obtained in food, but most of the necessary salt gain is accomplished by the gills. Freshwater teleosts' gills actively transport ions inward. Ions, which are in low concentration in the surrounding water, are taken up by energy-requiring active transport. Thus, active movement of ions in the gills of freshwater teleosts is in the opposite direction from that occurring in marine teleosts (see figure 13.13).

Because the adaptations of marine and freshwater teleosts to their environments are so different, it is remarkable that some teleosts actually move between the two environments during their normal life cycle. Salmon, for example, begin their lives in freshwater streams and rivers, move to the ocean for a period of time, and finally return to fresh water to breed. Eels, on the other hand, do just the opposite. They move from ocean to fresh water and back to the ocean. These fish must alter their gill and kidney functions in response to the sudden osmotic changes when they move from one environment to the other. Even their water drinking habits must be altered in response to environmental requirements.

While both marine and freshwater teleosts are adapted to solve their osmotic and ionic problems by complex energy-requiring responses of gills and kidneys, cartilaginous fish, such as sharks and rays, have an entirely different set of adaptations that allow them to evade osmotic problems of life in the marine environment. Body fluids of sharks and rays have much lower ionic concentrations than seawater (see figure 13.12), but these animals synthesize and retain enough urea in their blood and tissue fluids so that they are isotonic with seawater.

Biologists consider this strategy to be rather unusual for two reasons. First, since most aquatic animals excrete ammonia, urea excretion by sharks and rays is somewhat exceptional. Second, the high concentrations of urea found in sharks and rays would be very toxic to most other animals. Urea in high concentrations disrupts the structural integrity of protein molecules, thus breaking down structures made of protein molecules and interfering with the ability of enzymes to catalyze reactions. Clearly, the proteins of sharks and rays must be far less susceptible to urea damage than the proteins of most other animals.

Figure 13.14   The human urinary
system.

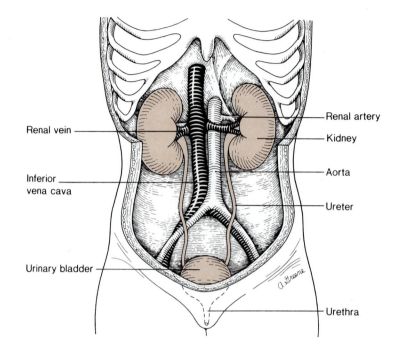

Renal vein

Inferior
vena cava

Urinary bladder

Renal artery

Kidney

Aorta

Ureter

Urethra

The Human Kidney and Fluid Homeostasis

The human kidney consists of a compact mass of individual functional tubular excretory units, the **nephrons.** Each of the about one million nephrons in an adult human kidney makes its own contribution to the total excretory function of the kidney. Human nephrons are similar to earthworm nephridia in that they are very closely associated with capillaries, and human nephrons function in a somewhat similar fashion. Fluid from the blood enters the excretory tubules. As this fluid passes through the tubules, some substances in it are reabsorbed for further use in the body, and cells in the tubules secrete other substances into the fluid for urinary excretion. In human nephrons, and those of other terrestrial vertebrates, water conservation is an important part of this processing of excretory fluid.

Structure of the Human Urinary System

Human kidneys are a pair of bean-shaped organs about 11 or 12 cm long that lie at the back of the abdominal cavity. Each kidney is connected to a duct, called the **ureter,** that carries urine from the kidney to the **urinary bladder,** where it is stored until it is voided from the body through

the single **urethra.** Each kidney receives its blood supply from a **renal** (kidney) **artery,** and blood leaves the kidney through a **renal vein.** (See figure 13.14.)

If a kidney is sectioned longitudinally, three major parts can be distinguished. The outer region, the **cortex,** has a somewhat granular appearance. The **medulla,** which lies inside the cortex, is arranged in a group of pyramid-shaped regions, each of which has a striped appearance. The innermost part of the kidney is a hollow chamber called the **pelvis.** Urine formed in the kidney collects in the pelvis on its way to the ureter. (See figure 13.15a.)

As mentioned earlier, the basic functional units of human kidneys are the one million or so nephrons that make up each kidney. Some nephrons are located primarily in the cortex, but the **juxtamedullary** (meaning "next to the medulla") **nephrons** lie mainly in the part of the cortex adjacent to the medulla, and a part of each of them extends down into the medulla (figure 13.15b).

Figure 13.16 shows more details of the anatomy of an individual nephron. Each nephron is essentially a rather complex tubule with an expanded, hollow end called **Bowman's capsule.** Enclosed within the Bowman's capsule is a knot of capillaries called the **glomerulus.** The remainder of the nephron consists of a tubule with three

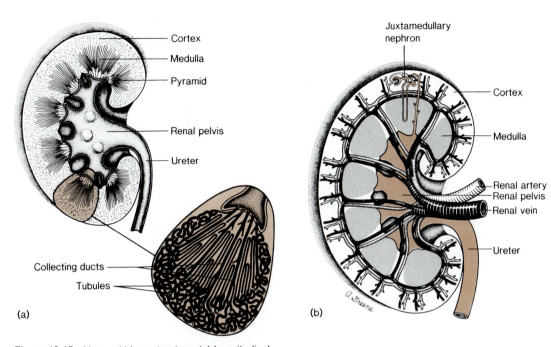

Figure 13.15 Human kidney structure. (a) Longitudinal section of the human kidney with an enlargement of one pyramid. (b) Internal details of the renal blood supply and the location of a juxtamedullary nephron.

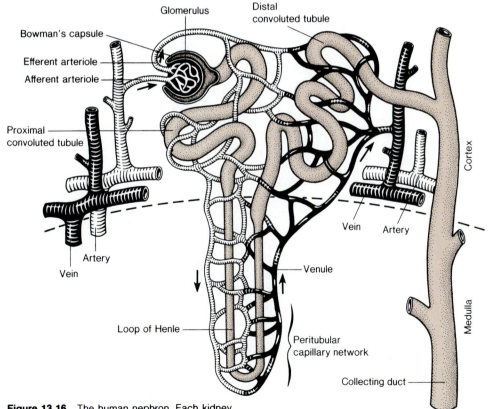

Figure 13.16 The human nephron. Each kidney contains over one million nephrons. The term proximal means nearer; distal means further. In this case proximal convoluted tubule is proximal (closer) to Bowman's capsule and distal convoluted tubule is further from it.

distinct regions. Adjacent to the Bowman's capsule is a coiled portion called the **proximal convoluted tubule.** The proximal convoluted tubule
leads to the **loop of Henle,** a narrowed portion
that plunges deep into the medulla. Another
coiled portion, called the **distal convoluted tubule,**
follows.

The end of each nephron drains into a **collecting duct** that receives fluid from several nephrons in an area of the kidney. Collecting ducts
transport fluid down through the medulla and
deliver urine to the pelvis. The loops of Henle
and the collecting ducts are the elements that
give the pyramids of the medulla their striped
appearance (see figure 13.15a).

Circulatory elements associated with the
nephrons are quite complex. Each nephron is supplied by an **afferent arteriole** that branches into
the capillary knot making up the glomerulus. The
glomerular capillaries drain into an **efferent arteriole,** which subsequently branches into a second capillary network around the tubular parts
of the nephron. These capillaries are called **peritubular** ("around the tubules") **capillaries.** This
network of capillaries around the tubules functions in a way similar to the network of capillaries
that surrounds the earthworm nephridium; they
pick up materials reabsorbed from the tubules
and deliver materials that are secreted into the

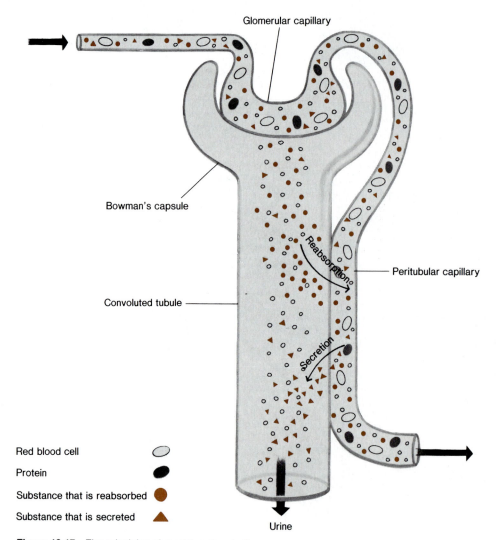

Figure 13.17   The principles of renal function. In the
glomerulus, water and solutes are filtered into the kidney
tubules while proteins and cells remain in the blood. As
the glomerular filtrate passes through the proximal and
distal convoluted tubules, solutes may be actively
reabsorbed from the filtrate or secreted into it.

tubules. The capillaries drain into venules, and these flow together into veins that carry blood to the single renal vein leaving the kidney.

Renal Function

Fluid from the blood is filtered through the walls of the capillaries of the glomerulus into the space inside Bowman's capsule, a process called **glomerular filtration**. This fluid (**glomerular filtrate**) is identical to the blood in composition except that it does not contain blood cells or plasma proteins. As the fluid is processed through the nephron, valuable substances, such as glucose, are actively reabsorbed and passed to capillaries.

This **tubular reabsorption** recovers a number of substances useful to the body. Certain other substances, delivered to the tubule area by the capillaries, are actively secreted by cells in the tubule walls. Through this **tubular secretion,** additional wastes are added to the fluid for eventual urinary excretion. (See figure 13.17.)

Glomerular Filtration

The glomerulus inside Bowman's capsule has a large surface area available for filtration because it consists of a great number of capillaries (figure 13.18). The walls of these capillaries are about 100 times more permeable to water, small solute

(a)

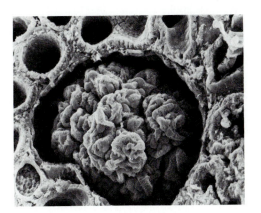

(b)

Figure 13.18 Structure of the glomerulus. (a) A diagram showing the glomerulus as a knot of capillaries inside Bowman's capsule. The juxtaglomerular apparatus is sensitive to fluid pressure. It responds to a decrease in sodium or blood pressure. (b) A scanning electron micrograph of a section of kidney cortex showing a glomerulus inside a Bowman's capsule. The holes surrounding the glomerulus are cross sections of tubules.

Proximal tubule

Glomerulus

Bowman's capsule

Juxtaglomerular apparatus

Efferent arteriole

Afferent arteriole

molecules, and ions than are the walls of most
capillaries elsewhere in the body. Because there
are numerous small pores or slits between the
cells that form the walls of glomerular capillaries,
water and some solutes readily pass through the
capillary walls and enter the space in Bowman's
capsule. But blood cells and larger molecules,
such as plasma protein molecules, cannot pass
through the pores and thus are held back and
retained in the blood. This selective passage of
materials through glomerular capillary walls is
glomerular filtration.

The driving force for glomerular filtration is
the blood pressure (hydrostatic pressure) inside
glomerular capillaries. Blood pressure is very effectively applied to the process of filtration because the efferent arteriole is smaller than the
afferent arteriole (figure 13.18a), and it offers
resistance to blood leaving the glomerulus. Blood
pressure acts to push fluid out through glomerular capillary walls because it is stronger than
the forces that oppose it (figure 13.19).

About 125 ml of fluid are filtered in the two
human kidneys every minute. This means that a
total of about 180 l of fluid leave a person's blood
and enter the Bowman's capsules each day. To
maintain this enormous normal glomerular filtration rate, a great deal of blood must pass
through the kidneys continuously. In fact, about
20 to 25 percent of total cardiac output passes
through the kidney circulation. Humans, however, do not produce and void 180 l of urine a
day. Human kidneys recover over 99 percent of
this glomerular filtrate and return it to the blood
leaving the kidney.

Tubular Reabsorption and Secretion

Both solutes and water are reabsorbed from the
nephron tubules into the capillaries that surround
the tubules. Most of this reabsorption takes place
in the proximal convoluted tubule, where active
transport mechanisms remove materials from the
filtrate inside the tubule. Some substances recovered by active transport are nutrients (glucose
and amino acids) and essential ions (Na^+, K^+,
Ca^{2+}, Mg^{2+}, HCO_3^- and HPO_4^{2-}). Other substances pass through the tubule walls by diffusion. Some substances, for example, chloride ions,

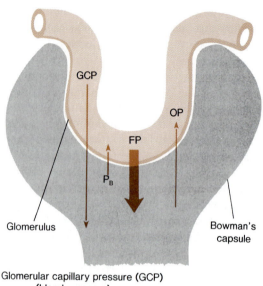

Glomerular capillary pressure (GCP) (blood pressure):	60 mm Hg
Pressure in Bowman's space (P_B):	—10 mm Hg
Net filtering pressure:	50 mm Hg
Osmotic pressure (OP):	—30 mm Hg
Effective filtration pressure (FP):	20 mm Hg

Figure 13.19 Diagrammatic representation of the
forces involved in glomerular filtration. Each force favors
fluid movement in the direction indicated by its arrow,
and lengths of arrows are proportional to the magnitude
of the forces operating in the glomerulus. Glomerular
capillary pressure (blood pressure) tends to force fluid
from the capillary to Bowman's capsule. Blood pressure
is opposed by a small fluid pressure in Bowman's
capsule (P_B) and by osmotic pressure factor (OP), which
is due to blood protein molecules that cannot pass out
of the capillaries into the Bowman's capsule space.

move across the tubule walls as a result of the active transport of other substances. Many of the sodium ions present in the filtrate are reabsorbed in the proximal tubule, mainly through operation of cellular sodium pumps. When sodium ions are reabsorbed, chloride ions follow because of electrical charge attraction; that is, the negatively charged chloride ions follow the positively charged sodium ions.

The reabsorption of materials by the proximal convoluted tubule is a vital function because it would be counterproductive for the body to lose all of the nutrients and ions that pass from the blood into the glomerular filtrate. A better picture of this reabsorption function can be obtained by considering the case of one nutrient, glucose.

Normally all of the glucose in the glomerular filtrate is reabsorbed in the proximal tubule unless the filtrate contains more than 180–200 mg of glucose per 100 ml. The filtrate would contain more glucose than that only if the blood glucose concentration were higher than that level. When the blood glucose level is higher, the active transport mechanism in the proximal convoluted tubule, even though it is working at its maximum capacity, cannot recover all of the glucose from the filtrate, and some of it passes out in the urine. Therefore, there is a **renal threshold** for glucose; that is, a blood glucose concentration above which glucose cannot be fully reabsorbed by the kidney, with the result that glucose can be detected in the urine. In a normal healthy person, the renal threshold for glucose usually is exceeded for only short periods following ingestion of a great deal of sugar. But blood glucose levels rise quite high in people with uncontrolled **diabetes mellitus** ("sugar diabetes"), and significant quantities of glucose regularly appear in their urine.

Because the kidney has a threshold for every substance that it reabsorbs, the blood levels of a number of substances automatically are kept within specific limits. If the plasma concentration of a particular substance exceeds its renal threshold, the excess is excreted in the urine. This arrangement is a very important homeostatic mechanism acting to regulate the content of the blood.

When sodium chloride, glucose, and other solutes are reabsorbed from the tubule, the fluid inside the tubule becomes more dilute than its surroundings. As a result, a large quantity of water leaves the tubule osmotically and is reabsorbed into the body. A great deal of fluid is reabsorbed in the proximal tubule, but the tubular fluid is still isosmotic with body fluids when it reaches the loop of Henle because both solutes and water have been reabsorbed.

Tubular secretion provides a means by which additional wastes can be added to the fluid as it passes through the tubules. Toxic substances, such as foreign acids and bases that have been absorbed in the gut, are eliminated by tubular secretion. The antibiotic penicillin and a number of other substances also are excreted in this way. But the process of tubular secretion is not limited to foreign substances. Many nitrogenous compounds produced in the body are expelled by active tubular secretion.

Production of Hyperosmotic Urine

A terrestrial animal, such as a human, normally has to conserve as much water as possible while at the same time excreting nitrogenous wastes and other materials that must be removed from the body. Human kidneys must be capable of producing urine that is hyperosmotic to body fluids if water conservation is to be adequate.

The human kidney does produce hyperosmotic urine and, if necessary, can eliminate body wastes in as little as 450 ml of urine a day. Most people with adequate access to drinking water, however, normally excrete several times that volume. Urine concentration results from reabsorption of water as fluid flows through the collecting ducts. This reabsorption of water is not by active transport; there is no cellular water "pump" in the human kidney. Instead, water is recovered because a very salty environment is maintained in the medulla of the kidney where the collecting ducts are located. Because this environment is hypertonic to the fluid in the collecting ducts, water tends to move out of the collecting ducts by osmosis. Thus, water is recovered by the body, and the urine that finally enters the renal pelvis is hyperosmotic to normal body fluids. The hypertonic environment maintained in the kidney medulla is the key to water conservation.

There is a gradient of increasing osmotic concentration from outside inward in the renal medulla. Maintenance of this gradient is the work of the juxtaglomerular nephrons, a special group consisting of about one-fifth of the nephrons. Juxtaglomerular nephrons lie in the cortex adjacent to the medulla with their loops of Henle penetrating deep into the medulla (see figure 13.15b).

These loops of Henle operate via a **countercurrent mechanism** that functions in maintenance of the osmotic gradient in the kidney. This countercurrent mechanism involves an exchange of material between two streams of fluid flowing past each other in opposite directions. The concentration of the material being exchanged decreases in one fluid stream as it increases in the other. (A countercurrent mechanism also operates in gas exchange in fish gills.)

The two streams in this countercurrent system are the two parts (**limbs**) of the loop of Henle. Fluid moves down toward the center of the kidney through the **descending limb** of the loop of Henle, and it flows in the opposite direction through the nearby **ascending limb.** What changes in the descending and ascending limbs is the concentration of Na^+ and Cl^- in the fluid inside them. This happens because there are important differences between the descending limb and the ascending limb of the loop. The descending limb does not actively transport salt. It is somewhat permeable to Na^+ and Cl^-, and it is very permeable to water. In contrast, the ascending limb actively transports sodium ions outward to the interstitial space, and chloride ions follow due to electrical attraction. Also, the ascending limb is practically impermeable to water. (See figure 13.20.)

What happens, then, is that the ascending limb actively pumps salt out of the fluid moving up through it. Ions from the salty environment around the loop of Henle diffuse into the descending limb, adding to the Na^+ and Cl^- already in the fluid in the lumen of the tubule, because these ions are more concentrated outside than inside

the tubule. At the same time, some water leaves the descending limb osmotically. The combination of salt gain and water loss makes the Na^+ and Cl^- content of the fluid in the tubule lumen at the bottom of the loop of Henle very high. But most of this salt is pumped out of the ascending limb, which at the same time is impermeable to water. So fluid reaching the distal convoluted tubule is not very salty at all. In fact, it actually is hypoosmotic to normal body fluids.

The result of all this activity is that a good deal of sodium chloride is retained near the bottom of the loop of Henle where it cycles. It is pumped out of the ascending limb into the interstitial fluid; then it diffuses into the descending limb and is carried around to the ascending limb, where it is pumped out again. Recycling sodium chloride permits the kidney to maintain a salty environment in the interstitial fluid in the area of the tips of the loops of Henle. (See figure 13.21.)

All of this sets the stage for the final phase of water reabsorption, which takes place from the collecting ducts. The collecting ducts can be made permeable to water so that as fluid in the ducts passes through the briny area in the renal medulla, water leaves osmotically through their walls. The urine delivered from the collecting ducts to the renal pelvis actually can be isosmotic with the briny interstitial fluid in the medulla. Thus, operating at peak efficiency, the human kidney can produce quite concentrated urine, about four times as osmotically concentrated as normal body fluids.

This last phase of water recovery is made possible by the functioning of the loop of Henle. Such efficient water recovery and production of concentrated urine are vital for all animals, including humans, that live in environments in which the water supply is limited. But human kidneys are not nearly as efficient as the kidneys of some other mammals that live with more severe permanent water shortages.

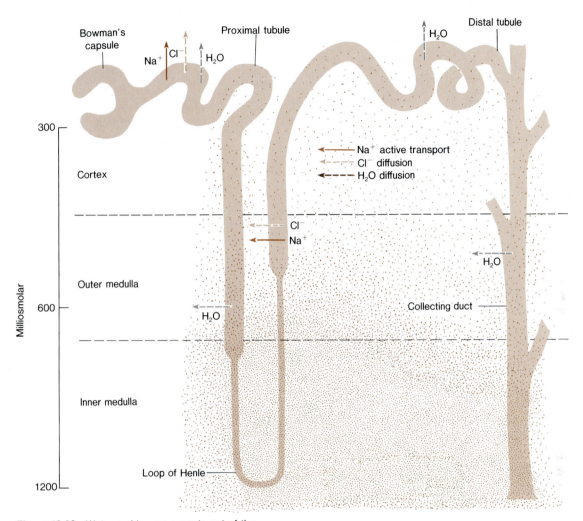

Figure 13.20 Water and ion movements out of the human nephron. The osmotic gradient in the kidney medulla is indicated by the osmotic concentration scale on the right and by the density of stippling, which increases as the renal pelvis is approached. Relative concentrations are shown in osmolarity, a measure of osmotically effective concentrations. The movements of sodium ions, chloride ions, and water are indicated by arrows of different colors and shades. Further details of the mechanism are shown in figure 13.21.

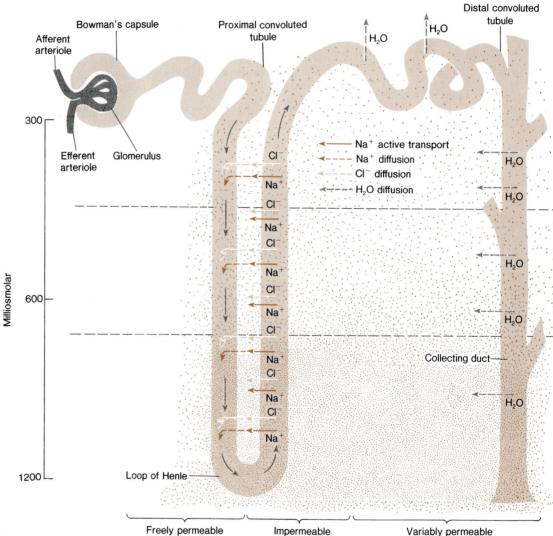

Figure 13.21 The countercurrent mechanism that
produces an osmotic concentration gradient in the
kidney. This gradient results from the specific activities
shown in figure 13.20 and the different permeabilities of
different parts of the nephron tubules. Sodium and
chloride ions pumped out of the ascending limb of the
loop of Henle diffuse into the descending limb. This
cycling of ions maintains the salty environment of the
interstitial fluid in the part of the medulla around the tips
of the loops of Henle, and there is a balance between
ions leaving the area via the peritubular capillaries and
ions coming down the tubule from Bowman's capsule.
Concentrations are shown in osmolarity, as in figure
13.20.

Box 13.2
The Kangaroo Rat:

A Mammal That Does Not Need to Drink

The kangaroo rat (box figure 13.2A), which gets its name from its habit of hopping on its hind legs, performs some remarkable feats of water balance maintenance in a very dry environment. This small rodent thrives in the desert regions of the southwestern United States, even in Death Valley where there is no drinking water available. How does this animal manage to live in its harsh, dry environment?

The kangaroo rat is behaviorally adapted to water conservation because it lives in burrows that are deep enough to be much cooler and more humid than the desert ground surface in the daytime. A nocturnal animal, the kangaroo rat stays in its burrow during the warmer hours of the day and ventures out at night in search of the seeds that make up its diet. This avoidance of daytime heat and dryness minimizes evaporative water loss.

Some animals living in waterless environments survive by eating succulent leaves that have a high water content, but the kangaroo rat eats mainly dry seeds. Water content in the seeds is so low that only 10 percent of the animal's daily water requirements can be supplied by the seeds. The remainder of its water gain is metabolic water formed as a result of oxidation of food material. Oxidation of one gram of carbohydrate yields 0.6 g of water, and oxidation of a gram of fat yields 1.1 g of water. The seeds that the kangaroo rat eats are high in fat and carbohydrate content and low in protein content. The low protein content of the seeds is important because using protein for energy requires more water for excretion of nitrogenous wastes.

Because the kangaroo rat's water supply is so meager, it must conserve water very efficiently. It reabsorbs so much water from its large intestine that its fecal material is almost completely dry. In addition, the kangaroo rat has exceptionally long loops of Henle in its kidneys. These enable the animal to produce urine that is over three times as concentrated as human urine. Only about one-fourth of a kangaroo rat's body water loss is due to use of water for excretion.

The efficiency of a kangaroo rat's kidney was demonstrated in experiments in which soybeans, which have a high protein content, were substituted for kangaroo rats' normal diet of high-fat, low-protein seeds. This created a need for more water to be used in nitrogenous waste excretion. If the rats were supplied with seawater to drink, however, they managed nicely. Humans cannot maintain water balance while drinking seawater because human kidneys do not produce urine that is concentrated enough to avoid extra water loss as a result of salt intake. But the more efficient water reabsorption mechanism of the kangaroo rat kidney makes it possible for the kangaroo rat to drink seawater as a supplemental water source.

Most of the total water loss from a kangaroo rat's body is by evaporation, mainly from its lungs. Even this loss is kept to a minimum because the animal's long nose allows exhaled air to be cooled. This exhaled air then carries out less precious body water than it would if it remained at body temperature.

Its minimal water needs and highly efficient methods of water conservation make the kangaroo rat remarkably well adapted to life in a dry environment.

Box figure 13.2A The kangaroo rat *(Dipodomys spectabilis).*

Regulation of Water and Sodium Balances

The human kidney can recover a great deal of water and eliminate wastes in a hyperosmotic urine. But what regulates the kidney so that it can change to meet either the need for water conservation or elimination of excess water?

The amount of water reabsorbed in the kidney is regulated by altering the water permeability of the collecting ducts. When collecting ducts are more permeable to water, more water is recovered and less eliminated. Thus, urine is more concentrated. When the collecting ducts are less permeable to water, less water is recovered and more eliminated. Urine then becomes more dilute, and the volume of urine produced increases.

One of the important factors regulating water recovery is a peptide hormone called **antidiuretic hormone (ADH).** ADH opposes diuresis, the increased discharge of urine. Specifically, ADH increases the water permeability of the collecting duct walls. When the circulating ADH level is high, more water is osmotically recovered and urine volume is decreased. A decrease in ADH concentration decreases the water permeability of the ducts and decreases water recovery so that urinary water elimination increases. ADH is produced by the hypothalamus and released into the blood from the posterior lobe of the pituitary gland (chapter 14). The rate of production and release of ADH is very responsive to changes in body water balance, which are detected by receptors that respond to blood pressure changes and osmotic changes in the blood. When the body is losing too much water, more ADH is produced and released. When excess water accumulates in the body, less ADH is produced and a larger volume of more dilute urine is excreted (figure 13.22).

This mechanism explains a disease called **diabetes insipidus,** which is caused by damage to the part of the hypothalamus that normally produces ADH. Such damage can occur, for example, when a tumor grows in the hypothalamus. The resulting ADH deficiency causes constant elimination of large quantities of dilute urine.

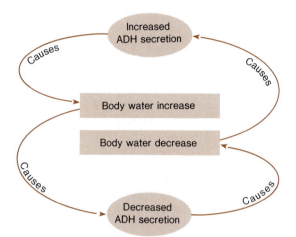

Figure 13.22 Regulation of body water content by antidiuretic hormone (ADH). Colored arrows indicate cause and effect relationships. This is a negative feedback relationship because increase in water content causes decrease in ADH secretion, while decrease in water content causes increase in ADH secretion.

A number of substances, called **diuretics,** increase urine volume either by acting directly on collecting duct walls or by inhibiting ADH production. Coffee, tea, and ethanol all act as diuretics.

Sodium ions (Na^+) are by far the most numerous positively charged ions in the extracellular fluids of the body. Because sodium ions are so numerous and because water moves osmotically in response to changes in osmotic concentration, control of the body's sodium content is closely related to regulation of water balance.

The sodium content of body fluids is regulated in several ways. One way is by increasing or decreasing sodium reabsorption from the fluid passing through kidney tubules. Sodium reabsorption by kidney tubules is controlled by the hormone **aldosterone,** a steroid hormone from the cortex of the adrenal gland. Aldosterone promotes sodium reabsorption, and the rate of aldosterone secretion goes up or down depending on the level of sodium ions in the blood.

Another control mechanism involved in sodium regulation depends on responses of the kidney itself to blood pressure changes. The **juxtaglomerular apparatus,** a small blood pressure sensor located along each afferent arteriole leading to a glomerulus in Bowman's capsule (see figure 13.18), responds to decreased blood pressure or decreased Na^+ concentration by releasing a substance called **renin** into the blood (figure 13.23).

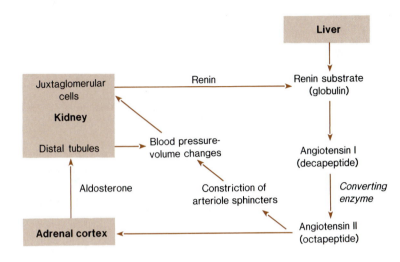

Figure 13.23 A diagram of the role of the renin-angiotensin in blood pressure and blood Na$^+$ regulation. Angiotensin I is a peptide that has ten amino acids; that is, it is a decapeptide. The active form, angiotensin II, has eight amino acids. Thus, the converting enzyme in the blood catalyzes the removal of two amino acids from the angiotensin I molecule.

Renin is an enzyme that is involved in conversion of a plasma protein into a substance called **angiotensin I,** which is then converted into an active form called **angiotensin II.** Angiotensin II causes blood vessel constriction, which raises blood pressure. Angiotensin II also stimulates increased aldosterone secretion, thereby promoting sodium and water reabsorption. This increasing fluid volume in the body also tends to increase blood pressure. Thus, the kidney itself is involved in maintenance of adequate blood pressure for normal kidney functioning.

The renin-angiotensin mechanism is a vital part of normal blood pressure regulation. It can, however, cause serious problems for kidney disease patients. Because damaged or diseased kidneys produce considerable quantities of renin, people suffering from kidney diseases often also have problems with hypertension (high blood pressure).

Maintenance of Constant Blood pH

Control of pH (hydrogen ion, H$^+$, concentration) in body fluids is critical to the maintenance of homeostasis. The pH of normal plasma is about pH 7.4, and the pH range compatible with life is fairly narrow—between pH 7.0 and 7.9.

The pH of the blood (and other body fluids) is stabilized in several ways. Acids or bases formed within cells are rapidly diluted in a large volume of extracellular fluid when they leave the cells. This large fluid volume tends to stabilize the pH. Also, excess acid can be taken up by buffers in the blood. Bicarbonate, phosphate, and proteins like hemoglobin are important plasma buffers.

The respiratory system also plays an important role in pH maintenance. Excess hydrogen ions (H$^+$) combine with bicarbonate to form carbonic acid. In the lungs, carbonic acid can then decompose to water and CO$_2$. Therefore, if a person breathes rapidly and loses more CO$_2$, the blood pH increases (becomes more alkaline) because hydrogen ions are used to form new carbonic acid as more CO$_2$ is breathed out:

$$H^+ + HCO_3^- \rightleftharpoons H_2CO_3 \rightleftharpoons$$
$$H_2O + CO_2 \longrightarrow \text{Elimination by lungs}$$

Conversely, a decrease in breathing rate leads to an increase in plasma CO$_2$ and a drop in blood pH. Because of the bicarbonate-carbonic acid system, blood pH is regulated to some extent through changes in the breathing rate.

Cells lining kidney nephron tubules also function in the stabilization of blood pH by adding bicarbonate ions to the blood. A tubule cell lining a nephron uses carbon dioxide to generate

Figure 13.24 Renal secretion of hydrogen ions (H^+).
Reactions occurring in cells that line a nephron tubule
are shown in one such cell. Carbonic anhydrase is an
enzyme that catalyzes the formation of carbonic acid.
Small circles indicate active transport mechanisms.
Hydrogen ions are actively transported into the anterior
of the tubule, while sodium is actively transported out
into the interstitial fluid. Broken arrows indicate
substances that diffuse in response to charge
differences created by the active transport processes.

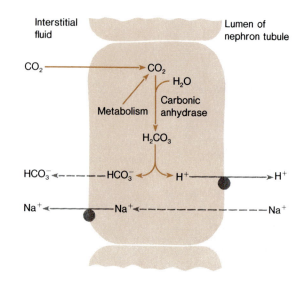

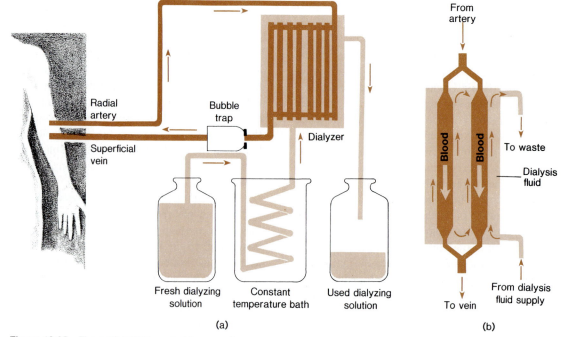

(a)   (b)

Figure 13.25 The artificial kidney and its operation.
(a) A schematic diagram depicting the machine's
construction and operation. (b) An enlarged view of a
parallel plate dialyzer unit. Blood and dialysis fluid flow
past each other in opposite directions (that is,
countercurrent flow). Body wastes diffuse from the blood
across the dialysis membrane to the dialysis fluid and
are carried away.

carbonic acid (figure 13.24). This CO_2 is obtained from the cell's own metabolism or from interstitial fluid near the cell. Once carbonic acid has been formed inside a tubule cell, the acid dissociates to bicarbonate and a hydrogen ion. The hydrogen ion is then actively transported into the urine inside the tubule. When this happens, a sodium ion leaves the urine and diffuses into the tubule cell to maintain electrical neutrality. On the other side of the tubule cell, the sodium ion is actively transported out into the interstitial fluid and, because of electrical attraction, is accompanied by a bicarbonate ion. In this way, the kidney excretes excess hydrogen ions and, at the same time, raises the bicarbonate concentration in the blood. These actions resist the acidification of body fluids.

Problems with Kidney Function

The great importance of the kidney for maintenance of body fluid homeostasis is vividly illustrated by the consequences of renal failure due to kidney disease. There are numerous kidney diseases, but acute and chronic glomerulonephritis probably are the most common.

Acute glomerulonephritis usually occurs about one to three weeks after an infection that occurred elsewhere in the body and involved a type of streptococcus bacteria; for example, following a strep throat infection. But the kidney problems are not due to direct effects of bacteria on the kidneys. During a streptococcus infection, the body produces specific antibody molecules that bind with certain molecules (antigens) of the bacterial cells. Antibodies and antigens combine during this normal body defense reaction to form antigen-antibody complexes that circulate in the blood. Antigen-antibody complexes reach the glomeruli, where they are trapped and cause irritation that leads to glomerular inflammation and damage. Such glomerular damage sometimes leads to blockage of the glomeruli so that no fluid moves into the tubules. Or it can cause increased porosity of the glomeruli so that blood cells and plasma proteins are no longer held back, but instead pass through the tubule system and out of the body in the urine.

Chronic glomerulonephritis is caused by several diseases, including streptococcus infections. The actual damage done to the glomeruli is very similar to that seen in the acute form. The most obvious difference between acute and chronic glo-

merulonephritis is time. Acute glomerulonephritis often subsides within as little as one week to ten days, whereas the chronic disease lasts for years.

Either disease can lead to partial or complete renal failure. Renal failure indicates severe kidney damage because the kidneys are able to handle normal waste excretion and other homeostatic functions with as few as one-third of the nephrons functioning normally. When kidney damage is so extensive that more than two-thirds of the nephrons are incapacitated, waste substances accumulate in the blood. This condition is called **uremia,** and it will lead to death if left uncorrected for long. In uremia, water and salt are retained so that generalized edema (fluid accumulation in body tissues) develops. Urea and other nitrogenous wastes build up in body fluids and cause serious damage. The ionic composition of body fluids is disturbed. For example, the potassium ion (K^+) concentration can rise to the point where it interferes with heart function and can even lead to heart failure. The pH of body fluids drops because the kidneys cannot excrete adequate quantitites of hydrogen ions and other acidic substances. The pH decrease is called **acidosis,** and it can cause a variety of problems, including loss of consciousness.

The Artificial Kidney

People suffering from either temporary or permanent renal failure can be treated with an artificial kidney machine. Blood circulates from an artery in the patient's arm to the kidney machine and returns to a vein. In this way, the blood continuously is circulated through a **dialyzing unit** in the artificial kidney. The dialyzing unit has very thin, selectively permeable membranes that separate blood flowing in one direction from a **dialysis fluid** that flows in the opposite direction. Dialysis fluid used in artificial kidneys contains concentrations of various substances that are the same as those in normal blood plasma, but it lacks such substances as urea, sulfate, and phosphate, which are abnormally concentrated in the blood of uremic patients (figure 13.25). The membranes separating blood and dialysis fluid are permeable to ions and small molecules such as urea, but are impermeable to blood cells and plasma proteins. Thus, as the blood and dialysis fluid flow through the dialyzing unit, the unwanted substances that diffuse out of the blood through the membranes are carried away by the dialysis fluid.

Dialysis speed and efficiency are improved by increasing the total membrane surface area in the dialyzing unit, and the units are designed to maximize membrane surface areas. For example, one type of dialyzing unit contains a stack of thin chambers, each of which has two compartments separated by membranes (figure 13.25b). Blood flows through one compartment while dialysis fluid flows in the opposite direction through the other. With efficient use of such thin chambers, some dialyzing units have total dialyzing surface areas that are as large as 20,000 cm², even though only a few hundred milliliters of blood are in the machine at any given time. The countercurrent flow through the system also makes waste removal more efficient.

Dialysis for several hours returns a uremic person's blood to its normal homeostatic balance and removes nitrogenous waste materials as well. Kidney machines have saved the lives of many people suffering short-term kidney failure resulting from disease or toxic responses, such as mercury poisoning. The machines also prolong the lives of victims of permanent renal failure. Unfortunately, artificial kidneys and extended dialysis treatment are very costly. Recent technical advances in dialyzing unit design, however, may make treatment available to more people who suffer from these disturbances of body fluid homeostasis.

Summary

All living cells must selectively exchange materials with their immediate environments. All unicellular organisms must be adapted to function in the changeable environments in which they live, but the cells of specialized multicellular organisms live in stable internal environments that are maintained through specialized activities of various organs and tissues. This specialized division of labor is especially important for maintaining the stability of the fluids in the several body fluid compartments of animals' bodies.

About two-thirds of body fluid is intracellular (inside body cells). The remaining third is extracellular fluid (interstitial fluid and blood plasma). Body fluid homeostasis requires exchanges of materials among these compartments and regulated exchanges between the blood plasma and the external environment. Kidneys and other excretory organs are particularly important sites of exchange between the internal fluids and the environment that surrounds the body.

Nitrogenous wastes, such as ammonia, result from the deamination of amino acids. Ammonia is toxic and must be excreted directly or after incorporation into other molecules, such as urea or uric acid. A good deal of water must be available for ammonia to be excreted directly. Organisms that must conserve water excrete urea or uric acid, even though these nitrogenous waste products require metabolic energy for their synthesis.

Water balance is vital to body fluid homeostasis for all organisms, and maintenance of water balance is closely related to regulation of ion balance in body fluids.

Marine invertebrates are isosmotic with seawater but still must regulate specific ions. Brackish water presents marine organisms with an osmotic water gain problem. Osmoconformers simply let their body fluids become diluted, but osmoregulators control the concentration of body fluids. Freshwater organisms require means of ridding their bodies of excess water, while terrestrial animals must vigorously conserve water to avoid dehydration.

Excretory structures are found even in the cells of protists such as *Paramecium*, where contractile vacuoles pump out excess water. Flame cells, nephridia, and Malpighian tubules are excretory structures in invertebrate animals. All of them function in waste excretion and maintenance of water and ion balance.

Fluid homeostasis in fish depends on the activity of kidneys and also on gill cells that can actively transport ions. Marine bony fish suffer from osmotic water loss and must drink seawater and excrete salts to maintain water balance. Freshwater fish suffer from osmotic water gain and must excrete copious amounts of urine while replacing lost body salts by gill uptake. Sharks and rays maintain and tolerate such a high urea content in their body fluids that they are isotonic with their marine environment.

The human urinary system consists of kidneys with tubular nephrons and a transport system to void urine to the outside.

Renal nephron function begins with pressure filtration of fluid from the blood in the glomerulus. The glomerular filtrate enters Bowman's capsule and moves on through the rest of the nephron tubule. Some materials are recovered from the filtrate by active transport mechanisms. Other substances are secreted into the fluid as it moves along the tubule.

Production of hyperosmotic urine is important for water balance maintenance in humans. Water is reabsorbed from several parts of the nephron, but the final phase of water reabsorption occurs in the collecting ducts. Water moves out of the collecting ducts osmotically because the countercurrent mechanism operating in the loops of Henle maintains a salty environment in the region through which collecting ducts pass.

The amount of water reabsorbed from the collecting ducts depends on the ducts' permeability, which is under the control of antidiuretic hormone (ADH). Conversely, the level of ADH in the blood depends on the body's water balance situation. Sodium reabsorption is promoted by the hormone aldosterone.

Kidney function is critically dependent on blood pressure, and the juxtaglomerular apparatus associated with each nephron responds to decreases in blood pressure or Na^+ concentration by secreting renin. Renin catalyzes the activation of angiotensin II, which in turn acts to raise blood pressure and increase aldosterone secretion.

Kidney tubule cells function in pH regulation by producing carbonic acid. When carbonic acid ionizes, the hydrogen ions produced are actively secreted into the tubule lumens for urinary excretion.

Glomerulonephritis is a kidney disease involving damage to the glomerulus. It can result in glomerular blockage and greatly diminished urine production, or a breakdown in filtration, which leads to urinary passage of blood cells and plasma proteins.

When two-thirds or more of the nephrons cease functioning, uremia develops, and excess water, salts, and urea are retained in the body. Acidosis may also occur in uremia.

Uremia can be treated with artificial kidney machines. These machines contain dialyzing units in which blood and dialysis fluid flow on opposite sides of selectively permeable membranes. Urea and other wastes diffuse from this blood to the dialysis fluid and are carried away, thus artificially helping to restore body fluid homeostasis.

Questions

1. Explain why peanut worms swell up and gain weight when placed in diluted seawater, but shore crabs show no similar weight gain.

2. What effect would you expect each of the following fluid environments to have on the rate that the contractile vacuoles inside a *Paramecium* cell contract: seawater, distilled water, a very dilute solution of potassium cyanide (KCN) in distilled water?

3. Describe three functional adjustments that must be made by a young salmon as it swims from a freshwater river out into the ocean.

4. Explain the role of the loop of Henle in maintaining a hypertonic environment around the collecting ducts of the kidney.

5. What effect would you expect elevated blood pressure to have on the volume of urine being produced? Why?

6. In terms of kidney function, how would you account for "beer diuresis," a condition experienced by many college students?

7. If a urine sample contains blood cells and plasma proteins, what kidney structures have been damaged? How did you reach your conclusion?

Suggested Readings

Books

Chapman, G. 1967. *The body fluids and their functions*. Studies in Biology no. 8. Baltimore: University Park Press.

Schmidt-Nielsen, K. 1979. *Animal physiology*. 2d ed. Cambridge, Mass.: Cambridge University Press.

Smith, H. W. 1961. *From fish to philosopher*. Garden City, N.Y.: Doubleday.

Articles

Crowe, J. H., and Cooper, A. F., Jr. December 1971. Cryptobiosis. *Scientific American* (offprint 1237).

Moffat, D. B. 1971. The control of water balance by the kidney. *Carolina Biology Readers* no. 14. Burlington, N.C.: Carolina Biological Supply Co.

Schmidt-Nielsen, K., and Schmidt-Nielsen, B. July 1953. The desert rat. *Scientific American* (offprint 1050).

Part **3**

Regulation in Living Organisms

Taken separately, the functions examined in Part 2 tell only part of the story of maintenance of organismic integrity because there also must be effective coordination of the many functional activities occurring concurrently in multicellular organisms. This coordination requires communication among body parts that are, in many cases, widely separated from one another. Furthermore, coordinating mechanisms must be flexible enough to permit organismic adjustment in response to changes in the external environment as well as changes in the age and condition of the organism itself.

In animals, chemical control by hormones provides relatively slower but more sustained regulation, while nervous mechanisms are involved in relatively faster but shorter-lived regulatory responses.

Hormonal regulation in plants functions mainly in control of growth and developmental responses to environmental factors, the most important of which is light. These responses are part of each plant's continuing interaction with its environment. In this interaction, various hormonal and timing mechanisms function to synchronize growth and other activities with favorable environmental conditions.

In chapters 14–18 hormonal and nervous control mechanisms in animals are examined, and some aspects of the structural and functional overlaps and interactions between these animal regulatory systems are explored. Discussion then turns to the functions of various plant hormones and the timing mechanisms that coordinate plants' lives with seasonal environmental changes.

Hormonal Regulation in Animals

Chapter Concepts

1. Throughout the animal kingdom, organisms' bodies are chemically regulated by hormones.

2. Hormones, in addition to nerve impulses, provide a means of communication within the organism. They help maintain homeostasis and aid in directing growth and development.

3. Neurosecretory substances that operate as hormones form an important link between the nervous and endocrine systems.

4. Many vertebrate endocrine glands are regulated by secretions from the anterior pituitary, which is in turn regulated by hormones released from the hypothalamus.

5. In humans, numerous hormones with complex interactions control metabolism and fluid and ionic balance.

6. Hormones may be regulated during their synthesis or by any of a number of mechanisms that affect the receptor molecules of target cells.

7. Protein and peptide hormones exercise their effects upon target cells by stimulating cyclic AMP synthesis, while steroid hormones act directly on the genome through a hormone-receptor complex.

8. Animals can chemically regulate one another when they release pheromones that affect other members of the same species.

In multicellular organisms, there is a division of labor among the cells, tissues, and organs of the body; each of the body regions performs specialized functions that help maintain a stable internal environment for all body cells. Obviously, all of these specialized activities must be coordinated if this stability is to be maintained. Coordinating mechanisms have thus evolved along with regional specialization and the division of labor in multicellular organisms.

Animals possess two major methods of internal coordination: nervous (neural) and hormonal control mechanisms. Neural control, which is achieved through electrochemical impulses, permits the rapid transmission of messages and causes quick but usually short-term responses. Hormonal control, on the other hand, is accomplished through specific chemical messengers carried in the blood. Hormonal control requires longer time periods both for the message to be transmitted and for its effects to take place.

It is impossible to discuss neural and hormonal systems entirely separately because there are some forms of regulation that do not fall neatly into either category. Instead of two separate coordinating systems with separate operating principles, there are really several kinds of coordinating mechanisms involving chemical substances originating from cells in both the nervous and endocrine systems. Furthermore, these different coordinating mechanisms may both act in the regulation of a single function in the body.

Ordinary nerve cells release chemicals called **neurotransmitters** that are discharged at the ends of the nerve cell processes (or axons) into tiny spaces between cells. There the neurotransmitters are involved in transporting nerve impulses from cell to cell, or in the stimulation of a specific action within an effector cell, such as the contraction of a muscle cell. Neurotransmitters are not released into the bloodstream, and thus they act only on cells that are a very short distance from the cells that release them.

Certain specialized nerve cells produce and release specialized substances that are carried in the blood and that have specific effects on **target cells** (responding cells) more or less remote from the nerve cells that produce them. These specialized nerve cells are called **neurosecretory cells** and the substances that they release are called **neurosecretions.** Some neurosecretory cells release their neurosecretions directly into the blood, while others pass their neurosecretions to storage

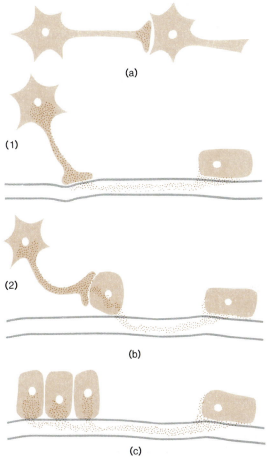

Figure 14.1 A comparison of different forms of chemical control. (a) A typical nerve cell discharges neurotransmitters into tiny spaces between cells. (b) Neurosecretory cells synthesize neurosecretions that are transmitted down axons and released either directly into the circulation (1) or into storage cells (2) for later release into the bloodstream. (c) A typical gland cell secretes its product directly into the circulation.

cells that later release them into the circulatory system (figure 14.1).

Finally, **hormones** are synthesized and released into the bloodstream by specialized glandular cells. Hormones have traditionally been defined as substances secreted by organs called **endocrine glands.** The word endocrine (from the Greek *endon,* meaning within) indicates that these glands secrete their products directly into the bloodstream, while **exocrine** (from the Greek *exo,* or outside) **glands,** such as sweat glands and salivary glands, secrete their products through ducts. It is now recognized that defining hormones only in terms of the familiar endocrine glands (such as the pituitary, thyroid, and adrenals) barely touches the surface of hormonal regulation in the body.

The term hormone comes from the Greek *hormon,* meaning to excite or set in motion. This term was applied to the chemical substances secreted by endocrine glands before it was known that there are inhibitory as well as excitatory hormones. Many other ideas about hormones have changed as hormone-producing cells have been discovered in other tissues. But the study of hormones began with early discoveries in endocrinology, or the study of hormone-producing tissues and their secretions.

Early Studies of Animal Hormones

Endocrinology has become a major branch of the field of biology only in the past century, but the actions of some endocrine glands have been widely known for many years. For example, the effects of castration (removal of the testes) on young male vertebrates have been known for several thousand years, and this knowledge has long been applied to prevent certain male animals from breeding or to raise animals that are more docile and that produce better meat. In fact, Aristotle, in the fourth century B.C., wrote an accurate, detailed description of the effects of castration on birds and on humans.

However, scientific experimentation on the functioning of endocrine glands did not actually begin until 1849, when A. A. Berthold transplanted testes back into the bodies of young castrated male chickens. Berthold found that this treatment restored the male secondary sex characteristics normally seen in roosters. They developed combs, gained the ability to crow, and began to exhibit the generally pugnacious behavior expected of roosters. These treated males looked and acted like roosters even when the transplanted testes were far from their normal location in the body and had none of the normal nerve connections of testes. Berthold concluded that testes must release some substance that causes the development of male characteristics, and that normal male development was not due to some influence of the testes transmitted through nerves to the central nervous system, as was earlier believed.

By 1904, the English physiologists Starling and Bayliss (who discovered the hormone secretin, which stimulates the release of pancreatic juice) were writing about an "endocrine system" and had introduced the term "hormone." Bayliss

defined a hormone as "a substance secreted by cells in one part of the body that passes to another part, where it is effective in very small concentrations in regulating the growth or activity of other cells." That operational definition of hormones is still useful today, although the term is now applied to a growing list of substances produced both by the traditionally recognized endocrine glands and by many other body tissues.

The standard techniques developed by early endocrinologists for identifying endocrine glands and for studying hormones involved several steps. Usually, a hormone was identified by (1) removing the structure that was thought to be an endocrine gland and noting the effects of its removal; (2) replacing the gland or treating the animal with a gland extract and noting the effect; and (3) attempting to isolate and purify the gland's secretion and to administer that substance to test animals. Early endocrinologists also carefully correlated disease symptoms with autopsies that showed the degeneration of possible endocrine glands. For example, in 1855 Thomas Addison described a set of symptoms—including poor appetite, low blood pressure, muscular weakness, and a characteristic skin discoloration—associated with the deterioration of the outer portion or cortex of the adrenal glands. This was the first of many clinical studies of the defects of endocrine glands, and the syndrome Addison described is still known as Addison's disease. With the discovery of insulin in 1922 by Banting, Best, and their colleagues, the door was opened for the treatment of a serious human hormonal disorder, and endocrinology was well on its way to becoming a modern science.

But thorough as they were, most of the early studies left many questions unanswered. It was simply impossible for test animals to survive the removal of certain organs; and not all hormone-producing tissues exist as well-defined, easily removable glands. Often it is hard to isolate hormones without changing them chemically and thereby altering their functions. And interactions between hormones complicate the problem of sorting out the actions of each particular hormone.

There are still many questions left to be answered in regard to hormonal regulation. How is the synthesis and release of hormones regulated? Why do hormones act on certain target cells, but not on other cells? What changes do hormones cause within target cells, and how do those

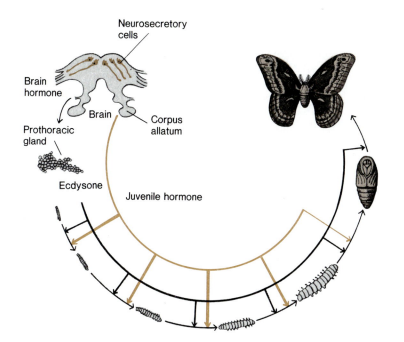

Figure 14.2 Hormonal control of molting and metamorphosis in a moth. Ecdysone and juvenile hormone together produce a larva-to-larva molt. In the last larval stage, the juvenile hormone level decreases, and pupation occurs.

changes cause the ultimate effects of the hormones? Before examining these questions further, it is necessary to identify some of the more important animal hormones.

Hormones in Invertebrates

One of the most important keys to the evolutionary success of any group of animals is the assurance that growth, maturation, and reproduction coincide with the most favorable seasons of the year so that the climate and food supply are optimal. It is not surprising, therefore, that hormones regulating growth and reproduction were probably the first hormones to appear during the course of animal evolution.

The first invertebrate hormones were likely neurosecretions, and most of the substances functioning as hormones in modern invertebrates are also neurosecretions. Only in a few of the more complex invertebrates (such as mollusks, arthropods, and echinoderms) have non-neurosecretory hormones been identified.

The principal hormones involved in the regulation of insect growth and maturation are undoubtedly the most thoroughly studied set of invertebrate hormones. In insects, as in all arthropods, neurosecretory hormones are important, but non-neurosecretory hormones also contribute to the regulation of insect development.

Hormones and Insect Development

Insects (and all other arthropods) have a rigid outer body covering called the **exoskeleton.** Because exoskeletons do not expand and thus do not allow for enlargement, insect growth is associated with the periodic shedding of the exoskeleton in a process known as **molting** (or **ecdysis**). Molting is followed by the expansion and hardening of a new, soft, larger exoskeleton that has developed underneath the old exoskeleton. Insect development therefore includes a series of molts and stages between molts, which are known as **instars.**

In insect development, **metamorphosis** is the change in body form that converts an immature individual into an adult (**imago**). Metamorphosis is either **complete,** involving **larval, pupal,** and **adult** stages, or **gradual,** involving a series of **nymphs.** (Nymphs generally resemble adults and become somewhat more adultlike at each molt until a fully developed adult emerges from the final molt.)

Several hormones regulate insect growth, molting, and metamorphosis; figure 14.2 summarizes the interactions of the principal hormones affecting the development of an insect undergoing complete metamorphosis. The neurosecretory cells of the brain produce a **brain hormone** that travels along axons to a structure called the corpus cardiacum, which is attached to the brain. From the corpus cardiacum, the

Figure 14.3 Experiments on the role of juvenile hormone in insect development. (a) Normal last (fifth) larval stage, pupa, and adult of the silkworm moth *Bombyx mori*. (b) Corpora allata, sources of juvenile hormone, were removed from this third larval stage. Instead of entering a larva-to-larva molt, it pupated and developed into a miniature adult. (c) Corpora allata from young larvae, which were still actively secreting juvenile hormone, were transplanted into a fifth larval stage. Instead of pupating, it entered a larva-to-larva molt and produced this large extra larval stage, which then pupated and developed into a giant adult.

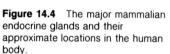

Figure 14.4 The major mammalian endocrine glands and their approximate locations in the human body.

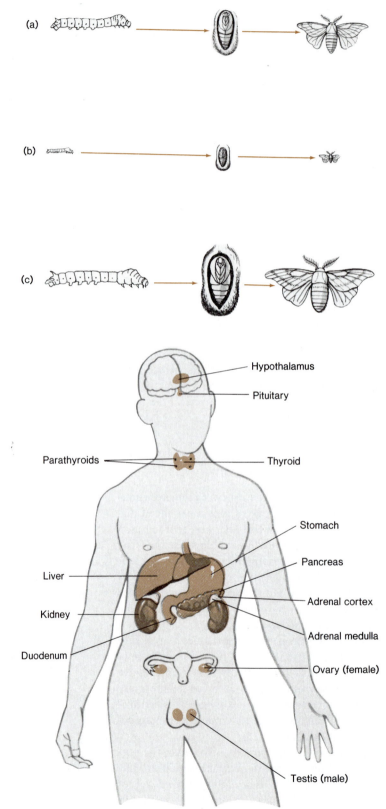

brain hormone is released into the blood and carried to the prothoracic gland located some distance from the brain. The brain hormone then stimulates the prothoracic gland to release its hormone, **ecdysone.** Ecdysone is also called the molting hormone or molt and maturation hormone because it promotes molting, and it also tends to spur the developmental process along toward pupation and adulthood. During the larval stages, the response to ecdysone is modified by the action of yet another hormone called **juvenile hormone.** Juvenile hormone does not alter the molt-promoting function of ecdysone, but it counters ecdysone's maturing effect. While juvenile hormone is secreted, the organism remains immature; larval molting leads simply to additional larval stages. But in the last larval stage, juvenile hormone secretion decreases dramatically, and the animal pupates and begins its transformation to the adult form.

It is interesting that the insect brain hormone, which is a neurosecretory hormone, is a peptide (a chain of amino acids) because hormones from the brain-pituitary complex of vertebrates, including humans, are also peptides. The insect hormone ecdysone is a steroid that is structurally similar to cholesterol, as are the steroid hormones of vertebrates. And juvenile hormone is a modified derivative of a fatty acid.

There have been many interesting experiments on the hormonal regulation of insect development. For example, removal of the corpora allata (the sources of juvenile hormone) from an early larval stage causes pupation and development of a miniature adult instead of the normal larva-to-larva molt (figure 14.3). Transplanting corpora allata from earlier larval stages into a final-stage larva induces an extra larva-to-larva molt. These giant larvae produce giant pupae and giant, abnormal adults.

The powerful effects of juvenile hormone on insect development make this hormone an interesting subject for insecticide research. If larval insects or the food they consume could be sprayed with juvenile hormone, the larvae might develop abnormally and die, or fail to reach an appropriate stage of their life cycle in time to survive the winter. Research on juvenile hormones as insecticides seems well worth the effort because the hormones are effective at very low concentrations and are naturally occurring, biodegradable compounds that should not pose the threat of cumulative toxicity that many insecticides do.

Because the juvenile hormones of different species differ, it might be possible to limit the effect to harmful insects without affecting many beneficial insects in the same environment. And it seems unlikely that the insects would develop resistance to juvenile hormone insecticides because their own development depends on sensitive, specific cellular responses to very small amounts of juvenile hormone.

From an evolutionary point of view, it is interesting that some plants produce compounds similar or identical to ecdysone or juvenile hormone. This apparently protects the plants by disrupting the development of insects that eat them.

Vertebrate Glands and Hormones

Although there has been a great deal of research on hormonal regulation in other vertebrates, especially in regard to hormonal control of reproduction, development, and growth, this discussion will concentrate on hormonal regulation in mammals. More is known about mammalian hormones than about the hormones of other vertebrates, and much of the study of mammalian hormones has focused on hormonal regulation in the human body.

Figure 14.4 is a diagram of a human body showing the locations of some of the major endocrine glands: the hypothalamus, the pituitary, the thyroid, the parathyroids, the pancreas, the kidneys, the adrenals, the ovaries, and the testes. Although they will not be discussed in this chapter, the pineal body and possibly the thymus, as well as some tissues in the stomach, intestine, and other organs all function as hormone sources in mammals. Because the total picture of mammalian hormonal regulation is very complex, only certain parts will be subject to closer scrutiny in this chapter.

Table 14.1 lists some of the major mammalian hormones along with their sources and the class of chemical compounds to which each belongs. The table omits some hormones that presently seem to be restricted in their effects, and some hormones, such as those involved in digestion, which are discussed in other chapters.

Table 14.1
Some Major Mammalian Hormones.

Source	Hormone	Class of Chemical Compound
Hypothalamus	Thyrotrophin-releasing hormone (TRH)	Peptide
	Corticotrophin (ACTH)-releasing factor (CRF)	Not determined (peptide?)
	Luteinizing hormone-releasing hormone/follicle stimulating hormone-releasing hormone (LH-RH/FSH-RH)	Peptide
	Somatostatin or growth hormone-release inhibiting hormone (GH-RIH)	Peptide
Hypothalamus, with storage in the posterior pituitary	Antidiuretic hormone (ADH) or vasopressin	Peptide
	Oxytocin	Peptide
Anterior pituitary	Growth hormone (GH) or somatotrophic hormone	Protein
	Thyroid-stimulating hormone (TSH) or thyrotrophin	Glycoprotein
	Adrenocorticotrophic hormone (ACTH)	Polypeptide (39 amino acids)
	Follicle-stimulating hormone (FSH)	Glycoprotein
	Luteinizing hormone (LH)	Glycoprotein
	Prolactin	Protein
Thyroid	Thyroxin and triiodothyronine	Iodinated amino acids
	Calcitonin	Polypeptide (32 amino acids)
Pancreas	Insulin	Polypeptide (51 amino acids)
	Glucagon	Polypeptide (29 amino acids)
Adrenal medulla	Epinephrine (adrenalin) and norepinephrine (noradrenalin)	Modified amino acids
Adrenal cortex	Glucocorticoids (cortisol, corticosterone, cortisone, and so on)	Steroids
	Mineralocorticoids (aldosterone and so on)	Steroids
Parathyroids	Parathyroid hormone	Polypeptide (32 amino acids)
Ovaries	Estrogens and progesterone	Steroids
Testes	Testosterone	Steroid

Figure 14.5 The location of the human pituitary gland below the brain. The posterior lobe has neural connections to the hypothalamus, but the anterior lobe does not. These two lobes are distinctly separate glands that differ structurally and functionally.

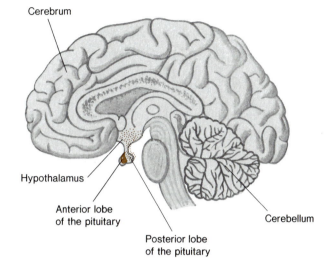

Cerebrum

Hypothalamus

Anterior lobe
of the pituitary

Posterior lobe
of the pituitary

Cerebellum

In considering mammalian hormones, three main topics will be explored: (1) the hormones of the pituitary gland and the relationship between the pituitary and the hypothalamus, (2) the hormones regulating metabolism, and (3) the hormones regulating ionic balance in the body. Grouping hormones in this way helps emphasize the role of hormonal regulation in the maintenance of homeostasis, or a stable internal environment, through the coordination of specialized body functions.

The Pituitary Gland

The pituitary gland, which in humans is about the size of a garden pea, is located below the brain and suspended on a stalk extending down from a part of the brain called the hypothalamus (figure 14.5). For many years, the pituitary was known as the "master gland" among the endocrine glands of the body. While this is true to the extent that pituitary hormones do regulate several other endocrine glands and their secretions, the pituitary does not function in a simple master-slave relationship with other glands.

The pituitary (also known as the hypophysis) is composed of two major regions, a **posterior lobe** and an **anterior lobe,** and, in some vertebrates, the **intermediate lobe,** which lies between the posterior and anterior lobes. The tissue of the posterior lobe resembles neural tissue, while the anterior lobe consists of specialized secretory cells. Although these two lobes lie close together in what appears to be a single structure,

they are in reality two distinctly different glands that secrete different hormones and that have different functions.

The Posterior Lobe

In early experiments on the posterior lobe of the pituitary, researchers prepared extracts of the lobe and injected them into test mammals. From these experiments they learned that an extract of the posterior lobe has four major effects on an experimental animal: blood pressure elevation, water retention by the kidneys, contraction of smooth muscles in the uterus, and ejection of milk from the mammary glands through contractions of the smooth muscles in the milk ducts. All of these effects strongly suggest that the posterior lobe is the source of one or several hormones.

After considerable research, it was established that the first two effects are caused mainly by **antidiuretic hormone** or **ADH,** also called **vasopressin.** The second two effects, involving the uterus and the mammary glands are attributed to a second hormone called **oxytocin.** Because of its powerful effect on uterine muscles, oxytocin is thought to be involved in normal childbirth, although it does not seem to be absolutely essential to that process. Nevertheless, oxytocin is a very reliable means of inducing labor and childbirth, for example, when a pregnancy has continued long past term. Oxytocin secretion is also a part of the reflex response of female mammals to suckling that results in milk being "let down" or released. The release of oxytocin in response to suckling is a good example of a **neuroendocrine**

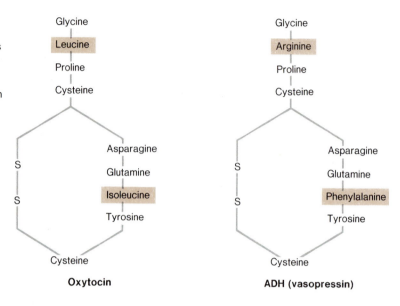

Figure 14.6 Structural formulas of oxytocin and ADH (vasopressin). Note that the amino acid sequences of the two hormones differ by only two amino acids (indicated by color shading). Each chain contains eight amino acids, counting cystine (which is cysteine-s-s-cysteine) as one amino acid. Vincent du Vigneaud received the Nobel Prize for chemistry in 1955 for his work in determining the structure of these molecules and synthesizing them in the laboratory; that was the first synthesis of a peptide hormone.

reflex. In such a reflex, sensory information received by the nervous system leads to the secretion of a hormone by an endocrine organ.

Both ADH and oxytocin are small peptides containing eight amino acids. It is not surprising that the two hormones overlap somewhat in function, because they are structurally very similar. Six of the eight amino acids in ADH and oxytocin are the same (figure 14.6). These two peptide hormones were originally thought to be synthesized and released by the posterior lobe of the pituitary, but more recent studies have clearly demonstrated that they are products of neurosecretory cells in the hypothalamus, and that they travel down axons to the posterior lobe, where they are stored until their release (figure 14.7). Thus, ADH and oxytocin are actually products of brain cells, and the posterior lobe of the pituitary is functionally similar to the insect corpus cardiacum in that both serve as release sites for hormones synthesized by neurosecretory cells of the brain.

The Anterior Lobe

The hormones of the anterior lobe have a wide range of effects on many processes in developing and adult mammals. Perhaps, however, the effects of **growth hormone** are best known because they are so obvious—and even dramatic, in some cases. If the secretion of growth hormone is insufficient during childhood and adolescence, the individual becomes a pituitary midget (figure 14.8), while excess secretion during the same period produces a pituitary giant. The oversecretion of growth hormone in an adult causes the toes, fingers, and face to resume growth. This results in a condition called **acromegaly,** which is characterized by distorted facial features, including an enlarged jaw and thick, heavy eyebrow ridges.

Growth hormone is also known as **somatotrophic hormone,** or **STH.** The suffix *trophic* comes from a Greek word meaning to feed or cause to grow, and is usually used to describe hormones that stimulate their target organs to secrete other hormones.[1] Two other trophic hormones from the anterior lobe are **thyrotrophic hormone,** also commonly called **thyroid stimulating hormone (TSH),** and **adrenocorticotrophic hormone (ACTH).** TSH stimulates the thyroid gland to produce its hormones, the major one thyroxin, while ACTH stimulates the adrenal cortex to produce a group of hormones known as the glucocorticoids. Proper regulation of the secretion of these hormones is vital to the normal metabolic functioning of the organism.

1. Many biologists instead use the suffix *tropic,* from a Greek word meaning "a turning," because these hormones give directions to their target organs.

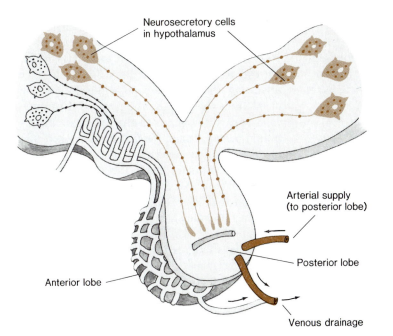

Figure 14.7 Relationship between the hypothalamus and the posterior lobe of the pituitary. Neurosecretory cells in the hypothalamus synthesize ADH and oxytocin. The hormones are transported down axons to the posterior lobe and released into the bloodstream there.

Neurosecretory cells in hypothalamus

Arterial supply (to posterior lobe)

Posterior lobe

Anterior lobe

Venous drainage

Figure 14.8 Result of a childhood growth hormone deficiency. "General" Tom Thumb (Charles Stratton) grew to be only about 84 cm tall. Tom Thumb, like other pituitary midgets, had normal body proportions despite his small size. He is shown here with P. T. Barnum, the showman who exhibited him.

Other trophic hormones from the anterior pituitary include the gonadotrophic hormones, primarily **follicle stimulating hormone (FSH)** and **luteinizing hormone (LH)**. FSH causes a female's eggs to mature by acting on the follicles in the ovaries. It also stimulates the ovaries to produce the female hormone **estrogen.** Luteinizing hormone causes the ovarian follicles to rupture, releasing eggs in a process called ovulation. LH also causes the ruptured, post-ovulatory follicle to change into a structure called the **corpus luteum,** which is the source of the hormone **progesterone.** In the male, FSH stimulates sperm production in the tubules of the testis, and LH stimulates other testis cells to produce the hormone **testosterone,** which is responsible for the development of a male's secondary sex characteristics.

The last anterior lobe hormone to be discussed here is **prolactin.** Interestingly, prolactin is produced in the anterior lobes of both female and male mammals, but its function is clearly understood only in females. Prolactin (along with other hormones such as thyroxin and ACTH) acts on the mammary glands, causing them to produce milk after they have been prepared to do so by estrogen and progesterone. The roles of prolactin and oxytocin differ in that the former is necessary for milk production, while the latter is necessary for milk ejection from the glands. More than thirty functions have been proposed for prolactin in male vertebrates, including the regulation of mating behavior, but these proposed functions, for the most part, are yet to be proven.

The Intermediate Lobe

Not all vertebrates have a distinguishable, functional intermediate pituitary lobe. Adult humans, for example, lack this third lobe. But in the vertebrates that do have it, the intermediate lobe is the primary source of a polypeptide hormone called **melanocyte stimulating hormone (MSH),** which is very similar in structure to ACTH. MSH causes the dispersal of pigment granules in the skin's pigment cells (or melanocytes), thereby causing the skin to darken (figure 14.9). This action of MSH is especially important in amphibians that are able to lighten and darken in response to changes in the color of their surroundings.

The Hypothalamus and the Pituitary

For years, certain facts have indicated that there might be influences from the brain on the pituitary. For example, it has long been known that stress can cause the pituitary to secrete more ACTH, thereby causing stimulation of the adrenal glands. But certainly the pituitary itself does not perceive and respond to stress. Similarly, in some species, the mere sight of a potential mate stimulates the release of pituitary gonadotrophins, which in turn stimulates the production of sex hormones by the gonads. Again, how does this happen?

Evidence points to the brain, and specifically to the hypothalamus, as the mediator between environmental stimuli and the pituitary. Thus, these responses are additional examples of neuroendocrine reflexes. The close relationship between the hypothalamus and the posterior pituitary has been established; the two structures are connected by neurosecretory fibers. But years of careful anatomical studies failed to show similar nervous connections between the hypothalamus and the anterior pituitary. However, in 1936, researchers discovered blood vessels extending from the hypothalamus through the pituitary stalk to the anterior portion of the pituitary (figure 14.10). This special network of capillaries and vessels has since been named the **hypothalamic-pituitary portal system.**

A portal system is a circulatory arrangement that begins in a capillary bed and leads not to vessels that carry blood directly toward the heart, but rather to a second capillary bed. Only after traversing this second capillary bed does the blood from the portal system proceed through veins toward the heart. In the hypothalamic-pituitary portal system, the first capillary bed is in the hypothalamus, and it is connected by vessels to a second capillary bed in the anterior lobe.

In 1945, G. W. Harris of Oxford University proposed that neurosecretory cells in the hypothalamus liberate substances into the blood vessels of the portal system and that these substances stimulate the anterior pituitary to secrete its hormones. The hypothetical substances were called **factors.** In 1955, scientists reported that extracts of the hypothalamus could, indeed, stimulate the anterior pituitary to secrete ACTH. This action was attributed to a small peptide called **corticotrophin releasing factor (CRF).** Since then, this hormone's activity has been well demonstrated. CRF seems to be one of at least nine different releasing and inhibiting factors from the hypothalamus that are themselves neurosecretions and that act upon the anterior pituitary via the portal system.

The effort to identify and synthesize the different factors from the hypothalamus that act upon the anterior pituitary has been led by two teams of researchers, although many other investigators have contributed significantly to the work. The leaders of these teams, Roger Guillemin and Andrew Schally, shared the Nobel Prize in 1977 for the isolation and characterization of hypothalamic factors with Rosalyn Yalow, who developed analytical techniques that have dramatically changed hormone research and diagnosis. (Incidentally, once a factor has been identified and synthesized, researchers prefer to call it a hormone.)

Thyrotrophin releasing hormone (TRH) was the first of the releasing factors from the hypothalamus to be isolated and purified. TRH is a small peptide composed of only three amino acids: histidine, proline and glutamic acid. It causes the anterior pituitary to secrete thyroid stimulating hormone (TSH), which, in turn, acts upon the thyroid to stimulate the release of thyroxin. TRH was isolated in 1969 by Schally's research team using pig hypothalami and by Guillemin's team using sheep brains. As has been the case throughout the history of endocrinology,

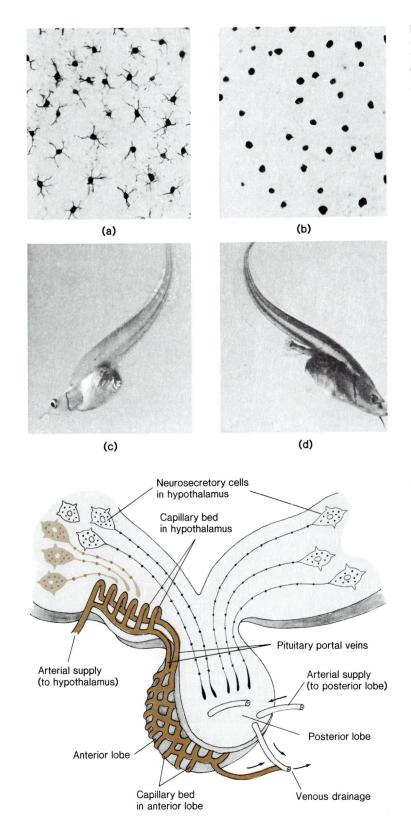

(a)

(b)

(c)

(d)

Figure 14.9 The effects of melanocyte stimulating hormone (MSH) in amphibians. (a) Frog *(Rana pipiens)* pigment cells stimulated by MSH. Pigment spreads inside the cells making the skin appear darker. (Magnification × 228) (b) In the absence of stimulation, pigment contracts into small dots inside cells. This makes the skin appear lighter. (Magnification × 228) (c) Larva of the clawed frog *Xenopus laevis* from which the pituitary gland has been removed. Note the absence of dark pigmentation. The pituitary is necessary for normal development of pigment cells. (d) Control *Xenopus* larva with its pituitary intact.

Neurosecretory cells
in hypothalamus

Capillary bed
in hypothalamus

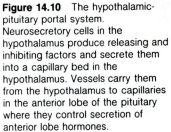

Pituitary portal veins

Arterial supply
(to hypothalamus)

Arterial supply
(to posterior lobe)

Posterior lobe

Anterior lobe

Capillary bed
in anterior lobe

Venous drainage

Figure 14.10 The hypothalamic-pituitary portal system. Neurosecretory cells in the hypothalamus produce releasing and inhibiting factors and secrete them into a capillary bed in the hypothalamus. Vessels carry them from the hypothalamus to capillaries in the anterior lobe of the pituitary where they control secretion of anterior lobe hormones.

this research was no easy task; Guillemin collected 500 tons of sheep brains to remove seven tons of hypothalami, and his team put in four years of intense work, to produce a single milligram of TRH!

In 1971, both Guillemin's and Schally's research groups determined the structure of a second releasing hormone—**luteinizing hormone-releasing hormone (LH-RH or LRF).** LH-RH is composed of a sequence of ten amino acids. As its name implies, it stimulates the anterior pituitary to release luteinizing hormone, the gonadotrophin necessary for ovulation.

For some time it was thought that there was a second gonadotrophin-releasing hormone, FSH-releasing hormone, but many biologists now believe that FSH-RH may be the same compound as LH-RH. Consequently, the proposed single hypothalamic hormone is sometimes designated by the name **LH-RH/FSH-RH.** There is tremendous interest in this releasing hormone because inhibiting its production or preventing it from acting can prevent ovulation in experimental animals. Thus, the potential seems to exist for developing an entirely new method of birth control.

Research on the regulation of growth hormone secretion has revealed a somewhat different pattern of control. In this case, there seem to be factors both stimulating and inhibiting the release of GH. Although the existence of a GH-releasing factor in the hypothalamus has been demonstrated, GH-RF has not yet been purified and isolated for study. However, a growth hormone-inhibiting hormone was found by Guillemin and colleagues in 1973 and by Schally and his group in 1976. The fourteen amino acid molecule, known as **growth hormone-release inhibiting hormone (GH-RIH)** or **somatostatin,** prevents the release of growth hormone from the pituitary.

Two other pituitary hormones, prolactin and melanocyte-stimulating hormone, also seem to be under this kind of dual control. Various studies have indicated the presence of both a prolactin-releasing factor (PRF) and a prolactin release-inhibiting factor (PRIF) in the hypothalamus. However, the chemical structures of the substances have not yet been determined. Because prolactin controls milk production and perhaps other aspects of reproduction, isolation and synthesis of PRF and PRIF could be useful in human and veterinary medicine.

The regulatory mechanisms involved in the hormonal control system thus operate at at least three levels of control: in the hypothalamus, in the pituitary, and in the other endocrine glands (figure 14.11).

Hormones Controlling Metabolism

Metabolism can be defined as the sum total of all the biochemical processes that take place within an organism, and the **metabolic rate** as the rate at which these reactions proceed. The individual rates of many of the reactions that contribute to the overall metabolic rate are affected by various hormones. For example, hormones affect the supplies of substrates (nutrients) that reach individual cells by controlling concentrations of substances such as glucose in the blood and tissue fluids of the body. Some hormones affect the rates at which substrates enter cells, while others affect reactions inside cells, either by controlling the synthesis of enzymes that catalyze the reactions or by affecting reaction rates directly. All of these mechanisms operate simultaneously and many interact with one another. Obviously, the hormonal regulation of metabolism is very complex.

Thyroid Hormones

Thyroid hormones, which are produced by the thyroid gland located at the base of the neck (see figure 14.4), have long been recognized as the principal hormones controlling metabolic rate. There are two major thyroid hormones, **thyroxin (T_4)** and **triiodothyronine (T_3).** Both molecules are synthesized from the amino acid tyrosine, and they differ only in that thyroxin contains four iodine atoms while triiodothyronine contains only three (figure 14.12). This difference is far from trivial, however. Although the two molecules have the same kind of effect on target cells, T_3 (which is secreted in much smaller quantities) is actually three to five times as active as T_4.

Both T_3 and T_4 are stored in the thyroid gland attached to large glycoprotein molecules until they are released into the bloodstream. These iodine-containing (iodinated) molecules are unique in the vertebrate body; no other iodinated compounds are known to be produced by vertebrates, or by any other animals, for that matter.

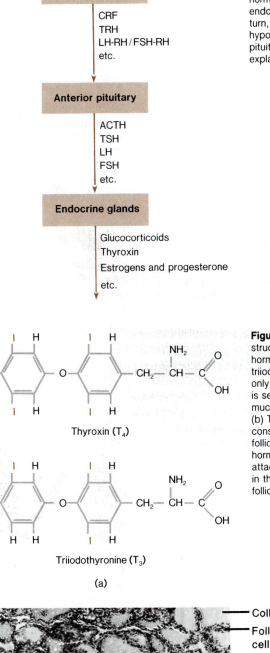

Hypothalamus

CRF
TRH
LH-RH/FSH-RH
etc.

Anterior pituitary

ACTH
TSH
LH
FSH
etc.

Endocrine glands

Glucocorticoids
Thyroxin
Estrogens and progesterone
etc.

Figure 14.11 Levels of control in mammalian hormonal systems. The hormones secreted by the various endocrine glands exert control, in turn, over the activity of the hypothalamus and the anterior pituitary. This relationship is explained further in figure 14.14.

Thyroxin (T_4)

Triiodothyronine (T_3)

(a)

Figure 14.12 The thyroid. (a) The structures of the two major thyroid hormones, thyroxin (T_4) and triiodothyronine (T_3). They differ by only one iodine atom, but T_3, which is secreted in smaller quantities, is much more active than T_4.
(b) Thyroid tissue. The thyroid consists of follicles, hollow balls of follicle cells, which secrete thyroid hormones. The hormones are stored attached to glycoprotein molecules in the colloid that fills the inside of follicles.

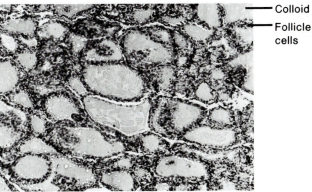

Colloid

Follicle cells

(b)

(a)

(b)

(c)

(d)

Figure 14.13 Metamorphosis in amphibians is a dramatic developmental process that is controlled by thyroid hormone. Stages in the metamorphosis of a tree frog are pictured here. Tadpoles (a) gain legs (b and c), lose their tails (d), and undergo other changes, including digestive tract reorganization, all in preparation for life as adult frogs. But if larval thyroids are inhibited, metamorphosis is prevented. The tadpoles continue to grow without making any of these changes. Experimentally, growth of thyroid-inhibited tadpoles has continued until they are several times normal size.

The thyroid hormones powerfully affect all body cells. T_4 and T_3 increase the metabolism and oxygen consumption of cells throughout the body. They increase rates of protein synthesis, including the synthesis of enzymes that catalyze metabolic reactions, and they stimulate carbohydrate metabolism by causing increased carbohydrate absorption and oxidation by cells.

Thyroid hormones are also instrumental in promoting normal growth and development largely through the stimulation of normal protein synthesis in developing vertebrates (figure 14.13). In humans, thyroid deficiency during infancy and childhood causes a dwarfed condition known as **cretinism,** in which the victim never matures sexually and suffers severe mental retardation. Fortunately, cretinism can be prevented by administering hormones to children with thyroid deficiencies.

As was previously mentioned, TSH from the pituitary's anterior lobe controls the production of thyroid hormones. TSH promotes the uptake of iodine by the thyroid, controls the rate of synthesis of the thyroid hormones, and regulates their storage and release into the bloodstream.

But the rate of TSH production also must be regulated so that only normal, adequate stimulation of the thyroid is achieved. TSH secretion is affected by the level of thyroid hormones in the blood. When thyroid hormone concentrations in the blood increase, TSH secretion temporarily slows down; when thyroid hormone concentrations in the blood decrease, TSH secretion temporarily increases. This relationship in which a substance (in this case, thyroid hormone) negatively affects the secretion of another substance that regulates its own production (TSH) is called a **negative feedback control mechanism** (figure 14.14). Many such feedback mechanisms control the secretion of various hormones.

The feedback control relationship between thyroid and pituitary secretion is mediated by the hypothalamus. Increased levels of T_4 and T_3 in the blood inhibit TRF production by the hypothalamus, while decreased T_4 and T_3 levels reduce inhibitions to TRF production and release, thereby causing increased TSH secretion.

The delicately balanced feedback control of thyroid secretion prevents excess thyroid secretion, (hyperthyroidism) which causes weight loss, elevated body temperature, profuse perspiration, and nervous irritability. On the other hand, thyroid deficiency (hypothyroidism) leads to a lowered metabolic rate, obesity, physical sluggishness, mental dullness, and abnormal skin and hair conditions.

Clear evidence that the normal feedback mechanism has gone awry is seen when a condition called **goiter** develops. Goiter is a tremendous expansion of the thyroid gland that is externally visible as a swelling at the base of the neck (figure 14.15). Low thyroid secretion rates, caused by hormone synthesis problems or by a deficiency in dietary iodine, result in increased TSH secretion through the normal feedback relationship. The excess growth of the thyroid tissue, producing a goiter, occurs in response to continuing TSH stimulation.

Hypothyroidism and goiter are rare in humans in seacoast areas because dietary iodine is usually adequate when a considerable amount of seafood is included in the diet. But chronic thyroid deficiency problems can develop inland, where dietary iodine levels are low. Fortunately, this problem can be prevented by adding very small quantities of iodine to table salt used in low-iodine areas.

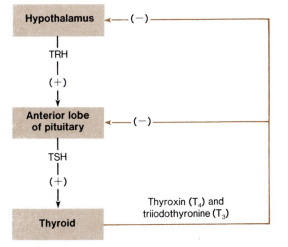

Figure 14.14 Negative feedback control mechanism that regulates secretion of the thyroid hormones. Increased thyroxine (T_4) and triiodothyronine (T_3) cause decreased TRH release from the hypothalamus and inhibit TSH release from the anterior lobe. When T_4 and T_3 in the blood decrease, inhibition of the hypothalamus and anterior lobe is relieved. More TSH is released, and the thyroid is stimulated to release more T_4 and T_3. Thus, the negative feedback system normally balances the secretions of all elements and keeps the concentration of T_4 and T_3 in the blood within the proper range. Negative feedback control relationships also function in regulation of secretion of many other hormones.

Figure 14.15 A woman suffering from goiter, an enlargement of the thyroid gland that often accompanies hypothyroidism. Continued TSH stimulation causes massive overgrowth of thyroid tissue.

Pancreatic Hormones

The control of blood sugar levels is one of the most important aspects of the hormonal regulation of metabolism. A readily available supply of energy is needed for all tissues in the organism, but some tissues, particularly the brain, have a high demand for glucose and no way to store it for any length of time. The task of storing glucose and making it available to other body tissues falls primarily to the liver, and, in large part, glucose storage and blood sugar levels are regulated by two major hormones produced in the pancreas, **insulin** and **glucagon.**

The cells producing insulin and glucagon are located in the **islets of Langerhans,** which are discrete patches of tissue scattered among the portions of the pancreas that secrete digestive enzymes (figure 14.16). In the human pancreas, the islets make up less than 2 percent of the tissue, and yet a human pancreas contains more than a million islets. The cells of the islets are glucose sensitive—they respond quickly to changes in blood glucose concentration. When glucose is abundant in the blood, insulin secretion is stimulated. When there is a decrease in blood glucose, glucagon secretion is stimulated.

To maintain a steady supply of sugar to body tissues, the liver stores glucose in the form of glycogen, a glucose polymer. After a meal, when blood glucose concentrations rise rapidly as sugar is absorbed from the digestive tract, the liver conserves some of the glucose by producing glycogen. The reactions involved in glycogen production are collectively known as **glycogenesis** ("glycogen formation"). Later, glycogen can be broken down or hydrolyzed to glucose molecules, which are gradually released into the bloodstream. Glycogen hydrolysis is called **glycogenolysis.** However, at times even the glycogen reserves in the liver may be depleted. In that event, more glucose can be manufactured from other substances, most notably by using the carbon chains of certain amino acids, in a set of processes known as **gluconeogenesis** ("new glucose formation").

Glucagon is a polypeptide composed of twenty-nine amino acids. It promotes glycogenolysis and stimulates gluconeogenesis in the liver. Thus, glucagon helps ensure a steady supply of glucose in the liver, causes the conversion of glycogen to glucose, and raises blood sugar levels.

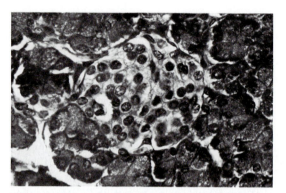

Figure 14.16 An islet of Langerhans in a section of pancreas tissue. The islets are separate areas of endocrine tissue surrounded by the tissue that secretes the digestive enzymes of the pancreas.

Insulin, a polypeptide composed of fifty-one amino acids, lowers blood sugar levels, mainly by promoting the uptake of glucose by body cells. Apparently, insulin acts on cell membranes to allow the faster, more efficient transport of glucose into the cell. In the presence of insulin, glucose enters the cell about twenty-five times faster than it does without the hormone. Insulin also promotes glycogen synthesis in the liver and muscle cells and depresses glycogenolysis and gluconeogenesis in the liver. Moreover, it inhibits the breakdown of fats and proteins for use in cellular oxidation, thereby promoting the use of glucose in cells' energy-yielding reactions. The net effect of all these actions is lowered blood glucose levels.

A deficiency in insulin secretion results in the condition called **diabetes mellitus.** In diabetes mellitus, glycogen synthesis in the liver and in muscles decreases, and the liver produces too much new glucose. At the same time, the insulin deficiency causes inefficient glucose absorption and utilization in the body cells. As a result, blood glucose levels rise so high that glucose is excreted in the urine. At the same time, more water is lost and the diabetic becomes dehydrated. Despite blood glucose excesses, body cells cannot meet their energy needs, and they therefore metabolize proteins and fats. These metabolic alterations quickly lead to a general weakness and susceptibility to disease.

Interest in insulin, and secondarily in glucagon, has been great ever since the discovery early in this century that diabetes mellitus is due primarily to insufficient insulin secretion by the pancreas. Now that the structure of the insulin molecule is understood and synthetic insulins can be manufactured, research and success in the treatment of diabetes are progressing rapidly. However, prolonged insulin therapy itself may cause problems for diabetes mellitus patients. Much further research is needed before the condition can be controlled without serious side effects.

In summary, glucagon and insulin are hormones that oppose and balance one another. Insulin promotes glycogen synthesis and stimulates the body cells to increase their glucose uptake, thus lowering blood sugar levels. Glucagon, on the other hand, promotes glycogenolysis and gluconeogenesis in the liver, thus raising blood sugar levels. A normal, adequate supply of glucose for body cells and tissues depends to a large extent on this antagonistic relationship between insulin and glucagon.

Adrenal Hormones

Some of the adrenal gland hormones also affect blood glucose levels and metabolism in general. The adrenal glands, one of which is located above each kidney, are really two glands in one. The two parts of an adrenal gland, the inner **medulla** and the outer **cortex,** are actually derived from different tissues in the embryo, and even though the two types of tissue grow together, they remain functionally distinct (figure 14.17).

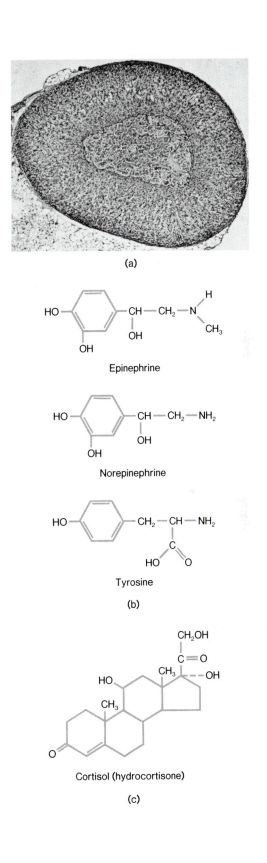

Figure 14.17 The adrenal gland. (a) Cross section of an adrenal gland of a mouse showing the inner medulla and outer cortex. Human adrenal glands are more flattened than this mouse adrenal because they lie against the surface of the kidney (see figure 14.4). (b) Structural formulas of the two hormones of the adrenal medulla, epinephrine (adrenalin) and norepinephrine (noradrenalin), and the amino acid tyrosine from which they are derived. (c) Cortisol (hydrocortisone), the principal glucocorticoid. Most of the carbons and hydrogens in the molecule are not shown because the adrenal cortical steriods share a common skeleton and differ only in side chains. But these side chain differences can result in very different functional properties (for example, compare the structure of cortisol with aldosterone, a mineralocorticoid shown in figure 14.19).

Table 14.2
Physiological Effects of Epinephrine Secretion.

Increased blood sugar level
Increased heart rate
Increased force of cardiac contractions
Increased respiratory rate
Increased blood flow to brain
Increased blood flow to liver and heart
Increased blood flow to skeletal muscle
Decreased blood flow to digestive tract
Decreased digestive function
Dilation of pupil of the eye
Dilation of bronchioles of the respiratory tract
Increased blood pressure

The adrenal medulla secretes the hormones **epinephrine** and **norepinephrine** (also called adrenalin and noradrenalin). The activities of the two hormones are similar, though not identical. The secretion of epinephrine causes an increase in blood sugar levels by stimulating liver glycogenolysis. Epinephrine also has many other effects not directly related to blood sugar (table 14.2). In general, epinephrine is secreted in response to stressful situations and prepares the organism physiologically to cope with such circumstances. The effects of the hormone are similar to the effects of stimulating the sympathetic nervous system, which also becomes highly active in emergency situations. This is perhaps not surprising, since the tissue from which the adrenal medulla develops originates from the embryonic nervous system.

The cortex of the adrenal gland secretes numerous hormones—perhaps as many as fifty different compounds—and all of them are steroids. Some of these steroids, the **glucocorticoids,** are involved in the regulation of blood sugar. The principal glucocorticoid is **cortisol (hydrocortisone).** The glucocorticoids raise blood sugar by inhibiting glucose uptake in various body cells, especially muscle and fat cells, and thus counteracting insulin. They also promote the release of amino acids from muscle and fatty acids and the release of glycerol from fat cells to be subsequently used for gluconeogenesis in the liver, and they stimulate the liver enzymes involved in gluconeogenesis. All of this tends to raise blood sugar levels.

Glucocorticoids are also involved in the body's adaptation to stress situations that continue longer than those met by the adrenal medulla and the sympathetic nervous system.

Unfortunately, if glucocorticoid secretion is increased for too long, it can cause gastrointestinal ulcers, brittle bones, high blood pressure, kidney problems, and increased susceptibility to certain diseases and infections. Clearly, although there are hormonal mechanisms that aid the body's adjustments to stress, these adjustments have definite limits. Similar limits apply to the use of glucocorticoids such as cortisone in the treatment of arthritis and other inflammatory diseases.

The control of blood sugar levels is actually even more complicated than has been indicated here and may involve, at least in a minor way, several more hormones. The complexity of blood sugar regulation indicates the importance of proper blood sugar balance for the overall well-being of the organism, particularly the brain.

Figure 14.18 summarizes the major controls governing blood sugar levels and glucose use by all body cells. Although many feedback mechanisms and other interconnecting loops have been omitted from the diagram, it is apparent that several pathways have evolved for the metabolic control of sugars. The controls for the metabolism of proteins and fats are intertwined with the controls for sugar metabolism. Imbalance in any one aspect of this finely tuned system for control of metabolic homeostasis can affect the whole organism.

Hormones Controlling Fluid and Ionic Balance

Animal blood and tissue fluids are essentially dilute salt solutions containing mostly sodium and chloride ions with smaller amounts of calcium, potassium, phosphate, magnesium, and other ions. Concentrations of these ions are kept under strict control; if the concentration of one or more exceeds or falls below the normal range, adverse physiological reactions occur, and death may result. Several hormones are involved in maintaining **ionic balance** in the body cells and in the tissue fluids bathing them.

The primary hormone that promotes sodium ion reabsorption is **aldosterone** (figure 14.19). Aldosterone is one of the steroid hormones of the adrenal cortex. It is not a glucocorticoid, but rather belongs to a second class of steroid hormones, the **mineralocorticoids.** As their name implies, the mineralocorticoids are involved in mineral regulation in the body.

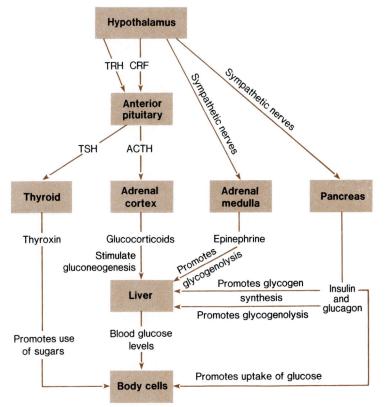

Figure 14.18 Some of the factors that control glucose supply to body cells, and the uptake and oxidation of glucose by cells.

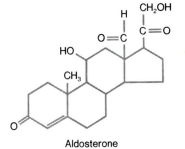

Aldosterone

Figure 14.19 Aldosterone, the principal member of the class of adrenal cortical hormones called mineralocorticoids, which are involved in ionic regulation in the body. Compare the structure of aldosterone with that of cortisol, a glucocorticoid, which is shown in figure 14.17c.

Control of Calcium Balance

Calcium is especially important to the functions of nerve and muscle tissue. Death due to calcium imbalance can occur very quickly as the result of a massive nervous disruption or disturbance of heart muscle contraction. Consequently, calcium levels in blood are closely regulated within a very narrow range. The **parathyroid glands** play a major role in this regulation.

Early endocrinologists assumed that the parathyroids were closely related to the thyroid in function because these small glands, usually four of them in humans, are located on the surface of the thyroid (see figure 14.4). Later they discovered that careful removal of the parathyroids from experimental animals caused very specific effects, including the lowering of blood calcium concentration and the disturbance of nerve and muscle function, which were different from the effects caused by thyroid removal. Further research established that the parathyroids produce a polypeptide hormone called **parathyroid hormone.** This hormone's major effect is to raise blood calcium levels and lower blood phosphate levels.

Parathyroid hormone achieves this effect in several ways. First it acts upon the bone cells, causing them to remove both calcium ions and phosphate ions from the matrix of the bone and transfer them into the bloodstream. Of course, this raises both calcium and phosphate concentrations in the blood, so the phosphate-lowering effect of parathyroid hormone depends on its action at another site, the kidney. The hormone stimulates kidney tubule cells to excrete phosphate from and retain calcium in the body.

For many years parathyroid hormone was thought to be the only hormone controlling calcium balance in the body. Scientists believed that after parathyroid hormone had raised blood calcium levels sufficiently, the high blood calcium acted through a feedback mechanism to shut down its own secretion until calcium levels in the blood dropped again. This may be true in part, but it is not the entire story. In the early 1960s a second calcium-regulating hormone was discovered and named **calcitonin.** Calcitonin is produced by a special group of cells in the thyroid, the C cells, which stain differently from the regular thyroid cells in histological studies. In other vertebrates, calcitonin is produced by separate glands, but in mammals, these specialized cells are incorporated into the thyroid gland during embryonic development.

Calcitonin is an **antagonist** to parathyroid hormone; that is, it has the opposite physiological effect. This hormone lowers the blood calcium level by inhibiting the resorption of bone by bone cells, thus decreasing the release of calcium from the bone. Calcitonin secretion is stimulated by a rise in blood calcium, and it acts quickly to lower the blood calcium to a proper level.

Yet a third hormone is now known to affect calcium and phosphate levels in body fluids. This recently discovered hormone, **1,25-dihydroxyvitamin D_3,** also known as 1,25-dihydroxycalciferol$_3$, is a derivative of a common form of vitamin D. Vitamin D_3 or calciferol is either ingested in the diet or synthesized in the skin from a steroid precursor under the influence of ultraviolet radiation. Vitamin D_3 is itself a **prohormone**; that is, it is a molecule that can be converted into an active hormone form. The first step in converting it into a hormone occurs when liver cells change the vitamin to 25-hydroxyvitamin D_3 (figure 14.20). This substance is then transported to the kidney, where enzymes in the mitochondria of the kidney cells change it to the active hormone, 1,25-dihydroxyvitamin D_3 (symbolized 1,25-$(OH)_2D_3$).

The major effect of this hormone is to stimulate the cells of the intestine to absorb more calcium, thus raising blood calcium levels. It is also believed to assist parathyroid hormone in the mobilization of calcium from the bone. Finally, 1,25-$(OH)_2D_3$ has been shown to stimulate phosphate absorption in intestinal cells, thus increasing the blood's phosphate level as well. The discovery of the conversion of vitamin D_3 to a hormone has done much to clarify the role of vitamin D in mineral metabolism.

What stimulates the synthesis of 1,25-dihydroxyvitamin D_3? As was previously stated, low levels of calcium in the blood cause the secretion of parathyroid hormone. Parathyroid hormone, in turn, stimulates kidney cells to synthesize and release 1,25-$(OH)_2D_3$. Low blood phosphate levels also seem to stimulate the final step in the synthesis of this hormone from vitamin D_3. Thus, 1,25-$(OH)_2D_3$ is synthesized through yet another feedback mechanism in which calcium and phosphate blood levels cause the synthesis and release of a hormone that, in turn, governs those levels.

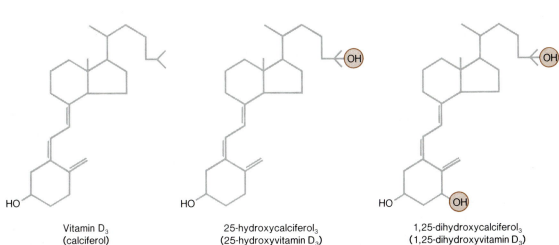

Vitamin D₃
(calciferol)

25-hydroxycalciferol₃
(25-hydroxyvitamin D₃)

1,25-dihydroxycalciferol₃
(1,25-dihydroxyvitamin D₃)

Figure 14.20 Steps in the conversion of vitamin D_3 into a molecule that is active as a hormone. The carbon and hydrogen skeleton of the molecule does not change during these steps so it is not shown in detail. Vitamin D_3, which is synthesized from precursors in the skin under the influence of ultraviolet radiation or ingested in the diet, is converted to 25-hydroxyvitamin D_3 in the liver and finally to 1,25-dihydroxyvitamin D_3 in the kidney. The latter is the active form of the hormone.

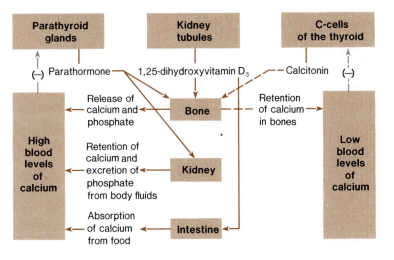

Figure 14.21 Feedback control diagram showing regulation of blood calcium and phosphate levels by parathyroid hormone, calcitonin, and 1,25-dihydroxyvitamin D_3.

Figure 14.21 summarizes the regulation of the secretion of parathyroid hormone, calcitonin, and 1,25-dihydroxyvitamin D_3.

Hormone Synthesis

Some of the most interesting recent discoveries in endocrinology have concerned the synthesis of polypeptide hormones. Polypeptide synthesis occurs in ribosome complexes on the rough endoplasmic reticulum of a cell. The coded information brought from DNA by messenger RNA directs the synthesis. The polypeptide produced could thus be called a **gene product** since it is made according to the specifications of the gene for that polypeptide. But when polypeptide hormone synthesis was studied using the sophisticated techniques of modern molecular biology, it soon became apparent that the gene product first produced in gland cells is *not* identical to the polypeptide hormone that the cells secrete. What are the differences between these gene products and the actual hormone molecule, and what is the significance of these differences?

Prohormones

A number of substances have been isolated and described as **prohormones,** or precursors in the biosynthesis of polypeptide hormones. Prohormones are large, physiologically inactive molecules that are converted to smaller, active

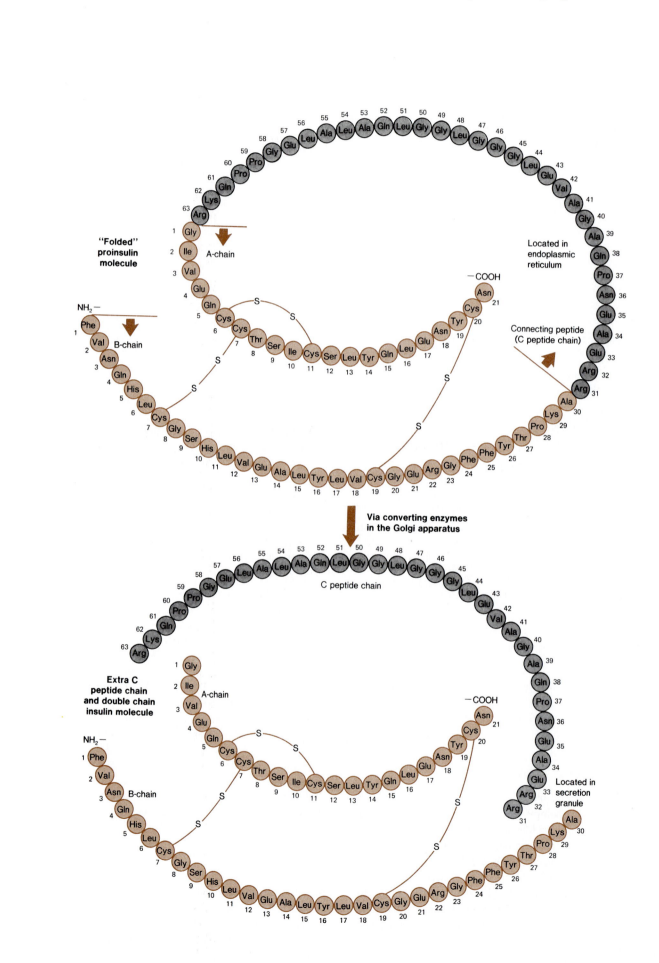

hormones through the enzymatic cleavage of specific peptide bonds. Anything affecting the enzyme activity that converts a prohormone to an active hormone exerts influence over the hormone synthesis and secretion of gland cells. Prohormones are relatively stable compounds, but they may exist for only fifteen or twenty minutes in intact cells before they are converted to active hormones.

A number of prohormones are now being studied, for example, the single-chain polypeptide **proinsulin,** which is larger than the insulin molecule itself. Proinsulin is converted to insulin in the Golgi regions of pancreas cells. The insulin is then stored in granules to await secretion (figure 14.22).

All propolypeptide hormones appear to be larger molecules than the hormones produced from them. **Proglucagon** is larger than glucagon. The **proparathyroid hormone** chain is several amino acids longer than the parathyroid hormone molecule. There are probably also *pro* forms for all of the anterior pituitary hormones, but they are especially difficult to extract and isolate because so many polypeptide hormones are produced in that gland.

Recently, even larger molecules have been found in biochemical studies of hormone synthesis involving cell-free extracts and messenger RNAs for specific hormones. These larger molecules are called **preprohormones.** The prepro-hormones are precursors of the prohormones, which are, in turn, precursors of hormones. Preprohormones may actually be the **initial gene products** in polypeptide hormone synthesis. Figure 14.23 outlines what is believed to be the synthesis process for an active hormone, although other intermediate steps may be possible.

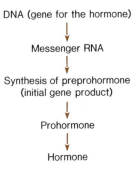

DNA (gene for the hormone)

↓

Messenger RNA

↓

Synthesis of preprohormone
(initial gene product)

↓

Prohormone

↓

Hormone

Figure 14.23 Steps in the synthesis of a polypeptide hormone. It is now thought that the preprohormone represents the direct product of gene expression and that the conversions to the prohormone and then to the active hormone forms are necessary subsequent steps. It is quite likely that each of these steps might be subject to complex control mechanisms.

Two preprohormones, **preproinsulin** and **pre-proparathyroid hormone,** are currently being studied intensely. Preproinsulin has been isolated in studies using rat insulin messenger RNA. The preproinsulin molecule is basically proinsulin with an extra 23 amino acids attached. Preparathyroid hormone has 115 amino acids, 25 more than the proparathyroid hormone molecule.

One of the main problems in preprohormone research is that preprohormones have not been found in studies using intact cells, only in cell-free biochemical systems. Preprohormones are probably extremely short-lived inside living cells; if they are converted rapidly to prohormones, they would be very difficult to extract and isolate from whole cells. Nonetheless, the concept of preprohormones as the initial gene product, or the first step in hormone synthesis, may prove vital in certain areas of genetic-engineering research. The goal of this research is to develop specialized microorganisms that can synthesize polypeptide hormones cheaply and in large quantities.

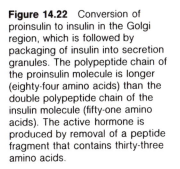

Figure 14.22 Conversion of proinsulin to insulin in the Golgi region, which is followed by packaging of insulin into secretion granules. The polypeptide chain of the proinsulin molecule is longer (eighty-four amino acids) than the double polypeptide chain of the insulin molecule (fifty-one amino acids). The active hormone is produced by removal of a peptide fragment that contains thirty-three amino acids.

Box 14.1
Peptides Everywhere:

The ''New'' Hormones

ADH and oxytocin, the peptide neurosecretory hormones that are released in the posterior pituitary, have been known to biologists for years. So has the hormone gastrin, which stimulates secretion by the glandular cells in the stomach. But in the last few years, sophisticated analytical techniques, many of them developed during the search for the hypothalamic releasing hormones, have revealed previously unknown hormones in the brain and other tissues. And perhaps an even more surprising discovery is that many familiar peptide hormones are produced in unexpected places.

For example, somatostatin, the growth hormone-release inhibiting hormone (GH-RIH), is a polypeptide composed of fourteen amino acids. It has been isolated from the hypothalamus, where it is synthesized, and from the anterior pituitary lobe, where it inhibits the release of growth hormone. But when somatostatin was applied to other tissues in experiments, it inhibited the secretion of other hormones as well; for example, somatostatin decreases glucagon and insulin secretion by the pancreas and gastrin secretion by the stomach. In fact, there are special cells in the pancreatic islets of Langerhans that actually secrete somatostatin, and the hormone has also been isolated from certain cells in the upper gastrointestinal tract. In both sites, somatostatin is probably involved in regulating hormone secretion.

An equally intriguing discovery concerns the hormone gastrin, which has long been known to regulate secretion of gastric juices in the stomach. Recently, a peptide very much like gastrin was isolated from brain tissue, but its function there is presently unknown.

Somatostatin and gastrin seem to be examples of hormones that ''plug in'' to different regulatory functions in different target tissues. Research on peptide hormones has also produced evidence of cases where two hormones do the same thing in the same place; in fact, some of the ''new'' peptide hormones overlap functions with other, more familiar hormones. For example, **somatomedin,** a peptide synthesized by the liver, causes many of the same effects as growth hormone, as well as some of the effects of insulin. Research on these apparent redundancies in hormonal regulation may help biologists to better understand both normal and abnormal endocrine functioning. For example, somatostatin and somatomedin may be the keys to discovery of new, more effective treatments for diabetes mellitus and other hormonally related metabolic disturbances.

Peptide research has also led biologists to new discoveries about the interaction of hormonal and nervous regulation. The powerful effects of opiates (opium, morphine, and related drugs) have been known for many years. Once the concept of receptor molecules was developed, it was

Hormone Receptors and Target Cells

How does a hormone interact with its target cells, and what determines which cells are target cells for a particular hormone and which cells are not? A target cell can be defined as a cell that has the appropriate receptors for a hormone and that responds physiologically to the hormone. A receptor is a molecule or molecular complex that is part of the target cell and that can specifically recognize a particular hormone and bind with it. Receptors for polypeptide and adrenal medulla hormones appear to be located on the cell surface,

that is, on the plasma membrane. Receptors for steroid hormones, on the other hand, appear to be located in the interior of the cell, in the cytoplasm. Because steroid hormones are relatively small molecules in contrast to polypeptide hormones, the steroids can enter the cell and even penetrate to the nucleus and bind with receptor molecules located inside the cell.

Nontarget cells presumably do not have the appropriate receptors for a particular hormone and, hence, do not respond to it. Different types of cells are assumed to have different hormone

hypothesized that the powerful pain-killing effects of the opiates must involve specific opiate receptors on cells in the nervous system. Such receptors were in fact detected, but this raised a new question: why would cells carry receptors for substances not normally found in the body? Perhaps, it was suggested, the opiates combine with receptors that normally bind naturally occurring regulatory molecules in the nervous system. Further research isolated small peptides called **enkephalins** (Greek for "in the head") that do bind to opiate receptors. Larger peptide molecules called **endorphins** ("internal morphine") that include the enkephalin peptide sequences were also found to bind with the opiate receptors.

The discovery of enkephalins, endorphins, and even larger polypeptides that include these smaller molecules within them has opened a new field of study in chemical regulation. It may now be possible to determine the basis for physical drug addiction. These peptides might even serve as nonaddictive pain killers. Nervous system disorders might also be treated with enkephalins, endorphins, or other peptide regulatory substances.

Science has barely scratched the surface in understanding the peptides that act as regulators throughout the body. The next few years should yield great progress and, very likely, more surprises.

hormone-receptor relationship can change with time and with changing conditions in the target cells. For example, in some target cells, as receptors for a particular hormone become occupied by hormone molecules, other receptors become less receptive to additional hormone molecules; that is, their affinity for the hormone decreases. This may be a homeostatic regulatory mechanism that controls excessive cell responses to hormonal stimulation when hormone concentrations are very high. In a sense, it is also a negative feedback mechanism because the cell responds to a hormone until a certain amount of the hormone has bonded, and then the cell's sensitivity to further stimulation decreases because the rest of the receptors do not bind so readily to additional hormone molecules. The total number of receptor molecules on target cell surfaces can also change with time. When the number of receptors decreases, a hormone seems to lose its effectiveness because target cells become less responsive to it.

A great deal of current research is devoted to the study of normal regulatory mechanisms and disease-related factors that affect a receptor's affinity for hormones, or that affect the number of receptor molecules per target cell. Another interesting research area has been opened by discoveries that the concentration of a particular hormone may affect not only the receptors for that hormone, but also those for other hormones. For example, when excessive amounts of thyroxin are present, it can cause a two- or three-fold increase in the number of receptors on heart cells for epinephrine and norepinephrine. This may help to explain why persons with excessive thyroid hormones can suffer the sort of heart palpitations that characteristically are caused by excess epinephrine and norepinephrine, even though these same patients do not produce excessive amounts of those substances.

Important as receptor-hormone binding is, it is only the first step in a chain of events that make up a target cell's reaction to a hormone. It is equally important to understand some of the other steps in these cellular response sequences and to consider some of the intracellular effects of hormones on their target cells.

receptors, and cells that respond to several hormones must have receptors for each one. An example of a type of cell that must have receptors for several different hormones is the liver cells that are involved in glucose storage and release (see figure 14.18).

The various interactions between hormones and receptors have several general characteristics. A receptor is very specific for the particular hormone that it binds and has a high affinity for it. A hormone binds very rapidly to its receptor, but the binding is usually reversible. Finally, the

Intracellular Actions of Hormones

Hormones exercise their effects on target cells in several major ways. Some hormones alter the transport of substances across the cell membrane. The prime example of this type of action is the hormone insulin, which increases the transport of glucose into cells by stimulating glucose transport systems in the plasma membrane. A second major mode of hormonal action is the alteration of enzyme activity to produce **cyclic AMP,** which can in turn alter numerous other enzyme systems inside the cell. Finally, a hormone can exert control by acting directly on the genome to stimulate the synthesis of messenger RNA, thus producing a particular protein or set of proteins as the final product.

Cyclic AMP, The Second Messenger

Cyclic AMP acts as a **second messenger** in responses of target cells to certain hormones. Earl W. Sutherland received the Nobel Prize in 1971 in recognition of his discovery of the importance of cyclic AMP in hormone activity. At the time, Sutherland was studying the effects of epinephrine on liver cells. Epinephrine stimulates the release of glucose into the blood by causing an increase in the rate at which glycogen is broken down in liver cells. It was known that the first step in the process is a cleavage of glycogen that yields glucose-1-phosphate and that this reaction is catalyzed by an enzyme called phosphorylase. The enzyme phosphorylase exists in two forms, an inactive and an active form. Epinephrine appeared to cause the conversion of inactive phosphorylase to active phosphorylase, thereby increasing the rate of glycogen breakdown and glucose release in liver cells.

However, there is a problem in interpreting the effect of epinephrine because although epinephrine binds to receptors on the surfaces of target cells, it does not enter the cytoplasm where phosphorylase is located. How, then, does this hormone exert its influence on the enzyme? Sutherland and his colleagues determined that epinephrine stimulation of liver cells causes an increase in the production of a small nucleotide, **cyclic adenosine monophosphate** (cAMP), in the cytoplasm. They demonstrated that this cyclic

AMP functions as a "second messenger" in stimulating phosphorylase activity. The hormone epinephrine is the first messenger because it brings a message to the plasma membrane of a target cell. There it stimulates production of the second messenger, cyclic AMP, which actually causes the response inside the target cell—in this case, activation of the enzyme phosphorylase.

Cyclic AMP is formed from ATP in a reaction catalyzed by the enzyme **adenylate cyclase.** In this reaction, ATP's two terminal phosphates are removed and the other phosphate group of the molecule is converted into a ring structure (figure 14.24). Adenylate cyclase is located in the plasma membrane, associated with epinephrine receptors that are situated in the membrane. As the hormone binds to the receptors, the activity of adenylate cyclase increases. The adenylate cyclase then produces more cyclic AMP from ATP, and the cyclic AMP stimulates the conversion of phosphorylase to its active form (actually through two intermediate, enzyme-catalyzed reactions). Activated phosphorylase then begins the process of glycogenolysis, finally resulting in the release of glucose molecules into the bloodstream, the ultimate effect of the hormone epinephrine (figure 14.25).

There is one final, critical stage in the cyclic AMP mechanism. No regulatory system can function efficiently if the regulator substance accumulates in excessive amounts; the regulator substance must be disposed of fairly quickly after it has had time to act. In the case of cyclic AMP, an enzyme called **phosphodiesterase** catalyzes the conversion of cyclic AMP to an inactive form of AMP. This reaction is just as important as the reaction catalyzed by adenylate cyclase involving cyclic AMP as a secondary messenger.

Sutherland also worked with the hormone glucagon, and other investigators have studied a series of hormones in relation to cyclic AMP (table 14.3). The result of all this research has been the further development of the concept of cyclic AMP as a second messenger. Hormone receptors, adenylate cyclase, and cyclic AMP are linked with the ultimate intracellular effects of a number of hormones. The enzymes activated by cyclic AMP and the end products are, of course, different for different hormones and their target cells, but cyclic AMP regulates several different processes in several kinds of cells.

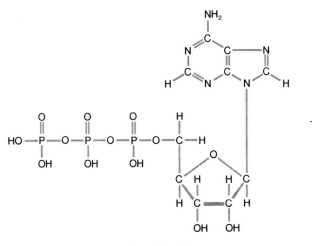
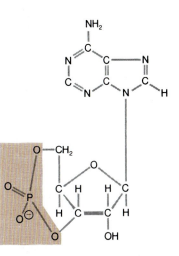

Adenosine triphosphate

Adenylate cyclase

Cyclic adenosine monophosphate

Figure 14.24 The enzyme adenylate cyclase catalyzes the production of cyclic AMP from ATP. Note that the phosphate group in the colored area is part of a ring structure with oxygens and carbons.

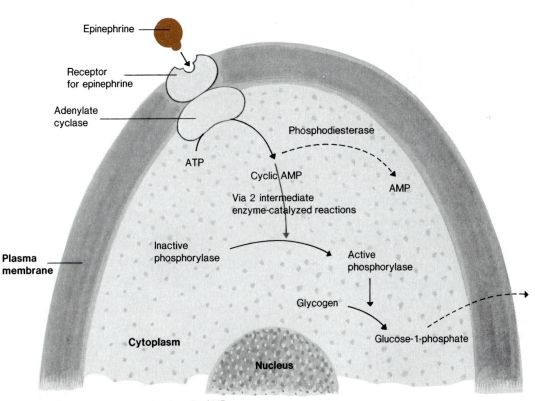

Figure 14.25 A model of the role of cyclic AMP as a second messenger in the effects of epinephrine on a liver cell. Cyclic AMP stimulates activation of phosphorylase (via several intermediate steps), thereby causing release of glucose into the bloodstream. Phosphodiesterase prevents accumulation of cyclic AMP by converting it to an inactive form of AMP.

Table 14.3
Hormone Actions in Which Cyclic AMP Serves As a Second Messenger.

Hormone	Target Cells	Action
Epinephrine	Liver cells	Stimulates glycogenolysis
Epinephrine	Muscle cells	Stimulates glycogenolysis
Glucagon	Fat cells	Breakdown of triglycerides to fatty acids
Parathyroid hormone	Bone cells	Resorption of calcium
Antidiuretic hormone	Kidney cells	Reabsorption of water from kidney filtrate
Thyroid-stimulating hormone	Thyroid cells	Stimulation of iodine uptake
Adreno-corticotrophic hormone	Cortex of adrenal gland	Stimulation of cortisol secretion
Luteinizing hormone	Corpus luteum cells of ovary	Stimulation of progesterone secretion

A variety of other substances besides hormones act on cells by affecting cyclic AMP levels. Some of them cause changes in adenylate cyclase activity, but a number of them act either by inhibiting or increasing phosphodiesterase activity, which also has an effect on cyclic AMP levels. Some very common substances are now known to affect phosphodiesterase activity. For example, caffeine inhibits the enzyme, while nicotine increases its activity. A cup of coffee and a cigarette must therefore send a very mixed signal to cells!

For a time after researchers had established cyclic AMP as a secondary messenger, endocrinologists thought that this might be the pathway by which all hormones act on target cells. However, although the hormones of the adrenal medulla (epinephrine and norepinephrine) and the polypeptide hormones act through cyclic AMP, the steroid hormones interact with cells through a different mechanism.

The Actions of Steroid Hormones on Genes

Steroid hormones are relatively small molecules (molecular weight about 300) that appear to enter the cell with ease. Once in the cell, a steroid hormone acts on the genetic material in a series of reactions. First the hormone binds to a high molecular-weight polypeptide receptor in the cytoplasm. The hormone-receptor complex is then activated; that is, it is transformed so that it acquires a high affinity for chromosomal material. This activation is probably caused by a conformational change in the hormone-receptor complex, perhaps by the attachment of an additional polypeptide subunit. The entire complex then moves into the nucleus of the cell.

Once it enters the nucleus, the hormone-receptor complex binds to specific parts of the chromatin, where it facilitates the transcription of the DNA in that region. In other words, the hormone-receptor complex activates certain genes. Messenger RNA is synthesized and performs its characteristic task in the cell, and an appropriate protein is synthesized. The steroid hormone thus induces gene activation, which leads to the cellular response characteristic of that hormone. All of this is summarized in figure 14.26.

Several steroid hormones cause the synthesis of messenger RNA and, presumably, the subsequent synthesis of proteins, which brings about the effects of the hormones in target cells. For example, estrogen and progesterone both cause the synthesis of messenger RNA in cells from reproductive structures such as the oviduct and uterus. Testosterone causes an increase in RNA synthesis in a variety of cells. Cortisone stimulates the synthesis of enzymes that catalyze the reactions in gluconeogenesis in liver cells, and aldosterone apparently stimulates the synthesis of proteins involved in the movement of sodium ions across the plasma membranes of its target cells.

Nonsteroid hormones may also act by stimulating messenger RNA synthesis; growth hormone, thyroxin, and insulin all appear to stimulate genes to produce certain proteins. Although this is certainly not the only mode of operation for such hormones, it certainly is a direct and seemingly effective way to achieve results in the target cell.

Not all of the loose ends have been tied together in this discussion of the hormone-receptor molecule mechanism. The thyroid hormones affect the growth, development, and metabolism of all cells, and yet the means by which they affect target cells is not at all clear. While polypeptide hormones characteristically bind with receptors in the plasma membrane, the smaller thyroid hormones, which also are synthesized from amino acids, pass through cell membranes and bind with receptors in the cytoplasm or nucleus of a target cell. Finally, some large polypeptide hormones

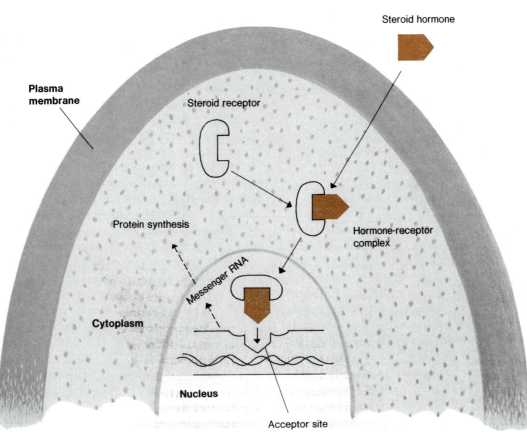

Figure 14.26 A model of the mechanism of action of a steroid hormone on genes in a target cell. The hormone passes through the plasma membrane and binds with a polypeptide receptor in the cytoplasm. The hormone-receptor complex is modified in some way that activates it, and then it enters the nucleus, where it facilitates transcription of specific genes. Much of this information about the action of steroid hormones was discovered by B. W. O'Malley and his colleagues.

that should not be able to pass through plasma membranes do, indeed, enter some cells. Various researchers have reported observing insulin, prolactin, parathyroid hormone, and gonadotrophins entering cells.

Prostaglandins

The **prostaglandins** are a family of twenty carbon polyunsaturated fatty acids formed in the plasma membranes of a wide variety of tissues. They are produced in small amounts, are extremely potent, and exist for only a short time.

The prostaglandins were first discovered in the 1930s and were named by the Swedish scientist Ulf S. von Euler. Von Euler first found prostaglandins in seminal fluid, where they are one hundred times more concentrated than they are in most body tissues. It was assumed that they are secreted by the prostate gland, hence the term *prostaglandin*. Actually, the name is somewhat of a misnomer because most of the prostaglandins in seminal fluid come from the seminal vesicles. Furthermore, prostaglandins have been found in virtually every tissue of the mammalian body.

There are many different prostaglandins, and, for convenience, they were originally grouped into five series that differ slightly in molecular structure. The series of prostaglandins are known as PGA, PGB, PGC, PGE, and PGF. Each series was further subdivided as more kinds of prostaglandins were identified; for example

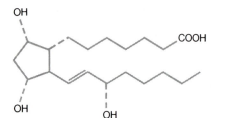

Figure 14.27 Molecular structure of one of the prostaglandins, PGE$_2$. The broken line denotes a bond that extends below the plane of the ring portion of the molecule. Only a skeleton of the molecule is shown without details of carbon and hydrogen atoms except for those that differ from other prostaglandin molecules.

PGE$_1$ and PGE$_2$ are chemically similar molecules. Figure 14.27 shows the structure of PGE$_2$, one of the better known prostaglandins.

The reported physiological effects of the prostaglandins are widespread and varied. Some of them even have opposing effects; for example, PGE$_2$ lowers blood pressure, but PGF$_2$ raises it. However, both PGE$_2$ and PGF$_2$ stimulate the contraction of the smooth muscle of the uterus, perhaps the most well-known effect of the prostaglandins. This uterine muscular response can make the careful administration of prostaglandins helpful in childbirth. Prostaglandins also are possible abortion-inducing agents, not only because they cause the uterus to contract but also because PGE$_2$ may cause a reduction in the secretion of progesterone, a hormone necessary for the maintenance of pregnancy. It is possible that some of the prostaglandins may prove useful for birth control in the future.

Prostaglandin research has already provided a possible beginning of an explanation of the way in which aspirin acts. Aspirin has been used as a drug for several hundred years without real understanding of what it does; in fact, some cynical medical scientists even doubted that aspirin did anything aside from making people think that they were going to feel better. It is now known that aspirin blocks the synthesis of prostaglandins, and this discovery seems to be a step toward explaining some of the effects of this most commonly used of all drugs.

Prostaglandins (and probably a chemically similar group of compounds, the thromboxanes) are released from tissues in response to a variety of stimuli, including hormones. Prostaglandins circulate in the blood and are like hormones in that they can exert their effects at distant locations, but prostaglandins have their principal effects within the cells where they are produced. Since hormonal stimulation of a cell sometimes results in increased prostaglandin synthesis as well as increased synthesis of cyclic AMP, prostaglandins may act as an intermediary between some hormones and cyclic AMP or may function independently as second messengers. No doubt many basic mechanisms of cell function will be clarified by research on prostaglandins and thromboxanes and on their interactions with other hormones and hormonelike substances.

Pheromones

In addition to the hormones secreted within an organism that affect its own physiological processes, there are other chemical regulatory compounds that are secreted by organisms into the environment. These compounds, called **pheromones,** apparently evolved as communication mechanisms within species. Pheromones may evoke behavioral, developmental, or reproductive responses within a population of organisms. Therefore, pheromones are sometimes called "social hormones."

A number of pheromones that affect behavior patterns have been intensively studied. Well-known examples include trail-marking behavior in ants and territory marking by dogs, deer, and antelope through pheromones secreted in urine or from scent glands. In these cases, the pheromone proclaims to other males of the species that the territory belongs to a particular individual.

An even more intriguing example of pheromones are the **sex attractants,** which have been most extensively studied in insects. The first sex attractant identified and synthesized was bombykol, which comes from females of the silkworm

Figure 14.28 The head of a male Polyphemus moth with its long, feathery antennae. The antennae bear receptors with which the moth can detect minute quantities of the sex attractant released by the female of the species.

moth *Bombyx mori*. Bombykol can attract male silkworm moths several kilometers away. Sex-attractant pheromones have been discovered in other moths and in many other insect species as well (figure 14.28). For example, the queen bee releases a pheromone during her nuptial flight that entices the drones to follow her and mate. Sex attractants have a very compelling effect on behavior.

Sex attractants have also been studied in mammals. In a number of mammalian species, females emit a special odor during their period of fertility and sexual receptiveness, which is called estrus. Normally docile, well-behaved male dogs will scratch doors, break leashes, or jump fences and roam great distances to reach a female in estrus once they detect her chemical signal.

Recently, a sex attractant was isolated from female rhesus monkeys. This focuses new attention on primate pheromones in general and human pheromones in particular. Whether or not sex attractants and other pheromones exist in humans remains a matter of some controversy. At any rate, modern men and women seem to prefer to wash off natural body chemicals and replace them with laboratory-compounded scents. Who knows what would happen if manufacturers of these scents should succeed in isolating, identifying, synthesizing, and marketing real human pheromones?

Summary

Multicellular organisms must have coordinating mechanisms for their cells, tissues, and organ systems. The endocrine and nervous systems are the two major methods of cellular communication within an animal. Both are necessary for the maintenance of homeostasis in the organism.

Endocrine glands are ductless glands that secrete substances called hormones. These chemicals serve as messengers carried in the bloodstream to the target cells, where the physiological effects directed by the hormone messengers occur.

The first hormones to develop during the evolution of the animal kingdom were probably hormones governing growth and reproduction. The first invertebrate hormones were most likely neurosecretions produced at the ends of nerve axons. Neurosecretory substances that operate as hormones have been observed in many invertebrate phyla.

Insect development is controlled by brain hormone from the neurosecretory cells in the brain, by ecdysone from the prothoracic gland, and by juvenile hormone from a portion of the brain known as the corpus allatum.

In humans and other vertebrates the major endocrine glands are the pituitary, the thyroid, the parathyroids, the pancreas, the adrenals, the ovaries, the testes, the hypothalamus of the brain, and the kidneys, but many other tissues also secrete chemical regulatory substances. The pituitary gland, located at the base of the brain, is composed of two major parts: the posterior lobe and the anterior lobe. Oxytocin and ADH, hormones of the posterior pituitary, are produced by neurosecretory cells in the hypothalamus and released into the blood from the posterior lobe. The anterior pituitary secretes at least six hormones, which are collectively known as trophic hormones. Anterior-lobe hormones stimulate the growth and development of the entire organism, secretion by other endocrine glands, and milk production by the mammary glands.

Hormone secretion by the anterior lobe is regulated by releasing hormones and release-inhibiting hormones from the hypothalamus. These small peptide messengers travel to the anterior pituitary from the hypothalamus via a special circulatory connection called the hypothalamic-pituitary portal system.

Metabolism, the sum of all the biochemical processes that take place within an organism, is under the control of several hormones, notably the thyroid hormones. These hormones stimulate oxygen consumption, carbohydrate oxidation, and protein synthesis. The regulation of carbohydrate metabolism is also under the control of insulin and glucagon from the pancreas, epinephrine and norepinephrine from the adrenal medulla, and glucocorticoids from the adrenal cortex. The complexity of the hormones governing blood sugar levels demonstrates how important finely tuned control of this parameter is to the overall well-being of the organism.

Three hormones are involved in the regulation of blood calcium levels: parathyroid hormone, thyrocalcitonin, and 1,25-dihydroxyvitamin D_3. The accurate regulation of blood calcium levels is extremely important because calcium imbalance can quickly prove fatal to the organism.

Research on hormone synthesis has led to the discovery of prohormones and preprohormones, the latter of which may be the initial gene product. The conversion of preprohormone to prohormone and then to the active hormone provides several levels of control in the synthesis and release of a hormone.

A receptor for a hormone is a molecule or molecular complex that is part of the cell and that can recognize a specific hormone, bind with it, and initiate the chain of events that lead to the cell's ultimate physiological response to the hormone.

Some hormones appear to affect their target cells by altering intracellular levels of a substance known as cyclic AMP. This chemical may then alter numerous other enzyme systems within the cell and produce the final physiological effect of the hormone. For this reason, cyclic AMP is sometimes known as the second messenger in hormone-target cell interactions. This seems to be particularly true for polypeptide and adrenal medulla hormones.

Steroid hormones evidently exert their effects upon target cells by entering the cell, combining with a polypeptide receptor, entering the nucleus of the cell, and stimulating messenger RNA synthesis. Steroid hormones thus cause protein synthesis, which leads to the cellular response characteristic of that particular hormone.

Prostaglandins are hormonelike substances found in cells' plasma membranes in a wide variety of tissues. Prostaglandins are extremely potent substances that affect many physiological functions. Some hormones may stimulate prostaglandin synthesis.

Pheromones are chemicals secreted by organisms into the environment that evoke behavioral, developmental, or reproductive responses in other members of the same species. Sometimes pheromones are referred to as "social hormones."

Questions

1. Why would it have been impossible for early endocrinologists to have determined the role of the kidneys in producing 1,25-dihydroxyvitamin D_3 using the standard classical methods for identifying an endocrine gland?

2. Describe the anatomical and functional connections between the hypothalamus of the brain and the posterior lobe of the pituitary.

3. Explain how the secretions of the anterior lobe of the pituitary are controlled by the hypothalamus.

4. American Indians called the famous mountain man Jim Bridger "Big Throat." Can you speculate on what Jim Bridger's problem might have been and on a possible environmental cause for that problem in a person who spent his life wandering the central part of the North American continent?

5. Why do the body cells of people suffering from diabetes mellitus oxidize vital proteins and fats when there is actually excess glucose available around them?

6. Describe how proper calcium levels are maintained in the blood of the human body. Why is this so critical?

7. In what way does cyclic AMP act as a second messenger in hormone action?

Suggested Readings

Books

Bentley, P. J. 1976. *Comparative vertebrate endocrinology*. Cambridge, Mass.: Cambridge University Press.

Turner, C. D., and Bagnara, J. R. 1976. *General endocrinology*. 6th ed. Philadelphia: W. B. Saunders.

Articles

Ashburner, M. 1980. Chromosomal action of ecdysone. *Nature* 285: 435.

Bloom, F. E. October 1981. Neuropeptides. *Scientific American* (offprint 1502).

Guillemin, R., and Burgus, R. November 1972. The hormones of the hypothalamus. *Scientific American* (offprint 1260).

Malkinson, A. M. 1975. Hormone action. *Outline Studies in Biology*. New York: John Wiley and Sons.

McEwen, B. S. July 1976. Interactions between hormones and nerve tissue. *Scientific American* (offprint 1341).

O'Malley, B. W., and Schroeder, W. T. February 1976. The receptors of steroid hormones. *Scientific American* (offprint 1334).

Pastan, I. August 1972. Cyclic AMP. *Scientific American* (offprint 1256).

Randle, P. J., and Denton, R. M. 1974. Hormones and cell metabolism. *Carolina Biology Readers* no. 79. Burlington, N.C.: Carolina Biological Supply Co.

Snyder, S. H. March 1977. Opiate receptors and internal opiates. *Scientific American* (offprint 1354).

Wigglesworth, V. B. 1974. Insect hormones. *Carolina Biology Readers* no. 70. Burlington, N.C.: Carolina Biological Supply Co.

Neurons
and Nervous
Systems

Chapter Concepts

1. The nervous system is specialized to provide coordination and communication in the animal body. This system senses environmental changes and organizes appropriate responses to them.

2. Neurons are the functional units of all nervous systems ranging from simple, diffuse networks of cells to highly specialized systems including large, complex brains.

3. The general cellular property of irritability is highly developed in neurons. Neurons respond electrically to certain stimuli.

4. Electrical changes in neurons result from orderly changes in the plasma membrane's permeability to specific ions.

5. Nerve impulses are electrical changes that sweep rapidly along neuron fibers.

6. Neurons communicate with one another by way of specialized connections called synapses.

7. Neurons are specifically arranged in functional pathways, the simplest of which are reflex arcs.

A complex division of labor is necessary to maintain homeostasis in multicellular organisms. Regulatory mechanisms coordinate the diverse, specialized activities of the various body parts. In animals, there are two major types of regulation: nervous or neural coordination and chemical coordination involving chemical messengers called hormones. Hormones are transported in the blood, and it may require minutes, hours, or even longer for a chemical message to produce an effect. Communication through the nervous system, on the other hand, is extremely rapid, requiring only thousandths of a second. Both coordinating systems are vital for maintaining homeostasis.

An animal's nervous system gathers information about both the external and internal environments. Most people are familiar with senses such as hearing, sight, and touch—senses that detect changes in the external environment. But there also are internal senses that monitor muscle tension, blood composition, and various other conditions in the body. The nervous system sorts, filters, and processes the information about these conditions that it constantly gathers. The information indicating trivial environmental changes is essentially ignored. Other information might be stored in the memory so that it can be retrieved later. Yet other information might elicit specific, immediate responses that are effected through the activities of muscles and glands. The nervous system also generates some spontaneous activities that do not depend directly on responses to environmental changes.

The next three chapters discuss the nervous system. In this chapter the structure and functions of nerve cells are explored, emphasizing the cellular, molecular, and electrical bases for these functions. The structural organization and function of nervous systems (which are, of course, composed of nerve cells) are also examined. Chapter 16 focuses on the complex organization of the vertebrate brain and the sensory mechanisms by which environmental information is collected and transferred to the brain. And chapter 17 discusses muscular structure and function and direction of muscular activity by the nervous system.

Neurons

Neurons, or nerve cells, are the functional units of any nervous system. Neurons vary greatly in appearance and function. Each neuron, however, is capable of receiving some form of input, responding to that input with an electrical change, and generating an output to one or several other cells.

Figure 15.1 shows an example of the type of neuron that directs the contractions of muscle cells involved in major body movements. Such a neuron consists of three basic parts. The **dendrites** are slender, branched extensions that constitute a major receptor area of the neuron; that is, they receive messages from other cells. The **nerve cell body** (or **soma**) contains the nucleus, mitochondria, ribosomes, endoplasmic reticulum, Golgi bodies, and other organelles common to all cells.

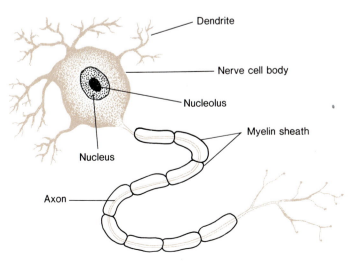

Figure 15.1 Structure of a vertebrate motor neuron. Note the branched dendrites and the long single axon that branches only near its tip. The myelin sheath is an insulating structure that is present around many neuron processes.

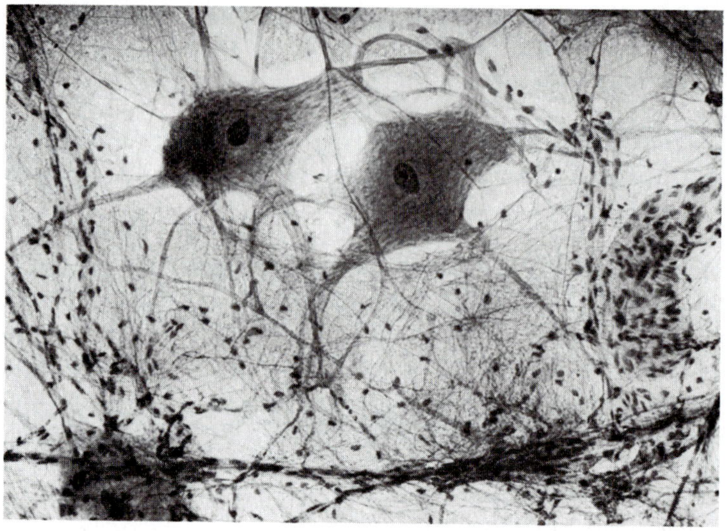

Figure 15.2 Photomicrograph of neurons from a vertebrate spinal cord. These cells with numerous processes extending from them are motor neurons.

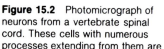

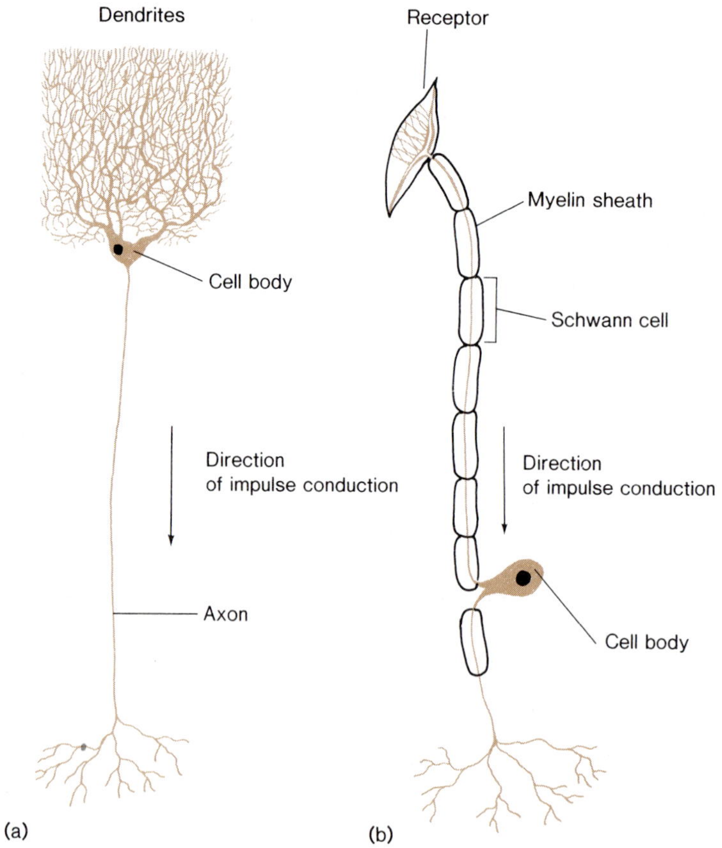

Figure 15.3 Examples of neuron diversity. (a) A brain cell (from the cortex of the cerebellum) with very highly branched dendrites. (b) A bipolar (sensory) neuron. The distinction between axons and dendrites that was made in motor neurons (figure 15.1) is not so obvious here because the neuron has only one long fiber.

Dendrites

Receptor

Myelin sheath

Schwann cell

Cell body

Direction of impulse conduction

Direction of impulse conduction

Axon

Cell body

(a)

(b)

Finally, the **axon** is a long process extending from the cell body that carries messages (nerve impulses) on to the next cell.

Figure 15.2 is a photomicrograph of neurons from a vertebrate spinal cord. The numerous processes seen in the figure are probably dendrites, since they show extensive treelike branching. But this is only one of many types of neurons, and dendritic branching is different in different types of neurons. Some large neurons in the brain, for example, have very intricately branched dendrites (figure 15.3a). In other neurons, it is impossible to distinguish between dendrites and axons because the cell has only a single branched fiber (figure 15.3b).

There are still other variations in neuron structure and an even greater array of variations in the specific functional characteristics of neurons. Many different types of neurons are incorporated into the nervous systems of several kinds of animals.

The Evolution of Nervous Systems

In single-celled organisms, or protists, the whole organism carries out sensory receiving and responding functions, but in animals, which are multicellular organisms, specialized cells receive sensory stimuli, process information about the stimuli, and direct organismic responses to them. Several broad categories of neurons can be recognized on the basis of their functional characteristics. Some are **sensory neurons** that either receive stimuli directly or respond to specific changes that occur in other specialized receptor cells. Others are **motor neurons** that direct the organism's response to stimuli. And still others are **interneurons** that act as intermediates between sensory and motor neurons. One important trend in the evolution of nervous systems has been an increase in the number of interneurons and in the complexity of their structural and functional relationships to sensory and motor neurons as well as to other interneurons.

It is necessary to exercise caution when drawing conclusions about evolution based on comparisons of various contemporary organisms. Animals living in the same era should never be thought of as evolutionary ancestors and descendants. But by comparing the nervous systems of various living animals, much can be learned about the fundamental functions of different nervous systems. These comparisons may also help scientists to visualize some of the steps that may have occurred during the evolution of the complex nervous systems existing in animals today.

Coelenterates such as *Hydra* have a very simple nervous system that consists of interneurons interwoven into a diffuse network, a **nerve net.** This nerve net extends through all parts of the body (figure 15.4a). Impulses can spread in all directions along this net from various points of origin. But this nerve net has no central processing unit, which severely limits its functions. Such a system allows for only a very limited repertoire of responses by the animal.

Several important advances in nervous system organization can be seen in members of the phylum Platyhelminthes (flatworms) such as *Planaria* (figure 15.4b). A flatworm's nervous system contains more neurons than a *Hydra's,* and the neurons are arranged differently. In a flatworm nervous system there is distinct centralization; many neurons are clustered together in specific locations. **Cephalization,** or the differentiation of a definite head region, is also seen in the flatworm's nervous system. In this head region, a concentration of neurons constituting a primitive brain is closely associated with sense organs located on the head. Two longitudinal nerve cords run through the body from the brain, and nerve cell fibers running through them connect the brain with other parts of the body.

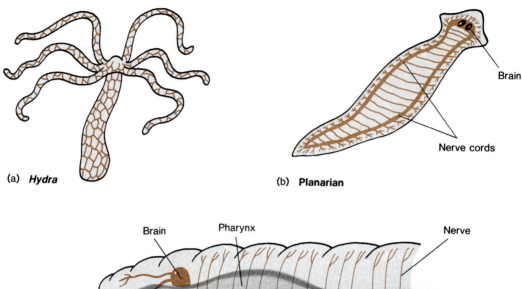

(a) *Hydra*

(b) **Planarian**

Brain

Nerve cords

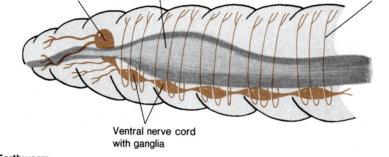

Brain   Pharynx   Nerve

Ventral nerve cord
with ganglia

(c)   **Earthworm**

Figure 15.4   Simple nervous systems. (a) The nerve net
of *Hydra,* a coelenterate. (b) Brain and paired nerve
cords of a planarian worm, a flatworm. (c) Brain, ventral
nerve cord, ganglia, and peripheral nerves of the
earthworm, an annelid (segmented) worm.

Annelids (segmented worms) such as the earthworm have an even more elaborate nervous system (figure 15.4c). In the earthworm, a concentration of nerve cells in the head constitutes the brain, and a large ventral nerve cord runs through the body. Along this nerve cord are clusters of nerve cell bodies called **ganglia** (singular: **ganglion**), arranged with one ganglion in each body segment. The brain, nerve cord, and ganglia make up the **central nervous system,** while bundles of neuron fibers called **nerves** make up the **peripheral nervous system.** These nerves, which contain both sensory and motor neuron fibers, connect all parts of the body with elements of the central nervous system. In such "mixed" nerves,

the sensory neurons that carry impulses toward the central nervous system are sometimes called **afferent neurons,** and the motor neurons that carry impulses away from the central nervous system are called **efferent neurons.**

Further evolutionary changes have involved vast increases in the numbers of neurons in nervous systems. This trend becomes particularly evident when vertebrate nervous systems are compared with invertebrate nervous systems. For example, an insect's entire nervous system may contain a total of about one million neurons, while a vertebrate nervous system may contain many thousand times that number. Similarly, some invertebrates have muscles that are innervated by a single axon from a single motor neuron, while a vertebrate muscle characteristically is innervated by many motor axons. But some of the most striking evolutionary advances in nervous systems have involved increasing numbers of interneurons. In vertebrates, huge numbers of interneurons are organized into brains that function as elaborate information-processing centers. Such centers can generate much more complex responses and behavior patterns than those of invertebrate animals.

Many other cells besides neurons are also present in a nervous system. In fact, neurons make up only about 10 percent of the cells in a vertebrate brain. The most abundant cells are called **glial cells** or **glia** (the word *glia* is derived from the Greek for "glue"). Obviously, early researchers thought glial cells helped hold the nervous system together. Although some glia help electrical signals to move quickly down a neuronal process, very little else is known about the role glial cells play in information processing. Further research may show that glial cells do, at least figuratively, help hold the nervous system together.

The Organization of the Human Nervous System

The overall design of the human nervous system (figure 15.5) is generally similar to the nervous systems of some of the other vertebrate animals. The brain and spinal cord constitute the **central nervous system (CNS),** while the **peripheral nervous system** consists of nerves containing sensory and motor neuron fibers. The sensory and motor fibers of the peripheral nervous system carry information to and from the central nervous system. The central nervous system decodes and evaluates the information it receives from the sensory (afferent) neurons and organizes appropriate responses to be carried back through the motor (efferent) neurons. The brain also serves as the site of learning and thought, which are the "higher" functions of the nervous system.

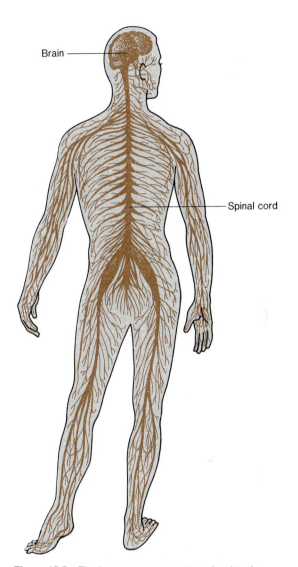

Brain

Spinal cord

Figure 15.5 The human nervous system showing the central nervous system (brain and spinal cord) and the nerves of the peripheral nervous system.

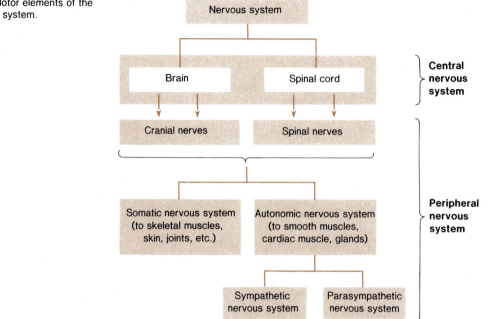

Figure 15.6 Motor elements of the human nervous system.

The motor portions of the peripheral nervous system can be divided into two categories called the **somatic motor** and **autonomic** or **visceral motor** divisions (figure 15.6). Interactions with the external world are mediated by the somatic motor division. These motor neurons lead to the skeletal muscles, which can function under voluntary control. Interactions with the body's internal environment are mediated by the autonomic division of the peripheral nervous system. Messages from the CNS that affect the activities of internal organs are carried by the motor neurons of the autonomic division.

Cellular Aspects of Neural Function

Each neuron in a nervous system is part of a functional array of neurons, and each plays a role in one or several neural circuits. Neuron participation in these circuits is made possible by the fundamental properties of neurons; neurons undergo specific electrical changes in response to appropriate inputs, and they generate outputs that communicate information to other cells.

Irritability and Membrane Potentials

A specific change in a neuron that occurs in response to an appropriate stimulus depends on the neuron's **irritability.** All living cells are irritable; that is, they can respond to stimuli. But irritability is especially well developed in neurons, making them extremely sensitive and responsive.

This quality of responsiveness is based on electrical phenomena that involve the plasma membranes of neurons. Unequal distributions of specific charged particles (ions) are maintained inside and outside the plasma membrane, and, as a result, there is an electrical charge difference between the inside and the outside of the neuron's plasma membrane. Ionic distribution and electrical charge differences across the plasma membranes of cells are not unique to neurons, but these phenomena function in a special way in neurons.

Table 15.1 lists some of the differences in the distribution of various ions on either side of a neuron's plasma membrane. Note that the extracellular fluid that bathes the neuron contains more sodium ions (Na^+) and chloride ions (Cl^-) than the neuron's interior, while the neuron's cytoplasm contains more potassium ions (K^+) and

Table 15.1
Concentrations of Ions Outside and Inside the
Plasma Membrane of a Cat Nerve Cell.

	Extracellular Fluid	Intracellular Fluid
Na^+	150 mM*	15 mM
K^+	5 mM	150 mM
Cl^-	125 mM	10 mM
	Fewer negatively charged proteins	More negatively charged proteins

*millimolar

Modified from A. J. Vander et al., *Human Physiology*, 3d ed.
(New York: McGraw-Hill, 1980).

more negatively charged protein molecules and phosphate ions than the cell's external environment. The net effect of these ion concentration differences is an electrical charge difference across the neuron's plasma membrane.

The electrical properties of neurons have been measured with microelectrodes, which are electrodes small enough to be inserted into individual nerve cells. The basic rules of electricity apply to neurons, as well as to other cells. If particles with opposite charges are separated, they will be attracted back together. This means that the particles have the potential to do work (work = force × distance). When positive and negative charges are separated by a plasma membrane, there is an **electrical potential** across the membrane. This potential is called a **membrane potential.** The unit of measurement for electrical potential is the **volt,** but electrical potentials across cell membranes are so small that they are measured in **millivolts** (thousandths of a volt).

The interior of a neuron is negative relative to its exterior. (This illustrates a fundamental principle of electricity: nothing is simply electrically positive or negative; it is electrically positive or negative relative to something else.) Another way of describing the electrical charge difference across the neuron's plasma membrane is to say that the membrane is **polarized**—it has both a positive pole or side and a negative pole or side. The electrical potentials across the polarized membranes of normal, resting neurons range between 50 and 100 mv, depending on the kind of neuron under observation. This normally maintained potential is called the **resting membrane potential,** or simply **resting potential.** The membrane potential of a neuron changes in a specific and characteristic way when that neuron responds to an appropriate input. This membrane potential change is called an **action potential.** An action potential is a short-lived, temporary event, but it is the key to a neuron's ability to conduct information over distances quickly. The **nerve impulses** that sweep along neuron fibers are based on action potentials.

The Resting Membrane Potential

Several properties of the plasma membrane establish and maintain a neuron's resting potential. Normally, there is a very small excess of positively charged particles outside the plasma membrane and a very small excess of negatively charged particles inside the cell. Despite the fact that the ion concentration differences shown in table 15.1 appear to be great, the total excess of negatively charged particles inside and positively charged particles outside the plasma membrane amount to only a tiny fraction of the total number of charged particles in each area. But even these tiny differences are significant because they account for the polarity of the membrane.

Plasma membranes, especially neuron plasma membranes, do not permit various ions to move with complete freedom into or out of cells, and plasma membranes are not equally permeable to all types of ions. Different ion-specific channels in the membrane provide protein "gates" that control the movement of ions through the membrane. Thus, plasma membranes are **selectively permeable;** some ions pass through more freely than others. The neuron's plasma membrane is quite impermeable to Na^+ but is 50 to 100 times more permeable to K^+, so K^+ leaks out through the plasma membrane fairly easily.

Another factor in the establishment and maintenance of the resting potential across the membrane is the presence of large, negatively charged molecules, mostly protein molecules, that are much more concentrated inside neurons than in the surrounding fluid. Because these molecules cannot pass through the plasma membrane at all, they constitute a source of negative charges confined in the interior of the cell.

Yet another factor is the presence of an ATP-driven "pump" mechanism in the membrane that works against concentration gradients by actively transporting K^+ to the inside of the cell and Na^+ out. This mechanism is called the **sodium-potassium** or **Na^+/K^+ ATPase pump** because energy for its ion-transporting work is obtained by ATP hydrolysis. The action of the sodium-potassium pump contributes to the differences in Na^+ and K^+ concentration inside and outside the cell (see table 15.1). In addition, the sodium-potassium pump is electrogenic; that is, it contributes directly to the charge difference across the membrane. It does so by sending 3 Na^+ out through the membrane for every 2 K^+ it takes in. Thus, the net effect is the active movement of an excess of positive ions out of the cell.

Sodium ions tend to diffuse from outside the cell (where they are in higher concentration) to the inside of the cell (where their concentration is lower). However, since the resting membrane is not very permeable to sodium ions, they cannot move freely along their concentration gradient. Potassium ions tend to diffuse from inside the cell, where their concentration is higher, to the extracellular fluid, where their concentration is lower. Because the membrane is more permeable to K^+, significant quantities of potassium ions "leak" out of the cell, diffusing down their concentration gradient. These potassium ions tend to draw negatively charged particles from inside the cell along with them, but a major group of negatively charged particles, the protein molecules, cannot pass through the membrane. Thus, the cell loses some of its positive charge as the lone potassium ions leak outward, unaccompanied by negatively charged particles (figure 15.7).

These factors—differential membrane permeability for Na^+ and K^+ and the resulting greater leakage of K^+ in response to its concentration gradient; the presence of large, negatively charged molecules inside the neuron; and the Na^+/K^+ ATPase pump—account for the resting membrane potential.

The Action Potential

A neuron's response to an appropriate input takes the form of an action potential. An action potential is a specific and characteristic set of changes in a membrane's electrical polarity and permeability characteristics.

An action potential involves a series of very rapid changes in the membrane potential (these changes can actually be completed within one millisecond). During an action potential, the potential across the membrane (inside relative to outside) changes from its normal -60 or -70 millivolts to $+30$ or even $+55$ millivolts and back again.

An action potential can be divided into three phases, the **rising** or **depolarization phase,** the **repolarization phase,** and the **hyperpolarization phase** (also called the hyperpolarizing after potential). Each of these electrical changes results from ion movements that take place as the plasma membrane undergoes a series of alterations in its permeability characteristics (figure 15.8).

During the rising phase the membrane becomes depolarized, until in the latter part of the phase, the inside of the cell actually becomes positive relative to the outside by as much as $+30$ to $+55$ millivolts. This change in membrane potential is due to a rapid influx of sodium ions caused by an abrupt increase in the membrane's permeability. During the rising phase, the membrane temporarily becomes up to about six hundred times more permeable to sodium ions.

The repolarization phase of the action potential is marked by the return of the membrane potential to its resting potential. The inside of the cell once again becomes negative relative to the outside, also as a result of permeability changes in the membrane. During this phase, the sodium permeability of the membrane returns to its regular low level, but the membrane's permeability to potassium ions suddenly increases. Potassium ions flow out of the cell in response to both the potassium concentration gradient and the excess of positive charges temporarily present inside the membrane. The outward flow of K^+ returns the membrane to its resting polarity.

But these electrical changes do not culminate in a direct return to the normal resting potential. During the hyperpolarization phase, the inside of the membrane becomes even more negative relative to the outside than it is in the resting state. This hyperpolarization results from the continuing flow of K^+ out of the cell. Soon, however, the membrane's potassium permeability returns to normal and the membrane potential returns to its resting level (-60 to -70 millivolts). Again, all three of these phases occur in about a thousandth of a second.

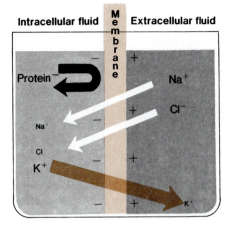

Intracellular fluid | Membrane | **Extracellular fluid**

Figure 15.7 Diagrammatic representation of permeability characteristics of the resting nerve cell membrane that contribute to the resting membrane potential. The intracellular fluid of the neuron is to the left of the neuron's plasma membrane, and the extracellular environment is to the right. The membrane is quite permeable to potassium ions (K⁺, colored arrow), less permeable to chloride ions (Cl⁻), and only slightly permeable to sodium ions (Na⁺). It is impermeable to negatively charged protein molecules, which are much more concentrated inside the cell than outside. Outward leakage of potassium ions is a major factor in maintenance of the resting membrane potential.

Figure 15.8 The action potential. Diagram shows timing of electrical changes measured in millivolts and relative permeabilities of membrane to Na⁺ and K⁺ in arbitrary units.

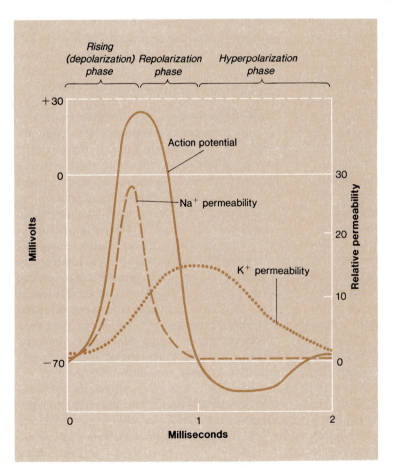

How does an action potential start, and how does the predictable series of events proceed in the same orderly fashion every time? An action potential is initiated by small electrical changes in a voltage-sensitive area of a cell. If the membrane in such an area becomes depolarized to a certain level called the **threshold potential,** an action potential is triggered. Once the threshold potential is reached, an action potential automatically results. The "size" of the action potential in no way depends on the strength of the stimulus causing depolarization; the action potential of a neuron is always the same, whenever it is triggered. Because the action potential will not occur if the threshold potential is not reached, and because the action potential proceeds in the same way every time it occurs, the action potential is an **all-or-none phenomenon.**

Permeability changes that occur in the membrane during the action potential involve changes in the gated channels through which ions move when they pass through the plasma membrane. Some of the gated channels through which ions move are sensitive to voltage changes (figure 15.9). When the threshold potential has been reached, voltage-sensitive gates open and close in a predictable and characteristic sequence that is completed within a millisecond.

From these descriptions of ion movements during action potentials, it might seem that massive quantities of ions move in an action potential and that their movement might abolish the concentration gradient across the membrane. This is not the case. It is estimated that perhaps only one out of every one hundred thousand potassium ions and a similarly low number of sodium ions move in or out of the cell to cause the action potential. Thus, concentration gradients are not abolished, and action potentials can follow one another at short intervals. But even with these low rates of ion exchange, a number of repeated action potentials would eventually abolish the Na^+ and K^+ concentration gradients across the membrane. This is another reason why the Na^+/K^+ ATPase pump is so important for neuron functioning. The pump is vital to the recovery of the ionic concentration gradients.

How rapidly can action potentials actually follow one another? The cell membrane is unresponsive for a certain length of time after an action potential has occurred. This time is called the **refractory period** and consists of two parts.

During the absolute refractory period, no stimulus, no matter how strong, can elicit another action potential. During the relative refractory period, a very strong stimulus can produce an action potential, but the strength of the stimulus must be considerably above the usual threshold level. The absolute refractory period lasts only a fraction of a millisecond, but the relative refractory period can last from 1 to 15 msec or longer. Although this limits the number of action potentials a cell can have, the fact that action potentials and refractory periods are completed in milliseconds means that many nerve cells can produce up to hundreds of action potentials per second.

Interestingly, local anesthetics seem to act by preventing action potentials in the nerve cells of the "deadened" area. Anesthetics such as novocaine and xylocaine prevent the increase in membrane permeability to sodium that is necessary for an action potential to occur.

Nerve Impulse Conduction

The action potential is a series of events occurring in a small area of a membrane. But action potentials form the basis of the nerve impulse, the passage of excitation from point to point along the fibers of neurons. How then are action potentials incorporated into nerve impulse conduction, which permits neurons to transmit information long distances in small fractions of a second?

The transmission of an impulse along the membrane of a neuron might be described as the conduction of an action potential. But a given action potential does not literally travel along a neuron's plasma membrane. Instead, an action potential is a localized event in one area of a membrane. In impulse conduction, each action potential triggers a new action potential right next to it in the plasma membrane. As each action potential generates a new one, the electrical change sweeps along the membrane. A nerve impulse is thus a chain reaction series of action potentials along a neuron's membrane.

Figure 15.10 pictures an impulse being transmitted down an axon. At any one point where depolarization (an action potential) occurs, the inside of the axon membrane, because of Na^+ influx, becomes positive and the outside negative. This triggers depolarization at the next point along the axon. The impulse thus travels

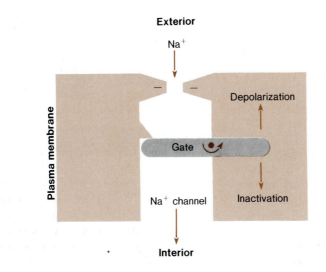

Figure 15.9 Hypothetical model of a voltage sensitive gate in a sodium channel through a neuron's plasma membrane. When threshold potential is reached, such gates open to permit the sodium influx that occurs during depolarization.

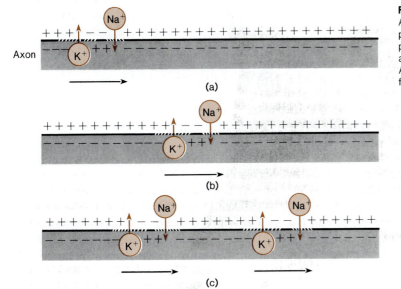

Figure 15.10 Impulse transmission. A nerve impulse is a self-propagating series of action potentials. (a and b) The progress of a single impulse along an axon. (c) A second impulse (left) following the first.

down the axon from the nerve cell body to the axon terminal. As the point of depolarization progresses down the axon, the membrane area just behind it becomes repolarized.

The spontaneous propagation of the action potential down the axon is due to changes in the sodium permeability of the membrane. The electrical influence of voltage changes that occur during an action potential extend far enough to affect the voltage-sensitive gates of nearby sodium channels. When those channels open, sodium ions enter, and a new action potential is underway. Thus, the nerve impulse is a self-propagating series of interdependent local membrane events. The nerve impulse, like the individual action potentials it is composed of, is an all-or-none phenomenon.

Several pioneers in the study of neuron functioning have made especially important contributions. A. L. Hodgkin and A. F. Huxley, English neurophysiologists, received the Nobel prize in 1963 (shared with John Eccles) for their work on action potentials and impulses. Their work verified a membrane-conduction hypothesis that Julius Bernstein of the University of Halle in Germany had proposed many years earlier. Hodgkin and Huxley and a group of researchers headed by K. S. Cole and H. J. Curtis at Woods Hole, Massachusetts, worked for many years on the giant axons of the squid *Loligo* (figure 15.11). Both teams used these especially large and accessible axons to measure electrical potentials across and ion fluxes through the membrane. Research has also been done with the large axons of lobsters, cockroaches, and annelid worms by other teams of investigators. This work has shed a great deal of light not only on how an action potential is generated, but also on how the nerve impulse is conducted down the nerve axon.

Myelin Sheaths and Conduction Velocities

The speed at which an impulse moves along an axon depends on the axon's diameter. The larger the axon's diameter, the faster the impulse is transmitted. The giant axons of squids and other invertebrates may be over 1,000 μm in diameter, and they conduct impulses very rapidly. Conduction velocities of up to 25 m per second have been recorded in squid axons at 15°C.

Figure 15.11 The squid *Loligo*. Giant axons from the squid have been used to study electrical events and permeability changes during nerve impulses.

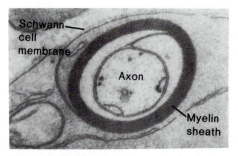

Figure 15.12 Electron micrograph showing a cross section of an axon enclosed by a myelin sheath. Schwann cells produce myelin sheaths.

A different mechanism has evolved in vertebrate nervous systems to facilitate rapid impulse conduction. This mechanism is associated with the **myelin sheath,** a wrapping of fatty material that encloses the axons of many vertebrate neurons (figure 15.12). Specialized cells called **Schwann cells** build the myelin sheath around axons during development. The longitudinal view of a myelinated axon in figure 15.13 shows that the myelin sheath is interrupted at intervals. These openings in the sheath, called **Nodes of Ranvier,** are unmyelinated portions of the axon located between Schwann cells.

Myelin insulates the axon's membrane and restricts ion flow so that the exchange of Na^+ and K^+ necessary for an action potential can occur only at the nodes. The current generated during an action potential at one node depolarizes the next node and sets off an action potential there. Thus, the impulse "jumps" from node to node. This phenomenon is called **saltatory** ("leaping") **conduction** (figure 15.13b). Conduction velocities of up to 120 m/sec are reached in myelinated axons of about 20 μm in diameter.

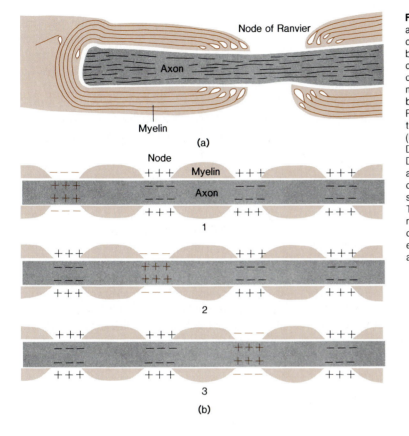

Figure 15.13 The myelin sheath and impulse conduction. (a) A diagram of myelin sheath structure based on electron microscope observations. Myelin sheath consists of layer upon layer of cell membrane material wrapped around the axon by a Schwann cell. Nodes of Ranvier are unmyelinated regions that interrupt the sheath. (b) Saltatory conduction. Depolarization occurs only at nodes. Depolarization at one node is an adequate electrical change to cause opening of voltage sensitive gates in sodium channels at the next node. Thus, depolarization "leaps" from node to node during impulse conduction. Repolarization occurs at each node as depolarization occurs at the next node.

The myelin sheath thus permits rapid impulse conduction in vertebrate neurons that are much smaller than the giant invertebrate axons described earlier.

Synaptic Transmission

For the nervous system to fulfill its functions of communication and coordination, it obviously must transmit impulses from nerve cell to nerve cell. Communication between neurons occurs at special junctions called **synapses.** Sir Charles Sherrington proposed the existence of the synapse (which means "to clasp") because of two major observations he and other neurophysiologists had made. First, when an impulse spreads from one nerve cell to another, it is always *unidirectional;* that is, the message can travel only one way. Biologists now know that this is because of different structural and functional characteristics of the neurons on each side of the synapse. Second,

the early neurophysiologists recorded a *delay* in the transmission of an impulse from one neuron to another. This led Sherrington and others to hypothesize that there was a gap between the two neurons, that their cell membranes were not actually touching, and that the message had to be carried across the gap by special chemicals called **neurotransmitters.**

Other investigators argued that the impulse transmission was an electrical phenomenon. The two opposing ideas came to be called the "sparks hypothesis" (electrical transmission) and the "soup hypothesis" (chemical transmission). As sometimes happens, both sides were partially correct. Most synapses are areas of chemical transmission, but **electrical synapses** have been found in the nervous systems of a number of invertebrates and at some locations in vertebrate brains.

In an electrical synapse, there is direct electrical coupling between the two neurons. Their membranes are fused in a special arrangement called a **gap junction,** an area in which the membranes are tightly fastened together and small channels provide a direct cytoplasmic connection between the cells. There is no delay in impulse conduction across electrical synapses, and pharmacological agents that block chemical synapses do not prevent impulse transmission across electrical synapses. Electrical junctions apparently synchronize the activities of populations of neurons very effectively in parts of some nervous systems. Electrical junctions also provide for very rapid impulse transmission in some invertebrate reflexes.

However, as was mentioned before, most synapses operate by chemical transmission. The cell sending the signal at a synaptic junction is called the **presynaptic neuron,** and the cell receiving the signal is called the **postsynaptic neuron.** A very small physical gap (only about 20 nm across) separates presynaptic and postsynaptic neurons. This gap is called the **synaptic cleft** and is bathed with extracellular fluid. A small molecule can diffuse across the cleft in about a tenth of a millisecond.

The presynaptic portion of a synapse is usually the end of an axon, called the **axon terminal** (or synaptic knob). The postsynaptic portion of a synapse is usually found on a dendrite or on the nerve cell body (figure 15.14). Occasionally, however, a synapse may lie between the axon terminal of one neuron and the axon of the second neuron, or even between a dendrite on one neuron and a dendrite on another neuron.

Although figure 15.14 shows several synapse arrangements, the illustration is actually quite simplified compared to situations that actually exist in nervous systems. A postsynaptic neuron may have thousands of synaptic junctions on its dendrites and cell body. Some neurons in the human spinal cord have as many as fifteen thousand synapses. In the brain, a single neuron may

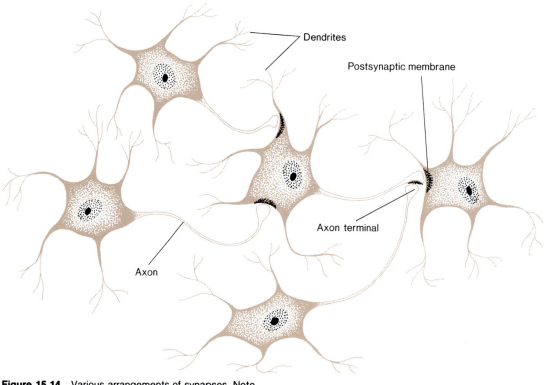

Figure 15.14 Various arrangements of synapses. Note the knob at the end of the axon called the terminal and the thickening of the membrane just under the synapse, the postsynaptic membrane. Usually, synapses are between axon terminals and dendrites or nerve cell bodies. A synapse can also form between the axon of one neuron and the axon of another. Even dendrite to dendrite synapses (not shown here) exist.

receive even more synapses—perhaps as many as one hundred thousand synapses per nerve cell. Since the brain contains perhaps as many as 10^{10} neurons, its complex circuitry contains an astronomical number of synapses.

In some cases, thousands of synapses from many different presynaptic neurons *converge* upon a single postsynaptic neuron. On the other hand, a single neuron may have several axon branches that *diverge* to form synapses with many other neurons. The properties of convergence and divergence (figure 15.15), along with the overwhelming number of synapses, contribute to the tremendous complexity of the central nervous system.

The Function of the Synapse

At the turn of this century, the Spanish histologist Santiago Ramón y Cajal, using special staining techniques, observed synapses in the central nervous system by means of light microscopy.

However, it was not until the 1950s, with the advent of the electron microscope, that the details of the structure of the synapse were elucidated.

The structure of a synapse, as magnified greatly by the electron microscope, is diagrammed in figure 15.16. The presynaptic terminal is typically shaped like a small knob. Its most obvious features are the small membrane sacs called **synaptic vesicles.** Each vesicle is about 20 nm in diameter and contains chemicals called **neurotransmitters.** These chemicals are released from the presynaptic terminal when an impulse arrives, and they then cross the synaptic cleft to the postsynaptic neuron.

Once neurotransmitter molecules have diffused across a synaptic cleft, they attach to special **receptor molecules** in the postsynaptic membrane. These receptors are large protein molecules embedded in the surface membrane of the postsynaptic cell and are similar to the hormone receptor molecules discussed in chapter 14.

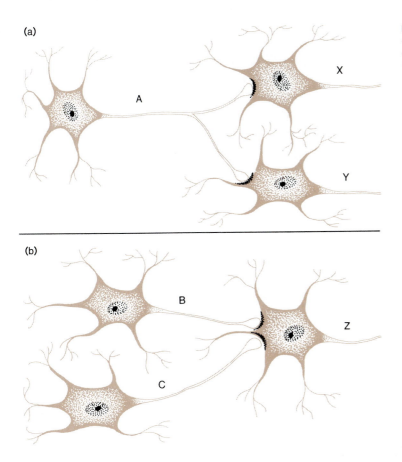

Figure 15.15 Simple models of divergence and convergence in the central nervous system. (a) Terminal branches of neuron A diverge to synapse with both neuron X and neuron Y. (b) Both neurons B and C synapse with neuron Z. This is convergence. In the central nervous system, divergent and convergent relationships may involve very large numbers of neurons.

The neurotransmitter molecules fit into the receptors' binding sites very specifically; a receptor for the neurotransmitter norepinephrine, for example, will not fit a molecule of the neurotransmitter acetylcholine.

The reason, then, that impulse transmission across a synapse is unidirectional is that only the presynaptic side has vesicles containing neurotransmitters, and the receptors for the neurotransmitters are present only on the postsynaptic membrane. The delay in the transmission of the impulse over the synapse, a matter of about 0.3 msec, is due to the time needed for the release of the neurotransmitter from the synaptic vesicles (about 0.2 msec) and diffusion of the transmitter across the synaptic cleft (less than 0.1 msec).

When the arrival of a nerve impulse transmitted down the presynaptic neuron causes the release of neurotransmitter into the synaptic cleft, the concentration of neurotransmitter within the cleft rises almost instantaneously from zero to a very high level. The release of the neurotransmitter molecules requires the presence of free calcium ions inside the axon terminal; the action potential triggers a flow of calcium into the terminal. The calcium then participates in some way in the fusion of the membranes of the synaptic vesicles with the presynaptic plasma membrane, leading to the release of their contents. This release through the plasma membrane is an example of **exocytosis.** Hundreds of synaptic vesicles fuse with the cell surface and empty

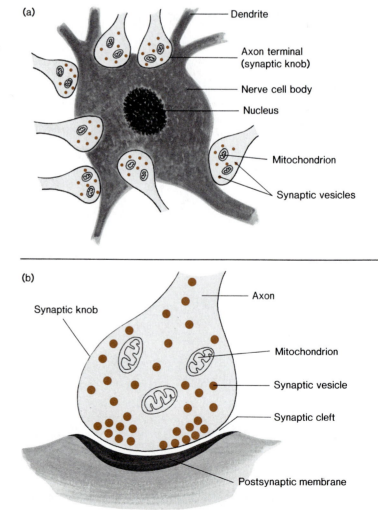

Figure 15.16 Synapse structure. (a) A nerve cell body with several axon terminals (synaptic knobs) on it. (b) Diagram of the structure of the synapse as revealed by the electron microscope.

(a)

Dendrite

Axon terminal (synaptic knob)

Nerve cell body

Nucleus

Mitochondrion

Synaptic vesicles

(b)

Synaptic knob

Axon

Mitochondrion

Synaptic vesicle

Synaptic cleft

Postsynaptic membrane

their contents into the synaptic cleft in a very short time (figure 15.17).

When neurotransmitter molecules attach to receptor molecules, specific changes in ion conductance occur in the postsynaptic membrane. The interaction of neurotransmitters and receptor molecules causes ion channels in the membrane to open. These particular channels are chemically gated rather than electrically gated. The ion flow through them causes the postsynaptic membrane to become partially depolarized. Changes in the polarization of the postsynaptic membrane are known generally as *postsynaptic potentials,* and this particular kind of postsynaptic potential, involving partial depolarization, is called an **excitatory postsynaptic potential (EPSP).** It is excitatory in the sense that the potential change

favors the initiation of an action potential, but not every EPSP leads to an action potential. EPSPs are **graded potentials** because the postsynaptic cell membrane is only partially depolarized and the degree of depolarization is variable. The size of the EPSP depends on how many chemically gated channels have been opened, which in turn depends on how much neurotransmitter was released into the synaptic cleft.

If postsynaptic potentials build up enough (that is, if the postsynaptic membrane becomes sufficiently depolarized), nearby electrically gated channels open and an action potential is generated. The postsynaptic membrane usually must be depolarized by 10 to 20 mv to reach threshold, so that an action potential is generated (figure 15.18a).

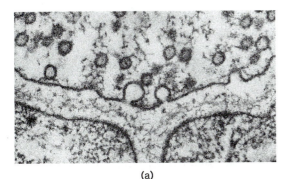

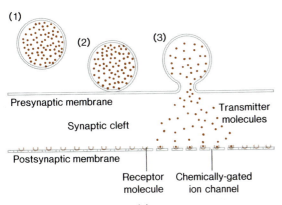

Figure 15.17 Synaptic transmission. (a) Electron micrograph showing synaptic vesicles apparently opening into the synaptic cleft. (Magnification × 104,000) (b) Proposed interpretation of (a). A synaptic vesicle (1) fuses with the presynaptic membrane of the synaptic knob (2) and releases its contents into synaptic cleft (3). Neurotransmitter molecules fuse with receptor molecules in the postsynaptic membrane. This causes opening of chemically gated ion channels and a change in membrane potential.

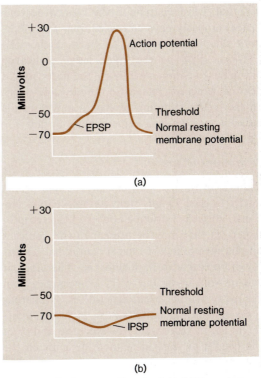

Figure 15.18 Postsynaptic potentials. (a) Record of potential across postsynaptic membrane. Excitatory postsynaptic potentials (EPSPs) cause partial depolarization of the membrane. If the depolarization reaches threshold (about −50 millivolts in this case), an action potential will be generated in the postsynaptic membrane. (b) Record of an inhibitory postsynaptic potential (IPSP) caused by an inhibitory synapse. The electrical potential across the membrane is greater than the resting membrane potential (further from threshold).

In addition to excitatory potentials, there are also postsynaptic potentials that *inhibit* the activity of postsynaptic neurons. For example, many inhibitory synapses exist in the cerebellum of the vertebrate brain. These inhibitory synapses filter out impulses that might cause excessive, spastic body movements.

A particular synapse can be either excitatory or inhibitory, but probably not both. Inhibitory synapses usually cause the postsynaptic membrane to increase in K^+ permeability, thereby causing the postsynaptic cell membrane to become hyperpolarized. A hyperpolarized membrane has a larger electrical potential across the membrane (as much as -90 mv) than the normal resting potential. This hyperpolarization results in an **inhibitory postsynaptic potential** (**IPSP;** figure 15.18b).

In the presence of an IPSP, it is much more difficult for an excitatory synapse on the same neuron to depolarize the cell and cause an action potential. Most neurons in the central nervous system probably have both excitatory and inhibitory synapses impinging on them. Either the EPSPs or IPSPs get the "upper hand" and cause or inhibit an action potential in the postsynaptic neuron. A neuron integrates the messages from all the synapses impinging upon it and responds either by passing on a nerve impulse or not.

Neurotransmitters

While a neuron uses only one kind of neurotransmitter to send signals from its presynaptic terminal, no such restriction applies to the receptor molecules that may be present in the neuron's plasma membrane. A typical neuron has a variety of receptors in its membrane. Therefore, it can respond in various ways to messages from the many neurons synapsing with it. Numerous types of neurotransmitters exist, each with a different set of responses. However, the ability of the neuron to respond to them depends on the types of receptor molecules that it possesses.

When a receptor molecule binds with the appropriate neurotransmitter, the result may be an increase *or* a decrease in the permeability of the postsynaptic membrane to any of the major ions (Na^+, K^+, Cl^-, or Ca^{+2}). This leads to excitatory or inhibitory potential changes in a small area of the membrane adjacent to the receptor molecule.

A number of chemicals have been demonstrated to act as neurotransmitters, and many others are thought to be likely candidates (figure 15.19). Neurotransmitter molecules usually have molecular weights of 200 or less. The most studied neurotransmitters include (1) **acetylcholine,** which was the first neurotransmitter to be discovered; (2) the **catecholamines,** such as dopamine and norepinephrine; and (3) **Gamma-amino butyric acid (GABA),** which appears to be the most abundant neurotransmitter in the brain. Several amino acids also function as neurotransmitters.

Establishing that a given chemical compound is a neurotransmitter is technically very difficult. It seems likely that a large number of compounds, some as yet unknown, act as neurotransmitters. Good examples of relatively recently discovered neurotransmitters are peptides such as the enkephalins and endorphins discussed in the last chapter (box 14.1). These molecules may act as both hormones and neurotransmitters, illustrating once again the overlapping functions of the endocrine and nervous systems.

The Destruction and Recycling of Neurotransmitters

Obviously, neurotransmitter molecules cannot remain permanently bound to receptor molecules if the postsynaptic neuron is to be capable of receiving subsequent messages. The binding of neurotransmitters to receptor molecules is, in fact, reversible, and the lifetime of a receptor-transmitter complex is extremely short. In the case of acetylcholine, for example, the neurotransmitter is degraded by enzymes to simpler chemicals that are unable to activate receptors.

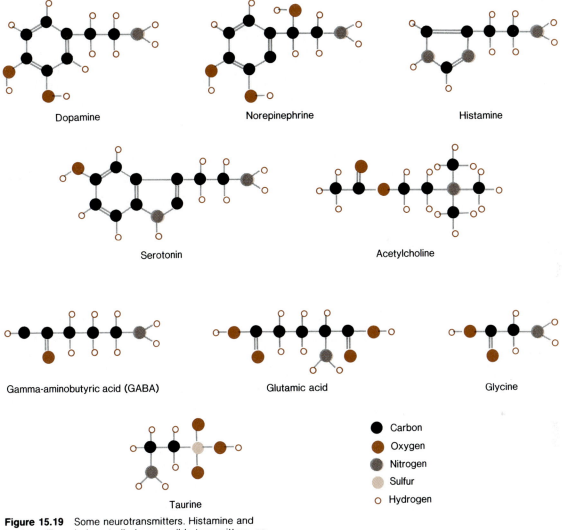

Figure 15.19 Some neurotransmitters. Histamine and taurine still are being studied as possible transmitters, as are many other substances not shown here.

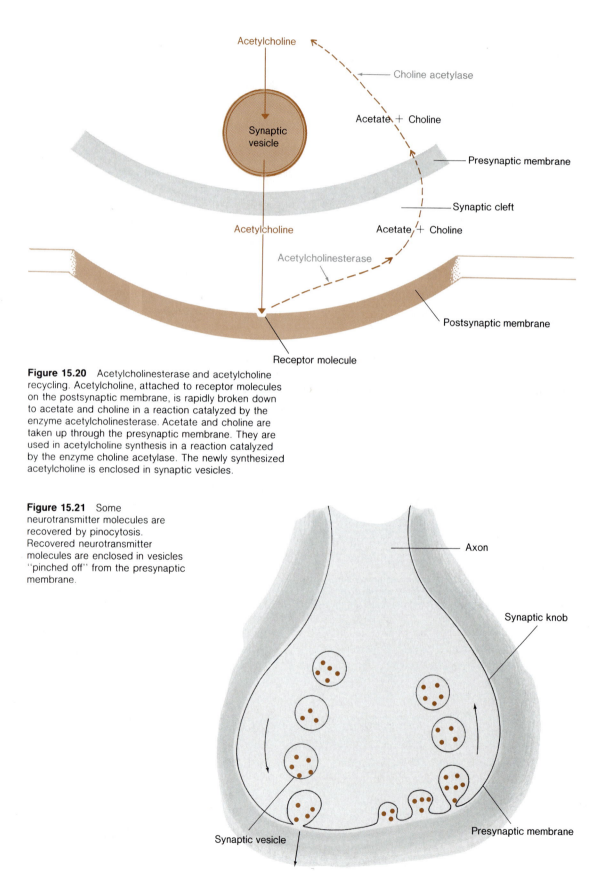

Figure 15.20 Acetylcholinesterase and acetylcholine recycling. Acetylcholine, attached to receptor molecules on the postsynaptic membrane, is rapidly broken down to acetate and choline in a reaction catalyzed by the enzyme acetylcholinesterase. Acetate and choline are taken up through the presynaptic membrane. They are used in acetylcholine synthesis in a reaction catalyzed by the enzyme choline acetylase. The newly synthesized acetylcholine is enclosed in synaptic vesicles.

Figure 15.21 Some neurotransmitter molecules are recovered by pinocytosis. Recovered neurotransmitter molecules are enclosed in vesicles "pinched off" from the presynaptic membrane.

The enzyme that breaks down acetylcholine is called **acetylcholinesterase.** It is one of the most effective enzymes known; a single acetylcholinesterase molecule can catalyze the breakdown of 25,000 molecules of acetylcholine in a single second. The breakdown products are then taken up into the presynaptic terminal and used in the synthesis of new acetylcholine molecules (figure 15.20).

With some neurotransmitters, specific uptake systems in the presynaptic membrane rapidly remove transmitter molecules from the synaptic cleft. Once transmitters are returned to the presynaptic neuron, they are either enclosed in synaptic vesicles for future use or are degraded by relatively slow-acting enzymes in the axon terminal. The enclosure of neurotransmitters in vesicles formed by pinching in the plasma membrane of the presynaptic terminal also helps to keep the surface membrane from growing larger (figure 15.21). This is because the formation of vesicles balances the increase in membrane surface caused by the fusion of membranes during the release of transmitters by exocytosis.

The Effects of Drugs at Synapses

Research on the chemistry of synaptic communication has provided not only a basic understanding of how two neurons interact but insights into how psychoactive drugs affect the nervous system. For example, drugs can cause changes in the synthesis of a specific neurotransmitter, in packaging transmitters into vesicles, in releasing transmitters, in receptor activation, or in the recovery of neurotransmitters or their breakdown products from the synaptic cleft.

Examples of the actions of particular drugs and pharmaceutical agents at the synapse include the discovery that Parkinson's disease can be treated by giving the patient L-DOPA, a precursor of the transmitter dopamine. Myasthenia gravis, a disorder characterized by skeletal muscle weakness and fatigue, involves a decrease in acetylcholine receptors and is treated by inhibiting acetylcholinesterase. Schizophrenia is treated by drugs that block dopamine receptors.

Tranquilizers such as valium apparently interact with GABA receptors. Stimulants such as cocaine and amphetamines block the recovery of catecholamine transmitters, leaving them in the synaptic cleft. Hallucinogens such as LSD and mescaline interact with the receptors for the neurotransmitter serotonin. The list could go on, but the point is that understanding the synapse enables biologists to understand better the actions of many of these drugs. Some of the drugs that act on synapses may provide new tools for treating neurological disorders and perhaps even various types of mental illnesses.

Reflexes

Not only are messages passed from nerve cell to nerve cell, but motor neurons, interneurons, and sensory neurons are linked together to carry messages through functional neural pathways. The simplest neural pathways are those involved in certain **reflexes,** which are automatic responses to specific stimuli. The messages involved in a reflex travel over a pathway called a **reflex arc** that links receptor and effector. Some of the simplest (but most important for survival) reflexes involve only the spinal cord and peripheral nerves, not the brain. Only one or a few interneurons may be involved; in fact, the simplest reflexes do not involve interneurons at all.

Almost everyone is familiar with the knee-jerk response that occurs when the leg is tapped lightly just below the kneecap, stretching the muscle. In a very short period, about 50 msec, the leg extends in response to the tapping. This response is so fast that it almost seems that the sensory information of the tapping is "reflected" back to the muscle; hence, the origin of the term *reflex.* Actually, the rapidity of this reflex response is made possible by the simplicity of the neuronal pathway that elicits the muscle contraction. Only one neuronal synapse is involved in the circuit that leads to the activation of the responding muscle. Since the reflex is stimulated by the stretching of a muscle and involves only

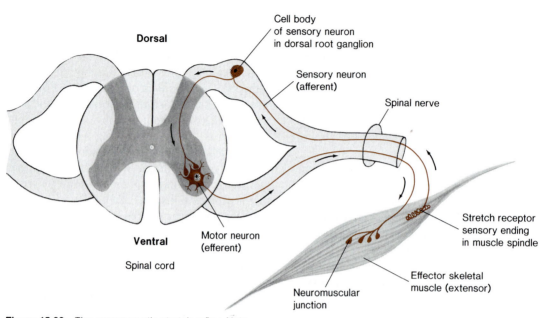

Figure 15.22   The monosynaptic stretch reflex. Note the sensory neuron coming into the dorsal side of the spinal cord. It synapses inside the spinal cord with the nerve cell body of the motor neuron. The axon of the motor neuron leaves the spinal cord on the ventral side and goes to the skeletal muscle. In humans, the dorsal or "back" side is the posterior surface of the body. The ventral side is the "front" or anterior aspect of the body.

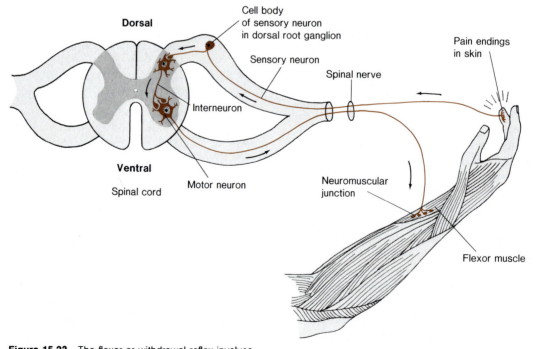

Figure 15.23   The flexor or withdrawal reflex involves three neurons and two synapses. Note the interneuron within the spinal cord. In this case the sensory receptors are just the naked (unmyelinated) ends of the sensory neuron in the skin. The flexor muscle bends the forearm, withdrawing the hand from the painful stimulus.

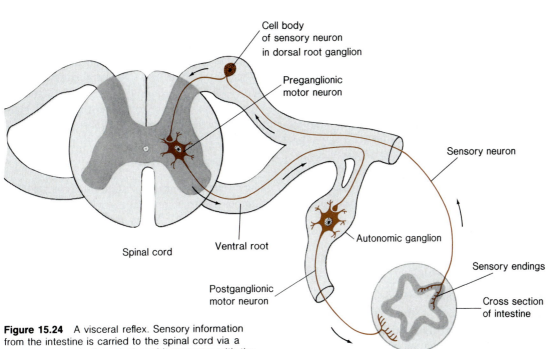

Figure 15.24 A visceral reflex. Sensory information from the intestine is carried to the spinal cord via a sensory neuron. In the spinal cord it synapses with the first motor neuron (the preganglionic neuron). This, in turn, synapses in an autonomic ganglion with the postganglionic motor neuron, which goes to the muscle in the wall of the intestine.

one neuron-neuron synapse, it is known as a **monosynaptic stretch reflex** (figure 15.22). Of course, many neurons and muscle fibers participate simultaneously in the stretch reflex, but for the purpose of clarity and simplicity, the reflex is illustrated by showing only one neuron chain.

A slightly more complex reflex involving more than one synapse (and hence called a *polysynaptic reflex*) is illustrated in figure 15.23. This particular example is a **flexor reflex** (also called a withdrawal reflex). If someone were to touch a hot stove, he would automatically flex his forearm, withdrawing his hand from the hot surface. This happens in a fraction of a second and does not require a conscious decision by the brain. Pain receptors in the skin are stimulated, and the message is relayed by sensory neurons to the spinal cord. There interneurons transmit impulses back to the appropriate motor neurons, and then impulses travel over the motor neurons to muscles. The proper muscles contract, the arm flexes, and the hand withdraws.

There are many other examples of reflexes involving special sensory receptors, sensory neurons, interneurons, motor neurons, and skeletal muscles acting as effectors. Another type of reflex involves the internal organs.

The Autonomic Nervous System

The skeletal muscles involved in the reflexes just described are under the control of the somatic motor division of the nervous system. Reflexes involving internal organs are called **visceral reflexes,** and their effectors are glands, the involuntary muscles of organs, or the heart. While somatic motor activities can generally be controlled voluntarily, the motor activities of the autonomic system, which controls effectors involved in visceral reflexes, are not normally under voluntary control.

The sensory neurons that function in visceral reflexes follow the same pathways as those of the somatic reflexes; that is, they enter the spinal cord on the dorsal side, and the nerve cell bodies of the sensory neurons are located in **dorsal root ganglia** (figure 15.24). The difference between somatic motor innervation and autonomic innervation is in the motor neurons of each system.

Somatic motor control depends on the axons of single motor neurons that reach from the central nervous system to the muscles being stimulated. But autonomic innervation involves two motor neurons arranged in sequence to transmit stimulation from the CNS to the effector. These two cells are called the **preganglionic motor neuron** and **postganglionic motor neuron**, respectively. The axon of the preganglionic motor neuron leaves the CNS, but it synapses with the postganglionic motor neuron in an **autonomic ganglion.** The axon of this second motor neuron reaches to the effector.

The Parasympathetic and Sympathetic Systems

A mammal's autonomic nervous system can be divided into two parts on the basis of structure and function; the **parasympathetic** nervous system and the **sympathetic** nervous system (figure 15.25).

The axons of the parasympathetic system (also called the *craniosacral system*) originate in the brain and in the sacral or lower end of the spinal cord. Axons from the cranial portion of the system innervate the iris of the eye, the salivary glands, the heart, respiratory passages, and the upper digestive tract. Axons from the sacral portion innervate the lower part of the digestive tract, the urinary bladder, and the reproductive organs.

As is indicated in figure 15.25, the ganglia of the parasympathetic system are located away from the spinal cord, within or very close to the organs innervated. Thus, the preganglionic axons are long and the postganglionic axons quite short. Parasympathetic postganglionic axons release the neurotransmitter acetylcholine at their terminals in or near the target organs.

The axons of the sympathetic system (also called the *thoracolumbar system*) originate in the thoracic and lumbar regions of the spinal cord. The sympathetic system innervates the same organs as the parasympathetic system, but the ganglia of the sympathetic system lie close to the spinal cord. Thus, the axons of the sympathetic preganglionic neurons are short and the axons of the postganglionic neurons are relatively long. Sympathetic postganglionic neurons release the neurotransmitter norepinephrine at their axon terminals in or near the target organs.

The difference in neurotransmitters released by postsynaptic neurons (that is, acetylcholine by the parasympathetic and norepinephrine by the sympathetic) is a key functional difference between the two systems. It is the basis of the antagonistic effects of parasympathetic and sympathetic neurons on target organs.

Autonomic Control

The autonomic nervous system innervates a large number of organs and has a wide variety of effects (table 15.2). Many organs are innervated by both sympathetic and parasympathetic neurons. Normally, sympathetic and parasympathetic neurons act antagonistically; one set of neurons *stimulates* the organ; the other *inhibits* it. For example, sympathetic nerves stimulate the heart to beat faster and stronger. Parasympathetic nerves cause the heart to contract more slowly and with less strength. On the other hand, the contractions (peristalsis) of the muscle in the digestive tract are speeded up by the parasympathetic system and slowed or even stopped by the sympathetic system.

The autonomic nervous system is essential to the maintenance of homeostasis, the dynamic, steady state of the organism's internal environment. In general, the parasympathetic system stimulates everyday activities and functions (digestion, urination, and so on). The parasympathetic thus might be regarded as the "housekeeper system." On the other hand, the sympathetic system promotes the ability of the body to meet emergencies, often called the "fight-or-flight" response. Stimulation by the sympathetic system, among other things, increases the heart rate, dilates the pupil of the eye (letting in more light), dilates the bronchioles of the respiratory tract (increasing gas-exchange efficiency), and inhibits digestion, which is not immediately essential during an emergency.

The Brain and Neural Reflexes

Although many somatic reflexes involve only the spinal cord, humans become conscious of what has transpired after the reflex has occurred and other interneurons have transmitted the information to the areas of the brain that are responsible for awareness.

Sympathetic   **Parasympathetic**

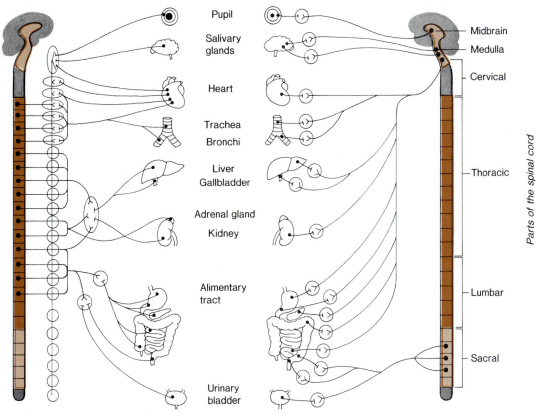

Figure 15.25 The sympathetic and parasympathetic divisions of the autonomic nervous system and the organs they innervate. Note the location of the sympathetic ganglia close to the spinal cord and the parasympathetic ganglia close to the target organs.

Table 15.2
Functions of the Autonomic Nervous System.*

Sympathetic Division	Parasympathetic Division
Dilates pupils of the eyes	Constricts pupils
Increases heart rate	Decreases heart rate
Increases strength of heart contractions	Decreases strength of heart contractions
Dilates bronchioles of respiratory system	Constricts bronchioles
Decreases peristalsis of digestive tract	Increases peristalsis
Probably inhibits the secretion of digestive enzymes	Stimulates the secretion of digestive enzymes
Relaxes urinary bladder	Contracts urinary bladder
Constricts blood vessels to genital organs	Dilates blood vessels to genital organs (causes erection)

*This list is not exhaustive. There are several other effects, particularly on the blood supply to various organs.

The brain can consciously override some reflexes. For example, the withdrawal reflex can be blocked by the cerebral cortex when necessary. When a patient allows someone to prick her finger with a lancet to obtain a blood sample, she is blocking this reflex.

Humans are never conscious of much of the sensory information that comes from the internal organs or of the functions of the autonomic nervous system. Information regarding changes in blood pressure, blood chemistry, urine formation by the kidneys, and other organ functions is constantly monitored by the brain at a level below consciousness, and the brain controls these functions through the sympathetic and parasympathetic nervous systems.

Summary

The nervous system provides a means of rapid communication in the animal body. It gathers information from both the external and internal environments, stores the information, and directs appropriate responses of the organism.

Neurons are the functional units of any nervous system. Neurons receive inputs, respond with electrical changes, and generate outputs to other cells. Many neurons have long fibrous processes along which nerve impulses are conducted.

The trend in the evolution of nervous systems has been toward larger numbers of neurons and greater centralization. Sensory neurons carry information toward the central nervous system, while motor neurons carry information away from the CNS toward effector cells. Interneurons are intermediate components involved in information processing.

The human central nervous system is made up of the brain and the spinal cord. The nerves of the peripheral nervous system connect to these two areas. Nerves contain sensory and motor neuron fibers. Somatic motor neurons innervate skeletal muscles. Autonomic neurons innervate internal effectors.

Neuron function in any nervous system depends on the irritability of the neurons, which is based on the neurons' electrical properties. Unequal concentrations of charged particles (ions) inside and outside a neuron's membrane result in a charge difference or electrical potential across the membrane. The resting membrane potential exists because of the selective permeability of the membrane and the action of the sodium-potassium ATPase pump.

The resting membrane is not very permeable to sodium ions, but is quite permeable to potassium ions. During an action potential, the membrane becomes much more permeable to sodium, which rushes inward along its concentration gradient. This results in a temporary reversal of the membrane's polarity. Then an outward flow of potassium restores the normal resting potential after a brief hyperpolarization.

An action potential is initiated as a result of small electrical changes in a voltage-sensitive area of a cell. If the threshold level is reached, an action potential occurs. The action potential is an all-or-none phenomenon; it proceeds in the same way each time it occurs. Following an action potential, there is a brief refractory period during which another action potential cannot be initiated.

The conduction of a nerve impulse involves a chain reaction series of action potentials sweeping along the neuron's plasma membrane. Each individual action potential influences nearby voltage-sensitive gates in sodium channels through the membrane. This triggers another action potential, which in turn triggers another, and so forth.

Impulse conduction velocities are proportional to the diameter of unmyelinated fibers—larger fibers conduct faster than small ones. Myelin sheaths are associated with rapid, saltatory impulse conduction in which depolarization occurs only at unmyelinated nodes.

Information is transmitted from neuron to neuron by way of synapses. In chemical synapses, neurotransmitters carry a message across the synaptic cleft. Some neurons are electrically coupled so that a nerve impulse can be conducted from cell to cell without a chemical intermediary.

In a chemical synapse, neurotransmitter molecules are released from synaptic vesicles through the presynaptic membrane. They then diffuse across the synaptic cleft and attach to receptor molecules in the postsynaptic membrane. Receptor molecules are associated with chemically gated ion channels so that their union with neurotransmitters can lead to ion flow and postsynaptic potentials. Some of these potentials are excitatory (EPSPs) and lead toward action potentials if the threshold is reached. Other postsynaptic potentials are inhibitory (IPSPs) and make it more difficult for the threshold to be reached. Most neurons in the central nervous system have both excitatory and inhibitory synapses impinging on them.

A number of chemicals act as neurotransmitters. A given neuron must have specific receptor molecules for a neurotransmitter if it is to respond to it. Neurotransmitters remain in the synaptic cleft for only a short time. Some are enzymatically degraded; others are actively recovered by the presynaptic cell.

The simplest neural pathways are reflexes that include one sensory cell and one motor cell in a reflex arc. More complex reflex arcs include one or several interneurons.

Visceral reflexes occur below the level of conscious control and operate by autonomic innervation of internal effectors. Autonomic innervation requires two motor neurons arranged in sequence. The axon of a preganglionic motor neuron reaches from the CNS to an autonomic ganglion, and the axon of a postganglionic motor neuron reaches to an effector.

The parasympathetic and sympathetic nervous systems usually act antagonistically; one stimulates a given process or activity while the other inhibits it. Generally, the parasympathetic system stimulates everyday activities in the body, and the sympathetic system stimulates activities needed in emergency situations.

Although humans are often aware of somatic reflexes, they are not aware of autonomic activity even though the brain continuously processes visceral reflexes.

Questions

1. How is the Na^+/K^+ ATPase pump electrogenic? (That is, how does it contribute to the resting membrane potential of a neuron?)
2. How does nerve-impulse conduction depend on voltage-sensitive (electrically gated) ion channels?
3. Why are chemically gated ion channels essential to synaptic functioning?
4. How is a neuron able to respond to several different neurotransmitters?
5. Explain why EPSPs can occur without the production of an action potential.
6. List three ways that autonomic innervation differs from somatic motor innervation.

Suggested Readings

Books

Bullock, T. H.; Orkand, R.; and Grinnell, A. 1977. *Introduction to nervous systems.* San Francisco: W. H. Freeman.

Miles, F. A. 1969. *Excitable cells.* London: William Heinemann Medical Books.

Usherwood, P. N. R. 1973. *Nervous systems.* Studies in Biology no. 36. Baltimore: University Park Press.

Articles

Gray, E. G. 1977. The synapse. 2d ed. *Carolina Biology Readers* no. 35. Burlington, N.C.: Carolina Biological Supply Co.

Iversen, L. I. September 1979. The chemistry of the brain. *Scientific American* (offprint 1441).

Jones, D. G. 1981. Ultrastructural approaches to the organization of central synapses. *American Scientist* 69:200.

Keynes, R. D. March 1979. Ion channels in the nerve-cell membrane. *Scientific American* (offprint 1423).

Patterson, P. H.; Potter, D. D.; and Furshpan, E. J. July 1978. The chemical differentiation of nerve cells. *Scientific American* (offprint 1393).

Schwartz, J. H. April 1980. The transport of substances in nerve cells. *Scientific American* (offprint 1468).

Shepherd, G. M. February 1978. Microcircuits in the nervous system. *Scientific American* (offprint 1380).

Stent, G. A., and Weisblat, D. A. January 1982. The development of a simple nervous system. *Scientific American.*

Stevens, C. F. September 1979. The neuron. *Scientific American* (offprint 1437).

The Brain and the Senses

Chapter Concepts

1. The central nervous system is composed of the brain and the spinal cord.
2. Sensory, association, and motor functions occur in specialized areas of the central nervous system.
3. Most neural functions involve two or more major regions of the central nervous system.
4. Sensory receptors respond to environmental changes with membrane potential changes. These changes cause nerve impulses to be transmitted to the central nervous system.
5. Each type of sensory receptor responds best to a specific kind of stimulus.

Of all scientific endeavors, one of the most fascinating is the study of the structure and function of the central nervous system. This awesome cellular circuitry, with billions of interconnected neurons, not only regulates the body's internal environment but allows humans to interact with the world in a rich and meaningful way.

The central nervous system is made up of the brain and the spinal cord. The spinal cord and the brain are actually continuous tissue, not separate organs. Together they look like a thick-walled tube that expands at one end. The brain is enclosed and protected by the bony cranium or skull. The spinal cord, which extends from the base of the skull, is protected by the vertebral column (figure 16.1).

Functionally, the spinal cord and brain play very different roles. The brain carries out the higher functions of the central nervous system. It decodes sensory information, evaluates it, and directs appropriate responses. The spinal cord acts as a liaison between many elements of the peripheral nervous system and the higher centers of the brain. It brings sensory information to the brain and carries motor commands from it. The spinal cord also generates some responses (reflexes) that are accomplished independently of the brain.

The Spinal Cord

Spinal nerves containing sensory and motor fibers are attached to the spinal cord. In humans there are thirty-one pairs of spinal nerves. Near the spinal cord, each spinal nerve splits into two branches—a **dorsal root** and a **ventral root,** carrying sensory and motor information, respectively. (There also are other nerves connecting to the central nervous system. The twelve pairs of **cranial nerves** connect directly to the brain.)

A cross-sectional slice of the spinal cord is shown in figure 16.2. As the figure shows, the spinal cord is divided into **gray matter** and **white matter.** The center of the cord is gray because it is composed mostly of cell bodies and unmyelinated fibers. The gray matter has small projections on opposite sides called the dorsal and ventral horns. The surrounding area is white because it is made up mostly of **tracts** of myelinated fibers. (Bundles of neuron fibers are called *nerves* in the peripheral nervous system and *tracts* when they run from one point to another within the central nervous system.) In the very center of the spinal cord is a hollow canal. This canal runs throughout the central nervous system. In the brain, the canal expands into four cavities called **ventricles** (see figure 16.1). The central canal of the spinal

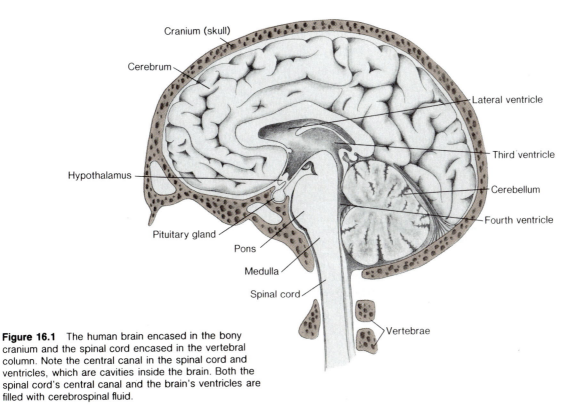

Figure 16.1 The human brain encased in the bony cranium and the spinal cord encased in the vertebral column. Note the central canal in the spinal cord and ventricles, which are cavities inside the brain. Both the spinal cord's central canal and the brain's ventricles are filled with cerebrospinal fluid.

Labels: Cranium (skull), Cerebrum, Hypothalamus, Pituitary gland, Pons, Medulla, Spinal cord, Vertebrae, Lateral ventricle, Third ventricle, Cerebellum, Fourth ventricle

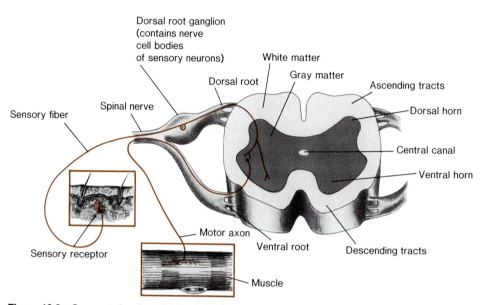

Dorsal root ganglion
(contains nerve
cell bodies
of sensory neurons)

White matter

Gray matter

Dorsal root

Ascending tracts

Sensory fiber

Spinal nerve

Dorsal horn

Central canal

Ventral horn

Sensory receptor

Motor axon

Ventral root

Descending tracts

Muscle

Figure 16.2 Cross section through the spinal cord showing white and gray matter. Note also the spinal nerve, dorsal root, and ventral root. Arrows indicate directions of nerve impulse conduction—toward the spinal cord via sensory fibers and away from the spinal cord via motor axons.

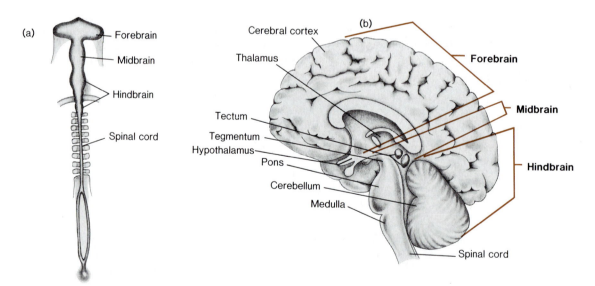

(a)

Forebrain

Midbrain

Hindbrain

Spinal cord

(b)

Cerebral cortex

Thalamus

Tectum

Tegmentum

Hypothalamus

Pons

Cerebellum

Medulla

Forebrain

Midbrain

Hindbrain

Spinal cord

Figure 16.3 Organization of the brain. (a) The major brain regions—forebrain, midbrain, and hindbrain—as they appear early in the development of a vertebrate embryo. (b) Structures present in the major brain regions of the fully developed adult brain. The hindbrain is composed of medulla, pons, and cerebellum. The tectum and tegmentum comprise the midbrain. The forebrain is composed of hypothalamus, thalamus, and cerebrum.

cord and the ventricles are filled with **cerebro-spinal fluid.** This fluid carries nutrients and chemical signals and also protects the central nervous system from damage by providing an inner cushion.

The dorsal and ventral sides of the spinal cord are specialized to handle sensory and motor information, respectively. Sensory information from the spinal nerves enters the spinal cord through the dorsal roots, while motor information from the spinal cord is sent to spinal nerves through the ventral roots. In the gray matter, dorsal cells function primarily in receiving sensory information, while ventral cells are primarily motor. Within the white matter of the spinal cord, ascending tracts of dorsal axons take information to the brain, while descending tracts in the ventral part of the cord primarily carry information down from the brain.

The Brain

Even without the aid of a microscope, it is obvious that the brain is a highly differentiated structure. The brain is not an amorphous collection of cells, but an anatomically specialized system containing discrete **nuclei** (clusters of nerve cell bodies) and tracts. It has characteristic grooves and ridges on its surface. Researchers have gained an elementary knowledge about the discrete parts of the brain and their functions.

Early in the development of a vertebrate embryo, three swellings arise in the developing brain region. These embryonic divisions form the basis for naming the general brain regions: **hindbrain, midbrain,** and **forebrain** (figure 16.3). The hindbrain is composed of the medulla, pons, and cerebellum; the midbrain is composed of the tectum and tegmentum; and the forebrain includes the hypothalamus, thalamus, and cerebral cortex.

The Hindbrain

The first structure within the brain, beginning just above the spinal cord, is the **medulla.** By virtue of its position, structure, and function, the medulla can almost be thought of as an extension of the spinal cord. All the ascending and descending tracts of the spinal cord pass through the medulla, and it also connects with some of the cranial nerves. Although a major function of this part of the brain is to relay signals to and from the forebrain, nuclei in the medulla play an important role in such visceral reflexes as control of the heart rate, digestion, and breathing—basic functions that are essential to life.

The **pons** is the next structure upward. Like the medulla, it contains ascending and descending tracts and receives input from cranial nerves. The pons has another important function: it funnels tracts into the cerebellum, the third major region of the hindbrain.

The **cerebellum** is devoted especially to the control of movement. It is a bulblike structure extending from the pons and contains two halves or **hemispheres.** The cortex, or outer portion, of the cerebellum is gray matter, which means that nerve cell bodies are located close to its outer surface. The cerebellum is one of three regions of the brain that have a gray cortex; in all other parts of the brain, fiber tracts lie near the outside and nerve cell bodies are confined to the interior, as in the spinal cord.

Underneath the cortex of the cerebellum are tracts of white matter and clusters of cell bodies in various nuclei. The cerebellum is responsible for fine tuning muscle activity and is especially important in controlling rapid movements. The cerebellum functions like a computer to keep track of movements initiated by the forebrain. It compares actual movements with desired movements and provides correctional signals as needed. To accomplish this, the cerebellum connects to cell groups in many parts of the brain and spinal cord and receives a rich supply of sensory signals. If the cerebellum is damaged, jerky body movements result because the corrections needed for coordination are no longer made.

The Midbrain

The **midbrain** is the small part of the central nervous system that contains the **tegmentum** (ventral portion of the midbrain) and the **tectum** (dorsal portion of the midbrain). The midbrain connects the hindbrain and the forebrain. All of the ascending and descending tracts between the hindbrain and the forebrain pass through the tegmentum. Together, the medulla, pons, and midbrain form the **brain stem,** linking the spinal cord and the forebrain.

The tectum, which has a gray cortex, receives sensory inputs from the eyes and ears. All auditory information enters the tectum before ascending to higher centers in the cerebrum. In lower vertebrates, all visual information is processed in the tectum, but in animals such as mammals, which have highly developed forebrains, this processing is done by visual centers in the cerebral cortex.

The Forebrain

The **forebrain,** the largest part of the human brain, handles the highest neurological functions. Sensory decoding, the initiation of movement, and the capacity for memory, language, and reasoning all reside in the forebrain. Some important derivatives of the embryonic forebrain are the **hypothalamus,** the **thalamus,** and the **cerebral cortex.** The forebrain also contains major portions of the limbic system, an important network of nuclei and structures that are linked together in a network of functional interactions.

Just above the midbrain is the hypothalamus, a tiny region containing several nuclei (clusters of cell bodies). Although the hypothalamus is small, weighing only about 4 g in humans, it fulfills several major functions. The hypothalamus is the central nervous system center responsible for regulating the autonomic nervous system. It controls the release of hormones from the pituitary gland. And vital controls over hunger and satiety, thirst, osmotic balance, body temperature, metabolic rate, and circulation also are under the influence of various nuclei in the hypothalamus.

The hypothalamus is also associated with the emotions, ranging from anger and hostility to pleasure. But the hypothalamus is not the only brain center associated with emotion. It is, in fact, part of a complex of interconnected brain parts that are involved with emotion. These parts are known collectively as the **limbic system.**

The limbic system is a network of nuclei linking parts of the forebrain and midbrain. It is believed that the limbic system contains the neuronal circuitry responsible for emotional and instinctive responses. The major structures within the limbic system are shown in figure 16.4. Functions of some of these structures are partially understood. Direct stimulation of the amygdala, for example, can provoke outbursts of extreme rage, while the removal of the amygdala produces passive, docile behavior in experimental animals. Another structure, the hippocampus, very likely plays an important role in memory processes. Major lesions in the human hippocampus have been known to block almost totally the ability to store new memories. An individual with hippocampal lesions may not even remember having read the beginning of a magazine by the time he or she reaches its end.

The **thalamus** is another major relay center of the brain. It is an oval-shaped structure deep within the forebrain and is made up of a large number of nuclei responsible for collecting, processing, and relaying sensory inputs to appropriate areas of the cerebral cortex.

The Cerebral Cortex

The forebrain is the largest and most striking part of the human brain, as it accounts for over half of the brain mass; and the **cerebral cortex** is by far the largest part of the forebrain. The cerebral cortex is a large mass of gray matter containing billions of nerve cell bodies. It is divided bilaterally into two **cerebral hemispheres.** The cerebral cortex is responsible for many of the sophisticated neurological capacities that characterize humans and other advanced vertebrates.

During the evolution of mammalian brains, the amount of gray cortex material has increased, along with the number of ridges (called **gyri;** singular: **gyrus**) and valleys (called **sulci;** singular: **sulcus**) in the cortex (figure 16.5). The surface of the cerebral cortex is not randomly folded; the locations of the gyri and sulci are always essentially the same in the brains of individuals of any given species. Because of this, the surface of the cortex can be mapped. In humans and other mammals, it has been divided into four major lobes: the **frontal, parietal, occipital,** and **temporal** lobes (figure 16.6). Different functions are associated with each lobe. For example, the frontal lobe controls motor functions and helps regulate and sequence behavior. The parietal lobe deals with somatosensory or touch inputs and spatial abilities. The occipital lobe handles visual functions. The temporal lobe processes both auditory and visual inputs and is associated with language. However, these are not the only functions associated with each of the lobes. These functions have been mapped only because they are fairly easy to identify. Much of the vast complexity of the cerebral cortex is yet to be unraveled.

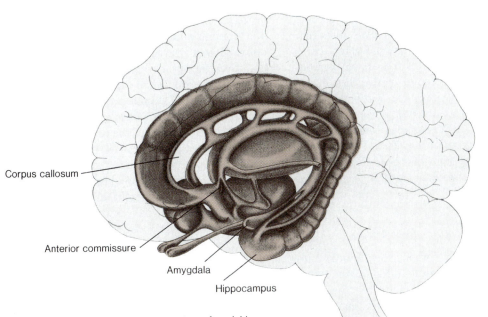

Corpus callosum

Anterior commissure

Amygdala

Hippocampus

Figure 16.4 The limbic system, a system of nuclei in the forebrain and brain stem.

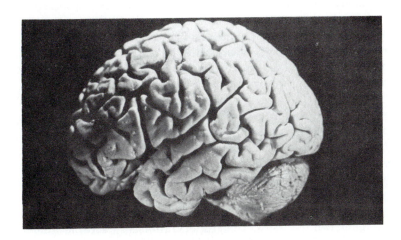

Figure 16.5 Photograph of the human brain showing the many convolutions of the cerebral cortex. Each ridge is a gyrus. Each groove is a sulcus. The surface area of the cortex is greatly increased by such folding.

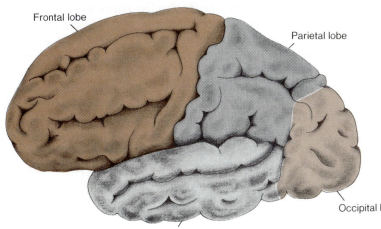

Frontal lobe

Parietal lobe

Occipital lobe

Temporal lobe

Figure 16.6 The lobes of the cerebral cortex.

Underneath the gray matter of the cortical surface run the billions of myelinated fibers that make up the white matter. Some tracts carry signals from other parts of the central nervous system toward the cerebral cortex, and others carry signals away from the cerebral cortex. These are known as ascending and descending tracts, respectively. Tracts also connect various regions within the cerebrum itself. For example, the right and left hemispheres of the cerebrum are connected by major transverse tracts called **commissures.**

Regional Specialization of the Cortex

Some regions of the cortex are specifically involved in receiving sensory information. Other regions are associated with motor functions, and yet other areas are not specifically affiliated with either function. These latter regions are called association areas.

One region of the parietal lobe, called the **somatosensory cortex,** handles information about the sense of touch. Other areas process visual and auditory information. These sensory-specific regions of the cerebral cortex are called **primary projection areas** and are named according to the senses with which they are affiliated (figure 16.7).

Within the somatosensory cortex are subdivisions that receive sensory information from specific body regions. Each of these body regions is called a **receptive field.** The receptive fields associated with different neurons in the somatosensory cortex can be determined by monitoring activity in single neurons or in groups of neurons when different parts of the body are stimulated. The somatosensory cortex can also be stimulated directly with electrical probes, and the subject will perceive certain parts of the body as having been touched. Both approaches furnish the same "map" of the body projected onto the cerebral cortex. Figure 16.8 shows the somatosensory homunculus, or miniature person, that symbolically illustrates what parts of the body map onto what parts of the somatosensory cortex. As the figure shows, some parts of the body are unequally represented in the projection area. Regions of the body with large numbers of sensory receptors have the greatest representation. The neural area serving the hand, for example, is as large as the area serving both torso and leg. In animals that rely less on forelimb or finger contacts and more,

perhaps, on the tactile stimulation of whiskers, the forelimb would have a smaller projection and the nose area a larger one.

Specific sensory information also goes to the **association areas** of the cerebral cortex. The association areas of human and primate brains are much larger than in other vertebrate animals. Each of the four lobes of the cerebral cortex contains one association area. In the primate brain, about 80 percent of the cortex is made up of association areas, and only about 20 percent by projection areas. Although they make up such a large and important part of the cerebral cortex, very little is known about the association areas. As suggested by their name, the association areas take specific information from the projection areas and *associate* the information into higher, more complex levels of consciousness. The association regions are linked to the projection areas by massive fiber tracts. They are connected with one another in each hemisphere by association bundles and are linked with areas on the other side of the brain by commissures.

The association areas are correlated with major neurological functions. The parietal association area, for example, plays a role in the subjective picture a human being has of his or her spatial environment and body image. Thus, visual, touch, and positional information all converge in this area of the brain. One way of discovering what different association areas do is to study neurological deficits caused by damage to specific areas. In humans, such deficits result from trauma such as strokes or gunshot wounds that damage small, localized areas. Certain lesions of the parietal association area, for example, cause what is known as visual agnosia. Agnosia is the loss of ability to recognize the significance of sensory stimuli. Patients with visual agnosia have normal eyesight, as judged by the ability to reach for objects or to avoid obstacles, but they cannot recognize objects by sight. However, if persons with visual agnosia can touch the object, they can identify it through their sense of touch. Other higher intellectual and cognitive functions of the nervous system also depend on the association areas. The capacity for speech is dependent on normal temporal lobe function, for example, and the organization of behavior patterns depends on the frontal lobe.

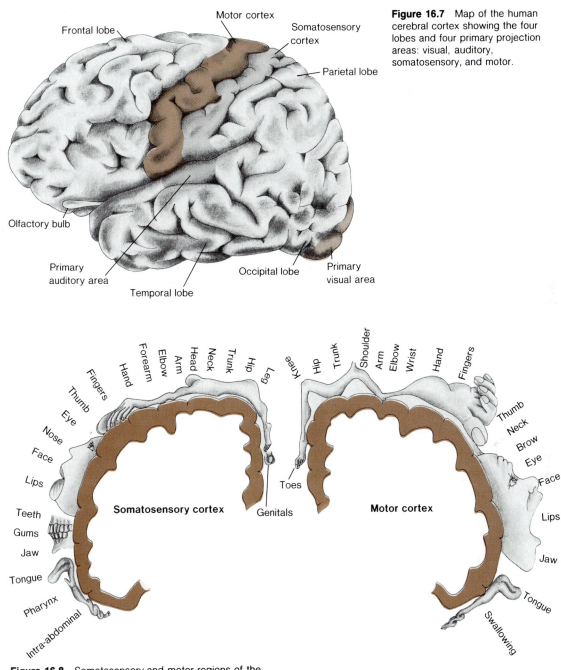

Frontal lobe

Motor cortex

Somatosensory cortex

Parietal lobe

Olfactory bulb

Primary auditory area

Temporal lobe

Occipital lobe

Primary visual area

Figure 16.7 Map of the human cerebral cortex showing the four lobes and four primary projection areas: visual, auditory, somatosensory, and motor.

Fingers
Thumb
Eye
Nose
Face
Lips
Teeth
Gums
Jaw
Tongue
Pharynx
Intra-abdominal

Hand
Forearm
Elbow
Arm
Head
Neck
Trunk
Hip
Leg

Knee
Hip
Trunk
Shoulder
Arm
Elbow
Wrist
Hand
Fingers

Thumb
Neck
Brow
Eye
Face
Lips
Jaw
Tongue
Swallowing

Somatosensory cortex

Toes

Genitals

Motor cortex

Figure 16.8 Somatosensory and motor regions of the cerebral cortex. The somatosensory projection area is located in the parietal lobe. The motor cortex is part of the frontal lobe (see figure 16.7). Only half of each cortical region is shown here, the left somatosensory area and the right motor cortex. Note the large proportion of both devoted to the face and hands and the relatively small areas devoted to the trunk and limbs.

Thus, sensory information is decoded and interpreted in the primary projection areas and association areas of the cerebral cortex. But axons from the sensory association areas also feed back to the limbic system, affecting emotion and motivation. And, of course, sensory information often triggers some type of appropriate motor response.

Motor Functions

The cortical area that controls movement, the **motor cortex,** is located in the frontal lobe of the cerebrum (figure 16.7). Together with the cerebellum and the descending tracts in the spinal cord, the motor cortex is responsible for the control and direction of body movement. As is the case with the somatosensory cortex, the motor cortex has been mapped for various body areas. Motor problems can be correlated with damage to specific cortical areas, and a general map of motor cortex functioning has been derived (figure 16.8).

Although body movements are generated in the motor cortex, the **cerebellum** influences posture and integrates signals necessary for rapid skilled movements. For example, perfect timing is necessary for smoothly coordinated muscular activity; movements are awkward and jerky if groups of muscles do not contract and relax at just the right instants. Coordinating this timing is the responsibility of the cerebellum. The neural activity involved is exceedingly fast and not under voluntary control. Lesions to the cerebellum cannot be compensated for by learned behavior, and cerebellar disorders result in uncoordinated muscular activity.

To coordinate activity, the cerebellum monitors vast amounts of relevant information. It receives information concerning the position and movement of muscles, their tension and frequency of contraction. It receives information about what muscles and limbs have just done and also what they are about to do. The cerebellum analyzes data from the cerebral cortex and from many of the body's sensory receptors: eyes, ears, touch, and receptors in the muscles and joints. The output of the cerebellum goes to the motor area of the cerebral cortex and to the motor centers of the brain stem. The cerebellum both influences the excitability of motor centers and modifies subsequent commands from the brain to the muscles.

The Reticular Formation

Stretching along the entire brain stem, from the spinal cord up into the beginning of the forebrain, is a diffuse collection of many nuclei with connecting fibers (axons) that together make up the **reticular formation** (figure 16.9). The reticular formation is more a network of small nuclei and fibers than a discrete structure. Together, these nuclei control the level of "arousal" of the cerebral cortex.

For the cortex to respond to incoming sensory information, it needs excitation from neurons in the reticular formation. The neurons that provide this excitation make up the **ascending reticular activating system (ARAS)** (figure 16.9). Many sensory fibers coming from the eyes, ears, and other external sensory receptors synapse in the reticular formation, which then evaluates their sensory signals and stimulates the cortex, if appropriate, to be ready to receive the incoming information. The drifting of an individual from full and undivided attention to drowsiness to deep sleep is determined by signals from this system.

If the ARAS of a sleeping animal is stimulated, the animal immediately awakens. Sustained wakefulness seems to require that the ARAS be continually active. It is kept active, apparently, by stimulation from sensory receptors and by neuronal feedback from the cerebral cortex. During sleep, the ARAS ceases to be active and, therefore, ceases to stimulate the cerebral cortex.

Special Senses

During the course of evolution, cells have evolved that can change or transduce thermal, mechanical, chemical, and electromagnetic energies into neuronal signals. Such energy-transducing cells are called **receptor cells.** There are several types of receptor cells, and cells of each type respond best to only one type of energy, called the cell's **adequate stimulus.** Light, for example, is the adequate stimulus for receptor cells in the eye.

While a receptor cell responds *best* to only one type of energy, it also can respond to other types. For example, the phosphenes, or flashes of light, that one sees when eyes are rubbed (or punched!) are caused by receptor cells in the eye that have been activated by the mechanical energy of pressure transmitted to them through eye

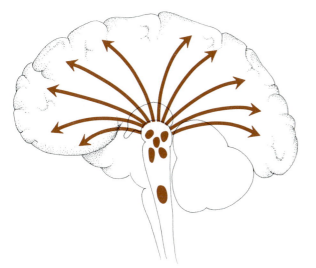

Figure 16.9 The reticular formation, consisting of several nuclei with connecting fibers, runs through the brain stem. Functionally, the reticular formation can be called the ARAS (ascending reticular activating system). It apparently directs the cerebral cortex to pay attention to some stimuli while ignoring others and helps maintain a conscious state. The ARAS is inactive during sleep.

fluid. However, the mechanical energy level required to trigger this sort of response is much higher than the energy level required for the eye's receptor cells to respond to their adequate stimulus, light.

Receptors usually are classified by the type of stimulus to which they respond. **Photoreceptors,** for example, respond to light. Some **thermal receptors** respond to heat, others to cold. **Mechanoreceptors,** such as touch receptors in the skin or the receptors in the ear that pick up vibrations in the air, respond to mechanical stimuli. **Chemoreceptors** in the tongue and the nasal passages are fundamental to the senses of taste and smell.

Most people are familiar with the five basic senses: touch, smell, taste, hearing, and sight. Actually, as shown in table 16.1, human beings have many more than five senses. Both internal and external environments are monitored, some at a conscious level and others below the level of consciousness.

How Receptors Function

Receptors respond to various types of energy, which they transduce into membrane electrical potential changes. Energy in the form of the proper stimulus is absorbed somewhere in the cell and changes in the cell's permeability occur in response to it. Changes in membrane permeability give rise to ion fluxes that lead to **receptor potentials.** If receptor potentials exceed threshold, action potentials are generated in sensory neurons.

Two general categories of receptors are separated on the basis of the identity of the specific responding cells functioning in the receptors. When a sensory neuron itself acts as a receptor cell, it is called a **primary receptor.** One type of primary receptor in humans is a pain receptor. Pain receptors are simply the unmyelinated ("naked") ends of the fibers of sensory neurons. Receptors that have a specialized type of receptor cell *in addition to* the sensory neuron are called **secondary receptors.** Examples of secondary receptors are the rods and cones in the retina of the eye, hair cells in the inner ear, and taste cells in taste buds.

Table 16.1
Types of Sensory Receptors.

Sense	Receptor	Location
Vision	Rods and cones	Retina of the eye
Hearing	Hair cells	Organ of Corti in the inner ear
Smell	Olfactory neurons	Lining of the nose
Taste	Taste receptor cells	Taste buds in the mouth
Rotational movement	Hair cells	Semicircular canals of the inner ear
Touch	Nerve endings	Skin
Heat	Nerve endings	Skin
Cold	Nerve endings	Skin
Pain	Naked nerve endings	Throughout body
Joint position and movement	Nerve endings	Joint
Muscle length	Muscle spindle	Muscles
Muscle tension	Golgi tendon organ	Tendons
Arterial blood pressure	Stretch receptors	Carotid sinus and aortic arch
Venous blood pressure	Stretch receptors	Venae cavae and right atrium
Inflation of lung	Stretch receptors	Lung tissue
Internal body temperature	Nerve endings	Hypothalamus of the brain
Osmotic pressure of blood	Osmoreceptor cells	Hypothalamus
Blood sugar level	Glucostatic cells	Hypothalamus
Blood oxygen levels	Chemoreceptor cells	Carotid and aortic arteries

Several general properties of sensory function apply to essentially all of the sensory receptors. One important property is the ability of the response to vary with the **intensity** of the stimulus. A sensory receptor responds to a more intense stimulus with a higher **frequency** of action potentials. Above a threshold stimulus level, the greater the intensity of the stimulus, the greater the receptor potential, and the greater the frequency of action potentials (and impulses) generated in the sensory neuron. The frequency of impulses becomes a code for the brain to determine the strength of the stimulus.

The brain also can distinguish the location and type of stimulus. This is possible because of the discrete sensory pathways to the brain and the discrete sensory projection areas in the cerebral cortex. The "map" of the somatosensory area of the cerebral cortex was discussed earlier in this chapter. Other senses also have their own pathways to their appropriate sensory projection areas in the cortex. Thus, the brain can discriminate the type of stimulus, the intensity of the stimulus, and the location from which the stimulus came.

Another interesting property of sensory reception is sensory adaptation. In sensory adaptation, the response of receptor cells declines, even though the stimulus is sustained at the same level of intensity. Some receptors adapt relatively quickly, while others adapt quite slowly. For example, touch receptors of the skin adapt quickly.

This is fortunate for individuals not wishing to be continually reminded of the shirts on their backs or the rings on their fingers. Pain receptors, however, adapt quite slowly. Slow adaptation to pain actually is a safety device. If pain were easy to ignore, its function as a warning system would be much less effective.

Sensory Receptors in the Skin

The skin has receptors for a number of senses, including touch-pressure, pain, warmth, and cold. Mechanoreceptors of the touch-pressure system have been studied in the greatest detail, while relatively little is known about the thermoreceptors for warmth and cold. Receptors for pain respond to stimuli that actually damage tissue. It is thought that pain receptors may, in fact, respond to chemicals liberated from damaged cells.

Sensory receptors of the skin, as well as those in other parts of the somatosensory system, actually are the ends of neuronal processes sent out from nerve cell bodies in the dorsal root ganglia. These sensory neuron ends may be encapsulated by specialized structures, or they may be naked nerve endings. As can be seen from figure 16.10, receptor terminals of the somatosensory system have a variety of shapes. The structure of the naked nerve endings of pain receptors clearly contrast with the encapsulated Meissner's corpuscle (a receptor for touch) and the Pacinian corpuscle (a receptor for pressure and vibration).

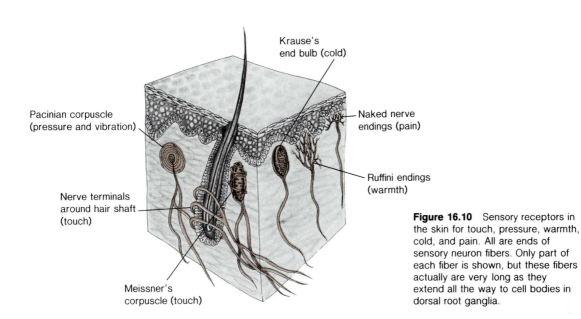

Krause's
end bulb (cold)

Pacinian corpuscle
(pressure and vibration)

Naked nerve
endings (pain)

Ruffini endings
(warmth)

Nerve terminals
around hair shaft
(touch)

Meissner's
corpuscle (touch)

Figure 16.10 Sensory receptors in the skin for touch, pressure, warmth, cold, and pain. All are ends of sensory neuron fibers. Only part of each fiber is shown, but these fibers actually are very long as they extend all the way to cell bodies in dorsal root ganglia.

Smell

The sense of smell, or **olfaction,** employs primary sensory neurons for capture and transduction of stimuli into electrical responses. Olfactory receptors are neurons embedded in a mucous membrane of the nasal passage. This membrane also contains non-neuronal cells, some of which secrete a watery fluid or mucus (figure 16.11).

The olfactory bipolar cells that are the actual chemoreceptors have what appear to be modified cilia extending into the layer of mucus on the surface of the membrane. Volatile chemicals dissolved in the watery mucus interact with the olfactory cilia, in some way initiating receptor potentials. Minute quantities of many different kinds of chemicals can initiate potentials in these receptors. At the other end of the neurons, the very short **olfactory nerves** extend toward the brain. The olfactory nerves reach the brain through tiny holes in the base of the skull, and they end in the nearby **olfactory bulbs.** From the olfactory bulbs, fibers form the **olfactory tracts,** which go directly to the olfactory region of the cerebral cortex.

Olfaction is unique among the specific senses in that it bypasses the thalamus, the major sensory relay station, and sends signals directly to its primary sensory region in the cortex. Olfaction is a much more important sense in many other mammals than it is in humans, and the olfactory bulbs of these mammals comprise a much larger fraction of the total brain mass. Although less

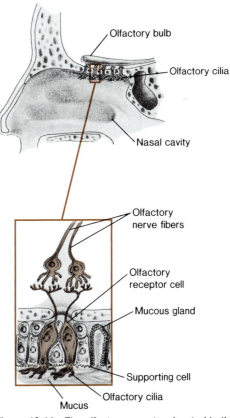

Olfactory bulb

Olfactory cilia

Nasal cavity

Olfactory
nerve fibers

Olfactory
receptor cell

Mucous gland

Supporting cell

Olfactory cilia

Mucus

Figure 16.11 The olfactory receptors located in the olfactory epithelium of the upper nasal passages. The olfactory receptors are neurons. Their cell bodies are embedded between supporting cells in the epithelium. Fibers project to the surface of the epithelium, where they bear special structures known as olfactory cilia. The cilia are the sites for interaction with the chemical stimuli.

important to humans, human olfaction nonetheless is quite acute and can be made even more so with practice and training. For example, trained perfumers can discriminate among hundreds of fragrances.

Taste

Taste "buds" in the tongue contain chemoreceptors specialized for registering sensations of sweet, salt, bitter, and sour. The sense of taste is neither so sensitive nor so diverse as the sense of smell. Indeed, a great deal of the "taste" of food actually comes from olfactory sensations received during eating, and the olfactory sense is dulled when smell is impaired, as it is by the common cold.

The structure of a taste bud is shown in figure 16.12. Receptors for taste are secondary sense cells, not neurons. Chemicals in food stimulate receptor potentials in the receptor cells, and receptor potentials in these receptor cells trigger the release of neurotransmitters that excite neurons surrounding the receptor cells. When activated by neurotransmitters from the receptors, axons of these neurons send impulses into the medulla. Messages then are relayed to the thalamus and ultimately to the cerebral cortex. It is interesting, however, that neither nerve impulses, nor even action potentials, develop in the secondary receptor cells that actually respond to chemicals. In this case, transduction from a stimulus (presence of a chemical) to neuronal information (impulses conducted toward the brain) is a several-step process.

Sense Organs of the Ear

The **ear** is considerably more complex than the other sensory structures discussed so far. It houses two quite different senses, **hearing** and **balance.** The organs for these senses are found in the inner part of the ear and are associated with several fluid-filled chambers. One of these chambers is the **cochlea,** which contains the **organ of Corti.** The organ of Corti contains specialized cells that initiate the auditory response. Three other chambers, the **semicircular canals,** respond to movements of the head and are responsible for the sense of balance. Receptors in both cases belong to the general receptor class called mechanoreceptors.

Hearing

The human ear has three general areas: the outer, middle, and inner ears (figure 16.13). The outer ear, which includes the **pinna** and the **auditory canal,** plays little direct role in the actual sensory functioning of the human ear, but some other animals, such as dogs, can direct their pinnae for better hearing.

The eardrum or **tympanic membrane** separates the outer ear and the middle ear. The transduction process for the sense of hearing begins in the eardrum, although neurons are not involved at that point. Sound waves elicit vibrations in the membrane, and induced vibrations occur at the same frequency as the sound. Within the middle ear, the vibrations of the eardrum are passed on to three miniature bones (**ossicles**). These three bones, commonly known as the **hammer (malleus), anvil (incus),** and **stirrup (stapes)** because of their shapes, cause the vibrations of the eardrum to be reproduced on another membrane located at the **oval window.** When the eardrum moves inward, the stirrup, connected to the oval window, also moves inward. The system is arranged so that vibrations of the large eardrum are fully transmitted to the much smaller oval window. This results in pressure increase that is needed to stimulate the movement of fluid within the inner ear. Movement of fluid in the inner ear triggers receptor potentials in the organ of Corti.

The middle ear has a reflex function that protects against sounds of too great intensity. Muscles in the ear contract to dampen the pressure exerted at the oval window, somewhat like the way the pupil of the eye constricts in bright light. The middle ear also is protected against rupture of the tympanic membrane, which could result from pressure differences between the middle ear and outer ear. The **eustachian tube,** which connects the middle ear cavity with the pharynx (throat), permits pressure equalization on the two sides of the membrane. Because of this passageway, for example, discomfort from pressure changes associated with airplane takeoffs and landings can be relieved by yawning and swallowing.

The oval window separates the middle ear from the inner ear. Vibrations in the oval window, initiated by pressure from the stirrup, induce movement of the fluid in the **cochlea.** The three-dimensional structure of the cochlea and its internal compartments looks something like a snail

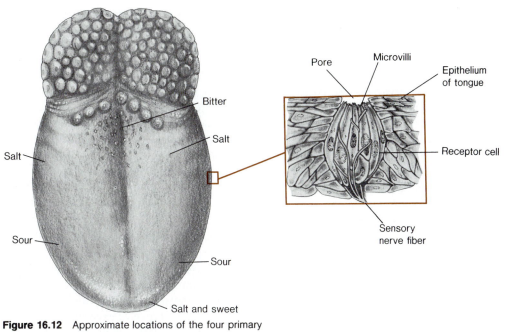

Figure 16.12 Approximate locations of the four primary tastes on the human tongue. The schematic representation of a taste bud on the right shows the receptor cells with their "hairlike" microvilli where the chemical stimuli contact the cells. There are also supporting cells around the taste bud. Sensory neurons synapse with the taste receptor cells and carry signals to the brain.

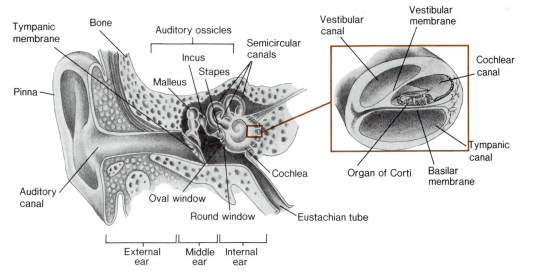

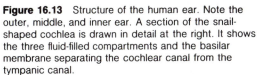

Figure 16.13 Structure of the human ear. Note the outer, middle, and inner ear. A section of the snail-shaped cochlea is drawn in detail at the right. It shows the three fluid-filled compartments and the basilar membrane separating the cochlear canal from the tympanic canal.

shell. Internally, the cochlea is divided by two membranes into three fluid-filled canals. The three canals, shown in cross section in figure 16.13, are called the **vestibular canal,** the **cochlear canal,** and the **tympanic canal.**

The **basilar membrane** separates the cochlear canal from the tympanic canal. The secondary sensory cells for hearing are located in the basilar membrane and are mechanoreceptors called **hair cells.** The **organ of Corti** includes the hair cells, together with nearby support cells, the basilar membrane, the auditory primary sensory neurons, and the overlying tectorial membrane. The organ of Corti is shown in figure 16.14.

Vibrations of the oval window cause movements of the fluids within the cochlea, and the moving fluid stimulates vibrations in the basilar and tectorial membranes. The tectorial membrane vibrates against the hair cells, setting up forces that bend the hairs. This bending motion causes receptor potentials in the hair cells, and changes in their membrane potentials lead to synaptic communication with the dendrites of auditory neurons localized at the base of the hair cells. This synaptic transmission from the hair cells generates action potentials and nerve impulses in the auditory neurons.

The movement of noncompressible fluid within the cochlea is possible because of the expandable membrane at the **round window** (figure 16.13). Stretched across the round window is a membrane that bulges into the middle ear as the membrane of the oval window bulges into the inner ear.

Although biologists understand how sound is captured, they have only a partial understanding of how **pitch** and **loudness** are discriminated. It is known that sounds of different pitches cause different parts of the basilar membrane to vibrate. Thus, low frequency (low pitch) vibrations stimulate the basilar membrane and hair cells near the apex (distal end) of the cochlea. High frequency (high pitch) vibrations stimulate the basilar membrane and hair cells near the base of the cochlea. Intermediate frequencies stimulate hair cells in intermediate regions of the cochlea. Sensory neurons from each of these regions in the cochlea lead to slightly different areas in the brain, and perception of pitch depends on which of these brain areas receives impulses from sensory neurons.

Loudness or intensity of sound, on the other hand, seems to be a function of the *number* of sensory neurons stimulated at a given time and the *frequency* of impulses that they carry. This probably is because louder sounds lead to greater vibrations in the cochlear fluid, greater vibrations of the basilar membrane, and more intense stimulation of hair cells. The brain interprets the increased number of nerve impulses received as a louder sound.

Balance

Three other fluid-filled chambers in the inner ear have similar properties to the cochlear receptor mechanisms, but their function is related to balance and movement rather than hearing. These chambers, shown in figure 16.13, are called the **semicircular canals.** Hair cells within these canals are stimulated by pressure exerted on them by the fluid that the canals contain. Pressure occurs because, as the head turns, the hair cells move with the head but the fluid lags slightly behind. This differential movement deforms the hair cells, causing receptor potentials. Receptor potentials in the hair cells cause synaptic transmission to sensory neurons. The sensory neurons respond with action potentials and nerve impulses that go to the cerebellum. The three semicircular canals are oriented so that movements in any direction are detected. The brain integrates the different impulses it receives from each of the three canals and, therefore, can determine the direction and rate of movement. Very rapid or prolonged movements of the head, such as a continual rocking motion in a boat, may cause uncomfortable side effects, typically vertigo or nausea. Seasickness is a reaction of the body to continual activation of receptors in the semicircular canals.

Vision

Figure 16.15 traces the visual pathway—from photoreceptors in the eye to the primary visual cortex in the occipital lobe. The reception of visual information, however, begins with cellular responses to light, which take place in the **retina,** the photosensitive (light sensitive) part of the eye.

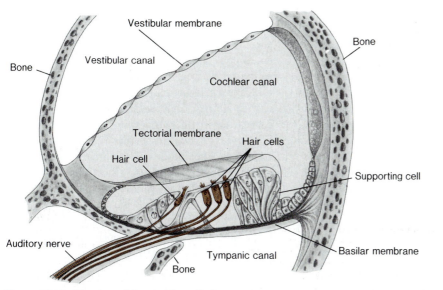

Figure 16.14   Structure of the cochlea with the organ of Corti. Note the hair cells, the sensory neurons with which they synapse, the basilar membrane, and the overlying tectorial membrane.

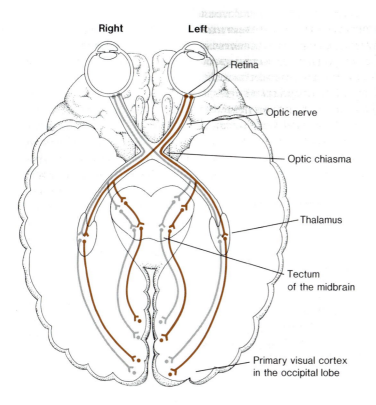

Figure 16.15   The visual pathway involves the retina of each eye, optic nerves, the optic chiasma (where axons from each eye cross to the other side of the brain), the thalamus, the midbrain, and the visual projection area in the occipital lobe of the cerebral cortex.

The retina is in the back of the eye, and the rest of the eye essentially is an optical instrument that focuses light on photoreceptors in the retina (figure 16.16).

Three external features of the eye are readily visible: the **sclera** (the white part of the eye), the **iris** (the colored part of the eye), and the **pupil** (an opening that appears as the dark center of the iris). Light passes into the eye through the pupil, and reflex-controlled smooth muscles in the iris increase or decrease the diameter of the pupil in response to the brightness of the light. The **cornea** is a piece of transparent tissue curving over the front of the eye. Between the cornea and the iris is a compartment filled with fluid called the **aqueous humor.** A larger fluid compartment, filled with **vitreous humor,** is between the lens and the retina, and gives the eye its shape.

Light is focused by the **cornea** and the **lens.** The cornea has a fixed shape and is a passive element in focusing light. The lens, on the other hand, changes shape when acted on by **ciliary muscles.** These changes alter the curvature of the lens and permit appropriate focusing on the retina of light coming from different distances.

Light passing the cornea and lens is refracted (bent). As a consequence, the actual image that arrives at the retina is inverted. It is upside down and right-left reversed. However, the brain's processing of sensory information inverts the images, thus compensating for the refraction and yielding a correct perception of the physical environment.

If the inside of the eye were white, like the outside, light would be reflected. Such reflections would obscure images and make it impossible to focus. As in a camera, however, the inner part of the eye is coated black. Black melanin pigment is produced by a layer of cells in the back of the eye, and the black surface reflects no light.

Events in the retina represent the real beginning of visual sensation. The retina is a thin sheet of tissue that contains three important layers of cells (figure 16.17). The layer at the back of the retina holds the **photoreceptor cells.** Photoreceptor cells are called **rods** and **cones** because of their shapes. Light passes not only through the cornea, lens, and humors, but also through several layers of neurons before reaching the rods and cones. The middle layer of the retina contains bipolar relay cells and also other cells that make complex connections among receptor cells and bipolar cells. The third layer, closest to the front of the eye, contains ganglion cells, whose fibers form the optic nerve. The optic nerve extends from the retina toward the brain (see figure 16.15).

The different shapes of the photoreceptor cells suggest that rods and cones have different functions. Nocturnal animals have a much higher ratio of rods to cones than do day-active animals. This suggests that rods play a role in *visual sensitivity* and *night vision* and that cones play a role in *sharpness* and *acuity* of vision. Other evidence indicates that this is true. For example, a bipolar cell usually is connected to only one or a few cones, whereas bipolar cells connected to rods usually synapse with several rods. Cones function well in bright light and permit sharply detailed vision. Rods function well in dim light and are extremely light sensitive, but if functioning alone yield only poorly defined images. Another important difference between rods and cones is that rods do not detect color differences while cones do. Thus, color vision depends on the functioning of cones. Humans and other primates may be the only mammals with color vision. Most other mammals probably see in shades of gray. Fish, some reptiles, and most birds also see in color.

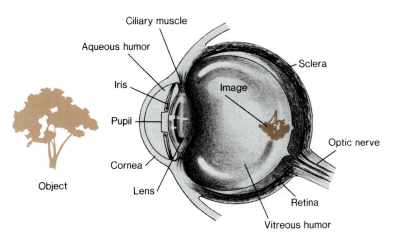

Ciliary muscle

Aqueous humor

Iris

Pupil

Cornea

Lens

Object

Sclera

Image

Optic nerve

Retina

Vitreous humor

Figure 16.16 A section of the human eye. The image is inverted on the retina.

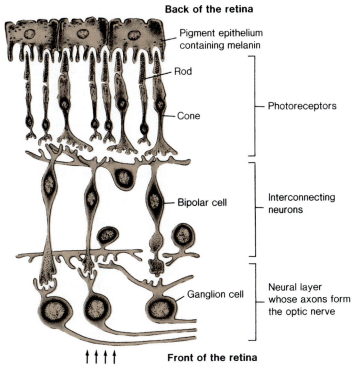

Back of the retina

Pigment epithelium containing melanin

Rod

Cone

Photoreceptors

Bipolar cell

Interconnecting neurons

Ganglion cell

Neural layer whose axons form the optic nerve

↑ ↑ ↑ ↑   **Front of the retina**
Incident light

Figure 16.17 Structure of the retina showing the three layers of cells. Light must pass through the ganglion cells and bipolar cells to get to the rods and cones at the back of the retina.

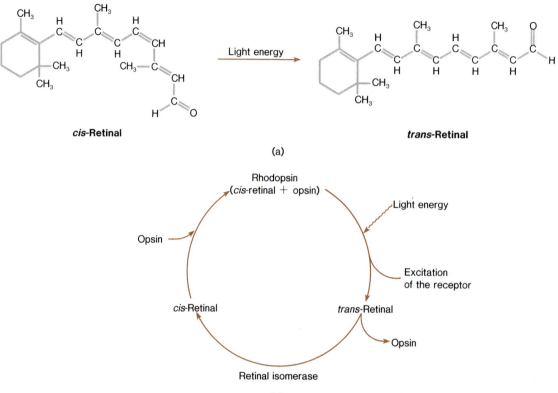

cis-Retinal

trans-Retinal

(a)

(b)

Figure 16.18   The chemistry of light reception by rods. (a) The *cis*- and *trans*- forms of retinal. Light causes a conformational change (isomerization) in retinal. (b) The rhodopsin cycle. The light-induced conformational change in retinal stimulates a change in the membrane potential of the rod. The opsin (protein part of the molecule) breaks off when the isomerization occurs. *Trans*-retinal is converted back to *cis*-retinal in a reaction catalyzed by the enzyme retinal isomerase. Then the *cis*-retinal joins with opsin to restore rhodopsin. A similar series of events occurs in the cones. This simple chemical cycle is the basis of humans' sense of vision.

The actual mechanism by which light causes electrical changes in photoreceptor cells is better understood for rods than for cones. Light reception in rods is based on a **photochemical reaction,** a light-induced change in a chemical substance. Rods contain a type of molecule that changes conformation (**isomerizes**) when struck by a single photon of light. This key molecule is a derivative of vitamin A called **retinal** (figure 16.18). Retinal is bound to a protein called **opsin** in a complex known as **rhodopsin.** Retinal changes from a bent (*cis*) conformation to a straight (*trans*) one when struck by light. This is the key event in photoreceptors' response to light. The conformational change in retinal that occurs when it is struck by a photon of light somehow causes a change in membrane potential (a receptor potential) in the rod cell.

Receptor potentials in rod cells cause the rods to release synaptic transmitters to the bipolar cells, which in turn release transmitters to the ganglion cells. Action potentials are produced in the ganglion cells, and nerve impulses are conducted toward visual centers in the brain.

In cone cells, slightly different protein molecules are involved, and the complexes are called **iodopsin.** Iodopsins appear to play a role in color vision, and several varieties exist that are specialized for detecting light of different wavelengths. Apparently, iodopsin also responds to photons of light with a conformational change in the molecule, which causes a membrane potential change in the cone, and eventually results in impulse transmission by ganglion cells connected to the cone.

The different iodopsins present in different cones are sensitive to different color wavelengths. Although a complete explanation of color vision has not yet been established, the trichromatic theory of color vision is widely accepted. This theory is based on fundamental properties of light and color. Any color can be obtained by an appropriate blending of three primary colors: red, blue, and green. It is no coincidence, then, that cone cells have three different pigments, one responding best to blue, one to green, and one to red. These cells respond *best* to the primary colors, but not exclusively to them. They also respond to light of other wavelengths, but not as well. In some way, the differential, selective responses of the various types of cone cells to light of different wavelengths are blended in the brain to create all of the sensations associated with the full range of colors perceived.

Summary

The spinal cord and brain make up the central nervous system. Spinal nerves containing sensory and motor fibers connect to the spinal cord. Cranial nerves connect to the brain.

The spinal cord has gray matter, composed mostly of nerve-cell bodies, in its center, and white matter made up of fiber tracts surrounds the gray matter. The ventricles in the brain and the hollow central canal in the spinal cord contain cerebrospinal fluid.

The forebrain, the midbrain, and the hindbrain are the three major regions in the embryonic brain. Each region produces a specific set of adult brain structures.

The medulla and pons, derivatives of the embryonic hindbrain, carry ascending and descending fiber tracts, and the medulla also conducts several important visceral reflex functions. The cerebellum (another hindbrain derivative) monitors body movements to keep them smooth and coordinated.

Midbrain derivatives in the adult brain are the tegmentum (a route for ascending and descending tracts) and the tectum (an area possessing a gray cortex that receives sensory input from the eyes and ears).

Important derivatives of the embryonic forebrain are the hypothalamus, the thalamus, and the cerebral cortex. The hypothalamus participates in control of temperature, metabolic rate adjustment, osmotic regulation, and other automatically regulated functions. The hypothalamus also interacts with the pituitary in hormonal regulation, and it is a part of the limbic system, a network of brain areas functioning in emotion and instinctive behavior. The thalamus is a relay center.

The cerebral cortex is divided bilaterally into two cerebral hemispheres that contain large quantities of gray cortex material arranged in gyri and sulci. The cerebral cortex is divided into four major lobes: the frontal, parietal, temporal, and occipital lobes. Each of these lobes is functionally specialized.

Various regions of the cortex are specialized as primary sensory projection areas, association areas, or motor areas. Within sensory projection areas are subdivisions that receive sensory input from specific body regions. Association areas associate sensory and motor functions and higher levels of awareness. Motor cortex areas control body movements, with the cerebellum playing a coordinating role.

The reticular formation is an activating system that stimulates the cerebral cortex to be ready to receive incoming sensory signals.

Sensory receptor cells transduce various types of energy into neuronal signals. Each type of receptor cell responds best to only one type of energy—its adequate stimulus. Receptor cells respond to stimuli with receptor potentials. Receptor cells are or are associated with neurons that carry impulses to the central nervous system via specific pathways.

Sensory receptors of the skin, as well as those in other parts of the somatosensory system, are ends of neuronal processes. Different groups of receptors are specialized to respond to touch-pressure, pain, and temperature changes.

Olfactory receptors also are parts of sensory neurons. They are embedded in nasal passages, where their olfactory cilia detect chemicals present in fluid on the surface of mucous membranes. Olfactory sensory information enters the cerebrum directly without passing through the thalamus, as other sensory information does.

Receptors for taste are secondary sense cells in the taste buds. Receptor potentials in these cells cause them to release neurotransmitters that cause impulse transmission in nerve cells that synapse with them.

The ear houses the senses of hearing and balance. Sound strikes the tympanic membrane as a series of vibrations. Vibrations are transmitted via the middle ear ossicles to the cochlea, where fluid movements trigger receptor potentials in hair cells of the organ of Corti. Hair cells communicate synaptically with neurons that conduct impulses to the brain.

Body movement and position are sensed by hair cells in the semicircular canals. The hair cells respond to fluid pressure changes inside the canals that result from head movement.

The eye is an optical device that functions to focus light on the retina, where photoreceptor cells are located. Rods are receptor cells that are very light sensitive and can function in dim light. Cones require bright light and are associated with sharp image formation. Cones also detect color differences.

When light strikes rhodopsin, its retinal portion isomerizes from its *cis* to its *trans* form. This isomerization causes a receptor potential in a rod cell. Rod cells communicate synaptically with bipolar cells which, in turn, communicate with the ganglion cells. Ganglion cells transmit impulses to the brain.

Cone cells contain different pigments that react to light of different wavelengths. Color vision results from reactions of light with one or several of these pigments.

Questions

1. Explain the distinction between gray matter and white matter. How are regions of the central nervous system that have a gray cortex organized? What regions in the human central nervous system have a gray cortex?

2. People who have suffered severe damage to a large portion of the cerebral cortex are mentally impaired, but they often remain alive. Significant damage to the brain stem, however, especially to the medulla, almost always causes death in a very short time. What aspects of brain functioning explain this difference?

3. What is the difference between primary receptors and secondary receptors?

4. Explain how intensity of stimulation is communicated.

5. Why is a blow to the eye often accompanied by what seems to be a bright flash of light?

6. How is vitamin A related to visual sensory function?

Suggested Readings

Books

Bullock, T. H.; Orkand, R.; and Grinnell, A. 1977. *Introduction to nervous systems.* San Francisco: W. H. Freeman.

Eccles, J. C. 1977. *The understanding of the brain.* New York: McGraw-Hill.

Hubbard, J. 1975. *The biological basis of mental activity.* Reading, Mass.: Addison-Wesley.

Articles

Amoore, J. E.; Johnston, J. W.; and Rubin, M. February 1964. The stereochemical theory of odor. *Scientific American* (offprint 297).

Friedmann, I. 1976. The mammalian ear. *Carolina Biology Readers* no. 73. Burlington, N.C.: Carolina Biological Supply Co.

Horridge, G. A. July 1977. The compound eye of insects. *Scientific American* (offprint 1364).

Hubel, D. H., and Wiesel, T. N. September 1979. Brain mechanisms of vision. *Scientific American* (offprint 1442).

Jerison, H. J. January 1976. Paleoneurology and the evolution of mind. *Scientific American* (offprint 568).

Nauta, W. J. H., and Feirtag, M. September 1979. The organization of the brain. *Scientific American* (offprint 1439).

Nicholls, J. G., and van Essen, D. January 1974. The nervous system of the leech. *Scientific American* (offprint 1287).

Parker, D. E. November 1980. The vestibular apparatus. *Scientific American* (offprint 1484).

Weale, R. A. 1974. The vertebrate eye. *Carolina Biology Readers* no. 71. Burlington, N.C.: Carolina Biological Supply Co.

Zwislocki, J. J. 1981. Sound analysis in the ear: A history of discoveries. *American Scientist* 69: 184.

Muscles and Movement

The contributions of the animal nervous system to the maintenance of homeostasis are varied and complex. Sensory elements of the nervous system transmit information to the central nervous system about conditions in the external environment around the body and in the body's internal environment. The central nervous system directs appropriate, adaptive responses to changes in either or both of these environments. Stimuli go out over motor elements of the nervous system to various **effectors,** which respond in specific ways to stimulation by the nervous system.

In chapter 15 some activities of the sympathetic and parasympathetic divisions of the autonomic nervous system were described. These motor elements direct the activities of many internal effectors. Some processes under the control of autonomic innervation include secretion by some glands, contraction of muscles in the walls of blood vessels and digestive organs, and the rate of heartbeat. These motor functions are automatic, and the activities of the internal effectors controlled by autonomic innervation are not subject to conscious control under normal circum-

stances. For example, it normally is not possible for a person to will a change in heart rate or the rate at which peristaltic waves of muscular contraction move along the digestive tract.

Another set of motor nerves controls the actions of **skeletal muscles,** the muscles attached to the supporting skeleton of the body. Skeletal muscles are responsible for major movements of body parts and of the whole body. The activities of this important group of effectors are under conscious, voluntary control. Vertebrates can make and act upon decisions about body movements that require skeletal muscle action.

Movement is not restricted to multicellular organisms that have muscles and skeletons. Single cells move by means of cilia, flagella, or cytoplasmic streaming. Some small animals move by means of coordinated action of many cilia on their body surfaces. Even parts of some plants can move, as a result of abrupt osmotic changes in strategically located cells (figure 17.1). But this chapter is focused on functions of skeletal muscles in vertebrate animals and comparisons with other types of muscles.

Figure 17.1 Movement in plants. (a) Photographs of the sensitive plant *Mimosa pudica* before (1) and after (2) the plant has been touched. Note how the leaflets fold up and the leaves droop in response to touch. (b) A specialized cluster of cells called the pulvinus, located at the base of the leaf stalk (1) is responsible for leaf drooping. In the resting condition, cells in the pulvinus are turgid and the leaf stalk stands erect (2). Following disturbance of the plant, cells in the bottom part of the pulvinus lose turgor and the leaf stalk droops (3).

Muscles, Skeletons, and Body Movement

Muscles are made up of contractile cells—cells that can shorten forcefully—and muscles exert force as a result of simultaneous shortening of many of these cells. To bring about body movements, however, the force of muscle contractions must be specifically directed against other body parts. In some animals, such as segmented worms (chapter 37), pressure of muscle contractions is applied to fluid-filled body compartments, and shape changes of those compartments are translated into whole body movements. But animals whose muscle contractions cause fluid pressure changes in such **hydrostatic skeletons** are not capable of such complex movements as animals whose movements are produced by muscles attached to and pulling against rigid supporting skeletons.

The animals with the most highly developed arrangements for body movement are the arthropods (insects and their allies) and the vertebrates. Both groups have skeletons made up of rigid supporting elements to which muscles are attached. Arthropods have an **exoskeleton,** an external skeleton that doubles as a hard, protective skin. Vertebrates have an **endoskeleton,** an internal skeleton that provides support from the inside.

Shortening is the only way that muscles can exert force; they cannot exert force by active stretching. Muscles lengthen by relaxing. This has important consequences for the design of the skeletal-muscular units that produce body movements. Movements usually are produced around **joints,** movable junctions between parts of the skeleton. Because muscles can exert force only by shortening, each joint is supplied with a double set of muscles (or groups of muscles) that function as an **antagonistic pair.** One muscle (or group of muscles) passively stretches while the other contracts. For example, in limbs, one muscle of an antagonistic pair brings the limb toward the body, bending the joint. This muscle is called a **flexor.** The other member of the antagonistic pair, called an **extensor,** straightens the joint. Figure 17.2 illustrates the arrangement of antagonistic muscle pairs and compares the actions of muscles in the limb of a vertebrate, where they are attached to elements of an endoskeleton, with the muscles in the limb of an arthropod, where they are attached to an exoskeleton.

Vertebrate Skeletons

The vertebrate skeleton provides rigid or semi-rigid support for the vertebrate body and the attachment points for the skeletal muscles that move the body. Many of the properties of the skeleton arise from properties of the cartilage and bone that make up vertebrate skeletons.

Cartilage and Bone

Cartilage and bone consist of cells and extracellular material that the cells produce. This extracellular material, called the **matrix,** makes cartilage and bone firm, rigid tissues, but differences in their matrix material give cartilage and bone somewhat different properties. **Cartilage** is a skeletal tissue that provides firm support with a considerable degree of flexibility, while bone provides much more rigid support (figure 17.3).

Cartilage makes up most of the skeleton during development of vertebrate embryos. Some vertebrates—notably sharks, skates, and rays—have a cartilaginous skeleton throughout life. But in most other vertebrates much of the cartilage of the embryonic skeleton gradually is replaced by bone during subsequent development. In adult humans, for example, cartilage remains only in such flexible structures as the ears and the nose tip, and it also provides a form of padding for the areas where bones lie adjacent to one another in movable joints.

Bone is a more rigid supporting tissue that occurs in two forms—spongy bone and compact bone (figure 17.4). **Spongy bone** is composed of a latticelike network of bars. The irregular spaces among the bony bars, as well as the hollow cavities inside many bones, contain **bone marrow.** Bone marrow is the site where replacement blood cells develop in adult vertebrates.

Compact bone is more solid than spongy bone and provides stronger support in such sites as the shafts of the longer bones of the body. Compact bone consists of structural units called **Haversian systems.** Each Haversian system has concentric layers of rigid bone matrix material that contains large amounts of calcium salts (mostly calcium phosphate with some calcium carbonate). These concentric layers are arranged around a central canal through which blood vessels and nerves pass. Materials are exchanged between bone cells located in small cavities in the matrix and the blood vessels in the central canal.

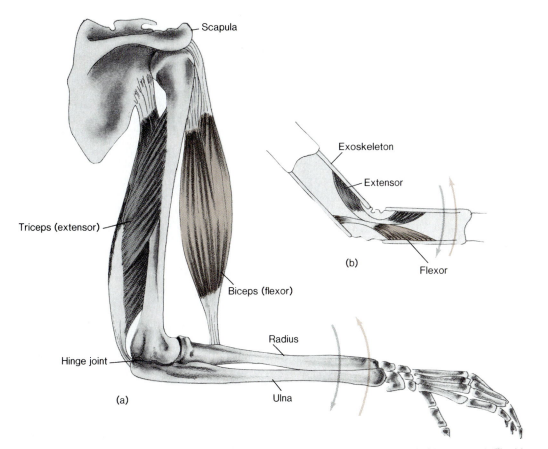

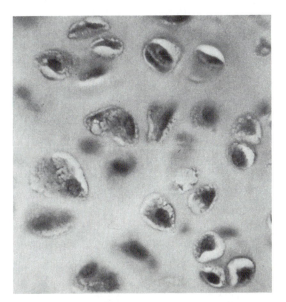

Figure 17.2 Antagonistic muscle pairs. Muscles can exert force only by shortening, and movable joints are supplied with double sets of muscles that work in opposite directions. (a) The human arm showing an example of antagonistic muscles attached to a vertebrate endoskeleton. The biceps is a flexor; it contracts and flexes the arm, that is, bends it at the elbow (colored arrow). The triceps passively stretches while the biceps contracts. When the triceps contracts, the lower arm is extended (gray arrow). The biceps passively stretches while the triceps contracts. (b) Muscles in an insect's leg as an example of antagonistic muscles attached to the inside of an arthropod exoskeleton. Note the relative positions of flexor and extensor when muscles are inside the skeleton as opposed to the vertebrate arrangement where muscles are attached to the outside of the skeleton.

Figure 17.3 Cartilage tissue. Cartilage cells are located in cavities called lacunae, which are scattered in the matrix material produced by the cells. Cartilage provides semi-rigid but flexible skeletal support because the matrix material is somewhat pliable.

Figure 17.4 Bone structure. The upper scanning electron micrograph shows spongy bone, which contains irregularly shaped bony bars and marrow spaces. (Magnification × 12) The lower scanning electron micrograph is a section of compact bone showing several Haversian systems. (Magnification × 147) Concentric layers of bone matrix surround a central canal (HC) in each Haversian system. In living bone, blood vessels and nerves pass through the Haversian canal and bone cells are located in lacunae (La).

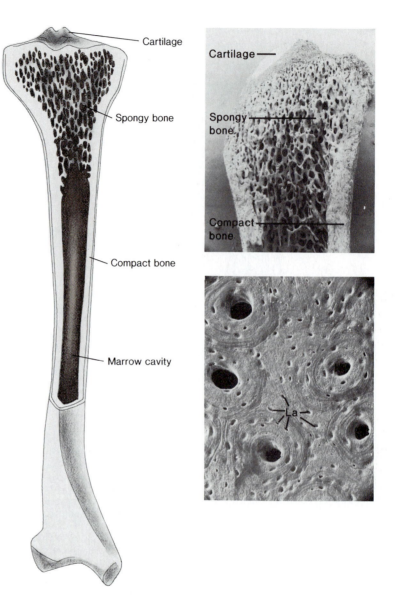

Cartilage

Spongy bone

Compact bone

Marrow cavity

Cartilage

Spongy bone

Compact bone

La

Nutrients, gases, and wastes diffuse through tiny passages that connect the cells' cavities with the central canal.

Bones examined in the laboratory reveal only the matrix material, which remains long after the death and degeneration of the bone cells and the blood vessels and nerves of the Haversian systems. Such dry, dead bones make it difficult to comprehend that during life, bone tissue is active and dynamic. For example, even adult bones can be reshaped by thickening and thinning of various bone parts in response to stresses and pressures applied during regular exercise. And bone cells are able to heal even badly broken bones.

The Human Skeleton

The human skeleton, like all vertebrate skeletons, can be divided into two parts for purposes of description (figure 17.5). The **axial skeleton** supports the main body axis—the head, neck, and trunk. It includes the **skull,** which surrounds and protects the brain; the **vertebrae,** a series of bones

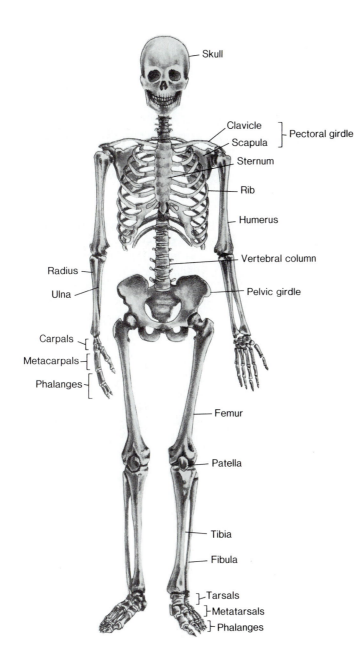

Figure 17.5 The human skeleton.

Skull

Clavicle
Scapula } Pectoral girdle

Sternum

Rib

Humerus

Vertebral column

Radius

Ulna

Pelvic girdle

Carpals

Metacarpals

Phalanges

Femur

Patella

Tibia

Fibula

Tarsals

Metatarsals

Phalanges

making up the **vertebral column** or backbone; and the **ribs,** curved bones that are attached to vertebrae and enclose and protect vital organs, such as the heart and lungs, located in the upper body trunk.

The **appendicular skeleton** includes the bones of the paired appendages (the arms and legs), as well as the bones of the pectoral and pelvic girdles that support the arms and legs, respectively, and connect them to the axial skeleton.

Some bones, such as those of the skull, are connected to one another by immovable joints so that they are permanently fixed in position relative to one another. But many other bones meet in movable joints that are held together by **ligaments.** Ligaments are tough, but flexible, connecting straps between bones. Movable joints are flexible enough to permit body movements in response to contractions of skeletal muscles.

Vertebrate Muscles

Vertebrate animals have three types of muscle tissue: **skeletal muscle, smooth muscle,** and **cardiac muscle.** Each of these three types of muscle tissue has different structural and functional characteristics.

Skeletal muscle tissue is the contractile tissue of the skeletal muscles involved in limb movements and other whole body movements. It consists of tubular, multinucleate cells, called **muscle fibers,** that are held together in bundles by connective tissue (figure 17.6a). Skeletal muscle tissue is also called **striated muscle** because of the alternating light and dark bands (**striations**) that cross its fibers. Nuclei in a skeletal muscle fiber lie near the edge of the tubular cell, just inside the plasma membrane, which is called the **sarcolemma.**

Smooth muscle tissue is found in many sites, including the walls of the digestive tract, the urinary bladder, the uterus, and encircling blood vessels, where smooth muscle contraction regulates blood flow to specific body regions. Smooth muscle tissue is different from skeletal muscle in that it contains individual contractile cells with *single* nuclei. Smooth muscle cells also do not have the obvious striations seen in skeletal muscle (figure 17.6b).

Cardiac muscle tissue is the specialized contractile tissue of the heart. Cardiac muscle resembles skeletal muscle in that it has obvious striations and tubular muscle fibers. Cardiac muscle fibers differ from skeletal muscle fibers in that they branch much more and are not multinucleate. Individual cells are separated by discs that appear under the light microscope as very prominent cross lines that are darker than the striations in the tubular fibers. Electron microscope studies have shown that each of these discs is the complex junction between the plasma membranes of adjacent cells.

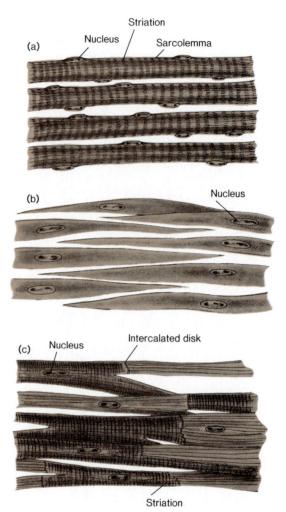

Figure 17.6 Vertebrate muscle fibers. (a) Skeletal (striated) muscle fibers. Note striations and the peripheral location of nuclei in the multinucleate fibers. (b) Smooth (visceral) muscle fibers. Smooth muscle fibers have a single nucleus and lack striations. (c) Cardiac muscle fibers. Note the branching of the fibers, the central position of the nuclei, and the presence of intercalated discs, which are the complex junctions between adjacent individual cells.

Many skeletal muscle fibers are innervated by motor neurons that are under conscious (voluntary) control. Smooth muscle tissue and cardiac muscle tissue are involved in movements in internal body structures that are not under voluntary control. Their contractions are controlled by nerves of the autonomic division of the nervous system.

Skeletal Muscle

Skeletal muscle is the most abundant muscle type in the body, constituting about 80 percent of the body's soft tissue.

A skeletal muscle is made up of many bundles of skeletal muscle fibers held together by connective tissue. All of these bundles are surrounded by a tough outer connective tissue sheath with a smooth surface that reduces friction between the muscle and adjacent structures during muscle contraction and relaxation.

Most skeletal muscles taper at their ends and are connected to bones by **tendons.** Figure 17.7a shows the **gastrocnemius muscle** in a frog's leg. The gastrocnemius muscle is attached by the Achilles tendon to a bone in the frog's foot. When the gastrocnemius muscle contracts, it produces a powerful straightening of the foot that is an important part of the frog's jump. The nerve innervating the frog gastrocnemius is the sciatic nerve, which passes through the muscles of the thigh (figure 17.7b).

The attachment of a muscle to a bone that moves is called the muscle's **insertion.** In the example, the insertion of the frog's gastrocnemius muscle is in the foot. The opposite end of a muscle, the end attached to a relatively stationary bone is called the **origin.** The relationships of a muscle's origin and insertion also can be seen in figure 17.2a where the origin of the biceps muscle is apparent in the shoulder area, and its insertion is on the radius bone of the forearm.

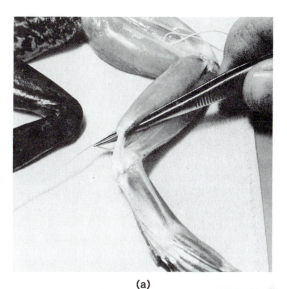

(a)

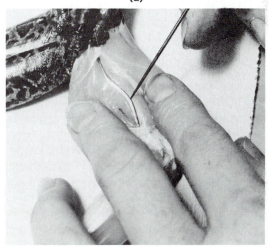

(b)

Figure 17.7 Muscles, tendons, and nerves. (a) Dissection of a frog's leg that shows the gastrocnemius muscle and the Achilles tendon that attaches it to a bone in the foot. The muscle shown here is being prepared for study with a kymograph apparatus (figure 17.8). (b) Dissection of a frog's thigh that exposes the sciatic nerve, which contains axons that provide motor innervation for the gastrocnemius muscle as well as those supplying other leg and foot muscles.

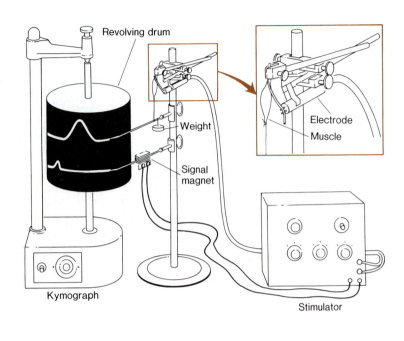

Figure 17.8 A frog gastrocnemius muscle and sciatic nerve attached to a kymograph apparatus. An electrode is shown here applied to the sciatic nerve. The electrode may instead be applied directly to the muscle, depending on the experiment being conducted. A stylus attached to the insertion end of the muscle traces a record of muscle movement on paper attached to the revolving drum of the kymograph. The stimulator sends a stimulus to the nerve and at the same time deflects the signal magnet, which makes a mark that provides a record of when the stimulus was applied. Muscle contraction raises the stylus aim, and relaxation lowers it, thus tracing out a record of the contraction on the kymograph drum. Electronic recorders often replace the kymograph apparatus in modern studies of muscle function.

Muscle Physiology

Many of the properties of vertebrate skeletal muscles have been studied using single muscles, such as the frog gastrocnemius muscle. A muscle selected for study is dissected and removed from the body. Then it is attached to a device known as a kymograph apparatus (figure 17.8) or to one of several kinds of electronic devices that have replaced kymographs in many laboratories. Often, the nerve that innervates the muscle is removed along with it. In this case, the sciatic nerve remains attached to the gastrocnemius muscle.

Once attached to the kymograph, a muscle can be stimulated electrically by applying an electric current either to the nerve or to the muscle itself. When the sciatic nerve is stimulated electrically, nerve impulses pass down some of the motor neuron axons, which are contained in the nerve, to the muscle fibers that each axon controls. A motor neuron, together with the group of muscle fibers innervated by its axon, are collectively known as a **motor unit.** Electrical stimulation of the sciatic nerve can cause contraction of muscle fibers in many motor units and result in contraction of the entire muscle.

The Simple Twitch

If a muscle is given a single stimulus, such as a very brief electrical shock applied to the nerve innervating it, the muscle responds with a single, quick contraction called a **simple twitch.**

Although a simple twitch occurs very rapidly, requiring only about 0.1 sec for a frog muscle and 0.05 sec for a human muscle, each twitch consists of three separate phases (figure 17.9). The first is the **latent period,** a very brief time interval between application of the stimulus and the beginning of muscle shortening. The latent period lasts only 0.005 sec in a frog muscle. Then the muscle shortens and does work during the **contraction period,** which lasts about 0.04 sec in a frog muscle. Finally, the muscle returns to its original length during the **relaxation period,** which lasts about 0.05 sec and thus is the longest of the three phases of the twitch.

While the timing of the phases in a simple twitch is the same in each twitch of a muscle, the strength (force of contraction) of twitches is variable. This variability arises from responses of the muscle fibers that make up muscles. Individual skeletal muscle fibers behave much like individual neurons in that they exhibit all-or-none responses to stimulation. To contract, an individual muscle fiber must receive an appropriate stimulus. Variations in the strength of a whole muscle's twitches depend on the number of muscle fibers

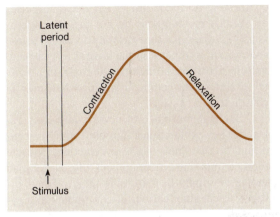

Figure 17.9 Drawing of a kymograph tracing of a simple twitch. The duration of the latent period actually is much shorter relative to the other periods than this sketch indicates.

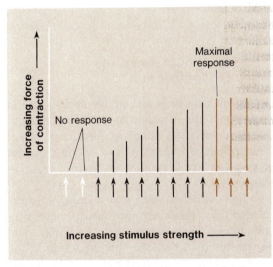

Figure 17.10 The relationship of increasing stimulus strength to force of contraction. Very weak stimuli (white arrows) do not activate enough motor units to cause a detectable contraction of the muscle. A threshold level stimulus gives a minimal contraction. Increasing stimulus strength increases contraction force until a maximal level is reached. Further increases in stimulus strength do not increase the force of contraction.

that are stimulated to contract. If only a few fibers contract, the muscle contraction is weak. If more fibers contract, the muscle contraction is stronger.

Variations in strength of muscle contractions can be demonstrated experimentally on the kymograph apparatus. A progressive increase in stimulus strength results in progressively stronger and stronger contractions as more and more motor axons are recruited and more and more muscle fibers contract. Eventually, however, a stimulus level is reached beyond which further increases in stimulus strength do not increase contraction strength. The stimulus level that produces the strongest obtainable twitch is called a maximal stimulus. Stimuli stronger than the maximal stimulus do not produce stronger contractions (figure 17.10). In fact, further increases in stimulus strength soon lead to damage of the muscle being studied.

How can results of experimental studies on simple muscle twitches be related to the question of what determines the strength of muscle contractions in the body? Normally, stimulus for a muscle fiber to contract is provided as a result of the arrival of an impulse (action potential) at the **neuromuscular junction,** the contact between a motor neuron's axon tip and an individual muscle fiber. The strength of the muscle contraction depends on the number of motor units in which motor neuron axons conduct impulses to the muscle fibers that they innervate. A second factor, not operative in simple muscle twitches produced in the laboratory, also influences muscle contraction strength. Motor axons carry series of impulses that cause repeated contractions of muscle fibers. The frequency of repetition of these impulses is a major factor in determining the strength of a whole muscle's contraction. The number of motor neurons conducting impulses and their impulse condition frequencies are determined by complex central nervous system activities that adjust the strength of muscle contractions to the work load at hand. For example, the same set of muscles is used to lift a piece of paper from the edge of a table or to lift the edge of the table itself. The nervous system adjusts the strength of the muscle contractions involved in the two jobs by varying the number of motor units activated and the frequency of impulse conduction by motor axons.

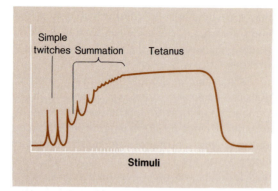

Simple
twitches Summation   Tetanus

Stimuli

Figure 17.11 Sketch of a kymograph record showing
summation and tetanus. Marks on the lower line indicate
stimuli given to the muscle. Individual simple twitches
appear as single sharp spikes because the drum is
rotated very slowly for this experiment. Increasing
stimulus frequency produces summation and, finally,
tetanus.

Summation and Tetanus

Simple muscle twitches studied under experimental conditions reveal some fundamental properties of muscle contraction, but normal body movements involve sustained muscle contractions that last much longer than the simple twitch obtained when a single, brief stimulus is applied to a muscle or the nerve that innervates the muscle.

The sustained muscle contractions involved in normal body movements can be closely replicated in the laboratory if a series of rapidly repeated stimuli are applied to the muscle. The kymograph record in figure 17.11 shows that the first few contractions represent simple twitches. But if stimuli are given at short enough intervals so that the muscle does not have time to relax between stimuli, a response called **summation** develops. Each contraction adds to the one that preceded it so that contraction strength is increased. If the frequency of stimuli is increased further, individual contractions blend into a single, sustained, forceful contraction called **tetanus** (not to be confused with tetany, the violent muscle spasms that result from parathyroid hormone deficiencies, or "lockjaw," which is caused by toxins produced in a bacterial infection). A tetanic contraction continues until stimulation stops or until the muscle becomes fatigued and is unable to respond to further stimulation.

Tetanic contraction seems to resemble the sustained muscle contractions involved in normal body movements. Those sustained contractions result from a series of rapidly repeated impulses conducted by sets of motor axons innervating the contracting muscles. But in the body, impulses on various motor axons do not arrive synchronously (at the same time) as they do when tetanic contraction is induced in the laboratory. Instead, muscle fibers are stimulated to contract at different times, thus helping to sustain whole muscle contraction.

Skeletal muscles are never completely relaxed. They exhibit **tonus** (muscle "tone"), a condition of sustained slight contraction. Muscle tonus is maintained throughout life, as long as the nerve supply to muscles is intact. Tonus involves alternating periods of contraction and relaxation in different groups of motor units within a muscle. Thus, maintenance of muscle tonus is a continuing form of energy-requiring work in each muscle, but it does not produce fatigue in any of the motor units because contraction of each is followed by rest.

The Molecular Basis of Muscle Contraction

Biologists have been seeking for many years to understand the mechanisms by which muscles contract and exert a pulling force, and rather detailed analysis of resting and working muscles was undertaken during the first half of this century.

Two proteins, **actin** and **myosin,** were extracted from muscle and identified as major constituents of the contractile cells. By implication, it was assumed that they played a role in the contraction process. Then in 1939, V. A. Engelhardt and M. N. Ljubimova, working in Moscow, demonstrated that the protein myosin functioned as an ATPase. That is, myosin acted as an enzyme that catalyzed the energy-yielding hydrolysis reaction in which the terminal phosphate is removed from ATP. This discovery directly connected an important energy-yielding reaction with one of the special proteins that characterize muscle cells, and it opened the way for the next major steps in analysis of the contraction process.

During the 1940s, Albert Szent-Györgyi extracted and purified actin and myosin, and then precipitated them as fibrous threads of pure actin or pure myosin. He also mixed actin and myosin in solution to form a complex called **actomyosin** that precipitates to form fibers just as actin or myosin alone do. Then Szent-Györgyi experimented with these protein fibers. He found that fibers of actomyosin, in a solution containing appropriate ions, actually shortened as he watched them, if he added ATP to the solution. Neither pure actin fibers nor pure myosin fibers would shorten in this way. Thus, Szent-Györgyi demonstrated that actin and myosin indeed are contractile proteins that can shorten physically, but that they do so only as actomyosin, a complex of the two proteins. He further showed that ATP can provide energy directly to this molecular contraction process.

Energy for Muscle Contraction

Analyses of resting and working muscles undertaken in the early 1900s had revealed that a number of chemical changes occur during muscle contraction. Contracting muscles consume oxygen and produce carbon dioxide; their glycogen content decreases, and their lactic acid content increases; their ATP concentration decreases, and ADP and inorganic phosphate concentrations increase; another phosphate-containing compound, **creatine phosphate,** also decreases (table 17.1). All of these changes were known to be involved in cellular energy conversions. But which of them provided energy directly to the contraction process?

Albert Szent-Györgyi had demonstrated that energy for the work of muscle contraction comes from ATP. However, only a very small supply of ATP is present in a muscle fiber at any given time, barely enough ATP for a few contractions. During exercise, a muscle must replenish its ATP supply while it is working.

One means of replenishing ATP seems to involve creatine phosphate, which can give up its phosphate to ADP, thus producing ATP (figure 17.12). The usual explanation of this reaction is that creatine phosphate serves as an energy reserve that extends the ability of a muscle to contract by allowing rapid replacement of ATP. The supply of creatine phosphate that is depleted during contraction is replaced while the muscle rests. Recently, it has been suggested, however, that creatine also may function as an "energy shuttle," accepting phosphate from ATP in one part of the muscle cell (the mitochondrion) and donating it to ADP near the contractile proteins. This proposed energy shuttle role will require more research in the next few years.

Table 17.1
Chemical Changes in Contracting Muscles.

Decrease during Muscle Work	Increase during Muscle Work
Oxygen	Carbon dioxide
Glycogen	Lactic acid
ATP	ADP and inorganic phosphate (P_i)
Creatine phosphate	Creatine

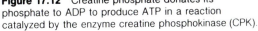

Figure 17.12 Creatine phosphate donates its phosphate to ADP to produce ATP in a reaction catalyzed by the enzyme creatine phosphokinase (CPK).

But even using up the creatine phosphate energy reserve only briefly extends the ability of a muscle to continue contracting actively. The ATP supply must be replenished by other means as the muscle continues to work.

The glycogen stored in muscle cells is hydrolyzed to release glucose that can be oxidized to provide energy for ATP production. Oxygen-requiring cell respiration is the most efficient means of producing ATP, and muscle fibers employ it, but its usefulness to a working muscle is limited by the amount of oxygen that can be delivered to the muscle. Physical training improves the efficiency of oxygen delivery to muscles, but even muscles in an individual in excellent physical condition suffer an oxygen delivery deficit as heavy muscle exercise continues.

Additional oxygen becomes available as the oxygen-binding compound myoglobin releases oxygen. Myoglobin, which is found in muscle cells, reversibly binds oxygen when plenty of oxygen is available and releases it when the oxygen content in muscles drops, as it does during exercise.

The oxygen made available from myoglobin helps for a short time, but ATP produced by oxygen-requiring respiration falls short of the muscle's needs because of the limits imposed by the available oxygen supply. How is the additional required ATP produced?

Muscle cells have the ability to carry on lactic fermentation (figure 17.13), which permits them to produce additional ATP anaerobically (without additional oxygen). As a result of this fermentation, lactic acid builds up in muscle fibers and in the fluid around them during heavy and extended muscle work, but an adequate supply of ATP to power continuing muscle contractions is maintained.

An accumulation of lactic acid in contracting muscles must be dealt with when vigorous contraction stops. The results of this temporary employment of an anaerobic metabolic pathway amount to an **oxygen debt,** a chemical deficit that must be made good when the muscle returns to rest. Some of the lactic acid built up in working muscles is exported to the liver, where it is converted into glycogen. The remainder of the lactic acid in the muscle fibers is converted back into pyruvic acid when an adequate supply of oxygen once again becomes available after the muscle has returned to rest. There are limits to this oxygen debt because excessive lactic acid accumulation contributes to muscle fatigue and can actually cause damaging pH changes in muscles. Fortunately, other functional mechanisms normally seem to protect muscles from dangerously excessive lactic-acid buildup.

The anaerobic metabolic capabilities in skeletal muscle fibers have great adaptive value. The ability to continue vigorous muscle contractions and to accumulate a temporary oxygen debt can be a matter of life and death for most organisms. A rabbit fleeing from a fox would not get far if its muscles could work only to the extent permitted by aerobic metabolism, which depends on the amount of oxygen delivered to the muscles.

The Sliding Filament Theory

Szent-Györgyi's demonstration that extracted and precipitated threads of actomyosin could shorten if supplied with ATP focused attention on the two contractile proteins actin and myosin and on the actual physical arrangement in muscle fibers that permits them to interact during contraction. Much of the present understanding of these relationships has resulted from the research of two English scientists: A. F. Huxley, who studied muscle contraction mainly with the light microscope, and H. E. Huxley, who studied the fine structure of muscles with the electron microscope.

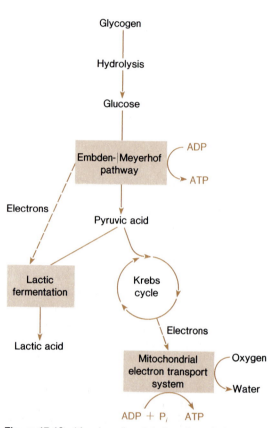

Figure 17.10 Muscle cell metabolism. In actively contracting muscles, ATP production by aerobic metabolic pathways (Krebs cycle and mitochondrial electron transport system) proceeds at the maximum rate permitted by delivery of oxygen to the muscle cells. Additional ATP production by the Embden-Meyerhof pathway is made possible by lactic fermentation (see chapter 7 for details). But supplementing ATP production in this way results in lactic acid buildup, an oxygen debt that must be repaid when the muscle returns to rest.

Each skeletal muscle fiber is striated (see figure 17.6); it has a pattern of alternating light and dark bands. The striation of whole fibers arises from the alternating light and dark bands of the many smaller, tubular **myofibrils** contained in each muscle fiber (figure 17.14). The myofibrils have dark bands, called **A bands,** that alternate with lighter bands, called **I bands.** Each A band has a region in its center that is lighter than the rest of the band. This lighter, central region is called the **H zone** (also known as the M band). The middle of each I band has a distinctive, thin, dark line called the **Z line** (figure 17.15). The banding patterns of myofibrils reflect their functional organization, and the portion of a myofibril running from one Z line to the next is a single contractile unit called a **sarcomere.**

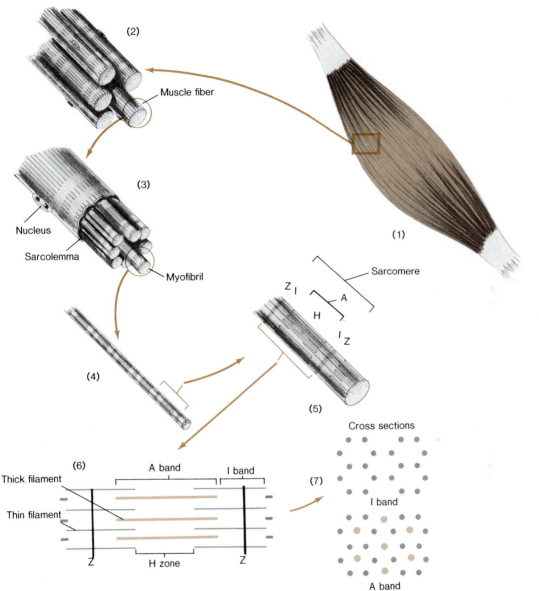

(2)

Muscle fiber

(3)

Nucleus

Sarcolemma

Myofibril

(4)

Sarcomere

Z I
A
H
I Z

(5)

(1)

Cross sections

(6)

Thick filament

A band   I band

Thin filament

(7)

I band

Z   H zone   Z

A band

Figure 17.14 The structure of skeletal muscle. Each muscle (1) contains many muscle fibers (2 and 3). Muscle fibers contain myofibrils (3 and 4), each of which consists of many sarcomeres (5). The banding of sarcomeres arises from the arrangement of thick and thin filaments inside them (6). The cross section sketches (7) show the very precise geometric arrangement of the thin and thick filaments in small areas of the I and A bands. Each sarcomere actually contains hundreds of thick and thin filaments.

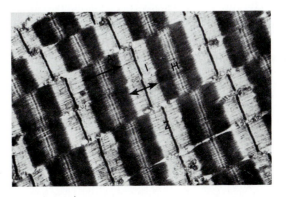

Figure 17.15 An electron micrograph of several myofibrils in a skeletal muscle fiber of a rabbit showing the appearance of the various regions represented diagrammatically in figure 17.14. A band (A), I band (I), H zone (H), and Z line (Z) are identified. (Magnification × 760)

Electron microscopy and biochemical analysis of myofibrils have shown that the banding patterns of myofibrils result from the arrangement of two types of filaments located inside the myofibrils. Relatively thicker filaments, which are made up of myosin, run through the A band. Relatively thinner filaments, which contain actin (and several other proteins), run through the I band and also overlap with the thick myosin filaments in part of the A band. The portions of the A band in which the thick and thin filaments overlap are darker than the H zone, which is the central portion of the A band that contains only the thick myosin filaments (see figure 17.14). The Z line is a structure to which the smaller actin-containing filaments are anchored.

The two Huxleys proposed a model for the molecular events in muscle contraction that has come to be called the **sliding filament theory.** The theory says that in response to a stimulus to contract, these filaments slide by one another and increase the amount by which they overlap. This sliding draws the two Z lines of each sarcomere closer together. The H zone in each sarcomere essentially disappears, and the I bands all along the sarcomere become narrower while the widths of the A bands remain constant (figure 17.16). Shortening of all of the sarcomeres within each myofibril shorten the entire myofibril, and simultaneous shortening of all of the myofibrils in a muscle fiber shortens the whole fiber. During relaxation, the filaments slide back to their original positions relative to one another, and all of the changes in myofibril and muscle fiber length are reversed.

Filament sliding involves temporary connections that form between myosin in the thick filaments and actin in the thin filaments. The connections are flexible, temporary cross-bridges that are established when the globular "heads" of myosin molecules attach to binding sites on actin molecules in the thin filaments (figure 17.17). Once a cross-bridge forms, it bends, thus exerting a pulling force on the thin filament that slides the thin filament by the thick filament.

The original binding between actin and myosin requires the presence of an ATP molecule. Hydrolysis of the ATP molecule occurs during cross-bridge binding and provides energy for the bending. Following ATP hydrolysis, the linkage between actin and myosin is released. The myosin "head" then returns to its prebending

position and is ready to form a new cross-bridge with an actin molecule further along the thin filament.

This cross-bridging cycle is repeated over and over by each of a huge number of myosin molecules during each contraction. At any given instant there are numerous links between myosin molecules and actin molecules, and filament sliding is the composite effect produced by bending of many of these cross-bridges.

Control of Muscle Contraction

As described earlier, motor units consist of a motor neuron and the several muscle fibers innervated by the branched end of the motor neuron's axon. Nerve impulses (action potentials) conducted by the motor neuron's axons cause muscle fibers to contract. But what occurs between the passage of an impulse down a motor axon and the filament sliding, which is the essence of muscle contraction?

The end of a motor axon forms a synapselike junction with an area of the sarcolemma, the muscle fiber's plasma membrane. Such junctions are called neuromuscular junctions. As in a synapse between nerve cells, a space separates the specialized end (often called the motor end plate) of the motor axon from the sarcolemma.

When an impulse passes down to the tip of a motor axon, acetylcholine is released from vesicles that open at the surface of the motor end plate. Acetylcholine released from the motor axon crosses the gap between the axon and the sarcolemma and binds with acetylcholine receptors in the sarcolemma. The sarcolemma, like the nerve cell membrane, normally is polarized; there is a resting potential across the membrane. Acetylcholine binding changes the permeability of the sarcolemma by opening chemically gated ion channels through it and initiates an action potential, which sets off a chain reaction series of action potentials that sweep along the sarcolemma. So far, these events closely parallel those seen in synapses between nerve cells. But how does a series of action potentials traversing the sarcolemma cause filament sliding in all parts of all myofibrils of the muscle fiber at precisely the same time?

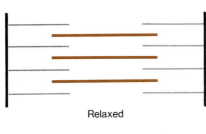

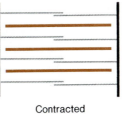

Relaxed

Contracted

Figure 17.16 Sketches showing how filaments in a sarcomere slide during muscle contraction.

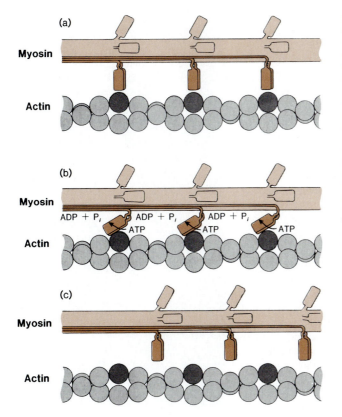

Figure 17.17 The proposed mechanism of filament sliding during contraction. (a) Thick filaments contain myosin molecules, which have globular double "heads" that form temporary connections with actin molecules. Thin filaments consist of two chains of spherical actin molecules that are intertwined in a double helix. (b) Myosin "heads" bind with actin molecules and then flex, thus exerting a sliding force on the thin filament. ATP is required for binding and flexing of the myosin "heads." (c) After flexing, the myosin heads release from actin, return to their original shape, and then bind with other actin molecules further along the filament.

Box 17.1
The First Site of Fatigue

When a muscle has contracted many times, it becomes fatigued. Its reserve of glycogen and its supply of creatine phosphate are depleted, and a great deal of lactic acid has accumulated. Animals feel fatigue, however, before these chemical changes, especially the lactic-acid buildup, threaten muscle fibers with permanent damage. Thus, a sense of fatigue normally causes an animal to rest and recover. But what happens if continued exercise is forced under extreme conditions?

The nerve of an isolated nerve-muscle preparation can be stimulated repeatedly until the muscle no longer contracts in response to further stimulation. What has failed to function? Electronic studies of axons under these conditions indicate that the axons still are able to conduct action potentials, and if electrodes are applied directly to the muscle, the muscle can contract strongly in response to direct stimulation. Thus, muscle fibers still are able to contract. These findings have led physiologists to conclude that during heavy exercise under experimental conditions, neuromuscular junctions are the initial sites of fatigue-induced failure.

It is assumed that fatigue-induced failure protects muscles from damage caused by excessive lactic-acid accumulation or other chemical damage due to massive fatigue. But it is not clear how these experimental findings relate to nerves and muscles in an intact normal body. Would muscles actually stop at some point if heavy muscular exercise continued too long? Do muscle cramps, which start much earlier in the fatigue process, normally help to keep motor units from reaching this "freeze-up" point? In the case of heavy exercise in humans, what is the role of determination ("willpower")? Do long-distance runners who "run through" cramps and muscle pain risk serious muscle damage (box figure 17.1A)? More research is needed to deal with these questions and to relate results of experimental studies of muscle functioning to general exercise physiology.

Box figure 17.1A An athlete displaying symptoms of muscle fatigue. Sebastian Coe crossing the finish line in a world record-setting performance.

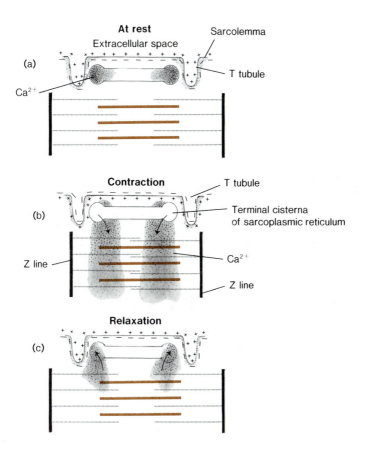

Figure 17.18 Simplified diagram of control of actin and myosin filament sliding by calcium release and recovery. (a) T tubules, which are continuous with the sarcolemma, actually branch extensively and penetrate all parts of the muscle fiber. (b) When action potentials pass down the T tubules, calcium is released from terminal cisternae, and filament sliding occurs. (c) Relaxation occurs when the terminal cisternae recover calcium ions. Cisternae membranes no longer permit calcium ions to flow outward, and cisternae quickly recover calcium by active transport inward across their membranes.

T Tubules and Cisternae

In the late 1940s, L. V. Heilbrunn and his colleagues injected a number of substances into muscle fibers and found that, of all those tested, only calcium salts caused fiber contraction. This discovery led to a proposal that permeability changes that occur in the sarcolemma during the action potential allow calcium ions (Ca^{2+}) to enter the cell and that these calcium ions diffuse inward and set off contraction of the contractile proteins.

This hypothesis was challenged as more research was done over the following years. One major problem is that very little calcium actually enters a muscle fiber as a result of action potentials in the sarcolemma. Another objection is that calcium ions moving inward by diffusion would reach outer myofibrils much sooner than the ones deep in the center of the cell, and yet, all of the myofibrils of a muscle fiber contract at exactly the same time. It seems, therefore, that if calcium ions actually do initiate contraction events in the myofibrils, some mechanism must permit simultaneous delivery of calcium ions throughout the interior of the muscle fiber.

Such a mechanism does exist. It is based on the functions of two networks of membranous structures that exist in the spaces among the myofibrils of the muscle fiber. One membranous network is the **sarcoplasmic reticulum,** the muscle fiber's endoplasmic reticulum (see chapter 3 for additional information on the endoplasmic reticulum). Expanded sacs of sarcoplasmic reticulum, called **terminal cisternae** (singular: **cisterna**), lie near the Z lines of the myofibrils. These cisternae contain a large quantity of calcium ions.

The second network of membranous structures is a system of hollow tubules whose walls are continuous with the sarcolemma. These tubules, which open to the outside of the cell, are called **T tubules** (for transverse tubules). They penetrate all parts of the cell and their tips come in contact with the cisternae of the sarcoplasmic reticulum (figure 17.18).

Each wave of action potentials passing along the sarcolemma results in depolarization of the membranes of all of the T tubules. When an electrical change in the membrane arrives at the tip of a T tubule near a terminal cisterna, it somehow triggers an abrupt change in the membrane of

Figure 17.19 Proposed relationships of tropomyosin, troponin, and calcium ions in the control of filament sliding. (a) Rodlike tropomyosin molecules lie along the actin chains. Globular troponin molecules are closely associated with tropomyosin molecules. (b) Details of the proposed arrangement of tropomyosin, troponin, and actin. Tropomyosin rods lie over the sites (cross-hatched) on actin molecules that combine with myosin cross-bridges. Troponin molecules have calcium binding sites. (c) When calcium combines with troponin, its confirmation is changed so that it moves tropomyosin and exposes sites on actin molecules that combine with myosin "heads." This permits the interactions between myosin and actin that cause filament sliding.

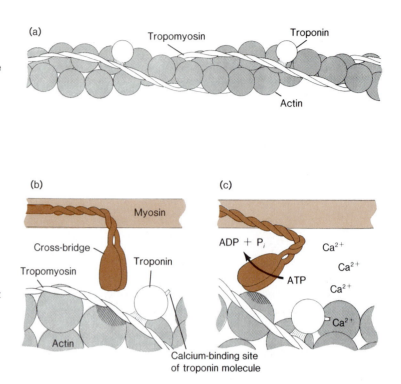

the cisterna, which then releases calcium ions. Because T tubules penetrate all parts of the muscle fiber, this calcium release is simultaneous in all parts of the fiber, and the calcium ions cause simultaneous contraction of all of the myofibrils.

After the calcium release, the membranes of cisternae again become relatively impermeable to calcium ions, an active transport system pumps calcium ions back into the cisternae, and contraction stops. This inward movement of calcium ions into the cisternae prepares the cell to respond to subsequent stimulations by the motor nerve.

Tropomyosin, Troponin, and Calcium

The mechanism by which calcium ions, which are released from the terminal cisternae, initiate filament sliding is now quite well understood. Two regulatory proteins, **troponin** and **tropomyosin,** which are present with actin in the thin filaments, are involved in regulation by calcium. Each thin filament is essentially a double helix; that is, two chains of linked actin molecules wind around each other (see figure 17.17). Actin molecules have combining sites for myosin "heads."

Tropomyosin molecules, along with troponin molecules, function to block the formation of linkages between myosin "heads" and actin molecules (figure 17.19).

Rodlike tropomyosin molecules lie nestled in the grooves between the chains of the double helix of the thin filament. At rest, in the absence of stimulation, tropomyosin occupies a key position, relative to the actin molecules, that precludes binding by myosin "heads."

Troponin molecules are globular protein molecules that are closely associated with tropomyosin, and their conformation affects the positioning of tropomyosin relative to the myosin-binding sites on actin molecules. When calcium ion concentration increases as a result of stimulation, troponin molecules bind calcium and, as a result, undergo conformational changes that shift the position of tropomyosin. This movement of tropomyosin presumably uncovers sites on the actin molecules and permits the linkage between actin and myosin that is essential for filament sliding and muscle contraction.

Contraction of Other Kinds of Muscles

Cardiac muscle and smooth muscle both have functional characteristics that distinguish them from skeletal muscle. Cardiac muscle requires at least ten times longer than skeletal muscle to contract and relax, and smooth muscle works even more slowly, requiring as much as a hundred times longer than skeletal muscle to contract and relax.

Tonus is different in smooth muscle because it is highly variable. Smooth muscle may remain slightly or strongly contracted for much longer periods than skeletal muscle.

Cardiac muscle differs sharply from skeletal muscle in that it has a long **refractory period,** the period following one stimulus during which it is unable to respond to another stimulus. Thus, cardiac muscle is not capable of tetanic contraction. Another interesting feature of vertebrate cardiac muscle is its inherent rhythmicity. Individual heartbeats are generated in the heart itself. Spontaneous depolarization of membranes in a pacemaker area results in transmission of impulses that stimulate contraction of muscle fibers in all parts of the heart (see chapter 12). This internal beat generation means that individual heartbeats do not depend at all upon neural stimulation from the central nervous system.

Despite these functional differences and differences in internal organization of individual muscle fibers, contractions of both smooth and cardiac muscle also depend on the same basic actomyosin sliding filament mechanism that operates in skeletal muscle.

In fact, muscles of invertebrate animals, which display a diverse array of functional characteristics that are very different in some cases from those associated with vertebrate muscles, also seem to contract by the same kind of molecular mechanisms, although molecular controls of contraction may differ considerably. For example, muscles of some invertebrates lack troponin; contraction and relaxation thus are regulated by other means. Nevertheless, cross-bridges between actin and myosin form the basis for filament sliding during contraction.

Even some other cellular movements, such as amoeboid movement (see chapter 3), involve these same contractile proteins. The ability of contractile proteins such as actin and myosin to convert chemical energy into physical movement at the molecular level forms the basis for many kinds of movement in living systems.

Summary

In response to sensory information about changes in the external environment or the internal environment of the body, the central nervous system sends stimuli over motor axons to various effectors.

Skeletal muscles exert force by contracting and pulling against rigid but jointed and movable skeletal elements. Because muscles exert force only by shortening, skeletal muscles usually are arranged in antagonistic pairs.

The two major skeletal tissues in vertebrates, cartilage and bone, consist of cells and considerable extracellular matrix material. Cartilage has some flexibility, but bone provides very rigid support.

The spaces among the bony bars of skeletal bone and the spaces in the cylindrical cavities of long bones contain bone marrow, which functions in replacement blood cell production.

Compact bone is made up of Haversian systems, which have concentric cylinders of bony matrix with scattered cells arranged around a central cavity that contains blood vessels and nerves.

Human skeletons have two main parts. The axial skeleton supports the head, neck, and trunk of the body. The appendicular skeleton includes the bones in the limbs and the bones that support the limbs.

Vertebrate animals have three types of muscle tissue. Skeletal muscle tissue makes up muscles responsible for limb movements and other whole body movements. Smooth muscle tissue is involuntary muscle located in the digestive tract and other internal organs. Cardiac muscle tissue is found in the heart. The three muscle types have different structural and functional characteristics.

Properties of skeletal muscles can be studied in isolated muscles or in nerve-muscle preparations, using an electrical stimulator and a kymograph apparatus.

A simple twitch—which includes a latent period, a contraction period, and a relaxation period—is a muscle's response to a single, brief stimulation.

The strength of whole muscle contractions depends on the number of fibers contracting and the frequency of stimuli conducted by the various motor axons.

Rapidly repeated stimuli produce a summation response, and a still higher frequency of stimulation produces tetanus, a smooth, sustained contraction. Laboratory-induced tetanic contractions resemble the contractions involved in normal body movements more closely than do simple twitches.

Muscle cells contain characteristic contractile proteins, actin and myosin, that interact to produce physical shortening. Energy for shortening is provided directly by ATP.

Creatine phosphate seems to provide an energy reserve that allows rapid replenishment of ATP in a working muscle. But continuing contraction soon uses up the creatine phosphate reserve and then exceeds the ability of aerobic respiration to supply ATP because oxygen delivery to the muscle is a limiting factor. The muscle then uses anaerobic lactic fermentation to supplement ATP production, but this causes lactic-acid buildup, and the muscle accumulates an oxygen debt that must be repaid when muscle work stops.

The physical basis for muscle contraction lies in the highly ordered arrangement of filaments of contractile proteins inside the myofibrils of muscle fibers. Thick myosin filaments overlap thin, actin-containing filaments in the darker A bands of myofibrils. The lighter I bands contain only thin filaments. During contraction, myosin molecules in the thick filaments form temporary, bendable cross-bridges with actin molecules in the thin filaments. Energy provided by ATP hydrolysis permits bending of these cross-bridges. The bending pulls the thin filaments along so that they overlap further with the thick filaments. Filament sliding shortens the myofibrils and whole muscle fibers. During relaxation, the filaments return to their original positions.

Acetylcholine released at the end of a motor axon triggers a wave of action potential that passes along the sarcolemma and the membranes of the T tubules that end near terminal cisternae of the sarcoplasmic reticulum. This triggers calcium ion release from the terminal cisternae and causes filament sliding.

Calcium ions bind with the regulatory protein troponin. This changes the conformation of troponin so that it shifts another regulatory protein, tropomyosin, away from the myosin-binding sites on actin molecules. This permits cross-bridge formation and filament sliding.

Cardiac and smooth muscle have functional properties that differ from those of skeletal muscle, but contraction of all three types of muscle is based on interactions of the contractile proteins actin and myosin. These contractile proteins also function in other types of movement in living cells.

Questions

1. What characteristic of muscle function necessitates the arrangement of skeletal muscles in antagonistic pairs?
2. Explain the mechanism by which increasing stimulus strength increases the force of whole muscle contractions.
3. How is muscle tonus maintained in skeletal muscles without development of muscle fatigue?
4. Why might you assume that Ca^{2+} ions had to be included in the medium that Szent-Györgyi used when he successfully induced shortening of actomyosin threads by adding ATP?
5. Explain why H zones seem to disappear in contracting myofibrils.
6. What is the functional significance of the fact that T tubules extend deep into muscle fibers and branch extensively?

Suggested Readings

Books

Carlson, F. D., and Wilkie, D. R. 1974. *Muscle physiology*. Englewood Cliffs, N.J.: Prentice-Hall.

Eckert, R., and Randall, D. 1978. *Animal physiology*. San Francisco: W. H. Freeman.

Wilkie, D. R. 1976. *Muscle*. 2d ed. Studies in Biology No. 11. Baltimore: University Park Press.

Articles

Buller, A. J. 1975. The contractile behaviour of mammalian skeletal muscle. *Carolina Biology Readers* no. 36. Burlington, N.C.: Carolina Biological Supply Co.

Cohen, C. November 1975. The protein switch of muscle contraction. *Scientific American* (offprint 1329).

Huxley, H. E. 1969. The mechanism of muscular contraction. *Science* 164: 1356.

Lazarides, E., and Revel, J. P. May 1979. The molecular basis of cell movement. *Scientific American* (offprint 1427).

Lester, H. A. February 1977. The response to acetylcholine. *Scientific American* (offprint 1352).

Murray, J. M., and Weber, A. February 1974. The cooperative action of muscle proteins. *Scientific American* (offprint 1290).

Plant Hormones and Plant Responses to the Environment

Chapter Concepts

1. A plant continually interacts with its environment.
2. Plant hormones mediate these interactions and cause specific developmental responses in plants.
3. Light is the dominant environmental factor to which plants respond. Light determines growth patterns of plants and coordinates the timing of important events in their lives.
4. Changes in the twenty-four hour cycle of alternating light and dark are photoperiodic stimuli that are particularly important in timing flowering and preparation for dormancy.
5. Internal clocks are necessary for time measurement in plant responses to day length changes.
6. Internal clocks also time daily rhythms in many physiological processes, but the nature of the basic cellular clock remains obscure.

Every phase of a terrestrial vascular plant's growth, function, and reproduction involves interactions with a changing environment. Because plants remain in a single place throughout their lives, they cannot avoid adverse changes in their environment by moving to more favorable locations. They are directly exposed to all of the seasonal changes in light, temperature, and available moisture, as well as to the possibility of being eaten by animals. They must adjust to seasonal changes in conditions around them, and occasionally, to the loss of parts of their bodies.

Plants develop in an organized way: flowering and seed production are completed during a specific part of the growing season, and then plants prepare for winter dormancy. How is all of this coordinated? How do seeds germinate at an appropriate time? What factors control shape, size, and organization of the plant body? How do whole fields of a given plant species flower during the same brief period, sometimes as short as only a few days out of an entire growing season (figure 18.1)? What factors cause plants to change their activities so that they are no longer vulnerable to harsh winter conditions?

In all of these orderly and properly timed processes, one or several plant hormones mediate plants' responses to such external factors as temperature, moisture, and especially, environmental light conditions. Throughout the year, changes in light conditions are the most reliable indicators of changing seasons. Seasonal day length and light intensity changes are the same every year, while temperature and moisture conditions can vary considerably. This reliability of seasonal light changes has probably been a major factor in evolution of plant responses to environmental change. Light is the dominant external factor affecting seasonal plant responses. Of course, light also is essential to plants as an energy source for photosynthesis, and there are numerous specific, hormone-mediated growth responses to light. Thus, the study of light's effects on plants is intertwined with the study of plant hormones.

Plant Reponses to Light

The three general categories of plant responses to light are phototropism, photomorphogenesis, and photoperiodism.

Figure 18.1 Events in the lives of plants are precisely regulated in a continuing interaction with the environment. This picture of a mountain meadow illustrates the simultaneous flowering of many individual plants that occurs during only a brief period out of the entire growing season.

Some plant responses depend on a *directional* light stimulus, and they develop a growth curvature that is specifically determined by the direction of the light striking the plant. This orientation of plant parts in response to a directional light stimulus is called **phototropism** ("light turning"). House plants that "turn toward the light" are good examples of phototropism.

Other plant responses are initiated by light stimuli that need not be either specifically directional or periodic. These responses fall in the category known as **photomorphogenesis**. For example, flowering plants grown from seeds in the dark are **etiolated**. That is, they have a pale, yellowish color rather than a bright green color because they have not produced mature chloroplasts with chlorophyll. For many of these plants, one brief exposure to light is enough to stimulate chloroplast maturation and the appearance of their normal green color. Photomorphogenic responses also affect seed germination, stem elongation, leaf unrolling, and many other structural and functional features in the life of plants.

Many plants respond to day length changes; that is, changes in the twenty-four-hour cycle of alternating light and darkness. This is called **photoperiodism**. Photoperiodic responses usually occur over a relatively longer period of time and involve qualitative changes in the plant (for example, from a nonflowering to a flowering condition).

Plant Hormones

A plant hormone is like an animal hormone in that it is produced in one part of the organism and transported to other parts of the organism, where it causes a response, even though it is present only in a very low concentration. But there also are some important differences between hormonal regulation in plants and animals.

Animal hormones are produced by specialized cells whose chief or only function is hormone production, and in many cases, hormone-producing cells are clustered in specific organs, the endocrine glands. Hormones from endocrine glands are carried throughout the body in the bloodstream, but they act on only one or a few kinds of cells, their target cells. While plant hormones also may be produced in localized parts of plant bodies, the cells that produce them are not specialized for hormone production alone. Hormone-producing cells of plants often are, for example, mitotically active cells in a meristematic region of the plant. Also, responses to plant hormones are not restricted to a few kinds of target cells; practically all plant cells respond in some way to each of the plant hormones, although some cells respond differently or more extensively than others.

Another important difference between hormonal regulation in plants and animals is the number of hormones produced. In mammals, where animal hormonal regulation is best understood, there are literally dozens of known hormones and probably additional ones yet to be discovered. Vascular plants have only about a half dozen major hormones, but this does not limit the number of types of responses caused by plant hormones. Different combinations of the plant hormones give different responses in many cases. Therefore, plant regulation depends heavily on changing balances of a few hormones, instead of specific actions by a larger number of individual hormones, as is the case in animal regulation.

Finally, the modes of cellular action of plant and animal hormones seem to be significantly different. Animal hormones bind with specific receptor molecules of target cells (page 412). Then the hormone-receptor complex causes a specific response of the target cell, either by direct action or by stimulating production of a second messenger, such as cyclic AMP (p. 414), which causes the specific response of the target cell. But it is questionable whether such specific hormone receptor molecules are involved in plant cells' responses to hormones. Some molecules in plant cell membranes do bind plant hormones, but it is not clear whether this binding is directly involved in normal hormone actions. Also, there is very little cyclic AMP in plant cells and no evidence, as yet, that cyclic AMP functions as it does in some animal target cells, or that there are other second messengers in plant cells. Rather, plant hormones appear to produce reactions in plant cells directly.

Almost all hormone effects in plants modify growth patterns; plant hormones are regulators of the lifelong growth of plants and are not as closely associated with short-term, reversible physiological adjustments as many animal hormones are.

Five major types of plant hormones have been identified: auxins, ethylene (a gas), cytokinins, gibberellins, and abscisic acid.

Auxins

Auxins were the first of the plant hormones to be identified. Their discovery resulted from their role in various tropisms (turning responses to environmental stimuli), especially phototropism. Auxins are now known to be involved in a great variety of plant cell responses.

Phototropism

Scientific analysis of phototropism began with the experiments of Charles Darwin and his son Francis around 1880. The Darwins used grass seedlings to investigate the bending of plants in response to unidirectional light. They found that seedlings failed to bend toward the light after their tips were cut off or covered with black caps. A seedling would bend toward the light, however, if its tip was exposed while the rest of the seedling was buried in fine black sand. The Darwins concluded that curvature of grass seedlings in response to light depended on some "influence" transmitted from the seedling tip to the rest of the seedling (figure 18.2b).

Work on phototropism then focused on this role of the seedling tip. Usually, oat (*Avena*) seedlings were used in the experiments. An oat seedling has a sheath, called the **coleoptile**, that encloses the first leaves, and the tip of this coleoptile is the key to the phototropic response of the young seedling (figure 18.2a).

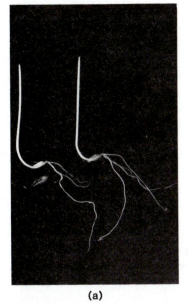

(a)

Figure 18.2 The coleoptile and the discovery of auxins. (a) A photograph of two oat (*Avena*) seedlings, one with its husk removed (left), showing the coleoptile and other seedling parts. The first leaves are rolled up inside the coleoptile. (b) A diagrammatic summary of some of the important experiments that demonstrated the existence of auxins. Both the Went and Paál experiments were conducted in the dark.

Darwin (1880)

Light

Intact seedling
(curvature)

Tip of coleoptile
excised
(no curvature)

Opaque cap
on tip
(no curvature)

Buried in fine black
sand but with extreme
tip left exposed
(curvature)

Boysen-Jensen (1913)

Tip removed

Tip replaced
with gelatin

Normal phototropic
curvature remained
possible

Paál (1919)

Tip removed

Tip replaced
on one side
of coleoptile stump

Growth curvature
developed without a
unilateral light stimulus

Went (1928)

Coleoptile tips
on agar

Tips discarded;
agar cut into
small blocks

Each agar
block placed on
one side of
coleoptile stump
in darkness

Coleoptile bent
in darkness;
angle of curvature
measured and is
proportional to number
of coleoptile tips
stood on agar
or to time of standing

(b)

P. Boysen-Jensen, working in Denmark about thirty years after the Darwins' work on phototropism, experimented with cutting off the tips of the oat coleoptiles. When he did this, the coleoptile stopped growing. Next he placed a piece of gelatin on the cut surface and set the cut tip on top of the gelatin. A short time later the coleoptile resumed growing, and in this condition it would also show the normal phototropic response to light from one side. Boysen-Jensen's experiments demonstrated that the influence of the coleoptile tip did not depend on the tip being in its normal place on the plant because the influence of the tip could pass through a gelatin block (figure 18.2b).

A few years later in 1918, Arpad Paàl in Hungary cut off coleoptile tips and then placed the tips to one side of the cut surfaces. When these plants were allowed to grow in the dark, Paàl observed a curvature that was very similar to the normal response of intact seedlings to unidirectional light (figure 18.2b). Thus, asymmetrical placement of the tip imitated the effect of light shining on the plant from one side. Paàl reasoned that the tip of a normal coleoptile produces a growth-promoting substance that travels downward and that light must make it travel asymmetrically, thus causing greater growth on the shaded side of the seedling.

Then in 1926 Frits Went's experiments in Holland demonstrated that Paàl's proposal was right: a substance produced in the coleoptile tip causes growth responses in the rest of the coleoptile. Went cut off coleoptile tips, placed them on an agar block, and left them for varying periods of time. He then cut up the agar and put pieces of it on "decapitated" coleoptiles that were kept in the dark. A piece of this agar placed squarely on top of a cut coleoptile caused the seedling to grow straight upward, but if Went placed the agar to one side of the top of the coleoptile, the seedling curved just as it did in Paàl's experiment with the tip itself. This seedling curvature response in the dark, and in the complete absence of the coleoptile tip, proved conclusively that the effect of the coleoptile depends not on the actual physical presence of the tip, but on a diffusible chemical substance that can accumulate in the agar (figure 18.2b).

Went also found that the angle of seedling curvature in the dark depended on the number of tips that had been on the agar and the length of time that they had spent on the agar. In other words, the degree of seedling curvature is proportional to the amount of growth substance in the agar. Thus, the degree of curvature can be used as a quantitative test for determination of the amount of growth substance present in a test sample. Such quantitative measurements employing specific biological responses are known as **bioassays**, and Went's coleoptile curvature bioassay (figure 18.3) and various other bioassays have been instrumental in the search for the plant hormones.

The growth hormone produced by oat coleoptile tips was given the name **auxin**, from the Greek word *auxein* meaning "to grow." Went's reliable bioassay provided a key tool for biologists seeking to determine the chemical nature of auxin.

Nature and Cellular Actions of Auxins

Early researchers thought that they were working with a single substance in their experiments on growing coleoptiles. But when they began to apply Frits Went's bioassay for auxin activity to substances extracted from a variety of organisms, they found that auxin could be isolated not only from plants, but also from a number of other seemingly unlikely sources, including various molds and yeasts, and even animals. In fact, the first auxin to be identified chemically was not isolated from plants, but from human urine! This auxin is **indoleacetic acid (IAA)**, a molecule whose molecular structure is very similar to that of the amino acid tryptophan (figure 18.4). Evidence now indicates that IAA is the principal naturally occurring auxin. Its isolation from plants, however, has been difficult because it is present and active in only extremely low concentrations inside plants.

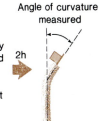

Cut coleoptile tip
transferred to agar
for known period

Coleoptile tips removed;
agar block cut
into small blocks;
similar agar blocks
prepared containing
known amounts
of authentic auxin

Agar block
containing
auxin placed
asymmetrically
on decapitated
coleoptile
in clamp,
seedling's first
leaf pulled up
slightly as
support

Angle of curvature
measured

2h

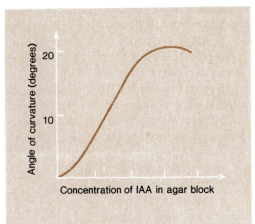

Figure 18.3 The oat (*Avena*) coleoptile curvature bioassay for auxin, which is derived from Frits Went's original study. The amount of auxin diffusing into an agar block can be estimated with reference to a standard curve (right) prepared from responses of coleoptiles to known concentrations of auxin. The tests are conducted in darkness.

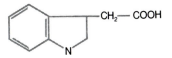

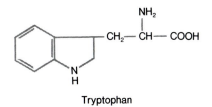

IAA

Tryptophan

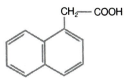

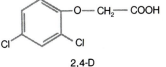

Naphthalenacetic acid

2,4-D

Figure 18.4 Indole acetic acid (IAA) is the principal naturally occurring auxin. The formula of the amino acid tryptophan is given to show its close similarity with IAA. Naphthaleneacetic acid and 2,4-dichlorophenoxyacetic acid (2,4-D) are synthetic auxins.

Other compounds isolated from plants also act as auxins, as do a number of laboratory-synthesized compounds. Some of these synthetic auxins are economically important. For example, **naphthalenacetic acid** is widely used in horticulture for plant propagation because it promotes root development on stem cuttings (figure 18.4). It also is used in orchards to prevent damaging early fruit drop. A commonly used **herbicide** (plant killer), **2,4-dichlorophenoxyacetic acid**, or **2,4-D**, is also a synthetic auxin (figure 18.4). It causes fatally deranged growth and metabolic responses in dicots at concentrations that do not harm monocots, such as lawn grasses. This selective killing of "broadleaf weeds" makes 2,4-D and similar synthetic auxins practical for weed control in lawns and especially in growing cereal grain fields. Although these synthetic auxins have been used for many years as weed killers, it still is not known why a given concentration of 2,4-D kills broadleaf plants and leaves grasses unharmed.

Normal responses of plants to auxin depend on cell elongation, but the cellular mechanisms by which auxin causes plant cells to elongate are not altogether clear. Cellular responses to auxin seem to occur in two phases. The first phase is a relatively fast response that takes place within ten to fifteen minutes after cells are exposed to appropriate quantities of auxin. A responding cell actively pumps hydrogen ions (H^+) out through its plasma membrane into the surrounding cell wall via an energy-requiring (using ATP) process. The cell wall contains cellulose fibers that are chemically cross-linked to form a firm, box-like structure just outside the plasma membrane. The hydrogen ions that are pumped out lower the pH in the cell wall, and the decreased pH activates enzymes that break down the cross-bridges linking the cellulose fibers. This weakens the wall and allows the normal turgor pressure of the plant cell to push the wall outward, thus expanding the cell.

A second, slower phase of the response to auxin takes thirty to forty-five minutes to begin. In this second phase, auxin stimulates transcription of part of the genome and the subsequent synthesis of certain enzymes involved in cellulose production.

The first, quicker cell elongation response phase is independent of the second phase. The first phase occurs even if the second phase is prevented by drugs, such as the antibiotic cycloheximide, that inhibit protein synthesis. It appears that the initial cell expansion phase stretches and thins the cell wall, and the second phase, which includes auxin-stimulated protein synthesis, leads to production of structural material that thickens cell walls to their prestretching condition, but further research is needed before this explanation of cellular responses to auxin can be fully accepted.

Cell Elongation and Auxin-Induced Responses

Greater cell elongation on the side of a plant away from a unidirectional light source causes the curvature response in phototropism (figure 18.5), and many years ago, it was proposed that cells elongate more on the shaded side because the auxin concentration is greater there (figure 18.5). How does the auxin concentration differential that causes the greater expansion on the shaded side come about?

In the 1920s, N. J. Cholodny and Frits Went proposed that a growth substance is laterally transported in response to a stimulus such as asymmetrical lighting. But this Cholodny-Went hypothesis fell into disfavor, first because other researchers proposed that light simply inhibited auxin synthesis on the lighted side of the plant and then later because of the discovery that bright light promotes the natural breakdown of auxin. This latter finding led to a new hypothesis—that the differential distribution of auxin in phototropic responses might be due to destruction of auxin on the lighted side rather than to some mechanism involving auxin transport to the shaded side. Light destruction of auxin, however,

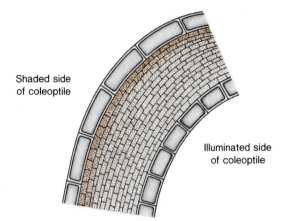

Figure 18.5 Greater cell elongation on the shaded side, a response to greater auxin concentration, causes curvature in phototropism.

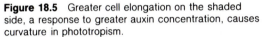

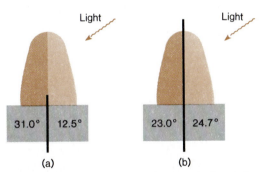

Figure 18.6 Evidence for lateral transport of auxin in corn (*Zea mays*) coleoptile tips in response to unilateral illumination. In (a), lateral transport of auxin is possible, but in (b) a thin glass barrier divides the coleoptile tip so that auxin cannot be transported laterally. The numbers represent degree of curvature of coleoptiles in a Went-type bioassay for auxin content in the agar blocks that were under the halves of the tip in each case. The experiment clearly showed that the asymmetrical distribution of auxin produced by unidirectional illlumination is due to lateral transport of auxin, not to light-induced destruction of auxin on the lighted side.

requires very bright light, and some phototropic responses do not require such bright light. A new set of experiments in the 1950s produced results that were much more in harmony with the Cholodny-Went hypothesis of lateral transport than with the hypothesis that light caused auxin destruction on the illuminated side.

W. R. Briggs and his colleagues at Stanford showed clearly that auxin produced in the coleoptile tip is transported to the shaded side of the coleoptile. Their key experiment compared corn coleoptile tips that were prepared in two slightly different ways. Some of the coleoptiles were set on agar blocks, and then the blocks and part of the coleoptiles were split by barriers that were left in place (figure 18.6a). Other coleoptiles on agar blocks were completely split with barriers, and again the barriers were left in place (figure 18.6b). Then unidirectional light was applied for three hours, and the pieces of agar block below the coleoptiles were removed for testing in a Went-type bioassay. The piece of agar below the shaded side of the partially split coleoptile showed three times as much auxin activity as the piece below the lighted side. But in the completely split coleoptile, where lateral transport was blocked, the auxin concentration in the two halves of the agar block was essentially equal.

These and other experiments demonstrated that auxin produced in the coleoptile tip is transported to the shaded side. This lateral transport of auxin, rather than light-induced destruction of auxin, establishes the auxin concentration difference, which in turn causes differential cell elongation and the resultant seedling curvature. Despite these clear demonstrations that auxin is transported laterally, important questions remain. For instance, very little is known about the actual mechanism by which such lateral transport takes place.

Light is not the only factor affecting the growth direction of seedlings. Growing seedlings also bend in response to gravity. Response to gravity, called **geotropism**, is very important in early seedling growth because it gets seedlings properly and uniformly oriented for growth, even though seeds are in a variety of orientations in the soil when seedling growth begins. In geotropism, stems curve upward and roots curve downward; thus, stems are **negatively geotropic** because they curve away from the earth's gravity center, and roots are **positively geotropic** because they curve toward it (figure 18.7).

The geotropic response of stems is fairly easily explained because it involves greater cell elongation on the lower side. This elongation apparently results from auxin accumulation due to auxin transport to the lower side in response to gravity. Thus, the general mechanism of stem curvature in geotropism is similiar to that seen in phototropism.

But the geotropic response of roots is not so simply explained. Auxin is transported to the lower side of a root in response to gravity, but the upper side of the root (where auxin concentration is lower) shows the greater cell elongation, with the result that the root turns downward in the soil. One explanation for the difference between stems and roots is that root cells are much more sensitive to auxin than are stem cells. An auxin concentration that stimulates stem elongation actually is high enough to inhibit root cell elongation (figure 18.8). Therefore, the upper side, the side *opposite* the area of high auxin concentration, actually elongates most because it has an auxin concentration in the range that favors root cell elongation. More recently, it has been reported that abscisic acid, an inhibitor, accumulates on the lower side of seedlings' roots, where it can directly inhibit cell elongation on the lower side. Finally, the gaseous hormone ethylene also may be involved in root geotropism, but its role remains to be determined. Thus, root geotropism appears to be one of many cases in which a plant growth response is controlled by interactive effects of several of the plant hormones.

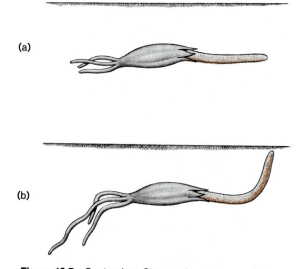

Figure 18.7 Geotropism. Stem and roots grow out of a seed (a) and respond to gravity (b) with stem growing up and roots growing down. Stem curvature results from differential auxin concentration. Root curvature involves several factors.

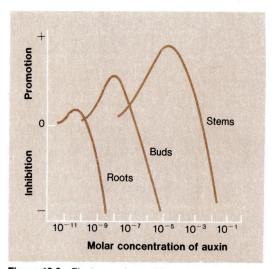

Figure 18.8 Plant parts have different sensitivities to auxin, and different concentrations promote optimum growth in each case. Concentrations above the optimum level actually inhibit growth. Note that roots show a much lower optimum than stems and that auxin levels that promote stem cell elongation strongly inhibit root growth. This difference is relevant for the explanation of root geotropism in the text.

(a)

(b)

Figure 18.9 General growth patterns of plants depend
on relative growth of main stem and lateral branches.
Compare the growth of trunk and branches and the
shapes of a poplar (a) and an American elm (b).

Apical Dominance and Other Auxin Effects

The general growth pattern of any dicot (broad-leaf) plant depends on the relative growth at the tip (apex) of the main stem and the growth of lateral buds, which produce branches. Plants with a dominant main stem and very little lateral growth have a straight, narrow shape, while plants with extensive lateral growth are bushy. This difference in growth pattern is illustrated quite dramatically in trees (figure 18.9).

Auxin produced in the apical meristem (the region of cell division at the stem tip) is mainly responsible for **apical dominance** (dominance of the main stem over lateral branches). Apical dominance is another example of the differential sensitivity of different plant parts to various auxin concentrations because the same auxin concentration that promotes cell elongation in the main stem simultaneously inhibits development of lateral buds (see figure 18.8). Clearly, the nature of the response to a given concentration of auxin depends on the type and condition of the responding tissue.

Removal of the stem tip (shoot apex) removes the source of the auxin that had maintained apical dominance and releases the lateral buds from inhibition. Lateral buds then grow actively and produce branches. This procedure is put to practical use in horticulture and everyday gardening when stem tips are "pinched off" to make plants such as *Coleus* develop more

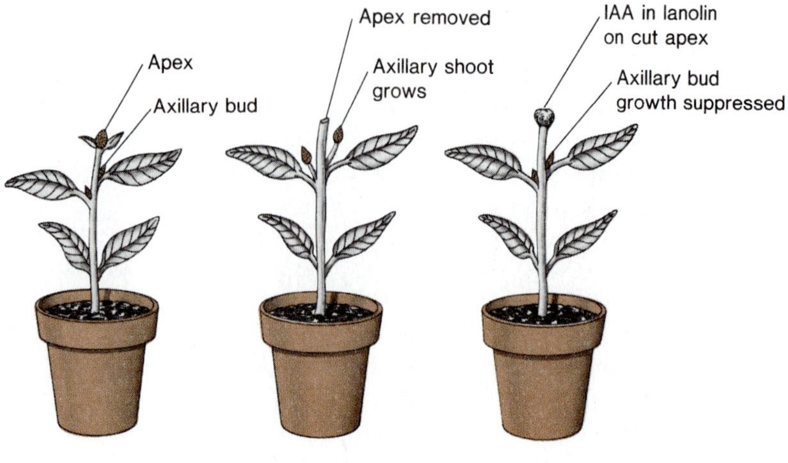

Figure 18.10 Experiments showing growth of lateral (axillary) buds when the stem apex is removed and role of auxin in apical dominance. IAA applied to cut apex inhibits growth of lateral buds.

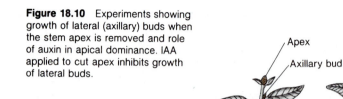

Apex

Axillary bud

Apex removed

Axillary shoot grows

IAA in lanolin on cut apex

Axillary bud growth suppressed

branches and a fuller growth pattern. Experimentally, apical dominance can be maintained following shoot apex removal if auxin is applied to the cut surface of the stem (figure 18.10).

Auxin also causes a variety of other responses, both alone and in conjunction with other hormones. For example, in many plants, fruits develop in response to auxin produced by embryos in developing seeds, and early removal of seeds causes abnormal fruit development (figure 18.11).

Ethylene

Ethylene is the only one of the currently known plant hormones that is a gas at normal environmental temperatures and pressures (figure 18.12). It has been known for many years that plants produce ethylene, and as early as 1935, there were suggestions that it might function as a hormone. However, the very small quantities of ethylene involved in plant responses were very difficult to detect and measure until quite recently.

(a)

(b)

(c)

(d)

Figure 18.11 Auxin released from seeds influences fruit growth. (a) Normal strawberry fruit. (b) Fruit of the same age from which all seeds had been removed. (c) Magnified view of a fruit with all seeds except one removed. There is a normal area of fruit development around this one seed. (d) A strawberry fruit that developed after all seeds were removed and replaced by a lanolin paste containing a synthetic auxin.

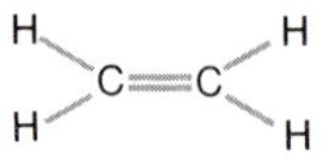

$$\begin{matrix} H & & & H \\ & \diagdown & & \diagup \\ & C & = & C \\ & \diagup & & \diagdown \\ H & & & H \end{matrix}$$

Ethylene

Figure 18.12 Structure of ethylene.

Ethylene is involved in flowering responses in at least some plants. It also is closely related to auxin effects because when auxin exceeds a certain concentration in tissues, the tissues begin to produce and release ethylene. This makes interpretation of responses to auxin more complex because ethylene is an antagonist of auxin in many processes; that is, ethylene causes responses opposite to those caused by auxin.

The best-known response to ethylene is fruit "ripening." K. V. Thimann rightly points out that "ripeness" is more of a subjective human judgment than a specific physiological state in fruits. Basically, fruits are said to be ripe when they are ready to be eaten. Ethylene promotes hydrolysis of starch, with a resulting increase in sugar concentration. Ethylene also stimulates production of **cellulase**, an enzyme that hydrolyzes cell wall cellulose and thus generally softens the fruit tissue during ripening.

Ethylene production is "contagious." That is, when a fruit releases ethylene, it stimulates other fruits to begin to produce it as well. Thus, the old saying, "one bad apple spoils the barrel," is accurate. To counteract this contagious ripening, apples now are stored in sealed compartments with a high concentration of atmospheric carbon dioxide because CO_2 is an antagonist of ethylene. On the other hand, ethylene can be used to stimulate ripening in fruits that are picked green and shipped long distances. Thus, ethylene is a natural fruit-ripening hormone that is economically important.

Cytokinins

In the early 1940s, biologists attempted to grow plant tissues in culture vessels using media containing auxin and all known plant nutrients, but they did not obtain completely satisfactory results. Something seemed to be missing because plant cells in these cultures would enlarge, often to spectacularly large sizes, but cell divisions were rare. Thus, a quest began for substances that would promote cell division in cultured plant cells. Eventually, by trial and error, it was discovered that coconut milk, which is a liquid endosperm (nutrient storage tissue), would greatly stimulate cell division in cultured cells when added to a culture medium. Other preparations, such as crude yeast extracts, also could provide the necessary stimulus for cell division.

However, coconut milk and yeast extracts are complex mixtures of many substances, and soon a search was underway for the identity of the actual division-promoting factors contained in these complex mixtures. Folke Skoog and his colleagues at the University of Wisconsin began this research by testing the effectiveness of nucleic acid extracts from various sources, and they found that DNA preparations from yeast promote cell division in cultured cells. But as they proceeded to test other nucleic acid preparations, they obtained some curious results. They got their best results with an old nucleic acid preparation that had been stored in the laboratory for some time. Old preparations, in fact, were very much more effective cell-division promoters than freshly prepared ones. Furthermore, they found that nucleic acid preparations sterilized by intense heating in an autoclave (a device that sterilizes with steam under high pressure) were much more effective than ones sterilized more "gently" by filtration.

These results strongly suggested that the actual division-promoting factors are similar to breakdown products produced when nucleic acids are stored for long periods or heated to high temperatures. This led Skoog and his colleagues to test a number of compounds that were chemically similar to components of nucleic acids, and they found that several of them were effective cell-division promoters. They called these compounds

cytokinins (from *cytokinesis*, meaning cell division) and named the active one that they isolated **kinetin** (figure 18.13). Although kinetin effectively promotes cell division in tissue cultures, it is not a naturally occurring compound. Years later, in 1964, D. S. Letham and his associates in New Zealand announced that they had isolated a natural cytokinin from *Zea mays* (corn), which they therefore named **zeatin**. Finally, in 1967, it was discovered that coconut milk contains zeatin as well as another structurally similar cytokinin called **zeatin riboside**. Thus, the mystery of coconut milk's cell division promoting effect on cultured plant cells was solved at last, some twenty-five years after the search began.

In cultured plant tissue, there is a complex interaction between auxin effects and cytokinin effects on cells (figure 18.14). When auxin and cytokinin concentrations are balanced, the tissue grows as an undifferentiated mass called a **callus**; when the auxin to cytokinin ratio is increased, roots develop; when the auxin to cytokinin ratio is decreased, shoots and leaves develop. Similarly, in intact plants, regulation of various growth processes involves changing ratios of auxin and cytokinin concentrations.

As with all plant hormones, cytokinins also are known to affect many different processes in plants. Cytokinins promote development of lateral buds. If kinetin is applied directly to lateral buds, they are released from apical dominance and begin to grow actively into lateral branches. Cytokinins are involved in breaking the dormancy of embryos in germinating seeds, and they promote flowering and fruit development in some plants. Cytokinins also retard senescence (aging) of leaves and other organs; commercially, cytokinins are sprayed on vegetables to keep them "fresh" during shipping and storage.

Gibberellins

Gibberellins were first found as a result of research on a rice plant disease called "foolish seedling disease." Affected rice seedlings become unusually tall, but they are spindly and weak, and usually break and fall over before they produce ripe rice grains. Japanese scientists learned that the disease is caused by a fungus, *Gibberella fujikuroi*. Then, in 1926, E. Kurosawa showed that a substance that caused excessive growth in rice plants could be extracted from the fungus,

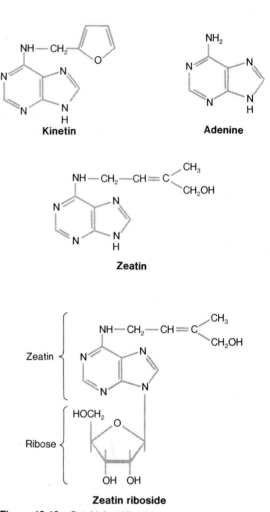

Figure 18.13 Cytokinins. Kinetin was isolated as a hormonally active fraction in breakdown products of nucleic acids. All cytokinins are structurally related to the purine, adenine. Zeatin is a natural cytokinin isolated from corn *(Zea mays)* seeds. Zeatin riboside occurs, along with zeatin, in coconut milk. Note that zeatin riboside has the same basic structure as zeatin but has a sugar (ribose) attached.

IAA CONCENTRATION IN MG./L.

Figure 18.14 Interactions between auxin (IAA) and cytokinin (kinetin) in cultures of tobacco tissue. When IAA and kinetin concentrations are balanced, cultures grow as undifferentiated calluses (middle rows). Higher IAA to kinetin ratios result in root growth (upper right). Lower IAA to kinetin ratios result in shoot growth (lower left).

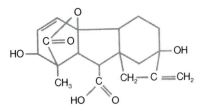

Gibberellic acid (Ga₃)

Figure 18.15 Structure of gibberellic acid (GA₃). Numbers in the abbreviations of gibberellins stand for the order in which the compounds were discovered.

Figure 18.16 Effect of a single application of gibberellic acid (GA), applied at the time of emergence of the first leaf, on genetic dwarf variety of corn (*Zea mays*). These plants are two weeks old. Note that GA has little effect on the length of normal plants.

or even from a culture medium in which the fungus had been grown. Because of World War II, biologists in other countries did not become aware of this work until the 1950s. Then biologists began to study the effects of this extracted substance, **gibberellic acid (GA)**, on vascular plants and to search for similar naturally occurring compounds that they assumed were produced in vascular plants. Eventually, chemically similar compounds that caused such growth responses were found in vascular plants, as well as in algae and in other fungi. Collectively, these compounds are known as **gibberellins**, and they are named according to their chemical structure and the order in which they were discovered. For example, the gibberellin most widely used in experimental work is a gibberellic acid that is abbreviated GA₃ (figure 18.15).

One of the most striking effects of gibberellic acid (GA) is the stem elongation that it causes in genetic dwarf varieties of certain plants. When GA is applied to such dwarf plants, they grow to normal size (figure 18.16). This response of dwarf plants is a useful bioassay for gibberellins, and it is quite specific because the dwarf plants do not elongate in response to auxins. The response also suggests a possible role for gibberellins in normal plant growth. It has been hypothesized that plants grow to their normal height because of the action of internally produced (endogenous) gibberellins and, therefore, that the dwarf mutants are short because they are unable to produce normal quantities of gibberellins. Reasonable and attractive as this hypothesis may be, it must be tested by further research before it can be accepted or rejected.

Another interesting effect of gibberellic acid is the response it causes in plants growing as **rosettes**. A rosette is a compact growth form where leaf attachments (nodes) are very close together because the stem length between leaf attachments is very small. Cabbage, for example, is a biennial plant that grows in a rosette form (the familiar cabbage "head") during its first growing season and then grows tall, flowers, and sets seeds during its second season. To grow tall during the second growing season, a cabbage plant normally must experience chilling during the winter between its first and second growing seasons. Thus, chilling is the normal stimulus for stem elongation. GA applied to a first-year plant, however, can replace chilling and cause a rapid internode elongation known as **bolting**. GA also causes bolting in other kinds of rosette plants that normally lengthen in response to photoperiodic stimuli (figure 18.17). But once again, the possible roles of endogenous gibberellins in normal second-year growth of biennials, such as cabbage, or in lengthening following photoperiodic stimulation are not established.

Gibberellins also stimulate leaf growth, cell division at the shoot apex, and the development of male flower parts. But the best understood action of endogenous gibberellins involves regulation of metabolic changes in germinating seeds. Embryos in seeds are in a metabolically inactive state called **dormancy**. At germination, the newly reactivated embryo resumes its growth, utilizing nutrients stored in the seed. In grass seeds, such as barley, starch is stored in the endosperm of the seed. This starch must be hydrolyzed in a reaction catalyzed by the enzyme α-amylase to make glucose available to the growing embryo. When germination begins in barley seeds, the embryo produces GA, which stimulates α-amylase synthesis by cells that surround the endosperm, and the necessary starch digestion begins. This reaction also is important in brewing because barley seeds are germinated in the malting process, an early step in beer production. GA is used in the brewing industry to promote this process.

Figure 18.17 Bolting caused by gibberellic acid. The cabbage plants on the left are untreated. Treatment with gibberellic acid has caused bolting and flowering in the plants on the right.

Abscisic Acid

The plant hormones discussed so far generally are stimulators of various processes in plants. Another group of plant hormones play very different roles in plants because they act basically as inhibitors. The most widely studied of these inhibitors is **abscisic acid** (**ABA**).

Over the years, researchers came to the conclusion that naturally occurring inhibitory substances are involved in plant dormancy. P. F. Wareing and his colleagues in Wales approached the question of inhibitors by studying the formation of dormant winter buds in several trees. As the days shorten in autumn, the shoot apex stops producing new leaves and switches over to the production of the bud scales that enclose and protect the bud during the winter. Wareing showed that this change in shoot apex activity is caused by a substance produced in senescing (aging) leaves and exported to the shoot. Because this regulatory substance is involved in preparation for winter dormancy, they named it *dormin*. At about the same time, Frederick Addicott and his colleagues in California isolated a substance that promotes the **abscission** (falling off) of cotton boles (fruits containing cotton seeds), and they named it *abscisin*. Finally, R. L. Wain and his colleagues in England found a substance that caused abscission of yellow lupine (*Lupinus luteus*) pods.

All of these discoveries were tied together when B. V. Milborrow and others showed that dormin, abscisin, and the lupine abscission substance were all one and the same substance. This single hormone was then renamed abscisic acid (figure 18.18).

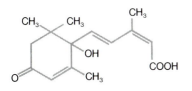

Abscisic acid

Figure 18.18 Structure of abscisic acid.

Abscisic acid (ABA) also appears to inhibit RNA and protein synthesis, thus maintaining plants' dormancy states. Naturally dormant plant material, such as seeds and potatoes, contain high concentrations of ABA, but the ABA concentrations fall as these plant materials approach the end of their dormancy period.

As mentioned earlier, abscisic acid seems to be a major factor in root geotropism. ABA accumulates on the lower side of a root, which is in a horizontal position, and inhibits cell growth on that side of the root. Greater growth on the other side then causes the curvature of the geotropic growth response.

Finally, ABA is involved in recovery from leaf wilting, which is the loss of turgor due to excess water depletion by transpiration. When a leaf wilts, its ABA content shoots up, and ABA causes stomata to close. This reduces water loss and gives the leaf a chance to recover as it accumulates water. This stomatal response involving ABA is the only firmly established exception to the statement that plant hormones generally are not involved in short-term, reversible, homeostatic physiological responses.

Photomorphogenesis

Some plant responses to light are initiated by light stimuli that need not be either specifically directional or periodic. This is called photomorphogenesis.

Etiolation

Seedlings grown in darkness have a peculiar spindly appearance and extremely long internodes (figure 18.19). No new leaves beyond those present in the seed are produced, the vascular tissue of the stem has a characteristically abnormal organization, and the plants have a very pale color because they have colorless **etioplasts** instead of chloroplasts. These symptoms, which develop in darkness, are caused by abnormally high auxin levels and excessive ethylene accumulation. Such dark-grown plants are said to be **etiolated**.

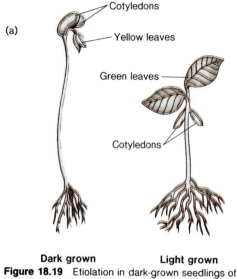

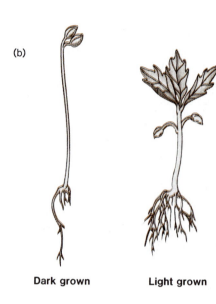

Figure 18.19 Etiolation in dark-grown seedlings of (a) bean and (b) mustard. Note the elongate stems in both and the small, colorless leaves of the bean. Leaves are lacking from the etiolated mustard seedling.

Early experiments on etiolation showed that even very dim, continuous light or a single, brief exposure to brighter light each day prevents etiolation of a growing plant. Later experiments determined the quantity and quality (wavelength) of light required to switch etiolated seedlings to normal growth. A dark-grown pea seedling, for example, switches to normal growth if exposed for only five minutes to red light (660 nm wavelength). But, interestingly, if such a red light exposure is followed by a five-minute exposure to far-red light (730 nm wavelength), etiolated growth continues. This rather intriguing result leads to a fundamental question: How does light exert control over plant growth processes?

Steps in Environmental Control

Three fundamental steps are involved when an environmental stimulus controls a growth response in a plant: the environmental stimulus must be perceived, the environmental stimulus must be converted into a biological stimulus, and that biological stimulus must exert control over aspects of the growth and differentiation of the plant. In the case of red-light effects, such as escape from etiolation, factors involved in the first two steps are quite well understood. A specific pigment (light-absorbing molecule) known

as **phytochrome** is responsible for stimulus perception and conversion in such reversible red-light/far-red-light effects.

Phytochrome

Phytochrome is involved in a number of plant responses to light, but the early history of phytochrome research was closely associated with the study of seed dormancy. Seeds of many plants remain more or less dormant as long as they are kept in the dark, and they germinate only after exposure to light. The seeds of one type of lettuce, the Grand Rapids variety, have been used most extensively for research on phytochrome. In 1935, Flint and McAlister, working at the United States Department of Agriculture, found that the wavelengths of light most effective in promoting lettuce seed germination were in the red region of the light spectrum, and they found that light in the far-red region of the spectrum depressed the germination rate, even below the low rate found in seeds kept in total darkness. Later, Borthwick and Hendricks, also at the Department of Agriculture, showed that if seeds stimulated by red light were then exposed to far-red light before germination had actually begun, germination was inhibited. Thus, far-red light reverses the effect of red light on Grand Rapids lettuce seeds (figure 18.20 and table 18.1). This led Borthwick and Hendricks to conclude that

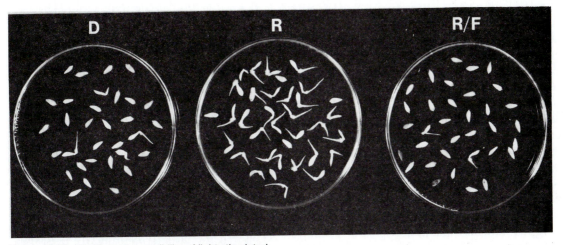

Figure 18.20 Red/far-red reversibility of light-stimulated germination of lettuce seeds. D: Seeds kept in dark. R: Received three minutes red light. R/F: received three minutes red light followed by three minutes far-red light.

germination is controlled by a reversible system, and they proposed that a single light-absorbing substance (pigment) is involved. They further proposed that this pigment exists in two forms, one that absorbs red light and one that absorbs far-red light.

Warren Butler and his colleagues demonstrated that the Borthwick/Hendricks hypothesis was correct, and they named the pigment phytochrome ("plant pigment"). Phytochrome exists in two forms that are interconvertible in a **photochemical** reaction. That is, the phytochrome molecule is reversibly converted from one form to the other upon the absorption of light energy. Red light (660 nm wavelength) converts the red-light-absorbing form of phytochrome, called P_r, to the far-red-light-absorbing form, called P_{fr} (figure 18.21). And P_{fr} is converted back to P_r by far-red light (730 nm wavelength). There also is another reaction system (nonphotochemical) that converts P_{fr} to P_r, but it works much more slowly than the photochemical reaction.

Would not sunlight, which contains a full-spectrum mixture of the various wavelengths, cause phytochrome reactions to occur in both directions? Sunlight does cause phytochrome reactions in both directions, but the relative light intensities are such that there is a net conversion of P_r to P_{fr} in sunlight. Thus, even though the nonphotochemical reversing process also is operating, increasing proportions of P_{fr} accumulate in plants during daytime hours. In darkness, only

Table 18.1
Data on Lettuce Seed Germination Percentages Following Exposure to Red and Far-Red Light.

Irradiation	Percentage Germination
Red	70
Red/Far-red	6
Red/Far-red/Red	74
Red/Far-red/Red/Far-red	6
Red/Far-red/Red/Far-red/Red	76
Red/Far-red/Red/Far-red/Red/Far-red	7
Red/Far-red/Red/Far-red/Red/Far-red/Red	81
Red/Far-red/Red/Far-red/Red/Far-red/Red/Far-red	7

From H. A. Borthwick et al., "A Reversible Photoreaction Controlling Seed Germination," *Proceedings of the National Academy of Sciences* 38(1952):662–66.

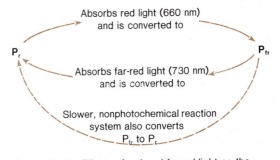

Absorbs red light (660 nm) and is converted to

P_r P_{fr}

Absorbs far-red light (730 nm) and is converted to

Slower, nonphotochemical reaction system also converts P_{fr} to P_r

Figure 18.21 Effects of red and far-red light on the phytochrome molecule.

the nonphotochemical reversing process operates, and the accumulated P_{fr} is gradually restored to the P_r form during the night. Therefore, there are recurring day-night fluctuations in the relative proportions of the forms of phytochrome molecules present in plants.

The phytochrome system, then, provides mechanisms for two of the three steps necessary for control of plant growth responses by an environmental stimulus. Phytochrome receives a light stimulus by absorbing light energy, and photoconversion of the phytochrome molecule converts the environmental stimulus into another form, a specific chemical compound. But it is much less clear how photoconversion of phytochrome actually exerts control over growth and differentiation. Changes in production of auxin and/or gibberellins or changes in cellular responsiveness to them must be involved, but it is not yet known how these are related to changes in phytochrome.

Photoperiodism

Knowledgeable gardeners in temperate zone areas can plan their flower gardens so that the advance of spring is marked by a very orderly series of floral displays. New blossoms appear on a regular basis as flowering border plants are followed by blooming vines, which, in turn, are superseded by flowering shrubs and trees. Considering the variability of springtime temperature and moisture conditions, how can gardens be programmed to provide a regular series of flowering events? Such programming is possible because many plants flower in response to photoperiodic stimuli, which, as mentioned earlier, are changes in the twenty-four hour cycle of alternating light and dark. Because different species respond to different day lengths, garden flowers can be selected so that they bloom in sequence as day lengths change.

Flower gardeners are simply taking advantage of a system, which has evolved in many species, that assures that all members of a species in a given area flower at the same time. The adaptive value of this precise photoperiodic timing of flowering is clear. It ensures simultaneous flowering and thus increases the opportunities for cross-pollination and for reproductive success.

Aspects of plant life other than flowering also are regulated photoperiodically. For example, plants must make structural and functional preparations for dormancy if they are to be ready to withstand the rigors of winter (figure 18.22). The photoperiodic stimuli of shorter days and longer nights in the fall stimulate plants to begin preparations for dormancy. The advantage of photoperiodic timing is that photoperiodic changes occur in the same regular fashion year after year. It is of great adaptive value for many species of plants that such critical parts of their lives as flowering and entry into dormancy are controlled by photoperiodic changes rather than by other, less regular, seasonal environmental changes, such as temperature and moisture changes.

Figure 18.22 A leafless, dormant tree in midwinter. Dormancy preparations are made in response to photoperiod changes that provide reliable information about the time of year.

Types of Photoperiodic Responses

Research on photoperiodism began in the 1920s when Garner and Allard at the United States Department of Agriculture made a series of soybean plantings over a period of several weeks. In late summer, they observed the flowering time of the plants in the various groups. Despite age differences due to different planting times, all of the soybeans flowered surprisingly close to the same time. Further study showed that all of the soybeans flowered in late summer in response to decreasing day length. Flowering was induced when days shortened below a critical length.

Soybeans and other plants that flower where day length decreases below a critical point are called **short-day plants**. Other plants, called **long-day plants**, show an opposite response: they flower when day length increases beyond a particular point. **Day-neutral plants** (for example, tomato) do not seem to be influenced by photoperiodism. They flower in response to other environmental factors or simply when they become mature. (See figure 18.23.)

Figure 18.23 The general categories of photoperiodic responses to the daily light-dark cycle in flowering plants. (a) Short-day plants. (b) Long-day plants. (c) Day-neutral plants.

(a)

Short-day plant

Shorter than
critical day length

Longer than
critical day length

(b)

Long-day plant

Longer than
critical day length

Shorter than
critical day length

(c)

Day-neutral plant
(flower after a certain period
of vegetative growth
or in response to
environmental factors other
than photoperiod)

Long-day length

Short-day length

Generally, in the temperate zone, short-day plants (for example, cocklebur, ragweed, and asters) grow and mature during spring and early summer and become ready to flower in response to the shortening day length of late summer and fall. Long-day plants (for example, radish, clover, and wheat) flower in response to day length increases in the spring and early summer.

The real distinction between short-day and long-day plants, however, is the way that each type of plant responds to its own **critical photoperiod**, a specific ratio of day and night length. Short-day plants flower when the light period becomes shorter than their critical photoperiod, however long the critical photoperiod might be. Long-day plants flower when the light period becomes longer than their critical photoperiod, however long that critical photoperiod might be. For example, *Xanthium* (cocklebur) is a short-day plant because it flowers when day length decreases below its critical photoperiod; this occurs in late summer. The critical daily light-dark cycle for *Xanthium* turns out to be fifteen hours of light and nine hours of darkness. Although this is not a particularly short day by everyday human standards, it is the critical photoperiod for this particular short-day plant.

Short Days/Long Nights

In the 1930s, K. C. Hamner and James Bonner experimented further with the relative roles of light and dark periods by testing the effects of a variety of light-dark combinations on the cocklebur *Xanthium*. As noted earlier, *Xanthium* normally flowers when it has fifteen hours of light and nine hours of darkness. But Hamner and Bonner found that flowering also occurred when they subjected *Xanthium* to artificial, experimentally lengthened daily cycles with light periods longer than fifteen hours *as long as the dark period was nine hours or more*. A critical dark period of longer than nine hours seemed to be the primary determining factor in the *Xanthium's* short-day response. Thus, short-day plants such as *Xanthium* might more properly be described as "long-night plants."

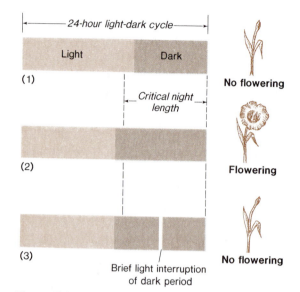

Figure 18.24 The importance of the dark period in the flowering response of short-day plants. The dark periods in cycles 2 and 3 are the same length, but the brief light interruption of cycle 3 prevents the flowering response.

Another experimental result underscored the importance of the dark period. If the dark period was interrupted by a short period of intense light, even a period as brief as one minute, short-day plants failed to flower (figure 18.24). This is further evidence that short-day plants respond to an uninterrupted dark period of a critical length.

Experiments with long-day plants have shown that night length also is critical in their photoperiodic responses. In the case of long-day plants, nights that are *shorter* than a particular critical length are necessary to cause flowering. Nights longer than the critical length inhibit the flowering of long-day plants, even if they are given very long days in artificial, experimentally lengthened daily cycles. If the longer than critical-length night is interrupted by a brief light period, however, long-day plants will flower. Clearly, the length of the *uninterrupted* night period is also a key factor in the photoperiodic responses of long-day plants.

Phytochrome and Photoperiodism

Night interruption experiments, especially with short-day plants, opened the way to making some connections between photoperiodic responses and the phytochrome system. Borthwick and Hendricks did a series of night-interruption experiments to test the effectiveness of light of various wavelengths in preventing short-day plants from flowering. They found that visible red light (with wavelengths between 620 and 680 nm) was most effective. Because other kinds of plant responses to red light usually are negated or reversed by far-red light, Borthwick and Hendricks also tested the effect of far-red light in these night-interruption experiments. They found that if a night interruption with red light was followed by a far-red light treatment, the plants responded as if their nights had not been interrupted at all (figure 18.25). This demonstrated very clearly that the phytochrome system is involved in photoperiodic responses.

Time Measurement in Photoperiodism

It was thought for some time that regular daily fluctuations in the two forms of phytochrome alone accounted for time measurement in photoperiodism. But there also seems to be time measurement by an "internal clock" mechanism that is separate from the phytochrome system. A short-day plant measures night length using its internal clock in a way that is as yet undetermined. If the uninterrupted dark period exceeds a critical length, the plant proceeds with processes leading to flowering. If night length is measured using an internal clock, what is the role of the phytochrome conversions? It seems that the far-red light-absorbing form of phytochrome (P_{fr}) inhibits many processes in plants, including some leading to the flowering response in short-day plants. Thus, a short-day plant, for example, measures a longer-than-critical dark period using its internal clock, but it is able to respond to the appropriately long dark period only because the slow nonphotochemical conversion of P_{fr} to P_r in darkness removes an inhibition. Thus, both the phytochrome system and an internal clock must be considered in descriptions of photoperiodic responses (figure 18.26).

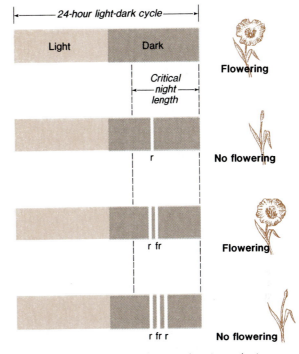

Figure 18.25 Night interruption experiments on short-day plants using a red light (r) interruption or combination of red light and far-red (fr) interruptions.

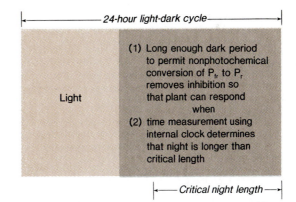

Figure 18.26 Two separate processes are involved in the response of a short-day plant when the dark period exceeds critical length.

Box 18.1
Florigen:

For more than one hundred years, biologists have searched for a "flowering hormone," a substance that causes the growth changes leading to flower development and subsequent reproductive events. Two discoveries, the discovery of photoperiodic induction of flowering followed by the discovery of the phytochrome system, have provided additional motivation for the search. But the flowering hormone has not, as yet, been isolated.

What evidence is there for a flowering hormone? The evidence has been accumulating since 1936 when M. H. Chailakhyan and his colleagues in Russia experimented with photoperiodic induction of flowering in chrysanthemums and found that a flowering stimulus appeared to be transmitted from one part of a plant to another. They removed the leaves from the upper part of chrysanthemums, which are short-day plants, and placed a light barrier between the upper and lower parts. Then they exposed the leafless upper part and the intact lower part to different photoperiods. When the upper (leafless) part was given long days and the lower part (with leaves) was given short days, the plant flowered. But the opposite treatment, in which the upper part received the short-day treatment, did not result in flowering (box figure 18.1A). Chailakhyan concluded that the leaves receive the photoperiodic

stimulus, and he proposed that leaves produce a chemical flowering substance that is transported through the plant. He suggested that the substance be called "florigen" (roughly meaning "flower maker").

Subsequent experiments by Hamner and Bonner extended and refined the concept of florigen. Cockleburs remain vegetative (nonflowering) as long as they are kept on long-day photoperiods. But Hamner and Bonner showed that if any part of a cocklebur, even a single leaf, is given short days, the plant flowers. Thus, the flowering stimulus can be transmitted from a single leaf to the whole plant. Another of Hamner and Bonner's experiments showed that the flowering stimulus can be transmitted from one plant to another. They induced a plant to flower by exposure to short days. Then they returned it to long-day photoperiods and grafted it to another plant that had been kept continuously in a long-day environment. This second plant also flowered even though it had not experienced inducing photoperiods. (See box figure 18.1B). Furthermore, grafting a photo-induced short-day plant to a *long-day plant* in a noninducing environment causes the long-day plant to flower. This indicates that the flowering substance is the same in both cases, even though the plants respond to different photoperiodic stimuli.

Other experiments have shown that the interruption of the phloem by girdling (removing a ring of stem tissue that includes phloem) prevents movement of the stimulus from one part of the plant to another. Since the phloem is the normal route of long-distance transport of organic molecules inside a plant, this is further evidence that a chemical flowering substance exists. There have also been reports that extracts of leaves from photo-induced plants applied to noninduced plants cause the noninduced plants to flower, but these results have been challenged because they are not always repeatable.

Not all biologists are convinced that there is a flowering hormone. They think that the flowering stimulus caused by photoperiodic induction might simply be change in the ratios of other hormones, auxins, cytokinins, or gibberellins.

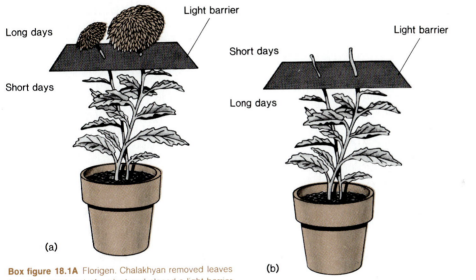

Box figure 18.1A Florigen. Chalakhyan removed leaves from the upper part of a plant and placed a light barrier between the upper and lower parts. (a) When leaves were exposed to the inducing (short-day) photoperiod, plants flowered. (b) Exposing the leafless upper part to short days had no effect.

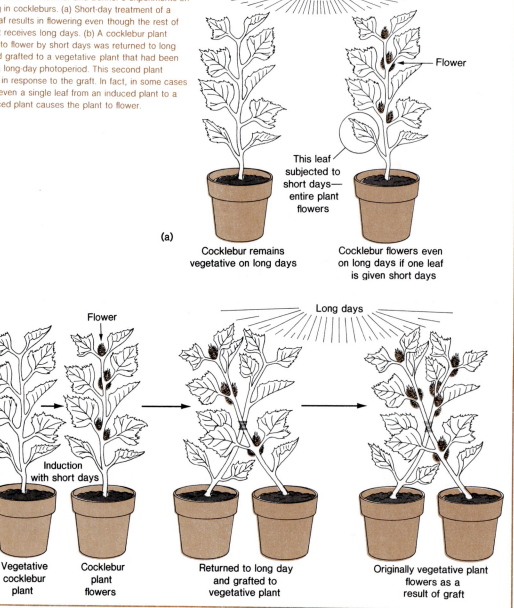

Box figure 18.1B Hamner and Bonner's experiments on flowering in cockleburs. (a) Short-day treatment of a single leaf results in flowering even though the rest of the plant receives long days. (b) A cocklebur plant induced to flower by short days was returned to long days and grafted to a vegetative plant that had been kept in a long-day photoperiod. This second plant flowered in response to the graft. In fact, in some cases grafting even a single leaf from an induced plant to a noninduced plant causes the plant to flower.

Long days

Flower

This leaf subjected to short days— entire plant flowers

(a)

Cocklebur remains vegetative on long days

Cocklebur flowers even on long days if one leaf is given short days

Flower

Long days

Induction with short days

(b)

Vegetative cocklebur plant

Cocklebur plant flowers

Returned to long day and grafted to vegetative plant

Originally vegetative plant flowers as a result of graft

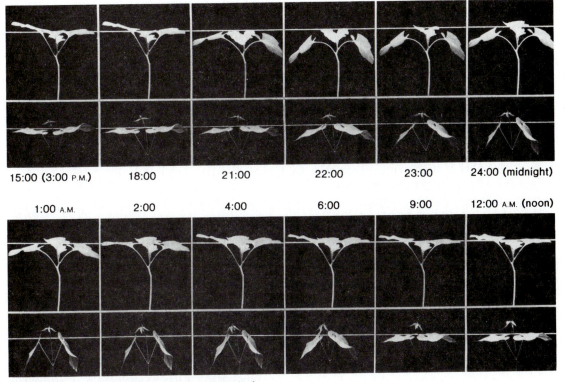

15:00 (3:00 P.M.) 18:00 21:00 22:00 23:00 24:00 (midnight)

1:00 A.M. 2:00 4:00 6:00 9:00 12:00 A.M. (noon)

Figure 18.27 A photographic record of rhythmic leaf movements in cocklebur (top row) and bean (second row). The plants were photographed at hourly intervals from noon to noon. Note that the bean leaves drop more sharply and later in the evening than the cocklebur leaves.

Internal Clocks

The nature of the internal clocks in living organisms is one of the most intriguing and puzzling questions in modern biology. One expression of the functioning of internal clocks is the obvious daily cyclic fluctuations (**rhythms**) seen in many physiological processes.

Daily Rhythms

Plants in nature show obvious daily cycles. For example, beans and many other plants spread their leaves to a horizontal position in the daytime and thereby expose a greater leaf surface area for light absorption. At night their leaves fold up or down in what has been called a sleep movement (figure 18.27). In addition to such externally obvious daily cycles or rhythms, plants have many subtle, rhythmic, internal physiological fluctuations. For example, some enzyme activity levels, certain ion concentrations in body fluids, and sensitivities to many drugs and other chemicals change throughout the day in a rhythmic fashion. All plants and all animals, including humans, have such daily rhythms. It is not surprising that life processes vary in a rhythmic manner with the time of the twenty-four-hour **solar day** because all organisms have evolved in a decidedly rhythmic world where light and darkness cycle with great regularity. Other physical factors, such as temperature and barometric pressure, also fluctuate, but not with nearly the same regularity as the daily light cycle.

What *is* surprising is that the solar-day rhythms of many living organisms continue even when the organisms are deprived of obvious external information about the time of day. The cycles continue when organisms are maintained in laboratories in constant light, constant temperature, and even constant barometric pressure. For example, bean plant sleep movements continue for days when a bean plant is left undisturbed under constant conditions (figure 18.28). Biologists have concluded that this persistence under constant conditions must mean that organisms possess internal clocks that time rhythmic processes even in the absence of obvious external cues about the time of day. This conclusion has inspired a number of studies of the properties of biological rhythms and the effects of various environmental factors on these rhythms.

Another surprising result in this research has come from the study of the effect of temperature on rhythms. In many cases, temperature change has very little effect on the rates at which rhythmic processes go through their daily cycles. This was unexpected because rates of many biological processes are strongly affected by temperature increases. Most biological rhythms, however, are remarkably insensitive to temperature change. This is demonstrated by the fact that the **period length**, which really is a measure of the rate at which a biological rhythm "runs," is quite temperature independent in most cases (figure 18.29). This relative temperature independence is a striking and commonly observed property of biological rhythms.

Another commonly observed feature of most rhythms in laboratory constant conditions, especially in constant light or constant darkness, is that period lengths usually are slightly longer or slightly shorter than exactly twenty-four hours. These are said to be *circadian* (literally "about a day") **rhythms** (figure 18.30). Sometimes, biologists use this name much more generally and speak of all clock-timed rhythmic phenomena as circadian rhythms or circadian clocks. But, technically, the term circadian describes only rhythms that have period lengths slightly different from twenty-four hours when measured under constant conditions. Circadian periods are seen only under laboratory conditions, where organisms are deprived of information about normal light/dark cycles. In nature, the daily light/dark cycle keeps period lengths of daily rhythms exactly twenty-four hours long. This external control by environmental factors that keeps rhythms precise is called **entrainment**, and the environmental factors that entrain rhythms are called **zeitgebers** (from the German *Zeitgeber*, meaning

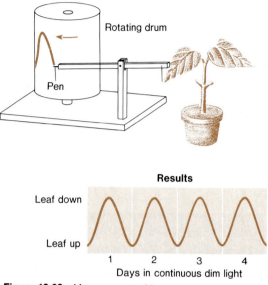

Figure 18.28 Measurement of bean plant sleep movements under constant laboratory conditions. Leaf is attached to a lever with a pen that traces on paper fastened to a rotating drum. Results of several days of recording are shown.

"time-giver," a name suggested by German students of biological rhythms). Thus, rhythms become circadian in the laboratory only when deprived of normal entrainment by environmental zeitgebers.

Rhythms and Clocks

Most or maybe all of the biological rhythms, such as plant sleep movements, that have been described and studied are overt, external expressions of timekeeping by an underlying timer (**clock**). In other words, rhythms represent the hands of the clock and are resetable just as the hands of an ordinary clock can be reset to any time setting. Even human jet lag following intercontinental travel disappears after a few days. There seems to be an underlying cellular clock, and the resetable rhythms (the hands) are linked to it. Environmental zeitgebers cause resetting of rhythms.

The interpretation of results of biological rhythms experiments is difficult because experimental results might relate to a property of only the clock's hands (observable, overt rhythms), the clock's basic timing mechanism, the linkage between the two, or some combination of all three. One certainty, though, is that there are some challenging unanswered questions about both the

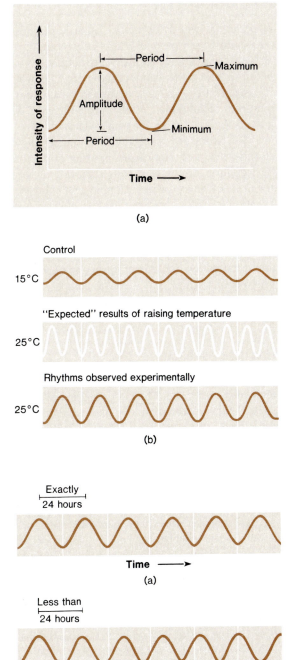

(a)

Control

15°C

"Expected" results of raising temperature

25°C

Rhythms observed experimentally

25°C

(b)

Figure 18.29 Biological rhythms and temperature. (a) Period of rhythm is the time from a point in one cycle to the same point in the next cycle. Amplitude is the difference between the high point and the low point within one cycle. (b) Temperature relationships. Increased temperature would be "expected" to change *both* period length and amplitude of rhythms. However, amplitude changes are as predicted, but period length changes little if at all in response to raising the temperature.

Exactly
24 hours

Time ⟶

(a)

Less than
24 hours

Time ⟶

(b)

Figure 18.30 Comparison between an exact twenty-four-hour daily rhythm and a circadian rhythm. (a) An exact twenty-four-hour solar day rhythm, the characteristic pattern of all daily rhythms under natural entrainment and a few rhythms even under laboratory constant conditions lacking normal entraining mechanisms. (b) A circadian rhythm with a period length less than twenty-four hours. This slightly exaggerated model shows how peaks of rhythmic activity occur earlier during each actual solar day. Such circadian rhythms are observed under laboratory constant conditions when organisms are deprived of normal entrainment, such as the daily light-dark cycle.

underlying cellular clock and its relationship to the overt rhythms that play such a large part in the lives of all plants and animals. Possibly the most fundamental of these questions involves the nature of the cellular clock (or clocks) and its location (or locations) in the organism.

Clock Timing: Internal or External?

There are two opposing, but not entirely dissimilar, viewpoints on the nature of the cellular clock. One hypothesis proposes that the cellular clock is an entirely *internal*, biochemical oscillator mechanism. The proposed cellular clock works like the pendulum that provides the time measurement in a pendulum clock (figure 18.31a). Of course, pendulum clock hands (rhythms) can be reset to any time on the face of the clock. The hypothesis also proposes that basic timing information comes from regular, cyclical changes in some cellular process, such as a complex enzyme reaction series, differences in membrane permeability, or possibly even repeated transcription of a segment of a DNA molecule.

The other hypothesis also suggests that the cellular clock is internal, but that its basic timing information comes from the external environment. This proposed cellular clock is like the motor of an electric clock (figure 18.31b). The hands (rhythms) of an electric clock also are resetable, but basic time measurement is by an electric motor driven by alternating current supplied from the *outside*. It is proposed that regular daily fluctuations of physical environmental forces, such as magnetic or electrostatic field intensities, cosmic radiation, or some combination of these or other physical forces, provide basic timing information that overrides all efforts of investigators to maintain constant laboratory conditions.

Either of these hypotheses can be used to explain almost every result of experiments on biological rhythms and almost every inferred property of biological clocks. It is difficult to conceive of experiments that would clearly distinguish between the two hypotheses regarding the nature of basic clock timing. All such studies are handicapped because it has proven difficult to learn about the clock mechanism when all experimental results actually give information about the behavior of the clock "hands," that is, the observable, overt rhythmic processes.

Season-Ending Processes

A period of winter dormancy is necessary for temperate zone perennial plants (plants that are active during several to many growing seasons) because these plants cannot survive freezing winter temperatures in their active, growing, summertime condition. What preparations do perennial plants make for winter dormancy?

Nutrients are transported to underground storage sites, and tough scales or other protective devices form around the buds that will initiate the next season's growth. Changing leaf color and eventual loss of leaves are the most obvious and familiar signs of approaching winter in plants. Leaves contain relatively soft tissue that would be very difficult to protect from winter damage, and preparation for winter dormancy includes **senescence** (aging) and eventual loss of leaves (**abscission**) with the formation of scar tissue over the points where the leaves detach.

Leaf Senescence

As the time of abscission approaches, leaves change dramatically. They lose protein and chlorophyll. This chlorophyll loss allows the autumn coloration caused by other pigments that have formed in the leaves to be seen. During early phases of senescence, leaf oxygen consumption rises sharply and reaches a peak level that is two or three times as great as the presenescence level. Oxygen consumption does fall off later, but this initial burst indicates that senescence is not just a slow, steady decline in activity, but a set of energy-requiring active metabolic changes. Hydrolytic enzymes actively break down components of leaf cells, and some essential components, such as mineral ions, are exported from the leaves to other parts of the plant. This prevents the loss of valuable plant resources that otherwise would fall with the leaves.

Hormonal control of leaf senescence is complex and not well understood, as yet. Experimentally, leaf senescence can be delayed by application of cytokinins, gibberellins, or auxin, or it can be promoted by application of ethylene or abscisic acid. Further research is needed, however, before the actual roles of these hormones in causing or preventing leaf senescence in intact plants can be clarified.

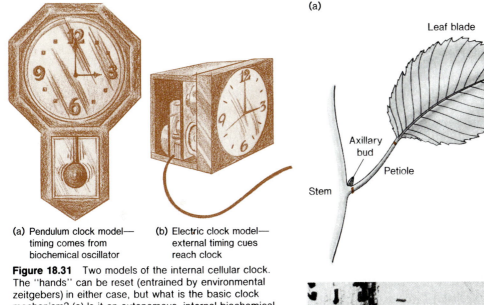

(a) Pendulum clock model—
timing comes from
biochemical oscillator

(b) Electric clock model—
external timing cues
reach clock

Figure 18.31 Two models of the internal cellular clock.
The "hands" can be reset (entrained by environmental
zeitgebers) in either case, but what is the basic clock
mechanism? (a) Is it an autonomous, internal biochemical
oscillator that measures time like a pendulum? This
hypothesis is supported by C. S. Pittendrigh, J. Aschoff,
and others. (b) Or is it like an electric motor clock,
where time measurement depends on alternating current
from the outside? This hypothesis is supported by F. A.
Brown, Jr. and others.

Abscission

While senescence changes most of the cells of a
leaf, abscission is due to changes in a very narrow
band of cells located, depending on the type of
plant, either at the base of the leaf petiole or at
the point where the blade of the leaf joins the
petiole. This band of cells forms an **abscission
zone**, the breaking point where the leaf eventually
detaches from the stem (figure 18.32). During
abscission, enzymes break down pectins, which
are important constituents of the material between cells, and cellulase affects the cellulose of
cell walls directly. These enzyme actions cause
the layers of the abscission zone to separate, and
the leaf falls. Corky material then covers the abscission zone and forms a **leaf scar**.

The control of leaf abscission probably involves a decrease in auxin production in leaves
because auxin prevents leaf abscission in most
plants. This can be demonstrated in simple experiments where the leaf blade is cut off, but the
petiole is left in place. Young and mature leaf
blades produce auxin. Thus, removal of the blade
deprives the petiole of its major auxin source.

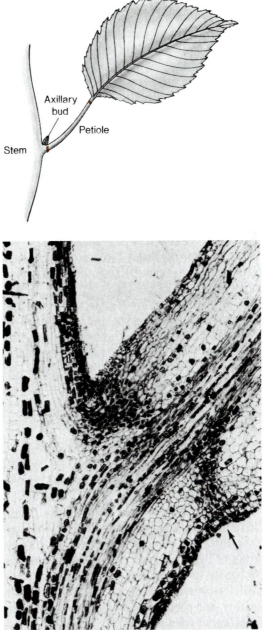

Figure 18.32 The abscission zone. (a) Abscission
zones (color) may be located at the base of a petiole, at
the junction of a petiole and a leaf, or even at several
different points. (b) Abscission zone at the base of the
petiole of a *Coleus* leaf.

Without further treatment, the petiole goes through normal abscission and falls. But if auxin is applied to the cut end, the petiole remains in place. Therefore, for some time it was thought that the control of abscission was fairly easily explained as the result of decreased auxin production in aging and senescing leaves. But even the role of auxin in leaf abscission now seems more complex. Cells of the abscission zone may respond to differences between the auxin concentration on the stem side and on the leaf-blade side of the abscission zone rather than simply responding to lower leaf auxin production.

Ethylene also is known to be a powerful abscission promoter, and senescing tissues produce ethylene. More might be said about possible roles of other hormones, but this is enough to indicate that control of abscission, like so many other processes in plants, probably involves complex interactions and changing balances of several plant hormones.

Thus, the whole life of a vascular plant, from seed germination to aging and senescence, involves continuing interactions with its environment, and plant hormones mediate the plant's responses at every step along the way.

Summary

A terrestrial vascular plant remains in one place and interacts with its changing environment throughout its life. These interactions are mediated by plant hormones.

Light is the dominant environmental factor affecting growth responses and seasonal changes in plants, and plants display several different kinds of light responses. They show phototropic responses to directional light, photomorphogenetic responses to changing light quantity and quality, and photoperiodic responses to changing light/dark cycles.

The cells that produce plant hormones are not specialized for hormone production alone. Also, responses to plant hormones are not restricted to a few kinds of target cells; practically all plant cells respond in some way to each plant hormone. Plant hormones act by directly affecting reactions in plant cells.

Auxin was discovered as a result of experiments on phototropic responses of grass seedlings. The factor produced in coleoptile tips that causes seedling elongation can be trapped in agar, and

the agar applied to the seedling has the same effect as the tip itself. This demonstrates that a diffusible chemical is involved.

The principal naturally occurring auxin is indoleacetic acid, but other substances, including some laboratory-synthesized substances, act as auxins. Auxin causes cell elongation first by weakening cross-linkages between cellulose fibers in the cell wall so that turgor pressure enlarges the cell. Then it stimulates synthesis of cellulose, which thickens cell walls to their prestretching condition.

Phototropic responses result from lateral transport of auxin to the darker side, where greater cell elongation occurs. Stem geotropism is similar in that auxin stimulates greater elongation on the lower side. Auxin also is responsible for apical dominance because auxin produced in the shoot apex inhibits the growth of lateral buds.

Ethylene, a gas, is an important fruit-ripening agent. It promotes hydrolysis of starch to sugar and stimulates cellulase activity, which results in fruit softening.

Cytokinins, such as kinetin, are cell-division promoters that act along with auxins in general regulation of growth responses. Zeatin and zeatin riboside are naturally occurring cytokinins.

Gibberellins, which were first extracted from a fungus, stimulate stem elongation in dwarf plants and probably play a role in stem elongation in normal plants as well. Gibberellic acid, a gibberellin, causes bolting in rosette plants. Gibberellic acid is important in seed germination because it stimulates synthesis of amylase, which hydrolyzes stored starch, thus making sugar available to the growing seedling.

Abscisic acid, an inhibitor, is involved in preparation for and maintenance of dormancy. Abscisic acid acts by inhibiting RNA and protein synthesis. It also promotes abscission of various fruits and seeds, and is involved in stomatal closure during wilting in some plants.

Many light responses, including escape from etiolation, are promoted by red light but opposed by far-red light. This generalization led to the discovery of phytochrome, a pigment that exists in two forms that are interconvertible in a photochemical reaction. Phytochrome is reversibly converted from one form to another upon the absorption of red or far-red light.

Photoperiodism is particularly important in timing of reproductive activity (flowering) and in onset of dormancy. A short-day plant responds positively when nights are longer than its particular critical length, while a long-day plant responds when nights are shorter than its particular critical length.

Photoperiodic responses involve time measurement that depends on an internal clock mechanism. The clock mechanism also times rhythmic daily changes in various physiological functions. But the nature of the underlying cellular clock mechanism remains obscure. One hypothesis is that it is a biochemical or biophysical oscillator system. Another hypothesis proposes that it is a responding system that receives timing information from subtle geophysical factors that are hard to exclude from any experimental situation.

Temperate zone plants prepare in several ways for winter dormancy, a state in which they can survive adverse winter conditions. Nutrients are stored underground, and protective scales form around delicate buds. But the most obvious sign of preparation for winter is leaf senescence and abscission. Leaves lose chlorophyll and protein, export various minerals to other parts of the plant, and develop abscission layers that are specific breakage points. These processes also are under hormonal control, as are all of the environmental responses in the lives of vascular plants.

Questions

1. Explain the adaptive value of using photoperiodic changes rather than seasonal temperature changes to time events in plants' lives.

2. Explain the Went bioassay for auxin and discuss the general usefulness of bioassays in hormone research.

3. Gibberellins are not required for the growth of the fungus that causes "foolish seedling disease" in rice. What adaptive value do you suppose gibberellin production has for the fungus?

4. Why does "one bad apple spoil the barrel"?

5. Propose a relationship between the phytochrome system and florigen production (the flowering stimulus) in cockleburs.

6. Why do you think that some short-day plants near the northern edge of their range are able to produce mature seeds and fruits during some growing seasons, but not during others?

Suggested Readings

Books

Bidwell, R. G. S. 1979. *Plant physiology.* 2d ed. New York: Macmillan.

Galston, A. W., and Davies, P. J. 1979. *Control mechanisms in plant development.* 3d ed. Englewood Cliffs, N.J.: Prentice-Hall.

Hill, T. A. 1973. *Endogenous plant growth substances.* Studies in Biology no. 40. Baltimore: University Park Press.

Kendrick, R. E., and Frankland, B. 1976. *Phytochrome and plant growth.* Studies in Biology no. 68. Baltimore: University Park Press.

Palmer, J. D. 1976. *An introduction to biological rhythms.* New York: Academic Press.

Salisbury, F. B., and Ross, C. W. 1978. *Plant physiology.* 2d ed. Belmont, Calif.: Wadsworth.

Thimann, K. V. 1977. *Hormone action in the whole life of plants.* Amherst, Mass.: University of Massachusetts Press.

Articles

Binkley, S. April 1979. A timekeeping enzyme in the pineal gland. *Scientific American* (offprint 1426).

Black, M. 1972. Control processes in germination and dormancy. *Carolina Biology Readers* no. 20. Burlington, N.C.: Carolina Biological Supply Co.

Cleland, C. F. 1978. The flowering enigma. *BioScience* 28:265.

Juniper, B. E. 1976. Geotropism. *Annual Review of Plant Physiology* 27:385.

Rubery, P. H. 1981. Auxin receptors. *Annual Review of Plant Physiology* 32:569.

Satter, R. L., and Galston, A. W. 1981. Mechanisms of control of leaf movements. *Annual Review of Plant Physiology* 32:83.

Part 4

Continuity of Life

The ability to reproduce is a fundamental property of living things. For each individual organism, reproduction is a biological admission that the mechanisms that maintain the integrity of each organism eventually will fail. This "failure" is evidenced by the fact that all multicellular organisms age and die. But for the species, reproduction provides a potential means of increasing numbers and results in continuing replacement of aging individuals with young, vigorous individuals. In addition to the general rejuvenating effect of reproduction for the species, sexual reproduction produces individuals possessing new genetic combinations that may be expressed as characteristics that make some of them more able to exploit various habitats successfully.

Reproductive processes involve specific activities of individual cells. Ordinary mitotic cell division is a fundamental reproductive strategy of unicellular organisms. In these organisms, one individual divides mitotically to produce two individuals.

The reproductive strategies of multicellular organisms also involve a return to the single cell level since multicellular organisms produce specialized individual reproductive cells such as eggs and sperm. Via these specialized cells, characteristics are transmitted from one generation to the next by specific genetic mechanisms. In sexually reproducing organisms, offspring resemble parents generally, but each individual offspring represents a new genetic combination, the expression of which yields a set of characters that is somewhat different from each of the parents.

Expression of the genetic complement of a new individual occurs through developmental processes. Repeated mitotic divisions produce a multicellular aggregate, and groups of cells differentiate as the various specialized functional parts of a multicellular body.

Developmental processes continue throughout life. Growth, maturation, and aging are all parts of development. Developmental processes are involved in continuing replacement of body cells, in healing of wounds and injuries, and in specific responses to infection and disease, all of which threaten the integrity of the organism. But developmental processes gone awry as abnormal growth can also threaten the organism.

In chapters 19–26 the cell division processes involved in reproduction and development, the mechanisms by which genetic information is transmitted and expressed, some details of reproductive processes in various organisms, and the nature of developmental processes that occur throughout the lives of organisms are examined.

Cell Division

Chapter Concepts

1. The cell cycle involves growth, duplication of genetic information, preparation for division, and division.

2. Mitosis is the process whereby the duplicated genetic information of a cell is precisely divided to produce two genetically identical nuclei.

3. The cytoplasm of a dividing cell is separated by cytokinesis, a process that differs in plant and animal cells.

4. A sexually reproducing organism must reduce its chromosome number by half at some point in the life cycle. This prevents a doubling of the normal chromosome number each time fertilization occurs.

5. Meiosis is a type of cell division by which this reduction in chromosome number occurs.

The idea that all organisms are composed of cells became quite widely accepted in the centuries following Robert Hooke's first description (in 1665) of his microscopic observation of small spaces in cork that he named "cells." But it was not until the 1830s that M. J. Schleiden and Theodor Schwann advanced the concept that *all* plants and animals are composed of cells and cell products.

Theories on how new cells originated were hotly debated in the decades following Schleiden and Schwann's work. Schleiden and Schwann, among others, thought that the nucleolus became a new nucleus and that this new nucleus then developed into a new cell. Other scientists thought that new cells either budded off preexisting cells or formed by an unknown process in the space between cells. Still others insisted that new cells formed when a cell divided into two equal halves.

Observations over the years finally confirmed that cells arise by a process of division, and this idea was well enough established in the 1850s for Rudolph Virchow to complete the formulation of the cell theory begun by Schleiden and Schwann. Schleiden and Schwann had said that all organisms are composed of cells and cell products. Virchow added that all cells are produced by the division of previously existing cells. This chapter explores how these cell divisions take place.

It is now known that genetic information is contained in the chromosomes. During cell division, this material is duplicated and distributed to the new cells with great precision, thus ensuring the continuity of genetic information from cell to cell. The process by which the genetic material is distributed is called **mitosis**. Mitosis results in two new nuclei, each genetically identical to the original nucleus. The cytoplasm is divided, more or less equally, by a separate process known as **cytokinesis**.

Any extensive growth depends on mitotic cell division because growth due to increase in the size of the component cells is limited. But cell division does not end when growth in body size slows or stops at maturity because cells produced by mitosis also serve as replacements for cells that die or that are lost to the organism. In humans, for example, the outer layers of the skin are constantly being sloughed off and replaced by division of cells lying beneath them.

In some types of organisms, specialized cells, resulting from mitotic cell division, may develop directly into new individuals that are genetically identical to the parental individual. This type of reproduction is called asexual reproduction (see chapter 23).

The other method by which organisms reproduce themselves is sexual reproduction. Whereas in asexual reproduction the genetic information for the new individual comes from a single cell, sexual reproduction requires a genetic contribution from two different cells. The two specialized reproductive cells, called **gametes**, combine to form a single cell called a **zygote**. The combining gametes may in some species be produced by the same individual, but in most species they must be produced by two different individuals. The fusion of two gametes to form a zygote is called **fertilization**. The zygote may then divide by mitosis, and the cells differentiate to form a mature multicellular organism.

Prior to fertilization, organisms that reproduce sexually must produce gametes with half the normal chromosome number of ordinary body cells. If this did not occur, the fusion of the two gametes would result in a zygote with twice the base chromosome number, and the number of chromosomes would continue to double with each generation. This necessary reduction in chromosome number is accomplished by a type of cell division called **meiosis**. The doubling of the chromosome number by fertilization and the halving of the chromosome number by meiosis ensure that the amount of genetic information is stable from generation to generation.

The Cell Cycle

In the 1870s, techniques of microscopy had advanced to the point where several scientists were able to provide detailed and accurate descriptions of the movement of chromosomes during mitosis. For a long time the emphasis in studies of cell division was on the process of mitosis itself. Because there was little visible activity in the remainder of the cell's life span, the period between the visible mitotic events was dismissed as a resting stage, an "interphase." It was not until much later, with the development of increasingly sophisticated biochemical techniques in the 1950s, that biologists were able to discover and describe many of the processes occurring in cells between mitotic divisions and to learn that the period between divisions is actually a very important and active time for the cell. The entire cycle—from the formation of a new cell until the cell divides—is called the **cell cycle**, and mitosis can be fully understood only in the context of this cell cycle.

Shortly after it was confirmed that DNA is the primary genetic material, new methods were developed to study the dynamics of DNA function and duplication throughout the cell cycle. Two major methods of studying DNA are Feulgen staining, which indicates the presence and relative amounts of DNA, and autoradiography (chapter 3), a process by which DNA can be radioactively labeled and then visualized on photographic film. A combination of the two techniques revealed that the DNA of a cell is replicated during a specific period of the so-called resting stage, frequently hours before it is separated during mitosis. The discovery of the timing of DNA replication led to a description of a cell cycle divided into four distinct, functional phases (figure 19.1).

The phase in which the DNA is synthesized is designated the S (for synthesis) phase. The period of time after a cell is produced by mitosis and before it enters the S phase is called the gap 1 (G_1) phase. (The G_1 phase is a "gap" only in the sense that it does not include major cell reproductive events involving DNA.) During this G_1 phase, growth and synthesis of many compounds other than DNA occur. After the genetic material is duplicated in the S phase, another gap occurs (G_2). Not much is known about the G_2 phase except that it apparently involves the synthesis of various enzymes and structural proteins necessary for mitosis. The fourth phase is the nuclear division process itself, mitosis (M phase).

The length of time spent in each stage of the cell cycle varies depending on the type of cell involved, and variations in temperature may increase or decrease the time spent in each phase. In human cells cultured at 37°C, the G_1 stage lasts for eight hours, the S phase lasts for six hours, the G_2 phase takes about 4½ hours, and mitosis is completed in one hour. In higher plants, the entire cycle lasts from ten to thirty hours,

with the M and S phases comparatively longer than in animals. Many protozoa, fungi, and most embryonic cells have much shorter cell cycles with greatly reduced G_1 phases. The length of the G_1 phase is apparently correlated with the amount of growth and biosynthesis necessary for the cell. Cells that reproduce very rapidly do not have an extended G_1 period.

The transition from the G_1 phase to the S phase is a critical point in the cell cycle. If the cell does not first enter the S phase, it will not divide. If the cell enters the S phase, it usually is committed to mitosis. This transition point is a key to the entire cycle. Important as this transition is, however, the stimulus that causes a cell to enter the S phase is unknown. A better understanding of what occurs at this point would offer clues for treatment and control of abnormal cell division. Cancer cells, for example, are cells that grow without restraint or control. They probably are derived from cells that are normally blocked from entering the S phase; that is, cells that normally do not divide. But some factor, external or internal, transforms the cells so that they continue to enter the S phase in each cell cycle and, thus, are committed to uncontrolled cell division (see chapter 26). Much of the basic research on cancer is now concentrated on the possibility of restoring the factor or factors that restrain cell division in normal cells.

Cell Division in Prokaryotes

Cell division in both prokaryotic cells (cells that lack nuclear membranes and membrane-bound organelles) and eukaryotic cells is similar in that in both cases duplicated genetic material is apportioned to each new cell. In prokaryotes, the process is simplified by the fact that the genetic information is not contained within a nucleus. In bacteria, for example, each copy of the replicated chromosome becomes anchored to a fold in the cell membrane (figure 19.2). The cell membrane elongates until the cell is approximately twice its original length. Then the membrane appears to pinch the cell into two segments. Each of these segments is a new cell, and each contains an identical copy of the circular bacterial chromosome. Genetic replication in prokaryotic cells is covered in more detail in chapter 22.

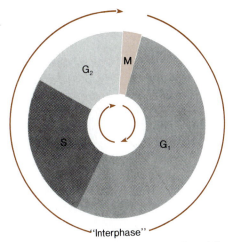

Figure 19.1 The cell cycle. Mitosis (M) and the synthesis of the genetic material (S) are separated by two gap phases (G₁ and G₂).

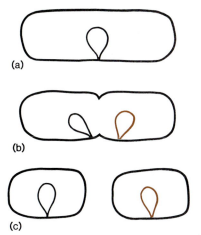

Figure 19.2 Cell division in bacteria. (a) A bacterium with its circular chromosome prior to division. (b) The cell begins to elongate. The replicated chromosomes are pulled apart by the elongation. (c) The elongated cell divides in the middle, resulting in two new cells, each with a copy of the chromosome.

Mitosis

Mitosis is the division of the eukaryotic nucleus that results in equal distribution of duplicated genetic information to the two new cells (called daughter cells). This division is very precise, and the end result of mitosis is two nuclei with genetic information identical to that of the parent cell's nucleus.

Stages of Mitosis

Mitosis is a continuous process that is arbitrarily divided into four stages for convenience of description. The first stage of mitosis is called **prophase**. A typical cell entering prophase from G_2 has an intact nuclear membrane and may have visible nucleoli, but the chromosomes themselves are not visible with a light microscope (figures 19.3a and 19.4). During the early stages of prophase, the chromosomes begin to condense. This apparent condensation is actually a coiling process that results in a shortening and thickening of each chromosome.

This coiling results in chromosomes that are easily visible with a light microscope after they have been stained. Chromosome condensation seems to be a necessary part of the mitotic process because chromosomes in their normally extended condition might become quite tangled during the mitotic separation. Chromosomes cannot remain permanently condensed, however, because genetic expression (the cell's use of information based on interpretation of the genetic code represented by the sequence of nucleotides in the DNA) is complicated when the chromosomes are condensed. Therefore, the chromosomes become coiled at the beginning and uncoiled at the end of each cell division.

In the cells of most organisms, the nuclear membrane begins to break down during prophase. This allows later for the chromosomes to be apportioned to what will become two new nuclei. Any nucleoli present in the dividing cell also disappear during the early part of prophase.

Figure 19.3 Mitosis. (a) A cell just prior to entering the M phase of the cell cycle. (b) Prophase. (c) Metaphase. (d) Anaphase. (e) Telophase and the beginning of cleavage.

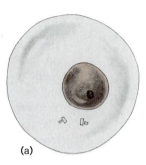

(a)

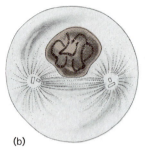

(b)

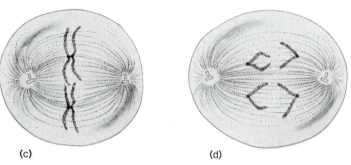

(c)

(d)

(e)

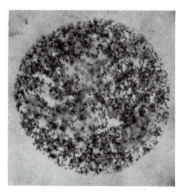

Figure 19.4 The nucleus of a peony cell prior to entering mitosis. Doubled chromosomes are not visible, and the nuclear envelope is intact. (Magnification $\times$ 1,046)

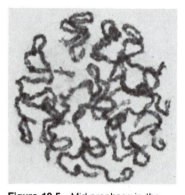

Figure 19.5 Mid-prophase in the peony. The chromosomes are now visible. (Magnification $\times$ 995)

By midprophase (figures 19.3b and 19.5), stained chromosomes are clearly visible under the light microscope as double-stranded structures. There are two strands at this time because the chromosomes were duplicated prior to mitosis, in the S phase of the cell cycle. Each strand of the doubled chromosome is called a **chromatid** (figure 19.6), and the two chromatids of each chromosome are joined in the regions of their centromeres. The **centromere** (or **kinetochore**) of each chromatid is a small platelike structure located on a constricted area of the chromatid.

Also during prophase the **spindle apparatus** begins to form. The spindle apparatus consists of a set of microtubules (see chapter 3 for a discussion of microtubules), which are associated with chromosome movement later in mitosis. Another obvious occurrence at this time is the movement of the **centrioles**, which were duplicated earlier during the S phase of the cell cycle. During prophase, one pair of centrioles moves to one side of the nucleus and the other moves to the opposite side. The centriole pairs are accompanied during this movement by a halo of microtubules called the **aster**, part of which remains around each of them in their new locations on opposite sides of

the nucleus. The spindle apparatus develops between these pairs of centrioles (figure 19.7). Despite the proximity of all these structures to one another, there does not seem to be any actual physical connection between the centrioles and the microtubules of either the aster or of the spindle apparatus.

Because of their location, centrioles were first thought to be responsible for the organization of microtubules into the spindle apparatus. But many organisms, including certain amoebae, the unicellular red algae, some gymnosperms (such as pines), and all angiosperms (flowering plants), do not even have centrioles. Another feature shared by the organisms that lack centrioles is that they all have eukaryotic cells but never produce flagellated or ciliated cells at any time in their lives. Cell biologists now believe that centrioles are, in fact, identical to basal bodies, the organelles that give rise to cilia and flagella (chapter 3). There is also experimental evidence that centrioles do not play a vital role in cell division because removal of centrioles does not prevent cells that normally have centrioles from forming the spindle apparatus and dividing.

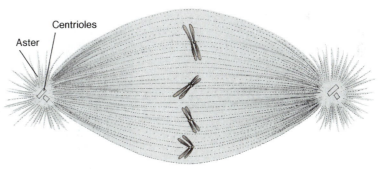

Figure 19.7 The spindle apparatus. Some of the microtubules are attached to the chromosome centromeres. Others overlap to extend from pole to pole. The centrioles are surrounded by short microtubules, forming the aster.

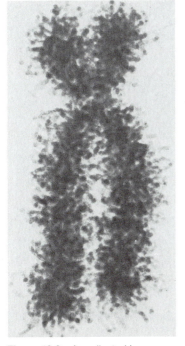

Figure 19.6 A replicated human chromosome. The two chromatids are connected in the region of their centromeres. (Magnification × 33,824)

Thus, it may be that under normal circumstances in those cells that do possess centrioles, the movement of the centrioles to the poles of the cell during prophase simply ensures that each new cell will receive a pair of centrioles.

During late prophase (sometimes called prometaphase), the chromosomes migrate toward the center of the cell. The nuclear membrane has completely disappeared by this point. Prophase ends when the chromosomes are lined up in the center of the cell.

The next phase of mitosis is a relatively short one called **metaphase**. During metaphase the chromosomes are lined up across the middle of the cell as if on a "plate" that is at right angles to the direction of the spindle microtubules (figures 19.3c and 19.8). The physical characteristics of the chromosomes are easily observed during this phase of mitosis. Figure 19.9 shows comparisons of metaphase chromosomes of different species. Metaphase chromosomes can also be studied to show differences among the individual chromosomes within a single cell.

Chromosomes differ in the relative position of their centromere and may also differ in size. If the centromere is located near the center, so that the "arms" of the chromatids are nearly equal in length, the chromosome is called **metacentric** (figure 19.10a). If the centromere is very near one end, the chromosome is called **acrocentric** (figure 19.10b). If the centromere is located at the very tip of the chromosome, the chromosome is called **telocentric** (figure 19.10c). Most cells are **diploid**, which means that they have two of each kind of chromosome. Members of these pairs of chromosomes are referred to as **homologous** chromosomes. Homologous chromosomes are similar in appearance, and the two chromosomes in a pair contain genetic information controlling the same characteristics.

The beginning of the next phase of mitosis, **anaphase** (figures 19.3d and 19.11), is marked by the separation of the two chromatids of each chromosome, so that each chromatid becomes an

Figure 19.8 Metaphase in the peony. The centromere of each duplicated chromosome is lined up on a "plate" across the middle of the cell. (Magnification × 850)

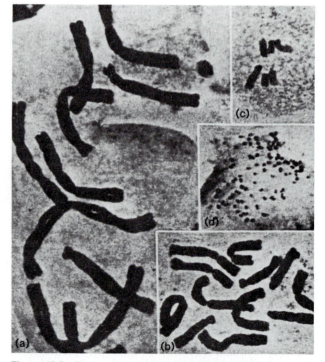

(a) (b) (c) (d)

Figure 19.9 These photographs, all taken through a microscope at the same magnification, show some of the differences that are found in chromosome sizes. (a) *Trillium erectum* (purple trillium). (b) *Podophyllum peltatum* (May-apple). (c) *Crepis capillaris* (hawk's beard). (d) *Sedum rupidragum* (stonecrop).

independent chromosome. During anaphase, one chromosome from each of these newly separated pairs of chromosomes moves to each end of the cell.

Neither the initial separation of the chromosomes during anaphase nor their subsequent movement toward the poles of the spindle is well understood, but the microtubules of the spindle seem almost certainly to be involved in the movement.

There are two types of microtubules in the spindle. One type stretches from the centromere of the chromosome to the area of the centriole; these are called centromere-to-pole microtubules. A second type of longer microtubules, not attached to the chromosomes, stretch further and overlap with one another to produce what appear to be pole-to-pole connections (see figure 19.7). During the anaphase movement of chromosomes, the entire spindle elongates, and this lengthening contributes to chromosomal movement. But it is

generally agreed that the major force involved in chromosomal movement is provided by actions of the centromere-to-pole microtubules that are attached to the centromere of each chromosome. As chromosomes move, chromosome arms lag as if the chromosome is being pulled by its centromere. This apparent pulling results in the appearance of characteristic shapes. Metacentric chromosomes assume a "V" shape and acrocentric a "J" shape as they move through the cytoplasm with the apex of the V or J being the centromere, which is at the point of spindle microtubule attachment.

The actual mechanism of chromosomal movement is still the subject of much investigation and some speculation. One mechanism proposed for chromosomal movement is the tubule assembly-disassembly mechanism. This hypothesis suggests that the centromere-to-pole microtubules are disassembled into their constituent subunits at each pole of the cell. This would shorten them and pull the attached chromosomes toward each pole. The released subunits are then reassembled in the spindle microtubules that are not attached to chromosomes but that overlap to stretch from pole to pole of the cell. This would pull the chromosomes toward the poles and, at the same time, add to the total length of the spindle.

The other proposed mechanism for chromosomal movement is called the sliding microtubule mechanism. This hypothesis suggests that the chromosomes move as chromosome-to-pole microtubules and pole-to-pole microtubules slide past each other. This sliding would be the result of chemical bonds forming, breaking, and reforming between the spindle fibers moving in opposite directions in a manner similar to the sliding of fibers in contracting muscle cells (chapter 17).

Some biologists think that there is yet a third possibility for chromosomal movement. They propose that actin-containing microfilaments might be involved in this movement, as they are in many other cell movements. Currently, there is some evidence favoring each of these hypotheses.

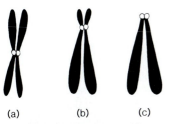

(a)   (b)   (c)

Figure 19.10 Three types of metaphase chromosomes. Chromatids are joined in the region of their centromeres in each case. (a) Metacentric. (b) Acrocentric. (c) Telocentric.

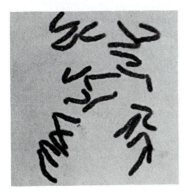

Figure 19.11 Anaphase in the peony. The centromeres have divided, and the chromatids have separated, becoming independent chromosomes. (Magnification × 850)

Figure 19.12 Early (a) and late (b) telophase in the peony. (Magnification × 1,445 and 1,598, respectively)

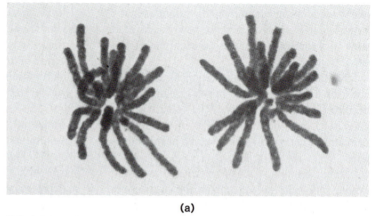

(a)

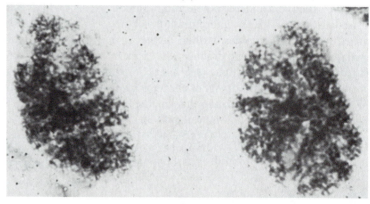

(b)

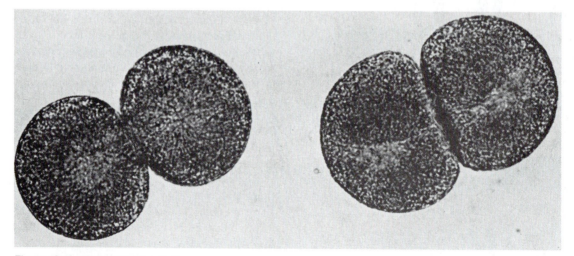

Figure 19.13 Cytokinesis in animal cells. These sea urchin eggs have been photographed in the process of furrowing or cleavage.

When the chromosomes arrive at each pole of the cell, **telophase**, the final phase of mitosis, begins (figures 19.3e and 19.12). During this phase the chromosomes begin to uncoil, and the two new nuclei organize. This nuclear reorganization involves formation of nuclear membranes and the reappearance of the nucleoli.

Cytokinesis

The discussion of cell division to this point has been limited to the activities of the nucleus. The rest of the material in the cell divides during a process called **cytokinesis**. Cytokinesis may or may not occur at the same time as mitosis. Animal cells divide by a process called furrowing or cleavage (figure 19.13). As the process begins, the cell membrane constricts across the middle of the cell due to the activity of microfilaments. The furrow progressively deepens until the cell is actually pinched into two new cells. This cleavage cuts through the cell exactly where the middle of the cleavage spindle was located. Thus, the plane of cleavage in a dividing animal cell can be predicted once the cleavage spindle forms because the cleavage will cut through the cell at right angles to the orientation of the spindle microtubules.

Cytokinesis in higher plants is a very different process because of the rigid cell wall that surrounds each plant cell. In plant cells, division of the cytoplasm occurs from the inside to the outside instead of outside to inside as in the furrowing of animal cells. A **cell plate** begins to develop in the center of the spindle at the end of mitosis. This plate forms by fusion of membranous spheres derived from the Golgi apparatus (figure 19.14). More and more spheres are added to the growing cell plate, and it increases in size until the cell is divided down the middle by two plasma membranes with a space between them. These membranes become continuous with the old plasma membrane, thus dividing the cell into two daughter cells. Contents of the vesicles that fused to form the cell plate are trapped between the plasma membranes of the two newly divided cells. These materials plus additional materials secreted by the two cells develop into new cell walls. At the end of the process, two plant cells, each surrounded by its own cell wall, exist where there had been one.

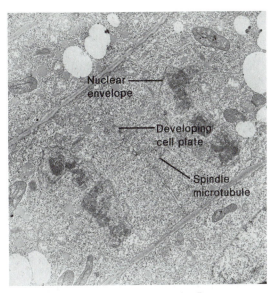

Figure 19.14 Cytokinesis in plant cells. Electron micrograph of cell plate formation in the root tip of a soybean. The cell plate forms by fusion of membranous spheres derived from the Golgi apparatus of the original cell. The process begins in the center of the spindle and spreads outward. Note the presence of some spindle microtubules and the re-forming nuclear envelopes of the daughter cells. (Magnification × 6,650)

Variations of Mitosis

As with most other biological processes, there are many variations of mitosis. In certain cells of the larval stages of some insects, the M phase is missing from the cell cycle so that mitosis does not take place at all. The S phase, however, does occur. Chromosomes replicate at each turn of the cycle, but the replicated strands are never separated. The resulting multistranded chromosomes are called **polytene** chromosomes (figure 19.15). These unique "giant" chromosomes and the information about developmental processes that has been gained by studying them are discussed further in chapter 25.

Another variation of mitosis occurs when the nuclear membrane remains intact throughout mitosis rather than breaking down in prophase. The chromosomes are replicated and separated but remain within the same nuclear envelope. The number of chromosomes doubles each time the cell goes through mitosis. This result is referred to as **endopolyploidy**. Some examples of endopolyploid cells are the liver cells of some animals, including human liver cells, and certain cells of tomato flowers.

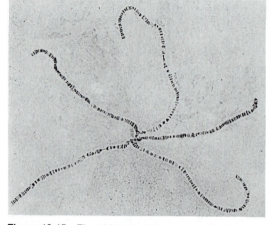

Figure 19.15 The polytene chromosomes of a larval *Drosophila* salivary gland cell. Polytene chromosomes are produced when there is repeated DNA replication that is not followed by mitosis.

A third variation of mitosis occurs when sister chromatids do not separate at the beginning of anaphase. When this happens, one new cell has an extra chromosome, and the other is missing a copy of that chromosome. This phenomenon is called **nondisjunction.**

Meiosis

In sexual reproduction, the zygote, the first cell of the new organism, is produced by fusion of two gametes. Each of the two gametes contributes half of the genetic information for the new organism. For this reason, the gametes must contain half the normal chromosome number of ordinary body cells. **Meiosis** is the process by which the number of chromosomes is halved. If meiosis did not occur, the total number of chromosomes would double every time two gametes fused to form a zygote.

Role of Meiosis in Life Cycles

Meiosis is found in all sexually reproducing organisms, but the timing of this process in the life cycle varies in different types of organisms. In most animals, reduction in the number of chromosomes from the diploid number to the necessary **haploid** number occurs during gamete production. The term haploid means that the cells have one of each "kind" of chromosome;

that is, only one member of each of the homologous pairs of chromosomes found in diploid cells. Haploidy is conventionally indicated by the designation N and diploidy by 2N.

In plants, meiosis occurs during the production of spores. These haploid (N) spores then develop into multicellular reproductive bodies made up of haploid cells. Gametes develop directly from some of those haploid cells. Thus, meiosis is not directly involved in gamete production in plants but occurs elsewhere in the plant life cycle. The role of meiosis in plant and animal reproduction is explored more fully in chapter 23.

Stages of Meiosis

Meiosis consists of two successive divisions (figure 19.16). The first division reduces the *actual number of chromosomes* to half. But at the end of this first division, the chromosomes are still doubled because they were replicated in the S phase preceding meiosis, and chromatids do not separate in the first division of meiosis. That is, at the end of the first division, each chromosome still consists of two chromatids attached in the areas of their centromeres. The second division resembles mitosis in that the centromeres separate and the chromatids part to become independent chromosomes. The second meiotic division has the effect of reducing the total DNA content of the nucleus to the haploid level.

As with mitosis, meiosis is divided into arbitrary stages for descriptive purposes. The first stage of the first division is called **prophase I.** During early prophase I (figure 19.17), the nuclear membrane and the nucleoli remain intact. The chromosomes appear as very long threads with beadlike thickened areas. Although these chromosomes have been replicated, they appear to be single strands; chromatids are not distinguishable at this stage.

Next, homologous chromosomes begin to come together and pair with each other. *This pairing process, called* **synapsis** *(figure 19.18), has no counterpart in mitosis.* The pairing between the homologous chromosomes is stabilized by a structure, called the **synaptonemal complex,**

Prophase I

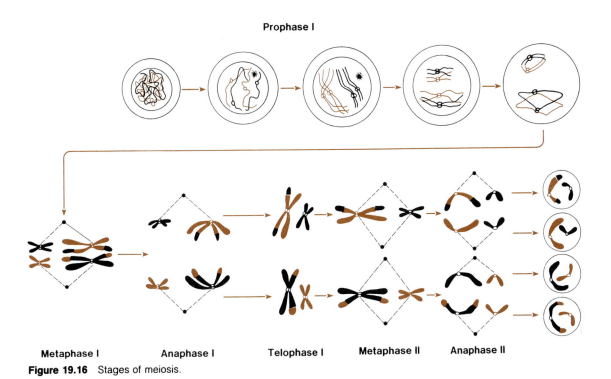

Metaphase I   Anaphase I   Telophase I   Metaphase II   Anaphase II

Figure 19.16   Stages of meiosis.

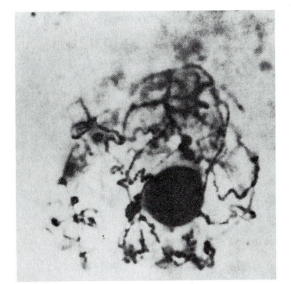

Figure 19.17   Early prophase I in corn (*Zea mays*). The large dark body is a nucleolus.

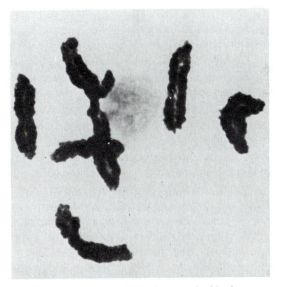

Figure 19.18   Synapsis. This photograph of barley chromosomes shows the pairing of the fourteen chromosomes into seven pairs of homologous chromosomes.

that appears in the space between each pair of chromosomes (figure 19.19). Each of these synaptic pairs, therefore, consists of two chromosomes (four chromatids) in close proximity to one another.

The chromosomes gradually become more condensed, and an important phenomenon called **crossing over** occurs. During crossing over, chromatids of the homologous chromosomes exchange segments (figure 19.20). This effectively redistributes genetic information among the paired homologous chromosomes and produces new combinations in the sets of genes borne by the various chromatids. After crossing over, there is a slight separation of the homologous chromosomes (figure 19.21). These chromosomes, however, remain attached at several points called **chiasmata** (singular: **chiasma**), which are the places where crossing over previously has occurred (figure 19.22). At the end of prophase I, the nuclear membrane breaks down, the nucleoli disappear, and the chromosome pairs move toward the center of the cell.

Metaphase I is similar to the metaphase of mitosis in that the chromosomes are lined up in the center of the cell (figure 19.23). The orientation of each chromosome, however, is different from that seen in mitosis because mitosis does not involve synapsis of homologous chromosomes. In mitotic metaphase, the centromeres of the chromosomes are lined up along the equatorial plane, while the arms of the chromosomes point in various directions (see figure 19.8). In the metaphase of the first meiotic division, however, the centromere regions of homologous chromosomes point to opposite poles of the cell.

Anaphase I begins when homologous chromosomes separate and begin to move toward each pole (figure 19.24). Sister chromatids do not part at this stage because there is no centromere separation in the first meiotic division. Because the chromosomes of each homologous pair separate and move in opposite directions, each daughter cell receives one chromosome from each homologous pair. Each of these chromosomes still consists of two chromatids and is called a **dyad.**

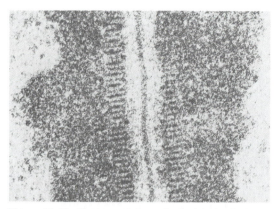

Figure 19.19 Electron micrograph of a longitudinal section through paired homologous chromosomes during prophase I of meiosis. The two wide dark areas are the two chromosomes. The synaptonemal complex consists of two striated (banded) elements separated by less dense areas and a dark line between the chromosomes. (Magnification ✕ 70,200)

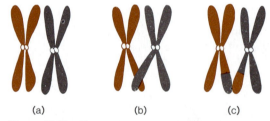

(a)   (b)   (c)

Figure 19.20 Chromosome segment exchange during crossing over. (a) Paired homologous chromosomes prior to crossing over. (b) Two of the chromatids exchanging segments. (c) The chromosomes after crossing over has been completed.

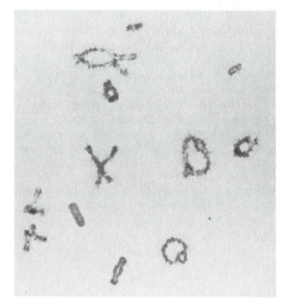

Figure 19.21 This photograph of a grasshopper cell shows how homologous chromosomes separate slightly near the end of prophase I but remain attached at chiasmata.

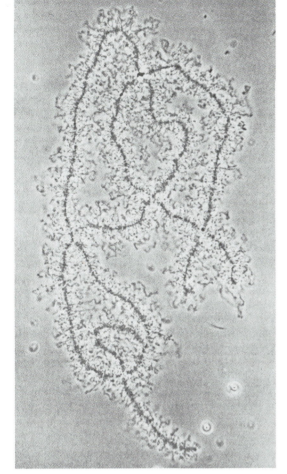

Figure 19.22 Three chiasmata hold the paired chromosomes together. These chromosomes are in a newt cell. (A newt is a tailed, aquatic amphibian.) (Magnification × 343)

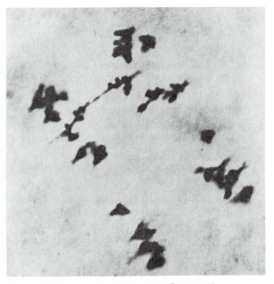

Figure 19.24 Anaphase I in corn (*Zea mays*).

At the end of the first meiotic division, the number of chromosomes has been reduced to half of the diploid number. But each chromosome consists of two chromatids; thus, the amount of DNA is still at the diploid level. One more division is needed to separate the chromatids and to complete the reduction of the DNA content to the haploid level.

In some species there is a **telophase I** stage at the end of the first division. In this stage, as in mitosis, the nuclear envelope reforms and the nucleoli reappear. The chromosome may also uncoil. In species where there is no telophase at the end of the first meiotic division, chromosomes which are grouped at anaphase I enter directly into the second division. In either case, there is *no* DNA replication (S phase) between the two meiotic divisions. In the second division of meiosis, the phases are referred to as **prophase II, metaphase II, anaphase II,** and **telophase II** (see figure 19.16). During anaphase II, centromere separation occurs, and the chromatids part to become independent chromosomes. At the end of telophase II, the final products of the two divisions of meiosis are four new cells. Each of these four cells is haploid; that is, it has half the diploid number of chromosomes and half the DNA content of diploid cells. Because of the crossing over that has occurred and because of the assorting of homologous chromosomes into different associations, the meiotic products are genetically different, not only from each other, but also from the original parental cell.

Figure 19.23 Metaphase I in corn (*Zea mays*).

Box 19.1
Independent Assortment of Chromosomes in Meiosis

All body cells are descended by mitotic cell divisions from a single cell, the zygote. Each cell has, just as the zygote had, a diploid (2N) number of chromosomes made up of a characteristic (N) number of homologous pairs. In each of these homologous pairs, one chromosome is a copy of a chromosome received from the individual's mother; for purposes of convenience, it can be identified as a "maternal chromosome." The other member of the homologous pair can be identified as a "paternal chromosome," as it is a replicate of a chromosome received from the individual's father. Of course, these "maternal" and "paternal" chromosomes differ from one another genetically.

Because the haploid cells produced by meiosis have only a single representative of each homologous pair of chromosomes, how are these different chromosomes distributed to daughter cells during the meiotic process? This is a key genetic question because the cells produced by meiosis become gametes directly (in animals) or develop into multicellular haploid structures that produce gametes (in plants).

The distribution of the members of any homologous pair of chromosomes to daughter cells during the first meiotic division is totally independent of the distributions of members of all other homologous pairs of chromosomes. *Each type of chromosome assorts independently of all others.*

What does this mean in terms of the meiotic process in cells? Box figure 19.1A diagrammatically shows some key parts of meiosis in an organism with a hypothetical diploid (2N) number of 4 and a haploid (N) number of 2. "Maternal" chromosomes are shown in color and "paternal" chromosomes are shown in black. Chromosomes are shown following chromosome replication and the synapsis of chromosomes that occurs in prophase I.

The key to understanding independent assortment of chromosomes is understanding the possible outcomes of anaphase I when members of homologous chromosome pairs separate from one another. As box figure 19.1A shows, the fate of each homologous pair during this separation is totally independent of the fate of each other homologous pair. It is a random process that can produce, in the example, four possible outcomes. As a result of this independent assortment, four chromosome combinations are possible for the haploid nuclei produced when chromatid separation in the second meiotic division is complete.

What about organisms with larger chromosome numbers? Meiosis in an organism with six chromosomes (2N = 6, N = 3) can produce twice as many different possible chromosome combinations as the example of an organism with a diploid number of 4. The number of possible meiotic outcomes doubles with each additional pair of homologous chromosomes. Another way of saying this is that the number of possible outcomes is equal to 2^N (2 to the Nth power), where N is equal to the number of homologous chromosome pairs (same as the haploid number, N). As was seen in the example, when N = 2, there are four (2^2) possible combinations. When N = 3, there are eight (2^3), when N = 4, there are sixteen (2^4), etc. In humans, where N = 23 (humans have forty-six chromosomes), the number of possible chromosome combinations produced by meiosis is a staggering 2^{23} or 8,338,608!

Independent assortment together with crossing over results in a great variety of genetic combinations in the haploid cells produced by meiotic cell division.

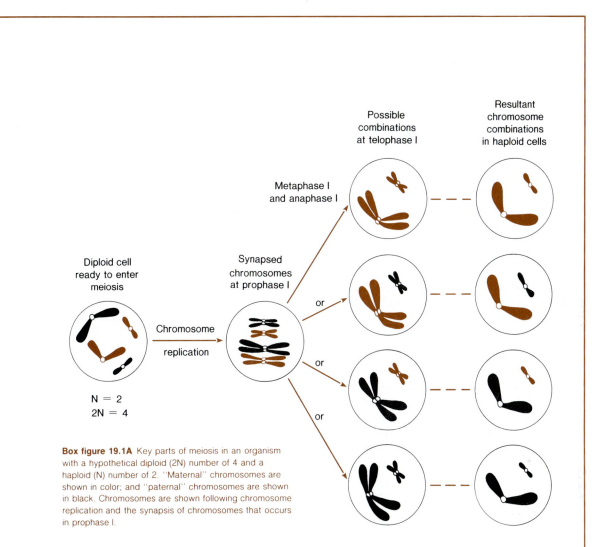

Box figure 19.1A Key parts of meiosis in an organism with a hypothetical diploid (2N) number of 4 and a haploid (N) number of 2. "Maternal" chromosomes are shown in color; and "paternal" chromosomes are shown in black. Chromosomes are shown following chromosome replication and the synapsis of chromosomes that occurs in prophase I.

Meiotic Variations and Errors

There are a number of interesting variations of meiosis. Some of these are associated with modifications of the process of sexual reproduction. For example, in several species of flowering plants, meiosis proceeds only as far as anaphase I. Then the two haploid nuclei recombine within a single nuclear membrane. Because of this modification in the meiotic process, egg cells that are eventually formed within such flowers are diploid. These eggs subsequently develop directly into embryos without the normally required union with a male gamete in fertilization.

A number of animal species also have the capacity to produce new individuals without fertilization, a process called **parthenogenesis.** In these species, diploidy is most commonly restored when two haploid nuclei combine during the second meiotic division in the development of the egg cell.

Occasionally, cells in which the diploid chromosome number is restored function as gametes. If these cells participate in fertilization, the resulting organisms have extra sets of chromosomes (they will be **polyploid**).

Errors in mitotic cell division ordinarily do not have serious consequences for organisms unless they occur early in development so that genetic or chromosomal problems are perpetuated in many body cells. But errors in meiosis are another matter because they result in genetic problems in gametes. These problems are perpetuated in all the cells of an individual produced as a result of fertilization involving such genetically abnormal gametes. A common error in meiosis is **nondisjunction,** which was described during the discussion of mitosis.

Nondisjunction can occur as a result of both chromosomes of a homologous pair moving to the same pole at anaphase I (figure 19.25). This means that, after meiosis is complete, some cells will have N + 1 chromosomes and some will have N − 1 chromosomes. If these cells ultimately participate in fertilization by combining with normal haploid gametes, the resulting organisms have either one extra chromosome (2N + 1) or are missing one chromosome (2N − 1). A chromosome number that varies from the normal chromosome number of the species by a small number of chromosomes is called **aneuploidy.** The most common of human birth

defects, Down's syndrome, is the result of aneuploidy. Affected individuals have an extra chromosome in one of the sets of homologous chromosome pairs (chromosome no. 21), and they suffer a number of physical problems as well as mental deficiencies that range from relatively mild impairment to severe retardation. Clearly, meiotic errors can have broad and significant consequences.

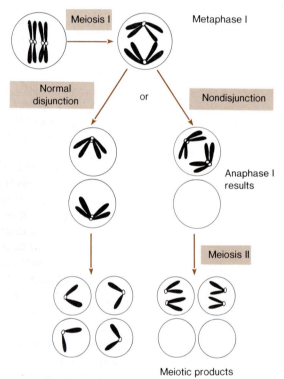

Figure 19.25 Nondisjunction in meiosis. Only the homologous pair of chromosomes showing nondisjunction are illustrated in this diagram.

A Comparison of Mitosis and Meiosis

Mitosis and meiosis are similar in that they are both forms of cell division, incorporating complex mechanisms that ensure that the genetic information is reliably passed to each new cell. The differences between these two methods of division involve their different roles in the life cycle of the organism.

The consequence of mitosis is simply production of new cells from previously existing cells. This is the method by which multicellular organisms grow and replace worn-out cells. Meiosis is necessary for the reduction of chromosome numbers in the life cycle of organisms that reproduce sexually.

Mitosis may occur in either haploid or diploid cells, and it does not result in any change in chromosome number. Meiosis always reduces a diploid cell to the haploid condition.

The process of mitosis takes place in one division with four stages. Meiosis requires two divisions; one to separate the homologous chromosomes and one to separate the replicated chromatids.

Finally, the products of the two processes differ. Mitosis produces two new cells, each of which has exactly the same genetic information as the parental cell. Meiosis produces four cells. These cells may all differ genetically from each other and from the parental cell.

Mitosis is the division of the eukaryotic nucleus so that two new nuclei are formed. For descriptive purposes, mitosis is divided into four stages. During the first stage, prophase, the chromosomes shorten and thicken, and the nuclear envelope and nucleoli disappear. The doubled chromosomes line up in the center of the cell during the second stage of mitosis, metaphase. Anaphase, the third stage, occurs when chromatids of each doubled chromosome move to opposite ends of the cell. Each chromatid becomes an independent chromosome at the time of separation. The fourth stage, during which the two new nuclear membranes are formed and the nucleoli reappear, is called telophase.

The cytoplasm of the cell is divided by cytokinesis. In most animal cells cytokinesis occurs by a constriction of the cell from the outside. The constriction continues until the cell appears to be pinched in two. This process is called furrowing or cleavage. Plant cells have a rigid cell wall, and direct cleavage is not possible. The cytoplasm of plant cells is divided by the formation of a cell plate across the center of the cell. This cell plate enlarges and materials are added until a new plasma membrane and cell wall separate the two nuclei formed by mitosis.

Meiosis produces cells that have half the diploid number of chromosomes. This reduction in chromosome number is accomplished in two successive divisions. The first division separates the pairs of homologous chromosomes. The two cells that are produced have the haploid number of chromosomes, but each of these chromosomes consists of two chromatids. The second division of meiosis separates these chromatids and produces cells that have a haploid (N) chromosome number of DNA content.

Summary

Cells arise from preexisting cells by a process of cell division. The life span of a cell from the time that it has been formed by cell division until it too divides can be described in terms of four distinct stages that make up the cell cycle. After division, a cell goes through a process of growth and maturation. This is the gap 1 (G_1) phase of the cell cycle. If the cell is going to divide, it moves into the synthesis (S) phase. During this phase the genetic material is replicated in preparation for division. After the S phase, the cell enters a gap 2 (G_2) phase. During this phase the cell synthesizes the necessary enzymes and structural proteins for division. The fourth phase (M) is nuclear division. It is accompanied by the division of the cytoplasm, cytokinesis.

Questions

1. When early microscopists studied cell division, they concentrated on the visible events of mitosis and named the period in which there was no visible activity between mitotic events the "interphase." They thought of the interphase as a "resting stage." Is the period between cell divisions (interphase) a resting stage? Justify your response.

2. Does a cell duplicate most of its component parts before or during mitosis? Explain.

3. What is a chromatid, and when does a chromatid become a chromosome?

4. What aspects of cell division are common to both plant and animal cells? What aspects are different?

5. Define synapsis and crossing over.

6. Explain why we say that at the end of the first meiotic division the chromosome number of the daughter nuclei is haploid, but their DNA content is diploid. How does this statement relate to the fact that there are two successive divisions in meiosis?

Suggested Readings

Books

Avers, C. J. 1976. *Cell biology*. New York: Van Nostrand.

DeRobertis, E. D. P., and DeRobertis, E. M. F., Jr. 1980. *Cell and molecular biology*. 7th ed. Philadelphia: Saunders College.

Dyson, R. D. 1978. *Cell biology: A molecular approach*. 2d ed. Boston: Allyn and Bacon.

Karp, G. 1979. *Cell biology*. New York: McGraw-Hill.

Kemp, R. 1970. *Cell division and heredity*. Studies in Biology no. 21. Baltimore: University Park Press.

Mitchison, J. M. 1971. *The biology of the cell cycle*. London: Cambridge University Press.

Swanson, C. P.; Merz, T.; and Young, W. J. 1967. *Cytogenetics*. Englewood Cliffs, N.J.: Prentice-Hall.

Wolfe, S. 1981. *Biology of the cell*. 2d ed. Belmont, Calif.: Wadsworth.

Articles

Beams, H. W., and Kessel, R. G. 1976. Cytokinesis: A comparative study of cytoplasmic division in animal cells. *American Scientist* 64:279.

John, B., and Lewis, K. R. 1973. The meiotic mechanism. *Carolina Biology Readers* no. 65. Burlington, N.C.: Carolina Biological Supply Co.

John, B., and Lewis, K. R. 1981. Somatic cell division. 2d ed. *Carolina Biology Readers* no. 26. Burlington, N.C.: Carolina Biological Supply Co.

Margolis, R. L.; Wilson, L.; and Kiefer, B. 1978. Mitotic mechanism based on intrinsic microtubule behavior. *Nature* 272: 450.

Mazia, D. January 1974. The cell cycle. *Scientific American* (offprint 1288).

Prescott, D. M. 1978. The reproduction of eukaryotic cells. *Carolina Biology Readers* no. 96. Burlington, N.C.: Carolina Biological Supply Co.

Sloboda, R. D. 1980. The role of microtubules in cell structure and cell division. *American Scientist* 68: 290.

Fundamentals of Genetics

In a sexually reproducing species, both parents contribute certain traits to their offspring. **Genetics** is the branch of biology that deals with this inheritance of characteristics. Although the concept of heredity has always been well known and its importance readily acknowledged, very little was known about the nature of the genetic process until 1865, when two of the basic laws of heredity were reported. Prior to that time, the prevailing theory of heredity was a blending theory. The hereditary material was thought of as something resembling a fluid, perhaps something in the blood of animals or the sap of plants. The traits of offspring were thought to be a combination of the parents' traits due to the blending of these fluids. The difficulty in determining the precise manner in which genetic information is transmitted from generation to generation was not due to a lack of study by scientists, but to the fact that most scientists were attempting to study simultaneously the transmission of a large number of characteristics. This made it much more difficult to explain what was happening.

Modern genetic analysis really began with a series of experiments presented orally in 1865 and published in 1866 by an Austrian monk named Gregor Mendel. Mendel's discoveries were the result of remarkably insightful analysis of simple but careful experiments. The significance of Mendel's conclusions is indicated by the fact that modern biologists refer to the study of certain types of inheritance patterns as **Mendelian genetics.**

Mendel's work went unappreciated until 1900 when it was rediscovered by three different geneticists—Hugo DeVries in Holland, Carl Correns in Germany, and Erich von Tschermak in Austria—working independently. Since that time, the science of genetics has expanded rapidly, and the last decades have seen especially significant advances in the understanding of the molecular aspects of genetics. Mendel's basic conclusions, however, remain valid.

Figure 20.1 Gregor Mendel.

The Work of Mendel

Gregor Mendel (figure 20.1) had two years of university training prior to entering the monastery. While at the university, he became familiar with the genetic theories of the time. He also was very aware of the interest of many scientists in plant hybridization (crosses between two differing types). Experiments conducted up to the time that Mendel began his work seemed to support the concept of blending inheritance.

Mendel began his own experiments around 1856 with the garden pea (*Pisum sativum*). This plant was well suited to Mendel's experiments for two reasons. First, the structure of the pea flower is such that it normally self-fertilizes; that is, the female gametes unite with male gametes from the same flower (figure 20.2). If a cross with a different plant is desired, the pollen-bearing structures (stamens) can be removed. Pollen from another plant, designated as the male parent, can then be placed on the appropriate part of the pea flower. Eventually, the male gamete contained in the pollen unites with the female gamete within the ovule. A second reason why the garden pea was well suited to Mendel's experiments is that genetic experiments require the testing of large numbers of individuals. Sufficient numbers of peas could be grown in the limited garden space available to Mendel at the monastery.

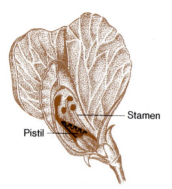

Figure 20.2 A cross section of a garden pea, *Pisum sativum*, flower. The petals that normally surround the reproductive organs have been cut away. These surrounding petals ensure self-fertilization under natural conditions. Pollen grains that later produce sperm nuclei are produced by the stamen. Haploid tissue that produces the egg develops in the pistil.

Table 20.1
The Seven Characteristics Used by Gregor Mendel in His Studies of the Garden Pea.

Seeds	Round vs. wrinkled
Seeds	Yellow vs. green
Flowers	Purple vs. white
Flowers	Axial vs. terminal
Pods	Inflated vs. pinched
Pods	Green vs. yellow
Stem	Tall vs. short

P₁
(first parental)   Tall × Short

F₁   Tall
                              (all plants)

                              │ Selfed

F₂   Tall (74%)   Short (26%)

Figure 20.3 A diagram of Mendel's experiments on the inheritance of stem length in pea plants.

Mendel was not the first to study hybridization in garden peas, but his approach was different. While others chose to examine the simultaneous inheritance of a large number of characteristics, Mendel chose, at the beginning of his experiments, to examine one characteristic at a time. He also chose characteristics with alternate forms that were easily recognized.

Mendel made a number of crosses involving each of seven different characteristics (table 20.1). In one experiment he crossed peas with tall stems and peas with short stems (figure 20.3). The plants he used for the first parents were plants derived from a number of generations of selfing (self-fertilization). Mendel considered these plants to be "true breeding" in the sense that the offspring of selfed tall plants always were all tall and the offspring of selfed short plants always were all short. Mendel crossed a true breeding tall plant with a true breeding short plant. He collected all the seeds (peas) resulting from this cross and planted them. The plants that developed from these seeds were all tall. The first generation of offspring in a genetic experiment such as this is called the **first filial** or the **F₁** generation. All of these tall plants were allowed to self, forming the **second filial** (**F₂**) generation, and they produced a total of 787 tall plants and 277 short plants. This works out to a ratio of about 3 tall:1 short. Mendel did the same type of crosses with each of the six other characteristics in table 20.1 and obtained approximately the same ratios each time.

Segregation

In interpreting these results, Mendel stated what is now called Mendel's **law of segregation.** Mendel concluded that characteristics such as tall or short stems are determined by discrete factors. These factors are now called **genes.** Mendel stated that *each organism contains two factors for each characteristic and that the factors segregate during the formation of gametes so that each gamete contains only one of each pair of factors.* Mendel was able to draw these conclusions from his experimental results even though he was unaware of the fact that his factors were located on chromosomes. He was also unaware of meiosis, which is responsible for the segregation of chromosomes (and therefore genes) prior to the formation of gametes.

Figure 20.4 shows how Mendel's theory can be used to interpret the results of his experiment on stem height in peas. Each pea plant has two copies of the gene controlling the height of the stem. In this example, each copy may be either of two possible forms. Alternate forms of a gene are called **alleles.** Thus, in this example there is an allele for tallness and an allele for shortness. The original parents used in Mendel's cross, referred to as the P_1 generation, were "true breeding." Therefore, the tall plants had two copies of the allele for tallness, and the short plants had two copies of the allele for shortness. When an organism has two of the same kind of alleles, it is **homozygous** for that gene. Because the first parents were homozygous, all gametes produced by the short plant contain the allele for shortness, and all gametes produced by the tall plant contain the allele for tallness. In the resulting F_1 generation, all the individuals have one allele for tallness and one for shortness. When an organism has two different alleles of a gene, it is **heterozygous** for that gene. Although the plants of the F_1 generation have one of each type of allele, they are all tall. When expression of only one allele in a heterozygous individual is observed, this allele is a **dominant** allele. The allele whose expression is masked in a heterozygote is a **recessive** allele. In modern genetic notation, the dominant allele is indicated with an uppercase letter and the recessive allele with a lowercase letter.

Genotype and Phenotype

It is obvious from these results that two organisms with different allele combinations for a trait may have the same outward appearance (*TT* and *Tt* peas are both tall). For this reason it is important to distinguish between the genetic constitution and the outward appearance of an organism. The genetic constitution of an organism is its **genotype.** The genotype is determined at fertilization and does not change during the lifetime of the organism. The **phenotype** of an organism is a set of observable characteristics of that organism. In the pea example, both *TT* and *Tt* are genotypes that give the tall phenotype. The distinction between genotype and phenotype must be kept in mind because while the genotype remains constant, the phenotype may vary over the lifetime of an individual.

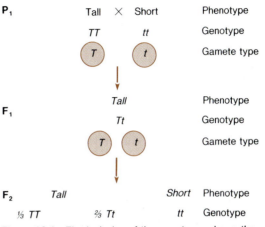

Figure 20.4   The inclusion of the genotypes shows the genetic basis for Mendel's results (figure 20.3).

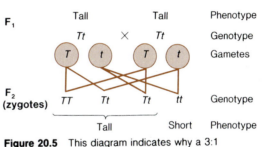
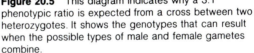

Figure 20.5   This diagram indicates why a 3:1 phenotypic ratio is expected from a cross between two heterozygotes. It shows the genotypes that can result when the possible types of male and female gametes combine.

The relationships between genotype and phenotype are complex because a phenotype is the result of the alleles present, possible interactions between different genes, and interactions between the genes and the environment. For this reason identical genotypes may have different phenotypes, and identical phenotypes may result from different genotypes.

When the heterozygous F_1 pea plants are selfed, the results are the same as those obtained when two heterozygous individuals (heterozygotes) are crossed. Approximately half of the gametes (both male and female) have the allele for tall stems and half have the allele for short stems. When these combine to form the F_2 zygotes (figure 20.5), the results are one-fourth homozygous tall plants, one-fourth homozygous short plants, and one-half heterozygotes. Because the heterozygotes give the tall phenotype, 75 percent of the F_2 generation are tall and 25 percent are short, thus producing the 3:1 phenotypic ratio.

Independent Assortment

Mendel next conducted experiments in which he traced the inheritance patterns of two different genes at the same time. He crossed plants producing round, yellow seeds with plants producing wrinkled, green seeds (figure 20.6). The round and yellow alleles are dominant; the wrinkled and green alleles are recessive. Both the seed shape and color traits had been previously shown by Mendel to obey the law of segregation.

When the two genes were traced simultaneously, the F_1 generation showed, as expected, that all individuals had the dominant characteristics—round and yellow seeds. Mendel realized that if the F_1 individuals were selfed, there were two possible outcomes for the F_2 generation.

If the two traits, seed shape and seed color, were inherited together, there would be just two kinds of seeds in the F_2 generation. The seeds would be either round and yellow, or wrinkled and green because these two sets of alleles would be inherited together. The two kinds would be in a 3:1 ratio with 75 percent of the seeds being round and yellow and the remaining 25 percent being wrinkled and green.

If the two traits were inherited independently of each other, however, the F_2 generation would show four possible combinations of seed color and shape: round and yellow (both dominant characteristics), round and green (dominant, recessive), wrinkled and yellow (recessive, dominant) and wrinkled and green (both recessive).

The F_2 generation conformed to the second of these hypothetical outcomes. That is, Mendel found all four different genetic combinations among the F_2 generation plants. The combinations occurred in a ratio of 9 round, yellow: 3 round, green: 3 wrinkled, yellow: 1 wrinkled, green.

The section on dihybrid crosses (page 554) explains why these particular ratios are obtained. For now, however, the important point is that the results of these experiments led Mendel to formulate a second genetic principle, called the **law of independent assortment.** This principle states that the members of one pair of genes segregate independently of other pairs.

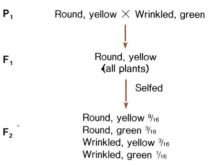

Figure 20.6 These results convinced Mendel that the genes for seed shape and color assort independently of each other.

These two laws, the law of segregation and the law of independent assortment, form the basis for modern genetic analysis, but when Mendel published his conclusions in 1866, no one realized their significance. One probable reason for this is that most investigators continued to study inheritance of large numbers of characteristics simultaneously. This led to such complex results that the underlying principles were virtually impossible to detect. These investigators thought that Mendel's experiments, based on the study of one or only a few traits, were a gross oversimplification of natural processes.

Mendel himself came to have some doubts about the generality of his laws. He attempted to verify his ideas with a native plant *Hieracium* (hawkweed). When he made the same type of crosses that he had made with garden peas, the results were totally different. In this case, Mendel's choice of an experimental organism was as unfortunate as his choice of the garden pea for his original experiments had been fortunate. What Mendel did not and could not know was that while *Hieracium* requires a pollen grain to develop a seed, the pollen grain is only needed to stimulate the development of the egg within an ovule. Fertilization does not occur. The only genetic information transmitted to the offspring in Mendel's experiments with hawkweed was from the female parent. No wonder Mendel was confused by the results.

In his later years at the small monastery, Mendel devoted most of his time to administrative battles with the local government, and he died in 1884 without the satisfaction of having the significance of his work understood and appreciated. In 1900 De Vries, Correns, and von Tschermak discovered that Mendel's early conclusions stated very clearly what they themselves were discovering from their own experiments.

Mendelian Genetics

The principles of Mendelian genetics—the law of segregation and the law of independent assortment—can also be applied to present-day genetic analysis of hereditary patterns.

Monohybrid Crosses

A **monohybrid cross** is a cross between parents that differ in a single gene. The individuals involved may actually differ in other characteristics, but the analysis is concerned with only one. Mendel's experiment with the tall and short pea plants is an example of a monohybrid cross. The law of segregation may be used to calculate the expected proportions of genotypes and phenotypes if the genotypes of the parents are known.

　　There are two ways to make this calculation. The first is to use a **Punnett square.** This simple method was introduced by a prominent poultry geneticist, R. C. Punnett, in the early 1900s. Figure 20.7 shows how the Mendelian cross between two types of peas can be diagrammed in a Punnett square. All the possible kinds of male gametes are placed across the top of the square, with the types of female gametes along the left side. The ways in which these gametes may combine are determined by filling in the cells of the square. The results show that the expected proportions of offspring are 3 tall: 1 short for the phenotypes, and 1 *TT:* 2 *Tt:* 1 *tt* for the genotypes.

　　These are only expected proportions, however, not absolute numbers. For example, if the cross shown in figure 20.7 produced 200 offspring, approximately 150 of them would be tall and approximately 50 would be short. In terms of the genotypes, approximately 50 of the plants would be *TT,* about 100 would be *Tt,* and the remaining 50 or so would be *tt.*

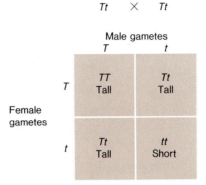

Figure 20.7 A Punnett square showing the results of a cross between two heterozygous pea plants.

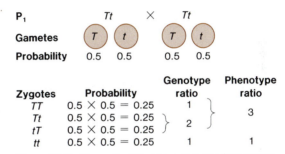

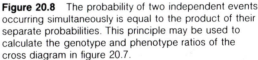

Figure 20.8 The probability of two independent events occurring simultaneously is equal to the product of their separate probabilities. This principle may be used to calculate the genotype and phenotype ratios of the cross diagram in figure 20.7.

　　Another method of calculating the expected ratios of offspring types is to multiply the probabilities of the formation of different types of gametes. When flipping a coin, the expectation is that the number of heads will approximately equal the number of tails. The probability of a head is considered to be 50 percent (0.5 in decimal form). Likewise, the probability of a *T* gamete from a *Tt* parent is 50 percent (0.5). The ratio rarely is exactly 50–50 but usually is close. The first step in estimating zygote genotype frequencies is to determine the probabilities for forming types of gametes for each parent. Then the probability of forming each type of zygote is obtained by multiplying the probabilities of forming the gametes involved. This is illustrated in figure 20.8. The probability of the zygote being a *TT* homozygote is determined by multiplying the probabilities of producing a *T* gamete from the male parent (0.5 in this example) and of producing a *T* gamete in the female parent (0.5).

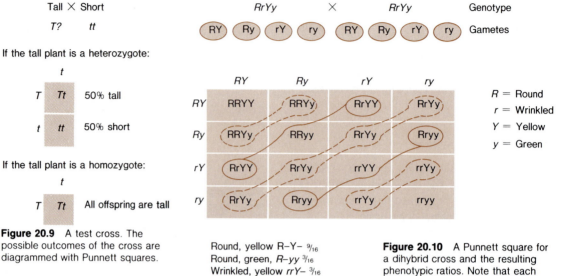

Tall × Short
T? tt
If the tall plant is a heterozygote:
t
T Tt 50% tall
t tt 50% short
If the tall plant is a homozygote:
t
T Tt All offspring are tall

Figure 20.9 A test cross. The possible outcomes of the cross are diagrammed with Punnett squares.

RrYy × RrYy Genotype
RY Ry rY ry RY Ry rY ry Gametes
RY Ry rY ry
RY RRYY RRYy RrYY RrYy
Ry RRYy RRyy RrYy Rryy
rY RrYY RrYy rrYY rrYy
ry RrYy Rryy rrYy rryy
R = Round
r = Wrinkled
Y = Yellow
y = Green

Round, yellow R–Y– 9/16
Round, green, *R–yy* 3/16
Wrinkled, yellow *rrY–* 3/16
Wrinkled, green *rryy* 1/16

Figure 20.10 A Punnett square for a dihybrid cross and the resulting phenotypic ratios. Note that each gamete has an allele for each gene being studied. The (–) symbol is used to indicate that, due to dominance, it does not matter which second allele is present.

The result is a proportion of 0.25 *TT* genotypes (0.5 × 0.5 = 0.25). The probabilities of the other possible genotypes are calculated in a similar manner. The sum of the decimal parts in such statements of probability, whether they are concerned with the gamete possibilities or the zygote genotype possibilities, always is 1.0 (100 percent).

To determine the types of gametes that an individual will produce, it is necessary to know the genotype, not just the phenotype, of that individual. This is because different genotypes may have the same phenotype. Mendel got around this problem by using plants that had been selfed for many generations. These "true breeding" plants were homozygous. If the genetic history of an individual is not known, the individual's genotype can still be determined by a genetic means that involves making a specific kind of cross, called a **test cross,** which takes advantage of certain predictions made by the law of segregation.

An example of a test cross is diagrammed in figure 20.9. In a test cross, the individual being tested for homozygosity or heterozygosity is crossed with an individual that is homozygous for the recessive characteristic. That is, an individual with an unknown genotype is crossed with an individual with a known genotype. If the tall plant in figure 20.9 is a heterozygote, half of the offspring will be tall, and half will be short. If the tall parent is a homozygote, all of the offspring will have at least one *T* allele and will exhibit the dominant characteristic. But if any short offspring are produced, the parent with the unknown genotype must be a heterozygote.

Dihybrid and Trihybrid Crosses

The same principles used for the analysis of monohybrid crosses can be used when analyzing the inheritance of the combinations of two or more different characteristics. Mendel's law of independent assortment states that the segregation of alleles of a single gene is independent of the segregation patterns of other traits. This makes it possible to predict the genotypes of the offspring of a cross involving more than one gene.

A **dihybrid cross** is a cross between individuals differing for two genes. Mendel's experiment with seed shape and color in peas is an example of a dihybrid cross. The Punnett square method of determining the genotypic and phenotypic ratios in the offspring is shown in figure 20.10. The method is basically the same as with the monohybrid cross. The major difference is that each gamete type now has one allele of each of two different genes rather than one.

Another means of working out the results of a dihybrid cross is by use of a dendrogram, a branching, treelike (dendro="tree") diagram. The basis for the construction of a dendrogram is probability. Rules of probability are applicable because, as Mendel showed, the different genes assort independently of each other. (There are, however, exceptions to the law of independent assortment and the reasons for, and consequences of, these exceptions are discussed in chapter 21.)

The dendrogram technique can be used, for example, to predict phenotypic ratios in a dihybrid cross (figure 20.11). The probability of producing offspring with round seeds is 3/4 (1/4 homozygous round + 2/4 heterozygous). The probability of the green phenotype is 1/4 because only the homozygous recessive individuals have this phenotype. Therefore, the probability of producing an individual with round, green seeds is 3/16 (3/4 × 1/4). The other probabilities can be worked out in the same way to show that a dihybrid cross between two heterozygous parents yields a ratio of 9 dominant, dominant: 3 dominant, recessive: 3 recessive, dominant: 1 recessive, recessive.

The dendrogram technique also can be used to predict genotypic ratios in the dihybrid cross shown in figure 20.11. The proportion of offspring that should be homozygous for both dominant characteristics can be calculated by determining the probability of producing a zygote that is homozygous for the round allele. That probability is 1/4 (25 percent) when both parents are heterozygous. The probability of producing an individual that is homozygous for the yellow allele is also 1/4. The probability, then, that an individual will be produced that is homozygous for both of those genes is the product of the independent probabilities. Therefore, one out of every sixteen offspring (1/4 × 1/4 = 1/16) should be homozygous for the alleles for both round seeds and yellow seeds.

A **trihybrid cross,** which involves three characteristics, follows the same principles as a dihybrid cross. However, using a Punnett square for a trihybrid cross becomes quite tedious because each trihybrid parent can produce eight different types of gametes (figure 20.12). The resulting Punnett square would have sixty-four cells! A dendrogram is a much more efficient method of predicting phenotypic and genotypic ratios in a trihybrid cross (figure 20.13).

Yellow (¾) = ⁹⁄₁₆ Round, yellow

Round (¾)

Green (¼) = ³⁄₁₆ Round, green

Yellow (¾) = ³⁄₁₆ Wrinkled, yellow

Wrinkled (¼)

Green (¼) = ¹⁄₁₆ Wrinkled, green

Figure 20.11 The dendrogram method of predicting phenotypic ratios from a dihybrid cross. The independent probabilities are indicated in parentheses, and the final ratios are the products of these probabilities.

Genotype	Aa Bb Cc
Gamete types	ABC
	ABc
	Abc
	AbC
	aBC
	abC
	aBc
	abc

Figure 20.12 An organism that is heterozygous for three genes (trihybrid) may form eight types of gametes.

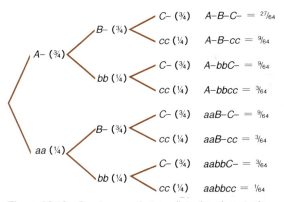

Figure 20.13 Dendrogram that predicts the phenotypic ratios resulting from a trihybrid cross. As with a dihybrid cross, the probability of a particular phenotype is equal to the product of the independent probabilities.

Multiple Alleles

All of the examples discussed so far have concerned genes that have two alleles. In some populations, however, there may be only one form of a gene for a particular trait. In other cases, a gene can have three or more alleles, although a diploid individual can have a maximum of two alleles per gene.

One of the best known examples of a three allele gene is the one that controls ABO blood types in humans. There are four possible ABO blood types or phenotypes: A, B, AB, and O. These phenotypes are produced from combinations of three different alleles: I^A, I^B, and i^O. The i^O allele is recessive, and the I^A and I^B alleles are both dominant. Figure 20.14 shows the different combinations of alleles that form the different phenotypes. Types O and AB are each produced from a single genotype, but types A and B may each be either homozygous or heterozygous.

It is possible, within limits, to predict the blood groups of children from those of their parents. This relationship is sometimes used, for example, to determine cases of disputed paternity. Paternity cannot be proven from blood types, but it is possible to eliminate certain individuals as possible fathers. For example, two parents with blood type O will have only type O children. Children whose mother was O and whose father was A could be neither B nor AB. On the other hand, a type A parent and a type B parent could possibly have children with any of the four blood types, if both of them were heterozygous; that is, if blood type A parent has the $I^A i^O$ genotype and blood type B parent has the $I^B i^O$ genotype (figure 20.15).

An important point to note from this example is that the addition of a third allele increases the number of possible genotypes from three to six. In general, if there are N alleles, there are N(N + 1)/2 genotypes (table 20.2).

Blood type (phenotype)	Possible genotypes
A	$I^A I^A$, $I^A i^O$
B	$I^B I^B$, $I^B i^O$
AB	$I^A I^B$
O	$i^O i^O$

Figure 20.14 How the three blood type alleles combine to form the four possible blood types.

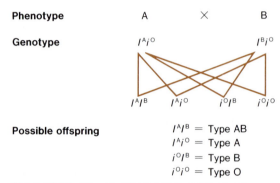

Phenotype	A	×	B
Genotype	$I^A i^O$		$I^B i^O$

	$I^A I^B$ $I^A i^O$ $i^O I^B$ $i^O i^O$

Possible offspring	$I^A I^B$ = Type AB
	$I^A i^O$ = Type A
	$i^O I^B$ = Type B
	$i^O i^O$ = Type O

Figure 20.15 How an offspring of A and B parents would have any of the four possible blood types if both parents happen to be heterozygous.

Table 20.2
Number of Different Genotypes Possible for Different Numbers of Alleles.

Alleles	Possible Number of Genotypes
1	1
2	3
3	6
4	10
5	15
N	$\dfrac{N(N + 1)}{2}$

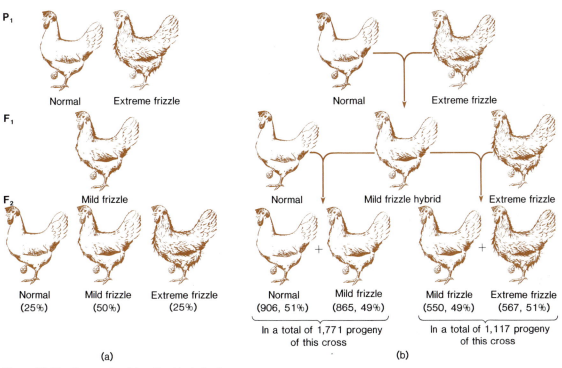

Figure 20.16 Crosses involving the frizzle feather abnormality demonstrate the phenomenon of incomplete dominance. (a) General pattern of inheritance. (b) Data from actual experimental crosses.

Other Inheritance Patterns

The pattern of human blood type genotypes and phenotypes, which was used in the previous section as an example of multiple alleles, differs in another important way from the inherited characteristics that Gregor Mendel studied. All of the characteristics of garden peas that Mendel reported studying display simple dominant-recessive relationships. Both individuals with the homozygous dominant genotype and individuals with the heterozygous genotype express the dominant phenotype, while individuals with the homozygous recessive genotype express the recessive phenotype. In other words, only two phenotypes are expressed. Human blood group inheritance, however, shows four phenotypes. Many other inherited traits also show several possible alternative phenotypes.

Intermediate Inheritance

Alleles are not always either dominant or recessive to each other. In chickens, for example, there is a mutation affecting feather structure that has the descriptive name "frizzle." This condition produces bristly feathers that wear off easily. Poultry geneticists have noted that some chickens have a more extreme form of the frizzle condition than others. If a chicken with extreme frizzle is mated with a normal feathered chicken (figure 20.16a), all of the F_1 offspring have a mild frizzle condition. If two of these mild frizzle chickens are mated to produce an F_2 generation, 25 percent of these offspring have extreme frizzle, 50 percent have mild frizzle, and the remaining 25 percent have normal feathers. With the segregation of parental types in the F_2 generation, it becomes obvious what has happened. The intermediate form occurs in the F_1 because neither of the two alleles exerts dominance, and an intermediate phenotype is produced. This phenomenon is called **incomplete dominance** (figure 20.16b).

Incomplete dominance also occurs in snap-dragons. If red-flowered snapdragons are crossed with white-flowered snapdragons, all of the off-spring have pink flowers. If these are selfed, the F_2 ratio is 25 percent red, 50 percent pink, and 25 percent white. Thus, the only way to maintain a garden with all pink snapdragons is to replant F_1 seeds each year.

Another type of intermediate inheritance occurs when the traits of both parents are dis-tinctly expressed in the F_1 as opposed to forming an intermediate phenotype. The genetic control of blood types in humans, which was described earlier, is an example of this. If one parent con-tributes an I^A allele and the other contributes an I^B allele, the child will have blood type AB. The A and B alleles are said to be **codominant.**

Modifier Genes and Epistatic Genes

Dominance occurs when one allele masks the expression of another allele of the same gene. But sometimes a gene affects the phenotypic expres-sion of an entirely different gene, a gene that is not one of its alleles. One example of this occurs in the genetic determination of coat pattern in cattle. The basic spotted pattern in Holstein cat-tle is determined by a single gene, but the relative amounts of black and white are determined by a different series of genes. Genes such as these are called **modifier genes.**

An **epistatic gene** is a gene that covers up the expression of another gene in the phenotype. This **epistasis** ("covering up") works much like ordinary dominance except that two different genes (genes that are not alleles) are involved instead of two alleles of a single gene. For ex-ample, there is a color gene in fowl designated *C*. A *cc* genotype is white; a *Cc* or *CC* genotype is expected to have color. Another gene, *I*, is epis-tatic over the *C* gene. A genotype of *II* or *Ii* produces a white fowl, no matter what the ge-notype is at the color gene. Thus, the fowl must have a homozygous recessive genotype *ii* if it is to have color.

The white leghorn fowl has the genotype *IICC*. The white wyandotte has the genotype *iicc*. Thus, the explanation for white feather color is different in the two breeds. The leghorn is white because of the effects of the *I* gene. The wyan-dotte is white because it is homozygous recessive (*cc*). To demonstrate the effects of epistasis, the results of a cross between a white leghorn and a

white wyandotte are diagrammed in figure 20.17. Fowl with color segregate out in the F_2 genera-tion.

Pleiotropy

Modifier genes and epistatic genes are examples of different genes affecting a single character. **Pleiotropy** is the name for an entirely different situation where a single gene affects several dif-ferent phenotypic characteristics. In humans, an example of pleiotropy is the gene which, in the homozygous recessive form, results in the disease phenylketonuria (PKU). Untreated individuals with this disease have unusual amounts of phen-ylalanine in the blood, and also lower IQ scores, somewhat larger heads, and lighter hair color. All of these characteristics are determined by a single gene.

There are many other examples of pleio-tropy. Blue-eyed white cats are generally deaf, minks with the colmira color gene are deep sleep-ers, and humans with Eddowe's syndrome have both brittle bones and a blue color to the normally

| P_1 | White leghorn $\times$ White wyandotte |
| | *IICC*   *iicc* |

| F_1 | White |
| | *IiCc* |

F_2

	IC	*Ic*	*iC*	*ic*
IC	*IICC*	*IICc*	*IiCC*	*IiCc*
Ic	*IICc*	*IIcc*	*IiCc*	*Iicc*
iC	*IiCC*	*IiCc*	Colored *iiCC*	Colored *iiCc*
ic	*IiCc*	*Iicc*	Colored *iiCc*	*iicc*

Figure 20.17 The colored fowl form segregates out in the F_2 generation of a cross between white leghorn and white wyandotte chickens. The I gene is epistatic over the C gene. The F_2 phenotypic ratio is thirteen white: three colored.

white sclerotic coat of the eye. An extreme example is a gene appropriately called polymorph (polymorph means "many form" or "many shape") in *Drosophila,* the fruit fly. This single gene affects eye color, body proportions, wing size, wing vein arrangement, body hairs, size and arrangement of bristles, shape of testes and ovaries, viability, rate of growth, and fertility.

The fact that pleiotropy is very common reflects the way in which genes act through effects on biochemical processes. An alteration in a single biochemical pathway, resulting from the expression of a single gene, may affect many developing characteristics.

Quantitative Genetics

The alternate traits discussed up to this point, such as tall vs. short plants, green vs. yellow seeds, A vs. B blood types, etc., are clearly different from each other. This kind of variation is called **discontinuous variation,** because the variants fall into discrete, nonoverlapping sets. With regard to other traits, however, organisms may exhibit **continuous variation** so that different individuals may vary from each other only slightly. Differences among phenotypes are *quantitative* rather than *qualitative,* and the study of the inheritance patterns of these traits is called **quantitative genetics.** This name is derived from the fact that, with regard to such traits, differences between individuals are quantitative—differences in degree or amount requiring precise measurement—rather than qualitative—requiring only simple observation for sorting of phenotypes into discrete groups.

Many sets of alleles may be involved in quantitative inheritance of certain characteristics, and the environment may play an especially large part in determining the phenotype. Because large numbers of genes may be involved, quantitative inheritance is sometimes called **multiple factor** or **polygenic inheritance.** Phenotypes of quantitatively inherited characteristics frequently are distributed in a bell-shaped or normal curve. Human height, for example, is a quantitatively inherited trait determined by many genes, and the distribution of human heights follows a normal distribution (figure 20.18).

| **Number of individuals** | 1 | 0 | 0 | 1 | 5 | 7 | 7 | 22 | 25 | 26 | 27 | 17 | 11 | 17 | 4 | 4 | 1 |

| **Height in inches** | 58 | 59 | 60 | 61 | 62 | 63 | 64 | 65 | 66 | 67 | 68 | 69 | 70 | 71 | 72 | 73 | 74 |

Figure 20.18 Height distribution in 175 men recruited for the Army around 1900.

Figure 20.19 An example of quantitative inheritance is this experiment showing the inheritance of ear length in corn. The height of the black bars indicates the number of ears at that length.

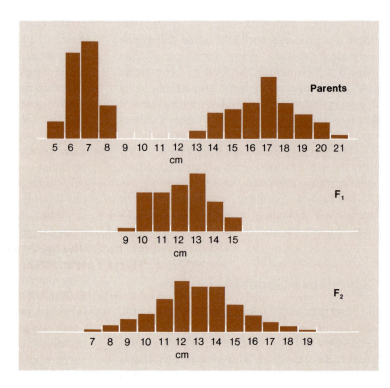

Another example of this kind of distribution was found in a study of the inheritance of ear length in corn (figure 20.19), which began with crosses between plants that produce short ears and plants that produce long ears. The bell-shaped distribution of offspring in the F_1 generation is retained in the F_2 generation. Very few of the parental phenotypes segregate out, as would be expected from the law of segregation if ear length were determined by a single gene pair.

This apparent conflict with Mendel's law can be resolved by looking at a hypothetical organism where the genotype $A^1A^1B^1B^1C^1C^1$ (three different genes) is 20 cm tall. The addition of an allele denoted by a "2" increases the height 1 cm. For example, the genotype $A^1A^2B^1B^2C^1C^2$ would be 23 cm tall, and the genotype $A^2A^2B^2B^2C^2C^2$ would be the maximum 26 cm tall. If a cross was made between an $A^1A^1B^1B^1C^1C^1$ individual and an $A^2A^2B^2B^2C^2C^2$ individual, and members of the F_1 generation were crossed with one another, the distribution of height in the F_2 generation would be as follows: 1/64, 26 cm; 6/64, 25 cm; 15/64, 24 cm; 20/64, 23 cm; 15/64, 22 cm; 6/64, 21 cm; 1/64, 20 cm. Only two out of sixty-four offspring are the parental types. This is due simply to the fact that three genes are involved. When still more genes are involved, even fewer offspring reflect the parental types.

Many of the geneticists of Mendel's time were unknowingly studying polygenic inheritance, and because quantitative inheritance patterns resemble a blending pattern, these scientists were unable to recognize the laws of segregation and independent assortment or to appreciate the significance of Mendel's work.

Many of the economically important characteristics of agricultural plants and animals are inherited in a quantitative fashion. Examples are milk production in dairy cattle and yield in corn. For this reason, those involved in breeding agricultural plants and animals must have a thorough understanding of quantitative genetics.

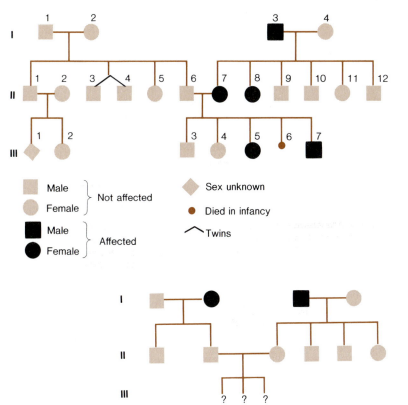

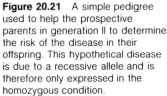

Figure 20.20 A typical human pedigree with the standard notation. For purposes of reference, Roman numerals to the left identify generations, and each individual within a generation has his or her own number. For example, II-7 is the seventh person in the second row.

Figure 20.21 A simple pedigree used to help the prospective parents in generation II to determine the risk of the disease in their offspring. This hypothetical disease is due to a recessive allele and is therefore only expressed in the homozygous condition.

Human Genetics and Genetic Counseling

Over the years, considerable progress has been made in the study of inheritance of many human characteristics. A list of Mendelian traits in humans compiled in 1978 includes 736 known dominants, 753 suspected dominants, 521 recessives, 596 suspected recessives, and 107 traits on the X chromosome (these are called X-linked genes and are discussed in the next chapter). Most of these identified traits are concerned with various diseases because there is much more medical incentive to study inherited diseases than the inheritance of various normal human characteristics.

Couples frequently are concerned with questions such as, "Will one of these genetic disorders, which we may display or may be present in a relative, or in previous children, also be likely to occur in future children?" Answers to such questions should be sought from a qualified genetic counselor. A genetic counselor (frequently a physician) is trained in genetic principles and is familiar with the inheritance patterns of human diseases.

The patterns of inheritance of a trait or characteristic in a particular family are commonly analyzed by drawing a **pedigree.** Figure 20.20 shows a sample pedigree and definitions of some of the symbols used. If the information is complete enough, a genetic counselor may be able to reconstruct the genetic history of a particular trait. The counselor can then inform the parents of the risk of having an offspring with the particular genetic defect. Once the risk has been determined, most genetic counselors encourage the prospective parents to decide for themselves whether or not to have children and, if so, how many to have.

For example, if each marriage partner has a parent with the same genetic disease, the genetic counselor is able to obtain enough information to construct the simple pedigree shown in figure 20.21. The pedigree shows that the partners are each heterozygous (assuming simple recessive Mendelian inheritance, as is the case with many genetic diseases). The probability that one child would be affected (homozygous recessive) is 1/4. The probability that each of three children would be affected is small $(1/4)^3 = 1/64$. But calculation of the probability that at least one of

Box 20.1
Marriage between Relatives

The risk of genetic defects in offspring is important not only when there is a history of genetic disease, but also when close relatives are considering marriage. The offspring of two individuals who are genetically similar show a decrease in the number of genes that are heterozygous and an increase in the number of genes that are homozygous. This increase in homozygosity may have serious consequences because most individuals contain at least a few recessive alleles that, if homozygous, would be harmful, perhaps lethal, to the individual. Because these alleles normally are not expressed in a heterozygous condition, their presence causes no harm to heterozygous carriers.

The increase in homozygosity in the offspring is proportional to how closely related the parents are. Mating between parents and children or between sisters and brothers is called incest. There is a taboo against incest in most societies, and most states and countries have laws forbidding this type of marriage. A more common type of marriage is between first cousins. The frequency of first cousin marriages varies from society to society. In the United States, almost half the states have laws forbidding such a marriage, but other states permit it. In Japan, first cousin marriages are encouraged and occur in up to 10 percent of the marriages in some areas.

Box table 20.1 shows some of the effects of the increased level of homozygosity due to first cousin marriages. The presence of harmful recessives is referred to as a type of **genetic load,** and unfortunately, the genetic loads of close relatives are more likely to include the same harmful recessive genes than are those of unrelated or more distantly related individuals. Clearly, there is a sound genetic basis for the ancient taboos against incest.

Box table 20.1
Mortality in Offspring of First Cousin Marriages Compared with That in Offspring of Unrelated Parents.

Trait	Unrelated	First Cousins
Stillbirths and neonatal deaths	0.044	0.111
Infant and juvenile deaths	0.089	0.156
Juvenile deaths	0.160	0.229
Postnatal deaths	0.024	0.081
Miscarriages	0.129	0.145

From H. E. Sulton, *An Introduction to Human Genetics* (New York: Holt, Rinehart and Winston, 1980). Used by permission.

three children would be affected yields a probability of considerably more than 1/2. Knowing all of this, the prospective parents must decide whether or not to have children.

Many pedigrees are much more complex than the example just given, and the risks are much more difficult to estimate. Sometimes, the risk must be estimated from past statistics rather than from the knowledge of the nature of inheritance of a particular trait. But some factors make the job of a genetic counselor easier. With some traits, heterozygotes can be detected by medical tests. For example, individuals heterozygous for Tay-Sachs disease or sickle cell anemia can be detected by appropriate tests. When such testing is available, risk to offspring of certain marriages can be much more precisely determined.

Heredity and Environment in Genetic Expression

One of the most basic, and also most often forgotten, principles of genetics is that the genotype and the environment interact to produce the phenotype. It is frequently quite difficult to determine how much of a phenotype is determined by heredity ("nature") and how much by environment ("nurture"). An example of the complex interaction of heredity and environment is seen in the determination of human height. Many genes affect height. Some genes involved are genes controlling growth hormone, digestive enzymes, rate of calcium deposition in the bones, and many others. In addition, environmental factors, such as the presence or absence of adequate nutrition during critical growing periods, may strongly influence an individual's height. A combination of all these factors ultimately determines how tall a person is.

Geneticists attempt to assess the relative contribution of genetic factors by the use of a concept called **heritability.** Heritability is the proportion of the phenotypic variation in a population that is due to genetic factors. If all the variation is due to genetic causes, the heritability is 1.0 (100 percent). If all of the observed phenotypic variation in a population is due to the environment, the heritability of that variation is 0.0. All traits demonstrate heritability values that fall between 0 and 1.0 (table 20.3).

Heritability is a concept that applies to populations, not to individuals. For example, the heritability of egg weight in the domestic chicken is 0.6. This does not mean that 60 percent of the weight is determined by the genes and 40 percent by the environment. It means that 60 percent of the total variation in egg weights among a group of chickens is due to genetic differences and the rest is due to environmental causes.

In economically important plants and animals, the heritabilities of different characteristics are estimated from careful breeding studies. In humans, of course, this cannot be done, but the heritabilities of some abnormalities have been estimated from the rate at which they occur among relatives of a person with the disease. Some examples of heritabilities that have been estimated in this manner are certain kinds of epilepsy (0.4), clubfoot (0.8), and harelip (0.7).

A better opportunity for the study of heritability in humans exists in the study of human twins. There are two kinds of twins: **monozygotic** (identical) and **dizygotic** (fraternal). Dizygotic twins are produced from two different zygotes and are therefore no more similar genetically than brothers and sisters of different ages. Monozygotic twins are produced from the same zygote and, as a result, are genetically identical. Usually, twins are raised in the same environment so that there is similarity in both the genetic and environmental components. But what happens when monozygotic twins are reared apart from each other? In these situations, the individuals have identical genetic components, but different environmental components. A comparison of the phenotypes of such twins provides an estimate of the relative genetic contribution to certain traits and therefore their heritability. Cases of identical twins raised separately, in different environments, are so rare, however, that progress is very slow in this research.

The heritability of human traits, such as intelligence, is very difficult to measure, and study of heritability in humans may sometimes lead to controversy.

Table 20.3

Examples of Heritability for Traits in Various Organisms (Estimated by Various Methods).

Trait	Heritability
Amount of white spotting in Frisian cattle	0.95
Slaughter weight in cattle	0.85
Plant height in corn	0.70
Root length in radishes	0.65
Egg weight in poultry	0.60
Thickness of back fat in pigs	0.55
Fleece weight in sheep	0.40
Ovarian response to gonadotropic hormone in rats	0.35
Milk production in cattle	0.30
Yield in corn	0.25
Egg production in poultry	0.20
Egg production in Drosophila	0.20
Ear length in corn	0.17
Litter size in mice	0.15
Conception rate in cattle	0.05

From Francisco Ayala and John Kiger, *Modern Genetics* (Menlo Park, Calif.: Benjamin-Cummings, 1980).

Summary

The modern science of genetics was founded by Gregor Mendel in the mid-1800s. From a series of experiments with garden peas, Mendel derived two basic laws of genetics that still are valid today.

Mendel's first law, the law of segregation, states that organisms contain two discrete factors (genes) for each characteristic and that these genes segregate so that a gamete has only one of the pair of alleles (forms of the gene) for each trait. Mendel's second law, the law of independent assortment, states that genes affecting different traits segregate independently of each other. These two laws can be used as a basis to predict the proportions of different types of offspring from a specified cross.

The types of genes present in an organism constitute that organism's genotype. Interaction of the genotype with the environment results in the outward appearance or phenotype of an individual.

Sometimes, one form of a gene, or allele, may mask the effect of another allele in the same organism. This effect, called dominance, means that organisms with different genotypes may have the same phenotype. In other cases, both alleles of a pair are clearly expressed in the phenotype, resulting in various types of intermediate inheritance. Modifier genes and epistatic genes sometimes affect the phenotypic expression of an entirely different gene. Pleiotropy is where single genes have multiple effects.

A number of genes, each having a small effect, can combine to form a particular characteristic. The study of this kind of inheritance is called quantitative genetics.

The application of genetic principles to human populations has resulted in genetic counseling services. Couples may, in many cases, obtain estimates of the risk of having offspring with certain genetic diseases.

Both genotype and environment contribute to an organism's phenotype. The relative contributions of each, however, are difficult to resolve.

Questions

1. Can you think of an experiment that would determine whether or not the differences between two populations of plants are due to genetic or strictly environmental influences? Remember that most plants can be propagated asexually and transplanted.

2. We have seen from the chapter on cell division that it is not the genes themselves that segregate during meiosis but the chromosomes on which the genes are located. If two genes were located next to each other on the same chromosome, do you think that they still would segregate independently? (We will discuss this question further in chapter 21.)

3. How many different alleles can be present for a single gene in a diploid individual? Is there any limit to the number of alleles that might be present in a population of individuals?

4. How many types of gametes may be formed in an organism with the genotype $A^1A^2B^1B^2C^1C^1$? A, B, and C are genes located on separate chromosomes.

5. In cattle, polled is dominant to horned. A polled bull is crossed with a horned cow. Of the four offspring, one is horned and the other three are polled. Is the bull homozygous or heterozygous for this gene?

6. In humans, assume that right-handedness is dominant to left-handedness. Explain how two right-handed people might have a left-handed child. If a right-handed person who had one left-handed parent married a left-handed person, what predictions could you make about their offspring?

7. Some genes are lethal when present in the homozygous condition. In chickens, when a gene known as creeper is present in the homozygous condition in a developing embryo, the spinal cord and vertebral column develop abnormally and the embryo dies inside the eggshell. When the creeper gene is present in the heterozygous condition, the chicken hatches, but it has skeletal abnormalities and walks with a peculiar creeping, stumbling gait. This set of characteristics gives the gene its name. Interpret the results of the following crosses:

Normal × Normal 96 normal
Normal × Creeper 51 normal, 48 creeper
Creeper × Creeper 54 creeper, 26 normal

8. In crosses between two crested ducks, only about three-quarters of the eggs hatch. The embryos in the remaining quarter of the eggs develop nearly to hatching and then die. Of the ducks that do hatch, about two-thirds are crested and one-third have no crest. What results would you expect from a cross between a crested and a noncrested duck?

9. A cross is made between two parents with genotypes *AaBB* and *aabb*. If there are thirty-two offspring, how many of them would be expected to exhibit both dominant characteristics?

10. In rabbits, black color is due to a dominant gene *B* and brown color to its recessive allele *b*. Short hair is dominant to the dominant gene *S* and long hair to its recessive allele *s*. In a cross between a homozygous black, long-haired rabbit and a brown, short-haired one, what would be the nature of the F_1 generation? Of the F_2 generation? If one of the F_1 rabbits was mated with a brown, long-haired rabbit, what kinds of offspring in what ratio would you expect?

11. In radish plants, the shape of the radish produced may be long, round, or oval. Crosses among plants that produced oval radishes yielded 121 plants that produced long radishes, 243 that produced oval radishes, and 119 that produced round radishes. What type of inheritance appears to be involved? What results would you expect from a long with long cross? A round with round cross?

12. A man of blood type A and a woman of blood type B produce a child of type O. What are the genotypes of the man, the woman, and the child? If the couple were to have other children, what possible blood types could be produced? Indicate genotypes as well as phenotypes.

13. Why do fewer parental phenotypes segregate out when several genes determine a particular characteristic?

14. Do you think the genetic risk to offspring of first cousin marriages is significant enough that such marriages should be legally discouraged? Justify your answer.

Suggested Readings

Books

Ayala, F. J., and Kiger, J. A., Jr. 1980. *Modern genetics*. Menlo Park, Calif.: Benjamin Cummings.

Crow, J. F. 1976. *Genetics notes*. 7th ed. Minneapolis: Burgess.

Gardner, E. J., and Snustad, D. P. 1981. *Principles of genetics*. 6th ed. New York: John Wiley.

King, R. C. 1974. *A dictionary of genetics*. 2d ed. London: Oxford University Press.

Singer, S. 1978. *Human genetics*. San Francisco: W. H. Freeman and Company.

Stern, C., and Sherwood, E. R. eds. 1966. *The origin of genetics: A Mendel source book*. San Francisco: W. H. Freeman and Company.

Sturtevant, A. H. 1965. *A history of genetics*. New York: Harper and Row.

Suzuki, D. T.; Griffiths, A. J. F.; and Lewontin, R. C. 1981. *An introduction to genetic analysis*. San Francisco: W. H. Freeman and Company.

Articles

Crow, J. F. February 1979. Genes that violate Mendel's rules. *Scientific American* (offprint 1418).

Fincham, J. R. S. 1971. Using fungi to study genetic recombination. *Carolina Biology Readers* no. 2. Burlington, N.C.: Carolina Biological Supply Co.

Chromosomes and Genes

Mendel's law of independent assortment states that genes segregate independently of each other during the formation of gametes. This, however, is not true of all genes. Mendel thought of genes as discrete factors within the cell. Actually, genes are organized into linear arrays on chromosomes. The fact that genes are located at specific sites on chromosomes has a number of very interesting and significant genetic consequences.

Chromosomes

When Gregor Mendel published his results in 1866, it was known that all living things are made up of cells. It was also known that plant and animal cell nuclei contain rodlike structures called **chromosomes.** But Mendel did not know enough about the behavior of chromosomes in dividing cells to realize the significance of chromosomes for his theories in inheritance.

Genes on Chromosomes

When Mendel's work was rediscovered in 1900, a great deal more was known about the behavior of chromosomes in dividing cells. In 1902, two years after Mendel's work was rediscovered, two investigators, Walter S. Sutton of the United States and Theodor Boveri of Germany, suggested independently that the genetic material is contained in the chromosomes. This idea has developed into the **chromosome theory of heredity.**

Sutton and Boveri had several reasons for believing that Mendel's independent factors, or genes, were located on chromosomes. They knew that within the nucleus, chromosomes exist in pairs. This coincides with Mendel's theory that organisms have two copies of each gene. According to the chromosome theory of heredity, one copy of each gene is located on each one of a pair of chromosomes. Thus, an individual with a heterozygous genotype has one allele on one chromosome of a given pair and another allele on the other. It had also been noted that gametes have only one of each pair of chromosomes. The separation of pairs of chromosomes during meiosis explains how the alleles of a gene segregate and then recombine in the formation of zygotes. Later in this chapter, some of the experimental evidence—especially the work of Thomas Hunt

Morgan and his colleagues—that proved that genes are arranged in linear arrays on chromosomes is examined.

Chromosome Structure

Prokaryotic cells and eukaryotic cells differ in the manner in which their genetic material is organized into chromosomes. The bacterial chromosome consists of a circular strand of DNA associated in an as yet undetermined way with protein molecules. The eukaryotic chromosome is more complex, and details of its structure currently are being clarified.

The chromosomes in most eukaryotic cells, when they are condensed during cell division, are rod shaped and stain quite strongly with certain dyes. Biologists call the stainable fibrous material that makes up chromosomes **chromatin.** Chromatin, however, is not a single substance, but a complex, highly organized set of substances. A chromatin fiber consists of a long molecule of DNA associated with basic protein molecules called histones, along with quantities of acidic and neutral nonhistone proteins, as well as several kinds of enzymes that are active in DNA and RNA synthesis.

The DNA and proteins in the chromatin are organized in a very specific way. When chromatin is isolated, spread out, and examined with the electron microscope, it appears to consist of a series of "beads" that are about 10 nm in diameter and arranged in a chain (figure 21.1a). Each of these "beads" is a basic structural unit of a chromosome called a **nucleosome.** Each nucleosome is an aggregate of eight histone molecules (two each of four kinds) with a specific length of DNA wrapped in a helical coil around the protein aggregate (figure 21.1b). This coiled wrapping of DNA around the proteins represents a second level of helical coiling in the DNA molecule, since the fundamental structure of DNA consists of two strands coiled around one another in a double helix. The helical coiling of DNA around the proteins in the nucleosome is called a "superhelix," to distinguish it from the familiar fundamental double helix structure of the DNA molecule.

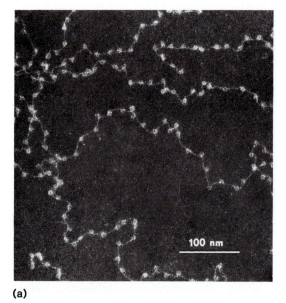

(a)

(b)

Figure 21.1 Nucleosomes. (a) This photograph shows uncoiled chromatin material from chicken red blood cells. The "beads" in the photograph are the nucleosomes and the "string" is the DNA between nucleosomes. (b) A drawing of a portion of a chromosome, showing the nucleosomes (spheres of aggregated histone molecules) around which the DNA (lines) is wrapped. The boxed area shows how the DNA is coiled around the nucleosomes and how it connects nucleosomes. The upper left portion of the diagram (outside the box) shows the coiled form in which the nucleosome chain sometimes is found. This coiled form seems to represent the condition of the chromatin in chromosomes when they are most condensed, as during metaphase of mitosis.

Nucleosome "beads" in chromatin appear to be held together by DNA which, in addition to coiling around each nucleosome, stretches from one nucleosome to the next. Finally, the nucleosome chain itself sometimes is found in a coil, called a solenoid, which is about 30 nm in diameter. It is not clear which, if either, of these arrangements represents the "native" state of chromatin inside a cell, nor is it clear how the arrangement of nucleosomes might change with changes in chromosome structure and function at different times in the cell cycle.

Number of Chromosomes

While the basic structural units in chromosomes, the nucleosomes, seem to be remarkably similar in all eukaryotic cells studied so far, there are major differences among the chromosome complements of various organisms. One of the most obvious differences is the number of chromosomes present.

Differences among Organisms

The number of chromosomes found in different organisms varies a great deal. The number ranges from 4 chromosomes found in one plant, *Haplopappus gracilis,* and in several insects and other invertebrate animals, to nearly 500 chromosomes in some ferns (table 21.1). The number of chromosomes present is usually, but not always, consistent within a species and is not correlated with either the size of the individual or the size of the cell, or with the evolutionary "advancement" of a species.

Polyploidy

Up to this point, this discussion has concerned mainly organisms that are diploid; that is, organisms that have two of each kind of chromosome. When an organism has more than two sets of chromosomes, the condition is called **polyploidy.** Polyploid organisms are named according to the number of sets of chromosomes that they have: triploids (3N) have three of each kind of chromosome, tetraploids (4N) have four sets of chromosomes, pentaploids (5N) have five sets, etc. Polyploids with uneven numbers of chromosome sets, such as triploids, are usually sterile because, with an uneven number, it is impossible to have an even distribution of chromosomes during meiosis.

Table 21.1
Chromosome Numbers of Some Common Plants and Animals.

Species	Common Name	Chromosome Number
Triticum aestivum	Wheat	42
Zea mays	Corn	20
Lactuca sativa	Lettuce	18
Lycopersicon esculentum	Tomato	12
Gossypium barbadense	New World cotton	52
Phaseolus vulgaris	Kidney beans	22
Glycine max	Soybeans	40
Pyrus communis	Apple	34
Ipomea batatus	Sweet potato	90
Ophioglossum vulgatum	Fern	480
Apis mellifera	Honeybee	16
Bombyx mori	Silkworm	56
Cyprinus carpio	Carp	100
Rana pipiens	Green frog	26
Alligator mississippiensis	Alligator	32
Anas platyrhynchos	Mallard duck	78
Bos taurus	Cattle	60
Equus caballus	Horse	64
Felis felis	Cat	38
Canis familiaris	Dog	78

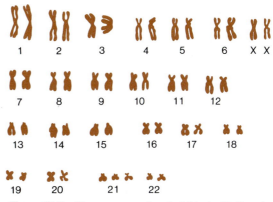

Figure 21.2 Chromosomes of an individual with Down's syndrome. Note the extra copy of chromosome no. 21.

Polyploidy is quite common in plants but relatively unusual in animals. This is because the polyploid condition usually causes relatively little physiological or developmental disturbance in plants, and polyploidy does not interfere with the reproductive process in plants as long as the number of chromosome sets is even (for example, 4N, 6N, 8N, etc.). In animals, where sex determination in many cases is based on a single pair of chromosomes, called sex chromosomes (which are covered later in the chapter), polyploidy disrupts normal reproductive processes. Many ornamental plants are bred as polyploids because they tend to have larger leaves and flowers. In some cases, triploids are specifically developed because these plants are sterile and will not put any energy into seed production. This may result in larger flowers or the development of a more edible seedless fruit. Because plants commonly can be propagated asexually from "cuttings" taken from plants with desirable characteristics, even plants with an uneven number of chromosome sets, which usually precludes sexual reproduction, can be used commercially.

Aneuploidy

While polyploidy involves the number of sets of chromosomes, **aneuploidy** is a chromosome number that varies from the normal number for the species by a small number of chromosomes. Usually, aneuploidy involves an excess or a deficiency of an individual chromosome in an otherwise normally diploid cell. If a diploid individual has one extra copy of a chromosome, the individual is called a **trisomic**. The condition itself is called **trisomy** and is designated as $2N+1$. If a diploid individual is missing one chromosome, the individual is called a **monosomic**. The condition is called **monosomy** and is indicated by $2N-1$.

As noted in chapter 19, aneuploidy usually arises because of nondisjunction of homologous chromosomes during cell division, and if it occurs during meiosis, it can have drastic consequences for development of a zygote produced as a result of fertilization involving a gamete with a chromosome excess or deficiency.

A well-known example of aneuploidy is a condition in humans known as **Down's syndrome.** This condition, sometimes referred to as "mongolism," results from the presence of an extra copy of the chromosome designated no. 21 (figure 21.2). The extra copy of the chromosome results from nondisjunction; either the duplicated pair of chromosomes fails to separate during the first division of meiosis or the chromatids of the duplicated chromosome fail to separate during the second division.

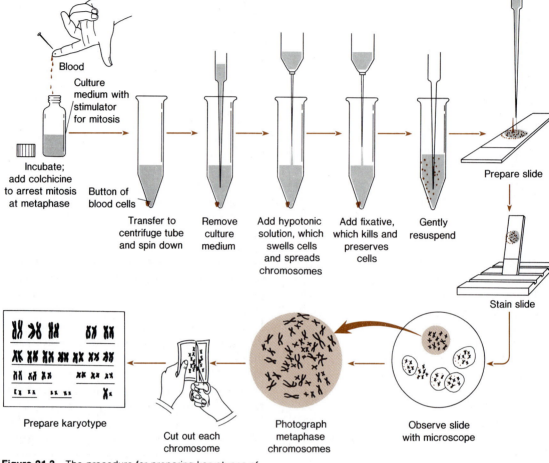

Figure 21.3 The procedure for preparing karyotypes of cultured human white blood cells.

Victims of Down's syndrome frequently have heart trouble and are almost always mentally retarded. Down's syndrome may be detected in early pregnancy by the use of a technique called **amniocentesis.** A long needle is inserted through the mother's abdomen into the uterus, and amniotic fluid is withdrawn. Cells from the amniotic fluid can then be cultured and examined for the presence of the extra no. 21 chromosome.

Human Chromosomes

Although it is now commonplace to see pictures of sets of human chromosomes such as that shown in figure 21.2, virtually all knowledge of human chromosomes has been obtained since the 1950s. Even the exact number of chromosomes in humans was not determined with precision until 1956. This seems extraordinary in light of

the fact that chromosomes in other organisms had been observed and described for over a hundred years. But the chromosomes of humans are quite difficult to prepare in a manner in which consistent chromosome counts can be made, and routine techniques for studying human chromosomes have been available for only a relatively short time.

Preparations of human chromosomes are usually made from certain types of white blood cells (lymphocytes). Figure 21.3 diagrams the procedure. When dividing cells are observed, metaphase chromosomes are photographed. The individual chromosomes are cut out of the photograph and lined up with one another in matching homologous pairs. This ordered arrangement of all the chromosomes in the nucleus is called a **karyotype.**

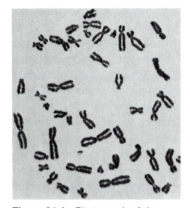

Figure 21.4 Photograph of the metaphase chromosomes of a human male.

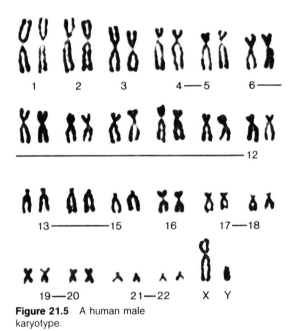

Figure 21.5 A human male karyotype.

Figure 21.4 is a photograph of the metaphase chromosomes of a human male. Figure 21.5 is a karyotype made from the chromosomes shown in such a photograph. Such karyotypes show the characteristic diploid chromosome number of human body cells, which is forty-six (twenty-three pairs). The last pair of chromosomes in the karyotype, labeled X and Y, are not identical. These are the **sex chromosomes.** The human male has the differing X and Y chromosomes in this pair, while the human female has two copies of the X chromosome.

The study of human chromosomes was greatly advanced in the 1970s with the introduction of new methods for chromosome staining. These new methods produce distinct patterns of dark and light stained bands. Figure 21.6 shows one such type of banding, referred to as G banding. The bands are called **G bands** because the stain used to prepare the chromosomes for examination is the Giemsa reagent. The visualization of this type of banding pattern is extremely important because it permits reliable identification of individual chromosomes. Many of the chromosomes are otherwise very similar and can be sorted only into groups with common general appearance when routine karyotyping procedures are used without Giemsa staining (see figure 21.5). Such staining also allows for the ready identification of structural abnormalities in the chromosomes and helps to detect changes in the number of chromosomes. The mapping of the locations of particular genes (p. 582) on specific chromosomes also is aided by the presence of G bands.

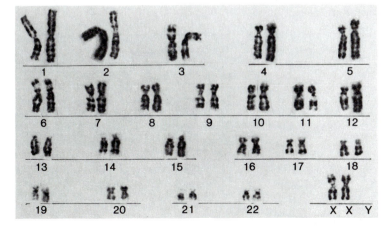

Figure 21.6 A karyotype of a normal human female, showing the different G-band patterns that permit reliable sorting of individual chromosomes. Note that the sex chromosome pair in females consists of two identical X chromosomes.

Sex Chromosomes

In the karyotype in figure 21.5, it can be seen that the human male has an X chromosome and a Y chromosome, and twenty-two other pairs of chromosomes. The karyotype in figure 21.6 shows that the genetic makeup of the human female differs from the human male only in that the female has a pair of X chromosomes and no Y chromosome. The X and Y chromosomes are called sex chromosomes. The chromosomes that are common to both males and females are called **autosomes.**

Sex Determination

The pair of sex chromosomes segregates at meiosis so that each gamete receives just one of the pair of sex chromosomes. In humans, each egg has an X chromosome, but a sperm cell can have either an X chromosome or a Y chromosome (figure 21.7). Therefore, the sex of the new individual is determined at the time of fertilization, and it depends on the type of sex chromosome present in the sperm. If the sperm cell has a Y chromosome, the zygote will become a male. If the sperm cell has an X chromosome, the zygote will become a female.

Many animals have the same sex determination system as humans. An organism with two Xs is a female, and an organism with an X and a Y is a male. But different sex determination patterns have also been observed. In grasshoppers there is no Y chromosome, so there are XX females and X– males. In birds, moths, and butterflies, the male has two X chromosomes, while the female has only one X chromosome. The Y chromosome may or may not be present, depending on the species. In ants, wasps, and bees, there is yet another pattern. Some eggs develop parthenogenetically (without fertilization) into males, while the fertilized eggs develop into females. Since there is no mechanism for return to the diploid number associated with parthenogenesis in these animals, as there is in some other animals that reproduce parthenogenetically, all cells in males are haploid (N), and all cells in females are diploid (2N). The development of different types of females, such as queens and workers in bees, is due to differences in nutrition, not to differences in chromosome complement.

Sex Linkage

Genes occupy specific sites on chromosomes, and the specific part of a chromosome that is the site of a particular gene is known as its **locus** (plural: **loci**). During this century, **chromosome mapping** studies (the determination of relative positions [loci] of individual genes on specific chromosomes) have yielded extensive maps of gene loci on chromosomes of many organisms, including humans. All of this work on determining the positions of gene loci on chromosomes began with a discovery that involved sex chromosomes.

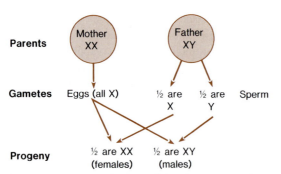

Figure 21.7   The segregation of sex chromosomes. The number of male zygotes produced should equal the number of female zygotes.

Box 21.1
Sex Ratios

The **sex ratio,** the ratio of males to females among offspring produced, is approximately one in many organisms with separate sexes. This is because of the segregation of the sex chromosomes in the organism with the XY genotype. The two kinds of gametes, X and Y, should be produced in equal proportions, and therefore, the sex ratio of the offspring should be roughly 1:1 (see figure 21.7).

The sex ratio in humans, however, shows that the number of males does not equal the number of females. Among caucasians in the United States, there are approximately 106 boys born for every 100 girls. This ratio, the ratio at the time of birth, is the secondary sex ratio. The sex ratio at the time of fertilization is the primary sex ratio. If there is a different survival rate between fertilization and birth for the two sexes, the primary and the secondary sex ratios may be different. Is the higher number of males at birth due to a higher death rate among female embyros? This apparently is not the case. Studies of unborn fetuses actually show an even higher ratio of males to females. In fact, it is estimated that the human primary sex ratio is about 114 males to 100 females.

It has been suggested that the sperm cells bearing the lighter Y chromosome are able to swim faster than those bearing the X chromosome, and that for this reason they are able to reach the egg sooner. Another possibility is that the physiological environment of the female reproductive tract favors the survival of more Y-bearing sperm than X-bearing sperm. The actual reason for the slightly distorted sex ratio in humans, however, remains unknown.

Even though there are more human males than females at birth, this ratio ultimately reverses itself because more males than females die at every age. Females begin to exceed males as early as age eighteen in some countries or as late as age fifty-five in others. The reasons for this differential survival also are unknown.

The sex ratio conceivably could be altered if parents were given the option of choosing the sex of an offspring. The technical methods of determining the sex of a fetus already exist. The sex of a fetus can be determined by amniocentesis followed by karyotyping. Selective abortion at this point could alter the sex ratio, although this approach is not likely to become either widely accepted because of the grave ethical problems that it raises, or readily available because of its expense. However, animal breeders now are able to separate X and Y sperm outside the body of the animal; either type can then be introduced into a female by artificial insemination. In years to come, many ethical and social problems could be raised by the introduction of parental choice regarding sex of human offspring.

Thomas Hunt Morgan and his colleagues at Columbia University were the first investigators to associate a specific gene with a specific chromosome when they published the results of the following experiment in 1910. In one of their many genetic experiments on the fruit fly (*Drosophila melanogaster*), they crossed female flies that had red eyes with males that had white eyes (figure 21.8). In the F₁ generation, all the flies had red eyes. This would be expected if red eyes were dominant to white. When crosses were made between members of the F₁ generation, a ratio of 75 percent flies with red eyes to 25 percent with white eyes was expected. The F₂ generation did have 3,470 flies with red eyes and 782 with white eyes, roughly a 4:1 ratio, a ratio not totally deviant from the expected ratio of eye colors. The experimenters noticed, however, that all the flies with white eyes were males, while there were both males and females among the red-eyed flies.

Because of this seemingly unusual result, further experiments were conducted. The experimenters crossed white-eyed males and red-eyed females from the F₂ generation with the expectation that possibly once again all of the white-eyed offspring would be male and the red-eyed offspring would be a mixture of males and females. What they found, however, was a mixture of red-eyed males and females and white-eyed males and females.

How could sex differences be taken into account in interpreting these results? The researchers knew that *Drosophila* males have an X and a Y chromosome, while the females have two X chromosomes. They therefore concluded that the gene for eye color is located on the X chromosome and that the Y chromosome bears no gene for eye color. Thus, female fruit flies have two genes for eye color while males have only one. Figure 21.9 illustrates this explanation of Morgan's results.

Any gene that, like the genes for red and white eye color in *Drosophila,* occurs on a sex chromosome, is called a **sex-linked gene.** The vast majority of sex-linked genes occur on the X chromosome, as opposed to the Y chromosome. In humans, for instance, very few genes occur on the Y chromosome. The most frequently cited example is a Y-chromosome gene with an allele

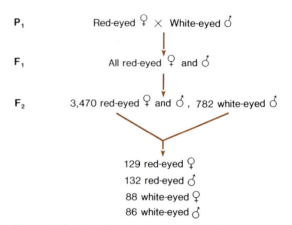

Figure 21.8 T. H. Morgan's experiments with *Drosophila* eye color.

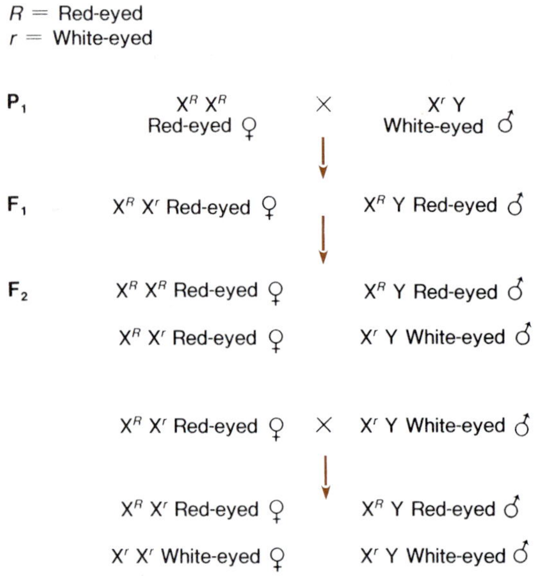

R = Red-eyed
r = White-eyed

Figure 21.9 How Morgan explained his results. The gene for eye color is found on the X chromosome and is therefore sex linked (or X-linked). The Y chromosome does not carry a gene for eye color.

for hairy ear rims (figure 21.10). Genetically, a gene on the Y chromosome is easy to detect because if the father has a particular allele, all of the sons and none of the daughters have it.

When analyzing any cross involving a sex-linked gene, it is important to keep track of the X and Y chromosomes and to treat the sexes separately (figure 21.11). In humans (and in *Drosophila* and other animals with the same sex chromosome pattern), all male offspring receive the Y chromosome from their father and thus

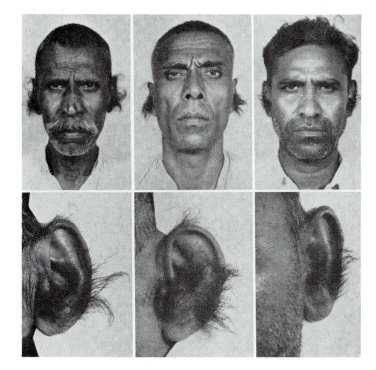

Figure 21.10 These three brothers have hairy ear rims. The gene for this condition is located on the Y chromosome.

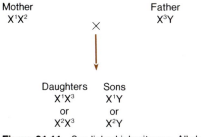

Mother X^1X^2 ×   Father X^3Y

Daughters   Sons
X^1X^3 X^1Y
or   or
X^2X^3 X^2Y

Figure 21.11 Sex-linked inheritance. All daughters receive the same X chromosome (X^3) from the father and either of two X chromosomes (X^1 or X^2) from the mother. All sons receive the Y chromosome from the father and one of the two X chromosomes from the mother.

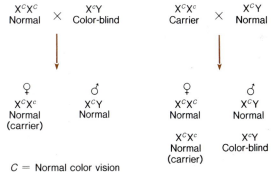

X^CX^C × X^cY
Normal   Color-blind

X^CX^c × X^CY
Carrier   Normal

♀   ♂
X^CX^c X^CY
Normal   Normal
(carrier)

♀   ♂
X^CX^C X^CY
Normal   Normal

X^CX^c X^cY
Normal   Color-blind
(carrier)

C = Normal color vision
c = Red-green color-blind

Figure 21.12 Color vision in humans as an example of sex-linked inheritance.

cannot receive any gene located on the father's X chromosome. All female offspring receive the father's X chromosome and one of the two maternal X chromosomes.

One example of a sex-linked gene in humans is the one responsible for the common red-green color blindness. The gene for this condition is carried on the X chromosome. The allele for color blindness is recessive to the allele for normal vision. If a male receives an X chromosome from his mother that has the allele for color blindness,

he will be color-blind because the Y chromosome does not carry an allele for normal vision. A female, because she has two X chromosomes, may be heterozygous for the alleles. She would have normal vision but would be a "carrier" of the allele for color blindness. Figure 21.12 shows a cross between a color-blind male and a female that is homozygous for normal vision. None of the male offspring can possibly inherit the allele for color blindness. All of the female offspring

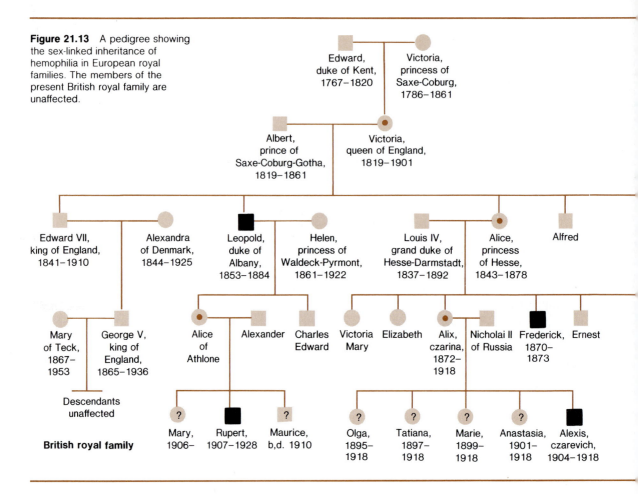

Figure 21.13 A pedigree showing the sex-linked inheritance of hemophilia in European royal families. The members of the present British royal family are unaffected.

are carriers because they all get the X chromosome with the allele for color blindness from their father and one of the two X chromosomes with the normal allele from their mother. If one of the carrier daughters marries a normal male, approximately half of the male offspring will have red-green color blindness. None of the female offspring will be color-blind, although they may be carriers. This is because half the males receive the allele for color blindness from their mother, and all of the females get a normal allele from their father. Again, the important points in working any problem concerning sex linkage are to diagram the crosses in terms of X and Y chromosomes and to look at the phenotypic ratios in both sexes in the offspring.

Another interesting example of a sex-linked trait in humans is a recessive allele that results in hemophilia, "bleeder's disease." This disease has been found in royal families in Europe that are related to Queen Victoria of Great Britain,

and it has resulted in or contributed to the early death of male members of several royal families (figure 21.13).

Sex Influence

Some genes not located on the X or Y chromosomes are expressed differently in the two sexes, due probably to the hormonal differences between the sexes. This is called **sex influence.** An example of a sex-influenced characteristic in humans is pattern baldness, which is far more common among men than women. The allele for pattern baldness is dominant in males but recessive in females (figure 21.14). Another example of sex-influenced inheritance in humans involves the comparative lengths of the index finger and the fourth finger of the hand. The index finger being equal to or longer than the fourth finger is dominant in females and recessive in males (figure 21.15).

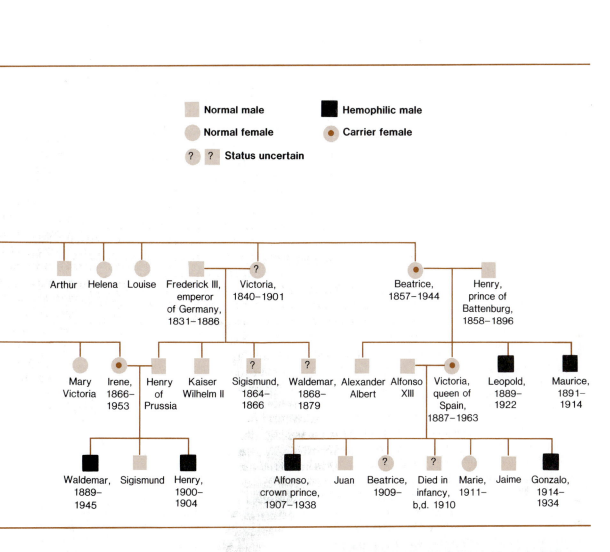

Normal male

Normal female

? ? Status uncertain

Hemophilic male

Carrier female

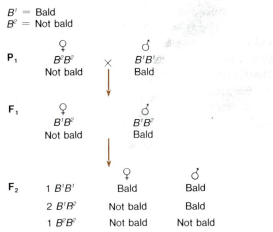

B^1 = Bald
B^2 = Not bald

	♀		♂
P_1	B^2B^2	×	B^1B^1
	Not bald		Bald

	♀		♂
F_1	B^1B^2		B^1B^2
	Not bald		Bald

		♀	♂
F_2	1 B^1B^1	Bald	Bald
	2 B^1B^2	Not bald	Bald
	1 B^2B^2	Not bald	Not bald

Figure 21.14 The inheritance of pattern baldness shows that it is a sex-influenced characteristic.

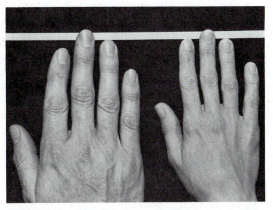

Figure 21.15 Another sex-influenced trait is the relative lengths of the index finger and fourth finger of the hand. Both individuals in the photographs are heterozygous for the trait. The hand on the right is that of a female, and the hand on the left is that of a male.

Chromosomal Abnormalities

Changes in chromosomal structure can have important genetic consequences. Heritable changes in the genetic material are called **mutations,** and changes in chromosome structure are called **chromosomal mutations** or **macromutations** to distinguish them from gene mutations, which are covered in chapter 22. Various types of chromosomal mutations are diagrammed in figure 21.16.

A **deletion** occurs when a segment of a chromosome is lost. Deletions usually are lethal when homozygous, although exceptions have been detected in corn and some other organisms. Even when heterozygous, a deletion often is expressed phenotypically. An example in humans is the **cri-du-chat syndrome** (cat's cry syndrome). This syndrome results from a deletion in the short arm of chromosome no. 5. Infants who are heterozygous for this deletion have a high-pitched mewing cry and usually have severe growth abnormalities and mental retardation.

A **duplication** is the presence of a chromosome segment more than once in the same chromosome. An example of the phenotypic expression of a duplication is the bar-eyed phenotype in *Drosophila*. This duplication reduces the number of facets in the insect's eye and results in development of small, bar-shaped eyes instead of normal round eyes.

Both deletions and duplications may arise following chromosome breakage. Chromosomes may be broken by radiation, various chemicals, or even viruses, and the way in which chromosome fragments rejoin or fail to rejoin following breakage results in the deletions or duplications.

An **inversion** is when a segment of a chromosome is turned around 180°. Inversions cause reproductive problems because lethal chromosome disorganizations often are produced by meiotic crossing over when the organism is heterozygous for the inversion.

Translocations result from the actual interchange of blocks of genes between two nonhomologous chromosomes. Translocation heterozygotes usually have reduced fertility, again due to problems with pairing during meiosis.

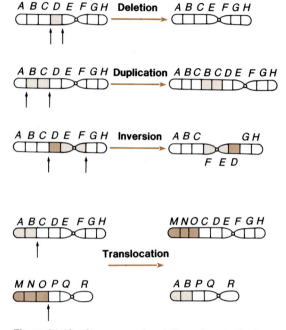

Figure 21.16 Chromosomal mutations. Arrows show points of chromosome breakage, and displaced chromosome fragments are colored. In deletions, the broken chromosome fragment does not reattach and is lost. In duplication, a broken segment from one chromosome attaches to its homologous chromosome. An inversion involves breakage and reattachment to the same chromosome in a reversed position. A translocation is a transfer of a chromosome fragment to a nonhomologous chromosome. Radiation damage to cells increases the frequency of these problems because radiation causes greatly increased chromosome breakage.

Chromosome Number Problems in Humans

Down's syndrome has already been discussed as an example of the effects of aneuploidy in humans. There are several other diseases that result from aneuploidy in the autosomes. For example, Patau's syndrome is due to an extra copy of chromosome no. 13. This syndrome is characterized by harelip and cleft palate, as well as other serious defects. Most infants with Patau's syndrome die during the first three months of life. Edward's syndrome results when an infant has an extra copy of chromosome no. 18. This disease is associated with many malformations of organ systems, and affected infants live an average of only six months.

There also are a number of abnormalities resulting from aneuploidy in the sex chromosomes. An individual with **Turner's syndrome** has only one X and no Y or second X chromosome.

People with this condition are sterile females with no ovaries and little development of the secondary sex characteristics. Usually, mental deficiency is not associated with this syndrome.

Metafemales have three or more X chromosomes. These females have only limited fertility and usually are mentally retarded.

An individual with **Klinefelter's syndrome** has a Y chromosome with two or more X chromosomes. Individuals with this chromosomal condition are sterile males with some tendency toward femaleness.

These human chromosomal abnormality studies provided important insights into the mechanism of sex determination in humans. Sex determination in *Drosophila* had been studied thoroughly during the early years of genetics before human chromosome studies began. Results of those studies indicated that sex is essentially determined in *Drosophila* by the number of X chromosomes present, whether or not a Y chromosome is present. Presence of only one X chromosome produces a male. Thus, either an XY or an XO (a lone X chromosome with no other sex chromosome) genotype produces a male. A *Drosophila* individual with an XXY genotype becomes an essentially normal female. These results indicate that the Y chromosome is a basically neutral factor in *Drosophila* sex determination.

When chromosomal studies began on human sex determination, it was assumed that the chromosomal mechanism of human sex determination might be similar to the *Drosophila* model because both have the X-Y sex chromosome systems. But this clearly is not the case. Turner's syndrome subjects have an XO genotype and an at least partly female phenotype. Klinefelter's syndrome subjects have an XXY genotype and a basically male phenotype. Clearly the Y chromosome is not a neutral factor in human sex determination, as it seems to be in *Drosophila*. Rather, it is a powerful male-producing factor. Among the few genes that the Y chromosome bears are genes whose expression results in development of the male phenotype.

Linkage and Chromosomal Mapping

During meiosis, the chromosomes assort at random into the gametes. The gametes then combine to form the zygotes of the next generation. Mendel's law of independent assortment states that each gene segregates independently of other genes. At the time, Mendel assumed that genes behaved as independent particles in cells. Genes, however, are not independent particles, but parts of linear arrays on chromosomes. It is actually the chromosomes that segregate independently during meiosis. Thus, if two gene loci are close together on the same chromosome, they tend to assort together during meiosis, and they do not obey Mendel's law of independent assortment. This tendency of certain genes to assort together is called **linkage.**

Linkage

An example of linkage is given in figure 21.17. If a test cross is made between an individual heterozygous at each of two genes and another individual that is homozygous recessive for both of those genes, each of the four possible genotypes is expected in the offspring in equal ratios. But in the example shown in figure 21.17, there is a strong deviation from this expected ratio.

This deviation from Mendelian expectations may be explained by assuming that the two gene loci involved are on the same chromosome and thus cannot assort independently of one another. That is, the genes are linked. But how can the particular ratio of genotypes found among the offspring be accounted for? The answer is that even when gene loci are linked because they are located on the same chromosome, linkage is not demonstrated 100 percent of the time because of the chromosome fragment exchange that occurs as a result of crossing over during meiosis.

AaBb × aabb

Genotypes	Expected	Observed
AaBb	25%	42%
Aabb	25%	8%
aaBb	25%	8%
aabb	25%	42%

Figure 21.17 Example of the kind of deviation from Mendelian predictions that might be seen when genes A and B are close together on the same chromosome. Genotypes are recognizable in this experiment since any offspring showing dominant phenotype must be heterozygous.

Box 21.2
Active Y Chromosomes and Inactive X Chromosomes:

The Functions of Sex Chromosomes in Body Cells

Although the Y chromosome in mammals carries few genes, it has a powerful effect on sex determination. One intriguing gene thought to be borne on Y chromosomes is identified as H-Y$^+$. If the H-Y$^+$ gene is not actually borne on the Y chromosome, possibly it is a gene on another chromosome whose expression is regulated by a gene that is located on the Y chromosome. In any event, this gene directs synthesis of a specific kind of protein molecule, **H-Y antigen,** which is present in the membranes of virtually all cells in male mammals but not in those of females. The H-Y antigen may play a role in determining that the gonads of a developing embryo become testes rather than ovaries. Once that determination is made, hormone differences cause other body parts to develop in a male direction (see chapter 25 for details of the process). In addition to a possible role in that first commitment to maleness in the developing embryo, the presence of the H-Y antigen may make male cells different from female cells for a lifetime, in ways that are as yet undetermined.

Another question about sex chromosomes that has interested biologists for some time is how males manage with only one "dose" of the genes present on the X chromosomes (sex-linked genes), while females apparently have a double "dose" of such X-linked genes. That question turned out to have a rather unexpected answer. Apparently, cells in females also have only one functional X chromosome. How does this come about?

Years ago, M. L. Barr observed a consistent difference between nondividing cells from female and male mammals, including humans. Females have a small, darkly staining mass of condensed chromatin present in their nuclei (box figure 21.2A). Males have no comparable spot of chromatin in their nuclei. This darkly staining spot in female nuclei is now called the **Barr body** or **sex chromatin body.**

In 1961, Mary Lyon, a British geneticist, proposed that the Barr body represents a condensed, inactive X chromosome whose genes are not expressed in cells of females, and that only genes on the other X chromosome are expressed. She further proposed that X chromosome inactivation occurs sometime during development and that once inactivated in a cell, a given X chromosome remains inactive in that cell and in all of its descendants.

This **Lyon hypothesis,** as it is called, has been tested and found to be valid. Early in development, X chromosomes are inactivated in cells of female embryos, but the inactivation is not a consistent inactivation of a single one of the two X chromosomes. That is, one X chromosome is inactivated in some cells while the other is inactivated in others. As Mary Lyon proposed, once inactivation of a particular X chromosome occurs in an embryonic cell, that X chromosome forms a Barr body in each and every cell descended from the original cell. Because many sex-linked genes are heterozygous in any given individual, this X chromosome inactivation has interesting genetic consequences for body cells. Some cells will have one allele of a gene being expressed while other cells will have another allele being expressed. The female body, therefore, is a mosaic with "patches" of genetically different cells. It has been demonstrated, for example, that blood cells from heterozygous females belong to two separate subpopulations that produce either one or the other of two different forms of certain enzymes, depending on which allele of sex-linked genes is being expressed. Cells from skin and other tissues have also been shown to be divided into two subpopulations with regard to expression of sex-linked genes. Thus, it now seems quite certain that female mammals, including human females, are genetic mosaics with reference to expression of sex-linked genes.

Having a mixture of cells with different X chromosomes inactivated allows heterozygous females to escape the effects of certain harmful, or even lethal, genes. Although harmful alleles may be the only ones expressed in some body cells, normal alleles are expressed in other body cells and "cancel out" the effects of the harmful alleles.

The connection between inactivated X chromosomes and Barr bodies also has implications for genetic testing. Human cells have one less Barr body than the number of X chromosomes they contain. Thus, normal males have no Barr bodies. Normal females have one. Metafemales may have two or more Barr bodies because they have three or more X chromosomes. This correlation between the numbers of X chromosomes and Barr bodies is useful in genetic screening because it provides a simple and direct means of detecting possible sex chromosome problems. Any suspected problems can then be pursued by preparation of a full karyotype.

Sex chromatin (Barr body) tests are sometimes used in other contexts as well. For example, they are sometimes applied to entrants in Olympic and other sporting events that are open only to one sex.

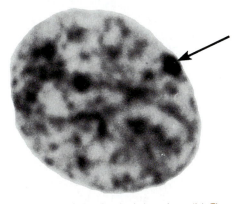

Box figure 21.2A The Barr body (sex chromatin). The inactivated X chromosome appears as a dense, darkly staining mass known as a Barr body in the nucleus of a cell from a female mammal. This nucleus is in a cell taken from the lining of a normal woman's mouth.

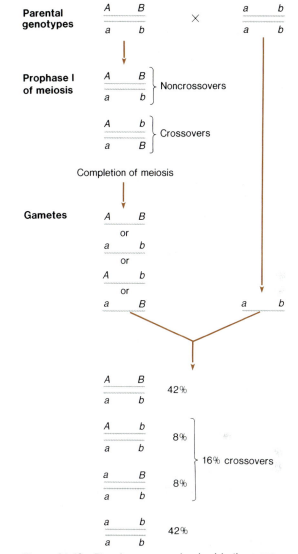

Figure 21.18 The chromosomes involved in the cross in figure 21.17. The linkage between genes A and B explains the observed ratio in the offspring.

Thus, both linkage and crossing over are involved. A diagrammatic representation of the chromosomes and alleles (figure 21.18) shows what happens. The homozygous recessive parent has the *a* and *b* allele on each chromosome of the pair. All gametes produced by that parent would contain a chromosome with *a* and *b* whether or not crossing over occurred. Given the ratios of genotypes in the offspring (figure 21.17), most of the gametes from the other parent had either *A* or *B* on the same chromosome, or *a* and *b* on the same chromosome. These would represent gametes resulting from a meiotic sequence during which no crossing over occurred between the two loci.

Cross-over events produce gametes in which *A* and *b* are on the same chromosome and *a* and *B* are on the homologous chromosome. The results show that a combined total of 16 percent of the offspring resulted from gametes formed after crossing over between the two genes in the heterozygous parent. The other 84 percent were formed by gametes resulting from meiosis during which no crossing over occurred between the two genes. When two or more gene loci are close enough on the same chromosome so that there is little crossing over between them, as the loci in this example are, Mendel's law of independent assortment does not hold.

Gregor Mendel reported experimental crosses involving seven characters of pea plants, and the genes determining each of these characters assorted independently of the genes for each of the other six characters. Independent assortment occurs optimally only if the gene pairs being studied are on separate chromosomes. This makes Mendel's reported results rather amazing because the pea *Pisum sativum* has seven pairs of chromosomes, and the gene pairs that he reported studying do occur one gene pair on each chromosome pair. This raises an interesting question about Mendel and his work. Was Mendel almost incredibly lucky in choosing seven characters to be studied genetically, or did he study other characters, which did not assort independently, and then choose to ignore them when he reported his results?

Chromosomal Mapping

The study of linkage and crossing over has been used to prepare **chromosome maps,** which are schematic representations of the relative positions of various gene loci on chromosomes.

In synapsis during meiosis, chromatids are breaking and rejoining all up and down their length. Sometimes broken fragments rejoin the same chromatid; sometimes they cross over to join a chromatid of the homologous chromosome in exchange for a broken piece of that other chromatid. The further apart two loci are, the greater the chance that one of these breaks will occur between them. This also means that the amount of crossing over shown between loci is proportional to the distance between them. That is, crossing over occurs more frequently between loci that are farther apart along the chromosome than between those that are closer together.

Working under these assumptions, it is possible to determine the arrangement of the different genes on a chromosome by examining their crossing over frequencies. For example, if there is a 10 percent crossing over between genes *C* and *A*, 27 percent between *C* and *B*, and 17 percent between *B* and *A*, there is only one possible arrangement of genes along the chromosome (figure 21.19). If another gene, *D*, is found to have 12 percent crossing over with *B* and 29 percent with *A*, its location can be added to the map (figure 21.20). By this method, a map of the entire chromosome can be made.

The distances between the genes along the chromosome are designated in terms of **map units.** One map unit is equal to 1 percent crossing over. Therefore, in the example, *C* is ten map units from *A*, *A* is seventeen map units from *B*, *C* is thirty-nine map units from *D*, etc.

The chromosome map (also called a linkage map) in figure 21.21 shows the relative locations and distances of many of the known genes on the chromosomes of the fruit fly *Drosophila melanogaster*.

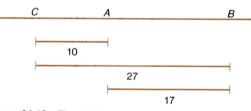

Figure 21.19 The relative positions of genes A, B, and C as determined by the analysis of the crossover percentages. The numbers given are the map units.

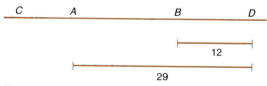

Figure 21.20 Another gene, D, is added to the map in figure 21.19.

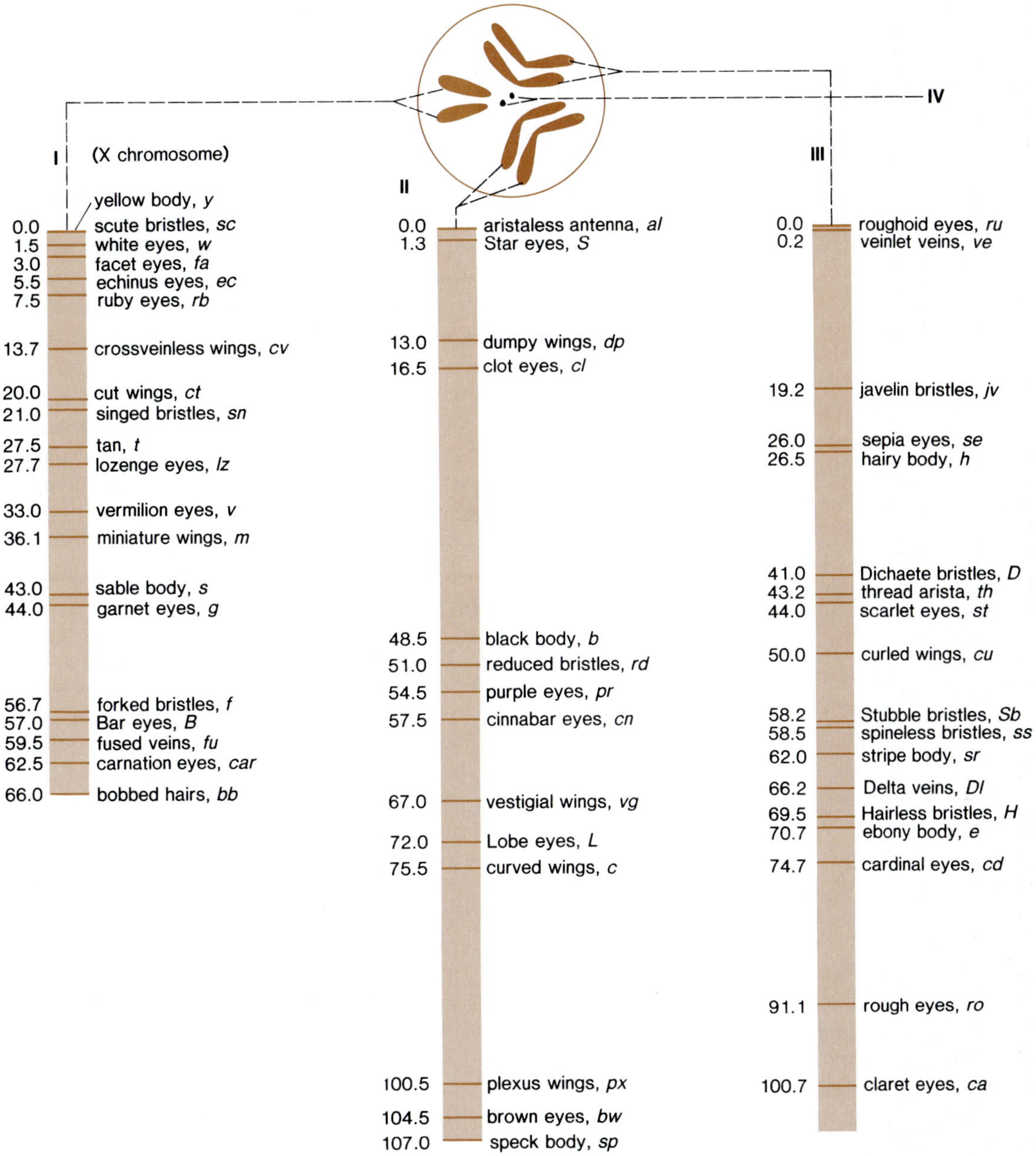

I (X chromosome)

	yellow body, *y*
0.0	scute bristles, *sc*
1.5	white eyes, *w*
3.0	facet eyes, *fa*
5.5	echinus eyes, *ec*
7.5	ruby eyes, *rb*
13.7	crossveinless wings, *cv*
20.0	cut wings, *ct*
21.0	singed bristles, *sn*
27.5	tan, *t*
27.7	lozenge eyes, *lz*
33.0	vermilion eyes, *v*
36.1	miniature wings, *m*
43.0	sable body, *s*
44.0	garnet eyes, *g*
56.7	forked bristles, *f*
57.0	Bar eyes, *B*
59.5	fused veins, *fu*
62.5	carnation eyes, *car*
66.0	bobbed hairs, *bb*

II

0.0	aristaless antenna, *al*
1.3	Star eyes, *S*
13.0	dumpy wings, *dp*
16.5	clot eyes, *cl*
48.5	black body, *b*
51.0	reduced bristles, *rd*
54.5	purple eyes, *pr*
57.5	cinnabar eyes, *cn*
67.0	vestigial wings, *vg*
72.0	Lobe eyes, *L*
75.5	curved wings, *c*
100.5	plexus wings, *px*
104.5	brown eyes, *bw*
107.0	speck body, *sp*

III

0.0	roughoid eyes, *ru*
0.2	veinlet veins, *ve*
19.2	javelin bristles, *jv*
26.0	sepia eyes, *se*
26.5	hairy body, *h*
41.0	Dichaete bristles, *D*
43.2	thread arista, *th*
44.0	scarlet eyes, *st*
50.0	curled wings, *cu*
58.2	Stubble bristles, *Sb*
58.5	spineless bristles, *ss*
62.0	stripe body, *sr*
66.2	Delta veins, *Dl*
69.5	Hairless bristles, *H*
70.7	ebony body, *e*
74.7	cardinal eyes, *cd*
91.1	rough eyes, *ro*
100.7	claret eyes, *ca*

IV

Figure 21.21 A chromosome map showing three of the four linkage groups of *Drosophila melanogaster*. The map of the tiny fourth chromosome is not included. The linkage groups are connected with the appropriate chromosomes (inset) by dashed lines. By convention, one end of each chromosome is designated as the zero end. Crossover map units are shown to the left of each map, and names and genetic symbols of various mutant alleles are shown to the right. When an entire chromosome is mapped, map units are determined from the gene closest to one end of the chromosome to the gene adjacent to it. Then the map units between the second gene and a third gene are determined, and so forth. In constructing a chromosome map, larger values are given as sums of shorter intervals, but there actually cannot be more than 50 percent crossing over between any given pair of genes on a single chromosome.

Extranuclear Inheritance

This chapter has concentrated on the relationships between chromosomes and the genes that they bear. But it is important to point out that this nuclear genetic material, organized in linear arrays of gene loci on chromosomes, does not represent the only repository of genetic information in eukaryotic cells.

Genetic information located outside the nucleus can affect the phenotype of a cell or an organism. For example, mitochondria contain DNA and functioning genes as do chloroplasts in plant cells. These genes direct synthesis of organelle-specific polypeptides, which are enzymes and constituents of membranes of the organelles. Mapping studies indicate that these extranuclear genes usually are part of a single circular molecule of DNA in each organelle.

The inheritance patterns of these genes differ from normal Mendelian inheritance. During the formation of gametes, so little cytoplasm is included in the male gametes, such as sperm cells, that most extranuclear genes, the genes contained within the mitochondria and chloroplasts, are transmitted through the female parent.

Summary

Genes are located in linear arrays on chromosomes.

Species differ widely in the number of chromosomes that they contain. The number of chromosomes in an organism may be increased or decreased. In polyploidy, the increase or decrease is in terms of entire sets of chromosomes. In aneuploidy, the increase or decrease is in terms of individual chromosomes.

Many organisms have sex chromosomes. In humans, an individual with two X chromosomes is a female, and an individual with an X chromosome and a Y chromosome is a male. Genes on these chromosomes are called sex-linked genes, and their patterns of inheritance are different than the patterns of inheritance of genes on autosomes. Sex-influenced inheritance is the differential expression of autosomal genes due to hormonal differences between the sexes, with one allele dominant in one sex and the other allele dominant in the other.

Changes in chromosomal structure include deletions, which occur when a segment of a chromosome is lost; duplications, which occur when a chromosome is repeated; inversions, which involve a reversal of part of a chromosome; and translocations, which involve an exchange of material between nonhomologous chromosomes.

Genes that are close together on the same chromosome deviate from Mendel's law of independent assortment. The amount of crossing over between any two genes on the same chromosome is proportional to the distance between them. This principle can be used to map genes on chromosomes.

Genes also are present in circular pieces of DNA in mitochondria and chloroplasts. These genes are usually inherited through the female parent.

Questions

1. Why are fewer instances of polyploidy found in organisms with sex chromosomes?
2. How does the visualization of bands in human chromosomes assist in the genetic mapping of these chromosomes?
3. If a heterozygous red-eyed female fruit fly is mated with a white-eyed male, what kinds of offspring will be produced and in what ratio?
4. Two normal-visioned parents produce a red-green color-blind son. What are the genotypes of the parents?
5. In cats, the genotype *BB* is yellow, *Bb* is calico, and *bb* is black. The gene is on the X chromosome. If a calico female is crossed with a black male, what types and frequencies of offspring would be expected? Could there ever be a calico male? Explain.
6. What is the difference between sex-influenced inheritance and sex linkage?
7. Why do you suppose that some chromosomal mutations cause difficulties during prophase I of meiosis?
8. How many Barr bodies would you expect to observe in the cells of individuals with each of these genotypes: XO, XYY, XXX, XXXY?

9. If there is 3 percent crossing over between genes 1 and 2, 12 percent crossing over between genes 2 and 3, and 15 percent crossing over between genes 1 and 3, what are the relative positions of these three genes along the chromosome?

10. Why are the extranuclear genes inherited almost exclusively from the female parent?

Suggested Readings

Books

Lewis, W. H. ed. 1980. *Polyploidy, biological relevance.* New York: Plenum Press.

Mange, A. P., and Mange, E. J. 1980. *Genetics: Human aspects.* Philadelphia: Saunders College.

Mittwoch, U. 1967. *Sex chromosomes.* New York: Academic Press.

Moore, D. M. 1976. *Plant cytogenetics.* New York: John Wiley.

Sutton, H. E. 1980. *An introduction to human genetics.* 3d ed. New York: Holt, Rinehart & Winston.

Articles

Fuchs, F. June 1980. Genetic amniocentesis. *Scientific American* (offprint 1471).

Jinks, J. L. 1976. Cytoplasmic inheritance. *Carolina Biology Readers* no. 72. Burlington, N.C.: Carolina Biological Supply Co.

Kornberg, R. D., and Klug, A. February 1981. The nucleosome. *Scientific American* (offprint 1490).

McKusick, V. A. April 1971. The mapping of human chromosomes. *Scientific American* (offprint 1220).

Mittwoch, U. July 1963. Sex differences in cells. *Scientific American* (offprint 161).

Upton, A. C. February 1982. The biological effects of low-level ionizing radiation. *Scientific American.*

Molecular Genetics

Chapter Concepts

1. Genetic information is coded in the DNA molecule.

2. In DNA replication, the two strands of the molecule are split apart, and these strands are used as templates to form new complementary strands.

3. Genetic information coded by the nucleotide sequence in DNA is transcribed into a complementary segment of RNA and then translated in the cytoplasm through the use of ribosomes into specific amino acid sequences in polypeptides.

4. Structural genes code information for synthesis of structural proteins or enzymes. Other genes, called regulatory genes, regulate the activity of structural genes.

5. Recombinant DNA techniques permit insertion of genes from one organism into the DNA of another organism.

6. Changes in the nucleotide sequence of DNA molecules are called point mutations. Such mutations can be random accidents in recombination, or they can be caused by various chemicals or by radiation.

Genes have been identified as the units of heredity. But what exactly are genes, and how do they work? How does the transmission of a part of a chromosome from one cell to another result in the transmission of genetic information? How is this genetic information stored, and how is it used by the organism? To answer these questions, genes and their actions must be studied at the level of the molecules involved. This area of study, molecular genetics, has expanded very rapidly, especially since the 1950s, due to the increasing sophistication of biochemical techniques and the subsequent increase in knowledge concerning molecular processes in living things.

DNA as the Genetic Material

The first task of molecular geneticists was to determine which molecular substance actually makes up the genetic material. After Sutton and Boveri suggested the chromosomal theory of inheritance in 1902 and T. H. Morgan and his colleagues followed with their proof that genes are arranged in linear arrays on chromosomes, scientists knew that the genetic material is contained within the chromosomes. Chemical analysis of chromosomes showed that they are made up of DNA and protein, and it then became a question of which of these substances was the genetic material.

At the time that this question was being argued most vigorously, the chemical structure of DNA was poorly understood. Many biologists thought that its chemical structure was too simple to encode genetic information. Proteins, on the other hand, were known to be quite complex. Biologists suspected that the sequence of amino acids in certain proteins in chromosomes formed a genetic code. They also thought that DNA might be a structural material that provided a supporting framework for the protein molecules that carried the genetic information.

A series of historically important studies in molecular genetics indicated, however, that DNA and not protein is the genetic material.

The first of these studies was published by Fred Griffith in 1928. Griffith's experiments involved the process of **transformation.** Transformation is a change in a bacterial strain that results when the bacteria receive genetic information from another bacterial strain.

Griffith's experiments were done with strains of the bacterium *Pneumococcus*. Some strains of *Pneumococcus* cause a fatal pneumonia in mammals. These pathogenic *Pneumococci* produce a polysaccharide capsule that protects the bacteria from the phagocytic cells that are part of the defense mechanisms of the infected animal. Other strains of *Pneumococcus* have lost the ability to synthesize this capsule. The two strains produce different types of colonies when grown in culture. Colonies made up of cells with capsules have a smooth appearance, and the cells are called strain-S cells. Colonies made up of cells of the strains without a capsule (strain-R cells) have a rough appearance. The strain-R cells do not cause pneumonia in animals that they infect; the strain-S cells do.

Griffith performed several experiments designed to reveal more about the differences between these two strains and the role of the capsule in infection of mammalian hosts. He found that living strain-S cells injected into a mouse invariably were virulent (disease-producing) and caused the mouse to die, while living strain-R cells were nonvirulent (relatively harmless). He then found, however, that heat-killed strain-S cells injected into a mouse were no longer virulent. The strain-S cells had to be alive for the strain to be virulent. In a later experiment, Griffith injected live strain-R bacteria (normally nonvirulent) together with heat-killed strain-S cells (figure 22.1). Surprisingly, the mice injected with this combination died of bacterial pneumonia. Somehow, the strain-R cells had been transformed from nonvirulent to virulent.

Further study of transformation in cultured bacteria showed that transformation does not depend on the living host animal. When heat-killed strain-S cells are added to a culture of living strain-R cells, some of the strain-R cells are transformed into capsule-producing cells that form smooth colonies. James Alloway demonstrated that even a cell-free extract of strain-S would transform strain-R cells into strain-S cells. Some chemical factor from the strain-S cells transforms living strain-R cells into strain-S cells. Furthermore, these transformed bacteria retain their capsule, colony, and virulence characteristics for generation after generation. Thus, transformation results in a stable genetic change.

After years of chemical investigation, O. T. Avery and his colleagues Colin MacLeod and Maclyn McCarty demonstrated conclusively that

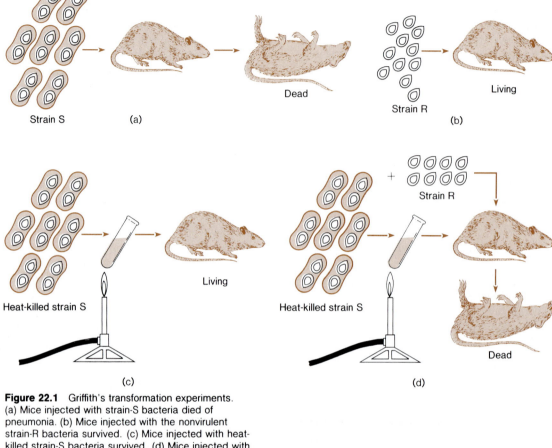

Figure 22.1 Griffith's transformation experiments. (a) Mice injected with strain-S bacteria died of pneumonia. (b) Mice injected with the nonvirulent strain-R bacteria survived. (c) Mice injected with heat-killed strain-S bacteria survived. (d) Mice injected with living strain-R bacteria and heat-killed strain-S bacteria contracted pneumonia and died. Furthermore, it was possible to extract strain-S bacteria from their blood.

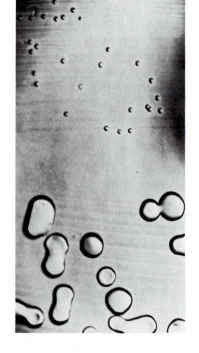

Figure 22.2 Two kinds of colonies of *Pneumococcus* growing on nutrient medium. The small colonies at the top are made up of strain-R cells. The large, glistening colonies on the bottom are made up of cells that have been transformed from strain-R to strain-S by DNA from strain-S cells.

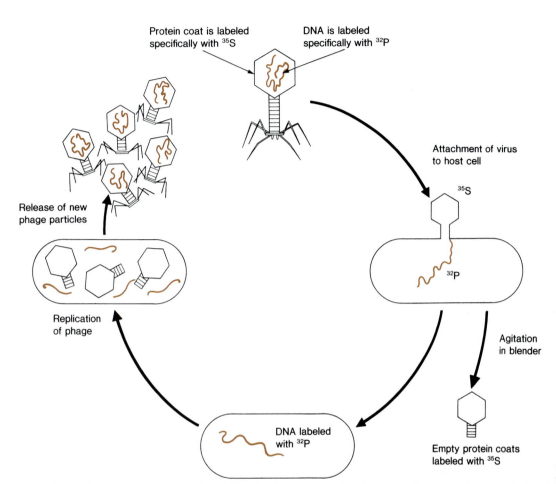

Protein coat is labeled
specifically with ^{35}S

DNA is labeled
specifically with ^{32}P

Attachment of virus
to host cell

^{35}S

^{32}P

Release of new
phage particles

Agitation
in blender

Replication
of phage

DNA labeled
with ^{32}P

Empty protein coats
labeled with ^{35}S

Figure 22.3   The Hershey-Chase experiment. Results of two experiments are combined in summary form in this diagram. Hershey and Chase labeled the protein coats of one batch of bacteriophage with ^{35}S. (Sulfur is found in protein but not in nucleic acids.) They allowed the phages to attach to bacteria for a time, and then they were agitated in a blender. Radioactivity measurements indicated that little radioactive material entered bacterial cells; the ^{35}S labeled protein remained outside. But the infection proceeded, and new virus particles were formed and released. Hershey and Chase labeled the DNA in another batch of phages with ^{32}P. (Phosphorus is an important constituent of nucleic acids but not of proteins.) Again, after a period of attachment, phages were separated from bacterial cells in a blender. This time the bulk of the radioactivity was found inside the host cells, indicating that it is DNA that enters host cells and infects them. These important experiments helped to prove that the primary genetic material is DNA, not protein.

DNA extracted from the heat-killed strain-S bacteria is the transforming factor (figure 22.2). If they treated the extract with enzymes that degrade DNA, the transforming activity was lost. Enzymes known to break down proteins or RNA had no such effect on the extract.

More evidence that DNA is the genetic material came from a series of experiments with bacteriophages published by A. D. Hershey and Martha Chase in 1952. A **bacteriophage** (also called **phage,** for short) is a virus that infects a bacterial cell and uses the cell's genetic machinery to produce more viruses. The T2 bacteriophage that Hershey and Chase used consists of a protein coat and a core of DNA. Hershey and Chase labeled the protein coat of one batch of bacteriophage with radioactive sulfur (^{35}S) and labeled the DNA of another batch of bacteriophage with radioactive phosphorus (^{32}P). The incorporation of radioactive elements into the two parts of the virus enabled Hershey and Chase to trace which material actually enters the bacterium. They found that DNA enters the cell and that the protein coat remains on the outside of the cell. Once the infection has been started, the empty protein coat can be removed from the cell without affecting the progress of the infection (figure 22.3). It is clear, then, that DNA is the substance inserted by the T2 bacteriophage to take over the genetic mechanisms of the cell.

The third major line of evidence that confirmed DNA as the genetic material came when the structure of the DNA molecule was finally revealed by Watson, Crick, Franklin, Chargaff, and Wilkins. (The story of their efforts and a description of the structure of the DNA molecule are given in chapter 2.) Because biologists now realized that DNA consists of two complementary strands of nucleotides, arranged in the now familiar double helix, it became obvious that each of these two strands could serve as a template for the synthesis of a precise copy of the partner strand. This would provide a mechanism for the duplication of genetic information during the S phase of the cell cycle. Also, it was pointed out that DNA molecules have a great deal of complexity and variety. The base pairing from one strand to another is restricted; that is, adenine (A) always pairs with thymine (T), and guanine (G) always pairs with cytosine (C). But there is no such structural restriction on the order of nucleotides in the molecule, and it was clear that their sequence could code genetic information. This work, published in a brief article in the journal *Nature* in 1953, marks the point at which it was finally confirmed to nearly everyone's satisfaction that the genetic code is carried by DNA molecules.

There are some exceptions to the rule that DNA is the primary genetic material. In some genetic systems, RNA is the primary genetic material, not DNA. Some viruses, for example, contain cores of RNA rather than DNA. But for the great majority of living things, DNA is the primary genetic material, and this examination of molecular genetics focuses on those organisms.

DNA Replication

As mentioned in the preceding section, discovery of DNA structure led investigators to predict a mechanism of DNA replication based on complementary base pairing. The proposed mechanism was that a double-stranded molecule of DNA would uncoil and that each single strand would serve as a template (pattern) for the formation of a new complementary second strand (figure 22.4). That is, where there is an A in the existing strand, a T is incorporated into the forming strand; where there is a C in the existing

strand, a G is incorporated into the forming strand, and so forth. Two different experiments confirm this pattern of replication of DNA molecules.

The first conclusive evidence was provided by an experiment by Matthew Meselson and Franklin Stahl in 1958. Meselson and Stahl were able to trace the mechanism of DNA replication by using strands of DNA with different densities. They produced DNA strands with different densities by supplying cultured bacteria with nucleotides (the building blocks of DNA strands) that contained different isotopes of nitrogen. Some bacterial cultures received nucleotides containing the normal (^{14}N) isotope, and others received the heavy (^{15}N) isotope. The density of the resulting DNA can be measured by centrifugation in a cesium chloride (CsCl) gradient. In this technique, DNA is mixed with a solution of CsCl and placed in centrifuge tubes. After the tubes have spun in the centrifuge at high speed for some time, the CsCl molecules become distributed in a gradient of increasing density toward the bottom of the tube. The DNA in the tube settles in a specific region of the gradient, where the density of the CsCl corresponds to the buoyant density of the DNA. Because the DNA can be visualized with ultraviolet light, the position of the DNA band or bands can be determined.

Figure 22.5 shows the experiment conducted by Meselson and Stahl. Bacterial cells that had been growing on the medium with nucleotides containing heavy nitrogen (^{15}N), which therefore have the high density DNA, are transferred to a medium with nucleotides containing light nitrogen (^{14}N). Any new DNA strands made in this medium should have a relatively lower density. After the bacterial cells are allowed to go through one cell cycle (one division), the measured DNA density is intermediate between the density of light DNA (DNA containing only nucleotides with ^{14}N) and heavy DNA (DNA containing only nucleotides with ^{15}N). This fits with the prediction that the two strands of DNA separate and that each serves as a template for formation of a new complementary strand. The original double strand with all heavy DNA has split into two strands. Each of those strands serves as the template for the formation of new strands, into which are incorporated nucleotides containing the light nitrogen. Two double-stranded DNA molecules

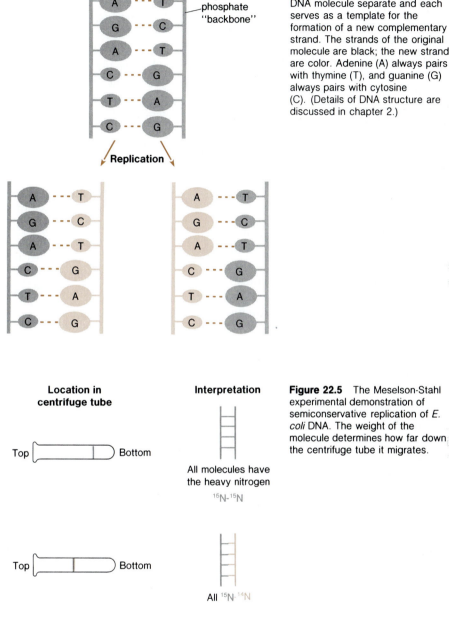

Figure 22.4 The two strands of a DNA molecule separate and each serves as a template for the formation of a new complementary strand. The strands of the original molecule are black; the new strands are color. Adenine (A) always pairs with thymine (T), and guanine (G) always pairs with cytosine (C). (Details of DNA structure are discussed in chapter 2.)

Sugar-phosphate "backbone"

Replication

Generations	Location in centrifuge tube	Interpretation

0   Top [ ] Bottom   All molecules have the heavy nitrogen $^{15}N\text{-}^{15}N$

1   Top [ ] Bottom   All $^{15}N\text{-}^{14}N$

2   Top [ ] Bottom   ½ $^{15}N\text{-}^{14}N$ ½ $^{14}N\text{-}^{14}N$

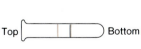

Figure 22.5 The Meselson-Stahl experimental demonstration of semiconservative replication of *E. coli* DNA. The weight of the molecule determines how far down the centrifuge tube it migrates.

result, each of which has a strand of light and a strand of heavy DNA. The density of the double strand therefore, is, intermediate between the density of DNA with two heavy strands and DNA with two light strands. After another division in the same medium, there are two bands in the CsCl gradient. The original heavy strands pair with new light strands, again producing DNA of intermediate density. The light strands from the previous generation serve as templates for assembly of new light strands, resulting in DNA molecules with light nitrogen in both strands. These molecules result in a low density band in the CsCl.

This type of replication, in which half of each new DNA molecule is carried over (conserved) from the previous generation and the other half is newly synthesized, is called **semiconservative replication.** Semiconservative replication was predicted after the publication of the Watson and Crick model of DNA structure and was therefore confirmed by Meselson and Stahl.

A different type of experiment by Arthur Kornberg showed clearly that a single strand of DNA can, by itself, act as a template for the formation of a new strand of nucleotides. Kornberg and his coworkers prepared a mixture of the four nucleotides found in DNA and **DNA polymerase,** an enzyme that catalyzes the formation of new strands of DNA. Then they added single strands of DNA. The result was new, double-stranded DNA. The length of the new DNA was determined by the length of the single-stranded DNA that was originally present in the mixture. When the nucleotide content of the new strand was analyzed, the new strand was found to be complementary to the original strand.

The Meselson-Stahl experiment and Kornberg's single-stranded DNA experiment were conclusive experimental demonstrations that DNA replication involves assembly of new strands complementary to each of the two separated strands of a previously existing DNA molecule.

Chromosome Replication

Any discussion of DNA replication must take into account the fact that DNA is in chromosomes. In bacteria, chromosomes are circular, which would seem to make separating and rejoining the DNA strands a problem. The linear chromosomes of eukaryotes present other problems.

One type of chromosome replication in bacteria is shown in figure 22.6. This process is known as "theta" (θ) replication because the characteristic shape of the nearly replicated chromosome resembles the Greek letter theta. Here, replication begins at a specific point by separation of the double helix. Synthesis of the new DNA strand begins along each strand and proceeds away from the separated region in opposite directions. The separation phenomenon and synthesis of the new complementary strands occur simultaneously until two new chromosomes are formed in a figure-eight-shaped structure. The process, which takes forty minutes to complete, is believed to be the most common way by which circular chromosomes of prokaryotic cells are replicated.

The bacterial chromosome also replicates by way of the rolling circle method illustrated in figure 22.7. The rolling circle method of DNA replication also is an important part of chromosome transfer in bacterial mating. In this form of replication, one strand is cut by an enzyme (part 2 in figure 22.7), and synthesis of a new complementary strand begins along the other (uncut) strand now exposed. The cut strand is displaced and "rolls" off the circular chromosome (part 3 in figure 22.7). Synthesis of a complementary strand also occurs along this cut strand. Finally, two new chromosomes are circularized (part 6 in figure 22.7).

In eukaryotic cells, replication of chromosomal DNA is more complex. Here, the chromosomes are much longer than bacterial chromosomes, yet the replication time is not so much greater as the size difference might suggest. For example, the human haploid nucleus has 1,000 times more DNA than does a bacterial cell, yet the replication time of DNA in human cells is only about twenty times longer than in bacterial cells. The explanation for this discrepancy is that DNA replication in eukaryotic cells is initiated at hundreds of sites, almost simultaneously, along each chromosome.

Figure 22.6 "Theta" (θ) replication of a bacterial chromosome. The two DNA strands separate at a specific point, and replication proceeds away from the separated region in both directions. The characteristic "theta" appearance is seen midway through replication. The helix uncoiling and recoiling involved in this replication pattern is very complex because the strands are literally whirling around one another, and special enzymatic mechanisms are required to accomplish it. Another set of enzymes is involved in closing and separating the two new circular chromosomes. All of these enzymes are in addition to DNA polymerase, which is required to join nucleotides together as new strands are formed.

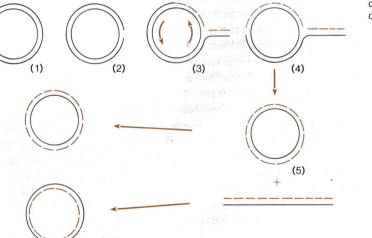

Figure 22.7 Rolling circle method of replication of circular chromosomes.

Figure 22.8 Replication of DNA in a eukaryotic chromosome. Nicks (N) in a single strand serve as swivel points as the double helix unwinds during replication. Replication begins simultaneously at a number of initiation sites (I) along the linear chromosome and proceeds in both directions from each initiation site. When the short, replicated segments have joined at common terminal points, the entire DNA molecule is replicated.

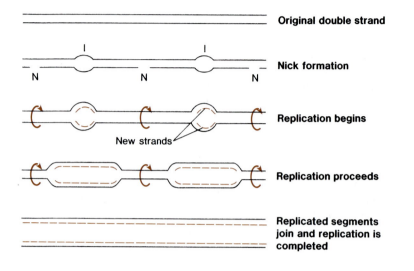

Original double strand

Nick formation

Replication begins

New strands

Replication proceeds

Replicated segments join and replication is completed

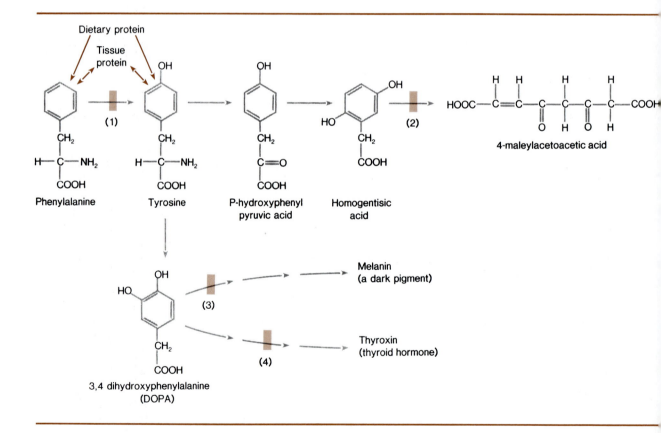

Dietary protein

Tissue protein

(1)

Phenylalanine

Tyrosine

P-hydroxyphenyl pyruvic acid

Homogentisic acid

(2)

4-maleylacetoacetic acid

3,4 dihydroxyphenylalanine (DOPA)

(3)

Melanin (a dark pigment)

(4)

Thyroxin (thyroid hormone)

A short segment from a eukaryotic cell chromosome showing two such sites is shown in figure 22.8. First, one strand of the double helix is broken or nicked by enzymes at various places (N) along the chromosome. These breaks or nicks serve as swivel points as the double helix unwinds at an initiation point (I). The original double helical molecule unwinds as its two strands separate. Complementary strand synthesis begins at these initiation points and moves along each strand. As new complementary strands form along each of the previously existing strands, two double helixes form where there had been one. Synthesis and unwinding occur simultaneously, followed by rewinding, up to the nicked region. When the nicks are rejoined, replication is complete.

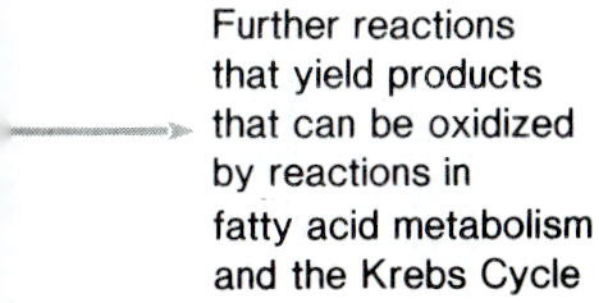

Further reactions
that yield products
that can be oxidized
by reactions in
fatty acid metabolism
and the Krebs Cycle

Figure 22.9 Alkaptonuria and related metabolic problems. The amino acids phenylalanine and tyrosine are obtained from dietary protein. Normally, they are used in a variety of reactions, each of which is catalyzed by a specific enzyme. The enzyme (homogentisic acid oxidase, site 2 in diagram) deficiency studied by Garrod is just one of several possible hereditary problems that can be associated with metabolism of these compounds. (1) Enzyme deficiency here leads to phenylketonuria (PKU). (2) Enzyme deficiency here leads to alkaptonuria. (3) Enzyme deficiency here leads to albinism. (4) Enzyme deficiency here leads to thyroid deficiency.

Chromosome replication is an extremely complex process involving many steps and enzymes. In addition to DNA polymerase (the enzyme that catalyzes the actual joining of nucleotides into strands), there are enzymes that make nicks in the sugar-phosphate backbones, enzymes that "heal" the nicks, and proteins that assist in the unwinding and rewinding processes. The description given here is greatly simplified, and additional details of the process may be obtained from some of the readings suggested at the end of this chapter.

Genes and Enzymes

Many types of genes function by providing coded information that directs synthesis of enzymes or other proteins. Enzymes are the catalysts involved in virtually all biochemical reactions in living cells. This relationship between genes and enzymes was first described by an English physician, Archibald Garrod, in the early 1900s.

In 1902 Garrod reported his findings on the disease **alkaptonuria.** Individuals with alkaptonuria usually have arthritis and characteristically produce urine that turns black when exposed to air. Garrod observed that people with alkaptonuria excreted, in their urine, all the homogentisic acid produced in their bodies. Homogentisic acid is produced in cells by reactions involving the amino acids tyrosine and phenylalanine. Normal healthy individuals can metabolize homogentisic acid; people with alkaptonuria cannot. Garrod hypothesized that people with alkaptonuria lack an enzyme necessary to catalyze the conversion of homogentisic acid to another compound. Because the disease is inherited in a simple Mendelian fashion and appeared to be controlled by a single recessive gene, Garrod proposed that there was a relationship between a gene and an enzyme. Figure 22.9 shows the reaction that is catalyzed by the enzyme missing in alkaptonuria. Garrod went on to propose similar mechanisms for three other human conditions. He referred to these biochemical hereditary diseases as "inborn errors of metabolism" and published a book with that title.

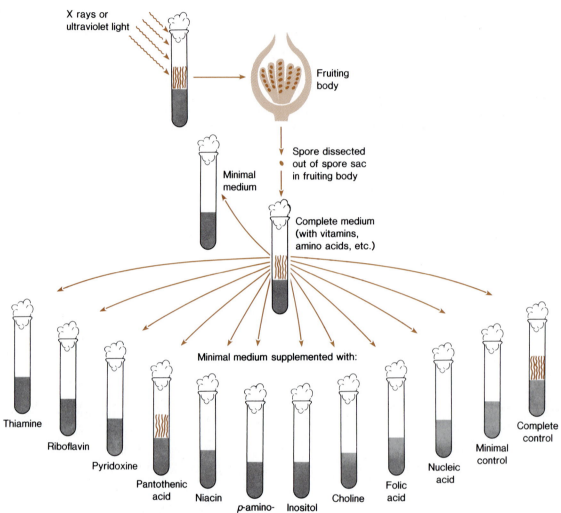

Figure 22.10 Beadle and Tatum's method for detecting nutritional mutants in *Neurospora crassa*. A single spore is placed in a complete medium that supplies a variety of nutrients, including vitamins, amino acids, and others. Then a sample of the growing culture is transferred to minimal medium (left), in which *Neurospora* normally can grow because it can synthesize all the compounds that it requires beginning with material present there. If the mold fails to grow in the minimal medium, it is a nutritional mutant. The nature of its deficiency can then be determined by transferring other samples from the culture in complete medium to various cultures containing minimal medium, supplemented with a different single nutrient in each case. In this example, the mutant strain is able to grow on minimal medium supplemented with pantothenic acid. This indicates that this strain lacks an enzyme involved in pantothenic acid synthesis.

It was many years, however, before Garrod's remarkably insightful hypothesis was verified experimentally. In 1951, G. W. Beadle and E. L. Tatum conducted a series of experiments on *Neurospora crassa,* the red bread mold, and were able to formulate the precise relationship between genes and enzymes. In their experiments, Beadle and Tatum used X rays to induce mutations in the mold. They isolated a series of mutant molds, each of which was unable to grow without the addition of a specific nutrient to the growth medium (figure 22.10). These were nutrients that normal molds were able to synthesize for themselves. In each case, however, the mutant lacked an enzyme that catalyzed a step in the synthesis of the nutrient. Different mutant strains required different nutrient supplements because they lacked different functional enzymes. Beadle and

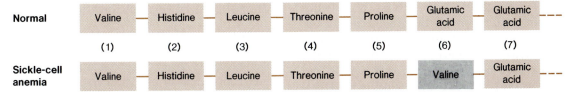

Normal	Valine	—	Histidine	—	Leucine	—	Threonine	—	Proline	—	Glutamic acid	—	Glutamic acid	- - -
	(1)		(2)		(3)		(4)		(5)		(6)		(7)	
Sickle-cell anemia	Valine	—	Histidine	—	Leucine	—	Threonine	—	Proline	—	Valine	—	Glutamic acid	- - -

Figure 22.11 The first seven amino acids of the β chain of human hemoglobin. The substitution of a single amino acid—valine substituted for glutamic acid—at the sixth position results in sickle cell anemia. This substitution of only one of the 146 amino acids in the polypeptide chain drastically alters properties of the hemoglobin that is formed by union of four polypeptides and the red blood cells that contain the altered hemoglobin.

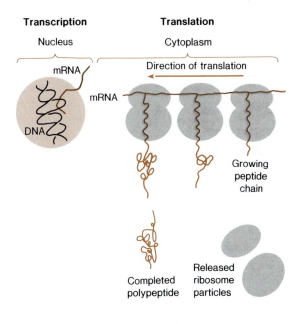

Transcription   Translation

Nucleus   Cytoplasm

mRNA

Direction of translation

mRNA

DNA

Growing peptide chain

Completed polypeptide

Released ribosome particles

Figure 22.12 Transcription and translation.

Tatum found a one-to-one correspondence between a genetic mutation and the lack of a specific enzyme required in a biochemical pathway. They proposed, therefore, that each gene specifies the synthesis of one enzyme. This is called the one gene/one enzyme hypothesis. It is now known that the relationship is more complex than that, but the work of Beadle and Tatum was clear proof of the relationship between genes and enzymes that Garrod had proposed so long before.

In 1957, V. M. Ingram reported that the hemoglobin in people with sickle-cell anemia differs from normal hemoglobin at a single amino acid in the β polypeptide chain (figure 22.11). (Hemoglobin is made up two α and two β polypeptide chains; see p. 55.) This indicated that genes actually specify the amino acid sequence of proteins and that they code specific information for assembly of all proteins (polypeptide chains), including both enzymes and other proteins. By renaming the one gene/one enzyme hypothesis the one gene/one polypeptide hypothesis, Ingram stressed that genes code for proteins that are single polypeptides and also for the individual polypeptides that aggregate in complex proteins such as hemoglobin. The genetic information for such complex proteins, therefore, is coded in not one gene, but several. The genes that code information for synthesis of polypeptide chains are called **structural genes.**

Genetic Expression

The process by which the information in DNA is expressed as a chain of amino acids arranged in a specific sequence in a polypeptide can be divided into two phases: transcription and translation (figure 22.12).

Transcription is the process of actually "reading and rewriting" the genetic code of the DNA molecule. In transcription, the sequence of nucleotides in a segment of DNA is transcribed into a complementary strand of RNA. In the process, the sequence of nucleotides in the DNA segment specifies the sequence of nucleotides in the RNA strand (table 22.1). The RNA strand produced is called **messenger RNA (mRNA).** It is a messenger in the sense that it carries a message about genetic information contained in nuclear DNA into the cytoplasm where the actual assembly of amino acids into polypeptides takes place.

The use of the information, now encoded in mRNA, to produce a polypeptide is called **translation.** The mRNA strand moves out into the cytoplasm, where it becomes associated with groups of ribosomes. Each ribosome moves along the mRNA strand and reads the genetic code. When it reads a three-base sequence along the mRNA strand (the code is a "three-letter" code, as shall be seen shortly), a specific **transfer RNA (tRNA),** bearing a specific amino acid, attaches to the ribosome. As the ribosome moves further along the mRNA strand, the tRNA leaves the ribosome, leaving behind its amino acid. The ribosome reads the next part of the coded information along the mRNA strand, binds another tRNA with its particular amino acid, and the process is repeated. Thus, sequentially coded information in mRNA specifies which tRNA molecules are bound and, therefore, what amino acids are added to the growing chain in what sequence. As each ribosome moves down the mRNA strand, it produces a chain of amino acids, a polypeptide. When the ribosome reaches the end of the mRNA strand or reaches a coded signal to stop, it breaks away and the completed polypeptide is released.

Mechanisms of Protein Synthesis

Transcription and translation are complex processes and not fully understood, but the molecular biology of genetic expression is one of the most active areas of biological research. In this section, the transcription and translation processes are explored in greater detail.

Transcription

The production of mRNA during transcription is somewhat similar to the replication of DNA. The formation of a mRNA strand requires the presence of an enzyme, **RNA polymerase,** that catalyzes the joining of nucleotides into a polynucleotide strand. One of the DNA strands acts as a template for the assembly of a complementary mRNA strand (figure 22.13). The transcript is made from only one of the two strands of DNA, and this strand is called the **sense strand.** The other DNA strand is called the **replication (or antisense) strand.** Although the DNA replication mechanism depends on the double-stranded nature of the molecule, only one strand's coded information is transcribed. In viruses, one of the two strands is transcribed in some regions of the chromosome, and the opposite strand is transcribed in other regions of the chromosome. This also has been demonstrated in certain chromosome areas of some types of eukaryotic cells.

Transfer RNA also is transcribed from the DNA. Molecules of tRNA are much smaller than those of mRNA, and they contain an arrangement of base sequences that results in complementary base pairing within the tRNA molecule. This pairing causes tRNA to have a shape that resembles a cloverleaf (figure 22.14). Each tRNA molecule attaches to one specific amino acid, and only to that kind of amino acid. A specific enzyme in each case functions to "charge" each kind of tRNA molecule with the appropriate amino acid. This specific charging of tRNA molecules is vital if assembly of normal protein molecules is to occur during the process of translation.

Table 22.1
Complementary Base Pairing between DNA and Messenger RNA.

DNA	RNA
Adenine	Uracil*
Cytosine	Guanine
Guanine	Cytosine
Thymine	Adenine

*Note that RNA contains uracil rather than thymine. Wherever an adenine appears in a DNA strand being transcribed, uracil is incorporated into the mRNA molecule produced.

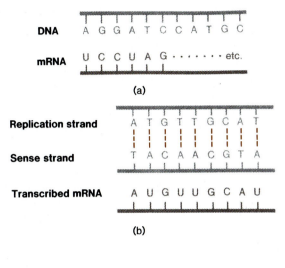

DNA

A G G A T C C A T G C

mRNA

U C C U A G · · · · · · · etc.

(a)

Replication strand

A T G T T G C A T

Sense strand

T A C A A C G T A

Transcribed mRNA

A U G U U G C A U

(b)

Figure 22.13 Transcription. (a) A representation of mRNA production. Transcription involves assembly of a strand of mRNA that is complementary to part of one of the strands of DNA. The enzyme RNA polymerase catalyzes the linking of nucleotides into a chain. Note the complementarity of the bases and that uracil (U) is present in RNA rather than thymine (T).
(b) Diagrammatic representation of the sense strand and replication strand in a segment of DNA. Only the sense strand is transcribed; mRNA produced during transcription is complementary to the sense strand. The replication strand is, of course, necessary for DNA replication.

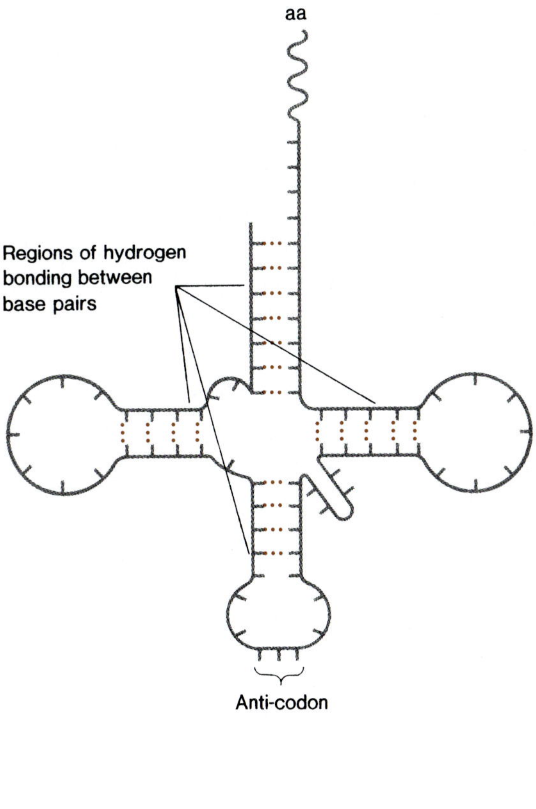

aa

Regions of hydrogen bonding between base pairs

Anti-codon

Figure 22.14 Transfer RNA (tRNA). Base sequences in tRNA are arranged so that complementary base pairing occurs within the molecule. Hydrogen bonds between base pairs stabilize the tRNA molecule in its characteristic cloverleaf shape. This "cloverleaf" actually twists into a more complex three-dimensional shape. The point of attachment of an amino acid is symbolized aa. Each type of tRNA molecule specifically attaches one and only one kind of amino acid, and has a specific base triplet (anti-codon) that is complementary to a specific codon in mRNA. The attachment of the appropriate amino acid to the correct tRNA molecule is catalyzed, in each case, by a specific enzyme.

Translation

Messenger RNA moves out into the cytoplasm and becomes associated with the ribosomes. Ribosomes become attached to the mRNA strand and move along the strand, "reading" its message. Usually, several ribosomes attach to the same mRNA strand so that a number of identical polypeptides are being assembled at the same time.

Each ribosome is made up of two subunits, one large and one small. In bacteria, these are designated 50S and 30S. S is the Svedberg unit, which is a measure of mass determined by the rate at which a particle in solution sediments in a centrifuge (see chapter 3). The larger the particle, the higher the number. The ribosomal subunits in eukaryotic cells are a little bit larger, consisting of 60S and 40S units. Both ribosomal subunits are made up of protein and nucleic acid. The nucleic acid is a third kind of RNA, called **ribosomal RNA (rRNA)**. The subunits are free in the cytoplasm while they are not directly involved in protein synthesis. When they attach to a mRNA molecule during protein synthesis, however, the two subunits join together to form a functional ribosome. After the polypeptide has been completed, the parts of the ribosome break apart again as the polypeptide is released.

The details of the translation process are illustrated in figure 22.15. Sequences of three nucleotides in mRNA are called **codons;** each codon specifies an amino acid. The three nucleotides on the tRNA that will base-pair with this codon are called the **anti-codon.** The specific pairing between each mRNA and the correct anti-codon of a tRNA molecule brings the correct amino acid into exactly the right place in a polypeptide. Obviously, each tRNA molecule must be charged with the appropriate amino acid if this is to occur correctly.

The mRNA transcript enters the 30S part of the ribosome. Then the 50S subunit joins the 30S-mRNA complex. There are two AA~tRNA (a tRNA charged with its Amino Acid) binding sites in the 50S part of the ribosome: the first site is the A (aminoacyl) site; the second, immediately adjacent, is the P (peptidyl) site. When the first codon of the mRNA appears at the A site, the correct AA~tRNA binds its anti-codon to the mRNA codon. The mRNA moves, sliding the bound AA~tRNA into the P site and exposing the next codon at the A site. The appropriate AA~tRNA then binds its anti-codon to this second codon of the mRNA. At this point the peptide bond forms between the two amino acids. A peptide bond is a covalent bond between the carboxyl carbon of one amino acid and the nitrogen of the amino group of the next amino acid. Then, the tRNA is disengaged from the first amino acid, and the tRNA falls away from the codon on the mRNA. The mRNA slides the second tRNA, together with the two amino acids now attached to it, into the P site. A dipeptide (two amino acid) chain has formed. The movement of the mRNA has exposed the next codon in the sequence at the A site. The appropriate AA~tRNA binds, a peptide bond forms, and a tripeptide (three amino acid) chain results. This process continues until the end of the message, at which time a newly completed polypeptide chain is released into the cytoplasm. This, then, completes a sequence of molecular processes involved in genetic expression in cells. The genetic message coded as a sequence of bases in DNA has been expressed, through the processes of transcription and translation, as a specific type of protein molecule consisting of a particular sequence of amino acids.

The Genetic Code

As explained in the previous section, codons and anti-codons are sequences of three bases in mRNA and tRNA respectively. The DNA molecule itself also has three-base sets that code information. But how was it actually proven that this code depends on base sequence, and how was the code "cracked"?

Figure 22.15 Details of the translation process. Initiation of translation always begins with an AUG codon. This codon apparently is in a special form because it specifies a tRNA bearing a modified (formylated) methionine molecule rather than the ordinary methionine that is coded for by AUG in other positions in mRNA. Termination of translation is brought about by one of the termination codons (UAA, UAG, or UGA). When a termination codon is present, a release factor joins site A instead of a tRNA. This leads to release of the polypeptide and separation of the ribosome subparticles. Note that tRNA molecules are drawn here in a form that symbolizes the three-dimensional shape that the "cloverleaf" shown in figure 22.14 assumes.

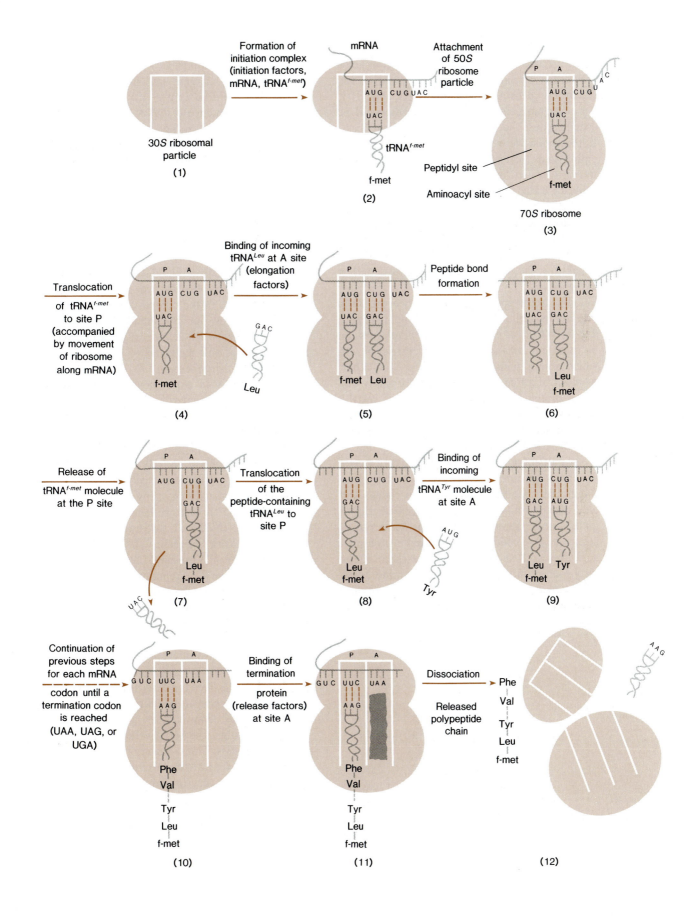

The first step involved some simple mathematical reasoning. Because twenty amino acids are formed in naturally occurring proteins, the codon has to be at least three bases long. If it were one base, only four amino acids could be coded because there are only four different bases. If the codon were two bases long, only sixteen amino acids (4^2) could be specified. If the codon were three bases long, sixty-four amino acids (4^3) could be coded. Therefore, three is the smallest number of bases that provides enough combinations to code for twenty different amino acids, and it has been demonstrated experimentally that the three-base (triplet) code, the simplest code possible, is the form of genetic coding found in living things.

The triplet code could be wholly or partially overlapping or it could be nonoverlapping (figure 22.16). Wholly or partially overlapping codes can be ruled out because they place restrictions on which amino acids can be next to each other. Analysis of amino acid sequences for many different proteins has shown that no such restrictions exist. All possible combinations of adjacent amino acids occur in protein molecules. Therefore, the triplet code must be nonoverlapping.

Could more than one three-base codon code for the same amino acid? With a three-base codon, there are sixty-four different codons and only twenty amino acids to be specified. Either there is more than one codon for some amino acids, or some codons do not code for an amino acid. Research shows that all but three of the sixty-four possible codons specify amino acids. Such a code, where more than one codon specifies the same amino acid, is said to be a **degenerate code.**

Which codons specify which particular amino acids was determined as a result of a series of discoveries. An enzyme was found by Severo Ochoa that could be used to link nucleotides together in the absence of DNA, thus producing artificial RNA. For example, uracil triphosphate (UTP) in the presence of this enzyme is converted into long chains of RNA that contain only uracil as a base (the composition of the RNA thus produced is described as poly U). It also was found that cell extracts can synthesize protein *in vitro* (in culture outside the cell) but that they produce protein for only a limited time. If more RNA is added, more protein is synthesized. This requirement of RNA addition suggested that RNA has a short life span, possibly because of RNA-digesting enzymes in the cytoplasm. It is, in fact, mRNA that runs out first in such cell extracts because messenger RNA generally is a relatively short-lived, unstable molecule.

Marshall Nirenberg, in 1961, used artificial messenger RNA in experiments with cell-free extracts. He found that when poly U is added to a cell extract that has used up its own mRNA, a particular type of new polypeptide is made. The single unusual polypeptide produced in the reaction mixture consists entirely of phenylalanine. Poly U directs production of polyphenylalanine, thus the codon specifying phenylalanine must be UUU. Rapidly, three large laboratories headed by Nirenberg (National Institutes of Health), Ochoa (New York University), and H. G. Khorana (University of Wisconsin) used similar methods to determine the remaining codons. This tremendous effort resulted in the construction of a simple but history-making RNA/amino acid dictionary (figure 22.17). The genetic code was "cracked!"

One of the most interesting outcomes of extensions of this code-cracking activity to a variety of organisms has been the determination that the genetic code appears to be universal. That is, in all living things, from bacteria and even viruses to human beings, the same codons specify the same amino acids. This discovery powerfully supports the idea that at the level of molecular genetics, there is a very fundamental unity of life processes that is indicative of a common evolutionary origin.

Wholly overlapping code

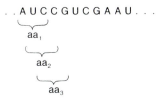

Partially overlapping code

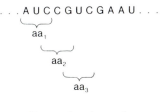

Nonoverlapping code

. . . A U C C G U C G A A U . . .

aa₁   aa₂   aa₃

Figure 22.16 Three models for a triplet nucleotide code using a hypothetical mRNA segment. The brackets show how the code might be read to specify three adjacent amino acids (aa) in a polypeptide being produced during translation. The code actually fits the third model. That is, the code is nonoverlapping.

First position

Second position

	U	C	A	G	
U	UUU ⎫ Phe UUC ⎬ UUA ⎫ Leu UUG ⎭	UCU ⎫ UCC ⎬ Ser UCA UCG ⎭	UAU ⎫ Tyr UAC ⎬ UAA ⎫ STOP UAG ⎭	UGU ⎫ Cys UGC ⎬ UGA STOP UGG Trp	U C A G
C	CUU ⎫ CUC ⎬ Leu CUA CUG ⎭	CCU ⎫ CCC ⎬ Pro CCA CCG ⎭	CAU ⎫ His CAC ⎬ CAA ⎫ Gln CAG ⎭	CGU ⎫ CGC ⎬ Arg CGA CGG ⎭	U C A G
A	AUU ⎫ AUC ⎬ Ile AUA ⎭ AUG Met	ACU ⎫ ACC ⎬ Thr ACA ACG ⎭	AAU ⎫ Asn AAC ⎬ AAA ⎫ Lys AAG ⎭	AGU ⎫ Ser AGC ⎬ AGA ⎫ Arg AGG ⎭	U C A G
G	GUU ⎫ GUC ⎬ Val GUA GUG ⎭	GCU ⎫ GCC ⎬ Ala GCA GCG ⎭	GAU ⎫ Asp GAC ⎬ GAA ⎫ Glu GAG ⎭	GGU ⎫ GGC ⎬ Gly GGA GGG ⎭	U C A G

Third position

Figure 22.17 The genetic code. This figure shows all sixty-four combinations of nitrogenous bases in mRNA and the amino acid for which each combination codes. Most of the combinations code for amino acids (note that there is more than one combination for most amino acids). There are three terminator codons (labelled "stop") that signal the end of a polypeptide. When a terminator codon is present, translation stops.

Box 22.1
Overlapping Genes in ØX174

A structural gene has been defined as a segment of DNA coding for a particular polypeptide. Also, the genetic code has been described as nonoverlapping. The assumption would be, then, that the genes themselves also are nonoverlapping. As with many generalizations in biology, however, exceptions have been found. One of them is in the bacteriophage ØX174.

The entire nucleotide sequence of the ØX174 DNA molecule was determined by Sanger and his coworkers in 1977, and the genes were mapped (box figure 22.1A). Gene E is found within gene D, and gene B is found within gene A.

It is not known whether or not overlapping genes such as these are a general phenomenon or whether they are restricted to very small segments of DNA, such as that found in ØX174 and a few other viruses. The presence of overlapping genes raises some interesting questions about gene expression.

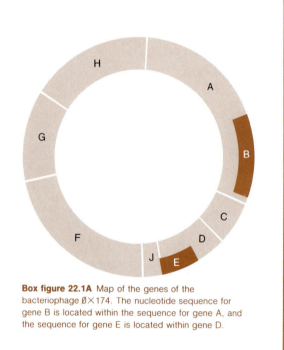

Box figure 22.1A Map of the genes of the bacteriophage Ø×174. The nucleotide sequence for gene B is located within the sequence for gene A, and the sequence for gene E is located within gene D.

Regulatory Genes

Thus far, genes that provide coded information for synthesis of proteins (polypeptides) have been discussed. These genes are called **structural genes** because they determine protein structure. There is strong evidence indicating that another class of genes, called **regulatory genes,** exists. Regulatory genes regulate the activity of structural genes. For example, it is known that specific mRNAs and their resulting proteins are synthesized at specific times in development. This fact implies that certain structural genes are somehow "turned on" and then "turned off" at specific times, apparently via the action of regulatory genes.

In bacteria, genetic elements that turn genes on and off have been described in great detail. These regulatory elements, together with the structural genes that they regulate, are called **operons.** The most thoroughly studied operon is the **lactose** or **lac operon.** François Jacob and

Jacques Monod in France proposed, based on their work with the lac operon, a general scheme for control of genetic expression that is now known as the Jacob-Monod model.

As shown in figure 22.18, there are five genetic units in the lac operon: two regulatory units (P and O) and three structural genes (Z, Y, and A). The P site, or promoter, is a segment of DNA to which RNA polymerase binds. Under appropriate conditions, RNA polymerase catalyzes transcription of mRNA from the structural genes Z, Y, and A. These genes code for enzymes that are involved with the breakdown of the disaccharide lactose. Another genetic unit, R, a regulator gene, is not a physical part of the operon but is involved in functional control of the operon. The regulator gene codes for production of a protein called the **repressor,** which has two binding sites, one for lactose and one for the operator gene O.

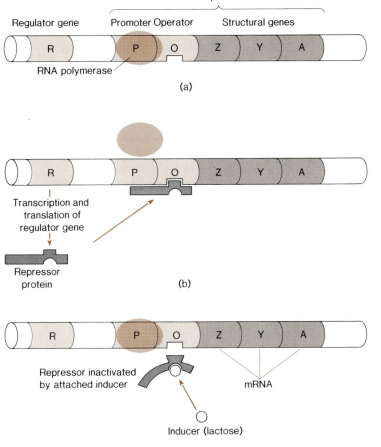

Operon

Regulator gene   Promoter Operator   Structural genes

R   P   O   Z   Y   A

RNA polymerase

(a)

R   P   O   Z   Y   A

Transcription and
translation of
regulator gene

Repressor
protein

(b)

R   P   O   Z   Y   A

Repressor inactivated
by attached inducer

mRNA

Inducer (lactose)

(c)

Figure 22.18 The Jacob-Monod model of structural gene regulation as illustrated by function of the lac operon of the bacterium *Escherichia coli.* (a) Elements of the operon with RNA polymerase in place on the promoter site of operon, as it must be for transcription to occur. Note that the regulator gene is not adjacent to the operon. (b) The lac operon in its repressed condition with the repressor molecule binding to the operator gene and preventing positioning of RNA polymerase on the promoter site. (c) The lac operon in its derepressed (induced) condition. The inducer (lactose) binds to its site on the repressor molecule, thus altering the repressor so that it no longer blocks positioning of RNA polymerase on the promoter site, and transcription of the structural genes proceeds.

The repressor is an **allosteric** protein molecule, a molecule in which the conformation of one binding site is affected by the condition of the other binding site. In this case, the lac operon repressor can bind with the operator (O) gene when it does not have its lactose binding site occupied. When its lactose binding site is occupied, the repressor no longer binds with the operator.

In the lac operon, when the repressor is binding the O gene, the polymerase cannot bind and function at the P site, and the structural genes Z, Y, and A are not transcribed; they produce no messenger RNA coding for enzyme synthesis. This is the situation when there are low levels of lactose in the medium. If lactose levels rise, lactose binding alters the conformation of the repressor so that it uncouples from the O gene, the polymerase binds at the P site, and the structural genes are transcribed (mRNA is produced). When the enzymes synthesized by translation of the messenger RNA have broken down all the lactose, the repressor returns to the O gene, shutting down enzyme synthesis. In this system the

presence of a low molecular weight molecule, lactose, can, in effect, turn on genes (induce transcription), whereas the absence of the molecule turns off the genes automatically. In the Jacob-Monod model, a substance that acts as lactose does is called an **inducer molecule.**

In addition to inducible genes, such as those of the lac operon, which normally are "off" and can be turned "on," there also are repressible genes, which normally are "on" and can be turned "off."

The value of gene regulation to the bacterial cell is that the enzymes are produced only when needed; thus, the cell's resources are conserved.

In eukaryotic organisms, similar regulatory systems probably exist, but the details have not been worked out as well. E. H. Davidson and R. J. Britten who, among others, have studied control of gene expression in eukaryotic cells, conclude that the control mechanisms involved are somewhat more complicated than those found in prokaryotic cells and described by the Jacob-Monod model. In fact, Donald Brown and others

Box 22.2
Split Genes

"What is true for *E. coli* is true for the elephant." These words became one of the popular mottos of molecular biology during the twenty years that followed the determination of the structure of DNA in 1953. Using prokaryotic organisms, especially the intestinal tract bacterium *Escherichia coli,* molecular biologists broke the genetic code and delved into the mechanisms by which genetic information, encoded in DNA molecules, is expressed through the mechanisms of transcription and translation.

Optimistic pronouncements about *E. coli* and elephants became commonplace when it was determined that the genetic code is universal. That is, the same three-nucleotide codons specify the same amino acids in all cells, from the prokaryotic cells of *E. coli* and other bacteria to the eukaryotic cells of complex plants and animals. It seemed quite logical to assume that similar parallels would be found in the organization of genes and in the mechanisms of genetic expression. This assumption was only slightly shaken when it was discovered that after transcription in eukaryotic cells, messenger RNA (mRNA) is modified in many cases by the addition of special end sequences of nucleotides. It was thought that possibly these posttranscriptional modifications had to do with transport out of the nucleus or with some special features of translation in eukaryotic cells.

But then, in the late 1970s, came the startling discovery that genes in eukaryotic cells are organized differently from genes in prokaryotic cells. In prokaryotic cells, the amino acid sequence in a polypeptide is a direct reflection of the codon sequence in the structural gene that codes for it. The codon sequence of the structural gene is transcribed in a mRNA molecule and translated directly into an amino acid sequence in a polypeptide.

This is not always the case in eukaryotic cells. Many genes in eukaryotic cells are split. Stretches of codons that specify parts of the amino acid sequence in the protein product are separated by intervening nucleotide sequences that are not translated. The codon-containing sequences are called **exons,** and the intervening sequences are called **introns.** An entire section of the DNA molecule, including both exons and introns, is transcribed, but only the exons are represented in the mRNA that is finally exported to the cytoplasm for translation. There is a splicing process by which the introns are cut out and the exons are connected in a single uninterrupted sequence.

The gene that codes for chicken ovalbumin, the major protein constituent in egg white, is an example of a split gene that has been studied extensively. The ovalbumin gene is 7,700 nucleotides long, almost seven times as long as the portion of mRNA (1,158 nucleotides) that is actually translated into the amino acid sequence that makes up ovalbumin. The ovalbumin gene has eight separate exons separated by introns.

Other protein-coding genes have been found that have even higher proportions of intron sequences, and split protein-coding genes have been found in many organisms. Split genes coding for ribosomal RNA and transfer RNA also have been found in eukaryotic cells.

Split genes seem to be widespread, possibly even commonplace in the genomes of eukaryotic cells, while they have not been observed in prokaryotic cells. Thus, in terms of gene organization, what is true for *E. coli* is not necessarily true for the elephant after all.

have discovered control regions located in the middle of structural genes in some eukaryotic cells. This is very different from the situation in bacteria, where control regions are separate from structural genes. The mechanisms involved in control of gene expression in eukaryotic cells currently are being studied in a number of laboratories.

One system that shows great promise for study is that of steroid hormone regulation of gene activity (chapter 14). The steroid hormone enters a cell of the target tissue. If a cell is a true target cell, it has a receptor protein that specifically binds the steroid, forming a complex. This hormone-receptor complex moves into the nucleus and binds to DNA. As a consequence, a specific mRNA and resulting protein are produced. Thus, if the hormone enters a cell, gene activity depends on the presence of the receptor protein. It is postulated that the hormone-receptor complex in some way regulates RNA polymerase activity. It would be interesting to see if the genes, which produce proteins in response to the steroid-receptor, can be cloned (this technique is discussed in the next section), and whether the isolated genes still respond to the steroid-receptor complex.

Recombinant DNA

A method has been developed recently for inserting a small segment of DNA from one species into the genome of another unrelated species. This method is called the **recombinant DNA** (or gene-splicing) **technique.**

Recombinant DNA technology is based on two sets of discoveries in molecular genetics. The first of these was that there is extrachromosomal DNA in bacterial cells. (Eukaryotic cells also carry extrachromosomal genetic information in mitochondria and chloroplasts.) One form of extrachromosomal DNA in bacteria is the **plasmid,** a relatively small circular DNA molecule that can replicate inside the bacterial cell and that also can be transferred to other bacterial cells. Resistance to certain antibiotics, for instance, is passed from cell to cell by plasmid transfer.

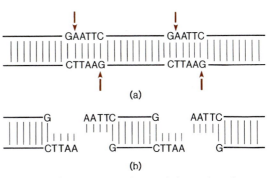

Figure 22.19 The action of a restriction endonuclease. A restriction endonuclease recognizes and attaches to a specific sequence of nucleotides in DNA. It then nicks (cuts) the single strands at specific points, indicated here with colored arrows (a). This leaves DNA fragments that have free single-stranded ends (b).

The second major finding that made recombinant DNA technology possible was the discovery of enzymes known as **restriction endonucleases.** These enzymes have the capacity of recognizing and binding to a specific pattern of base sequences in a DNA molecule, and cleaving the molecule in a very characteristic fashion (figure 22.19). When a restriction endonuclease cuts (nicks) the two strands of a DNA molecule in this way, it leaves fragments that have overlapping free ends. DNA from most species of organisms has sites that can be split by the same restriction endonucleases that nick bacterial plasmids.

How, then, does the recombinant DNA technique work? The actual methods used were devised by Stanley Cohen, Herbert Boyer, and others. Specific plasmids containing a single restriction endonuclease site are harvested from their host cells and cut open with restriction endonuclease (figure 22.20). The foreign DNA to be inserted is extracted and purified. Then it is cut up into many fragments using the same restriction endonuclease (part 2 in figure 22.20). A specific fragment, known to contain the desired gene or genes, is selected according to weight. Because the overlapping ends cut by a restriction endonuclease are complementary, a specific recognition and association can take place between the cut ends of the foreign DNA fragment and the cut ends of the plasmid. The sugar-phosphate backbones of the double helix are joined using an enzyme called a **ligase** (part 4 in figure 22.20),

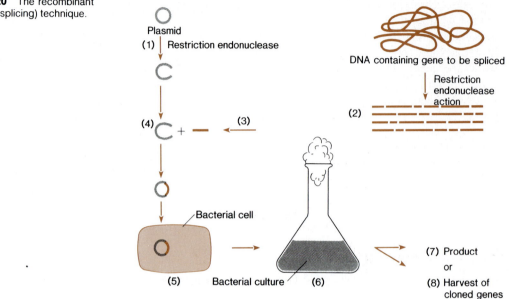

Figure 22.20 The recombinant DNA (gene-splicing) technique.

thus producing the normal closed circle of the plasmid into which the foreign gene is now incorporated. The recombinant plasmid composed of the plasmid DNA and the foreign DNA fragment (that is, recombinant DNA) is introduced into a bacterial cell (part 5 in figure 22.20). Cells containing the recombinant plasmid are allowed to reproduce until there are large numbers of cells (part 6 in figure 22.20), thus producing many more copies of the recombinant DNA. The many exact copies of the plasmid make up a clone of recombinant DNA. Certain strains of plasmids replicate many times inside each cell, further amplifying the number of **cloned genes.** Figure 22.21 shows electron micrographs of several steps in the recombinant DNA technique.

Either the products of the expression of the cloned genes or the cloned genes themselves then can be isolated. For gene products to be isolated, the gene must be transcribed in the bacterial cell, and translatable mRNA must be produced. Normal transcription can take place if the proper genetic control elements are present either in the plasmid or in the recombinant DNA. If these requirements are met, it is possible to design bacteria that "manufacture" a specific polypeptide product.

The cloned gene itself can be isolated by harvesting the plasmids from the cells, purifying them, and using the same restriction endonuclease to cut the recombinant DNA from the plasmid. The fragment, now in numerous copies, can be purified by the same technique used earlier.

One of the most important reasons for cloning genes is economic. For example, the protein hormone insulin, used for treatment of diabetes, is currently isolated, in an extremely expensive operation, from the pancreas of animals. But now it has become possible to produce insulin using recombinant DNA techniques. Copies of the human insulin gene have been obtained. The gene is inserted into a plasmid, and the plasmid is introduced into bacteria that can be grown in large quantities; transcription and translation occur in bacteria containing the plasmid. Thus, bacteria can now be made to synthesize human proinsulin, the larger polypeptide precursor of insulin (see page 410). This proinsulin can be converted into active insulin and purified (figure 22.22).

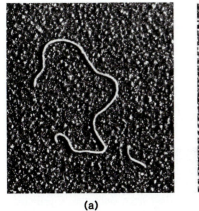 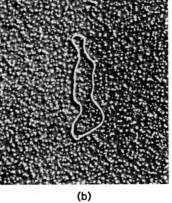 

(a)   (b)   (c)

Figure 22.21 Electron micrographs taken at stages during recombinant DNA experiments. (a) Bacterial plasmid ring has been nicked by restriction endonuclease. Fragment of foreign DNA to be inserted is at lower right. (b) Plasmid after incorporation of foreign DNA has been incorporated. (c) Plasmid adjacent to bacterial cell into which it can be introduced.

Figure 22.22 Crystals of purified, bacterially produced human insulin. These represent a payoff of practical application of the recombinant DNA technique.

In July of 1980 seventeen healthy volunteers in London and Liverpool were given small doses of bacterially produced insulin. The injected hormone caused lowered blood glucose levels. Clinical trials have proceeded, and there is every expectation that insulin produced by recombinant DNA techniques will be widely used in medicine and that it will be cheaper to produce insulin in this way than it is to extract insulin from animal tissues obtained in slaughterhouses. Furthermore, this insulin should be less likely to provoke immune responses in patients than are the animal insulin preparations now being used.

It also seems very likely that other polypeptides that are medically important will soon be in commercial production using these techniques. Some of these polypeptides, such as human growth hormone, can be obtained only from human tissues and thus have been so scarce that supplies have fallen far short of medical needs. It is hoped that polypeptide production using recombinant DNA techniques will make needed treatments available to many people previously not able to obtain them.

Gene cloning has also found valuable applications in genetic research. In detailed study of genes, such as determination of base sequences, large quantities of purified genes are required. Already, base sequences of certain cloned genes have been determined, and it looks as if the technique will be even more useful in the future.

Recombinant DNA and gene-cloning experiments may, however, have potential dangers. The possibility of scientists creating a "monster" strain that might escape from the lab and wreak havoc to life caused considerable public discussion of this work in the mid-1970s. Since *Escherichia coli* (an inhabitant of the human colon) is the major host culture organism used in recombinant DNA studies, there has been concern that dangerous genes or groups of genes might somehow be transferred to humans. For example, what dangers would be in store if *E. coli* containing cloned genes of a cancer-causing virus were to escape the lab and infect humans? Would it cause a cancer epidemic? As a result of these fears, federal guidelines have been devised to limit research in this area and permit dangerous experiments to be carried out only in carefully constructed fail-safe laboratories. In addition,

host cells and plasmid vectors have been constructed such that they will survive only under very specialized culture conditions, greatly limiting the chance of an escape into the environment or infection of humans. Furthermore, a series of experiments have been done to alleviate the fear of a cancer epidemic. In one such experiment, various parts of tumor-causing viruses were cloned in *E. coli,* and the cells containing the recombinant plasmids were used to infect the intestinal tracts of mice. No increase in the incidence of cancer over the normal incidence found in control animals has been detected. Other experiments on potential dangers have also yielded negative results, and research using recombinant DNA techniques is proceeding with caution.

Looking further into the future, there is the distinct possibility that gene-splicing techniques may be modified for use in treatment of hereditary diseases in what amounts to "gene therapy." It is also possible that these techniques may allow desirable traits to be put together in useful new combinations in agriculturally important plants and animals.

Mutations

The concept of mutation was introduced in chapter 21 with a discussion of stable heritable changes in chromosome structure that are transmitted to all of the cells descended from any cell in which such a change occurs. Such changes are chromosomal mutations (or macromutations).

Much smaller changes in the genetic material are called **point mutations.** Macromutations are large, in most cases involve breaks in the chromosome, and frequently can be detected using the light microscope. But point mutations generally involve a single base change in a DNA molecule and, as a result, are not microscopically detectable.

Thus, the definition of mutation can be broadened to include any relatively stable change in the genetic material of a cell that is transmitted to all cells descended from it. Although mutation is defined to include both general types of genetic change, biologists often tend to use the word mutation to mean point mutation because point mutations are the more common genetic changes.

There are several kinds of **mutagens** (agents that cause mutation). A number of chemicals are **mutagenic** (mutation causing), as are ultraviolet light and various forms of ionizing radiation (**X** rays, gamma rays, etc.).

Although point mutations—changes in single bases in the long sequences of bases in DNA molecules—may seem small in chemical terms, they certainly are not biologically unimportant. The effect of a single-base change can be dramatic, and it may spell the difference between life and death for an organism whose cells all contain an inherited point mutation.

Types of Point Mutations

There are several classes of point mutations. The first is the **missense** mutation. A missense mutation is a single-base substitution that changes a codon specifying one amino acid to a codon specifying another amino acid. The sickle-cell hemoglobin mutation is a classical example of a missense mutation. The change from the glutamic acid codon to a valine codon can be achieved by a single-base substitution: GAG→GUG (see figure 22.17).

When a gene containing a missense mutation is expressed, there is an amino acid substitution in the polypeptide formed (valine substituted for glutamic acid in the sickle-cell hemoglobin mutation, for example). How can substitution of a single amino acid among many in a polypeptide have such drastic consequences? If the substituted amino acid has a different net charge from the original, the resulting protein may have quite different properties. For example, glutamic acid has a free carboxyl group (giving the molecule a net negative charge), whereas valine has no net charge. If the amino acid substituted is at a critical region, the molecule could be nonfunctional. In the case of the sickle-cell hemoglobin, the substitution does occur at a critical position, and properties of the entire hemoglobin molecule are abnormal as a result. The structure of the molecule is unstable in solutions of low oxygen concentration. Structural changes in the molecules are related to a structural change in red blood cells (that is, assumption of the sickle shape) that interferes with circulation through small blood vessels and causes a host of problems in body tissues. This whole chain of events is initiated by a missense mutation, a single base substitution in a DNA molecule.

There are other missense mutations in the hemoglobin gene and in other genes that do not adversely affect the function of the molecules. Occasionally, a missense mutation actually improves the function of the molecule produced as a result of expression of the altered gene. Such mutations are very important in terms of evolution because they provide a source of new characteristics.

The second class of point mutations is the **frameshift** mutation. Messenger RNA transcribed from DNA is a sequence of codons in which each set of three bases constitutes a separate reading unit or frame. If an extra base is inserted into the base sequence in DNA, all of the codon frames after the insertion are changed or shifted to codons calling for different amino acids. This drastic change is transcribed into mRNA and translated into an altered polypeptide. Figure 22.23 shows a segment of an mRNA molecule that illustrates the effect of a frameshift. By inserting a guanine between the second and third codon, the third and fourth amino acids and all subsequent amino acids in the polypeptide produced during translation will be changed. If such a mutation occurs anywhere but at the very end of the gene (the last codons transcribed and translated), a completely nonfunctional protein results. The deletion of a base has a similar frameshift effect. But if a base is added close to the region of the deleted base, the reading frame returns to the normal sequence after the second frameshift mutation.

Before	AUG	CCA	UAC	UGG	
	Methionine	Proline	Tyrosine	Tryptophan	

After	AUG	CCA	(G)UA	CUG	G—
	Methionine	Proline	Valine	Leucine	

Figure 22.23 A segment of mRNA showing the effect of frame-shift mutation resulting from insertion of an extra base into the DNA strand from which the mRNA was transcribed. During translation, the mRNA transcribed from a gene that is frame-shifted will direct incorporation of incorrect amino acids because all codons beyond the addition point (G) are altered.

Often, a frameshift results in the production of a **nonsense** mutation. There is a codon that indicates where the reading of the mRNA begins (see figure 22.14) and codons (terminators) to indicate where it stops. No AA ~ tRNA attaches at a point in an mRNA molecule where there is a terminator codon. Because this interrupts assembly of the polypeptide chain, a terminator codon functions as a period in the genetic code. The three terminator codons do not code for any amino acids and for this reason are called nonsense codons (that is, no sense). If a frameshift or a base substitution produces a nonsense codon (for example, the mRNA transcribed might contain UAA instead of AAA), then during translation, a peptide bond does not form between the amino acids called for by the codons straddling the nonsense mutation. As a result, the translation terminates prematurely, producing a shortened polypeptide. If a nonsense mutation occurs near the end of the message, less harm is done, since almost all of the protein has been formed. However, if the mutation occurs earlier in the message, translation does not yield a functional polypeptide.

Causes of Mutations

Some mutations arise accidentally as a result of errors in replication of the DNA. One reason for such errors is that the four bases are somewhat unstable and can exist in separate states or chemical configurations. This shift from one state to another is called **tautomerization.** Normally, the bases in DNA are found in their more stable forms. But if a base undergoes a tautomeric shift to its less stable form just at the time of replication, a "mismatched" base pair can be formed. The error introduced into one of the strands is then perpetuated during subsequent replications, and there is a stable change in the base sequence of all progeny molecules.

Chemical Mutagens

Some mutations are caused by chemical mutagens. An example of a chemical mutagen is the very reactive chemical nitrous acid (HNO_2). Nitrous acid converts adenine into a compound (hypoxanthine) that resembles guanine. As a result of HNO_2 treatment, adenines are converted into guaninelike compounds that pair with cytosine during DNA replication. This change is permanent because progeny strands now have C-G base pairs instead of A-T pairs.

Certain other mutagenic compounds act by being incorporated into the DNA during replication. They are chemically similar to the normal bases and consequently substitute for a normal base. However, once incorporated into newly synthesized DNA, they are more unstable than normal bases and tend to form base pairs with the wrong bases. For example, the mutagen 5-bromouracil is similar to thymine and consequently would be expected to pair with adenine. However, 5-bromouracil is unstable and readily shifts into a tautomeric form that pairs more easily with guanine. Once again, the result is a stable, heritable change in the genome.

A third type of chemical mutagen is the frameshift mutagen. Here a normal base is added to or subtracted from an existing sequence of bases in the DNA during replication. The effect of the frameshift mutation was discussed in the previous section.

Physical Mutagens

Types of electromagnetic radiation also cause mutation. Although wavelengths of the electromagnetic spectrum longer than those of visible light, such as infrared and microwaves, also are believed to be mutagenic, the best understood types of mutagenic radiation are ultraviolet light, X rays, and gamma rays.

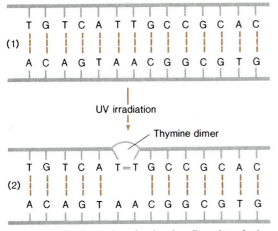

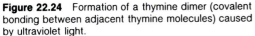

Figure 22.24 Formation of a thymine dimer (covalent bonding between adjacent thymine molecules) caused by ultraviolet light.

Ultraviolet light is most mutagenic at the wavelength that is absorbed most by DNA—260 nm. Ultraviolet light has its mutagenic effect on DNA through the process of excitation. An atom in a DNA molecule becomes excited by ultraviolet light, which results in a rearrangement of electrons. Excited, the atom is chemically reactive, allowing it to bind covalently with another atom. One frequent effect of ultraviolet light is the formation of a thymine dimer, where a thymine molecule bonds to an adjoining thymine (figure 22.24). Thymine dimers do not pair well with adenine but can form base pairs with guanine. Thus, ultraviolet light can cause a T-A pair to become a C-G pair.

If thymine dimers do occur, most cells possess the ability to repair the damage by enzymatically breaking the dimer with a specific endonuclease enzyme (figure 22.25). Humans that have the genetic disease *xeroderma pigmentosum* are very sensitive to ultraviolet light, have a high degree of freckling, and often suffer from skin cancer. Cells in such individuals are unable to repair thymine dimers, apparently because they lack the functional endonuclease.

The other important type of mutagenic radiation is ionizing radiation, such as X rays, gamma rays, and high energy particles, such as protons and electrons. This type of radiation is called ionizing radiation because when it hits a molecule directly, electrons are dislodged completely. Such a hit can result in a broken covalent bond, and an ion pair forms at the break point. The ions at a break point are chemically reactive, and if it is a base in DNA that is ionized, a point mutation may result. If an ion pair is produced in the sugar-phosphate backbone of the double helix, a break in the helix may occur, resulting in a broken chromosome.

All living things have been subjected to naturally occurring physical mutagens in the environment throughout the history of life on earth. But modern use of radioactive materials in industrial, medical, and energy conversion processes has significantly increased human exposure to ionizing radiation. It is not yet known how serious the long-term effects of this increased exposure will be, but increased exposure to ionizing radiation together with increased exposure to many potentially mutagenic chemicals does increase the frequency of alterations in the heritable set of chemical instructions for life known as the genetic code.

Summary

At the molecular level, genetic information is contained within the DNA molecule or, in the case of a few viruses, an RNA molecule. DNA is double stranded with complementary base pairing, and it goes through a semiconservative form of replication.

Structural genes code for the formation of polypeptides. The genetic code is transmitted to the cytoplasm of the cell by mRNA. Ribosomes read the code, and tRNA molecules bring the appropriate amino acid to be attached along the developing polypeptide chain. The genetic code itself is determined by the sequence of nitrogenous bases in the DNA. It is a nonoverlapping, triplet code.

Regulatory genes regulate the activity of structural genes.

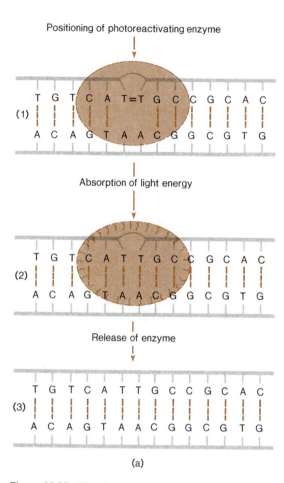

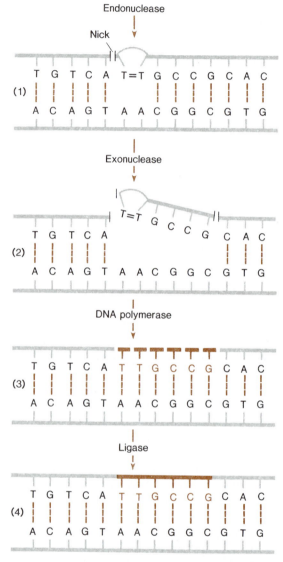

(a)

(b)

Figure 22.25 Thymine dimer repair. (a) Some cells possess a photoreactivating enzyme that locates itself on the DNA molecule over the dimer. The enzyme absorbs visible light in the blue part of the spectrum. This light absorption provides energy that enables the enzyme to split the covalent bonds of the dimer and allows the thymines to assume their normal configuration. Photoreactivating enzyme is present in cells of many organisms but may not be present in cells of mammals. (b) Another repair mechanism involves four enzymes acting in sequence. An endonuclease nicks the strand adjacent to the dimer. Then an exonuclease removes the affected thymidine plus four or five additional nucleotides from the exposed strand. Next, DNA polymerase catalyzes addition of nucleotides complementary to those exposed in the other strand. Finally, DNA ligase seals the newly added nucleotides into the now-corrected strand. (c) Xeroderma pigmentosum, an inherited disease in which cells are less efficient at repairing damage to DNA caused by exposure to ultraviolet light. The basis for the deficiency seems to be less efficient functioning of endonuclease enzymes. This boy suffers from the extensive skin tumors that characteristically develop in the disease.

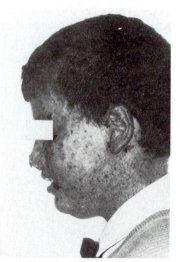

(c)

The use of enzymes called restriction endonucleases has led to the development of a technology for transferring genes from one organism to another. Pieces of DNA can be removed from one organism and spliced into the DNA of another. This is called the recombinant DNA technique.

A number of chemical and physical factors can produce changes, called point mutations, in the base sequence of DNA. A point mutation can cause a change in the code because of insertion of a single different amino acid in the polypeptide, a change that results in production of a completely abnormal polypeptide chain, or a chain that is terminated far short of completion.

Most cells have repair mechanisms that correct some of the DNA alterations caused by certain mutagenic agents.

Exposure to mutagenic agents has increased with development of modern industrialized societies.

Questions

1. Why did the structure of DNA have to be known before it was finally accepted as the hereditary material?
2. How many nucleotides would be necessary in a length of DNA to code for a polypeptide fifty amino acids long? (Remember that a polypeptide must be terminated.)
3. Take the following hypothetical stretch of DNA and diagram the base sequence in the mRNA transcribed from it and the anticodon sequences of the tRNAs involved in translation. Then refer to the mRNA genetic code in figure 22.17 to determine what amino acids this DNA would ultimately code for.
G C G G T G C A C T T T
4. Why must gene activity be regulated? (Remember that all cells contain the same genetic information.)
5. The ends of DNA molecules cleaved by restriction endonucleases are sometimes described as being "sticky." How do you interpret that description?
6. Distinguish between a missense and a nonsense mutation.
7. Why is it possible that some point mutations do not result in the insertion of a different amino acid in the polypeptide produced when the gene is expressed?

Suggested Readings

Books

Goodenough, U. 1978. *Genetics.* 2d ed. New York: Holt, Rinehart, and Winston.

Kornberg, A. 1980. *DNA replication.* San Francisco: W. H. Freeman.

Stent, G. S., and Calendar, R. 1978. *Molecular genetics: An introductory narrative.* 2d ed. San Francisco: W. H. Freeman.

Strickberger, M. W. 1976. *Genetics.* 2d ed. New York: Macmillan.

Suzuki, D. T.; Griffiths, A. J. F.; and Lewontin, R. C. 1981. *An introduction to genetic analysis.* 2d ed. San Francisco: W. H. Freeman.

Watson, J. D. 1968. *The double helix.* Norton Critical Reader, Stent, G. S. (ed.), 1980. New York: W. W. Norton.

Watson, J. D. 1976. *Molecular biology of the gene.* 3d ed. Menlo Park, Calif.: W. A. Benjamin.

Articles

Bauer, W. R.; Crick, F. H. C.; and White, J. H. July 1980. Supercoiled DNA. *Scientific American* (offprint 1474).

Chambon, P. May 1981. Split genes. *Scientific American* (offprint 1496).

Cocking, E. C.; Davey, M. R.; Pental, D.; and Power, J. B. 1981. Aspects of plant genetic manipulation. *Nature* 293:265.

Cohen, S. N., and Shapiro, J. A. February 1980. Transposable genetic elements. *Scientific American* (offprint 1460).

Gilbert, W., and Villa-Komaroff, L. April 1980. Useful proteins from recombinant bacteria. *Scientific American* (offprint 1466).

Hopwood, D. A. September 1981. The genetic programming of industrial microorganisms. *Scientific American.*

Lake, J. A. August 1981. The ribosome. *Scientific American* (offprint 1501).

Maniatis, T., and Ptashne, M. January 1976. A DNA operator-repressor system. *Scientific American* (offprint 1333).

Novick, R. P. December 1980. Plasmids. *Scientific American* (offprint 1486).

Rich, A., and Kim, S. H. January 1978. The three-dimensional structure of transfer RNA. *Scientific American* (offprint 1377).

Sutherland, B. M. 1981. Photoreactivation. *BioScience* 31:439.

Travers, A. A. 1974. Transcription of DNA. *Carolina Biology Readers* no. 75. Burlington, N.C.: Carolina Biological Supply Co.

Watson, J. D., and Crick, F. H. C. 1953. Molecular structure of nucleic acids: A structure of deoxyribose nucleic acid. *Nature* 171:737.

Principles
of Reproduction
and Development

A fundamental characteristic of all living things is the capacity for **reproduction.** This ability to produce offspring is vital to the continuation of species. In each species, new, vigorous, young organisms replace the aging and dying organisms, and the cycle of life continues.

Humans have a rather simple life cycle. Mature adults produce specialized reproductive cells, called **gametes.** These gametes—relatively large, nonmotile **eggs** in females and small, motile **sperm** in males—join to produce **zygotes,** each of which is a single cell that begins the life of a new individual organism.

But many other types of reproductive patterns are found among the numerous species of multicellular organisms. Many of these diverse reproductive strategies involve one form or another of **asexual reproduction,** the production of new individual organisms without the fusion of gametes.

Asexual Reproduction

The most common and direct form of asexual reproduction is mitotic cell division, which is the means by which unicellular organisms multiply. Reproduction by cell division confers a type of immortality on unicellular organisms that has no counterpart among multicellular organisms. While it is true that the life of each individual unicellular organism does end when it divides, the organism does not die, as all multicellular organisms eventually must. One individual simply divides to produce two new individuals. Genetically, of course, each of these new individuals is identical to the original individual (cell).

Some multicellular organisms also reproduce asexually by mitotic cell division. In algae, a series of cell divisions leads to production and release of a number of specialized reproductive cells called **zoospores,** each of which can develop into a new individual (figure 23.1).

Figure 23.1 Asexual reproduction in the green alga *Ulothrix.* Mitotic divisions produce flagellated swimming spores (zoospores), each of which can develop into a cell capable of starting a new filament.

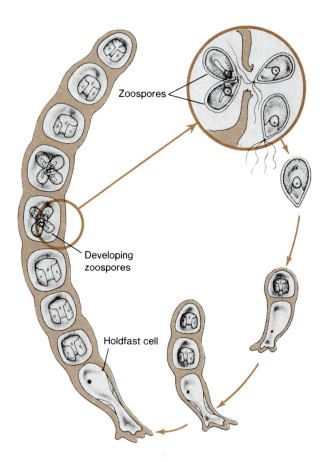

Zoospores

Developing zoospores

Holdfast cell

Other multicellular organisms reproduce by fragmenting or subdividing their bodies. New individuals grow from multicellular pieces of the original individual (figure 23.2). In genetic terms, this has the same result as reproduction by mitotic cell division because cells of the new individual produced by these fragmentation processes are genetically identical to the cells of the original parent individual.

Vegetative reproduction in plants is another kind of asexual reproduction. A new individual starts when some plant part, which is not a specialized reproductive structure, simply begins to grow into a new plant. The rooting response of strawberry plant runners in a spreading strawberry bed is a familiar example of the proliferation of individuals by vegetative reproduction (figure 23.3). Vegetative reproduction is economically important because an individual plant with especially desirable characteristics easily can be propagated. Small pieces (cuttings), especially pieces of stem, are treated so that they are induced to produce roots and develop into complete new individuals. Often, a number of plants can be grown from cuttings of one parent plant. Each of these plants has the same desirable traits of the original individual because all of them are genetically identical to the parent plant. Such groups of organisms are called **clones.** Members of a clone are derived directly from a single parent organism and are genetically identical with it and with one another.

Asexual reproduction also has some economic applications in animals. For example, sponge producers chop sponges into small pieces and place them where each piece can grow into a new sponge mass. Oyster fishermen, however, who chopped starfish to pieces to keep the starfish from preying on oysters, have found that asexual reproductive capacities in animals can have a detrimental economic effect. Chopping a starfish to pieces does not kill a starfish at all. Instead, each of the fragments grows into a complete starfish, and the total number of starfish actually increases in the process.

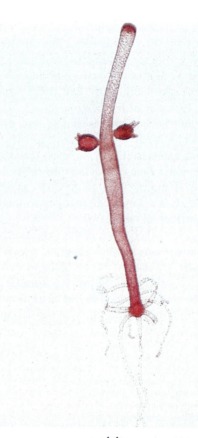

(a)

(b)

Figure 23.2 Asexual reproduction by fragmentation or direct outgrowth from the body. (a) Bud formation in *Hydra*. The bud eventually breaks off to become an independent, new individual. (b) Gemmae cups on the liverwort *Marchantia*. Multicellular bodies (gemmae) produced in these cups can develop directly into new plants when they are washed out on the ground.

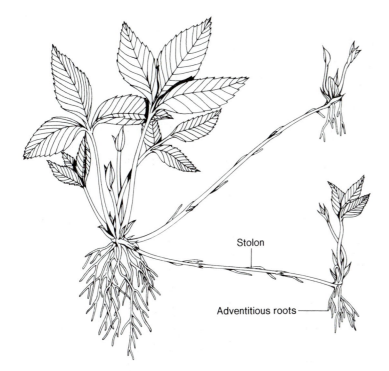

Figure 23.3 Rooting by strawberry plant stolons (runners) produces a new plant wherever adventitious roots are put down.

Stolon

Adventitious roots

Why Sexual Reproduction?

Despite the many reproductive mechanisms for the direct production of new individuals from single parents by asexual reproduction, sexual reproduction, which requires the presence of two parents together, is the dominant reproductive pattern in multicellular organisms.

One hypothesis for the prevalence of sexual reproduction is that it serves as a dispersal mechanism. For example, many marine animals produce countless thousands or even millions of eggs that, after fertilization, can be scattered over great distances to a variety of habitats by tides or currents. Larvae hatched from many eggs can swim considerable distances or be swept along by the same forces that disperse eggs. Asexual reproduction, however, also can result in the production of numerous offspring and thus function in dispersal in some cases. For this reason, dispersal does not seem to be the key adaptive advantage of sexual reproduction.

The most important advantages of sexual reproduction involve genetic and evolutionary considerations.

Genetically, sexual reproduction is associated with the diploid condition in ordinary body cells. This diploid condition masks the effects of harmful mutant genes, many of which are recessive. Many organisms that reproduce primarily asexually are haploid. Thus, harmful recessive genes are expressed, and they kill or weaken the individuals in which they occur.

For the evolutionary process, sexual reproduction is a source of variability. Sexual unions of reproductive cells from parents with different genetic makeups result in new genetic combinations. Offspring develop new combinations of characteristics that differ from the characteristics of either of their individual parents. Some of the individuals possessing these new combinations of characteristics may be better suited than others to live in either existing or newly encountered environments. Because these better-suited individuals are more likely to succeed in reproducing themselves than are other individuals, their genes and characteristics are likely to increase in frequency in the population. Sexual reproduction thus provides raw material for the evolutionary process in the form of new combinations of genes and the characteristics the genes produce.

In organisms that reproduce exclusively by asexual processes, mutation is the only source of new genetic variants that can be tested against diverse environments, but mutation frequencies usually are quite low.

Sexual Reproduction in Animals

Sexual reproduction is conducted at the single-cell level. A single egg and a single sperm combine to produce a zygote, the cell that carries the genetic design for a new individual. Therefore, in each generation, there is a return to the single-cell level as mature multicellular animals produce gametes involved in beginning the next generation. The process provides for continuation and renewal of the species as aging and dying individuals are replaced by new, young individuals.

In view of this, it might be tempting to think of development and maintenance of a body simply as a means to bridge the gap from one generation of gametes to the next generation of gametes. And, in fact, this idea of the body being simply a carrier of gametes has been emphasized at times during the history of biology. August Weismann, a German biologist who worked late in the last century and early in this century, proposed that gametes and body cells are two almost totally separate entities. He suggested that gametes belong to a **germ cell line** (also called germ plasm) that is continuous from generation to generation. And he said that body cells are part of a **somatic cell line** (also called somatoplasm) built up in each generation to house the germ cell line temporarily. But today adult bodies, and the gametes and developmental stages that connect adult generations, are thought of as parts of a life cycle that goes on generation after generation.

Reproductive Cycles

The majority of nonparasitic animals reproduce during only a relatively short time out of each year. Their reproductive activities are part of annual reproductive cycles that, in many animals, include only one breeding period each year. During that one period, the majority of individuals in a species come into breeding condition simultaneously, thus increasing the chances that fertilization will occur and that the offspring produced will encounter favorable conditions for their development and growth (figure 23.4).

Whatever reproductive strategies various species use, coordination of breeding periods is important for reproductive success in all cases. For example, because many marine invertebrate animals shed their gametes directly into the water, fertilization depends largely on random meetings of egg and sperm in the ocean. Part of the randomness of such patterns is reduced by the production and release of huge quantities of gametes (a female oyster may release sixty million eggs in one season). But even such prolific gamete production would do little to guarantee fertilization if individual animals shed gametes haphazardly on their own time schedules. Definite annual cycles ensure that many individual males and females shed gametes during the same time period.

Timing of reproductive periods is related to total reproductive strategy. Many animals produce and shed eggs that contain relatively small quantities of yolk or other stored nutrient material. Such eggs usually develop quickly into free-living individuals that can begin feeding. In many cases, this free-living form is a **larva**, an individual that is structurally different from the adults of the species, but capable of independent existence. Following a period of independent life, which may involve a progression through several larval stages, a marked structural modification known as **metamorphosis** converts the larva into an adult. Reproductive cycles in such larva-producing animals usually are timed so that breeding occurs when food will be immediately available for the larvae.

In mammals, the developing offspring is retained inside the female's body for a long period of time. Breeding in wild mammals is timed so that the end of the period of internal development (**gestation**) comes when conditions are favorable for the appearance of the newborn animal (figure 23.4b). Thus, many temperate zone mammals breed in the fall, when it would be difficult for young offspring to begin an independent life, but deliver their young in the spring, when conditions are becoming most favorable for young individuals.

(a)

Figure 23.4 Annual reproductive cycles are timed so that members of animal populations come into breeding conditions simultaneously at times that are favorable for reproductive success. (a) Amplexus, simultaneous gamete shedding by female and male frogs. Frogs breed when conditions are favorable for larval feeding and growth.
(b) Female white-tailed deer with her young fawn. Offspring of many wild mammals are born in favorable seasons after an extended period of internal development (gestation).

(b)

Timing Cues for Reproduction

The breeding period can be timed in several ways so that reproduction occurs when environmental conditions are favorable. Some animals remain in a reproductively ready state until temperature or moisture conditions change appropriately. Thus, the onset of favorable conditions directly triggers the reproductive response.

A good example of this is the flurry of reproductive activity in the temporary ponds created by occasional rainfall in deserts and other arid regions. Many of the aquatic invertebrates living in such environments can enter **diapause**, a dormant state in which metabolism is reduced and resistance to adverse environmental conditions is greatly increased. Diapause permits survival through long periods of drought or other environmental stresses. But when conditions change (for example, when a temporary pond forms), the animals quickly become active, breed, and offspring develop to a stage capable of entering diapause, all within a very short time. Thus, by the time the pond dries up, a new generation of diapausing individuals is ready to lie in wait for the next brief period of active life.

While such immediate responses to abrupt environmental changes are adaptive for many animals, the reproductive cycles of a very large percentage of other animals, both vertebrates and invertebrates, are timed by responses to day length changes (photoperiodic stimuli). Reproductive activities of many plants also are timed by photoperiodic changes (p. 513). Year after year, day length changes remain the most reliable indicators of the passage of the seasons. And response to photoperiodic changes permits organisms to anticipate the onset of favorable temperature, moisture, and food supply conditions locally, or even in distant places, as migratory animals must do. Use of reliably repeated seasonal time cues is especially important for migratory animals that must prepare to leave their wintering grounds and set off on their annual trek, even though they have no information about conditions on distant breeding grounds.

Some animals are not reproductively seasonal at all. For example, the human menstrual cycle continues summer and winter, year after year, as do the estrus cycles of some of our domesticated vertebrate animals.

Gamete Formation

Gametes are highly specialized cells that are adapted for their reproductive functions. The characteristic structural and functional specializations of eggs and sperm develop during the process of **gametogenesis** (gamete formation).

Sperm cells are relatively small cells that are specialized for motility, while egg cells are larger, nonmotile cells that, in many species, contain considerable quantities of stored materials that are used during the early development of the zygote. Both eggs and sperm also are specialized for the complex cellular interactions of the fertilization process itself.

In animals, gametogenesis involves more than the structural and functional development of gametes. The genetically important process of meiosis is integrally involved. Animal body cells normally have a diploid (2N) chromosome number, and each species of animals has a characteristic diploid chromosome number that remains the same from generation to generation. Because fertilization involves a fusion of two cells and brings together the cells' sets of chromosomes, some mechanism must reduce the chromosome number or it would double every generation. Meiosis is the process that reduces the chromosome number, and in animals, meiosis occurs during gamete formation. Thus, gametes are haploid (N), and gamete fusion during fertilization establishes the diploid chromosome number in the zygote. All body cells are mitotic descendants of the zygote, and thus they bear the diploid chromosome number. Therefore, the gametes are the only haploid cells in the animal life cycle (figure 23.5).

Figure 23.5 Generalized life history of an animal.

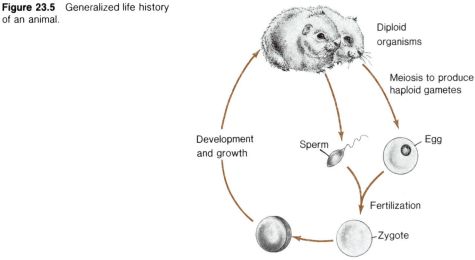

Diploid organisms

Meiosis to produce haploid gametes

Development and growth

Sperm

Egg

Fertilization

Zygote

Most animals have specialized reproductive organs, called gonads, within which meiosis and the structural and functional development of gametes take place. Sperm develop in the **testes** (singular: **testis**) of males, and eggs develop in the **ovaries** of females.

Spermatogenesis

A population of specialized cells called **spermatogonia** (singular: **spermatogonium**) are the ancestors of sperm cells that develop in the testes. In sperm formation (**spermatogenesis**), spermatogonia grow to become **primary spermatocytes.** These larger cells enter meiosis. The cells produced by the first of the two meiotic divisions are known as **secondary spermatocytes.** Each of these divides again in the second meiotic division, and at the completion of meiosis, there are four haploid cells, called **spermatids,** for each spermatogonium that began the process (figure 23.6a). Then the spermatids, which are fairly ordinary looking cells, proceed to develop into functional sperm cells.

Conversion of a spermatid into a sperm cell involves several important changes in the cell (figure 23.6b). The nuclear genetic material (chromatin) of the spermatid becomes condensed and is enclosed in the rounded **head** of the developing sperm. Much of the cytoplasm of the spermatid is lost as the sperm assumes a streamlined shape, an adaptation for swimming efficiency.

But mitochondria are retained, and they become tightly packed (and sometimes highly modified) in the **middle piece** of the sperm cell. Respiration in the mitochondria of the middle piece provides energy for the beating **tail** that propels the sperm when it begins to swim. The sperm cell tail, which has the same general organization found in flagella of all kinds (p. 90), propels the sperm cell. Energy for swimming is limited, however, because metabolic reserves of sperm cells are very limited, and the active, functional life of animal sperm cells is short. Sperm structure and function reflect an evolutionary compromise between weight conservation and streamlining on the one hand and nutrient reserves to provide energy for swimming on the other.

Each sperm cell has two centrioles (figure 23.6c). One of the centrioles is located near the chromatin in the sperm head, and the other lies at the base of the tail. Like the flagella and cilia of other motile cells, the sperm tail has a basal body, and this second centriole serves that function. Finally, a caplike covering, the **acrosome,** develops over the front tip of the sperm's head. The acrosome's functions differ in various species, but generally it functions in the fertilization process. Acrosomes release enzymes that help sperm to reach the egg cell plasma membrane. In sperm of some species, proximity to an egg causes a visible **acrosomal reaction,** in which a tubular extension, the **acrosomal process,** extends out from the front tip of the sperm. (The acrosomal process is discussed in more detail later in this chapter.)

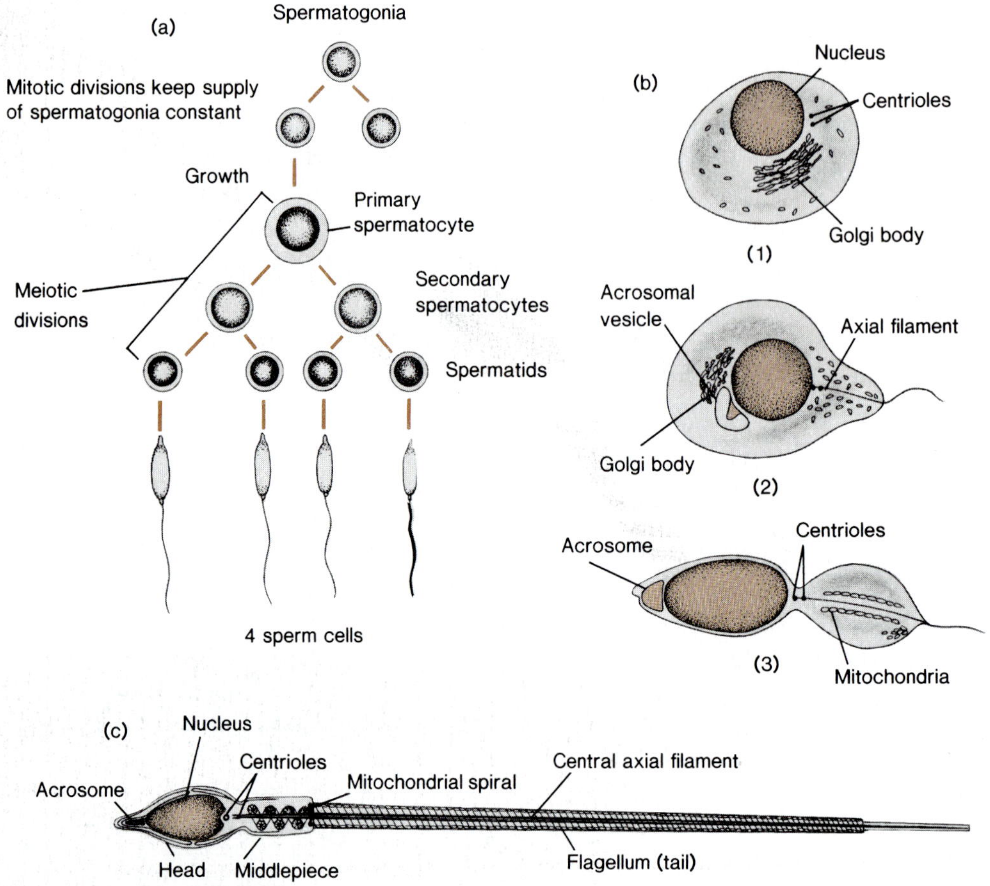

Figure 23.6 Spermatogenesis.
(a) Spermatogonium grows to
become primary spermatocyte.
Meiotic divisions produce four
spermatids, which become sperm
cells. (b) Stages in the structural
modification of a spermatid to
produce a functional sperm cell.
(c) An animal sperm with details
discovered with the electron
microscope.

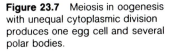

Figure 23.7 Meiosis in oogenesis
with unequal cytoplasmic division
produces one egg cell and several
polar bodies.

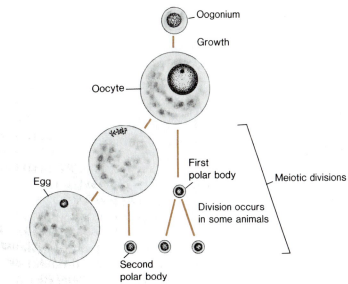

Oogonium

Growth

Oocyte

First
polar body

Meiotic divisions

Division occurs
in some animals

Egg

Second
polar body

Oogenesis

Meiosis in egg formation (**oogenesis**) is somewhat different from meiosis in spermatogenesis because the cytoplasmic divisions during oogenesis are very unequal. At each division, virtually all of the cytoplasm goes to one of the daughter cells, while the other daughter cell receives a nucleus and only a minimal amount of cytoplasm. These tiny cells, produced during oogenesis, are called **polar bodies** (figure 23.7). As a result of these unequal divisions, only one functional egg cell is produced from each **oogonium** (egg-producing cell) that grows into an oocyte and enters meiosis. The egg cell receives virtually all of the cytoplasm originally contained in the rather large oogonium as well as additional cytoplasm synthesized during pauses in the meiotic process. Cytoplasmic materials, which are to be used during early development of the zygote, are concentrated in one cell, and the polar bodies become reproductively nonfunctional by-products of the meiotic divisions in oogenesis. In contrast, four functional sperm cells eventually are produced as a result of meiotic divisions of one spermatogonium.

While spermatogenesis in the majority of animals is direct and may require only a matter of days or weeks, the meiotic divisions in oogenesis may include pauses over long portions of the individual female's total lifetime. For example, meiotic divisions of prospective egg cells begin in the ovaries of various young female vertebrate animals even before they are hatched (or born). Meiosis begins, but then is arrested in the prophase of the first meiotic division until the animal reaches maturity. Meiosis is resumed only at the time that eggs are being prepared for release from the ovary (**ovulation**). In human females the meiotic pause in some cells can last forty years or more.

Sexual Reproduction in Echinoderms

To illustrate some additional principles of sexual reproduction in animals, the reproduction and development of one group of animals, echinoderms, is discussed in detail. Emphasis is on sea urchins and sand dollars, which are commonly studied echinoderms.

In many animals (for example, humans) there is pronounced **sexual dimorphism;** that is, males and females are recognizably different in appearance and, in many species, in size as well.

Figure 23.8 (a) Adult sea urchin *Strongylocentrotus purpuratus*. (b) Location of sea urchin gonads and genital pores through which gametes are shed.

(a)

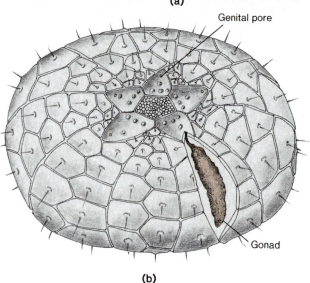

Genital pore

Gonad

(b)

But male and female echinoderms look alike. Thus, the only really effective means of sex determination in sea urchins, for example, is examination of the gametes that each animal sheds. Internally, a male sea urchin has a ring of testes whose very short ducts open to the outside of the body by way of **genital pores** (figure 23.8). Sperm are discharged through the genital pores directly into the water around the body. Female urchins also have a ring of gonads, and their ovaries discharge large numbers of eggs directly into the water through ducts and genital pores similar to those of the males. This shedding into the water makes fertilization a chancy proposition because there seems to be no specific mechanism directing eggs and sperm toward each other.

Sea urchins do tend to live in clusters, and they do release large numbers of gametes, but still, it is a big ocean! Thus, as with so many other animals, it is significant that most species of sea urchins have definite reproductive cycles, and individuals of a species in any given area generally become reproductively ready ("ripe") and shed gametes at the same time of year.

Fertilization: First Interactions

When sea urchin eggs and sperm come close together in the water, a series of prefertilization interactions begins. Each sea urchin egg is covered by a jelly coat that is "sticky" for sperm cells. Although there is no compelling evidence that eggs attract sperm cells from a distance,

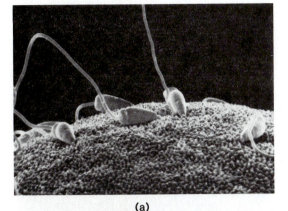

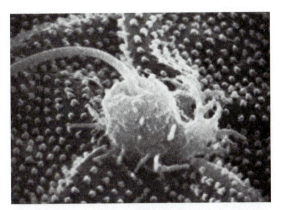

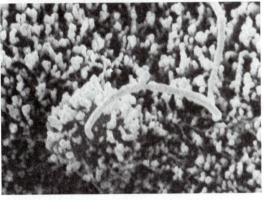

Figure 23.9 Scanning electron micrographs of early stages of fertilization. (a) Sperm attached to membranes on the surface of a sea urchin egg. (Magnification × 1,920) (b,c,d) Stages of sperm penetration through membranes and into the egg. The fertilization membrane was removed before micrographs (c) and (d) were made. (Magnification × 10,000, 4,355, and 6,750, respectively)

when sperm do make contact by random movement, they tend to adhere to the jelly coat (figure 23.9). This "stickiness" is species-specific because sperm tend to stick much more firmly to the jelly around eggs of their own species.

Another kind of interaction between egg and sperm before they actually touch one another is the acrosomal reaction of the sperm cell, which was mentioned earlier. When a sperm gets close to an egg, the acrosome releases enzymes, called **sperm lysins,** that dissolve holes in the material that surrounds the egg so that the sperm can reach the egg cell surface. In sea urchins and many other species, the acrosomal reaction has a second part. The acrosome reorganizes and sends out the acrosomal process.

As the sperm moves through the egg's jelly coat, aided by the action of sperm lysins, the acrosomal process is the first part of the sperm actually to contact the **vitelline membrane,** a membrane that lies just outside the egg cell's plasma membrane. On the vitelline membrane are specific receptor molecules, much like those that bind hormone molecules to the plasma membranes of target cells, that bind to molecules on the acrosome surface. Following this binding, the acrosome penetrates the vitelline membrane and fuses with the egg's plasma membrane, beginning an elaborate membrane fusion and reorganization process that permits the sperm to enter the egg (figure 23.10).

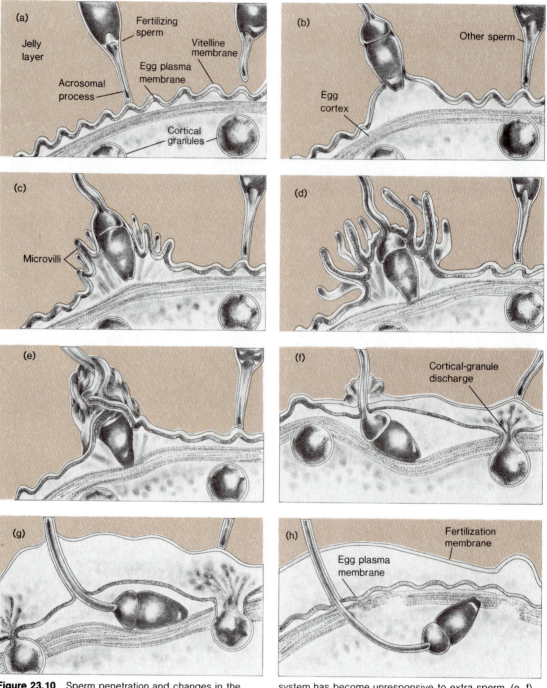

Figure 23.10 Sperm penetration and changes in the egg cortex. (a) Acrosomal process of sperm making initial contact with the vitelline membrane. (b, c, d) Filament penetrates vitelline membrane and makes contact with plasma membrane. Then an opening into the interior develops, and projections of the egg surface (microvilli) rise up around sperm. Other sperm are attached to sites on the vitelline membrane, but the system has become unresponsive to extra sperm. (e, f) Sperm moves into egg, cortical granules begin to discharge, and the vitelline membrane begins to lift away from the plasma membrane. (g, h) The vitelline membrane continues to raise off the surface of the egg. It is thickened by some of the proteins released by the cortical granules and transformed into the fertilization membrane.

But this is true of only the first sperm to contact the egg. After the initial contact, the egg surface quickly changes so that it is no longer responsive to contacts with additional sperm cells. A wave of negativity begins at the point of first sperm contact and spreads out rapidly over the entire surface in a matter of seconds. Nonresponsiveness to other sperm apparently serves as one of several blocks that prevent **polyspermy,** the entry of more than one sperm into a single egg. This is important because polyspermy, when it occurs, usually causes abnormal or arrested development in sea urchin eggs and other types of eggs that are relatively small and contain little yolk. An additional block to polyspermy is development of a tough protective envelope, the **fertilization membrane,** which forms within a minute or so following the initial sperm contact. It is produced as granules in the **cortex** of the egg (the area just under the egg's plasma membrane) open to the outside and release their contents into the space between the plasma membrane and the vitelline membrane lying just over it. Cortical granule contents include enzymes that loosen the vitelline membrane's contact with the egg and alter the vitelline membrane so that extra sperm previously attached to its surface come loose.

The fertilization membrane develops from the vitelline membrane that is elevated off the egg's surface and thickened. The vitelline membrane is elevated because material from the cortical granules causes water to enter the space between it and the plasma membrane by osmosis. This accumulating water expands the vitelline membrane, thus lifting it off the plasma membrane of the egg. Then, structural proteins from the granules unite with the vitelline membrane to thicken it into the fully developed fertilization membrane. The cortical response and membrane elevation begin where the first sperm makes contact, and then, spread peripherally from the point of sperm contact until the fertilization membrane is elevated and thickened over the egg's entire surface. The fully developed fertilization membrane is a virtually impenetrable barrier to sperm cells and is the final permanent block to polyspermy.

Egg Activation

Fertilization is often defined only in genetic terms because the fusion of the haploid sets of chromosomes in the gametes to produce the diploid zygote is emphasized. It should be apparent by now, though, that the egg-sperm interactions of fertilization begin long before the union of chromosomes, in fact, even before the gametes touch one another. Interactions continue through the time that egg and sperm cooperate to accomplish the sperm's entry into the egg. The egg does not passively wait for the motile sperm to burrow its way in; rather, the egg actively participates in the entry process once it has been contacted by the sperm.

Active participation in sperm entry is only the beginning of egg **activation** during fertilization, because activation involves many physical and biochemical changes in the egg cell. At the time of fertilization, the egg is a complex developmental system, poised and waiting for fertilization to stimulate it to develop further. Ordinarily, the stimulus that causes egg activation is contact with a sperm, but egg activation is not entirely dependent on sperm contact. It is possible experimentally to induce **parthenogenesis** (development of an egg without fertilization). Parthenogenesis occurs naturally in some species of animals, but it also can be induced in the eggs of other species. Various physical (for example, chilling, heating, pressure shocks) or chemical (for example, treatment with weak acids or bases) treatments of egg cells can set off some of the postfertilization reactions of egg cells. Sometimes, a biological agent is used, as in the case of frog eggs, which can be activated by pricking them with a needle that has frog blood on it. In some animal species, artificially activated eggs develop quite extensively. Over the years, observation of these responses to artificial activation agents and the ability of eggs to begin development without contact with sperm has led biologists to conclude that egg cells contain all of the information needed to direct early development, and modern biochemical studies have verified this conclusion. The egg seems to exert major control over early development; genetic information from the sperm becomes essential only after development is well under way. What, then, are some of these early activation responses of the egg cell?

One postfertilization response is the activation of DNA synthesis. Following sperm entry into the egg, the **pronuclei** (haploid nuclei of egg and sperm) migrate toward the center of the egg cell where they join on the mitotic spindle which has been assembled there in preparation for the first of the repeated cell divisions of the zygote. DNA synthesis has not occurred for some time in the egg and sperm nuclei, but now it must occur to complete the DNA replication and chromosome duplication that precede this first division and each of the subsequent cleavage divisions.

Protein synthesis also must become activated to meet the enzyme and structural protein needs of the new cells being produced by cleavage divisions. Furthermore, the cell division process itself requires synthesis of specific proteins, especially **tubulin** (a protein used to construct the microtubules of the mitotic spindle for each of the cell divisions), as well as histones, and other protein components of the chromosomes produced during these cell divisions.

Study of this early protein synthesis has revealed some intriguing facts about the messenger RNA (mRNA) that directs it. Although there is a striking increase in protein synthesis following fertilization, only a small amount of mRNA synthesis occurs in the early embryo. Much of the mRNA used for early protein synthesis is synthesized and stored in the oocyte before ovulation. Messenger RNA usually is thought of as a relatively unstable and short-lived molecule, but this mRNA is stored in a stable, inactive form and becomes involved in protein synthesis only after fertilization. Because this mRNA involved in early protein synthesis is produced by transcription of genetic information of only the oocyte, it is called **maternal mRNA,** to distinguish it from the mRNA produced and used directly when genetic information of both maternal and paternal origin is transcribed in the zygote (figure 23.11). In this and other ways, there is a distinct maternal dominance in the control of early development.

For many years, biologists have been seeking the "key reactions" of fertilization that actually initiate the egg's response. Increases in DNA and protein synthesis, which were identified as major responses to fertilization, do not begin for several minutes following sperm contact. What takes place in the meantime?

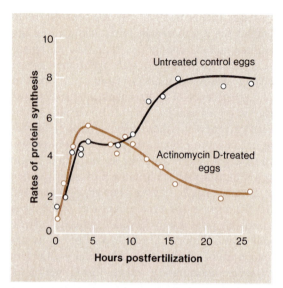

Figure 23.11 Experimental results that indicate the role of maternal mRNA in protein synthesis during early development. Protein synthesis increases in untreated eggs (black line) following fertilization. Actinomycin D, an antibiotic, inhibits messenger RNA synthesis. But actinomycin D-treated eggs (colored line) still show the initial postfertilization protein synthesis increase. This indicates that messenger RNA, already present before fertilization, is being used in early protein synthesis. Only later does the inhibition of new mRNA synthesis in the treated embryos result in deficient protein synthesis.

Synthesis and other cell activities in early development require energy, and, after fertilization, the metabolic rate of sea urchin eggs (and the eggs of many other animals as well) increases significantly. In fact, Otto Warburg discovered this metabolic rate increase in 1908, long before much was known about the biochemical changes in the egg following fertilization. When Warburg measured sea urchin eggs' oxygen consumption and found a sixfold increase following fertilization, he proposed that this metabolic acceleration was the "essence of fertilization." He said that before fertilization, eggs are metabolically depressed. He suggested that fertilization removes some sort of metabolic inhibition and that all subsequent development is set in motion by the metabolic rate increase following fertilization. This simple and direct hypothesis has not emerged as a general principle, however, because it has not withstood the test of further experiments. There is no change in the oxygen consumption of some other animals' eggs following fertilization, and the metabolic rates of some kinds of eggs (for example, eggs of some marine worms) actually *decrease* after fertilization!

Other changes initiated by sperm contact, however, may be basic to the activation process. For example, ion concentrations in the egg cell change. Shortly after sperm contact, a membrane permeability change permits sodium ions (Na^+) to enter the cell from the surrounding medium. This temporary permeability change occurs at the same time that the egg is becoming unresponsive to any sperm cells other than the first one that contacted it. As sodium ions are entering the cell, hydrogen ions (H^+) leave the cell, and this hydrogen ion for sodium ion exchange temporarily increases the pH inside the egg cell. This temporary pH change may play a role in activating enzymes involved in synthetic processes and other biochemical changes that follow fertilization. Free calcium ions (Ca^{2+}) also increase inside the cell, and the increased calcium concentration is somehow related to physical changes at the egg's surface, such as the elevation of the fertilization membrane.

Studies of fertilization interactions in various animals produce results that are of interest in themselves, but that also have potential for practical application in fertility control. Solutions to some infertility problems in humans and in economically important animals might emerge if more of the subtleties of fertilization responses were better understood. Such knowledge also might aid development of desirable new birth control techniques.

Postfertilization Development

The echinoderm zygote (like the zygotes of all other animals) carries a genetic blueprint specifying the characteristics of the adult organism that will develop. It also carries developmental information, a set of genetic instructions for the step-by-step construction of the new individual. Many of the developmental processes directed by these instructions are one-time events in the life history of the individual, but each must be completed successfully if the new individual is to develop normally.

Between zygote and body are many steps. The zygote usually is a relatively large single cell, while the body of a multicellular organism is an aggregate of many smaller cells arranged very precisely in specific spatial relationships. In various body areas, different populations of cells are structurally and functionally specialized (**differentiated**) to carry out the diverse functions needed in a multicellular body with functional "division of labor."

The postfertilization developmental program converts the unicellular zygote into a highly organized multicellular aggregate with definite regional specializations. This conversion begins with a series of mitotic cell divisions known as **cleavage.** Cleavage produces a cluster of smaller cells. Then the general structure of the body is organized by a series of cell movements and the segregation of groups of cells. These processes are known collectively as **gastrulation.** Finally, cells become differentiated and organized into functional groups to carry out specific functions in localized body areas. These processes are illustrated with photographs of developing sand dollar embryos.

Cleavage Cleavage is the process that converts the unicellular sand dollar zygote into a multicellular aggregate. During cleavage, each mitotic cell division produces two cells that divide again, and their daughter cells divide again, and so forth, all without growing between divisions. Cleavage divisions in echinoderm development, especially the early ones, occur in an orderly, predictable fashion. The first two cleavage divisions are longitudinal, but the third cleavage division cuts across horizontally at right angles to the first two. All of these cells divide, and the cycle of repeated cell divisions continues until a spherical mass of cells known as the **morula** stage of development is formed (figure 23.12).

As cleavage divisions continue in the morula, cells begin to move apart so that spaces appear among the cells in the center of the mass. Cells keep pulling away from the central area until, near the end of the period of most active cleavage divisions, the cells have organized themselves into a single-layered, hollow ball that surrounds a fluid-filled cavity known as the **blastocoel.** This hollow-sphere embryo, which develops at the end of cleavage, is called a **blastula** embryo.

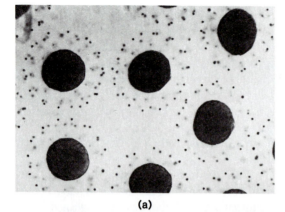

(a)

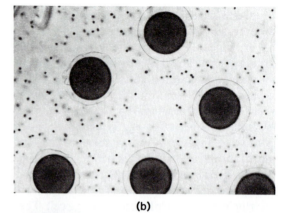

(b)

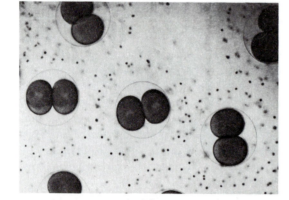

(c)

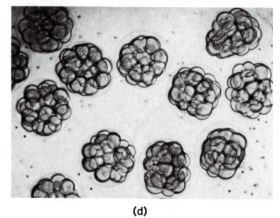

(d)

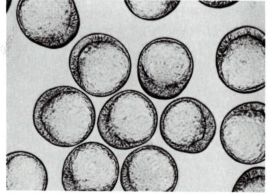

(e)

Figure 23.12 Cleavage in the sand dollar.
(a) Unfertilized eggs. (b) Fertilized eggs. Fertilization
membranes are complete. (c) First cleavage. (d) Sixteen
to thirty-two cell stage morula. (e) Blastula stage
embryos.

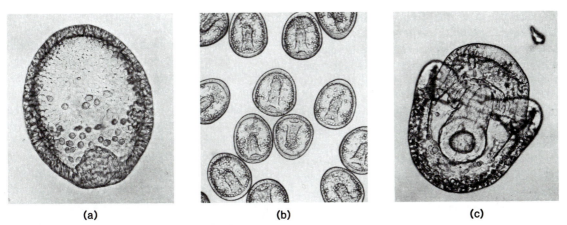

(a)   (b)   (c)

Figure 23.13   Gastrulation. (a) Sand dollar blastula with
primary mesenchyme cells. (b) Mid-gastrula stage of
sand dollar. (c) Early pluteus larva of sand dollar.

Development of the hollow blastula embryo
requires that the embryo's cells behave and move
in a specific way. This is only the first of several
cell movement patterns that are essential to the
orderly conversion of a loosely organized aggregate of cells into a complexly organized body.
These specific cell movements are one-time events
in the life history of the individual and must be
conducted precisely if development is to proceed
normally. These changes in the embryo's organization, which depend on cell movements and
cell shape changes, are part of **morphogenesis**
(meaning "development of form"), the progressive development of pattern and form of the developing body.

The mitotic divisions of cleavage thus subdivide the cytoplasm of the egg cell into smaller
units, the cleavage cells. And specific cell movement organizes the cells into a hollow sphere, the
blastula. Through all of this, some of the basic
organization of the egg cytoplasm is preserved;
that is, the various areas of egg cytoplasm relative to one another are not displaced. For example, a pigmented band present in an egg cell
is represented by a pigmented band in the blastula. It is likely that possession of specific kinds
of cytoplasm, inherited during cleavage from specific areas of the egg cell, may determine the
subsequent fates of cells in various locations in
the wall of the blastula.

Gastrulation   The next steps in organizing the
echinoderm embryo's body involve additional cell
movements and shape changes. Collectively,
these processes, known as **gastrulation,** convert
the simple, single-layered sphere of the blastula
into a several-layered body with structurally and
functionally specialized areas. Gastrulation in
echinoderms begins when one side of the blastula
wall sinks in to form a pit. This process can be
visualized by imagining the indentation formed
if a finger is pushed into one side of a balloon or
a soft, hollow rubber ball. To cause this shape
change, all of the cells in the area must change
their shapes at the same time in a display of
collective cell behavior (figure 23.13).

A second aspect of gastrulation, this one involving individual cell behavior, becomes apparent even as this sinking process is beginning.
Some of the cells in the depressed area pull loose
from their positions in the wall of the blastula
and come to lie free inside the blastocoel cavity.
Soon, these cells move out to strategic locations
in the embryo, where they position themselves
and produce the skeletal spicules, which support
the body of the developing larva. The cells that
set off individually on this specialized developmental course are known as **primary mesenchyme
cells.**

Sinking in of the blastula wall continues until the pit resembles the finger of a glove. This lengthening hollow structure is the developing gut, the future digestive tract. Cells at the tip of the developing gut send out long processes that contact the inside wall of the blastula and adhere to it. Then the processes contract, producing a cellular pulling force that draws along the developing gut. Finally, the gut contacts another point in the wall of the still hollow embryo. There, a fusion with the wall and a breakthrough opens the gut to the outside. This newly produced opening becomes the **mouth,** and the opening into the original pit on the side where the sinking process began becomes the **anus,** the posterior exit from the larval digestive tract. Thus, as a result of gastrulation, the hollow sphere of the blastula stage embryo has been converted into a body that has several layers and includes a complete gut, beginning with a mouth and ending with an anus.

Cell Differentiation and Body Organization

The fundamental body layers, or **germ layers,** that characterize the basic body plan of the more complex, multicellular animals are seen in fairly simple form in the echinoderm embryo (figure 23.14).

Cells remaining on the surface after gastrulation make up a body surface layer known as **ectoderm.** Ectodermal cells make up the "skin" of the echinoderm embryo, and they become ciliated, thus permitting the embryo to swim.

Gut cells constitute the innermost body layer, the **endoderm.** In animals, in general, endoderm is the layer that produces the digestive system lining and derivatives of the lining. The larval echinoderm gut quickly becomes functional, and the larva begins to feed on small particles from the surrounding water.

Tissue between ectoderm and endoderm forms the middle body layer, the **mesoderm.** The mesoderm of the echinoderm embryo develops from primary mesenchyme cells and some additional tissue that also separates from the developing gut. Mesodermal cells produce the larval skeleton, and mesodermal tissue lines the body cavity or **coelom,** the space between internal organs and the outer body wall. Eventually, in the adult echinoderm, mesoderm produces a number of structures that make up a considerable part of the body mass.

Vertebrate Development

The same three basic body layers (germ layers) that develop in echinoderms—the ectoderm, endoderm, and mesoderm—also develop in vertebrate embryos. In vertebrates, the outer (epidermal) layer of the skin is derived from the ectoderm of the embryo. The ectoderm also gives rise to the nervous system and the sense organs. As in echinoderms, the endoderm (the innermost body layer) of vertebrate embryos produces the lining of the digestive tract (gut) and various digestive tract derivatives. The middle (mesodermal) layer of the vertebrate embryo produces the skeletal, muscular, circulatory, excretory, and reproductive systems.

Despite some general organizational similarities, there are some noteworthy differences between early developmental processes in vertebrate embryos and those in early echinoderm development. These differences are apparent beginning with cleavage and gastrulation. An important factor in these differences is the relationship between the amount of stored nutrient material (yolk) in eggs and the eggs' developmental patterns.

Amphibian Development

Amphibian eggs contain considerable yolk, much of it concentrated in one hemisphere of the egg in a yolky area known as the **vegetal pole.** The opposite side of the egg, called the **animal pole,** is darkly pigmented and has much less yolk. The yolky cytoplasm in the vegetal pole side of the egg is heavier than the animal pole cytoplasm, and when amphibian eggs are free to rotate in their jelly, the vegetal pole becomes oriented downward in response to gravity (figure 23.15).

When an amphibian zygote divides, the cleavage furrow that cuts the cytoplasm in half moves quickly through the animal pole but is slowed by the bulky yolk in the vegetal pole. Even as the first division is being completed, a second set of cleavage furrows begins to divide each of the first two cells to produce four cells. Then, as in sea urchins, the division plane changes at the third cleavage (figure 23.16). This third division yields two distinctly different sets of cells—four smaller, darkly pigmented cells lying above four larger, yolky cells. As cleavage continues, the smaller animal pole cells divide more rapidly, and they produce a population of uniformly shaped,

Figure 23.14 A fully developed
pluteus larva. The mouth has
formed, and three basic body layers
have differentiated. Skeletal spicules
are produced by mesoderm (primary
mesenchyme cells).

Skin (ectoderm)

Gut (endoderm)

Mouth

Skeleton
(produced by mesoderm)

Anus

Figure 23.15 Frog (*Rana
temporaria*) zygotes surrounded by
their jelly coats.

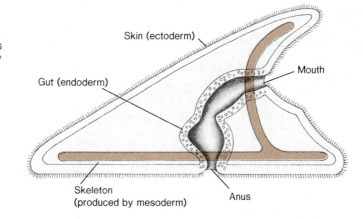

(a)

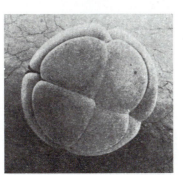

(b)

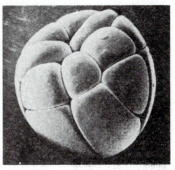

(c)

Figure 23.16 Scanning electron
micrographs of frog (*Rana pipiens*)
embryo cleavage stages. (a) The
first cleavage division. The cleavage
furrow cuts through the animal pole
(upper left) quickly and then slows in
the yolky vegetal pole.
(Magnification × 29) (b) The eight-
cell stage at the end of the third
cleavage viewed from above. Four
smaller cells lie above four larger,
yolky cells. (Magnification × 32)
(c) The sixteen-cell stage showing
the size difference between animal
pole cells (upper left) and yolky
vegetal pole cells (lower right).
(Magnification × 32)

small pigmented cells. The larger vegetal pole cells divide more slowly and produce a population of larger, less regularly shaped cells. Differences between animal and vegetal pole portions of the embryo also are apparent when the embryonic cells organize into a blastula because the blastocoel is displaced toward the pigmented animal pole (figure 23.17).

Gastrulation in the amphibian embryo involves extensive movement by all of the cells in the blastula. Smaller animal pole cells move downward to envelop the larger, yolky cells of the vegetal pole. Some of the animal pole cells from the surface of the embryo turn inward and move to the interior through a structure called the **blastopore**. Once surface cells have moved down into the blastopore and reached the interior of the embryo, they turn, move away from the blastopore, and organize the mesoderm of the embryo. Cells that remain at the surface after these movements produce the ectoderm. Yolky vegetal pole cells rearrange themselves to produce the gut; that is, they become the endoderm.

Bird Development

The avian (bird) egg proper is what is called "yolk" in everyday terminology. The albumen ("egg white"), the shell, and the adjacent shell membranes are accessory structures that surround and protect the embryo, which develops inside them.

Patterns of cleavage and gastrulation in birds are very different from those of amphibians. The large mass of yolk is not cleaved during development of the avian zygote. In fact, it may be physically impossible for ordinary cell division processes to cleave that much yolk. Cleavage is restricted to a small quantity of developmentally active cytoplasm on one side of the yolk. The cleavage divisions there produce a flat disc of cells called the **blastoderm** (figure 23.18).

The cells in the **blastoderm** then sort into two distinct layers—an upper **epiblast** and a lower **hypoblast.** Subsequent gastrulation movements involve certain cells of the epiblast that migrate toward and down into a shallow depression called the **primitive streak.** After moving

Figure 23.17 Amphibian gastrulation. (a) A section cut through a frog blastula-stage embryo showing the location of the blastocoel, which is displaced toward the animal pole side. (b) A section of a frog embryo at the beginning of gastrulation. Small animal pole cells move downward to surround and enclose the yolky vegetal pole cells. Some animal pole cells leave the surface and migrate to the interior through the blastopore (colored arrow). When they reach the interior, they migrate away from the blastopore and organize the embryonic mesoderm. (c) A scanning electron micrograph showing the surface of a frog embryo at the time that cells begin to move into the blastopore. The first part of the blastopore to form is called the dorsal lip (DL). (Magnification × 35) (d) Scanning electron micrograph of a later stage of frog gastrulation. As gastrulation proceeds, the blastopore becomes circular with cells migrating inward all around it. Small animal pole cells have completely enclosed the yolky vegetal pole cells except for the yolk plug (YP), made up of vegetal pole cells that lie in the center of the blastopore. (Magnification × 31)

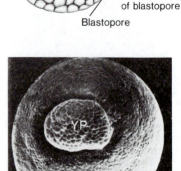

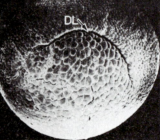

(c)   (d)

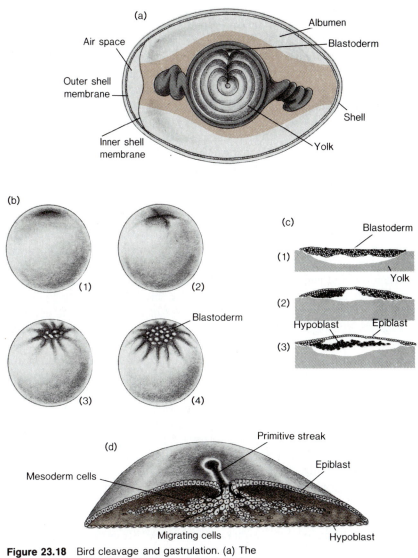

Figure 23.18 Bird cleavage and gastrulation. (a) The avian (bird) egg. The "yolk" is the egg proper, and development is restricted to an area of active cytoplasm at one side of the egg cell. Albumen, shell membranes, and shell are accessory structures that enclose and protect the developing embryo. (b) Cleavage divisions (1–4) in birds produce an embryonic disc (the blastoderm) at one side of the yolky egg. (c) Cells of the blastoderm (1–3) sort into two separate layers, an upper epiblast and a lower hypoblast. (d) Some of the cells of the epiblast migrate toward, down into, and away from the primitive streak to become the mesoderm of the avian embryo.

downward into the space between the epiblast and hypoblast, many of these migrating cells move outward to produce the mesoderm of the avian embryo, which is a new, third layer located between the two previously existing layers.

Epiblast cells that remain in the top layer constitute the ectoderm, while the endoderm of the avian embryo is made up of the original hypoblast plus some cells that come down from the primitive streak to join it.

Nervous System Development

Development of the nervous system is a good example of morphogenesis of an organ system in a developing vertebrate embryo. The morphogenetic processes involved in establishment of the nervous system are called **neurulation,** and neurulation is very similar in amphibian and bird embryos (and in other vertebrate embryos, as well). Amphibian and bird embryos are organized quite differently at the time that neurulation takes place, however. The amphibian embryo at the end of gastrulation is oblong with a basically tubular internal body organization, while the avian embryo consists of several virtually flat layers (figure 23.19).

Neurulation in both amphibian and avian embryos is a process in which the ectoderm thickens along what will be the dorsal midline of the body to form a **neural plate.** Cells in the plate change shape due to contractions of microfilaments that are strategically located inside them. The collective action of these cell shape changes cause the neural plate to roll up first into a pair of neural folds and then into a **neural tube.** The cell shape changes play a major role in the morphogenetic process.

The entire central nervous system (brain and spinal cord) of a vertebrate embryo develops from the neural tube.

Further Development of Flat Embryos

Eventually the flat, three-layered avian embryo rolls up to produce a tubular body form that is similar to that of amphibian embryos and that characterizes all vertebrate bodies. But during all of these changes, the large yolk mass of the avian egg is not incorporated into the embryo. Rather,

a structure known as the yolk sac, which is attached to the embryo, encloses the yolk and absorbs nutrients from it. These absorbed substances are transported through blood vessels of the yolk sac into the embryo, where they are used to support the embryo's growth and development.

The basic pattern of development in reptiles, which also have large yolky eggs, is very similar to that of birds. Cleavage is restricted to a small area of active cytoplasm, and the embryo goes through a stage when it is a circular disc consisting of relatively flat layers.

Mammals' eggs contain very little yolk, and the zygotes of mammals divide completely during cleavage. But the mammalian embryo still passes through a stage when it is a circular disc with relatively flat layers, and it assumes a tubular form after passing through this flat stage. Thus, the general pattern of early embryonic development in mammals closely resembles the pattern seen in reptiles and birds, whose eggs have huge quantities of yolk, despite the fact that the great majority of modern mammals produce eggs that contain very little yolk. This developmental similarity with reptiles and birds strongly suggests that mammals are descendants of animals that had much larger eggs containing much more yolk than do the eggs of present-day mammals.

Parthenogenesis

The echinoderm reproductive pattern represents a rather average animal life cycle and typifies the reproductive patterns seen in the majority of animals. But not all animals fit this pattern. One element that makes some animals' reproductive patterns exceptional is development by parthenogenesis. Artificial parthenogenesis (that is, experimental treatments that induce eggs to develop without fertilization) has already been discussed. But parthenogenesis also occurs *naturally* in many animals, and it is a regular part of the reproductive strategy of some animals.

For example, male honeybees develop by parthenogenesis. The queen bee receives a lifetime supply of sperm cells during her nuptial flight. She stores these sperm and controls fertilization of the eggs that she lays. When she permits fertilization, the eggs that she lays develop into females, workers, and the few young queens.

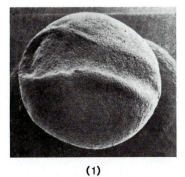
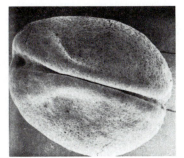

(1)

(2)

(a)

Figure 23.19 Nervous system
development. (a) Scanning electron
micrographs of frog embryos. In (1),
the outlines of the neural plate can
be seen as the edges of the plate
begin to roll up as neural folds.
(Magnification × 23) In (2), neural
folds approach one another as
neural tube formation is nearly
completed. (Magnification × 30)
(b) Scanning electron micrograph of
part of the developing nervous
system of a chick embryo showing
neural folds. (Magnification × 120)
(c) A series of sections showing the
process of neural tube formation in
the chick embryo. (d) Diagrams
showing how the contraction of
microfilaments strategically located
in cells brings about shape changes
in individual cells, and in entire
sheets of cells, such as those
involved in nervous system
development.

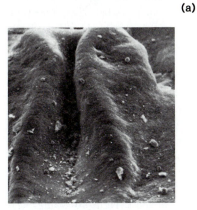

(b)

(c)

Neural plate

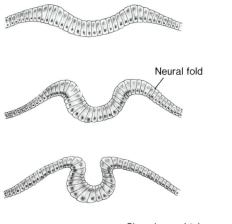

Neural fold

(d)

(1)

(2)

Contracted

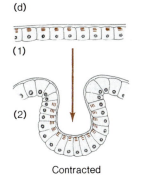

Closed neural tube

Skin ectoderm

When she withholds sperm, thus preventing fertilization, the eggs develop into males (drones) (figure 23.20). These male bees retain the haploid chromosome number of the egg cell in all of their body cells, including the testis cells that eventually produce gametes. Therefore, meiosis is not required for sperm production because testis cells that are already haploid can be converted directly into sperm cells.

Parthenogenetic development, however, does not necessarily lead to a general haploid condition in all body cells as it does in honeybees. In some other parthenogenetically reproducing species, the second polar body is not formed, and two haploid sets of chromosomes are retained. In other animals, there is an initial chromosome duplication at the beginning of cleavage, and this chromosome duplication is not followed by mitosis. Either of these methods provides the new individual with a diploid chromosome number, but of course, it is a double set of maternal chromosomes.

Some organisms, such as the common, small aquatic animals known as rotifers, produce eggs that develop directly by parthenogenesis at certain times of the year and then switch over to ordinary sexual reproduction with fertilization at other times of the year. Still other animals, such as some gastrotrichs (tiny creatures similar to rotifers), have no males at all and exist exclusively as populations of parthenogenetically reproducing females.

Life Cycle Complexity in Animals

Many animals have complex life cycles that include several different forms of reproduction. This is particularly true of some internal parasites, whose life cycles include both sexual and asexual reproductive processes. The oriental human liver fluke *Opisthorchis (Clonorchis) sinensis)* is an excellent example of an organism that exploits diverse reproductive possibilities within a complex life cycle (figure 23.21a).

Adult flukes live in bile passages in the human liver and are **hermaphroditic;** that is, each individual has complete male and female reproductive systems. An individual fluke can produce fertile eggs by itself, although sperm exchanges do take place when several animals are together. Adult flukes can live for a number of years (five to twenty) in the bile ducts, and they can release hundreds or even thousands of eggs each day.

The eggs pass through the common bile duct from the liver to the small intestine and eventually move through the digestive tract and exit with the feces.

If human feces bearing eggs enter water, it becomes possible for the eggs to be eaten by an appropriate type of snail (figure 23.21b). Once the egg has been eaten by a snail, a small larva, the *miracidium,* hatches and burrows into the body of the snail, where it forms a cyst. A series of asexual reproductions takes place within this cyst. Eventually, a large number of another type of larva, the *cercaria,* are formed, and these ultimately break out of the body of the snail into the water.

Cercaria larvae have a short life span, but if they contact an appropriate fish during this brief period, they burrow into the fish's body and form another larval stage. This last larval stage can remain in an inactive condition imbedded as a cyst in the muscles of the fish's body for long periods of time.

Finally, if the fish is eaten raw or poorly cooked by a human, the cycle is complete. The cyst opens, and the worm finds its way up the common bile duct into the liver and becomes sexually mature in the bile passages. Soon, this newly matured adult begins to produce and release large numbers of eggs.

The impressive number of eggs that can be produced by a single liver fluke only partially describes the fluke's reproductive power. This already large reproductive potential is multiplied by the asexual reproduction that occurs within the snail's body. It has been estimated that the hatching of a single egg in a snail's body can ultimately lead to the release of as many as 300 cercaria larvae.

The tremendous reproductive potential of the liver fluke is essential, however, because there is a large margin of error involved in the transfers from human to snail, from snail to fish, and fish to human again. At each step in the chain of transfers, the majority of individuals are lost. The huge numbers of individuals involved help to ensure some successful passages from host to host so that the life cycle continues. This strategy of producing many offspring, thus increasing the odds that some will survive and mature, is common not only among parasitic animals, such as the human liver fluke, but also in many other animals that provide their offspring with little or no parental care or attention.

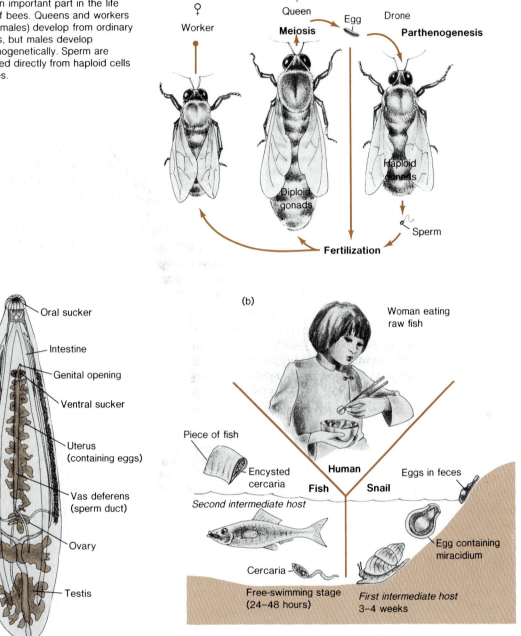

Figure 23.20 Parthenogenesis
plays an important part in the life
cycle of bees. Queens and workers
(also females) develop from ordinary
zygotes, but males develop
parthenogenetically. Sperm are
produced directly from haploid cells
in testes.

(a)

Oral sucker

Intestine

Genital opening

Ventral sucker

Uterus
(containing eggs)

Vas deferens
(sperm duct)

Ovary

Testis

Figure 23.21 (a) The human liver fluke *Clonorchis
sinensis.* Each individual has both female and male
reproductive structures. The fluke lives attached to the
wall of a liver bile passage and releases hundreds of
eggs each day. (b) Simplified life history of the human
liver fluke.

Sexual Reproduction and Plant Life Cycles

Plant life cycles are fundamentally different from the majority of animal life cycles because, in plants, meiosis usually is not directly involved in gamete production. Meiosis produces spores, and these haploid spores develop into haploid bodies (that is, multicellular structures consisting entirely of haploid cells). Then, gametes are produced by direct differentiation of already haploid cells within specialized areas of these bodies. Gametes fuse to produce diploid zygotes, which grow into diploid bodies (plants whose body cells are all diploid). The life cycle is completed when meiosis, occurring in specialized areas of these diploid bodies, produces haploid spores (figure 23.22).

Development of haploid spores into haploid bodies means that plants have two body forms (the haploid body and the diploid body) that alternate with one another in plant life histories. This is called **alternation of generations.** The diploid, spore-producing plant body is known as the **sporophyte generation,** and the haploid, gamete-producing plant body is known as the **gametophyte generation.**

While this general life history description can apply to all multicellular plants, there are many variations on the basic plan. Variations in reproductive strategies and details of life histories are related to habitats and general biology of the various kinds of plants (chapter 36). The highly specialized flowering vascular plants, the **angiosperms,** are the focus in this section.

Flowers

The majority of angiosperms are terrestrial plants, and their reproductive strategies are tied to their habitat. The delicate, unicellular reproductive units (spores and gametes) of angiosperms are enclosed and protected inside **flowers.**

A flower is a specialized shoot with a cluster of highly modified leaves, and it forms a protective envelope around the areas where cellular reproductive events take place. The modified leaves are arranged in concentric rings, or **whorls,** which are attached to a modified stem tip, the **receptacle.** The modified leaves in the lowest, outermost whorl are called **sepals,** and the next whorl inward consists of **petals** (figure 23.23). Sepals frequently are green and quite similar to ordinary foliage leaves, while petals often are large and colorful. Sepals and petals, while not directly involved in reproductive processes, attract insect or bird pollinators in those plant species that depend on this process for reproduction.

The whorl inside the petals consists of **stamens,** and the innermost whorl is the **pistil.** Stamens are very highly modified and not very leaflike in appearance, while the pistil of most flowers consists of several very highly modified leaves that are fused into a single unit. Many, but certainly not all, flowers contain both stamens and pistils. In some plants, separate flowers with stamens but no pistils (**staminate flowers**) and flowers with pistils but no stamens (**pistillate flowers**) are borne on the same plant body. In other plants, staminate and pistillate flowers are borne on entirely separate plants.

Spores and Gametophytes

Flowering plants produce two different kinds of spores. **Megaspores** develop into female gametophytes within the pistil, and **microspores** become modified into **pollen grains.** Pollen grains actually are very small male gametophytes that are released and carried by the wind or by animal pollinators.

The enlarged base of the pistil, the **ovary,** contains from one to many **ovules** depending on the plant species. Each ovule is the site of production of a functional megaspore. Actual megaspore production begins with a **megaspore mother cell** that divides meiotically to produce four haploid **megaspores.** Only one megaspore eventually produces a female gametophyte structure, while the other three degenerate without developing further.

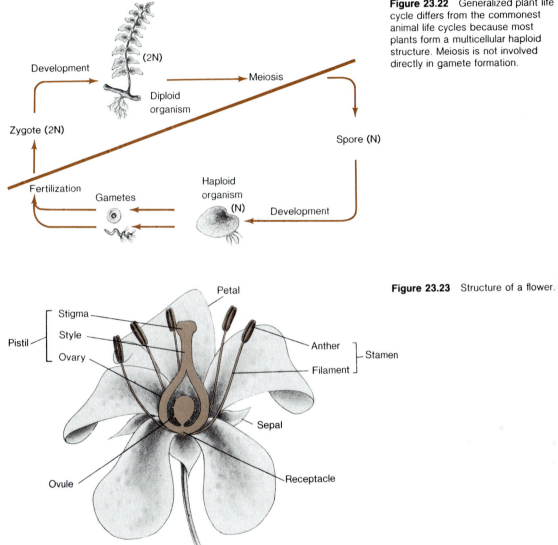

Figure 23.22   Generalized plant life cycle differs from the commonest animal life cycles because most plants form a multicellular haploid structure. Meiosis is not involved directly in gamete formation.

Figure 23.23   Structure of a flower.

Female gametophyte (**megagametophyte**) development also takes place inside the ovule where the megaspore was produced (figure 23.24). The functional megaspore cell expands, and three successive mitotic divisions produce eight nuclei in this one expanded cell. Three nuclei are clustered at each end of the cell, and the other two nuclei, the **polar nuclei,** are located in the middle of the cell. Cell walls develop around the nuclei at the ends to form clusters of three small cells. The polar nuclei are left in the middle in a larger, seventh cell. This seven-celled structure is called the **embryo sac,** and it is the fully developed female gametophyte. Thus, the haploid generation develops completely while still enclosed inside the structure that originally produced the megaspore. Clearly, the haploid generation is a very reduced and inconspicuous part of the flowering plant's life history.

One of the cells at one end of the embryo sac becomes the functional egg cell. The other two cells at the egg-cell end and the three cells at the opposite end all degenerate without playing any obvious role in reproduction. The large cell in the middle of the embryo sac with its two polar nuclei is involved in the formation of an important accessory nutrient storage structure in the seed called the **endosperm.**

Microspores are produced in pollen sacs in the **anther,** which sits atop the **filament** at the tip of the stamen (figure 23.24). Specialized **microspore mother cells** enter meiosis, and four microspores are produced per microspore mother cell. External ornamentation develops on the cell wall as the microspore becomes a functional pollen grain. Internally, the haploid nucleus of the developing pollen grain divides mitotically to produce two nuclei, the **tube nucleus** and the **generative nucleus.** Pollen grains are released in this condition, and they develop further only if **pollination** occurs; that is, if they land on the appropriate part of the pistil. There they germinate and continue male gametophyte development.

Pollination and Fertilization

Pollen grains land on the **stigma,** which is the sticky upper tip of the pistil. Contact with the stigma induces pollen grains to germinate and begin a characteristic growth response. A long outgrowth from the pollen grain, the **pollen tube,** grows down through the **style** and into the ovary.

During pollen tube growth, the two nuclei of the pollen grain enter the tube. The tube nucleus remains near the tip of the growing tube. The generative nucleus divides mitotically to produce two haploid **sperm nuclei.** The sperm nuclei are located just back from the tube tip, which enters the ovule through an opening called the **micropyle** and approaches the embryo sac.

Flowering plant fertilization actually involves two separate nuclear fusions. One of the two sperm brought into the embryo sac by pollen tube growth fuses with the egg cell to produce the diploid zygote, while the other sperm fuses with the two polar nuclei. This latter fusion brings together three haploid nuclei (the sperm nucleus and the two polar nuclei) and produces a triploid (3N) nucleus known as the primary endosperm nucleus.

Early Zygote Development

The primary endosperm nucleus divides repeatedly to establish a multicellular endosperm, which stores nutrients in the developing seed. The extent of endosperm growth and the endosperm's relative importance for food storage differ among angiosperms, but in many, the endosperm is the main storage site in the seed.

Zygote development begins with a transverse cell division that produces two cells, a terminal cell and a larger basal cell near the micropyle. Several sequential transverse cell divisions follow an initial division of the basal cell. These divisions produce a linear chain of cells, the **suspensor,** which is an embryo attachment structure.

Terminal cell divisions occur in several planes so that a flat plate of cells, the **embryo,** is produced. Further specifically oriented cell divisions and regional specializations carry development of the embryo body to the point where it is suspended until the seed containing it germinates.

The embryo body consists of several major areas (figure 23.25). A **hypocotyl,** attached at its end to the suspensor, develops into the lower part of the stem and the underground parts of the plant. There is a definite *polarity* in the development of the embryo; that is, the embryo is oriented in a specific way relative to the ovule. The hypocotyl is the end of the embryo that lies toward the suspensor (and, therefore, the micropyle) end of the ovule. At the opposite end of the embryo body, seed leaves, or **cotyledons,** develop.

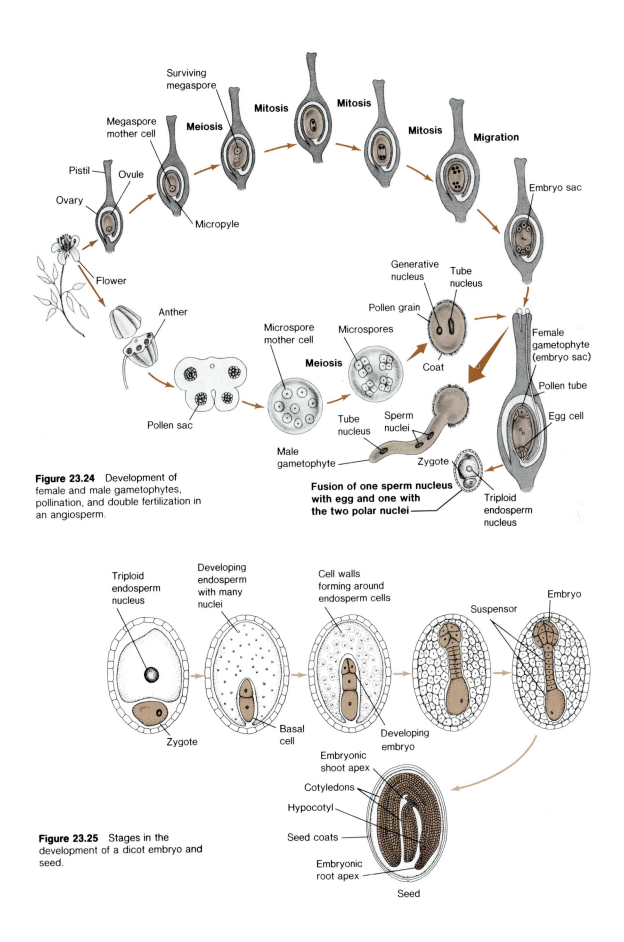

Figure 23.24 Development of female and male gametophytes, pollination, and double fertilization in an angiosperm.

Figure 23.25 Stages in the development of a dicot embryo and seed.

Some angiosperms (the dicots) have two cotyledons and some (the monocots) have only one. As the description "seed leaves" implies, the cotyledons are temporary structures and are lost in postgermination growth. In some angiosperms, especially in those species having a large endosperm, the cotyledons are relatively small. But in other species, the cotyledons may be very large and serve as the major nutrient storage sites in the seed. The large, fleshy cotyledons of bean seeds are familiar to anyone who has seen beans split open while they were being soaked or cooked. Near the base of the cotyledons is the **shoot apex.** While the shoot apex may seem a fairly inconspicuous part of the embryo, it is actually the future growing source of practically all of the permanent aboveground parts of the plant.

The sites of primary growth of the stem and root are established in the embryo. **Meristematic regions** (centers of continuing cell division activity) are located in the shoot apex and also in the **root apex** at the tip of the hypocotyl. When the seed germinates, active proliferation in these regions lengthen the shoot and root of the growing seedling.

Seeds

The embryo, packed in with the endosperm, is surrounded by ovule tissue. Parts of the ovule harden to form tough, protective **seed coats.** Most seeds then dry out until their water content falls to very low levels (5 to 20 percent), and they maintain only a minimal maintenance level of metabolism. In many species, the remainder of the ovary develops into a **fruit** around the seed.

Embryos within seeds develop further only after seed **germination.** Germination occurs when a dry seed, in a favorable site, takes up water and swells. Then, metabolic activities of the embryo increase and rapid growth resumes. The seedling grows out of the seed and develops into a mature, diploid plant body (sporophyte).

Some kinds of seeds do not respond directly to conditions that generally induce seed germination because they are **dormant;** that is, they are not physiologically responsive even when subjected to conditions that are fully appropriate for germination. Such dormant seeds require exposure to some specific environmental stimulus (in many cases, the required stimulus is a period of chilling) before they mature to the point where they can germinate.

Summary

All organisms have the capacity for reproduction. Many organisms can reproduce asexually, but the majority of multicellular organisms reproduce sexually with specialized gametes fusing in fertilization to produce zygotes. Because asexual reproduction produces new individuals directly from one parent organism, new genetic combinations are not introduced into the population, as they are in sexual reproduction.

A zygote carries the blueprint for a new individual, including developmental instructions that may be used only once during the life of the new individual. When an individual matures, gametes are produced and the whole process begins again. The certainty that this cycle will be repeated again and again led to development of the concept of the body (somatic line) as bearer and releaser of gametes (germ line) that are continuous from generation to generation. Both lines are now accepted as part of the generation-to-generation cycle of life.

Most organisms reproduce seasonally. Their periods of reproductive activity are timed to occur when reproductive success is most likely.

Gametes are highly specialized cells that are well adapted for their reproductive functions. During animal gametogenesis, meiosis reduces the chromosome complement to the haploid number characteristic of gametes. Fertilization restores the diploid number in the zygote.

Sperm cells are specialized for motility. They lose most of their cytoplasm, become streamlined in shape, and develop motile tails. They also develop an acrosome, which functions in fertilization.

Unequal cytoplasmic divisions during meiosis in oogenesis concentrate cytoplasmic materials in the future egg cell. These materials are used during the early development of the zygote.

Animal fertilization is a complex set of interactions between egg and sperm. Interactions begin even before physical contact between the cells and continue as membrane fusion permits sperm entry into the egg. In most animal species, the egg becomes unresponsive to contact with additional sperm once it has been contacted by one sperm. In sea urchins, polyspermy is further inhibited by a fertilization membrane that develops from the elevated and thickened vitelline membrane.

Fertilization is much more than sperm entry and subsequent nuclear union. Fertilization causes metabolic activation of the egg, which initiates synthesis of nucleic acids and proteins required during cleavage and subsequent development. However, the earliest fertilization responses of the egg seem to be ionic changes, which precede changes in synthesis activity.

Zygotes, single relatively large cells, are converted into multicellular embryos by cleavage, a series of mitotic divisions without intervening growth. These cells become organized according to a basic body plan during gastrulation. Finally, cells become structurally and functionally differentiated to perform diverse, specialized functions in the body.

Some animals, such as honeybees, reproduce by parthenogenesis as well as by sexual reproduction with gamete fusion. In bees, haploid males are produced by parthenogenesis. In some other animals, generation after generation of females develop parthenogenetically, and males appear only rarely, if at all.

Parasitic animals often have both sexual and asexual reproductive processes in their life cycles, which enhances their total reproductive potential. This increased reproductive power is adaptive for parasites, who must successfully transfer from one host to another so that the life cycle continues.

Plant life cycles include two separate generations—a haploid gametophyte generation and a diploid sporophyte generation. In flowering plants, gametophytes are reduced and inconspicuous, but several-celled structures. Gametes are produced by differentiation of already haploid cells of the gametophytes.

Flowers are clusters of highly modified leaves that protect reproductive structures. Meiosis occurs in the development of spores. Megaspores develop into embryo sacs (female gametophytes), and microspores develop into pollen grains. Pollen grains complete development into male gametophytes if pollination occurs; that is, if they land on an appropriate pistil. Then a pollen tube carries two sperm nuclei to the embryo sac where double fertilization occurs, producing a zygote and a primary endosperm cell.

Mitotic divisions of the zygote produce an embryo that is enclosed, along with the storage tissue of the endosperm, inside seed coats. Following seed maturation, embryo development pauses until germination of the seed. At that time, growth resumes and proceeds toward development of a mature sporophyte.

Questions

1. Some biologists say that, from a species point of view, reproduction is the ultimate homeostatic mechanism. Can you explain that statement?

2. Discuss advantages of sexual and asexual reproduction in terms of reproductive success of individual organisms and in evolutionary terms.

3. Fertilization is important genetically because it brings together haploid paternal and maternal chromosome sets, but fertilization interactions begin long before fusion of the pronuclei. Discuss the "early" interactions between egg and sperm in animals, and the concept of egg activation.

4. How is the pattern of division without intervening growth related to the basic function of cleavage in animals?

5. Contrast animal and plant life cycles generally with reference to relationships of meiosis and gamete formation. How is spermatogenesis in male bees different from the "average" situation in animals?

Suggested Readings

Books

Cohen, J. 1977. *Reproduction*. London: Butterworths.

Johnson, L. G., and Volpe, E. P. 1973. *Patterns and experiments in developmental biology.* Dubuque, Iowa: Wm. C. Brown Company Publishers.

Karp, G., and Berrill, N. J. 1981. *Development*. 2d ed. New York: McGraw-Hill.

Raven, P. H.; Evert, R. F.; and Curtis, H. 1981. *Biology of plants*. 3d ed. New York: Worth Publishers.

Saunders, J. W., Jr. 1982. *Developmental Biology*. New York: Macmillan.

Stern, K. R. 1982. *Introductory plant biology*. 2d ed. Dubuque, Iowa: Wm. C. Brown Company Publishers.

Articles

Epel, D. 1980. Fertilisation. *Endeavour* 4:26.

Fawcett, D. W. 1979. The cell biology of gametogenesis in the male. *Perspectives in Biology and Medicine*, part 2, p. 556.

Human Reproduction and Development

Chapter Concepts

1. Human reproductive systems are well adapted for reproduction in the terrestrial environment. Internal fertilization and extended internal development shelter gametes and developing offspring from the environment.

2. Complex hormonal mechanisms control reproductive processes and pregnancy responses.

3. Embryonic development establishes basic body organization, and growth and maturation proceed during fetal development.

4. The placental relationship sustains the developing infant throughout its development in the uterus.

5. After birth, a human infant is physiologically independent but still weak and helpless.

6. Fertility regulation and the increasing incidence of sexually transmitted diseases are worldwide social problems associated with human sexuality and reproduction.

The basic design of the human reproductive system is a result of adaptation to life in a terrestrial environment. As with vascular plants and other land-dwelling animals, the human reproductive system is specialized so that delicate reproductive cells and small, fragile, developing individuals are enclosed and protected within moist internal body environments. For internal fertilization and development, sperm must be delivered to the inside of the female body. This internal delivery is achieved when the male **penis** deposits fluid containing sperm cells in the moist internal environment of the female **vagina.** Should fertilization occur, the developing zygote is maintained inside the female reproductive tract, and development proceeds inside the **uterus,** an organ highly specialized for maintenance of the developing embryo. The embryo becomes embedded in the uterine wall in a process known as **implantation,** and a **placenta** then develops in the implantation site. The placenta is a highly specialized composite organ, made up of both maternal and embryonic tissues, that functions in exchange of materials between the circulatory systems of mother and offspring. This intimate functional connection between the uterus and the developing individual continues through embryonic development and the long **fetal period** of growth and maturation until the time of birth.

The Male Reproductive System

The male gonads are paired testes that are suspended in a saclike structure, the **scrotum.** Testes have a dual function; in addition to producing sperm, they synthesize and release male sex hormones, thus serving as endocrine organs.

Sperm production occurs in the seminiferous tubules of the testes. Each testis contains about one thousand of these small, highly coiled tubules. The total length of the seminiferous tubules in an average human testis is estimated to be as much as 250 m. Because sperm production is continuous in healthy adult males, all stages of spermatogenesis are present at any given time and can be observed in a cross section of a seminiferous tubule. Spermatogenesis begins in the outer part of the seminiferous tubule, and as the

process proceeds, the developing cells move from the periphery toward the center of the tubule. Mature sperm cells become detached and lie in the central cavity (**lumen**) of the tubule (figure 24.1).

Male sex hormones (chiefly, testosterone) are produced by the **interstitial cells** scattered in the spaces among the seminiferous tubules (figure 24.2). Testosterone is responsible for development of **secondary sex characteristics,** such as general male body form and muscle development, male hair distribution, and voice deepening in maturing males. If the testes are surgically removed (an operation called castration) before the time of sexual maturation (**puberty**), male secondary sex characteristics do not develop.

Two hormones from the anterior lobe of the pituitary gland regulate testis functions. **Follicle-stimulating hormone (FSH)** controls spermatogenesis, while **luteinizing hormone (LH)** stimulates testosterone production by interstitial cells. These hormones, along with several other pituitary hormones, are classified as trophic hormones (see chapter 14) because they stimulate secretion of other glands. Because they are trophic hormones associated with the gonads, they are called **gonadotrophic hormones** or **gonadotrophins.**

As with other anterior pituitary hormones, gonadotrophin secretion is controlled by releasing factors produced in the hypothalamus. Maturation of this control system is involved in the activation of reproductive function at the time of puberty. In adult human males, this hypothalamus-pituitary link is part of a feedback relationship that controls and balances the level of testosterone in the blood (figure 24.3). This functional connection between hypothalamus and pituitary also accounts for the ability of psychological factors, such as visual sexual stimuli, to influence blood testosterone levels and, thereby, change the general level of sexual responsiveness.

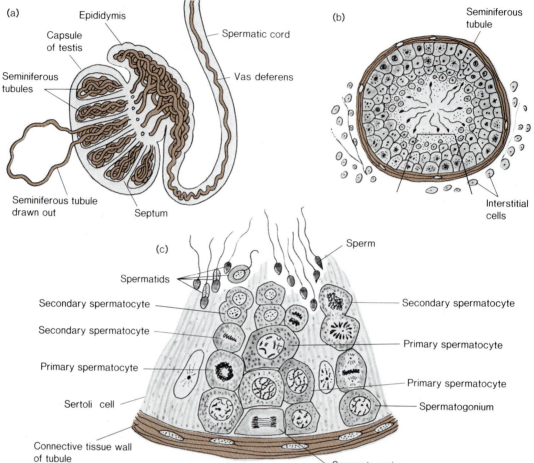

(a)

Epididymis

Capsule
of testis

Spermatic cord

Seminiferous
tubules

Vas deferens

Seminiferous tubule
drawn out

Septum

(b)

Seminiferous
tubule

Interstitial
cells

(c)

Sperm

Spermatids

Secondary spermatocyte

Secondary spermatocyte

Secondary spermatocyte

Primary spermatocyte

Primary spermatocyte

Sertoli cell

Primary spermatocyte

Spermatogonium

Connective tissue wall
of tubule

Spermatogonium

Figure 24.1 Human testis organization and sperm
production. (a) Schematic diagram of testis organization.
Packed-in, coiled seminiferous tubules produce sperm,
which are transported to the epididymis and through the
vas deferens. (b) A diagrammatic cross section showing
how sperm production proceeds as cells move from the
periphery toward the lumen in the center of a tubule.
Interstitial cells produce male sex hormones. (c) Stages
of spermatogenesis in one small area (outlined in b).
Mature sperm become detached in the lumen and are
moved away toward the epididymis. Sertoli cells function
in support and maintenance of developing sperm.

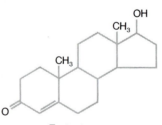

Testosterone

Figure 24.2 The structure of the steroid hormone
testosterone. Carbons and hydrogens of the basic
steroid molecular skeleton are not shown. Testosterone
is the main male sex hormone (androgen), and it causes
development of male secondary sex characteristics.

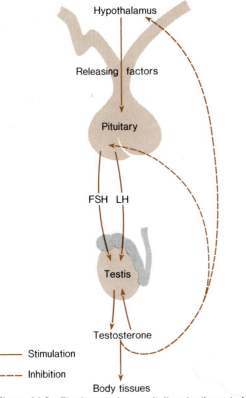

Figure 24.3 The hypothalamus-pituitary-testis control relationship (see chapter 14). Testosterone acts on various body tissues and balances the amount of releasing hormones being sent to the pituitary. This, in turn, affects gonadotrophin secretion by the pituitary. FSH controls spermatogenesis, and LH regulates the amount of testosterone produced.

Sperm Transport and Seminal Fluid

Sperm are moved along the seminiferous tubules, through small collecting ducts (the **vasa efferentia**), to the **epididymis,** a long, coiled tube lying on the surface of the testis. Here the sperm are stored until they are transported through the reproductive tract. Sperm are propelled through the reproductive tract by peristaltic waves of muscular contractions that move through the system. These contractions are part of **ejaculation,** a reflex response to sexual intercourse or other sexual stimulation. The contractions move sperm out of the epididymis and on through a tubular transport tract beginning with the **vas deferens** (figure 24.4). The vas deferens runs out of the scrotum, passes into the body cavity, loops across the surface of the bladder, and then turns down toward the **prostate gland,** which surrounds the urethra just below the point where the urethra drains the urinary bladder. At the lower end of each vas deferens, a glandular **seminal vesicle** adds an alkaline mucous secretion to the sperm. The seminal vesicles have been mistakenly identified as sperm storage organs, but their function actually is secretory. Sperm, with this added secretion, then move into the urethra at a junction contained within the body of the prostate gland. At this point, the thinner, milky-colored secretion of the prostate is added to the sperm. From here

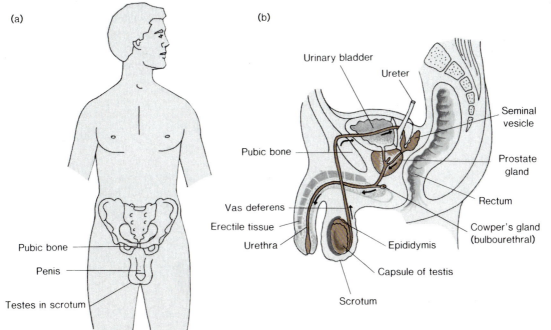

Figure 24.4 The human male reproductive system.
(a) Frontal view showing location of reproductive organs.

(b) Relationships of male reproductive structures and the pathway of sperm transport (indicated by arrows).

on, the excretory and reproductive systems share a common duct, the urethra. Seminal fluid moves from the prostate area through the urethra toward the base of the penis. There, a third pair of glands, **Cowper's** (or **bulbourethral) glands,** add more thick, mucous material to the seminal fluid. The **seminal fluid,** or **semen** (suspended sperm cells plus the various glandular secretions) accumulates in the urethra at the base of the penis. During ejaculation, it is forcefully ejected through the penis in a series of pulses. If sexual intercourse is occurring at the time of ejaculation, the seminal fluid is deposited in the upper portion of the female's vagina.

Seminal fluid is more than a simple suspending liquid for sperm. It contains buffers that partially neutralize the normally acidic environment in the vagina. Glucose and fructose in the seminal fluid can be used as a source of energy by sperm cells, which contain very limited amounts of nutrients. Seminal fluid also serves as a lubricant for the pathways through which the sperm swim. The average total volume of seminal fluid in each ejaculation is approximately 3 to 4 ml.

The penis is a cylindrical organ that hangs in front of the scrotum (figure 24.5). Spongy **erectile tissue** containing distensible blood spaces extends through the shaft of the penis. During sexual arousal, nervous reflexes cause an increase in arterial blood flow to the penis. This increased blood flow fills the blood spaces in the erectile tissue, and the penis, which is normally limp (flaccid), stiffens and increases in size. These changes are called **erection.** The erect penis can be inserted into the female vagina and can deliver seminal fluid into the upper portion of the vagina during ejaculation. In many male mammals, a bone in the penis gives it a permanent partial rigidity, but the copulatory function of the human penis depends entirely on erection by filling of the blood spaces in the erectile tissue.

The Female Reproductive System

Human ovaries are paired oval organs, about 3 cm long, located in the lower part of the abdominal cavity, where they are held in place by ligaments. Like the testes, the ovaries have a dual function; in addition to producing and releasing egg cells, they secrete two kinds of steroid hormones—**estrogens** and **progesterone**—and thus also function as endocrine glands (figure 24.6). Female body form and hair distribution, breast development, and other female secondary sex characteristics develop in response to ovarian hormones.

Every twenty-eight to thirty days, on the average, an egg is released from one of the ovaries, by a process known as **ovulation.** Ovulation involves a small rupture in the ovarian wall, which accounts for the slight pain that some women feel at the time of ovulation. Ovulation releases the egg into the abdominal cavity. The expanded end of an **oviduct,** or **Fallopian tube** (figure 24.7a), has beating cilia that move fluid

Figure 24.5 The male reproductive system showing the location of the penis and scrotum before (light gray outline) and during erection.

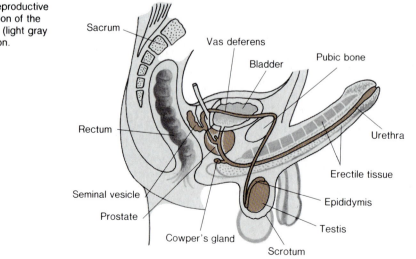

Sacrum
Vas deferens
Bladder
Pubic bone
Rectum
Urethra
Seminal vesicle
Erectile tissue
Prostate
Epididymis
Cowper's gland
Testis
Scrotum

Figure 24.6 The structures of two steroid hormones (estradiol, which is a common estrogen, and progesterone) that are produced by human ovaries. Carbons and hydrogens of the basic steroid molecular skeleton are not shown. Despite similarities in structure, these hormones have quite different effects on human tissues and very different effects from testosterone, which also has a similar structure (figure 24.2). This illustrates the high degree of molecular specificity in hormone effects on target tissues.

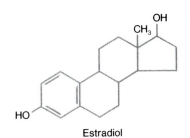

Estradiol

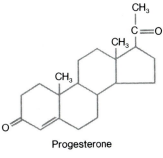

Progesterone

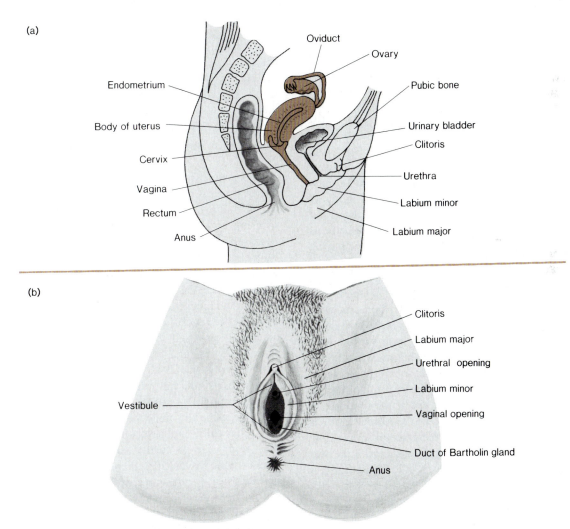

(a)

(b)

Figure 24.7 The human female reproductive system.
(a) Positions of female reproductive structures.
(b) Frontal view of external parts of the female reproductive system.

from the abdominal cavity along into the oviduct opening, the **ostium.** Usually, this movement of fluid carries the egg along so that the egg enters the oviduct. Occasionally, however, an egg fails to follow this normal route and escapes into the body cavity. This ordinarily causes no problem because the unfertilized egg degenerates and is picked up by phagocytic cells. But if fertilization occurs in the body cavity, as it does on rare occasions, a developing embryo may implant on the surface of one of the abdominal organs, where it can cause a dangerous cyst that requires surgical treatment.

Following normal entry into the ostium, the egg moves along through the thin-walled oviduct toward the upper corner of the **uterus,** which is a thick-walled, muscular, pear-shaped organ in the center of the lower portion of the abdominal cavity. The uterus lies just above and back of the urinary bladder. The larger, upper part of the uterus is called the **body.** Its muscular walls are capable of tremendous expansion in response to growth of a fetus during pregnancy. At its lower end, the uterus tapers down to the **cervix,** a muscular extension that protrudes into the **vagina.** A strong muscular ring surrounds the small cervical opening through which sperm enter the uterus from the vagina, where they are deposited during sexual intercourse. The vagina is an extended, thin-walled tubular canal connecting the cervical area with the outside of the body. It opens into the vestibule, a space that is partially enclosed by a pair of thickened skin folds, the **labia majora,** that contain fatty tissue and have surface hair (figure 24.7b). Within the labia majora are another pair of folds, the pink, membranous **labia minora.** At the point where the labia minora come together near the anterior end of the vestibule is a small knoblike structure, the **clitoris,** which comes from the same developmental source that produces the penis in male embryos. Like the penis, the clitoris contains erectile tissue that fills with blood during times of sexual arousal. Because the clitoris is richly supplied with nerve endings, it is very sensitive to the touch and is involved in female sexual responsiveness.

Copulation

Copulation (sexual intercourse) in humans normally is preceded by courtship behavior, or foreplay, which sexually arouses and prepares both partners. There is a great deal of variability in courtship behavior among human cultural groups, and even among individuals within cultural groups, but the most general pattern is pleasurable mutual stimulation by both partners. Physical signs of arousal include erection of the male penis and, in many cases, swelling of the female clitoris and tissue around it. In the female, blood vessel dilation also produces vasocongestion and swelling of tissues in the vaginal and vestibular regions. Fluid secreted from the cervix lubricates the vagina, and **Bartholin's glands** produce fluid that lubricates the vestibule area. Some lubricating fluid also oozes out of the erect penis if sexual arousal continues for a period of time. All of these secretions together permit smooth entry of the penis into the vagina (figure 24.8a).

Repeated stimulation of the penis, especially its sensitive tip, as it rubs against the walls of the vagina and the vestibule increases sexual tension to a higher level, known as the **plateau phase.** Male sexual climax, or **orgasm,** follows. In males, ejaculation, which is reproductively essential, is an integral part of orgasm. Orgasm is a complex of pleasurable sensations that includes involuntary muscular contractions in the abdominal area and other parts of the body, as well as the ejaculation process itself. During orgasm, blood pressure, heart rate, and breathing rate all increase. Following orgasm, all of these return toward their normal levels, muscles relax, and the penis softens and eventually returns to its flaccid condition. After orgasm, the male becomes very relaxed both physically and psychologically and is not very responsive to further sexual stimulation. Penis erection is usually impossible during this unresponsive, or **refractory,** period. The length of the refractory period varies among individuals and may even be quite different for the same person at different times.

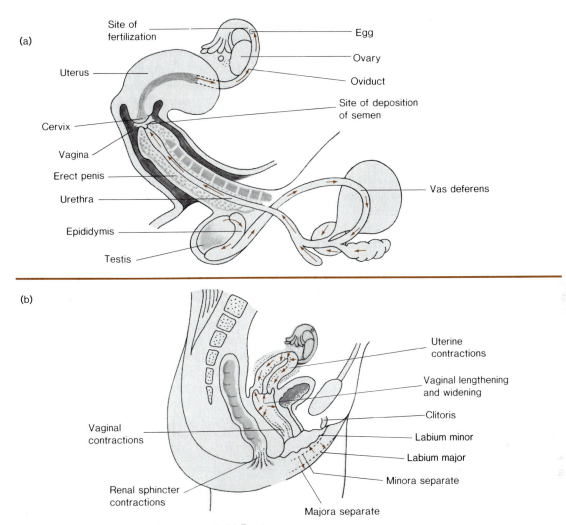

Figure 24.8 Copulation (sexual intercourse). (a) Erect penis in vagina during copulation. Colored arrows indicate the path of sperm transport through the male reproductive system. (b) Contractions and spontaneous movements of female reproductive structures during orgasm.

Female orgasm differs fundamentally from male orgasm in that it is not essential for basic reproductive function. Fertilization does not depend on occurrence of female orgasm. Female orgasm also differs from male orgasm in that women are capable of several orgasms following one another almost immediately. In fact, repeated orgasms may blend together so that one more-or-less continuous orgasm can be maintained for a considerable time. Female orgasm involves muscular contractions of several reproductive structures, as well as tensing of other body muscles (figure 24.8b). Breast enlargement, nipple erection, and skin flushing, which began during the arousal and plateau phases, continue.

Orgasm brings women a strong sense of tension relief, and after orgasm, women experience a feeling of general comfort and relaxation.

Complex psychological and physiological factors are involved in the production of female orgasms. Researchers and interviewers disagree on some points, but there is general consensus that the clitoris is a focal point for receiving stimulation leading to orgasm. Some women reach orgasm as a result of general stimulation during copulation, but clearly, other women require more directly focused clitoral stimulation to achieve orgasm. Even positive clitoral stimulation may be ineffective if negative psychological factors intervene. Female orgasm appears to be a distinctly human phenomenon that may have evolved along with essentially continuous female sexual responsiveness. Both of these special human female traits appear to play a role in maintenance of the long-term pair bonds characteristic of human mating systems.

The Menstrual Cycle

Changes in the uterine lining, the **endometrium,** are coordinated with egg maturation and ovulation so that the uterus is prepared to accept an implanting zygote. The timing is such that the uterus is fully ready for implantation just as the zygote is ready to implant, if fertilization and early development have proceeded on schedule. Cyclical changes in hormone levels time these processes during the human female **menstrual cycle.** The name menstrual cycle is derived from **menstruation,** the periodic shedding of blood and tissue from the endometrium. Because this shed material flows out through the vagina, it is the most obvious external sign of cyclical changes. But, actually, the uterine cycle runs concurrently with an ovarian cycle, and the two cycles are intimately related to one another.

The ovarian cycle involves maturation of the future egg and development of the **follicle,** a specialized area of ovarian tissue around the maturing egg (figure 24.9). Following ovulation, the follicle changes and develops into another type of tissue, the **corpus luteum** (literally "yellow body" because the cow's corpus luteum, which was studied first, is yellow; the human corpus luteum actually is cream colored). Changes during the uterine cycle are responses to hormones secreted by the ovary during the various phases of its cycle, and ovarian hormone production, in turn, is regulated by a feedback relationship involving the hypothalamus-pituitary axis (see chapter 14).

The day on which menstruation begins commonly is called "day one" of the menstrual cycle. During the early days of the cycle, **follicle stimulating hormone** (FSH) from the pituitary causes a follicle containing a prospective egg cell to grow and proceed toward ovulation. Growing ovarian follicles secrete increasing amounts of estrogens (primarily estradiol) that stimulate growth in the uterus that rapidly replaces layers of the endometrium that were shed during menstruation. This phase of endometrial repair lasts for eight to ten days after the end of menstruation.

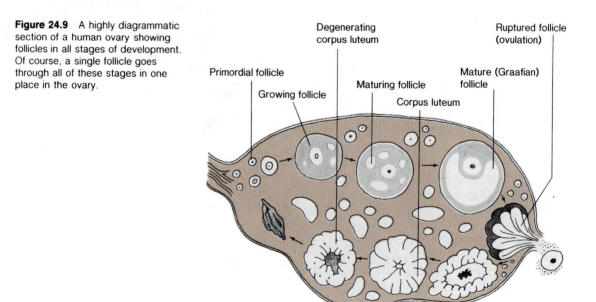

Figure 24.9 A highly diagrammatic section of a human ovary showing follicles in all stages of development. Of course, a single follicle goes through all of these stages in one place in the ovary.

Primordial follicle

Growing follicle

Degenerating corpus luteum

Maturing follicle

Corpus luteum

Mature (Graafian) follicle

Ruptured follicle (ovulation)

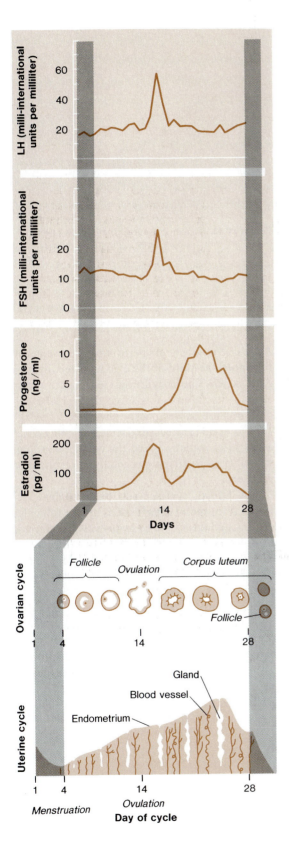

At about the midpoint of the twenty-eight or twenty-nine day menstrual cycle, the pituitary gland produces and releases a surge of **luteinizing hormone (LH)**. This LH surge sets in motion preovulation and ovulation changes in the follicle. Continued LH stimulation causes the postovulation follicle to develop into the corpus luteum and to continue hormone secretion. The maturing corpus luteum secretes both estrogens and increasing quantities of another steroid hormone, progesterone. Progesterone acts on the uterus by stimulating vascularization, glandular development, and glycogen accumulation in the endometrium. All of these progesterone-induced changes are necessary to make the uterus ready for the potential arrival of an implanting zygote (figure 24.10).

In a normal menstrual cycle, when a pregnancy has not begun, the hypothalamus and then the pituitary respond to the high levels of steroid hormones put into circulation by the active corpus luteum. This feedback response causes a decrease in LH production, and the resultant falling LH levels cause the corpus luteum to begin to degenerate by about the twenty-second or twenty-third day. As the corpus luteum is degenerating, its hormone output decreases.

Maintenance of the fully developed uterine endometrium depends on the relatively high progesterone levels present while the corpus luteum is most active. Thus, when circulating progesterone levels decrease as a result of corpus luteum regression, the uterine linings start to degenerate and slough off, and menstruation begins. Before long, however, the hypothalamus and pituitary respond to decreased circulating estrogen: FSH production rises again, follicle growth and maturation are stimulated, and the whole complex cycle starts over.

Menstruation marks the end of the time when the uterine endometrium could accept an implanting zygote, and it is a clear signal that pregnancy has not begun during that particular menstrual cycle.

Figure 24.10 A composite diagram showing the time relationships of pituitary hormone secretion, the ovarian cycle, ovarian hormone secretion, and the condition of the uterine lining during the human menstrual cycle. Note that different units and different scales are used for the various hormones. Pay special attention to relative hormone levels at different points in the cycle.

The beginning of a pregnancy, of course, greatly modifies hormonal and uterine events. In pregnancy, the corpus luteum is maintained in response to a hormonal influence from the developing embryo, and it continues production of adequate levels of progesterone to keep the uterine lining intact.

The Oocyte and Ovulation

Up to this point the term "egg cell" has been used somewhat loosely, but now clarification becomes important. Meiotic divisions of prospective egg cells begin in the ovaries of human female fetuses. Thus, a human female has initiated the process of egg production even before she is born. Some 400,000 prospective egg cells enter the first meiotic prophase and become arrested at that stage in the two fetal ovaries of each human female.

It is not clear how the 400 or so cells that proceed to ovulation during the reproductive period of an average adult woman are selected from this much larger population. But during each menstrual cycle, one (rarely two or more) prospective egg resumes meiosis and proceeds as far as metaphase of the second meiotic division. The cell, now properly called a secondary oocyte, pauses again and remains at that meiotic stage through the time of ovulation. Meiosis is completed only as part of the cell's fertilization **activation reactions** (chapter 23). Without sperm cell contact, meiosis proceeds no further, and the oocyte degenerates and is phagocytized somewhere in the reproductive tract.

Ovulation occurs in response to the midcycle LH surge, and it involves a rupture of the ovarian surface over the mature (**Graafian**) follicle (figures 24.11 and also 24.9). Fluid accumulated inside the mature follicle escapes and carries the oocyte with it. The oocyte, as it leaves the ovary, is enclosed by two covering membranes and several layers of cells that surround them. Immediately over the surface of the oocyte is the vitelline membrane, which is secreted by the oocyte itself. Just outside the vitelline membrane is another covering layer, the **zona pellucida,** which is produced by cells in the follicle. Around the zona pellucida are several layers of follicular cells, which accompany and enclose the oocyte and its membranes.

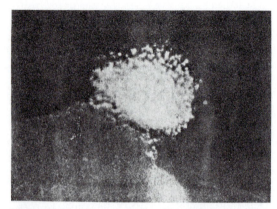

Figure 24.11 Human ovulation. The surface of the ovary swells and breaks open. Fluid from the follicle pours out, and the oocyte emerges surrounded by the zona pellucida and several layers of follicle cells.

Fertilization and Early Development

Adequate numbers of sperm must be present in the upper one-third of the oviduct if fertilization is to occur. The oocyte must encounter sperm early in its three-day passage through the oviduct because as yet poorly understood "aging" changes seem to affect the oocyte beginning about a day after ovulation, and the oocyte soon loses its ability to participate normally in fertilization reactions.

A human sperm lysin, **hyaluronidase,** released from sperm cells, loosens up the follicular cells around the oocyte by hydrolyzing **hyaluronic acid,** a mucopolysaccharide that cements the cells together. Dispersion of these cells is necessary if sperm are to gain access to the egg surface. The need for adequate quantities of hyaluronidase may explain why a large number of sperm must be present in the oviduct for fertilization to occur, even though only one sperm actually interacts with the oocyte in the fertilization process.

Contact with a sperm cell activates the oocyte, and it resumes the second meiotic division, which has been arrested since before ovulation.

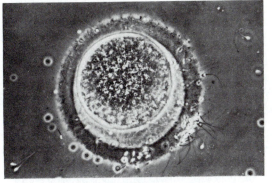

(a)

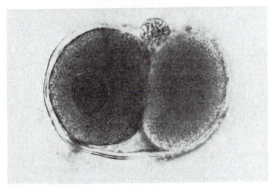

(b)

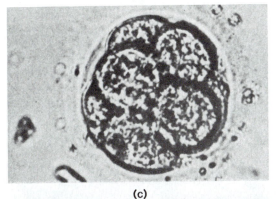

(c)

(d)

The second polar body is extruded, leaving the functional egg with a haploid nucleus (**pronucleus**). Then egg and sperm pronuclei migrate into the center of the cell where they meet. Chromosomes of both pronuclei replicate, condense in the manner that always precedes mitosis, and become arranged on a mitotic spindle that has formed in the center of the egg. Then the first of the mitotic cell divisions of cleavage begins (figure 24.12). This first cleavage division is completed within about a day after fertilization, and subsequent divisions occur at intervals of from eight to ten hours as the early embryo moves down through the oviduct toward the uterus (figure 24.13). Passage of the embryo through the oviduct takes from three to three and one-half days. During this time, the follicular cells around the developing embryo, already loosened by the action of hyaluronidase, are lost completely. When the cleaving cells reach the uterus, they are enclosed and held together by the zona pellucida.

Implantation

While the embryo lies free in the uterine cavity, further cell divisions produce a rather loose aggregate of much smaller cells, the **morula.** During the next two or three days, morula cells segregate themselves into two distinctly different groups as they organize the characteristic structure of the **blastocyst** stage of development (figure 24.14). The blastocyst consists of a hollow sphere of small, flattened cells, the **trophoblast,** surrounding a fluid-filled cavity. A mass of larger, rounded cells, the **inner cell mass,** is situated to one side of the cavity.

Figure 24.12 Human fertilization and early development of the human embryo. (a) A human oocyte with its enclosing layers. Note sperm cells around it and in the follicle cells that surround the oocyte. (b) A human embryo at the two-cell stage. (c) The six-cell stage. Cell divisions are not synchronous so there is not a regular progression of two, four, eight, sixteen cells, and so on. (d) A human embryo at the morula stage of development.

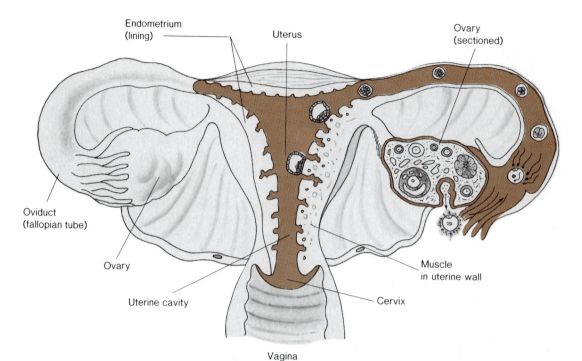

Endometrium
(lining)

Uterus

Ovary
(sectioned)

Oviduct
(fallopian tube)

Ovary

Uterine cavity

Muscle
in uterine wall

Cervix

Vagina

Figure 24.13 Human female reproductive tract with the uterus and one oviduct opened up to show progress of the developing embryo at various stages. The embryo normally becomes implanted in the lining of the uterus, as shown here. Implantation in other areas of the endometrium causes pregnancy complications.

Figure 24.14 The blastocyst and implantation. (a) A human blastocyst. Note the inner cell mass and the individual trophoblast (developing chorion) cells. (b) A section of a monkey blastocyst attached to the endometrium which shows how the blastocyst approaches and enters the endometrium inner cell mass side first. (c) Section of a human implantation site showing the implanted blastocyst and the glands and blood vessels of the fully developed endometrium. (d) A surface view of an implantation site twelve days after fertilization. Uterine tissue grows over the surface and heals the point of entry.

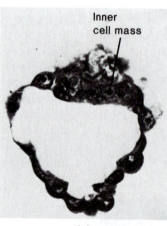

Inner
cell mass

(a)

(b)

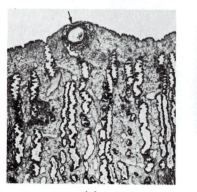

(c)

(d)

This segregation into two populations of cells represents an interesting degree of early cell specialization. Cells of the trophoblast layer play no part in the formation of the embryo itself but produce an extra-embryonic membrane, the **chorion**, and contribute to placenta development. The body of the developing embryo is produced entirely within the inner cell mass.

The blastocyst stage is a critical point in human development because the blastocyst must implant in the wall of the uterus. At least 25 percent and possibly as many as 40 percent or more of all developing blastocysts fail to implant in the wall of the uterus. In the absence of implantation, there is no sign of pregnancy, and the embryo simply dies and is lost. Some of these implantation failures undoubtedly involve developmental abnormalities that produce blastocysts incapable of implanting. In other cases, there may be timing disturbances such that the blastocyst contacts the lining of the uterus either before or after the time that the uterus lining is receptive to an implanting blastocyst. Sometimes, the blastocyst simply is lost because it never establishes a proper physical contact with the endometrium.

Normally, the inner cell mass side of an implanting blastocyst contacts the endometrium first. The initial contact involves a specialized tissue on the surface of the trophoblast. This specialized surface tissue is syncytial; that is, it is a multinucleate tissue not divided into individual cells. Properties of this syncytial tissue are very important because, in a sense, implantation of the human blastocyst into the uterine wall is the acceptance of a "graft" of foreign tissue. A better understanding of this special tolerance of the endometrium for an implanting blastocyst could provide information that would be helpful in improving transplantation and organ grafting techniques. Also, more precise information about the initial contact between blastocyst and uterine lining might be used to develop new and physiologically safer birth control techniques.

Enzymes secreted by trophoblast cells erode uterine tissue as the blastocyst sinks into the endometrium. It is fairly accurate to say that the blastocyst digests its way into the uterine wall.

In fact, materials from eroded uterine tissue that are absorbed into blastocyst cells probably serve as an important nutrient source for the blastocyst during implantation. After the blastocyst has entered the endometrium, uterine tissue grows over the surface and heals the implantation site.

Hormones and Pregnancy

During a normal menstrual cycle, a chain of events involving lowered LH production, regression of the corpus luteum, and the consequent decrease in circulating progesterone levels leads to menstruation. If implantation occurs and pregnancy is to proceed, this normal sequence must be interrupted.

An implanted blastocyst produces a hormone that prevents menstruation. The developing chorion (which is produced by the trophoblast and surrounds the developing embryo) functions as an endocrine organ because it begins very early to produce a hormone called **chorionic gonadotrophin** (abbreviated **HCG** for human chorionic gonadotrophin). Functionally, HCG replaces LH and keeps the corpus luteum active. The resulting continued progesterone secretion by the corpus luteum keeps the uterine lining intact. Technically, chorionic gonadotrophin might be classified as a pheromone (see chapter 14) because it is a chemical regulator released by one individual (in this case the embryo) that has a specific influence on another individual (the mother). Clearly, early pregnancy involves a hormonal interaction between embryo and mother. Later, the interaction becomes even more complex when the placenta, a composite organ formed from tissues of both individuals, begins to produce hormones.

Chorionic gonadotrophin is produced in such large quantities that pregnant women excrete considerable amounts of it in their urine. Thus, chemical pregnancy tests can be based on detection of urinary chorionic gonadotrophin (figure 24.15). Unlike many other human hormones, HCG is so readily available that it is routinely used in many teaching and research laboratories. It is even used to induce ovulation in female frogs to obtain frog eggs for embryological studies.

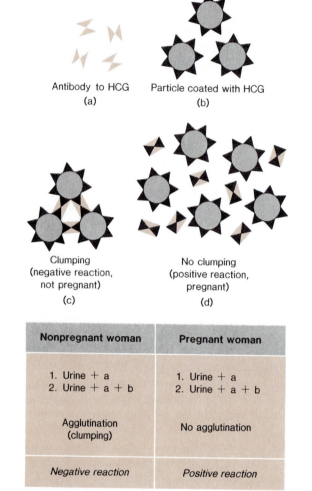

Antibody to HCG
(a)

Particle coated with HCG
(b)

Clumping
(negative reaction,
not pregnant)
(c)

No clumping
(positive reaction,
pregnant)
(d)

Nonpregnant woman	Pregnant woman
1. Urine + a 2. Urine + a + b	1. Urine + a 2. Urine + a + b
Agglutination (clumping)	No agglutination
Negative reaction	*Positive reaction*

Figure 24.15 The basis for the commonly used chemical pregnancy tests. (a) Antibodies to human chorionic gonadotrophin (HCG) are obtained by immunizing animals against HCG. This is one reactive agent in the test. (b) Inert particles coated with HCG are the second reactive agent for the test. (c) If these two are mixed together, a reaction causes the particles to clump. (d) Urine from a woman being tested is first mixed with the antibody to HCG. If the woman is pregnant and HCG is present in her urine, it will combine with the antibodies. Thus, mixing this already bound antibody with the particles coated with HCG in the second step cannot cause clumping. Such chemical tests have replaced older tests that depended on reproductive tract responses of rabbits or frogs to injected urine or blood samples. Now, much more sensitive tests known as radioreceptor assays (developed by B. B. Saxena and his colleagues) permit detection of much smaller quantities of HCG and can determine the presence of a developing embryo even before implantation.

Occasionally, corpus luteum activity may decrease due to illness of the mother or a failure of the chorion to continue adequate HCG production. The resultant drop in progesterone level initiates menstruation-like breakdown of the endometrium and loss (**miscarriage**) of the implanted embryo. Miscarriage during the first month or so of pregnancy may result in a blood flow somewhat heavier and longer than normal menstruation, but the embryo is still so small that the entire brief and abruptly terminated pregnancy simply may be mistaken for a somewhat delayed menstrual period.

In a normal pregnancy, chorionic gonadotrophin maintains vital steroid hormone production by the corpus luteum until the ninth or tenth week of pregnancy. By then, the placenta itself is well established as a steroid-producing organ. It produces both progesterone and estrogens, and these placental steroids replace ovarian hormones in the maintenance of the uterine linings during the remainder of pregnancy. In fact, beyond a certain point, even surgical removal of the ovaries would not terminate pregnancy because placental hormones alone are adequate to maintain pregnancy even in the complete absence of the ovary. Obviously, the placenta becomes a very important endocrine organ during pregnancy.

The Major Stages of Prenatal Development

Development during the first two weeks following fertilization includes cleavage, blastocyst development, implantation, and very early postimplantation development. From the third through the eighth weeks, basic body organization is established. **Rudiments** (the first visible evidences of development) of all major organ systems are produced, and by the end of this period, the body is recognizably human. From the ninth week until birth is the **fetal period.** The **fetus** grows larger, and cell and tissue differentiation take place in the rudimentary organs originally laid down during the embryonic period. Various organs mature enough to permit termination of functional dependence on the placenta by the time of birth. The total period of time involved in all of these stages—from the time of fertilization until birth—is called the **gestation period.**

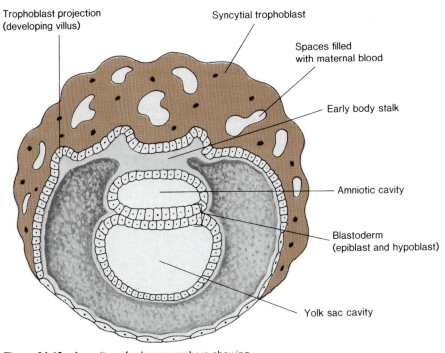

Figure 24.16 A section of a human embryo showing the yolk sac and the amniotic cavity. The blastoderm is located in the area where they are in contact with one another. These derivatives of the inner cell mass are attached to the inside of the chorion by the small **body stalk (connecting stalk)**. The body stalk becomes a major part of the umbilical cord.

The Embryo during Early Pregnancy

During the first days following implantation, organization of the embryo's body and some extraembryonic structures closely associated with the embryo's body proceeds within the inner cell mass, while the chorion (a product of trophoblast development) is involved in establishing the important embryonic connections with uterine tissue.

Cells of the inner cell mass reorganize during implantation. This cell activity establishes a flat, round, two-layered disc known as the **blastoderm (embryonic disc)**, which develops into the embryo. Actually, the blastoderm is the flattened area of contact between two hollow spheres of cells formed from the originally solid ball of inner cell mass cells. One of these spheres of cells, the **yolk sac,** is so named because of its obvious similarity to yolk-digesting structures in other vertebrates, such as chick embryos. But the human yolk sac does not function in yolk digestion because the small amount of yolk originally present in the human egg is used up well before the yolk sac forms. One side of the yolk sac is one of the layers of the blastoderm. That layer, the **hypoblast** ("lower layer"), is pressed against a second layer, the **epiblast** ("upper layer"), which lies at the bottom of another spherical space, the **amniotic cavity** (figure 24.16).

All parts of the embryo are produced from this flat disc after it is converted into the tubular body of the embryo. But even before the disc begins to change shape by folding, the process of gastrulation adds a third layer to the original two. Mechanics of mesoderm formation in the human embryo are virtually identical to those described for bird and reptile embryos (p. 636). Some cells leave the epiblast, migrate downward through a primitive streak, and assume a new position between the epiblast and hypoblast. Here they establish a new, third layer that lies between the two previously existing layers. This new, third layer is the mesoderm of the human embryo. The three primary body layers (germ layers) of the human embryo can now be identified. From top (amnion side) to bottom (yolk-sac side), they are ectoderm, mesoderm, and endoderm.

Later, the edges of the blastoderm fold under so that the flat disc rolls up into a tubular three-layered body. This body folding brings the germ layers into their permanent relationships with one another: ectoderm on the outside, endoderm on the inside, and mesoderm between the other two. As development proceeds, primary functional parts of major organ systems develop from each of the germ layers. For example, skin develops from ectoderm, gut linings develop from endoderm, and mesoderm produces skeleton, muscles, body cavity linings, and the circulatory system. Table 24.1 is a more detailed list of structures derived from the various germ layers.

Placenta Development and Function

At the same time that the embryo's body and structures closely associated with the embryo's body are being organized within the inner cell mass, a placental relationship is developing between the embryo and its mother.

A series of fingerlike outgrowths, the **villi** (singular: **villus**), develop over the chorion's surface (figure 24.17a). Villi on one side of the chorion are destined to grow, branch elaborately, and participate in development of the placenta, while those scattered over the remainder of the chorion regress and later disappear. Capillary beds develop inside the branching villi in the placenta area, and the villi receive a blood supply from vessels growing out from the embryo's body. These are called umbilical vessels because they pass through the **umbilical cord,** which develops from the **body stalk,** a restricted, narrow connection between the embryo and the developing placenta (figure 24.17b). Two **umbilical arteries** carry blood to the villi, and one **umbilical vein** carries blood back to the body from the placenta (figure 24.17c).

Chorionic tissue grows against and fuses with uterine tissue so that the placenta is literally a single organ constructed from tissues of the mother and the embryo. A very special circulatory arrangement develops in the spaces among the chorionic villi. Maternal arteries open up directly into these **intervillus** spaces (figure 24.17d).

Table 24.1
Germ Layer Sources of the Major Functional Parts of Various Body Tissues, Organs, and Systems.

Ectoderm
Nervous tissue
Epidermis of skin

Mesoderm
Dermis of skin
Skeleton
Muscle
Circulatory system
Excretory system
Reproductive system
Connective tissue

Endoderm
Digestive system linings
Digestive glands
Lung and respiratory tract linings

Maternal blood flows through these spaces, completely bathing the villi, before draining into veins that carry blood away from the placenta. This open circulation of maternal blood around the villi makes exchange by diffusion between the two circulations quite efficient.

Several layers of tissue normally separate the maternal bloodstream and the embryonic bloodstream. These layers are known collectively as the **placental barrier.** Properties of the placental barrier are of great interest to human embryologists because the barrier is selective in allowing substances to cross from one side to the other. Some drugs and other chemicals cross; others do not. Some everyday substances, such as ethanol (alcohol), that cross the placental barrier are mildly toxic to adults but *very* toxic to developing embryos and fetuses. Physicians now recognize a fetal alcohol syndrome characterized by reduced growth, possible mental impairment, and other symptoms. The syndrome often is seen in infants born to mothers who drink more than a minimal amount of ethanol during pregnancy. Bacteria cannot cross the placental barrier, but viruses can. Thus, viral diseases, such as measles, can be transmitted from mother to child across the placental barrier. Nutrients and oxygen cross from the maternal side of the barrier to the fetal

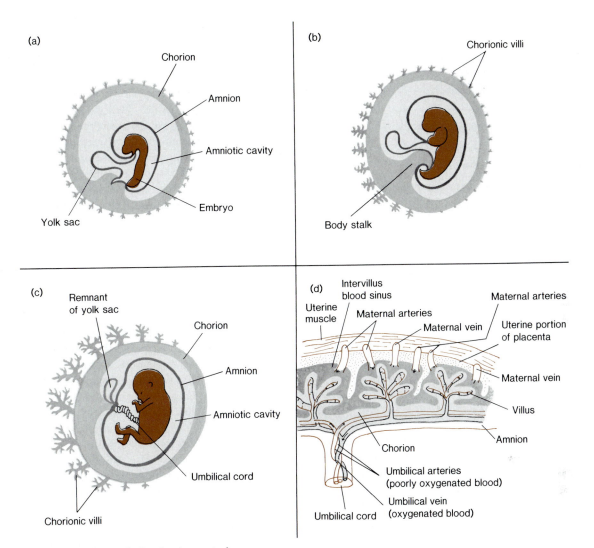

Figure 24.17 Stages in the development of
extraembryonic membranes and the placenta in humans.
(a) Syncytial portion of trophoblast (figure 24.16) and
uterine tissue not shown. Note villi projecting on surface
of chorion. Chorion has developed from trophoblast.
(b) Body stalk is producing umbilical cord. Note the
regression of the chorionic villi in the region furthest from
the developing umbilical cord. (c) Final relationships of
embryo and extraembryonic structure. The umbilical cord
with umbilical blood vessels is well developed and
serves as the route for exchange between the fetus and
its mother. (d) Circulatory relationships within the
placenta. Maternal blood pours into open spaces and
bathes the villi.

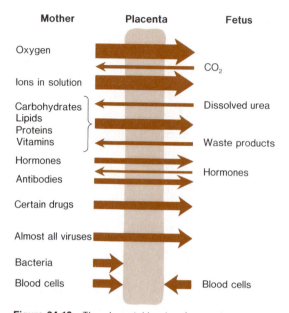

Figure 24.18 The placental barrier. Arrows that end at the barrier indicate that the substance normally does not cross the placental barrier. When the barrier is intact, there is no direct mixing between the two bloodstreams. Small breaks in the barrier apparently occur in a significant percentage of pregnancies without causing major problems.

side, while carbon dioxide and nitrogenous wastes pass in the other direction. Maternal hormones and placental hormones also can pass from one side of the barrier to the other. All of these placental functions are essential to the welfare of the embryo and fetus, and they are maintained until birth (figure 24.18).

The Amnion and Its Functions

The **amnion** (or amniotic sac) originally encloses the amniotic cavity, the space above the embryo's body. Later, this amniotic cavity spreads and enlarges to enclose and line the whole space within which the developing fetus remains throughout development.

Amniotic fluid secreted by amnion cells fills the cavity and lubricates its surfaces. This lubrication prevents delicate tissues from sticking to one another and cushions the developing fetus by absorbing shocks when the mother moves quickly or is bumped. In later stages of development, the fetus drinks and inhales amniotic fluid; thus, amniotic fluid bathes the linings of mouth and nasal passages as well as the skin.

The types and quantities of chemicals diffusing from the fetus into the amniotic fluid change characteristically as development proceeds, and cells sloughed off body surfaces float in the amniotic fluid. Therefore, samples of amniotic fluid can provide valuable information about both the progress of development and the genetic and biochemical characteristics of cells from the fetus. Amniotic fluid samples can be withdrawn for analysis using a needle inserted directly into the amniotic cavity. This technique is called **amniocentesis.** Amniocentesis is used to monitor the progress of difficult pregnancies and to detect potential genetic problems or biochemical deficiencies early so that appropriate treatments can begin as soon as possible.

The First Month

As explained earlier, during the first week, cleavage of the original single cell produces the morula, a cluster of loosely organized cells. Reorganization of the morula cells establishes the blastocyst, which begins to implant in the uterine wall near the end of the first week.

Implantation continues during the second week, and uterine lining cells heal over the surface of the implantation site. Reorganization of the inner cell mass produces the two-layered embryonic disc. At the very end of the second week, the primitive streak develops, and mesoderm development begins.

Mesoderm formation continues during the third week, and the nervous system begins to develop as a pair of thickened ridges of ectoderm that rise up, meet, and fuse with each other into a hollow tube (figure 24.19a). This establishment of future brain and spinal cord occurs relatively early by comparison with other major organ systems, and the brain and other head structures continue to be disproportionately large and precociously developed throughout embryonic and fetal development. Paired chunks of mesoderm tissue, called somites, appear alongside the developing nervous system. Somites later give rise to vertebrae, muscles, parts of the ribs, and the dermis of the skin. The first evidence of heart formation is a pair of separate tubes that move together and fuse into a single tubular heart during this time.

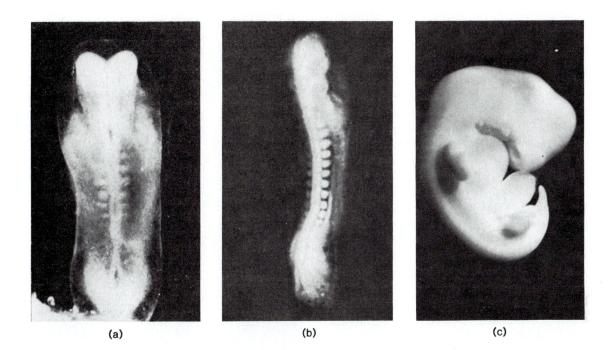

(a)   (b)   (c)

Despite the embryo's rather extensive development up to this point, the mother's menstruation has been delayed by only about a week.
Even though the embryo already has a developing
heart and central nervous system, the mother is
not even certain that she is pregnant!

During the fourth week, the embryo develops the beginnings of a digestive system (figure
24.19b). Hollow, tubular parts of the gut form
at the head and tail ends of the body. Strong,
coordinated heart beating initiates circulation
through the embryonic body and through the umbilical vessels supplying the placenta. The first
circulating blood cells are produced in special
areas of the yolk sac known as **blood islands.**
Thus, the first blood cells that develop come from
a source outside the body of the embryo itself.
Only later on, during the second month, does
blood cell formation begin inside the body, when
the liver becomes the second temporary source of
blood cells. Finally, the bone marrow matures
and takes over as the third, and permanent,
source of blood cells.

At the end of the first month, the embryo is
still only about 5 or 6 mm in length, and the total
diameter of the chorion enclosing the whole developing system is only about 3 cm (figure
24.19c).

Figure 24.19 Embryonic development. (a) Embryo near
end of third week of development. Ectodermal folds
(neural folds) are closing to produce brain and spinal
cord. Somites are forming. (b) A twenty-four-day human
embryo. Note that body folds have produced a tubular
body from the previously flat disc. (c) A thirty-four-day
human embryo.

The Second Month

During the second month, rudiments of all the rest of the major body organs are produced in the embryo, which assumes an increasingly human appearance. The relatively simple digestive tube now lengthens, and liver, gallbladder, and pancreas form as outgrowths from the embryonic gut. In addition, lung rudiments begin their development as pouches growing out near the anterior end of the gut. Paddlelike **limb buds** grow out from the sides of the body and gradually transform in shape into recognizable arms and legs. Eventually, fingers and toes are produced, and the complex skeleton and musculature of the limbs develops. In the head, eye and ear development proceeds, and the face differentiates. The embryo now looks like a human being and can easily be distinguished from other vertebrate embryos, which earlier it resembled very closely (figure 24.20).

Excretory and reproductive system elements develop extensively during the second month. At first, however, male embryos cannot be distinguished from female embryos because the developing reproductive system at this stage contains rudiments of both female and male structures. This "indifferent condition" of reproductive system development persists until about the seventh week of development, when signs of sex determination begin to appear.

The establishment of human appearance externally and all of the major organ rudiments internally marks the end of the embryonic period of development and the beginning of the fetal period. This is an important milestone in the development of the individual, and it has broader implications for the pregnancy, as well. The fetus generally is much less sensitive to harmful external influences than the embryo. Embryonic development includes a whole series of **sensitivity periods** during which delicately balanced developmental processes involved in organ formation are subject to disruption by chemical agents or disease. Harmful agents still can have damaging effects on the fetus, but they are much less likely to cause gross developmental disturbances.

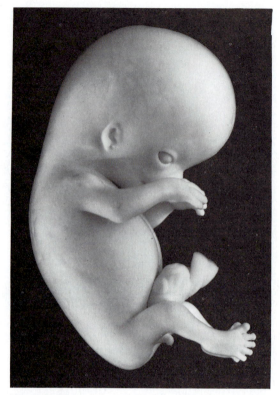

Figure 24.20   A fifty-six-day human fetus. From the end of the eighth week on, it is proper to call the developing individual a fetus because rudiments of all major organs and systems are present. The fetus is clearly human in general body form.

The Third Month

The third month of pregnancy is a relatively quiet period of development because the mother does not experience new symptoms, and fetal development involves mainly growth of structures already present. One notable feature, though, is that reproductive system development proceeds to the point that the sex of the fetus is externally apparent. Overall fetal growth is considerable because the fetus grows in crown (of the head) to rump length from 2.5 cm at the beginning of the month to about 6 cm at the end of the month. Its weight increases from less than 2 gm to about 12 gm. While the body and limbs may move slightly, the fetus is so small that the movements are wholly undetectable to the mother.

Four to Six Months (The Second Trimester)

During the second third of pregnancy (the second **trimester**), rapid fetal growth continues. For the first time, the mother can feel fetal movements. This "quickening" when the mother first feels movements within the uterus usually occurs late in the fourth month or during the fifth month. **Ossification** (calcified bone development) of the previously cartilaginous skeleton proceeds more rapidly during the second trimester than in the first, and general body form becomes increasingly more human. Hair develops on the head, and a very fine hair, the **lanugo,** forms a downy covering over the body. Waxy, almost cheeselike, secretions cover the skin surface and apparently protect it during its long immersion in the watery amniotic fluid. The fetal brain develops impressively during the second trimester, and the characteristic ridges and furrows appear in the surface of the cerebral hemispheres. Sense organs become functional, and the fetus shows reflexes that clearly indicate that it is becoming responsive to changes in its environment. The fetal heartbeat can be detected fairly readily with a stethoscope because of its distinctive sound and relatively rapid rate (120 to 150 beats per minute).

Especially during the sixth month, the fetus gains weight rapidly, and by the end of the second trimester, the average fetal weight is about 630 gm. Still, this is less than 20 percent of normal average birth weight. Babies born prematurely at the end of the second trimester sometimes survive, but they invariably require intensive care under strictly controlled conditions. Infants born that early very frequently suffer from respiratory distress due to lung immaturity, and they must be carefully protected from environmental stresses, such as temperature changes.

The Final Trimester

During the last trimester, fetal growth continues with an average weight gain of about 25 gm per day. The mother's uterus continues to expand to contain the growing fetus and its membranes. At the end of gestation, the uterus weighs more than twenty times as much as it did before pregnancy began. The uterus also continues to move higher in the abdominal cavity, as it has been doing through much of pregnancy, until just a couple of weeks before birth, when it settles down toward the pelvic area again.

Brain development and peripheral nervous system differentiation speed up during the final trimester. Nutritional research indicates that this is a critical period in brain development because normal nervous system development is very dependent on the mother's nutritional status. Protein deficiencies in the mother's diet, for example, can have adverse effects on nervous system development that may even limit the infant's future mental capacities.

Within the ovaries of female fetuses, primary oocytes enter prophase of their first meiotic division and then pause until after puberty. Testes of male fetuses begin to move during the seventh month, and they descend out of the abdominal cavity to their final location in the scrotum. This descent is necessary for eventual normal reproductive functioning after puberty. Testes that remain in the abdominal area (a condition known as **cryptorchidism;** literally, "hidden testes") are too warm to produce normal, functional sperm cells because normal spermatogenesis occurs only at the slightly lower temperatures of the scrotum.

Other systems progress toward their full-term condition. Especially because of rapid lung maturation, survival chances of prematurely born infants increase dramatically during the final trimester.

Maternal antibodies cross the placental barrier and enter the fetal circulation so that the newborn infant (**neonate**) carries immunities to bacteria, viruses, and other foreign materials. This passive immunity, "borrowed" from its mother, serves the infant only temporarily, and its own immune system begins to produce antibodies actively shortly after birth.

Birth

Human birth (**parturition**) occurs after an extended period of rhythmic uterine muscular contractions known as **labor.** Birth normally takes place about 280 days after the beginning of the last regular menstrual period or about 266 days after fertilization.

Factors determining the exact time of birth are obscure, but several items, including changes in the condition of the uterus, may be involved. Recent research indicates that there may be a "readiness hormone," that is, some chemical factor (possibly a pheromone?) that the fetus releases as a signal to the mother when the fetus has reached an appropriate stage of maturation.

Before the time of birth, hormone levels in the mother's bloodstream change. Progesterone production decreases during the last two months of pregnancy (progesterone's role has been to keep the uterine lining intact). Estrogen production increases gradually throughout the same period and then shows a sharp increase just before birth (estrogens increase the irritability and contractility of the uterine muscles). But the most direct hormonal stimulus probably comes from **oxytocin,** an octapeptide hormone from the pituitary's posterior lobe, that directly stimulates contractions of uterine muscles. The cause of the increase in oxytocin level and whether the increase is absolutely essential for normal birth are not known, but oxytocin's effect on uterine muscles is reliable enough that the hormone is used routinely to induce labor artificially when the birth process must be hurried along.

Labor is a progressive three-stage process that leads to the delivery of the fetus and the afterbirth (the placenta and the fetal membranes).

Uterine contractions during early labor usually are somewhat irregular and may be mistaken for the gas pains or other intestinal discomforts of late pregnancy. But when the contractions become regular and occur at intervals of twenty minutes or less, they signal the onset of labor, although they may still stop and resume some time later. Usually, the fetal membranes rupture early in labor, allowing the amniotic fluid to flow out through the vagina.

The first stage of labor involves important changes in the cervical portion of the uterus. A mucous plug that has blocked the uterine opening during pregnancy comes loose and is shed as the cervix begins to change shape. The cervix shortens and flattens so that it does not protrude into the vagina so far. The tiny uterine canal expands to a diameter of about 10 cm. By the end of this first stage of labor, uterine contractions have become much more forceful and frequent.

During the second stage of labor, expansion of the uterine canal is completed, and the infant is delivered through the opened uterine canal and the vagina (figure 24.21).

Continued uterine contractions, during the third stage of labor, dislodge the placenta, and normally the placenta is expelled through the birth canal about twenty minutes after delivery of the infant. The placenta is pulled loose from the uterine wall as a unit. Thus, in the separation of the human placenta, there is considerably more bleeding than there is at delivery of the placenta in many other mammals, where there are looser bonds between fetal and maternal tissue in the placenta.

Newborn infants (neonates) must make rapid respiratory and circulatory adjustments during the transition from complete dependence on the placenta to independent life outside the uterus. Circulation to the placenta normally shuts down after delivery (in attended births, the umbilical cord is clamped and cut shortly after birth) (figure 24.22). This shutdown deprives the neonate of placental gas exchange and leads to a falling oxygen concentration and a rising CO_2 concentration in its blood. Brain respiratory centers respond to this increasing level of CO_2 by sending impulses to chest and abdominal muscles, thereby stimulating the contractions needed to initiate breathing if the infant has not already gasped or begun to cry in response to pressure changes or other stimuli experienced during and after delivery. Because rapid circulatory changes direct increased quantities of blood to the newly expanded lungs, the neonate is able to make the transition to independent gas exchange.

(a)

(b)

(c)

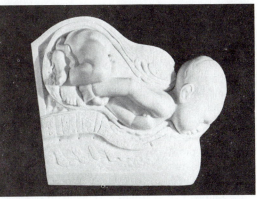

(d)

Figure 24.21 Models of the birth process. (a) Cutaway
side view of a fetus in the uterus near the end of
pregnancy. (b) Position of fetus at beginning of labor.
(c) Early in the second stage of labor as the infant
moves into the birth canal. (d) Later in the second
stage. The head has emerged.

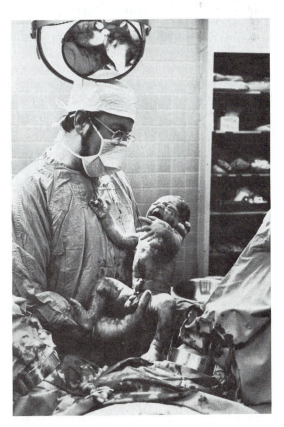

Figure 24.22 A newborn human infant (neonate) just
after delivery.

Box 24.1
Infertility, Sperm Storage, and "Test-Tube Babies"

On a personal level, infertility can represent as urgent a problem as excessive population growth can on an international scale. A significant percentage of couples wishing to have children are unable to do so. Some of this infertility is irreversible, but research into human reproductive physiology also has yielded some solutions.

One cause of male infertility is production of inadequate numbers of sperm. The average human male ejaculation produces 3 or 4 ml of semen and, with 120 million sperm per ml, a total sperm count of nearly 500 million. Because only one sperm can fuse with an egg in fertilization, 500 million seems to be an excessive number. But, in fact, when a man's sperm count is as low as 40 to 50 million per ml, he is likely to be infertile. Sperm storage technologies have provided help in such cases because semen from several ejaculations over a period of time can be stored, and the sperm can be concentrated and combined. This preparation can then be introduced into the vagina (by **artificial insemination**) at an appropriate point in the menstrual cycle, and pregnancy sometimes results. Of course, this technique does not help in cases where no sperm are produced, where sperm counts are extremely low, or where structurally abnormal sperm are produced.

A common cause of female sterility is blockage of the oviducts. This blockage prevents eggs from moving down and eliminates any possibility of fertilization. A technique for treating this condition was pioneered in England by Patrick Steptoe and Robert Edwards. In this technique, women are given pituitary hormone injections (the same treatment given other women who fail to ovulate because of pituitary gonadotrophin deficiencies), which cause maturation of one or several follicles. An oocyte is then removed through a small abdominal incision and placed in culture medium. Sperm are added, and fertilization occurs. After a blastocyst develops in culture, it is introduced into the uterine cavity at what would be the normal time of implantation if fertilization had occurred in the oviduct and the zygote had developed normally on its way down the oviduct to the uterus. If implantation occurs, development can proceed normally in some cases.

Moral problems are raised by such medical treatment. Though scientists tend to dismiss the concept, many people feel that it is not right to overcome "natural" infertility.

Also, during the early experiments that laid the groundwork for these treatments, human zygotes were cultured and discarded. Some people question whether this can ever be morally justified. And yet, the techniques that resulted from these experiments permit the couples involved to become parents, which they could not have done without treatment. The borderlines are rather fuzzy when the issue involves what is "natural" and what is "not natural" in human medicine.

But, still, a newborn human infant is helpless and immature in comparison to other neo-
natal mammals. Baby pigs or calves, for example,
are standing up, walking around, and attempting
to feed literally within minutes following birth.
Human neonates are weak; they are only poorly
able to regulate their body temperatures and are
incapable of making many types of precise physiological adjustments. They are completely dependent on parental care. The evolution of
concentrated parental care in a sheltered environment has had to parallel the evolution of humans' developmental timetable.

Regulating Human Fertility

No one knows how many people there are. Estimates of world population are only estimates and
crude ones at that. Clearly, though, world population is growing at an alarming rate, and predictions of a world population of 6,000 million or
more by the year 2000 and a population of two
and one-half times that by 2050 are chilling.
While population growth rates are falling toward
zero in the United States and several other industrialized countries, this is not the case in many
underdeveloped and developing countries. Population growth either causes or aggravates practically all international social and political
problems today. Effective means of dealing with
population growth problems depend on development and distribution of safe and biologically effective birth control techniques.

In addition to being an international social
problem, regulation of fertility is a pressing individual concern. Family size can be an urgent
personal and financial problem, and effective
family planning depends on adequate birth control technology and information.

Birth Control

The prevention of pregnancy is called **contraception** (literally "against conception," though not
all forms of birth control actually prevent fertilization). Contraceptive techniques are evaluated
on the basis of several criteria. Are they effective
in reliably preventing pregnancy? Are they reversible so that when couples do wish to have
children, they can stop using the technique and
successfully initiate a pregnancy? Are they safe
and relatively free from physiological side ef-

fects? Are they acceptable on personal and social
grounds; that is, do they interfere with sexual
enjoyment for one or both partners, or do they
place an unfair burden of responsibility for contraception on one member of the pair? The birth
control techniques currently in use all fall short
of these criteria for one reason or another. Thus,
the available techniques must be judged on their
relative merits and weighed against the chances
of unwanted pregnancy.

Birth control techniques fall into the following categories: (1) abstinence from intercourse,
especially during the portion of the menstrual
cycle when conception might occur; (2) suppression
of egg or sperm production or release; (3) prevention of contact between egg and sperm by use
of physical barriers; (4) prevention of blastocyst
implantation; and (5) abortion (termination of
pregnancy by removal of the embryo after it has
implanted in the uterine wall). Techniques from
all of these categories are used to control human
fertility, but their desirability and effectiveness
vary considerably (see table 24.2).

Abstinence

Usually, total abstinence is not psychologically
acceptable, so abstinence is used in the context
of the so-called "rhythm" method of birth control. Couples refrain from intercourse during
what is judged to be the fertile period, that is, for
several days around the time of ovulation in the
middle of the menstrual cycle. But the method
has marked shortcomings because variations in
the length of menstrual cycles and days of ovulation are very common.

Suppression of Gamete Production

The most popular birth control technique in industrialized countries is the use of hormonal contraceptive pills that prevent ovulation. Characteristically, these pills contain small quantities
of estrogens and larger quantities of a progester-
onelike compound. These hormones suppress pituitary release of FSH and LH by interfering
with the normal feedback relationship between
the ovary and the hypothalamus and pituitary,
thus preventing ovulation. Hormonal contraceptive pills also affect the remainder of the reproductive tract. They change the lining and motility
of the oviducts. They also change the way in
which the uterine endometrium grows and dif-

Table 24.2
Summary Information about Various Birth Control Techniques. Effectiveness is presented as the average number of women who become pregnant in a population of 100 sexually active women using the technique for one year. Undesirable side effects occur in only a small percentage of individuals in the case of most techniques.

Method	Effectiveness (Pregnancies Per 100 Women in One Year)	Required Medical Services	Possible Undesirable Side Effects
Tubal ligation	0.04	Surgical procedure	Infection during surgery
Oral pill (21-day administration	0.07	Prescription, regular physical exams	Early: water retention, breast tenderness, nausea Late: blood clots, hypertension
Vasectomy	0.15	Surgical procedure	Infection, possibly an autoimmune reaction
Intrauterine device (IUD)	1.5–3.4	Insertion	Menstrual discomfort, increased menstrual flow, uterine infection
Diaphragm*	12.0	Sizing, instruction	None known
Condom	14.0	None	Decreased sensation and sexual pleasure
Withdrawal	18.0	None	Decreased sexual pleasure
Spermicides alone	20.0	None	Usually none, occasional irritation
Abstinence during fertile period (rhythm method)†	24.0	None, physician counseling recommended	Psychological and physical frustration
Morning-after pill‡		Prescription	Breast swelling, nausea, water retention, cancer

*Effectiveness improved when used in combination with spermicidal jelly
†Effectiveness improved when monitored with cyclical temperature changes
‡Not frequently used as a form of birth control anymore because DES, the drug in the morning-after pill, is considered to be quite dangerous

ferentiates so that it probably would not be capable of accepting a normal implantation even if a blastocyst should reach it.

The "pill," as it is commonly called, is extremely effective in preventing pregnancy. It is taken from the fifth to the twenty-fifth days after the onset of the last menstruation. When a woman stops taking the pill on the twenty-fifth day, the endometrium begins to slough off in what approximates normal menstruation. But taking the pill on a daily basis, with no break, prevents menstruation completely. Some women may find this convenient or desirable, but they should be advised that physiologists have little idea of the long-term consequences of this abuse, which obliterates the natural cycle of menstruation.

Some unpleasant side effects of birth control pills experienced by certain women are nausea, dizziness, headache, and vomiting; these and other symptoms similar to those of early pregnancy can make the pills undesirable for some women. But far more serious are problems of increased tendency toward blood clot (**thrombus**) formation within the circulatory system of some women who take hormonal contraceptive pills. Occasionally, these clots come loose and circulate, and such a circulating clot (**embolus**) can lodge in a blood vessel somewhere else in the body and block a vital circulatory route. These problems must be weighed against the effectiveness and convenience of the pill, as well as the fact that the immediate risks of taking the pill are smaller than the risks that would accompany all of the pregnancies that would occur in the same population in the absence of contraception.

One unknown factor is the lifelong effect of this type of extended endocrine therapy, and further study is needed in this area. Research currently is being done on hormonal contraceptives that contain much lower doses of hormones and might therefore cause fewer side effects. Another possibility for the future is development of implantable capsules that would release small quantities of hormones over a long period of time and relieve the necessity of taking pills daily.

Research on suppression of sperm production is far behind research on suppression of ovulation. Some compounds (for example, diamines) do cause reversible suppression of spermatogenesis, but the range of their possible side effects has not been completely explored. Already it is known that men taking these compounds must avoid any form of alcohol consumption because alcohol causes an unpleasant or even dangerous reaction.

Prevention of Egg-Sperm Contact

Some very old techniques and some newer surgical techniques fall into this category. Physical barriers in the vagina that prevent sperm from moving into the uterus and oviduct have been used for years. The most widely used current version of these devices is the **diaphragm.** Diaphrams are latex domes that are inserted into the upper part of the vagina to cover the cervix. They usually are used with a **spermicidal** ("sperm-killing") **jelly** that is smeared around the edge of the diaphragm. This combination is quite effective.

Another physical barrier device is the **condom,** a rubber or plastic sheath that is pulled over the erect penis just before it is inserted into the vagina. The condom traps semen and quite effectively blocks sperm transfer into the vagina.

Surgical techniques can prevent egg-sperm contact by blocking gamete transport in the reproductive system. **Vasectomy,** the cutting and ligature (tying off) of the vas deferens is the male version of this procedure, and **tubal ligation,** the cutting and ligature of the oviduct, is the female version (figure 24.23). Although microsurgical procedures can reverse some vasectomies and tubal ligations, both of these techniques should be considered irreversible.

Both tubal ligations and vasectomies are done routinely, but each involves an element of risk. Tubal ligation does not interfere with normal ovulation or other ovarian functions, but the procedure requires opening the abdominal cavity, which always entails some risk, no matter how small the incision. Risks of infection in vasectomy are much lower, and the procedure is simpler, but some long-term effects are still unknown. There seems to be no natural feedback mechanism to shut down sperm production when the transport system is blocked. Some studies with experimental animals after vasectomies have shown phagocytic cell responses in testes that lead to destruction of all sperm and sperm-producing cells in seminiferous tubules. Any such reaction against a body tissue by the body's own defense mechanisms must be regarded with caution (chapter 26). The extent of such reactions,

(a)

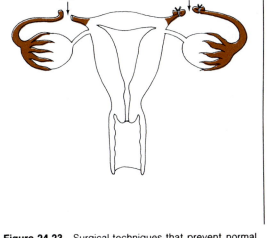

Oviducts

Cut   Tied

(b)

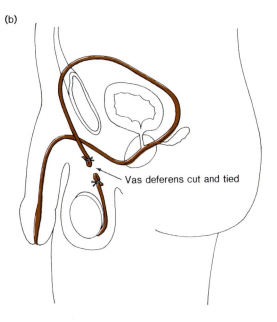

Vas deferens cut and tied

Figure 24.23 Surgical techniques that prevent normal movement of gametes. (a) Tubal ligation. (b) Vasectomy.

if any, in human testes is not known. Thus, both tubal ligations and vasectomies entail some established risks and some possible risks, which must be weighed against their nearly 100 percent effectiveness in preventing pregnancy.

Prevention of Blastocyst Implantation

Intrauterine devices (IUDs) are placed in the lumen of the uterus, where they prevent implantation of blastocysts, apparently by causing slight inflammatory responses in the endometrium. Current intrauterine devices include pieces of plastic in various shapes and small copper devices, which now seem more effective (figure 24.24). Intrauterine devices are inserted through the cervical opening and left in place in the uterus. They have the advantages of not requiring some action before or during sexual activity and not involving continuing hormone therapy. But they cause excessive menstrual bleeding and cramps in some women and produce uterine damage in a small percentage of cases. Sometimes, they are spontaneously expelled from the uterus, and pregnancy occurs before the loss is detected. Properly inserted IUDs that remain in place are very successful in preventing pregnancy and are cheap enough to provide hope for effective pregnancy control even in developing countries.

Blastocyst implantation also can be prevented by a hormonal treatment. The so-called "morning-after" pill actually is a large dose of an estrogenic substance such as diethylstilbestrol (DES). DES is a synthetic, nonsteroidal compound that produces abnormal growth of the endometrium, making normal implantation impossible. This treatment can be used in an emergency to prevent pregnancy resulting after an incident of unplanned or forced intercourse without contraceptive protection. But it must be given for several days after intercourse, and its side effects, such as nausea, vomiting, and heavy vaginal bleeding, are unpleasant. If DES somehow is taken mistakenly by a pregnant woman during a specific sensitivity period, it has an important effect on a developing female baby. Daughters of women treated with DES during pregnancy show a markedly increased incidence of certain types of cancer of the reproductive tract. This tendency does not express itself until these women reach their twenties or thirties, and only in the last few years has it been statistically

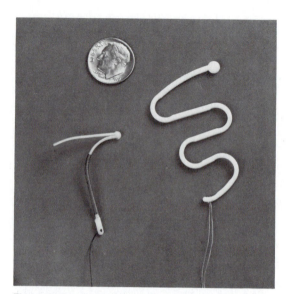

Figure 24.24 Two types of intrauterine devices: the Copper 7 (left) and the Lippes loop (right). Apparently, there is no special shape requirement. An effective IUD must remain in place and induce the necessary reaction of the uterine endometrium.

connected with DES treatments of these women's mothers. The ideal morning-after pill of the future, if it is developed, will not be DES, with its carcinogenic (cancer-causing) potential.

Abortion

Artificially induced abortions provide means of terminating pregnancy after implantation, but abortion is a controversial method of regulating fertility in most societies because of moral and legal concerns.

During the first three months of pregnancy, abortion can be induced by dilating the cervical opening and scraping the uterine lining, including the implantation site. An alternative is an aspiration technique that removes the embryo by sucking it out through a cannula. These techniques disturb the endometrium and lead to menstruationlike bleeding. Another technique leading to the same result is the injection of hypertonic saline into the amniotic cavity. This kills the embryo and sets off an expulsion response in the uterine lining.

New techniques for induction of abortion are being developed using some of the **prostaglandins,** a group of seemingly ubiquitous chemical messengers (chapter 14). Prostaglandins induce uterine contractions that lead to menstruationlike bleeding. It has been proposed that a monthly vaginal prostaglandin insert might replace other forms of birth control. The insert would be used to induce menstruation each month, and the users would have no idea whether or not pregnancy had begun. Clearly, this sort of technique will also raise moral questions.

Sexually Transmitted Diseases

Another social problem related to human reproductive function is that of **sexually transmitted (venereal) diseases.** A dozen or more diseases fit this general categorization, and the causative agents come from a number of taxonomic groups.

Gonorrhea, a bacterial disease, is the best-known sexually transmitted disease (figure 24.25a). In men, gonorrhea is usually readily detected because pus is discharged from the penis, and burning sensations during urination develop within a few days after infection. But in women, the infection concentrates in the cervical canal and produces very mild symptoms, if any at all. Thus, women may unknowingly develop extended infections in the reproductive tract. The most serious result of such extended infections in women is infection and inflammation of the oviducts, which can lead to partial or complete blockage of the ducts and eventually cause sterility. Gonorrhea routinely has been treated with large doses of penicillin, but penicillin-resistant strains are becoming more common, and treatment is more complex than it once was.

Syphilis, a spirochete disease, has a rather long incubation period (figure 24.25b). On the average, the first symptom, a hard painful ulcer called a **chancre,** develops at the site of infection after about three weeks. Even if the disease is left untreated, the chancre disappears after a short time, and no more sign of disease is seen until two to four months later. At this time the disease enters its secondary stage as a generalized skin rash, and infections of various organs sometimes develop. Then the disease goes into a latent period. This latent period can last throughout life, but in some people the disease can enter a tertiary phase that produces severe nervous system or circulatory system damage and even death. Syphilis also usually yields to penicillin treatment, but as in the case of gonorrhea, increasingly common antibiotic resistance is a growing problem.

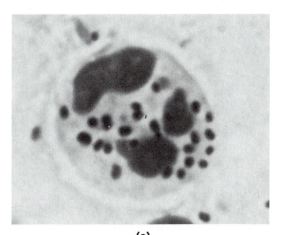

(a)

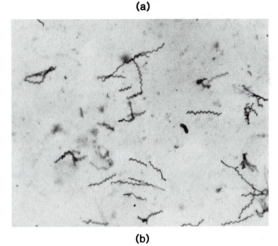

(b)

Figure 24.25 Organisms that cause two common venereal diseases. (a) Human leukocyte containing *Neisseria gonorrhoeae*. This cell was found in a gram-stained smear of urethral discharge. Note that the small, dark bacterial cells occur in pairs. (Magnification $\times$ 2,655) (b) The spirochete *Treponema pallidum*, which causes syphilis seen in an immunofluorescent smear. (Magnification $\times$ 580)

Several virus diseases are sexually transmitted. The best known of these usually is caused by the **Type 2 herpes simplex virus,** which is very similar to the virus that causes cold sores and fever blisters on the lips (Type 1 herpes simplex). Recently, however, genital infections with Type 1 herpes simplex also have become quite common. The incubation period is variable, but between two and twenty days after infection, blisters develop on the affected area. After the blisters rupture, the painful surface ulcers left by the blisters heal, and the disease becomes dormant. But as with cold sores, these blisters can reoccur repeatedly at variable intervals. Generally, the disease symptoms are more distressing for women than men, and the disease probably is more serious for women because there seems to be a higher-than-average incidence of cervical cancer among women infected with herpes simplex virus.

A variety of other diseases caused by viruses, bacteria, yeast, protozoa, and even small lice also are sexually transmitted. One of these, **nongonococcal urethritis (NGU)**, is so common as to demand special mention. NGU's symptoms are very similar to those of gonorrhea, but the gonorrhea-type bacterium (gonococcus) is not found in pus discharged from infected individuals. Therefore, NGU often escapes diagnosis as a sexually transmitted disease. Treatment of NGU is progressing now that physicians suspect NGU when gonorrhea symptoms are present but no ordinary gonococcus bacteria are found.

Because it is becoming more and more common for organisms that cause sexually transmitted diseases to be resistant to antibiotic treatment, interest in immunization programs is increasing. Perplexing social problems would be raised, however, by the development and experimental testing of vaccines for sexually transmitted diseases. And the publicity campaigns that would accompany these general vaccination programs undoubtedly would inspire much public comment.

Summary

The basic design of the human reproductive system is an adaptation to life in a terrestrial environment. Fertilization takes place inside the female reproductive tract, and there is a long gestation period of sheltered and physiologically supported development.

Testes produce sperm and testosterone, the male sex hormone responsible for development of secondary sex characteristics. Sperm are produced in seminiferous tubules and transported to the epididymis, where they remain until ejaculation carries them through the vas deferens and urethra, which opens to the outside at the tip of the penis. During this passage, the secretions of several glands are added to the sperm to produce seminal fluid (semen).

Ovaries produce oocytes and two kinds of steroid hormones—estrogens and progesterone. At ovulation, an oocyte is released and carried into the oviduct. Fertilization occurs if the oocyte encounters viable sperm during the first part of its journey to the uterus.

Copulation in humans involves transfer of seminal fluid into the vagina. Male orgasm is essential for reproductive success, but female orgasm is not. Female orgasm seems rather to be an adaptation that has evolved along with continuous female sexual responsiveness, which is rare among mammals.

Timing of reproductive events in the female is accomplished through the menstrual cycle. Pituitary hormones regulate maturation and ovulation of the oocyte, as well as hormone production by the ovary. Ovarian hormones, in turn, control cyclical events in the uterus so that the endometrium is fully prepared for implantation at the time that an embryo could arrive.

Meiosis in the ovary extends over many years with one of the arrested oocytes resuming its development each month. Meiosis is completed, however, only as part of the activation reaction in fertilization. Fertilization is made possible by sperm hyaluronidase that loosens the follicular cells around the oocyte, thus giving the sperm access to the oocyte.

Pronuclear fusion follows sperm entry, and development begins in the oviduct. The blastocyst stage, which is reached in the uterus, is capable of implanting in the uterine endometrium.

During early pregnancy, chorionic gonadotrophin stimulates continued progesterone production by the corpus luteum, and this, in turn, keeps the endometrium intact and prevents menstruation. Failure of this mechanism leads to miscarriage.

During most of gestation, the placenta is the site where materials are exchanged selectively between mother and infant across a placental barrier. The placenta also functions as a hormone-producing organ.

The developing infant is enclosed and protected within the amnion and bathed with amniotic fluid throughout development. The infant's condition can be examined by analysis of amniotic fluid, removed by amniocentesis.

Development of the embryo proceeds during and following implantation. Rudiments of all body organs are produced during the embryonic period, which ends at eight weeks. All body parts differentiate and grow during the long fetal period, which occupies the remainder of gestation, so that all systems are capable of functioning independently of placental support by the time of birth.

Near the end of gestation, during the final trimester, the fetus gains weight rapidly, and maturation of the respiratory system, the nervous system, and other systems accelerates so that chances of survival, in case of premature birth, increase dramatically.

Birth occurs about 266 days after fertilization. Labor is initiated by hormonal changes that are not completely understood. Uterine contractions expel the infant through an expanded cervix and vagina. Continued contractions result in expulsion of the afterbirth. Human neonates are relatively weak and helpless compared to other newborn mammals.

Regulation of human fertility is an important social and scientific concern. A variety of birth control techniques are in use, and each has advantages and disadvantages. Hormonal suppression of ovulation via the birth control pill is the most widely used technique in industrialized societies.

Another social concern associated with human reproduction is the increasing incidence of sexually transmitted diseases. Antibiotic resistance to bacterial venereal diseases is becoming a common problem. Venereal diseases caused by viruses are increasingly prevalent.

In the area of human reproduction, it is impossible to separate scientific and medical problems from personal and societal problems. Human sexuality is much more than just a reproductive capacity; it is a powerful psychological and social force in human life.

Questions

1. Can you suggest a hormonal basis for the physical and psychological letdown that some women experience just before menstruation? (Hint: Progesterone has powerful general metabolic effects.)

2. Discuss the adaptive significance of human female orgasm and speculate on the selective pressures involved in its evolution.

3. Fertilization involves one egg and one sperm, and yet human males with sperm counts below 50 million per ml are likely to be infertile. Suggest a possible explanation.

4. Distinguish clearly between the terms "oocyte" and "egg" in the context of human meiosis and fertilization.

5. Discuss implantation as a critical point in the continuing development of an embryo and as an example of "tolerance" in terms of the mother's normal defense mechanisms.

6. Many older legal systems make a distinction between developing humans during the first two months of pregnancy and those that have developed for more than two months. What developmental factors can be correlated with these traditional viewpoints?

Suggested Readings

Books

Langman, J. 1981. *Medical embryology.* Baltimore: Williams and Wilkins.

Mader, S.S. 1980. *Human reproductive biology.* Dubuque, Iowa: Wm. C. Brown Company Publishers.

Moore, K.L. 1977. *The developing human: Clinically oriented embryology.* 2d ed. Philadelphia: Saunders.

Nilsson, L. 1973. *Behold man: A photographic journey of discovery inside the body.* Boston: Little, Brown.

Articles

Beaconsfield, P.; Birdwood, G.; and Beaconsfield, R. August 1980. The placenta. *Scientific American* (offprint 1478).

Edwards, R. G. 1981. Test-tube babies, 1981. *Nature* 293: 253.

Epstein, C.J., and Golbus, M.S. 1977. Prenatal diagnosis of genetic diseases. *American Scientist* 65: 703.

Hart, G. 1977. Human sexual behavior. *Carolina Biology Readers* no. 94. Burlington, N.C.: Carolina Biological Supply Co.

Hart, G. 1977. Sexually transmitted diseases. *Carolina Biology Readers* no. 95. Burlington, N.C.: Carolina Biological Supply Co.

Rhodes, P. 1976. Birth control. 2d ed. *Carolina Biology Readers* no. 4. Burlington, N.C.: Carolina Biological Supply Co.

Segal, S.J. September 1974. The physiology of human reproduction. *Scientific American.*

Differentiation

Chapter Concepts

1. Cells with the same nuclear genetic makeup differentiate in many directions during development of a multicellular organism.

2. Principles of differentiation often can be studied using model systems. Results of these studies contribute to formulation of developmental generalizations.

3. Determination, the commitment to differentiate, can be imposed on cells by cytoplasmic factors, inductive relationships, hormones, and positional information.

4. Cellular responses to determination result in specific gene activations that lead to synthesis of specific types of proteins.

5. Differentiation, based on specific molecular syntheses, is expressed at all levels of organization.

6. Developmental specializations establish each individual as a biochemically unique entity.

Few natural phenomena are so impressive as the normal development of a complex, multicellular organism from a single cell, the zygote. Early in development, this single cell gives rise to a population of cells through a series of mitotic cell divisions. All of the cells in this population seem very similar to one another. In fact, they all carry identical sets of nuclear genetic information because they are all direct mitotic descendants of the same original cell. But before long, differences appear among the cells located in different parts of the developing body. Cells become arranged in specific, orderly relationships to other cells, and different groups of cells in various areas become structurally and functionally specialized (differentiated) to carry out their individual and group functions (figure 25.1).

Biologists have long been aware that this process of differentiation is a very orderly progression that is repeated exactly each time that an individual of any given species develops. Karl Ernst von Baer summarized this idea long ago (1828) in his book on animal development where he said that the more general characteristics of an animal group appear earlier in development

and more specific characteristics appear later. For example, all vertebrate embryos develop brains, spinal cords, the beginnings of vertebral columns, and other features common to all vertebrates early in development. Only later do recognizably different body shapes and special features, such as the hair of mammals or feathers of birds, appear. Still later, the specific features that separate closely related species develop, and finally, features that distinguish single individuals from one another emerge. Today developmental events are described in cellular and genetic terms, and the development of functional capabilities as well as structural features is emphasized, but the development of any organism is still viewed as a series of events leading from the more general to the more specific.

The blueprints for this progressive specialization are laid out in a complex genetic program for development. All parts of this genetic program must be expressed correctly and in the proper sequence if the precisely patterned relationships and regional specializations of body parts are to develop normally.

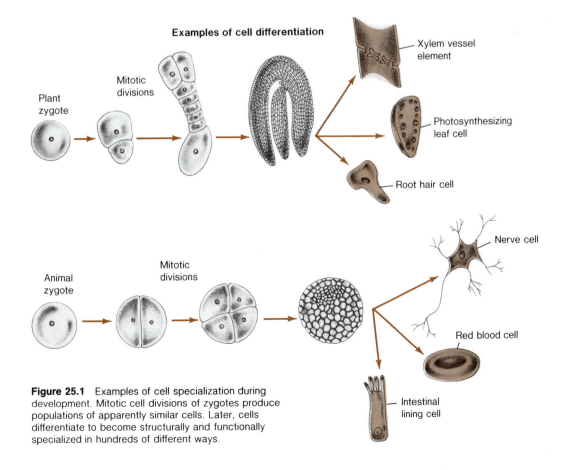

Figure 25.1   Examples of cell specialization during development. Mitotic cell divisions of zygotes produce populations of apparently similar cells. Later, cells differentiate to become structurally and functionally specialized in hundreds of different ways.

In this chapter, some current views of the mechanisms by which cells become organized into bodies and develop their specific functional specializations, under the direction of the genetic program for development, are examined. Study of the genetic program for development and its expression during development is one of the most active areas of modern biological research. Better comprehension of the processes involved will add to the understanding of normal development and provide clues for the prevention of abnormal development.

The Problem of Cell Differentiation

If all of the cells in the body of a multicellular organism receive a complete and accurate copy of the zygote's nuclear genetic information transmitted through a series of mitotic cell divisions, how do cells become specialized in so many different ways? Cell **differentiation** (specialization) must result from different uses of this common store of genetic information by various cells in different parts of the body.

Some parts of the genetic information are used by practically all cells because there are some "housekeeping" activities, such as cell respiration and protein synthesis, that all cells have in common. Genetic information that directs the synthesis of the enzyme systems needed for these processes must be actively expressed in all cells. But how do cells become structurally and functionally specialized for the division of labor that occurs in complex, multicellular organisms?

The special structural proteins and enzyme systems found only in certain specialized cells sometimes are called "specific" proteins, and each kind of specialized cell has its own particular set of specific proteins. Thus, cell differentiation depends on production of different sets of specific proteins by different groups of cells.

To produce these special sets of specific proteins, cells must actively express specific parts of their total store of genetic information (genome), while other parts are kept inactive (repressed) (figure 25.2). For example, some portions of the genome are used to produce one type of specialized cell (for example, a brain cell), while different portions of the genome are used to produce another type of specialized cell (for example, a kidney cell).

Cell differentiation, then, results from differential gene activation, and the genetic program for development directs a series of precisely timed and positioned sets of these differential gene activations.

Preformation and Epigenesis

During the seventeenth and eighteenth centuries, many biologists believed that eggs contained miniature bodies and that the seminal fluid was the "vital stimulus" that caused the unfolding and expansion of these transparent, preformed body parts. However, this **preformation theory** was complicated by the discovery of sperm cells. Once biologists had discarded the idea that sperm might just be parasites in the seminal fluid, some of them began to argue that the preformed body was in the sperm rather than in the egg (figure 25.3). They proposed that the preformed body contained in the sperm could develop only when the sperm entered the hospitable environment of the egg.

But then, in 1759, Caspar Friedrich Wolff published his detailed observations of chick embryo development and concluded that there was no evidence whatsoever of a preformed body. He said that the body was assembled from simpler, less organized material. He reported seeing "granules" (cells or nuclei?) that organized layers that folded into a body. This hypothesis of progressive organization of a body from less organized material was called the **theory of epigenesis.**

K. E. von Baer's book on animal development, mentioned earlier, was another powerful blow against the preformation theory because von Baer's careful observations of progressive differentiation revealed no evidence at all that a preformed body was present at the beginning of development.

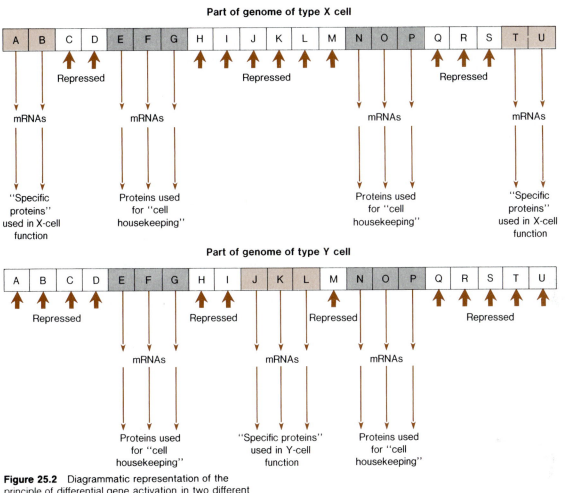

Part of genome of type X cell

| A | B | C | D | E | F | G | H | I | J | K | L | M | N | O | P | Q | R | S | T | U |

Repressed   Repressed   Repressed

mRNAs   mRNAs   mRNAs   mRNAs

"Specific proteins" used in X-cell function   Proteins used for "cell housekeeping"   Proteins used for "cell housekeeping"   "Specific proteins" used in X-cell function

Part of genome of type Y cell

| A | B | C | D | E | F | G | H | I | J | K | L | M | N | O | P | Q | R | S | T | U |

Repressed   Repressed   Repressed   Repressed

mRNAs   mRNAs   mRNAs

Proteins used for "cell housekeeping"   "Specific proteins" used in Y-cell function   Proteins used for "cell housekeeping"

Figure 25.2 Diagrammatic representation of the principle of differential gene activation in two different hypothetical cell types. "Housekeeping" activities are common to all cell types. "Specific proteins" are used for specific functions of specialized cell types. Different parts of the genome are expressed in different types of cells.

Figure 25.3 Seventeenth-century biologists proposed that preformed bodies existed inside sperm cells. They made sketches such as this and called the miniature body a homunculus ("little man"). Absurd arguments even arose over the possibility that bodies of all future generations were still smaller and were inside the homunculus.

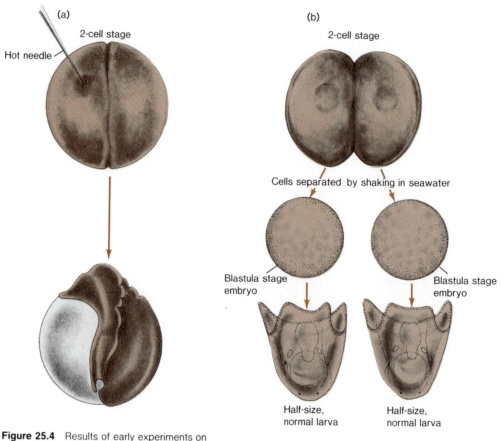

Figure 25.4 Results of early experiments on developmental potential of embryonic cells. (a) Roux's experiment. One frog embryo cell was killed with a hot needle. The surviving cell produced a half-embryo. (b) Driesch's experiment. Driesch separated the first two cells of a sea urchin embryo and found that each can produce a small, but normal and complete, embryo.

The First Experiments

Many biologists, however, were not convinced by Wolff and von Baer's conclusion that development was epigenetic. These biologists argued that Wolff's granules might represent a preformed set of determinants for all the body parts and that cleavage divisions might simply separate and properly position these determinants. To test this idea, Wilhelm Roux destroyed one of the first two cells of a frog embryo with a red-hot needle. The surviving cell formed only a half body. Roux concluded that the surviving cell had received only half of the determinants of the egg and that this accounted for the abnormal development (figure 25.4a). Hans Spemann later showed, however, that the degenerating dead cell very likely inhibited the surviving cell's further development.

When Hans Driesch completely separated the cells of sea urchin embryos, each of the first two cells developed into normal, but half-size embryos (figure 25.4b). Driesch concluded that early development was not controlled by rigid preformation involving specific determinants but that there was flexibility, which permitted adjustment of developmental processes. Years later, Hans Spemann and others showed that amphibian embryos could also show the same flexibility. If cells were completely separated at the two-cell stage without damage, each cell could develop into a half-size embryo and, eventually, a miniature tadpole.

Box 25.1
Flexibility in Early Development:

Two Parents, Four Parents,
or Six Parents!

Hans Driesch showed long ago that the first cleavage cells of sea urchins have considerable developmental flexibility. If left to develop normally, the first two cleavage cells produce one pluteus larva. If separated experimentally, the first two cleavage cells each develops into a half-size, but normal pluteus. Since Driesch's time, biologists have demonstrated that the early cleavage cells of a number of organisms (but by no means all of those tested) have this ability to regulate their development following separation, just as Driesch's sea urchin cleavage cells did.

But what about developmental flexibility in the other direction? Instead of two embryos from one, can one embryo be obtained from two or more? Beatrice Mintz and her colleagues removed the zona pellucida, the membrane that encloses mammalian embryos during early development, from cleaving mouse embryos that had been taken from their mothers' oviducts. Two embryos at the eight-cell stage were placed together and allowed to fuse into a single cluster of cells. After fusion, the resulting embryo (for such a cell cluster does organize a single embryo) was placed in the uterus of a foster mother, where development proceeded normally. Mice that develop from two embryos are called tetraparental mice (mice having four parents).

If the two embryos that are fused into one have different coat color genotypes, the results of fusion can be observed simply by examining the hair pattern of the mice after birth. The mice's coats are a mosaic of patches of the two colors. This is clear proof that cells of both embryos survive the fusion technique and participate in development of a single mouse.

In 1978, Clement Markert and Robert Petters fused three mouse embryos produced by three sets of parents, each of which had a different coat color genotype. These fused embryos were transferred to foster mothers, and some of the mice that were born had coats that were mosaics of three different coat colors. Thus, cells from all three of the original embryos participated in the development of a single fused embryo. Mice that develop from three embryos are called hexaparental (six-parent) mice (box figure 25.1A).

Beyond being an important demonstration of the flexibility of early development, these embryo-fusion techniques produce animals that are valuable subjects for research on other biological phenomena, such as the functioning of the immune system.

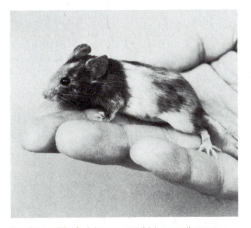

Box figure 25.1A A hexaparental (six-parent) mouse. This tricolored mouse was produced by aggregating three eight-cell embryos, each of which had a different coat color genotype. The mouse's coat has a mosaic of white, yellow, and black patches.

Nuclear Genetic Equivalence

If each of the first cells of an embryo can go on by itself to produce a small, but normal and complete body, each of them obviously contains a complete, accurate, and usable copy of the genetic information in the zygote; that is, each cell has the information required to direct the entire range of normal developmental processes. Might one explanation for cell differentiation be that later in development nuclei of some cells lose that ability to express this full range of developmental potential? In other words, is the key to cell differentiation a progressive series of restrictions of nuclear genetic capacity that leaves each nucleus capable of expressing only part of its original total genetic potential, only that part needed for development and function of a single cell type?

Nuclear Transplantation

Robert Briggs and T. J. King set out to test this hypothesis by transplanting nuclei taken from various embryonic cells of the frog *Rana pipiens* into eggs whose nuclei had been removed (**enucleated eggs**). These combinations tested the capacity of nuclei from cells at advanced stages of development to interact with mature egg cytoplasm to produce the full range of normal developmental processes (figure 25.5). Briggs and

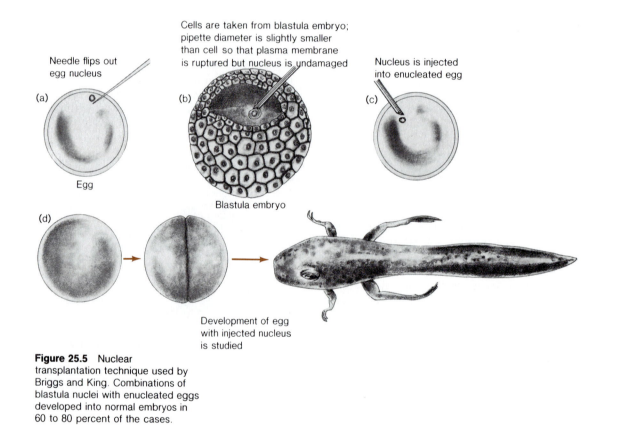

Needle flips out egg nucleus

(a)

Egg

Cells are taken from blastula embryo; pipette diameter is slightly smaller than cell so that plasma membrane is ruptured but nucleus is undamaged

(b)

Blastula embryo

Nucleus is injected into enucleated egg

(c)

(d)

Development of egg with injected nucleus is studied

Figure 25.5 Nuclear transplantation technique used by Briggs and King. Combinations of blastula nuclei with enucleated eggs developed into normal embryos in 60 to 80 percent of the cases.

King found that nuclei from cells of blastula-stage embryos could interact with enucleated eggs to direct normal development. Thus, even in blastula embryos, which have 8,000 to 16,000 cells, it appears that nuclei have not lost any of their developmental capacity; they still are able to direct the entire range of developmental processes—from egg to tadpole and even on through metamorphosis to adulthood. Briggs and King found in further experiments, however, that if nuclei from cells from still later stages of development were transplanted into enucleated eggs, very few normal tadpoles developed. At the time, these results led Briggs and King to conclude that as development proceeds beyond a certain point,

nuclei of embryonic cells become restricted in their developmental potential, and that this nuclear restriction might be one mechanism by which cell differentiation takes place.

However, J. B. Gurdon obtained very different results when he did two successive nuclear transplantations using nuclei of the clawed frog *Xenopus*. Gurdon transplanted nuclei from advanced embryos into enucleated eggs. After these combinations reached the blastula stage, he transplanted nuclei from cells of these blastulae into other enucleated eggs (figure 25.6). This several-step process overcame a problem of basic timing incompatibility. Nuclei from slower-dividing cells of older embryos are not immediately

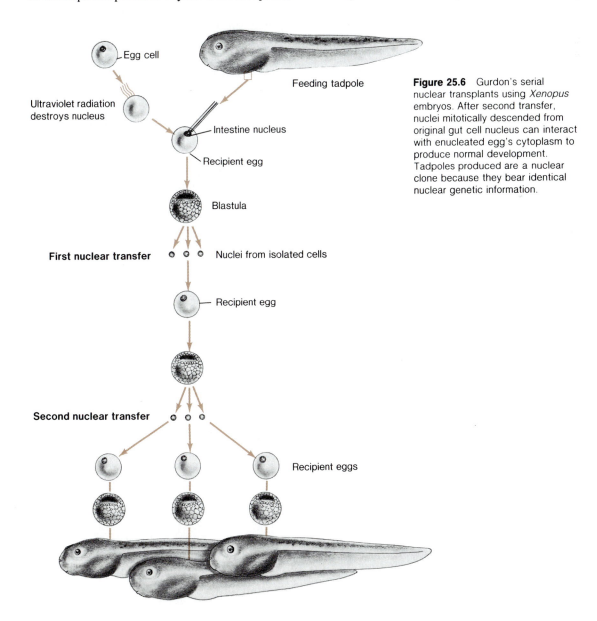

Figure 25.6 Gurdon's serial nuclear transplants using *Xenopus* embryos. After second transfer, nuclei mitotically descended from original gut cell nucleus can interact with enucleated egg's cytoplasm to produce normal development. Tadpoles produced are a nuclear clone because they bear identical nuclear genetic information.

Egg cell

Feeding tadpole

Ultraviolet radiation destroys nucleus

Intestine nucleus

Recipient egg

Blastula

First nuclear transfer   Nuclei from isolated cells

Recipient egg

Second nuclear transfer   Nuclei from isolated cells

Recipient eggs

compatible with enucleated eggs, which are geared up for the relatively rapid divisions of early cleavage. Using a serial transplantation technique, Gurdon found that even nuclei originally transplanted from differentiated gut cells of feeding tadpoles produced normal development. Thus, when tested under appropriate experimental conditions, nuclei from even differentiated cells of advanced embryos do not show signs of restricted developmental capacity. Eventually, Gurdon and his colleagues cultured adult frog skin cells and transplanted their nuclei into enucleated eggs. Even these combinations produced normal embryos (figure 25.7).

This work has attracted widespread interest because it seems to open the way for cloning of adult animals. Even mammals, including humans, possibly could be cloned. Experiments with mammals would pose great technical difficulties because mammalian eggs are much smaller than amphibian's eggs and their nuclei are more sensitive to handling. But there seems to be no theoretical barrier to nuclear transplantation and even cloning in mammals if the technical problems can be solved.

Totipotency of Plant Cells

Nuclei of even adult animal cells can express a wide range of developmental potentials, but testing their potential requires that they be placed in appropriate new cytoplasmic environments (enucleated eggs). Genetic potentials of plant cell nuclei, however, can be tested much more directly.

Many plants can be propagated from cuttings of stems or leaves that can take root and produce whole growing plants. In plant tissue culture studies, whole plants can be grown from small clusters of cells broken off the masses (**calluses**) of undifferentiated tissue that can be started from tissue taken from almost any part of a vascular plant. Thus, under appropriate conditions, whole plant bodies grow from cells originally descended from only one part of a plant. This is strong evidence that no permanent and irreversible restriction on the genetic capacity of plant cells occurs during development.

If small clusters of cells can grow into a whole plant, what is the developmental potential of a single cell? F. C. Steward and his colleagues broke up callus cultures of carrot root phloem tissue into small pieces, some of which contained a few cells and some of which probably were individual cells. When these pieces were placed in a medium containing coconut milk, some began to divide and produced cell clusters that resembled early embryos. These clusters (**embryoids**) developed into little plantlets that were transferred to solid cultures and grown into whole plants (figure 25.8) Then, Vasil and Hildebrandt showed unequivocally that single tobacco plant cells, isolated from calluses, could divide to produce embryoids that eventually grew into completely normal tobacco plants.

Since then, embryoids have been obtained in cultures of cells from mature tissues of many species of plants. It appears that it is possible to grow clones of whole plants from individual body cells of many species. This ability of single plant cells to develop into whole plant bodies is known as **totipotency.**

As a result of these studies on totipotency, it is clear that nuclei of some differentiated plant cells retain the ability to express all of the genetic information needed to direct development of a complete body, just as animal nuclei can. An important difference between animal and plant cells, however, is that intact individual plant cells can retreat from a commitment to being part of a differentiated multicellular body and divide to produce a number of cells that can each develop an entire new body. Totipotency of this sort has not been demonstrated for whole cells of animals.

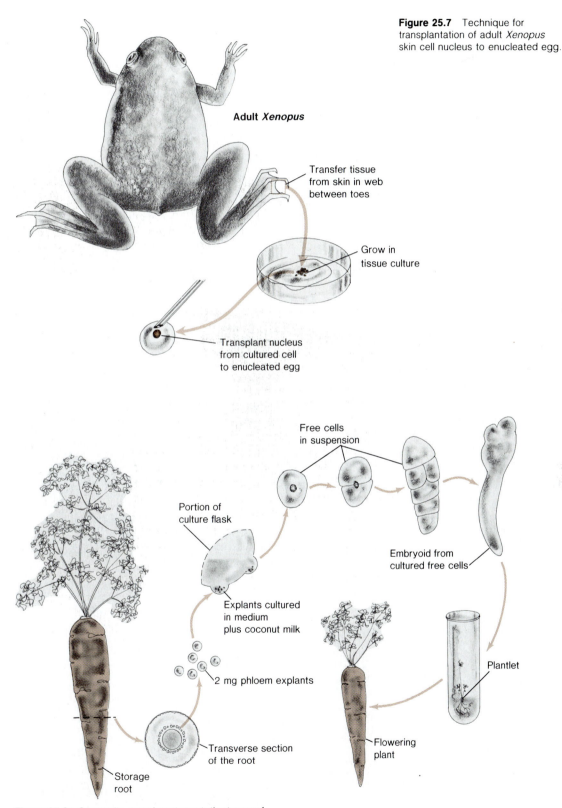

Figure 25.7 Technique for transplantation of adult *Xenopus* skin cell nucleus to enucleated egg.

Adult *Xenopus*

Transfer tissue from skin in web between toes

Grow in tissue culture

Transplant nucleus from cultured cell to enucleated egg

Free cells in suspension

Portion of culture flask

Embryoid from cultured free cells

Explants cultured in medium plus coconut milk

Plantlet

2 mg phloem explants

Transverse section of the root

Flowering plant

Storage root

Figure 25.8 Steward's experiments on totipotency of carrot cells. Embryoids develop from small clusters of cells or single cells broken off cultured phloem tissue. Some embryoids can go on to produce whole normal plants.

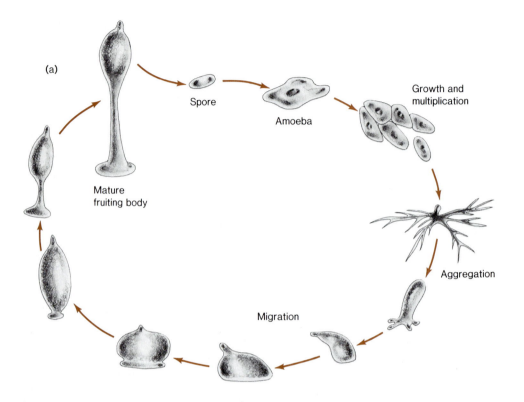

(a)

Spore

Amoeba

Growth and
multiplication

Mature
fruiting body

Aggregation

Migration

Figure 25.9 Slime molds. (a) Life
cycle of the cellular slime mold
Dictyostelium discoideum. Different
parts of the cycle are drawn to
different scales. (b) Individual
amoeboid cells moving toward an
aggregation center. (Magnification
× 389) (c) Streams of cells moving
toward aggregation centers.
(Magnification × 12) (d) Migrating
grex. (e) Scanning electron
micrograph of a mature fruiting
body. (Magnification × 149)

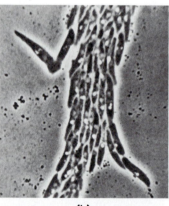

(b)

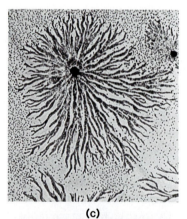

(c)

(d)

(e)

Slime Molds: A Model of Differentiation

During development of multicellular organisms, cells with a common genetic makeup respond to the external influences of many types and differentiate to produce the many specialized types of body cells. It is helpful to examine developmental control in a more simply organized system before considering the cell interactions that control differentiation in multicellular plants and animals. Developmental biologists often resort to the study of such model systems because information obtained about them can help to define basic developmental problems more clearly and prompt better and more precise questions about the development of complex plant and animal bodies. One widely studied model system for developmental analysis is the life cycle of cellular slime molds, such as *Dictyostelium discoideum,* a slime mold that has been studied in detail by K. B. Raper, J. T. Bonner, and others.

Individual amoeboid *Dictyostelium* cells feed on bacteria, grow, and divide. During this period of feeding and cell multiplication, individual amoebae show no tendency to come together in clusters; in fact, they repel one another when they get close together. This goes on for generations, but when food runs out, a change occurs. Amoebae swarm together and form a multicellular mass, the **grex.** Grex cells secrete a slimy sheath around themselves, and the grex moves around in a sluglike fashion. Eventually, the grex develops into a **fruiting body** consisting of a stalk holding up a rounded, spore-forming body. Spores with tough protective coverings and an ability to withstand dryness and other conditions that would kill the amoebae are produced in this rounded, spore-forming body. If the spores later are distributed in favorable locations, they germinate to release amoebae and start the whole cycle anew (figure 25.9).

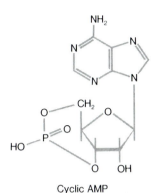

Cyclic AMP

Figure 25.10 Structure of cyclic adenosine monophosphate (cAMP), the aggregation-promoting substance of cellular slime molds. Cyclic AMP also functions as a second messenger in animal cell responses to some hormones (chapter 14).

How does all of this represent a model developmental system? Developmental changes in cellular slime molds parallel (in a simpler, more readily observable form) changes in developing cells of more complex, multicellular organisms. A population of identical, unspecialized cells goes through a series of specific differentiations, which result in aggregation, cellular cooperation within the moving grex, and finally, in development of a precisely patterned fruiting body.

The first change, initiation of aggregation, is a response to a chemical aggregation promoter, cyclic adenosine monophosphate (cAMP), secreted by cells forming an aggregation center (figure 25.10). Waves of cAMP secretion spreading from these centers cause amoebae to migrate toward the source and to develop specific cell surface adhesion sites that cause them to stick to one another. Cyclic AMP also is an important regulator substance in animals, where it serves as a secondary chemical messenger in animal cell responses to several hormones (chapter 14).

After the grex has migrated for a time, it produces the fruiting body. Cells from the front one-third of the grex form the basal disk and the stalk, and secrete cellulose cell walls that help to hold the stalk upright (see figure 25.9a). Cells from the back two-thirds of the grex form spores in a rounded mass at the top of the stalk.

and spore-forming area, and cellulose is produced by stalk cells. Analysis of the various cellular developmental changes, the grex pattern formation, the fruiting body production, and the controls on these processes have contributed to the understanding of cell interactions in development of all multicellular organisms.

Determination and Differentiation

Differentiation is the structural and functional specialization of cells during development. But what are the control mechanisms that cause cells to differentiate? Observable changes that signify cell differentiation take time to develop; thus, cells clearly must have made a commitment to a specific course of development some time before those changes actually occur. This commitment to undertake a particular differentiative pathway is called **determination.**

Some of the most challenging questions in developmental biology are concerned with determination and differentiation of cells in developing multicellular organisms. What kinds of influences cause cells to become determined? What specific cellular responses characterize cells that become developmentally determined? Is determination reversible? If so, for how long is it reversible? These are complex questions for which there are only partial answers, at best.

Cytoplasmic Factors and Differentiation

During a cell division, special parts of the cell's cytoplasm may pass to only one of the two daughter cells. The cell receiving the special cytoplasmic material may develop quite differently from another cell that does not receive it. For example, near the surface of the yolkiest part of a frog's egg is a special cytoplasm area that contains material long ago named the "germinal plasm." This material becomes localized in a cluster of relatively large cells which, at the blastula stage of development, are found among cells destined to become part of the digestive tract. But the cells containing the germinal plasm migrate out of the developing gut area into the area of the developing gonads. These special cells, the **primordial germ cells,** settle in the gonads and become the ancestors of all of the eggs or sperm ever produced by the gonads (figure 25.12).

Slime trail

Figure 25.11 Results of cutting a *Dictyostelium* grex in half. Each half forms a fruiting body with stalk and spores.

A definite pattern of anterior cells destined to form stalk and posterior cells destined to produce spores is well established within the migrating grex, but the pattern is adjustable. Experimental analysis has shown that individual grex cells are not irreversibly committed to their normal fate. If the grex is cut into parts, there is internal reorganization so that each part forms a small, but normal fruiting body (figure 25.11). Whatever the size of the grex piece, the front one-third of its cells forms stalk, and the rear two-thirds forms spores. Thus, even within a subdivided grex, there is still a definite **polarity,** an end-to-end difference in developmental tendency. The front of the grex piece forms stalk, and the rear forms spores, no matter what their fate would have been in the whole, intact grex.

As a model system, cellular slime molds show several interesting properties. Feeding amoebae change into aggregating cells that show specific movement behavior and develop new cell surface properties. Within the grex, a definite pattern of polarity develops, and cellular commitments are expressed during development of the fruiting body. Cell movements shape the stalk

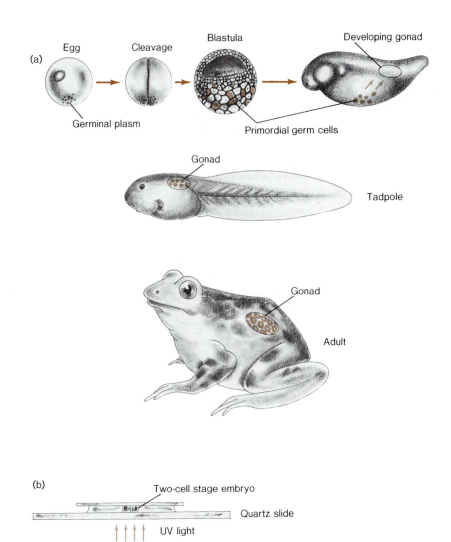

(b)

(c)

Figure 25.12 "Germinal plasm" and primordial germ cells. (a) "Germinal plasm" in the egg becomes localized in primordial germ cells that colonize the developing gonad. (b) Two-cell stage is pressed against a quartz slide, which allows ultraviolet light to reach the area containing "germinal plasm." (c) E. P. Volpe and S. Curtis surgically paired these two tadpoles so that their circulatory systems were connected early in development to test the possibility that frog primordial germ cells migrate through the blood vessels. Before they were surgically joined, the individual on the right was treated with ultraviolet light as in (b), but the one on the left was not. If primordial germ cells migrated through blood vessels, germ cells from the unirradiated embryo should reach the gonads of both individuals, but they did not. Note primordial germ cells (arrow) in the untreated individual and the complete absence of germ cells in the irradiated individual. This experiment clearly demonstrates the effects of ultraviolet treatment of the "germinal plasm" and strongly indicates that frog primordial germ cells do not migrate to the gonads via the circulatory system.

If ultraviolet light is applied to the germinal plasm while it is near the surface of the frog egg during the first cleavage division, no primordial germ cells reach the developing gonads. Therefore, an ultraviolet light-sensitive cytoplasmic material (probably containing nucleic acid) present in one area of the egg cell causes the cells that receive it to become primordial germ cells. If this germinal plasm is inactivated, the primordial germ cells do not develop normally.

The germinal plasm of frog eggs is only one example of many special cytoplasmic substances involved in differentiation. While the effects of such a special cytoplasmic factor can be described, how did only that one area of the cytoplasm in any cell originally come to contain the developmentally important factor? This question is particularly interesting when the cell being considered is an egg because the cytoplasmic organization of a developing oocyte is directed not only by expression of its own genetic information but also by influences exerted by follicular cells that surround the developing oocyte in the ovary. Which of these influences causes a particular cytoplasmic substance such as the germinal plasm to be located in a particular part of the oocyte? Answers to such questions will emerge only after additional research on oogenesis.

Induction

Many cell differentiations during animal development occur in response to **induction,** the critical influence of one group of cells on another group of cells that results in altering the responding cells' path of development.

For example, formation of the lens of the vertebrate eye involves an induction. The eyes begin to develop as bulging lateral outgrowths from the embryonic brain. When one of these **optic vesicles** touches the skin ectoderm, the contacted ectoderm thickens into a **lens placode.** As eye development proceeds, the lens placode develops into the rounded lens of the eye. Lens cells produce large quantities of special proteins (crystallins) that give the lens its light-refracting properties. Thus, induction causes specific structural and biochemical changes (synthesis of the crystallins) in the responding ectodermal cells (figure 25.13).

Other nearby ectodermal cells normally show no sign of lens development and become ordinary epidermal cells of the skin. If the optic vesicle is removed before the induction occurs, the cells that normally would be induced to form the lens form epidermis also. Furthermore, prospective lens and prospective epidermal portions of the ectoderm can be exchanged before induction occurs, and development of both tissues proceeds normally. In other words, other ectodermal cells are competent to produce the lens if they are moved into place to receive the inductive influence at the proper time. Induction by the optic vesicle, then, is necessary for conversion of embryonic skin ectoderm into lens tissue, and only those portions of the ectoderm that come under the inductive influence of optic vesicles produce lens tissue.

Much of the normal expression of the genetic program for animal development depends on series of such inductive relationships. Tissue A induces nearby cells to differentiate into Tissue B. Then Tissue B induces nearby cells to differentiate into Tissue C, and so forth.

For example, the chain reaction inductions involved in eye development begin with a primary induction in which mesoderm along the midline of the body induces overlying ectoderm to differentiate into the nervous system. Next, optic vesicles induce lens development. Then, the eye with

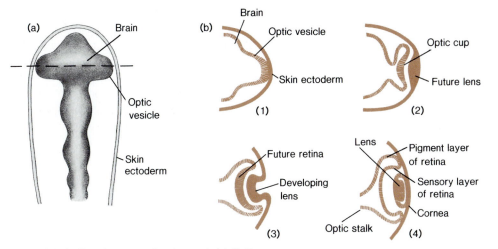

Figure 25.13 Vertebrate eye development. (a) Optic vesicles bulge from side of brain to contact skin ectoderm. Broken line shows where cut is made to show cross sections in (b). (b) Four stages in development of the lens from the area of skin ectoderm contacted by the optic vesicle. The completed lens separates from the skin ectoderm and sits in the opening (pupil) of the optic cup.

its lens in place induces the ectoderm that lies over it to differentiate into the transparent cornea of the eye. Failure of one of the first inductions in such a reaction series can cause drastically abnormal development. Should induction of brain differentiation fail, no eye development whatsoever occurs. The portions of the ectoderm that normally would respond to inductions to form lens and cornea simply produce epidermal tissue.

Even after years of study, the actual mechanisms of inductive interactions are still fairly obscure, but some inductions seem to involve release of specific molecules from inducing tissue. These "inducer molecules" probably bind with receptor molecules on the surface of responding cells in much the same way that hormones bind with membrane-bound receptors of target cells (chapter 14). This complex of inducer molecule and receptor molecule probably then enters the cell and causes genetic activation leading to differentiation.

Hormonal Control of Differentiation

Inductive interactions involve chemical influences that, at most, pass only short distances between adjacent groups of cells. Hormones act as chemical messengers that travel further and have more general and widespread effects in developing bodies. Hormones are powerful regulators of developmental processes.

Because growth with addition of new tissue continues throughout the lives of many plants, hormones continue to control growth and differentiation even in mature plant bodies. The hormonal regulation of plant development in that general, lifetime context is discussed in chapter 18. Hormonal regulation of development in animals, however, is most apparent during embryonic and prematuration phases of life. For example, hormones can control dramatic changes in body form (**metamorphosis**). The familiar conversion of tadpole to frog is caused by the thyroid hormone thyroxin (figure 25.14). If thyroid secretion is inhibited in tadpoles, metamorphosis does not occur. Insect metamorphosis also is hormonally controlled. The egg-larva-pupa-adult sequence is precisely regulated by a group of interacting hormones (p. 391).

Figure 25.14 Metamorphosis in the frog *Rana pipiens*.
(a) Premetamorphic tadpole. (b) Prometamorphic tadpole
(with growth of hindlimbs). (c) Onset of metamorphic
climax (eruption of forelimbs, degeneration of the tail).
(d, e) Later stages showing gradual appearance of
froglike features. The molecular structure of the iodine-
containing thyroid hormone, thyroxine, also is shown.
Metamorphosis can be accelerated experimentally by
adding thyroid hormone or iodine to the water around
tadpoles. But if antithyroid drugs are added to the
water, metamorphosis does not occur. In such
experiments it sometimes is possible to raise giant
tadpoles that are several times as large as normal
tadpoles.

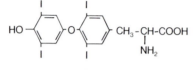

Thyroxine

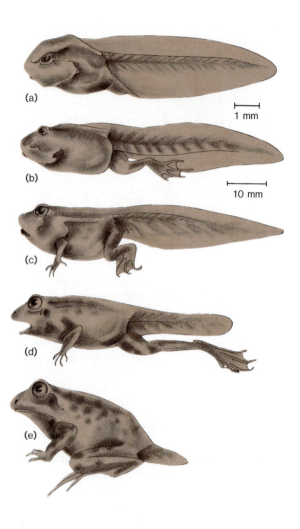

(a)

1 mm

(b)

10 mm

(c)

(d)

(e)

Another good example of operation of hormonal control in animal development is seen in mammalian reproductive system development.

Hormones and Development of the Mammalian Reproductive System

Early in development, the reproductive systems of vertebrate embryos contain double sets of ducts that represent the beginnings of both male and female reproductive systems. In humans, for example, this "indifferent" stage of reproductive system development is reached about seven weeks after fertilization. The **Wolffian ducts** are the rudiments of the vas deferens, and the **Müllerian ducts** are the rudiments of the oviducts, the uterus, and the vagina. If the cells of the developing gonad contain two X chromosomes, the system develops in a female direction. The Wolffian ducts degenerate, while the Müllerian ducts persist and develop. An XY sex chromosome makeup produces an opposite result (figure 25.15).

Expression of the genetic sex determination mechanism depends on the small amounts of sex hormones produced by the developing gonads. The process seems fairly simple and direct in some vertebrates; that is, male sex hormones cause development of a male duct system, and female hormones cause development of a female duct system. But the situation is somewhat different in mammals.

Experiments with tissue cultured outside the embryonic body indicate that the developing mammalian reproductive system is delicately balanced between the male and female directions of development and is very responsive to the presence or absence of male sex hormone produced

Figure 25.15 Transition from indifferent stage with a double-duct system to male or female reproductive systems during mammalian development. Mesonephros is a temporary embryonic kidney in mammals. Metanephric kidney is the permanent functional kidney in mammals.

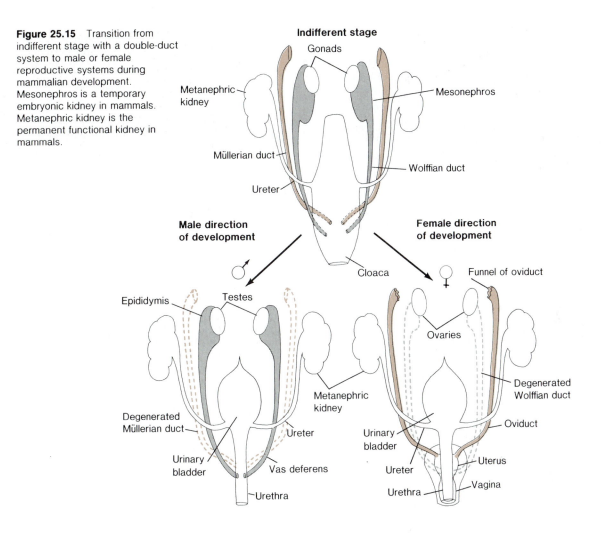

by the developing gonad. Male sex hormone causes development of a male duct system; *absence* of male sex hormone (not production of female sex hormones) results in female duct system development. This **monohormonic theory** of mammalian sex determination fits the experimental data and makes sense when the environment of the developing mammal (the mother's body) is considered. It probably is a good thing that control of this delicately balanced system does not depend on the small amounts of hormones produced by the developing female gonad because they might not make much difference in a developmental environment that is overwhelmingly female, at least in terms of hormones. But the presence or absence of male sex hormones is a reliable developmental switch that can be used to direct development beyond the indifferent stage.

Pattern Analysis in Development

During the normal development of many structures, very precisely ordered patterns of differentiation take place. Cells that are very close to one another can develop in very different ways to produce the specifically arranged parts of complex structures. Cells develop as if they were responding to very specific positional information.

Such precise pattern formations in development are being investigated, and new ideas about control of patterns are emerging, but these are complex research problems. The following examination of a relatively simple development pattern in a developing fern is a good introduction to pattern formation.

One- and Two-Dimensional Growth

A fern spore germinates and develops into a small, green, heart-shaped thallus that bears egg- and sperm-producing organs. Germination of the spore releases a single cell that divides to produce several green cells in an end-to-end chain called a **protonema.** Then the cell division plane changes, and the terminal cell of the chain divides to produce two side-by-side cells. This pattern change from one-dimensional to two-dimensional growth marks the beginning of the formation of the plate of cells making up the thallus (figure 25.16).

Possibly, the cells somehow record their developmental history so that they are influenced by the number of cell divisions it took to produce them. At a certain "count," the change in division plane is triggered in the terminal cell. This cell has a developmental history (a specific number of cell divisions) and positional information about its place in the pattern (it is contacted by another cell on only one side).

This normal developmental pattern is susceptible to change by external influences. For example, if kept under red light, protonemal cells grow long and thin, and one-dimensional growth

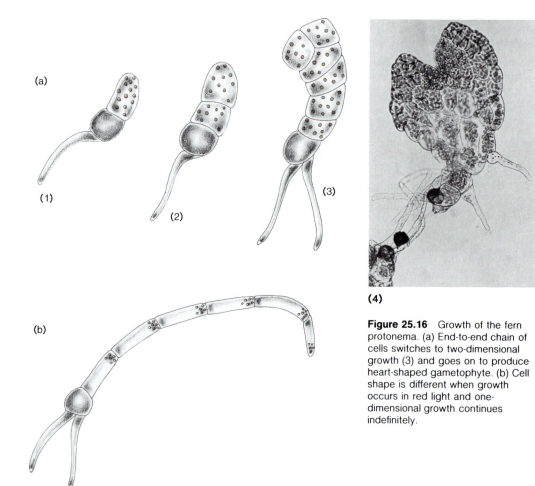

(4)

Figure 25.16 Growth of the fern protonema. (a) End-to-end chain of cells switches to two-dimensional growth (3) and goes on to produce heart-shaped gametophyte. (b) Cell shape is different when growth occurs in red light and one-dimensional growth continues indefinitely.

continues indefinitely. Interfering with RNA synthesis also prevents the pattern change. This indicates that the changes require specific genetic expression, but the control of that genetic expression under normal circumstances is not yet clear.

Chick Embryo Limbs

The idea of control of differentiation by an interaction of developmental history and specific positional information also can be applied to the problem of vertebrate limb development. In a vertebrate forelimb, for example, skeletal elements of the upper arm, the forearm, and the wrist and hand all must develop in the proper places and in the proper relationships to one another. Then, a complex of muscles, blood vessels, and other tissues must develop in proper associations with the limb skeleton. This discussion will concentrate on the chick embryo's wing, which is the most thoroughly studied example of vertebrate limb development.

Each limb begins its development as a small bulge, the **limb bud,** on the side of the embryonic body (figure 25.17). A limb bud consists of a core of mesoderm enclosed in a jacket of skin ectoderm. One strip of this ectodermal jacket is distinctly thickened to produce the **apical ectodermal ridge.** The apical ectodermal ridge (AER) seems very important for normal limb development because surgical removal of the ridge terminates the outward growth of the bud. An incomplete limb produced by very early removal of the AER contains only **proximal** limb structures, that is, limb elements closer to the midline of the body. Slightly later removal of the AER allows development of a longer limb with more **distal** (further from the body's midline) skeletal parts present. Thus, there seems to be a progressive laying down of the limb's normal proximal to distal pattern that depends on the presence of the apical ectodermal ridge.

In the mesoderm, just under the apical ectodermal ridge, is an area where many cell divisions take place. These divisions of mesodermal cells produce the bulk of the cells that are added to the length of the wing bud during its outgrowth. The active division zone is pushed further

and further distally as it leaves behind a lengthening limb. And the cells left behind produce specific elements of the limb. But what makes cells at a given place form part of a humerus, as opposed to a radius or ulna, or an even more distal element (figure 25.17c)? What gives the cells positional information about their place in the proximal to distal pattern of the developing limb?

Lewis Wolpert and his colleagues in London have advanced a hypothesis for limb pattern formation that suggests partial answers to these questions. Their hypothesis is based on experiments on developing chick embryo wings. They propose that the apical ectodermal ridge together with the mesodermal area where cell divisions occur beneath it constitute a "progress zone" in limb development and that this progress zone is the key to understanding how limb cells receive information about proximal to distal position in the developing limb. They suggest that the cells left behind by the progress zone early in the limb's development form proximal structures (for example, the humerus). Cells left behind later form more distal structures (for example, radius and ulna), and so forth as development continues. Somehow, cells receive information about how long a time they spent in the progress zone before being left behind to differentiate. Possibly, as suggested in the case of fern cells, these cells also "count" cell divisions somehow.

Wolpert and his colleagues tested this hypothesis. They found that if a progress zone is taken from a wing bud at an early stage of its development and substituted for the progress zone of a more advanced wing bud, the "young" progress zone proceeds with development according to its own developmental timer. Specifically, the "young" progress zone adds elements to the limb that repeat parts of the limb that already are there. They also found that substituting a progress zone from a more advanced wing bud for that of a younger wing bud results in a gap in development; that is, some elements are missing (figure 25.18).

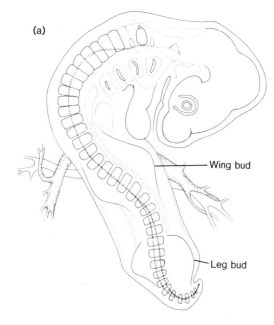

(a)

Wing bud

Leg bud

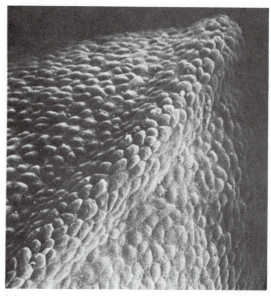

(b)

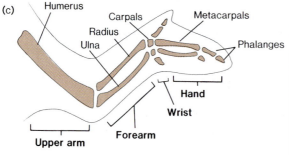

(c)

Humerus

Carpals

Metacarpals

Radius

Ulna

Phalanges

Hand

Wrist

Forearm

Upper arm

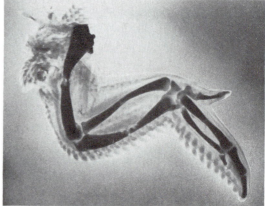

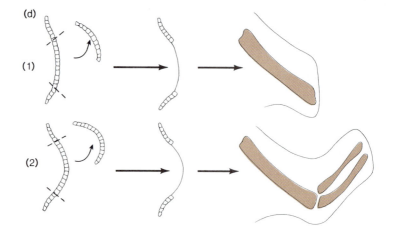

(d)

(1)

(2)

Figure 25.17   Chick embryo limb bud development. (a) Location of wing and leg buds of a chick embryo after about three days of incubation. (b) Scanning electron micrograph of the apical ectodermal ridge on a chick embryo's wing bud. (Magnification × 824) (c) The finished product of normal limb development. A simplified outline sketch of a bird wing to show the general pattern of relationships of parts of a vertebrate limb. Arm terminology for areas is used because it is familiar. (d) Result of surgical removal of apical ectodermal ridge. (1) Early removal. (2) Later removal. Results generalized from experiments of J. W. Saunders, Jr. and Edgar Zwilling.

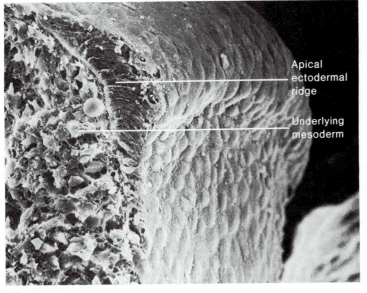

(a)

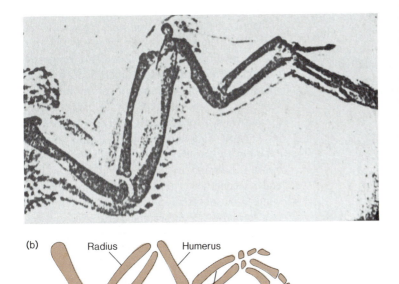

(b)

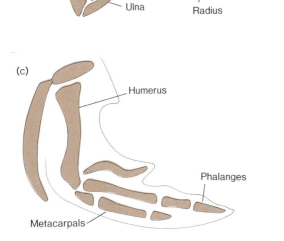

(c)

Figure 25.18 Experiments of J. H. Lewis, D. Summerbell, and L. Wolpert in which progress zones (apical ecotodermal ridge plus underlying mesoderm) were exchanged between wing buds at different stages of development. (a) The progress zone of a chick embryo limb bud. This scanning electron micrograph shows the apical ectodermal ridge (note that cells in the ridge are taller than other ectodermal cells covering the rest of the bud) and the underlying mesoderm cells. The sphere beneath the ridge is a preparation artifact. (b) Photograph and sketch of the composite wing produced when a progress zone of a younger wing bud is substituted for the progress zone of a more advanced wing bud. The host developed a normal humerus, radius, and ulna. Then the younger progress zone added another humerus, radius, and ulna, as well as distal elements. (c) A tracing of a composite wing produced in the opposite kind of experiment when a progress zone from a more advanced wing bud is substituted for that of a younger wing bud. A humerus was formed from host tissue, and the more mature progress zone added distal elements, but the radius and ulna are missing. This sketch includes several shoulder elements not shown in other figures.

Other general hypotheses for wing bud development might be consistent with these and other experimental data that have been obtained, but the Wolpert hypothesis regarding proximal to distal axis positional information in wing bud pattern formation is one of the most provocative.

What about pattern formation *within* each segment of the wing bud? Some cells form bone, others form muscles, still others blood vessels or connective tissues. Seemingly similar cells that are located very close to one another differentiate in very different directions. In fact, some cells even have degeneration and death as their normal developmental fate. One rather surprising element of the normal pattern of limb differentiation is development of **necrotic zones,** areas containing dying and degenerating cells (figure 25.19). The **morphogenetic cell death** that occurs in these necrotic zones is a normal part of limb development. Some of the necrotic zones produce the spaces that separate the digits in the originally paddlelike distal portion of the limb bud. Other necrotic zones help to give the limb its general shape. In these cases, death is a normal part of the developmental program of differentiation of populations of embryonic cells, and death of these cells is a necessary step in the shaping of a developing limb.

For all of this to happen, cells must receive fairly accurate information about their exact position in a limb, more than simply how far along a proximal to distal axis they are located. How do they receive this information? In the case of wing buds, Wolpert and his colleagues propose that proximal to distal axis positional information might also be available to cells within each segment of the limb. They suggest that this information is in the form of some chemical substance that is produced on one side of the bud and diffuses across the bud, and that different concentrations of the substance cause cells to differentiate into very different cell types.

The French Flag Problem

Lewis Wolpert has proposed an idealized system, which he calls the "French flag problem," for discussing the general concept of positional information. A French flag has three color bands: blue, white, and red. Wolpert supposes that an imaginary row of cells can produce these colors and that the cells in this row can neither divide nor change position. How can they be made to form a French flag? This is basically the same question asked about cells in many developing structures. How are these cells notified to produce different types of specialized tissues, depending on their location, and thus to assemble the normal pattern of tissues within a developing structure? One hypothesis holds that positional information could depend on a specific substance that diffuses from its place of production, called its "source." Then, if there is a place where this substance is actively destroyed (called its "sink"), a gradient of decreasing concentration exists between the source and the sink. Wolpert supposes that the cells in his imaginary row of cells are prepared, as a result of their developmental history, to respond to different concentrations of this diffusible substance in three different ways. Concentrations greater than a specific critical level (a **threshold** level) cause cells to differentiate in one way (for example, to become blue). At concentrations between this first threshold (Threshold B, for blue) and a second, lower threshold (Threshold W, for white), the substance causes cells to differentiate in a second way—to become white. And concentrations lower than this second threshold level cause cells to become red. Figure 25.20 shows how this hypothetical mechanism would work to produce elements of the French flag in the proper pattern.

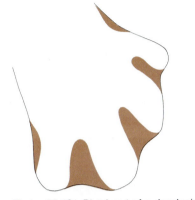

Figure 25.19 Distal part of a developing chick embryo's leg bud showing necrotic zones (color) where morphogenetic cell death is occurring. A leg is shown rather than a wing because it is easier to visualize the role of morphogenetic cell death in separating the toes. The distal end of the bud initially is paddlelike, and morphogenetic cell death separates the toes. Morphogenetic cell death is likewise involved in separating the digits of both hands and feet during human development.

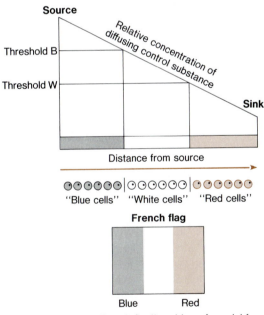

French flag

Figure 25.20 The "French flag" problem. A model for the action of a hypothetical diffusing control substance that is produced at a source and destroyed at a sink. A pattern can be generated if cells respond differently to different ranges of concentrations of the substance.

Such a simple model could explain differentiation of three cell types within a previously homogeneous population. It is hypothesized that such diffusible control substances, called morphogens, do exist and function during development, but their existence has not yet been demonstrated. Another possibility is that there are several simultaneous morphogen gradients at various orientations to one another. They could control development of very complex patterns.

Cell Responses during Differentiation

Cells descended from the zygote by mitotic cell divisions produce the specialized parts of multicellular bodies. Although all the cells in an embryo receive the same set of nuclear genetic information, they differentiate into the various body cell types in response to such controls as differential distribution of cytoplasmic factors, inductions, hormones, and positional information—all of which affect the differentiative pathways that cells pursue. Now that some of these controls of differentiation have been identified, it is time to look at the characteristics and properties of differentiating cells.

The idea of differential gene activation has been used to explain diverse specializations of cells with the same genetic makeup. In most cases, gene activation is assessed in terms of the products of that activation: the beginning of synthesis of a specific protein, or better yet, the beginning of production of specific messenger RNA. But has gene activation ever been directly observed? Again, this question can be approached via a model system, that is, observations of certain processes in a selected developing organism. These observations yield partial answers that sometimes can be extrapolated to other instances in which what is happening is not so clearly seen.

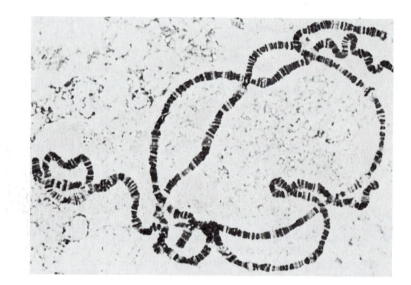

Figure 25.21 Giant polytene chromosomes from a *Drosophila* larval salivary gland cell stained to show banding pattern.

Chromosome Puffing

One model system for studying differential gene activation during development involves special chromosomes in the cells of certain tissues of some insects. These cells grow very large, and as they grow, their genetic material replicates repeatedly. Up to ten sequential replications produce nuclei that are 1024N (normal diploid body cells are 2N). This repeated replication without mitosis produces giant **polytene** (multistrand) **chromosomes** because the replicated DNA and associated chromosomal protein of each chromosome remain together in one structural unit. Each polytene chromosome, when it is stained for microscopic examination, has characteristic patterns of light and dark bands (figure 25.21). Genetic analysis has correlated various bands with mapped locations of genes on the chromosome. Thus, specific bands are considered to be sites of genes (gene loci).

Genetic expression in polytene chromosomes involves a specific change in chromosome organization. The many strands of a particular region of a chromosome loosen up and loop out. This local expansion produces a **puff** on the chromosome. Function of the puffs can be demonstrated by supplying radioactively labeled uridine (a component of RNA, but not DNA) to cells with polytene chromosomes. Radioactivity accumulates selectively on the chromosome puffs (figure 25.22). This result implicates the puffs as sites of genetic transcription (messenger RNA synthesis).

Because the gene loci and sites of genetic transcription can be seen directly, these giant polytene chromosomes are remarkably useful model systems for studying gene activation during development.

If it is assumed that some parts of the genome must be used for activities common to all cells, then all cells containing polytene chromosomes would have a number of puff locations in common. It also would be expected that certain chromosome puffs would represent parts of the genome being expressed to produce proteins unique to individual cell types. These latter sets of puffs would be different in the cells of each organ where polytene chromosomes are found. Both of these expectations have been justified in studies on polytene chromosomes. A number of common puff locations are found in all types of cells, and smaller sets of puff locations are found in certain cell types but not in others. These latter sets are called organ-specific puff patterns.

Not surprisingly, as development proceeds, puff patterns change in the chromosomes of cells within a single organ. For example, as the pupal case develops and the insect enters the pupal stage of development (pupation), some puffs disappear and other puffs appear for the first time. Puffing patterns during this larva to pupa transition have been studied extensively in polytene chromosomes of salivary gland cells of *Drosophila* (figure 25.23). The precise timing of appearance and disappearance of the puffs has made the larva to pupa transition in *Drosophila* and other insects a very useful system for experimental analysis.

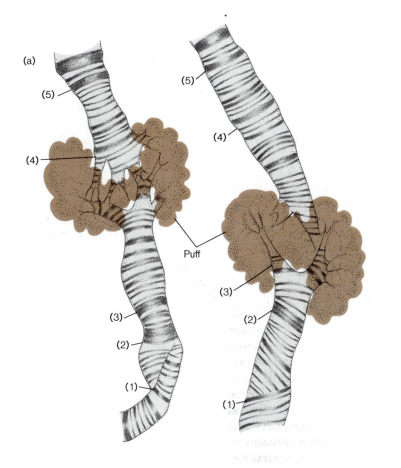

(a)

(5)

(4)

(3)

(2)

(1)

(5)

(4)

(3)

(2)

(1)

Puff

Figure 25.22 Puffing in insect polytene chromosomes. (a) A short section of the same chromosome taken from two different cells of the fly *Trichocladius*. Puffs are regions in which the chromosome strands loosen up and loop out. The bands are numbered for reference to show that different parts of the chromosome are puffed in different cells. This indicates that different parts of the genome are being transcribed in each cell.
(b) Autoradiogram of polytene chromosomes from the salivary gland of the larva of the midge *Chironomus tentans*. Radioactively labeled uridine, indicated by the black spots, accumulates selectively on puffed part of the chromosome, indicating that puffs are sites of RNA synthesis. For more details of the techniques of autoradiography used to prepare autoradiograms, see chapter 3.

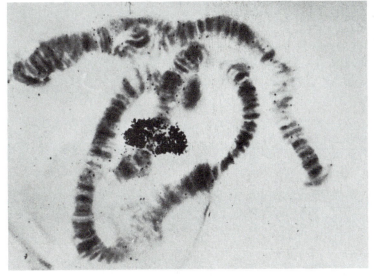

(b)

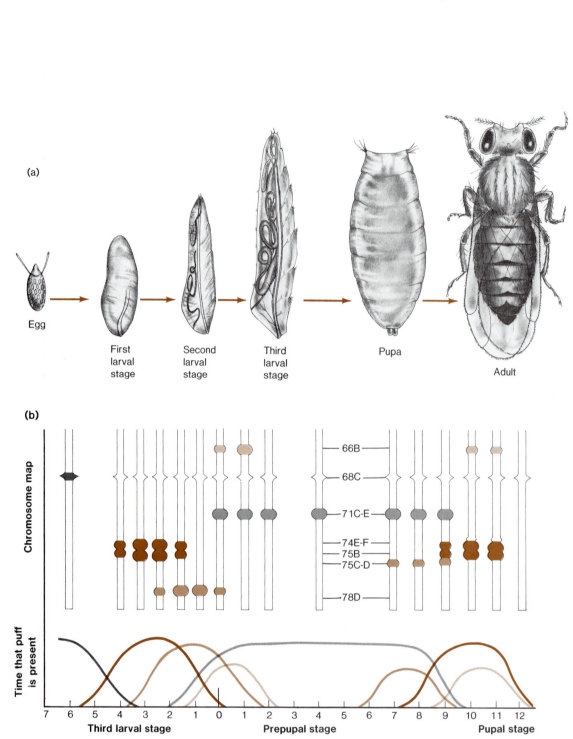

Figure 25.23 Developmental changes in puff patterns occur during stages in development of the fruit fly *Drosophila melanogaster*. (a) The familiar life history of *Drosophila melanogaster*—egg, larva, pupa, and adult. *Drosophila* actually goes through three larval stages before pupation. (b) Puffs appear and disappear at specific times during development. Here are the puffs in one part of a chromosome (chromosome arm 111L) from a *Drosophila* salivary gland around the time of pupation. This is a clear demonstration that normal development does depend on a complex series of precisely regulated gene activations.

Wolfgang Beermann, Ulrich Clever, and others have used this system to study the action of the molting hormone ecdysone (chapter 14), which promotes pupation. If larvae are given ecdysone injections, the characteristic prepupation puff patterns appear at an earlier than normal time. Clearly, the hormone causes changes in gene activation, as evidenced by changing puff patterns. Further analysis has suggested that ecdysone may act by changing plasma membrane permeability because changes in cellular ionic composition, especially changes in K^+ and Na^+ ion levels, in cells seem to be steps in the chromosome puffing response to ecdysone.

Isozymes

Genetic transcription as evidenced by insect chromosome puffing is only one step in the flow of genetic information in cells. The end result of cellular genetic expression is protein synthesis (translation). A great many biochemical studies of differentiation have charted the appearance and accumulation of cell type-specific proteins, such as hemoglobin, in developing red blood cells, and the contractile proteins in developing muscle cells.

But one of the most profitable approaches to the study of protein synthesis in differentiating cells has been the study of proteins that have several molecular forms, such as the enzyme lactate dehydrogenase (LDH). LDH, which catalyzes the reversible interconversion of lactate and pyruvate, is found in cells of practically all vertebrate tissues. The different molecular forms of LDH (**isoenzymes**, or simply **isozymes**) all catalyze the same reaction but have different kinetic properties. LDH has five isozymes that can be separated due to their different electrophoretic mobilities (figure 25.24). That is, the five isozymes differ in their migration in a starch gel through which there is an electrical voltage gradient because the different isozymes have different electrical charges.

Each functional LDH unit contains four subunits. Thus, LDH is an oligomeric protein (chapter 2). The five isozymes represent various combinations of two types of subunits, called A and B. LDH-1 contains four identical B subunits; LDH-5 contains four identical A subunits; and LDH-2, LDH-3, and LDH-4 contain mixtures of A and B subunits (see figure 25.24c).

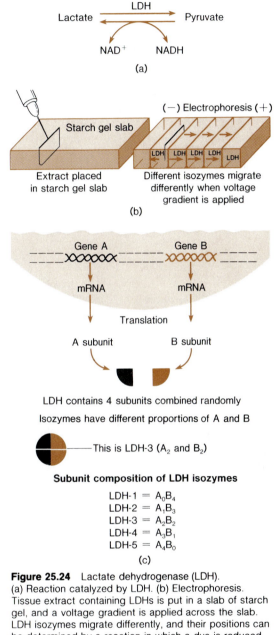

Subunit composition of LDH isozymes

LDH-1 $= A_0B_4$
LDH-2 $= A_1B_3$
LDH-3 $= A_2B_2$
LDH-4 $= A_3B_1$
LDH-5 $= A_4B_0$

(c)

Figure 25.24 Lactate dehydrogenase (LDH). (a) Reaction catalyzed by LDH. (b) Electrophoresis. Tissue extract containing LDHs is put in a slab of starch gel, and a voltage gradient is applied across the slab. LDH isozymes migrate differently, and their positions can be determined by a reaction in which a dye is reduced (and thereby becomes colored) while lactate is being converted to pyruvate. (c) Subunits of LDH. A and B subunits are polypeptides coded by separate genes.

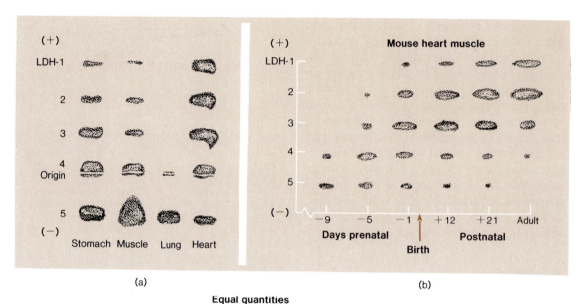

Figure 25.25 LDH isozyme patterns. (a) LDH isozyme patterns of four tissues from adult rat. (b) Developmental changes in LDH isozyme patterns in mouse heart from a predominance of LDH-5 to a predominance of LDH-1. (c) Schematic diagrams of an experiment showing how pattern seen in mouse one day before birth in (b) would arise from equal quantities of A and B subunits combining randomly.

Subunits A and B are independently synthesized products, the expression of separate genes. When they are together, the subunits combine spontaneously and randomly to produce LDH. Thus, cells containing a preponderance of the LDH-1 isozyme must be synthesizing mostly B subunits, while cells containing largely LDH-5 must be producing mostly A subunits. Differences in LDH isozyme pattern reflect differences in synthesis of the two kinds of subunits. Each tissue has its own characteristic pattern of LDH isozymes.

During development of some tissues, there is progressive change from predominance of one isozyme type to predominance of another (figure 25.25). For example, the embryonic mouse heart contains mainly LDH-5 when it first forms, but

fully developed adult heart tissue contains mainly LDH-1. This changing isozyme pattern must be based on a change in genetic expression, that is, a change from synthesizing mostly A subunits in the embryonic heart to synthesizing mostly B subunits in the adult heart. The pattern of isozymes in the one-day-old mouse heart is especially interesting (figure 25.25c). At about that point, approximately equal quantities of A and B subunits are being produced because the one-day-old pattern is exactly what would be expected as a result of random combinations of equal quantities of subunits A and B.

Thus, through studies such as this LDH isozyme research, biologists learn about differentiation by studying the end products of genetic expression: protein molecules.

Differentiation of the Cell Surface

Differentiated cells need specific enzyme systems to conduct specialized kinds of reactions. They also need to develop specific cell surface characteristics that permit them to receive and respond to chemical messages from their environment and to interact with cells around them. For example, cell responses to a number of hormones require that hormone molecules bind to specific receptor molecules at the cell surface (chapter 14). During development, cells must synthesize and position receptors if they are to be target cells capable of responding to a particular hormone. It now seems very likely that some hormone-related metabolic problems are not actually deficiencies in hormone production but are instead problems involving the quantity or specificity of cell surface hormone receptor molecules.

When hypothetical inductor molecules or hypothetical diffusing position indicators (morphogens) in developmental pattern formation are discussed, it is also necessary to hypothesize that responding cells have receptors. This means that "developmental history," as it was discussed earlier, may involve production of receptor molecules that make possible the proper responses to developmental stimuli when cells receive them. This also raises the possibility that developmental defects in which inductions fail to occur might be due, among other possibilities, to either a failure of inducing cells to act or a failure in the development of appropriate receptor molecules on the surfaces of potential responding cells.

Cell surface properties are also important in making and maintaining the contacts that hold cells together in very specific ways in developing tissues. Long before the discovery of the glycocalyx of the cell surface, with its specific array of glycoprotein recognition factors (chapter 2), developmental biologists were aware that cell surfaces had tissue-specific features. Johannes Holtfreter dissected pieces of tissue from amphibian embryos and separated the cells from one another by raising the pH of the medium around them. Upon return to normal pH, these **dissociated** cells moved around, contacted one another, and **reaggregated** (clumped together). When Holtfreter and his colleagues mixed dissociated cells taken from two types of tissues, the cells reaggregated into mixed aggregates, but then they sorted out within the clusters. Holtfreter said that cells have a selective affinity for cells of their own type (figure 25.26). Biologists today

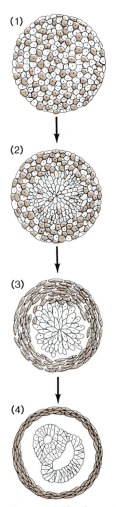

(1)

(2)

(3)

(4)

Figure 25.26 An example of results of Holtfreter's experiments on selective affinities of cells from different embryonic tissues. Dissociated cells from prospective nervous system and epidermis areas reaggregate to form a mixed ball of cells (1). Cells move and sort out by types within the aggregate (2) and (3). Epidermal cells form a thin-spread covering, and neural cells form hollow masses resembling developing nervous systems within normal embryos (4).

would say that cells have cell surface recognition sites for cells of their own type.

A. A. Moscona, Malcolm Steinberg, and others have dissociated cells of chick and mouse embryos by removing calcium and magnesium ions from the medium around the embryos. When these cells are returned to a medium containing calcium and magnesium, they reaggregate and then sort within the aggregates according to tissue types. Calcium and magnesium seem generally to cross-link proteins or glycoproteins at the cell surface, but the problems of specific sorting and selective affinity require more detailed explanations (figure 25.27).

Steinberg says that once cells come in contact through random movements, the sorting specificity depends on differences in quantities and distribution patterns of binding factors on cell surfaces. But Moscona contends that binding depends on tissue-specific binding molecules (**ligands**) that are released from cells and that selectively hold together cells of the same type.

The role of these factors in organizing cells into tissues during normal development is not clear. Of course, dissociation and reaggregation of embryonic cells that already are in contact with each other is a strictly experimental test of cell activities. But migrating cells such as the primordial germ cells of the frog embryo "dissociate" from the embryonic endoderm, migrate through the body, and "aggregate" with cells in the developing gonad (see figure 25.12). It is tempting to propose that these changes in association involve stage-specific changes in cell surface recognition factors, which are similar to those evidenced in the reaggregation studies of Holtfreter, Steinberg, Moscona, and others.

The permanent associations of cells in tissues involve some special junctions among cells, which develop after cells have become arranged in tissues. Some areas of very close junctions of plasma membranes called **desmosomes** bind cells very firmly to one another. Another special kind of junction, the **gap junction,** actually provides channels connecting the cytoplasm of adjacent cells (figure 25.28). Before gap junctions were observed directly, W. R. Loewenstein had proposed that they must exist. He found that, in

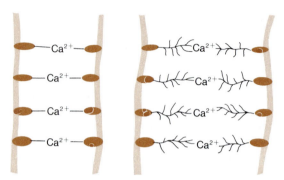

Figure 25.27 Role of Ca^{2+} in cell adhesion. The hypothesis is that Ca^{2+} forms bridges between cell surfaces by binding to anionic groups in plasma membrane proteins (left) or glycoproteins (right). Selective recognition and adherence between cells might then depend on quantities and distribution patterns of these anionic groups, as Steinberg suggests.

certain instances, inserting an electrode into a single cell and shifting the voltage across the cell membrane caused similar transmembrane voltage changes in adjacent cells, indicating that there is electrical coupling among the cells. This could happen only if there were low-resistance cytoplasmic channels between the cells because intact plasma membranes have high resistance to current flow and would have blocked the passage of current from one cell to the next.

Gap junction connections are physiologically significant, for example, in nervous system functioning, and they may also be very important developmentally in organizing what were previously separate, individual cells into integrated multicellular tissues in which cells cooperate in functional activities. The gap junctions may permit direct cytoplasmic passage of developmental regulators from cell to cell, and they may also be important in establishing positional relationships among cells. Electrical coupling indicating gap junction development appears at a specific time during development, and development of gap junctions is now recognized as a normal step in cell differentiation. It is interesting that cancer cells lack electrical coupling and thus may be incapable of receiving the signals that pass between normal cells via gap junctions.

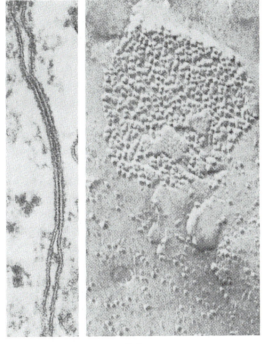

(a)

(b)

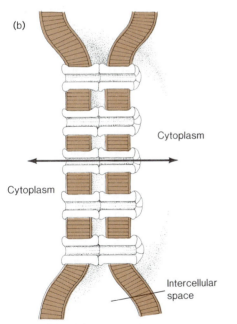

Cytoplasm

Cytoplasm

Intercellular space

Figure 25.28 (a) Thin section showing a gap junction between two membranes. The gap junction is the area where the membranes are very close together in the upper half of the photo on the left. Freeze fracture image of a gap junction (right photo). Note that gap junctions are constructed of clusters of small particles. (b) A hypothetical model showing how the small particles in a gap junction may be organized to make small channels connecting the cytoplasm of adjacent cells.

Human Biochemical Individuality

K. E. von Baer long ago described development as a process that proceeds from the more general to the more specific. He emphasized that characteristics that identify specific individuals within a species are the last to emerge during development. But von Baer was discussing morphological (structural) characteristics that are subject to direct visual observation. In this chapter, cellular and molecular aspects of differentiation have been emphasized. Several developmental processes common to large groups of organisms, as well as some processes (for example, changes in LDH isozyme patterns) that characterize development of all members of a species, have been considered. But what about development of real individuality at the cellular and molecular level?

Differences in certain cell surfaces in humans are examples of differentiation that produces biochemical individuality. The familiar ABO blood type distinctions that group people into rather large biochemical categories have been known for years. The ABO groups depend on differences in specific molecules (antigens) located on the surfaces of red blood cells. More recently, other complex cell surface recognition factors have been identified. One set of these factors make up the **human leucocyte antigen (HLA)** system, a set of antigens that occur on the surfaces of white blood cells. The HLA system depends on the expression of four gene loci, each of which can bear any one of five or more alleles that determine which antigens will be present on cell surfaces (table 25.1). Thus, the number of possible HLA system combinations is staggering. HLA typing is not just an interesting way to determine biochemical individuality among human beings. It also is of practical medical importance in tissue matching for organ transplants. These specific cell surface factors are important in the processes by which organisms recognize "self" and "nonself," and thus must be considered in judging whether foreign cells and tissues will be accepted or rejected. Cell surface recognition factors are the focus of much of the research in modern immunology, which is explored further in the next chapter.

Table 25.1
The Human Leukocyte Antigen System.

Locus*	Number of Alleles †
HLA-D	6
HLA-B	8
LHA-C	5
HLA-A	8

*These four loci lie on human chromosome 6, with the D locus being closest to the centromere.

†Several more alleles at each locus are being studied, and the number of well-established alleles will be larger by the time this book is published. Since each person's cells contain a maternal and a paternal chromosome 6 and several alleles are possible for each locus, there is a very low probability that any randomly selected pair of subjects would be HLA-identical. The HLA system and other antigen series are the bases of the biochemical individuality of human beings.

Other biochemical factors, some of which are known and some that remain to be discovered, add to the complexity of individual variability. In addition to easily observed physical and emotional differences among people, it can safely be concluded that the completion of the genetic program for development makes each person (except identical twins) a unique individual even at the cellular and biochemical level.

Summary

All cells in a multicellular organism are descended from a single cell, the zygote, by mitosis. They all carry the same set of nuclear genetic information, but they become structurally and functionally specialized in many different ways. This differentiation depends on synthesis of specific proteins that are required for cells' specialized functions.

Throughout history, many explanations for development and differentiation have been advanced. The preformation theory suggested that development is simply an unfolding of a small, preformed body that is already present in one of the gametes. An alternative theory is epigenesis, the idea that a body becomes organized progressively from simpler elements.

Early experiments on developing organisms were designed to test these ideas, and successful separation of cleavage cells, each of which developed a complete normal embryo, was powerful evidence against the idea of preformation.

Modern experiments have tested the developmental potential of the nuclei of developing cells by means of nuclear transplantation experiments. Under proper circumstances, nuclei from cells at any developmental stage (even adult cells) can interact with an enucleated egg to produce a normal embryo.

Plant cell nuclei also retain full developmental potentials. Isolated cells derived from fully developed plants can develop into embryoids that grow into new individuals.

The problem of differentiation can be approached through the study of model systems that do not include large numbers of differentiated cell types. Cellular slime mold development is such a model system. Individual amoebae aggregate, form a migrating grex, and then differentiate as the specialized cells of the fruiting body.

Differentiation is preceded by determination, the commitment to differentiate. A variety of influences lead to determination. Inheritance of a special kind of cytoplasm may result in determination. In inductive relationships, groups of cells critically influence the future development of other nearby group of cells. Hormones provide more general influences leading to determination.

Differentiation in response to specific positional information establishes definite patterns in some developing systems. Positional information may be derived from a timer that counts mitotic divisions, or possibly, from varying concentrations of diffusible morphogens.

Cellular responses to determination involve specific gene activations, such as those that can be visualized in the giant polytene chromosomes of some insects. This activation leads to synthesis of specific messenger RNAs, and their translation into specific proteins. Protein production may change with developmental stages, as the LDH isozyme patterns do in tissues of some animals.

A step above protein synthesis in complexity is assembly of aggregates of molecules, such as the plasma membranes of cells. Cell surface receptor molecules and the specialized contact areas that bind cells together and interconnect their cytoplasms electrically are features of surfaces of differentiated plasma membranes. Cells seem to recognize surfaces of other cells even after being subjected experimentally to dissociation and reaggregation.

The cell surface is a key area where biochemical individuality is expressed. Cell surfaces are specialized end products of development that contain unique sets of specific proteins in each and every individual organism.

Questions

1. Differentiating cells are characterized by selective gene activation. Explain this statement.

2. Distinguish between the terms determination and differentiation.

3. What problems do you see in applying Wolpert's "French flag" idea to slime mold development?

4. In cattle, membranes around developing twin embryos may fuse so that the circulatory systems of the embryos are connected. Often, twin births in cattle produce one male calf and a calf that resembles a male superficially but is genetically female (XX) with abnormal internal reproductive structures. These abnormal calves are called freemartins. Use what you know about reproductive system development in mammals to suggest an explanation for freemartin development in cattle.

5. How might you determine experimentally when cells in an area of a limb that is destined to become a necrotic zone actually become determined (in the developmental sense) to die?

6. What could you do to produce a chick embryo wing with phalanges attached directly to the shoulder?

Suggested Readings

Books

Browder, L. W. 1980. *Developmental biology.* Philadelphia: Saunders College.

Karp, G., and Berrill, N. J. 1981. *Development,* 2d ed. New York: McGraw-Hill.

Wessells, Norman K. 1977. *Tissue interactions and development.* Menlo Park, Calif.: Benjamin/Cummings.

Articles

Bonner, J. T. December 1978. The life cycle of cellular slime molds. *Natural History.*

DeRobertis, E. M., and Gurdon, J. B. December 1979. Gene transplantation and the analysis of development. *Scientific American* (offprint 1454).

García-Bellido, A.; Lawrence, P. A.; and Morata, G. July 1979. Compartments in animal development. *Scientific American* (offprint 1432).

Gierer, A. December 1974. Hydra as a model for the development of biological form. *Scientific American* (offprint 1309).

Gurdon, J. B. 1978. Gene expression during cell differentiation. 2d ed. *Carolina Biology Readers* no. 25. Burlington, N.C.: Carolina Biological Supply Co.

Northcote, D. H. 1974. Differentiation in higher plants. *Carolina Biology Readers* no. 44. Burlington, N.C.: Carolina Biological Supply Co.

Patterson, P. H.; Potter, D. D.; and Furshpan, E. J. July 1978. The chemical differentiation of nerve cells. *Scientific American* (offprint 1393).

Shepard, J. F. May 1982. The regeneration of potato plants from leaf-cell protoplasts. *Scientific American.*

Wolpert, L. October 1978. Pattern formation in biological development. *Scientific American* (offprint 1409).

Lifelong Developmental Change

How long does development last? One answer might be that, in animals, development lasts until birth (or hatching, as the case may be). But a moment's thought dismisses that answer as inadequate. For example, human development clearly continues through the teenage years, and even twenty-year-olds are not finished products who will not change further. Development continues throughout adulthood and even in old age. (Aging is not a process found only in very old people. The onset of some biochemical changes, which become more obvious parts of the aging process later, can first be detected in humans before age twenty!) Some early developmental changes are sweeping and dramatic, while later ones, such as aging, are much more gradual. But if development is defined in terms of continuing changes in the structure and functioning of organisms, then development can be considered a lifelong process.

Some developmental changes are important for maintenance of homeostasis in organisms. For example, development of resistance to infection by disease-causing (**pathogenic**) organisms is essential to well-being. There are many types of resistance mechanisms; some are general, while others involve very specific cell interactions with the invading foreign cells. In vertebrate animals, specific resistance mechanisms are centered around **immune responses,** which are based on a series of cell differentiations that take place throughout life.

Homeostasis also depends on continuing replacement of worn-out and lost cells, as well as repair of injuries. Thus, cell division and differentiation of replacement cells are lifetime processes. For example, in the human body, blood cells and gut-lining cells, as well as many other types of cells, are replaced throughout life. All of these repair and replacement processes normally are regulated so that the number of cells produced is adequate for replacement and no more.

When the mechanisms controlling normal cell division fail, abnormal growth of cell populations threatens the well-being and even the lives of organisms. Abnormal growth is considered in this chapter as an aspect of development. With advancing age, both resistance and repair mechanisms lose efficiency. Control over cell division also breaks down more frequently so that the incidence of abnormal growth increases with age.

This cycle of life, from vigorous and efficient activity to eventual aging and death, is all part of lifelong developmental change.

Passive and General Responses to Foreign Organisms

Some defenses against disease-causing organisms are simply passive barriers that keep potentially dangerous organisms away from parts of the body where the organisms might establish themselves and do damage. Other defenses are active but not biochemically specific. They involve cell activities that are effective against a variety of foreign cells but that do not depend on specific recognition of a single type of foreign cell.

Surface Barriers

Land-dwelling plants usually have waterproof, tough, and therefore, relatively impenetrable surfaces. Corky bark, waxy leaf cuticles, and other hard outer surfaces are barriers against excessive water loss from plants' bodies to the environment. But they also prevent entry of some of the pathogenic bacteria and fungi that can grow and thrive when they reach plants' internal tissue environments.

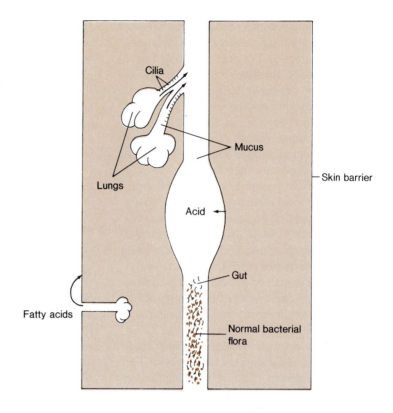

Figure 26.1 Surface barriers as passive defense against infection in humans. Skin is relatively impermeable, has fatty acids on its surface, and has a rather low pH. Digestive system linings are barriers, stomach acid kills some bacteria, and the normal bacterial flora of the intestine inhibits growth of other organisms. Cilia along air passages move materials away from lungs, which are more susceptible to infection. Mucus secretion helps to keep bacteria away from surfaces, which they might otherwise penetrate.

Animal body surfaces also serve as barriers to entry of microorganisms (figure 26.1). This is important because body surfaces are often densely populated with microorganisms. For example, various areas of human skin normally carry large numbers of bacteria. This **normal flora** of the skin usually causes no special problems, even though there may be as many as 1.5 to 2 million bacteria per square centimeter of surface, in addition to an assortment of yeasts and other fungi. Animal body surfaces prevent these microorganisms, some of which are potentially pathogenic, from entering the internal body environment.

Antibiotics

An **antibiotic** is a chemical substance that is produced by a microorganism and that can kill or inhibit growth of other microorganisms. Although interest tends to focus on those antibiotics that can be extracted for use in treating human diseases or diseases of domestic animals, what is the significance of antibiotic production for the

producing organisms themselves? The adaptive value of antibiotic production by soil organisms and other microbes is that the antibiotics inhibit other organisms that compete for resources in the immediate environment.

Antibiotic production is an important normal protective mechanism for many organisms, and literally thousands of antibiotics have been discovered. Only a few of them are valuable as medicine, however, because to be safe and effective they must be selectively toxic. That is, they must be much more toxic to disease-causing microorganisms than they are to animal cells. Antibiotics that are very toxic or even lethal to the organisms being treated obviously have very little medical application even though they might effectively inhibit growth of pathogenic organisms.

The environment-clearing effect of antibiotics is the basis for the continuing search for new antibiotics. Potential antibiotic producers are isolated and grown in pure culture. Then discs of the culture medium containing the organisms are transferred to culture dishes seeded with bacteria. If the organism produces an antibiotic effective

(a)

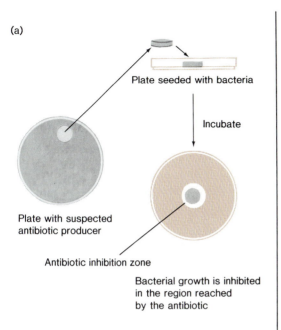

Plate seeded with bacteria

Incubate

Plate with suspected
antibiotic producer

Antibiotic inhibition zone

Bacterial growth is inhibited
in the region reached
by the antibiotic

(b)

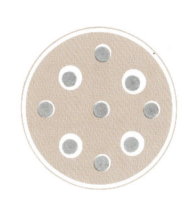

Figure 26.2 Laboratory tests for antibiotic production.
(a) A disc of culture medium containing an organism
being tested for antibiotic production is transferred to a
plate seeded with bacteria. If the antibiotic is released,
bacteria fail to grow in a zone around the disc. (b) A
simultaneous test of nine different possible antibiotic
producers against a particular type of bacteria. Note the
inhibition zones around four of the discs.

against the test bacterium, an inhibition zone
develops around the disc (figure 26.2). Some antibiotics are effective against only a limited number of other organisms, but others, the **broad-
spectrum antibiotics,** are effective against a wide
range of organisms.

Chemical Inhibitors in Plants

Some plants' cells normally synthesize organic
compounds that interfere with the metabolism of
invading microorganisms. These substances provide protection against microorganisms throughout a plant's life and, in many cases, for years
after a plant's death. Treating exposed wood with
creosote, one such natural product, protects
against rot after the wood being treated has lost
its own inhibitors.

Other inhibiting compounds, called **phytoalexins,** are produced in some plants only when
an invading bacterium or fungus is present. Once
phytoalexin production has begun in some plant
tissues, growth of invading organisms is very
strongly inhibited (figure 26.3).

Interferon

Interferon is an antiviral substance produced by
animal cells when the cells are infected by certain
types of virus. Interferon was discovered as a result of studies on the **interference phenomenon,**
the observation that infection by one virus makes
cells resistant to infection by additional viruses.

When viruses infect cells, they subvert the
cells' synthetic apparatus and cause the cells to
make more virus particles. When these new virus
particles are assembled in a host cell, the cell
bursts and releases them. These many new particles can infect additional cells, where the multiplication cycle is repeated. If this process
continues unchecked, huge numbers of cells soon
are destroyed, with disastrous consequences for
the organism.

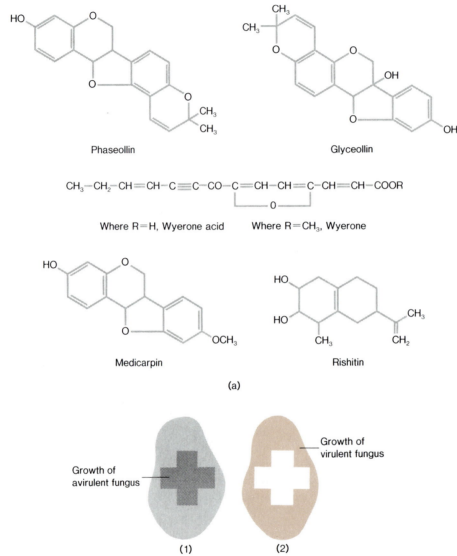

Figure 26.3 Phytoalexins and plant resistance to infection. (a) Molecular structure of some phytoalexins. Phaseolin is produced by beans of the genus *Phaseolus*. Soybeans *(Glycine max)* produce glyceollin. Broadbeans *(Vicia faba)* produce wyerone acid and wyerone. Clovers *(Trifolium)* and alfalfa *(Medicago sativa)* produce medicarpin. The tubers of potatoes *(Solanum tuberosum)* produce rishitin. (b) Diagram of an experiment by Müller and Börger in 1941. Areas of cut surfaces (in gray) of potato tubers were first inoculated with an avirulent (weakly infective) form of a blight fungus that grew only a little. Later, the entire surfaces were inoculated with a virulent form that grew everywhere (colored) but in the previously inoculated areas. The "protection" came from phytoalexin produced in response to the first inoculation.

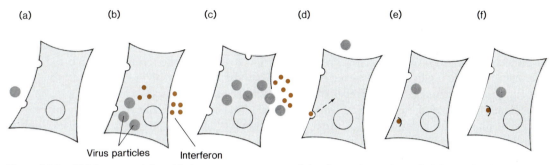

Figure 26.4 Alick Isaacs and Jean Lindemann discovered interferon in 1957. They showed that a material left in culture medium after viruses and virus-infected cells were removed from it would protect a new batch of healthy cells from virus infection (a). Since then, it has been shown that cells infected by virus particles produce and release interferon (b) as they also produce new virus particles. These first cells rupture (lyse) and release new virus particles (c). Interferon then prevents virus multiplication in other cells (d–f).

Interferon breaks this chain of cell infection and reinfection. While infected cells are synthesizing components of new virus particles, they also synthesize interferon (a small protein with a molecular weight of about 20,000), which they release into their environment. Interferon subsequently protects other cells against virus infection and thus breaks the chain of infection (figure 26.4).

Appropriate stimulation, especially exposure to **double stranded RNA** (which is present in animal cells only during virus infection), can induce interferon production by most vertebrate cells. Interestingly, interferon is not specific for types of viruses, but for the host cells. In response to flu viruses, interferon produced by chickens, for example, inhibits multiplication of other types of viruses in chicken cells. But it has little or no effect on multiplication of viruses—even flu viruses—in the cells of other vertebrate animals.

The host cell specificity of interferon relates to its mode of action. Interferon released from infected cells binds with specific cell surface receptor molecules on other cells. These interferon-receptor complexes enter the cells and induce them to synthesize an enzyme that interferes with translation during protein synthesis. The complex also activates a ribonuclease that degrades viral RNA so that it cannot be translated into protein. This inhibition of protein synthesis by two different mechanisms prevents viral replication needed for infection to spread from cell to cell.

Interferon, thus, is a specific protein synthesized in response to viral infection, using genetic information of the host cell (genetic information for human interferon synthesis is coded on chromosome number 5). Its host cell specificity

relates to a need to combine with a specific membrane receptor and to interact very specifically with the host cell's own protein synthesizing mechanisms.

Interferon might seem to be the ultimate weapon against virus infection, and it could well become that in the future. But for now there are several difficulties in its use, though it has been used experimentally in treatment of chronic serum hepatitis (hepatitis B). Interferon is difficult to purify, and until very recently, it had to be produced by cells of the species in which it is to be used. That is, human cells had to be used to produce human interferon. For many years, Karl Cantell and his colleagues in Finland produced most of the world's supply of human interferon using a painstaking technique that involved isolating and culturing human leukocytes, infecting them with viruses, and collecting the interferon that they produced. This work required a large, continuing supply of white blood cells, and the Finnish Red Cross supplied Cantell daily with the leukocytes extracted from several hundred liters of blood. Possibly, in the future, new purification methods will make interferon's use more practical. It is also likely that genetically engineered bacteria can be made to synthesize human interferon in large quantities because the human interferon gene has been isolated and cloned. Large-scale production of interferon will accelerate research and clinical trials on interferon's antiviral actions.

Interferon might hold promise of being more than a new and powerful treatment for virus diseases. It also is a cell division inhibitor in normal cells, and especially, in tumor cells. The prospect of an anticancer drug that might be effective against a variety of abnormal growths has spurred a new wave of intensive research on interferon production, interferon's effects on abnormally growing cells, and on the possible virus connections of various types of abnormal growth. Some very optimistic predictions have been made about future roles for interferon in cancer treatment. Even if not all of these predicted benefits materialize, major pharmaceutical firms are preparing for large-scale interferon production, and interferon research will be one of the most active and interesting areas of biomedical investigation for years to come.

Phagocytosis

Phagocytosis is the process by which amoeboid cells in animal bodies engulf and destroy microorganisms, other cells, and various foreign particles. Phagocytosis has been recognized as a general defense mechanism for about 100 years, ever since Elie Metchnikoff proposed his theory of phagocytosis. Metchnikoff had known that cells accumulated around splinters in human skin. One afternoon, as he was watching motile cells within transparent starfish larvae, it occurred to him that amoeboid cells might protect organisms against harmful invaders. He set out to test this hypothesis by picking rose thorns from his garden and inserting them into the starfish larvae. The next morning he found that the intruding thorn tips were surrounded by clusters of the amoeboid cells. Metchnikoff extended his work by observing amoeboid cell clustering in response to yeast infection in the common water flea *Daphnia* and by studying infections in other organisms. He faced opposition to his theory of phagocytosis from others who believed that resistance to infection was strictly humoral, that is, due to chemicals in the fluid portion of the blood. But Metchnikoff stuck by his theory as he sought to correlate the roles of cellular and humoral resistance, and in 1908 he shared the Nobel Prize in recognition of his work.

It is now clear that, at least in vertebrate animals, phagocytosis and humoral immunity are closely interrelated. As shall be seen later in this chapter, the activities of phagocytic cells are enhanced and focused as a result of antigen-antibody reactions.

There are several important types of human phagocytic cells. **Polymorphonuclear leukocytes,** sometimes also called **granulocytes** because they contain distinctive granules, are found mainly in the bloodstream (figure 26.5a). They generally are involved in relatively quick, focused, short-term responses to infection. **Monocytes** are cells that regularly move freely between blood vessels and tissue spaces (figure 26.5b). In the tissue spaces they sometimes enlarge and remain as **macrophages.** Macrophages generally are responsible for long-term phagocytic activities, such as cleaning up dead microorganisms and host cells at the end of a successful battle against infection.

Once ingested by phagocytic cells, microbes are held inside vesicles that fuse with lysosomes. The lysosomes supply a variety of factors, including proteolytic enzymes, that digest the microbes. After this process is completed, the vesicles are expelled from the phagocytic cells.

Complement

Blood serum enzymes known collectively as **complement** or the **complement system** attack bacterial cells or other foreign agents. Complement is a series of proteins, designated C1, C2, C3, through C9, that work against a variety of invading cells and are always present in the bloodstream, but they are normally inactive until an antigen-antibody reaction occurs on the surface of a foreign cell. Thus, the nonspecific complement system depends on specific recognition for its activation.

The complement system's attacks on foreign cells depend on a series of reactions of the complement proteins. C1 combines with the antigen-antibody complex and becomes able to interact with the C2 through C5 proteins in a set of reactions (figure 26.6a). The reacted forms of C3 and C5 serve as chemical attractants to phagocytic cells. This part of the complement reaction sequence, which makes foreign cells more susceptible to phagocytosis, is called **opsonization.**

The terminal series of reactions from C5 to C9 results in cell lysis, the destruction of the integrity of the cell membrane. The reacted forms of C8 and C9 cause holes to develop in the cell surface. Cytoplasm leaks out, and the cell dies (figure 26.6b).

The complement system, therefore, is a set of general defenses that is effective against a wide range of foreign cells. For its initial activation, however, it depends on an interaction with very specific antigen-antibody reactions of the immune response.

(a)

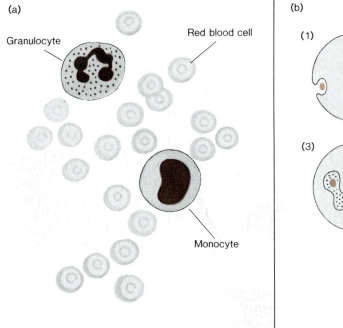

(b)

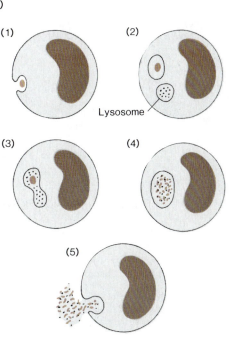

Figure 26.5 Phagocytic cells: (a) Two types of phagocytic cells. This granulocyte (polymorphonuclear leukocyte) is actually a neutrophil (chapter 12). Red blood cells average 7 μm in diameter so they can be used for a size comparison. (b) Phagocytosis by a monocyte. Bacterium (color) is digested by material from a lysosome. Remains are expelled from cell (5).

Figure 26.6 Complement.
(a) Antibody binding on the cell surface makes possible the chain of complement factor activations. C1 interacts with C2, C3, C4, and C5 in a complex set of reactions that yields active forms of C3 and C5, which promote destruction of the foreign cell by phagocytosis. Further reactions yield reacted forms of C8 and C9, which cause holes to develop in the foreign cell's surface. (b) Multiple lesions in a bacterial (*Escherichia coli*) cell wall caused by complement activation. (Magnification × 176,000)

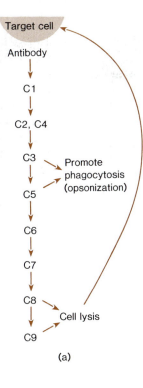

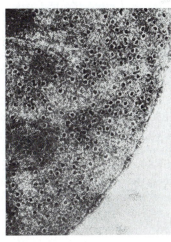

(b)

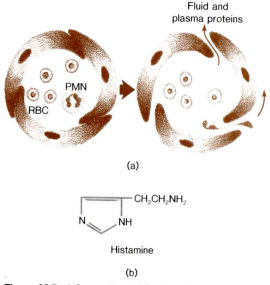

Figure 26.7 Inflammation. (a) During inflammation, cells in vessel walls shrink, allowing easier movement of phagocytic cells (such as the polymorphonuclear, PMN, leukocytes) into tissue, as well as increased movement of fluid and plasma proteins. (b) Histamine, a powerful inflammation promoter.

Inflammation

Inflammation is a tissue and blood vessel response to infection or injury that is externally visible as a swollen, warm, and often painful area. Inflammation involves movement of fluid, plasma proteins, and especially phagocytic cells out of the blood vessels into tissue spaces. This movement is facilitated because chemical regulators cause contractions of cells in the walls of blood vessels, which make passage of materials easier (figure 26.7a).

One of the best known of these chemical regulators is **histamine,** a substance released from granules in **mast cells** (figure 26.7b). These responses are complex and are involved in several different types of normal reactions (**hypersensitivity responses**) to foreign material so inflammatory responses are discussed again later. It is worth mentioning here, however, that some **allergy reactions** can involve such excessive and widespread expression of inflammatory-type responses that extreme drops in arterial blood pressure, bronchial spasms, and even death can result.

Specific Recognition and Defense Mechanisms

It has been known for centuries that once people have recovered from certain diseases, they are not likely to get them again; they have become resistant (**immune**) to subsequent infection. Despite earlier attempts to induce immunity, it was not until Edward Jenner did his work in the late eighteenth century that it was not necessary to have a disease to acquire immunity to the disease. Jenner observed that farm workers who had been exposed to cows with cowpox did not contract smallpox, which at the time was a common and extremely serious disease. Jenner purposely infected people with cowpox pus by scratching it into their skin. This treatment caused a very mild disease with only one pock in humans, but such **vaccinated** (from the Latin *vacca,* meaning cow) people did not catch smallpox. After Jenner did his work, Louis Pasteur and a long line of other workers developed vaccinations for many other infectious diseases, until now some diseases occur only rarely in developed areas of the world. Interestingly, in the late 1970s, smallpox, the first disease to be prevented by vaccination, became the first disease to be proclaimed completely eradicated by the World Health Organization.

Specific immunities to diseases are only part of a much broader system of specific chemical recognition that normally makes distinctions between parts of the body ("self") and foreign cells and their cell products ("nonself"). For example, certain cells of vertebrate animals recognize foreign proteins and carbohydrates, especially those on the surfaces of microbial cells, and respond with defense reactions aimed at removing or chemically neutralizing them. Substances causing such specific responses are called **antigens.**

Specific immune responses fall into two categories that are not really independent of one another. One category involves direct attacks by certain white blood cells (**lymphocytes**) on foreign antigens that they specifically recognize. This is **cell-mediated immunity**.

The second response category involves other lymphocytes, which specialize to become **plasma cells.** Plasma cells produce **antibodies** (protein molecules that combine specifically with antigens). This **humoral antibody synthesis** is important in body defense in general and absolutely vital for acquired immunity to specific disease-causing microbes.

Because of their obvious relevance to health problems, vertebrate immune mechanisms have been studied most intensively, but comparative immunology is an active research field, and work on other animals has helped to clarify some principles of immunology that apply to vertebrates as well.

More recently, there has been increased study of the role of specific recognition mechanisms in plant resistance to infection. This field of research is in its infancy, but is enormously important because of the impact of plant diseases on food and fiber production. Further research on phytoalexin (antimicrobial agents produced by plant tissues in response to infection) production now indicates that recognition of specific substances in cell walls of microbes elicits phytoalexin production. For example, Peter Albersheim has shown that the presence of certain polysaccharide molecules from a fungus parasite's cell walls causes soybeans to produce the phytoalexin glyceollin, whether the molecules are presented on fungal cells or in cell-free extracts of fungal cultures. Thus, specific recognition mechanisms are involved in disease resistance in both plants and animals.

B Cells, T Cells, and the "New" Cellular Immunology

Clarification of understanding of the two-part nature of the specific immune responses (cell-mediated immunity and humoral antibody synthesis) began with Bruce Glick's research in the late 1950s. Glick was studying the physiological role of the **bursa of Fabricius,** a saclike structure just off the cloaca in birds (figure 26.8). He was bursectomizing (removing the bursas from) young chickens to test the importance of the bursa for normal growth and sexual maturation. When a colleague asked Glick for some chickens for a laboratory demonstration of antibody production, Glick supplied him with some of the bursectomized chickens. They were surprised to find that, in response to antigen injection, bursectomized chickens produce antibodies only poorly or not at all.

Figure 26.8   Location of the thymus (in the neck) and the bursa of Fabricius (near the cloaca) of a young chicken.

This result, along with Noel Warner's work on thymectomized chickens (chickens whose thymuses were surgically removed), led Robert Good and Max Cooper to do a series of experiments in which they surgically removed either the thymus or bursa from young chickens, did various replacement therapies, and studied the chickens' immune responses. All of this eventually proved that the thymus and the bursa each are responsible for production of a distinct group of small lymphocytes. The two groups of lymphocytes thus came to be called **T cells** and **B cells.** T cells are lymphocytes responsible for direct cell attacks on foreign antigens (cell-mediated immunity, or **CMI**). B cells are lymphocytes that can mature into the plasma cells that produce humoral antibodies.

Once the two-part nature of specific immune responses in birds was clarified, experiments could be performed on mammals with the same principles in mind. It soon became clear that specific immune responses of mammals, including humans, show a similar two-partedness. In birds,

Figure 26.9 Processing of stem cells to produce immunocompetent T cells and B cells.

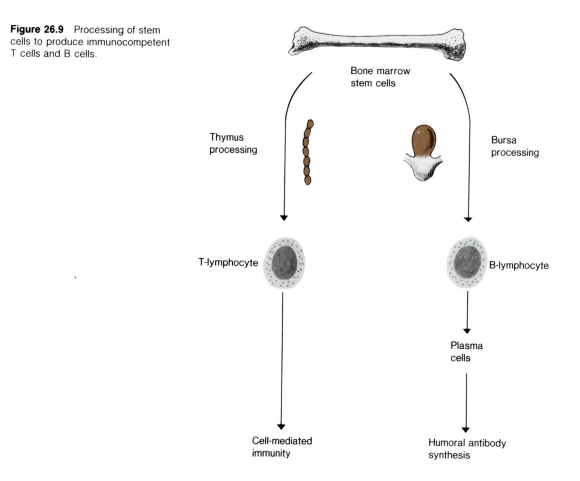

lymphocytes become T cells because they pass through the thymus, and B cells develop under the influence of the Bursa of Fabricius (figure 26.9). In mammals, T cell development seems the same. But since mammals do not have a Bursa of Fabricius, immunologists must speak of a hypothetical "bursa-equivalent" when discussing B cell differentiation in mammals. This clarification of development of the two parts of the immune system has promoted understanding of some human congenital immune deficiencies, which actually are rather common birth defects, and has led to research on cell replacement therapies for treatment of those children who suffer from these birth defects (table 26.1).

Cell-Mediated Immunity

T cells produce responses collectively known as cell-mediated immunity (CMI). CMI is also known by an older name—delayed hypersensitivity—because the responses take from one to several days to develop as opposed to immediate hypersensitivity, which is caused by circulating antibodies and develops in minutes. CMI is involved in many allergic reactions to bacteria, viruses, and fungi, as well as in contact dermatitis (skin irritation) resulting from sensitization to certain chemicals. It also causes the hardening and reddening reaction of skin in the Mantoux test for tuberculin sensitivity and is responsible for rejection of transplanted tissue.

A cell-mediated immune response begins when some of the T cells become activated by a specific foreign antigen. The compact chromatin of their nuclei spreads, and the cells enlarge and divide mitotically. A portion of the stimulated T cells develop cytotoxic powers and become "killer cells" that attack and kill foreign cells directly, and a separate population of T cells release soluble factors that mediate the hypersensitivity response that follows (figure 26.10).

Table 26.1
Some Congenital Immune Deficiencies in Humans

Condition	Symptoms	Cause
Congenital hypogammaglobulinemia	Inability to produce adequate specific humoral antibodies	Problem with development of "bursa-equivalent"
DiGeorge syndrome	T cell deficiency and very poor cell-mediated immunity	Lack of thymus or a poorly developed thymus
Stem cell (Swiss-type) deficiency	Both antibody production and cell-mediated immunity impaired	Apparent absence of stem cells (ancestors of both B cells and T cells)

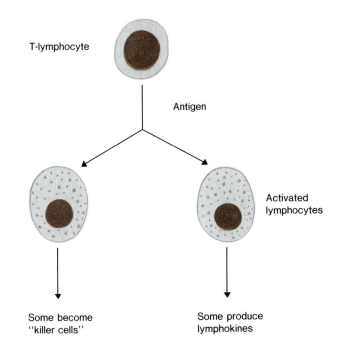

T-lymphocyte

Antigen

Activated lymphocytes

Some become "killer cells"

Some produce lymphokines

Figure 26.10 T cells divide after antigen stimulation and specialize for roles in cell-mediated immunity.

The soluble factors include **macrophage migration inhibition factor,** which tends to keep macrophages (large phagocytic cells) in the area, and **monocyte chemotactic factor,** which attracts monocytes to the area. Other factors activate macrophages and make them more active ("angry") and more efficient in ingesting and killing bacteria. These soluble factors collectively are known as **lymphokines.**

Humoral Immunity

B cells are responsible for humoral antibody production. In response to a specific foreign antigen, certain B cells develop into plasma cells that produce and release an antibody that binds specifically with the antigen. Other B cells become **memory cells,** which somehow retain a special responsiveness to the specific antigen so that when the system is challenged again by the same antigen, antibody production is much quicker and more efficient then it was the first time (figure 26.11). Thus, "memory" accounts for the difference in humoral antibody production between **primary response** (following initial exposure to an antigen) and **secondary response** (following subsequent exposures to the same antigen). This difference is the basis for acquired immunity to infectious microbes.

Interactions among Defense Mechanisms

Although individual defense mechanisms are described separately, it is important to realize that these mechanisms are an interlocking network that depend on and reinforce one another. Activation of the complement system depends on specific antibody binding to foreign antigens, such as those present on the surface of bacterial cells. In addition, phagocytosis by polymorphonuclear leukocytes is stimulated by antigen-antibody complexes on cell surfaces; specific antibodies seem to "mark" foreign cells for destruction by phagocytic cells.

In addition to these examples of activation of nonspecific defense mechanisms by specific immune responses, there are important interactions in the specific immune responses themselves. Macrophages "present" antigens to B cells in the beginning of the B cell response to specific antigens. And organisms with severe T cell deficiencies show weakened antibody production responses to certain foreign antigens. This is in-

terpreted as meaning that some T cells "help" B cells to become plasma cells and produce specific antibodies. Control of the amount and rate of antibody synthesis also involves T cells because there are other T cells that are "suppressors" of B cell activity.

Structure and Function of Immunoglobulins

Antibodies are **immunoglobulins,** a class of protein molecules present in blood, tissue fluids, and certain secretions, such as mucus, tears, and saliva. They are identified by letters: IgG (for immunoglobulin G), IgM, IgA, IgD, and IgE. IgA and IgD are involved in defense mechanisms in the digestive tract and are found mainly in intestinal fluids of normal individuals. Increased amounts of IgD are produced in people suffering from multiple myeloma, a malignant disease of the lymphoid tissues. No beneficial aspects of IgE antibodies have yet been discovered, but IgE antibodies are involved in some of the most unpleasant symptoms of allergy reactions because IgE strongly stimulates mast cells to release histamine. IgM antibodies appear early in the response to an antigen and later are replaced by IgG antibodies. About 80 to 85 percent of the circulating antibodies belong to the IgG class. These circulating immunoglobulins are found in what has classically been called the gamma globulin fraction of blood serum.

Structure of IgG Molecules

R. R. Porter and Gerald Edelman shared a Nobel Prize in 1972 for their work on the structure of antibodies. They determined that the IgG antibody consists of four polypeptide chains. Two of these chains contain 210 to 230 amino acids apiece, depending on the source of the IgG, and are called **light chains.** The other two contain 420 to 440 amino acids apiece, again depending upon the source, and are called **heavy chains.** IgG with its two heavy and two light chains has a total molecular weight of about 150,000. The molecule can be visualized as two heavy chains forming a Y shape, with a light chain located along each side of the arm of the Y (figure 26.12). Disulfide bonds hold the chains in place relative to one another, but the antibody is flexible at two hinge areas of the heavy chains so that the arms of the Y can open as wide as 180°.

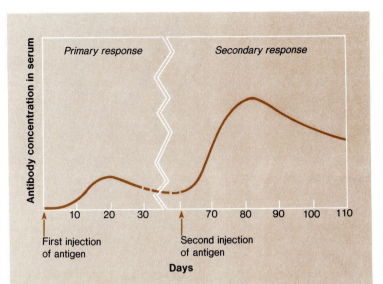

Figure 26.11 Time course of two exposures to an antigen showing the difference between primary and secondary responses.

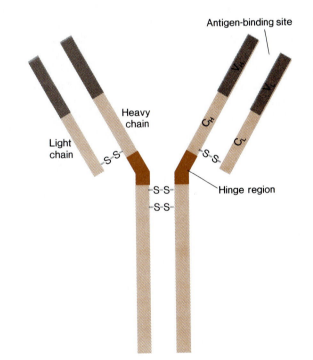

Figure 26.12 The structure of IgG. V_H and V_L are the variable regions of heavy and light chains respectively. C_H and C_L are the constant regions. Each IgG has two antigen-binding sites. Hinge allows molecular flexibility during antigen binding.

The sequences of amino acids in the heavy and light chains have been studied in detail, and it turns out that part of every heavy chain in every IgG molecule always contains the same amino acids arranged in the same sequence, as does part of every light chain in every IgG molecule. These regions of the polypeptides are called **constant regions.** The ends of heavy and light chains toward the tips of the arms of the Y, however, contain different amino acids in different IgG molecules and are known as **variable regions.**

It is not surprising that the actual antigen-binding sites of antibody molecules also turn out to be at the tips of the arms of the Y. To specifically bind different antigens, antibody molecules' combining sites have to be in the parts of the antibodies where the molecules differ from one another. Thus, antigen-binding specificity is based on three-dimensional differences in binding sites that arise as a result of differences in the amino acid composition of variable regions of the polypeptide chains. That single statement seems to explain antigen-binding specificity. However, it is estimated that an individual organism is able to make antibodies for at least one million different antigens. How can this enormous range of antibody diversity be accounted for?

Antibody Diversity

Because the variable regions of antibody polypeptide chains are important in antigen-binding specificity, they have been studied intensively. Elvin Kabat and T. T. Wu found that, within the variable regions, amino acid differences are more likely to occur in some portions of the chain (called the **hypervariable regions**) than in other portions (called the **framework regions**) (figure 26.13). On the basis of that discovery, Kabat and Wu predicted that the hypervariable regions of the chains would be important parts of the actual antigen-binding sites of antibodies. If this were the case, amino acid differences in the hypervariable regions of the polypeptide chains would be responsible for the subtle differences in shape and binding site characteristics that permit binding with different antigens. This prediction about the relationship between hypervariable regions and the binding sites was confirmed when three-dimensional structures of antibodies were studied by X-ray crystallography.

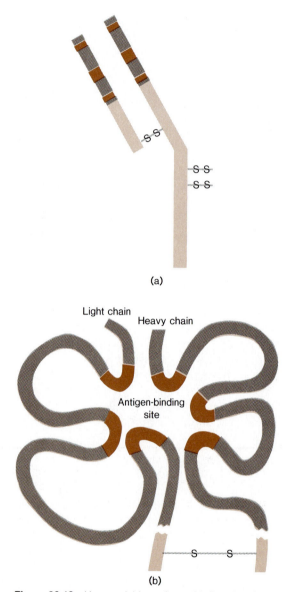

(a)

(b)

Figure 26.13 Hypervariable regions of IgG and antigen-binding site. (a) The hypervariable regions of a light chain and a heavy chain are shown in color. (b) Diagrammatic representation of the three-dimensional folding of light and heavy chains to show how hypervariable regions (color) actually make up the antigen-binding site.

Even though this work gives a good idea of how many different specific binding sites can be produced, there is still a question about how the necessary polypeptide chain diversity arises.

Possibly, there are many different genes for the variable regions of the light and heavy chains, and each of these genes codes for a different chain of amino acids. Any one of these variable genes (called V genes) could combine with the gene for the constant region (called the C gene). Thus, there would be a two gene/one polypeptide situation in antibody synthesis.

It is interesting that in the mouse, at least, C genes and V genes are separate in embryonic cells. Later, in maturing cells, they appear to be joined. Therefore, there is a gene shifting before messenger RNA is transcribed from them (figure 26.14). This results in various C-V combinations in different developing cells, which then are maintained in those cells and their descendants. Although the number of different V genes is not known, there almost certainly are not enough to produce enough different polypeptide chains to account for all the different kinds of antibodies that can be produced. There must be additional genetic mechanisms to provide diversity in antibody molecule polypeptide chains.

One hypothesis is that there is a period in the development of B cells during which replication of the V genes, especially the parts coding for hypervariable regions of polypeptide chains, is very error prone. Thus, a number of copying mistakes produce many different gene versions during cell multiplication. This hypothesis presumes a temporary "relaxation" of the normally very accurate copying mechanisms involved in DNA replication. Following this "relaxation" period, replication is presumed to become accurate once again. Thus, following this period, many subpopulations of cells with different V genes are maintained among the B lymphocytes.

Antibody genes

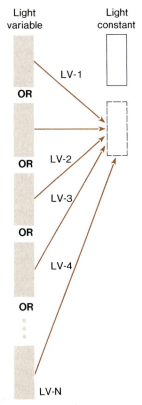

Figure 26.14 Diagram of the "gene shifting" believed to occur during differentiation of antibody-producing cells. Example shows only genes coding for the light chain, but genes for the heavy chain seem to behave in the same way. Any one of several genes (left) coding for different versions of the variable portion of the light chain may combine with the gene for the constant portion of the chain (right). Once this "gene shifting" has occurred, the genome stabilizes, and all descendants of each of the resulting cells produce identical light chains.

Although much has yet to be learned about the mechanisms by which B lymphocytes differentiate as future antibody-producing cells, it is known that normal individuals do possess the potential to produce antibodies for a million or so different antigens. How does that antibody production response actually work? How does the presence of an antigen set off production of a specific antibody?

Clonal Selection and Antibody Production

In the last century, Paul Ehrlich proposed that the body had preformed antibodies whose production was increased by the entry of antigen. But, over the years, discovery of the very wide range of antibody-forming capabilities, including the ability to synthesize antibodies even against substances not normally encountered in nature, made it difficult to continue to accept Ehrlich's idea.

Thus, an alternative, the **instructive model** of antibody synthesis, was devised. This hypothesis stated that an antigen serves as a template or model around which a standard, unfolded protein is molded into an appropriate complementary shape. Then disulfide bonds, hydrogen bonds, and other forces bind the molecule in that shape, leaving the antibody with a specific combining site for that antigen after it separates from the antigen that served as a model. Later, however, it was discovered that proteins fold spontaneously into their normal secondary and tertiary structures, which are determined by the amino acid sequences of the proteins, not by some outside directing force. This spontaneous folding of proteins was not compatible with the idea of specifically directed folding of an uncommitted polypeptide chain around an antigen template.

F. M. Barnet, G. J. V. Nossal, and others then developed a new hypothesis that resembles Ehrlich's much earlier model. This widely accepted model, called the **clonal selection model,** states that there are many homogeneous subpopulations of B cells, each of which contain cells that produce one antibody with unique, specific, antigen-binding characteristics. Each group of B cells may be descendants from a single cell (or a small number of identical cells) that became stabilized as a producer of IgG molecules with a set of particular variable region amino acid sequences. Thus, these are **clones** of B cells, and all the members of each clone produce the same antibody. It is an important part of this hypothesis that these antibody-producing clones arise by genetic means and are produced before the organism ever has been exposed to the antigens that its antibodies can bind. In other words, subpopulations of B cells capable of producing all possible antibodies ever to be produced are present among the circulating B cells.

Antibody production results when an antigen selectively stimulates the group of B cells bearing appropriate antibody molecules (figure 26.15). These B cells divide to produce a larger population and differentiate into either plasma cells or memory cells. The plasma cells produce and release immunoglobulin molecules of their own unique type, which are, of course, complimentary to the stimulating antigen molecules. Memory cells, however, do not release immunoglobulin molecules; rather, they are prepared for subsequent exposures to the same antigen and production of a secondary antibody response when next that antigen enters the body.

Transplantation and Tolerance

Cells, tissue, or an organ grafted from one adult animal (the **donor**) to another (the **host**) usually grow successfully for only a short time. Then a **rejection response** sets in, and **allografts** (grafts between individuals who are not genetically identical) are destroyed and sloughed off within about ten days in mammals (figure 26.16). The rejection is the work of T cell lymphocytes and the macrophages they stimulate in cell-mediated immunity. Something similar to the clonal selection described in antibody production responses occurs in graft rejection. But individual T cells seem to be able to respond to several types of foreign antigens, so their responses may not be quite so highly specific as the response of a clone of B cells to only one specific foreign antigen. There also is a memory component in graft rejection because a second graft from the same donor is rejected more quickly than the first (within six days in mammals). This is called a **second set reaction** as opposed to the **first set reaction** to the first graft from the donor.

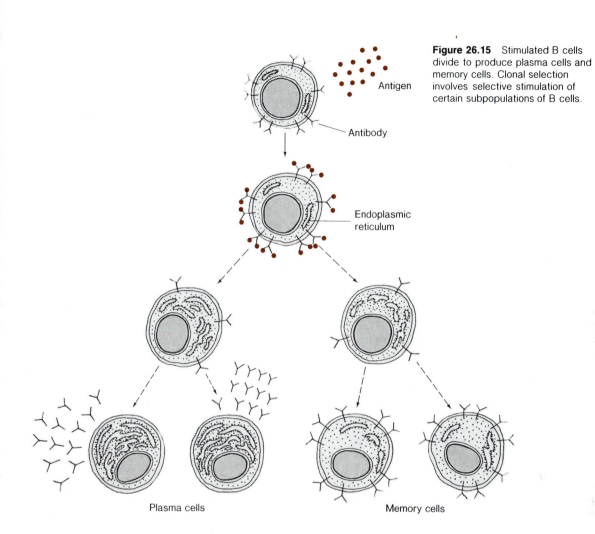

Figure 26.15 Stimulated B cells divide to produce plasma cells and memory cells. Clonal selection involves selective stimulation of certain subpopulations of B cells.

Antigen

Antibody

Endoplasmic reticulum

Plasma cells

Memory cells

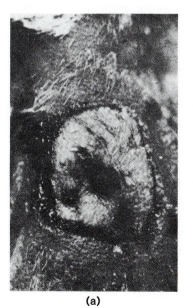

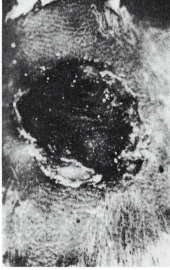

(a)

(b)

Figure 26.16 Rejection of a skin graft from one strain of mouse to another. (a) Discolored areas show drying of rejected and separated parts of graft after ten days. (b) Scabby surface indicates total rejection and sloughing of the graft after thirteen days.

In contrast with the allograft situation, **autografts** (tissue taken from a donor and grafted back into the donor's body) and **isografts** (grafts between genetically identical individuals, such as identical twins) are not rejected. This **tolerance** is built in during differentiation of the cells involved in specific immune responses. During development, clones of lymphocytes that are responsive to the organism's own antigens are eliminated or fail to develop; there is a sort of developmental "paralysis" of these clones. It also is possible to induce tolerance to another individual's cells by injecting donor cells into a developing host before the capability for specific immune responses becomes established.

Graft acceptance or rejection depends mainly on cell surface antigens known as **histocompatibility antigens.** Tissue matching in humans depends especially on a set of histocompatibility antigens known as the **HLA** (for human leukocyte antigen) system, which is the major histocompatibility complex in humans (see p. 711). In addition, the ABO group provides strong transplantation antigens. When the best possible match of HLA and other antigens has been made between prospective donor and prospective graft recipient, it is usually still necessary to give **immunosuppressive drugs** to prevent rejection, and this intentional weakening of immune responses places the transplant recipient at a much greater risk from infectious disease. Sometimes, popular articles about organ transplantation describe specific immune responses as if they were simply a nuisance factor preventing medical progress. Yet, these specific immune responses are essential to the normal protection of organismic integrity, and any reduction in their efficiency, either as a result of disease or medical treatment, is a serious matter.

Autoimmune Diseases

Usually, the body has appropriate mechanisms to prevent formation of **autoantibodies,** antibodies capable of reacting with "self" components. But when tolerance of some body component fails, **autoimmune disease** can result, and components of the immune system then attack the body in various ways. Autoimmune diseases may be organ-specific diseases, such as Hashimoto's thyroiditis, which severely impairs thyroid function, or glomerulonephritis, a degenerative disease of the kidney. Or they may be nonorgan specific, such as rheumatoid arthritis, which can attack connective tissues in a number of body areas at the same time.

In general, treatment of autoimmune diseases presents some of the same problems as prevention of graft rejection because long-term treatment with strong, specific, immune suppressive drugs is very hazardous. Thus, in rheumatoid arthritis, for example, steroids are used for partial immune suppression despite their undesirable side effects, and antiinflammatory drugs, such as aspirin, are used to relieve some symptoms. But even with these somewhat milder treatments, side effects of the therapy compound the problems of people suffering from autoimmune diseases. The development of newer, nonsteroid treatments would be greatly advanced if more could be learned about the causes of these malfunctions of the immune responses.

Cell Replacement and Body Repair

The ability of groups of B cells to multiply in response to antigen stimulation is just one example of a lifelong developmental process involving cell division that contributes to maintenance of homeostasis. Additional examples of such homeostasis-related developmental processes are continuous cell replacement and the repair of damage to body parts.

Cell division does not end in the animal body just because all tissues and organs have attained their adult cell number and size. Cell division must continue for normal cell replacement in many parts of the body throughout life. For example, it is estimated that about two million worn-out red blood cells are removed from circulation per second. This means that, on the average, about two million cell divisions must be completed in the bone marrow every second to replace those cells. Cells lining parts of the vertebrate intestine live only a few days and must be replaced constantly. Skin cells are constantly dying and sloughing off the epidermal surface. They must be replaced by cell divisions taking place deeper in the epidermis.

Figure 26.17 Outcome of a normal replacement cell division. One cell differentiates to become a functional cell; the other retains division potential. Thus, the number of dividing cells does not change.

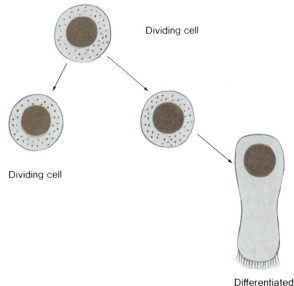

Dividing cell

Dividing cell

Differentiated functional cell

All of these cell replacement processes are regulated. Cell division rates normally are adequate to provide constant cell replacement and no more; on the average, each cell division produces one cell that differentiates to become a functional replacement for a lost cell, while the other remains a dividing cell (figure 26.17). This balances the replacement process so that both the number of functional cells and the number of dividing cells remain constant.

Replacement processes are flexible, and replacement rates can be increased enormously in some cases. For example, blood loss results in a rapid stimulation of cell division in bone marrow, but the process clearly is a controlled one because cell division returns to the normal rate when the lost cells have been replaced.

Other types of cells are permanently arrested and never multiply during adult life. In mammals, nerve cells, voluntary muscle cells, and heart muscle cells do not divide. Repair of damage involving cell death in such tissues results in the formation of **scar tissue.** Scar tissue is a connective tissue containing a network of tough, fibrous, protein (collagen) strands. It holds the tissue together, but it does not replace lost functional cells. Thus, even relatively minor damage to these tissues causes significant reduction in the number of functional cells in the organ.

Other types of cells normally do not divide but can divide under certain conditions. For example, bone cells (osteocytes) normally do not divide. But, if a bone is broken, osteocytes in the vicinity of the break begin to divide within a few hours and to participate in the repair process. Once damage is repaired, the bone cells return to their nondividing condition.

Thus, there are clear differences in the responses of various tissues to injury. Wound healing ends with scar formation in some tissues but proceeds to functional cell replacement in other tissues. Finally, there is a fascinating extension of repair processes called **regeneration,** which results in complete replacement of lost body parts.

Regeneration

While all organisms must have wound-healing abilities if they are to survive even the most minor injuries, the ability to regenerate lost body parts varies greatly among groups of animals. Two groups with very different regenerative capacities are considered here to illustrate some of the principles of regeneration.

Planarian worms—small, free-living members of the phylum Platyhelminthes—have extensive regenerative ability. When a planarian worm is cut in half, regeneration begins at each cut surface with formation of a **regeneration blastema.** The blastema is a growth center in which cell division continues until an adequate population of replacement cells is produced. Even as divisions are continuing, progressive differentiation produces the organized, specific body parts. When the process is complete, cell divisions cease, differentiation is completed, and two complete and normal worms have been produced from the one original worm (figure 26.18).

In planarian regeneration, many of the cells responsible for blastema formation seem to come from a special class of body cells called **neoblasts.** Neoblasts are unspecialized cells that are similar to embryonic cells. They migrate into a wounded area to become part of the blastema. Other cells in the blastema may be contributed by dedifferentiation of specialized cells in the area of the cut surface. In dedifferentiation, cells revert to an essentially embryonic state by losing their structural and functional specializations, and beginning to divide.

It appears that planarian cells can migrate from quite far away and then successfully participate in a regeneration of a body part different from the part in which they were originally situated. This has been demonstrated in experiments in which part of an irradiated area of the body is cut away. Irradiation kills or inactivates cells in the immediate vicinity of the wound so that they cannot participate in regeneration. Regeneration occurs, however, after a slight delay, apparently due to the activities of neoblasts that migrate in from unirradiated parts of the body (figure 26.19).

Regenerative abilities vary greatly among vertebrate animals, with the greatest regenerative capacity found in urodele (tailed) amphibians, where entire limbs can be regenerated when they are lost either in larval or adult stages. Urodele amphibian limb regeneration has been studied as a model of regeneration in vertebrates.

Limb amputation in urodele amphibians is followed by blastema formation and regrowth that continues as long as needed to replace the lost parts. There is evidence that cells do not migrate from any distance to form a blastema; therefore, it is not likely that reserve cells play a role in urodele regeneration. The blastema forms by dedifferentiation of limb cells near the cut surface (figure 26.20).

It is not clear if the dedifferentiation that occurs during urodele limb regeneration is a return to a completely uncommitted state or if dedifferentiated muscle cells are ancestors of only regenerated muscle cells, bone cells of bone cells, and so on. Moreover, the control mechanisms that turn off cell divisions when regeneration is complete are not understood. It also is not clear why the regenerative ability of anurans (frogs and toads) is different. Frog tadpoles can regenerate lost limbs, but adult frogs simply form scar tissue on the cut surface of lost limbs and do not regenerate.

It is known that the nerve supply to the cut tip is an important factor in the regeneration of amphibian limbs. Marcus Singer and his colleagues have shown that the ratio of the total cross-sectional area of axons (nerve cell processes) to the cross-sectional area of the limb is critical for regeneration. In salamanders, if nerves leading to a part of a limb are severed before the part is amputated, regeneration does not take place. Adult frogs (and other nonregenerating vertebrates, as well) have less than what Singer estimates to be the critical nerve supply ratios required for regeneration.

But another factor may hold the key to future research on limb regeneration and possibly to regeneration of other body parts. Through the years, biologists occasionally have suggested that electrical currents are somehow involved in regenerative and other developmental processes, but some of these suggestions have been vague and they often have been conflicting. Now, new techniques for measurement of electrical currents in biological systems have been developed and used by L. F. Jaffe, R. B. Borgens, J. W. Vanable, Jr., and their colleagues. They find that there is an electrical current that moves out

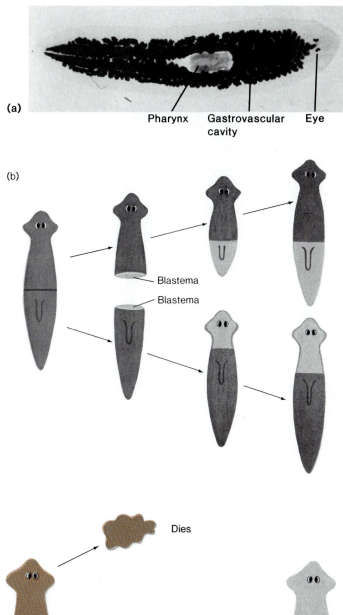

(a)

Pharynx　　Gastrovascular　　Eye
　　　　　　cavity

Figure 26.18 Planarian regeneration. (a) Photograph of a planarian worm with its gastrovascular cavity stained. (b) Regeneration of a planarian cut in half to produce two individuals.

(b)

Blastema

Blastema

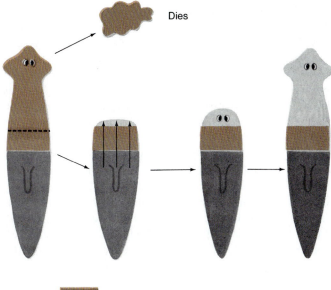

Dies

Irradiated portion

Figure 26.19 Regeneration in a planarian when cut is made through an X-rayed part of the body. Regeneration blastema forms slowly because neoblasts migrate from unirradiated part of the body.

Figure 26.20 Amphibian limb regeneration. (a) Stages in regeneration of limbs of a urodele amphibian (a newt) after amputation through lower part (left) and upper part (right) of forelimb. Numbers indicate days after amputation. (b) Results of amputation through irradiated (in circles) and shielded (unirradiated) parts of limbs. Regeneration occurs when cut is through any unirradiated part of limb (line (2)), but does not occur when cut is through irradiated area (lines (1) and (3)). This indicates that the blastema forms only from cells in the immediate vicinity of the cut and that cells do not migrate in from a distance to form a blastema, as they can in planarians.

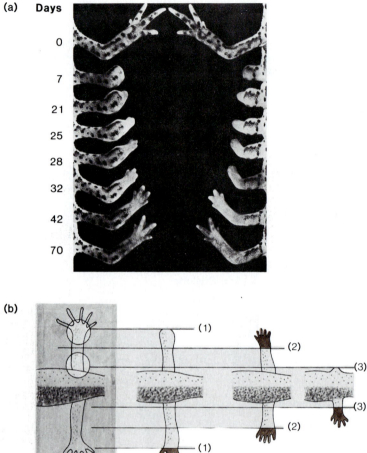

through the cut surface of a regenerating salamander limb and back in through the skin well above the level of the cut (figure 26.21). The current flow involves the positive charges of moving sodium ions (Na$^+$). The current is generated by the skin, which actively transports sodium ions (Na$^+$) inward, driving positive current outward from the cut surface. When these biologists compared these electrical currents in regenerating salamander limbs with currents in and around nonregenerating adult frog limbs, they found significant differences, and experimental reduction of the currents in salamander limbs interfered with their regeneration. Possibly, the electrical currents are crucial for stimulation of cell activities required for regeneration. The roles of electrical currents in regeneration (and other developmental processes) clearly need further investigation.

Regeneration in humans and other mammals is limited essentially to the liver and the skin. Attempts to induce regeneration in other structures, such as whole or partial limbs, have been unsuccessful. When it was discovered that adult frogs, which normally do not regenerate, could at least be made to begin to regenerate by repeated reinjury of the cut surface, similar experiments were tried with mammals, but they proved unsuccessful.

Young children (younger than age eleven), however, actually can regenerate fingertips, including even normal fingernails. But this regeneration occurs only if the cut surface of the finger is *not* closed surgically by drawing the skin over the wound. It is intriguing to consider the possibility of electrical current involvement in this process and to ask what changes occur so that all humans over eleven years old have lost this particular regenerative capacity. Possibly, further research will help to clarify the constraints on

Figure 26.21 Electrical currents in regenerating salamander limbs that can be measured with sensitive vibrating electrodes. The current is generated by the skin, which actively pumps sodium ions (Na⁺) inwards, driving positive charge out through the cut surface. The current flow may stimulate cell activities that are vital for regeneration.

2 mm

mammalian regeneration, but the long-cherished dream of regrowing new human body parts to replace lost or badly damaged ones still seems far away.

Cancer and Other Neoplastic Growth

Actively dividing populations of cells are necessary for normal continuing cell replacement processes in the body, and the actual number of dividing cells varies now and then to meet changing needs, but it remains relatively stable over the long term. Occasionally, however, body cells become **transformed** so that they divide in an uncontrolled manner. Then, normal regulation of cell division fails, and instead of contributing to maintenance of homeostasis, unregulated cell divisions disrupt homeostasis and threaten life itself. Transformed cells give rise to populations of cells committed to repeated cell divisions (figure 26.22). Thus, both the cell population size and the number of dividing cells increases. These characteristics distinguish the growth of **neoplasias** (literally, "new growths") from regulated cell divisions in normal replacement processes. Such neoplastic growths often are called **tumors,** especially when they produce a definite mass or nodule of tissue.

Tumors may be **benign**; that is, their cells remain within their site of origin, especially when they are enclosed in a capsule of connective tissue. But the term benign can be misleading. It definitely does not mean harmless. Benign tumors can grow until they severely interfere with normal function simply by their bulk, by diversion of circulatory supply, or by other means. Tumors

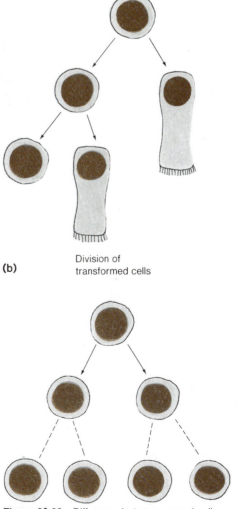

Figure 26.22 Difference between normal cell replacement (a) and cell division of transformed cells in neoplasia (b). In neoplasia, the number of dividing cells increases continually.

Figure 26.23 Metastasis. Cells from primary tumor spread through blood vessels and lymphatic vessels and establish secondary tumors in other body organs.

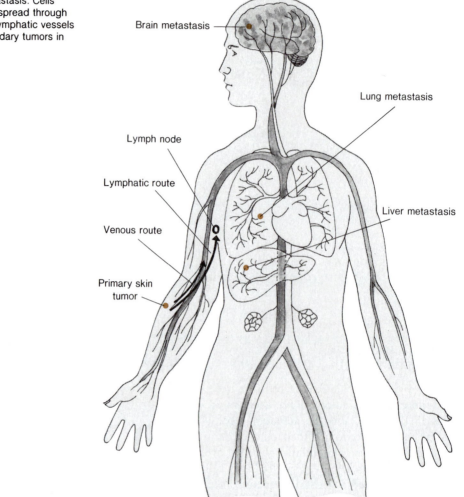

Brain metastasis

Lung metastasis

Lymph node

Lymphatic route

Venous route

Liver metastasis

Primary skin tumor

stimulate circulatory development so that they become well supplied with blood vessels (**vascularized**). As a result, tumors receive a good blood supply that assures nutrient availability, waste removal, and gas exchange, often at the expense of other body parts.

Other neoplastic growths are **malignant.** They are much more likely to cause death because their cells are invasive and they **metastasize;** that is, they spread to new locations in other parts of the body where they establish colonies that grow into additional tumors (figure 26.23). Medically, the primary tumor often can be treated quite effectively by surgery, local radiation, chemotherapy, or some combination of the three. But the colonies established by metastasizing cells are harder to detect and treat successfully. Malignant neoplastic growths are called **cancers.**

Virtually all body tissues and organs can be sites for neoplastic growths, and tumors are classified according to the tissue of origin. **Carcinomas** arise in epithelia (the lining and covering tissues), such as gut, skin, and glandular tissues. **Sarcomas** are sometimes called solid-tissue tumors because they arise in mesodermal tissues, such as bone, muscle, and connective tissue. Finally, the neoplastic growths of blood-forming tissues make up a special class called **lymphomas** or **leukemias.** Some other specialized names are applied to neoplasias with special characteristics, such as the term **melanoma** for heavily pigmented tumors.

Neoplastic Cell Characteristics

Popular descriptions of tumors often include the phrase "rapid and uncontrolled cell division." Cell division rates in some neoplasms may be

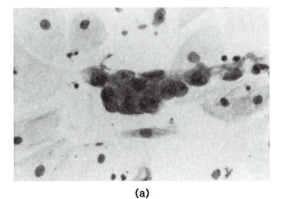

(a)

(b)

Figure 26.24 Neoplastic cells and neoplastic tissue. (a) Photomicrograph of a Pap smear, a sample of cells that were scraped off a human cervix. The more darkly stained cells clumped in the center are neoplastic cells. Note that they have larger nuclei and much less cytoplasm than the normal cells around them. (Magnification × 450) (b) Photomicrograph of a section of tissue from a human colon (large intestine). The left part of the picture contains normal colon tissue with prominent glands separated from other tissue beneath them by a definite boundary layer. The disorganized abnormal tissue in the right half of the picture is characteristic of colon cancer. Note the absence of functional glandular tissue and the lack of separation between tissue regions. (Magnification × 30)

very rapid, but others have cell division rates slower than those in some normal cell replacement processes. An unregulated commitment to cell division leading to a continually increasing population of dividing cells is the one most obvious and devastating property of neoplastic cells.

Appearance of Neoplastic Cells

Microscopically, neoplastic tissues usually look different from tissues around them because they do not contain structurally differentiated cells (figure 26.24). Tumor cells have a low cytoplasm to nucleus ratio, and this unspecialized appearance makes many tumor cells resemble embryonic cells structurally. (Differentiated cells have relatively more cytoplasm than undifferentiated cells.) Mitotic figures are common in sections of tumor tissue because of the high proportion of dividing cells.

Other Embryonic Parallels

Many tumor cells are also metabolically unspecialized. They contain much lower levels of distinctive specialized enzymes associated with specialized functions than are found in normal cells in their tissue of origin. Different types of tumor cells tend to resemble one another more than they resemble normal cells. But that is not to say that all kinds of tumor cells are metabolically exactly alike. If they were, the search for treatments for neoplastic growths might be greatly simplified.

Another interesting parallel with embryonic cells is that some kinds of tumors bear certain cell surface antigens normally found only on proliferating embryonic cells. One of these is **carcinoembryonic antigen (CEA),** which is found both in fetal gut tissue and in the blood of patients with advanced colon cancer and some other diseases. Another of these, **alphafetoprotein,** is present both in the fetal liver and in the blood of patients with certain kinds of tumors, including liver cancer. These data support the idea that transformation to the neoplastic state may involve activation of sets of genes normally expressed only during embryonic and fetal development.

Tumor Cell Surfaces

Other cell surface properties of tumor cells also deserve attention. Several cell surface properties of tumor cells have been studied in culture vessels (*in vitro*). Normal cells show **contact inhibition** in culture. When they encounter other cells, they stop migrating and also are inhibited from dividing. They then form orderly monolayers of cells in culture vessels. Tumor cells are not contact inhibited. They move over each other and continue to divide to form dense heaps of cells. These properties quite probably are related to tumor cells' tendency to invade other tissues during metastasis.

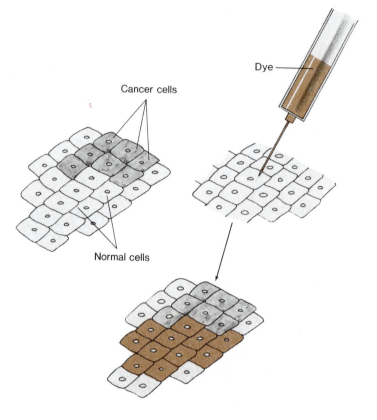

Figure 26.25 Gap junctions between normal cells permit electrical coupling and passage of small molecules directly from cell to cell (see p. 710). Dye experimentally injected into cells of normal tissue spreads through gap junctions but does not enter tumor cells, which lack gap junctions.

Dye

Cancer cells

Normal cells

Tumor cells lack gap junctions, the connections among cells that produce electrical coupling and provide channels for diffusion of low molecular weight materials from cell to cell (figure 26.25). Some researchers propose that the lack of gap junctions at least partly explains the absence of normal controls over cell division.

Tumor cells *in vitro* are easily agglutinated (clumped together) by plant **lectins,** such as **concanavalin A (Con A).** Lectins are proteins or glycoproteins, extracted from various organisms, that can bind to specific molecules on animal cell surfaces. Somehow, transformation produces exposure of lectin binding sites.

Chromosomal Abnormalities

Many, though not all, types of tumor cells have abnormal chromosomal complements. Some have broken or missing chromosomes. Others are **hyperdiploid;** that is, they have extra copies of many chromosomes. Yet others are **polyploid** with complete extra sets of chromosomes. One form of leukemia definitely is related to chromosomal breakage; however, the relationship between specific chromosomal abnormality and other tumor

types is not so consistent. It is possible that at least some of the observed chromosomal abnormalities arise as properties of transformed cells and that the chromosomal changes may not be part of the transformation process in those cases. It is worth noting in this connection that mitotic abnormalities also occur frequently in rapidly proliferating populations of normal cells in culture.

Neoplastic Transformation

Transformed cells continue to divide repeatedly. This indicates a lesion in the control over cell division because cell proliferation normally is regulated both in developing and adult organisms.

At a certain point during the G_1 phase of the cell cycle, a commitment is made to replicate DNA, and once replication is complete, a cell will divide (figure 26.26). In the course of normal development, cells drop out of repeated cell cycling. They either enter a prolonged G_1 phase and do not commit themselves to replication and division, or they leave the cell cycle completely to enter what is called the G_0 phase of specialized function. For transformed cells, the controls that keep normal cells in prolonged G_1 or shift them to G_0 simply do not operate.

One important hypothesis concerning replication of transformed cells says that this permanent loss of control over cell division is a "somatic mutation"; that is, an inheritable genetic change that is transmitted from cell to cell during divisions. The somatic mutation hypothesis also is supported by several general observations: (1) the widespread occurrence of chromosome damage in transformed cells, (2) the **carcinogenic** (cancer-causing) potential of radiation and mutagenic (mutation-causing) chemicals, and (3) the apparent permanent loss of specialized functional characteristics.

Transformation of normal to neoplastic cells occurs at all stages of life, but neoplasms appear more frequently in aging individuals. Genetic changes may accumulate with age, or the normal genetic repair mechanisms may become less efficient in aging individuals' cells. Or, aging changes in the immune system might weaken its effectiveness if it is involved in prevention of neoplastic growth, as the immune surveillance hypothesis suggests.

Immune Surveillance

At the turn of the century, Paul Ehrlich and others suggested that the immune system normally recognizes tumor cells as nonself because of their abnormal properties and destroys them. It is now known that specific cell surface differences can target tumor cells for destruction by the immune system. The modern version of this old hypothesis states that transformed cells are being produced continually, and that tumors develop only when transformed cells slip through this immune surveillance. But there is considerable controversy about this hypothesis. Robert Good and others report that a great many cancer patients, upon thorough examination, do have immune deficiencies or immune system problems of one sort or another. But it is not clear which came first—the cancer or the immune deficiency.

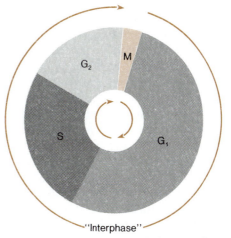

Figure 26.26 The cell cycle. During normal development and cell replacement processes, some cells leave the cell cycle (enter G_0 phase) and irreversibly differentiate for specialized function, or enter a prolonged G_1 phase and do not commit themselves to further replication and division. Transformed cells, however, continue to divide repeatedly.

Other studies have examined the incidence of neoplasia in patients with abnormal immune responses or purposely altered immune responses (for example, kidney transplant recipients). A cautious, and possibly too early, interpretation is that the incidence of most major types of cancer is not increased in patients with immune problems. On the other hand, experimental animals with suppressed immune responses are much more susceptible to tumors that are known to be virus-induced. And human warts, which can be classified as virus-induced benign tumors, are made worse during immune suppression. The full story is not yet in on the immune surveillance hypothesis.

Etiology of Neoplastic Growths

The properties of transformed cells that can be described directly were examined in the previous section. A more complex problem—**etiology** (study of causes) of transformation to the neoplastic state—is now explored. The facts are not very clear here, and interpretations are conflicting at almost every point. The problem can be illustrated by examining some specific cases.

Crown Gall Tumor

Crown gall tumor in plants is an especially useful subject for transformation studies because the inducing mechanism has been clearly described and because transformation involves an obvious defect in cell control mechanisms. In nature, crown gall tumors develop at the crowns (junction of root and stem) of plants (figure 26.27). Transformation requires two steps: a wound and an infection by the bacterium *Agrobacterium tumefaciens*. The wound causes cell division, which makes cells competent to respond to the tumor-inducing principle supplied by the bacterium. Transformed cells are stable and continue to divide, so that instead of development of the usual callus (a mass of disorganized tissue) that stops when the wound is healed, a large tumor develops. This occurs because transformation involves removal of the normal repression of synthesis of the hormones cytokinin and auxin in the cells of the healing wound. Once the cells begin to secrete these hormones, autonomous growth continues and produces a tumor.

Normal plant cells grown *in vitro* require added hormones if they are to divide and grow, but transformed cells from a crown gall grow in a simple culture medium containing inorganic salts and sucrose because they produce the hormones required to stimulate their own growth.

Some researchers believe that the bacterium carries a virus, or at least some viral genetic information, into the plant and that a virus is responsible for transformation of the cells. Whether or not this is the case, the important principle illustrated by crown gall is that transformation results from stabilized activation of genetic information normally present in the plant cells themselves.

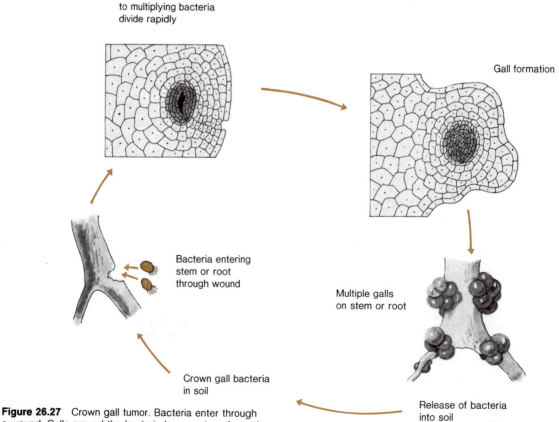

Cells adjacent to multiplying bacteria divide rapidly

Gall formation

Bacteria entering stem or root through wound

Multiple galls on stem or root

Crown gall bacteria in soil

Release of bacteria into soil

Figure 26.27 Crown gall tumor. Bacteria enter through a wound. Cells around the bacteria become transformed, and the tumor grows autonomously. Scales change in all these sketches, and bacteria, of course, are relatively much smaller than they appear here.

Viruses and Animal Tumors

In 1910, Peyton Rous showed that viruses extracted from chicken sarcomas could induce tumor formation when reinjected into tumor-free animals. This **Rous sarcoma virus** is one of the most intensively studied of **oncogenic** (tumor-causing) **viruses.** Rous sarcoma virus is a small virus containing a single strand of RNA surrounded by a protein coat, and it is one of several small RNA viruses associated with animal tumors.

Small DNA-containing viruses are associated with other animal tumors, such as the polyoma (meaning many tumors) virus that causes a wide range of tumors in many species of rodents. The adenoviruses, such as the one causing adenocarcinoma in frogs, are larger oncogenic DNA viruses (figure 26.28).

Possible roles of viruses in tumor induction should be approached with general patterns of virus infection in mind. Viruses affect cells in several ways. In one type of infection, viruses enter cells and are replicated directly. Then the infected cells rupture to release the new virus particles. In another pattern, virus genetic information is incorporated into the host cell genome, where it is replicated at each division and carried through generations of cell reproductions only to become active and cause virus replication much later.

Tumor viruses operate in the latter pattern; that is, virus genetic material is incorporated into the host cell genome during cell divisions that viruses stimulate. Incorporation is direct in the case of DNA viruses, but a special step is involved in the case of RNA viruses. First, DNA complementary to the virus RNA must be synthesized. A special enzyme, RNA-dependent DNA polymerase (**reverse transcriptase**), is responsible for this synthesis (figure 26.29). Then, the new DNA is inserted into the host cell genome. Transformed cells thus contain viral genetic information in

Figure 26.28 Tumor growing on the tail of a tadpole after it received a small graft of tissue from a virus-induced adenocarcinoma of an adult frog kidney. The oncogenic virus that induces this tumor is a large DNA virus.

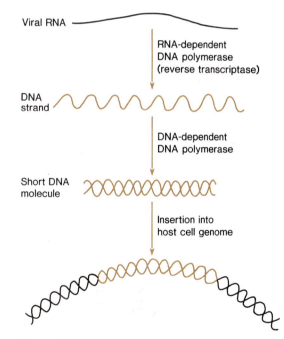

Figure 26.29 Steps in incorporation of genetic information of an oncogenic RNA virus into host cell genome.

their genomes, but it is unlikely that expression of the relatively small amount of viral genetic material alone accounts for the transformed state. Rather, presence of the viral genome releases the host cell from normal restraints, and it behaves as a transformed cell. In fact, some biologists go so far as to say that transcription of viral genetic material is not needed for the expression of the transformed state. The virus only causes the transformation.

Another interesting hypothesis states, in simplified form, that tumor viruses arose in the first place by incorporating a part of the genome of "spontaneously" transformed cells. The viruses cause transformation because they have become accidental transmitters of faulty host cell genetic information that induces the transformed state when they infect normal cells.

Unfortunately, the twin hopes of discovery of a common virus transformation mechanism and possible subsequent development of an immunization against oncogenic viruses have faded considerably because there may be several or even many quite different models for virus involvement in transformation. Nevertheless, experimental study of virus-induced transformation has greatly expanded the knowledge of the cell biology of transformation.

Viruses and Human Tumors

Some benign human tumors, such as skin warts, are caused by viruses. Other tumors contain viruses and viruslike particles, but these viruses do not cause transformation of human cells *in vitro.* Thus, ultimate proof that they are oncogenic viruses depends on ethically unacceptable experiments on human subjects. Some of the excitement about the discovery of virus particles in human tumor cells faded anyway when electron microscopists reported sighting similar viruslike particles in a variety of normal cells, especially embryonic cells.

The human tumor most closely associated with virus induction is **Burkitt's lymphoma,** which affects children in certain portions of Africa. The associated virus is called **Epstein-Barr virus.** It is very similar to the common **herpes viruses,** one of which causes cold sores. Epstein-Barr virus is found in many humans around the world, but Burkitt's lymphoma occurs mainly in areas of southern Africa with certain rainfall and temperature patterns (figure 26.30). Its absence

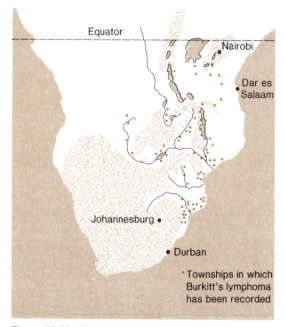

Figure 26.30 Distribution of Burkitt's lymphoma. This rare children's cancer occurs in southern Africa but is seldom found in cooler areas where temperatures fall below about 18°C (64°F) (shaded areas). Some interaction between the Epstein-Barr virus and chronic malaria may cause the disease.

in nonmalaria areas indicates some complex interaction between chronic malaria and the viral infection, if the virus does actually cause Burkitt's lymphoma.

Even if one or several viruses is successfully linked with human tumors, it still must be determined how virus infection leads to transformation in some individuals and not in others. The role of environmental factors in transformation is critically important and must be studied carefully.

Natural Carcinogens

The somatic mutation hypothesis must be considered when discussing natural carcinogens in the environment. Chronic exposure to physical agents such as certain parts of the spectrum of sunlight and other natural background radiation does induce mutations, some of which may not be repaired by normal repair mechanisms. Some chemical constituents of the normal diet or their breakdown products may also be mutagenic. According to the somatic mutation hypothesis, some of these mutation events might cause cell transformations. Increasing incidence of neoplasia with advancing age would then be correlated with accumulation of this natural damage or with reduction of the ability to repair it.

Carcinogens in an Industrial Society

Persistent irritation contributes to transformation, and some irritants, such as asbestos dust, are widespread in industrial societies. Other airborne particles, such as coal dust and ashes, cause irritation and contain carcinogenic chemicals as well. Other carcinogenic compounds are some common organic solvents, additives to fuels, and the combustion products of a number of normally harmless organic compounds (for example, charred foods). The chemical industry adds many newly synthesized compounds to their production lists every year, and many of them prove to be carcinogenic.

Screening of literally hundreds of chemical compounds is difficult because the standard animal tests are expensive and time consuming. There also are uncertainties because some neoplasias do not appear until long after the cellular insult that caused them. Some progress has been made in this testing because of the close correlation between causing mutations and being carcinogenic. The degree of mutagenic potential is easier to measure, and it gives an estimate, at least, of carcinogenic potential. Bacterial tests of mutagenesis, such as the very important **Ames test,** are very useful in identifying suspected carcinogens, which can then be tested further.

Cancer Therapy and the Cancer Cure

Considerable progress has been made in the detection and treatment of tumors over the past quarter century. But there have been no dramatic breakthroughs to compare, for example, with the revolution that antibiotics caused in the treatment of bacterial infections.

For now, the best hope for a reduction in the incidence of tumors seems to involve controlling environmental insults and reducing dietary intake of known carcinogens. The best example of potential for control using currently available knowledge is in the case of lung cancer, which is a high incidence human neoplasia (figure 26.31).

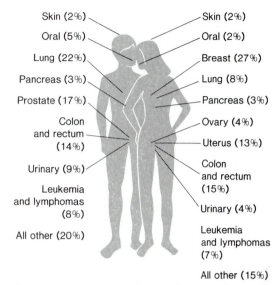

Figure 26.31 Incidence of cancer in men and women (excluding nonmelanoma skin cancer and some carcinomas). The incidence of lung cancer in women increased dramatically during the 1970s. If that increase continues, it may soon erase the sex difference in lung cancer incidence.

Lung cancer is so closely correlated with cigarette smoking that only the most resolutely closed-minded individuals could doubt that cigarette smoking increases the incidence of lung cancer. Yet cigarette smoking, especially among young people, has increased rather than decreased since discovery of the relationship.

One suggested explanation for this failure to heed warnings about dietary intake of carcinogens and environmental insults to the body is that somehow many people have embraced the optimistic idea that a general "cure" for cancer is just around the corner and that it will be available by the time they need it. Cancer researchers, on the other hand, tend to view neoplastic growth as a complex of problems in cell and developmental biology, and they are much less optimistic about finding a generally effective cure anytime in the immediately foreseeable future. This difference in understanding of the problem is unfortunate. People could become so disillusioned with apparent lack of progress that they insist upon further diversion of public support to treatment and detection, and away from crucial fundamental research. It is hoped that scientists also will do more educating so that cancer prevention becomes a still greater part of the environmentalist movement.

Aging

Aging is a continuation of lifelong developmental processes. The outward signs of aging may be obvious, but understanding of the underlying bases for aging changes is poor. **Gerontology,** the study and analysis of aging, is in its infancy as a science. Thus, most aging studies still are largely concerned with assembling enough facts about aging processes so that better experimental and analytical approaches can be organized.

Aging in Populations

Each species has its own characteristic, genetically determined range of life spans. Illness and accidents cause death of some young individuals, but progressive aging changes decrease the **vitality** of all individuals. Vitality can be defined simply as the ability to sustain life, and this vitality loss eventually results in a marked increase in the death rate of a population made up of aging individuals (figure 26.32).

It is important to understand that aging changes leading eventually to death are normal developmental processes. While the average human life expectancy has been increased, for example, by eliminating many early deaths due to infectious diseases, the maximum expected life span itself has not been altered. Still more could

be done to increase the average life span by eliminating such major causes of human death as cancer and heart disease, but the expected increase (possibly ten to twelve years) would again be due to an increase in the number of individuals living out their genetically determined expected life spans, not to any change in the expected maximum life span.

Expected life spans vary among individuals in a species, but each species has its approximate maximum expected life span. The maximum recorded life spans for some species of mammals are given in table 26.2.

From an evolutionary viewpoint, there appears to be no selection for longevity, because there is no particular biological advantage in a long postreproductive life. In fact, it may be a disadvantage for a population to have many individuals who compete for resources with those who currently are reproducing.

The human postreproductive life span is relatively long by comparison with other animals, but humans have a genetic limit just as other animals do. Thus, if there are important reasons for extending human life beyond the genetically determined maximum, the mechanisms by which the genetic limits are expressed would need to be discovered and manipulated.

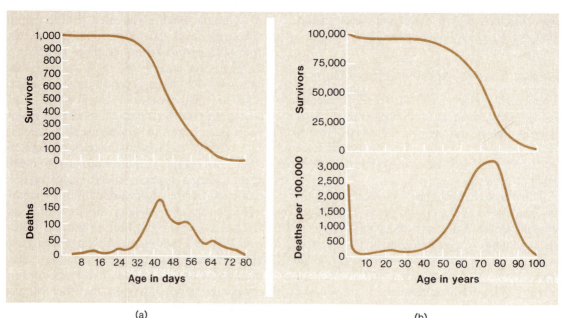

(a)   (b)

Figure 26.32 Survival curves and distribution of ages at death. (a) Male *Drosophila melanogaster*. (b) Human males.

Table 26.2
Maximum Age Attained for Selected Mammals.

Species	Common Name	Maximum Life Span (in Years)
Mus musculus	House mouse	3.5
Canis familiaris	Dog	20
Equus caballus	Domestic horse	46
Elephas maximus	Indian elephant	70
Pan troglodytes	Chimpanzee	44.5
Pongo pygmaeus	Orangutan	50+
Homo sapiens	Human	118+

Table 26.3
Functional Losses Accompanying Aging.

Type of Decline or Loss	Remaining Functions or Tissues* (Percentages)
Weight of brain	56
Number of spinal nerve fibers	63
Velocity of nerve impulse	90
Flow of blood to brain	80
Adjustment to normal blood pH after displacement	17
Output of heart at rest	70
Number of glomeruli in kidney	56
Glomerular filtration rate	69
Number of taste buds	36
Maximum oxygen uptake during exercise	40
Maximum breathing capacity	43
Hand grip	55
Basal metabolic rate	84
Water content of body	82
Maximum work rate for short burst	40

From E. Peter Volpe, *Man, Nature, and Society*, 2d ed. (Dubuque, Iowa: Wm. C. Brown Company Publishers, 1979). Used by permission.

*Figures are the approximate percentages of functions or tissues remaining in the average 75-year-old male, taking the value found in the average 30-year-old male as 100 percent (based on studies by Nathan W. Shock).

Functional Aging Changes

Many functional and structural features of organisms show signs of loss or decreased function with advancing age, and these changes, singly and in various interactive combinations, weaken an organism's homeostatic responses (table 26.3). Specific immune resistance mechanisms also decline with age so that organisms become more susceptible to infection. But even in the absence of accident or disease, the general physiological decline eventually reaches a point where some vital function passes a critical low point.

Both extracellular and cellular changes in tissues are being studied to identify these functional aging changes.

Cellular Aging

Tissues without significant cell turnover and replacement lose cells with aging. The nervous system, for example, is made up entirely of postmitotic cells and suffers a progressive loss in cell numbers throughout life (figure 26.33).

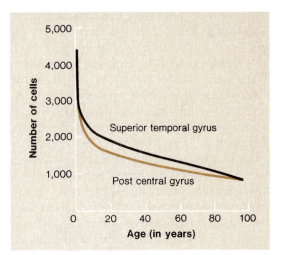

Figure 26.33 Cell counts for two areas of the human cerebral cortex. There is no cell replacement in nervous tissue so cell deaths result in permanent reduction of the total number of cells.

Several possible causes of cell death during aging have been suggested. Certain metabolic wastes, known as **aging pigments,** accumulate in aging cells. The aging pigments are mainly cell membrane breakdown products that are stored rather than released, and their accumulation may eventually lead to cell death.

Somatic mutation, the accumulation of DNA damage, is not readily detectable in nondividing cells, but it does cause disturbances in cell functions if it leads to faulty synthesis of enzyme molecules that are less efficient catalysts of cellular reactions.

A limit on the cell division potential may be important in the aging of tissues that do have cell replacement. Evidence for existence of such a limit comes from a discovery of Leonard Hayflick and Paul Morehead. They cultured and recultured human embryonic cells, always providing the cells with optimum growth conditions. They found that instead of growing and dividing indefinitely, the cells went through about fifty generations of divisions and then died, even though they were still maintained under the same favorable conditions. Similar results have been found with a variety of normal cell lines in culture. Tumor cells do not show this limitation, by the way, as they reproduce indefinitely under favorable culture conditions. Hayflick has proposed that such a limit on the number of possible divisions determines the life span of strains of somatic cells within organisms. Cells within aging organisms simply run out of division potential and eventually cell replacement becomes inadequate to maintain important functions at essential levels. This hypothesis describes, but still does not explain, a cellular mechanism. Furthermore, many researchers are cautious about making the extrapolation from studies on cultured cells to behavior of cells inside the intact aging organism. But the idea is used in development of another general hypothesis of aging: the immunological hypothesis.

The Immunological Hypothesis of Aging

The immunological hypothesis of aging is intriguing because it suggests how cellular aging might cause a wide spectrum of age-related problems. The life cycle of the thymus also attracts interest to this hypothesis because the thymus reaches a maximum weight sometime before puberty and then progressively atrophies (gets smaller) until it virtually disappears later in life. The hypothesis says that thymus atrophy is correlated with a decreasing role in differentiation of T cells. A similar decline in B cell production is proposed as well. The argument is that stem cell production in the bone marrow slows down as strains of cells run out of cell division capability.

F. M. Burnet and other supporters of the hypothesis suggest that this explains decreasing resistance to infection with advancing age. It also might be an explanation for the increasing incidence of neoplastic growth in older individuals. Finally, loss of balanced control between the parts of the specific immune response may increase risk of autoimmune diseases. All of those problems are seen in aging individuals, but the immune hypothesis is by no means established as the fundamental cellular cause of aging.

Extracellular Aging

Collagen fibers (figure 26.34), which are important permanent components of tissues because they form an extracellular framework around tissue cells, show significant aging changes. Collagen fibers consist of many tropocollagen subunits (chapter 2), and each of those tropocollagen subunits is made up of three polypeptide chains. In newly formed collagen, these components are linked by noncovalent bonds, but with advancing age, increasing numbers of stronger covalent cross-linkages form among the collagen components. Physically, the collagen becomes less elastic (more rigid) as this cross-linking increases, and loss of collagen elasticity changes the physical structure of tissues.

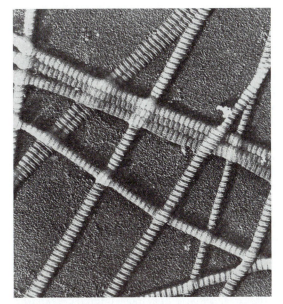

Figure 26.34 Collagen fibers. With increasing age, the number of covalent cross-linkages in collagen increases. This makes the collagen less elastic and changes the physical characteristics of the tissues in which the collagen is found. (Magnification X 25,194)

Another aging change in the extracellular environment is a decrease in the water-rich portion of the extracellular space. In a kind of progressive dehydration, fibrous material fills more of the spaces among cells, and as a result of this change, diffusion of materials to and from cells slows down.

Future Prospects

It is difficult to distinguish primary aging changes (the fundamental changes that make cells, tissues, and organisms "old") from the secondary changes that occur as a result of the primary changes. Until the primary changes are identified and understood, scientists can only work to increase the number of people who live out more of their expected life span in good health by treating diseases of the aged more effectively. But, if the primary cellular aging processes are identified and the life span is expanded, a new dilemma of sorts will have to be faced. Aging and death have always made room in populations for new, young individuals. Aging and death are part of humans' natural biological heritage and, from an evolutionary point of view, are a positive force for the welfare and survival of the species.

Everyone individually rebels at the notion of aging and death, and longs, along with people down through the ages, for indefinite extension of youth and vigor. For the overall good of the species, humans may have to find ways to accept this somewhat depressing individual burden and balance it with a broader view of the long-term importance of the continuing cycle of human life.

Summary

Developmental change continues throughout life.

Several of the developmental processes that occur even after maturity contribute to maintenance of homeostasis. These are processes involved in disease resistance and in wound healing and regeneration.

Some of the specializations involved in resistance to infection and disease are general and become functional before contact with specific invading organisms. Both plants and animals have relatively impermeable body surfaces that bar entry by microorganisms. Some microorganisms produce antibiotics that suppress growth of other nearby microorganisms. Plants produce phytoalexins, which also are inhibitors of microbial growth. Animal cells produce and release an antiviral substance called interferon when they are infected by viruses. Interferon prevents virus replication in other cells and thus breaks the cycle of infection, replication, and virus release by cell lysis.

Phagocytosis is the process by which specialized cells engulf and destroy invading microorganisms, other cells, and various foreign particles. The ingested materials are digested inside vesicles and then expelled from the phagocytic cells.

The complement system is a set of nonspecific antibacterial factors in the blood that become activated in response to specific recognition of foreign organisms.

Inflammation reactions provide easier access for phagocytic cells and complement factors to infected areas.

Specific immune responses fall into two categories. Cell-mediated immunity is based on direct attacks by T cells on foreign cells that are recognized by their surface antigens. Humoral immunity is based on certain B cells that, in response to specific foreign antigens, develop into plasma cells that produce and release specific antibodies (immunoglobulins).

Each immunoglobulin G molecule has two heavy chains and two light chains. The actual antigen-binding sites of the molecules are in the variable regions near the ends of the arms of the Y-shaped molecule. Diversity in antibody-producing cells is believed to arise during development, so that there are many small populations of B cells, each of which produces its own specific antibody, before antigen exposure ever occurs. Entrance of an appropriate antigen stimulates cells in one of these clones to multiply and differentiate either as plasma cells that produce and release antibody, or as memory cells.

Allografts (grafts between individuals who are not genetically identical) result in a rejection response. Autografts (tissue taken from a donor and grafted back into the donor's body) and isografts (grafts between genetically identical individuals) are not rejected. Graft acceptance or rejection depends mainly on cell surface antigens known as histocompatibility antigens.

Autoimmune diseases develop when the immune system mistakenly begins to recognize some body cells as foreign and attacks them.

Cell replacement continues throughout life because cells in a number of tissues wear out or are lost as a result of damage, but the cell divisions involved in these normal replacement processes are regulated.

Regeneration is complete replacement of lost body parts. Regrowth begins with a blastema made up of dedifferentiated cells that proliferate to replace lost tissue. Electrical currents in regenerating structures may stimulate cellular activity, and differences in this electrical activity could be the basis for differences in regenerative capacity.

Transformed cells are different from cells involved in all normal developmental processes because their divisions are not regulated. Transformed cells have relatively little cytoplasm, and many of them have surface antigens that resemble those on embryonic cells. Transformed cells also lack gap junctions, and chromosomal abnormalities are common.

Transformation can occur at any stage of life, but neoplasms appear more frequently in aging individuals. The immune surveillance hypothesis proposes that transformed cells should be recognized as nonself and destroyed, but that failures of the immune system allow them to live and divide. Such failures may become more common with advancing age.

Plant tumors can be transmitted by bacteria, and some animal tumors clearly are virus induced. But no such definite connections have been made for human neoplasms, although viruses are very closely associated with several human tumors.

People in industrialized societies are exposed to many carcinogenic compounds, both by choice, as in cigarette smoking, and by accident. Reduction of exposure to carcinogens in the human environment remains the most effective available means of combating cancer because expectations of a general cancer cure may not be realized for a long time, if ever.

Aging changes continue throughout adult life, and they progressively weaken homeostatic responses. The cellular basis for aging may be accumulation of somatic mutations or cell wastes, or a decrease in the ability to replace cells because of loss of division potential. Immune system changes may lead to increased incidence of disease and, possibly, of neoplastic growth. Irreversible extracellular changes alter the structural and functional characteristics of aging tissues. Aging changes reduce the efficiency of the mechanisms that function to maintain homeostasis in individual organisms, but primary aging changes (the fundamental changes that make cells, tissues, and organisms "old") cannot yet be distinguished from the secondary changes that occur as a result of the primary changes. It is necessary to view aging and death biologically from a species point of view because they are part of a cycle of life that results in replacement of older individuals by new, young members of the species.

Questions

1. What is the adaptive significance of antibiotic production for microorganisms?

2. Explain why interferon produced by cultured cells of other animals is not effective in treatment of human virus diseases.

3. If testosterone is injected into chicken eggs at an appropriate stage of incubation, the embryos that develop either lack a Bursa of Fabricius or have a very poorly developed one. What effect does this treatment have on immune system functioning of the chick after hatching?

4. Propose a hypothetical explanation for the experimental observation that one treatment that can induce some beginnings of regenerative responses in amputated frog limbs, which normally do not regenerate at all, is to immerse the amputated limb in a sodium chloride solution.

5. List several characteristics that distinguish transformed cells from normal cells, including general appearance, membrane organization, and regulation of cell division.

6. How does the increased incidence of neoplastic growths in older individuals provide indirect support for the immune surveillance hypothesis of neoplastic transformation and growth?

Suggested Readings

Books

Cairns, J. 1978. *Cancer: Science and society*. San Francisco: W. H. Freeman.

Deverall, B. J. 1977. *Defense mechanisms of plants*. London: Cambridge University Press.

Hammond, S. M., and Lambert, P. A. 1978. *Antibiotics and antimicrobial action*. Studies in Biology no. 90. Baltimore: University Park Press.

Lamb, M. J. 1977. *Biology of ageing*. New York: John Wiley.

Roitt, I. M. 1980. *Essential immunology*. 4th ed. Oxford: Blackwell Scientific Publications.

Articles

Borgens, R. B.; Vanable, J. W. Jr.; and Jaffe, L. F. 1979. Bioelectricity and regeneration. *BioScience* 29: 468.

Cairns, J. 1981. The origin of human cancers. *Nature* 289: 353.

Devoret, R. August 1979. Bacterial tests for potential carcinogens. *Scientific American* (offprint 1433).

Hayflick, L. January 1980. The cell biology of human aging. *Scientific American* (offprint 1457).

Henle, W.; Henle, G.; and Lennette, E. T. July 1979. The Epstein-Barr virus. *Scientific American* (offprint 1431).

Leder, P. May 1982. The genetics of antibody diversity. *Scientific American*.

Milstein, C. October 1980. Monoclonal antibodies. *Scientific American* (offprint 1479).

Nicolson, G. L. March 1979. Cancer metastasis. *Scientific American* (offprint 1422).

Porter, R. R. 1976. The chemical aspects of immunology. *Carolina Biology Readers* no. 85. Burlington, N.C.: Carolina Biological Supply Co.

Rose, N. R. February 1981. Autoimmune diseases. *Scientific American* (offprint 1491).

Talmage, D. W. 1979. Recognition and memory in the cells of the immune system. *American Scientist* 67: 173.

Willoughby, D. A. 1978. Inflammation. *Endeavor* 2: 57.

Part 5

Evolution

Conceptions of present-day organisms are enhanced by the recognition that all living things are both products of and participants in a continuing evolutionary process.

Scientific perceptions of the evolutionary process developed relatively slowly until Charles Darwin proposed his theory of natural selection. His proposals regarding the causes of evolutionary change had an accelerating effect on the study of evolution.

Current concepts of evolution still rest on principles of natural selection, but modern evolutionary theories have incorporated principles of population genetics that have been developed since Darwin's time. And modern evolutionary research incorporates techniques of the physical sciences that make possible determination of the ages of fossil organisms and detailed molecular evolutionary studies of contemporary organisms.

Exciting progress also is being made in the investigation of human evolution as newly discovered human fossils provide more information on the origins of modern humans.

In chapters 27–29 the origins of evolutionary theory, some modern concepts of evolution, and the status of the study of human evolution are examined.

27

Development of Evolutionary Theory

The idea that living things have arisen by evolution, a process of change through time, is an old one, dating at least to the time of ancient Greece, but this concept was not generally accepted as a valid explanation of the natural world until the latter part of the nineteenth century. Up to and including the early decades of the nineteenth century the common belief was that all species of plants and animals had been specially created at one time and that no new species had been formed since. Further, it was believed that variation was not the rule in nature but that each species had been created according to an ideal plan called an archetype and that each individual of a species strove toward the perfection of form represented by that archetype. The earth itself was thought to be quite young—only five or six thousand years old—and to have remained unchanged since the time of its origin. These concepts were generally accepted by the scientific community of the early 1800s, although a few scientists, such as Buffon, Lamarck, and Diderot, had suggested that evolutionary processes occurred.

It was Charles Darwin, however, who provided the foundation upon which the modern **theory of organic evolution,** the origin of living things by evolutionary processes, has been built. In 1859 Darwin published *On the Origin of Species,* a thoughtful exposition of his views on the evolution of life on earth. In the more than 100 years that have elapsed since the publication of *On the Origin of Species,* the theory of organic evolution has gained almost unanimous acceptance among scientists and has been greatly strengthened by evidence obtained from a wide variety of scientific disciplines, especially in the areas of genetics and molecular genetics.

Figure 27.1 Charles Darwin.

Charles Darwin and the *Beagle* Voyage

Charles Darwin was born in England into a wealthy family (figure 27.1). His father was a prominent physician, and his mother's maiden name was Wedgwood, from the family of renowned china manufacturers. Later, Darwin himself was to marry his cousin Emma Wedgwood. All his life Darwin loved nature. As a young man, he was an ardent naturalist, walking the fields and woodlands of the English countryside, studying plants and animals, and observing rock formations. This interest continued into his college years for, although his father wanted him to become a physician, Darwin preferred to associate himself with professors of biology and geology. When it became obvious that Darwin would not become a doctor, his father tried to guide him into a career in the clergy. This profession held little more interest for Darwin than did medicine.

Figure 27.2 The route followed by H.M.S. *Beagle* on its surveying trip around the world (1831–1836).

Finally, in 1831, Darwin was offered the opportunity to serve as ship's naturalist on the H. M. S. *Beagle,* a ship commissioned to spend five years on a survey mission around the world. Darwin saw in this post an opportunity to see more of the natural world than was afforded many people of his time. In December of 1831 Darwin embarked on a voyage that was to change the history of biological thought (figure 27.2).

Darwin began his voyage believing in the special creation of species and the short history of the earth. He returned with the dawning belief that species had arisen by organic evolution and that the age of the earth must be measured on a much longer time scale than was generally accepted.

Lyell, Hutton, and Uniformitarianism

Darwin was not a good sailor and spent much of his time in his bunk suffering from seasickness. During the early part of the voyage, he occupied much of his time reading the first volume of Charles Lyell's *Principles of Geology,* in which Lyell attempted to formalize James Hutton's uniformitarian theory of geology. Hutton believed that the earth was not a static, unchanging sphere but that the present features of the earth's surface were the result of a continuous cycle of erosion and uplift. Hutton saw evidence that the earth's continents had been continually worn away by the agents of erosion and that weathered rock debris was transported by rivers to the oceans, where the loose sediments were deposited

in thick layers that were eventually converted into sedimentary rocks. Hutton proposed that the forces of erosion were sufficient to, in time, reduce all continental surfaces to sea level. That this had not occurred indicated to Hutton that "subterranean forces" had operated to uplift sedimentary rocks from below sea level to form new land surfaces. Hutton's concept of uniformitarianism was that present forces had acted at the same rate throughout the earth's history to produce similar geological events. Hutton's general ideas about continuing geological change are still accepted today, although modern geologists realize that rates of change have not always been the same.

Having studied geology, Darwin understood Lyell's arguments and realized the implications that uniformitarian theory held for biological thought. If the earth itself was dynamic and present features were the result of a long process of gradual change, could it not be possible that the biological world was also dynamic rather than static?

Fossils of South America

Darwin found the first evidence of change in the biological world in South America. Along the east coast of Argentina, in the mud and silt of river deposits, Darwin discovered a large number of fossil bones literally sticking out of the loose sediment. Among these bones he identified the remains of a giant ground sloth, a huge hippopotamuslike animal, and a species of horse. All three of the fossil forms represented **extinct species,** species not represented by any living individuals. That some species had become extinct suggested to Darwin that not all species were immutable (unchanging). But, even more unsettling to Darwin was the close resemblance of these fossil forms to living species of sloths, hippopotamuses, and horses. Was it possible, Darwin wondered, that these fossil forms might have been ancestral to species found on earth today? If so, the implication was that new species appear on earth as a result of organic evolution. Darwin later wrote that his first ideas of evolution began with his observations of these South American fossils.

Galápagos Islands

From Argentina the *Beagle* slowly progressed southward, rounded Cape Horn, and worked its way northward along the coast of South America to a tiny group of islands called the Galápagos. It was in these islands that Darwin was to discover a veritable laboratory of evolution. The Galápagos were pushed up out of the sea by volcanic eruptions, and they are isolated by some 950 km of ocean from the South American mainland. Ancestors of the plants and animals that inhabit the islands were introduced from the South American mainland—brought to the islands by wind currents or on floating debris (rafting). The colonization of the islands by plants and animals was a slow process, but the weathered volcanic rock provided a good substrate for the growth of some plant species, and as seeds germinated and vegetation spread, food and shelter were available for animal species.

Adaptive Radiation of the Tortoises and Finches

When Darwin arrived in the Galápagos in 1835, he found a variety of plants and animals, many of which are found nowhere else in the world (figure 27.3). Among the most interesting of the Galápagos creatures were the giant tortoises that gave the islands their name (*galápago* in Spanish means tortoise). Darwin's interest in these animals was aroused when the governor of the islands chanced to remark that he could tell from which island a tortoise had come by observing the shape of its shell. Upon close inspection, Darwin noticed that not only did the shape of the

(a)

(b)

(c)

Figure 27.3 Many species of plants and animals are found only in the Galápagos islands. (a) Drab-colored marine iguanas climbing over rocks on Narborough (Fernandino) island. Marine iguanas are found only in the Galápagos. (b) The colorful iguana found only on Hood (Española) island. Note the large claws on its feet, which are well adapted to a life spent clinging to rocks along the shoreline. Galápagos iguanas eat seaweed exposed at low tide. (c) The rare flightless cormorant found only on two of the Galápagos islands.

shell vary from one island to another but that other features of the tortoise varied as well. For example, the necks of the tortoises living in dry areas were longer than the necks of the tortoises living in moist areas (figure 27.4). Why, Darwin wondered, should so many distinct forms of the tortoise appear in such a limited geographical area?

Darwin hypothesized that very few tortoises could have arrived at the Galápagos from the mainland, but for those few that had survived the arduous journey, adjustment to conditions on the islands was not difficult. The tortoises found little competition for food, and there were no natural predators. Under such conditions, the tortoises became established on the islands and gradually increased in numbers, spreading out to initiate tortoise colonies wherever conditions were favorable. Because the physical conditions for survival varied from island to island, the tortoise population of each island developed, over time, individual characteristics that distinguished it from the tortoise populations inhabiting neighboring islands. Long-necked tortoises inhabited dry areas where food was scarce since the longer neck was helpful in reaching high-growing foliage. In moist regions with relatively abundant foliage, short-necked tortoises fared well. Thus, Darwin concluded that the major factor in development of interisland variation among the tortoises was the environmental variation that existed from one island to the next.

Darwin also observed a wide range of variation among a group of small, drab birds that lived on the Galápagos Islands. These relatively inconspicuous birds have gained fame in the annals of biology as "Darwin's finches." Like the tortoises, finches were introduced into the Galápagos from the mainland of South America, either blown to the islands by strong offshore winds or rafted on floating vegetation. Also like the tortoises, finches found little competition for food and no natural predators. Under these conditions, the finch population became established.

The original finches were seed eaters, and for a time there was plenty of food and they thrived. But the very factors that led to the initial success of the finches also led to a major problem. Finches are prolific breeders, and with an initial abundance of food and no predators to control population size, the population increased to the point that competition for food and nesting sites became severe. During this period of population increase, the finches spread to all of the various islands.

On the different islands, the finches underwent evolutionary modification in several directions. Although the finches were primarily seed eaters, they would also eat insects, and those finches that could not obtain sufficient seed to stay alive were compelled to turn to this alternate food source. By doing so, they decreased their competition with the seed eaters and increased their chances for survival and reproductive success. Over time, finches evolved that were insect eaters rather than seed eaters. An important ingredient in the success of this transition was the lack of competition from other bird species for the insect food supply. In most ecological situations, there are bird species that feed specifically on seeds, on insects, on fruits, on the larvae found under the bark of trees, and so on. Under such conditions it would be very difficult for members of a seed-eating species to make the transition to insect eating. But because the Galápagos finches had little competition from other bird species over long periods of time, such evolutionary transitions were possible. Some of the finches became specialized to other habits, such as cactus eating.

After periods of isolation on separate islands, various groups of finches came to be very different from one another. And these differences involved more than variations in food habit and appearance. These groups of birds, which were descended from common ancestors, also had changed genetically during the period of geographic isolation. They could no longer interbreed when they encountered one another later. They had become reproductively isolated from one another.

(a)

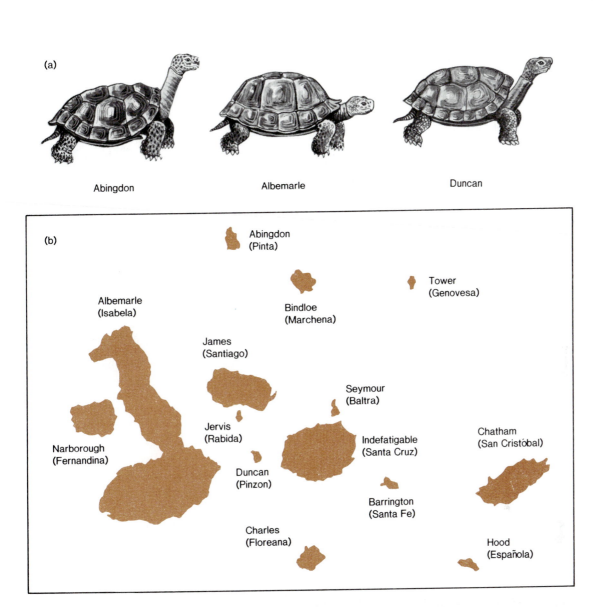

Abingdon   Albemarle   Duncan

(b)

Abingdon
(Pinta)

Tower
(Genovesa)

Bindloe
(Marchena)

Albemarle
(Isabela)

James
(Santiago)

Seymour
(Baltra)

Jervis
(Rabida)

Indefatigable
(Santa Cruz)

Chatham
(San Cristóbal)

Narborough
(Fernandina)

Duncan
(Pinzon)

Barrington
(Santa Fe)

Charles
(Floreana)

Hood
(Española)

Figure 27.4 Tortoises of the Galápagos islands. (a) Sketches of Galápagos tortoises from three different islands illustrating neck differences. Longer-necked species live in relatively dry areas and feed on the high-growing vegetation of cacti. Shorter-necked species live in moister regions with more abundant ground vegetation. (b) Map of the Galápagos islands with the islands' English names first and the current Spanish names of the islands in parentheses. (c) Photographs of tortoise shells illustrating the different shapes that the islands' governor pointed out to Darwin. The round shell on the left comes from James (Santiago) island and the other comes from Hood (Española) island.

(c)

Figure 27.5 Some species of Galápagos finches. All of the species are descended from a common ancestor, but they differ markedly in appearance, habitat occupied, and food sources. Ground finches (lower right) eat cactus flesh; warbler finches (upper left) eat insects; others eat seeds or have mixed diets. The most remarkable of the Galápagos finches is the woodpecker finch (lower left). It pecks tree bark as a woodpecker does, but it lacks the characteristically long woodpecker tongue. Instead, it uses cactus spines as tools to draw insects out of the holes that it makes.

Original ancestor

So varied were the unoccupied habitats of the Galápagos and so malleable was the finch population that by the time Darwin arrived in 1835 there were thirteen distinct species of finches occupying the Galápagos islands and one species on Cocos Island, which is located 700 km to the northeast. Each of these species was adapted to a different life-style. In the Galápagos, finches fill the ecological roles of warblers, parrots, woodpeckers, and other bird species that would normally occupy these niches in other geographical areas. Darwin found distinct differences among the thirteen Galápagos finch species in such traits as plumage, nesting sites, body size, and size and shape of the beak (figure 27.5).

The development of the tortoise and finch populations of the Galápagos islands are examples of a process modern biologists call **adaptive radiation.** Adaptive radiation occurs when members of a single taxonomic group enter environmental niches in which there is little initial competition for resources. Under such conditions, there is often an initial period of adjustment, followed by a rapid increase in population size, and a period of intense competition for resources. This competition may result in the fragmentation of the population into a number of new species, each

of which occupies a new and distinct niche, as in the example of Darwin's finches. More details regarding specific mechanisms of species formation are discussed in chapter 28.

Profoundly influenced by his observations of the tortoises and the finches of the Galápagos islands, Darwin later deduced that variation is an integral part of the biological world and that these factors—environmental variation and biological competition—are involved in the initiation of adaptive changes among the individuals of species. This led him to the unavoidable conclusion that species are not immutable and that new species are formed through the gradual modification of existing traits in response to the pressure of both the environment and biological competition.

Darwin had come to accept the idea of organic evolution—the concept that all species of plants and animals have arisen through descent with modification from previously existing species. A corollary of species formation through organic evolution is that of species extinction, for not every species passes the test of adaptation. That is, not every species continues to interact successfully with its environment, especially when there are changes in the environment.

Change is the rule of nature, and the history of the earth has been filled with many geological and climatological changes. Thus, a species that was very well adapted to the conditions that existed during one period of earth history could become extinct because of changes in world temperature, sea level, rainfall patterns, and so on. In the history of life many more species have become extinct than now inhabit the earth.

The Theory of Organic Evolution

Upon his return to England in 1836, Darwin married Emma Wedgwood and, after living in London for a short time, moved to a country house near the village of Downe in Kent. There he spent the next twenty years in research and reading, seeking evidence in support of the idea that life evolves, that living species are the product of a series of gradual changes that have occurred through a long period of geological time.

Darwin was convinced that variation was a natural part of any species. He was also convinced that, in the natural range of variation that appeared in any population, some of the variants were better suited to survival in a given set of environmental conditions than others. If the environment changed, so did the conditions for survival. Thus, the environment exerted an influence on variants that appeared in any species: adaptive traits (traits that permit successful interaction with the environment) tended to be amplified and nonadaptive traits tended to disappear. Darwin saw the end result of adaptive change as the eventual formation of a new species. Species not adapted to changed environmental conditions became extinct. This was, in essence, the theory of organic evolution: that every species of plant and animal, living or extinct, was the modified descendant of a previously existing species. The concept of organic evolution was not original with Charles Darwin, by any means, but he stated it clearly and directly, and systematically gathered a large quantity of evidence in support of the idea.

Darwin realized that this theory required a mechanism that would explain the ultimate source of biological variation and the means by which heritable traits were passed from generation to generation. He also realized that he needed a better explanation for the mechanism by which new species were formed. Although

Gregor Mendel was a contemporary of Darwin's, the gene theory of inheritance was not developed until the early years of the twentieth century. Thus, Darwin never succeeded in explaining biological variation. Lacking an understanding of the mechanism of genetics, Darwin was left with such vague and inadequate concepts as the "blending" theory of inheritance and Lamarck's theory of acquired characteristics. The blending theory of inheritance attributed the inheritance of traits to a mixing, or blending, of traits found in the parents. The theory was wholly inadequate to explain the wide range of variation found in any species and could not explain how some traits became more pronounced rather than being diluted generation by generation. Lamarck's theory, though later proven wrong, offered some possibilities for Darwin.

Lamarck and Inheritance of Acquired Characteristics

The theory of inheritance of acquired characteristics was developed by Jean Baptiste de Lamarck as part of his theory of organic evolution published in his *Philosophie Zoologique* in 1809. Lamarck accepted variation as part of the natural world and believed that the environment played a role in the origin of new species. He also emphasized the gradualness of evolutionary change and the tendency of living organisms to be adapted to their environments. However, Lamarck thought that new traits were acquired by an organism in response to a *need* imposed by the environment and that evolution proceeded toward a perfect form.

Lamarck stated that traits could be acquired through the use or disuse of an organ; that is, a body part not used by an organism atrophies and is lost to future generations, while a body part used extensively is amplified. For example, the Lamarckian explanation for the long neck of the giraffe is that elongation occurred because the ancestors of the giraffe constantly reached into the trees to feed on high-growing vegetation. This repeated stretching of the neck over time resulted in the modern giraffe with an exceptionally long neck. A similar argument would be that a son born into a family of weight lifters should have large muscles because his father, grandfather, and great grandfather had developed bulging biceps lifting weights.

Lamarck's theory of inheritance of acquired characteristics never gained wide acceptance in scientific circles, partly because it was teleological, assuming a final goal toward which the process of evolution was directed. Although his theory was an attempt to explain how traits are gained and lost, it lacked a basic mechanism of inheritance. When such a mechanism became known with the rediscovery of Mendel's principles of inheritance, it became apparent that acquired characteristics, which are not genetically based, cannot be inherited.

Artificial Selection

Although Darwin was not able to provide an adequate explanation for biological variation because he lacked knowledge of mechanisms of inheritance, he did develop a clear explanation of a mechanism by which evolution of new species can occur.

Observation of a commonplace activity that had been practiced by the human race for thousands of years provided a clue for Darwin's explanation. For centuries humans have selectively bred fruits and vegetables, cattle, chickens, and dogs in an effort to obtain greater yields of corn, a better milk cow, more productive hens, or a dog more adept at sheepherding or rabbit hunting. These practical breeders observed the variation appearing in their crops and herds and selected for breeding those individuals that best represented the properties they desired in future generations (figure 27.6). The success of this **artificial selection** was undeniable, and Darwin saw in artificial selection a model for change in the natural world. What puzzled him was how selection worked in nature. The consciousness of the human mind directed artificial selection, but what force directed selection in nature?

The Theory of Natural Selection

An essay on human population written by Thomas Malthus gave Darwin a clue to the solution to this problem. Malthus stated that human populations have a general tendency to increase in size, but that such factors as fire, war, plague, and famine serve to keep the human population of the earth more or less in check. If it were not for these factors of population control, the human population of the earth would increase explosively (a prediction that agricultural, medical, and technological advances of the last century have proven to be accurate).

By applying the Malthusian doctrine to the natural world, Darwin reasoned that every species produced many more young than could survive to reproductive maturity. Of the many new individuals produced by all species in nature each year, most are eaten by predators, succumb to disease, or are unable to obtain an adequate food supply or habitat in which to live. There is a constant struggle for survival. Only the best adapted members of any species live to reproduce, and only the members of any generation that succeed in reproducing contribute to the evolutionary future of the species. Darwin recognized that the same factors he had encountered in the Galápagos islands operated throughout the biological world: nature in the form of environmental variation and biological competition operates as the selective factor in natural populations.

From these hypotheses Darwin derived his major theory, the **theory of natural selection,** to explain how evolution takes place. The theory of natural selection can be summarized as follows: All populations of organisms possess the potential to increase at a very rapid (geometric) rate, but this rate of increase is very seldom observed among natural populations. Instead, the size of populations remains almost the same from one year to the next. Thus, Darwin reasoned, there is an intense, constant struggle for food, water, and other resources among the many young born every generation. And those individuals with the most adaptive traits are the most likely to survive and reproduce. Thus, generation by generation, adaptive traits become more frequent in the population, while nonadaptive traits tend to be eliminated.

Darwin's theory of natural selection also explains why some species perish while others survive because survival is predicated on the adaptability of the species as a whole. Under changing environmental conditions, species with the greatest amount of biological variability are more likely to survive than species that are highly specialized to a given set of environmental conditions. These highly specialized species may have lost the flexibility required to survive in the face of changing conditions.

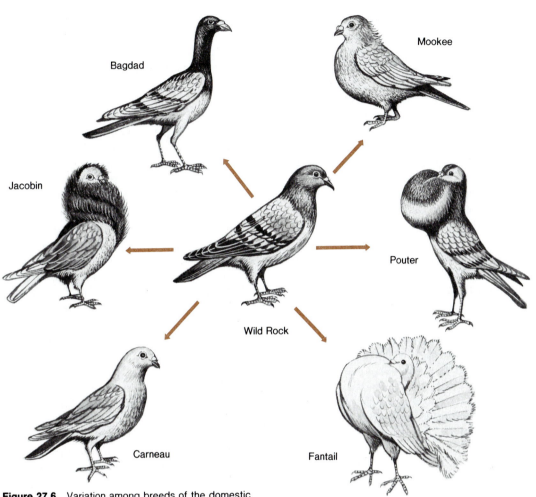

Figure 27.6 Variation among breeds of the domestic pigeon. Darwin saw artificial selection, as practiced in the breeding of pigeons and agricultural plants and animals, as a model for evolutionary change in nature.

Over long periods of time, natural selection produces species of organisms that are finely adapted to the set of environmental conditions in which they must compete. Each species is not only adapted to its physical environment but to its biological interaction with other species.

Alfred Russel Wallace

Early in 1858 Charles Darwin received an essay from the young English biologist Alfred Russell Wallace titled "On the Tendency of Varieties to Depart Indefinitely from the Original Type." Wallace wanted Darwin to review the essay and to decide whether it merited further critical reading by Charles Lyell. Darwin was startled to find in Wallace's short essay the basic elements of his own theory of natural selection. So similar were Wallace's ideas to his own that Darwin felt that

he should defer publication of his own work in favor of that of Wallace. However, both Lyell and the botanist Sir Joseph Hooker knew of Darwin's long efforts in the development of his theory, and they also knew that he had set out his own ideas on the origin of species and natural selection in an abstract sent to Lyell in 1844 and in a letter to the famed American botanist Asa Gray in 1857. They persuaded Darwin to have these statements of his theory, along with Wallace's paper, presented to a meeting of the Linnean Society in the summer of 1858. The arrival of Wallace's paper prodded Darwin into proceeding with a much more complete statement of his ideas. Darwin's major exposition of his ideas of organic evolution and natural selection, *On the Origin of Species,* was published in November of 1859.

The singular coincidence of the similarity of Wallace's concept of species formation with that of Darwin was heightened by the fact that Wallace had taken an exploratory journey up the Amazon River Valley, had extensive field experience in the East Indies, and had formulated his ideas as a direct result of reading Malthus' essay on population. Nevertheless, Wallace himself maintained Darwin's primacy in the development and formalization of the theories of organic evolution and natural selection. Not only had Darwin's work preceded that of Wallace in a chronological sense; it also was of considerably greater breadth and depth.

Darwin and Wallace disagreed on the importance of natural selection in the process of organic evolution. Wallace believed that every trait of an organism was the product of natural selection and therefore must possess an adaptive function. Darwin, on the other hand, maintained that selection was not the only means of species modification and that organisms are complex systems in which certain specific features may have adaptive value while others are nonadaptive. Darwin did maintain, however, that natural selection is the major force in directing the process of organic evolution.

Evolutionary Studies after Darwin

It is an historical fact that new scientific theories stimulate a great deal of controversy, with the beneficial effect that a period of intense, intellectual activity ensues during which scientists with opposing beliefs attempt to seek evidence to support their points of view. The theory of organic evolution and Darwin's theory of natural selection initiated such a period of debate, and the search for data that would either support or refute these theories began. The weight of scientific evidence over the past 100 plus years is strongly supportive of Darwin's basic precepts, and such areas of biological specialization as biogeography, paleontology, comparative anatomy, comparative embryology, genetics, and molecular biology have provided especially strong support.

Biogeography

Biogeography is the study of the geographic distribution of plants and animals. It attempts to explain why species are distributed as they are and what their pattern of distribution reveals about prehistoric habitats and their climates. As Darwin toured the world on the H.M.S. *Beagle*, he discovered that plant and animal species are not generally distributed as broadly as are their potential habitats. Studies in biogeography since the time of Darwin have confirmed this over and over again. For example, plants of the cactus family *(Cactaceae)* are found in the deserts of southwestern North America and the high deserts of the Andes, but nowhere else. Comparable arid habitats in Africa are occupied by members of the spurge family *(Euphorbiaceae)*. In fact, each major geographic region of the world has its own characteristic **flora** and **fauna** (figure 27.7). In addition, the fossil record of each region reflects a historical sequence of biological events that is distinct from that of all other regions. Within each sequence, the most recent fossil forms closely resemble the living species found in that region.

Such observational evidence supports the theory of organic evolution by reinforcing the concept that the environment is the major force that has molded modern species, adapting the life forms of each geographic region into species that are suited for survival within the topographic and climatic condition of their surroundings. The fossil record is evidence that each region has a distinct evolutionary history, which explains why species are not distributed as widely as their potential habitats.

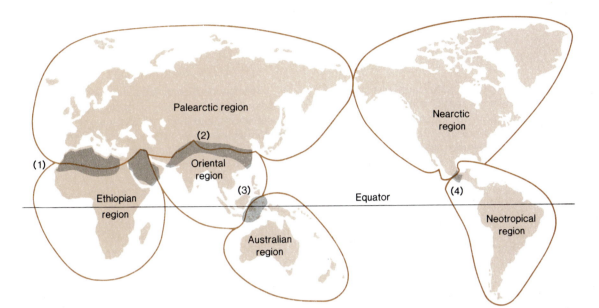

Figure 27.7 Major biogeographical regions of the world. Some natural barriers that separate the regions are indicated in gray. (1) The Sahara and Arabian deserts. (2) Very high mountain ranges, including the Himalayas and Nan Ling mountains. (3) Deep water marine channels among islands of the Malay Archipelago. (A. R. Wallace recognized and wrote about this barrier, which has been called Wallace's line.) (4) The transition between highlands in southern Mexico and the lowland tropics of Central America.

Paleontology

The only set of observations relating to the history of life on earth is contained in the fossil record, a somewhat sketchy collection of the remains of extinct forms. Incomplete as it is, the fossil record does permit modern scientists to consider events and processes that have taken place through the long eons life has occupied this planet. The study of fossils and the fossil record is called **paleontology.**

Fossil Formation

One reason for the incomplete nature of the fossil record is that very few organisms have become fossilized, and many of those that were fossilized have been destroyed by various geological processes.

The natural fate of the great majority of dead organisms is that they are devoured by scavengers and decomposed by bacteria and fungi.

Bones left exposed to the atmosphere are reduced to dust by the combined action of water, sun, and wind. The organisms preserved in the fossil record are those few that have been buried by loose sediments very soon after their death. This happens most often underwater and least often in dry upland regions. For this reason, the fossil record is biased in favor of aquatic organisms and organisms that lived near water.

When an organism is buried quickly, it is not found by scavengers; nor is it weathered by sun, wind, and rain. Because oxygen is sealed out, decomposition is retarded. Under these conditions the organism can become fossilized. To become fossilized, the organism must generally possess hard body parts, such as the woody tissue of plants, an external skeleton (shell), or an internal skeleton such as vertebrate animals have. Soft body tissue seldom survives the rigors of burial in coarse sediments, and organisms without hard body parts are poorly represented in the fossil record.

Box 27.1
Continental Drift

The concept of a static earth was changed by Hutton's concept of uniform change through time. Hutton's proposal that the continents are continually worn away by the processes of erosion has been substantiated by geologists, as has his hypothesis that the accumulated debris of the erosion cycle is compacted and cemented into sedimentary rocks. However, not even Hutton envisioned an earth as dynamic as that postulated by modern geologists.

Early in this century Alfred Wegener in Germany proposed that at one time in the past all the continents of the earth were united in one large continental mass that he called Pangea. Wegener believed Pangea began breaking apart millions of years ago, undergoing a series of changes that have resulted in the configurations and positions of the continents found today. Although Wegener presented a variety of geological, paleontological, and climatic data in support of his theory, it was not widely accepted by geologists until the second half of this century. It was given strong support by H. H. Hess of Princeton University in the early 1960s, and since that time, a great deal of evidence has been brought forward in support of the concept, which is now called the theory of **continental drift.**

According to this theory, the continents are thin plates of lower density rock that float atop a denser rock that forms a "skin" around the earth. As the light continental plates move over the denser layer beneath, they are subjected to stresses that cause cracking and crumbling of the plates— stresses that geologists believe are responsible for faulting, earthquakes, volcanic activity, and mountain building. The relatively new subdiscipline of geology that is concerned with these matters is called **plate tectonics.**

The theory of continental drift has been substantiated by a variety of geological observations, and from their studies, geologists have been able to piece together the following story of the earth's history: About 225 million years ago all of the continental plates were united in a supercontinent called Pangea. Then Pangea began breaking up. It first split into two huge continental masses—Laurasia and Gondwana. Gradually, Laurasia and Gondwana broke up into the continental plates familiar today. About 65 million years ago the last of the previously existing plates separated, and since that time, the plates have been drifting slowly to the positions in which they are now found (box figure 27.1A).

The fossil record reflects these geological events. During the time when the continents were united as Pangea, many species enjoyed worldwide distribution—a phenomenon hard to explain if the continents were separated by ocean barriers as they are today. Even during the existence of Laurasia and Gondwana, species could be very widely distributed. But as the continents separated and drifted apart, the oceans became physical barriers to plant and animal distribution, and the flora and fauna of each continent became isolated. The plant and animal species of the individual plates have developed in geographic isolation, and over the many millenia, each continent has developed its own species. This sequence of events is reflected in the unique fossil record of each continental region and also in the distinct species of each biogeographical region.

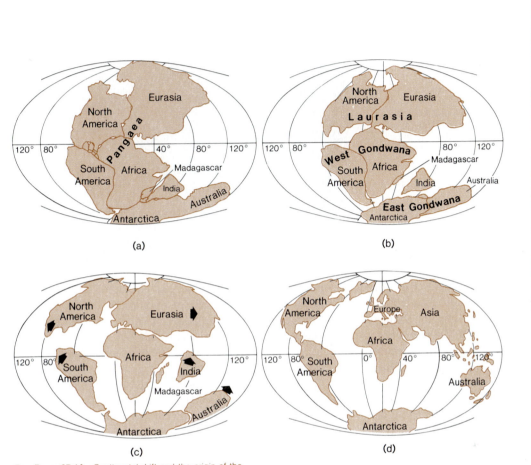

(a)

(b)

(c)

(d)

Box figure 27.1A   Continental drift and the origin of the continents. (a) About 225 million years ago, early in the Mesozoic era, all of the earth's land existed as a single massive continent, which geologists have named Pangea. (b) About 135 million years ago, early in the Cretaceous period, Pangea had broken up into a northern supercontinent called Laurasia and a southern supercontinent called Gondwana, which itself split into east and west parts. (c) About 65 million years ago, Laurasia remained a single continent, but Gondwana's breakup was complete, and its parts were drifting apart toward their present locations (arrows indicate direction of motion of the various plates). (d) The present arrangement of the continents.

Figure 27.8 A trilobite fossil from central Utah. Trilobites have been extinct for millions of years and are only known from the fossil record, but trilobites were once very numerous in terms of numbers of species and numbers of individuals. While the fossil record often is described as being fragmentary or incomplete, a sufficient number of fossils have been found to permit identification of nearly 4,000 species of trilobites and allow study of juvenile forms of some of those species.

Actual fossilization can occur in one of several ways. Water circulating in the sediments may slowly dissolve the water-soluble calcium of shell or bone and leave another mineral deposited in its place. In this process, called mineralization, a durable fossil replica of the original material remains in the sedimentary rocks.

Fossilization also occurs when the original shell or bone is completely dissolved away and removed from the sediment layers by groundwater, leaving behind a perfect mold of the organism. At times, a cast is formed when minerals are later deposited in the hollow chamber of the mold. Both casts and molds may retain a surprising amount of surface detail.

Other fossils include footprints or the skin impressions of organisms made in wet mud that later hardens to form a shale.

Interpretation of the Fossil Record

Even though knowledge of the past is limited to those relatively few fossils that have been unearthed and studied, paleontologists have done a remarkable job of reconstructing the history of life. From these reconstructions, certain definite trends can be observed. The first life forms to appear in the fossil record were primitive single-celled organisms resembling modern bacteria. In time, more complex cellular forms appeared, followed by multicellular plants and animals. The increase in the complexity of life forms is seen as evidence of an evolutionary sequence—the development of new life strategies continually being put to the test of natural selection. Successful innovations—nervous tissue, circulatory systems, endo- and exoskeletons, sexual reproduction—all provided adaptive advantages or they would not have persisted. Throughout the long period of trial and error, the great majority of species failed to meet the challenge of adaptation and became extinct (figure 27.8).

In addition to a general increase in structural complexity, the fossil record shows clear evidence of gradual change within single lines of descent. The classic example of such change is the evolution of horses. The fossil record of horses covers a period of some 60 million years (figure 27.9). The earliest fossil horse was the dog-sized *Hyracotherium* (popularly known as *Eohippus*), a foot-high animal with four toes on its front feet and three on its hind feet. Based on the structure of its teeth, *Hyracotherium* is thought to have fed on forest underbrush. Following *Hyracotherium,* there appeared a number of species of horses, not all on the main line of evolution to the modern horse *(Equus)*. From the variety of horses appearing in the fossil record, certain trends in horse evolution are apparent. First, the molar teeth gradually enlarged and became flatter and higher crowned, thus making horses better suited for feeding on grasses. Second, the leg bones became fused, providing increased strength in running. And third, the digits of both the forelimb and hindlimb were gradually reduced in number. In the modern horse one digit remains, and the hoof is formed from the nail of this digit. The fusion of the leg bones and reduction of digits were accompanied by an overall elongation of the limbs, which provided greater speed and strength

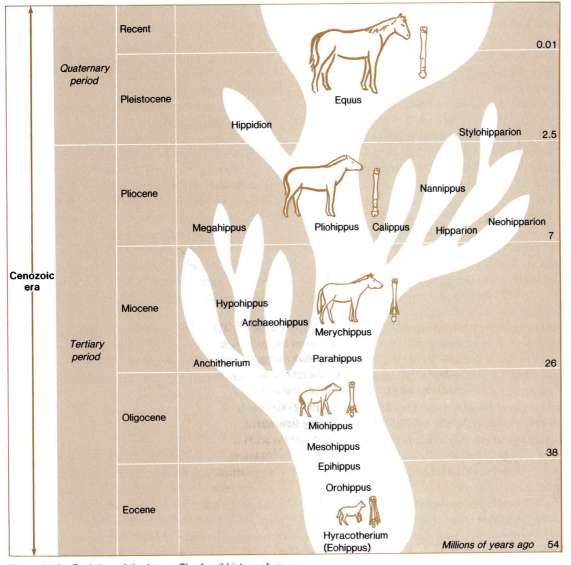

Figure 27.9 Evolution of the horse. The fossil history of horses dates back to early Cenozoic times, about 60 million years ago. (The Cenozoic era is divided into periods, which are subdivided into epochs—see figure 27.12 for more information on the geologic time scale.) The modern horse has evolved from the dog-sized *Hyracotherium* (also called *Eohippus*) via a number of intermediate forms. Note the gradual reduction in the number of digits of the limbs. A number of side branches of horse evolution have ended in extinction.

when running over rough ground. In short, the horse became specialized to the life of a grassland grazer. These changes were gradual, and they occurred in a continuous sequence of events, which are depicted in the fossil history of the equine group.

Adaptation to a particular life-style involves a variety of individual traits. In the horse, for example, tooth size and shape, the elongation and fusion of limb bones, increase in body size, neck extension, and change in skull structure are just a few of the changes that occurred during the transition to a different life-style. This illustrates that evolutionary changes occur in conjunction with one another, as *sets* of traits rather than independent traits. It should be noted that the modern horse is a highly specialized animal, molded by natural selection to life on open grasslands. Such specialized animals lack the genetic flexibility to meet changing conditions and are likely candidates for extinction if their environment is modified significantly.

The fossil record clearly indicates that the major groups of organisms appeared on earth in a sequential pattern. The reasonable assumption is that the older groups gave rise to the younger groups. However, observational evidence is required to support this assumption, and there are some "missing links"—forms intermediate between major groups of organisms that are missing from the fossil record.

However, a number of such intermediate forms have been found. For example, among the vertebrate animals, fossils of intermediate forms support the idea that each of the major groups of vertebrate animals have evolved from older forms. A well-known example of such a transitional form is the fossil species *Archeopteryx,* a pigeon-sized animal that possessed true wings with feathers, the beak of a bird, and a skeletal structure that partially resembled that of the modern bird. But *Archeopteryx* had reptilian claws, and its skeleton, while birdlike, was persistently reptilian in most aspects (figure 27.10).

Another fossil form that had characteristics intermediate between major vertebrate groups is *Seymouria,* an amphibian whose skeletal structure shows features that are a mixture of reptilian and amphibian characteristics. In fact, the skeleton of *Seymouria* is so intermediate in form that it has alternately been classified as a reptile and an amphibian by paleontologists.

These specific forms that have been found may not be the precise species that gave rise to the birds or reptiles, but they reflect periods during which conditions favored change, and one result of this change was the formation of entirely new types of organisms.

It should be mentioned at this point that there is some debate as to the tempo of evolutionary change. Many evolutionists believe that the gradual transition in form represented by the fossil record of the horse is typical of evolutionary modification. If this view is correct, then a large number of intermediate forms should be found in the fossil record. However, other evolutionists believe that there have been long periods of **stasis** in the fossil records, periods during which species displayed little morphological change. According to this theory, new species form very rapidly, during periods of environmental stress. Such a relatively rapid production of new forms would leave few transitional forms in the fossil record (see the discussion of **punctuated equilibrium,** chapter 28). This would make it highly unlikely that fossil "missing links" between certain groups would ever be found.

Relative Dating Techniques

The effective use of the fossil record to represent sequential changes that have occurred in the history of life on earth requires that actual dates be assigned to the fossil organisms comprising this record. Paleontologists use two types of dating to determine the age of fossil-bearing rock layers—relative dating and absolute dating. Absolute dating is the method of choice but can only be used when radioactive materials are present in the rock sequence to be dated. Since this is not always the case, relative dating techniques must often be used.

Relative dating techniques are not new. In the seventeenth century Nicolaus Steno established the principle of **superposition** for the relative dating of sedimentary rock layers. This principle states that in any undisturbed sequence of sedimentary rocks, the bottommost layer is the oldest, and the topmost layer is the youngest. This interpretation follows from the manner in which sedimentary layers are formed. When a river reaches the ocean, it deposits its sediment load on the ocean floor, and over a number of years, hundreds of meters of sediments can accumulate. Eventually, these sediments may be

(a)

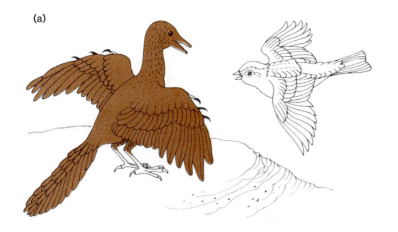

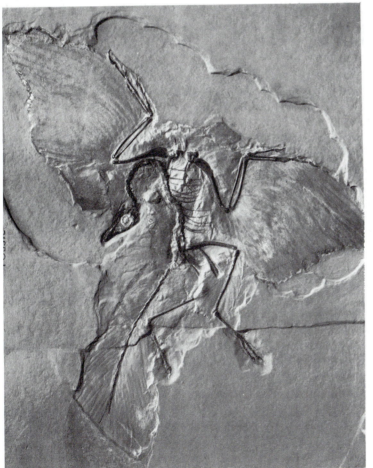

(b)

Figure 27.10 (a) Comparison of a modern bird and a representation of *Archeopteryx* based on fossils. Like modern birds, *Archeopteryx* had feathers on its body and wings, but it had jaws with teeth, clawed digits at the margins of its wings, and various skeletal features that were intermediate between those of reptiles and modern birds. (b) A fossil of *Archeopteryx*.

converted into sedimentary rock, and the oldest sediments (and hence the oldest rock) is at the bottom of the rock sequence. Therefore, in any undisturbed sequence of sedimentary rock, fossils found in the bottommost layer are older than fossils found in the layers above.

A second method of determining the relative age of fossils was developed by William Smith, a geologist, surveyor, and engineer, who traveled extensively through the British Isles in the late 1700s, constructing roads and canals. Smith's interest in geology was a practical one, since he was concerned with the physical properties of the rock layers he encountered in his work. At that time there were no accurate geological maps of the British Isles, and Smith had no way of knowing whether sandstone he encountered in Wales was of the same origin, and thus possessed the same physical properties, as sandstone found in the east of England. Such knowledge would make the job of the road engineer much simpler, and Smith searched for a method to correlate strata from one geographic region with another.

Smith was aware that many of the sedimentary strata of Britain contained fossils, and he also noted that in some sequences of sedimentary rock, the kinds of fossils in the bottom layers differed from those in higher layers. Because he had encountered similar fossil groupings in sedimentary layers occurring in widely separated parts of Britain, Smith wondered if all rocks containing similar fossils were of the same geological age. If so, then the fossil assemblage of a particular rock layer would provide the means of correlating the age of rock layers occurring in widely separated geographical regions.

Smith's idea proved to be a fruitful one and became known as the principle of **fossil correlation,** a valuable tool used by geologists to this day.

These two principles—superposition and fossil correlation—were the only dating techniques available to geologists before the twentieth century. They were used to painstakingly piece together the fossil record of life in the form of a set of correlations between contents of fossil-bearing rock layers (strata) and their relative ages. No one area on earth contains a continuous sequence of rock strata from beginning of the earth to the present. Neither does any one location contain a continuous sequence of fossils. Instead, a short period of earth history is represented in an outcrop in Australia, another in Africa, still another in Europe or North America, and so forth. The task confronting geologists and paleontologists was to place these scattered rock layers in the proper time sequence.

Superposition was used to determine the relative ages of layers within a particular rock formation, and if similar fossils were found in the layers of two separate regions, then it was assumed that those layers were formed during the same geological period. By carefully examining sedimentary layers from all parts of the world and correlating the fossils they contained, a continuous sequence of rock strata and fossil life was attained. This record became known as the geological time scale (table 27.1). A major shortcoming of this system, however, was that there was no way of assigning an absolute age to any single rock layer or fossil. With the discovery of radioactivity and the characteristics of radioactive decay, however, a means was devised to assign specific dates to certain rock layers.

Absolute Dating

Radioactive atoms have unstable atomic nuclei. When these nuclei break down, they emit characteristic particles or rays. The end result of this radioactive decay is that another kind of atom is formed. For example, atoms of ^{238}uranium (^{238}U) break down, through several intermediate stages, to form atoms of ^{206}lead (^{206}Pb). If a rock sample contained 100 percent uranium atoms and no lead atoms, all of the ^{238}U atoms would eventually be converted to ^{206}Pb, but this process would take a very long time. In fact, it takes 5,500 million years for one-half of the ^{238}U to change to ^{206}Pb. This period of time, called the **half-life,** is the time interval commonly used in radiometric dating.

Every radioactive substance has its own unique half-life. Radioactive potassium (^{40}K) has a half-life of 1,300 million years (breaking down to form ^{40}argon (^{40}Ar)). If a rock sample contains ^{40}K and ^{40}Ar, an exact measurement of the proportion of ^{40}K to ^{40}Ar yields the absolute age of that rock sample. For example, if the rock sample originally contained 1,000 grams of potassium, after 1,300 million years it would contain 500 grams of ^{40}K and 500 grams of ^{40}Ar. If a rock sample contained 250 grams of ^{40}K and 750 grams of ^{40}Ar, it would have passed through two half-lives and would be 2,600 million years old.

Table 27.1
The Geological Time Scale. Major Divisions of Geological Time with Some of the Major Evolutionary Events of Each Geological Period.

Era	Period	Millions of Years Ago	Major Biological Events	
			Plants	*Animals*
Cenozoic	Quaternary		Rise of herbaceous plants	Age of humans
		2.5		
	Tertiary		Dominance of the angiosperms	First hominids Rise of modern forms Mammals and insects dominate the land
		65		
Mesozoic	Cretaceous		Spread of angiosperms Decline of gymnosperms	Extinction of the dinosaurs
		130		
	Jurassic		First flowering plants (angiosperms)	Age of dinosaurs First mammals and first birds
		180		

	Triassic		Land plants dominated by
			  gymnosperms
				First appearance of the
				  dinosaurs

		230		
Paleozoic	Permian		Land covered by forests of primitive vascular plants	Expansion of the reptiles Decline of the amphibians Extinction of the trilobites
		280		
	Carboniferous		Land covered by forests of coal-forming plants	Age of amphibians First appearance of reptiles
		350		

	Devonian		Expansion of primitive
			  vascular plants over land
				Fishes dominate the seas
				First insects
				First amphibians move
				  onto land

		400		
	Silurian		First appearance of primitive vascular plants on land	Expansion of the fishes
		435		
	Ordovician		Marine algae	Invertebrates dominate the seas First fishes (jawless)
		500		
	Cambrian		Primitive marine algae	Age of invertebrates Trilobites abundant
		600		
Precambrian			*Aquatic Life Only* Origin of the invertebrates Origin of complex (eukaryotic) cells Origin of photosynthetic organisms Origin of primitive (prokaryotic) cells Origin of life	
		4,600		

There are several problems with radiometric dating. One is the possibility that in the thousands or millions of years a rock has existed, some of the radioactive isotope or its decay product may have been weathered from the rock. In such a case an erroneous date would be obtained. Another problem is that the best samples of radioactive materials for dating are found in rocks of volcanic origin (igneous rocks). Fossils are very rarely found in igneous rocks because the heat of the molten material from which these rocks are formed destroys organic matter rather than preserving it. For this reason, the dating of fossils often depends on finding an igneous rock layer in close association with a sedimentary rock formation. When this occurs, an absolute date can be assigned to a particular fossil species found in one location, and then fossil correlation can be used to assign an absolute age to all strata containing the same species.

Carbon dating is a means of dating fossils that still contain some carbon material from the living organism. The radioactive isotope ^{14}carbon (^{14}C) has a half-life of 5,760 years, and it breaks down to form a stable isotope of nitrogen (^{14}N). ^{14}Carbon is formed in the earth's atmosphere through the action of cosmic radiation from the sun on nonradioactive nitrogen, which is converted to ^{14}C. The ^{14}C in the atmosphere, in the form of ^{14}CO$_2$, is incorporated into organic molecules in living tissue through the process of photosynthesis (see chapter 6), and when plants are eaten by animals, the ^{14}C is incorporated into the animals' tissues. When an organism dies, carbon exchange with the environment ceases. Any subsequent change in the ratio of ^{14}C to stable carbon isotopes in its tissues is due to radioactive decay of ^{14}C to nitrogen.

The assumption is made that the rate of carbon-14 formation in past times was similar to that occurring today and that the ratio of ^{14}C to stable isotopes of carbon (mainly ^{12}C) incorporated into living material was the same in the past as it is now. Then the emissions that are characteristic of ^{14}C (beta particles) being released from a sample can be counted, the total carbon content of the sample determined, and the age of the sample calculated. Age is a function of the ratio of ^{14}C to total carbon content; the smaller the ratio of ^{14}C to total carbon content, the older the sample.

One problem associated with the use of the ^{14}C technique is that the relatively short half-life of ^{14}C (about 5,760 years) make this technique valid only for fossil materials dating back to about 50,000 years ago. Also, it is believed that the influx of cosmic radiation has varied from time to time in earth history. Thus the rate of ^{14}C formation from nitrogen may not have been consistent, and the photosynthetic uptake of ^{14}C by plants may also have varied.

Overall, the use of radiometric dating has proven to be a valuable tool to both the geologist and the paleontologist. The age of the earth has been estimated at 4,500 to 5,000 million years, and reasonably accurate dates have been assigned to many of the major events of earth history and to the majority of fossil species appearing in the fossil record. The oldest known fossils are primitive bacterialike organisms and date to approximately 3,000 to 3,500 million years ago.

Comparative Anatomy

Another invaluable instrument in paleontological interpretation is **comparative anatomy.** Comparative anatomists try to find similarities and differences among the fundamental structures of living organisms. Their studies have revealed certain basic plans upon which groups of organisms are structured. For example, all vertebrate animals have a central skeletal axis composed of the skull and vertebral column. A rib cage, made up of ribs attached to vertebrae, protects the heart and lungs. Many vertebrates also have paired appendages. The basic plans of the vertebrates' circulatory, respiratory, digestive, and other systems have similarities. There are, however, important variations on these basic plans that distinguish the major groups of vertebrate animals from one another. These similarities and variations are the areas of interest for the comparative anatomist.

Since all species of organisms have arisen from previously existing species, it follows that those species that share a recent common ancestor possess a large number of gene alleles in common. Over time, the number of genetic differences increases, and the species become genetically and morphologically more dissimilar. All of the similarities and differences studied by comparative anatomists are products of gene action and thus yield information about the heredity of organisms and species.

Box 27.2
Stable Isotopes and Dinosaurs' Diets

Studies of radioactive isotopes have been used for some time to date fossils, but more recently developed techniques for study of naturally occurring, stable (nonradioactive) isotopes have opened the door for other kinds of fossil research.

Instruments known as mass spectrometers can separate quantitatively the common stable carbon isotope (^{12}carbon) from the much less abundant stable carbon isotope (^{13}carbon), thus permitting calculation of the relative abundance of the two isotopes in organisms from which samples are taken. This technique can be very informative because of the way in which the two isotopes enter living material. During photosynthesis, there is discrimination in favor of $^{12}CO_2$ as opposed to $^{13}CO_2$ by the enzyme systems that incorporate carbon dioxide into organic molecules. But the degree of discrimination is different in the two major photosynthetic pathways for carbon fixation, the C_3 and C_4 pathways (chapter 6). Organic material produced by C_3 plants has a different $^{13}C/^{12}C$ ratio than material produced by C_4 plants.

Furthermore, the $^{13}C/^{12}C$ ratios in plant material seem to be perpetuated in the cells and tissues of the animals that eat the plants. It, therefore, is possible to use $^{13}C/^{12}C$ ratios in tissue or even in feces to determine whether an animal has been eating mainly plants that use the C_3 photosynthetic pathway, plants that use the C_4 pathway, or a mixture of the two.

A number of researchers are using these techniques to study dietary patterns in modern animals, but other geologists and biologists, including M. J. DeNiro, S. Epstein, and L. L. Tieszen, are using these same techniques on fossil organisms. There is, for example, enough carbon preserved in some animal fossils to allow determination of $^{13}C/^{12}C$ ratios for those organisms. Such data can indicate something about food preferences of the organisms or about the food sources available to them. Although results of these studies can be difficult to interpret, biologists are able to partially analyze the diets of long dead animals, including even members of extinct species.

Fundamental similarities in structure that have arisen through descent from a common genetic ancestor are called **homologies.** The classic example of homology is the vertebrate forelimb (figure 27.11). All vertebrates are thought to have evolved from a common ancestor, and despite the fact that bird, whale, and human forelimbs serve different functions, they all possess the same fundamental bone structure. While comparative anatomists can compare many anatomical features of contemporary animals, the study of comparative skeletal anatomy is most important to the vertebrate paleontologist because fossil evidences of anatomy are almost entirely limited to skeletal material.

Analogy is another important concept derived from comparative anatomy. Analogous structures are similar in function and appearance but differ in their fundamental structural plan. The wings of birds and insects are analogous structures. In both cases, the organs are used for flying, and each has the broad, flattened shape

that is well adapted to this function. However, the two kinds of wings are structured differently. The insect wing is a stiffened membrane supported by hard chitinous veins, while the bird wing has an internal skeleton covered by skin and feathers.

Analogous structures differ in basic form because they have arisen separately in the evolution of life forms, and the different kinds of organisms possessing them do not derive from the same genetic ancestor. The ability to fly evolved completely independently in birds and in insects. Superficial similarities in the structure of their wings are due to the fact that a broad, flat surface is a prerequisite for flight. Thus, the ability to fly has appeared several times in the history of life, apparently because flight provides a real advantage to an organism: it permits the animal to flee from threatening conditions, to seek food over a broad range, and to migrate to avoid harsh climatic conditions. Analogous structures illustrate

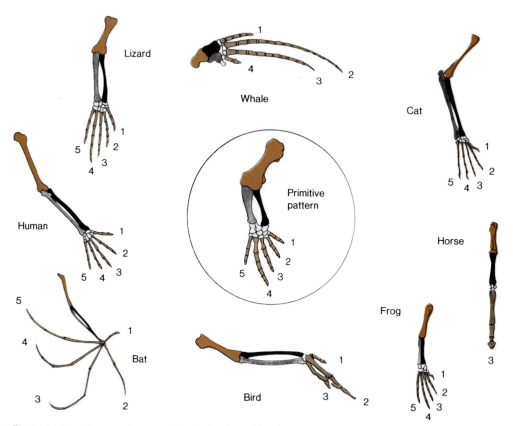

Figure 27.11 Homology in the vertebrate forelimb. All of these forelimbs are built on a fundamentally similar framework of the same set of homologous bones. This indicates that they share a common ancestry even though the limbs now look very different from one another and are adapted to a variety of functions.

Homologous bones are indicated as follows: humerus (upper arm)—color; radius (forearm)—black; ulna (forearm)—gray; carpals (wrist)—white; metacarpals (palm) and phalanges (digits)—heavy stippling. The digits are numbered, beginning with the first digit (thumb).

Figure 27.12 Comparison of a bird wing, which has feathers attached to skin covering a skeletal framework, and the stiff, membranous insect wing as an example of analogy. Structures that have evolved from very different evolutionary precursors and have different basic structures serve the same function. This independent evolution of wings as an adaptation for flight is an example of convergent evolution.

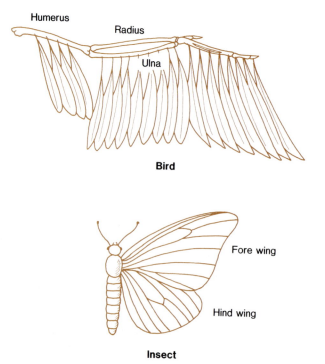

convergent evolution—evolution toward a common adaptation to a similar life-style (figure 27.12).

Comparative Embryology

Studies of the developmental stages of vertebrate animals reveal similarities that are not apparent in the adult stage. Early embryonic stages of fish, reptiles, birds, and mammals are surprisingly similar. In the past, these similarities were interpreted to mean that, among vertebrate animals, "ontogeny recapitulates phylogeny." That is, the stages of embryonic development (ontogeny) of an organism repeat the evolutionary history of the species (phylogeny). This notion has been discarded in modern evolutionary theory because it is obvious that embryos do not, in fact, pass through stages that resemble adult ancestors. Instead, developing embryos of each major vertebrate group pass through some stages that resemble embryos of the other vertebrates.

Vestigial Structures

Vestigial structures are those that appear in a rudimentary form in some organisms but in a fully developed, functional form in other, closely related animals; or they are poorly developed structures thought to have been fully functional in ancestors of the modern form. Some examples of vestigial organs are: the rudimentary, functionless eyes of cave-dwelling animals; the wisdom teeth of humans; and the presence of three to five caudal vertebrae in humans (remnants of the tail structure found in most mammals but absent in higher primates).

One explanation for the decline of organs whose adaptive value to the organism has diminished is that the development and maintenance of body organs requires energy. If a particular organ has no function, animals that use less energy to produce and maintain the unneeded structure have a selective advantage because they have more energy to carry on other, more adaptive functions. Thus, natural selection favors those individuals in which the unneeded organ is reduced. For example, there is selection against full expression of the genes for eye development in cave-dwelling animals. Very slowly, over long periods of time, natural selection may completely remove such genes from populations. However,

it appears that this process may be quite slow and incomplete and that suppression of gene expression may occur long before genes actually are eliminated from the genome. A good example of this is the discovery that chickens retain genes for development of tooth dentine, even though these genes are never expressed in modern chickens, which do not have teeth.

Molecular Evolution

Organisms possessing homologous structures are assumed to have descended from a common genetic ancestor. Therefore, genetic similarities should exist among these organisms. For example, the forelimbs of all terrestrial vertebrates are considered to be homologous because they possess common skeletal elements—the hereditary legacy of the early amphibians, in which the basic vertebrate forelimb structure evolved.

In addition to homologies found in such anatomical structures as the vertebrate forelimb, biochemists have found many homologies at the molecular level. Virtually all living things possess the same genetic material (DNA), with information spelled out in the same genetic code, and conduct energy conversions involving the same molecule (ATP). Many other molecules, including a number of enzymes, are found widely distributed among living forms.

The fact that practically all living things possess DNA as genetic material is interpreted as meaning that this trait developed very early in the evolution of life and has been passed down through the millenia to all organisms that have followed. The DNA method of hereditary transmission and genetic control is efficient and effective, and thus there seems to be little selective pressure favoring change in fundamental hereditary mechanisms.

Protein homologies have been of special interest to the biochemist because proteins are direct products of gene expression. Modern laboratory techniques permit biochemists to ascertain the exact amino acid sequence of protein molecules, and from this information, the makeup of genes can be inferred. Because proteins are the products of gene expression, it can be inferred that two organisms that possess similar proteins must also possess similar genes. This information is a valuable tool in the study of evolutionary relationships because possession of similar genes is indicative of common ancestry.

At the biochemical level, the evolutionary process, as revealed by study of protein molecules, has been very conservative in some cases and much less conservative in others. The amino acid sequences of certain histones (proteins closely associated with DNA in chromosomes) have been highly conserved. The amino acid sequences of one of them, histone IV, are virtually identical in molecules extracted from a wide range of organisms. For example, amino acid sequences in histones from such divergent sources as pea plants and cattle differ by only two out of the 102 amino acids in the molecules. Apparently, the proper functioning of histones depends on maintenance of a very specific amino acid sequence in the molecules, and there has been strong selective pressure against change in histone amino acid sequences. Amino acid substitutions that may have appeared in the past as a result of mutation probably were eliminated by natural selection.

Other proteins have not been so highly conserved, however, and it is possible to use amino acid sequences of such proteins to determine the closeness of evolutionary relationships among different species. If, through protein analysis, it is determined that the amino acid sequences in proteins of two species are very similar, it is assumed that these two species have only recently evolved from a common genetic ancestor. If, on the other hand, the proteins have very different amino acid sequences, it is assumed that the two species are only distantly related (that is, they have been separated from a common genetic ancestor for a long period of time).

These assumptions are possible because gene mutations that change DNA base coding sequences are expressed as variations in the structure of proteins (changes in amino acid sequences) and because gene mutations occur at relatively constant rates. Thus, the accumulation of protein dissimilarities through gene mutation has been used to estimate when ancestors of present-day species diverged from one another. Data from studies on cytochrome *c*, an electron carrier molecule from mitochondria, illustrate this principle. The amino acid sequences in cytochrome *c* from humans and chimpanzees are identical, indicating relatively recent divergence of ancestors. Humans and chimpanzees differ from rhesus monkeys by only one amino acid out of the 104 that constitute cytochrome *c*, but they differ from dogs by eight amino acids, from rattlesnakes by 12, and from dogfish sharks by 24.

Using mutation rates of various genes, each of which is different, it has been possible to calculate the actual time since divergence. It is interesting to note that, in general, phylogenies (lines of evolutionary descent) developed using this "molecular clock" agree very well with most phylogenies developed by paleontologists using the fossil record and radiometric dating techniques.

Summary

The theory of organic evolution states that all species of plants and animals are the modified descendants of previously existing forms. Charles Darwin formalized the theory of organic evolution in 1859 in his book *On the Origin of Species*. Darwin's work was based on a long period of observation, experimentation, and contemplation. His ideas about evolution began during his five-year tour of duty as ship's naturalist aboard the H.M.S. *Beagle*. The fossils of South America and the great biological variation he observed in the Galápagos islands convinced Darwin that nature molds living forms to their environment and that modern species must be the modified descendants of extinct life forms.

Darwin also envisioned the process by which organic evolution took place—a process he called natural selection. Each species produces more young than can survive, and among the range of biological variants appearing in any species populations, some individuals are better adapted than are others. These better-adapted individuals make up the greater percentage of the breeding population of any generation, and thus it is likely that succeeding generations will contain a high percentage of individuals with the adaptive traits.

Darwin's theory generated a great deal of interest in evolution, among scientists and nonscientists alike, and it has stimulated much research on evolutionary processes in the years since its publication. The bulk of the scientific evidence gathered over the years has strongly supported the theory of organic evolution and Darwin's theory of natural selection. Support has come from a variety of scientific disciplines, including biogeography, paleontology, comparative anatomy, comparative embryology, genetics, and molecular biology.

Biogeography is the study of geographic distribution of plants and animals. Each major region of the world has its own characteristic flora and fauna. Biogeographic evidence indicates that different environments have exerted different selective pressures during the evolution of the living things that occupy them.

Paleontology is the study of fossils and the fossil record. Although vast quantities of fossils have been found, the fossil record is incomplete probably because of the large element of chance involved in the various means by which fossils form. The fossil record can be used to reconstruct a general history of life on earth and to trace development within certain lines of descent. Fossils also document the past existence of transitional and intermediate forms.

Effective use of the fossil record requires dating of the fossil organisms that constitute the record. Relative dating can be done using the principle of superposition and the fossil correlation of strata. Absolute dating is based on known rates of radioactive decay of various unstable isotopes that are found in fossil organisms or in layers near them. Uranium-lead dating and potassium-argon dating permit age determinations ranging back thousands of millions of years. Carbon dating permits dating of more recent fossils that still contain carbon from the organism. Effective carbon dating is valid only back to about 50,000 years ago.

Comparative anatomy, comparative embryology, and the study of vestigial structures all provide data on evolutionary relationships. Comparative anatomy is especially useful for study of evolution of skeletal structures, which are well represented in the fossil record.

Studies of molecular evolution are based on comparisons of molecular structures, mainly amino acid sequences in proteins, found in present-day organisms. Amino acid sequences are most similar in species that have relatively recently evolved from a common ancestor and much less similar in species that have been separated from a common genetic ancestor for a long period. Results of this research have largely confirmed phylogenetic relationships suggested by other types of research on the evolutionary process.

Questions

1. What is the theory of uniformitarianism? How did it influence Darwin's thinking about the history of life on earth?
2. What is adaptive radiation? Under what conditions does it occur?
3. Describe Lamarck's theory of inheritance of acquired characteristics and explain why it is not generally accepted by scientists.
4. What is Darwin's theory of natural selection? What is the essential difference between artificial selection and natural selection?
5. Explain why the fossil record does not represent a complete history of life on earth.
6. List some fundamental differences between uranium-lead dating and ^{14}carbon dating.
7. What is the difference between homology and analogy?

Suggested Readings

Books

Darwin, C. 1859. *On the origin of species*. Reprint. 1975. London: Cambridge University Press.

de Beer, G. 1965. *Charles Darwin: A scientific biography*. Garden City, N.Y.: Anchor Books, Doubleday Book Co.

Stebbins, G. L. 1977. *Processes of organic evolution*. 3d ed. Englewood Cliffs, N.J.: Prentice-Hall.

Articles

Grant, P. R. 1981. Speciation and the adaptive radiation of Darwin's finches. *American Scientist* 69:653.

Lack, D. April 1953. Darwin's finches. *Scientific American* (offprint 22).

Mayr, E. September 1978. Evolution. *Scientific American* (offprint 1400).

Ostrum, J. H. 1979. Bird flight: How did it begin? *American Scientist* 67:46.

Valentine, J. W. September 1978. The evolution of multicellular plants and animals. *Scientific American* (offprint 1403).

Modern Concepts
of Evolution

The modern theory of organic evolution is a union, or synthesis, of the theories of organic evolution and natural selection with modern concepts of genetics. Modern genetics has provided information that was not available to Darwin and his contemporaries about sources of variation in populations and about the mechanisms by which those variations are passed from generation to generation. This knowledge has helped to explain how natural selection can function as a causative mechanism that brings about evolutionary change.

The primary source of variation is the genetic segregation and recombination (see chapter 19) that occurs in sexually reproducing organisms. Another important source of variation is gene mutation, which represents the only source of *new* genetic material in a species. Natural selection operates on genotypic variation that is exposed to the environment as phenotypes. Under any set of environmental conditions, some phenotypes are better adapted than others, and this situation results in a differential survival of phenotypes. Even though natural selection operates on phenotypes, gene alleles confer adaptive traits on individual organisms and, in a very real sense, genes are being selected. Natural selection can best be explained in terms of its effect on the genetic composition of a population. In fact, one definition of organic evolution is any change in the frequency with which various gene alleles occur in a population.

Population Genetics

A **population** is an interbreeding group of individuals that occupy a specific geographical area. Because populations may be quite large, numbering in the thousands or millions of organisms, the fate of an individual organism has little influence on the fate of the population as a whole. Thus, modern biologists study changes taking place in an entire population, and evolutionary studies concentrate not on the genotypes of individual organisms, but on the **gene pool,** the sum total of all genes available for reproduction in a population. The science of population genetics has grown up as a mathematical study of the events taking place in gene pools, and **population genetics** forms a fundamental part of modern evolutionary thought.

The Hardy-Weinberg Law

The basis for study of population genetics is a simple set of mathematical concepts called the **Hardy-Weinberg law** (named for G. H. Hardy, an Englishman, and W. Weinberg, a German, who independently developed and proposed these concepts in 1908). The Hardy-Weinberg law is a mathematical model of a gene pool that is at equilibrium under certain idealized conditions, which are considered later. The law shows how genotype and phenotype stability can be achieved in a population. It examines the frequency of occurrence of alleles of various genes in the total gene pool of a population, but it is not concerned with genotypes or phenotypes of specific individual organisms within the population.

Although any individual may carry only two alleles of a given gene, many different alleles of that gene may be present in a population. However, a simplified hypothetical example can best be used to illustrate the Hardy-Weinberg law. A gene that has only two alleles, a dominant allele *A* and a recessive allele *a,* present in an entire population of sexually reproducing organisms is used in this example. In this situation, three genotypes—*AA, Aa,* and *aa*—would be possible in the population. Two of those genotypes, *AA* and *Aa,* would produce the same phenotype.

The Hardy-Weinberg law examines population gene pools, not individual genotypes, and is concerned with frequencies of occurrence of alleles in the total gene pool. The frequencies of the hypothetical alleles *A* and *a* in the example have been arbitrarily set at 0.6 and 0.4, respectively. Since the totals of allele frequencies always equal 1.0 (0.6 + 0.4 = 1.0 in this example), allele *A* makes up 60 percent of the total of the two alleles in the population, while allele *a* constitutes 40 percent of the total. (Decimal frequencies and percentages, of course, are interchangeable.)

If allele *A* has a frequency of 0.6, it follows that 60 percent of all sperm cells and all egg cells produced in the population carry allele *A*. Similarly, allele *a,* which has a frequency of 0.4, is represented in 40 percent of all gametes produced in the population. These gamete allele frequencies can then be used in a Punnett square to determine the outcome of mating in this hypothetical population (figure 28.1).

Because allele *A* is dominant, 84 percent of the offspring (all those with genotypes *AA* or *Aa*) produced by mating in this population display the phenotype produced by expression of the dominant allele, while only 16 percent display the phenotype produced by expression of the recessive allele.

Will the more frequently occurring allele, allele *A* (which is also dominant in this case), tend to occur even more frequently in subsequent generations, at the expense of allele *a?* Do less frequently occurring alleles tend to become lost from gene pools over long periods of time? The answer to these questions provided by the Hardy-Weinberg law is that if the population meets a particular set of qualifications, there will be no change in relative frequencies of alleles in populations. Even very rare alleles can continue to occur at constant frequencies generation after generation and are by no means doomed to disappear from the gene pool simply because they are rare.

For example, the summary of allelic frequencies in the gene pool of the generation produced by the first hypothetical mating in figure 28.1 shows that the relative allele frequencies are the same in the offspring generation as in the parental generation. Thus, generation after generation of reproduction under these same conditions does not change the relative frequencies of these two alleles *A* and *a* in this gene pool.

The Hardy-Weinberg law provides a general mathematical statement of this type of equilibrium in the relative frequencies of alleles. In this general statement, the frequency of one allele is represented by the letter *p* and the frequency of the other by the letter *q*, and $p + q = 1$. These symbolic gene frequencies can be inserted into a Punnett square as gamete frequencies, or the same results can be obtained algebraically (figure 28.2). The results represent expansion of the binomial expression $(p + q)^2$. Because the sum of

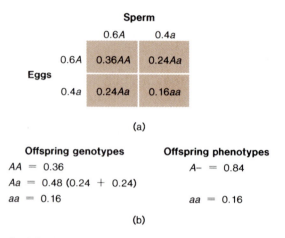

(a)

Offspring genotypes

AA = 0.36
Aa = 0.48 (0.24 + 0.24)
aa = 0.16

Offspring phenotypes

A– = 0.84

aa = 0.16

(b)

Contributions to frequency of alleles in gene pool in offspring generation

	Allele *A*	Allele *a*
AA individuals	0.36	0
Aa individuals	0.24	0.24
aa individuals	0	0.16
Totals	0.60	0.40

(c)

Figure 28.1 Allele frequencies and the Hardy-Weinberg law. (a) A Punnett square illustrating the effects of totally random mating in a parental population in whose gene pool allele *A* has a frequency of 0.6 and allele *a* a frequency of 0.4. (b) Determination of offspring phenotypes in the offspring generation. A simple dominant-recessive relationship between these alleles is assumed. (c) Analysis of allele frequencies in the gene pool of the offspring generation, demonstrating the equilibrium in relative allele frequencies described by the Hardy-Weinberg law.

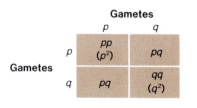

$$(p + q)^2 = p^2 + 2pq + q^2 = 1$$

Figure 28.2 Demonstration of the general Hardy-Weinberg model of allele frequency equilibrium by use of a Punnett square and by algebraic expansion of the binomial expression $(p + q)^2$.

genotypes in any gene pool is 1, the equation that serves as a model of the equilibrium described by the Hardy-Weinberg law is:

$$p^2 + 2pq + q^2 = 1$$

This can be illustrated further by substituting the frequencies of the two alleles (0.6 and 0.4) from the hypothetical gene pool for p and q respectively:

$$p^2 + 2pq + q^2 = 1$$
$$(0.6)(0.6) + 2(0.6)(0.4) + (0.4)(0.4) = 1$$
$$0.36 + 0.48 + 0.16 = 1$$

Thus, the terms of this equation indicate frequencies of genotypes that are identical with those obtained by the Punnett square method in figure 28.1:

$$p^2 = \text{Frequency of } AA = 0.36$$
$$2pq = \text{Frequency of } Aa = 0.48$$
$$q^2 = \text{Frequency of } aa = 0.16$$

Calculations become more complex in situations where more than two alleles are present, but the Hardy-Weinberg law applies equally well in those situations.

The Hardy-Weinberg law is more than a description of equilibrium in situations where relative allele frequencies are known. The Hardy-Weinberg equations also permit other kinds of calculations, such as calculation of the frequencies of the alleles and the other genotypes when the frequency of one genotype is known, although such calculations must be based on the assumption that the population is at genetic equilibrium, which often is not the case.

In the case of cystic fibrosis, an inherited human disorder caused by expression of a recessive gene, the homozygous recessive genotype that results in development of the condition occurs in about one out of every thousand people. If this frequency (0.001) represents q^2 in the Hardy-Weinberg equation, the other frequencies can be calculated. The frequency (q) of the recessive allele for cystic fibrosis in the gene pool equals $\sqrt{q^2}$. In this case, $q = \sqrt{0.001} = 0.03$ (approximately). The frequency (p) of the dominant allele therefore equals $1.0 - 0.03$, or 0.97. According to the equation, the frequency of the heterozygous genotype is $2pq$. In this case, $2pq = 2(0.97)(0.03)$, or about 0.06. Thus, about 6 percent of all people are heterozygous carriers of the recessive allele for cystic fibrosis.

Gene Frequencies and Evolution

The Hardy-Weinberg law is a mathematical model of genetic equilibrium in a population, a model that demonstrates that the genetic recombination that occurs in sexual reproduction as a result of meiosis and fertilization does not by itself produce changes in gene frequencies. But the Hardy-Weinberg model applies only under the following certain conditions:

1. The population must be large enough so that changes in gene frequency do not occur as a result of chance alone.

2. In-migration or out-migration of individuals must not produce any change in the gene pool of the population.

3. Mutations must not occur, or there must be mutational equilibrium, where genetic changes in one direction are balanced by an equal number of changes in the opposite direction.

4. Reproduction must be totally random; every individual in the population must have an equal opportunity for reproductive success. Every individual must have an equal chance of mating with any other individual in the population, all matings must produce the same number of offspring, and all offspring must similarly have equal opportunity for reproductive success.

No natural population meets all of these conditions. It may seem rather startling, after having developed the concepts of the Hardy-Weinberg law, to say that the conditions under which such allele frequency equilibrium is maintained from generation to generation do not exist in nature. But this seeming paradox is really a key to understanding the modern synthetic theory of evolution. The very fact that gene frequencies in the gene pools in natural populations *do not* remain in equilibrium from generation to generation indicates that these natural populations are evolving. Evolution involves changes in gene frequencies.

What are some forces that bring about these evolutionary changes in gene frequencies? Some of these forces can be identified by reexamining the four conditions just mentioned for maintenance of equilibrium in gene pools.

Condition one implies that there can be changes in gene frequency in small populations due to chance occurrences. **Genetic drift**, which is change in gene frequency that is not closely tied to the adaptiveness of the traits involved, can occur in relatively small populations.

Condition two is sometimes met in natural populations but is not met in other cases. Obviously, migration of individuals from one population to another can and does change gene frequencies in gene pools.

Condition three is never met in natural populations because there is always mutation. And even if mutational changes are reversed as frequently as they occur, the condition still is not met. Mutational change is not an isolated entity. The importance of mutation can properly be assessed only in the context of the major factor that affects condition four. That factor is natural selection.

The requirement for random reproduction is never met in natural populations. Many animals, for example, show mating preferences based on phenotypic expressions of size, coloration, and behavioral patterns. Different individuals do not produce equal numbers of offspring. And some offspring have definite selective advantages over others and are more likely to contribute to gene pools of future generations. Thus, natural selection always affects gene pools. Natural selection is the most important factor in bringing about evolutionary change (changes in gene frequencies in gene pools of natural populations). Every other factor that causes evolutionary change must be viewed in the context of natural selection because, as Charles Darwin rightly surmised, it is the major driving force that causes evolutionary change in living things.

Each of the change-inducing factors—genetic drift, migration, mutation, and most importantly, natural selection—are now examined in detail.

Genetic Drift

In large populations, chance events have little effect on gene frequencies, but in small populations, chance may play a much larger role. Genetic drift can occur in small, isolated populations, especially populations with only 10 to 100 breeding-age individuals. Fluctuations due to chance factors often result in the loss of certain

alleles from such small gene pools, while other alleles that have been "fixed" in the gene pool become much more common. These changes occur independently of natural selection and even can result in the loss from a gene pool of alleles that actually would confer a selective advantage. As a result, small populations where genetic drift has occurred tend to have a higher degree of homozygosity than larger populations, which are more variable.

A special case of genetic drift is the **founder effect.** In nature, a few members of a parent population may migrate to a new area and establish a small, interbreeding population (called a **deme**). Because a deme is established by only a few members of the parent population, it is highly unlikely that the gene frequencies of these few individuals are the same as those of the entire population from which they were drawn. It therefore follows that the gene pool of these individuals' descendants will reflect the gene frequencies of the founder organisms rather than those of the entire population.

A commonly used example of the founder effect in human populations is a religious sect descended from the Old German Baptist Brethren. In the United States, they are called the Dunkers (from the German *dunken,* "to dip"). The Dunkers immigrated to the United States from Wittgenstein, Germany, in the early eighteenth century. Twenty-eight individuals founded the community in Pennsylvania in 1723, and approximately 200 additional members joined them a few years later. Since their arrival in the United States, the Dunkers have maintained strict marriage customs that prohibit marriage outside the sect. Thus, the Dunkers have remained an essentially isolated gene pool. Unlike some other religious sects that also have restricted social contacts with nonmembers, the Dunkers do permit the use of modern medical care and use modern technology. Thus, they probably are subject to the same selection pressures as members of the general population of their geographic area.

Under these conditions—that is, a small initial population, genetic isolation from other populations, and no obvious difference in selective pressure—significant differences in gene frequencies between the Dunkers and the United States population as a whole can be ascribed to genetic drift through the founder effect.

Table 28.1
Blood Types in Populations (Percentages)

	A	B	AB	O
United States	40	11	4	45
Dunkers	59	3	2	36
Western Germany	45	10	5	41

Comparisons of the frequencies of genes for several traits, including those expressed in the ABO blood types were conducted by Bentley Glass and his associates at Johns Hopkins University. Three populations were compared: the Dunker population, the population of the United States, and the population of western Germany, from which the original Dunkers emigrated to the United States (table 28.1). The data show that the frequencies of the genes conferring the ABO blood types differ significantly from the Dunker population to the other two studied. Blood type A occurs more frequently among the Dunkers than in the other two populations. The O type is somewhat rarer among the Dunkers, and the B and AB types occur very infrequently. In fact, it was estimated in this study that the I^B gene, which must be expressed to produce either type B or AB, had been virtually lost from the Dunker population because most of the individuals with B or AB blood types were not born in the community but had entered it as converts through marriage to community members.

Dunkers do not differ from the surrounding population, however, in frequency of the Rh blood groups (Rh positive and Rh negative).

Certain directly observable physical traits also were examined (figure 28.3). Frequencies of some traits, such as middigital hair patterns, distal hyperextensibility of the thumb, and attached earlobes were significantly lower among the Dunkers than in the surrounding population. On the other hand, Dunkers have essentially the same incidence of left-handedness as that found in the surrounding population.

Apparently, the special set of gene frequencies found among the Dunkers can be attributed to the founder effect and genetic drift.

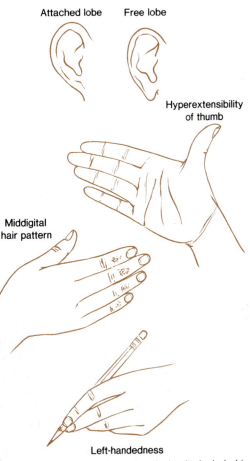

Figure 28.3 Some inheritable physical traits included in the study of gene frequencies, conducted by Bentley Glass and his co-workers, in the Dunker population in Pennsylvania. Traits studied included nature of earlobes, hyperextensibility of the thumb, middigital hair pattern, and left- versus right-handedness. Frequencies of some of these traits and ABO blood types among Dunkers differed significantly from those in the surrounding population. Apparently, these gene frequency differences have arisen through genetic drift.

Migration

Unless there are complete barriers to movement of individuals from one population to another, **gene flow** (gene migration) occurs as a result of migration and interbreeding. Gene flow is a factor that adds to the variability of the majority of natural populations, and it upsets the type of equilibria described by the Hardy-Weinberg law. Of course, the fate of an allele newly introduced into a population depends on natural selection.

Some natural populations, however, do live in complete isolation from other populations and are not subject to the effects of gene flow. Thus, depending on the situation, the relative importance of gene flow ranges from significant effects on gene pool in some cases to no effect at all in others.

Mutation

Mutations can occur as modifications of chromosome structure or, more frequently, as specific changes in individual genes, which result from base substitutions in nucleic acid molecules (chapter 22). Genes have their own characteristic mutation rates, and it is estimated that these rates range from about one mutation for every one hundred thousand (10^5) replications to one mutation per one hundred million (10^8) replications.

Mutation rates attracted the attention of researchers when it was learned that mutation rates of different alleles for the same character rarely are in equilibrium. That is, the rate of forward mutation, which is mutation from the more common allele to the less common allele, seldom is the same as the rate of back mutation, which is mutation in the reverse direction. The difference between the two is a mutation pressure that tends to produce a very slow change in allele frequencies.

When such differences in mutation rates were first discovered, many biologists believed that mutation pressure was important in determining the direction of evolutionary change. Modern evolutionists, however, do not. It would take an enormous amount of time for mutation pressure alone to cause significant changes in allele frequencies. Furthermore, mutations are random events affecting individual genes, but they occur within the context of a whole constellation of genes in a gene pool, and natural selection acts on phenotypes that represent composites produced by expression of these many genes.

The real importance of mutation is that it is the only mechanism by which *new* genetic material enters the gene pool, as some mutations produce new alleles not previously present in the gene pool.

It is important to note, however, that it is much more likely that a new allele produced as a result of mutation will prove detrimental rather than beneficial when the gene is expressed. This is understandable since the gene pool of any population is the product of a long period of natural selection during which genes producing more adaptive traits have increased in frequency and those producing less adaptive traits have decreased in frequency. Under such conditions, it therefore is highly unlikely that a completely random mutational event will produce an allele that is more adaptive than existing ones.

This is not to say, however, that even potentially detrimental new alleles will quickly be eliminated by natural selection, or that such alleles will permanently remain detrimental under all conditions. An important consideration in assessing the interaction of mutation and selection is the fact that the majority of mutational events produce alleles that are recessive to the original allele. Such recessive mutant alleles are not expressed as phenotypes until they occur in the homozygous condition. Thus, new recessive mutant alleles are not immediately exposed to the effects of natural selection. In other words, heterozygosity functions to perpetuate the existence of recessive mutant alleles in populations, even though they would have a detrimental effect if they were expressed as phenotypes.

Maintenance of recessive alleles in a gene pool is important because this reservoir of genetic variability may prove advantageous should environmental conditions change. Although specific alleles may be classified as beneficial or detrimental under a given set of circumstances, these classifications are only meaningful in terms of a current situation. Should environmental conditions change, the adaptive value of a specific allele may also change.

An example of a detrimental mutant gene that can become beneficial under different conditions is seen in the water flea *Daphnia* (figure 28.4). Normally, *Daphnia* thrives at temperatures around 20°C but cannot survive temperatures of 27°C or more. There is, however, a mutant strain of *Daphnia* that requires temperatures between 25°C and 30°C, and cannot live at temperatures around 20°C. Thus, as is often the case, the adaptive value of this mutant condition is entirely dependent on environmental conditions.

Figure 28.4 The microscopic water flea *Daphnia*, which normally lives and thrives at temperatures around 20°C and dies when temperatures rise above 27°C. There is a mutant strain of *Daphnia* that lives at temperatures between 25°C and 30°C, and cannot live at 20°C.

Sickle Cell Anemia

The human condition known as sickle-cell anemia is a good example of the difference in survival value of a specific allele under varying environmental conditions.

Sickle-cell anemia is an inherited disease that results from the formation of abnormal hemoglobin molecules in the red blood cells of afflicted individuals. The abnormal sickle-cell hemoglobin differs from normal hemoglobin by substitution of a single amino acid in a polypeptide. This substitution is the result of a single base substitution in the DNA of the hemoglobin gene (see p. 611). This base substitution is the only difference between the two hemoglobin gene alleles. Expression of the *S* allele produces normal hemoglobin, and expression of the *s* allele produces abnormal hemoglobin. Individuals with the *SS* genotype have 100 percent normal hemoglobin in their erythrocytes, individuals with the *ss* genotype have 100 percent abnormal hemoglobin in their red blood cells, and individuals with the *Ss* genotype have approximately 50 percent normal and 50 percent abnormal hemoglobin in their red blood cells (figure 28.5).

Hemoglobin electrophoretic pattern

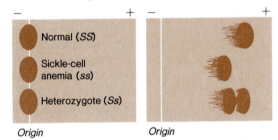

Figure 28.5 Electrophoretic patterns of hemoglobins. Electrophoresis of hemoglobins extracted from red blood cells demonstrates differences between hemoglobin from a normal individual and a person suffering from sickle-cell anemia. It also demonstrates that both types of hemoglobin are produced in heterozygous individuals.

Individuals with the *ss* genotype suffer from a condition known as sickle-cell anemia. When they are physically active, their red blood cells collapse, and the cells assume a characteristic shape resembling a sickle (figure 28.6). Physiologically, the sickling of the red blood cells is a serious problem because sickled cells can block small blood vessels and cut off circulation to a tissue in a particular region of the body. Such crises can result in tissue necrosis (death). Individuals with the *ss* genotype are unable to exercise and are generally less vigorous and more susceptible to diseases than are individuals with the *SS* genotype. A large number of *ss* individuals die before maturity, and very few of them live to age thirty. Heterozygous *(Ss)* individuals are generally intermediate in trait expression; they sometimes experience some problems when they exercise vigorously, but they are generally healthier than *ss* individuals.

It might be expected that the *s* gene would have a very low frequency in the gene pool. Many *ss* individuals die before reaching reproductive maturity, either directly from sickle-cell anemia or from other diseases that affect them more severely because of the weakening effect of sickle-cell anemia. Heterozygous *(Ss)* individuals also are seriously enough affected at times to be at a selective disadvantage to individuals who are homozygous for the normal hemoglobin allele *(SS)*. However, when geneticists studied populations of black Africans, in which the disease is most common, they found that the recessive allele had a surprisingly high frequency (0.2 to as high as 0.4 in a few areas).

Further research showed that another important factor affects the frequencies of these alleles. In malarial regions, the heterozygous *(Ss)* genotype had an adaptive advantage over both the *SS* and the *ss* genotypes (figure 28.7). Because the malarial parasite, which enters red blood cells, does not inhabit cells that contain the abnormal form of hemoglobin produced by expression of the *s* gene, individuals with the *SS* genotype suffer a higher death rate from malaria than do individuals of the *Ss* genotype. Therefore,

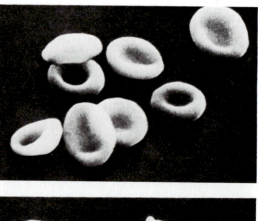

Figure 28.6 Electron micrographs of normal red blood cells (top) and sickled red blood cells (bottom). (Magnification × 4,350 and 3,375, respectively)

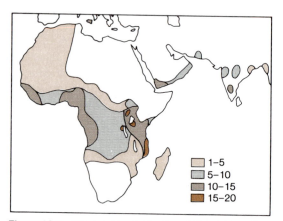

1–5
5–10
10–15
15–20

Figure 28.7 Geographic distribution of the sickle-cell condition, shown as percentages of the population afflicted with sickle-cell anemia. The frequency of the *s* gene is highest in those parts of the world in which falciparum malaria is endemic.

in malarial regions, individuals with the *Ss* genotype enjoy selective advantage over both the *SS* and the *ss* genotypes. And since a relatively high proportion of *Ss* individuals reproduce under these conditions, the frequency of the recessive gene remains greater than would be expected in the absence of the malaria factor. As might be expected, among American black populations living in nonmalarial regions, the frequency of the *s* allele is lower than in black populations inhabiting malarial regions.

The sickle-cell anemia example is an illustration of an evolutionary phenomenon called **balanced polymorphism.** The name describes a situation in which several very different phenotypic expressions are maintained in a population without one increasing in frequency at the expense of the others. In the case of the sickle-cell phenomenon in Africa, heterozygotes have a strong selective advantage over either of the two types of homozygotes. Even though both homozygotes are selected against, neither gene is eliminated because reproduction of the favored heterozygous individuals contributes both genes in equal quantities to subsequent generations. Such heterozygote superiority strongly favors balanced polymorphism.

The selective forces at work in other examples of balanced polymorphism are not always so evident. Human ABO blood types are one example. The relative frequencies of the alleles are quite stable in populations, implying that the positive and negative selective forces acting on individuals with different phenotypes (blood types) must be in balance, thus producing little selective pressure for change in allele frequencies.

Natural Selection

Natural selection is the major force causing changes in gene frequencies within gene pools.

Directional Selection

When the environment changes or when organisms migrate into new environments, natural selection operates to select those alleles that confer traits that are adaptive under the new set of environmental conditions. Because these changes represent a progressive adaptation to a changing environment, this type of selection is called **directional selection.** Directional selection acts for or against individuals possessing phenotypes that exhibit one of the more extreme expressions of a characteristic.

Directional selection is clearly illustrated by a phenomenon known as **industrial melanism,** which refers to a progressive change in the average color of moths that has occurred in industrial areas, particularly in Britain and continental Europe. The best-documented example of industrial melanism occurred in the population of peppered moths *(Biston betularia)* in England between the mid-1800s and the early 1900s.

In the mid-1800s the peppered moth population was almost entirely made up of a light-colored body phase with only the rare appearance of a dark-bodied (melanic) form. Biologists assumed that the greater survival of the lighter-colored moths was due to the fact that predatory birds had a harder time seeing these moths against light-colored vegetation, while the melanic forms stood out sharply and were quickly attacked and eaten.

Genetically, body color in the peppered moth can be treated as a simple dominant-recessive gene relationship with the gene for dark body color being dominant to the gene for light body color. In 1850, the moth population consisted of approximately 99 percent light-colored individuals and 1 percent dark individuals. During the latter half of the nineteenth century, the Industrial Revolution came to England, and great quantities of soft coal were burned to fuel the fires of industry. Burning soft coal produced a large quantity of air pollution and soot from smokestacks, which settled over areas surrounding industrial centers and darkened the foliage.

Figure 28.8 Light and dark forms of the peppered moth *Biston betularia*. Both forms are shown on a soot-darkened oak tree in Birmingham, England (a), and on a lichen-coated tree in an unpolluted region (b).

(a)   (b)

As the foliage became progressively darker, the adaptive advantage of the light-colored phase of the peppered moth was lost, and natural selection now favored the darker, melanic form. Against the soot-darkened tree trunks, the light-colored moths now stood out clearly, while the melanic form enjoyed the advantage of protective coloration (figure 28.8). Because of the darkening of the habitat, and the change in selective advantage, by 1900 the peppered moth population of industrialized areas of England was 90 percent melanic and 10 percent light-colored. Obviously, the frequency of the dominant allele increased over this period of time, while the frequency of the recessive allele decreased dramatically. Industrial melanism clearly shows directional selection in which there is environmental change accompanied by differential selection of phenotypes, resulting in changes in gene frequencies.

The selecting factor involved predatory birds that feed on the moths. To provide experimental evidence that birds prey on the peppered moths and that body color of the moths influences prey selection, H. B. D. Kettlewell of Oxford University reared moths of both body colors and released equal numbers of the two color phases in two areas: (1) the nonindustrial, unpolluted Dorset area, and (2) the highly industrial and highly polluted Birmingham area. Each released moth was marked so that when Kettlewell later recaptured his moths, he could identify those he had released as compared with wild moths that had inadvertently been captured.

Kettlewell's recapture data showed that from the unpolluted Dorset area he recaptured 14.6 percent of the light-colored moths and 4.7 percent of the melanic form. From the polluted area around Birmingham 27.5 percent of the melanic form were recaptured as compared with 13 percent of the light-colored moths. Kettlewell observed the taking of moths by birds from blinds set up in each area and also recorded the selective predation of moths on film. Kettlewell's experiment demonstrates that body color of the moth is a factor in survival with the selective advantage of body color varying with the color of the background vegetation.

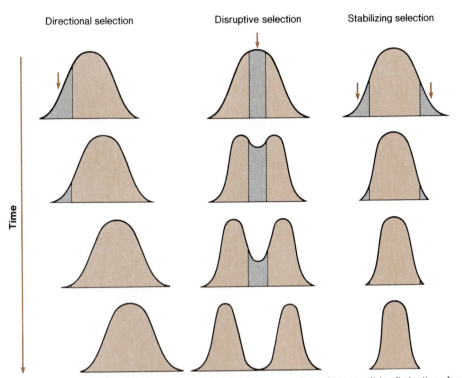

Directional selection   Disruptive selection   Stabilizing selection

Time

Figure 28.9   Three different types of selection. Each set of curves symbolizes the status of a particular set of characteristics in a population. Horizontal axes indicate the ranges of variability in the populations and vertical axes the number of individuals distributed in different parts of the ranges. Directional selection acts against individuals possessing one of the phenotypic extremes.

It tends to result in elimination of one phenotype and proportional increase in others. Disruptive selection favors two different extreme phenotypes and acts against intermediate ones. It produces two divergent subpopulations with very different gene pools. Stabilizing selection operates when conditions remain stable for long periods, and it tends to make the gene pool more homogeneous.

Other Types of Selection

Directional selection is the driving force that produces changes in gene frequencies when the environment is changing in such a way that phenotypes representing one extreme of phenotypic variation in a population are better adapted to new conditions than are other phenotypes. Directional selection generally results in change in one direction in response to steadily changing environmental conditions. But natural selection can have other effects on gene pools in other situations (figure 28.9).

A second type of selection effect is **disruptive selection.** In disruptive selection, both categories of phenotypic extremes are favored over the average phenotype in a population. This happens when a population previously exposed to a homogeneous environment becomes exposed to very different conditions in different parts of its area.

A population being subjected to disruptive selection tends to become divided into two contrasting subpopulations.

Another type of selection is **stabilizing selection,** which functions when the conditions under which a population is living remain constant over a period of time. Stabilizing selection tends to focus adaptation to existing conditions by selecting against phenotypic variation. This helps to conserve the adaptive fit of the population to its environment by selecting against phenotypes produced by expression of new genetic combinations produced in sexual reproduction or expression of new alleles produced by random mutational events. Thus, selection is not always an agent of change moving populations in new directions. Stabilizing selection plays an important conservative role in populations that are well adapted to stable environmental conditions.

Adaptations

All biology texts, including this one, are books about adaptations because all characteristics of living things must be examined in the context of adaptation. An **adaptation** can be defined as a characteristic of an organism that increases the organism's fitness for life in its environment. And fitness is a measure of the likelihood that an organism will live and succeed in reproducing. Fitness, therefore, is a measure of the odds that an organism will make a genetic contribution to the next generation.

Adaptations range from molecular specializations of enzyme systems to the complex behavioral rituals involved in reproductive processes in many animals. The precisely regulated plant responses to photoperiod change that coordinate reproductive timing in plants (chapter 18), the complex role of stomata in carbon dioxide uptake and control of water loss in leaves during photosynthesis in vascular plants (chapter 6), fungi that trap worms and digest them (chapter 10), and the highly efficient water-conserving capacities of mammalian kidneys (chapter 13) are all adaptations. These are only a few examples drawn from a vast array. Each living thing possesses a battery of adaptations that contribute to its fitness. The complexity and diversity of adaptation can be further illustrated by examining some examples of special types of adaptations, such as the adaptations involved in certain interactions between organisms of different species.

(a)

(b)

Figure 28.10 Cryptic coloration. For cryptic coloration to provide effective concealment, cryptically colored organisms must be behaviorally adapted to remain absolutely motionless when danger threatens. (a) A leaf katydid from Brazil. (b) A stonefish.

Cryptic Coloration and Mimicry

One of the most urgent and continuing threats to the survival and eventual reproductive success of most organisms is predation, and some adaptations associated with predator avoidance are truly remarkable. Some organisms have **cryptic coloration** ("hidden coloration"), which makes them virtually undetectable when they are in position against their normal background (figure 28.10). Certain animals even undergo complex color changes that improve their "match" with the background when they move from one location to another.

A related but somewhat different kind of adaptation that also functions in avoiding predation is **mimicry.** Some organisms, instead of being hidden from the eyes of potential predators, present a showy but misleading appearance that resembles some other organism. For example, a relatively harmless animal's appearance may mimic that of an organism that predators avoid (figure 28.11). The similarity between the mimic and the model organism that it resembles is so close in some cases that predators avoid both model and mimic to nearly the same extent.

(a)

Figure 28.11   Mimicry. (a) A lacewing (below) that mimics a wasp (above). The lacewing presents no threat to predators, but is avoided along with the stinging wasp. (b) A robber fly (left), a mimic of the bumblebee (right).

(b)

Coevolution

Other sets of intriguing adaptations are seen in organisms involved in **coevolution.** Coevolution is a situation in which the mutual evolutionary interaction between species is so intense that each exerts a strong selective influence on the other.

Some examples of adaptations resulting from coevolution have already been described. For example, rumen microorganisms and their grazing hosts (chapter 10) are products of a long coevolutionary process. In chapter 30, the striking interplay between the echo system that bats use to locate their prey and the defensive measures used by the nocturnal moths that bats often hunt is examined. The adaptations of both bats and moths have arisen through coevolution. Their ultrasonic struggle is played out every summer evening as bats scream through the nights, making sounds that would sound louder than a passing freight train if only human ears were attuned to the frequencies of their cries. And moths respond to these ultrasounds with diving, fluttering, evasive maneuvers.

Perhaps the best-known products of coevolution are the relationships between certain flowering plants and the animals upon which they depend for pollen transport from flower to flower. Adaptations of the flowers have evolved along with adaptations of the pollinators.

Flowers pollinated by bees usually are showy and bright, and they are always open during daytime hours, when worker bees do their foraging. Bee-pollinated flowers tend to be blue or yellow, but usually are not red, since bees are blind to red colors. They often have sweet, aromatic odors because bees depend on a well-developed sense of smell to locate nectar-containing flowers. Bee-pollinated flowers also have petal arrangements that provide bees a place to land before they push into the flower in search of nectar (figure 28.12).

Hummingbirds, on the other hand, see red well but blue only poorly. Because hummingbirds have a poor sense of smell, aroma is not an important factor in attracting them to flowers. Hummingbirds, like bees, forage in the daytime, but they require no landing perch. A hummingbird hovers in front of a flower and inserts its long beak and tongue, which is specialized for sucking, into the nectar-containing part of the flower. Characteristically, hummingbird-pollinated flowers usually are red or yellow and often are odorless.

Moths, such as the sphinx moth, also hover as they feed on nectar from flowers, but they feed during evening or nighttime hours. Thus, the flowers that moths pollinate are ones that remain open at night, and in most cases are light colored.

Even behavioral adaptations of pollinating animals show evidence of the coevolutionary process. Bees, for example, tend to feed "faithfully" on one kind of flower at a time. This is adaptive for bees because plants of many species flower practically in unison, and by tending to seek more flowers like those in which they have found nectar, the bees' search for nectar is more efficient. If bees tend to move from plant to plant of the same species, pollination is effectively accomplished. This systematic feeding pattern is not limited to bees. It is fascinating to watch a hovering sphinx moth move systematically from flower to flower in a flower bed. It pauses before a blossom long enough to unroll its long sucking proboscis and extend it deep into the flower to feed. Then it rerolls its proboscis and moves on, repeating the process again and again until every flower has been visited.

Figure 28.12 Pollen that sticks to a bee as it pushes into a flower in search of nectar is carried to other flowers that the bee visits subsequently.

Species and Speciation

Any change in gene frequencies is considered to be evolutionary change. The changes considered thus far are changes that occur within the gene pools of populations and constitute what is called **microevolution.** But on occasion sufficient genetic change has accumulated so that a new species has been formed. This is called **macroevolution.** The species concept and the mechanisms by which macroevolutionary change (the production of new species) takes place are discussed in this section.

Species Defined

A **species** is a population of organisms that may display a range of phenotypic (and genotypic) variation, but that still represents a biological entity distinct from all other populations. One of the definitions of a species is based on the degree of similarity among species members and their differences from members of other species. This is a morphological definition of a species that says rather simply that all members of a species are more like one another than they are like members of any other species. In such a scheme it is sometimes necessary to weigh some characteristics more heavily than others, as in the classification of human ancestors in which brain size, facial structure, and tooth characteristics are considered to be traits of prime importance.

From an evolutionary perspective, all members of a species are expected to have similar traits because they share a gene pool that has a history distinct from that of any other species, a gene pool that is reproductively isolated from all other gene pools. The concept of **reproductive isolation** leads to a second definition of a species as an interbreeding population that possesses a genetic identity because members of the species do not crossbreed with the members of any other species. Reproductive isolation is the major working criterion for species definition, but it does have certain shortcomings.

First of all, it generally is not possible to test for crossbreeding among all members of any two natural populations, and it is therefore impossible to state with absolute certainty that complete reproductive isolation does exist.

Second, an occasional mating does occur between members of two species, as in the 1975 mating and production of an offspring by a siamang and a gibbon at Atlanta's Grant Park Zoo, which was an example of crossbreeding above the species level. An important consideration in such matings is whether or not the offspring is fertile because matings that produce only sterile offspring have no evolutionary future and are considered to fall within the definition of reproductive isolation.

A third problem with the reproductive isolation criterion is that it cannot be applied to species known only from the fossil record. Certainly, no test for cross-fertility can be conducted on these organisms, and studies of similarity and difference, especially based on comparative anatomy, remain the major criteria used for species definition.

A final shortcoming of the reproductive isolation criterion is that it cannot be used to define organisms that reproduce asexually. Despite these reservations, however, reproductive isolation remains the major criterion used in defining species.

It may not be obligatory that reproductive isolation be 100 percent because species represent a wide range of genotypes. Since closely related species have diverged from a common gene pool, it may be possible that some organisms among the two species might retain the ability to mate successfully. But this question is at the crux of what is sometimes called the "species problem." How much interbreeding is "permitted" between supposedly separate species before they are no longer truly reproductively isolated? It should be clear by now that the criteria currently used for species definition may not be altogether clear-cut, nor are biologists in total agreement on how the criteria should be applied.

Box 28.1
Birth of the "Siabon"

In 1979, R. H. Myers and D. A. Shafer reported the birth of a hybrid between two species of lesser apes, a male gibbon (*Hylobates moloch*) and a female siamang (*Symphalangus syndactylus*). The female infant has been called a "siabon." This mating occurred in captivity in Atlanta's Grant Park Zoo, where the two apes were housed together. In nature these two animals occupy the same geographic area of Southeast Asia and feed on the same plants, although the gibbon feeds primarily on fruits and nuts, while the siamang eats leaves and shoots. But when members of the two species encounter one another, the smaller gibbon either withdraws immediately or is chased away by the male siamang. Thus, under natural conditions there is a social/behavioral barrier between the two species, and it is unlikely that interspecific breeding occurs in the wild. Nevertheless, a successful mating between the two species did take place in captivity.

This successful mating between two species that probably never mate in nature raises an interesting genetic question about how species are formed. Although the gibbon and the siamang are members of different genera, scientists have concluded, on the basis of protein analysis, that they are genetically very similar. But, despite the high degree of genetic similarity indicated by these biochemical data, the two species show striking structural differences in their chromosomes. Structural changes in chromosomes (see chapter 21) take place when a small piece of a chromosome is broken off and reattached in reverse order (inversion), or when a piece of one chromosome breaks off and reattaches to a second nonhomologous chromosome (translocation). Such structural changes can interfere with the proper pairing of chromosomes during meiosis, and this, in turn, can prevent the formation of viable gametes. The gibbon and the siamang diverged from a common stock some 15 million years ago and in that period of time have accumulated a large number of structural changes in the chromosome sets of each species. These structural changes may prove to be sufficient to render the siabon (the hybrid offspring) infertile because of an inability to produce viable egg cells. In that case the siabon hybrid would be incapable of mating successfully with either parent type or a male siabon, if one should be produced. If offspring cannot be produced, the gibbon and the siamang still would meet the test of reproductive isolation because the siabon would have no evolutionary future.

In this example, it is a change in chromosome structure that is considered an isolating mechanism rather than changes in gene frequency. Biologists think that such **chromosomal evolution** has taken place often in the history of life and that structural changes in chromosomes have been an important factor in evolutionary change.

Speciation

A very important mechanism involved in species formation is **geographic isolation,** which occurs when natural barriers, such as deserts, rivers, mountains, or oceans, form and physically separate a species into two or more **allopatric** populations (populations that are separated by physical barriers). An important corollary of allopatry is that gene flow between the separated populations is prevented.

Figure 28.13 hypothetically illustrates the way in which geographic isolation into allopatric populations can lead to speciation. In stage 1, species A is continuously distributed over a region that, except for a small desert, is uniformly cool and moist. At this stage there is a large population with random mating and no interruption of gene flow. Under such conditions, stabilizing selection would exist, and there would be little or no change in gene frequencies from one generation to the next. Then the climate of the region begins to change, such that a dry wind starts to blow across the desert, causing an extension of very dry conditions and a considerable warming of another portion of the region. Under these conditions, the homogeneity of species A is disrupted, and in stage 2 there appear local vari-

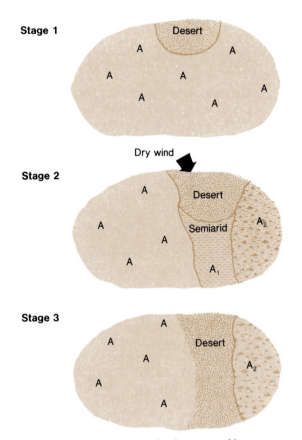

Stage 1

Stage 2

Stage 3

Figure 28.13 Species formation by geographic isolation. *Stage 1:* Uniform environmental conditions. Stabilizing selection operates in the gene pool of species A. *Stage 2:* Three separate environments have been produced. These are: cool, semiarid, and warmer. There are three demes (A, A_1, and A_2), that is, populations adapted to local environmental conditions. *Stage 3:* The expansion of the desert now represents a geographic barrier to gene flow between A and A_2. In time, microevolutionary changes in these two populations may produce reproductive isolation, meaning that new species have been formed.

are separate populations that experience differences in mutation and selection pressure. It also is possible that the founder effect may contribute to differences between the two populations. In time, sufficient variation may accumulate in the two gene pools for reproductive isolation to occur. At this point, new species will have evolved. Even if those new species would come back together, they would not be able to interbreed. They are reproductively isolated.

Specific mechanisms develop within geographically isolated populations of organisms that prevent successful interbreeding should individuals of the two new species later encounter one another. These **reproductive isolating mechanisms** are the result of the accumulated genetic differences within the two gene pools.

In animals, premating isolating mechanisms operate to prevent mating between members of isolated populations. One example of a premating isolating mechanism is **mechanical isolation.** There may be differences in the reproductive structures of members of the two populations that make copulation difficult, if not impossible. Another example of premating isolation is **habitat isolation,** which occurs when populations occupy different specific habitats within the broad range occupied by the two species, and individuals seldom venture out of their habitat. **Seasonal isolation** is a third premating mechanism. In many geographically isolated populations the time of sexual activity is variable, with individual groups becoming sexually active at different times of the year. A final premating isolating mechanism is **behavioral isolation.** The various songs, calls, display patterns, and courtship "dances" of birds and other animal species attract mates of that species but also eliminate members of other species as possible mates, thus preventing gene flow between populations.

Prefertilization reproductive isolation mechanisms also function in plants because pollen tubes often fail to grow when pollen falls on the stigmas (chapter 23) of flowers of other species. Cross-pollinations among such species cannot result in fertilization.

Postmating isolating mechanisms operate after copulation has taken place. These include **gamete mortality, hybrid inviability,** and **hybrid sterility.** In matings between geographically isolated populations of animals with internal fertilization, sperm produced by males of one population may not be able to survive in the female

ants of the species (demes)—subpopulations that are adapted to local environmental conditions. Demes possess slight genetic variation but are not so divergent that gene flow cannot occur. In stage 2, the A deme is essentially like the original population; A_1 is adapted to dry, semiarid conditions; and A_2 is adapted to a warmer environment.

By stage 3 the desert has extended and become a physical barrier that has completely separated A from A_2, stopping gene flow. The two populational units are now allopatric; they are separated by a physical barrier, the desert. The A_1 deme has disappeared because its habitat has disappeared. From this point onward, A and A_2

reproductive tract of the other species, or the zygote may not even be able to begin its development. Hybrid inviability refers to the death of hybrid offspring anytime before reproductive maturity. Hybrid sterility is when offspring are healthy and survive to reproductive age, but are incapable of producing offspring when mated with either parent type or with another hybrid. The commonly used example of hybrid sterility is the mule, the hybrid offspring of the horse and the donkey. The mule is a strong animal, well suited to certain work tasks, but sterile. In hybrid inviability and hybrid sterility, members of two populations may mate successfully and produce offspring, but in neither case is a new genetic line of descent established, and thus there is no effective gene flow between the two populations. These isolating mechanisms function because of either significant genetic or chromosomal differences between the parents.

In addition to allopatric speciation, new species might be formed through a process called **sympatric speciation** (that is, the origin of different species from populations that occupy the same geographic area). No geographic barriers are involved, and sympatric speciation would be a slower process, resulting from more subtle selective pressures than in the case of allopatric speciation. But some biologists are debating whether sympatric speciation actually occurs. If it does take place, however, it may be that changes in gene frequency result, because no habitat is entirely homogeneous, and at times small units of a population become highly specialized to local environmental conditions. For example, mountain species are subjected to a wide variation in soil type, soil moisture, temperature, and so on within fairly limited geographic areas, and they often have local demes, which are adapted to growth in very different conditions. The question of whether or not enough differences to produce reproductive isolation can accumulate between demes whose members have the opportunity for interbreeding is still a point of argument among students of evolution. The debated question, put another way, is, can speciation actually occur among truly sympatric populations?

Box 28.2
Evolutionary Controversies

The controversy between the phyletic gradualism and punctuated equilibrium viewpoints (p. 800) is by no means the first such controversy in evolutionary theory. Like many other scientific fields, evolutionary theory has had its share of controversy and disagreement. Even Darwin and Wallace, who together formulated the theory of natural selection, came to disagree about the actual role of natural selection in the evolutionary process.

Wallace favored a very strict application of the concepts of natural selection. He and others thought that each and every characteristic of every organism was a product of natural selection. Darwin, however, was convinced that natural selection was the main factor, but not the only factor, involved in evolutionary change. He viewed an organism as an integrated whole and thought that while natural selection might result in adaptive changes in one part, other parts might change in ways that had neither positive nor negative effects on the odds of reproductive success, which is the final test of adaptation according to the theory of natural selection.

The recent discussion about punctuated equilibrium must be viewed in the context of the "modern synthesis" of evolutionary theory, which developed in the first half of this century. This modern synthesis developed following the discovery of some of the basic mechanisms of inheritance and the growth of the science of population genetics. At the heart of the modern synthesis is the idea of long periods of accumulating gradual change, usually associated with geographic isolation, and the eventual emergence of new and reproductively isolated species. The concept of punctuated equilibrium challenges this central idea of the modern synthesis. It proposes instead that evolution of any particular line is characterized by long periods of relative stability punctuated by periods of abrupt change, change that occurs in periods of time that are relatively short compared to the enormity of the geologic time scale.

Disagreements such as the one that exists among evolutionary theorists regarding phyletic gradualism and punctuated equilibrium are not unusual happenings in science. Unfortunately, however, people who do not think that life on earth has an evolutionary history cite such controversies as indications of some fundamental weakness in the theory of organic evolution. Such controversies are inherent in the nature of science and characterize most, if not all, fields of scientific endeavor. This vigorous debate is by no means a weakness in the theory of organic evolution. Rather it is a strength, an expected characteristic of an active field of scientific endeavor. Scientific progress is made through careful examination that leads to acceptance or rejection of competing conceptual schemes. The theory of organic evolution remains one of the central unifying themes of modern biology, and it provides one of the basic conceptual frameworks for the interpretation of biological phenomena.

Views of the history of life on earth, possibly more than any other set of scientific concepts, engender emotional responses and controversy of another sort. The great majority of scientists think that available evidence indicates a very long history of life on earth that is measured in thousands of millions of years and characterized by evolutionary descent. But some individuals believe that the history of life on earth is much shorter, possibly as short as 10,000 years, and that it is characterized by a series of divine creation events. Their view of life is essentially compatible with literal interpretation of the biblical creation story of the Judeo-Christian religious tradition.

Other scientists, however, do not find the idea of a long evolutionary process incompatible with their religious faith and experience. They do not feel that their faith is compromised because they interpret a creation story in terms of modern scientific understanding. They recognize that the biblical creation story was written in a form compatible with the experience of people living several thousand years ago. Possibly their view of faith and life can best be summarized with the words that Charles Darwin used following his summary of the theory of natural selection at the very end of *The Origin of Species:* "There is a grandeur in this view of life, with its several powers, having been originally breathed by the Creator into a few forms or into one; and that, whilst this planet has gone cycling on according to the fixed law of gravity, from so simple a beginning endless forms most beautiful and wonderful have been, and are being evolved."

The Tempo of Evolution

In recent years a major debate has developed over the question of the tempo of species formation. The classic concept of evolutionists has been that species are formed as a result of a slow, gradual accumulation of adaptive traits through a selective process that requires many thousands or even millions of years for the formation of new species. This is the concept of **phyletic gradualism.** Recently, however, a number of evolutionists, notably S. J. Gould and Niles Eldredge, have advanced the hypothesis of **punctuated equilibrium**, which states that evolution is "concentrated in very rapid events of speciation." They argue that throughout the greater part of its existence, a species displays very little morphological (or genetic) change. This period of comparative equilibrium may be quite suddenly interrupted by environmental events that result in the geographic isolation of population units in which new species are produced within a relatively short period of time. Proponents of punctuated equilibrium claim that the fossil record supports their view of the history of life because it shows long periods during which species remained essentially intact. They also point out the lack of transitional forms between taxonomic groups. Gradualists claim that transitional forms are missing because relatively few organisms have been preserved in the fossil record, and it is not likely that, by chance, a large number of intermediate forms would be found. The supporters of punctuated equilibrium claim that transitional forms are not found because they never existed—that species formation occurs so rapidly that the long sequence of intermediate types predicted by the gradualistic theory simply were not a part of species formation.

The debate between the phyletic gradualists and the punctuationists is far from settled, with each side seeking evidence that supports its point of view. Unfortunately, some observers outside the field of biology misinterpret such arguments and conclude that biologists "are beginning to have doubts about evolution." These biologists are not expressing doubts about the overwhelming evidence indicating the occurrence of an evolutionary process. Rather, they are arguing about the way that the evolutionary process proceeds and the mechanisms by which new species arise. If nothing else, this debate will stimulate biologists to reexamine existing data and to seek new information regarding the process of evolution.

Summary

Modern biologists explain the process of organic evolution in terms of changes in gene frequencies occurring in gene pools. The Hardy-Weinberg law provides a model of genetic equilibrium against which evolutionary changes in gene frequency can be measured. Gene frequency changes may be caused by genetic drift, migration, and mutation, but the major force is natural selection.

Each organism possesses a set of adaptations to its environment. Some adaptations are particularly striking because of their involvement in interactions between organisms of different species. Cryptic coloration and mimicry are adaptations that function in avoiding predation. In coevolution there is an intense mutual evolutionary interaction between species in which each exerts a strong selective influence on the other.

A species is a population of organisms that is reproductively isolated from other members of other populations. Species may also be defined on the basis of structural similarities and differences. Geographic isolation is a major factor in species formation. A population is divided into smaller units by the establishment of physical barriers, and the resultant allopatric populations accumulate changes in gene frequencies within their individual gene pools until reproductive isolation is established—at which point new species have been formed. It is also possible that species may be formed sympatrically, as a result of local habitat variation.

There is a major debate among modern students of evolution concerning the tempo of species formation. Does speciation occur as a result of long-term, gradual, steady accumulation of adaptive traits? Or are there relatively short periods of rapid formation of new species separated by longer periods of comparative equilibrium?

Questions

1. Why does the synthetic theory of organic evolution deal with populations rather than individual organisms?

2. What is a gene pool? What factors can cause changes in gene frequencies within a gene pool?

3. Using either the Punnett square method or the algebraic method, demonstrate the genetic equilibrium described in the Hardy-Weinberg law for a gene pool in which the alleles B and b have frequencies of 0.1 and 0.9 respectively. Assume that allele B is dominant to allele b.

4. What is the difference between stabilizing and directional selection?

5. What is reproductive isolation? How effective is its use in the definition of a species?

6. What are the basic differences between allopatric speciation and sympatric speciation?

7. Explain how both proponents of phyletic gradualism and punctuated equilibrium can argue that data from the fossil record support their viewpoints even though the two hypotheses are very different.

Suggested Readings

Books

Dobzhansky, T.; Ayala, F. J.; Stebbins, G. L.; and Valentine, J. W. 1977. *Evolution*. San Francisco: W. H. Freeman.

Edwards, K. J. R. 1978. *Evolution in modern biology*. Studies in Biology no. 87. Baltimore: University Park Press.

Mettler, L. E., and Gregg, T. G. 1969. *Population genetics and evolution*. Englewood Cliffs, N.J.: Prentice-Hall.

Simpson, G. G. 1967. *The meaning of evolution*. Rev. ed. New Haven, Conn.: Yale University Press.

Articles

Ayala, F. J. September 1978. The mechanisms of evolution. *Scientific American* (offprint 1407).

Dickerson, R. E. March 1980. Cytochrome *c* and the evolution of energy metabolism. *Scientific American* (offprint 1464).

Friedman, M. J., and Trager, W. March 1981. The biochemistry of resistance to malaria. *Scientific American* (offprint 1493).

Jones, J. S. 1981. An uncensored page of fossil history. *Nature* 293:427.

Lewontin, R. C. September 1978. Adaptation. *Scientific American* (offprint 1408).

Rensberger, B. April 1982. Evolution since Darwin. *Science 82*.

Human Evolution

In the introduction to his book *The Descent of Man*, Charles Darwin wrote: "During many years I collected notes on the origin or descent of man, without any intention of publishing on the subject, but rather with the determination not to publish, as I thought that I should thus add to the prejudices against my views." At the time that he wrote these words, Darwin was keenly sensitive to the continuing storm of controversy that publication of his *Origin of Species* had stirred some years before, and he felt that dealing explicitly with the matter of human evolution would once again fan the flames of emotional response. Nevertheless, he thought that his ideas about evolution could logically be extended to include humans, and he proceeded to present the evidence available at that time. His work did cause further controversy as many of Darwin's contemporaries generally objected to the idea of evolution and were especially hostile to the notion that humans had descended from other earlier organisms and had not always been as they are now.

To a large extent, controversy about the evolution of human beings has passed; at least it is not a significant issue among biologists. But controversy of another kind still surrounds the study of human evolution. Today paleontologists exchange heated arguments regarding interpretation of the fossil record of human evolution. For years, there have been disagreements and debates about various fossils assumed to be or to resemble our evolutionary ancestors. These fossils often were named without adequate concern for their possible relationships to one another or for their place in the general evolutionary history of human beings and humans' closest relatives in the animal kingdom, the great apes.

In recent years, this taxonomic tangle has been somewhat straightened out, and names of fossil organisms now better reflect their relationships, but controversy still remains over interpretation of evolutionary lines of descent. Views on these evolutionary matters are subject to revision as new fossils are discovered and new insights are gained.

Human evolutionary ancestry is represented in the fossil record by a frustratingly discontinuous series of potential ancestral organisms. Some major trends of human evolutionary history are apparent, but details can be filled in only by further discoveries in paleontology.

Before these various fossil organisms can be examined and their places in humans' evolutionary ancestry considered, a brief survey of humans' relationships to present-day organisms is necessary.

Taxonomic Relationships

Humans are mammals. That is, human beings belong to a group of vertebrate animals that are included in a taxonomic category called the **class Mammalia** (table 29.1). Mammals are homeothermic ("warm-blooded") animals that have body hair, feed their young milk produced by mammary glands, and possess a muscular diaphragm that separates their body cavity into thoracic and abdominal portions and functions in breathing movements.

The great majority of modern mammals are **viviparous.** Their offspring develop inside a specialized portion of the female reproductive tract, the uterus; and nutrients, gases, and other materials are exchanged between the bloodstreams of mother and developing offspring within a highly specialized organ known as the placenta. Generally speaking, placental mammals also provide their offspring, after birth, with parental care for a more extended time than do most other vertebrate animals.

Table 29.1
Examples of Taxonomic Categories.

Categories	Human	Domestic Dog
Kingdom	Animalia	Animalia
Phylum	Chordata	Chordata
Subphylum	Vertebrata	Vertebrata
Class	Mammalia	Mammalia
Order	Primates	Carnivora
Family	Hominidae	Canidae
Genus	*Homo*	*Canis*
Species	*sapiens*	*familiaris*

The earliest fossil mammals are found in rocks of the Jurassic period of the Mesozoic era. These early mammals were small and inconspicuous compared to the reptiles that dominated that period of the earth's history. The beginning of the rise of modern mammalian forms coincided with the mass extinction of the ruling dinosaurs near the end of the Mesozoic era. During the early stages of the Cenozoic era, the mammals began an adaptive radiation that has resulted in their achieving dominance among terrestrial vertebrates.

Primate Evolution

Humans are members of a group of mammals classified in the **order Primates** (table 29.2). The primates first evolved as **arboreal** (tree-dwelling) animals, although not all modern primates live in the trees. Most of the characteristics of primates can be viewed as adaptations to arboreal habitats.

Primates retain five functional digits on their hands and feet, while many other modern mammals have undergone evolutionary reduction in digit number from the ancestral vertebrate five-digit conditions. The digits of primates are very mobile, and in many primates, thumbs are **opposable;** that is, they close to meet fingertips and function efficiently in grasping. The tips of primates' digits have nails that protect them instead of claws that extend beyond the digits, as many other mammals have, and primates' digits have pads that are very sensitive to touch.

Table 29.2
Classification of the Primates.

Order Primates
    Suborder Prosimii (lemurs and tarsiers)
    Suborder Anthropoidea
        Superfamily Ceboidea
            Family Cebidae (New World monkeys)
        Superfamily Cercopithecoidea
            Family Cercopithecoidae (Old World
                monkeys and baboons)
        Superfamily Hominoidea
            Family Pongidae (apes)
            Family Hominidae (humans)

Other parts of primate skeletons also reflect arboreal adaptations. Primates' shoulder joints permit much more extensive forelimb rotation than the shoulders of other mammals allow, and their elbow joints permit some rotational movement as well.

Other adaptations to life among the tree branches include relatively short snouts and eyes placed at the front of the head. This results in excellent binocular stereoscopic (three-dimensional) vision that permits primates to make very accurate judgment about distance and position, judgments that are essential for animals that swing and leap from branch to branch. Paralleling this development of greater visual acuity has been evolutionary elaboration of anterior portions of the brain, especially the cerebral cortex. Compared with many other mammals, however, some primates have a relatively weak sense of smell.

Most primates have a pair of mammary glands and produce only one offspring per pregnancy, whereas a great many other mammals produce litters, batches of several offspring born at the same time. Primates have relatively long infancies and develop strong, long-lasting mother-infant bonds.

The primates have evolved from a group of small insect-eating animals that entered the trees long ago and there escaped predators on the ground and found new food sources. The earliest known occurrence of fossil primates is a few teeth found in Montana, in rocks from the Cretaceous period of the Mesozoic era. Teeth are important indicators of an animal's diet, and these fossil teeth indicate that the earliest primates were insect eaters. Fossil materials from the earliest primates are scarce, but it generally is thought that they may have been quite similar to the living tree shrews of Southeast Asia (figure 29.1). By the time that the Cenozoic era began, 65 million years ago, a variety of small primitive primates had spread through the forests of Europe and North America. These primates, however, still had clawed feet and climbed trees by digging claws into the wood, as all tree-climbing animals except modern primates still do. By 50 million years ago, primates that grasped limbs with their digits, as modern primates do, had evolved. They had nails, and their eyes faced forward rather than to the sides, as the eyes of their shrewlike ancestors had.

Figure 29.1 A tree shrew. Modern tree shrews have many characteristics that are intermediate between those of modern insectivores (shrews and moles) and primates. Tree shrews may resemble the ancient ancestors of modern primates.

Table 29.3
Periods and Epochs of the Cenozoic Era.

Era	Period	Epoch	Millions of Years before Present*
	Quaternary	Recent	0.01
		Pleistocene	2.5
Cenozoic	Tertiary	Pliocene	7
		Miocene	26
		Oligocene	38
		Eocene	54
		Paleocene	65
Mesozoic			230

*These are approximate dates of the beginnings of these intervals.

By about 30 million years ago, the distribution of the evolving primates had changed considerably. Climates had become cooler and drier, and the lush forests in which primates had thrived disappeared from many parts of the world. This shrinkage of the semitropical forests restricted the primates to part of Asia and the southern continents, Africa and South America, where their evolution continued.

The Primate order is divided into two suborders, the Prosimii and the Anthropoidea (see table 29.2). The suborder Anthropoidea includes monkeys, apes, and humans, and the evolution of that group is emphasized, but it is also important to discuss the prosimians, which arose earlier in the evolution of the primates.

Because the major events in primate evolution occurred during the Cenozoic era, the names and approximate dates of these various geologic time intervals are presented in table 29.3.

Prosimians

The oldest known prosimian fossils are of Paleocene age and show a great deal of similarity with the present-day tree shrews (see figure 29.1). Late in the Paleocene epoch and throughout the Eocene epoch, lemurs were common small primates in Europe and North America (figure 29.2a). Tarsier fossils are common in strata of the Tertiary period in northern continents (figure 29.2b). But modern prosimians—which include lemurs, tarsiers, and other primitive primates with such exotic names as lorises, pottos, aye-ayes, and galagos—are very restricted in their ranges. Today, in fact, lemurs are found only on the island of Madagascar, and tarsiers occur only in the Philippines and the East Indies.

Monkeys

By the Oligocene epoch, which began about 38 million years ago, the earliest members of the suborder Anthropoidea had diverged from the prosimians, and the two major lines of primate evolution were established (figure 29.3). Several lines of divergence soon developed among the primitive anthropoids. One group of anthropoids included ancestors of the **New World monkeys,** and the other group included ancestors of **Old World monkeys,** apes, and humans.

New World and Old World monkeys differ in several ways, the most obvious of which is that New World monkeys have a long **prehensile** (grasping) **tail** that they regularly use almost as a fifth limb while swinging through the trees. Old World monkeys do not have prehensile tails. New World monkeys are sometimes called **platyrrhines** (flat-nosed) because their nostrils are directed outwards and separated by a broad, flat partition. Old World monkeys, as well as apes and humans, are called **catarrhines** (downward-nosed) because their nostrils are closer together and are directed downward. Many of the Old World monkeys have brightly colored areas on their buttocks (ischial callosities), but New World monkeys do not.

(a)

(b)

Figure 29.2 Some living prosimians. (a) A ring-tailed lemur. (b) A tarsier.

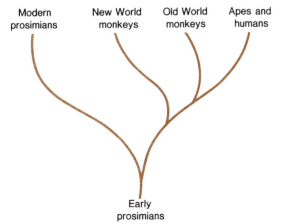

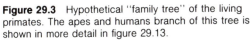

Figure 29.3 Hypothetical "family tree" of the living primates. The apes and humans branch of this tree is shown in more detail in figure 29.13.

New World monkeys are found in Central and South America. Some well-known New World monkeys are the spider monkey (figure 29.4a), the capuchin ("organ-grinder's monkey"), howler monkeys, and squirrel monkeys. Old World monkeys are found in Africa and southern Asia. Some Old World monkeys, such as the colobus monkeys and mona monkeys are arboreal, but others are ground-dwelling descendants of arboreal ancestors. These ground-dwelling monkeys include the colorful mandrill, the familiar baboon (figure 29.4b), and the macaques (figure 29.4c), some of which are very widely used in medical research. All of the ground-dwelling monkeys walk on all fours.

(a)   (b)   (c)

Figure 29.4 Monkeys and baboons. (a) The spider monkey, a New World monkey. Note that it uses its prehensile tail to grasp as it moves through the branches. (b) A baboon, a ground-dwelling Old World monkey. (c) Macaques.

Dryopithecines

During the Miocene epoch, which began about 25 million years ago, apes became abundant and widely distributed in Africa, Europe, and Asia. Among these Miocene apes, members of the genus *Dryopithecus* are of particular interest because they are thought to be the ancestors of modern chimpanzees, gorillas, orangutans, and humans.

Dryopithecines were forest dwellers who probably spent most of their time in the trees, but their foot skeletons indicate that they also may have spent some time on the ground. When they did walk on the ground, however, they undoubtedly walked on all fours virtually all the time, using the knuckles of their hands to support part of their weight. Such "knuckle-walking" is clearly illustrated by modern apes (figure 29.5). The dryopithecine skull had a low, rounded cranium and **supraorbital ridges** (bony ridges that protrude above the eyes). The dryopithecine face and jaws projected forward.

One species, *Dryopithecus major,* appears to have been closely related to the modern gorilla. While not so large as the gorilla, *D. major* was **sexually dimorphic** (that is, males and females had some different general body features), a condition found only in the gorilla among the living pongids (apes).

Climatic changes during the Miocene epoch had important effects on the evolution of the dryopithecines. The climate became progressively cooler and drier. The lush forests where the dryopithecines thrived dwindled, and vast **savannas** (grasslands with occasional clumps of trees) spread (figure 29.6). Some late Miocene relatives of the dryopithecines, however, were not restricted to the shrinking forests. These primates were able to live on the savannas, or at least at the edges of the savannas. Among these late Miocene primates, which came out of the trees, stood partially upright, and moved onto the savannas, were the ancestors of humans.

The Origin of the Hominids

Every successful species is adapted to the environment in which it lives. Chimpanzees, for example, are adapted to existence in the more open areas of a tropical rain forest. Within this habitat,

Figure 29.5   A chimpanzee "knuckle-walking."

Figure 29.6   An African savanna, part of Serengeti plain in Tanzania. During the late Miocene epoch, the climate became cooler and drier, lush forest dwindled, and savannas spread.

they live primarily on the forest floor but are also very comfortable in the trees, easily swinging from branch to branch as they feed, and sleeping in the trees at night. Early hominids, on the other hand, were adapted to life on the open savanna and inhabited the grassland areas. What evolutionary steps were involved in the transition of the hominid line from forest dwellers to grassland dwellers?

The early Miocene climate was warm and moist, and much of the earth was covered with dense forest—an environment very suitable for the arboreal primates. However, as the forests withdrew during the cooling and drying that occurred later in the Miocene, broad areas of forest were replaced by bushy savannas.

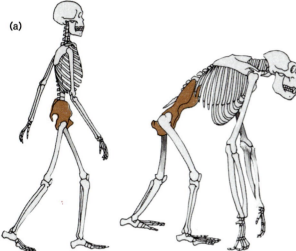

(a)

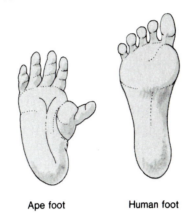

(b)

Ape foot   Human foot

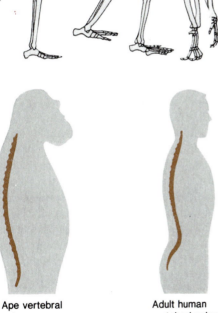

(c)

Ape vertebral
column

Adult human
vertebral column

Figure 29.7 Adaptations for erect posture.
(a) Comparison of the skeleton of a "knuckle-walking"
ape (a gorilla) with a human skeleton. Note differences in
proportions of hindlimbs and forelimbs, differences in
shape of the pelvis (color), and differences in the angle
and position of attachment of the vertebral column to
the skull (see also figure 29.8). (b) Comparison of an ape
foot and a human foot. (c) Comparison of an ape's
vertebral column and a human vertebral column. Note
the greater curvature in the human vertebral column.

Erect posture and **bipedal locomotion** (walk-ing on two legs) were two important requirements for success on the savanna. Erect posture pro-vided several advantages: (1) the ability to see predators in time to flee to safety; (2) freeing of the hands from the necessity to support the body, permitting their use for carrying food or using tools; and (3) enhancement of the ability to spot prey. Bipedal locomotion, an adjunct of erect pos-ture, permitted efficient movement while both hands were free to perform other functions.

The transition to erect posture required a number of individual adaptive changes (figure 29.7). The opposable toe that provides gripping ability to the rear feet of tree dwellers makes walking on land awkward. Thus, those primates that first ventured onto the savannas and had the rear toe more in line with the other digits were favored. The elongation and straightening of the

hindlimb was also important for erect posture and bipedal movement. Figure 29.7 illustrates the differences in proportion of hindlimb to fore-limb, and in the straightness of the limbs, that exist between modern apes and humans. Another modification of the skeleton was a more curved spine than is found in the apes. The curved hom-inid spine permits a better weight distribution and improved balance in the erect-standing ani-mal. Also, the pelvis of humans is shorter and wider than that of apes. Finally, the transition to erect posture involved a change in position of the **foramen magnum,** the hole in the skull through which the spinal cord passes from the vertebral column to join the brain. In apes, this opening is well to the rear of the skull, which permits the head of the ape to face forward when knuckle-walking. In humans, with erect posture, the fo-ramen is almost directly in the bottom center of the skull.

There is considerable argument among biologists concerning the time sequence in which adaptations to life on the savannas developed. Was it, as some have concluded, a gradual process that occurred as the savannas spread and the forests receded? Or were there primates that already possessed these characteristics, which would prove to be adaptive for life on the savannas, before the climatic changes set in? In other words, were some Miocene primates preadapted for the move onto the savannas? If this were the case, what other selective forces that operated in the forest environment favored development of these traits?

As with several other aspects of human evolution, there are as yet no unequivocal answers to these questions. There is, however, a fossil representative of the organisms that lived on the edge of the savannas. But, as is so often the case, study of this fossil creature, *Ramapithecus,* raises almost as many questions as it answers.

Ramapithecus

Fossils of the genus *Ramapithecus* are found in the late Miocene strata in India and Africa. This is a tantalizing group of fossils, known only from a few specimens of teeth, jaws, and parts of the face because no parts of the body skeleton have been discovered. Even so, the small amount of fossil material found has led a large number of researchers to consider *Ramapithecus* either an immediate ancestor of the hominids, or quite possibly, a member of the family Hominidae.

Fossil evidence of *Ramapithecus* has been found in strata from 14 to 10 million years ago. Evidence found in rocks containing the Indian specimens suggest that *Ramapithecus* lived in a forest environment with slow-moving rivers flowing down shallow gradients. Toward the end of the period that *Ramapithecus* lived in this area, the environment underwent a change, with vast forest areas gradually becoming grasslands. The African deposits, while younger than the Indian, suggest a similar sequence of events.

To understand why *Ramapithecus* might be called a hominid rather than a pongid, it is necessary to know what characteristics are used by paleontologists in their evaluation of primate fossils. Some of the major points of comparison between the skulls of apes and humans are shown in figure 29.8. Apes have pronounced supraorbital ridges (bony crests above their eyes). In apes, the plane of the face projects forward, forming a muzzle, while in humans the plane is flat. The reason for the pronounced muzzle of the apes is the greater number and larger size of their teeth. Apes' canine teeth are much larger than the adjacent teeth, while the canines of humans are approximately the same size as the adjacent teeth. The two sides of the ape jaw are roughly parallel, giving a rectangular shape to the **dental arcade.** In humans, the tooth pattern curves gently and continuously, giving a broad U shape, or parabolic curve, to the arcade.

The face of *Ramapithecus* was flatter than the faces of the apes, and its teeth were more like those of the hominids, with much reduced canines. Due to the fragmentary nature of the fossil evidence, the shape of the dental arcade is uncertain. One specimen appears to be parabolic, while another is V-shaped, that is, intermediate between rectangular and parabolic. Another dental evidence that qualifies *Ramapithecus* as an early hominid (or direct ancestor) is the pattern of wear found on the molar teeth, which is very close to that observed in humans and not at all like that found in the pongids. The molars of *Ramapithecus* exhibit differential wear; that is, the first molar is more worn than the second, which is, in turn, more worn than the third. This pattern indicates delayed eruption of the teeth and suggests that physical maturity was delayed in *Ramapithecus,* a trait more characteristic of hominids than of apes.

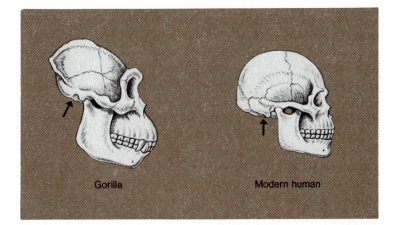

Figure 29.8 Comparison of an ape (gorilla) skull and a modern human skull. Ape skulls have prominent supraorbital ridges, which human skulls lack. The ape face projects forward, forming a muzzle, while the human face is flat. Apes have larger, heavier teeth and have especially pronounced canine teeth. The gorilla's brain capacity is only about 450 cc while that of this modern human skull is about 1,500 cc. Note also the difference in the position of the foramen magnum (arrows).

Gorilla

Modern human

But several important questions have arisen concerning the status of *Ramapithecus* in relation to evolution of the hominids. One of the most distressing problems in the paleontological study of hominid evolution is the very long gap in the fossil record between *Ramapithecus* and the next known fossil hominids, which come from strata that are nearly 10 million years more recent than the date of *Ramapithecus*. Could there have been conditions during that long period of time that made fossil formation unlikely? Or, are there undiscovered hominid fossils formed during this long time interval that are waiting in the rocks for paleontologists who look in the right places?

Other questions about *Ramapithecus* have been raised by recent biochemical projections of evolutionary rates. These studies, which compare amino acid sequences in proteins of living primates, are interpreted as indicating that hominids and pongids did not diverge until much later than the time of *Ramapithecus*. But paleontologists still think that the divergence must have occurred at about the time of *Ramapithecus* for there to have been enough time for the development of characteristics of the next more recent of the known hominid fossils.

Although some puzzling questions concerning *Ramapithecus* and its relationship to the evolution of the hominids remain, many biologists think that *Ramapithecus* should be classified as a hominid and that it is a representative of the organisms that made the important transition from the forests to the savannas.

The Australopithecines

A small skull discovered in 1924 in a limestone quarry in South Africa was sent to the anatomist Raymond Dart in Johannesburg for study. Dart named the skull *Australopithecus africanus* (southern ape of Africa) and proclaimed it to be a form intermediate between ape and human. At first, Dart's pronouncement created a great deal of controversy, but as more and more of these skulls were found in South and in East Africa, it became clear that *A. africanus* was a true, upright-walking member of the family Hominidae (table 29.4). Many physical characteristics of *A. africanus* are transitional between apes and humans. The brain capacity of *A. africanus* varies from just over 300 cc (cubic centimeters) to about 600 cc. The foramen magnum is farther forward than in the apes but not centered as in

Table 29.4
A Classification of the Family Hominidae.

Family Hominidae
    Genus *Australopithecus*
        A. afarensis
        A. africanus
        A. robustus
    Genus *Homo*
        H. habilis
        H. erectus
        H. sapiens neanderthalensis
        H. sapiens sapiens

Figure 29.9 Dental arcades of a (a) chimpanzee, (b) an australopithecine, and (c) a modern human. The two sides of the chimpanzee's jaw are roughly parallel, giving its dental arcade a rectangular shape. Its canine teeth are much larger than adjacent teeth, as they are in the jaws of all pongids (apes). The human jaw and the australopithecine jaws both curve gently to give a parabolic dental arcade. Note, however, that the large grinding molars (three rear teeth on each side) of australopithecines are much sturdier and broader than human molars.

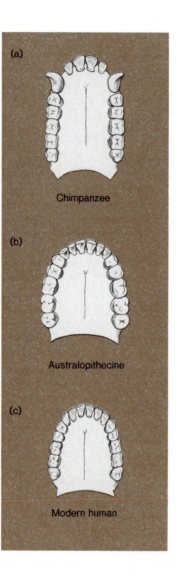

(a)

Chimpanzee

(b)

Australopithecine

(c)

Modern human

the humans. The dental arcade is generally U-shaped, and the canines are reduced (figure 29.9). The face is flatter than that of the apes, and the cranium is higher and rounder. The pelvis is shorter than in the apes, and the spine is S-shaped. The leg and pelvis characteristics indicate that locomotion was bipedal.

Another species of *Australopithecus, A. robustus,* has also been found in South Africa (figure 29.10). This hominid was larger than *A. africanus,* and its large jaw and strong teeth suggest that its diet was of rougher vegetation. Both species lived during the period from about 4 million years ago to about 1 million years ago, although *A. africanus* is a somewhat older species and may actually have given rise to *A. robustus.* Because of their different feeding habits, it seems unlikely that the two species competed for food or for living space.

Another important hominid fossil was found in 1974 in the Afar region of Ethiopia (figure 29.11) by Donald Johanson, who discovered a group of surprisingly complete specimens that he and his colleagues have named *Australopithecus afarensis.* Skeletal evidence indicates that this hominid walked fully erect some 3 million years ago. However, the skull capacity of *A. afarensis* is only 500 cc, and its dentition is persistently primitive; that is, the dental arcade is rectangular, and the canines are long. Some biologists have suggested that *A. afarensis* was ancestral to both *A. africanus* and *A. robustus* and that *A. africanus* gave rise, in turn, to *Homo,* the genus of modern humans. Johanson, however, thinks that *A. afarensis* may have given rise to two separate lines, one of which led to the other australopithecines and the other to the genus *Homo.*

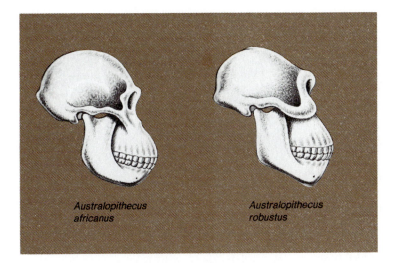

Figure 29.10 Skulls of *Australopithecus africanus* and *Australopithecus robustus*.

Australopithecus africanus

Australopithecus robustus

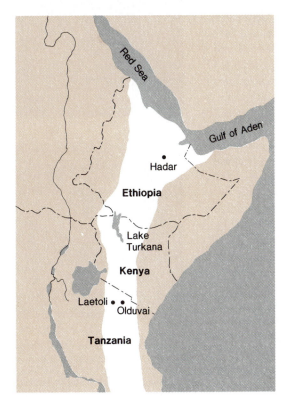

Red Sea

Gulf of Aden

• Hadar

Ethiopia

Lake Turkana

Kenya

Laetoli • • Olduvai

Tanzania

Figure 29.11 Location of the Rift valley of Africa. The Rift valley, which varies in width, lies within the lighter area of this map. It is part of a long line of depressions running down from Turkey, through the Jordan River and the Dead Sea area of Israel and the Red Sea, and on through eastern Africa. The Rift apparently has been produced by movement of continental plates. Several very important sites of hominid fossil discovery are located in or near the African Rift Valley.

There are, however, other fossils and still other interpretations. One of them comes from the "first family" of hominid paleontology, the Leakeys (figure 29.12), who have hunted hominid fossils at Olduvai Gorge and other sites in the Rift Valley of Africa for many years (see figure 29.11).

In addition to their many other important discoveries, Louis and Mary Leakey found an especially interesting hominid skull in 1964. The Leakeys considered this 2-million-year-old skull to be much more advanced than any of its australopithecine contemporaries, and they named it *Homo habilis* ("handy man", or "man who is able to do or make"). The Leakeys have maintained that the genus *Homo* is much older than previously thought, essentially as old as the australopithecines and, therefore, not descended from them. But there is much disagreement over this conclusion. Even though the skull of *Homo habilis* does have a comparatively large (650 cc) cranial capacity (table 29.5), some researchers prefer to classify it with the australopithecines. Thus, there are several suggested "family trees" of the hominids (figure 29.13).

Arguments about relationships of the hominid fossils are far from settled. Perhaps because of frustration with this situation, when Richard Leakey (son of Mary and Louis Leakey) discovered an exceptionally interesting fragmented skull in the Lake Turkana region of Kenya, he did not designate a specific classification but simply called it by its museum number. This "skull 1470," as it is called, had a cranial capacity of 800 cc, and it has been dated at 2.6 million years ago. Because of the fragmented condition of this skull, reconstruction was difficult, and there remains some uncertainty as to its original shape. There are also some disagreements about the dating of the skull. But its large cranial capacity (see table 29.5) and very early date support the Leakey hypothesis of the early origin of the genus *Homo*.

Another very significant set of finds from eastern Africa, especially from Olduvai Gorge, has been a number of small stones dating to some 2.5 million years ago that appear to have been intentionally chipped so that they could be used for cutting, pounding, and scraping; that is, they were made and used as tools (figure 29.14). At a slightly later date, there is indisputable evidence that rocks were worked and used as tools—large cobbles that were hit against one another

(a)

(b)

(c)

Figure 29.12   The search for the early hominids. (a) Mary and Louis Leakey at work in Olduvai Gorge. Louis Leakey died in 1975, but Mary Leakey has continued her work on fossil hominids, as has their son Richard. (b) Photograph of the Leakeys' excavations in Olduvai Gorge, one of the most important sources of early hominid fossils. (c) Richard Leakey and Donald Johanson examine fossils at the Kenya National Museum in Nairobi, Kenya.

Table 29.5
Cranial Capacity of Hominids.

Name	Cranial Capacity (in cc)
Australopithecus afarensis	500
A. africanus	300–600
A. robustus	300–600
Homo habilis	650
H. erectus	750–1,300
H. sapiens	1,400–1,700

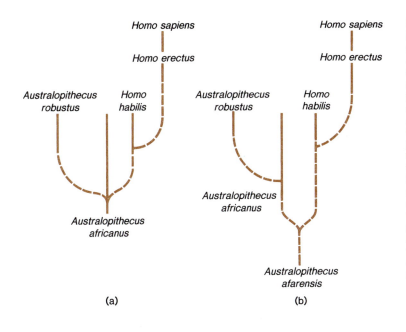

(a)   (b)

Figure 29.13 "Family trees" of the hominids. (a) A "standard" version of hominid relationships with *Australopithecus africanus,* or a similar form, giving rise to both the australopithecine and *Homo* lines. (b) A version that includes Johanson's views regarding *A. afarensis* as an ancestor of all of the hominids. There are several points of conflict. Many researchers would include *Homo habilis* in the main line of evolution of modern humans. And the Leakey hypothesis (not shown here), for which there is considerable evidence, proposes that the divergence between australopithecines and the genus *Homo* occurred much earlier and that the two groups have descended as separate evolutionary lines.

Figure 29.14 Examples of the earliest human tools, in the style of those found by the Leakeys in Olduvai Gorge. These simple tools probably were used as choppers in food preparation.

to produce sharp cutting edges. Who made these tools? Picking up and using objects as tools is common among primates. Chimpanzees and gorillas, for example, pick up and use objects as tools or weapons. But who consciously set out to make these tools more than 2 million years ago? One standard answer is that the australopithecines made tools. But another answer might be that while australopithecines may have found and used various objects as tools, the design and fabrication of tools is and has been a human enterprise that was limited at that time to the work of *Homo habilis* or possibly other members of this genus. Here again, as is often the case, there are only partial answers for questions about human evolution.

The Genus Homo

The earliest fossils that are accepted by all paleontologists as being members of the genus *Homo* are from a series of finds in Africa, Asia, and Europe. Collectively, these fossils are called *Homo erectus*.

Homo Erectus

The oldest of the *Homo erectus* fossils dates to about 1.5 million years ago and the youngest to about 300,000 years ago. The leg, pelvis, and associated structures indicate that members of this species were bipedal and fully erect in posture. The cranial capacity was large, 750 to 1,300 cc, as compared with 1,400 to 1,700 cc for modern humans. *Homo erectus* differed from modern humans in that it retained some primitive skull features: the face was more projecting than that of modern humans, and the cranial bones were thicker. The teeth were large, and the lower jaw sloped back so that there was no distinct chin (figure 29.15).

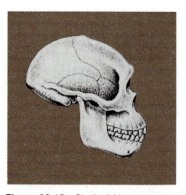

Figure 29.15 Skull of *Homo erectus. Homo erectus* was a widely distributed species, and many important human fossil finds (including "Java man" and "Peking man") are now assigned to this species. Earlier, the species was called *Pithecanthropus erectus*, but now there is general agreement that it belongs in the genus *Homo*.

The oldest fossil remains of *H. erectus* have been found in Africa and younger ones in other parts of the world, a pattern that suggests that the species originated in Africa and migrated outwards to other suitable habitats—an event that took place in a relatively short period of time. This migration occurred at a time when continental environmental conditions would have placed severe selective pressures on the species. The rather rapid development of typically human characteristics during this period may have been strongly influenced by these conditions since the more ingenious and dexterous individuals would have been more likely to survive and reproduce.

Two sites explored near Torralba, Spain, provide evidence that these early humans were big game hunters. The bones of 150 animals have been found here, including elephants, deer, horses, wild cattle, and rhinoceroses. There is evidence to suggest that these early humans set fire to the grass to force the large animals into a swampy region, where they could be more easily killed. The tools found at Torralba are much more elaborate than the older, more primitive tools found at Olduvai. Tools of *Homo erectus* included heavy, wedge-shaped choppers, small hand axes, and various small tools, such as scrapers, borers, and engravers.

Another site, found during the excavation of an apartment complex at Terra Amata in Nice, France, has provided some insight into the daily life of *Homo erectus.* This site, dating to about 400,000 years ago, contained several temporary shelters of about five meters by nine meters each. The shelters, built at different times, contained hearths, animal bones, and a fragment of a wooden bowl. Evidence clearly indicates that these humans used fire for cooking. Near the shelters were found several pieces of the natural pigment red ocher, sharpened to a point as though it had been used for body adornment. It seems likely that these were temporary shelters, used by tribe members as they followed the annual migration of the big game animals they hunted. These evidences indicate that *Homo erectus* was a social animal, and that cooperative effort and some form of social structure contributed to group survival.

The earliest *Homo erectus* skulls had an average cranial capacity of about 940 cc, while the later ones averaged about 1,000 cc. A small difference, but one that reflects the overall tendency toward increase of brain size in human evolution (see table 29.5). In general, increase in brain size is correlated with increased intelligence.

Homo Sapiens

Modern humans, *Homo sapiens,* display a number of advances in skull structure over *H. erectus.* These changes include an increase in cranial capacity from an average of 1,000 cc to an average of about 1,500 cc; a decrease in the size of the teeth and jaws; flattening of the plane of the face; rounding of the cranium; and lightening of the cranial bones. How long ago did *Homo sapiens* emerge? Two specimens of early *H. sapiens,* one from Swanscombe, England (dated 250,000 years ago), and another from Steinheim, Germany (dated 200,000 years ago), are somewhat transitional in their skull features but are classified as *H. sapiens.*

About 100,000 years ago a distinctive group of humans called **Neanderthals** emerged. The name Neanderthal comes from the Neander Valley in Germany, where the first fossil of this type was unearthed. Actually, the Neanderthals were widespread geographically, with fossils being found in much of southern Europe, Asia (including China, Java, and Sumatra), and in northern, eastern, and southern Africa. During the first half of the tenure of the Neanderthals, the climate was temperate, but later a new glacial period developed, reaching its maximum late in the Neanderthal period, which lasted until about 40,000 years ago.

There is extensive evidence that the Neanderthals carried out ritualistic practices, which is strongly suggestive of abstract thought and, perhaps, even a concept of religion and life after death. The arrangement of bear bones and rocks in many caves suggests that a bear cult existed among some Neanderthals, and they also were known to bury their dead. A Neanderthal skeleton found in Shanidar Cave in Iraq was buried some 60,000 years ago amidst a profusion of flowers.

Usually, the Neanderthals are classified as *Homo sapiens neanderthalensis,* a variety of modern humans. The distinction is based on skeletal characteristics, including some persistently primitive skull features (a protruding forehead, no chin, and a flattish skull) (figure 29.16). But the Neanderthal brain capacity was essentially the same as that of modern humans. Some researchers suggest that the Neanderthals gave rise to modern humans *(Homo sapiens sapiens);* others believe that the Neanderthals were modern humans and that they simply displayed a normal degree of structural variation.

Figure 29.16 Early humans.
(a) Restored skulls of *Homo erectus,*
a Neanderthal man, and a Cro-
Magnon man. There are significant
differences in shape and size of
craniums, supraorbital ridges, and
sizes and shapes of chins and jaws.
(b) Artistic reconstruction of possible
facial features of the three types.

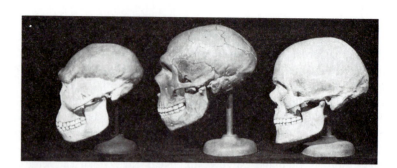

(a)

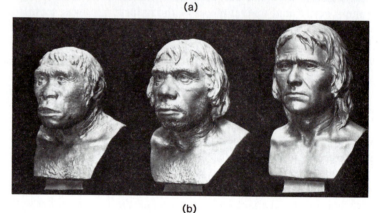

(b)

Fossils of modern humans appear in the fossil record near the end of the Neanderthals' time. These earliest modern humans have been designated as **Cro-Magnon** people and are classified as *Homo sapiens sapiens* (as are present-day humans) to distinguish them from the Neanderthals. After a brief period of coexistence with the Cro-Magnon people, the Neanderthals disappeared, and from about 40,000 years ago onward, no more Neanderthal fossils occur in the fossil record.

The rapid (in terms of geological time) disappearance of the Neanderthals is somewhat of a mystery. Were they eliminated as a result of general competition with the Cro-Magnons, or were they actually annihilated in warfare with the Cro-Magnons? Some biologists think that the two groups interbred to such an extent that distinctive Neanderthal characteristics were no longer recognizable, and the Neanderthals were simply incorporated into the general *Homo sapiens* population.

About 40,000 years ago, a new tool industry spread through the human population. The major characteristic of this industry was the production of the blade, a tool with roughly parallel sides (figure 29.17). Many other tools also were made

Figure 29.17 Blades produced by Cro-Magnon people. Some Cro-Magnon blades were as much as 30 cm long and only 1/2 cm thick. The blades represent the pinnacle of stone tool manufacture. (Compare them with the crude choppers in figure 29.14.) The next major step in human tool making occurred with the advent of metal working.

from a variety of materials, including bone, antler, wood, and ivory. These increasingly complex tools are physical evidences of the growth and spread of human culture. People taught other people how to use and make tools. This information (and very likely a great deal of information about many other subjects) was transmitted by personal communication. There was now a human **culture,** a body of information transmitted from generation to generation by means that did not depend on genetic mechanisms. Human evolution had entered a new and important phase, a phase involving cultural evolution, that has continued at an accelerating pace to the present time.

Cultural Evolution

In several important ways the ancient hominids and the living great apes differ very little from modern humans. Humans, chimps, and gorillas possess over 90 percent of their genes in common; differences in skeletal anatomy of apes, early hominids, and modern humans are more a matter of degree than of any absolute differences in morphology; and the roots of human social behavior can be found in other primates' behavior. The most significant differences between modern humans and ancient hominids—or humans and apes for that matter—are the increased intelligence and cultural advancement of modern humans. Culture is passed from one generation to the next through teaching and example, and it is dynamic since it is continually modified as it is transmitted. Cultures do not evolve in the biological sense, but they do display an overall progressive change with occasional large-scale advances that take place in relatively short periods of time.

The earliest cultural stage of the hominid line was that of the hunter-gatherer, a stage in which hominids first learned to hunt small animals and later the larger herd animals. The second major cultural stage was agriculture, a change in life-style that greatly altered not only the means by which humans provided for their physical existence but also marked a fundamental change in structure and values of human societies. The agricultural period extended from approximately 12,000 or 15,000 years ago to about 200 years ago, a period during which there was relatively little change in agricultural practices or in the basic values of the agriculturalist.

However, a trend was taking place throughout this period, a trend that was alien to the agricultural existence, that is, the growth of cities. The cities became more and more the home of the merchant, the artisan, and the ruling class. Then, about 200 years ago, a third major stage of human cultural development began with the initiation of the Industrial Revolution, a technological and economic movement that was to have a profound influence not only on the growth and structure of cities, but on all areas of human endeavor.

Each of these phases of cultural evolution is now briefly examined.

The Hunter-Gatherers

From their earliest beginnings, humans have been hunter-gatherers, and some contemporary peoples, such as the !Kung of the Kalahari and the Bushmen of Australia, still make their living in this way. In such cultures, the men wander afield to hunt for game, and the women and children gather fruits, nuts, herbs, and rootstalks around the campsite. This type of existence places a premium on communal effort. A single man is neither strong enough nor fast enough to track down and kill most game animals, but cooperative hunting efforts can be quite effective. Also, an individual woman cannot forage for food, care for an infant, and perform the myriad tasks required around the homesite. But, a group of women of different ages can share the work effectively. In the hunter-gatherer society, the women and children provide over two-thirds of the food for the tribe, but game is the favorite food, and hunting forms the central theme of these tribes in a ritualistic sense.

The hunting-gathering existence requires intelligence, cooperative effort, and a ruggedness of disposition. Members of these tribes spend, on the average, twelve to twenty hours per week working, with the remainder of their time spent around the campsite indulging in such communal activities as playing music and dancing, telling stories, reciting tribal histories, and training children in toolmaking and usage. Members of hunting-gathering tribes are generally well fed, the adults are resistant to endemic diseases, and epidemics are rare except when a disease is introduced from outside the tribe.

The making and use of tools is important to the hunter-gatherer. Their tools include digging sticks to unearth edible roots, spears for hunting, scrapers to clean hides, and a variety of tools for cooking, clothesmaking, and so on. For the most part, early tools were items associated with obtaining and preparation of food, or the making of clothing. But early humans made objects for other purposes as well. Other **artifacts** (objects fashioned by humans) associated with the hunter-gatherer stage of human development include bone whistles, necklaces, sculptured figurines, knives, spears, scrapers, awls, and cave paintings. Some of these obviously were used in ritualistic practices, perhaps early efforts at obtaining some magical control over nature. Many cave paintings (see page 3) suggest that the artists were attempting to guarantee success in the hunt, and some female figurines strongly suggest that they were fertility symbols used to guarantee the production of offspring. Burial of the dead dates to about 60,000 years ago and suggests a belief in an afterlife because flowers, food, and implements were sometimes interred with the dead.

The decline of the hunter-gatherer culture began some 20,000 to 15,000 years ago, brought about by a diminishment in the number of large game animals upon which they depended, a catastrophic event in the fortunes of these people. It is not possible to look back into prehistoric times and provide a strict cause-effect relationship for events that occurred so long ago, but many experts believe that the ancient hunter-gatherers were quite possibly the authors of their own decline. The tribesmen had become efficient hunters, and possibly they began to kill many more animals than were needed for food and clothing. The decline of the game herds coincided with a period during which the human population of the earth was increasing rapidly. It has been estimated that 30,000 years ago the human population of the earth was between 1 and 3 million. By 12,000 years ago, the population had increased to 6 to 10 million. The combination of decreased food supply and increased population provided a strong impetus for people to find a new way of obtaining food.

Agriculture

Agriculture appeared independently in several places in Eurasia and the New World some 12,000 to 9,000 years ago, and there is even evidence that people of the Nile Valley were milling grain as early as 15,000 years ago. It is believed that the first steps toward an agricultural existence were taken by some ancient hunter-gatherer tribes who settled close to a natural supply of grain or rootstalk and gave up following the declining herds. At first these people probably harvested crops without seeding for new ones and depended on natural processes to provide a new crop each season. This probably worked well in tropical or semitropical environments, where there is always some variety of fruit, nut, or seed available, and many rootstalks grow year-round. These people probably continued to supplement their vegetable diet with whatever game was present in the immediate area. However, as populations continued to grow, shortages of food supply would force some members of the tribe to move to a new area. Possibly, these early migrants carried with them seeds and cuttings that they hoped could be used to produce the same types of foods they were used to eating. Slowly, a variety of agricultural practices grew up that helped in the tenuous processes of making a living from the earth, and a new set of tools was fashioned to aid in the arduous labor involved in farming. Along with these changes in the way these humans obtained their food came changes in their social life. Humans were no longer nomadic; instead they were tied to a specific geographic location. Their horizons were narrower, and rather than viewing land as a base for the migrating herds, they now thought of land ownership and means of protecting it. The need to defend a territory was introduced into the human consciousness, and warfare between tribes increased as a result.

The hours of labor were longer for the agricultural people than they were for their hunter-gatherer ancestors. Sharing among tribe members decreased as individuals spent most of their time working their own land. And, while in good times food was more abundant, a single crop failure could create widespread famine.

On the more positive side, the life of the farmer was more settled and stable than that of the hunter-gatherer. Housing was permanent and generally of better quality. In fact, the larger families of the farmer were made possible by the increased amount of food and the more settled conditions of life. The growth of agriculture also was important to the growth of cities, most of which began as regional trading centers.

By about 9,000 years ago, agricultural practices were well developed, and in the Middle East, a number of animals, including sheep, cattle, goats, pigs, and dogs, had been domesticated. These animals not only were an important source of food and clothing; some of them were also used as draft animals. A fundamental difference between the development of agriculture in the Middle East and the New World was an almost complete lack of draft animals in the New World.

An entirely new array of tools accompanied the development of agriculture, and the origin of metallurgy some 8,000 years ago gave great impetus to the making of tools for agriculture and warfare. Once humans learned to produce temperatures hot enough to melt and cast copper, the use of other metals, such as lead, silver, and iron, followed. Copper itself is too soft to hold a sharp edge, but bronze, an alloy of copper and tin, proved invaluable in the making of a wide variety of cutting implements.

As populations grew, again people were forced to migrate to new areas, seeking new lands to farm. As they entered new areas, they cut down trees, and often they burned off the surface vegetation, preparatory to planting their first crops. But much of the farmland obtained in this way was not really suitable for long-term agricultural usage. Some experts believe that the spread of this cut-and-burn agriculture has been responsible for the spread of nonarable lands, even the formation of deserts. The cutting of native forests and the subsequent overfarming of these lands has definitely been responsible for the spread of deserts in India within historic times. The desertification of the Fertile Crescent of the Tigris and Euphrates Rivers may well have resulted from a similar sequence of events. And there is good evidence that the ancestors of the Egyptians lived in lush forests where vast expanses of the Sahara spread today. Human agricultural activities caused large-scale changes on the earth. Clearly, a new relationship between organism and environment was evolving.

The Industrial Revolution

The third major cultural stage of human development is in large part a product of the urbanization and specialization that has always been an integral part of civilizations. The earliest cities originated as trade centers and grew as artisans and shopkeepers formed a new class of human industry. The long history of toolmaking, practical chemistry, and artisanship reached its culmination in the eighteenth century in the initiation of the Industrial Revolution which, from its inception, has made it possible to produce many articles cheaply and efficiently, placing many objects within the reach of the general public that formerly were reserved for the rich. The Industrial Revolution had the side effect of increasing the growth of the cities by making farming more efficient and thus reducing the need for farm laborers while at the same time providing many jobs in urban areas for those displaced from the farm. The initial result of this mass movement from an agricultural society to a manufacturing society was chaotic and tragic because people in early industrialized societies often were forced to live and work under miserable conditions. Also, industrialization in some parts of the world has been accomplished using resources drawn from many other parts of the world. As a result of this unequal consumption of the earth's resources, the relative abundance of the Industrial Revolution has not been shared equally by all humans.

Another problem associated with the Industrial Revolution is environmental degradation. Some early human activities had adverse effects on the environment, such as the decline and extinction of some species, and the overfarming and desertification of large land areas. But present human culture has the power for disruption and destruction of the environment that far exceeds that of any other stage of human development. The burgeoning world population is placing an ever greater strain on natural resources, and the waste products of industry are poisoning the environment. The very future survival of human beings, and the many other organisms that human activities affect, may depend on entry into a new cultural stage, one in which the emphasis is placed on the ability of humans to live in harmony with their environment. It seems that if this does not occur, the quality of human life must inevitably decline in the future.

Summary

The study of human evolution has produced considerable controversy, including substantial disagreement about interpretation of the fossil record.

Humans are primates. Primates evolved as arboreal animals, and primate characteristics generally reflect adaptations for arboreal life. The oldest primates are the prosimians. Modern prosimians include lemurs and tarsiers. Monkeys, apes, and humans belong to the suborder Anthropoidea. Monkeys have diverged into New World monkeys, which have prehensile tails, and Old World monkeys.

The dryopithecines were ancestors of apes and humans. They were forest dwellers who may have spent some of their time on the ground. Climatic changes that led to forest shrinkage and produced vast savannas provided habitats for ground-dwelling hominid ancestors such as *Ramapithecus*. Following *Ramapithecus,* there is a several million year break in the fossil record of hominid evolution.

The australopithecines, which lived from about 4 million years ago to about 1 million years ago, definitely were hominids, and some of them were on, or near to, the line of human ancestry. *Homo habilis* was a contemporary of the late australopithecines, and its early emergence may indicate very early divergence of australopithecines and the genus *Homo*. Hominids made simple tools more than 2 million years ago.

Homo erectus emerged about 1.5 million years ago and spread rapidly over large areas of Africa, Europe, and Asia. *H. erectus* made more elaborate tools, hunted in groups, built shelters, and apparently had some sort of social structure.

Modern humans, *Homo sapiens,* emerged between 250,000 and 200,000 years ago. From 100,000 years ago until about 40,000 years ago, Neanderthal humans lived in much of Europe and Asia. They were replaced (absorbed?) by modern humans, who have been dominant for the past 40,000 years.

Recent human evolution has been marked by rapid cultural evolution. Humans have progressed from hunter-gatherer societies to agricultural societies to the Industrial Revolution and the social development that has followed it.

Questions

1. What human characteristics result in humans being classified as mammals?
2. What are some adaptive advantages of erect posture and bipedal locomotion?
3. Why is it useful for development of an understanding of human evolution to study chimpanzees, gorillas, and baboons in the wild?
4. Explain the fundamental disagreement between the Leakey hypothesis and the views of some other paleontologists regarding the early history of the genus *Homo*.
5. What are some differences between the skulls of *Australopithecus* and modern humans? between *H. erectus* and modern humans?
6. What reasons might be suggested for the decline of the hunter-gatherer culture?
7. What was the probable relationship between population size and the beginning of agriculture?

Suggested Readings

Books

Leakey, R. E. 1981. *The making of mankind.* New
   York: E. P. Dutton.
Leakey, R., and Lewin, R. 1977. *Origins.* New
   York: E. P. Dutton.
Pfeiffer, J. E. 1978. *The emergence of man.* 3d ed.
   New York: Harper and Row.
Pilbeam, D. 1972. *The ascent of man: An
   introduction to human evolution.* New York:
   Macmillan.

Articles

Day, M. H. 1977. The fossil history of man. 2d ed.
   Carolina Biology Readers no. 32. Burlington,
   N.C.: Carolina Biological Supply Co.
Hay, R. L. and Leakey, M. D. February 1982. The
   fossil footprints of Laetoli. *Scientific American.*
Isaac, G. April 1978. The food-sharing behavior of
   protohuman hominids. *Scientific American*
   (offprint 706).
Johanson, D. C., and Edey, M. A. 1981. Lucy: A 3.5
   million-year-old woman shakes man's family
   tree. *Science 81* (March).
Napier, J. R. 1977. Primates and their adaptations.
   2d ed. *Carolina Biology Readers* no. 28.
   Burlington, N.C.: Carolina Biological Supply
   Co.
Simons, E. L. May 1977. Ramapithecus. *Scientific
   American* (offprint 695).
Trinkaus, E., and Howells, W. W. December 1979.
   The Neanderthals. *Scientific American* (offprint
   722).
Walker, A., and Leakey, R. August 1978. The
   hominids of East Turkana. *Scientific American.*
Washburn, S. L. September 1978. The evolution of
   man. *Scientific American* (offprint 1406).

Part 6

Behavior and Ecology

Organisms continually interact with their physical and biological environments. Interactions such as selective exchanges of materials with the environment and the effects of light and temperature on organisms have been discussed in previous chapters. In chapters 30–33 some additional interactions between organism and environment, especially interactions among various living things, are examined.

Animals, for example, demonstrate behavioral responses to information about their environment that they receive through their senses. They interact with members of their own species in very specific ways, especially in the context of reproductive activity. Animals also respond to members of other species. Often, these interactions between species take the form of predator-prey contacts where the outcome of the contact determines whether the organism obtains food or does not obtain food, or whether the organism becomes food for another animal or successfully avoids becoming food.

There are many other complex relationships among organisms, and living things can only be understood in the context of their membership in and interactive contributions to populations, communities, and ecosystems.

Animal Behavior

Chapter Concepts

1. Behavior allows animals to respond rapidly to the problems they encounter.
2. Organisms are genetically programmed through natural selection to behave in an adaptive way.
3. In some cases, it is possible to demonstrate directly that a behavior is under genetic control.
4. In some species, there is selection for rigid, relatively unvarying responses to a given situation, while in others selection produces behavior that is highly modifiable through experience (learning).
5. The expression of certain behavior patterns requires that an organism be at a given state of morphological and physiological maturation.
6. Hormones control behavior by altering the internal motivational states of organisms.
7. It is possible to gain an understanding of a wide variety of behavior by asking two basic questions: (a) Of what benefit is the behavior to the organism? (b) How might the behavior have evolved?
8. In recent years, sociobiology has provided a new way of looking at the genetics and evolution of social behavior, but application of sociobiological principles to human behavior has caused controversy and mixed reactions.

Walking through a gull colony can be an exciting if somewhat baffling experience. Gulls fly overhead in all directions, calling, swooping down, sometimes buffeting the intruder's head with their feet. Gulls and their young are in constant danger from predators. Crows, humans, and small rodents steal their eggs, larger gulls eat their chicks, and foxes attack them while they are sitting on their nests. Thus, it is not surprising that gulls attack humans just as they do any other potential marauders that wander near their nests (figure 30.1).

But if one hides in a blind, the gulls soon calm down and return to their normal activities. Some individuals stand alert by their nests or raise their heads in the raucous long-call so characteristic of their species. At some nests, pairs are courting, jerking their heads up and down in a chokinglike motion, while at other nests the birds are fighting with their neighbors. Other gulls are settled on their nests, incubating their eggs, while their mates stand vigilantly nearby.

Gulls' nests are made of a few stones and some grass scraped together into a shallow cup. Although the construction appears haphazard, the nests are not scattered randomly around the gull colony. Like houses in a suburban development, there is a regular spacing between neighboring nests. Each gull staunchly defends the area (**territory**) around its nest, allowing only its mate into the territory. The female lays several mottled, greenish-brown eggs that are well camouflaged against the nest background. The two gulls take turns incubating the eggs for three weeks, and then they collaborate in the demanding task of feeding the young.

Shortly after the time of hatching, parent black-headed gulls systematically pick up empty eggshells from around the nest and remove them. If one of these discarded shells falls by the nest of a neighbor, the neighboring gull, in turn, also removes it.

Figure 30.1 A black-headed gull.

Many people might have ignored this apparently trivial piece of behavior, but it intrigued Nobel prize-winning **ethologist** Niko Tinbergen. Ethologists such as Tinbergen analyze behavior patterns in the field and are particularly interested in the adaptive significance of behavior. Tinbergen and his associates knew that not all birds do such meticulous housekeeping around their nests. For example, ducks, hawks, and even some closely related cliff-nesting gulls called kittiwakes simply trample the broken eggshells into their nests, or sometimes they eat the shells. Why should the black-headed gull and other ground-nesting gulls take such pains to remove the shells?

There seemed to be several possibilities. For one, the sharp edges of the broken shell might be harmful to the chick or uncomfortable for the brooding parent. But sharp eggshell edges are presumably just as sharp for the kittiwakes, yet kittiwakes do not systematically remove eggshells from their nests. Another possibility was that the behavior might have something to do with reducing the risk of predation on the chicks. Only the outside of the gull's eggshell is camouflaged. Eggshells are basically white because they contain a great deal of calcium, and the pigment is added only to the outside after most of the shell is formed. The empty shell with its jagged edge and white interior might flash like a beacon to any passing predators, especially to crows flying overhead. As long as the empty shells are in or near the nest of a gull that nests on flat ground, the tiny chicks could be in danger.

Since kittiwakes nest on steep cliff-edges, they suffer much less predation than gulls that nest on flat ground (figure 30.2). Thus, kittiwakes may not need to remove the eggshells because crows do not maraude their nests anyway.

Tinbergen and his colleagues did a series of experiments to test these ideas about predation and eggshells. But before they attacked the problem directly, they made some preliminary observations. They observed that the gulls also removed a wide assortment of highly visible items from their nests. The gulls removed bottle caps, aluminum foil, film boxes, clam shells, and just about anything else that had been carried to the nests from the garbage dumps where the gulls feed.

The scientists wondered whether the gulls were selectively picking out and removing the most conspicuous objects, as the hypothesis would predict. So they put dummy plaster eggshells in assorted colors, sizes, and shapes into the nests of wild gulls. The scientists then hid in a blind nearby and observed gulls' responses to the shells. When a gull returned and settled down on its nest, it pecked at the dummy eggshell for a few moments and then picked it up and discarded it from the nest. The more the dummy looked like the conspicuous inside of the gull's egg, the more likely the gull was to pick it up and carry it from the nest. But if the dummy eggshell closely resembled the camouflaged color of the outside of the gull's own eggs, the gull was less likely to remove it from the nest. Not surprisingly, the gulls seldom removed anything that looked like an unhatched egg.

Thus, the results of these experiments tended to confirm the idea that the more conspicuous items that did not resemble intact gull eggs were more likely to be removed, but, actually, how conspicuous were these shells to real would-be predators?

This was a more difficult question to answer. It would have been useful to introduce shells into ordinary nests containing eggs and see if nests with empty shells were more likely to be attacked by predators. But, of course, the parent gulls prevented this experiment because they would never allow the shells to stay near the nests long enough for the test to take place. To perform this experiment, Tinbergen and his colleagues built artificial nests in various places around the edge of the gull colony and placed a clutch of eggs in each mock nest.

Figure 30.2 Kittiwakes nest on steep cliffs. Thus, they suffer less predation than other gulls that nest on flat ground.

Eggs in nest ←— Distance —→ Broken eggshell	Percentage of eggs destroyed by predators
←15 cm→	42
←— 100 cm —→	32
←— 200 cm —→	21

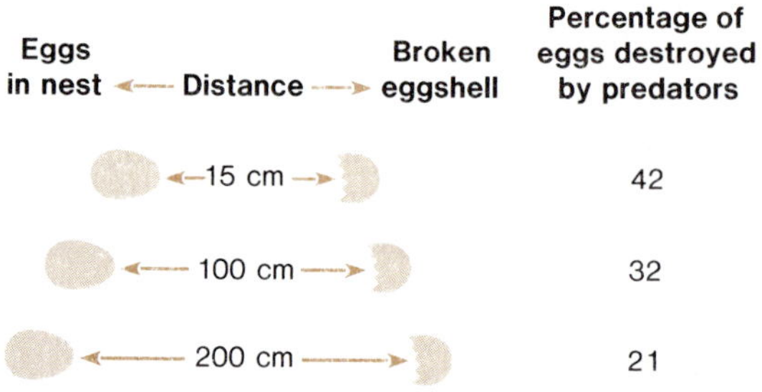

Figure 30.3 In an experiment with mock black-headed gull nests, predation was more likely if eggshell fragments were located close to the nest.

Before doing the experiment with empty shells, they had to confirm first that unhatched gull eggs are really camouflaged. So, in some of their mock nests they put white hen's eggs, and in other nests they put gull's eggs. The result was not quite what the scientists had expected. All eggs in all of their experimental nests were destroyed by crows! Undaunted, Tinbergen realized that he had made it much too easy for the crows. To make it more difficult to find the eggs, he put some grass over the gull's and hen's eggs in the artificial nests. Now he did find a difference. The white hen's eggs were much more likely to be eaten than the mottled, camouflaged gull's eggs.

Now Tinbergen was ready to do the key experiment. What was the effect of having empty shells near the nest? He and his colleagues put broken eggshells at varying distances from mock nests with gull's eggs in them. Exactly as their hypothesis had predicted, the nearer the eggshells were to the nest, the more likely it was that the eggs in the nest would be eaten (figure 30.3). It seems, then, that a parent gull saves its chicks from being eaten by removing the empty eggshells from around the nest.

But one puzzling feature of the behavior remained. Despite the great risk from predation, the parent gulls did not remove the eggshells immediately after their chicks hatched. If the nests were so predictably destroyed when eggshells were present, why did the gulls wait several hours after the chick had hatched before removing the broken shell? Tinbergen made a fascinating, if somewhat gruesome, observation. When parent gulls flew up and left a chick that had not yet dried out after hatching, the chick was very likely to be eaten by a neighboring gull. The neighbor simply walked into the territory and gobbled up the helpless chick, swallowing it whole. But after a chick had time to dry out and gain a little strength, it could run and hide in the grass if the parents flew up from the nest, and thus could escape cannibalistic neighbors. Tinbergen reasoned that there were really two factors affecting eggshell removal. The parents need to remove the eggshells quickly, but as long as the chicks are still wet, it is better that the parents not leave the nest, even for the few minutes required to carry off a broken eggshell.

While it appears that parent gulls are considering the welfare of their offspring and planning their actions carefully, there is no reason to believe that they are thinking out the consequences of their behavior. Removing eggshells does protect the young, but as far as the parent is concerned, it is simply responding mechanically and automatically to a **stimulus** in the environment. It is as if gulls are programmed by their genes to behave in this manner when they encounter eggshells or other conspicuous objects in or around their nests. In evolutionary terms, the complex combination of genes that causes gulls to remove objects from their nests is more likely to survive (and thus be passed on to future generations) than other genetic combinations that produce other behavior patterns. If a mutant gull that did not remove eggshells appeared, chances are that all of its eggs would be eaten, and the mutant genes would not be passed on to another generation of gulls. There is **natural selection** for eggshell removal behavior, and **selection pressure** resulting from predation maintains the genes for the behavior in the population.

Simultaneously, however, there is also selection against the gulls leaving wet chicks unattended, even for a few moments. Hence, gulls with genes that program them to remove eggshells immediately are less likely to raise their young than gulls with genes that program them to wait before removing eggshells. This is a good example of the delicate balance between complex selective pressures that results in the evolution of the finest details in the timing and form of behavioral responses.

To discuss how a trait such as eggshell removal behavior might have evolved, one important assumption had to be made. That assumption was that the trait was **inherited,** that is, that the offspring of parents with the behavioral trait would be more likely to show the behavior than offspring of parents without the trait. Somehow, the behavior must be programmed by the genes, and those genes must be passed in an essentially unaltered form to the next generation, whose individual members are then similarly programmed. But how do genes actually program such complex behavior?

Genes act directly by controlling protein synthesis and thereby exert their effect through control of development and of physiological processes. However, such control is too slow and too indirect to be involved in animals' responses to immediate problems. Therefore, animals' genes act to build bodies with nervous systems that make possible quick behavioral responses to changes in the environment. Plants also respond to changes in their environment, but by and large these are slow alterations in patterns of growth, or changes in turgor pressure that regulate water loss.

Thus, animal behavior is an extraordinary evolutionary invention. **Behavior,** muscular contractions under nervous control, allows animals to respond quickly and adaptively. Animal behavior is not limited to automatic stereotyped responses that are genetically determined in advance. Genes may also program animals to be able to **learn** to respond in particularly adaptive ways.

Genes and the behavior they program are preserved because the individuals in which they occur are more likely to survive and pass their genes on to the next generation.

Genes and Behavior

The idea that genes control behavior needs to be examined more closely. Behavior patterns that clearly are inherited have been described for a number of organisms, ranging from unicellular organisms such as the protozoan *Paramecium* to complex animals such as birds.

Swimming in *Paramecium*

The single-celled *Paramecium* is covered with tiny hairlike cilia. A *Paramecium* moves through the water in a spiral fashion, propelled by its beating cilia. When it hits something, it backs up and moves away. Much of the detailed mechanism of this **avoidance response** is now understood. Running into an object causes a depolarization of the plasma membrane at the point of contact. This depolarization results from a sudden influx of calcium ions into the cell. The depolarization spreads quickly over the rest of the cell's plasma membrane. This process is very similar to the way that impulses pass through the nerve cells of animals (see p. 432). The depolarization of the *Paramecium's* plasma membrane causes a reversal in the direction of movement of the cilia, and the *Paramecium* backs up (figure 30.4). Within a few seconds, the membrane becomes repolarized again by active transport of ions back across the membrane, and the *Paramecium* resumes its forward path.

A number of intriguing *Paramecium* mutants that differ in their swimming behavior have been found in the laboratory. One of these is a mutant named "paranoic." When a "paranoic" *Paramecium* runs into an object, it backs up as usual but continues swimming backwards much longer than a normal *Paramecium* would. This abnormal behavior apparently results from a failure in the normal process of plasma membrane repolarization. Because of a problem in the structure of the cell's plasma membrane, the sudden influx of ions during depolarization is not turned off properly, and the cell remains depolarized much longer than normal. This "mistake" in membrane structure is known to be caused by a change in a single gene.

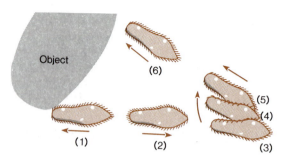

Figure 30.4 Avoiding response of *Paramecium*. When the organism swims into an object (1), it backs up by momentarily reversing the direction of its beating cilia (2). After the *Paramecium* is positioned at a different angle (3–5), the cilia resume their normal movement and the organism once again moves forward (6).

Thus, a change in a gene results in a change in the electrical properties of the plasma membrane, which, in turn, has a direct effect on the behavior of the organism.

Nest Building in Lovebirds

Lovebirds are small, green and pink African parrots that nest in tree hollows. There are several closely related species of lovebirds of the genus *Agapornis*. The species differ in the way they build their nests. The female of one species picks up a large leaf (or in the laboratory, a piece of paper) with her bill, perforates it with a series of bites along its length, and then cuts out long strips. She carries the strips to the nest in her bill and weaves them in with others to make a deep cup. The female of another species cuts somewhat shorter strips in a similar manner, but then she carries them to the nest in a very unusual way. She picks up the strip in her bill and inserts the strip deep into her rump feathers (figure 30.5). In this way she can carry several of these short strips with each trip to the nest, while females of the first species can carry only one of the longer strips at a time.

(a)

Figure 30.5 Nest-building behavior in lovebirds. (a) Fischer's lovebirds carry strips of nest material in their bills, as do most other birds. (b) Peach-faced lovebirds tuck strips of nest material into their rump feathers before flying back to the nest.

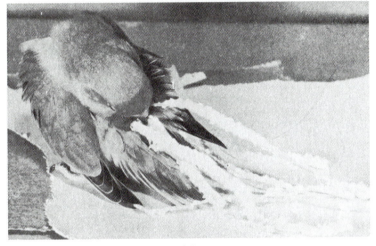

(b)

Crosses between these two species result in daughters that show intermediate behavior. When first beginning to nest, a hybrid female acts completely confused. She cuts strips of intermediate length and then attempts to tuck the strips into her rump feathers. She does not push the strips far enough into the feathers, however, so that when she walks or flies, the strips always come out. (See figure 30.6.)

Such hybrid behavior clearly indicates that the differences in the way the two species carry nesting material is the result of differences in their genotypes, and thus this is another illustration of how genes control behavior.

But this study also illustrates another important point. Eventually, the hybrid bird's behavior changes so that she is no longer so inefficient. With time, she learns to fly to the nest with the cut strip in her bill and does not even try to tuck it into her feathers. But even after three years, a hybrid female still makes at least a furtive turn of her head toward her rump before flying off to the nest with the strip in her bill.

The improvement in the hybrid female lovebird's ability to carry nesting material suggests that there is an interaction between what the animal is programmed to do and the results of actually doing it. The fact that differences in the behavior of individuals are due to differences in their genotype is not incompatible with the fact that behavior is influenced by the individual's experience with the environment in which it develops and lives. Adaptive behavior that results from an organism's experience with its environment is known as **learned behavior.**

Figure 30.6 Inheritance of nest-building behavior in lovebirds. The peach-faced lovebird cuts a strip and then tucks it into her rump feathers (a). Fisher's lovebird simply cuts a strip and carries it to the nest in her bill (b). The daughters of a cross between the two species at first act completely confused (c). They succeed in carrying the strip to the nest only when they carry it in the bill like Fisher's lovebird. The hybrid (d) eventually learns always to carry the strip in her bill but not without a turn of the head as though she were about to tuck it in her feathers.

Learned Components in Behavior

As was seen in the case of the hybrid female lovebird, basic inherited behavior patterns are modified by learning. When biologists analyze behavior, it is quite often difficult or impossible to distinguish clearly between innate and learned components of behavior. Some behavior patterns that have recognizable learned components are now examined.

Gull Chick Pecking

Parent gulls have the arduous task of trying to keep up with the feeding demands of their chicks. The parent feeds at nearby dumps, behind fishing boats, or on marshes, and carries the food back to the nest in its crop (a sac in the first part of the bird's digestive tract). When an adult lands at the nest, the chick pecks at the parent's bill (figure 30.7). This behavior causes the parent to regurgitate the food it is carrying. All chicks, even newly hatched ones, peck at the parent's bill. This is an example of an innate, or what is sometimes called an **instinctive behavioral pattern.** The young gulls respond automatically to a **visual communication signal.**

This pecking behavior can be studied outside of the nest because newly hatched gull chicks will peck at a wooden model of a gull's head (figure 30.8). Different rates of pecking at gull heads of different shapes and colors are reliable measures of how like the parent the chick finds various models. After initial testing, the chicks can then be returned to their nest and tested again when they are a little older. Newly hatched chicks are not very accurate in their pecking, but their aim improves markedly after only a couple of days in the nest.

The development and improvement in pecking behavior has been thoroughly studied in the American laughing gull. Laughing gull chicks improve in pecking accuracy, and they also develop the ability to identify details of their parents' bills specifically. In addition, they add a head-turning motion to their pecking.

Improvement in pecking accuracy apparently results from the fact that the chicks undergo morphological and physiological maturation (for example, they become more steady on their feet), and some improvement comes from the experience of pecking. By a kind of learning known as **trial and error,** the chicks apparently *learn* how far to stand from the parent's bill in order neither to fall short nor to overshoot.

Figure 30.7 Gull chick pecks at the parent's bill for food.

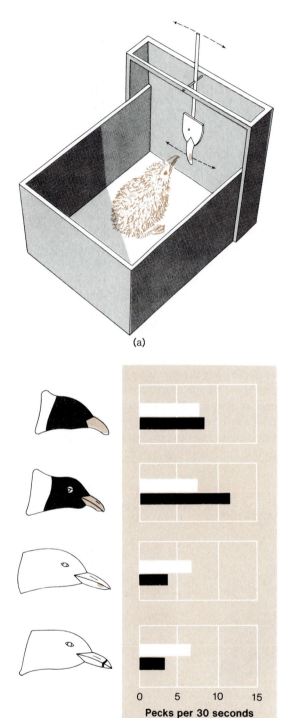

(a)

(b)

Figure 30.8 Pecking behavior in gull chicks. (a) Herring gull chick pecking at a model of its parent's bill. (b) Responses of **newly hatched** (white bars) and **older** (black bars) laughing gull chicks to various models. Note that newly hatched chicks do not discriminate among the very different models (white bars of similar length), while older chicks prefer models that resemble the parent birds (black bars). The model that most closely resembles the parent laughing gull is second from the top.

Newly hatched laughing gull chicks peck at almost any object that is long and thin (like a bill) with red on it—they do not even discriminate between the adult laughing gull's black head and red bill and an adult herring gull's white head and yellow bill with a red spot at the tip. Week-old chicks, on the other hand, show a very different pattern. They are much more sensitive to small differences in shape and detail of the parent's head and beak. The older the chicks become, the more discriminating they are, until eventually they peck only at heads that look like their own parents' heads.

The newly hatched gull chicks show only a simple reaching out and jabbing motion when they peck, whereas week-old chicks show an anticipatory turning of the head when reaching up toward the parent's bill.

Gull chick pecking behavior was once regarded as a classic example of an innate and **fixed action pattern.** Young chicks need no prior experience to show the behavior, clumsy though they may be. They respond automatically by pecking at a specific stimulus. However, the realization that chicks got better at pecking as they grew older cast doubt on whether this behavior could be regarded as strictly innate. Not only did chicks become more coordinated, but even the form of the behavior and especially the stimulus that elicited the pecking changed as the chicks grew older.

It has become increasingly obvious to biologists who study animal behavior that there is no clear distinction between the innate (instinctive or genetically determined) and learned components of behavior. Thus, newer discoveries in animal behavior research do not "resolve" the old "nature-nurture" controversy. Rather, they eliminate the controversy entirely.

All behavior at each stage of development is the result of an interaction between the effects of the animal's genes and the effects of its environment. Behavior cannot be due to genetic factors alone. At the least, the expression of a given behavior requires a proper developmental environment and the existence of stimulus objects. Similarly, no behavior can be attributed solely to environmental factors. At the least, a multicellular animal must have sensory receptors, motor neurons, and a brain that detects and processes the relevant information, as well as muscles that actually perform the behavior. Thus, virtually all behaviors are mixtures of innate components and learned components.

Learning to Follow

Sometimes it seems very logical to think that certain behavior patterns would be either almost completely innate or else entirely learned. For example, an animal's ability to recognize members of its own species and to distinguish them from other species might be expected to be innate, that is, purely instinctive. But is it, in fact, innate?

Once young birds such as ducklings have dried off and rested from the arduous task of hatching, they begin to take notice of moving objects in their environment and follow whatever moves away from them. It helps if the moving object quacks like their mother. The famous Austrian ethologist Konrad Lorenz discovered that the process of recognizing and following members of their own species, or **imprinting,** as it is called, occurs very rapidly and that it occurs during a definite **critical period** after hatching.

Lorenz raised ducklings and goslings and allowed them to follow him as he walked along quacking (figure 30.9). When he then introduced the young birds to adult members of their own species, the ducklings and goslings ran from the adults and hurried back to Lorenz. Later in life, these imprinted ducks often paid little attention to members of their own species, preferring the companionship of humans. As adults, they sometimes even tried to mate with humans!

Many birds and mammals do not somehow "know" how to recognize a member of their own species because such knowledge is not innate. Instead, many animals depend on the generally reliable method of *learning* early in life to make this distinction. Birds, therefore, define a member of their own species as that which looks like what they learned to follow just after hatching. Normally, birds do encounter their mothers during the critical period. However, the imprinting mechanism works just as effectively in experiments where the moving object is not the mother bird, but a biologist interested in animal behavior.

Figure 30.9 Imprinted goslings follow Konrad Lorenz.

The Development of the Chaffinch Song

On a beautiful spring day, the melodious songs of birds might seem to be emotional outbursts of pure joy. But birds' songs actually are specific signals among members of a species, and their songs play important roles in the birds' behavior and life. The songs usually proclaim the singers to be sexually active males with territories. Generally speaking, the effect of the singing is that other males of the species are repelled, and females are attracted to the territory.

Do birds know instinctively how to sing the songs that communicate these important messages? Members of some species, such as chickens and doves, can utter their species-typical calls without having to learn them, but this is not the case for many of the true songbirds.

Many male birds *learn* their species-typical songs from other males that are around their nest when they are juveniles. For example, if a young adult chaffinch is captured before he starts to sing an adult song, and is isolated from all sounds, his song still develops exactly as it would in the wild (figure 30.10a). The young bird's song sounds just like the songs of neighbor males that were around his nest when he was younger. He has a "memory" about the sorts of vocalizations adult male chaffinches were uttering.

However, if very young nestlings are taken from the nest and isolated from all other sounds, the results are different. These birds eventually sing an adult song that has normal duration and pitch, but their song lacks the phrasing and complexity so characteristic of the adult chaffinch song (figure 30.10b). If these isolated birds are put with adult male singing "tutors," however, the young birds learn to imitate the pattern of the tutor. This tutoring is effective, though, only up until the end of the first year, after which the song has crystallized and does not change. It appears that during the first year, male chaffinches listen to and remember the song of other males, although they sing a song that is generally correct in form even when no models are present. This suggests that the basic pattern of the bird's song is programmed in the nervous system, and the flair and nuances are added later. But this is not quite the full story.

If young chaffinches are deafened as very young birds in the nest, their song is totally different from normal birds. They lack both the appropriate basic pattern of adult song as well as the right phrasing and complexity. Their songs are more like the food-begging cries of the nestling (figure 30.10c). Deafening prevents the bird from being able to hear itself sing; that is, it prevents **auditory feedback.** The inability of birds deafened at an early age to develop anything even resembling normal adult song, suggests that the basic pattern of the chaffinch song is programmed in the nervous system, but that the bird must be able to hear itself sing because the song is self-taught. The animal apparently has a kind of **template** (pattern) of what the song should sound like, and he practices to develop a basic song using that pattern. If he cannot hear himself, he has no way of matching the sounds he is uttering with his template.

But there is one more fascinating feature of chaffinch song. If the bird is deafened after the normal song has crystallized, his song remains essentially normal. This means that although the bird must hear himself to learn his song, he does not need auditory feedback to sing it. The crystallization of song is apparently accomplished by the use of a memory of the song, almost like a tape-recording. (This is not true of all species; in some species, the song deteriorates at any stage if the bird is deafened.)

In chaffinches, then, the birds "know" what the basic song should sound like, but they must be able to practice in order to teach themselves to sing. Superimposed on the basic pattern are the nuances and detailed phrasing that the bird learns by listening to neighboring males at an early age. Not surprisingly, this method of song acquisition results in all the males in one localized area tending to sing alike. There are regional dialects in the song of chaffinches, as there are in the songs of many other songbirds.

Thus, development of normal chaffinch singing results from a complex combination of an innate song pattern, song development using auditory feedback, and phrasing learned early in life from adult birds singing near the nest. This is a good, and quite typical, example of the interaction of innate and learned components in the development of behavioral patterns.

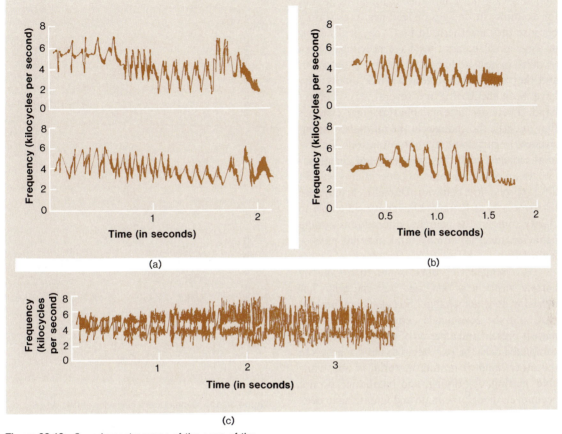

Figure 30.10 Sound spectrograms of the song of the chaffinch (*Fringilla coelebs*) showing the change in frequency (kilocycles per second) on the vertical axis with time (in seconds) on the horizontal axis. (a) Normal songs of a wild-caught male. (b) Songs of two males raised in auditory isolation from a very young age. (c) Song of a male deafened at a very young age.

Hormones and Behavior

Most temperate zone birds do not sing all year; they sing only in the spring. The expression of singing behavior, like the expression of many other behaviors, depends on **internal motivational state,** the set of conditions that exist in an animal's body at any given time. In birds, motivational states are closely related to changes that occur at different times in the quantities of circulating hormones in the birds' bodies.

A good example of the powerful effect of hormones on motivational state is seen in the reproductive behavior of ring doves.

If a pair of experienced ring doves are placed together in a cage, they immediately launch into a fairly predictable cycle of reproductive behavior that lasts about forty-five days. The male begins strutting around, bowing and cooing, in front of the female, much like the behavior seen in street pigeons. Then, the pair works together in selecting a nesting site, and they build a nest out of twigs and straw.

During the course of nest building, they copulate, and seven to eleven days after the beginning of courtship the female lays two eggs. When the chicks hatch, both parents feed them with **crop milk** that is produced by the lining of the crop. After ten days or so, the chicks begin to feed more and more on their own, and the parents become less willing to feed them. When the young are between two and three weeks old, the adults begin a new cycle by courting again.

This orderly sequence of reproductive events is controlled by a complex set of factors. During the cycle, the birds are not simply responding to changing external conditions. Ring doves do not, for example, sit on eggs or feed nestlings simply because they are there. In fact, they will not sit on eggs at all unless they have previously been involved in nest-building and courtship behavior, and they are unable to feed a chick unless they have been incubating for several days beforehand. There are internal physiological changes that underlie the changes in the animals' responsiveness to particular stimulus situations. Behavioral biologists call these **motivational changes.**

In an extensive series of experiments on ring dove reproductive behavior, Daniel Lehrman and his associates at Rutgers University were able to clarify some of the complex interactions among behavior, external stimuli, and internal physiological state.

If a female ring dove is put in a cage by herself with a nest bowl and nest material, her behavior does not change. She does not build a nest, and she does not lay eggs. Only after the male is present and a period of courtship has been completed does the pair begin to build a nest. On the other hand, if nesting material is not available, mating, egg laying, and incubation do not occur even though the male and female have been together for a period of time and are courting. If the pair is given nesting material a week after courting started, the two will build a nest very quickly, copulate, and the female will lay eggs, all within a couple of days, rather than the usual period of about a week. Clearly then, the presence of the male and the actual building of a nest are essential for inducing normal reproductive behavior in the female (figure 30.11).

The physiological changes that underlie these behavioral responses are due to changing hormone levels. As the male courts the female, the level of the female hormone estrogen increases (figure 30.12). It appears that the male's bowing-cooing courtship display actually stimulates his mate in such a way that her ovaries begin to develop and produce more estrogen. As estrogen production increases, she spends increasing amounts of time around the nest. After several days of ovarian development, she ovulates and lays her eggs. However, when no nesting material is present, estrogen levels remain high and the female is ready to ovulate, but she does not ovulate until there is a proper nest in which to put her eggs.

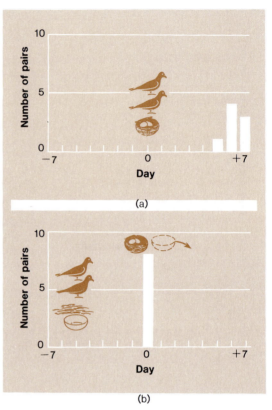

(a)

(b)

Figure 30.11 Reproductive behavior of ring doves under laboratory conditions. Vertical bars indicate the number of pairs that begin sitting on eggs each day. (a) Pairs consisting of males and females that had not been together before were placed in cages containing a nest and eggs (day 0 on the graph). Eight such pairs were used, and they spent from five to seven days (as indicated by the bars) in courtship and nest-building behavior before finally sitting on the eggs. Lehrman and his associates concluded from these results that courtship is necessary for nest building and nest building is necessary for incubation. (b) In a second experiment, pairs of doves were placed in cages containing a nest bowl and nesting materials and were permitted to complete a seven-day period of courting and nest building. At the end of the seven days (day 0 on the graph) the nest bowl and nest were replaced with a nest containing eggs. All eight pairs of doves tested responded immediately (as indicated by the single long bar) by sitting on the eggs, demonstrating once again that presence of the mate and nest building have a dramatic effect on incubation readiness.

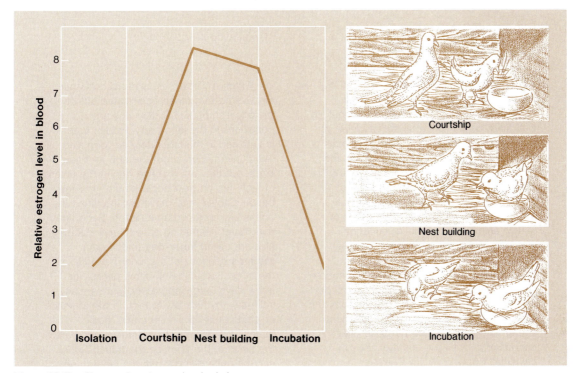

Figure 30.12 Changes in estrogen levels during different stages in the reproductive cycle of the ring dove. If a female ring dove is experimentally injected with estrogen, the cycle can be sped up.

It is clear, then, that the changes in the behavior and motivation of the pair are tightly linked to one another, to specific external stimuli in the environment, and to their internal physiological states. Complex mechanisms ensure that the behavior of an animal and its internal state are coordinated, for example, that a nest is ready when the female ring dove is ready to lay eggs. These coordinating mechanisms also ensure that the two birds cooperate with each other and that their activities take into account conditions in the external world.

The Adaptive Significance of Behavior

Behavior has adaptive significance; that is, a behavior evolves through time because it confers some benefit on the organism exhibiting it. In this section the adaptive significance of several behavior patterns is examined.

Communication in Bees

Honeybees **communicate** information about new food sources by completing a movement known as the "waggle" dance. In the waggle dance a worker who has located a new food source moves around in a figure-eight pattern. When she is on the straight run between the loops of the eight, she moves her abdomen from side to side, hence the name "waggle" dance. The direction of the straight run in the worker's dance tells other workers the exact direction of the food source, while the tempo of the dance communicates information about the food's distance away from the hive.

The significance of the waggle dance was first discovered by the Austrian ethologist and Nobel prize winner Karl von Frisch. He observed that when a worker returns to the hive, she often does the waggle dance, sometimes on a horizontal surface outside the hive or sometimes on the comb inside the hive. This dancing attracts the attention of fellow workers, who follow the dancer's movements for a few minutes and then fly out to the same feeding area that she has been

Box 30.1
Solar-Day and Lunar-Day Rhythms

Modification of the bee dance through the course of the day is only one of many behavioral patterns that change rhythmically with time. In fact, the total activity level of most animals is rhythmic. That is, animals are active at certain times of the day and inactive at others. Bats, rats, and moths are **nocturnal** (active at night), while dogs, butterflies, frogs, and lizards tend to be active in the daytime. Thus, members of certain animal species never encounter members of certain other species, even though they may live in the same area. Orlando Park, one of the pioneers in modern research on daily rhythms in organisms, put it this way: ''Two organisms may occupy the same physical space and yet never meet one another face to face because their periods of activity and inactivity do not coincide.''

As was pointed out in chapter 18, solar-day (twenty-four-hour) rhythms are very widespread among living things. Many such solar-day rhythms continue even when organisms are placed under constant conditions in the laboratory, where they are deprived of obvious information about the time of day.

However, solar-day rhythms are not the only rhythms expressed in animal behavior. Many animals that live in the intertidal zones along ocean shores display rhythmic changes in activity that correlate with the ebb and flow of tides.

Fiddler crabs of the genus *Uca,* for example, become active and feed during low tide periods, but plug their burrows from the inside and remain hidden during high tide periods (Box figure 30.1A). Since tides occur mainly in response to variations in the moon's gravitational pull on the earth during the 24.8-hour lunar day, tidal ebbs and flows occur about fifty minutes later each day than they did on the preceding day. Because fiddler crabs emerge from their burrows and forage on exposed mud flats during each low tide, their activity periods must begin fifty minutes later each day.

Frank A. Brown, Jr. and his colleagues discovered that these tide-related activity periods also persist under constant conditions in the laboratory. Fiddler crabs maintained in the laboratory are deprived of direct information about the ebb and flow of tides, but they are inactive during the times of high tide on their home beach, and they become active and move around when the tide is low on their home beach. Such tidal rhythms of alternating activity and inactivity also are called **lunar-day** (moon day) **rhythms** because of the lunar periodism of the tides. Some tidal rhythms persist for many days in the laboratory, just as many solar-day rhythms do. In fact, in the absence of a tide table for fiddler crabs' home beach, it is possible to determine accurately the times of the tides on the beach (which may be quite distant from the laboratory) simply by observing the crabs' activity cycles!

Daily rhythms, and the underlying cellular clocks that time the rhythms, are important adaptations because rhythmic changes in physiology and behavior help to ensure continuing successful interactions with a fundamentally rhythmic environment.

Box figure 30.1A Fiddler crabs (genus *Uca*) show a definite tidal rhythm. Here fiddler crabs are moving over a muddy beach at low tide. During high tide periods, fiddler crabs remain inactive in their burrows.

using. In this way, foraging bees can quickly recruit other bees to harvest particularly rich sources of food.

If the dance is completed outside the hive, the direction of the straight run in the worker's dance tells other workers the exact direction of the food source. The other workers need only fly in the same direction as the straight run, using the sun as a compass (figure 30.13a).

This pattern of bee flight, which is specifically oriented with reference to a light source (in this case, the sun), is an example of a **phototaxis.** A **taxis** is a body movement in which an animal assumes a specific orientation to a stimulus source. The term phototaxis is used for movement oriented to a light stimulus.

If the dance is done inside the hive, the workers cannot see the sun and so cannot use it directly as a compass. In a clever series of experiments, von Frisch demonstrated that foraging bees use a symbolic representation for the direction of the sun. If the forager dances the straight run part of the dance directly up the vertical comb surface, the food source is on a straight line from the hive toward the sun (figure 30.13b). If the dance is

straight down the comb, then workers know they can find food by flying away from the sun. If the forager dances 20° to the left of straight up, then the workers know to fly 20° to the left of the direction of the sun, and so forth. The direction to the food source is symbolized by the dancer's movement with respect to gravity (called a **geotaxis**).

Even though the sun moves across the sky at about 15° per hour, the bees can compensate for this even in the darkened interior of the hive because they possess an accurate **biological clock** (see p. 519). A dancing bee inside a dark hive gradually changes the direction of its dance with the passage of time. If at 9 A.M. a bee is dancing straight up the comb, indicating that a food source is directly toward the sun, then an hour later she will be dancing 15° to the left of vertical to indicate the same food source. In fact, the straight run of the dance rotates on the surface of the comb like the hands of a clock (only counterclockwise). Smell and visual cues also are used by bees, but there is enough information in the dance to get a recruited worker very close to the food source.

Figure 30.13 The waggle dance of the honeybee. When the bee moves up the straight run of the dance, she vibrates her abdomen back and forth (waggles). At the conclusion of the straight run, she circles back to the same starting position. The nestmates gather information about the location of a food source from the straight run. If the bee performs the dance outside the hive (a), the straight run points directly toward the food source. If she performs the dance inside the hive (b), she orients herself using gravity as a reference, and the point directly overhead represents the direction of the sun. For example, in (b), because the food source is X degrees to the right of the direction of the sun, the straight run is X degrees to the right of straight up.

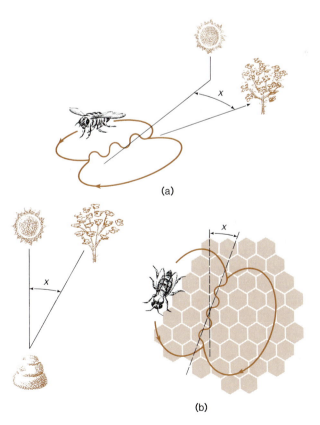

Because individual worker bees receive specific directions to good food sources through the waggle dance, they do not have to waste energy exploring until they encounter the sources on their own. This complex system of communication and location finding is adaptive because it permits more workers to reach food sources more quickly, and thus, it enhances the food-gathering efficiency of the hive.

This efficient communication system is also used when bees **swarm.** Swarming is a process by which a queen and a large number of worker bees leave an established hive and start a new hive. The bees fly out and settle on a tree branch or another surface in a mass. Then individual workers fly out and scout the area for appropriate new hive sites (figure 30.14).

When a scout bee locates a potential hive site, it returns to the swarm and does a waggle dance to communicate the distance and direction to the site. This allows other bees to find and explore the site also. When scouts return from very desirable sites, they dance vigorously. Vigorous dancing recruits other bees, who also examine the site. If they also find it favorable, they also return and dance vigorously. Eventually, a "decision" is reached, and one site is chosen from among the sites being explored.

Again, use of the waggle dance is adaptive. Bees in swarms are vulnerable because they are exposed to predators and the weather. Thus, it is important that the new hive be established as quickly as possible, and communication by the waggle dance facilitates hive site location and selection.

The waggle dance is a form of symbolic communication because each dance is really a miniaturized, symbolic version of the journey to a specific point. It is difficult to reconstruct the evolution of a complex behavior such as the waggle dance, but it is easy to understand that selection pressure has favored the evolution of behavior that increases the efficiency of such vital processes as food gathering and location of appropriate living sites.

Figure 30.14 A swarm of honeybees on a fire hydrant in San Francisco. Swarming bees assemble in masses (usually on tree branches), while individual workers scout for a suitable new hive site.

The Seeing Ear

How do animals that are active in almost total darkness find their way? If they are not only active but also are trying to catch elusive and fast-moving prey, they require a very accurate system for finding their way around. One solution has been the evolution of **sonar systems,** where animals make loud noises and listen to the returning echoes. Using such a system requires that an animal be able to overcome many technical problems, but this is exactly what bats, dolphins, and a few other organisms have done in a variety of intriguing ways.

Donald Griffin clearly demonstrated that bats **echolocate.** He allowed them to fly in a room that was specially equipped with fine wires running from the ceiling to the floor. He soon learned that even blindfolded bats could find their way around these wires in total darkness.

However, plugging a bat's ears or covering its mouth seriously affected its performance. Such a bat was very reluctant to fly at all. It would hang on its perch and groom its ears and mouth vigorously, trying to remove the plugs. If the bat was forced to fly, it flew "blindly," running into wires and crashing into walls.

Although these facts strongly suggested that the animal was using sound to find its way, it was puzzling to researchers because bats rarely make audible sounds. However, improved technology and the invention of what is now called a "bat detector" solved the mystery. Bats produce a more or less constant stream of clicks, but these clicks are all outside the range of human hearing; that is, they are in the **ultrasound** range. Humans can hear sounds to a maximum **frequency (pitch)** of about 20 KHz (20,000 cycles per second). This means that humans can hear only the lowest sounds that bats occasionally make because bats' ultrasonic cries usually fall between 30 and 50 KHz. It has taken years of research to find out how bats actually manage to locate objects and capture moving prey with the echoes of these ultrasonic cries (figure 30.15).

The echo of a bat's ultrasonic cry is much softer than the cry itself. There are two reasons for this. First, the sound spreads out widely from the bat's mouth, like the waves from a stone dropped in a pond. Second, some of the energy that is put into the sound is lost simply in doing the work of moving air molecules around. Unfortunately for bats, more energy is lost in this way as the sound becomes higher in frequency. A 10 KHz sound (in the range of human speech) is half as loud by the time it is 15 m from the mouth, but a 50 KHz sound (a bat's call) is half as loud only 10 m from the mouth. Some bats solve this problem of sound **attenuation** (becoming softer) with "megaphones," bizarre leaflike structures on the ends of their noses, which incidentally give them a more frightening appearance (figure 30.16). Bats belonging to these

Figure 30.15 A greater horseshoe bat, *Rhinolophus ferrum-equinum*, catching a moth with the aid of its left wing.

Figure 30.16 Outlines of the heads of five bats. (a) African yellow-winged bat. (b) Horseshoe bat. (c) Funnel-eared bat. (d) Mouse-eared bat. (e) Free-tailed bat. Two of these (a and b) emit sounds through their nostrils, and they have nose-leaves that act as a megaphone to reduce the attenuation of their ultrasonic cries. The others emit sounds through their mouths.

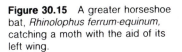
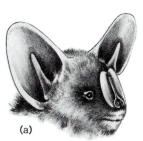

(a)

(b)

(c)

(d)

(e)

Box 30.2
Animal Compasses

One of the most remarkable aspects of animal behavior is that animals have what might be called a sense of place. Foraging animals move out, sometimes over great distances, from their homes and are able to find their way back. Some of these animals might be using landmarks to find their way along familiar paths. But many animals demonstrate much more impressive abilities to find their way from place to place, often traversing hundreds or even thousands of kilometers in the process of migration.

For example, young green sea turtles are found on the east coast of South America, where they feed and grow. Every two or three years adult females set off swimming eastward across the Atlantic against the prevailing current. Their sea journey covers more than 2,200 km and ends on the beaches of tiny Ascension Island, which is only about 10 km across, where they lay their eggs. They leave the eggs buried in the sand and head out to sea on their return voyage. When hatchling turtles dig their way out of the sand, they scramble across the beach to the water. They apparently are carried westward by the current to the coast of South America, where the females live until they set out across the Atlantic on their own journey to Ascension Island.

Green sea turtles, however, do not hold the record for long-distance migration. Many common birds migrate across large parts of a continent or even from continent to continent. Probably the champion long-distance migrant is the arctic tern. In the autumn, the artic terns that nest in the North American arctic regions migrate across the Atlantic and travel southward along the west coasts of Europe and Africa until they reach the tip of South Africa. Then they cross the South Atlantic to Antarctica, where they live along the shore during the Antarctic summer. In this journey they cover a distance of about 18,000 km.

Many other birds' long-distance migrations are spectacular because they involve nonstop flights requiring great endurance. For example, a flock of blue geese and snow geese was observed to migrate from James Bay, Canada, to Louisiana apparently without stopping. They covered this distance of 2,700 km in sixty hours.

Some species of birds migrate in flocks in which young birds and mature adults travel together, and thus young birds travel with individuals who have made the trip before. In other species, however, older birds migrate first, leaving young birds to make their first migratory journey without the company of experienced adults.

Birds also can find their way on long journeys that are not part of their normal migrations. For example, a Manx shearwater was captured on the west coast of England, banded, and transported by plane to Boston, Massachusetts, where it was released. The shearwater returned to its nest in England just twelve and one-half days later after crossing the Atlantic Ocean, a distance of about 5,300 km.

How do birds find their way on these long journeys? The use of familiar landmarks might be a factor near the beginning and end of a journey, but long-distance navigational capabilities must exist as well. It is well established that various birds use the sun and/or the stars for orientation. Such celestial navigation requires an accurate time sense because animals using the sun as a compass, for example, must continually change their orientation relative to the sun to stay on course in a single direction. Experiments have shown that birds do indeed have and use accurate time information in their celestial navigation.

However, the migratory flights of many birds do not stop on overcast days, when celestial navigation is not possible. Birds continue their journeys, sometimes actually flying through clouds as they go. Early in the study of migration, some biologists suggested that migrating animals might use a magnetic sense for navigation, but the idea was dismissed because it was assumed that it was not possible for animals to sense the low energy levels of the earth's magnetic field.

During the 1960s, Frank A. Brown, Jr. and his colleagues presented very persuasive data indicating that some animals show orientational responses to weak magnetic fields. But again, despite the strong evidence presented, the idea of a magnetic sense for navigation was rejected by the majority of biologists for the same reasons given years before.

Finally, during the 1970s, the idea that animals can sense the earth's magnetic field and can use it in orientation and navigation became widely accepted. Some of the most convincing evidence came from studies of homing pigeons. Homing pigeons were carried some distance from their home loft and released to find their way home. Observers then recorded their vanishing bearing—the direction in which they disappeared as they flew away from the release point. The birds tended to orient themselves so that they disappeared in the direction of their home loft whether the day was clear or cloudy.

W. T. Keeton, Charles Walcott, and others have demonstrated, however, that a pigeon's homing ability on a cloudy day depends on its being able to sense the earth's magnetic field. A pigeon's orientation is confused on cloudy days if it has a small bar magnet attached to its back or if it carries a small battery-powered coil that induces a uniform magnetic field in its head.

It appears that use of a magnetic compass for orientation functions as a backup to the sun compass, which cannot be used on cloudy days. Magnets attached to the pigeons do not disturb the pigeons' orientation on clear days.

Since the demonstration of a magnetic sense that can be used for orientation, some biologists have proposed that a magnetic sense may also be involved in animals' strong sense of place. Local and regional anomalies in the earth's magnetic field might actually provide "map" information to animals, but this proposal requires much more testing and evaluation.

How do organisms actually perceive magnetic fields? This question has been only partially answered. Some organisms possess grains of magnetite (Fe_3O_4). For example, bacteria that show specific orientation in magnetic fields contain chains of magnetite particles. Magnetite particles ("magnetosomes") also have been reported in a number of animals, including tuna, dolphins, whales, pigeons, and green turtles. In animals, movement of magnetite particles in response to magnetic fields might be detected by adjacent sensory receptors.

W. T. Keeton, one of the leading researchers in this field, suggested that one definite conclusion can be drawn from the work done so far. It is apparent that organisms can respond to relatively weak magnetic fields. It is now important to determine how organisms respond and to investigate the adaptive significance of the responses.

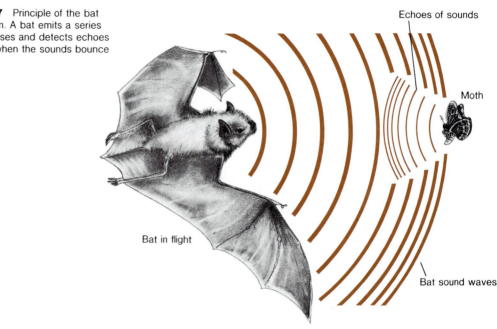

Figure 30.17 Principle of the bat sonar system. A bat emits a series of sound pulses and detects echoes that return when the sounds bounce off objects.

Echoes of sounds

Moth

Bat in flight

Bat sound waves

species emit ultrasounds through their nostrils rather than through their mouths, like other bats. The nose-leaves focus or beam the sound forward rather than allowing it to spread out.

Since making a loud cry is so important for getting a good echo, and high-pitched sounds attenuate more rapidly, it would seem to be more efficient for bats to scream at a slightly lower frequency. Although ultrasound has the disadvantage of greater attenuation, it also has certain important advantages. The principle of the sonar system is that a sound bounces off an object and returns as an echo to the bat's ears (figure 30.17). Sound with very high frequency (pitch) has short **wavelengths,** while low-frequency sounds have long wavelengths. The chance that a sound wave will actually hit an object and bounce back is affected by the size of the object and the length of the sound wave. Shorter wavelength (higher frequency) sounds reflect off small objects much more reliably than longer wavelength sounds. Thus, if bats are to detect small insects, they must use high-frequency sounds that reflect off very small objects reliably.

Bats must cry very loudly if they are to receive loud echoes from objects at any distance. Bat cries have been measured at intensities of 60 to 100 dynes per cm^2. If these sounds fell within the range of human hearing, bats would sound like jet airplanes screaming through the night. These sounds are so loud, in fact, that researchers wondered why bats do not deafen themselves.

Bat ears have to be enormously sensitive to hear the faint echoes returning from their calls. How can they possibly hear these echoes when they are making such loud noises? Some bats can temporarily shut off their hearing mechanism. They have a muscle with which they decouple the ear ossicles while a sound pulse is going out and then recouple in time for the echo. Other bats can broadcast and listen at the same time. For reasons having to do with the physics of sounds produced by moving objects, the echoes returning to a flying bat are slightly higher in pitch than those the bat sends out. This means that the bat can be somewhat deaf to the pitch of the sound it actually produces but have very sensitive hearing for the somewhat higher pitch of the returning echoes.

Bats must not only be able to hear the returning echoes; they must be able to detect the distance and direction of objects from the reflected echoes. A bat's cry moves out from its mouth at 340 m/sec, so if, for example, a tree is 100 m away, the echo returns in just over half a second. Since distance is measured by the delay between the time when the sound was made and the time the sound returned to the ear, the bat must have an extremely accurate sense of timing that can measure differences of very small fractions of seconds.

If a bat's cries were not sent out often enough, a bat would get only a jerky, disjointed picture of its surroundings, rather like a slowly flickering motion picture. Such a system would not allow the animal to respond quickly to objects in its path or to catch flying insects. Therefore, the cruising bat sends out a very rapid stream of short clicklike pulses at a rate of about 15 per second, incidentally about the same shutter speed as a silent motion picture. When it is zeroing in on an insect and really good resolution is required, the bat can speed up its clicks to 200 per second, the equivalent of very high speed motion picture photography.

When a bat is sending out a rapid series of sounds, echoes are returning while subsequent clicks are being produced. Furthermore, an echo from the beginning of the click bouncing off a distant object may return at the same moment as an echo from the end of the click returning from a nearer object.

To get around these problems, bats make the clicks very short, each one lasting only 5 to 10 msec (thousandths of a second). Some bats also use frequency modulation within each click (FM radio operates on the same principle). The click begins at a high frequency and swoops down to a lower frequency. The bat then "knows" that if the echo is higher frequency, it is a reflection from the beginning of a click, and if it is lower frequency, it is a reflection from the end of a click.

The processing and integrating of this complex pattern of sounds returning at differing times and at different pitches might seem impossible. Yet recognizing patterns and depth by a sonar system is no more complex than what humans do every day when processing visual information. Recognizing shapes and patterns, colors, shades, sizes, distances, and movements with the human eye is an even more elaborate and complex (and no better understood) phenomena than a bat's sonar system.

Bats' favorite prey have evolved ways of avoiding capture as a result of detection by bats' echo systems. Many of the common night-flying (noctuid) moths have ears that are extremely sensitive to ultrasound. When moths hear batlike sounds, they respond quickly. If one jingles a set of keys (this makes some sounds in the ultrasonic range) near a noctuid moth that is clinging to a wall, the moth lets go and drops to the ground. It was just such a chance observation that first puzzled Kenneth Roeder, a biologist from Tufts University.

For several summers, Roeder spent every warm evening in his floodlit backyard playing sounds he could not hear to moths that happened to be passing by. Their behavior was striking. When a moth was 20 to 30 m from the loudspeaker, it turned and flew rapidly in the opposite direction. But when it was very close to the sound, it folded its wings and dropped to the ground, or flew hard toward the ground in a power dive, or executed a series of loops and dives that generally took it downward. It was completely unpredictable as to which one of these three courses of action the moth would take.

By watching interactions between moths and local bats, Roeder and his associates quickly accounted for the different kinds of moth behavior. A moth that is 20 m away from a bat is still "invisible" to the bat; that is, it is out of sonar range. The moth's turning away behavior is very adaptive, for it gets completely out of the bat's range before it is even detected. However, when a moth is only 10 m or so away (within sonar range), it is in acute danger because bats can fly faster than moths. Only by executing evasive action does the moth escape. The moth does not always dive to the ground because the bat would soon predict the moth's behavior and intercept at a point below the place where the moth was first

Figure 30.18 Tracks of bats intersecting those made by moths tossed in the air. (a) The bat enters from the left; the moth dives and turns sharply, escaping capture. (b) The bat enters from the right; the moth track begins below that of the bat and terminates on it, indicating capture.

(a)

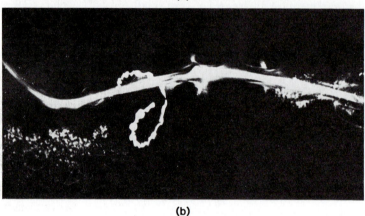

(b)

detected. Clearly, variable evasive maneuvers are adaptive for the moth, or for any other animal escaping a predator, for that matter. Therefore, there is selection pressure for moths to behave in an erratic fashion, sometimes diving and sometimes going in for aerobatics (figure 30.18). In experiments, moths that did not take evasive action were seven times more likely to be captured than those that did.

Behavior of predators and their prey usually include measures and countermeasures, such as those seen in bats and moths. In most cases, a balance has been reached through the course of evolution. The capture behavior of the predators is efficient enough so that a stable population of predators can obtain food. But the evasion and escape behavior of prey organisms are efficient enough so that a stable living and reproducing population of the prey organisms can be maintained.

The Courtship of the Stickleback

Several examples of adaptive behavior used for gathering food have been examined. Behavioral mechanisms also are involved in various other phases of animals' lives, such as reproductive behavior.

For example, the behavior that brings male and female three-spined sticklebacks together at appropriate times for reproduction is very complex and interesting. This small, freshwater fish is found in ponds or slowly moving streams throughout most of Europe and along both coasts of North America.

The male three-spined stickleback defends a territory to which the female comes to lay her eggs and is also the sole guardian of the eggs. A male competes with other males for possession of a territory with a good nesting site. Then he builds a nest by carrying mouthfuls of sand away from the nest site until a small pit is formed in the sandy bottom. Next he collects bits of vegetation in his mouth and spits them into the pit.

Figure 30.19 Male stickleback selects a nest site, excavates it by removing sand and gravel, and then constructs a nest over it. The nest is made of bits of vegetation, which the male glues together with a secretion from his kidney. The completed nest contains a tunnel in which the female lays her eggs.

When he has a little pile, the male glues the bits together with a secretion from his kidney. He consolidates the mass of nest material by pushing it together with his snout and picking up bits of sand and spitting them onto the vegetation. Finally, he forcefully wriggles and burrows his way through the pile of glued vegetation, making a tunnel (figure 30.19).

The male vigorously defends his nest site from other stickleback males and other intruders, such as invertebrates or fish that are likely to eat eggs. A male stickleback recognizes another male by the bright red color of his belly (this is another example of a visual communication signal). When a male sees a spot of red in his territory, he charges out to expel it. He may bite his rival or carry him bodily from his territory. It is easy to fool these aggressive little fish with a piece of cardboard of almost any shape, as long as it has red along the bottom side (figure 30.20a). One of the reasons for this fierce defense is that neighboring males are very disruptive to the nests. They steal nest material, destroy nests, or may even eat the eggs.

If an intruder does not flee after the initial rush, the two males may circle, or they may stand on their heads in the water, jerking up and down rapidly in gestures of **threat** (figure 30.20b), rather like a dog baring its teeth in a snarl or a man shaking his fist in an opponent's face.

When two males' territories are side by side, fighting may take on a rather bizarre appearance. Each male is **dominant** in his own territory, but **subordinate** when in the territory of another. The aggressiveness of the male decreases as he gets farther from his nest. Therefore, a stickleback that is fearlessly pursuing a rival suddenly turns tail and flees when he discovers that he has crossed the territory boundary. These territorial males seem almost as if they are attached to their nests by an elastic band that allows them to move only so far from the nest before they are compelled to come racing back again.

When a female enters the male stickleback's territory, his behavior becomes very different. He recognizes a female ready to lay eggs by her swollen belly. (Again, it is easy to fool him for he will respond to any object of about the right size and

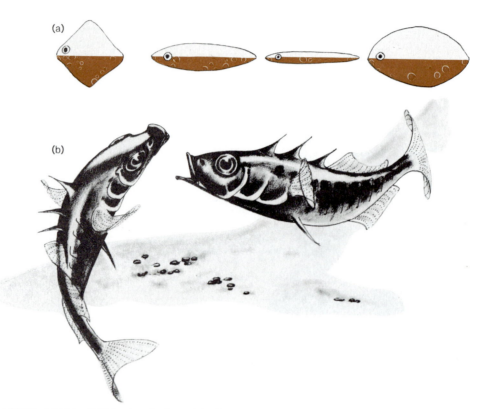

Figure 30.20 (a) Male sticklebacks attack models of various shapes such as these, so long as the ''belly'' is red. (b) Two male three-spined sticklebacks fighting.

shape.) He launches into an elaborate course of zigzag swimming and jumps in front of the gravid female, interrupted now and then by a quick dash back to his nest. If she approaches, which she usually does not do right away, he swims toward the nest, leading her. If she stops, he begins to zigzag again. If she follows him further, he leads her to his nest and sticks his snout into the entrance. The female may then enter and wriggle into the nest tunnel. If the female does enter the nest, the male nuzzles her tail, and as he does so, she spawns her eggs and then swims out the other side of the tunnel. The male then immediately enters and sheds sperm on the newly laid eggs (figure 30.21).

The female stickleback may break off this **chain of behavior** at any point and swim from the territory. This only causes a renewed course of zigzag swimming by the male. Neighboring males also may disrupt the process. The neighbor may court the female and attempt to lure her off to his own nest. He may even sneak in behind a spawning female and shed sperm on eggs in another male's nest!

After fertilization, the male's behavior changes dramatically. He chases the female and drives her from his territory. Then he pushes and flattens the egg mass to the floor of his nest and repairs any damage to the nest. He also lengthens the nest so that the next batch of eggs that is laid will not overlap with the first. Within an hour he again courts any females who come into his territory.

As the embryos develop, however, they require more and more of his time, and he is less likely to court. He swims near the nest and alternately beats his two pectoral fins, forcing a current of water through the nest (figure 30.22). This fanning behavior is essential for the successful development of the eggs because it makes adequate gas exchange possible. During this period of tending the young, the male stickleback loses much of his bright red coloration. Loss of the male's conspicuous coloration probably further protects the nest from predators.

When the young fish begin to hatch, the male makes holes in his nest so they can get out, but he continues to hover over them for several

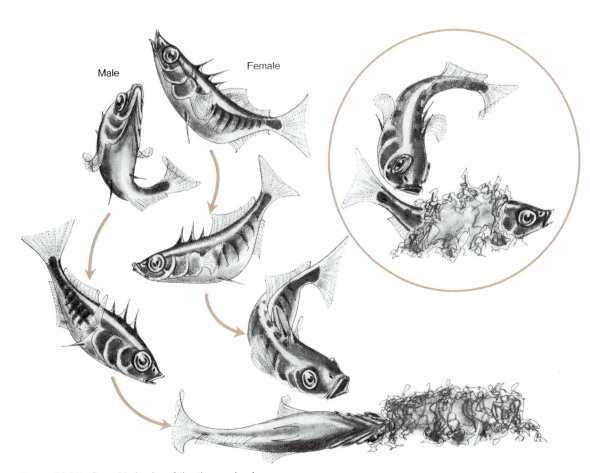

Male   Female

Figure 30.21 Sexual behavior of the three-spined stickleback. When a female enters a male's territory, he courts her with zigzag swimming. If she responds, he leads her to his nest, where he adopts a special posture with his snout in the entrance. Then if the female enters the nest, the male nuzzles her tail, and as he does so, she spawns her eggs (inset). After she spawns, the female swims out of the nest and the male enters and sheds sperm on the eggs.

Figure 30.22 A male stickleback beats his pectoral fins and creates a current of water that flows through the nest.

days. If any stray away from the nest, he sucks them up into his mouth and spits them back into the nest. Finally, when they are large enough, they swim off, or the male simply deserts them and the nest.

The reproductive behavior of this little fish is one of the most thoroughly studied patterns of animal behavior, and it illustrates a number of important concepts of ethology. First, the male fish is not responding generally to the female or a rival male. Only a specific part of the overall stimulus (for example, the red belly of another male) elicits a response. A specific stimulus such as another male's red belly is called a **sign stimulus,** and it is said to **release** the male's **aggressive behavior.** Sign stimuli are found throughout the animal kingdom. For example, the parent gull's bill stimulates begging movements in the newly hatched chick.

Second, there are progressive changes in the male's **motivation** or likelihood to behave in a certain way, which in turn are functions of such variables as his reproductive state, whether or not his nest contains eggs, and even the developmental stage of the eggs. In part, these outcomes are the result of hormonal changes in the male, but they also result from changes in the central nervous system that alter his responsiveness to particular stimuli in the environment. For example, just prior to mating he vigorously courts the female, and just afterwards he chases her from the territory as though she were a trespasser. One might expect that this change in his behavior results from the male having just ejaculated, but this is not the case. In fact, it has been shown experimentally that it actually is the presence of freshly laid eggs in his nest that causes him to drive away all other fish and to tend the eggs closely. Understanding the bases of motivational changes such as these is an important part of understanding and predicting animal behavior.

Courtship in the Bowerbird

In many animals, copulation is preceded by elaborate patterns of **courtship behavior.** In some species, males are normally aggressive, and courtship helps to resolve the conflict between the drive to be hostile and the drive to mate. In such species, courtship behavior often has both **aggressive** and **appeasing** components. An example

of this would be the zigzag swimming behavior of courting male sticklebacks.

In other species, courtship behavior may be necessary to resolve the conflict involved in animals touching each other, as they obviously must do during mating. This is not a trivial problem because many adult animals respond very negatively to being touched by any other animal. This general response is understandable when one considers that, in most contexts, being touched can well mean being captured and killed. The negative response is so strong that it includes even members of the animal's own species. Even birds assume regular spacing when perched on a tree limb or an electrical wire. For such animals, courtship behavior is necessary to resolve this conflict if they are to come into close contact for mating and nesting or other parental care activities.

Another important function of courtship behavior is that it helps ensure that males attract females of their own species. This is obviously adaptive, and a variety of behavioral mechanisms that attract females have evolved. Many of the mechanisms involve specific responses of females to combinations of coloration and display by males. But the matter can be more complex, as in cases where male behavior involves objects that males collect and display for females.

Male bowerbirds go to particularly great lengths to attract females. These chickenlike birds are found in the forests of Australia and New Guinea, where the males defend territories. Unlike the males of many other species in which there is no paternal care of the young, the male bowerbird is often drably colored, much like the female. However, he makes up for it in the most surprising way. He constructs a "bower."

The male satin bowerbird builds a walled avenue of sticks (figure 30.23). He collects distinctively colored objects and displays them at one end of the bower. Satin bowerbird males specialize in anything that is blue, including blue parrot feathers, blue berries, and freshly picked bluebells and other flowers. When a male has his territory near human habitation, he may collect and display blue crockery, rags, bus tickets, candy wrappers, and so forth. These males also make a paste of blue fruit pulp and saliva, which they paint on the inner walls of their bowers. They do this with the aid of a piece of bark that they use like a paintbrush.

Figure 30.23 A male satin bowerbird courts a female who is sitting in his bower. The ground in front of the bower is decorated with shells, feathers, and flowers.

Other species of bowerbird choose totally different kinds of display objects. The spotted bowerbird, for example, chooses white objects, such as bleached bones, dried white snail shells, and so forth. When they are nesting near humans, male spotted bowerbirds collect an astonishing array of miscellaneous objects, including silverware, coins, keys, and jewelry. One of these birds even hopped through an open window and stole a gentleman's glass eye from his bedside!

When a female comes to a male's bower, he displays by picking up feathers or flowers or others objects in his bill. He makes a rhythmic whirring, not unlike the sound of a mechanical toy. He arches his tail, stiffens his wings, and hops around the bower with his display objects. The rather shy female usually leaves before courtship is consummated, but occasionally a female crouches down and copulates with the male in the bower.

In these species, females apparently choose males on the basis of their ability to assemble materials in a bower. Here the forces of sexual selection focus on the display objects that the males collect, rather than on display plumage. During the courtship dance, the male bowerbird

holds objects much the same way other males display a bright crest or brilliant tail feathers. The display objects he holds are essentially externalized secondary sexual characteristics.

Foster Parents

Many kinds of animals expend a great deal of energy on the care and feeding of their young offspring. In some cases, paying attention to young offspring or gathering food for them can expose parents to predators. But, despite these disadvantages, behavior that assures care of helpless offspring is adaptive because it is essential for reproductive success. Thus, it is an important part of the behavior of many adult animals. Behavior patterns have evolved, however, that allow some animals to avoid the burdens of parental care. A few animals trick other animals into caring for their young.

Some birds, known as **brood parasites,** lay their eggs in the nests of other species with the result that birds of the **host species** take care of the parasite's eggs and nestlings, along with their own. Included among the brood parasite species are many cuckoos, one species of duck, and the

American cowbirds. The behavior of brood parasites is a lesson in deception.

Female cuckoos spend most of their time patiently watching the nest-building activities of prospective host species. Their egg-laying is carefully timed to coincide with the host's egg-laying. When the female cuckoo is ready to lay an egg, she quietly flies to the host's nest, moves one of the host's eggs, slips onto the nest, and lays one of her own in its place. In a matter of seconds she is gone again, taking a stolen egg with her. Each female cuckoo lays her eggs in the nest of one particular species of host, and it is thought that she chooses the same species by which she herself was raised. The eggs of different European cuckoos closely resemble the eggs of the host species, so that the cuckoos that parasitize one sort of nest have eggs that look very different from those that parasitize another. Obviously, eggs that are less likely to arouse the host's suspicions are more likely to be accepted and cared for.

The incubation period of the cuckoo egg is slightly shorter than that of the host's eggs. For this reason the nestling cuckoo generally hatches first. It then engages in an extraordinary behavior. A day after hatching the apparently helpless cuckoo becomes totally intolerant of anything else in the vicinity and begins the strenuous task of evicting everything else from the nest. Supporting an egg or nestling on its back and holding it with its wings, the young cuckoo pushes it along the rim of the nest and out over the edge (figure 30.24a). After this, the foster parents devote all of their attention to the growing cuckoo. The baby cuckoo grows quickly, and in some cases, soon becomes larger than its foster parents (figure 30.24b). They have to work furiously to feed their insatiable adopted nestling.

There is an evolutionary "race" between the cuckoo and its host. Selection favors hosts that avoid taking care of cuckoo young. Hosts often desert nests that have been disturbed by cuckoos

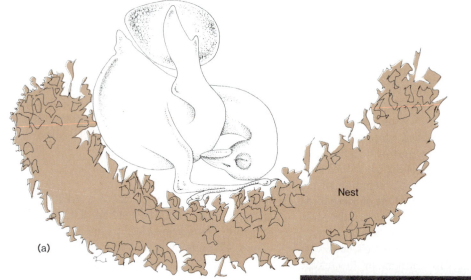

Nest

(a)

Figure 30.24 Brood parasitism by cuckoos. (a) A newly hatched cuckoo pushes its host's eggs out of the nest. (b) A hedge sparrow feeds a young cuckoo that is much larger than its foster parent.

(b)

or remove eggs that do not look like their own. Hosts mob cuckoos and try to chase them away whenever they see them. There is continuing selection pressure for hosts to become better at distinguishing cuckoo eggs. But there is also selection pressure for cuckoos to become more successful at getting hosts to accept their eggs. This is a balanced relationship, and neither host nor parasite really wins the "struggle."

Sociobiology: A New Look at the Evolution of Social Behavior

During the 1970s, a new emphasis developed in behavioral research. This new emphasis has been named **sociobiology,** and it focuses on certain genetic and evolutionary explanations for social behaviors. Advocates of sociobiology maintain that many complex social behaviors, including human social behaviors, could be better understood if they were reexamined in the context of sociobiology. Because many biologists disagree with some of its basic assumptions, a great deal of controversy surrounds sociobiology.

This discussion of sociobiology begins with an example of a sociobiological analysis and interpretation of a behavior pattern.

Helpers at the Nest

Mexican jays are sociable birds who live in the Arizona mountains in permanent flocks of five to fifteen birds. Only one or two pairs of birds in the flock build nests at any one time. All the eggs in a given nest are laid by one female and incubated by a single pair of jays. Thus, jays' nesting behavior is similar to that of most other species of birds. The unusual thing about the jays is that after the nestlings have hatched, all the birds in the flock feed the young, even the parents from the other active nest! It is a sort of bird commune. Although the nestlings' own parents supply more food than other individuals in the flock, they provide only about half the food eaten by the hungry nestlings (figure 30.25).

Such cooperative behavior in caring for offspring seems unusual, although it is by no means unique. Many female mammals, such as lions and wild pigs, are known to nurse another female's offspring now and then, and African hunting dogs will help feed another female's pups.

Figure 30.25   A Mexican jay feeding young in a nest.

Such apparently unselfish interest in the welfare of others is called altruism, and behavioral biologists call this **altruistic behavior.** But how did altruistic behavior evolve? How could there ever be selection pressure for taking care of another individual's offspring with the same care that would be given to one's own?

It appears to be very difficult for young Mexican jays to establish themselves in a new flock. Therefore, they usually remain in the same flock throughout their unusually long lifetime. Young jays do not begin breeding until they are three years old, so a flock usually contains several of these young birds. Because most jays remain within the same flock, most of the individuals in a flock are likely to be very close relatives. The flock actually is an extended family unit made up of several generations of uncles, aunts, parents, siblings, cousins, and so forth, joined by an occasional outsider. Thus, if an animal helps another member of the flock, it is almost certainly helping a close relative. Many examples of altruistic behavior actually turn out, like this one, to be cases of animals helping relatives. But the question still remains, how could such behavior evolve?

In the early 1960s, W. D. Hamilton, then a graduate student at London University, developed a hypothesis that has generated much interest and controversy among behaviorists. Hamilton pointed out that a parent caring for its young is just a special case of the more general phenomenon of animals helping genetic relatives. Close genetic relatives, in addition to parents and offspring, have genes in common. Thus, an animal showing altruistic behavior toward a close relative is promoting survival and transmission of at least some genes that are the same as its own, and sociobiologists propose that there is selection pressure for such altruistic behavior toward relatives.

Does Hamilton's hypothesis provide any insight into the evolution of helpers at the nest in Mexican jays? Since members of a flock are almost always close relatives, helping to feed another individual's nestlings means that a jay is helping a close relative. Since individuals often cannot breed for several years, the best way to promote one's own genes is by helping relatives in the flock that are able to breed and transmit those same genes.

This is not to suggest that animals "know" in any conscious sense who their relatives are. However, animals who behave in this way have more surviving relatives that share and transmit the same genes, including genes for altruistic behavior, than those who do not. By behaving altruistically they are really doing little more than advancing their own genetic self-interests. Incidentally, they may even be helping their own future reproductive prospects by gaining valuable experience in raising young.

Sociobiology and Human Behavior

Sociobiology assumes that individuals within social groups behave in ways that improve the chances for passing on their own genes. Put another way, sociobiologists think that individuals in a social group have been programmed through evolution to behave as self-serving opportunists. Even altruistic behavior can be self-serving in an evolutionary sense if the altruism is directed toward close relatives with whom the altruists have genes in common. Because the adaptiveness of behavior is measured ultimately in terms of effects of the behavior on successful transmission of genes, some sociobiologists go so far as to say that the basic "selfish" units are genes rather than individuals.

Sociobiologists insist that these ideas apply to human behavior as well as to the behavior of other animals, and it is on this point that sociobiology has generated the greatest amount of controversy. People who disagree with some of its basic assumptions argue that sociobiology as applied to human beings promotes a kind of **biological determinism.** They feel that sociobiology leads to the conclusion that much of human behavior, including some of its most negative aspects, is at least in part genetically determined and thus is inevitable and unalterable. Opponents of sociobiology feel that this determinism can lead people to an acceptance of the social status quo that can condone racism, sexism, and class determinism. They argue that culture and the social environment are of primary importance in shaping human behavior.

Sociobiologists answer this criticism by saying that it is not dehumanizing to investigate the roles of genes in establishing human behavior. They think that the prospect of discovering that human behavioral patterns are more a part of human biological nature than originally thought should not make people uneasy. They argue that it does not detract from one's humanity to discover that humans are different from one another in terms of behavior and that some of these differences have genetic roots deep in humans' evolutionary past.

In a very real sense, this argument is another version of the old nature-nurture controversy, which has been a part of behavioral research throughout its history. Even so, the questions raised by the sociobiologists are intriguing, and the continued development of sociobiology is anticipated with a great deal of interest.

Summary

Behavior allows animals to respond to environmental stimuli quickly and adaptively. Adaptive behavior increases the probability that animals, or their offspring, live to reproduce successfully so that their genes are perpetuated in future generations. Thus, there is selection pressure for behavior such as removal of broken eggshells from gulls' nests.

The innate (genetically determined) nature of certain behavior is clear. Behavioral mutants show altered behavior patterns that are inherited by offspring of the mutant organisms. The genetic basis of behavior is also demonstrated when hybrid individuals show behavior patterns that are intermediate between the two different behavior patterns of their parents. For example, female lovebirds that are hybrids between two species, each of which shows a different nest-building behavior, show a nest-building behavior that is intermediate between those of their parents.

Learned behavior results from an organism's experience with its environment, and the influence of learning on behavior patterns begins very early in life. Even obviously instinctive behavior patterns, such as the pecking response in the feeding behavior of gull chicks, is modified by learning through experience. All behavior results from a combination of animals' genes and interactions with their environments (learning).

The learned component in behavior is sometimes surprising. Many animals do not instinctively know the difference between members of their own species and animals belonging to other species. Rather, they learn to recognize members of their own species and distinguish them from other organisms because they become imprinted on their parents at an early critical age. Experimental substitution of another animal during that critical period alters the behavior of the mistakenly imprinted animal throughout its life.

Many behavior patterns have a definite instinctive component, but they require one or several kinds of learning through sensory input for the definitive pattern to emerge. Chaffinches sing a roughly correct adult song if they can hear themselves sing during the learning period. But they do not learn proper adult song phrasing and complexity unless they hear the song sung by an adult before they are one year old.

Behavioral expression depends on internal motivational state. Physiological changes, such as changing hormone levels during reproductive cycles, result in changing motivation. Other motivational changes depend on completion of specific sequences of behavioral expressions. Thus, normal expression of complex chains of behavior depends on stimuli being presented in the normal sequence.

Adaptive behavior patterns play roles in diverse aspects of animals' lives, including such vital activities as obtaining food and finding proper living places. Bee communication through the waggle dance increases the efficiency of food gathering and facilitates location of new hive sites during swarming. Bats have highly specialized sensory and behavioral mechanisms that permit them to fly safely at night and to locate and capture flying prey.

Courtship behavior is adaptive because it brings animals together with prospective mates of their own species and resolves potential behavioral conflicts between males and females. In many cases, courtship behavior also is part of a chain of behavior leading to parental behavior that provides care for helpless offspring. Sometimes, courtship behavior involves elaborate rituals using display objects rather than body displays. For example, male bowerbirds collect and display an array of inanimate objects to attract a female.

Some adaptive behavior patterns are exceptional or unusual. For example, brood parasites such as cuckoos shift a large part of the burden of their reproduction to members of other species, whom the cuckoos trick into incubating cuckoo eggs and feeding young nestling cuckoos.

Sociobiology focuses on genetic and evolutionary explanations for social behavior. Among other things, sociobiology attempts to explain how there might be selection pressure for even seemingly altruistic (unselfish) behavior. Altruism directed toward close relatives promotes survival and transmission of at least some genes that are the same as those of the altruist.

Use of the concepts of sociobiology as explanations for human social behavior has resulted in much controversy. Critics of the approach fear that sociobiology applied to human behavior promotes acceptance of biological determinism that discourages attempts to deal with human behavioral problems through improvement of social environments. Sociobiologists maintain that they are simply taking an honest, open approach to the evolution and genetics of human behavior and that this approach can only increase human self-understanding.

Questions

1. Explain how genes control behavior.
2. What evidence is there that nest-building behavior in some lovebirds has a strong genetic component?
3. How does imprinting affect the behavior of ducks, geese, and some other animals throughout their lifetimes?
4. Discuss the development of chaffinch singing as an example of the interaction of genes and learning in the development of a behavior.
5. What do behavioral biologists mean by motivational state?
6. How does the waggle dance of bees communicate information about direction and distance?
7. Describe at least one positive way and one negative way that the use of ultrasound affects the efficiency of the bat sonar system.
8. Courtship behavior produces several important results. List and describe three of those results.
9. Although brood parasitism seems to be highly adaptive for some animals, there are only a few species that use it as part of their reproductive behavior. Why do you suppose that this is true?
10. Explain the sociobiological reasoning leading to the conclusion that altruistic behavior can be adaptive.

Suggested Readings

Books

Alcock, J. 1975. *Animal behavior: An evolutionary approach*. Sunderland, Mass.: Sinauer Associates.

Catchpole, C. K. 1980. *Vocal communication in birds*. Studies in Biology no. 115. Baltimore: University Park Press.

Dawkins, R. 1976. *The selfish gene*. New York: Oxford University Press.

Lorenz, K. 1961. *King Solomon's ring*. New York: Crowell.

Lorenz, K. 1981. *The foundations of ethology*. New York: Springer-Verlag.

Manning, A. 1979. *An introduction to animal behavior*. 3d ed. New York: Addison-Wesley.

Palmer, J. D. 1976. *An introduction to biological rhythms*. New York: Academic Press.

Wilson, E. O. 1980. *Sociobiology: The abridged edition*. Cambridge, Mass.: Belknap Press.

Articles

Bekoff, M., and Wells, M. C. April 1980. The social ecology of coyotes. *Scientific American* (offprint 726).

Brines, M. L., and Gould, J. L. 1979. Bees have rules. *Science* 206:571.

Crews, D. August 1979. The hormonal control of behavior in a lizard. *Scientific American* (offprint 1435).

Dilger, W. C. January 1962. The behavior of lovebirds. *Scientific American* (offprint 1049).

Fenton, M. B., and Fullard, J. H. 1981. Moth hearing and the feeding strategies of bats. *American Scientist* 69:266.

Gould, J. L. 1980. The case for magnetic sensitivity in birds and bees (such as it is). *American Scientist* 68:256.

Hailman, J. P. December 1969. How an instinct is learned. *Scientific American* (offprint 1165).

Keeton, W. T. December 1974. The mystery of pigeon homing. *Scientific American* (offprint 1311).

Maugh, T. H., II. 1982. Magnetic navigation an attractive possibility. *Science* 215:1492.

Silver, R. 1978. The parental behavior of ring doves. *American Scientist* 66:209.

Silver, R. 1978. The parental behavior of ring doves. *American Scientist* 66:209.

Smith, J. M. September 1978. The evolution of behavior. *Scientific American* (offprint 1405).

Population Ecology

Chapter Concepts

1. A population is a group of individuals of the same species that interbreed and occupy a given area at a given time.
2. Populations are dispersed in time and space.
3. When resources in an environment are nonlimiting, populations grow exponentially, but as resources run out, population growth slows and eventually stops.
4. Population size may be influenced by forces within or outside the population.
5. Populations may be divided, based on reproductive strategies, into two groups consisting of opportunistic and equilibrium species.
6. Two species that have the same ecological requirements cannot coexist indefinitely.
7. Predator and prey systems are the outcomes of coevolution.
8. Parasitism and mutualism are two symbiotic relationships that exist between pairs of species.

Ecology has been traditionally defined as the study of the relationship of an organism to its environment. The word ecology comes from the Greek root *Oikos,* which means something like "house," or "household." Implicit is the idea that the members of a household interact to form a functional unit. In recent years ecologists have become especially interested in coming to a better understanding of functional units of organisms as they occur in specific places at specific times. Groups of organisms together with their physical environment comprise **ecosystems.**

Ecology is a complex and encompassing subject. Plants and animals do not exist by themselves but interact with both the physical and biological aspects of their environment. The **abiotic** or nonliving environment includes such physical factors as moisture, temperature, light quality and quantity, mineral nutrients (especially for plants), and the terrestrial and aquatic medium on or in which animals and plants live. The biological environment of an organism includes individuals of its own kind and individuals of other species.

One approach to ecology is to examine the interactions that exist between a single kind of organism and its physical environment. Such interactions are treated elsewhere in this book (for example, see chapters 7 and 13). In another approach the unit of special focus is the population or community or the total ecosystem. In this chapter ecology is dealt with at the population level.

The Population

A **population** is a group of individuals of the same species that interbreed and occupy a given area at a given time. Populations have certain properties and characteristics and respond in a predictable way to the selective pressures of competition, predation, and resource availability.

Because a population consists of many individuals giving birth and dying over a period of time, a population has a birth rate and a death rate. Because a population usually consists of individuals of different ages, it has an age structure. And because individuals occupy different points in time and space, a population has **dispersion** in time and space.

Figure 31.1 The distribution of human populations is often clumped or aggregated, as is clear from this aerial view of human settlements.

Population Patterns in Space

Individuals in a population are scattered across the space they occupy. The pattern of cities, suburbs, towns, and farms across the landscape when viewed from the air is an example of how human populations are dispersed (figure 31.1).

The most common type of spacing observed among organisms is **clumped** (figure 31.2). Clumping may result because individuals of many species tend to form social groups. Examples include elk and caribou, which live in herds of varying sizes. Clumping also results because environments often are not uniformly suitable. In such environments organisms occupy only patches of favorable habitat. Colonially nesting birds, such as gulls and swallows, congregate on suitable nesting sites. Human communities are located along rivers, in valleys, at the junction of major highways and railroads, and at other such areas where transportation and other needed resources are available. Plants also exhibit clumped distribution. Seedlings may grow near the parent plant because of the limited dispersal of seeds; other clumped distributions result from vegetative reproduction by plants.

In most clumped distributions, however, organisms tend to maintain some spacing between individuals. Most animals appear to possess some minimal distance within which they do not tolerate other individuals. Plants and animals that remain attached in one place require minimal distances for proper growth and development.

Rarely found in nature is **random** distribution, in which the location of one individual has no influence on the location of another (see figure 31.2). In this situation individuals exhibit no tendency to attract or repel, and the habitat is so uniform that it plays no part in determining the location of individuals. Some invertebrates of the forest floor and some forest trees are randomly distributed.

Population Patterns in Time

Populations are also variable in time. The populations of birds in a woodland in the northern United States in summer are somewhat different from those found in the same woodland during winter. Robins, wood thrushes, and vireos are present in the summer but leave for warmer places in the winter. During the winter populations of pine siskins and evening grosbeaks may inhabit the woodland.

Plant populations also change seasonally. Early spring flowers in the woodland, such as large-flowered trillium, yellow violet, and spring beauty, are replaced in the fall by white wood aster, gray-stemmed goldenrod, and white snakeroot. Among plants, such changes usually occur because different species under the influence of changing day length become dormant at different times. Individuals of dormant herbaceous species spend the nongrowing season as seeds or bulbs, or as rosettes or buds buried beneath the leaves.

Since animals are mobile, the seasonality of some animal species is due to local and long-range migratory movements. However, many of the smaller species, like insects, enter a state of dormancy, spending the winter in the adult or some subadult stage.

Population Density

Another important dispersion characteristic of a population is its **density.** Density is the number of organisms per unit area of the earth's surface. A referral to 25 deer or 250 people per square kilometer or 150 mice per hectare is in terms of

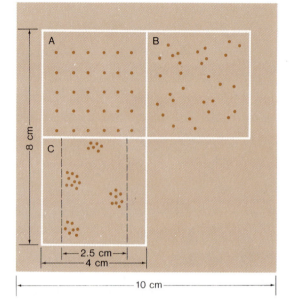

Figure 31.2 Population dispersion and density. If the large square is considered as the population boundary, the entire population shows a clumped distribution. Subpopulation A is uniformly distributed; B is random; and C is clumped or aggregated. Crude density, to a large degree, is dependent on how the boundaries are drawn around a population. If the large square is considered to be the population boundary, the crude density is 0.9 dots per cm². If the lines drawn about the three groups are considered the population boundaries, crude density is 1.9 dots per cm². In the clumped population (C), the lines can be drawn even tighter to produce a crude density of 3 dots per cm². To be meaningful statements about density, estimates should include information about population dispersal.

Trees growing close together, for example, often have stunted growth or are crowded out by more vigorous individuals.

Less common than clumping is **uniform** distribution. In uniform distribution individuals of a population are more or less evenly spaced. Animals that defend a territory tend to be uniformly spaced because they divide the available habitat among them. In desert communities or in dense pure stands of trees, plants tend toward uniform spacing because of competitive elimination or because older plants produce toxins that inhibit the establishment of their own seedlings within the spread of their crown or the reach of their roots.

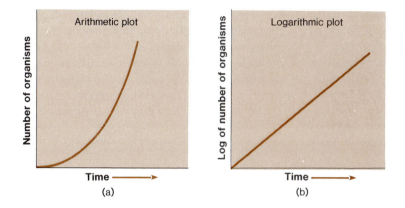

Figure 31.3 The exponential growth curve plotted (a) arithmetically and (b) logarithmically.

crude density. Such numbers tell very little about the nature of that density. Are the 250 people scattered over the square kilometer or are they living in a compact settlement? If the distribution is clumped, the density is much greater from the organism's point of view. Thus, density data may be influenced by the way boundaries are placed around a population. If the boundaries are drawn only around the areas the plants or animals actually occupy, the density per unit space is increased greatly (see figure 31.2).

Because organisms are confined to suitable habitat patches within any given area, density should be expressed in terms of **ecological density,** the number of organisms per unit of occupied habitat. Thus, crude densities can be very misleading unless something is known about the dispersion of the population and how the boundaries about the population are circumscribed.

Population Growth

The change in density of a population over time often reflects change in population size. What stimulates the growth of a community, city, town, or suburban area over a span of a few years? Is it increased births, decreased deaths, or increased movement of people into the area? Very likely all of these to some degree are responsible.

If a small number of reproducing individuals of some species were introduced into a vacant habitat with an abundance of food and other resources, the population would grow in a predictable manner. As long as the total environment continued to be optimal, the population would grow **geometrically (exponentially).** The number of individuals added to the population would become larger for each successive increment in time. This becomes apparent from a graph, using an arithmetic scale, of population size against time (figure 31.3). Such a graph shows that the population grows slowly at first, and then the curve becomes steeper because increasingly large numbers of individuals are added to the population. If the same data are plotted using a logarithmic scale for population size, exponential growth plots as a straight line.

The equation that describes exponential growth in a population with a stable and stationary age structure is:

$$\frac{dN}{dt} = r_m N$$

N is the number of individuals in a population at any particular time; $\frac{dN}{dt}$ is simply calculus notation for rate of change in the population over time; and r_m is the rate of increase, per individual, of the population in an optimal environment.

The value of r_m is basically a function of birth rates and death rates in the population (r_m = births − deaths). Since by definition r_m measures the rate of increase for a population growing under optimal conditions, birth rates theoretically are as high as they could be and death rates are as low as they could be. Thus, r_m describes an innate maximum rate at which a species can increase and is a measure of that species' **biotic potential.**

There have been many instances of exponential growth in real-life populations. These have occurred, primarily, when individuals of a species have been introduced into areas or habitats (or laboratory flasks) that are nonlimiting. Two examples of this are the starling and the house mouse, whose populations increased dramatically following their introduction into the United States.

Even though a population is not increasing at a rate predicted by its biotic potential (that is, at r_m), it may still be increasing exponentially. This, in fact, is the usual case for natural populations because environments are seldom nonlimiting. The rate of increase for such populations is something less than r_m and is usually described with the use of the symbol *r*.

As a population that initially grows in exponential fashion increases, resources have to be shared by an increasing number of individuals, and eventually the environment becomes limiting. The population now begins to increase at a rate less than that predicted by its biotic potential. The environment exerts its depressing effect on animal populations by influencing such things as the number of litters produced per year, the number of young produced per litter, the age at which reproduction first occurs, and the length of time during which individuals are fertile. As a group, the factors that serve to depress population growth are known as **environmental resistance.** Thus, a compromise is struck between maximum reproductive capability and environmental conditions that restrain actual growth.

When population size is plotted against time for a population in a limiting environment, the resulting curve is **S-shaped** and is called a **logistic** curve (figure 31.4). It is really a modification of the exponential growth curve and includes a factor to account for limitation by the environment. The equation that describes the logistic curve is:

$$\frac{dN}{dt} = r_m N \left(\frac{K-N}{K}\right)$$

K is the asymptote or upper level of the S-shaped growth curve. This equation represents the maximum number of organisms a particular environment can support on a continuing basis and is known as the **carrying capacity.** When the value for N is low, the value of $\frac{K-N}{K}$ approaches 1, and the equation becomes exponential:

$$\frac{dN}{dt} = r_m N$$

When the value of N is high, the value of $\frac{K-N}{K}$ approaches 0, and the population growth rate $\left(\frac{dN}{dt}\right)$ also approaches 0.

Carrying capacity varies from year to year. During favorable periods, when resources such as space or food may be more than adequate, a population may increase its size. In other periods, the environment may be less favorable, and the population declines in size. As a result, populations tend to fluctuate from year to year (figure 31.5). The mean of the series of ups and downs over time represents the carrying capacity.

In some cases, when a population exceeds a resource such as food, the population undergoes a sharp decline. Such population growth curves are sometimes described as **J-shaped.** An example is the reindeer herd introduced on St. Matthews Island in Alaska. The population experienced a rapid expansion, and then declined just as quickly as the food resources became depleted (figure 31.6).

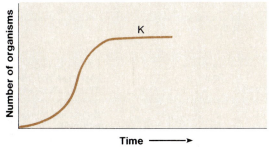

Figure 31.4 The logistic or S-shaped growth curve. The population initially grows exponentially and then begins to slow down as resources become limiting. Population size remains stable when the carrying capacity, K, of the environment is reached.

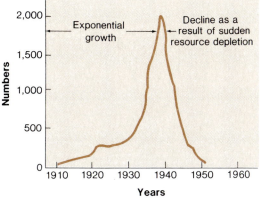

Figure 31.5 Population fluctuations in a bobwhite quail population. Open bars represent spring population sizes, and colored bars represent fall population sizes. Note that quail populations increase from spring to fall and decrease from fall to spring. But spring populations are relatively constant from year to year, reflecting the wintertime carrying capacity of the habitat and the territorial behavior of the species.

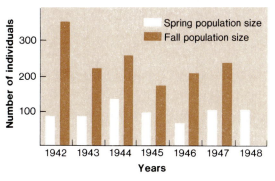

Figure 31.6 The St. Matthews Island reindeer herd grew exponentially for several decades and then underwent a sharp decline as a result of overgrazing the available range.

Mortality and Survivorship

Rates of increase in population size are a function of additions by birth and losses by death. The rate of increase sometimes increases not because of a growing birth rate but because of a decline in the death rate. This decline in deaths is one reason for the explosive growth of the human population in certain countries of the world.

Individuals that make up a population have a limited life span, and death is the means by which most organisms are lost to a population. (The other means is emigration.) The number of individuals in a population that are dying per unit of time per unit of population is the **death rate.** The death rate is usually determined by dividing the number dying during a given time span by the number alive at the beginning of the time period.

Because living members of the population are more important to the population than dead ones, another way of looking at mortality is survivorship. A **life table** summarizes statistics of life and death in a population by age group (table 31.1). Life tables, devised by actuaries for use by life insurance companies, consist of five columns: units of age (x), survivorship (l_x), mortality (d_x), mortality rate (q_x), and life expectancy (e_x), the average time left to an individual.

A survivorship curve can be drawn from a life table by plotting the survivorship data in the l_x column against time. This survivorship curve represents a group of individuals born at a given time in the population. Such a group is known as a **cohort.** In theory there are three types of survivorship curves (figure 31.7). A type I curve shows that most individuals in the cohort live out their allotted life span and then die of old age. Under very favorable environmental conditions, certain mammals and certain grasses tend to approach this curve. The type II curve results when a cohort of organisms die at a constant rate through time. Such curves are most typical of birds, rodents, and some perennial plants. Type III curves are evidence of organisms experiencing high mortality rates early in life. This curve is typical of fish, many invertebrates, and some plants. Most real-life survivorship curves are modifications of one of these three types of curves (figure 31.8).

Table 31.1
Life Tables for Cohorts of 100,000 White Male and Female Humans in the United States, as of 1972.

Age (years)	Survivorship (Number Alive at Beginning of Age Interval)	Mortality (Number Dying during Age Interval)	Mortality Rate	Life Expectation (Average Number of Years Left to an Individual at Beginning of Age Interval)
x	l_x	d_x	q_x	e_x
White, Male				
0–1	100,000	1,824	0.0182	68.3
1–5	98,176	323	.0033	68.5
5–10	97,853	220	.0023	64.7
10–15	97,633	238	.0024	59.9
15–20	97,395	735	.0075	55.0
20–25	96,660	927	.0096	50.4
25–30	95,733	799	.0083	45.9
30–35	94,934	845	.0089	41.3
35–40	94,089	1,154	.0123	36.6
40–45	92,935	1,840	.0198	32.0
45–50	91,095	3,013	.0331	27.6
50–55	88,082	4,542	.0516	23.5
55–60	83,540	6,913	.0827	19.6
60–65	76,627	9,618	.1255	16.1
65–70	67,009	12,014	.1793	13.1
70–75	54,995	14,160	.2575	10.4
75–80	40,835	14,798	.3624	8.1
80–85	26,037	12,354	.4745	6.3
85 and over	13,683	13,683	1.0000	4.7
White, Female				
0–1	100,000	1,370	0.0137	75.9
1–5	98,630	252	.0025	75.9
5–10	98,378	164	.0017	72.1
10–15	98,214	141	.0014	67.2
15–20	98,073	285	.0029	62.3
20–25	97,788	314	.0032	57.5
25–30	97,474	339	.0035	52.7
30–35	97,135	461	.0047	47.9
35–40	96,674	687	.0071	43.1
40–45	95,987	1,093	.0114	38.4
45–50	94,894	1,710	.0180	33.8
50–55	93,184	2,425	.0260	29.3
55–60	90,759	3,649	.0402	25.1
60–65	87,110	5,096	.0585	21.0
65–70	82,014	7,370	.0899	17.1
70–75	74,644	10,784	.1445	13.6
75–80	63,860	15,078	.2361	10.4
80–85	48,782	17,237	.3534	7.8
85 and over	31,545	31,545	1.0000	5.7

*From *Vital Statistics of the United States 1972*, vol. 2, part A, Table 5–1, U.S. Department of Health, Education, and Welfare, Public Health Service.

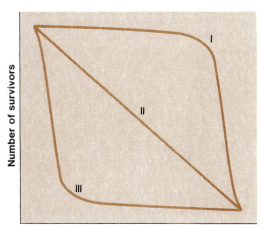

Age

Figure 31.7 The three basic types of survivorship curves. Curve I is for organisms living out the full physiological life span of the species. Curve II is for organisms in which the rate of mortality is fairly constant at all age levels. Curve III is for organisms with a high mortality early in life.

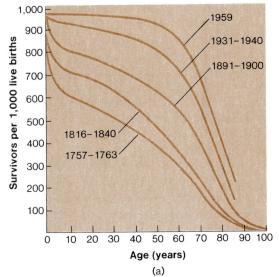

(a)

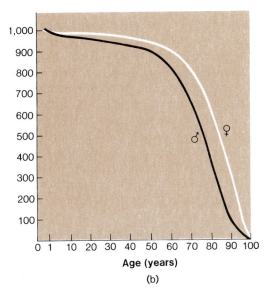

(b)

Figure 31.8 (a) Survivorship curves for the population of Sweden over several centuries. Note that as health conditions improved and the standard of living increased over the years, the survivorship curves began to approach the physiological life span, changing from slightly concave to convex. (b) Survivorship curves for white males and females in the United States (1972) drawn from life table data. Note the greater survivorship of women over men through all age classes.

Age Structure

One characteristic of some populations is the range in ages of individuals in those populations. This age distribution of individuals occurs only in populations of plants and animals that have overlapping generations.

The age structure of a population can best be visualized if the proportions of individuals in each age class are plotted as a bar graph. This produces an age pyramid (figure 31.9). The shape of this pyramid can be used to make predictions about future changes in population size.

All age classes in the population can be divided into three main groups: prereproductive, reproductive, and postreproductive. The proportions in each are a function of the birth rate and the death rate. If the birth rate is high and mortality is low, most of the individuals in the population will fall in the prereproductive and early reproductive classes (see figure 31.9). This results in a pyramid with a broad base and a narrow top—indicating very small old age or postreproductive age classes. Because large numbers of prereproductive individuals will later be entering the reproductive age classes, such pyramids suggest that the population will grow rapidly. If the pyramid shows that the numbers of individuals in each of the age classes are about the same, except for the oldest age classes, the population probably will not experience significant growth. Finally, if the number of individuals in the prereproductive age classes is small and most of the individuals are in the midreproductive and postreproductive age classes, the population will probably decline. A similar situation exists if a large number of individuals are in the reproductive age groups and few are in the prereproductive and postreproductive age groups. In this last situation, since few individuals are being born to enter the reproductive age classes, postreproductive age classes will continue to increase in relative size, and the population will decline.

Age structure is related to the concepts of a stable population and a stationary population. In a **stable population** the age structure does not change; the relative proportions of individuals in each of the age classes remains the same. A stable population may be increasing, decreasing, or remaining at the same level as far as its total size is concerned. A **stationary population** is one that remains the same size over a period of time. The number of individuals added by births and immigration must be equal to the number of individuals removed by death and emigration. Thus, a stationary population's rate of increase is zero.

Population Regulation

Monitoring of population size in some animals, such as the meadow vole, cottontail rabbit, or robin, shows that in some years the species are quite abundant and in other years relatively scarce. Over a long period of time, however, the average size of the population changes little (see figure 31.5). In general, population size is a result of the interaction between biotic potential and environmental resistance, but how are populations actually regulated?

This question has intrigued ecologists for years. Many population ecologists have argued that factors within the population related to density are of primary importance in regulating population numbers. They argue that if the population goes much over the carrying capacity, certain **density-dependent** mechanisms operate to decrease the birth rate or increase mortality, thereby slowing the population's growth. If the population falls below the carrying capacity, density-dependent factors function to decrease mortality and increase the birth rate. Other population ecologists have argued that populations are controlled or influenced by mechanisms outside of the population that do not necessarily relate to population density. These **density-independent** mechanisms exert their influence in a manner that is independent of the size of the population. Today most population ecologists agree that the fluctuations of populations of most organisms result from the interaction of both types of mechanisms. These mechanisms are perhaps best discussed as influences imposed from within and from outside of the population.

Regulatory Forces within a Population

Several density-dependent regulatory forces operate within a population, including intraspecific competition, stress, and dispersal.

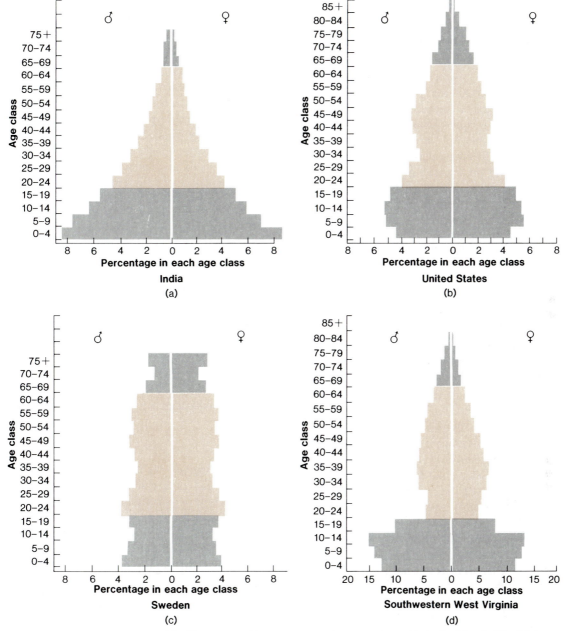

Figure 31.9 Types of age pyramids. All age pyramids are divided into dependency periods, rather than reproductive periods. Dependency periods, an economic rather than biological division, have more meaning for human populations. The divisions are 0–20, young dependents; 21–64, working period; 65+ retired dependency period. (a) The age pyramid for India (1970) is one of a rapidly expanding population. Note the broad base of young that will enter the reproductive age class (15–45). (b) Age pyramid for the United States (1970). The pyramid shows a constricted shape. The youngest period or class is no longer numerically the largest. This type of age structure reflects declining fertility. (c) Age pyramid for Sweden (1970). The population is nearly equally distributed over all age classes. This pyramid is typical of a population that is either aging and possibly declining or that has achieved zero population growth. (d) Age pyramid for a coal mining region in Appalachia (1960). In this pyramid the middle is pinched, reflecting a heavy emigration of younger age classes as they enter the economically active period. This emigration resulted in a high ratio of dependent young and old to the economically active. It also reflects a declining population.

Intraspecific Competition

As a population increases, a larger number of individuals require a share of resources such as food and space, and these become short in supply relative to the number of individuals seeking them. Members of the population increasingly compete for these resources. Since competition in this case is among individuals of the same species, it is known as **intraspecific competition.** If the population continues to grow, it may eventually exceed the capacity of the environment to support it. At that point, intense competition for the limiting resource may increase the mortality rate or decrease the birth rate. Intense competition will have its largest negative effect on the young, and their loss will reduce the number that will be added to the reproductive age classes. Declining birth rates and increased mortality drop the population back to the carrying capacity of the environment.

Competition may take two forms. In **scramble competition,** all individuals of a population have access to the resource. The resource, in effect, is subdivided among all members of the population. Each individual uses the resource without regard to other individuals. While each gets a share, usually this amount under crowded conditions is insufficient for adequate growth, successful reproduction, or even survival.

The English animal ecologist A. H. Nicholson demonstrated the nature of scramble competition in laboratory experiments with populations of blowflies. The blowfly life cycle has four stages: egg, larva, pupa, and adult. The larvae live on decaying meat, and the adults feed on sugar-rich foods. The size and vigor of the larval population influence the size of the adult population. The larvae need a certain level of food intake to grow large enough to pupate. The larger the larvae, the larger the adults (within limits of course). Larger adults lay more eggs.

Nicholson raised blowflies in closed containers in which he could regulate the amount of food given over daily intervals. In one experiment he rationed food to the larvae at a constant daily level but provided an unlimited amount of food to the adults. When larval density was high and the food supply low, few larvae reached a sufficiently large size to pupate, and the adults they produced were small. Small adults laid a small number of eggs, which, in turn, resulted in few larvae. A small larval population now had access to an abundance of food, grew to large size, pupated, and developed into large adults that laid many eggs. Thus, the population size from one generation to the next was influenced by the amount of food available to the larvae.

Contest competition results in more economical use of resources. In contest competition individuals compete indirectly for resources through social dominance and territoriality. The successful individuals are assured a supply of the limited resource, while the unsuccessful are denied access to the resource. Thus, the deleterious effects of resource shortage are confined to a certain fraction of the population, and the chance of large oscillations in a population is reduced.

Stress

Overcrowding and the resulting competition for resources and rank can result in stress among individuals in a population. In vertebrate animals, stress operates through certain physiological pathways often involving the anterior pituitary gland and its influence on other endocrine glands. Stress may result in lowered resistance to disease, failure of females to reproduce, failure of females to nurse and care for the young properly, and failure of the young to develop normally. Such symptoms of stress are most easily observed in laboratory or enclosed populations of animals. They are more difficult to demonstrate or prove in the wild.

Dispersal

In many animal populations, some individuals are behaviorally predisposed to leave and seek residence elsewhere. This tendency increases as populations become larger, possibly because of overcrowding and increased aggressiveness among individuals. Such dispersal constitutes one mechanism by which population size is regulated before the carrying capacity is reached. In such cases, enough animals leave so that the population of an area rarely reaches a high density. This was demonstrated by Charles Krebs and his associates, who held populations of meadow mice in enclosed areas with natural vegetation. In one area, mice were not disturbed. In two others, mice were removed by trapping. Mice from the first area moved into the removal areas, especially in fall and winter. The movement was most pronounced when the undisturbed population was increasing, whereas few mice moved when that population was low. In other experiments, Krebs found that in areas from which mice could not leave, population density reached a level nearly three times higher than that of natural populations in which dispersal was possible. These studies suggest that dispersal is an important means of reducing population size and controlling density. The odds are great that many of the dispersing animals will succumb to disease, predation, exposure, and starvation. A few, however, will be successful in establishing populations elsewhere.

Regulatory Forces Outside a Population

Other forces that regulate population size are density-independent. Examples of these forces are the weather and human activity.

Weather

One of the most significant influences on population growth and decline is weather. A long, cold winter with deep snow can cause death among terrestrial organisms by chilling and starvation. A long dry spell can dry up ponds and marshes and diminish the flow of streams. Under such conditions, fish and aquatic invertebrates perish, and waterfowl fail to nest successfully because of the lack of nesting cover and food. Muskrats, normally protected by water in the marshes, are exposed to predation by foxes and mink. Drought in winter or summer can kill stands of trees and shrubs. Late-spring frosts can eliminate the flowers of fruit- and nut-bearing trees, causing a shortage of food for squirrels, mice, and other animals dependent on such seed crops for winter food. A cold, wet spring can reduce the nesting success of rabbits and birds because the young drown or are chilled. On the other hand, highly favorable conditions of moisture and temperature stimulate population growth. Factors such as weather exert their influence on population growth regardless of population size.

An example of the close association between weather and population growth in some species is the relationship between reproductive success and winter rains in desert-dwelling rodents and quail in the southwestern United States. Winter rains stimulate the germination of herbaceous annual plants. These plants then provide the water and nutrients required by lactating female rodents and the vitamin E necessary to stimulate reproductive activity and successful nesting in quail. If the rains fail, herbaceous plants do not become available, and rodents and quail produce only a few young. If dry conditions persist for several years, quail may remain in flocks and fail to nest.

Figure 31.10 Cutting of a forest can affect some species of animals adversely and benefit others. This block clear-cutting in a western coniferous forest will result in patches of trees of different ages, which in time will increase the diversity of animal life. In the cutover areas, species of the mature forest will disappear and be replaced by animals of the shrubland.

Human Activity

Native populations of plants and animals can be either adversely or favorably affected by land management decisions of humans. Decisions to build a dam, construct a highway, drain a marsh, or cut down a forest have had a major impact on plant and animal populations (figure 31.10). Such human activities have destroyed habitats of many plants and animals and created habitats for some others. The destruction of forests and native prairies has brought some species, such as the ivory-billed woodpecker and Attwater's prairie chicken, to the edge of extinction. Drainage of potholes in prairie regions of the Midwest has greatly reduced the breeding populations of waterfowl. Widespread use of pesticides has lowered the breeding success of many birds by interfering with calcium metabolism, which results in the production of soft-shelled eggs. Accumulation of pesticides in tissues also affects survival of young birds and fish. On the other hand, the clearing of forests and subsequent planting of crops have extended the range of coyotes and horned larks.

Strategies for Growth

Another way to come to an understanding of population growth in a particular species is to examine growth characteristics in terms of the kind of environment the species inhabits. Environments range from being unoccupied, unstable, and unpredictable to occupied, predictable, and stable. Species that typically occupy the first kind of environment are said to be **opportunistic species,** while those in the second are known as **equilibrium species.**

Ragweeds in gardens and along roadsides and tent caterpillars in cherry and apple trees are opportunistic species. Opportunistic species usually are relatively short-lived, small in size, produce a large number of offspring usually in a single reproductive effort, and expend minimal energy on each individual produced. Among opportunistic animal species, for example, the eggs or young receive little or no parental care. Opportunistic species often possess efficient dispersal mechanisms that enable them to reach and exploit new food sources or newly opened habitats that are relatively free of competitors. The strategy of opportunistic species is to produce large numbers of offspring, a few individuals of which quickly disperse to colonize "open" habitats. From an evolutionary point of view, such species have undergone selection to maximize their intrinsic rates of natural increase and, for that reason, are said to be *r*-**selected** (p. 864).

Equilibrium species, such as oak trees and whales, are relatively large, long-lived, and slow to mature. They produce a small number of well-developed offspring and expend considerable energy on each individual produced. Among animals of equilibrium species, the young receive parental care; in plants of equilibrium species, the seeds contain a large amount of stored energy to ensure a vigorous start for seedlings. Equilibrium species are strong competitors and once established can dominate or exclude opportunistic species. They are habitat specialists rather than colonizers. Mortality is mostly density dependent, and population growth curves are logistic in form. Such populations are mostly stable and are at or near the carrying capacity of the environment. Equilibrium species are said to be *K*-**selected** (p. 864).

Interspecific Competition

In the same way that individuals within a population of a single species may compete with one another for a certain resource in short supply, populations of different species may compete for limited resources. Competition among individuals of different species is known as **interspecific competition.**

Two mathematicians, A. J. Lotka and V. Volterra, used a mathematical model to demonstrate that two species competing for the same resources cannot both survive. To develop this model, the two mathematicians modified the logistic equation for each population by including a factor that would account for the depressing effect of one species on the population growth of the other.

Lotka and Volterra suggested four theoretical outcomes to competition between two species. Under certain conditions species 1 will outcompete and completely replace species 2. Under a different set of conditions species 2 will completely replace species 1. In both cases neither species can coexist with the other. A third outcome is one in which an unstable equilibrium is established. Both species compete for the same resource, but which one will "win" depends on initial densities and chance fluctuations in population size. The fourth outcome is a stable equilibrium in which both species coexist but at a lower level than they would if the other species were not present. In this situation both species are weak competitors whose population growth is influenced more by competition among members of their own population than by competition with members of other species.

The theoretical work of Lotka and Volterra prompted G. F. Gause to undertake experimental studies with laboratory populations of protozoans in the genus *Paramecium.* Among his experiments were those involving *P. caudatum* and *P. aurelia.* These species have similar food requirements and therefore are strong competitors. When grown separately in culture, both *P. caudatum* and *P. aurelia* increased and eventually reached the carrying capacity of the culture medium. When the two species were grown together, *P. aurelia* eventually replaced *P. caudatum* (figure 31.11).

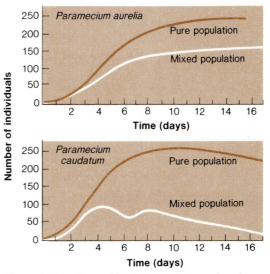

Figure 31.11 Competition between two species of *Paramecium.* When grown alone in pure culture, *Paramecium caudatum* and *Paramecium aurelia* exhibit sigmoid growth. When the two species are grown together in mixed culture, *P. aurelia* is the better competitor and its population increases. *P. caudatum* cannot grow in the presence of a growing population of *P. aurelia,* and it dies out.

In another experiment Gause raised *P. bursaria* (rather than *P. caudatum*) with *P. aurelia.* Although *P. aurelia* outcompetes *P. bursaria, P. bursaria* was able to survive because it occupied the lower part of the cultural medium, where *P. aurelia* did not occur. In this case, coexistence was possible because *P. bursaria* occupied a kind of spatial refuge.

The outcomes of Gause's experiments led to the formation of **Gause's principle,** which states that two species with exactly the same ecological requirements cannot coexist indefinitely. This idea has also come to be known as the **competitive exclusion principle.**

A corollary of the competitive exclusion principle is that coexisting species necessarily have ecological requirements that are different in some way. That is not to say that coexisting species never compete for the same resource. They often do, and competition becomes especially sharp when that resource is in short supply. Competition for a single resource sometimes becomes especially intense when an exotic species is introduced into a habitat. For example, the introduced starling competes strongly with bluebirds and flickers for nesting sites. Such competition occurs much less often among coexisting native species.

Over time, native species have evolved means of partitioning or dividing up resources. For example, they feed in different areas or utilize different food types or occupy different parts of the habitat (figure 31.12).

Robert MacArthur demonstrated how five species of North American warblers partition resources in their northern spruce forest habitat (figure 31.13). He divided spruce trees into zones and recorded the length of time each species spent in the different zones. In this way he was able to determine where each species did most of its feeding. He discovered that different species used different parts of the tree canopy, although some overlap did occur. This partitioning of resources permitted all five species to feed in the same forest without seriously competing with each other.

Competitive exclusion can be more easily demonstrated among plants and sessile animals. Among such organisms an individual species must compete, be eliminated, or exist in an area to which the other is poorly adapted. One good example of such competition involves two species of barnacles studied by Joseph Connell on the Scottish coast. *Chthamalus stellatus* lives on the high part of the intertidal zone, while *Balanus balanoides* lives on the lower part (figure 31.14). Free-swimming planktonic larvae of both species are able to attach themselves to rocks at any point in the intertidal zone and after doing so develop into the sessile adult form. In the lower zone, faster-growing *Balanus* individuals either force *Chthamalus* individually off the rocks or grow over them. *Balanus,* however, is not as resistant to drying out as *Chthamalus* and for that reason does not survive well in the upper intertidal zone, thus allowing *Chthamalus* to grow there.

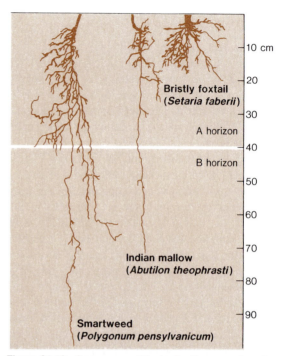

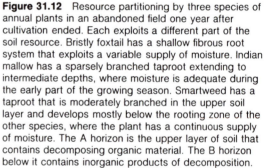

Figure 31.12   Resource partitioning by three species of annual plants in an abandoned field one year after cultivation ended. Each exploits a different part of the soil resource. Bristly foxtail has a shallow fibrous root system that exploits a variable supply of moisture. Indian mallow has a sparsely branched taproot extending to intermediate depths, where moisture is adequate during the early part of the growing season. Smartweed has a taproot that is moderately branched in the upper soil layer and develops mostly below the rooting zone of the other species, where the plant has a continuous supply of moisture. The A horizon is the upper layer of soil that contains decomposing organic material. The B horizon below it contains inorganic products of decomposition.

(a) Bay-breasted warbler

(b) Cape May warbler

(c) Blackburnian warbler

(d) Black-throated green warbler

(e) Yellow-rumped warbler

Figure 31.13 Resource partitioning among five species of coexisting warblers. The time each species spent in various portions of spruce trees was determined. Each species spent more than half its time in the stippled gray zones.

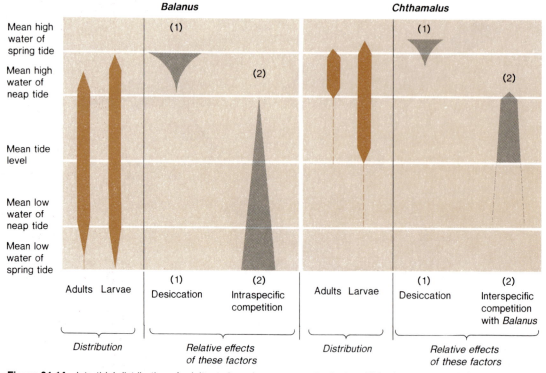

Figure 31.14 Intertidal distribution of adults and newly settled larvae of the barnacles *Balanus* and *Chthamalus* as influenced by interspecific competition and dessication. The following conclusions can be drawn: (1) *Balanus* can be found everywhere in the intertidal zone except the upper part where *Chthamalus* predominates; (2) in the upper part of the intertidal zone *Balanus* is less resistant to dessication than is *Chthamalus*; (3) the distribution of adult *Chthamalus* below the upper part of the intertidal zone is restricted by interspecific competition with *Balanus*.

Predation

All organisms are faced with the need to extract energy from their environments. Because for many organisms, available energy occurs in the form of other organisms, energy acquisition involves a series of interactions between prey organisms and their predators. Predators are under constant selective pressure to improve their ability to capture food, while prey organisms continually evolve to become more successful at preventing capture. Predation can be defined simply as one living organism feeding on another. Examples include a cougar feeding on a deer or a robin feeding on an earthworm (figure 31.15).

Predators can influence the number of prey organisms. While predation from an individual prey organism's point of view is always harmful, predation when considered from a population point of view may, within limits, be to the population's advantage.

Lotka and Volterra examined the effects of predation on both predator and prey. They independently developed mathematical models that predict that as a predator population increases, its prey population declines. Eventually, the prey population decreases to such a low number that the predator population also declines sharply because of lack of food. At this point the trend is reversed, and prey numbers increase with the result that predators also begin to increase. Prey increase the fastest when both predator and prey populations are at their lowest. Predators increase most rapidly when the prey population is at its highest. These interactions produce oscillations, or cycles, in both populations, the one lagging behind the other (figure 31.16).

(a)

(b)

(c)

Figure 31.15 A predator is any organism that feeds on another organism. Examples of some predators are (a) a white-tailed deer feeding (preying) on pine trees; (b) a barn owl feeding on a lizard; and (c) a mountain lion burying its kill.

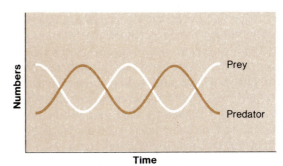

Figure 31.16 A visualization of the Lotka-Volterra predation theory. The equations generate cycles between predator and prey. As the predator population increases, the prey population declines. The decline in the number of prey eventually causes a decline in the number of predators. The decline in predators allows the prey population to recover.

Russian biologist G. F. Gause raised laboratory populations of two species of protozoans, a predator *Didinium* and its prey *Paramecium.* He discovered that the predator soon eliminated its prey, and the predator then perished from starvation. Gause was able to generate oscillations only by periodically introducing into the culture new individuals of both species. Gause also discovered that if he provided a refuge for the prey, in this case sediment on the bottom of the culture tubes, part of the prey population was able to survive. Under these conditions the predators died of starvation, after which the remaining prey repopulated the medium.

Carl Huffacker experimented with a large, complex, patchy laboratory environment to see if it would prevent a predator from eliminating its prey. He introduced two species of mites, one of which preys on the other, onto trays containing an array of oranges. The oranges constituted food for the prey species. Among the oranges he dispersed a number of rubber balls of the same size. By increasing or decreasing the number of oranges and rubber balls, by covering parts of oranges with paper and sealing wax, and by establishing barriers of petroleum jelly across which the mites could not move, Huffacker could manipulate the complexity of the environment.

He found that the predator population increased as the prey population increased; that when food for the prey was not well dispersed (so that the prey population was highly clumped), the predator eliminated the prey; and that when the food (oranges) was widely dispersed across the tray (so that prey organisms were also widely dispersed), both predator and prey fluctuated as the Lotka-Volterra model predicted (figure 31.17). However, after several oscillations both predator and prey died out. Like Gause, Huffacker discovered that a predator population could not be maintained for a great length of time without adding prey from time to time.

While predators have evolved strategies to secure a maximum amount of food with a minimal expenditure of energy, prey organisms have evolved strategies to escape predation. Natural selection operates constantly to refine these strategies. Such refinements are coupled; that is, as predators' capture efficiency increases, there is a corresponding increase in the ability of prey to escape capture. If this were not the case, prey would be hunted to extinction, or predators would be eliminated by starvation. When two species interact so closely that evolutionary changes in one influence the direction of evolutionary changes in the other, the process is called coevolution (see p. 793).

Predators are constantly improving their hunting techniques. Individual predators that are slow, lack strength, or fail to react quickly to prey movements are eliminated from the population. Through evolutionary times predators have acquired such "tools" as claws, talons, sharp beaks, and shearing teeth that enable them to capture, hold onto, kill, and process their prey. Some predators have improved their hunting success by acquiring coloration that permits them to blend into their surroundings. Some vertebrate predators have color lines leading from the eye that seem to function as aiming devices (figure 31.18). Circles and lines about the eyes of some predatory birds, mammals, fish, and reptiles also may reduce glare about the eyes.

Color patterns in some prey species match those of their surroundings and make it possible for such species to escape detection by predators. In some animals such concealing patterns involve **counter-shading,** in which coloration is darker above than below. Other prey species resemble some background object. Katydids look much like green leaves, the walking stick resembles a twig, and some moths may look like bark on a tree (figure 31.19).

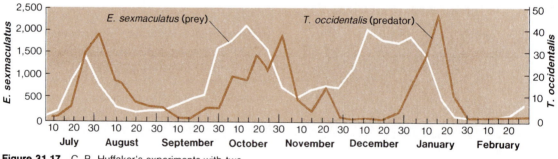

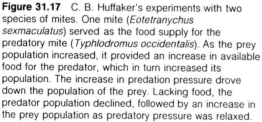

Figure 31.17 C. B. Huffaker's experiments with two species of mites. One mite (*Eotetranychus sexmaculatus*) served as the food supply for the predatory mite (*Typhlodromus occidentalis*). As the prey population increased, it provided an increase in available food for the predator, which in turn increased its population. The increase in predation pressure drove down the population of the prey. Lacking food, the predator population declined, followed by an increase in the prey population as predatory pressure was relaxed.

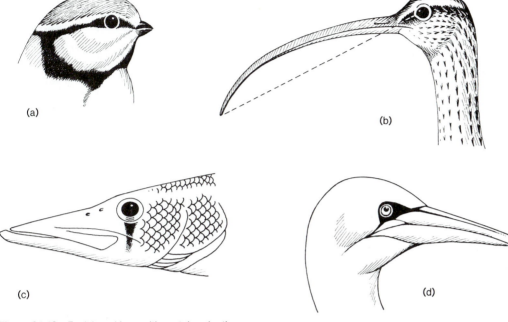

(a)

(b)

(c)

(d)

Figure 31.18 Facial markings with certain adaptive functions. (a) This simple eye line of the partially insectivorous blue tit is an example of the most common type of eye line in vertebrates. It serves as a sighting and aiming device. (b) The eye line in long-billed curlews is directed forward from the center of the pupil to the top of the bill and aids in securing prey. (c) The teardrop mark of the pickerel is associated with downward dashes at its prey. (d) The facial patch in the gannet reduces glare from the open sea.

Figure 31.19 The broad-winged katydid, an insect of the treetops, is an example of cryptic coloration and shape. Both its green color and its body shape and posture give the katydid the appearance of a leaf.

Some species show **disruptive coloration** (figure 31.20). Lines and marks tend to break up the outline of the animal or concentrate the attention of the predator on some appendage or less vulnerable part. For example, when startled, the white-tailed deer flashes its tail and in this way distracts the predator's attention away from the main body region (figure 31.21). White tailfeathers in birds function in a similar manner.

Some prey species avoid predation because there is something about them that is noxious to predators. An example is the monarch butterfly, which lays its eggs on the leaves of the milkweed, a plant rich in cardiac glycosides. Cardiac glycosides are powerful poisons used in minute quantities to treat heart disease in humans. They also activate nerve centers that cause vomiting. The green caterpillar of the monarch consumes the leaves of milkweed, detoxifies the poison, and incorporates the glycoside into its own tissue, where the glycoside remains through the monarch's pupal stage and into the adult stage. Thus, the monarch gains a chemical defense that causes it to be vomited and rejected by the predator. Of course, some individual monarchs are going to be sacrificed during the time the predator learns to avoid the prey, but the monarch butterfly population as a whole escapes any significant predation.

Unpalatable prey are often highly colored, which serves to warn potential predators that such organisms possess obnoxious qualities and should be avoided. Once a naive predator has experienced such prey, it associates the bitter taste or unpleasant experience with the color pattern and avoids the animal. This is advantageous both to the predator and to the prey. The prey is attacked less frequently, and predators don't waste time catching unpalatable prey.

Another strategy for escaping predation is to closely resemble a species that is protected by some defense mechanism. Such mimics, known as **Batesian mimics,** gain protection by looking and behaving like their models. An example of a mimic is the viceroy butterfly, a palatable species that feeds on willows and other nonpoisonous plants. The viceroy closely resembles the monarch butterfly, which as discussed earlier is unpalatable to predators (figure 31.22).

Figure 31.20   Orange and black stripes break up the body outline of this tiger, making it more difficult to locate in the grass. This is an example of disruptive coloration.

Figure 31.21   Distraction display. When startled, the white-tailed deer quickly flashes its tail, distracting any potential predators away from vulnerable parts of the animal.

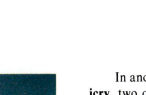

Figure 31.22 Batesian mimicry. The unpalatable monarch butterfly (above) and its mimic, the viceroy butterfly (below).

Figure 31.23 Thorns of an African acacia tree. Many plants have adaptations such as thorns or spines that discourage grazing herbivores.

In another type of mimicry, **Mullerian mimicry,** two or more protected species mimic each other. Such mimicry gives survival advantage to each species involved because the predator soon learns to associate a particular pattern with distastefulness or danger without trying all of the species involved. An example of such mimicry can be found among the wasps with yellow and black bands, many of which have potent stings.

Just as prey animals have developed defenses against their predators, plants have evolved defenses against herbivores. One strategy is to decrease the nutritional value of plant tissues. In some plants a high cellulose content in the leaves serves to reduce both palatability and digestibility. Oaks and other plants possess certain chemical substances, such as tannin. When tannin is combined with plant protein, the protein cannot be digested by the herbivore. Tannin also imparts an unsatisfactory taste to leaves and twigs and in this way discourages herbivores from consuming them. Other plants possess toxic chemicals, such as nicotine, that disrupt the normal function of the nervous system. A number of plants develop spines and thorns that discourage grazing animals (figure 31.23). Thistles in pastures are avoided by cattle, and spiny cactuses are avoided by desert herbivores. In many plant species tough seed coats discourage predation. Another method of seed defense involves predator satiation. Plants may produce such an abundance of seeds over a wide area that seed predators are unable to utilize the whole crop and some seeds escape and germinate.

Can Predators Regulate Prey?

In some situations and under certain conditions, predators may control prey populations. Specific examples are hard to find because predation is only one of a number of factors, including food supply, cover, and competition, that regulate populations.

Some of the more clear-cut examples can be grouped under the heading of biological control and involve specialized predators. One such example is the cottony-cushion scale, a small sap-sucking insect that was accidently introduced into California citrus groves from Australia sometime before 1872. Entomologists, seeking some means of control, identified its natural enemy in Australia. It happened to be a predaceous ladybird beetle, a number of which were imported from Australia and reared in California. In 1899 over 10,000 ladybird beetles were released in the citrus groves, and within a year the cottony-cushion scale was virtually eliminated. The scale was held to a very low level until the orchards were sprayed with DDT, which killed the beetle, permitting the scale to increase its population size.

Examples involving large predators are not so clear-cut. In general, evidence seems to indicate that a large predator, such as a wolf, can limit prey, such as deer or moose, only when the ratio of predators to prey is relatively high. Although such predators do sometimes take healthy animals, the majority of individuals preyed upon are very young, very old, or diseased. Populations of deer, moose, and other hoofed mammals in presettlement North America probably were limited by predation. Postsettlement human activity, such as cutting of forests and removal of large predators, has confounded the impact of predation on such populations today.

Parasitism

Parasitism is another kind of interaction that exists between individuals of different species. Parasitism is one of several kinds of symbiotic relationships. (In general, relationships in which one organism lives in intimate association with another are known as **symbiotic** relationships.) Parasites, like predators, gain their livelihood from other organisms. They are different from predators in that they live in close association with their prey (**host**). Parasites are smaller than their hosts, they live in or on their hosts, and they utilize only a portion of the host's energy content. Parasites usually do not kill their hosts. To do so would mean that the parasite itself could no longer survive. Thus, parasites and their hosts have through time coevolved a mutual tolerance for each other.

There are exceptions, however, and they include certain parasitic flies and wasps that lay their eggs on the larvae of other insects. When the eggs hatch, the parasitic larvae attach themselves to the body of the host, suck out the body fluids, and thereby bring about the eventual death of the host. By the time the larvae are ready to pupate, the host is dead.

Some parasites, known as **ectoparasites,** live on the body surface of their hosts. Ectoparasites, examples of which are lice, fleas, and ticks, usually possess some kind of specialized organ by which they remain attached to their hosts. Other parasites, known as **endoparasites,** live within the bodies of their hosts. They typically are found in the digestive tract, the blood stream, or the lungs. Some parasites utilize one host throughout their life cycle, while others require more than one host. One example of a parasite that requires more than one host is the human liver fluke (p. 640).

Mutualism

Another important symbiotic relationship between two organisms is **mutualism.** In mutualism, both organisms benefit from the association. Some mutualistic relationships are so close that neither organism can survive without the other. Such mutualism is said to be **obligatory.**

An example of obligatory mutualism is the relationship between the swollen-thorn acacias and certain species of ants in Central America. One in particular involves the bullhorn acacia and ants in the species *Pseudomyrmex ferruginae*. The large hornlike thorns of this acacia have a tough, woody covering and a pithy interior. The queen ant makes room for her brood by boring a hole into the base of a thorn and cleaning out the pith. As the colony grows, an increasing number of thorns on the plant become filled with ants. Eventually, the ant colony grows quite large, numbering in the tens of thousands of individuals. In addition to finding shelter in the acacia, the ants also obtain food from nectaries at the bases of leaves and in nodules at the tips of the leaves. In return, the ants bite and sting invertebrate and vertebrate herbivores when they attempt to feed on the acacia. The relationship is such that apparently neither ant nor acacia can survive without the other. Other species of acacias harbor different species of ants, but these relationships are not obligatory.

Many examples of the coevolution of nonobligatory mutualistic associations are emerging as studies of relationships between species become more numerous. One such mutualistic interaction involves the pinyon jay and the pinyon pine in the southwestern United States (figure 31.24).

Pinyon pine is an irregular producer of cones and seeds. Its cones take three years to mature, and production is often determined by weather conditions. Over its range, pinyon pine populations produce seeds in synchrony. This synchrony appears to be an evolutionary response to seed predation by both invertebrate and vertebrate seed eaters, including the pinyon jay. During poor seed years, wandering flocks of jays locate and utilize those few stands of pine that are producing seed. Because most of the seeds of such pines are eaten, selection has operated through time to reduce the number of pine trees that produce seed out of synchrony. Since abundant food is not available during poor seed years, the buildup of populations of seed predators, like the pinyon jay, is prevented. When a good seed year does arrive, low populations of seed-consuming animals are swamped by an overabundance of seeds, and many seeds escape predation and germinate. During these same years of seed abundance, flocks of pinyon jays store seeds for use in winter and early spring. Flocks of jays can store enormous quantities of seeds in open sites favorable for seed germination. Among the millions of seeds carried away and stored, a sufficient number escape consumption and germinate to start new stands of pinyon pine. In effect, pinyon jays plant pinyon pines.

The irregular but synchronized production of seed by pinyon pine results in synchronized large-scale colonial breeding of pinyon jays in areas where pinyon pine nuts are maturing. Thus, the pinyon jay and pinyon pine interact in a mutualistic fashion. The jays provide the pine a primary means of seed dissemination, while the pine furnishes the jay an abundant and highly nutritious source of food that stimulates very successful reproduction.

Figure 31.24 A pinyon jay. The pinyon jay utilizes irregular but heavy seed crops of pinyon pine, which stimulates colonial nesting and a high rate of reproduction. The pinyon pine, in turn, depends on the jay for the dissemination of its seeds to sites favorable for germination. Thus, the pinyon pine and the pinyon jay have a mutualistic association.

Summary

A population is a group of interbreeding organisms occupying a given area at a given time. Populations are characteristically dispersed in time and space.

Under ideal conditions populations grow geometrically. However, as environmental resources become limiting, growth eventually slows down and in stable environments population size usually comes to fluctuate around the carrying capacity of the environment. This produces an S-shaped growth curve. In some cases, particularly in unstable environments, a population may grow to exceed the carrying capacity and then sharply decline as some critical resource suddenly becomes depleted. Such populations have J-shaped growth curves.

The size of a population appears to be influenced by forces from within and also from outside the population. A major influence from within a population is competition, which may be either of the scramble type or the contest type. Forces from outside the population include the effects of weather and such human activities as land clearing and the spraying of pesticides.

Species adapted to compete well in occupied, stable environments are known as equilibrium species, while those adapted to produce large numbers of offspring in unstable, unpredictable environments are known as opportunistic species.

Several types of interactions occur between populations of different species. One of these is interspecific competition. Studies of interspecific competition have led to the important theoretical conclusion that coexistence is impossible for two species that have exactly the same ecological requirements.

Another interaction between species is predation. Population ecologists have used a number of approaches to come to a better understanding of predator-prey relationships. They include mathematical modeling, experiments involving laboratory populations, and studies of predators and prey in their natural environments. Predators and their prey have coevolved. As predators evolved ways of improving hunting efficiency, prey species evolved ways of avoiding predation.

Prey strategies include counter-shading, disruptive coloration, mimicry, chemical defense, and in some plants, the presence of thorns and spines. Some plants have evolved means of using herbivores to disperse seeds.

Parasitism is a relationship in which one organism derives its nourishment from a host organism usually without killing the host. Successful parasitism, in which the host and the parasite possess some mutual tolerance, is another example of coevolution.

Another important relationship between organisms is mutualism, an association from which both organisms benefit. Some mutualistic associations are obligatory: neither organism can survive without the other. Other associations are nonobligatory: both organisms can do without the other, but both benefit from the relationship.

Questions

1. What is a population and what are its attributes?
2. When can a population grow exponentially?
3. Distinguish between a stable population and a stationary population.
4. Describe the basic types of survivorship curves and indicate the significance of each.
5. Distinguish between contest and scramble competition, and explain how each might influence population size.
6. Distinguish between opportunistic and equilibrium species.
7. What is interspecific competition, and how does it relate to the competitive exclusion principle?
8. What is parasitism? Mutualism?

Suggested Readings

Books

Billings, W. D. 1970. *Plants, man and the ecosystem.* 2d ed. Belmont, Calif.: Wadsworth.

Emmel, T. C. 1976. *Population biology.* New York: Harper and Row.

Etherington, J. R. 1978. *Plant physiological ecology.* Studies in Biology no. 98. Baltimore: University Park Press.

Giesel, J. T. 1974. *The biology and adaptability of natural populations.* St. Louis: Mosby.

Hazen, W. E. 1975. *Readings in populations and community ecology.* 3d ed. Philadelphia: Saunders.

Kormondy, E. J. 1976. *Concepts of ecology.* 2d ed. Englewood Cliffs, N.J.: Prentice-Hall.

McNaughton, S. J., and Wolf, L. L. 1979. *General ecology.* 2d ed. New York: Holt, Rinehart & Winston.

Odum, E. P. 1971. *Fundamentals of ecology.* Philadelphia: Saunders.

Smith, R. L. 1980. *Ecology and field biology.* 3d ed. New York: Harper and Row.

Wilson, E. O., and Bossert, W. H. 1971. *A primer of population biology.* Stamford, Conn.: Sinauer Associates.

Articles

Bekoff, M., and Wells, M. C. April, 1980. The social ecology of coyotes. *Scientific American* (offprint 726).

Bell, R. H. V. July 1971. A grazing ecosystem of the Serengeti. *Scientific American* (offprint 1228).

de Beer, G. 1972. Adaptation. *Carolina Biology Readers* no. 33. Burlington, N.C.: Carolina Biological Supply Co.

Myers, J. H., and Krebs, C. J. June 1974. Population cycles in rodents. *Scientific American* (offprint 1296).

32

Community Ecology and Ecosystems

Chapter 31 focused on individual populations and the way size is regulated in such populations. This chapter examines *all* of the populations that inhabit a given area at a given time. Such interacting populations of different species form a **community.** The ways in which communities of organisms interact with the abiotic (nonliving) components of their environments also are explored in this chapter. A community of organisms together with its abiotic environment comprise an ecological system or **ecosystem.**

Community Structure

A grassy field is obviously quite different from a forest. Even a casual observer can point out that the field is dominated by grasses while the forest is dominated by trees. A careful study of these two communities also would reveal other important differences in community structure. The animal life of the grassland includes relatively fewer species than that found in the forest, and these species spend most of their time close to the ground surface. For example, many bird species are ground nesters and feed on the ground or in vegetation close to the ground. Similarly, grassland mammals live on or below the ground. In the forest, where there are relatively more species, some birds and mammals live in and on the ground, while other forms live in the shrubby understory, in the lower canopy, or in the tops of the trees.

An important difference, then, has to do with the way the two communities are stratified. The grassland essentially has two strata: a layer of litter and a layer of herbaceous vegetation. A well-developed forest may have as many as five or six strata, including a layer of litter at the soil surface, a layer of ground or herbaceous vegetation, a layer of woody shrubs, a layer of understory trees, and a canopy layer (figure 32.1). Stratification in any habitat is important because as the number of strata or "floors" increases, so does the number of microhabitats available to organisms. Stratification also influences such aspects of the physical environment of the community as light, temperature, humidity, and wind.

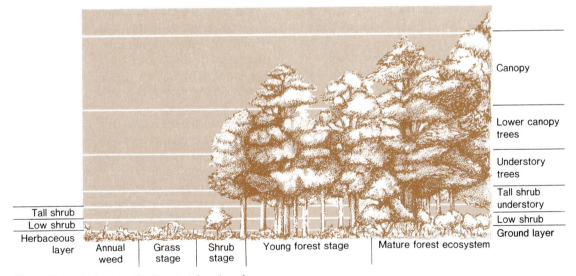

Figure 32.1 Vertical stratification as a function of developmental stage in terrestrial vegetation. Note that only two layers exist in the annual weed stage, while as many as six can be identified in the mature forest ecosystem.

Dominance and Diversity

One of the concepts that is useful in making distinctions between one community and another is that of **dominant species.** A dominant species is one that exerts more control over the character of the community than do the other species. Dominant species are usually characterized by greater numbers or **biomass** (living weight) than other species in the community. For example, in the northern coniferous forest, spruce and fir dominate the community. They severely reduce the amount of light that reaches the forest floor, produce a litter that is not easily decomposable, and create acidic conditions that influence mineral weathering and thereby affect the character of the soil. Only those species of plants and animals able to tolerate or adapt to the conditions created by spruce and fir can inhabit those forests. Other communities are dominated by animal species. Examples include oyster and mussel beds and coral reefs. In some communities dominance results from an interaction between species, often between animal and plant species. Animals such as cattle, buffalo, or prairie dogs may inhibit the growth of certain plant species by selective grazing or by disturbing the soil. Their activity stimulates the growth of other plant species that maintain dominance as long as grazing continues.

Another important community concept is that of **species diversity.** Species diversity as calculated by ecologists considers both the number of species present (**richness**) and the relative abundance of each species (**evenness**). The greater the number of species and the more evenly the individuals are distributed among the species, the greater the species diversity.

Species dominance and species diversity are related. Species diversity is low in communities in which a single species expresses dominance. In communities in which no one species is truly dominant, the total number of individuals is equally distributed among all species, and consequently species diversity is high.

This relationship between dominance and diversity is illustrated by data on trees over six inches in diameter in two Appalachian hardwood stands (table 32.1). Stand A has a lower species diversity because over one-half of all individuals

Table 32.1

A Comparison of Species Richness and Evenness in Two Appalachian Hardwood Stands.

Species	Stand A Number of Individuals	Stand B Number of Individuals
White oak	59	11
Red oak	7	17
Black oak	1	18
Chestnut oak	1	16
Scarlet oak	1	5
Red maple	19	22
Sugar maple	2	0
Cucumber magnolia	6	0
Black cherry	3	6
Black gum	1	0
Sassafras	0	5
Total individuals	100	100
Total species	10	8

in the stand are of one species. In stand B diversity is higher because individuals are more uniformly distributed among the species present. In this example evenness is more important than richness in determining diversity; stand B actually has fewer species than stand A.

It has been a axiom among ecologists that as species diversity increases so does community stability. A stable community is able to return to its original condition after having been disturbed in some way. It has been argued that this relationship exists in part because increased species diversity means that there are alternative pathways by which individuals in the community can obtain energy and nutrients. Or, to put this another way, a large number of species means that if one species declines or disappears, its place and function can be assumed in part by another. Thus, in general, communities consisting of many species do not exhibit the pronounced fluctuations in population size that are characteristic of communities composed of few species. More recently, however, ecologists have come to recognize that in some ecological situations increased diversity does not necessarily result in increased stability. It may be that increased community stability is more dependent on the number of well-adapted species than on the total number of species.

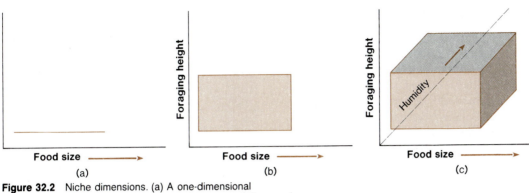

Figure 32.2 Niche dimensions. (a) A one-dimensional niche, the range of food size that a species can exploit. (b) A two-dimensional niche, in which a second axis, the height of foliage over which the species forages, is included. The enclosed space represents two dimensions in the ecological niche of the species. (c) The two-dimensional niche has been expanded to include a third axis representing humidity. The enclosed volume now represents a three-dimensional fundamental niche.

The Niche

No two species in a community have exactly the same ecological requirements. Each species is unique in the way it interacts with other species and with the physical environment. All of this is caught up in the term **niche.** A species' niche describes the unique structural and functional role the species plays in its natural community. A species' niche describes the species' way of life.

The concept of niche was formalized by the California ornithologist Joseph Grinnell early in the present century. He considered the niche to be a habitat space, the smallest subdivision of a habitat that can be occupied by one species. He further stated that "no two species of bird or mammal will be found to occupy exactly the same niche." In his 1927 book *Animal Ecology,* the English ecologist Charles Elton described the niche as an animal's place in the community. He looked upon the role of the animal in its community as its "profession." Elton thought of the niche largely in terms of how animals utilized food resources. Later, the American ecologist G. E. Hutchinson visualized the niche as an abstract multidimensional hypervolume. The axes of this hypervolume are described by all the environmental variables important to a particular species. Hutchinson's view of the niche can be examined by considering a limited number of environmental variables. For example, an important environmental variable for a species might

be food item size, with the range of food items sizes that could potentially be utilized plotted as a line in one dimension (figure 32.2a). A second variable might be foraging height, plotted as a line in a second dimension (figure 32.2b). These lines would then enclose a rectangle that would represent a two-dimensional niche. This can be taken one step further by adding a third dimension for another variable—humidity. The three lines now describe a three-dimensional niche (figure 32.2c).

G. E. Hutchinson used the term **fundamental niche** to describe the multi-dimensional niche. The fundamental niche includes the full range of environmental conditions under which an organism can live in the absence of interference from other species. Because most species occur together with other species, however, the fundamental niches of the species in a community often overlap. When this takes place, species compete for that part of the environmental resource represented by the overlap. The outcome of this competition is that either the region of overlap is incorporated into the niche of one species or the overlap area is divided between them. In either case, the fundamental niche is reduced to the **realized niche,** that part of the fundamental niche actually occupied under natural conditions.

Succession

Natural communities are constantly changing. Old fields become forests. Ponds fill in to become meadows. In mature forests, trees die and are replaced by others. As the vegetational components of a community change, so do its animal components. Communities undergo a predictable series of changes in structure and function that eventually result in the emergence of a relatively stable community known as a **climax community.** The process by which a climax community is achieved is called **succession.**

Ecologists distinguish between primary succession and secondary succession. **Primary succession** takes place on areas that are devoid of soil and have not previously supported a community. Primary succession occurs on sand dunes (figure 32.3), lava flows, raw glacial till, and rock outcrops. **Secondary succession** takes place on areas that were previously occupied by life and where soil is present. Secondary succession occurs on such disturbed areas as abandoned agricultural fields (figure 32.4) and burned-over or cut-over land.

Primary Succession

An isolated sandy beach by the ocean often provides a firsthand look at primary succession. The vegetation changes as one walks into the dunes beyond the beach. Surface temperatures on the sand dunes during the summer are high, and moisture is in short supply. These conditions, among others, make it difficult for organisms to colonize the dunes. One of the most successful pioneering plants on the dunes is marram, or beach grass. Once this grass becomes established and stabilizes the dunes against movement by the wind, associated plants such as beach pea invade the area. These are followed by mat-forming shrubs, such as poison ivy, beach heath, and rose. Later, the shrubs may be succeeded by trees—first pines and then oaks. As soil fertility and moisture availability increase, oaks may be replaced by maples.

Figure 32.3 Succession on a sand dune. The beach grass shown here is a pioneer species. Shrubs and trees replace the grass on older dunes after the dunes become stabilized and some organic matter accumulates.

Figure 32.4 Secondary succession on an old-field site. The shrubs and invading trees indicate that this field is moving toward a forest community climax.

In primary succession, changes in vegetation come about in part because the plants themselves make the environment more favorable, thus permitting the invasion of other species. The pioneer species build up organic matter, shade the ground from the sun, reduce evaporation, and improve moisture conditions. In this way grass species pave the way for shrubs, and shrubs improve conditions for trees. When pines come in, they eventually shade out the shrubs. But pine seedlings are unable to grow in the shade of mature pines, and eventually pines give way to oaks.

Secondary Succession

A study of old-field succession in North Carolina serves as an example of secondary succession. When cornfields are abandoned after harvest in the fall, annual crabgrass, an opportunistic species, takes over. The seeds of another plant species, horseweed, also germinate and by early winter form rosettes in the disturbed soil. The following spring, horseweed gets a headstart on crabgrass and crowds it out. But during that summer another plant, the white aster, germinates and begins to crowd out the horseweed. Dying horseweed opens up sites for the establishment of broomsedge, a bunchgrass. Eventually, broomsedge dominates the field. The growth form of broomsedge is such that open ground exists between individual plants. These moist, lightly shaded, and plant-free spots provide an ideal place for pine seeds to germinate. If a seed source is near, a pine stand develops and shades out the broomsedge. Since pine seedlings cannot grow in the shade, the pines eventually die and are replaced by climax species of oaks and hickories.

In secondary succession the site is colonized by short-lived, opportunistic species whose seeds are already present in the soil or are carried to the site by wind and animals. Opportunistic annual plant species are able to dominate the site quickly because they are adapted to severe environmental conditions and their seeds, given the proper conditions, germinate quickly and the resulting seedlings grow rapidly. Because annual plants die each year and need to grow anew from seeds in the spring, they get off to a slow start

each new year. This gives certain other longer-lived opportunistic biennial and perennial plant species a head start because they carry roots and growing tissue over the winter. These plants, especially perennials, may claim the site for an indefinite period of time. But as individuals in the population die, their place may be taken by still longer-lived plants of different species that have been growing slowly in the understory. These plants then take over the ground and prevent other species from developing until some of the dominant residents are damaged or killed. Eventually, the site is claimed by long-lived species whose own seedlings or those of associated species can replace older individuals that die. At this point succession will have reached the relatively stable climax condition.

It would be a mistake to think of climax communities as static and unchanging. Although a climax community such as a beech-maple forest is relatively permanent and self-sustaining, there are small-scale changes occurring all the time. For example, trees die and often are replaced by species other than their own kind (figure 32.5). Thus, changes in species composition occur in patches across the community.

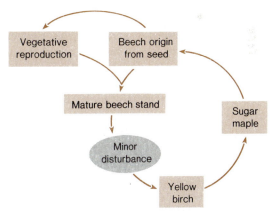

Figure 32.5 Pathway of replacement after minor disturbance to a climax beech forest community in northeastern North America. If a minor disturbance such as a windstorm removes mature individuals from a beech stand; the first species to colonize the newly exposed soil is yellow birch. As time continues, sugar maple becomes established under the yellow birch, and eventually it, in turn, is replaced by seedlings and vegetative groves of beech. Presumably, the site will continue to be dominated by beech until the next minor disturbance occurs.

Communities sometimes undergo disturbances that slow down the process of succession. One example is marsh succession in the prairie region of the midwestern United States. This aquatic succession typically proceeds from open water to submerged aquatic vegetation, to floating vegetation, to emergent vegetation, and finally to a climax terrestrial community. But in many marsh communities this progression is retarded by cycles of wet and dry years. During dry years, the pond dries up, and vegetation is eliminated. Organic matter decays rapidly, releasing nutrients. When the rains return and the marsh basin once again contains water, the vegetation at first is relatively sparse, consisting of annuals and some perennials. But eventually, perennial emergents such as sedges and cattails take over again. Thus, cyclic changes in water availability tend to retard the progess of succession and ensure the long-term existence of the marsh.

A number of community characteristics change predictably as a community undergoes succession. One change in a community moving toward a forest climax involves vertical structure, or stratification (see figure 32.1). Early successional grass stages have two layers, a litter layer and a foliage layer. During the shrub stage the shrub canopy becomes an additional layer. The mature forest stage may have as many as five or six layers, as described earlier in this chapter. Other changes involving community productivity and species diversity also occur. As succession continues, biomass increases. Although gross production (rate at which organic matter is produced) among the various successional stages remains about the same, net production (rate of storage of organic material not used by heterotrophs) declines because more energy goes into the maintenance of the increased biomass. Species diversity is higher in later stages of succession in part because of the greater number of microhabitats that exist when vegetation is more highly stratified.

The direction that succession takes is affected by a number of physical influences, such as soil, slope, nutrient and moisture availability, fire, and grazing. Of major importance is fire. Fire is a natural disturbance that sets back succession, shapes the character of the community, and influences the community's future composition (figure 32.6). Some species, such as pine, birch, and aspen, depend on fire for their persistence, while other species are eliminated by fire. Some communities, such as jack pine forests and chaparral, are fire-controlled. Their growth is renewed by periodic burning. If fire is prevented in a fire-adapted community, for example, a jack pine community, the pine may be replaced by competitively superior deciduous trees. Or fuel in the form of wood and litter may build up so that when a fire does occur, the heat becomes so intense that it is destructive even to the fire-controlled community. Because fire can be used selectively, foresters use it as a management tool to maintain desirable species.

Grazing and browsing by domestic and wild animals can arrest succession or influence the species composition of a community (figure 32.7). In parts of eastern North America with high populations of white-tailed deer, there are areas where the deer have eliminated sprouts and seedlings of certain species (pin cherry, black cherry, and sugar maple), while ignoring others (beech, birch, and certain oaks). As a result, deer influence the species composition of the forest. In other areas deer inhibit forest reproduction on cutover sites. In these areas woody vegetation is replaced by grasses and other herbaceous plants.

The Ecosystem

All of the organisms in a given area, together with the abiotic (nonliving) components of their environment, comprise an **ecosystem.** The term ecosystem was coined in 1935 by the British plant ecologist A. G. Tansley.

Ecosystem Structure

Ecosystems have both structure and function. Ecosystem structure is determined by the components that make up the system, while function is determined by the manner in which these components interact in a complementary way. All ecosystems possess both biotic and abiotic structural components. Biotic components include all of the organisms in the system, while abiotic components include such things as soil, water, light, inorganic nutrients, and weather variables.

The biotic components of an ecosystem can be categorized on the basis of how they obtain energy and nutrients. The **producer** organisms in any ecosystem are chiefly green plants that use the process of photosynthesis to convert radiant energy from the sun into the chemical energy of carbohydrates, fats, and proteins. In terrestrial ecosystems the producers are predominately herbaceous and woody plants, while in freshwater and marine ecosystems the dominant producers are various species of algae. From a nutritional point of view, producer organisms are **autotrophic.** That is, they can synthesize organic compounds using only inorganic raw materials and energy from an external source (usually sunlight).

Consumer organisms obtain energy and nutrients by ingesting producer organisms or other consumer organisms. **Primary consumers,** or **herbivores,** eat green plants, while **secondary consumers,** or **carnivores,** eat primary consumers. Both types of consumers show **heterotrophic** nutrition since they must obtain organic nutrients by ingesting other organisms.

Figure 32.6 A burn area seven years after a forest fire. Fire can be a powerful influence on vegetational development because it influences species composition and shapes the character of the community.

Figure 32.7 Example of the effects of overgrazing on a community. Wild burros have overgrazed the area to the left of this fence in Death Valley National Monument.

Decomposer organisms obtain their energy and nutrients by breaking down dead organic matter (figure 32.8). These heterotrophic decomposer organisms are mostly bacteria and fungi. In the soil bacteria grow as individuals or as small microcolonies on the surface of soil particles. Fungi, on the other hand, grow actively through soil containing organic material. For this reason fungi, rather than bacteria, are the major decomposers of dead organisms.

Many fungal and bacterial decomposers are specialists. Some groups, especially bacteria, quickly utilize any amino acids, peptides, and low molecular weight carbohydrates that are present. But much of the organic matter must be broken down before it can be utilized, a process accomplished by other species of microorganisms. Thus, one group of decomposers follows another as each exploits what it can from the food base. As a result of this sequential processing, organic material eventually is broken down into simple inorganic (mineral) substances. In this way mineral nutrients become available once again to producer organisms.

Previous use of the word *consumer* implied that the material being eaten was alive. In any ecosystem there are heterotrophic organisms that ingest large masses of dead organic matter, thereby reducing the masses to smaller chunks that are more susceptible to the decomposing action of bacteria and fungi. Dead plant and animal tissue is called **detritus,** and organisms that ingest this material are called **detritivores.** Earthworms and soil arthropods are detritivores.

The importance of detritivores in decomposition has been demonstrated experimentally. Litter bag experiments in which leaf litter is placed in bags with mesh too fine to allow soil arthropods to enter demonstrate that microorganisms can be relatively ineffective in decomposing detrital material if there has not been prior conditioning by detritivores.

Figure 32.9 shows the relationships that exist between biotic and abiotic components of an ecosystem.

Figure 32.8 The presence of decomposer fungi is confirmed by the development of the mushroom, which is the fruiting body of the fungus. The mushroom may be consumed by deer, squirrels, or certain insects, thus tying the decomposer food chain to the grazing food chain.

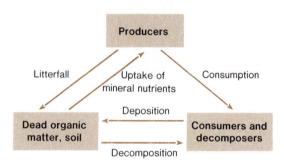

Figure 32.9 A simplified concept of the ecosystem showing the major components and the processes by which these components are interrelated. Producers obtain mineral nutrients from decomposed organic matter, are eaten by consumers, and contribute to the pool of dead organic matter. Consumers eat producers (and other consumers) and contribute to the pool of dead organic matter. Decomposers break down dead organic matter and contribute to the pool of dead organic matter.

Energy Flow in Ecosystems

Energy flow is a one-way process; that is, as the energy available in the chemical bonds of organic molecules moves through an ecosystem, it is gradually degraded to a nonuseable form. What this means, effectively, is that ecosystems are unable to function unless there is a constant energy input from an external source.

Food Chains and Food Webs

One way to follow energy flow through an ecosystem is to identify the sequences of organisms through which the energy moves. Such a sequence of organisms is known as a **food chain.** A simple food chain involves a producer and a primary consumer (herbivore):

Sunlight $\longrightarrow$ Plants $\longrightarrow$ Meadow mouse

A longer chain would involve a secondary consumer (carnivore):

Sunlight $\longrightarrow$ Plants $\longrightarrow$
Meadow mouse $\longrightarrow$ Weasel

A still longer food chain in the same ecosystem might be as follows:

Sunlight $\longrightarrow$ Plants $\longrightarrow$
Grasshopper $\longrightarrow$ Meadowlark $\longrightarrow$
Cooper's hawk (**tertiary consumer** or **top carnivore**)

Each step involves a transfer of energy from one feeding group, or link, to another. The type of food chain described is called a **grazing food chain** because it involves the consumption of green plants by herbivores, which in turn support additional feeding links of carnivores and top carnivores.

Not all of the biomass produced in an ecosystem is channeled into grazing food chains. In fact, most plants and animals die without being eaten, thus contributing to **detritus-based food chains.** The two types of food chains in ecosystems can be illustrated as follows:

Grasshopper $\longrightarrow$ Meadowlark $\longrightarrow$ Cooper's hawk
$\uparrow$ (Grazing food chain)

Sunlight $\longrightarrow$ Plants

$\downarrow$

Fungi and bacteria $\longrightarrow$ Mites $\longrightarrow$ Spiders
(Detritus-based food chain)

At any point the grazing food chain may be tied into the detrital food chain. This happens when a consumer in the grazing food chain feeds on some organism in the detrital food chain, as when a meadowlark eats a spider.

An important characteristic of food chains is that there is less energy available at successively higher feeding levels. A major portion of the energy incorporated at a given feeding level is not utilized, that is, not eaten, by organisms of the next feeding level. This energy (organic matter) eventually becomes part of the detritus. A significant portion of the energy that is consumed by organisms in the next feeding level fails to be assimilated and is egested as feces. Of the portion that is assimilated, some is lost to the system as the heat of respiration. Heat is a by-product of the metabolic processes involved in the breaking down of chemical bonds in food and their rearrangement into new bonds. Energy that is assimilated but not lost during respiration is incorporated into the tissues of organisms.

On the average, in grazing food chains only about 10 percent of the energy ingested by the organisms at a particular feeding level is incorporated into the tissues of those organisms. This unavoidable inefficiency in energy transfer between feeding levels explains why food chains are relatively short—at the most four or five links—and why mice are more common than weasels, foxes, or hawks. This also explains why more people could be fed on grain than on meat from grain-consuming animals. See figure 32.10.

A food chain represents just one path of energy flow through an ecosystem. But any ecosystem consists of numerous food chains, each linked to others to form complex **food webs** (figure 32.11).

Trophic Levels and Pyramids

If all the links in a food web are brought together according to the feeding level involved, or in other words, if all organisms in each link are grouped according to their general source of nutrition, the food web can be converted into feeding or **trophic levels.**

Biologists have agreed to assign producer organisms to the first trophic level, primary consumers (herbivores) to the second trophic level, the secondary consumers (carnivores) to the third trophic level, and tertiary consumers (top carnivores) to the fourth trophic level (figure 32.12).

The trophic structure of an ecosystem can be summarized in the form of ecological pyramids (figure 32.13). The base of the pyramid represents the producer trophic level; the apex is the tertiary or some higher level consumer; other consumer trophic levels are in between.

There are three kinds of pyramids. One is based on the numbers of organisms at each trophic level and consequently is known as a **pyramid of numbers.** For most ecosystems, pyramids of numbers are right side up because numbers of organisms decrease at successively higher trophic levels. However, there are some ecological systems for which the pyramid of numbers is inverted. An example would be the system including a single tree and all of its insect predators. In this case the part of the pyramid representing the producer trophic level that is the base would be smaller than the part representing the consumer trophic level.

A second type of pyramid is the **pyramid of biomass.** Biomass is the weight of living material.

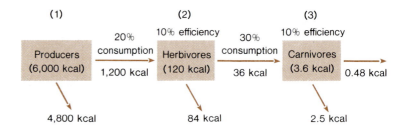

Figure 32.10 A hypothetical diagram showing energy relationships in a grazing food chain. For the herbivore feeding level, the relationships are as follows: (1) 1,200 kcal (20 percent of the 6,000 kcal available at the producer levels) are consumed; (2) herbivores are only 10 percent efficient in incorporating this energy into tissue; that is, only 120 kcal of the 1,200 consumed are incorporated into herbivore tissue; (3) although not indicated in the diagram, the rest of the 1,200 kcal is used to support the metabolism of the herbivores (and is lost as the heat of respiration) or passes unabsorbed through the digestive tracts of the herbivores. The diagram also indicates that only 36 kcal or 30 percent of the 120 kcal available at the herbivore trophic level is consumed by organisms in the next trophic level (carnivores). The rest of the 120 kcal, that is, 84 kcal, becomes part of the detritus.

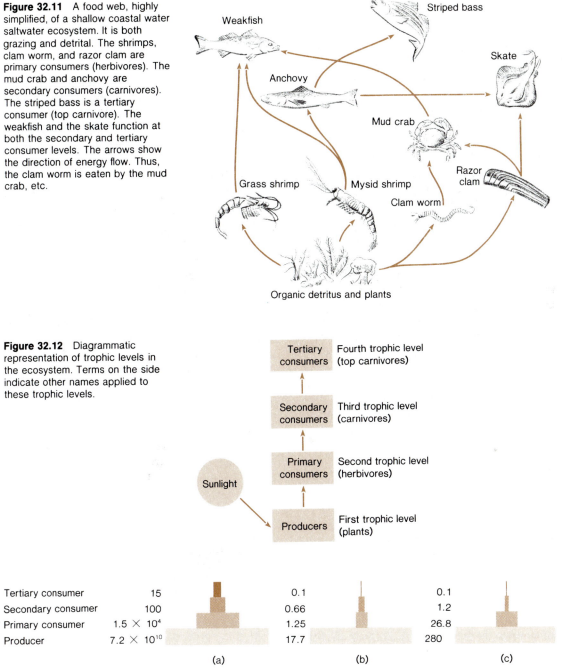

Figure 32.11 A food web, highly simplified, of a shallow coastal water saltwater ecosystem. It is both grazing and detrital. The shrimps, clam worm, and razor clam are primary consumers (herbivores). The mud crab and anchovy are secondary consumers (carnivores). The striped bass is a tertiary consumer (top carnivore). The weakfish and the skate function at both the secondary and tertiary consumer levels. The arrows show the direction of energy flow. Thus, the clam worm is eaten by the mud crab, etc.

Figure 32.12 Diagrammatic representation of trophic levels in the ecosystem. Terms on the side indicate other names applied to these trophic levels.

	Tertiary consumers	Fourth trophic level (top carnivores)
	Secondary consumers	Third trophic level (carnivores)
	Primary consumers	Second trophic level (herbivores)
Sunlight	Producers	First trophic level (plants)

	(a)	(b)	(c)
Tertiary consumer	15	0.1	0.1
Secondary consumer	100	0.66	1.2
Primary consumer	1.5×10^4	1.25	26.8
Producer	7.2×10^{10}	17.7	280

Figure 32.13 Ecological pyramids based on data for an experimental pond. (a) Pyramid of numbers (individuals/m²). (b) Pyramid of biomass (dry g/m²). (c) Pyramid of energy (dry mg/m²/day). Tertiary consumers were feeding at both primary and secondary consumer levels.

Weighing plants in sample plots in a field or forest at the end of the growing season gives an estimate of the biomass of plants in that area. Multiplying the number of animals by the average weight per individual gives an estimate of biomass for each of the consumer trophic levels. In general, the biomass of producers is much greater than the biomass of herbivores feeding on the producers, and the biomass of herbivores is much greater than the biomass of carnivores. However, this is not always the case. In some aquatic ecosystems, the pyramid of biomass may be inverted because of the rapid turnover of the small, short-lived, single-celled algae that dominate the producer trophic level. At a particular point in time in such systems consumer biomass may exceed producer biomass. If, however, biomass is measured over a period of time rather than at a single point in time, producer biomass will be greater than consumer biomass.

A third kind of pyramid is the **pyramid of energy.** The pyramid of energy is always upright because, for reasons discussed earlier, a given trophic level has a smaller energy content than does the trophic level immediately below it. The pyramid of energy is more meaningful than pyramids of numbers or biomass when studying trophic relations in ecosystems, but the information necessary to construct the pyramid is more difficult to obtain.

Primary Productivity

In general, **productivity** refers to the rate at which energy is incorporated into organic material. Since productivity is a rate function, it is always expressed on a per unit time basis. The terms production and productivity are used interchangeably.

Through the process of photosynthesis green plants utilize the sun's energy to convert carbon dioxide and water into carbohydrates. The total amount of energy incorporated by the producer organisms during photosynthesis is called **gross primary productivity.** Use of the word "primary" indicates that the reference is to productivity at the producer trophic level.

Some of this energy is utilized by the producers themselves during respiration to support their own metabolism. **Net primary productivity** is the rate at which energy left after respiration is stored by producer organisms. Arithmetically,

net primary productivity is equal to gross primary productivity minus respiration rate. Gross and net primary productivities are most often expressed as dry organic matter ($g/m^2/yr$) or calories (kilocalories/m^2/yr).

Net primary production accumulates over time in the form of leaves, twigs, bark, wood, roots, fruits, flowers, and seeds. The less energy plants require for their own metabolic needs, the more energy is left to accumulate as net production. If the energy intake (gross primary production) of plants just equals the amount lost as respiration, the plant does not accumulate net production. It is difficult to measure net primary productivity, in part, because during the measurement period some of the plant material produced is eaten by heterotrophs. Net primary production should not be confused with biomass. Biomass is the weight of living material present at a given point in time, while productivity is measured over a period of time.

Figure 32.14 compares primary productivities in different ecosystems. Least productive are open oceans and deserts. Productivity of open oceans is limited by a lack of nutrients because light energy is absorbed by water before it can reach the nutrient-rich bottom waters. Deserts have low primary production because of a lack of water, even though they may be nutrient rich. Grasslands, coastal waters, shallow lakes, temperate forests, and most agricultural systems have a net primary production ranging between 1,000 and 10,000 kcal/m^2/yr. The most productive ecosystems are estuaries, marshes, coral reefs, moist tropical forests, and systems under intensive agriculture. Their net primary production ranges between 10,000 and 20,000 kcal/m^2/yr. These ecosystems have available to them an abundance of nutrients and moisture and a long, favorable, growing season.

Secondary Productivity

Net primary production represents the amount of energy potentially available to consumers. Although some of this organic matter is ingested quickly by primary consumers, the greater portion of it is not utilized immediately. Some of it is unavailable or inaccessible—for example, most net primary production stored as wood in trees becomes available only when the tree dies and falls. Some of it is unpalatable or indigestible.

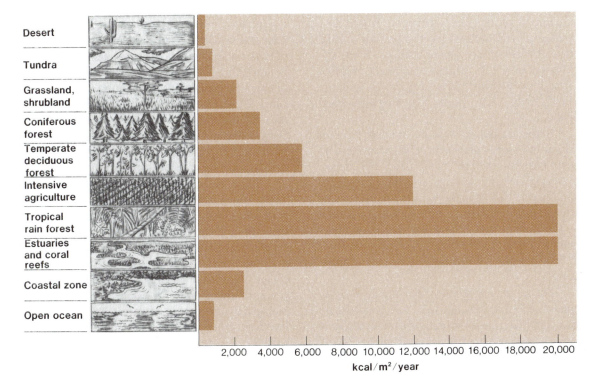

Figure 32.14 Gross primary productivity (kcal/m²/year) of major world ecosystems.

Some net production remains unconsumed simply because there is more net production available than can be immediately used by herbivores. Eventually, though, almost all net production is processed by consumers and/or decomposers.

A large portion of the net primary production that is actually eaten by consumers passes through the gut only partially digested. Grasshoppers, for example, assimilate only 30 percent of the food they consume; the rest is lost as feces and becomes part of the detritus. Organic matter that is actually assimilated is used for maintenance, growth, and reproduction. Most of the energy that goes into maintenance is used to meet the everyday demands of metabolism and tissue repair and is eventually lost as heat of respiration. In some consumers the costs of maintenance are high; in others, energy costs are relatively low.

The assimilated energy left over after maintenance goes into body growth and reproduction. The organic matter represented by growth and reproduction is **secondary production.** More generally, secondary production refers to the rate at which heterotrophs in an ecosystem store energy.

The amount of secondary production varies widely among consumers. Homeotherms use about 98 percent of the energy assimilated in metabolism, and only 2 percent goes into secondary production. Among heterotherms, especially invertebrates, about 75 percent of the assimilated energy goes to metabolism and 25 percent to secondary production.

Biogeochemical Cycles

Energy, entering as sunlight and leaving as heat, is continually lost from an ecosystem. By contrast, inorganic nutrients—the individual chemical elements out of which organisms are made—continually cycle through the abiotic and biotic components of the ecosystem. Because there is no input from outside the ecosystem, the thirty to forty elements essential to life must be used over and over. Some of these elements are abundant and easily recycled; others are less available to living organisms. For example, the activities of humans tend to send some of these elements, such as phosphorus, into deep sediments where they become unavailable for very long periods of time while the immediate supply declines. Since the pathways by which inorganic nutrients circulate through ecosystems involve organisms and the "earth," they are known as **biogeochemical cycles.**

Inorganic nutrients begin moving through the various biotic components of an ecosystem by being incorporated into living tissue. For plants, sources of inorganic nutrients are the atmosphere and the soil. For animals, sources are plants and other animals. Nutrient availability depends both on the amount tied up in plant and animal biomass and the rate at which the organic material in this biomass is decomposed (the rate of turnover).

There are mechanisms existing in ecosystems that ensure that nutrients will be available on a continuing basis. Many of these mechanisms are related to adaptive strategies of plants. Some plants have short life spans, possess low biomass, reproduce quickly, die, and decompose. Such plants take up nutrients quickly and also return nutrients to the system in a very short time. This provides a small but continuous supply of nutrients through the spring and summer. Other plants are longer-lived and tend to accumulate nutrients. Some of these store nutrients for a relatively long period of time in woody trunks and limbs and thus impede the flow of nutrients. Yet, many of these plants, especially deciduous trees, partition nutrient reserves between long-lived and short-lived parts. Although there is much nutrient storage in wood tissue, another portion is concentrated in leaves and twigs that are recycled annually.

Variations in the concentration of nutrients by different plant species also contribute to the regulation of nutrient flow. Some evergreen species, such as rhododendron and pine, retain their leaves (and the nutrients they contain) for several years and recycle them over a longer period of time. Deciduous trees recycle their leaves annually, losing them in the fall.

Among decomposers there is also controlled release of nutrients. Actively growing populations of bacteria and fungi hold considerable quantities of nutrients in their biomass and thus make them temporarily unavailable for use by other organisms. While this process, called **immobilization,** may result in nutrient deficiencies in plants under certain circumstances, it reduces the loss of nutrients due to leaching. In this way nutrients are held in short-term reserve in the soil.

Some biogeochemical cycles are gaseous, while others are sedimentary. Elements whose main reservoirs exist in the atmosphere or ocean move through **gaseous cycles.** By this definition, four critical elements—oxygen, carbon, nitrogen, and hydrogen—are involved in mostly gaseous cycles. These four elements constitute about 97 percent of living matter. The thirty-six or so other biologically important elements follow a **sedimentary cycle** in which the main reservoir is the earth's crust. They initially become available as a result of rock weathering and are taken up by organisms in ionic form.

Gaseous Cycles

Chapters 6 and 7 discussed how carbon moves through the biotic components of an ecosystem, and chapter 8 examined the nitrogen cycle. The **carbon cycle,** considered on a global scale, requires further discussion. The global carbon cycle involves an exchange of CO_2 between the atmosphere and the main carbon reservoir, the ocean. An equilibrium in the CO_2 concentration is maintained by exchanges at the interface between the two. Other exchanges of CO_2 also take place within the aquatic system. Excess carbon dioxide may combine with water to form carbonates and bicarbonates. Carbonates are not very soluble and precipitate out in bottom sediments, from which they return to the system slowly. These carbonates serve to buffer the CO_2 concentration in the water. If CO_2 becomes depleted, carbonates may be converted into bicarbonates and ultimately to CO_2 and water, thus increasing the CO_2 concentration in the water.

Some carbon is incorporated into forest vegetation biomass and may remain out of circulation for hundreds of years or even much longer. Incomplete decomposition of organic matter in wet environments results in the accumulation of peat. Such accumulation during the Carboniferous period resulted in great stores of such fossil fuels as coal, oil, and gas.

The total carbon pool, estimated at 55×10^{12} metric tons, is distributed among organic and inorganic forms. Fossil carbon accounts for 22 percent of the total world carbon. The oceans contain 70 percent of the inorganic carbon, mostly in the form of bicarbonate and carbonate ions, and an additional 3 percent in dead organic matter and phytoplankton. Terrestrial ecosystems hold only 4 percent of the total carbon; forests are the main reservoirs of terrestrial carbon. Only about 1 percent of the total world carbon is in the atmosphere. This is the amount circulated and utilized in photosynthesis.

However, because of the rapid burning of fossil fuels, the clearing of forests, and other land use, the amount of CO_2 in the atmosphere has been increasing at an exponential rate since the Industrial Revolution. Atmospheric concentrations have risen from an estimated 260 to 300 parts per million (ppm) 100 years ago to 330 ppm today.

Increased CO_2 in the atmosphere may have a significant impact on world climate. The CO_2 in the atmosphere transmits shortwave radiation from the sun to the earth but absorbs the longer-wave radiation that passes from the earth back toward outer space. Some of this absorbed heat energy is re-radiated towards the earth. Thus, CO_2 makes a contribution to the **greenhouse effect** of the atmosphere, and an increase in CO_2 concentration could result in increased warming of the earth. This could, in turn, melt the polar caps, cause the oceans to rise, increase evaporation, and decrease oceanic circulation. However, other changes in the atmosphere could produce a cooling effect. An example is an increase in the number of particles in the atmosphere that block incoming solar radiation. Climatologists are not yet able to predict the net effect of these changes on mean earth air temperatures.

Sedimentary Cycle

The calcium cycle is an example of a sedimentary cycle (figure 32.15). It begins in the soil, where calcium is taken up by plants and deposited in leaves, twigs, stems, and trunks. Rain dripping from the foliage dissolves some of the calcium from the leaves and carries it back to the soil,

Figure 32.15 The calcium cycle.

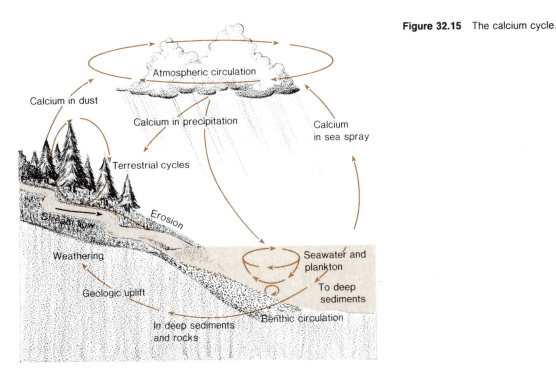

where it is quickly taken up by the plants again. Herbivores obtain their supply of calcium from plants, and carnivorous animals obtain their calcium from herbivores and other animals that they consume. Some animals have other means of obtaining calcium. For example, birds ingest calcium directly by picking up limestone particles from the soil and roadside, and rodents supplement their intake of calcium by gnawing on bones and shed antlers of deer. The death and decay of plants and animals return calcium to the soil where it is mineralized and taken up again by plants.

Only a part of the calcium and other nutrients taken up by plants is returned to the short-term cycle each year. For example, in the forest ecosystem, a substantial amount of calcium is stored in woody tissue, where it remains until the tree falls to the ground and decays or until the tree is reduced to ashes and mineral elements by fire. Some calcium may be lost to the ecosystem by movement of rainwater through the soil. Water carries calcium and other nutrients in solution through the soil to the groundwater and eventually to streams, lakes, and oceans. There some of the calcium is recycled through phytoplankton, zooplankton, fish, and other aquatic and marine organisms. A portion of this calcium is returned to land by way of sea spray. Another portion becomes incorporated in the bottom sediments of lakes and ocean where it is removed from circulation for extremely long periods of time. Many other biologically important elements, such as phosphorus, also are involved in complex sedimentary cycles.

Summary

All of the populations that inhabit a given area at a given time make up a community. Communities can be characterized in terms of structure, dominant species, and species diversity. Vertical community structure involves the way in which the plants in the community are layered. Dominant species are those that exert more control over the character of the community than do other species. Species diversity relates to both the number of species in a community and the evenness with which individuals are apportioned among the species. Each species in a community occupies a niche. A species' niche describes the unique structural and functional role the species plays in its natural community.

Communities undergo predictable changes through time in a process called succession. Primary succession occurs on sites that have not previously supported life, while secondary succession occurs on sites where life has occurred and where there is a soil base. Succession reaches a climax or stable stage when the vegetation dominating the site tends to replace itself and species composition tends to remain somewhat the same.

The community of an area, together with the physical environment in which it is found and with which it interacts, make up an ecological system or ecosystem. From a functional point of view, the major living (biotic) components of an ecosystem are the producers, consumers, and decomposers. Basic to an understanding of how ecosystems function is an understanding of the way energy and mineral nutrients move through an ecosystem. The flow of energy through an ecosystem is a one-way process because eventually all energy is dissipated as heat. Nutrients, however, circulate through an ecosystem, and they can be used over and over.

Questions

1. What is a community, and what are some of its characteristics?
2. What is meant by dominance? diversity? community stability?
3. Define the niche. What is the difference between the fundamental and realized niche?
4. What is succession? Describe the difference between primary succession and secondary succession.
5. Describe some mechanisms by which succession takes place.
6. What is meant by the climax vegetation?
7. How does fire influence succession?
8. What is an ecosystem? What are its basic components?
9. What is a food web? A food chain? A trophic level?
10. Distinguish among the three types of ecological pyramids.
11. Define gross primary productivity, net primary productivity, and secondary productivity.
12. What are the two types of biogeochemical cycles? Distinguish between the carbon cycle and the calcium cycle with emphasis on cycling mechanisms and major reservoirs of the nutrient.

Suggested Readings

Books

Cole, G. A. 1979. *Textbook of limnology.* 2d ed. St. Louis: Mosby.

Hazen, W. E. 1975. *Readings in population and community ecology,* 3d ed. Philadelphia: W. B. Saunders.

Whittaker. R. H. 1975. *Communities and ecosystems.* 2d ed. New York: Macmillan.

Articles

Brill, W. J. July 1977. A grazing system of the Serengeti. *Scientific American* (offprint 922).

The biosphere. September 1970. *Scientific American* (entire issue).

Gose, J. R.; Holmes, R. T.; Likens, G. E.; and Bormann, F. H. March 1978. The flow of energy in a forest ecosystem. *Scientific American* (offprint 1384).

Horn, H. S. May 1975. Forest succession. *Scientific American* (offprint 1321).

Likens, G. E.; Wright, R. F.; Galloway, J. N.; and Butler, T. J. October 1979. Acid rain. *Scientific American* (offprint 941).

Woodwell, G. M. January 1978. The carbon dioxide question. *Scientific American* (offprint 1376).

Ecosystems
of the World

The earth supports many types of ecosystems that collectively make up the **ecosphere** (figure 33.1). The ecosphere in turn functions as one huge ecosystem. Ecosystems are either water-based (**aquatic**) or land-based (**terrestrial**).

Aquatic Ecosystems

Aquatic ecosystems may be fresh water or salt water (marine). Freshwater ecosystems include streams, rivers, lakes, swamps, marshes, and bogs. Marine ecosystems include saline areas (like the Great Salt Lake and associated saline marshes), salt marshes, estuaries, sandy and rocky ocean shores, shallow seas, and open oceans.

Freshwater Ecosystems

Freshwater ecosystems may be **lotic** or **lentic.** Lotic systems are characterized by flowing water and include brooks, streams, and rivers (figure 33.2). Differences among lotic systems involve size, current velocity, water depth, and type of bottom (mud, salt, stone, or gravel). Lotic systems are inhabited by organisms well adapted to maintaining their position in flowing water. For example, fish in mountain streams, such as trout, have a streamlined shape. Invertebrates, such as caddis flies and mayflies, live beneath stones or attach themselves to stones by means of sticky surfaces, hooks, and suckers. Algae, mainly diatoms, grow tightly attached to the surface of rocks and may be encased in a slippery, gelatinous sheath.

The most important primary producers of lotic systems are algae, but in most streams the major energy source is detrital material carried in from surrounding terrestrial ecosystems. Many of the stream inhabitants feed on **drift,** which consists of organisms and detrital material carried downstream by the current. Others feed on leaves and other organic matter carried into the stream from vegetation along the banks.

In more arid regions, lotic systems apparently depend more on primary production for energy than on detrital material. As streams reach lower elevations and surface gradients become less steep, the water becomes deeper and slower, and the stream or river takes on some aspects of lakes. Phytoplankton (small floating plants) and rooted aquatics may become the dominant producers. Free-swimming invertebrates replace stone-dwelling insects, and bass replace trout.

Lakes and ponds are lentic systems and are characterized by still water, vertical stratification, and horizontal zonation (figure 33.3).

Vertical stratification of temperature, light, and oxygen can influence the distribution of organisms in lentic systems. For example, temperature strongly affects the structure of lakes and ponds in north temperate regions. During the summer, intense solar radiation is absorbed by surface waters so that a layer of warmer, lighter water "floats" on top of heavier, cooler water. Since the warm upper layer does not mix with the colder, more viscous, lower layer, a zone of steep temperature decline, called the **thermocline,** develops between the two layers (figure 33.4). This thermocline marks the boundary between the upper warm layer (**epilimnion**) and the deeper cold layer (**hypolimnion**). The upper, well-lighted layer, where most of the phytoplankton grows, is well aerated because of oxygen production by plants and mixing by the wind. The lower waters, however, may be deficient in oxygen because of bacterial decomposition. Because sediments and organic matter accumulate on the bottom, the deeper waters are relatively high in nutrients, while the upper waters become depleted of nutrients because of uptake by phytoplankton during the growing season.

During the fall, the upper waters cool, become more dense, and sink to the bottom, while the lighter, relatively warmer water rises to the top, producing the fall overturn. This cooling continues until temperature is uniform throughout the water column. At this point the wind aids in the circulation of water throughout the lake or pond basin.

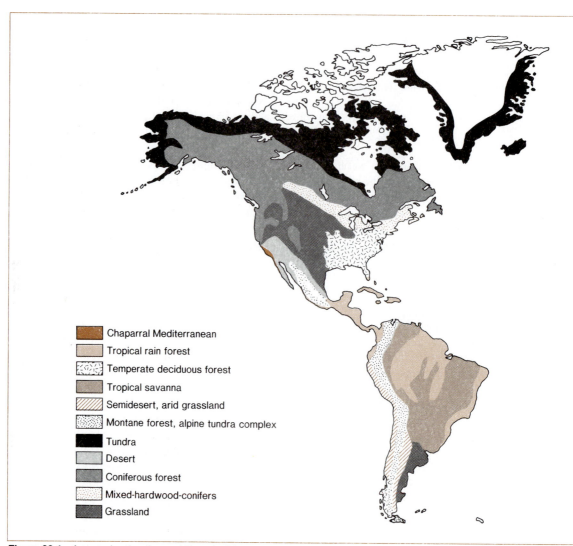

Figure 33.1 A map showing the world's major biomes. The map is a generalized one, and indicated biome boundaries are only approximately correct.

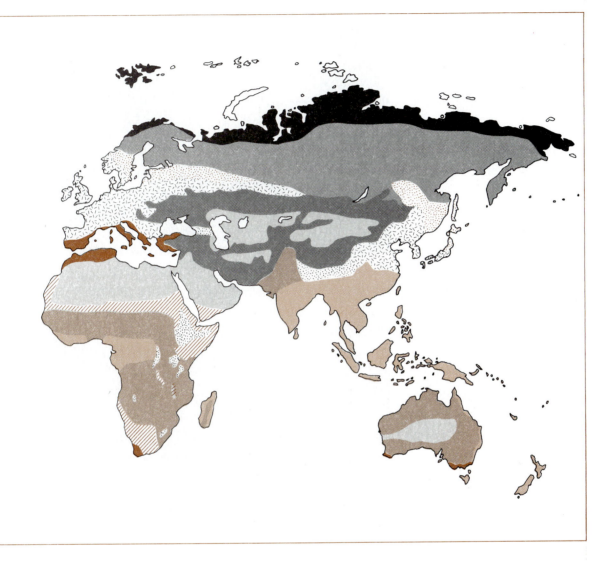

Figure 33.2 The stream ecosystem largely depends on inputs of energy and nutrients from the watershed it drains and from streamside vegetation. Two basic stream habitats include the riffles, where swift water flows over stones and gravel, and pools, which serve as catchment basins.

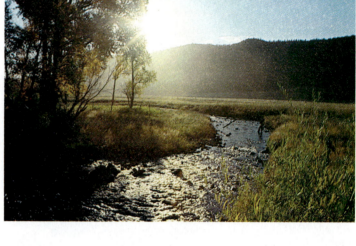

Figure 33.3 Lakes, like streams, depend in part on the watersheds that drain into them. Lakes that receive rich nutrient inputs from surrounding watersheds are eutrophic, while lakes in nutrient poor watersheds are oligotrophic. Additional nutrients are supplied to lakes from the emergent marshy vegetation along the shores.

Figure 33.4 Physical and biological zonation of a pond. The littoral zone includes the emergent and floating and submerged vegetation; the open water zone is dominated by phytoplankton. The profundal zone includes deep waters where light does not penetrate. The epilimnion is an upper layer of warm water separated from the deeper, colder hypolimnion by a zone of steep temperature decline called the thermocline.

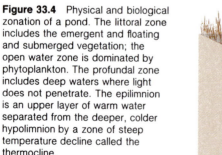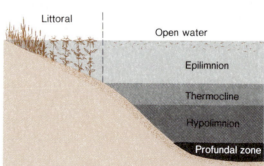

As winter approaches, the water in lakes and ponds cools. When water reaches its freezing point, ice formation begins at the top of the water because ice is less dense than water. Forming ice floats on the more dense cold water (water actually reaches its maximum density at 4°C). This freezing pattern leaves unfrozen water beneath the ice, which has an insulating effect that retards further cooling. This permits aquatic organisms to live through the winter in the water beneath the surface ice.

In spring, after the ice has melted, the surface water warms to 4°C and sinks, while lighter, warmer water rises to the top, producing the spring overturn.

This vertical stratification and seasonal change in temperature in a pond or lake basin influence the seasonal distribution of fish and other aquatic life in the lake basin. For example, cold-water fish move to the deeper water in summer and inhabit the upper water in winter.

Horizontal zonation is another characteristic of lentic systems. Horizontal zones in lakes and ponds include open water dominated by phytoplankton and a **littoral** zone, near the shore, that is dominated by floating and emergent vegetation rooted in the bottom. Characteristic bottom-dwelling (**benthic**) organisms occur in the "ooze" beneath each of these zones. Because of the wide range of habitats in lakes and ponds, aquatic life is diverse, ranging from phytoplankton, zooplankton (small floating animals), and fish of open water to organisms adapted to live in the littoral vegetation. Organisms of the littoral zone include *Hydra,* dragonflies, frogs, blackbirds, and muskrats.

Although lakes commonly are regarded as self-contained ecosystems, they are strongly influenced by the input of nutrients from surrounding watersheds through precipitation, drainage of surface water, and leaching of nutrients from soil, groundwater, and detrital material. The open water of the lake is dominated by phytoplankton and grazing zooplankton and therefore is characterized by grazing food chains. If the littoral zone is well developed, it becomes a source of considerable detrital material, so that detrital food chains become very important in the functioning of lakes, including open-water portions. Thus, nutrients move from littoral plants to bottom sediments and then after decomposition become available to the phytoplankton of open-water areas. At the same time, much of the fine organic matter not completely decomposed is consumed by the zooplankton.

Lakes and ponds often are classified by their nutrient status. **Oligotrophic** (nutrient-poor) lakes are characterized by low organic matter, low nutrient release from bottom sediments, and low productivity of phytoplankton. Such lakes are usually situated in nutrient-poor watersheds. **Eutrophic** (nutrient-rich) lakes are characterized by high organic matter, high release of nutrients from bottom sediments, high productivity of phytoplankton, and often a well-developed littoral zone. Such lakes are usually situated in either naturally nutrient-rich watersheds or in watersheds affected by agriculture or urban and suburban human settlements. Oligotrophic lakes can become eutrophic through large inputs of nutrients. This is usually the fate of oligotrophic lakes when they become polluted by sewage and other wastes.

Wetlands are another kind of lentic ecosystem and occur in areas where the water table is at or above the level of the ground most of the year. In some cases, wetlands occur where lake and pond basins have filled in with sediments. Wetlands include marshes, swamps, and bogs. **Marshes** are wet "grasslands" dominated by emergent vegetation, such as cattails, bulrushes,

Figure 33.5 A marsh is a wet grassland. Emergent vegetation includes species of sedges, cattails, and bulrushes. To be a suitable habitat for wildlife, the marsh should consist of patches of emergent vegetation interspersed with areas of open water.

Figure 33.6 A Florida swamp. A swamp is a wooded wetland. Swamps may be forested swamps, dominated by such species as swamp oak and cypress, or they may be shrub swamps.

and sedges (figure 33.5). **Swamps** are wetlands dominated by woody vegetation, such as button-bush, willow, swamp oak, and cypress (figure 33.6). **Bogs** have wet mats of vegetation, often dominated by *Sphagnum* moss (figure 33.7). Acidic decaying material beneath the surface forms peat.

Marine Ecosystems

Marine ecosystems begin at the edges of oceans, where there is a transition between fresh water and salt water. Areas where fresh water meets the sea are called **estuaries.**

Estuaries are characterized by salinities that are intermediate between those of fresh water and salt water. Because the interaction between outflowing river water and inflowing tidal salt water tends to create vertical water movements that keep nutrients in circulation, estuaries are highly productive, and serve as nursery grounds for a variety of organisms, including molluscs and fish eaten by humans (figure 33.8).

Salt marshes are associated with estuaries and are dominated by salt marsh cordgrass, a plant that is highly adapted to a saline environment. In addition to being an important wildlife habitat, salt marshes contribute nutrients and detrital material to estuaries.

Because of incoming and outgoing tides, sandy beaches and rocky shores constitute rather severe environments for organisms. For almost twelve hours of each twenty-four-hour period, organisms of the seashore are exposed to the heat and drying effect of the sun. During the remaining time, they are under water. Many of the organisms on a sandy beach remain below the sand, becoming active and rising to the surface only during the period of high tide (figure 33.9). On

Figure 33.7 A bog is a small body of water dominated by a wet mat of *Sphagnum* moss. Bogs often develop around lakes, but sometimes they develop on upland sites in which case the sphagnum moss holds water to create a perched water table.

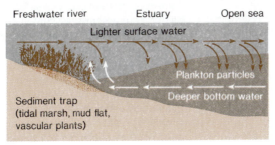

Figure 33.8 Mechanisms by which nutrients are conserved and retained in an estuary. Lighter, fresh water from rivers flows seaward on top of denser seawater flowing in from the estuary. Living and dead plankton carried seaward settle to the bottom and are carried toward land along with their nutrients. Many of these nutrients are taken up by plants in tidal marshes and mud flats, which act as nutrient filters. Tidal marshes, for example, tend to trap particulate nitrogen and phosphorus and convert them to inorganic chemical forms which are then exported back to the open water of the estuary.

Figure 33.9 A long stretch of sandy beach may appear to be a lifeless place, but its teeming life is buried beneath the sand and becomes active only at high tide. During low tide, sandpipers, plovers, and gulls hunt the sandy beach for food located just beneath the surface.

Box 33.1
Life Around Deep-Ocean Warm Springs

Benthic animals are scarce on the seafloor beneath the open ocean. The few animals that do live in the perpetual darkness of the ocean depths must obtain nutrients either by ingesting material that sinks from areas near the surface, where there is light enough to permit photosynthetic production, or by eating other organisms that consume the material that falls from above.

However, there are some places on the deep ocean floor where these rules do not apply. In 1977 scientists aboard the deep-diving research submarine *Alvin* of the Woods Hole (Massachusetts) Oceanographic Institution discovered some areas of the ocean floor 2,500 to 2,600 m below the surface in the eastern part of the Pacific Ocean that contain surprisingly dense populations of relatively large animals. These areas occur in volcanic zones along rifts where the seafloor is spreading tectonically (see p. 766). Water falls into volcanic hot spots in the rift, where it is heated, and it then emerges from hydrothermal vents. Water emerging from the vents is much warmer (20 to 22° C) than the cold water (2° C) normally found on the ocean floor.

Huge clams (many of which are up to 25 cm long) and mussels, crabs, polychaete worms, and fish have been discovered around hydrothermal vents along two spreading zones, the Galápagos Rift and the East Pacific Rise (box figure 33.1A). Some of the species found in these areas have not been observed elsewhere. Probably the most unusual animals in the hydrothermal vent communities are huge tube worms (*Riftia pachyptila*) that belong to the phylum Pogonophora, a relatively obscure group of marine worms. *Riftia* individuals live in tubes that they construct near vents, and they grow up to 3 m long and 10 cm in circumference (box figure 33.1B). One of the most remarkable characteristics of these worms is that they do not possess mouths or digestive tracts.

How do all the animals in these dense communities obtain nutrients? One suggestion is that the warm water rising from the vents sets up local convection circuits that sweep nutrients inward from large areas of the surrounding seafloor, thus concentrating nutrients around the vents. This may indeed be the case, but there is another intriguing factor in the nutrient equation of deep-ocean hydrothermal vent ecosystems.

The warm water in the vent areas contains high densities of bacteria. These bacteria are autotrophic; that is, they use carbon dioxide as a precursor in synthesis of reduced organic compounds. But they do not require light energy to accomplish this. In complete darkness, they utilize the hydrogen sulfide (H_2S), which is present in the vented water, in chemosynthesis. H_2S serves as a source of electrons for ATP-generating oxidation-reduction reactions and for the reduction reactions involved in CO_2 incorporation. Thus, a supply of H_2S coming out of the vents makes possible considerable primary production in an environment where photosynthesis is impossible.

These bacteria, therefore, are the primary producers in a food chain that is not based on photosynthesis. Filter-feeding animals such as clams and mussels can trap and digest the bacteria that grow in the water around them. Crabs and fish can live as scavengers, eating small animals or the debris from larger animals after their death. But how does the large, gutless worm *Riftia* manage nutritionally?

Pogonophoran worms are generally thought to live by actively transporting organic material from the seawater around them inward across body surfaces. *Riftia* may do this, but it also appears to utilize another nutritional strategy. Biochemical analysis of parts of the *Riftia* body has demonstrated the presence of enzymes that catalyze reactions in which ATP is produced during the oxidation of reduced sulfur compounds. Analysis has also revealed some enzymes that catalyze reactions in the Calvin cycle, a set of reactions by which carbon dioxide is incorporated into reduced carbon compounds (see chapter 6). Some

biologists think that symbiotic bacteria living in some tissues of *Riftia* produce these enzymes, but others think that there is at least a possibility that the worm itself might produce these enzymes. If this were the case, it would mean that these strange animals could be classified as autotrophs. While it seems more logical to assume that the enzymes do belong to symbiotic bacteria, this is just one of a number of questions that remain to be answered concerning the organisms that live around deep-ocean hydrothermal vents in the dark depths of the sea.

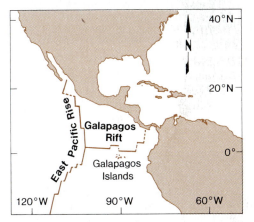

Box figure 33.1A Location of the spreading zones in the East Pacific along which deep-ocean hydrothermal vents occur. The "oases" of abundant life around the vents have been given such names as Dandelions, the Rose Garden, and the Garden of Eden.

Box figure 33.1B Giant tube worms, *Riftia pachyptila*, living near a deep-ocean hydrothermal vent. These large worms (up to 3 m long) have no digestive tract.

rocky shores, organisms such as rockweed, barnacles, and periwinkles are tightly attached to the rocks to withstand the action of waves (figure 33.10). During low tides, many of these organisms close their shells to prevent drying out. At high tide they open their shells to feed on detrital material and small forms of life carried by the water.

The open ocean consists of zones similar to those of lakes, but the zones have greater depth (figure 33.11). Oceans are relatively unproductive because there is little circulation between nutrient-containing bottom waters and light-absorbing upper waters. However, certain areas of shallow water over the continental shelf are highly productive. High productivity also is found in regions where cold and warm currents meet to produce upwellings that bring deep, nutrient-rich, bottom waters to the surface. This stimulates high phytoplankton production, which in turn supports dense populations of fish and fish-eating seabirds.

Oceans also exhibit horizontal zonation (figure 33.12). The open sea is the **pelagic** zone, the shallow seas over continental shelves make up the **neritic** zone, and coastal waters are in the littoral zone.

Figure 33.10 The rocky seashore, especially along the northeastern coast of North America, shows strong zonation of life exposed at low tide. The dark zone consists mainly of blue-green algae. The zone below it contains barnacles. Below the barnacles is a zone dominated by rockweeds *(Fucus)*.

Terrestrial Ecosystems

Major terrestrial ecosystems are categorized on the basis of the climax community types they support. These community types are called **biomes.**

Tundra

Arctic Tundra

Lying largely north of latitude 60°N and encircling the top of the earth is a vast, treeless region called the **arctic tundra** (figure 33.13). The arctic tundra biome is dominated by mosses, lichens, sedges, grasses, and low-growing shrubs. The growing season is short (sixty days or less); the winters are extremely cold; and the precipitation, falling mostly as rain in summer and fall, is usually less than 25 cm a year. However, because of poor drainage, low temperatures that reduce evaporation, and the presence of a permanently frozen soil (permafrost) a short distance below the surface, the land is wet and covered with numerous ponds, lakes, and bogs.

Figure 33.11 The open sea, which covers most of the globe, is dominated by phytoplankton, which in turn support all other life in the seas. Although the ocean from the surface may seem to be uniform, it is really a patchy environment. Some areas, such as those over continental shelves and in regions of upwelling, may be highly productive. Other areas of the seas may support relatively few life forms.

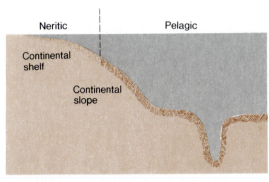

Figure 33.12 Diagrammatic section of an ocean showing the major life zones.

Figure 33.13 The arctic tundra. The name tundra comes from the Finnish *tunturi*, meaning treeless plain. It is a land of bogs, sedge marshes, and lakes. Frost molds its landscape, and permafrost, a frozen layer of soil, impedes drainage and keeps the land constantly wet, even though precipitation is low. This photo shows part of an arctic caribou herd during its annual migration across the tundra.

Net annual primary productivity in the arctic tundra is low because the growing season is short and nutrient availability is poor. Nutrient levels are low because decomposition is extremely slow in soils that are very cold most of the year. In much of the arctic tundra the most important herbivores are lemmings, small rodents related to meadow mice. Lemmings consume, on the average, about 3 percent of the aboveground biomass, but when their population peaks, they consume much more plant material. Consumption by lemmings is heaviest under the snow in winter. In other parts of the tundra, musk ox and caribou are the major herbivores. The herbivores support a number of carnivores, including snowy owls, foxes, and weasels.

Alpine Tundra

Above the timberline in the higher mountains of the world is the **alpine tundra** (figure 33.14). The alpine tundra is similar to the arctic tundra but lacks a permafrost or has only a poorly developed one. Alpine tundra, which shares few species with the arctic tundra, has moderate precipitation, widely fluctuating temperatures, and high ultraviolet radiation. The vegetation includes more grasses and sedges and fewer shrubs and lichens than the arctic tundra.

Coniferous Forest or Taiga

South of the tundra lies a worldwide belt of **coniferous forest (taiga)** that has southward extensions at higher elevations in the mountains (figure 33.15). The vegetation is dominated by spruces, firs, and pines. The climate is characterized by cool summers, cold winters, and a growing season of about 130 days. Precipitation ranges between 40 and 100 cm per year with much of it coming as heavy snow. The soils are acid and infertile and have a thick litter of slowly decomposing needles. Much of the precipitation moves through the soil, carrying with it important nutrients. Because of poor evaporation, this movement is not balanced by an upward movement of nutrients. Although the growing season may be short and nutrients limiting, annual net production is rather high, between 2,000 and 3,000 kcal/m². This is true, in part, because the evergreen nature of the needles permits photosynthesis on any favorable day during the year.

Figure 33.14 The alpine tundra is a cold land of strong winds, snow, and widely fluctuating temperatures. It is a land of rock-strewn slopes, small bogs, alpine meadows, and shrubby thickets. Plant life is dominated mostly by cushion and mat-forming plants.

Figure 33.15 The coniferous forest, dominated by spruce and fir, forms a belt around the world below the tundra. It is also called the taiga.

Many of the coniferous tree species that comprise the taiga are accumulator species; that is, they remove large quantities of elements from the soil, especially calcium, nitrogen, phosphorus, and potassium. This can upset the nutrient balance and impoverish the system because nutrients are not recycled until the tree dies or is consumed by a forest fire. Nutrient accumulation occurs, in part, because the needles are held on the tree for a long period of time and not dropped seasonally like the leaves of deciduous trees. Important in nutrient recycling in coniferous forests are the **mycorrhizae,** which are symbiotic associations between fungi and the roots of coniferous trees. Mycorrhizae extend into the fermenting litter and speed the transfer of nutrients from the fermenting litter to the plant. Most of the litter is

decomposed by fungi, which are resistant to leaching. These fungi act as a nutrient sink, holding nutrients in place to be used eventually by the trees.

Compared to temperate and tropical biomes, the coniferous forest has relatively few consumer species. However, the number of individuals within each consumer species tends to be large. Insects such as the spruce budworm and larch sawfly are grazing herbivores whose populations can reach outbreak proportions and defoliate trees over large areas. Larger grazing herbivores include the snowshoe hare, which is preyed upon by the lynx, and moose and woodland caribou, which are prey of wolves.

Figure 33.16 A temperate deciduous forest in eastern North America.

Temperate Deciduous Forest

South of the coniferous forest lies the **temperate deciduous forest** (figure 33.16). Deciduous forest is found in areas of moderate climate with a well-defined winter and summer season and relatively high precipitation (75 to 150 cm/yr). The growing season, which lasts from the last frost of spring to the first frost of fall, ranges between 140 and 300 days. The soil is relatively fertile and mildly acid, and litter decomposes quite rapidly. The dominant trees are **deciduous;** that is, they lose their leaves in fall. Although different areas have different dominant tree species, vertical stratification is well-developed throughout the deciduous forest biome. Net primary production ranges from 4,000 to 8,000 kcal/m^2 a year, of which about two-thirds is stored in wood.

Production stored as wood forms a nutrient pool unavailable for short-term cycling. Some short-term nutrient cycling does take place through the death and decomposition of roots and litterfall. Nutrient cycling in deciduous forests is strongly influenced by the geological nature of the site on which the forest grows. For example, forests growing in soils over nutrient-poor bedrock store and cycle smaller quantities of nutrients than do forests growing in soils over rocks rich in calcium and magnesium.

The deciduous forest supports a high diversity of consumer organisms, the most important of which are insects. Insect populations seldom reach outbreak proportions because target tree species seldom occur in the extensive single-species stands that make such outbreaks possible.

The notable exception is the gypsy moth, which was accidentally introduced into northeastern United States and feeds on a wide variety of hardwood trees, especially oaks. A major large herbivore is the white-tailed deer. In areas of high population, the white-tailed deer can influence the structure and development of a forest by over-utilizing certain species of forest tree seedlings and sprouts. Because of human activity, deer predators like the wolf and the mountain lion have disappeared from most areas.

Scrub Forest

In regions characterized by hot, dry summers and mild, damp winters, the dominant vegetation is **scrub forest,** or as it is known in North America, **chaparral** (figure 33.17). Chaparral is characterized by low, shrubby vegetation with tough, waxy-coated leaves that are resistant to drought. Net production averages about 3,000 kcal/m² a year with much of the production tied up in woody growth and litter. Periodic fires that sweep through the scrub forest are important in nutrient cycling and in stimulating new vegetative growth. Chaparral occurs in southwestern United States and Mexico and in parts of southern Europe and southern Australia.

Tropical Rain Forest

Tropical rain forest is located in the equatorial regions of Central America, central and northern South America, western Africa, and Indonesia (figure 33.18). Tropical rain forest occurs in areas where average temperatures are high (between 20 and 25°C) and precipitation is heavy (in excess of 200 cm per year). Temperatures fluctuate little, and the rainfall is evenly distributed through the year. The forest is dominated by broadleaf nondeciduous trees, which are part of a very rich diversity of plant species. The forest is highly stratified in the canopy, but because of dense shade the understory is poorly developed.

Figure 33.17   Scrub forests occur in the semiarid areas of western North America, the Mediterranean region, Australia, South America, western India, and Central Asia where summers are hot and winters wet. Scrub forests are communities of broadleaf shrubs and dwarf trees not over eight feet tall. In North America these shrubland communities are called chaparral. They are highly susceptible to fire because they possess volatile and flammable compounds in their leaves. Chapparral is a fire-dominated ecosystem. Burning renews growth and recycles nutrients.

Figure 33.18 The tropical rain forest forms a worldwide belt around the equator. The largest continuous rain forest is found in the Amazon Basin, where its existence is threatened by land clearing. The tree species number in the thousands, climbing plants are common, and the interior is dark and moist.

Figure 33.19 Tropical savannas are found over large areas in the interior of continents. The best known are the African savannas, characterized by scattered, flat-topped *Acacia* trees.

Decay of organic matter is rapid, and little litter accumulates on the forest floor. Nutrients are cycled directly from the litter to the plants by way of mycorrhizae. Thus, most nutrients are stored in trees with little nutrient reserves in the soil. Although the soil itself is relatively infertile, net production is very high because of high temperatures, a yearlong growing season, and the rapid recycling of nutrients from the litter. Even though the soil is poorly suited for agriculture, the tropical rain forest is being converted to agricultural uses and may be destroyed by the turn of the century.

At the edge of the tropical rain forest is the **tropical seasonal forest.** Not as lush as the rain forest, the tropical seasonal forest grows in regions where rainfall is seasonal with pronounced wet and dry seasons. Many trees in tropical seasonal forests are deciduous and lose their leaves during the dry season.

Tropical Savanna

Also associated with tropical regions is the **tropical savanna,** which is dominated by tall grasses and also supports scattered trees (figure 33.19). In tropical Africa these trees are flattopped, thorny *Acacia* trees, while in South America they are palms. Tropical savanna occurs in regions where temperatures are high and where rainy and dry seasons are pronounced. During the rainy season, precipitation ranges between 90 and 150 cm. Savannas are fire-dominated ecosystems to which grasses and a few tree species are well adapted. The heavy rains of the wet season cause extensive leaching and result in nutrient-poor soils. Annual net production, however, ranges between 800 and 8,000 kcal/m². African savanna supports a rich diversity of grazing animals whose reproductive season and migrations are correlated with the rainy season. Tropical savanna has been highly disturbed by human activity, particularly agricultural development and overgrazing by domestic animals. The large herds of native grazing animals that once occupied the savanna are rapidly disappearing.

Grassland

Grassland is found on the plains areas of North America, Eurasia, and South America and supports a variety of bunch and sod-forming grasses (figure 33.20). Rainfall is variable, and in many areas could support tree growth. Tree encroachment is halted by fires and periodic drought. Compared to forest ecosystems, grassland exhibits little biomass accumulation, and two-thirds of the biomass is below ground. Most of the annual aboveground and below-ground production dies each year. Thus, the turnover of biomass and nutrients is rapid; a complete turnover occurs in approximately three years. Because of relatively low rainfall and high evaporation rates, leaching is not excessive, and nutrients tend to accumulate in the lower part of the soil. At the same time, organic matter accumulates in the upper soil. The result is very fertile soil that has been exploited worldwide for agriculture.

Grassland, which evolved under a combination of grazing pressure and fire, supports large populations of herbivores. In North America, the native grassland once supported millions of bison, and the African plains are dominated by a large assemblage of grazing animals, including many species of antelope, the wildebeest, zebra, and giraffe. These herbivores support a number of large predators, including lions and hunting dogs in Africa, and at one time the wolf in North America. Also common are insect herbivores, notably grasshoppers and ants. On occasion, grasshopper populations can reach outbreak proportions and consume vast quantities of plant material. In North America and much of Africa large native herbivores have been replaced by cattle, sheep, and goats, and in many places overgrazing has resulted in the deterioration of the grassland. In North America along an east to west gradient of decreasing moisture there occur tall-grass, mid-grass, and short-grass grassland. Because of the highly fertile soils, most native temperate grassland of the world has been exploited for agriculture and has been largely transformed into grainland.

Figure 33.20 A preserved area of native prairie in South Dakota in autumn. Before settlement by European settlers, the interior of the North American continent supported one of the great grasslands of the world. The grassland varied from east to west across the plains: eastern tall-grass prairie gave way progressively to mixed prairie and short-grass plains. These plains once supported vast herds of bison. Today much of the native grassland is under cultivation.

Deserts

Desert forms a worldwide belt at about 30°N and 30°S latitude (figure 33.21). Desert is associated with the rainshadows of mountains, coasts next to cold ocean currents, and deep interiors of continents. It is characterized by rainfall of less than 25 cm per year, low humidity, a high evaporation rate, and high daytime temperatures and low nighttime temperatures. Desert plants are widely scattered shrubs with short, waxy, drought-resistant leaves. The soil is high in inorganic salts and low in organic matter. Because of lack of water, and in some areas, cold temperatures, annual net production is very low (below 200 kcal/m²).

Desert may be hot or cold. The North American hot desert is found in southwestern United States and Mexico. It is dominated by cactus and desert shrubs. Northern cold desert is dominated by sagebrush.

Even desert is subject to extensive human disturbance. Irrigation has made it possible to utilize desert for agriculture. But a price is paid for such agriculture. Water for irrigation often is pumped to the surface from deep wells and essentially is not replaced. Irrigation also results in the accumulation of salts in surface soils as a result of the evaporation of irrigation water.

Figure 33.21 Deserts, defined as land areas where evaporation exceeds rainfall, occur in two distinct belts about the earth, one near the Tropic of Cancer, the other near the Tropic of Capricorn. In North America there are two types of desert. One is the cool desert of central and northwestern North America, which lies in the rain shadow of the Sierra Nevada and Cascade Mountains. This desert of the Great Basin is dominated by sagebrush. The other is the hot desert of the southwest in Arizona (pictured here), New Mexico, and Southern California. This desert is dominated by cacti, creosote bush, paloverde, and ocotillo.

Summary

The earth supports a number of recognizable ecosystem types. Some of these are aquatic (water-based) systems, while others are terrestrial (land-based) systems.

Aquatic ecosystems may be fresh water or salt water (marine). Lotic freshwater ecosystems are characterized by flowing water and include brooks, streams, and rivers. Lentic freshwater systems are characterized by still water and include lakes, ponds, and marshes. In many lentic systems there is both vertical stratification and horizontal zonation. Marine ecosystems include salt marshes, estuaries, rocky shores, sandy shores, and the open sea. Salt content increases in marine ecosystems in the direction of the open sea.

Major terrestrial ecosystems are categorized on the basis of the climax community types they support. These community types reflect differences in climate and soil and are called biomes.

North of latitude 60°N is a vast, treeless region called the arctic tundra. To the south of the tundra is a worldwide belt of coniferous forest called the taiga. Still further south in areas of moderate climate is the temperate deciduous forest. Regions of the world with hot, dry summers and mild, wet winters support the scrub forest biome type. The tropical rain forest occurs in equatorial regions with high temperatures and heavy rainfall. The tropical savanna also is found in tropical regions and is an extensive area of tall grass with scattered trees. Grassland is found in temperate areas where rainfall is limited and variable, and evaporation is high. Desert is located at about 30°N and 30°S latitude, where rainfall is very low and daytime temperatures are very high.

Questions

1. Distinguish between aquatic and terrestrial ecosystems.
2. What are some characteristics of a lotic freshwater ecosystem? of a lentic freshwater ecosystem?
3. How does vertical temperature stratification develop in temperate lakes?
4. Distinguish between oligotrophic and eutrophic lakes.
5. What is an estuary?
6. Explain why environmental conditions are rather severe on rocky shores and sandy beaches.
7. How are oceans zoned in the horizontal direction? What are some characteristics of these zones?
8. What is meant by the term biome?
9. Characterize each of the following as to location, general climate conditions, and climax communities: arctic tundra, alpine tundra, taiga, temperate deciduous forest, scrub forest, tropical rain forest, tropical savanna, grassland, desert.

Suggested Readings

Books

Krutch, J. W. 1960. *The desert year.* New York: Viking Press.

Odum, E. P. 1971. *Fundamentals of ecology.* 3d ed. Philadelphia: W.B. Saunders.

Shelford, V. E. 1963. *The ecology of North America.* Urbana, Ill.: University of Illinois Press.

Smith, R. L. 1980. *Ecology and field biology.* 3d ed. New York: Harper & Row.

Weaver, J. E. 1954. *North American prairie.* Lincoln, Neb.: Johnsen Publishing.

Whittaker, R. H. 1975. *Communities and ecosystems.* 2d ed. New York: Macmillan.

Articles

Cooper, C. F. April 1961. The ecology of fire. *Scientific American* (offprint 1099).

Enright, J. T.; Newman, W. A.; Hessler, R. A.; and McGowan, J. A. 1981. Deep-ocean hydrothermal vent communities. *Nature* 289:219.

Jannasch, H. W., and Wirsen, C. O. 1979. Chemosynthetic primary production at East Pacific seafloor spreading centers. *BioScience* 29:592.

Richards, P. W. December 1973. The tropical rain forest. *Scientific American* (offprint 1286).

Part 7

The Diversity of Life

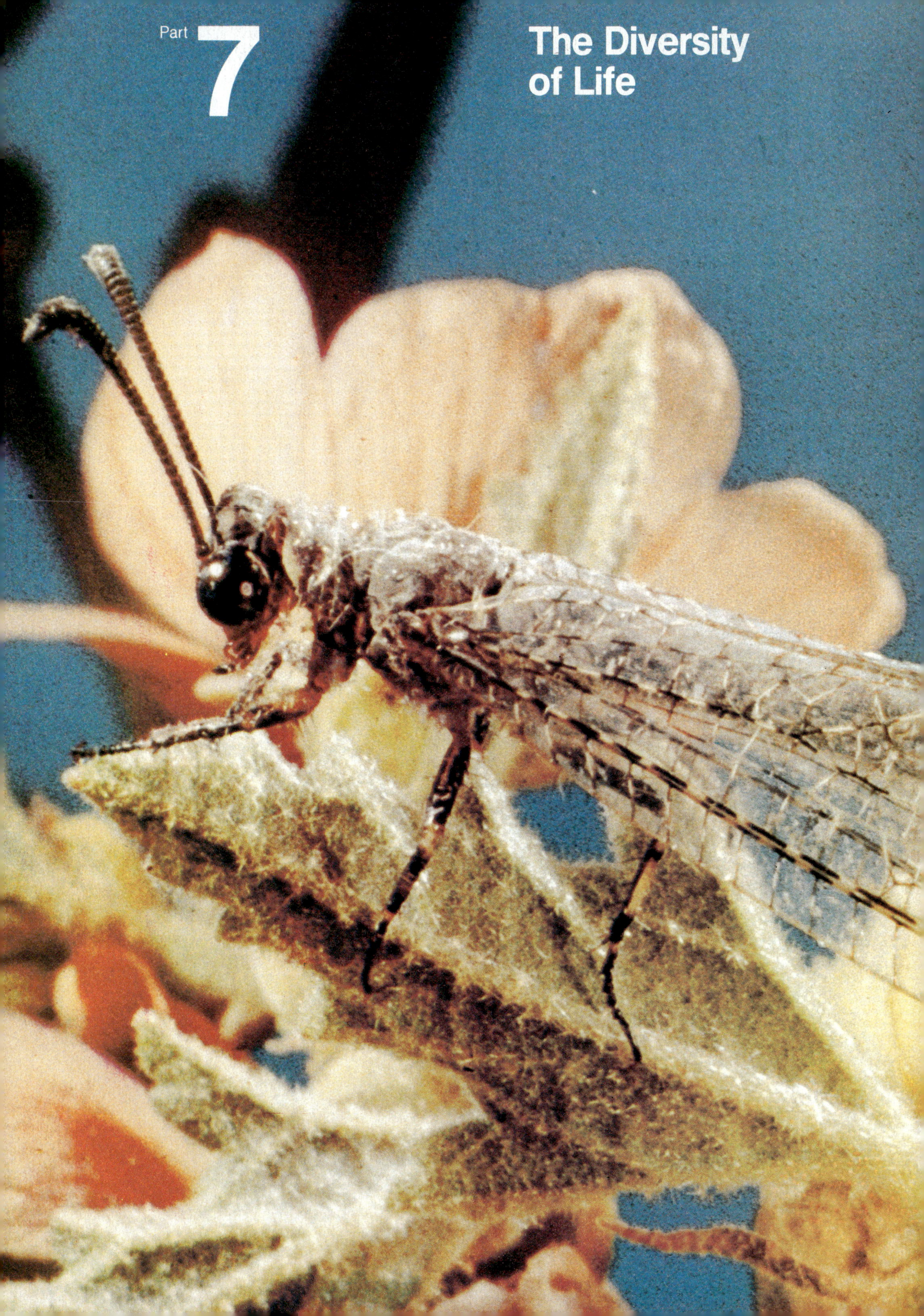

The most striking characteristic of the world of life is its diversity. Organisms of innumerable types possess a multitude of specializations that permit various living things to interact successfully with virtually every type of environment on earth. How do biologists make sense of this sometimes bewildering diversity?

On the basis of certain ranges of shared characteristics, biologists divide living things into categories (taxonomic groups). No single classification scheme is ideal, and biologists continually argue about what methods of categorization should be used to classify organisms. Generally accepted taxonomic schemes, however, are essential tools for biologists.

Taxonomic grouping facilitates recognition and study of organisms and makes it possible for scientists of many lands to communicate clearly and directly with one another about the organisms they are studying. The systematic process of classifying organisms into taxonomic groups also aids in the study of evolutionary relationships.

In chapters 34–37 the fantastic diversity of life is sampled within the context of one of the most widely used general taxonomic schemes.

Taxonomy Principles and Survey of Monera

Humans are constantly prodded by curiosity and a desire to understand the world about them. Undoubtedly, even before the beginnings of recorded history, people were observing and mentally cataloging plants and animals. This process of naming living things and relating them to each other in some sort of organized taxonomic system is of great importance in biology because it allows scientists to communicate with universally understood names. Such a system also facilitates the efficient organization of a great deal of information.

Until about 300 years ago, attention was exclusively centered on larger organisms. Scientists were not even aware of the existence of microorganisms. With the advent of the microscope, the discipline of microbiology arose and has become increasingly active over the past 300 years. Microorganisms are of extreme importance in both negative and positive senses. On the one hand, they can cause disease and destroy crops; on the other, they are responsible for recycling nutrients within ecosystems by acting as decomposers and nitrogen fixers. Moreover, research concerning their metabolism and genetics has contributed significantly to the development of biochemistry and molecular biology.

This survey of the living world begins with a brief consideration of the principles of taxonomy, and the rest of the chapter is devoted to two important and fascinating groups of microorganisms, the viruses and bacteria.

Principles of Classification

A fundamental principle of biology is that organisms possess a biological individuality. Even identical twins, though they are alike genotypically, often are slightly different phenotypically. For this reason, the task of organizing or grouping individual organisms is an extremely difficult one.

Nevertheless, some biologists specialize in taxonomy. Taxonomists (from the Greek *taxis* = arrangement) describe and name organisms,

arranging them in a system of classification that takes their similarities and differences into account. Such a task is enormous. Life arose at least 2,000 million years ago, and organisms have been evolving ever since with increasing complexity and diversity. At the present time, approximately 400,000 plant species and 1,200,000 animal species have been described, with perhaps many more yet to be found.

Another basic concept of biology is adaptation. Environmental pressures (through natural selection) determine which organisms live to reproduce in a particular environment. Thus, "selected" reproducing organisms determine what genotypes are passed on to new generations (see chapter 28). These genotypes are expressed as phenotypes that may be described or measured. As environments change or evolve, so do genotypes (and phenotypes) of organisms. A good classification system should accurately reflect the evolutionary history of organisms. Two organisms that are classified as members of the same genus, for example, should be very closely related, from an evolutionary standpoint, and consequently have many characteristics in common. If two organisms are in the same genus, knowledge of the characteristics of one should enable a biologist to predict many characteristics possessed by the other organism without ever having seen it.

The first recorded attempts to categorize organisms systematically were made by Greek philosophers. Aristotle (384–322 B.C.) grouped plants into herbs, shrubs, and trees. He used medicinal value and size as criteria. In general, early taxonomic schemes were based on concrete and practical criteria, such as size, habitat, medicinal value, and ability to harm humans.

Modern classification has stemmed from the work of John Ray (1627–1705), an English naturalist. He developed a classification scheme for his *Historia Plantarum* and was the first to outline the species concept, that is, that there were small groups of organisms with similar morphological characteristics that could be distinguished from other closely related organisms. This scheme was followed for the most part by Carolus Linnaeus (1707–1778), a Swedish physician and botanist who is considered to be the father of modern taxonomy.

Linnaeus changed the course of systematics in 1758 with the publication of the tenth edition of his *Systema Naturae.* Rather than using a polynomial ("many name") system that included long, descriptive phrases to identify organisms, as others had done previously, Linnaeus used only two names to identify a species, and his binomial system is still in use. He recognized four **taxa** (groupings) of organisms: class, order, genus, and species. Organisms were placed in these taxa mainly on the basis of structural similarities and differences. Modern taxonomists have added only two major taxa to his scheme: phylum and family. Two aspects of Linnaeus's system—nomenclature and the classification hierarchy—are now considered in detail.

Nomenclature

Because science is a truly international enterprise, scientists of all nations must be able to communicate with one another. The use of universally understood names for organisms plays an essential role in making this communication possible. The use of common names for organisms does not work because common names vary considerably, even within an individual country. For example, the sparrow *Passer domesticus* is known variously as the English sparrow, the house sparrow, or the common sparrow. The European white water lily actually has 245 different common names distributed among four languages.

To be useful, a scientific naming system (nomenclature system) must have at least three properties. Each organism must have a unique name. This name must be universal; it must be used by scientists regardless of their native tongue. Finally, the name must be as unchanging as possible. Once an organism has been properly named, its name should not be changed without an excellent reason.

Before Linnaeus's time there was little order or consistency in the way species were named. Names ranged in length from one to many words and sometimes even took the form of descriptive phrases. Linnaeus introduced the **binomial** ("two-name") system. Each organism is given a two-word name that is always in Latin or latinized.

The first word is the name of the genus to which the organism belongs; the second word indicates the particular species. The initial letter of the genus name is capitalized, and the name is italicized in print or underlined when written or typed. A good example of the binomial system is provided by *Paramecium,* the common ciliate protozoan encountered at one time or another by nearly all biology students. *Paramecium* is the genus name, and within this genus are a number of species—*Paramecium aurelia, Paramecium bursaria,* and so forth.

To ensure the uniqueness, universality, and stability of each name, all new names must meet the criteria set down in one of three special international codes: the International Rules of Botanical Nomenclature, the International Code of Zoological Nomenclature, and the International Bacteriological Code of Nomenclature.

The Hierarchy of Taxa

It is not sufficient to simply name each organism, even if this is done properly. So many species inhabit the earth that the situation would be chaotic without further grouping or classification of these species. A hierarchy of taxa (groups of organisms) has been developed to organize species in a coherent, meaningful way. The smallest unit in this hierarchy is the **species.** A species has normally been defined as a population of morphologically similar organisms that can sexually interbreed but are reproductively isolated from other organisms. This definition is not adequate for organisms such as bacteria that generally lack sexual reproduction. In these cases, individuals are placed in the same species on the basis of overall similarity.

Each species is precisely located within the living world by placement in a succession of ever broader taxonomic ranks. The major taxonomic ranks used, in decreasing order of breadth, are: kingdom, phylum or division, class, order, family, genus, and species. Usually, every rank in the hierarchy contains several groups in the level below it (for example, a particular family normally is composed of several genera, and each genus may have several species within it). Part of such a hierarchical classification system is illustrated in figure 34.1.

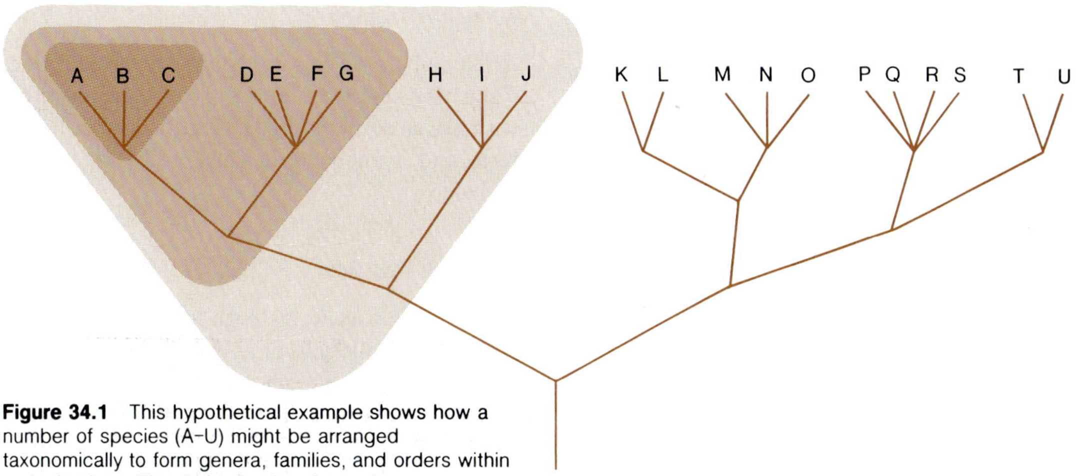

Figure 34.1 This hypothetical example shows how a number of species (A–U) might be arranged taxonomically to form genera, families, and orders within a single class. The darkest colored area is a genus containing three species. The family containing this genus is composed of two genera and seven species; the order, of two families and three genera (the second family has only one genus).

The Kingdoms of Organisms

There are several taxonomic systems that divide living things into kingdoms in various ways. Two of these systems are the two-kingdom system and the five-kingdom system.

The Two-Kingdom System

For centuries the living world was divided into two kingdoms—plants and animals. Plants stay in one place and carry out photosynthesis. In contrast, animals are motile and ingest their food. For years this two-kingdom system seemed adequate. Trees, flowers, mosses, and seaweeds are clearly plants. Fungi are at least plantlike—non-motile and "rooted" to solid objects. Cougars, cod, caterpillars, and clams all move and eat other organisms (in one way or another). Problems arose, however, when scientists attempted to incorporate microorganisms into this scheme.

There are at least four major difficulties with the two-kingdom system. The first difficulty lies in the nature of bacteria and cyanobacteria (blue-green algae). In recent years it has been recognized that these groups, called **prokaryotes,** are radically different from other living organisms, the **eukaryotes.** Prokaryotes are, for the most part, single prokaryotic cells that lack a true membrane-bound nucleus, histone-complexed DNA, mitochondria, chloroplasts, endoplasmic

reticulum, a Golgi apparatus, mitosis, sexual reproduction with meiosis, and so forth. Eukaryotes have eukaryotic cells that possess all these organelles and processes, especially a membrane-bound nucleus. There are greater differences between prokaryotes and eukaryotes than between plants and animals, yet this distinction is not recognized in the two-kingdom system.

The second difficulty with the two-kingdom system becomes apparent upon closer inspection of unicellular eukaryotic microorganisms. Many of these microorganisms do not really "fit" in either the plant or animal kingdom because they have properties intermediate between those of plants and animals. *Euglena* and its close relatives provide a good example. *Euglena* has both plant and animal characteristics: it carries out photosynthesis, but is also quite motile and lacks a rigid cell wall. Furthermore, heat treatment of *Euglena* can cause it to permanently lose its chloroplasts and become heterotrophic. Other euglenoid (*Euglena*-like) flagellates, such as *Peranema,* lack chloroplasts and obtain their nutrients by ingesting bacteria, algae, and even *Euglena.* A large number of species simply do not readily fit the usual plant and animal categories.

The third difficulty with the two-kingdom system is similar to the second. Fungi, originally associated with the plant kingdom, do not really "fit" into the plant kingdom. Higher fungi are composed of a mass of tubes or **hyphae.** This network of hyphae, called a **mycelium,** is quite different from the plant body plan; so also are fungal reproductive structures. In addition, fungi do not carry out photosynthesis but absorb their nutrients in a dissolved form. These and other differences sharply distinguish fungi from true plants.

Finally, the two-kingdom system implies that there are two major modes of nutrient procurement—photosynthesis and ingestion. As just noted, however, nutrients can be procured in a third way. Fungi secrete digestive enzymes, which then degrade food molecules extracellularly. The smaller nutrient molecules (sugars, amino acids, fatty acids, and so on) are subsequently absorbed. This process is quite different from the ingestion practiced by animals. In ingestion, solid material is taken into the body and digested internally, either in food vacuoles or a digestive tract. These major differences in the mode of nutrient acquisition mean that fungi can be nonmotile, while animals often have to move about in search of suitable food. Thus, most animals must be motile and also be able to sense their surroundings and prey; fungi do not.

The Five-Kingdom System

The five-kingdom system developed primarily by R. H. Whittaker of Cornell University is used in this text. Living organisms are distributed into five kingdoms on the basis of level of organizational complexity and mode of nutrition (figure 34.2). The five kingdoms are Monera, Protista, Fungi, Animalia, and Plantae.

Kingdom Monera

All members of this kingdom are prokaryotic. They may be either autotrophic or heterotrophic with the absorptive mode of nutrition. All bacteria and cyanobacteria (blue-green algae) are included.

Kingdom Protista

This kingdom contains organisms that exist as single eukaryotic cells or as colonies of eukaryotic cells. Sometimes, simple multicellular organisms without true tissues also are included among the protists. Nutrition may be ingestive, absorptive, or photosynthetic. The line dividing this kingdom from the other three kingdoms of organisms with eukaryotic cells is sometimes vague. For example, the chlorophyta (green algae) may be divided between the protist and plant kingdoms, or placed completely in either.

Kingdom Fungi

The fungi are mainly multinucleate organisms with walled mycelial organization, although some are unicellular. Most are nonmotile. Nutrition is absorptive.

Kingdom Animalia

Animals are multicellular with wall-less eukaryotic cells and are usually motile. Most have ingestive nutrition; phagocytosis and pinocytosis are present.

Kingdom Plantae

Plants are multicellular with walled eukaryotic cells, and are usually nonmotile. They are primarily photosynthetic and possess chloroplasts.

In this chapter and the chapters that follow, each of these kingdoms is considered in more detail, and the distinctions among them should become even clearer.

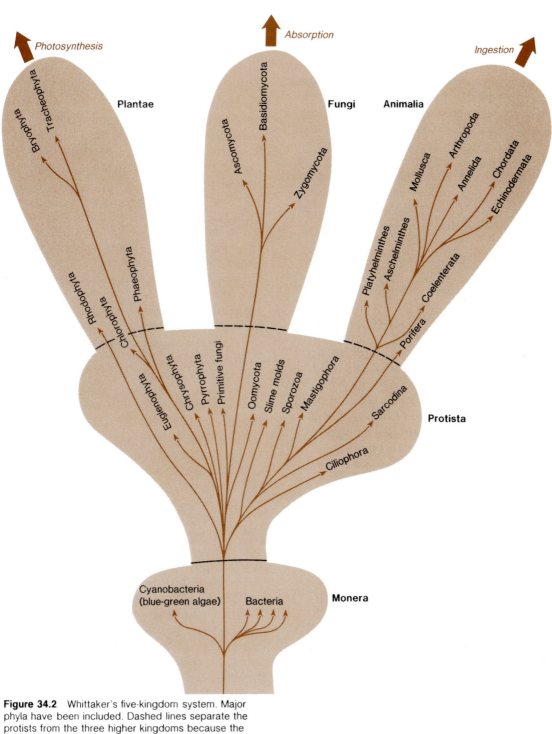

Figure 34.2 Whittaker's five-kingdom system. Major
phyla have been included. Dashed lines separate the
protists from the three higher kingdoms because the
boundaries are indistinct. For example, chlorphyta may
be found in both the Plantae and the Protista. The
system is based on three levels of organization—
prokaryotic; eukaryotic unicellular; and colonial,
eukaryotic multicellular. The three higher kingdoms are
distinguished from each other primarily on the basis of
nutritional mode. The arrows indicate evolutionary
relationships only in a general, simplified way.

Viruses

Viruses, in a strict sense, are nonliving particles. They lack multiple enzyme metabolic pathways and therefore cannot generate ATP or independently synthesize their own constituents. Although some viruses are more morphologically complex, most viruses consist basically of nucleic acid enclosed in a protein coat. Such simple entities can only be reproduced within living cells. Nevertheless, a portion of this chapter is devoted to viruses because viruses are so important to biology and often have reproductive cycles intimately related to those of bacteria.

Discovery of Viruses

By the beginning of this century, diseases caused by microorganisms were well known, and the disease agents themselves were being isolated in suitable culture media. The term "virus," as it was originally used, denoted any infectious agent causing disease. As such, pathogenic (disease-producing) bacteria, fungi, and protozoa were included among the "viruses." In time, however, the term virus became associated only with those submicroscopic agents that could not clearly be identified as bacterial, fungal, or protistan.

At first, size was the primary reason for differentiating viruses from known microorganisms. The studies of Dimitri Ivanowsky, a Russian biologist, gave the first hint of the presence of viruses in 1892. He worked with tobacco mosaic disease, a condition in which leaves of affected tobacco plants become wrinkled and mottled in appearance. Ivanowsky was able to transmit the disease to healthy plants by rubbing them with juice extracted from diseased plants. While attempting to isolate these "bacteria," he passed the infective extract through a finely meshed porcelain filter. To his surprise, the filtrate was still infective. Thus, the disease-producing agents of tobacco mosaic were smaller than any known bacteria. Any disease-producing agents that could pass through such a filter subsequently were known as filterable viruses and later, simply **viruses**.

Although considered by many to be very small bacteria, these viruses continued to puzzle biologists because they could not be cultured in media normally used for bacteria. Moreover, the viruses retained their infectivity even when precipitated by alcohol. In 1935, W. M. Stanley demonstrated that viruses were remarkably different from bacteria and any other living cells for that matter. He was able to crystallize infectious tobacco mosaic virus and show that it was mostly protein in composition.

The Structure of Viruses

The virus particle is referred to as a **virion.** The simplest virions are composed of a protein coat, the **capsid,** enclosing nucleic acid (figure 34.3). This complex is often referred to as the **nucleocapsid.** In some viruses, the nucleocapsid is surrounded by a membraneous lipoprotein **envelope.**

Careful examination of virus capsids in the electron microscope has shown that they are built from protein subunits, the **capsomers** (figure 34.4). These capsomers can generate three capsid types—helical capsids, icosahedral capsids (twenty-sided figures with each side an equilateral triangle), and complex capsids. Regardless of the shape, inside the protein capsid is nucleic acid. The nucleic acid carries genetic information, while the capsid protects the nucleic acid and aids in its transmission between cells.

Although living cells contain both DNA and RNA, virions have either DNA or RNA, but not both. The nucleic acids can range in size from a few thousand nucleotide base pairs (fewer than five genes) up to around 250,000 base pairs (several hundred genes). The variety of nucleic acid structure in the virus genome is remarkable. Depending on the type of virus, the nucleic acid can be single- or double-stranded. In most instances, each virion has a single nucleic acid molecule. However, some RNA viruses have their genome in pieces (the influenza virus has eight or nine pieces of RNA). As shall be seen shortly, viruses with different kinds of genetic material are reproduced differently.

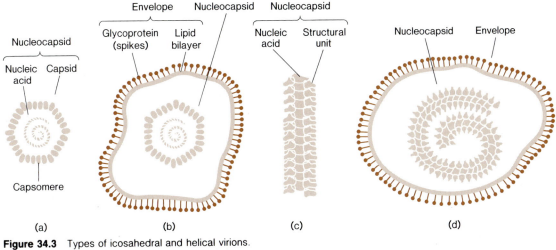

Figure 34.3 Types of icosahedral and helical virions.
(a) Naked icosahedral. (b) Enveloped icosahedral.
(c) Naked helical. (d) Enveloped helical.

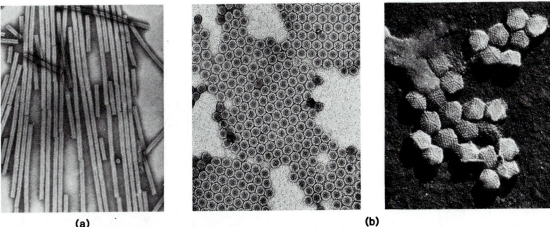

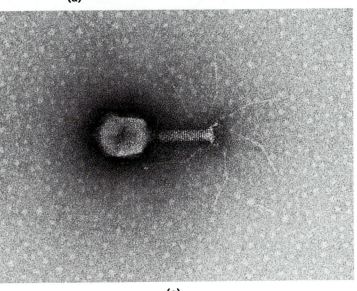

Figure 34.4 Electron micrographs
of different virion morphological
types. (a) Tobacco mosaic virus
(helical). (Magnification × 66,000)
(b) Icosahedral virions. On the left is
a crystalline array of particles of
Brome mosaic virus. On the right
are freeze-dried adenovirus particles.
(Magnification × 75,000 and 81,200
respectively) (c) T4 bacteriophage
(complex). Note head, tail, and tail
fibers extending from the base plate.
(Magnification × 140,400)

Viral Reproduction

Probably the best studied virus reproductive cycle is that of the large, complex viruses that infect the bacterium *Escherichia coli*. Since these viruses attack bacteria, they are referred to as **bacteriophages,** or **phages** for short. Most is known about the reproduction of T-even phages (figure 34.5).

The phage's life cycle begins with a chance collision between the virus and an *E. coli* cell (figure 34.6). The virion attaches to the bacterium at a specific receptor site on the surface. The tail base plate and its fibers are involved in this attachment. Next, a phage enzyme digests a portion of the cell wall; the tail sheath contracts, pushing the core into the bacterium; and virus DNA is injected. The virus genetic material takes control of the host cell within a few minutes and initiates the destruction of host DNA and inhibition of the synthesis of host cell proteins.

Viral messenger RNA is synthesized, and it directs the production of virus proteins and any enzymes required for the manufacture of new virions. The host bacterium provides all the required energy and building blocks for virion synthesis. As soon as all virion components have been prepared, the virus particles are assembled in a complicated, but orderly, sequence. Finally, the mature virions—about 100 per bacterial cell infected by T2 phages—are released when an enzyme called a lysozyme synthesized under virus direction disrupts the cell wall. The new virions can now attack neighboring *E. coli* cells.

The reproductive cycle just described is called a **lytic cycle** because the host cell is lysed (broken open) by the phage. In contrast, some bacteriophages do not immediately destroy their hosts but instead reproduce in synchrony with the bacterium to generate infected bacterial progeny. This relationship between a phage and its host is

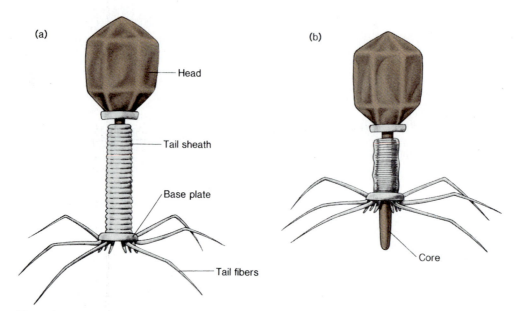

(a)

— Head

— Tail sheath

— Base plate

— Tail fibers

(b)

—Core

Figure 34.5   The morphology of a T-even bacteriophage. In (a) the tail sheath is relaxed; in (b) it has contracted. The head contains double-stranded DNA.

called **lysogeny.** Infected bacterial cells capable of producing phage virions are said to be **lysogenic,** and viruses that can enter into this relationship are **temperate phages.** The way in which this takes place is illustrated in figure 34.7. When a lysogenic phage such as the phage lambda (λ) infects a bacterium, it directs the synthesis of a special repressor protein. If this repressor reaches a high concentration fairly quickly, it inhibits phage reproduction. Instead of lysing the *E. coli* cell, the lambda DNA is integrated into the bacterial DNA and becomes a **prophage.** The prophage is replicated along with the bacterial chromosome, and all the lysogenic cell's progeny will carry the prophage but not be lysed. If lysogenic cells are exposed to harsh environmental factors (such as ultraviolet light), repressor levels drop and **induction** occurs. As a result of induction, the prophage leaves the bacterial chromosome, virus nucleic acids and proteins are

synthesized, new virions are constructed, and the host cell lyses. Thus, the same virus can be responsible for both lytic and lysogenic cycles.

Before leaving the topic of viral reproduction, it must be emphasized that viruses do not always mimic the behavior of T-even bacteriophages. For example, many animal viruses seem to be taken into cells by phagocytosis and their capsid removed by host cell digestive enzymes in a process referred to as **uncoating.** The mechanism of virus release after virus reproduction also varies. Some bacterial viruses and a number of animal viruses are released by the host cell without lysis. Indeed, the host cell plasma membrane may be used in the formation of the virion envelope when the virions are released by a kind of cell budding process.

It was mentioned earlier that a number of viruses differ from living cells in that they have RNA as their primary genetic material. Some

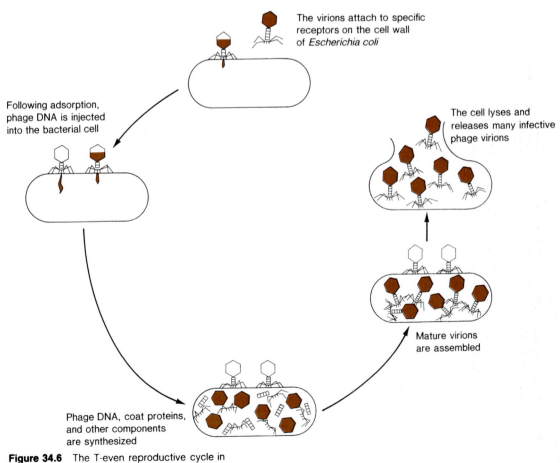

The virions attach to specific receptors on the cell wall of *Escherichia coli*

Following adsorption, phage DNA is injected into the bacterial cell

The cell lyses and releases many infective phage virions

Mature virions are assembled

Phage DNA, coat proteins, and other components are synthesized

Figure 34.6 The T-even reproductive cycle in *Escherichia coli.*

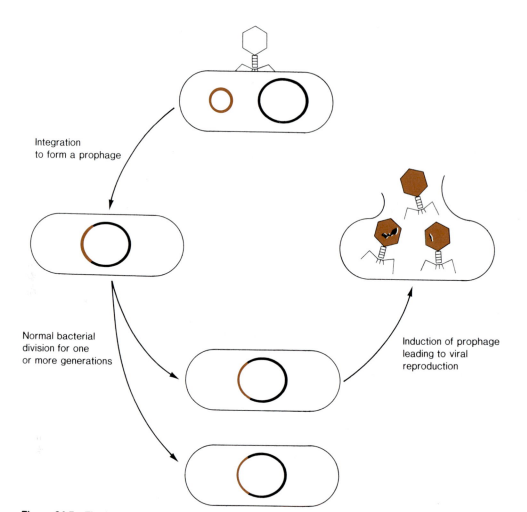

Figure 34.7   The lysogenic cycle of a bacteriophage.
The bacterial chromosome is in black; the phage
chromosome is in color.

Integration
to form a prophage

Normal bacterial
division for one
or more generations

Induction of prophage
leading to viral
reproduction

viruses have double-stranded RNA, but they
function much as if they had DNA. A special
RNA transcriptase copies one strand to produce
mRNA, which can then direct protein synthesis
or be copied to produce new double-stranded
RNA. Thus, genetic expression depends on
RNA→RNA information flow, not the
DNA→RNA flow, which is more familiar. The
primary genetic material in some other viruses
is single-stranded RNA. These single-stranded
RNA viruses operate in a different way. A **RNA
replicase** (coded for by the virus genome) copies
the single strand to form a double-stranded RNA
replicative form. This replicative form, which is
present only during virus reproduction, then directs the synthesis of single-stranded RNA for
new virions. Polio virus reproduces in this way.

A radically different approach is taken by
some RNA tumor viruses (retroviruses). A vi-
rion-associated **reverse transcriptase** can copy the
virus single-stranded RNA to form a complementary DNA copy (see figure 26.29). This
DNA is used as a template to produce double-
stranded DNA. The double-stranded DNA may
direct the synthesis of new virus RNA, just as
any DNA is transcribed to form mRNA, but the
DNA may also be incorporated into the host genome and replicate with it. This integration of viral
DNA can result in tumor formation in a number
of animals. Therefore, many scientists believe
that viruses also may cause some human malig-
nancies. Although a direct causative link between
viruses and cancer has not yet been established,
there is some evidence in favor of such a link for
human leukemias, lymphomas, and breast cancers.

Host Resistance

Animals can produce antibody molecules in response to infection (see chapter 26 for more details). Virus infections trigger the immune response, and it is particularly effective in preventing reinfection of an individual. However, the rise in antibody levels in response to a new infection is relatively slow. Thus, a second defense mechanism is employed in resisting virus attacks. Infected animal cells can synthesize and release one or more types of small proteins (about 20,000 molecular weight in humans) called **interferons** (see p. 717). Interferons bind to neighboring, uninfected cells and make them resistant to virus infection. They probably act by stimulating the synthesis of an antiviral polypeptide that can interfere with transcription of the viral genome. This either completely blocks or greatly reduces virion production and release. In contrast with antibody synthesis, interferon release is rapid. Interferon production peaks within three to five days after the onset of influenza; antibodies are not manufactured in large amounts for at least seven days. Needless to say, there is great interest in the future manufacture of interferon for possible use in treatment of viral diseases.

Viral Diseases

Many important human diseases result from viral infections. Virus pathogens have high tissue specificity: they can multiply only in particular tissues. This specificity appears to be due to the presence of viral receptors on the surface of target organ cells and plays a major role in determining the nature of the disease. Some viruses attack the respiratory tract (influenza and rhinoviruses), while others grow in the skin (rubella, smallpox, measles), the liver (hepatitis), the central nervous system (rabies, polio), and the parotid salivary glands (mumps). In addition, viruses inflict extensive damage on other animals (swine influenza, rabies, hog cholera) and plants (mosaic diseases, leaf curls).

Cell and tissue damage can result in a number of ways. Some viruses, particularly enveloped viruses, may possess antigens that stimulate the host to form antibodies against some of its own constituents. A number of virus-induced proteins are actually toxic for the cells and cause morphological damage or block cellular biosynthetic processes. Lysosomes may be disrupted so that the cells digest themselves. Virions can even directly disrupt cell structure through the formation of **inclusion bodies.** These are large collections of incomplete or mature virions that accumulate in the nucleus or the cytoplasm. The destruction of a particular tissue may result from the simultaneous operation of more than one of these processes.

Viroids

It has recently been discovered that a number of plant diseases—potato spindle-tuber disease, exocortosis of citrus trees, chrysanthemum stunt disease, and others—are caused by a class of infectious agents called **viroids**. These are very short strands of RNA that can be transmitted between plants through mechanical means or by way of pollen. Viroids are smaller even than the smallest viruses, which can possess fewer than five genes.

The potato spindle-tuber disease agent has been most intensely studied. Its RNA is about 130,000 molecular weight, much smaller than the nucleic acids of viruses (figure 34.8). It can exist as either a linear RNA strand or a closed circle collapsed into a rodlike shape due to intrastrand base pairing.

Viroids are found mainly in the nucleus of infected cells. They do not serve as messengers to direct protein synthesis, and it is not yet clear how they are reproduced or cause disease symptoms.

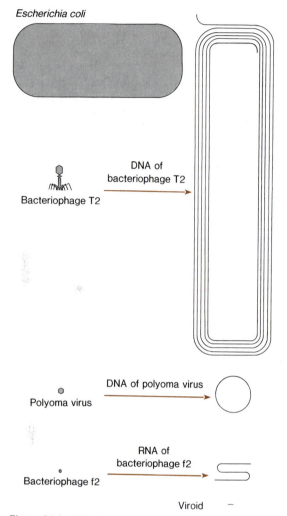

Escherichia coli

DNA of
bacteriophage T2

Bacteriophage T2

DNA of polyoma virus

Polyoma virus

RNA of
bacteriophage f2

Bacteriophage f2

Viroid   —

Figure 34.8 This illustration compares *Escherichia coli*, several viruses, and the potato spindle-tuber viroid with respect to size and the amount of nucleic acid in the genome. All dimensions are enlarged about 40,000 times. Bacteriophage f2 is one of the smallest known viruses (the capsid is about 20–25nm in diameter).

Although the viroids so far discovered cause plant diseases, there is a possibility that similar agents are responsible for animal diseases as well. There are a number of nonbacterial, neurological diseases (scrapie of sheep and goats; kuru and Creutzfeldt-Jakob disease in humans) in which it has not yet been possible to isolate viruses from diseased victims. Some scientists believe that infectious nucleic acids might be involved. Although it is too soon to tell, this is an intriguing possibility and is currently being vigorously pursued in several laboratories.

The Kingdom Monera (Prokaryotae)

The Bacteria

Bacteria were discovered by Antony van Leeuwenhoek, who published drawings of a variety of bacteria in 1684. Bacteria are all prokaryotes. Nevertheless, they vary enormously in size and complexity. Bacterial cells can range in size from as small as spheres 0.2 or 0.3 μm in diameter up to bacteria hundreds of micrometers in length. The average cell is around 1 μm to 5 μm in size. There is a similar variation in the morphology and arrangement of cells (figure 34.9). A bacterium may be spherical (a **coccus**), rod-shaped (a **bacillus**), or helical (a **spirillum**). The rods and cocci can also associate to form clusters and linear filaments. For example, some cocci form chains of cells (streptococci) or grapelike clusters (staphylococci). Bacteria may have stalks, buds, or even fruiting structures (figure 34.10). In fact, a large group of bacteria, the actinomycetes, closely resemble eukaryotic fungi in their morphology. Incidentally, actinomycetes produce many valuable antibiotics.

In view of this extraordinary diversity, which is even more pronounced when bacterial physiology is considered, it is not surprising that microbiologists disagree on bacterial classification. The standard work in the area, *Bergey's Manual of Determinative Bacteriology*, divides the bacteria into nineteen different groups. More recently, Gibbons and Murray have proposed that the kingdom Monera or Prokaryotae be divided into four divisions on the basis of cell wall structure.

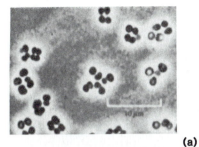
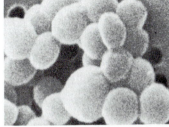

(a)

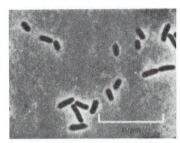
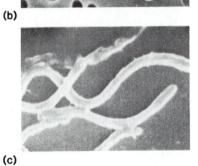

(b)

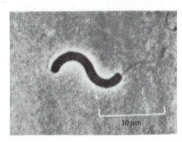

(c)

Figure 34.9 The three most common bacterial shapes— (a) cocci, (b) bacilli, and (c) spirilla. The lefthand pictures were made with a phase contrast microscope. The righthand pictures were obtained with a scanning electron microscope. (Magnifications of scanning electron micrographs × 20,000, 13,000, and 5,000, respectively)

(a)

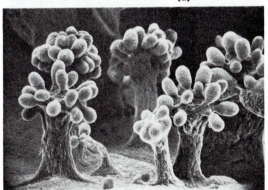

(b)

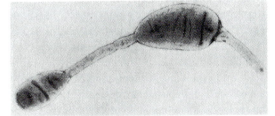

(c)

Figure 34.10 Electron micrographs showing some of the variety in bacterial morphology and reproduction. (a) *Caulobacter crescentus* cells. The cell on the left is preparing to divide. The cells on the right are the two types that are produced by cell division: a stalked, nonmotile cell, and a flagellated, motile swarmer cell. (Magnification × 6,030) (b) The fruiting bodies of the myxobacterium *Chondromyces crocatus*. (Magnification × 262) (c) *Hyphomicrobium* producing a bud at the end of a hypha.

Figure 34.11   A typical Gram-
positive bacterium (*Bacillus
megaterium*) undergoing division.
(Magnification × 61,500)

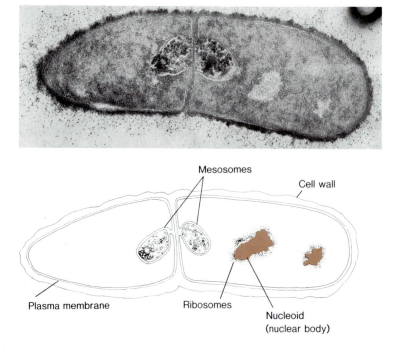

Mesosomes

Cell wall

Plasma membrane   Ribosomes

Nucleoid
(nuclear body)

Bacterial Morphology

In distinct contrast with the variety in bacterial
size, shape, and life-style, the internal structure
of bacteria is moderately uniform (figure 34.11).
Bacterial cytoplasm is bounded by a normal-
looking **plasma membrane** (cell membrane). As
with eukaryotic cells, the plasma membrane
serves as a mechanical boundary for the cell and
a permeability barrier. It contains a number of
nutrient transport systems. Because bacteria lack
mitochondria and chloroplasts, enzymes and
electron carriers involved in respiration and photosynthesis are located in the plasma membrane.
Sometimes the plasma membrane extends into
the cell interior and provides additional membrane surface area for use in metabolic activities.

Most bacteria have a **cell wall** outside their
plasma membrane. This wall gives the cell its
shape and rigidity. It is chemically different from
eukaryotic cell walls. Almost all bacterial cell
walls contain a rigid **peptidoglycan** layer composed of N-acetylglucosamine, N-acetylmuramic
acid, and amino acids. Peptidoglycan is actually
a meshwork of polysaccharide chains cross-linked
by polypeptide chains. N-acetylmuramic acid
and diaminopimelic acid, present in some bacterial cells, are never found in eukaryotic cell
walls. D-amino acids also are found in bacteria,
but not in eukaryotic cells.

Bacteria can be divided into major divisions
on the basis of staining properties. **Gram-positive
bacteria** stain purple with the Gram stain and
have walls with a thick peptidoglycan layer.
Gram-negative bacteria have a multilayered wall
with a thin peptidoglycan layer and appear pink
or red after the Gram staining. Their outer layer
is membraneous and gives some Gram-negative
pathogens their toxic properties.

Penicillin acts by blocking peptidoglycan
synthesis; and bacterial cells affected by penicillin lyse because they lack the osmotic protection
provided by a cell wall.

It is important to note, however, that not all
bacteria have cell walls. The **mycoplasmas** are
bacteria that lack walls, yet do not lyse in dilute
media. Some mycoplasmas are human pathogens, while others are free-living and can be isolated from the soil, sewage, and other environments.

Bacterial cytoplasm appears homogeneous
under the electron microscope. The more distinctive structures are ribosomes, the nuclear area or
nucleoid, and the mesosome. Since there is no
endoplasmic reticulum, the **ribosomes** are scattered free in the cytoplasm. Bacterial ribosomes
are smaller and lighter than eukaryotic cytoplasmic ribosomes. Antibiotics like erythromy-
cin, streptomycin, tetracycline, and chloramphenicol destroy bacterial pathogens by acting on

bacterial ribosomes and thus inhibiting bacterial protein synthesis without inhibiting protein synthesis in the eukaryotic cells of the organisms being treated for a bacterial infection.

Bacteria do not have true nuclei as do eukaryotes. Their DNA is present as a single circular strand, about 1 mm long, associated with protein (but not histones). It is packed together in a discrete part of the bacterial cytoplasm, the **nucleoid.** Rapidly dividing bacteria may have more than one nucleoid at a time.

Many bacteria also have large plasma membrane infoldings called **mesosomes.** Their function is still not clear. Many microbiologists believe that they may be involved in transverse septum formation during cell division. They may also pull the daughter chromosomes apart into separate cells during cell division.

Bacteria can have a number of structures outside the cell wall. Some bacteria secrete a shiny or gelatinous compact layer outside the cell wall. This **capsule** is composed of polysaccharides or polypeptides and protects against phagocytosis by host white blood cells. For example, *Streptococcus pneumoniae* can cause pneumonia only when it possesses a capsule (see p. 587). The **pilus** is another external structure. Pili are short, slender tubes formed of protein subunits that extend from the cell wall. One type of pilus, the sex pilus, is involved in bacterial mating or conjugation. Other pili help the bacterium adhere to objects. There is evidence that some pathogens attach to susceptible host cells by use of pili; and vaccines against these particular bacterial pathogens are being prepared with the use of their pili.

A large variety of bacteria are motile, and most of these employ **flagella** to propel themselves. The bacterial flagellum is quite different from eukaryotic flagella (figure 34.12). They are thin (about 20 nm) protein strands made of **flagellin** subunits. At the base of the flagellum is a curved hook joined to a **basal body.** The basal body contains rings that interact with the cell wall and plasma membrane. It is thought that the basal body rotates, somewhat like an electric motor, and this twists the rigid flagellum and thus moves the bacterium at speeds up to about 80–90 μm per second. Bacteria can even direct their movement towards nutrients or away from harmful chemicals. This **chemotaxis** is made possible by special chemoreceptors in the cell membrane.

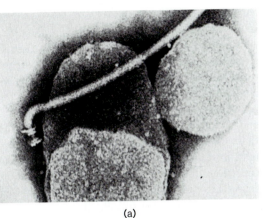

(a)

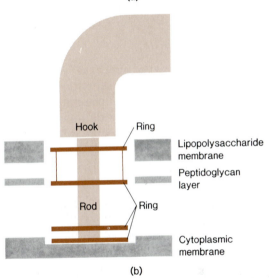

(b)

Figure 34.12 Flagellar structure. (a) An electron micrograph of a negatively stained flagellum showing the basal body and curved hook region. This picture also contains other fragments from the cell that was lysed to release this flagellum. (Magnification × 136,800) (b) The relationship of the basal body to the membrane of a Gram-negative bacterium.

The bacterial **endospore** has a thick wall of several layers (figure 34.13). Inside are DNA, ribosomes, and other cytoplasmic constituents. Endospores are the most resistant living structures known. For example, many can survive several hours in boiling water. Endospores are produced by some of the Gram-positive bacteria, and they serve as a mechanism of survival for those bacteria that can form them. When a bacterium is subjected to unfavorable environmental conditions, it responds by developing a dormant spore that may remain viable for years. When exposed to the proper stimuli, the spore germinates and produces a new vegetative cell.

Bacterial Reproduction

Bacteria normally reproduce asexually by regular cell division (**binary fission**). After the DNA geome has been replicated, the plasma membrane pushes inward to form a central transverse septum. Next, the cell wall grows inward within the transverse septum and eventually divides the cell in two. The newly duplicated DNA genomes are attached to the plasma membrane or mesosome. Since the older cell walls are elongating as the new walls are being formed, the two chromosomes are separated and pulled into the daughter cells by cell wall and plasma membrane expansion.

Although it takes a period of hours to days for most eukaryotic cells to reproduce, bacteria are able to carry out binary fission at much more rapid rates. *E. coli* populations double about once every twenty minutes under optimal culture conditions. If this were to continue for a day or so, the mass of *E. coli* formed would be many times the weight of the earth! Obviously, such maximal growth rates are not sustained for very long, but decrease due to exhaustion of nutrient supplies or accumulation of toxic wastes. In nature, where growth conditions are not so favorable, a population can take days to double in number. For example, *E. coli* has a doubling time of twelve hours even in the intestinal tract (a fairly nutrient-rich environment). This is much slower than the twenty-minute doubling time seen in culture media.

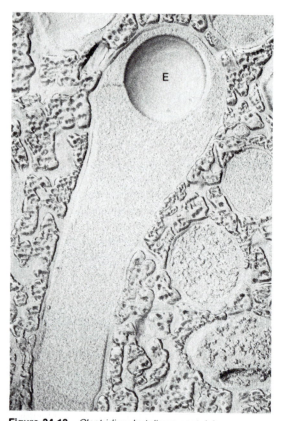

Figure 34.13 *Clostridium botulinum* containing an endospore (E). This organism causes the form of food poisoning known as botulism. (Magnification × 37,400)

One of the major differences between prokaryotes and eukaryotes is that eukaryotes regularly carry out organized genetic recombination as part of their sexual life cycle; prokaryotes do not. Although prokaryotes do not have this ability, they can achieve partial genetic recombination or gene exchange under certain conditions. There are three major ways in which recombination can come about through the transfer of a DNA fragment from a donor cell to a recipient.

The simplest recombination process is **transformation**. The study of transformation provided the best early evidence for the role of DNA as genetic material (chapter 22). When a bacterium dies, fragments of its DNA may be released intact. If this fragment contacts a competent member of a transformable species, it can be bound to the cell and taken inside. Not all bacteria can be transformed, only competent strains of particular species, such as *Streptococcus pneumoniae* and *Escherichia coli*. After the DNA fragment has been taken up, part of it is incorporated into the genome of the recipient. In this way, a number of genes can be transferred. It has been shown that this recombination mechanism functions in nature.

As discussed earlier, bacteriophages can infect bacterial cells and either lyse or lysogenize them. This phenomenon makes possible a second mechanism of gene exchange, **transduction.** Transduction (virus-mediated genetic transfer) can occur in two ways (figure 34.14). It sometimes happens that when virus capsids are assembled around virus DNA, a fragment of host DNA is mistakenly incorporated into a newly formed virion instead. The defective virion can subsequently infect a recipient bacterium. In this way, any cluster of genes may be transferred to a recipient bacterium during the course of a lytic cycle or after a lysogenic bacterium has been induced to form viruses.

Temperate viruses also can transfer specific sets of genes between bacteria by a second type of transduction mechanism (figure 34.14b). When a prophage has been induced to detach from the host chromosome and begin a lytic cycle, it may exchange some of its DNA for that of the bacterium. The resulting defective virus DNA contains bacterial genes from the region adjacent to the original prophage location. These bacterial genes are then transferred to a recipient bacterium when it is infected by the defective phage particle. Since it appears that most bacteria in natural environments are lysogenic for one or more viruses, these two transduction processes may be of great importance in ensuring genetic exchange within bacterial populations.

It is also possible for bacteria to exchange genetic material during direct cell-to-cell contact. This process is called bacterial **conjugation** and can be used to transfer either a plasmid or a portion of the bacterial chromosome. A **plasmid** is an extrachromosomal circular DNA segment that can reproduce independently of the bacterial chromosome (see p. 607).

Many plasmids also can be integrated into the chromosome and replicate with it (a situation very similar to that seen with temperate viruses). *E. coli* and a number of other bacteria may contain a special plasmid, the **F factor,** that codes for the **sex pilus** and any other molecules required for conjugation. If a cell lacks the F factor, it can only receive DNA during conjugation and is called **F⁻**. A cell possessing the F factor has a sex pilus and can donate DNA. However, its behavior depends on whether the F factor is free of the bacterial chromosome or integrated into it.

The simplest situation is when the F factor is separate from the chromosome (figure 34.15a). The **F⁺ cell** contacts the F⁻ cell using its sex pilus, and a conjugation bridge is formed. The F factor DNA replicates, and one copy is transferred to the F⁻ cell, thereby changing it to F⁺. In this way, a plasmid may spread throughout a bacterial population, but with little or no transfer of bacterial genes.

Sometimes an F factor is incorporated into the bacterial chromosome. In this case conjugation results in very efficient transfer of bacterial genes, and thus the cells are referred to as **Hfr cells** (for "high frequency of recombination"). After the conjugation bridge has formed between a Hfr cell and a F⁻ recipient, one strand of the Hfr chromosome breaks next to the F factor and is transferred into the F⁻ cell while being replicated (figure 34.15b). Since the end of the chromosome farthest away from the F factor is transferred first, these matings result in efficient gene transfer, but few conversions of the recipient to F⁺ or Hfr. The conjugation bridge usually breaks before transfer is complete, and the F factor thus is not transferred. After the transfer, part of the DNA from the Hfr donor is incorporated into the recipient chromosome, and the remainder is destroyed.

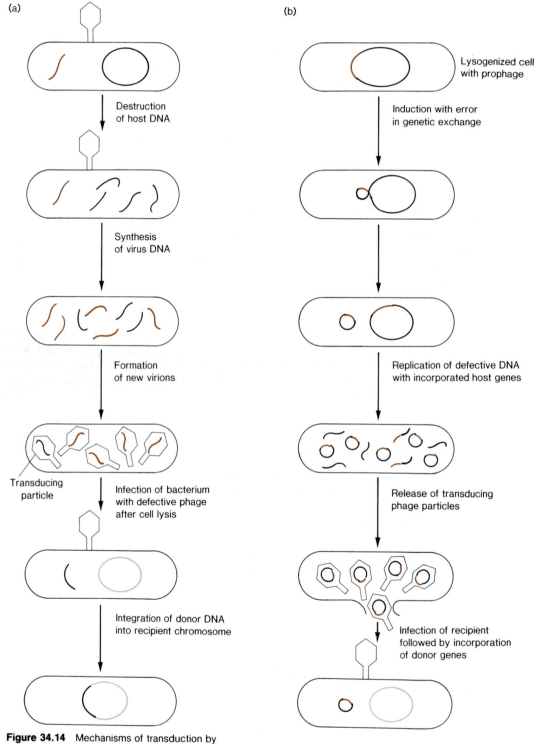

Figure 34.14 Mechanisms of transduction by bacteriophages. (a) Generalized transduction in which any cluster of genes may be transferred. (b) Specialized transduction in which only genes next to a prophage are transferred after the prophage has been induced to begin a lytic cycle.

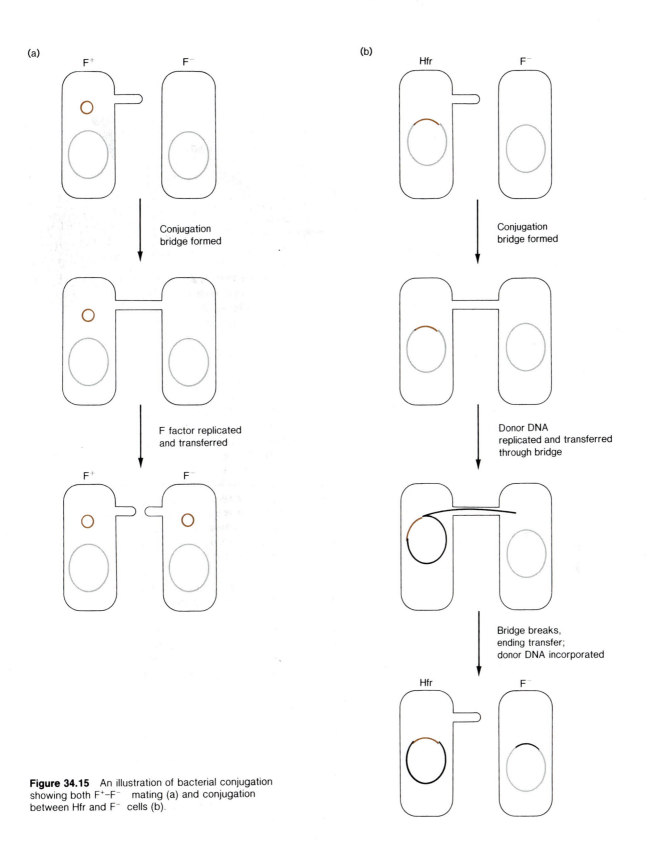

Figure 34.15 An illustration of bacterial conjugation
showing both F⁺–F⁻ mating (a) and conjugation
between Hfr and F⁻ cells (b).

Bacterial Nutrition and Metabolism

One of the most remarkable qualities of bacteria is their vast metabolic diversity. There probably is no natural organic molecule that cannot be degraded by at least one bacterial species. Indeed, some bacteria are chemical omnivores. *Pseudomonas multivorans,* for example, can use over ninety different organic molecules as its source of carbon and energy. Bacteria that are quite fastidious also can be isolated (methane-oxidizing bacteria, for example, only metabolize methane, methanol, and a few related substrates). This metabolic diversity and flexibility makes bacteria an extremely important component of the ecosystem. They play a critical part in the decomposition and recycling of organic matter, including many manmade materials.

The majority of bacteria are chemoheterotrophs. Most require oxygen for growth; that is, they are **aerobic.** Oxygen is used for such purposes as accepting electrons from the electron transport chain during aerobic respiration. Many bacteria possess the Embden-Meyerhof pathway, the Krebs cycle, and an electron transport chain (although the electron transport chain differs from that seen in most eukaryotes).

Other bacteria are **facultative anaerobes.** They do not require oxygen but grow better when it is present. *E. coli* is a well-known facultative bacterium. It grows adequately under anaerobic conditions in the human intestine but flourishes when incubated in the aerobic atmosphere of an incubator. Under anaerobic conditions, the bacterium switches to the process of **fermentation** for its source of energy. That is, it oxidizes a sugar such as glucose to pyruvate and produces ATP. To eliminate the excess electrons thus generated, it reduces pyruvate to lactate (see chapter 7 for more details on respiration and fermentation). Here an organic molecule, not oxygen, is serving as the electron acceptor. Many bacteria are exceptionally active fermenters and can produce a dozen or more products.

Anaerobic bacteria not only do not require oxygen for growth; they are actually killed by it. They lack special enzymes that protect aerobic and facultative organisms from toxic oxygen derivatives like hydrogen peroxide. Members of the genus *Clostridium*—the causative agents of botulism, tetanus, and other serious diseases—are obligate anaerobes. That is, they can grow only under anaerobic conditions.

Identification of bacteria has always been a difficult task because many lack complex, distinctive morphological features. Therefore, microbiologists have made extensive use of physiological tests in the identification of bacteria, particularly of potential pathogens. In practice, the nutrient sources used and products released by growing bacteria are identified by use of color reactions (figure 34.16). Frequently, it is possible to incorporate the appropriate indicator directly in the growth medium. A hospital laboratory, for example, often can make a tentative identification of a type of bacteria within a few hours using commercially available color test systems.

Animals like humans must oxidize organic molecules such as sugars and fatty acids to obtain ATP. Bacteria are not always thus limited. **Chemoautotrophs** can extract electrons from hydrogen sulfide, elemental sulfur, iron, and other inorganic substances and use these electrons to generate ATP by electron transport. They then use the ATP to provide energy for synthesis of organic compounds from simple, inorganic precursors. These bacteria are quite important ecologically because they oxidize large quantities of sulfur and ferrous iron to sulfuric acid and insoluble ferric iron. They can render a mine drainage stream (which contains large quantities of sulfur and iron) barren and lifeless through their metabolic activities.

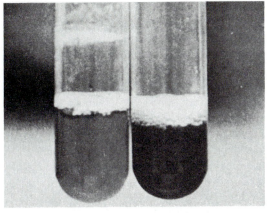

(a)

(b)

Figure 34.16 Some physiological tests used in
bacterial identification. Results of such tests usually are
"read" as color changes in the medium. Two tests used
to distinguish certain bacillus bacteria, the Voges-
Proskauer test (a) and the nitrate reduction test (b), are
shown. In each case the tube on the left shows a
positive result, and the tube on the right is negative.

Purple and green bacteria resemble plants
in being **photoautotrophs,** using light as their energy source and producing reduced carbon compounds using carbon dioxide as a precursor. Of
course, they lack the chloroplasts possessed by
eukaryotes and have their photosynthetic membranes spread throughout the cytoplasm. Their
distinctive colors show that their chlorophylls and
other photosynthetic pigments differ from those
of plants. Plant chloroplasts have two functional
photosystems and yield oxygen from water oxidation (see chapter 6). In contrast, photosynthetic bacteria have only one photosystem
(Photosystem I) and do not oxidize water to O_2.
Instead, they use light to produce ATP and extract electrons from inorganic molecules like H_2S
in order to reduce NAD^+ or $NADP^+$. As a result,
granules of elemental sulfur are products of photosynthesis instead of O_2. Purple and green bacteria are usually found a few meters below the
surface of lakes, where there is sufficient light
and also a supply of H_2S.

Bacteria As Pathogens

When most people think of bacteria, they think
of "germs" and disease. This is quite natural.
Bacterial pathogens have had an incalculable effect on humans. In fact, modern microbiology
developed in the nineteenth century in response
to the need for disease control.

One of the foundations for **the germ theory
of disease** resulted from the work of Louis Pasteur on spontaneous generation, the notion that
living things spontaneously arise from nonliving
material whenever proper conditions exist. During the course of his studies, which indicated that
spontaneous generation does not occur, Pasteur
showed that the air was filled with microorganisms that could cause decay unless removed or
destroyed. The British surgeon Joseph Lister was
immensely impressed by Pasteur's work. He applied Pasteur's principles to the protection of
wounds from airborne microbes by sterilizing
surgical instruments, dressings, and the operating
room. His success provided indirect evidence in
support of the belief that germs might cause disease.

The germ theory was placed on a firm basis by the German physician Robert Koch. In 1876, Koch published an exhaustive study of anthrax, a disease of cattle that could be tramsmitted to humans. He managed to prove in a systematic way that anthrax was due to the spore-forming bacterium *Bacillus anthracis*. His approach and method of proof can be expressed in terms of the criteria known as **Koch's postulates.** Koch maintained that one could prove a bacterium caused a particular disease if four conditions were met:

1. The microorganism must be present in all diseased animals and no healthy ones.
2. The organism must be isolated from a diseased animal and grown in pure culture.
3. When the isolated microorganisms are injected into healthy animals, precisely the same disease must result.
4. One must be able to reisolate the suspected pathogen from the experimentally infected host and show it to be the same as the one first isolated.

Koch's work not only established rigorous criteria for the study of disease agents; it also stimulated the use of **pure cultures** (cultures containing only one kind of bacterium) and the laboratory cultivation of bacteria. Furthermore, Koch pioneered the use of solid nutrient media, particularly agar. This made the isolation of single colonies derived from individual bacteria easier. A mixture of bacteria could simply be streaked out on agar to such an extent that each cell reproduced to form an isolated colony.

These technical and conceptual breakthroughs led to a vigorous spurt of activity in medical microbiology beginning in the 1870s. Bacterial pathogens for many of the most deadly diseases were discovered during the next thirty years: anthrax, gonorrhea, typhoid fever, tuberculosis, cholera, diphtheria, tetanus, pneumonia, meningitis, gas gangrene, plague, botulism, dysentery, whooping cough, and many more. These discoveries made possible the subsequent progress in disease control that has transformed human lives.

Chapter 26 focused on how the body protects itself from foreign invaders. Here the focus is on some of the ways in which bacteria can harm their hosts. The ability of a bacterial parasite to cause disease is called **pathogenicity.** The relative degree of pathogenicity, or **virulence,** is fundamentally a function of two major factors—invasiveness and toxigenicity. **Invasiveness** refers to the ability of a pathogen to proliferate in its host. **Toxigenicity** is the capability of producing a chemical substance or **toxin** that can damage the host and lead to disease. A pathogen may be virulent either because of high invasiveness or great toxigenicity. An excellent example of invasiveness is *Streptococcus pneumoniae.* This bacteria does not appear to produce a toxin, yet it reaches such a high population level in the lungs that the lungs fill with serum and white blood cells in response to its presence. In contrast, *Clostridium tetani* rarely leaves the wound in which it lives. Nevertheless, its toxin is so potent that the victim may well die of paralysis, which is caused by the toxin.

There are two types of bacterial toxins, each with different effects. **Exotoxins** are protein toxins that are usually released by living Gram-positive bacteria. Botulism, tetanus, gas gangrene, diphtheria, and cholera are all caused by exotoxins, each acting in a different fashion. For example, diphtheria toxin inhibits protein synthesis; one of the gas gangrene toxins attacks the host cell membrane phospholipid, lecithin; and cholera toxin disrupts ion transport in the intestinal epithelium by raising cyclic AMP levels and thus interferes with the ability of the host organism to maintain normal ionic composition of body fluids and normal body fluid volume. **Endotoxins** are derived from the outer portion of the Gram-negative bacterial cell wall. They are very complex in composition—containing lipid, polysaccharide, and protein—and normally seem to be released only upon bacterial cell death. Endotoxins cause generalized host responses, including shock, diarrhea, hemorrhages, and fever.

The disease process is very complex and depends on the interaction of host and pathogen. A bacterial pathogen may have its most powerful effect on the host by triggering allergic responses rather than by toxin production; that is, the host is harmed by its own defense mechanisms.

Beneficial Activities of Bacteria

Pathogenic bacteria have such widespread impact that all bacteria automatically are considered dangerous and harmful. In truth, beneficial bacteria vastly outnumber the pathogens and are so important that the ecosystem could not function without them. For example, bacteria play an indispensable role in both the carbon (chapter 32) and nitrogen (chapter 8) cycles. They help degrade dead organic material, mostly soil humus, to CO_2. This makes the carbon available for photosynthetic incorporation by plants. Organic nitrogen in decaying material is released as ammonium ions by bacterial activity. Nitrifying bacteria can oxidize this ammonia to nitrate, a nitrogen source readily used by plants.

One of the most important ecological contributions of bacteria is the **fixation of nitrogen,** the utilization of N_2 (atmospheric molecular nitrogen) as a nitrogen source. Nitrogen limits productivity in many environments because most organisms cannot use molecular nitrogen. Around 85 percent of nitrogen fixation is biological; the remainder results from lightning or industrial activity. *Rhizobium* is a bacterium that fixes nitrogen when in the root nodules of legumes such as soybeans, clover, and alfalfa (see p. 224). Several free-living bacterial species also fix nitrogen and thus contribute to soil fertility.

Bacteria in human bodies are often beneficial rather than harmful. Their importance has become obvious from studies on **germ-free animals** grown in special isolation units. These germ-free animals do not have well-developed immune systems and therefore are very susceptible to pathogens. The absence of normal bacteria also lowers resistance because an invading pathogen does not encounter any competition from indigenous nonpathogenic bacteria. Intestinal bacteria also seem to be necessary for proper intestinal development. The intestinal walls of germ-free animals are thin and underdeveloped. Finally, intestinal bacteria produce and release vitamins like vitamin K, which their host can absorb.

Of course, bacteria are indispensable to many industries. Lactic acid bacteria are used by the dairy industry in the manufacture of yogurt, cottage cheese, and cheese. The distinctive flavors of cheddar, Swiss, Parmesan, and Limburger cheeses are the result of bacterial products. Vinegars are made by allowing acetic acid bacteria to oxidize the alcohol in wine, apple cider, or malt. Bacteria also are used to synthesize amino acids (glutamic acid, lysine), organic acids (lactic acid, butyric acid), steroids, and enzymes. With the exception of penicillin and ampicillin, most important antibiotics are produced by actinomycetes.

The future of industrial microbiology looks even more promising. It is quite likely that methane for use as fuel can be manufactured from agricultural organic wastes by methanogenic bacteria. Several bacteria already are being grown for use as biological insecticides. A toxin from *Bacillus thuringensis* kills a dozen or more common insect pests (cabbage worm, gypsy moth, etc.). More intensive use of biological control agents is a certainty in the future.

Finally, some countries are already growing bacteria for use as a protein source. The utilization of bacteria as a food source is attractive because bacteria can grow rapidly on many indigestible materials, such as wastes from the petroleum industry.

Figure 34.17 An electron micrograph of the cyanobacterium *Anabaena azollae*. Note the prokaryotic structure and the extensive photosynthetic membranes (lamellae). Polyhedral bodies are structures commonly found in cyanobacterial cells that seem to be involved in some photosynthetic functions.

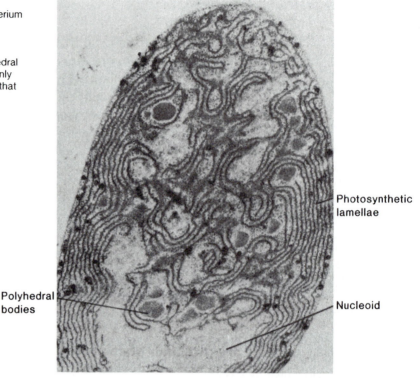

Photosynthetic lamellae

Polyhedral bodies

Nucleoid

Cyanobacteria

The cyanobacteria (blue-green algae) are a large, complex group of prokaryotic microorganisms that are similar to eukaryotic algae in terms of photosynthesis and general appearance. However, it is probably best to call them bacteria since they are genuine prokaryotes; that is, their cells lack a true nucleus and other complex, membraneous organelles (figure 34.17). Even their cell walls are similar to those of Gram-negative bacteria.

The group is very diverse morphologically, ranging from single cells to colonies and filamentous forms (figure 34.18). A number of species coat themselves with a gelatinous sheath. There is similar diversity in size. The smallest organisms are around 1 µm in diameter. *Oscillatoria princeps* can actually reach a diameter of 60 µm, a giant among prokaryotes.

Although a number of cyanobacteria are nonmotile, many possess gliding motility. When a cell or filament is in contact with a solid surface, it glides along by some unknown mechanism (cyanobacteria lack flagella). Some even rotate and flex as they move. Phototaxis and chemotaxis have also been observed.

Reproduction is asexual by binary fission, as with other prokaryotes. Sometimes, filamentous forms break off fragments (**hormogonia**), which then glide away as new individuals. Cyanobacteria may also form thick-walled resting spores called **akinetes** (figure 34.18). Akinetes survive periods of drying or cold and germinate under more favorable conditions.

Cyanobacteria are photosynthetic and usually have very simple nutritional requirements—CO_2, nitrate or ammonia, and some inorganic ions. Their photosynthetic apparatus, even though it is not contained in a chloroplast, is very similar to that of red algae, rather than the systems of green and purple bacteria. They have chlorophyll a and the accessory pigments **phycocyanin** (blue in color) and **phycoerythrin** (red). Because of these pigments, blue-green bacteria are not always blue-green but may even be red or brown. Unlike other photosynthetic bacteria, cyanobacteria do possess Photosystem II and produce O_2 photosynthetically. Many also have **gas vesicles** that make them able to float close to the surface, where light intensity is greater.

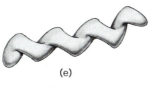

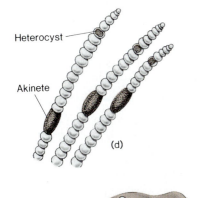

Heterocyst

Akinete

Heterocyst

Figure 34.18   Common cyanobacteria (blue-green
algae). (a) *Lyngbya*. (b) *Gomphosphaeria*.
(c) *Chamaesiphon*. (d) *Anabaena*. (e) *Spirulina*.
(f) *Oscillatoria*. (g) *Aphanocapsa*. (h) *Gloeocapsa*.
(i) *Nostoc*. (j) *Merismopedia*.

A number of cyanobacteria can fix nitrogen and possess a unique structure, the **heterocyst** (figures 34.18 and 34.19). Heterocysts are specialized, thick-walled cells that have lost their nuclei and been transformed into nitrogen fixation centers. **Nitrogenase,** the enzyme responsible for N_2 fixation, is inactivated by oxygen. Therefore, when the heterocyst develops, Photosystem II (the photosystem required for O_2 generation) is lost. The mature heterocyst still contains Photosystem I and can generate ATP, but without producing O_2. It seems that the heterocyst provides just the proper anaerobic environment for nitrogen fixation. Heterocysts develop when the organism lacks a source of nitrogen other than N_2. Cyanobacteria are such efficient fixers that in certain rice-growing regions of the world, Southeast Asia in particular, nitrogen fertilizers are unnecessary because cyanobacteria abound on the surface water of paddy fields. Consequently, rice may be grown on the same land year after year without the addition of fertilizers.

Cyanobacteria, with their resilience and simple nutritional requirements, are capable of colonizing areas unable to sustain eukaryotic algae (for example, in hot springs and deserts) and are found in almost all environments. It is presumed that these organisms were the first colonizers of land during the course of evolution. Once established, their mass could have provided a physical as well as chemical substrate for the ultimate attachment and growth of more complex plants.

Cyanobacteria are also participants in a variety of symbiotic associations. They are associated with liverworts and ferns, and with corals and other invertebrates. One of the most interesting of these associations is the **lichen** (a symbiotic association of a fungus with either green algae or cyanobacteria). In a lichen, the cyanobacterium provides organic nutrients for the fungus, while the latter protects the cyanobacterium and furnishes the inorganic nutrients required by its photosynthetic partner.

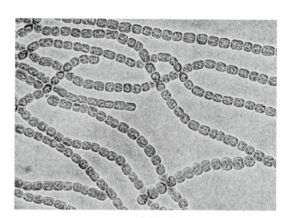

(a)

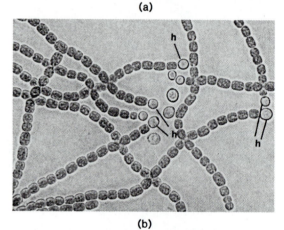

(b)

Figure 34.19 The effect of changing the nitrogen source on *Anabaena*. (a) Filaments without heterocysts formed in presence of ammonia. (Magnification × 437) (b) Filaments form heterocysts (h) when grown with N_2 as the nitrogen source. (Magnification × 437)

Although exceedingly beneficial and ecologically important, cyanobacteria also may become a nuisance. It must be noted, however, that they are aided and abetted in this regard by the ignorance and thoughtlessness of humans. These bacteria thrive in environments high in phosphates and nitrates. If care is not taken in the disposal of industrial, agricultural, and human wastes, phosphates and nitrates drain into lakes and ponds, resulting in a "bloom" of cyanobacteria. The surface of the water becomes turbid. Light available for photosynthesis by other aquatic algae is reduced, and the toxic by-products of the cyanobacteria can kill fish.

Summary

Taxonomy is concerned with naming organisms
and arranging them in a system of classification
based on their similarities and probable evolutionary relationships. A species is the basic taxonomic unit and is represented by a Latin
binomial (two-word name), the genus name, and
the species name. Together, the binomial represents a species. Species are arranged within a
succession of ever broader taxonomic ranks. The
major taxonomic ranks used, in decreasing order
of breadth, are: kingdom, phylum, class, order,
family, genus, and species.

The classification system used in this book
was developed by R. H. Whittaker and others to
handle the problem of "borderline" species—
those with common animal and plant features.
It is referred to as the five-kingdom system and
uses cellular structure, general level of organizational complexity, and modes of nutrition (photosynthetic, absorptive, ingestive) to group organisms into one of five kingdoms: Monera,
Protista, Fungi, Animalia, and Plantae.

Viruses are included in this chapter, although they are not considered to be living in the
usual sense of the word. However, they do contain
nucleic acids and protein, the basic chemicals of
living systems. To reproduce, viruses enter host
cells, taking over the metabolic machinery of the
cell. In so doing, the viruses cause the cells to
replicate viral nucleic acids and synthesize virus
capsids. In the process, cells are often destroyed,
and viral diseases can develop.

Bacteria and cyanobacteria are true monerans in that they are unicellular prokaryotes.
Prokaryotic cells differ from eukaryotic cells in
that the former do not possess a true nucleus,
have their DNA as a circular molecule not bound
to histone proteins, and lack membrane-bound
organelles, such as mitochondria and chloroplasts. Reproduction is asexual, usually by binary
fission. In bacteria, genetic recombination can
sometimes occur through transformation, transduction, and conjugation. Cyanobacteria differ
from other photosynthetic bacteria because they
possess chlorophyll a and use water as an electron
donor (with the release of oxygen as a by-product
of photosynthesis).

Bacteria and cyanobacteria are of great importance to human beings and the ecosystem as
a whole. They are critical components of the nitrogen and carbon cycles, since they are involved
in nitrogen fixation, nitrification, and the decomposition of dead organic matter. They are useful
industrially in the manufacture of food, antibiotics, and other important products. Some bacteria
are pathogenic for humans and the plants and
animals on which humans depend.

Questions

1. Briefly describe the five-kingdom system
   and its advantages over the older two-
   kingdom system.
2. What is a virus? Write a description of the
   major features of virus morphology.
3. How do T-even bacteriophages reproduce
   themselves? What is a lysogenic life cycle,
   and how does it operate?
4. How do RNA viruses reproduce?
5. What are viroids?
6. In what ways do bacteria differ from
   eukaryotic microorganisms?
7. Describe and contrast transformation,
   transduction, and conjugation. What is a
   plasmid? an Hfr strain?
8. Define the following terms: aerobe,
   facultative anaerobe, anaerobe,
   fermentation, aerobic respiration,
   chemoautotroph.
9. How does bacterial photosynthesis differ
   from that of plants?
10. What are Koch's postulates and what is
    their importance?
11. Define pathogenicity and virulence. What
    factors determine virulence? Characterize
    exotoxins and endotoxins.
12. Describe four major ways in which bacteria
    benefit humans or the human ecosystem.
13. What are cyanobacteria, and how do they
    differ from other bacteria? Describe briefly
    their morphology and reproduction.
14. What are heterocysts, and what is their
    significance?

Suggested Readings

Books

Brock, T. D. 1979. *Biology of microorganisms.* 3d ed. Englewood Cliffs, N.J.: Prentice-Hall.

Luria, S. E.; Darnell, J. E.; Baltimore, D.; and Campbell, A. 1978. *General virology.* 3d ed. New York: Wiley.

Savory, T. 1962. *Naming the living world.* New York: Wiley.

Stanier, R. Y.; Adelberg, E. A; and Ingraham, J. L. 1976. *The microbial world.* 4th ed. Englewood Cliffs, N.J.: Prentice-Hall.

Articles

Burke, D. C. April 1977. The status of interferon. *Scientific American* (offprint 1356).

Butler, P. J. G., and Klug, A. November 1978. The assembly of a virus. *Scientific American* (offprint 1412).

Costerton, J. W.; Geesey, G. G.; and Cheng, K. J. January 1978. How bacteria stick. *Scientific American* (offprint 1379).

Demain, A. L., and Solomon, N. A. September 1981. Industrial Microbiology. *Scientific American.*

Diener, T. O. January 1981. Viroids. *Scientific American* (offprint 1488).

Gilbert, W., and Villa-Komaroff, L. April 1980. Useful proteins from recombinant bacteria. *Scientific American* (offprint 1466).

Gould, S. J. The five kingdoms. *Natural History* 85(6):30.

Sanders, F. K. 1981. Interferons: An example of communication. *Carolina Biology Readers* no. 88. Burlington, N.C.: Carolina Biological Supply Co.

Walsby, A. E. August 1977. The gas vacuoles of blue-green algae. *Scientific American* (offprint 1367).

35

Survey of Protists and Fungi

Chapter Concepts

1. Eukaryotic cells may have arisen from an endosymbiotic association between anaerobic amoeboid prokaryotes and aerobic bacteria and photosynthetic bacteria.

2. Members of the protist kingdom are mainly unicellular or colonial eukaryotes, although some may be primitive multicellular organisms.

3. Despite their relatively simple structure, many protists are adapted to life in variable environments.

4. Both autotrophs and heterotrophs are found among the protists.

5. Many protists are major primary producers and of great ecological importance. A number also cause serious diseases in humans, their crops, and their domesticated animals.

6. Fungi are multicellular heterotrophic eukaryotes that absorb nutrients after secreting extracellular digestive enzymes. Because of their nutritional mode, they are either saprophytic or parasitic.

In many ways, the protists are the most heterogeneous and fascinating of the five kingdoms. This kingdom contains unicellular and colonial eukaryotic organisms, and even some simple multicellular forms (figure 34.2). These broad and somewhat vague boundaries encompass a tremendous diversity of organisms.

Table 35.1 lists the protist groups considered in this chapter. Only the most common and important major protist groups are listed, and there is one major omission from the table. As mentioned in chapter 34, the green algae (chlorophyta) contain unicellular, colonial, and multicellular forms. Biologists have called these algae both protists and plants. Even though some chlorophyta are clearly protist in terms of organizational complexity, a discussion of the green algae is deferred to chapter 36 ("Survey of Plants"). Distribution of protists among the categories protozoa, unicellular algae, and funguslike protists is not as clear and unambiguous as the table implies. For example, protozoologists would consider euglenoids, dinoflagellates, and slime molds to be protozoa. Despite these difficulties and limitations, this text covers the protists from the perspective of table 35.1, keeping in mind that this classification system mirrors reality only imperfectly.

This chapter also includes an introduction to the kingdom Fungi. Fungi are eukaryotic multicellular heterotrophs. They do not actively move about as do protozoa and animals, but they act as important decomposers in the ecosystem.

Despite the diversity of organisms to be covered in this chapter, there is a unifying theme. The majority of these organisms must cope with their environment without the benefit of the complex tissues and organ systems possessed by plants and animals. Indeed, these organisms frequently consist of a single cell. The apparent simplicity of protists and many fungi is deceptive, for they have very complex ultrastructure and can adjust to a changing environment with impressive ease.

The Origin of the Eukaryotic Cell

All of the organisms covered in this chapter and in chapters 36 and 37 are eukaryotic (see p. 93). The radical dissimilarity between prokaryotes and eukaryotes has provoked a great deal of research and speculation among biologists as to how eukaryotic cells might have arisen in the first place.

Probably the majority of biologists believe the **endosymbiotic theory** to be correct, at least in broad outline. **Symbiosis** is an intimate and long-term association between two organisms of different species. For example, algae such as dinoflagellates and diatoms can be found living within a variety of invertebrates. In the past few years, it has become clear that symbiotic associations are widespread and important. If symbiotic associations are prevalent today, such associations might have originated and been favored by natural selection early in the development of life. Thus, the mitochondria and chloroplasts of today may have been the endosymbionts of yesterday. This concept is the foundation of the endosymbiotic theory suggested by Lynn Margulis and others to explain the evolution of eukaryotic cells.

Table 35.1
The Kingdom Protista.

Protozoa

Mastigophora	Flagellated protozoa
Sarcodina	Pseudopodial protozoa
Sporozoa	Spore-forming protozoa
Ciliophora	Ciliated protozoa

Unicellular Algae

Euglenophyta	*Euglena* and relatives
Pyrrophyta	Dinoflagellates
Chrysophyta	Diatoms and others

Funguslike Protists

Chytridiomycota	Chytrids with a single, posterior flagellum
Oomycota	Water molds with biflagellate zoospores
Gymnomycota	Slime molds

The fossil record indicates that prokaryotes may have arisen about 3,000 million years ago, while eukaryotes are thought to have first appeared late in the Precambrian period, about 1,500 million years ago. During the 1,500 million years between the appearance of prokaryotes and the origin of the eukaryotes, contact between different prokaryotes must have taken place frequently. The first step in the evolution of the eukaryotic cell is thought to have occurred when a large anaerobic amoeboid prokaryote ingested a smaller aerobic bacterium and stabilized its prey as an endosymbiont rather than digesting it (figure 35.1). The plasma membrane of the aerobic bacterium eventually became the inner mitochondrial membrane. Possession of such an endosymbiont conferred the advantage of aerobic respiration on its host. Flagella might have arisen through the ingestion of prokaryotes similar to spirochetes. Ingestion of prokaryotes somewhat resembling present-day cyanobacteria (blue-green algae) could lead to the endosymbiotic development of chloroplasts in a fashion similar to formation of mitochondria from aerobic, nonphotosynthetic bacteria.

Margulis and others have assembled a considerable amount of indirect evidence in support of their theory:

1. Mitochondria and chloroplasts are similar to bacteria and cyanobacteria in morphology and size.

Figure 35.1 The endosymbiotic theory of how eukaryotic cells first arose.

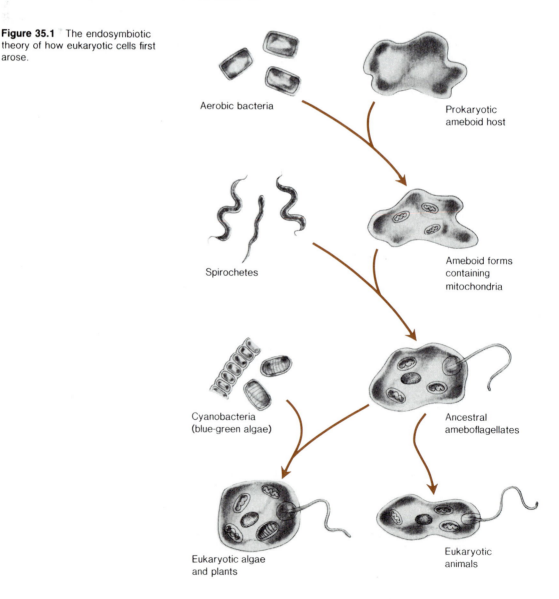

Aerobic bacteria

Prokaryotic ameboid host

Spirochetes

Ameboid forms containing mitochondria

Cyanobacteria (blue-green algae)

Ancestral ameboflagellates

Eukaryotic algae and plants

Eukaryotic animals

2. These organelles are self-replicating and contain DNA that resembles prokaryotic cells' DNA more closely than that of eukaryotic cells.

3. Ribosomes in these organelles are smaller than those in the cytoplasm of eukaryotic cells and are similar in size to the ribosomes of prokaryotic cells.

4. Protein synthesis by bacterial ribosomes and the ribosomes of mitochondria and chloroplasts is inhibited by chloramphenicol, but not by the drug cycloheximide. In contrast, eukaryotic cytoplasmic ribosomes are inhibited by cycloheximide rather than chloramphenicol.

Proponents of the theory explain the observations that DNA of mitochondria and chloroplasts is much smaller than bacterial DNA and that some organellar proteins can be coded for by nuclear genes with the hypothesis that the endosymbionts must have transferred some of their genes to the host nucleus and thus relinquished their independence for the sake of the symbiotic relationship.

There are some difficulties with the endosymbiotic theory, and a number of alternative theories for the origin of the eukaryotic cell have been proposed. A representative example of these is described in figure 35.2. In brief, it has been proposed that the prokaryotic cell membrane invaginated to enclose copies of the genome, thereby forming several double-walled entities within a single cell. These entities could then have evolved into the eukaryotic nucleus, mitochondrion, and chloroplast.

Regardless of the exact mechanism involved, the emergence of the eukaryotic cell during the Precambrian period led to a dramatic increase in the complexity and diversity of living organisms. At first, organisms were capable of existing only as independent single cells. Later, some evolved into multicellular organisms in which different cells became specialized for a variety of different functions. In this process, the organism became better fitted, in some ways, to deal with environmental changes. But while cells in multicellular organisms gained the cooperation and support of their fellows, they sacrificed their independence and could no longer survive outside the body of the organism to which they belonged.

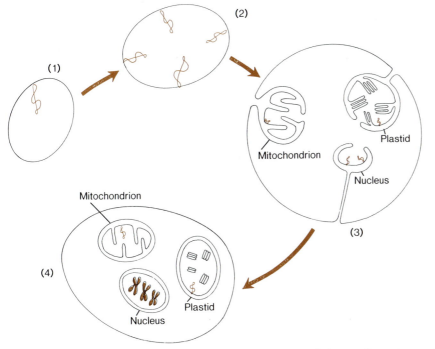

Figure 35.2 The origin of the eukaryotic cell may involve membrane invagination to encompass separate genomes in double-membrane boundaries. A prokaryotic cell (1) duplicates its genome (2). This is followed by invagination of the plasma membrane to form double-walled organelles and separate the individual genomes from one another (3). The nuclear genome enlarges while the organellar genomes lose many of their genes, and a modern eukaryotic cell results (4).

Box 35.1
The Origin of Life

Not only is there considerable interest in the origin of the first eukaryotic cells, but a number of scientists are actively studying those processes thought to be involved in the origin of life itself. This field is very speculative by nature, since it obviously is impossible to observe or to describe precisely the original events, and many of the assumptions and theories about how the first life might have arisen are tentative, and even controversial.

The earth probably formed by condensation of dust and gases approximately 4,500 to 5,000 million years ago. The lighter gases, such as hydrogen and helium, were lost as the planet formed because it was too small to hold them by gravity. It is thought likely that a primitive atmosphere was produced later by volcanic action, but there is considerable disagreement about the exact nature of this primitive atmosphere. Harold Urey proposed in the early 1950s that the atmosphere was strongly reducing due to the presence of hydrogen. He felt that it was composed of hydrogen, water, ammonia, and methane. More recent geochemical studies, however, favor the existence of a mildly reducing atmosphere consisting mostly of water, nitrogen gas, and carbon dioxide, with small amounts of hydrogen and carbon monoxide also present.

As early as the 1920s, the Soviet biochemist A. I. Oparin proposed that organic molecules could be produced from the inorganic constituents of the primitive atmosphere in the presence of an energy source, such as lightning or sunlight. In 1953, Stanley Miller provided support for Oparin's ideas through an ingenious experiment (box figure 35.1A). Miller placed a mixture resembling the primitive atmosphere (methane, ammonia, hydrogen, and water) in a closed reaction vessel at 80°C. He then exposed it to electrical spark discharges for a week or more. At the end of this period, Miller discovered that a variety of amino acids and organic acids had been produced.

(a)

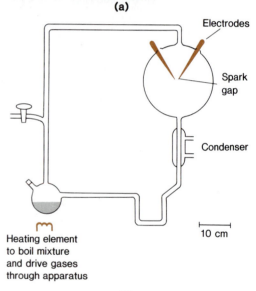

Electrodes

Spark gap

Condenser

10 cm

Heating element
to boil mixture
and drive gases
through apparatus

(b)

Box figure 35.1A   The spark-discharge apparatus for synthesizing organic molecules from gases thought to be present in the primitive atmosphere. (a) Dr. Stanley Miller and his apparatus. (b) A diagram of the apparatus.

A number of scientists have conducted similar experiments since the mid 1950s and have shown that a wide variety of biological molecules can be generated using several energy sources (box table 35.1). Similar molecules also arise if the less reducing atmosphere proposed recently by geochemists is employed. Either of at least two prevalent energy sources, ultraviolet light or electrical discharges, can promote these syntheses.

The results are consistent with Oparin's original proposal that early oceans might have contained considerable numbers of organic molecules. Indeed, the early oceans might have resembled a very dilute broth or soup. The formation of this broth constitutes the first phase of what has been called **chemical evolution.**

The next phase would involve the joining of molecular building blocks to yield polymers, such as polypeptides, polynucleotides, and polysaccharides. Such molecules are relatively unstable but might have been synthesized under gentle conditions through the action of chemical condensing agents like polyphosphates. Polyphosphates are readily formed and are capable of aiding in the formation of complex polymers, such as peptides and polynucleotides. It has also been suggested that reacting molecules became concentrated on the surfaces of clay particles. This concentrating effect would favor biopolymer formation.

Macromolecules could have subsequently coalesced to form small droplets containing the polymers in an aqueous matrix. These **coacervate**

Box table 35.1

Some of the Organic Compounds Produced under Experimental Conditions that Simulate Primitive Atmospheric Conditions.

Compound	Energy Source
Amino Acids Glycine, alanine, aspartic acid, glutamic acid, lysine, serine, valine, arginine, histidine, etc.	Electric sparking Ultraviolet light Heat
Peptides	Electric sparking Ultraviolet light Heat
Purines, Pyrimidines, Nucleosides, and Nucleotides	Ultraviolet light Heat
Polynucleotides	Heat
Sugars and Polysaccharides	Ultraviolet light Heat
Organic Acids Acetic, propionic, lactic, succinic, malic	Electric sparking Ultraviolet light
Simple Molecules Formaldehyde, HCN, pyrophosphate	Electric sparking Ultraviolet light

droplets, as Oparin called them, could have trapped primitive catalysts or enzymes and thus acquired a very rudimentary metabolism. S. W. Fox has heated mixtures of amino acids at 130°C to 180°C to form amino acid polymers, which he calls **proteinoids.** When the hot solution is cooled, these proteinoid mixtures form small, cell-like structures, which Fox has named **microspheres.** Microspheres seem to possess a selectively permeable surface and show cell-like behavior. Fox has speculated that such microspheres might be able to incorporate or develop self-replicating polynucleotides and gradually evolve into protocells.

Some biologists, Francis Crick for instance, disagree with Oparin and Fox. They believe that chemical evolution may have proceeded by the initial development of ribonucleic acid genetic material, rather than protein coacervates or microspheres. The most primitive nucleic acids might have reproduced themselves and subsequently acquired the ability to direct the synthesis of peptides by the use of condensing agents like polyphosphates.

Eventually, in one way or another, biological evolution began with the formation of the first true cells. These must certainly have been fermentative, heterotrophic forms, living at the expense of reduced organic molecules available in the ''soup'' that surrounded them. Eventually, as nutrients were depleted, the first autotrophs capable of incorporating CO_2 would have arisen. But these first autotrophs were not photosynthetic organisms. They were chemoautotrophs; that is, they derived energy for their synthetic activities not from light, but from chemical compounds available in their environment. These first autotrophs would have had a selective advantage over their heterotrophic predecessors because they would have been able to synthesize organic material from CO_2, rather than having to depend on the dwindling supply of complex nutrient molecules. This would have been followed by the development of the ability to trap light energy and use it in the process of photosynthesis. With the origin of photosynthetic cells that began to use water as an electron donor and produce oxygen, the primitive reducing environment would have been altered so drastically that the spontaneous development of molecules would have ceased. As oxygen levels increased, cells capable of respiratory metabolism—a much more efficient mode of energy conversion than fermentation—must have developed and flourished as a result of the cells' ability to make effective use of oxygen.

More recently, Berkner and Marshall have attributed the great increase in living forms that took place at the beginning of the Cambrian period, some 600 million years ago, to increases in oxygen concentration. Between the time at which photosynthetic autotrophs arose (around 3,000 million years ago) and the beginning of the Cambrian period, the oxygen level was too low to screen out ultraviolet radiation through the formation of an ozone layer. Early organisms could have lived only deep in the oceans, where they were shielded from the intense radiation striking the earth's surface. When atmospheric oxygen, and the resulting ozone, reached sufficiently protective levels at the beginning of the Cambrian period, life could spread to shallower waters. Finally, about 420 million years ago, the oxygen level rose high enough to screen out ultraviolet radiation and permit colonization of the land.

The Protozoa

The **protozoa** are such a diverse, complex group that it is difficult to devise a suitable definition for them. Most protozoologists consider protozoa to be organisms consisting of one cell in which the organelles take the functional role of the organ systems seen in animals and plants. This definition of protozoa is very broad, and most protozoologists (as mentioned previously) would include the Euglenophyta, Pyrrophyta, and Gymnomycota (table 35.1) among the protozoa. Even though this is a quite legitimate viewpoint, the protozoa are defined in this text as heterotrophic single-celled organisms without fungus-like fruiting bodies. The group is very large, with over 65,000 known species and more being discovered regularly.

Protozoa are found in a wide variety of environments. The majority are free-living and inhabit freshwater or marine environments. There are, however, a number of terrestrial protozoa, and even beach sand has its own special protozoan population. There are also a number of parasitic protozoa. Some very widespread diseases such as malaria are caused by protozoa.

Despite their ability to flourish in a variety of environments, protozoa are limited by their need for a moist habitat. Cells in multicellular organisms have an advantage over protozoa in being bathed by body fluids held constant by homeostatic mechanisms. Protozoa do not have this protection and must deal with their environment directly. If survival becomes difficult, frequently the best they can do is form a dormant cyst and wait until more favorable conditions once again arise.

Protozoa are the generalists among cells. Unlike multicellular organisms, which have cells that can specialize and do not have to move or reproduce themselves, protozoa must be able to move, reproduce, and respond to their environment in sophisticated ways, as well as carry out the metabolic and physiological functions seen in all cells. They truly are complex, sophisticated organisms, even though they are only single cells.

All normal physiological functions are carried out by these single-celled organisms. Protozoa take in food by simple diffusion and by active transport, through pinocytosis, and by phagocytosis (chapter 10). Usually, food is digested in food vacuoles. Gas exchange occurs through the cell membrane. Freshwater protozoa eliminate excess water acquired from their hypotonic environment by use of contractile vacuoles (see pp. 364–65). Marine and parasitic protozoa usually do not have active contractile vacuoles. Protozoa eliminate nitrogen wastes by simple diffusion of ammonia through the plasma membrane to the outside. Movement of protozoa can be due to flagella, pseudopodia of various types, or cilia (see chapter 3). In fact, protozoa are capable of fairly sophisticated responses to environmental cues (p. 830). Protozoa also show considerable variety in reproduction. Asexual reproduction is the rule. However, many protozoa can also reproduce sexually during some part of their life cycles.

The protozoa are such a diverse assemblage that their taxonomy is exceptionally complex. The problem is made even more difficult by the fact that protozoan classification is constantly changing due to the rapid accumulation of new observations, particularly as a result of the use of the electron microscope. Protozoa are divided among seven different phyla in the latest classification. For the sake of simplicity, however, this text uses a more traditional system of four groups: Mastigophora, Sarcodina, Sporozoa, and Ciliophora (table 35.1). These four groups are distinguished from one another on the basis of their locomotor organelles and their modes of reproduction.

Mastigophora

Evolutionarily, mastigophores, or **zooflagellates,** as they are often called to distinguish these heterotrophs from the unicellular autotrophic green algae (**phytoflagellates**), are considered to be quite ancient. It is very likely that the other groups of protozoa arose from the flagellates.

Figure 35.3 (a) A stained blood smear from a patient suffering with African sleeping sickness showing trypanosomes among the blood cells. Note the darkly stained kinetoplast in each trypanosome and the white blood cell in the center of the picture. (Magnification × 1000) (b) Structure of the intermediate bloodstream form of *T. rhodesiense* as revealed by the electron microscope. The pellicle is a flexible, proteinaceous covering over the cell.

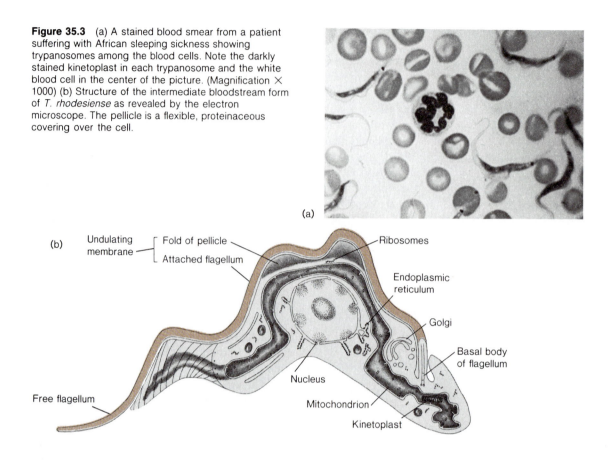

(a)

(b)

Undulating membrane — Fold of pellicle — Attached flagellum

Ribosomes

Endoplasmic reticulum

Golgi

Basal body of flagellum

Nucleus

Mitochondrion

Kinetoplast

Free flagellum

Zooflagellates are characterized by the presence of one to many flagella. Some also are capable of forming pseudopodia. The flagella have the characteristic 9 + 2 microtubular structure associated with eukaryotic cells. Most are uninucleate, but members of one major group, the kinetoplastids, possess an extranuclear source of DNA, a mitochondrial structure called the **kinetoplast** (figure 35.3). Asexual reproduction by way of longitudinal binary fission is most common, although sexual reproduction is known for a few multiflagellate forms, including certain symbionts *(Trichonympha)* in the digestive tract of termites. The majority of the zooflagellates are symbiotic and are given a great deal of attention because of their medical importance.

Many zooflagellates are important human parasites. **Parasitism** is defined as a symbiotic relationship in which the parasite derives benefit from the host at the host's expense. The parasite is dependent on the relationship, while the host might be far better off by the absence of the parasite. This is certainly true of the **trypanosomes,** for they cause severe human diseases, particularly in Africa (figure 35.3).

One major trypanosome pathogen *(Trypanosoma cruzi),* the causative agent of Chagas' disease, is found in South America. Charles Darwin may have acquired Chagas' disease when in South America as a crew member on the H.M.S. *Beagle.* He reported being bitten many times by bugs now known to transmit the disease organisms. Later, at home in England, he developed characteristic symptoms of that disease.

The majority of species of trypanosomes live in the blood of a variety of vertebrates, including humans. One of them, *Trypanosoma rhodesiense,* can cause African sleeping sickness. Developing first in the circulatory system, *T. rhodesiense* eventually invades the cerebrospinal fluid. The symptomatic sleeping sickness results as a function of the host's immune response. White blood cells accumulate around the outside of capillaries in the brain, a condition called perivascular cuffing. The presence of large numbers of these leukocytes decreases the lumen of the capillaries, reducing the blood supply to the brain. Decreased oxygen supply to the brain results in the lethargy symptomatic of *T. rhodesiense* infections. Eventually, the victim dies. A

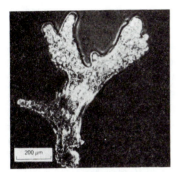

(a)

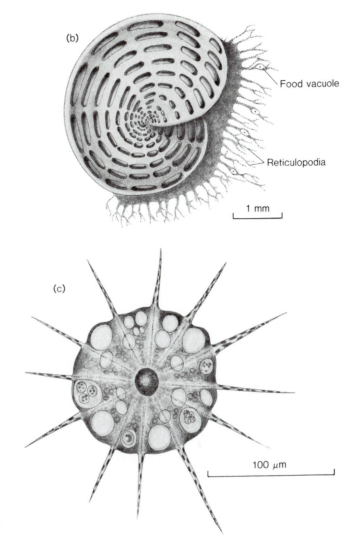

(b)

Food vacuole

Reticulopodia

1 mm

(c)

100 µm

Figure 35.4 Three different pseudopodial types. (a) *Amoeba proteus* with lobopodia. (b) A foraminiferan feeding with a reticulopodial net. (c) *Actinophrys sol* with axopodia.

related hemoflagellate, *T. brucei,* causes ngana in cattle, a disease that can inflict severe economic loss on farmers.

Both *T. rhodesiense* and *T. brucei* are transmitted by the bite of a species of the tse-tse fly *Glossina.* The tse-tse fly becomes infected by a diseased host when it takes a blood meal. The trypanosomes multiply and change into different life cycle forms within the fly's intestine; they then migrate to the fly's salivary glands, multiply further, and develop into infective forms. When the fly now feeds on a person, the trypanosomes are transmitted in its saliva. Sleeping sickness and ngana make large areas of Africa virtually uninhabitable. The only potentially effective control measure seems to be eradication of *Glossina.*

Sarcodina

Sarcodines are capable of movement and feeding by extension of pseudopodia; a few also can form flagella. Pseudopodial morphology is used to classify the various groups. Temporary pseudopodia having blunted and rounded tips and possessing a clear ectoplasmic anterior area are called **lobopodia** (figure 35.4). **Reticulopodia** are slender and filamentous in form, and usually fuse, forming a cytoplasmic net to enmesh food. **Axopodia** are supported by axial filaments of microtubules, about which the cytoplasm of the pseudopodium flows. Individual axopodia are very slender and may radiate from the surface of the sarcodine with great uniformity.

The majority of species are marine, comprising the major groups **radiolarians** and **foraminiferans.** Radiolarians are pelagic, floating on or near the surface of the open sea, while foraminiferans are most often found in the ooze of the ocean floor. Other species of sarcodines are freshwater amoebae living in the silt of streams, the ooze of ponds, moss and sphagnum, or in moist soil. Parasitic species may be found in the alimentary tracts and associated structures of vertebrates and invertebrates. A pathogenic species parasitizing humans, *Entamoeba histolytica,* causes amoebic dysentery.

Feeding is by phagocytosis or pinocytosis. Sarcodines have a varied diet, ranging from algae and bacteria to protozoa of all kinds. Sarcodines with lobopodia can simply surround and engulf their prey. Others with reticulopodia and axopodia can trap their prey with a sticky pseudopodial surface and then draw the victims inside.

One of the most common and popular sarcodines (at least in student labs) is *Amoeba proteus,* a large (up to 600μm) freshwater carnivore usually found in slow-moving or still-water ponds, often on the underside of decaying leaves (figure 35.5). It moves by extending lobopodia. Some amoebae are protected by a loose-fitting **test** or **shell.** For example, *Arcella* looks much like a cream-colored teacup turned upside down (figure 35.6). The amoeba extends its pseudopodia from the test aperture to feed or creep along.

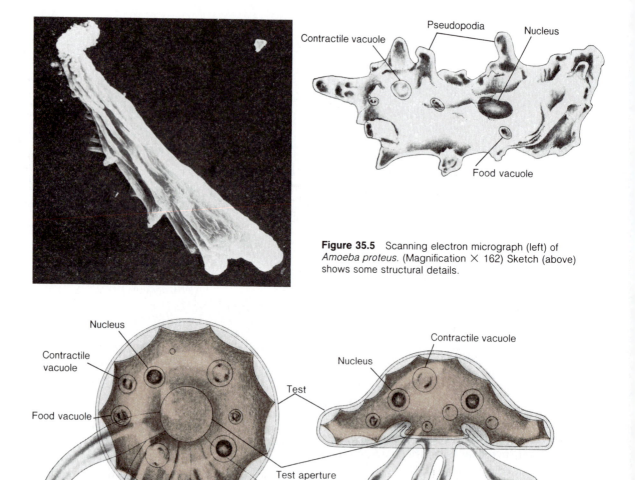

Figure 35.5   Scanning electron micrograph (left) of *Amoeba proteus.* (Magnification × 162) Sketch (above) shows some structural details.

Figure 35.6   Top and side views of *Arcella.*

Foraminiferans are important marine sarcodines that form calcium carbonate tests (figure 35.7). The organisms extend through holes in their tests, called **foramina,** and feed with a reticulopodial net. Upon death, these protozoa contribute to the mud on the ocean floor. They have been so populous through the ages that their bodies have formed such massive deposits as the White Cliffs of Dover and the limestone used to build the Egyptian pyramids. Radiolarians also contribute to the bottom ooze of the oceans. These marine forms float near the ocean surface despite the fact that they have internal skeletons composed of silica or strontium sulfate (figure 35.8). Their skeletons are intricate, exquisite, and of almost infinite variety. These skeletons sink to the bottom after the death of the cells.

Sporozoa

Sporozoans are parasitic protozoa with complex life cycles that almost always involve the formation of infective spores at some point. Some have been a scourge of humans and their domestic animals for longer than recorded history. Hippocrates recognized the relationship between swamps (where mosquitos breed) and the symptoms of the disease now known as malaria. Coccidiosis, a sporozoan disease of domestic animals, causes economic loss to farmers; and a related coccidian, *Toxoplasma gondii,* carried by cats, can be a dangerous parasite in humans, especially children, since it invades many tissues.

The most important human parasite among the sporozoa is *Plasmodium,* the causative agent

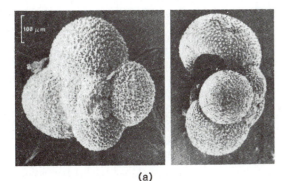

(a)   (b)

Figure 35.7 Scanning electron micrographs of two foraminiferan shells. (a) *Globigerina bulloides.* (b) *Elphidium crispum.*

Figure 35.8 Skeletons from a variety of radiolarians.

of **malaria.** The life cycle of *Plasmodium vivax,* one of the species responsible for this devastating disease, is shown in figure 35.9. The infection is initiated when a female *Anopheles* mosquito feeds. The mosquito injects a small amount of saliva containing an anticoagulant. If the mosquito is infected, small haploid **sporozoites** may be injected too. Initially, sporozoites become established in hepatic cells of the host and undergo multiple fission (schizogony), their products being called **merozoites.** This pre-erythrocytic phase within the host lasts about ten days. Merozoites next invade erythrocytes and repeat this asexual process. The erythrocytic phase is cyclic and repeats approximately every forty-eight hours in the most common form of malaria. Thus, every forty-eight hours a large number of the victim's erythrocytes simultaneously rupture and release merozoites that can invade uninfected erythrocytes. This sudden release of toxins and cell debris triggers an attack of the chills and fever characteristic of malaria. Occasionally, merozoites differentiate into **macrogametocytes** and **microgametocytes.** When taken up by a mosquito while feeding on the infected human host, the macrogametocytes and microgametocytes develop into female and male gametes. Fertilization results in a diploid zygote, the **ookinete.** The ookinete migrates to a position on the outside of the mosquito's gut wall and forms an **oocyst.** Once the zygote is established on the mosquito's gut wall, meiosis takes place, followed by the asexual formation of numerous sporozoites that migrate to the mosquito's salivary glands. This process is known as sporogony. The cycle is now complete, with only the bite of the infected mosquito needed to initiate a new infection in a human host.

Malaria is still a major killer of humankind, despite medical advances, and may actually be making a recovery. Over a thousand million people live in malarious areas, and millions are infected each year. Generally, efforts at control have involved reduction of the mosquito population through massive spraying campaigns. The current resurgence of malaria is due to a number of causes—an increase in insecticide prices, development of insecticide-resistant strains of mosquitos, and parasite resistance to antimalarial drugs. William Trager at Rockefeller University managed to cultivate malarial parasites in tissue culture. This remarkable achievement may make possible the production of vaccines capable of protecting humans. Effective protection against malaria would drastically reduce the level of human suffering, but at the same time probably would result in a further increase in the rate of population growth in areas of the world already suffering problems due to overpopulation.

Ciliophora

The ciliate protozoa are the most complex of the protozoan groups. They all possess cilia at some stage in their development. Each ciliate has two types of nuclei, a macronucleus and micronucleus. Sexual reproduction takes place through a unique process called conjugation. Asexual reproduction is by way of transverse binary fission, which may occur independently of conjugation.

Cilia function in locomotion, feeding, and attachment. They may be independent, or they may form more complex organelles, such as a **cirrus** (a tuft of fused cilia) or blocklike clumps of cilia (**membranelles**) that are in or near the oral

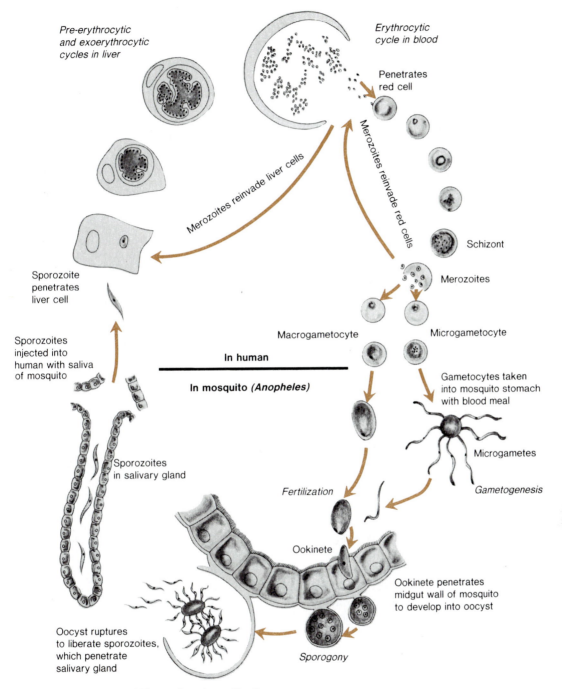

Figure 35.9 Life cycle of *Plasmodium vivax* with all stages to approximately the same scale.

Pre-erythrocytic and exoerythrocytic cycles in liver

Erythrocytic cycle in blood

Penetrates red cell

Merozoites reinvade liver cells

Merozoites reinvade red cells

Schizont

Merozoites

Sporozoite penetrates liver cell

Macrogametocyte

Microgametocyte

In human

Sporozoites injected into human with saliva of mosquito

In mosquito (Anopheles)

Gametocytes taken into mosquito stomach with blood meal

Microgametes

Sporozoites in salivary gland

Gametogenesis

Fertilization

Ookinete

Ookinete penetrates midgut wall of mosquito to develop into oocyst

Oocyst ruptures to liberate sporozoites, which penetrate salivary gland

Sporogony

cavity of many ciliates and aid in feeding. The beetle-shaped *Euplotes,* for example, crawls about its environment using cirri as "legs" (figure 35.10). Cilia are usually arranged in rows, each cilium being derived from a basal granule or **kinetosome,** which is structurally the same as a centriole. The linear orientation of these kinetosomes and associated fibrils is known as the **infraciliature** and is demonstrable even in those few forms in which cilia are absent. Although the functional role of the components in the infraciliature is not clear, ciliates are capable of coordinated movement and complex behavior (see chapter 30).

A description of ciliate morphology should include the oral system, since it is a most conspicuous feature of many of these protozoans. A **cytostome** (cell mouth) used in feeding is often highly modified by the presence of ciliary membranelles. A cell anus (**cytopyge** or **cytoproct**) is also present, usually in a fixed position. Protozoan digestion is a highly organized process, as can be seen in *Paramecium.* In *Paramecium,* food is sorted out and swept to the cytostome by cilia in the **oral groove,** and food vacuoles are formed (figure 35.11). Shortly after the vacuoles arise, lysosomes fuse with the vacuoles and empty digestive enzymes into them. The vacuoles also quickly become acid (pH values as low as 1.4 to 3.4 may be reached). Later, after the prey has been killed and digested, the vacuoles turn alkaline, and soluble nutrients are absorbed. Indigestible residue is eliminated at the cytoproct.

Ciliates are unique in having two types of nuclei. The **macronucleus** is polyploid (it may have even thousands of gene copies in some ciliates!) and is responsible for regulating normal physiological and metabolic functions. The **micronucleus** is diploid and serves as a genetic repository. It is active during reproduction and can give rise to macronuclei. The numbers of each nuclear type may vary from species to species. For example, *Stentor* has a chainlike macronucleus with over twenty macronuclear lobes and as many as fourteen micronuclei (see figure 35.12b).

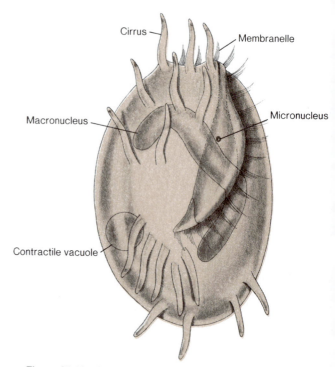

Figure 35.10 A ventral view of *Euplotes patella* showing its cirri and membranelles.

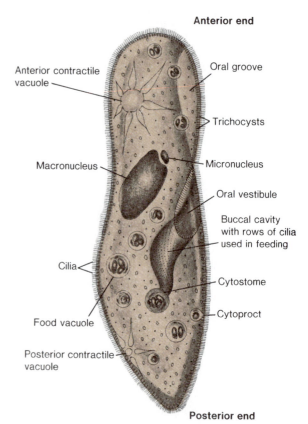

Figure 35.11 The principal structures seen in *Paramecium caudatum* with the light microscope.

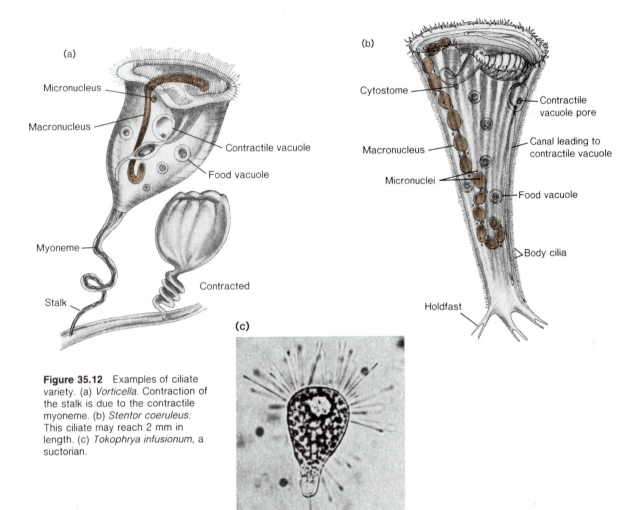

Figure 35.12 Examples of ciliate variety. (a) *Vorticella*. Contraction of the stalk is due to the contractile myoneme. (b) *Stentor coeruleus*. This ciliate may reach 2 mm in length. (c) *Tokophrya infusionum*, a suctorian.

Sexual reproduction is through a process called **conjugation.** Two organisms of the proper mating types join together at their oral regions and temporarily fuse. The micronuclei divide meiotically to produce haploid nuclei, while the macronuclei usually degenerate. Next, all micronuclei but two in each partner are destroyed. One of the two remaining micronuclei in each partner crosses over to the other ciliate and fuses with that ciliate's stationary nucleus (the female nucleus) to create a new recombinant diploid micronucleus. As a result of this exchange, each partner receives a set of chromosomes from the other. The two organisms then separate. Their diploid micronuclei divide, and at least one micronucleus in each organism develops into a new macronucleus.

The complexity, beauty, and variety of the ciliates are almost overwhelming. A colony of *Vorticella* looks much like a bouquet of beautiful flowers until they are disturbed and retract into a compact cluster (figure 35.12a). *Stentor coeruleus,* a favorite subject for research in regeneration, resembles a giant blue vase decorated with stripes when viewed under the microscope (figure 35.12b). The barrel-shaped *Didinium* (see chapter 10) can gobble up a much larger *Paramecium* much like a snake swallowing a rabbit. Suctorians have even stranger feeding habits. *Tokophrya* is covered with tentacles and rests quietly on its stalk until a hapless victim blunders into it and is promptly paralyzed. *Tokophrya* then uses its tentacles like straws and sucks its prey dry (figure 35.12c).

Unicellular Algae

The unicellular algae consist of three major groups: the Euglenophyta, Pyrrophyta, and Chrysophyta. These groups are all photosynthetic or are closely related to photosynthetic species.

Euglenophyta

The euglenoids are a small group of unicellular flagellates that are either autotrophic—with chlorophylls a and b in their chloroplasts—or closely related to the photosynthetic forms. Commonly found in fresh as well as brackish water, these organisms are usually biflagellate, one flagellum being shorter than the other (figure 35.13). Even in uniflagellate forms, there are always two basal granules. Euglenoids lack a cellulose cell wall, but are bounded by a flexible **pellicle** composed of proteinaceous strips joined together. A **reservoir** or **gullet** contains the base of the flagellum. A water expulsion vesicle is in the adjacent cytoplasm and empties into the reservoir. A **stigma** formed from granules of a red pigment is also usually located near the base of the flagellum. Carbohydrate is stored in the form of **paramylum,** a glucose polymer somewhat different from starch and glycogen. A special structure within the chloroplast, the pyrenoid, is involved in paramylum synthesis.

Reproduction is usually asexual by means of longitudinal binary fission, and it occurs only during darkness. The nucleus behaves unusually during mitosis since neither the prominent nucleolus nor the nuclear envelope disappear. The chromosomes remain condensed during interphase.

As might be expected from the fact that they are photosynthetic, most euglenoids are attracted by light as long as it is not too bright. A light-sensitive flagellar swelling lies at the base of one flagellum. The stigma shades this photoreceptor

Figure 35.13 *Euglena spirogyra.*

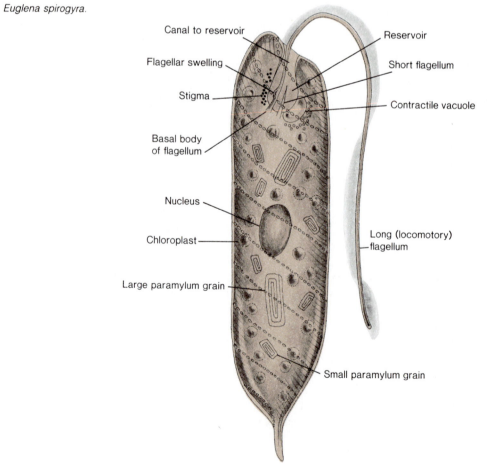

when it comes between the flagellum and a light source. This allows the organism to judge the direction from which light is coming and respond accordingly.

These flagellates are very flexible nutritionally. Most are photosynthetic but do require one or two vitamins. They grow in the dark, however, if provided an adequate supply of nutrients. It is possible to destroy the chloroplasts by heating *Euglena* at 35°C for a few generations or treating it with streptomycin. *Euglena* is permanently bleached by these treatments and behaves like a typical heterotroph.

Pyrrophyta

The Pyrrophyta are mostly **dinoflagellates.** Dinoflagellates are among the most numerous of marine organisms, making up a significant portion of the ocean plankton, and are among the most important primary producers of the food supply of the sea. Only the diatoms surpass them

in this regard. Approximately 1,000 species of dinoflagellates are known.

The dinoflagellate cell can be either naked (unarmored) or covered by a cellulose envelope (**theca**). The theca may be smooth or divided into thecal plates. Thecal plates are formed by the deposition of cellulose within a membranous covering. Dinoflagellates possess two flagella. One flagellum circles the cell, while the other extends posteriorly. The circling or **transverse flagellum** is located within a transverse groove (the **annulus**), while the **longitudinal flagellum** originates from a posterior groove (the **sulcus**) (figure 35.14).

Most species of dinoflagellates are autotrophic, with chlorophylls a and c in their yellowish or brownish chloroplasts. Their reserve material is either starch or oil. A few dinoflagellates do ingest particulate material. *Ceratium* (figure 35.15) actually extends pseudopods between its thecal plates to capture food.

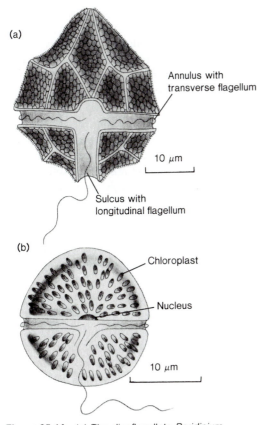

Figure 35.14 (a) The dinoflagellate *Peridinium tabulatum,* which has prominent thecal plates. (b) The unarmored dinoflagellate *Gymnodinium neglectum.* Note the lack of thecal plates.

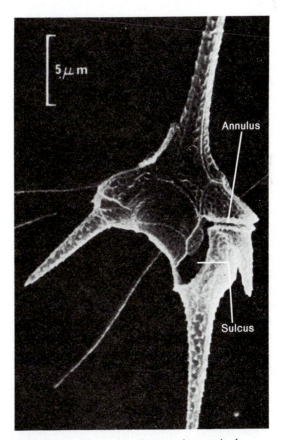

Figure 35.15 A scanning electron micrograph of *Ceratium hirudinella.* Note the thecal plates, the annulus or equatorial girdle, and the large sulcus where food is ingested.

The dinoflagellate nucleus is unique and thought to be quite primitive. The chromosomes resemble prokaryotic genetic material because they lack histones. They also lack centromeres and remain permanently condensed. The nuclear envelope and nucleolus remain intact during mitosis, and a spindle is not formed. Dinoflagellates reproduce by binary fission. In armored species, the daughter cells each receive a share of the thecal plates of the original cell and produce new plates to complete a set of plates on each daughter cell.

Dinoflagellates can multiply rapidly under appropriate conditions, and enormous numbers can be generated in a short period of time. Certain pigmented species can discolor the water, leading to a condition called the **red tide.** These bursts of reproductive activity, or **blooms,** can have an economic impact on the region in which they occur. Toxic by-products can affect aquatic organisms, causing widespread fish kills. The same toxicity causes a condition known as paralytic shellfish poisoning (PSP). PSP is widespread throughout the world and occurs during the months of May through October. Various species of *Gonyaulax* and *Gymnodinium* are the primary culprits. Their metabolic by-products accumulate in the tissues of shellfish. Their toxin—saxitoxin—is a nitrogenous substance capable of paralyzing striated muscles by inhibiting the movement of sodium ions.

However, dinoflagellates are more important for their beneficial effects. In addition to being major primary producers in the oceans, they are also among the most successful symbionts in the living world. They live within a wide variety of marine protozoa and in some invertebrates. The host provides protection, carbon dioxide, ammonia, and salts of various kinds. The dinoflagellates (or **zooxanthellae,** as they are often called when symbionts) furnish their hosts with photosynthetic products, such as sugars and amino acids. For example, coral usually contains large numbers of dinoflagellate symbionts and grows ten times faster in the light when photosynthesis is active than in the dark.

Chrysophyta

Chrysophyta are yellow-green to golden-brown algae that have chlorophylls a and c and store their excess carbon in the form of the glucose

Figure 35.16 A scanning electron micrograph of the diatom *Triceratium* showing wall perforations in its frustule.

polymer **chrysolaminarin (leucosin)** or oils. The color of these algae is usually due to the accessory pigment fucoxanthin in their chloroplasts. The most important group of organisms in this division are the **diatoms.** This is such a large and important group that it is often given phylum rank and called the Bacillariophyta.

Diatoms are intriguing silica-clad organisms found in both fresh water and marine habitats. Their remains accumulate as sediment on ocean floors. This **diatomaceous earth,** as it is called, is mined and used commercially in the making of insulation, as an abrasive in toothpaste and silver polish, and in water-filtering systems. From an ecological standpoint, diatoms are considered to be the most important primary producers in the marine ecosystem because of their enormous numbers. As such, they represent an important source of food for many animals.

The diatoms possess cell walls consisting of two **valves (frustules)** in which polymerized silica is embedded. The valves are of unequal size, the larger fitting over the smaller like the lid of a box, and are beautifully decorated with delicate, complex designs (figure 35.16). The region of overlap, called the **girdle,** is less siliceous than the rest of the valve, permitting some flexibility. The organism is typically uninucleate and differs from many other unicellular algae in being diploid. The adult cells lack flagella but can still glide along surfaces by means that are not yet understood.

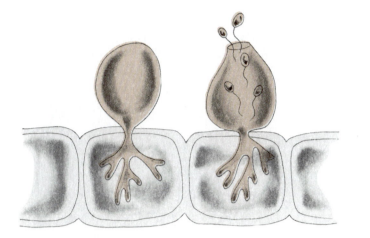

Figure 35.17 The thallus of a typical chytrid parasite. Rhizoids penetrate the host cells. The chytrid on the right is releasing zoospores that will eventually develop into new adults.

Diatom reproduction is both asexual and sexual. In asexual reproduction, each daughter cell receives a frustule after mitosis and manufactures a new valve to fit inside it. Thus, one daughter is slightly smaller than the parent, and the population slowly decreases in size. Eventually, the diatoms carry out sexual reproduction and return to their original maximum size as a result.

Funguslike Protists

A number of organisms, traditionally classified as fungi, can be placed with the protists because of their simple organization. In fact, the slime molds might even be considered protozoa with some justification. At least three protist groups can be classified as funguslike: the chytrids (Chytridiomycota), the water molds (Oomycota), and the slime molds (Gymnomycota).

Chytridiomycota

The chytrids are normally aquatic, although some live in moist soil. Many have very simple morphology, often consisting of a spherical cell penetrating a host with colorless, rootlike **rhizoids** (figure 35.17). Their cell walls contain chitin. All produce motile cells with a single, posterior flagellum.

Many of the aquatic chytrids are parasitic on algae. Some parasitize aquatic plants and animals. Other species are saprophytic and grow on decaying plant and animal remains. Species of *Allomyces* and *Blastocladiella* are valuable research organisms in the study of morphogenesis. On the whole, chytrids seem to have little direct effect on human welfare.

Oomycota

The water molds are very common aquatic forms and are often found as cottony masses on sick or dead insects and fish. They are also widespread in the soil. Their structure usually consists of a network of branching tubes called **hyphae** (singular: **hypha**). The whole mass of hyphae, the **mycelium,** lacks cross walls dividing hyphae into individual cells; that is, the mycelium is **syncytial** (also called **coenocytic**). The cell walls contain cellulose and other glucose polymers. Water molds produce biflagellated zoospores at some point in their life cycle.

Most members of the Oomycota are harmless saprophytes. That is, they live on decaying organic matter. However, several parasitic forms are of great economic importance, and one has shaped history. *Phytophthora infestans* causes potato blight and was responsible for the 1840s potato famine in Ireland. The entire potato crop was destroyed in one week during the summer of 1846. The mass migration of the Irish to the United States before the turn of the century was at least partially due to this organism. Interest in protists as disease agents was also stimulated by the loss of French wine grapes to *Plasmopara viticola,* the causative agent of grape downy mildew. This disaster led to the development of the first fungicide, the Bordeaux mixture, a combination of copper sulfate and lime. *Saprolegnia,* another water mold, can parasitize fish. Mycelia begin to form over fins and eventually can cover an entire fish, paralyzing and killing it.

Gymnomycota

The slime molds are an unusual and varied collection of amoeboid organisms that possess both plant and animal characteristics. One of the two best-known slime mold groups, the **cellular slime molds** (Arasiomycetes), has been very intensely studied by developmental biologists and is discussed at length in the chapter on differentiation (chapter 25). Therefore, attention here is focused on the other major group, the plasmodial slime molds.

There are 450 to 500 species of **plasmodial slime molds** or **true slime molds** (Myxomycetes), and they are found worldwide in moist, dark areas, such as under the bark of decaying trees or in layers of decomposing leaves. The active, vegetative form of the organism is a **plasmodium** that moves over surfaces, feeding on bacteria, protozoa, and other organisms (figure 35.18). As it grows, its diploid nuclei synchronously undergo mitosis. Plasmodial contents are constantly mixed by rapid cytoplasmic streaming. Although a number of plasmodia are small, many can become fairly large (30 cm or more across) and quite colorful. The sudden appearance of a large, bright yellow plasmodium on someone's lawn has more than once caused a panic in a neighborhood or community.

Figure 35.18 Plasmodium of the slime mold *Physarum.*

Figure 35.19 Life cycle of the plasmodial slime molds.

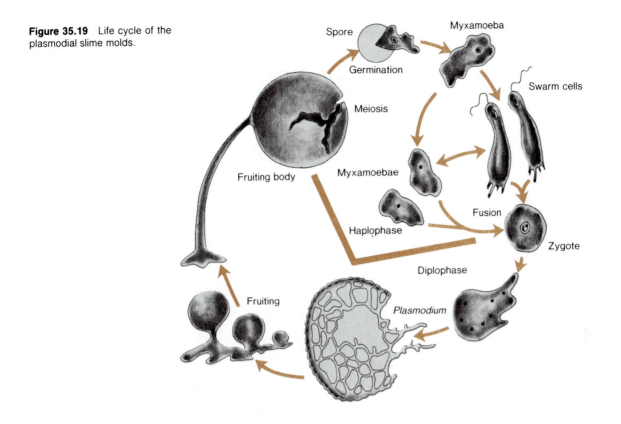

When the plasmodium has matured, it creeps out into a lighted area (light is often required for fruiting) and forms **fruiting bodies** (figure 35.19). Meiosis takes place as the spores develop within the **sporangium** (the part of the fruiting body that contains the spores) so that the mature spores are haploid. The spores can survive for years under unfavorable environmental conditions and then germinate in the presence of moisture to release **myxamoebae** or flagellated **swarm cells.** These haploid forms feed and reproduce. Eventually, they fuse to form a diploid zygote. The zygote then feeds, grows, and multiplies

its nuclei through synchronous mitotic divisions. Finally, a mature plasmodium develops, and the life cycle has come full turn.

Although plasmodial slime molds have little direct economic importance, they are interesting in their own right. Slime mold fruiting bodies, though only a few millimeters tall, are beautiful and essential in slime mold identification (figure 35.20). The slime mold *Physarum polycephalum* and its close relatives can be cultured in the laboratory and are proving very useful for research in cell physiology and molecular biology.

Figure 35.20 Plasmodial slime mold sporangia. (a) *Trichia.* (b) *Hemitrichia.* (c) *Stemonitis.*

(a)

(b)

(c)

The Fungi

Long grouped with plants, the fungi are now recognized as a unique group and placed in a separate kingdom. The majority are nonmotile, coenocytic organisms composed of masses of tubes or filaments, the hyphae. Collectively, the hyphae of a single organism are called a mycelium. Hyphal walls are semirigid and usually formed from chitin. Sometimes, hyphae lack cross walls or septa except where a reproductive organ is formed. Even the more advanced fungi have septa with pores or perforations so that the hyphal cells are not completely separated from one another.

Fungi are heterotrophic, but since hyphae have rigid walls, fungi cannot actively engulf particulate food by phagocytosis. They must rely on dissolved inorganic and organic food materials. Thus, they secrete digestive enzymes and then absorb the soluble digestion products. Because of this nutritional strategy, fungi are particularly important as decomposers, aiding in decomposition of dead matter and the subsequent recycling of inorganic and organic molecules in an ecosystem.

Reproduction can be either sexual or asexual. Although asexual reproduction can result from fragmentation of hyphae, asexual spores are frequently formed. Asexual spores of some of the lower fungi develop on specialized hyphae, **sporangiophores,** within a saclike structure, the sporangium. In the more complex fungi, spores develop in the tip of specialized hyphae, **conidiophores,** and are referred to as **conidia** (from the Greek, meaning "dust").

Sexual reproduction is accomplished in several ways, and the pattern of sexual reproduction differs among the major groups of fungi.

Although fungal taxonomy is not in a settled state, at least four divisions or subdivisions of the true fungi or Eumycota are usually recognized: the Zygomycota, Ascomycota, Basidiomycota, and Deuteromycota.

Zygomycota

Members of the Zygomycota division are widespread and include some of the familiar molds often found growing on foodstuffs such as bread and fruit. They are terrestrial and lack flagellated spores or gametes. Sexual reproduction involves the fusion of gametangia with subsequent formation of a resistant **zygospore.**

A commonly encountered member of the division is *Rhizopus stolonifer,* the black bread mold, so-called because mature sporangia turn black. The life history of *R. stolonifer* begins when a spore germinates on bread (figure 35.21). Hyphae, called **stolons,** extend laterally over the surface of the bread, growing with amazing rapidity. Specialized hyphae called **rhizoids** extend from the stolons. Rhizoids penetrate the bread and act in a dual capacity: they serve as anchors, and at the same time secrete digestive enzymes and absorb organic by-products. Additional hyphae rise into the air as sporangiophores, supporting sporangia. Mature sporangia eventually burst, releasing spores that are dispersed by air currents.

Sexual reproduction involves specialized gametangia that are indistinguishable as to sex. When hyphae of different mating strains (usually referred to as plus and minus) approach each other, swellings called **progametangia** form, come into contact, and develop into **gametangia** by formation of septa. The walls between the gametangia dissolve, and the nuclei fuse to form diploid zygote nuclei. The zygote wall then thickens and blackens. The resulting **zygospore** remains dormant for several months. Meiosis takes place upon germination. One or more haploid sporangia are immediately produced and release spores that germinate after dissemination. The adult fungus in this life cycle is haploid; only the zygospore is diploid.

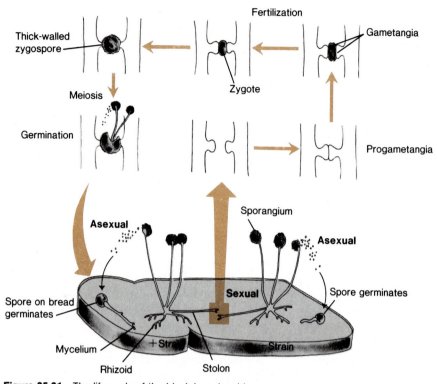

Figure 35.21 The life cycle of the black bread mold, *Rhizopus stolonifer*. Both asexual and sexual processes are illustrated.

Ascomycota

The Ascomycota is the largest and most varied of fungal divisions and includes the unicelled yeasts, as well as multicellular forms with extensive mycelia. Filamentous sac fungi, as the Ascomycota are sometimes called, have septate, chitinous hyphae.

Asexual reproduction usually occurs by way of hyphal budding. The tips of specialized hyphae are pinched off to form chains of spores called conidia. Fragmentation is also possible in this group: new individuals can arise from bits of hyphae.

Sexual reproduction in the Ascomycota is distinctive and characteristic for the group. It always involves the formation of an **ascus,** a saclike structure containing **ascospores** produced by **karyogamy** (fusion of two nuclei) and meiosis. Most often, an ascus contains eight ascospores, although the number can vary with species. The life cycle of a typical filamentous Ascomycota is shown in figure 35.22. The male organ (antheridium) joins with the female gametangium (the ascogonium) of a compatible mating type. After fusion, the two types of nuclei pair up and disperse into separate hyphal cells. Since each of these cells contains two nuclei, one from each mating type, they are referred to as **dikaryotic.**

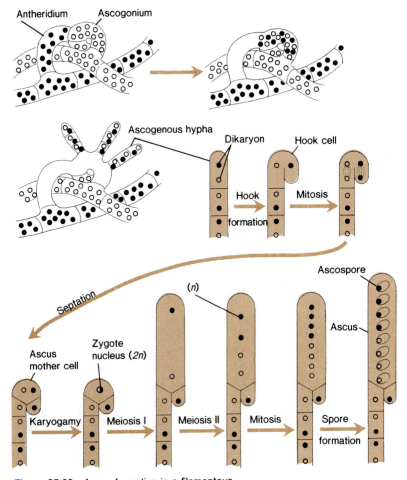

Figure 35.22 Ascus formation in a filamentous Ascomycota. The nuclei of the two mating types are represented by filled and unfilled circles.

The dikaryotic phase is very short-lived. Soon the nuclei in the tip cell fuse to form a diploid zygote nucleus, followed immediately by meiosis and mitosis. During this later phase, the ascus mother cell expands to form an ascus with eight haploid ascospores. The asci of many Ascomycota are contained within a large fruiting structure called an **ascocarp.**

Single-celled Ascomycota are called **yeasts.** They may reproduce asexually by budding or sexually when a yeast cell develops into an ascus.

Yeasts are of great economic importance. They tend to be found in environments with high sugar content, such as flowers and fruit. Wild yeasts associated with grapes, for example, are important in wine making. The yeasts ferment sugars to ethanol and CO_2. Baker's yeast, *Saccharomyces cerevisiae,* is responsible for the rising of dough during baking as a result of CO_2 production.

Certain of the Ascomycota (for example, morels and truffles) are edible and highly prized as a food delicacy. Morels develop large, visible, fleshy ascocarps (figure 35.23.) Truffles, in contrast, form subterranean ascocarps that must be dug from the ground.

Figure 35.23 *Morchella esculenta,* a common, edible morel that fruits in the spring.

Not all Ascomycota are of benefit to humans. A number are plant parasites and responsible for a variety of serious diseases, such as apple scab, powdery mildews, and Dutch elm disease. Possibly the most well known of these parasites is *Claviceps purpurea,* the causative agent of ergot disease in rye plants. Although not particularly harmful to rye, it causes severe illness in domestic cattle and humans who consume infected grain. The condition is called ergotism and is accompanied by muscular spasms, paralysis, and convulsions. Extracts of ergot are of medical importance because they are being used to stimulate certain muscle contractions and also as blood vessel constrictors. They are also used in the synthesis of the psychedelic drug LSD (lysergic acid diethylamide).

Basidiomycota

If anything, the Basidiomycota or club fungi are even more familiar than the yeasts and molds. Basidiomycota fruiting bodies are often seen thrusting through the surface of a yard or decorating the leafy floor of a forest. The Basidiomycota division is a very large one with a varied and interesting membership, including mushrooms, puffballs, shelf or bracket fungi, rusts, and smuts. Many of the fruiting bodies or **basidiocarps** are quite beautiful (figure 35.24). They are distinguished by the formation during sexual reproduction of a swollen cell, the **basidium,** at the tip of hyphae.

(a)

(b)

(c)

(d)

Figure 35.24 Representative Basidiomycota.
(a) *Pleurotus ostreatus*, a common, gilled mushroom with
a 2 to 30 cm cap that grows on hardwoods and
conifers. (b) *Polyporus sulphureus*, a bracket fungus with
large caps (5 to 25 cm or more) that can cause rot in
trees. (c) *Cantharellus cinnabarinus*, a funnel-shaped, red
chanterelle mushroom with exposed gills. (d) *Calvatia
gigantea*, one of the largest puffballs, 20 to 50 cm
across.

The life cycle of mushrooms is outlined in figure 35.25. A **basidiospore,** under suitable conditions, germinates to produce a **monokaryotic mycelium** (one with a single nucleus in each cell). This mycelium grows and spreads within the soil. When it meets another monokaryotic mycelium of a different mating type, the two may join to initiate a new **dikaryotic mycelium,** every cell of which contains two nuclei, one of each mating type. A mushroom dikaryotic mycelium can sometimes grow outward for hundreds of years in an ever-expanding ring (the older mycelium in the center of the circle ages and dies even though the outer ring is still flourishing).

Eventually, the mycelium is stimulated to form basidiocarps. First, a solid mass of hyphae called a **button** forms. This then pushes through the surface and develops into a basidiocarp. The mushroom cap supports a large number of plate-like **gills.** The surface of these gills is coated with basidia. The two nuclei in the tip cell of each basidium fuse to form a diploid zygote nucleus that immediately undergoes meiosis to form four haploid nuclei. These nuclei subsequently push their way through tiny projections (the sterigma) on the basidium and into the maturing basidiospores. A large mushroom can produce and release many million basidiospores, which are then dispersed by the wind.

Many mushrooms are considered gourmet delicacies and have been prized since the time of the Roman Empire. The best-known edible mushroom is *Agaricus campestris,* one of the few gilled mushrooms that can be cultivated commercially (figure 35.26). About 65,000 tons of mushrooms are produced annually in this country. Unfortunately, some mushrooms (particularly members of the genus *Amanita*) are extremely poisonous and have killed unwary mushroom hunters.

Some Basidiomycota, the rusts and smuts, are virulent plant pathogens and inflict extensive damage on agricultural crops. The wheat rust, *Puccinia graminis,* has been a severe problem for farmers. New rust-resistant strains of wheat must be bred continually to block or minimize outbreaks of new *Puccinia* strains.

Deuteromycota

Much of the taxonomy of the fungi is based on patterns of sexual reproduction. When a fungus species lacks sexual reproduction or has not yet been observed reproducing sexually, it is placed in the Deuteromycota or Fungi Imperfecti (only "perfect" fungi have sexual reproduction) division. The Deuteromycota is a large, heterogeneous group. Its members usually appear related to the Ascomycota or Basidiomycota and are transferred to the appropriate division if their mode of sexual reproduction is established.

Many economically important fungi are found in this division. *Penicillium* molds are used in penicillin production and in the cheese industry, where they give cheeses like Roquefort and Gorgonzola their distinctive flavors. *Aspergillus* is employed by industry in the manufacture of citric and gluconic acids.

Other imperfect fungi are important disease agents. *Candida albicans* grows on mucous membranes of the mouth and throat, causing the disease called "thrush." The skin can be affected by fungi causing ringworm and the related "athlete's foot." Some species of *Aspergillus* produce **mycotoxins** when growing in stored foodstuffs. These mycotoxins are quite poisonous and render the food unusable.

Lichens

Lichens are a classic example of a symbiotic association. They are a combination of a green alga or blue-green alga (cyanobacterium) with a fungus, usually a member of the Ascomycota, although some Basidiomycota and Deuteromycota can be lichenized. The photosynthetic partner supplies the food, while the fungus protects the alga and absorbs water and minerals for both. If the two symbionts are isolated in the laboratory, both grow satisfactorily. However, the fungus does not seem able to survive independently in nature, and so lichens are identified according to the fungal partner present. There are around 20,000 known species of lichens, many with interesting shapes and beautiful coloring from fungal pigments (figure 35.27).

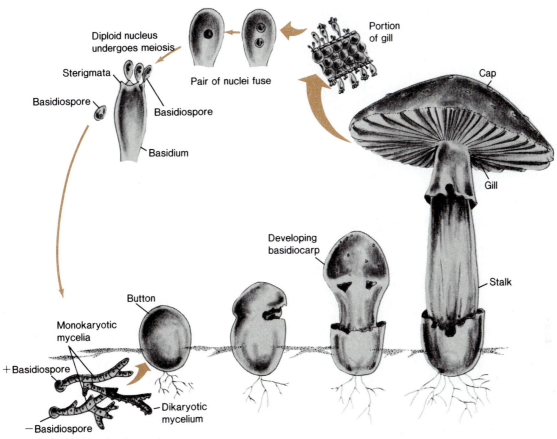

Diploid nucleus
undergoes meiosis

Portion
of gill

Pair of nuclei fuse

Sterigmata

Basidiospore

Basidiospore

Basidium

Cap

Gill

Developing
basidiocarp

Stalk

Button

Monokaryotic
mycelia

+Basidiospore

Dikaryotic
mycelium

−Basidiospore

Figure 35.25   Life cycle of a mushroom.

Figure 35.26   *Agaricus campestris*, the meadow
mushroom. Common edible mushroom often found
growing on lawns.

(a)

(b)

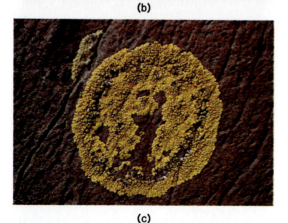

(c)

Figure 35.27 Typical lichens of the three major types. (a) A foliose lichen (leaflike) growing on a log. (b) A fruticose lichen (shrubby or hairlike). (c) Crustose (crustlike) lichens growing on a rock.

Lichens grow very slowly (a centimeter or less per year) but are remarkably resistant to environmental extremes. They are found on solid objects like rocks and trees in virtually all environments—from the desert to the arctic. They simply remain dormant when conditions become too adverse. Lichens produce acids while growing and thereby aid in breaking down rock in the initial stages of soil formation. They are used as food by many animals, including large mammals like reindeer. Interestingly, they are very sensitive to air pollution and are being employed as pollution indicators in some places.

Summary

The protists are unicellular and colonial eukaryotes. A few simple multicellular forms are included in this kingdom by some taxonomists. The fungi are eukaryotic multicellular heterotrophs.

The eukaryotic cell is thought to have arisen from the prokaryotes around 1,500 million years ago. The majority of biologists believe that eukaryotes arose through endosymbiosis when anaerobic prokaryotes ingested aerobic bacteria and photosynthetic bacteria. These bacteria survived and eventually evolved into present-day mitochondria and chloroplasts. Other biologists believe that plasma membranes invaginated to enclose genetic material and form the nucleus, mitochondria, and chloroplasts.

The protists encompass three major varieties of organisms—protozoa, unicellular algae, and funguslike protists. The protozoa are single-celled or colonial heterotrophs without funguslike fruiting bodies. This collection of organisms is very large and can be divided into four groups—Mastigophora, Sarcodina, Sporozoa, and Ciliophora—on the basis of locomotor organelles and reproductive mechanisms. Protozoa are found in almost all environments, and a number of them are serious parasites, causing diseases like African sleeping sickness and malaria.

The unicellular algae consist of three major groups: the Euglenophyta, Pyrrophyta, and Chrysophyta. They are all photosynthetic or closely related to photosynthetic species. The dinoflagellates and diatoms are particulary significant since these two groups are the most important primary producers in the oceans. Dinoflagellates are also responsible for toxic red tides that result in large fish kills.

The chytrids (Chytridiomycota), water molds (Oomycota), and slime molds (Gymnomycota) bear some resemblance to the fungi but may be considered protists. Most are harmless decay organisms, but a few cause serious diseases (for example, the Irish potato blight).

Fungi are multicellular heterotrophic eukaryotes that secrete digestive enzymes and absorb the resulting soluble nutrients. As such, they act as important decomposers in the ecosystem. There are four major divisions of the fungi: the Zygomycota, Ascomycota, Basidiomycota, and Deuteromycota. These are distinguished primarily on the basis of their mechanisms of sexual reproduction.

Fungi can cause diseases in both plants and animals. Many fungi produce toxic compounds and are poisonous if eaten. In spite of this, some fruiting bodies are considered gourmet delicacies. Yeasts are employed in baking and the production of alcohol, while filamentous fungi are used to manufacture antibiotics, cheese, and other products.

Lichens are the result of an association between fungi and green algae or cyanobacteria. They serve as a food source for animals living in barren regions and initiate the colonization of rocky, sterile areas.

Questions

1. Briefly describe the endosymbiotic theory for the origin of eukaryotic cells. What alternate hypothesis has been proposed?
2. What are protozoa? Describe the major characteristics of the Mastigophora, Sarcodina, Sporozoa, and Ciliophora.
3. Briefly describe African sleeping sickness and malaria in terms of the nature of the pathogen, the way in which the disease is caused, and the route of transmission.
4. What is a pseudopodium? Characterize the different types of pseudopodia. What are foraminiferans and radiolarians?
5. How do the Ciliophora catch and digest their prey?
6. Describe the nuclear types seen in the Ciliophora. How do the ciliates reproduce sexually?
7. Characterize the three major divisions of the unicellular algae with respect to morphology, reproduction, ecology, and importance.
8. Describe and contrast the Chytridiomycota, Oomycota, and Gymnomycota.
9. What are the fungi? What are the most important properties of the four major divisions of fungi?
10. Describe the Zygomycota, Ascomycota, and Basidiomycota life cycles.
11. What are lichens? How does each member of this symbiotic association contribute to the welfare of the whole?

Suggested Readings

Books

Alexopoulos, C. J., and Mims, C. W. 1979. *Introductory mycology*. 3d ed. New York: Wiley.

Bold, H. C., and Wynne, M. J. 1978. *Introduction to the algae: Structure and reproduction*. Englewood Cliffs, N.J.: Prentice-Hall.

Curtis, H. 1968. *The marvelous animals*. Garden City, N.Y.: Natural History Press.

Noble, E. R., and Noble, G. A. 1976. *Parasitology: The biology of animal parasites*. 4th ed. Philadelphia: Lea and Febiger.

Sleigh, M. A. 1973. *The biology of protozoa*. London: Arnold.

Trainor, F. R. 1978. *Introductory phycology*. New York: Wiley.

Articles

Amadjian, V. 1982. The nature of lichens. *Natural History* 91(3):30.

Dickerson, R. E. September 1981. Chemical evolution and the origin of life. *Scientific American* (offprint 1401).

Leedale, G. F. 1971. The euglenoids. *Carolina Biology Readers* no. 5. Burlington, N.C.: Carolina Biological Supply Co.

Margulis, L. August 1971. Symbiosis and evolution. *Scientific American* (offprint 1230).

Smith, D. C. 1973. The lichen symbiosis. *Carolina Biology Readers* no. 42. Burlington, N.C.: Carolina Biological Supply Co.

Trager, W., and Jenson, J. B. 1978. Cultivation of malarial parasites. *Nature* 273:621.

Survey of Plants

The plant kingdom includes a diverse collection of multicellular photosynthesizing organisms. Most terrestrial (land-dwelling) members of the kingdom are vascular plants with highly specialized body parts: roots, stems, and leaves. Other plants are simpler. For example, **algae** are photosynthesizing organisms, living in aquatic environments, that have relatively unspecialized structure. Algae are difficult to fit neatly into the five-kingdom scheme. Several groups of *unicellular* algae are included in the kingdom Protista in chapter 35, and now the green algae, the brown algae, and the red algae, which, for the most part, are *multicellular* organisms, are included in the kingdom Plantae in this chapter (figure 36.1).

This problem is complicated still more because some of the green algae are single-celled and thus do not fit the basic definition of plants as "multicellular photosynthesizing organisms." However, the green algae are biochemically similar to other plant kingdom members, especially in terms of pigments. Further, most biologists think that there is an evolutionary reason for including green algae in the plant kingdom because they regard ancient green algae as the most likely ancestor of the more complex modern green plants.

In this chapter, structural and functional adaptations of various plant groups—from the simply organized green algae to the huge vascular plants with highly specialized body parts—are compared.

The reproductive patterns of various groups of plants also are examined in this chapter. Most plant life cycles include two phases: haploid plant bodies, called **gametophytes,** which produce and release gametes; and diploid plant bodies, called **sporophytes,** which produce spores by meiosis. These haploid and diploid bodies alternate in plants' life histories (**alternation of generations**), but the relative prominence of these two reproductive patterns is different in various groups of plants (figure 36.2).

While many comparisons of plants' body organizations and reproductive patterns will be made in this chapter and evolutionary relationships will be discussed, it is important to avoid regarding any group of present-day organisms as

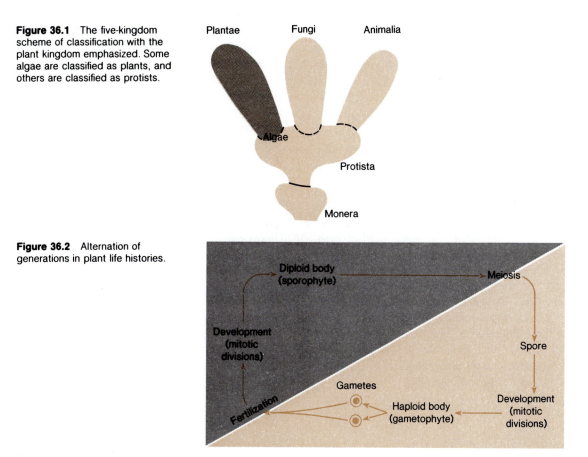

Figure 36.1 The five-kingdom scheme of classification with the plant kingdom emphasized. Some algae are classified as plants, and others are classified as protists.

Plantae   Fungi   Animalia

Algae

Protista

Monera

Figure 36.2 Alternation of generations in plant life histories.

Diploid body (sporophyte) → Meiosis

Development (mitotic divisions)

Spore

Fertilization

Gametes

Haploid body (gametophyte) ← Development (mitotic divisions)

Table 36.1
Divisions of Algae Included in the Plant Kingdom.

Group	Common Name	Approximate Number of Living Species
Division Chlorophyta	Green algae	7,000
Division Phaeophyta	Brown algae	1,000
Division Rhodophyta	Red algae	4,000

ancestors of any other group of present-day organisms. Contemporary organisms have not descended from one another. By studying the body organizations and reproductive patterns of various groups of living plants, however, it is possible to surmise some things about the possible course of plant evolution.

Botanists use the term "division" as an equivalent for the term "phylum" used in the animal kingdom. The International Code of Botanical Nomenclature (naming) specifies that the term "division" should be used to designate the largest taxonomic groups of plants. This text abides by that rule.

The survey of plants begins by examining the three divisions of algae included in the plant kingdom (table 36.1).

Division Chlorophyta

The division Chlorophyta is a diverse group of about 7,000 species of green algae. Some are unicellular; others are hollow balls of cells, filaments of end-to-end cells, or broad, flat sheets of cells. Some are **coenocytic (syncytial);** that is, there are many nuclei scattered in a large mass of cytoplasm inside a single plasma membrane.

The majority of green algae live in fresh water, but some live in moist terrestrial environments, and many are marine.

Green algae possess chlorophylls a and b, the same chlorophylls found in all of the terrestrial plants. Other algae and the photosynthetic protists also possess chlorophyll a, but most of them do not have chlorophyll b.

Unicellular Green Algae

Some unicellular green algae, such as **desmids,** display very ornate cellular organization, but many others are rather simply organized single cells. A good example is the simple unicellular flagellated green alga *Chlorella,* which was used by Melvin Calvin in his classical studies of photosynthesis (see chapter 6). *Chlorella* also has been studied for its potential as an aquatic crop plant for food production.

Another flagellated unicellular alga, *Chlamydomonas,* also is representative of the unicellular algae. *Chlamydomonas* is found in pools, lakes, and even in damp soil. The ordinary **vegetative** (nonreproducing) *Chlamydomonas* cell has a cellulose cell wall and a pair of flagella. Each cell has a single, large, cup-shaped chloroplast, which contains lamellae in an arrangement something like that of vascular plant cells' chloroplasts. The chloroplast contains a conspicuous **pyrenoid,** which is the site of starch production. The **stigma,** or **"eyespot,"** is a modified, carotenoid-containing portion of the chloroplast. It lies near the base of the flagella, as do two small **contractile vacuoles** that discharge rhythmically and expel excess water from the cell.

Asexual reproduction in *Chlamydomonas* is quite simple. A cell withdraws its flagella, its nucleus divides by mitosis, and cytoplasmic division produces two cells still contained within the original cell wall. Division stops at this point in some species and the daughter cells are released directly, but in other species, additional divisions may produce four or even more cells. Each daughter cell secretes a wall around itself and develops flagella. Then the parent cell wall ruptures, releasing these small cells, called **zoospores.** Each zoospore is a smaller copy of the parent cell. Subsequently, the zoospores grow to mature size, and division occurs again (figure 36.3).

In sexual reproduction, vegetative cells either convert directly into gametes, or they divide mitotically to produce a number of gametes. The gametes have flagella and resemble vegetative cells.

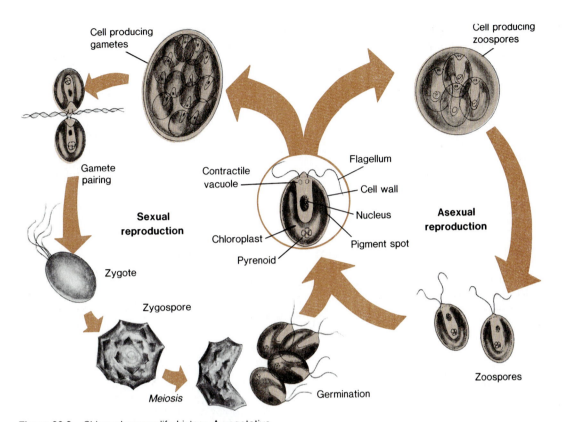

Figure 36.3 *Chlamydomonas* life history. A vegetative cell is shown in the center. Note that first contact between gametes is made by flagella. Gametes must be of opposite mating types to fuse. The only diploid *Chlamydomonas* cell is the zygote, which develops a thick wall to become a resistant **zygospore.** The haploid portion of the life history is in color.

Gametes of different **mating types** fuse in **syngamy** (fertilization). The different mating types are morphologically identical, but gametes do not fuse with gametes of their own mating type. The initial contact during syngamy is between flagella, and the mating types are recognized by differences in flagellar surfaces. After flagellar adhesion, the gametes gradually fuse. Later, the zygote loses its four flagella (two from each gamete), secretes a thick wall, and becomes dormant. The resistant, dormant cell, called a **zygospore,** can withstand adverse conditions, such as winter or a dried-up pond.

When conditions are suitable again, the resistant zygospore germinates. Meiosis occurs, and four haploid, flagellated cells emerge through the ruptured wall. Each of these cells matures into an ordinary vegetative cell, and the sexual cycle is complete (figure 36.3).

Although *Chlamydomonas* has mating types, the gametes are identical in appearance. Thus, they are classified as **isogametes** (iso = the same), and this pattern of reproduction is called **isogamy.**

The only diploid cell in the life history of *Chlamydomonas* is the zygote. All gametes, zoospores, and vegetative cells are haploid.

Motile Colonial Algae

One line of multicellular green algae is called the **volvocine line.** The name comes from *Volvox,* the most complex member of the group. The volvocine line consists of colonial forms made up of cells that individually resemble *Chlamydomonas* cells. The simplest form is *Gonium.* Each colony of *Gonium* consists of from four to thirty-two cells, depending on the species, embedded in a jelly matrix and arranged in a flat or slightly curved disc. Cytoplasmic threads connect the cells. These threads may be involved in coordinating cells because the flagella of all of the cells beat so that the colony can swim as a unit. Each cell of *Gonium* can give rise to a complete new colony.

Members of the genus *Pandorina* form slightly more complex colonies than those of *Gonium*. A *Pandorina* colony is a hollow oval sphere with sixteen to thirty-two cells arranged in a single layer, their flagella pointing outward. The colony shows regional specialization because one end is always anterior and one is always posterior when the colony swims. Cells in the anterior end have larger eyespots. In asexual reproduction, each cell divides to produce a daughter colony, and then the parent colony breaks open to release all the new colonies.

Volvox is considered to be the most complex genus in this line of colonial green algae. *Volvox* colonies have thousands of cells whose flagella beat in a very coordinated fashion. Colonies have regional specialization and show some complex responses, such as positive orientation toward moderate light, but movement away from very strong light. Only cells in one part of the colony can produce daughter colonies. Daughter colonies remain inside the parent colony for some time until the parent colony breaks apart and releases them. Daughter colonies are very commonly seen inside *Volvox* colonies. See figure 36.4.

There clearly is interdependence among cells in these colonial algae because individual cells of *Pandorina* or *Volvox* colonies cannot survive if isolated from the colonies, and fragmented colonies also die.

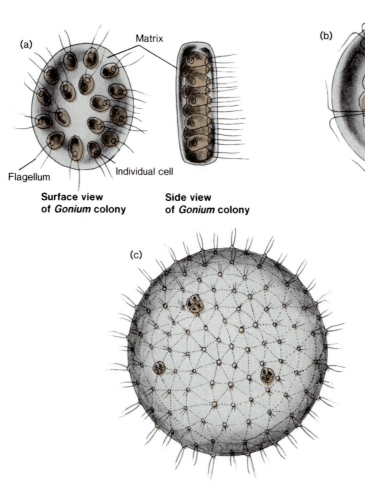

(a) Matrix

Flagellum   Individual cell

**Surface view
of *Gonium* colony**

**Side view
of *Gonium* colony**

(b)

(c)

Figure 36.4 The volvocine line.
(a) *Gonium* is a flat colony of cells embedded in a jelly matrix.
(b) *Pandorina* is an oval colony.
(c) The *Volvox* colony is a large sphere with thousands of cells embedded in a jelly matrix and connected by cytoplasmic strands. Note the daughter colonies inside the *Volvox* colony.

Filamentous Green Algae

Filaments (end-to-end chains of cells) form because cell divisions occur in only one plane and cells remain attached to one another after divisions. In some species of filamentous green algae, a specialized **holdfast cell** anchors the end of each filament to some surface. In other species, filaments float free in the water.

There are several different patterns of sexual reproduction among the filamentous green algae, and the patterns are best illustrated by examining the reproduction of several representative species.

In the genus *Ulothrix,* reproductive patterns are actually very similar to those seen in *Chlamydomonas* except, of course, that they occur in cells of a filament rather than in free-swimming individual cells. Any cell in the filament may function as a zoospore-producing structure. The

cell divides to produce four to eight zoospores. Each of them develops four flagella. After they swim about for a short time, the zoospores settle down and begin to divide to produce new filaments.

Ulothrix sexual reproduction occurs when filament cells produce several small gametes, each bearing two flagella. The gametes are released into the water, where fusion occurs. *Ulothrix* gametes are isogametes; that is, they are all identical in appearance. But there are mating types in *Ulothrix* just as there are in *Chlamydomonas.* The zygote forms a resistant, thick wall and then enters a resting condition, in which it can survive adverse conditions. Later, the zygote divides by meiosis to produce four haploid zoospores. Each of them can develop into a new *Ulothrix* filament (figure 36.5).

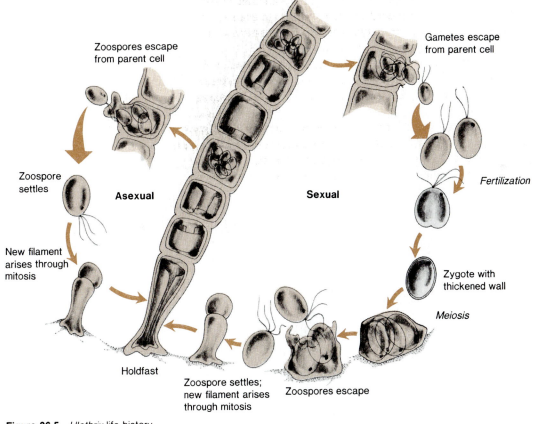

Zoospores escape from parent cell

Gametes escape from parent cell

Zoospore settles

Asexual

Sexual

Fertilization

New filament arises through mitosis

Zygote with thickened wall

Meiosis

Holdfast

Zoospore settles; new filament arises through mitosis

Zoospores escape

Figure 36.5 *Ulothrix* life history. The haploid portion is in color.

Spirogyra is a very common, freshwater, filamentous green alga. It is often an important part of "pond scum" when algal blooms cover the water surface. The most conspicuous features of *Spirogyra* cells are one or more large helical chloroplasts with numerous pyrenoids.

Sexual reproduction in *Spirogyra* occurs by a special fusion process called **conjugation.** Usually in the autumn, two filaments come to lie side by side. Protuberances that form on the cells enlarge, meet, and fuse to form **conjugation tubes.** Once cells are paired in this way, one cell pulls away from its cell wall, rounds up, squeezes through the conjugation tube, and fuses with the other cell. Thus, individual vegetative cells function as gametes. This happens all along the paired filaments and results in the production of zygotes in one filament and a series of empty cell walls in the other filament. Each zygote forms a thick, resistant wall. Before germination, meiosis occurs, and four haploid nuclei are produced. Three of them degenerate. The fourth becomes the functional nucleus of the cell and grows out of the broken zygote wall to begin development of a new filament (figure 36.6).

Just as in *Chlamydomonas* and *Ulothrix*, all *Spirogyra* vegetative cells are haploid. The zygote is the only diploid cell present in *Spirogyra's* entire life history.

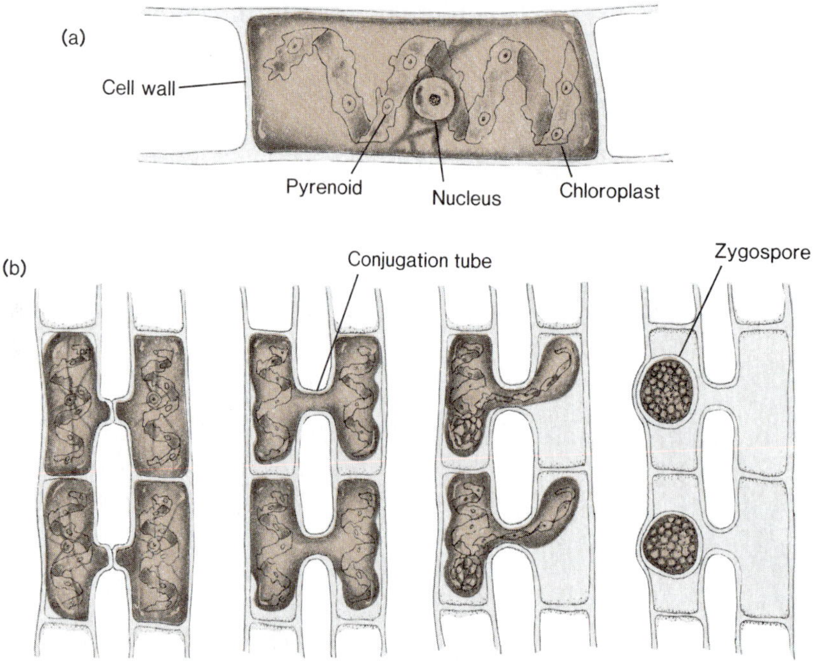

(a)

Cell wall

Pyrenoid   Nucleus   Chloroplast

(b)

Conjugation tube   Zygospore

Figure 36.6 *Spirogyra* conjugation. (a) Vegetative cell. (b) Stages of conjugation. (c) Photograph of the final stages of conjugation. Zygotes form in one filament of each pair, leaving empty cell walls in the other filament.

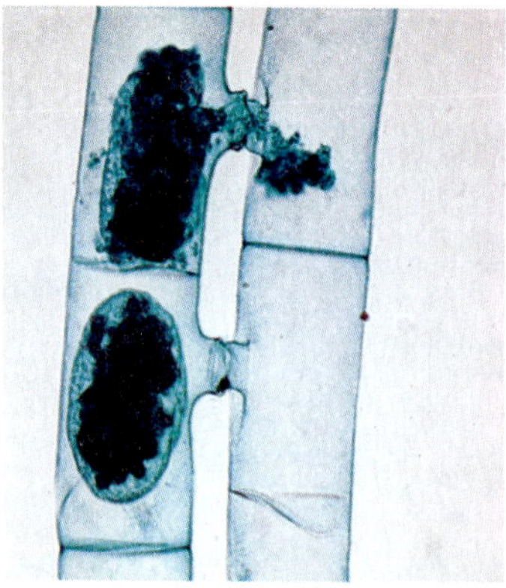

(c)

Oedogonium, another common freshwater alga, has a markedly different reproductive pattern. *Oedogonium* has two distinctly different types of gametes. Small, motile sperm cells fuse with large, nonmotile egg cells. Sexual reproduction involving fusion of such unlike gametes is called **heterogamy,** from **heterogametes** (hetero = different). Any vegetative cell can differentiate into an egg-forming cell (**oögonium**) or divide to produce several small sperm-forming cells (**antheridia;** singular: **antheridium**). An antheridium produces two swimming sperm cells, each of which has a circle of flagella.

Sperm are released and swim to eggs. Apparently, they are attracted by some chemical released by eggs. As in other filamentous algae, the zygote secretes a thick wall and can withstand adverse conditions. The zygote eventually undergoes meiosis to produce four haploid zoospores. Each of them can initiate formation of a new *Oedogonium* filament (figure 36.7).

Reproductive Patterns in Green Algae

Heterogamy with well-defined eggs and sperm is considered a more advanced form of reproduction than isogamy. A large, nonmotile egg can contain much more nutrient material than can be carried by a swimming gamete. This stored nutrient is then available to support growth of the zygote.

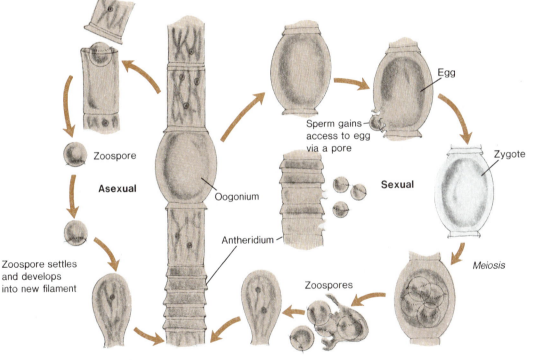

Zoospore

Asexual

Zoospore settles
and develops
into new filament

Oogonium

Antheridium

Egg

Sperm gains
access to egg
via a pore

Sexual

Zygote

Meiosis

Zoospores

Figure 36.7   *Oedogonium* life history.

In *Oedogonium,* only the sperm must be able to swim and expend energy on movement, as opposed to the *Ulothrix* situation, where both gametes must be equally active.

In the filamentous algae, only one diploid cell is present in the entire life history. Thus, alternation of generations is limited because there is no diploid plant body. In the green alga *Ulva* ("sea lettuce"), there is a different pattern.

Ulva grows in a simple, flat, sheetlike form called a **thallus** (plural, **thalli**). There are actually two different kinds of thalli in *Ulva*. One kind grows by mitosis from a spore and consists entirely of haploid cells. Gametes produced by these haploid bodies (gametophytes) fuse to produce zygotes. The zygote divides by mitosis and grows into a thallus made up of diploid cells. This is a marked difference from the filamentous green algae, where the zygote was the only diploid cell and it divided meiotically to produce haploid spores. Eventually, specialized cells of *Ulva's* diploid body (sporophyte) divide by meiosis to produce haploid zoospores that grow into haploid thalli. This life cycle pattern with multicellular haploid bodies alternating with multicellular diploid bodies is a clearcut alternation of generations (figure 36.8). Other variations on this life cycle pattern are seen in the terrestrial plants.

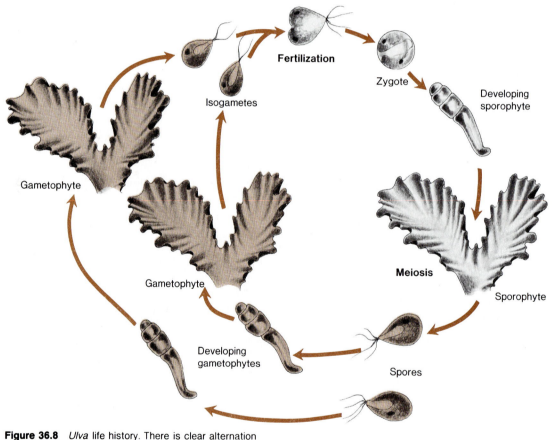

Figure 36.8 *Ulva* life history. There is clear alternation of generations between gametophyte (haploid) thalli and sporophyte (diploid) thalli.

Division Phaeophyta

The brown algae make up the division Phaeophyta (*phaeo*=brown). The group includes many of the plants that are commonly called seaweeds. The approximately 1,000 species of brown algae are all multicellular, and the great majority of them are marine. Brown algae range from small plants with simple branched filaments to large plants with thalli that may be between 50 and 100 m long. Large brown algae, known as kelp, are common in the intertidal zone, and they spread over large areas of rocky shorelines in cooler regions of the world. In deeper water, giant kelp such as *Macrocystis* and *Nereocystis* often form spectacular underwater forests (figure 36.9).

Intertidal algae are pounded by waves as the tide comes in and exposed to drying at low tide. Brown algae can withstand these conditions because their cell walls contain a mucilaginous, hydrophilic material that helps to prevent drying. They are firmly anchored by **holdfasts,** and broad flattened **blades** are connected to the holdfasts by **stipes.** Some have air bladders that buoy up the blades.

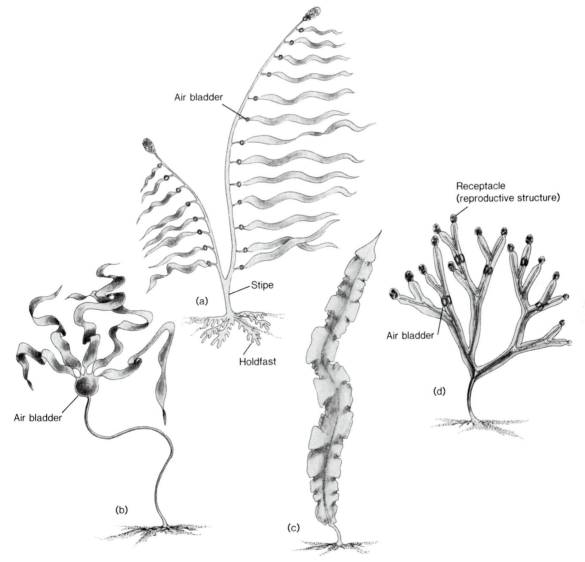

Figure 36.9 Representative brown algae. These are not drawn to the same scale because *Macrocystis* (a) and *Nereocystis* (b) are giant kelp, which grow larger than *Laminaria* (c). *Fucus* (d) ("rockweed") is smaller.

While most of the brown algae grow attached to substrates along shorelines, one genus, *Sargassum*, is a free-floating form that makes up much of the dense floating mat of algae accumulated in the Sargasso sea. The Sargasso sea is a huge (more than 5 million square km) eddy in the Atlantic ocean between the West Indies and the Azores.

Brown algae provide food and habitat for many marine animals, and some kelps are used as human food in several parts of the world. Brown algae also have been processed for fertilizers. Several kelp, such as *Macrocystis* and *Laminaria,* are economically important because a colloidal carbohydrate called algin can be extracted from their cell walls. Algin is a pectinlike material that is added to ice cream, sherbet, cream cheese, and other products to give them a stable, smooth consistency. It also is used in the manufacture of adhesives and many other industrial products.

Biochemically, brown algae are quite different from green algae. They possess chlorophylls a and c, as well as a special carotenoid pigment called **fucoxanthin,** which largely accounts for their brown color. Brown algae store nutrients as a unique polysaccharide called **laminarin,** which is a soluble glucose polymer, or as fat, but not as starch, which is the common storage compound found in green algae and the more complex green plants.

Division Rhodophyta

The division Rhodophyta, the red algae, includes approximately 4,000 species. The red algae contain red pigment called **phycoerythrin,** one of two types of **phycobilins** that they possess. The other phycobilin is a blue pigment called **phycocyanin.** These pigments are accessory pigments to chlorophyll a and are involved in light energy absorption for photosynthesis. Chlorophyll d also has been extracted from some red algae. Because of various mixtures of pigments, "red" algae range in color from red to almost black.

The presence of phycobilins probably explains how some red algae can grow even in water deeper than 100 m. Wavelengths of light that penetrate to such depths are not absorbed by chlorophyll a. Thus, the accessory pigments absorb light energy and pass energy to chlorophyll a. Other algae lacking these accessory pigments cannot grow at such great depths.

Red algae generally are much more common in deeper and warmer waters than the brown algae. A few red algae are up to a meter long, but most are smaller than that. The lacy, delicate bodies of most red algae would not fare well in the intertidal zone (figure 36.10).

Some tropical red algae, the **coralline** algae, have a flatter growth form and are able to accumulate calcium from seawater and deposit it in their bodies as calcium carbonate. In some cases, coralline algae may contribute as much to growth of coral reefs as the coral animals themselves do.

Red algae are economically important because valuable colloidal substances can be extracted from the outer layer of their cell walls. The best known of these compounds is **agar.** Agar is used in culture media for bacteria and fungi. It and other colloids from red algae are used in many foods to produce a smooth consistency and to help retain moistness. Soup is made from red algae in Scotland, the Orient, and elsewhere in the world, and more exotic uses of red algae as a food source also are known.

Table 36.2 summarizes this section on the three divisions of algae included in the plant kingdom by comparing the pigments found in each division and each division's type of nutrient storage.

Table 36.2
Comparison of the Divisions of Multicellular Algae.

Division	Pigments	Nutrient storage
Chlorophyta (green algae)	Chlorophyll a and b Carotenoids	Starch
Phaeophyta (brown algae)	Chlorophyll a and c Carotenoids (especially fucoxanthin)	Laminarin, fat
Rhodophyta (red algae)	Chlorophyll a Phycobilins, carotenoids Chlorophyll d in some	Floridean starch

(a)

(b)

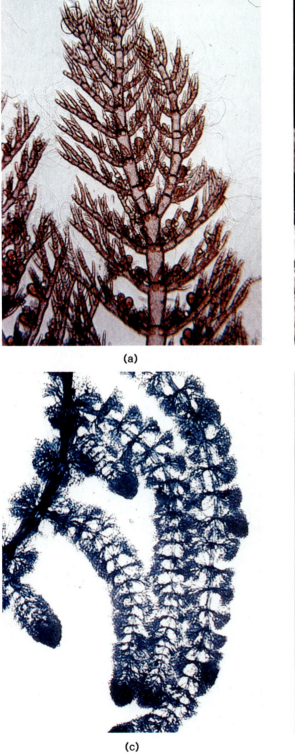

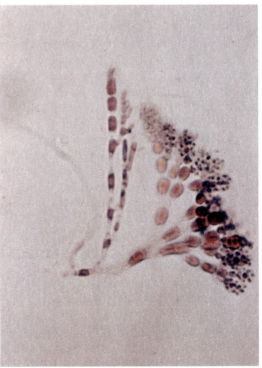

(c)

(d)

Figure 36.10 Some red algae. Red algae are smaller
and more delicate than the large brown algae.
(a) *Antithamnium.* (b) *Corallina.* (c) *Batrachospermum.*
(d) *Nemalion.*

Terrestrial Plants

Terrestrial (land-dwelling) plants encounter opportunities and problems different from those of aquatic plants.

One advantage of terrestrial plants is that the light supply for photosynthesis is much better on land. Even clear water filters light, and turbid water greatly reduces the light energy reaching aquatic plants. Also, carbon dioxide and oxygen are available in higher concentrations in air than in water, and gases diffuse more readily through air.

The principal problem of terrestrial plants is the constant threat of losing excessive amounts of water to the environment. Mineral nutrient supply is a related problem. Ions are present in the water that continually bathes aquatic plants, but terrestrial plants must get mineral nutrients from the soil. Thus, terrestrial plants must have means of preventing excessive water loss and of obtaining an adequate supply of water and minerals from the soil.

Water loss from terrestrial plants is reduced by waxy cuticles that provide a waterproof barrier over exposed surfaces. Nonvascular terrestrial plants grow low to the surface and usually are restricted to fairly sheltered and moist environments. But vascular plants can absorb water and minerals from the soil and transport them through roots and stems to large aboveground bodies. Thus, terrestrial plants are adapted to meet the everyday problems encountered in terrestrial environments, which are more stressing and changeable than aquatic environments.

Terrestrial plants, however, face another important problem. Spores, gametes, and developing zygotes of terrestrial plants are small and fragile. They can stand very little water loss or environmental stress of any kind. Different land plants have different solutions for the problems of spore and gamete transport.

The zygotes of terrestrial plants develop into multicellular **embryos.** All terrestrial plants protect developing embryos inside multicellular reproductive organs. Each reproductive organ is surrounded by a protective outer layer of sterile (nonreproductive) tissue. Thus, as a multicellular embryo develops from the zygote, it is enclosed and protected inside a female reproductive organ. The embryo receives water and nutrients from the parent plant and, therefore, is dependent on the parent plant. Because of these characteristics, the terrestrial plants are known collectively as **embryophytes** ("embryo plants").

While not actually a taxonomic designation, the terms "embryophyta" and "embryophyte" are used in discussing terrestrial plants in general. Some algae do live in terrestrial environments, but they are severely restricted to especially moist environments. All of the widely distributed and successfully adapted terrestrial plants are embryophytes.

Division Bryophyta

Mosses, liverworts, and their relatives make up the division Bryophyta. They are all relatively small terrestrial plants that grow in moist places. The bryophytes have waxy cuticles that control water loss, but they do not have vascular tissue to transport the soil's water and minerals through their bodies. Thus, their body sizes are limited because they are dependent on diffusion of materials.

Mosses

Mosses are the most familiar members of the division Bryophyta. About 14,500 species of mosses make up the class Musci. The small "leafy" bodies of moss, which grow densely in soft mats, are gametophyte generation members. Though it is tempting to identify the parts of moss gametophytes as roots, stems, and leaves, it is not appropriate to do so. Those names are used only for plant structures containing vascular tissues, and moss plants do not have vascular tissue.

The densely concentrated growth habit of mosses helps to compensate for the lack of transport tissue, because water becomes trapped and held in spaces among the crowded bodies of the moss plants. This provides a water source close to aboveground body parts.

Rhizoids, simple anchoring structures, absorb mineral nutrients from the soil, but some plant parts are quite a distance away from the rhizoids, and materials move slowly by diffusion. Some soil minerals are dissolved in the water found in spaces among the plants, and nutrients can diffuse into nearby plants from this water as well.

When the environment becomes dry, moss plants soon suffer from water deficiency. Some species can become dormant and revive when water becomes available again, but water supply limits the distribution of mosses.

Bryophytes depend on external surface water for sexual reproduction because they produce swimming sperm cells. In some cases, splashing raindrops or a heavy dew may provide enough water to make reproduction possible, but water is absolutely essential. This water requirement for reproduction is another limiting factor in the distribution of bryophytes.

In bryophytes, the gametophyte generation is larger and more conspicuous than the sporophyte. Sex organs develop at the tips of gametophytes. In some mosses, **antheridia** (sperm-producing structures) and **archegonia** (multicellular egg-producing structures with sterile cells around the egg) are borne on the same plant, but in others there are separate male and female plants.

Motile sperm swim from the antheridium to the archegonium. Fertilization occurs inside the archegonium, and the zygote develops into an embryo sporophyte in the archegonium's enclosed protected environment. Eventually, the sporophyte grows up out of the archegonium, but it remains attached by a foot that anchors it to the gametophyte plant. Moss sporophytes may have chlorophyll and be able to photosynthesize, but they are still dependent on water and minerals that diffuse in from the gametophyte plant. Thus, the sporophyte generation remains as a dependent attachment on the gametophyte and does not have an independent life of its own.

At the tip of a stalk (**seta**) extending up from the foot, a **sporangium** develops. In the sporangium, meiosis occurs, and haploid spores are produced. In some species of moss, a hoodlike covering of archegonium tissue, the **calyptra,** is carried along upward by the growing sporophyte. The calyptra comes off before spores are mature. A lid, the **operculum,** falls off the capsule, and mature spores escape (figure 36.11).

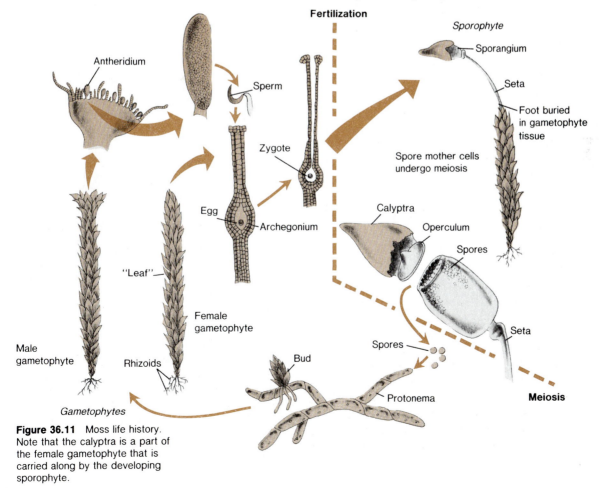

Figure 36.11 Moss life history. Note that the calyptra is a part of the female gametophyte that is carried along by the developing sporophyte.

Spore release is controlled by a ring of teeth. The teeth expand and stretch over the opening when it is wet, but curl up and free the opening when it is dry. This appears to be a mechanism that allows spore release when spores are more likely to be distributed by wind.

When a spore lands in an appropriate site, it germinates. A single row of cells grows out and then branches. This algalike structure is called a **protonema.** After about three days of growth under favorable conditions, "buds" appear at intervals along the protonema. Each of these sends down rhizoids and grows up into a gametophyte plant. This completes the moss life cycle.

Liverworts

About 9,500 species of liverworts (wort = plant or herb) make up the class Hepaticae. Most liverworts are small, flattened green plants that do not resemble mosses superficially, but they do have a reproductive cycle similar to mosses. Also, like mosses, they have rhizoids and lack vascular tissue. The flat, lobed body of liverworts is called a thallus.

Marchantia is a very common liverwort that is often found on damp soil, such as sheltered areas around buildings where water runs off roofs. The *Marchantia* thallus is a gametophyte body (figure 36.12a). It branches dichotomously (in twos) as it grows. Rhizoids grow down into the soil from the bottom of the *Marchantia* thallus. Pores in the upper epidermis permit CO_2 diffusion into air chambers occupied by chlorophyll-containing cells. Several layers of nonphotosynthetic cells lie below the air chambers, and many of those cells contain starch-storing bodies. Some contain mucilage, which tends to absorb and hold water.

Marchantia reproduces asexually by producing multicellular structures known as gemmae inside gemmae cups (see page 618). Each gemma can grow into a new thallus if water washes it into an appropriate place (figure 36.12b).

Sexual reproduction in *Marchantia* is very similar to that of mosses. Male and female reproductive organs develop on upright stalks of separate gametophytes. Antheridia develop on a

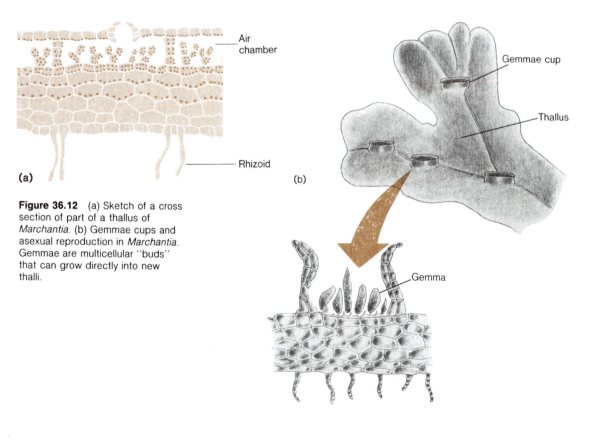

(a)

(b)

Figure 36.12 (a) Sketch of a cross section of part of a thallus of *Marchantia*. (b) Gemmae cups and asexual reproduction in *Marchantia*. Gemmae are multicellular "buds" that can grow directly into new thalli.

flattened disc, while archegonia develop suspended underneath fingerlike structures that hang from the top of a female stalk. Sperm swim to and enter the archegonium, where fertilization occurs.

The zygote develops into an embryo within the archegonium. It eventually outgrows the archegonium and hangs down from it. The mature *Marchantia* sporophyte, which remains attached to and dependent on the gametophyte plant, then develops. Meiosis occurs within the sporangium (capsule) of the sporophyte. As spores mature, the capsule opens, and **elaters** flip spores out of the capsule. Some of the spores, carried by air currents, land in favorable locations and grow into new gametophyte plants (figure 36.13).

In liverworts, as in mosses, sexual reproduction requires water for sperm to swim in, and the sporophyte (diploid generation) is a dependent attachment on the body of the female gametophyte. In vascular plants, there is a very different situation. Water is not a requirement for sexual reproduction, and the sporophyte generation is much more prominent.

Division Tracheophyta

The division Tracheophyta is a large and diverse group that includes the most complex and advanced plants. The primary distinguishing characteristic of tracheophytes is the presence of vascular tissues. **Xylem** conducts water and minerals up from the soil, and **phloem** transports nutrients and other materials from one part of the plant body to another. Because they have vascular tissues, the specialized body parts of tracheophytes can properly be called roots, stems, and leaves.

Tracheophyte reproduction involves a conspicuous alternation of generations, but the sporophyte clearly is the dominant generation. Sporophytes contain vascular tissues and specialized body organs. Gametophytes are small and relatively inconspicuous. The gametophytes of some tracheophytes live as independent plants, but the most complex tracheophytes have very reduced gametophyte generations that live dependently on sporophyte generation plants. Gametophytes lack vascular tissue.

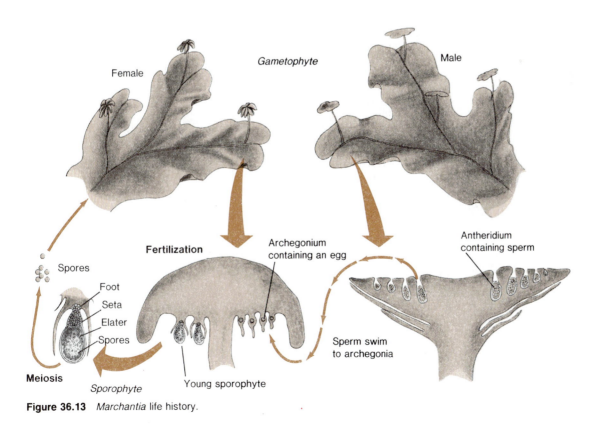

Figure 36.13 *Marchantia* life history.

Tracheophytes are believed to be descended from ancient filamentous green algae, which probably had dominant gametophyte generations. Since the vascular tissues and specialized body parts of sporophyte generation plants are important adaptations for success as terrestrial plants, development of a dominant sporophyte generation has been an important trend in the evolution of tracheophytes. Some possible evolutionary trends can be surmised by comparing some living tracheophytes, but it is necessary to guard against the temptation to think of living plants as ancestors of other living plants.

Subdivision Psilopsida

The psilopsids are the simplest known vascular plants. Most members of the group are extinct, and these fossil plants are known collectively as **psilophytes** (figure 36.14a). The **"whisk ferns,"** which do not particularly resemble common ferns but do look something like small green whisk brooms, may be living members of this group. But this is not certain since *Psilotum,* the most common living genus, is the subject of some disagreement among botanists. Traditionally, *Psilotum* has been included in this subdivision along with the primitive psilophytes that flourished about 400 million years ago. But David W. Bierhorst and others firmly maintain that *Psilotum* is either a very primitive fern genus or a degenerate genus descended from more complex fern ancestors.

At any rate, *Psilotum* is discussed here because it is one of the most primitive genera of living tracheophytes. *Psilotum* sporophytes are simple, dichotomously branched stems without leaves. The stems, which commonly grow to about 30 cm, but occasionally grow to a height of 1 m or more, are green and photosynthesize.

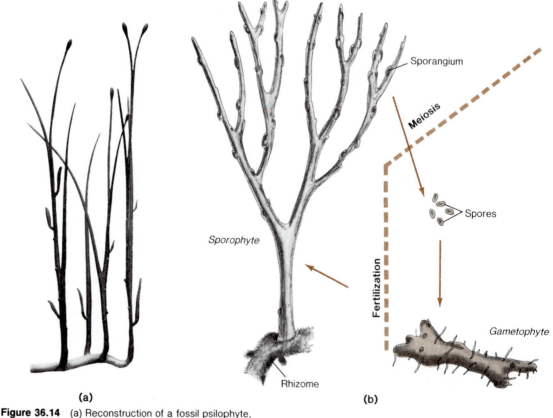

(a) **(b)**

Figure 36.14 (a) Reconstruction of a fossil psilophyte, *Rhynia gynne-vaughani.* These plants grew to a height of about 20 cm. Fossils of these primitive vascular plants are from the Devonian age (350–400 million years ago). (b) *Psilotum.* Underground gametophytes are relatively smaller than shown in this sketch.

They lack true roots but have underground stems (called **rhizomes**) that bear rhizoids. They have no cambium and thus do not show secondary growth.

Small sporangia produce spores by meiosis. The spores give rise to very small underground gametophytes, each of which has both antheridia and archegonia randomly distributed over its surface. Following fertilization, the zygote develops into a sporophyte body (figure 36.14b).

Subdivision Lycopsida

Plants belonging to the subdivision Lycopsida were widespread during the Carboniferous period (more than 280 million years ago), and some of them were treelike forms. Modern members of the group are small, inconspicuous plants called club mosses, spike mosses (actually, neither of them are mosses at all), and quillworts (figure 36.15).

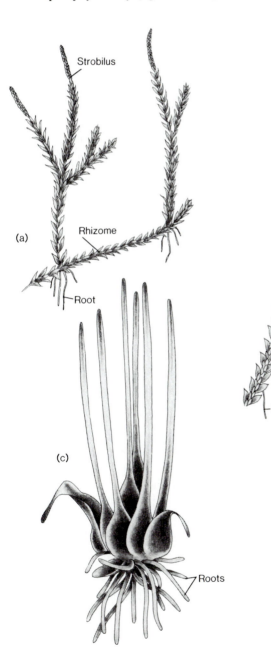

Figure 36.15 Lycopsids. (a) Club moss *(Lycopodium)* is sometimes called "ground pine" because it is green year round. This evergreen condition has made club moss popular as a holiday decoration and has led to depletion of some populations formerly found in the eastern United States. (b) Spike moss. (c) Quillwort.

The lycopsids have true roots and leaves and thus are more complexly organized than the psilopsids. Lycopsid sporangia are borne on modified sporangium-bearing leaves called **sporophylls.** In some species, the sporophylls are arranged in terminal structures called **strobili** (singular: **strobilus**), which resemble pinecones.

Spike mosses, such as *Selaginella,* have a more complex reproductive pattern than other members of the group because they produce two different types of spores, which develop into two separate types of gametophytes. Tiny **microspores** grow into male gametophytes; the larger **megaspores** produce female gametophytes. Plants that produce two separate kinds of spores are called **heterosporous** (hetero = different), as opposed to **homosporous** plants, which produce a single type of spore. The heterosporous pattern is considered to be more advanced in evolutionary terms.

Subdivision Sphenopsida

Like the subdivisions Psilopsida and Lycopsida, the subdivision Sphenopsida includes more fossil plants than living ones. Some of the fossil sphenopsids were giant treelike plants that were up to 13 m tall, and their bodies were transformed into important components of coal.

Today there is only one surviving genus of sphenopsids, the genus *Equisetum,* and most of the twenty-five or so living species are less than 1 m tall, although a few grow to be as much as 4.5 m tall.

Equisetum is commonly called "horsetail" because of the appearance of some stems, or "scouring rush" because the plants have been used for scouring and cleaning metal (figure 36.16). They are useful for that because of silica deposits in stem cell walls that make the stems very hard.

Equisetum has true roots, stems, and leaves, but the roots come off rhizomes. The stems are hollow and appear jointed because tiny, scalelike leaves are arranged in circles at regular intervals along the stem. Where branches are present, they also arise in circles at these definite points, the **nodes,** which are separated by stretches of plain stem, the **internodes.**

Strobili develop at the tips of special (fertile) shoots. Each strobilus contains many sporangia, which produce and release spores. Spores produce

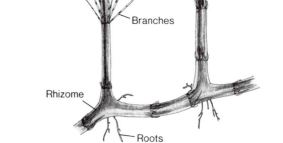

Figure 36.16 *Equisetum,* commonly called "horsetail" or "scouring rush." Strobilus is a cluster of sporangia.

small gametophytes that bear both antheridia and archegonia.

Subdivision Pteropsida

The subdivision Pteropsida is a large group that includes the dominant land plants. It is composed of three classes of familiar plants: the **ferns,** class Filicineae; the **conifers,** class Gymnospermae; and the **flowering plants,** class Angiospermae. The sporophytes of all of these plants have well-developed roots and stems. But their leaves clearly distinguish them from other vascular plant subdivisions. They have larger leaves with many veins instead of small scalelike leaves with only one vein. Well-developed vascular and supporting tissues permit many of the pteropsids to grow to very large sizes.

The gametophytes of all pteropsids are small and inconspicuous. Fern gametophytes are small but independent green plants, while conifers and flowering plants have tiny, dependent gametophytes that sometimes consist of only a very few cells.

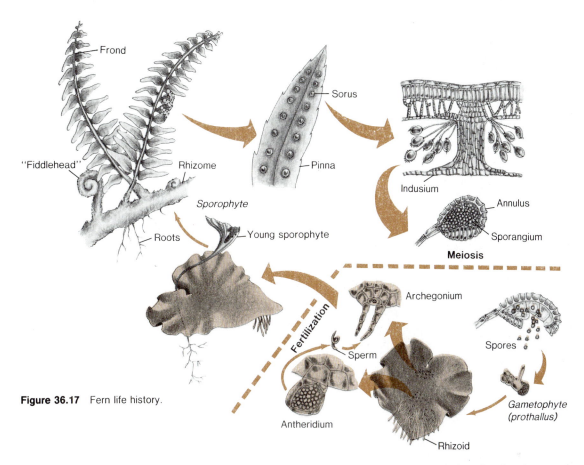

Figure 36.17 Fern life history.

Class Filicineae

The class Filicineae includes about 11,000 known species of ferns that vary in size from tiny water ferns only a centimeter in diameter to the tree ferns of the tropics, which may be 25 m tall in some cases. The familiar leafy fern plant is the sporophyte generation, and the entire aboveground growth consists of leaves, commonly called fronds. Both fronds and roots grow out of the rhizome, the underground stem. Young fronds grow in a curled-up form called "fiddleheads" out of rhizomes. Then they unroll to produce mature fronds.

Sporangia develop in clusters called **sori** (singular: **sorus**) on the undersurface of fronds. In many species, each sorus is protected by a flap of tissue called the **indusium.** Each individual sporangium is attached by a stalk. A band of thickened cells known as the **annulus** functions in expelling mature spores from the sporangium. The annulus snaps in response to moisture changes and flings spores out.

Spores are carried by the wind, and a few end up in appropriate wet habitats. They ger-

minate and produce a chain of cells, the protonema. After several linear divisions, the end cell divides transversely and begins development of the heart-shaped **prothallus** (plural: **prothalli**), the gametophyte body. A prothallus is a thin plate of cells with rhizoids extending down from its lower surface. Antheridia form among the rhizoids. Archegonia develop close to the notch of the "heart," and each of them produces a single egg.

Spiral-shaped sperm swim to the archegonium, where fertilization occurs. Thus, external water is required for sexual reproduction in ferns.

The zygote begins its development inside the archegonium, but soon the embryo outgrows the space available there. The young sporophyte becomes visible as a distinctive first leaf appears above the prothallus and the sporophyte's roots develop below it. Often, gametophyte and sporophyte tissue are distinctly different shades of green. This young sporophyte grows and develops into a mature sporophyte, the familiar fern plant (figure 36.17).

The requirement of surface water for sexual reproduction does restrict the distribution of some ferns, but other ferns can grow in fairly dry habitats once they become established. For example, the common bracken fern *Pteridium aquilinum* spreads over meadow and fenceline areas by vegetative reproduction. The rhizomes spread out, and fiddleheads grow up to produce fronds in new areas.

Seed Plants

The major groups of seed plants share two key characteristics. The first of these is the formation of **seeds,** protective structures that enclose the sporophyte embryo during a dormant stage. A seed includes the sporophyte embryo, a reserve of nutrients stored in nutritive tissue, and a tough, protective **seed coat.** Seeds are resistant to adverse conditions, such as dryness or temperature extremes. They provide for wide dispersal because seeds can be spread into new areas and they begin to grow when conditions become suitable.

The second key characteristic of seed plants is **pollination,** a process by which male gametes are brought to eggs without a requirement for surface water.

Seed plants are heterosporous; they produce two different kinds of spores. Megaspores develop into female gametophytes, and microspores develop into **pollen** (immature male gametophytes). Grains of pollen are transferred by wind or by insects to the vicinity of the developing female gametophyte. Then a **pollen tube** grows out to carry the male gametes to the egg.

Pollination and seed production have been critically important for the success of seed plants as terrestrial organisms.

The two groups of seed plants are the gymnosperms ("naked seeds") and the angiosperms ("enclosed seeds"). The names come from different relationships of the seeds to the structures where they are produced. Gymnosperm seeds are produced exposed on the surface of the sporophylls that make up cones, while angiosperm seeds usually are enclosed by a fruit produced from part of a flower (figure 36.18).

Class Gymnospermae The gymnosperms include only about 700 living species, but some of them—especially conifers, such as pine, spruce, cedar, and fir—have huge numbers of individuals and cover large areas of the earth's surface. Other, less familiar plants that are very different from the conifers have long been grouped with them. Many biologists think, however, that some of these less familiar plants, such as the cycads, the maidenhair tree *(Ginkgo biloba)*, and the gnetophytes, should be put in separate classes entirely (figure 36.19).

Cycads were very abundant during the Mesozoic era, but many of the 100 or so living species that remain are near extinction. These tropical or subtropical plants are often mistaken for ferns or miniature palm trees. Microspore- and megaspore-producing cones are borne on separate plants. Sometimes, cones can be as much as a meter long and covered with wooly hairs.

Ginkgo biloba, the maidenhair tree, is the only surviving species of a once widespread group. In fact, ginkgo trees probably do not now occur naturally anywhere in the world. All surviving ginkgos are cultivated, and the ginkgo is a popular tree in some areas of the United States. Pollen-producing and egg-producing structures grow on separate trees, and because the outer covering of seeds has a foul, rancid odor, the two are seldom grown together in parks or along streets. In fact, "male" trees, vegetatively propagated from cuttings, are used most commonly as ornamental trees. One special reproductive feature of ginkgos is that flagellated sperm develop within their pollen tubes.

(a)

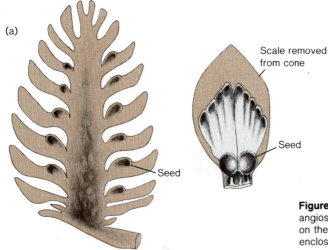

Scale removed
from cone

Seed

Seed

(b)

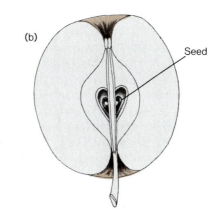

Seed

Figure 36.18 Comparison of gymnosperm (pine) and angiosperm (apple) seeds. Pine seeds (a) are ''naked'' on the surface of a cone scale. Apple seeds (b) are enclosed in a fruit.

(a)

(b)

(c)

Figure 36.19 Gymnosperms. (a) A cycad *(Encephalartos transvenosus)* growing in South Africa. (b) The maidenhair tree *(Ginkgo biloba)*. (c) *Welwitschia mirabilis*, a gymnosperm that grows in very harsh desert environments in Africa. The leaves of these plants are about 2 m long.

The **gnetophytes** are unique among gymnosperms because their xylems have both tracheids and vessels. Other gymnosperms have only tracheids in their xylem. One group of gnetophytes, the "joint firs" (genus *Ephedra*), are rather obscure, shrubby plants. One peculiar genus of gnetophyte is *Welwitschia,* which grows in the extremely dry deserts of southwestern Africa. Its short stem grows as a large, shallow cup, and leaves grow out from it. Leaves grow from their bases and continually replace tissue that is worn off at the tip because they flap around in the wind. *Welwitschia* plants may live to be 100 years old, so they are obviously well adapted to their harsh environment.

The conifers are a widely distributed and economically important group of plants. They produce much of the wood used for building and paper, and they also produce many other valuable products, such as those extracted from **resins.** Resins contain waxy substances dissolved in the liquid solvent **turpentine.**

The oldest and the largest trees in the world are conifers (figure 36.20). Bristlecone pines (*Pinus aristata*) in the Nevada mountains are known to be more than 4,500 years old, and a number of coastal redwood (*Sequoia sempervirens*) trees in California are 2,000 years old and over 90 m tall. Though not so tall as the coastal redwoods, the giant sequoias (*Sequoiadendron giganteum*) have the greatest mass of any living trees. Some of the famous "big trees" of California, which belong to this species, have trunks that are up to 10 m in diameter.

Although there is some variation among conifers, the pine life cycle is a good example of a conifer life history. Pines are heterosporous, and the two types of spores develop in separate types of cones. Cones are made up of highly modified sporophylls called **scales,** and sporangia are borne on the scales.

Microspores are produced in male cones that develop in clusters near the tips of branches. Male cones are usually not more than 1 or 2 cm long. Two **microsporangia** (microspore-producing sporangia) are located on each scale in a male cone. Within microsporangia, **microspore mother cells** divide by meiosis to produce microspores. Microspores develop into pollen grains that have a thickened, protective coat and a pair of flattened, winglike structures on their sides. Inside each pollen grain, two mitotic divisions produce four haploid cells. Two of the haploid cells are small and soon degenerate. The other two, the **tube cell** and the **generative cell,** become involved in further development after pollination. Pine trees release so much pollen that during the pollen season, everything in the area around pine trees can be covered with a dusting of yellow, powdery, pine pollen.

Megaspores are produced inside ovules that occur in pairs on the upper surface of female cone scales. Each ovule is surrounded by a thick, layered **integument,** which has an opening called the micropyle at one end. A single **megaspore mother cell** inside each ovule divides meiotically to produce a row of four megaspores. Three of the four degenerate, and the fourth develops slowly into a female gametophyte that contains several thousand cells. The mature female gametophyte produces several (two to six) archegonia at the end near the micropyle. A single large egg develops inside each archegonium. These events, from megaspore production to completion of female gametophyte development, require more than a year. During all of this time, the developing female gametophyte is enclosed in and dependent on the sporophyte plant body of the pine tree.

Pollination occurs during the first spring of a female cone's life, while the female gametophyte is still developing. The scales of the green cone separate, and pollen grains fall between the scales, where they become trapped in sticky material near the micropyles. After pollination, scales grow together, closing the cone again, and the sticky material shrinks back, drawing the pollen grains into the micropyle. The tube cell of the pollen grain develops a **pollen tube,** which slowly grows into the sporangium, and the generative cell enters the tube. The generative cell divides to produce two cells, one of which divides again to produce two sperm. A pollen grain with its pollen tube and two nonflagellated sperm is a fully developed pine male gametophyte.

The pollen tube reaches the archegonium and discharges sperm about fifteen months after pollination. One sperm unites with the egg to form a zygote. Clearly, pollination and fertilization are completely separate events in pine reproduction. Fertilization occurs in several archegonia in each ovule, and several zygotes begin to develop, but normally only one embryo completes its development as the seed is formed.

(a)

(b)

Figure 36.20 Conifers. (a) A giant sequoia *(Sequoiadendron).* (b) Bristlecone pine *(Pinus aristata).* Some living bristlecone pines are known to be over 4,000 years old. What may have been the oldest living thing, a 4,900 year old bristlecone pine, unfortunately was cut in 1965 to determine its age. (c) Pine needle cross section. The hypodermis below the epidermis is a further barrier to surface water loss. Injury to the needle causes resin ducts to release resin, which closes wounds.

Epidermis
(with cuticle)

Photosynthesizing cells   Vascular bundle

Resin duct   Hypodermis

Stoma

(c)

Seed development proceeds with accumulation of nutrients in the gametophyte tissue around the embryo. An integument layer hardens to form the **seed coat,** and a thin membranous layer of the cone scale becomes the seed "wing." Finally, in the third season of their lives, pine cones, by now woody and hard, open to release their seeds. Seeds germinate under appropriate conditions to begin the growth of new pine trees, the sporophyte plants, and the cycle is completed (figure 36.21).

The reproductive pattern of conifers has several important advantages over reproduction in other plants considered so far. These differences make the conifers better adapted for terrestrial life. Transfer of pollen grains and growth of the pollen tube eliminate the requirement of surface water for swimming sperm. Enclosure of the dependent female gametophyte inside a cone protects it during its development and shelters the developing zygote as well. Finally, the embryo is protected by the seed and provided with a store of nutrients that support development for the first period of its growth following germination. All of these factors increase chances for reproductive success.

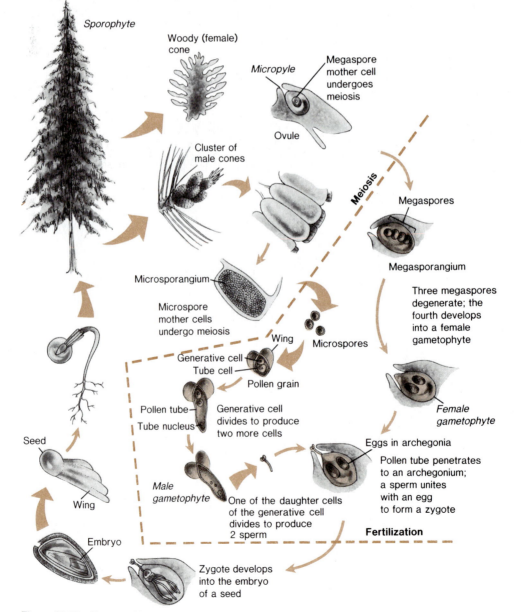

Figure 36.21 Pine tree life history.

Class Angiospermae The angiosperms, or flowering plants, are an exceptionally large and successful group of plants. At least 250,000 species of trees, shrubs, vines, and herbs belong to this class. Angiosperms range in size from tiny pond surface plants, which are only 0.5 mm in diameter, to very large trees (figure 36.22). The oldest fossils definitely recognized as angiosperms come from the Cretaceous period, which began only about 130 million years ago. Yet during this relatively short evolutionary history, the angiosperms have diversified and multiplied tremendously.

Angiosperms get their name from one of their most important characteristics. Angiosperm means "enclosed seed," and it describes the arrangement in which ovules are enclosed within an **ovary** that is situated at the base of the **pistil** of the flower. The ovary provides protection for the delicate developing female gametophytes inside the ovules and, later, for developing embryos and the seeds that enclose them.

Ovaries also produce fruits, either alone or together with other adjacent flower tissue. Fruits provide protection for seeds and a mechanism for their dispersal. Animals carry fruit and thereby transport seeds to locations often far from the plant that produced them.

Angiosperms have well-developed vascular and supporting tissues that make them very well adapted for terrestrial life. Their xylem tissue is different from that of other vascular plant groups because they have xylem vessels as well as tracheids. Other vascular plants, including virtually all gymnosperms, have only tracheids in their xylem.

The angiosperms are divided into two subclasses: the Dicotyledonae and the Monocotyledonae. Dicots and monocots differ in several important ways (see table 36.3). Common names for dicot families include many familiar plant groups, such as the buttercup, mustard, maple, cactus, carnation, pea, and rose families. The rose family, for example, includes roses, apples, plums, pears, cherries, peaches, strawberries, raspberries, and a number of other shrubs. Monocot families include lilies, palms, orchids, irises, and grasses. The grass family, for example, includes wheat, rice, corn (maize), and other agriculturally important plants.

Flower

Figure 36.22 Scanning electron micrograph of a duckweed *(Lemna paucicosta)*. Other pond surface angiosperms are even smaller. (Magnification × 21)

Table 36.3
Differences between Monocots and Dicots.

Monocots	Dicots
Embryo has one cotyledon (seed leaf).	Embryo has two cotyledons (seed leaves).
Leaves have parallel veins.	Leaves have nets of veins (branched and rebranched).
Leaf edges smooth.	Leaf edges usually lobed or indented.
Flower parts in threes or multiples of threes.	Flower parts in fours or fives or multiples of them.
Stem vascular bundles are scattered.	Stem vascular tissue is solid mass in center or ring of bundles between cortex and pith.
Cambium absent.	Cambium present.

Reproduction in angiosperms is generally similar to gymnosperms in that the gametophyte generation members are small and inconspicuous. But angiosperm pollen and ovules are produced in flowers, rather than in cones.

A flower is a cluster of modified leaves on a specialized stem tip, the **receptacle. Sepals** and **petals,** the outside parts, are not directly involved in reproductive processes, but they are important because they enclose and protect reproductive parts and because they attract insects to insect-pollinated plants.

Inside the petals are **stamens,** and at the center of a flower is the **pistil.** Stamens bear **anthers,** the sites of microspore production. The pistil is a complex, multi-part structure that consists of **stigma, style,** and **ovary.** Usually, several to many ovules develop inside the ovary. **Perfect** flowers have all of these elements, while **imperfect** flowers lack either stamens or pistils.

Details of reproductive events in a representative dicot with perfect flowers are discussed in chapter 23. Reproductive events in *Zea mays,* a representative monocot, are described here.

Corn (*Zea mays*) is called "maize" by Native Americans and by most English-speaking people outside of the United States. "Corn" to them means cereal grain generally and often barley in particular.

Zea mays has imperfect flowers. Staminate (pollen-producing) flowers occur in clusters (the tassel) at the top of the shoot. Pistillate flowers are borne in clusters lower on the plant. Drawn-out stigmas of many pistillate flowers make up the familiar bundles of corn "silk."

Microspores are produced by meiosis in the anthers. The microspores develop into pollen grains, which are released and carried by wind. Self-pollination, cross-pollination between plants, or both may occur. In production of hybrid seed corn, cross-pollination is assured by removal of tassels (staminate flowers) from one of the two genetic lines being crossed to produce the hybrid. "Detasseling" in seed corn fields is a summer job for students in a number of areas of the United States.

Meiosis in the ovule produces four haploid (*N*) megaspores, three of which degenerate. Mitotic nuclear divisions in the remaining megaspore produce a multinucleate cell, the **embryo sac.** Then cell walls partition the embryo sac. An egg along with two synergid cells, which are not involved in subsequent reproductive events, are at one end, and a cluster of reproductively nonfunctional antipodal cells are at the other end. A pair of nuclei, the polar nuclei, remain together in the middle of the embryo sac. This group of cells is the fully developed female gametophyte. It is still completely enclosed in the ovule, ovary, and flower of the sporophyte plant.

When a pollen tube grows into the ovule, it delivers two sperm nuclei. One fuses with the egg nucleus to produce the zygote. The other sperm nucleus fuses with the two polar nuclei to produce a triploid (*3N*) **primary endosperm nucleus.** This double fertilization is characteristic of angiosperms in general. The zygote develops into an embryo sporophyte plant, and the endosperm develops into a nutrient storage tissue (figure 36.23).

In addition to the seed, consisting of embryo, endosperm, and hardened ovule tissue, the ovary also contributes a hard covering to the corn **grain.** Botanically, this hard, dry ovary tissue is classified as a fruit. Grains of wheat, rice, barley, and oats all are fruits as well, even though they do not fit the familiar notion that a fruit should contain moist, fleshy tissue. A fruit is any covering that forms over seeds and that is derived from the ovary, or the ovary and other adjacent tissue.

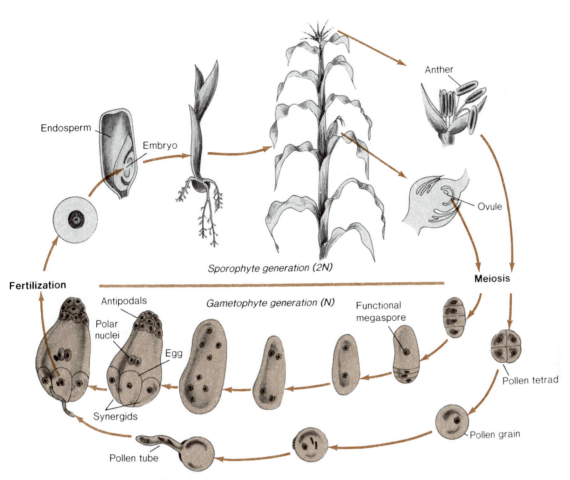

Figure 36.23   (Corn) *Zea mays* life history.

Summary

The roles of gametophyte and sporophyte generations in life cycles in various plant groups is related to the relative success of the groups as terrestrial organisms (figure 36.24). Mosses and liverworts, with their dominant gametophytes and small, dependent sporophytes, have limited distributions determined by external water supply.

While ferns have well-developed sporophyte bodies with vascular tissues, they still require very wet conditions for the growth of their small, independent gametophytes and for fertilization with swimming sperm.

The entire lives of gymnosperms and angiosperms can be lived on dry land because the large, dominant sporophyte is well adapted to terrestrial life. Small dependent gametophytes and delicate spores, gametes, zygotes, and embryos are enclosed within protective coverings that are parts of the sporophyte plant.

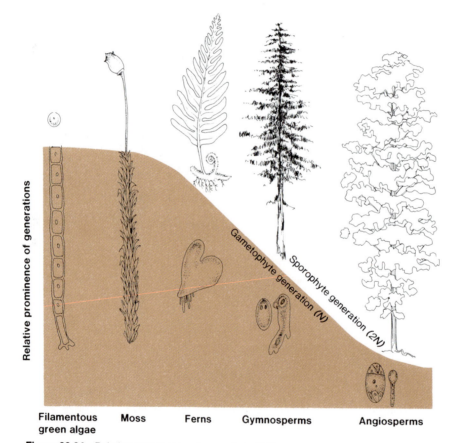

Figure 36.24 Relative prominence of sporophyte (2N) and gametophyte (N) generations. Some algae (for example, *Ulva*) do have separate gametophyte and sporophyte thalli, but generally gametophyte is more prominent in algae. Gametophyte is greatly reduced in vascular plants.

Questions

1. Explain how classification of the green algae illustrates the problems inherent in assigning organisms to the categories of a classification scheme.

2. It is often implied that "lower" plants (nonvascular plants) do not attain large size. Give an example that contradicts that implication.

3. What features of the division Bryophyta prevent its members from spreading widely in terrestrial environments?

4. Discuss the importance of vascular tissue as an adaptation for terrestrial life in plants.

5. Explain how the flowers and seeds of angiosperms function as adaptations for terrestrial life.

Suggested Readings

Books

Bold, H. C., and Wynne, M. J. 1978. *Introduction to the algae: Structure and reproduction.* Englewood Cliffs, N.J.: Prentice-Hall.

Raven, P. H.; Evert, R. F.; and Curtis, H. 1981. *Biology of plants.* 3d ed. New York: Worth Publishers.

Stern, K. R. 1982. *Introductory plant biology.* 2d ed. Dubuque, Iowa: Wm. C. Brown Company Publishers.

Trainor, F. R. 1978. *Introductory phycology.* New York: Wiley.

Weir, T. E.; Stocking, C. R.; and Barbour, M. G. 1974. *Botany: An introduction to plant biology.* 5th ed. New York: Wiley.

Articles

Beadle, G. W. January 1980. The ancestry of corn. *Scientific American* (offprint 1458).

Mulcahy, D. L. 1981. Rise of the angiosperms. *Natural History* 90(9):30.

Rickson, F. 1974. Plant diversity: Organisms. Biocore Unit XV. New York: McGraw-Hill.

Sporne, K. R. 1971. The mysterious origin of flowering plants. *Carolina Biology Readers* no. 3. Burlington, N.C.: Carolina Biological Supply Co.

Watson, E. J. 1972. Mosses. *Carolina Biology Readers* no. 29. Burlington, N.C.: Carolina Biological Supply Co.

Survey of Animals

Biologists classify more than one million species of organisms as **animals,** and within this huge array there is a very wide range of diversity. What common characteristics led biologists to establish a **Kingdom Animalia** that includes organisms ranging from sponges to elephants? What characteristics are shared by jellyfish and human beings?

All animals are multicellular and heterotrophic. That is, animals have bodies that consist of aggregates of many cells, and they must take in organic compounds from their environments to obtain chemical energy. The ultimate source of organic compounds (reduced carbon compounds) is photosynthetic activity in autotrophic organisms, and animals can obtain nutrients by consuming autotrophic organisms directly or by consuming other animals that have eaten autotrophic organisms. For example, lions eat zebras that have eaten plants.

In addition to this basic requirement for external sources of reduced carbon compounds, animals also must obtain usable nitrogen compounds (for amino acid, protein, and nucleic acid synthesis), certain ions, water, and oxygen from their environments. They must eliminate various metabolic wastes: carbon dioxide, nitrogenous wastes from amino acid metabolism, and in many cases, excess water.

In animals that do not have thick bodies, gas exchange and waste elimination depend on direct interactions between individual body cells and the animals' environments. But even these animals have specific body cells that are specialized to obtain organic nutrients for all of the body cells.

In fact, virtually all animals, except some parasites living inside the bodies of other organisms, have internal **digestive cavities** lined with cells that function specifically in obtaining nutrients. Thus, animal bodies are hollow and have at least two body layers, an external covering layer (**ectoderm**) and a layer of cells lining a digestive cavity (**endoderm**). All but the simplest animals also have a third middle body layer (**mesoderm**).

Most free-living (nonparasitic) animals move about actively and consume nutrients as they find them, but other animals remain in one place, usually attached to a surface. These **sessile** (attached) animals are all aquatic and obtain nutrients from water that is moved over body surfaces.

An important principle of biology is the close correlation between structure and function in bodies of organisms. Some of the most obvious correlations in animals are between body organization and the method of nutrient procurement. More complex structure-function correlations are found in animals that have distinct regional body specializations for other functions, such as gas exchange, waste elimination, movement, internal regulation (nervous and hormonal), circulation, and reproduction.

Over one million animal species have been identified. The **vertebrate** group of animals (animals with backbones) to which humans and domesticated mammals and birds belong is only one subphylum within the phylum Chordata. There are only a little over 50,000 known species of vertebrate animals, and more than 30,000 of these are fish. The great majority of the more than 1 million known animal species are **invertebrates** (animals without backbones). This survey of structural and functional diversity in the animal kingdom begins by considering eight "major" phyla of invertebrate animals as well as several interesting "minor" groups of invertebrates.

Phylum Porifera (The Sponges)

Sponges are very simply organized animals. They consist of loose aggregates of cells that have very little functional specialization. The majority of sponges are marine, but there are a few small freshwater sponges.

Adult sponges are sessile; that is, they remain attached to surfaces. They obtain nutrients from a stream of water that they continually move through their bodies. Water enters the central cavity, the **spongocoel,** through pores in the body wall (figure 37.1). The pores are passageways through special **pore cells.** This characteristic gives the phylum its name, as Porifera means "pore bearer." Water moves through the spongocoel and out through a single opening called the **osculum.**

A number of special flagellated cells, called **choanocytes** (or **collar cells**), in the lining of the spongocoel move considerable quantities of water through sponges; simple sponges only 10 cm tall are estimated to filter as much as 100 l of water each day. Choanocytes trap food particles from

Figure 37.1 Phylum Porifera. (a) Representative sponges. (b) A simple sponge on the left and a diagram of the body organization of such a sponge shown in longitudinal section on the right. Flagella of choanocytes cause water movement through the sponge. Water passes through pores, which are enclosed in porocytes (pore cells), into the spongocoel and then out through the osculum. (c) (facing page) Sponge cell types.

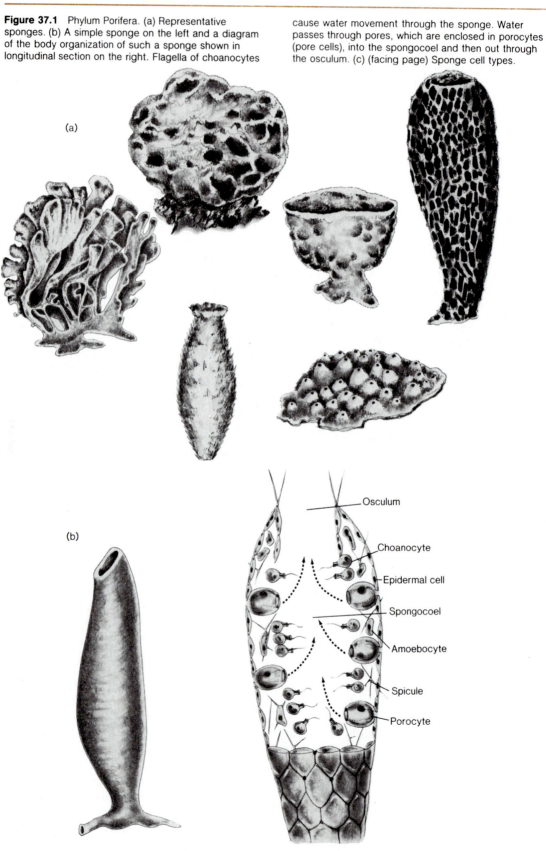

(a)

(b)

Osculum

Choanocyte

Epidermal cell

Spongocoel

Amoebocyte

Spicule

Porocyte

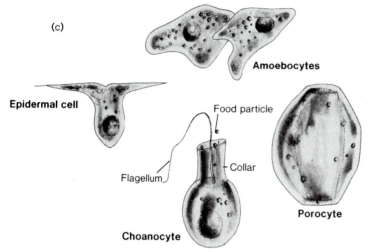

(c)

Amoebocytes

Epidermal cell

Food particle

Flagellum

Collar

Choanocyte

Porocyte

the water by phagocytosis. Part of this trapped food is used by the choanocytes, and part of it is transferred to wandering cells called **amoebocytes** that transport nutrients to other body cells.

The sponge body consists of two cellular layers: the layer containing choanocytes that lines the spongocoel and an outer covering layer of flattened **epidermal cells.** Between these two layers is a space containing a jellylike material and a few scattered cells. Some of these cells are the wandering amoebocytes; others are cells that produce supporting (skeletal) structures.

Some sponges are supported by **spicules,** which are crystalline structures made of calcium carbonate or silica (glasslike) compounds (figure 37.2). Other sponges have skeletons made of **spongin,** a fibrous protein. Commercial sponges are prepared by beating spongin-containing sponges until all of the living cells are removed and just the skeleton remains. Natural sponges, however, have largely been replaced by manufactured artificial sponges.

Sponges reproduce sexually. Eggs and sperm are released into the water, where fertilization occurs. The zygote develops into a ciliated larva that swims freely for a time before it settles, attaches, and develops into a small sponge.

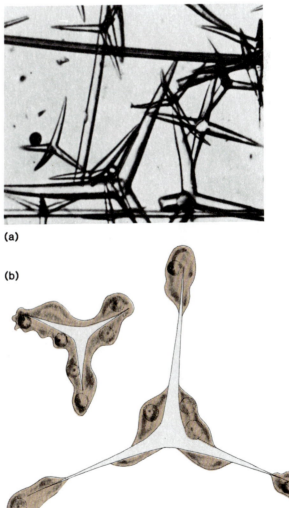

(a)

(b)

Figure 37.2 Sponge spicules. (a) Photomicrograph of spicules. (b) Two steps in the process of spicule formation. Several cells cooperate in an organized fashion and secrete the material making up the spicule.

Many sponges are simple and small with pores leading directly from the outside water into the spongocoel. Other sponges are more complex as they have canals leading to pores that are located internally (figure 37.3). Sponges of some species develop into large masses containing many individual sponges. These masses often have many canals and a number of spongocoels, each with its own osculum. Commercial sponge growers chop such masses into small pieces and separate them. The fragments grow to produce many sponges.

A dramatic demonstration of the flexibility of sponge body organization was reported by H. V. Wilson in 1907. He cut sponges into small pieces and squeezed the pieces inside cheesecloth bags so that individual cells or small clusters of cells were **dissociated** (separated) from each other. These dissociated cells, if left undisturbed, moved about and reaggregated into small clusters. Eventually, cells within each of these clusters organized themselves into a small but normally functional sponge.

Figure 37.3 Canal systems in sponges. In the simplest sponges (a), water flows in directly through pores. Other types of sponges have thicker bodies with networks of canals leading to internal pores (b and c).

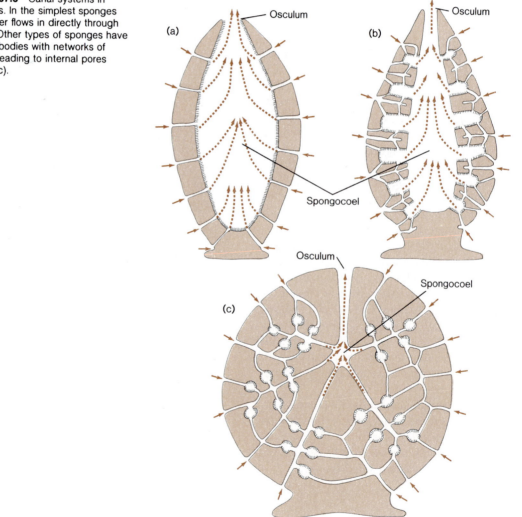

Such disruptive treatment would kill other animals. Even if their isolated cells survived and reaggregated, the cells would not reorganize whole bodies in the same way that sponge cells do. For this reason, biologists believe that some fundamental differences exist between the relationships among the cells making up sponges and the cells making up other animals.

Phylum Coelenterata

Coelenterates are a large group of simply organized aquatic animals. Some, such as the **hydras,** which are familiar to many biology students, live in fresh water, but the great majority of coelenterates are marine organisms. The marine coelenterates include **jellyfish, sea anemones, corals,** the **Portuguese man-of-war,** and others.

All coelenterates have the same basic body plan. Their bodies have well-developed ectodermal and endodermal layers, which are called the **epidermis** and **gastrodermis,** respectively. Between these layers is a jellylike layer, the **mesoglea,** that contains only a few scattered cells. The fibrous processes of cells that make up a rather extensive **nerve net** also run through the mesoglea.

Coelenterates have a single opening, a mouth, leading to an internal cavity called the **gastrovascular** cavity because it has both digestive ("gastro") and transport ("vascular") functions.

Coelenterates are **radially symmetrical;** that is, their bodies are arranged around a central axis so that splitting them lengthwise along any plane that passes through the central axis results in the separation of two equal halves (figure 37.4). Most other animals are **bilaterally symmetrical;** they have definite anterior and posterior ends, and right and left halves. There is only one midline plane that separates them into equal halves.

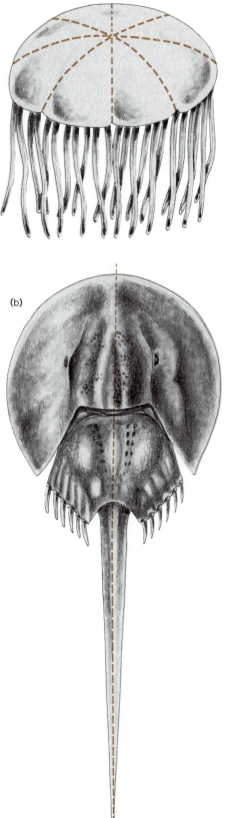

(a)

(b)

Figure 37.4 Types of animal symmetry. (a) A radically symmetrical animal can be sliced on any of several or many planes along the main body axis to yield two equal, mirror-image halves. (b) This sketch of a horseshoe crab (an arthropod) shows that there is only one plane along which a bilaterally symmetrical animal can be sliced to yield two equal, mirror-image halves.

The basic coelenterate body plan has two expressions. The **medusa** has a rounded, somewhat flattened, jellyfish form and swims about with its mouth directed downward. A **polyp** generally is sessile and is more slender and elongate than a medusa, and a polyp's mouth usually is directed upward. The majority of polyps live attached to surfaces (figure 37.5). Most coelenterates have rather complex life cycles that include both polyp and medusa stages.

Coelenterates are carnivores. They use their **tentacles** to maneuver other animals through their open mouths into their gastrovascular cavities. Coelenterates can capture free-swimming prey because the epidermis of the tentacles has numerous cells containing stinging devices called **nematocysts** (figure 37.6). Nematocysts contain coiled fibers that discharge on contact. Some nematocyst fibers are simply threads that tangle prey, but others are hollow and have sharp, pointed tips that penetrate and inject a paralyzing substance into prey organisms.

Class Hydrozoa

Freshwater hydras are the most familiar hydrozoan coelenterates, and they illustrate some important principles of coelenterate biology, such as methods of prey capture and feeding behavior (figure 37.7). But hydras differ from the average coelenterate because their life histories do not include medusa stages.

There is considerable cell specialization in hydras, but practically all of it is directed toward food gathering and digestive functions. Thus, some specialized cells obtain nutrients used by the whole organism, while other functions, such as waste excretion and gas exchange, are accomplished by every individual cell for itself. Each individual cell exchanges materials with water around the body or water inside the gastrovascular cavity.

Cells in the epidermis are specialized for the first stages of nutrient procurement. The epidermis contains **sensory cells,** nematocysts, and **epitheliomuscular cells.** Epitheliomuscular cells

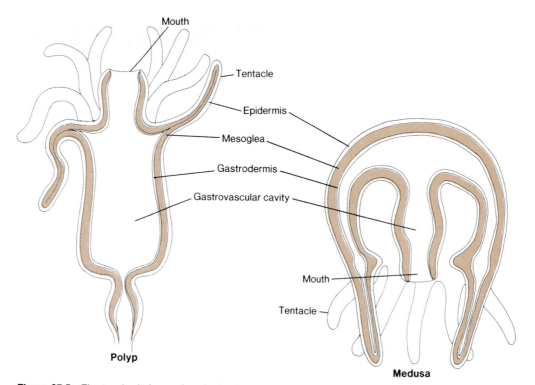

Figure 37.5 The two body forms of coelenterates. Polyps are attached to surfaces; medusae are free-swimming.

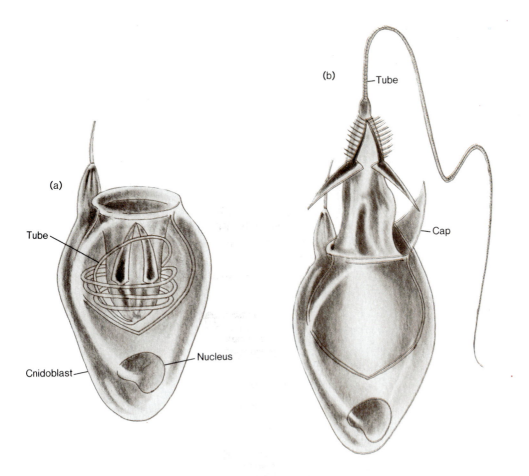

(a)

Tube

Cnidoblast

Nucleus

(b)

Tube

Cap

Figure 37.6 Nematocysts. (a) A nematocyst is contained within a specialized cell, the **cnidoblast.** (b) When a nematocyst is discharged, a cap opens and the tube shoots out. (c) Electron micrograph of an undischarged hydra nematocyst.

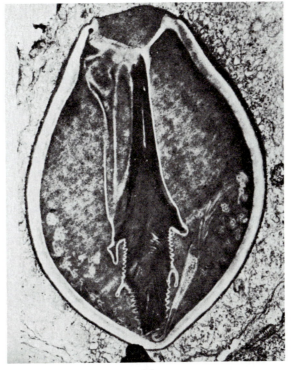

(c)

Figure 37.7 Body organization of a hydra.

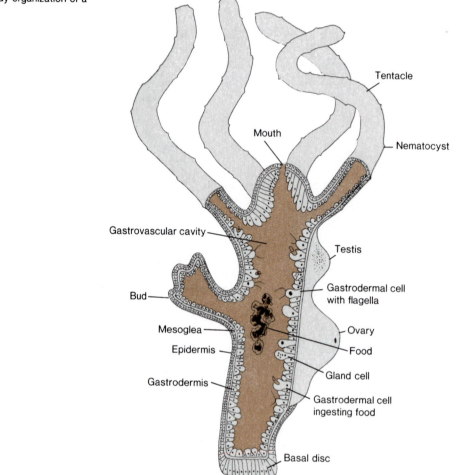

- Tentacle
- Mouth
- Nematocyst
- Gastrovascular cavity
- Testis
- Gastrodermal cell with flagella
- Bud
- Ovary
- Mesoglea
- Food
- Epidermis
- Gland cell
- Gastrodermis
- Gastrodermal cell ingesting food
- Basal disc

contract under control of the nerve fibers of the nerve net and provide the coordinated body movements needed for feeding and other activities. The gastrodermis also contains contracting cells, as well as **gland cells** that secrete digestive enzymes into the gastrovascular cavity.

After food is captured by the tentacles and pushed into the gastrovascular cavity, enzymes secreted into the cavity begin digestion. Then gastrodermis cells phagocytize chunks of this partially digested food material and incorporate them into vacuoles, where digestion continues. Thus, digestion in hydras (and other coelenterates as well) is a combination of **extracellular** digestion in the gastrovascular cavity and **intracellular** digestion inside vacuoles in gastrodermis cells.

Hydras can reproduce asexually and sexually. They reproduce asexually by forming **buds,** small outgrowths containing the body layers and

an extension of the gastrovascular cavity (figure 37.8). After tentacles and a mouth form, the bud pinches off and becomes an independent hydra. Small, unspecialized **interstitial** cells are important in bud growth. They also function throughout the life of the hydra in replacement of lost body parts or individual cells because they can differentiate and replace any of the cell types in the body.

Sexual reproduction only occurs at certain times in response to environmental stimuli, especially increased CO_2 concentration in the water. **Spermaries** that develop on the outside of the upper half of the body produce sperm cells. The female gonad, an **ovary,** develops lower on the body. The egg, which is produced in the ovary, remains attached to the adult body, and fertilization and early development occur there. There is a pause in development during which the embryo becomes **dormant** (physiologically inactive) and is surrounded by a hard, protective

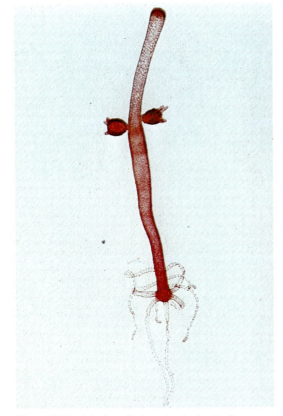

Figure 37.8 A hydra with two developing buds.

shell. Such embryos survive even when the adult bodies die and degenerate. Embryos live through winter (or other unfavorable periods) in this condition and then emerge in the spring to produce new polyps.

Obelia, a colonial, marine hydrozoan, is a more typical member of Class Hydrozoa than the hydras because *Obelia* has both polyp and medusa stages in its life cycle (figure 37.9). *Obelia* forms a colony of polyps, all of which are attached by hollow stems that connect the gastrovascular cavities of all polyps in the colony. These connections are important because some polyps are **feeding polyps** while others are **reproductive polyps** that lack tentacles. Thus, feeding polyps gather food used by all polyps in the colony.

Small, translucent medusae bud off the reproductive polyps and drift away in water currents. They produce gametes that are released into the water, where fertilization occurs. Each zygote develops into a small, solid, ciliated larva called a **planula.** Eventually, the planula settles

to the bottom, attaches, and develops into a little polyp that starts a new colony.

One of the most unusual hydrozoans is the Portuguese man-of-war *Physalia* (figure 37.10). *Physalia* looks as if it might be an odd-shaped medusa, but actually it is a colony of polyps. One polyp is specialized as a gas-filled float that provides buoyancy to keep the colony afloat. Other polyps are specialized for feeding or for reproduction. The Portuguese man-of-war also has fighting polyps armed with numerous nematocysts. Swimmers who accidentally encounter Portuguese men-of-war can receive painful, sometimes even very serious, injuries from these fighting polyps.

Class Scyphozoa

Because the medusa stage is the dominant phase of scyphozoan life histories, the members of this class sometimes are called the "true jellyfish." The scyphozoan planula attaches and develops into a relatively small polyp stage that buds off new medusae (figure 37.11).

Class Anthozoa

Anthozoans include the sea anemones and the corals. These animals have no medusa stages. Some species have solitary polyps, and others are colonial. Because their gastrovascular cavities are divided by partitions, they have greatly increased surface areas for digestion and absorption.

Sea anemones are fleshy animals, some of which are large enough to capture and digest fish or crabs (figure 37.12a). Anemones often live along coasts in areas between the high and low tide water marks (the intertidal zone).

Corals live as solitary individuals in cold-water areas, but tropical corals form large, spreading colonies. Corals secrete hard, limy skeletons that remain in place even after the polyps die and degenerate. Each coral polyp is rather small and inconspicuous, but new generations of corals grow on the skeletal remains of past generations. In this way, massive deposits formed by corals and the coralline algae that live with them build up over the years (figure 37.12b). Coral islands and various reefs and atolls are prominent oceanic features, especially in the South Pacific ocean. These coral formations provide sheltered habitats within which a great variety of organisms live and thrive.

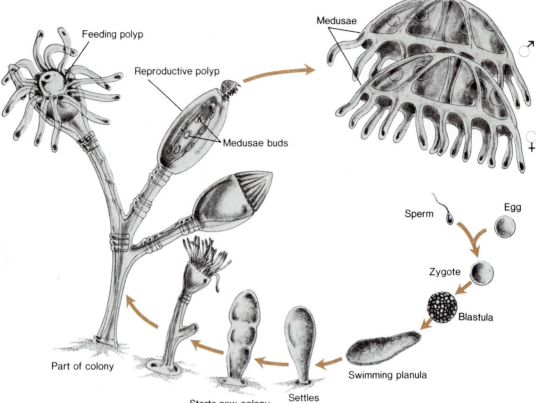

Feeding polyp

Reproductive polyp

Medusae buds

Medusae

Sperm

Egg

Zygote

Blastula

Swimming planula

Settles

Starts new colony
by asexual budding

Part of colony

Figure 37.9   The life cycle of *Obelia*, a member of the
class Hydrozoa. The polyp stage is the larger, more
prominent stage of the life cycle.

Figure 37.10   A Portuguese man-of-
war *(Physalia).*

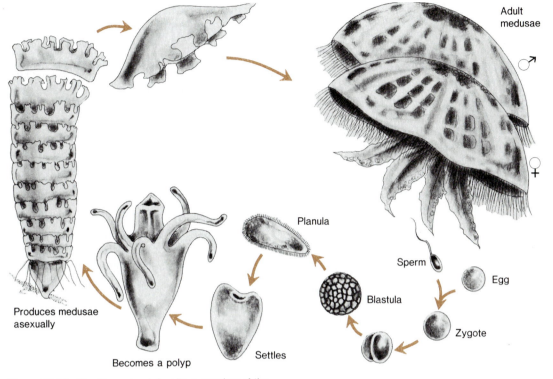

Figure 37.11 The life cycle of *Aurelia*, a member of the class Scyphozoa. The medusa stage is the larger, more prominent stage of the life cycle.

(a)

Figure 37.12 Class Anthozoa. (a) Sea anemones *(Metridium)*. (b) A collection of various corals. (1) Star coral. (2) Brain coral. (3) Staghorn coral. (4) Mushroom gill coral.

(b)

Phylum Platyhelminthes (Flatworms)

Flatworms are the most simply organized of the bilaterally symmetrical animals, but their elongate, flattened bodies have a number of features that are more complex than those seen in coelenterates. Flatworms have a solidly packed mesodermal layer between their ectoderm and endoderm. This mesoderm contains well-developed reproductive organs and an excretory system. The presence of definite **organs** is an important feature of flatworm body organization. Organs are functionally specialized structures consisting of several tissue types.

Flatworm body organization is more complex than that of coelenterates, yet flatworms are primitive in the sense that their digestive tracts have only one opening, a mouth, and they have no specialized circulatory or respiratory structures. Thus, gases have to be exchanged between all individual body cells and the environment. Oxygen and carbon dioxide diffuse through the body surface or through the lining of the gastrovascular cavity.

There are three classes of flatworms. The **class Turbellaria** includes free-living animals such as the freshwater **planarians.** Members of the **class Trematoda** (flukes) and the **class Cestoda** (tapeworms) are all parasites.

Class Turbellaria

Most of the members of class Turbellaria are marine organisms, but the freshwater planarians of the genus *Dugesia* are commonly studied representatives of the group.

Planarian worms move about along slimy mucous trails that they secrete as they go. They are propelled by ciliated cells on their body surfaces.

Planaria feed on dead organisms or on small, living animals. A planarian has an eversible **pharynx,** which is extended during feeding and retracted at other times. The pharynx is quite muscular and can produce such a strong sucking action that it tears off food chunks, which are then drawn into the gastrovascular cavity. Planarian digestion is a combination of extracellular and intracellular digestion. Undigested material must be eliminated through the mouth (figure 37.13).

Bilaterally symmetrical animals, such as planarians, that move about actively show **cephalization;** that is, they have definite **heads** with concentrations of sensory structures on the anterior ends of their bodies. On their heads, planarians have pigmented **eyespots** that are sensitive to light intensity changes and **chemoreceptors** that detect potential food sources. The planarian head area contains a concentration of nerve cells that functions as a primitive brain. The brain is connected with the rest of the body by a pair of nerve cords (figure 37.14a).

A network of **flame cells, excretory canals,** and **excretory pores** function as an excretory system (figure 37.14b). This system serves mainly to eliminate excess water that enters the body osmotically.

Planarians have tremendous powers of regeneration and actually can reproduce asexually by constricting their bodies and splitting in half. After splitting occurs, the posterior half grows a new anterior region, and the anterior half grows a new posterior region. Experiments on planarian regeneration have yielded a great deal of basic information about regeneration, and many students have done their first regeneration experiments by cutting away parts of planarians with ordinary single-edge razor blades.

Sexual reproduction in planaria involves a mutual exchange of sperm between a pair of planarian worms. Planarians are **hermaphroditic;** that is, each individual has both male and female reproductive organs. There are no larval stages. Zygotes develop into small worms that hatch and grow directly to become adults.

(a)

Figure 37.13 (a) Living planaria. (b) A planarian with its pharynx extended in feeding position. (c) Organization of the planarian digestive (gastrovascular) cavity. The mouth is the only opening into the cavity. Small branches of the cavity penetrate much of the body and distribute material to body tissues.

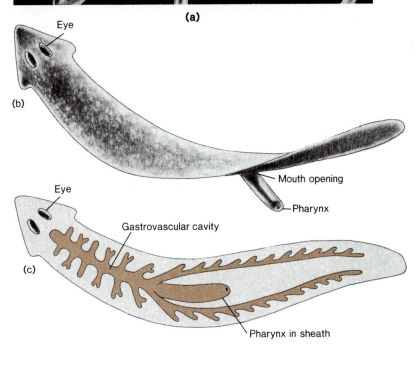

(b)

Eye

Mouth opening

Pharynx

(c)

Eye

Gastrovascular cavity

Pharynx in sheath

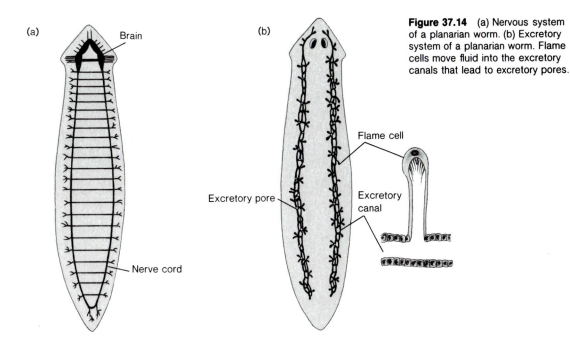

(a)

Brain

Nerve cord

(b)

Excretory pore

Flame cell

Excretory canal

Figure 37.14 (a) Nervous system of a planarian worm. (b) Excretory system of a planarian worm. Flame cells move fluid into the excretory canals that lead to excretory pores.

Class Trematoda (Flukes)

Flukes are **parasites:** they derive nourishment from a living **host** organism. A few flukes are **ectoparasites** (external parasites), but the majority are **endoparasites,** living inside the bodies of their hosts.

Flukes are socially and economically important because they infect agricultural animals and because some of the most widespread human parasitic diseases result from fluke infection.

An adult fluke feeds on host tissues or body fluids by using its muscular **pharynx** to suck material into its digestive cavity.

Flukes have highly specialized reproductive systems, and much of the space inside a fluke's body is occupied by reproductive structures. Most species of flukes are hermaphroditic and can produce zygotes either by self-fertilization or cross-fertilization. A fluke can produce huge numbers of eggs. For example, the Chinese liver fluke *Opisthorchis (Clonorchis) sinensis* produces literally hundreds of eggs every day during its lifetime in the human liver, where it can remain active for five, ten, or even twenty years (figure 37.15).

Most flukes have complex life cycles that involve several hosts. For example, humans or other mammals are **definitive** hosts of the liver flukes, such as *Opisthorchis;* that is, sexually reproducing adult flukes live in mammalian livers. Larval stages of flukes live in **intermediate hosts**—snails and then fish in the case of the liver flukes (see chapter 23 for a more extensive discussion of the human liver fluke life cycle). Because transferring from host to host is very risky and many individuals are lost, survival and success of the species depends on massive reproductive rates that increase the odds that at least some individuals will complete the entire life cycle.

Human blood flukes of the genus *Schistosoma* are among the most important of human **helminth** (worm) parasites. It is estimated that very nearly half of all of the people living in the more than seventy tropical and subtropical nations and islands where schistosomes occur are infected by these flukes. *Schistosoma mansoni* is the most widely distributed of the schistosome flukes and causes major public health problems in many nations (figure 37.16).

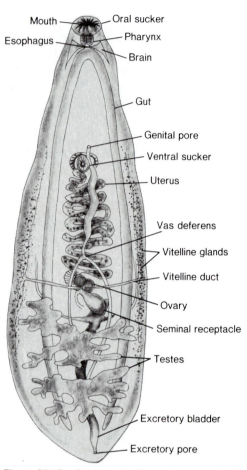

Mouth — Oral sucker
Esophagus — Pharynx
Brain
Gut
Genital pore
Ventral sucker
Uterus
Vas deferens
Vitelline glands
Vitelline duct
Ovary
Seminal receptacle
Testes
Excretory bladder
Excretory pore

Figure 37.15 *Opisthorchis (Clonorchis) sinensis,* the Chinese liver fluke. Each worm contains a complete set of male and female reproductive organs, and one individual worm can produce hundreds of fertile eggs every day for as long as twenty years.

Figure 37.16 Distribution of three species of human schistosome flukes: *Schistosoma mansoni* (a), *Schistosoma japonicum* (b), and *Schistosoma haematobium* (c). *S. mansoni* and *S. japonicum* adults live in intestinal vessels, and their eggs leave the body with feces. *S. haematobium* adults live in vessels of the urinary bladder, and their eggs leave the body with urine.

Figure 37.17 Scanning electron micrograph of a pair of *Schistosoma mansoni* adults copulating. The smaller female lies in a groove along the larger male's body.

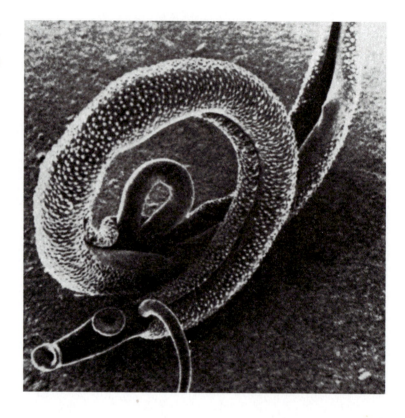

Adult *Schistosoma mansoni* live in blood vessels of the human digestive tract. Schistosome flukes have separate sexes, and the males and females live together in more or less permanent copulatory contact (figure 37.17). Some eggs laid in the blood vessels penetrate the intestinal walls and leave the body with feces, but other eggs accumulate and cause reactions that are responsible for some of the symptoms of **schistosomiasis.** The symptoms include dysentery, anemia, general weakness, and greatly reduced resistance to other infections.

Eggs that reach water hatch and release free-swimming **miracidium larvae.** The miracidium finds and penetrates the body of a snail, usually a member of the genus *Biompholaria.* In the snail's body, it transforms into a **mother sporocyst.** Each mother sporocyst produces many **daughter sporocysts,** each of which in turn produces huge numbers of **cercaria larvae.** These escape from the snail and become free swimming. In one study, scientists observed that more than 200,000 cercaria larvae escaped from a snail infected by one miracidium (figure 37.18).

When a cercaria encounters the skin of a human or another mammal in the water, it actively penetrates the skin (figure 37.19). Once inside the body, the cercaria loses its tail, migrates through the body for a time, and eventually reaches the intestinal vessels at maturity, where the cycle begins again.

Schistosome transmission depends on human feces reaching water, subsequent infection of the appropriate snail intermediate host, and contact between the cercaria and human skin in the water. Rice paddy farming, which is the heart of agriculture in many tropical countries, creates a nearly perfect environment for transmission of these flukes because human feces are used for fertilizer and people work in the water tending the rice plants.

Unfortunately, large modern irrigation projects, such as those made possible by building of the Aswan Dam in Egypt, increase problems with schistosomiasis at the same time that they bring new land into agricultural production. Thus, public health problems must be considered as part of plans for economic development.

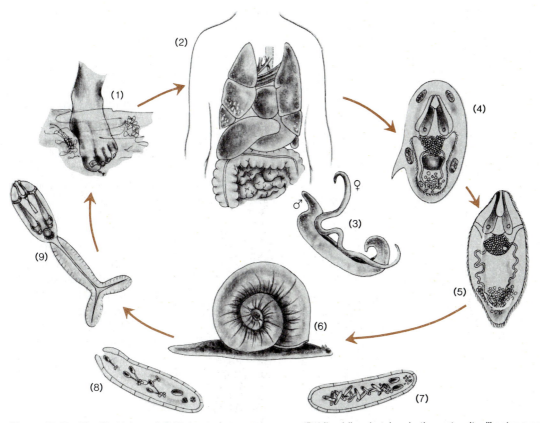

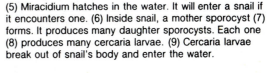

Figure 37.18 The life history of *Schistosoma mansoni*. All of the sketches are drawn to different scales. (1) Cercaria penetrate skin in water. (2) Adult worms live in blood vessels of intestines. (3) Copulating worms. (4) Eggshell containing developing miracidium.

(5) Miracidium hatches in the water. It will enter a snail if it encounters one. (6) Inside snail, a mother sporocyst (7) forms. It produces many daughter sporocysts. Each one (8) produces many cercaria larvae. (9) Cercaria larvae break out of snail's body and enter the water.

Figure 37.19 Scanning electron micrograph of a typical cercaria larva. This specimen is the cercaria of one of the species of *Trichobilharzia*, which parasitizes ducks. This species cannot develop to adulthood in the human body, but the cercaria can penetrate the skin and cause a form of dermatitis called "swimmer's itch." (Magnification X 162)

Class Cestoda (Tapeworms)

Tapeworms are very long, flattened worms that live as adults in the intestines of vertebrate animals, where they attach to the intestinal wall.

Tapeworms have a series of similar body subunits called **proglottids** (figure 37.20). The anterior tip of a tapeworm, the **scolex,** has hooks and suckers that permit firm anchoring in the host's intestinal wall. At the end of a narrow neck, new proglottids are produced continually. These new proglottids push older, maturing proglottids away from the neck. When a proglottid is mature, it contains excretory organs and a complete set of male and female reproductive organs, but no digestive structures. Tapeworms live by absorbing digested nutrients directly from the gut cavities of their hosts.

Sexual reproduction occurs either by self-fertilization or by cross-fertilization between proglottids of different worms. Each proglottid produces thousands (more than 100,000 in some cases) of eggs and stores them in the uterus, which expands until it fills the entire proglottid and crowds all other structures. Such "ripe" proglottids detach and pass out with the feces, scattering zygotes on the ground.

If a zygote is swallowed by an appropriate animal, a larva hatches and enters blood or lymph vessels. The larva is carried by the circulation and eventually enters a muscle, where it forms a hard-walled cyst called a **bladder worm** or **cysticercus.** This muscle cyst will pass to the definitive host if the flesh containing it is eaten raw or poorly cooked. Once the muscle cyst is inside the definitive host, the cyst wall is digested, and the bladder worm develops into a new tapeworm that attaches to the intestinal wall.

Beef tapeworm is rare in the United States, but pork tapeworm *(Taenia solium)* is common enough that great care should be taken to make certain that all pork is thoroughly cooked before it is eaten. Fish tapeworms also are relatively common in the United States and Canada.

Tapeworms generally grow to lengths of from 6 to 8 m in human intestines, and tapeworms approaching 20 m have been found. Tapeworms cause problems for their hosts because they absorb significant quantities of nutrients, excrete toxic wastes, and can sometimes interfere with passage of food through the gut.

Flatworm Parasitism

Flukes and tapeworms clearly illustrate several of the adaptations generally found in endoparasites.

Parasites must make their way from one host to another, but this transmission is seldom direct. Fluke and tapeworm life cycles, like those of many other endoparasites, involve intermediate hosts. While these complex life cycles aid in distributing the parasites, the several transmission steps involved make the odds against successful transfer of a single individual parasite great.

Because of the risks associated with transfer from host to host, the enormous reproductive potentials of endoparasites are important adaptations for parasitism and usually are associated with reduction of body structures not involved in reproduction. Adult tapeworm proglottids, for example, are essentially reproductive sacs. All other functional systems are absent or very reduced. Thus, little metabolic energy is spent on any activity except reproduction.

Massive sexual reproduction is only part of the reproductive strategy of flukes and tapeworms. Asexual reproduction by larvae greatly increases the numbers of individuals. An infected snail can release literally thousands of cercaria larvae for each miracidium that successfully enters its body.

In addition to overcoming the odds against transmission, parasitic adaptations must provide adequate protection against host defense mechanisms. Digestive enzymes kill many organisms that enter vertebrate digestive tracts, but flukes and tapeworms are unaffected by these enzymes. Tapeworms also are remarkably tolerant to the pH changes that occur in the intestinal tract.

Specific vertebrate host defense responses are circumvented in several ways. Flatworm parasites have surface properties that do not elicit strong, immediate immune responses. In fact, *Schistosoma mansoni* larvae actually incorporate host red blood cell antigens into their membranes after they burrow into the body. Thus, they "confuse" the immune system's ability to distinguish self from nonself.

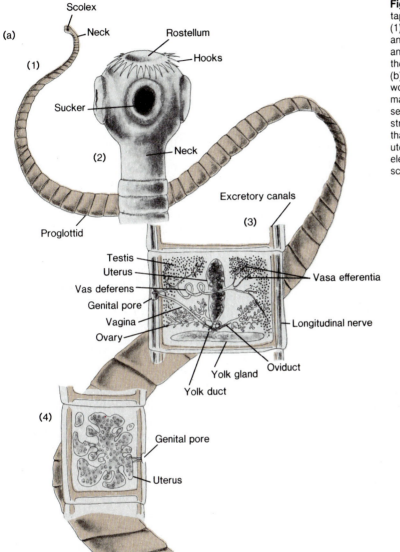

(a)

(1)

Scolex

Neck

Rostellum

Hooks

Sucker

Neck

(2)

Proglottid

Excretory canals

(3)

Testis

Uterus

Vas deferens

Genital pore

Vagina

Ovary

Vasa efferentia

Longitudinal nerve

Oviduct

Yolk gland

Yolk duct

(4)

Genital pore

Uterus

Figure 37.20 (a) The pork tapeworm *Taenia solium*. (1) Tapeworm showing scolex, neck, and proglottids. (2) Details of scolex and neck. Compare the sketch and the scanning electron micrograph in (b). Suckers and hooks attach the worm to the intestinal wall. (3) A mature proglottid with a complete set of female and male reproductive structures. (4) "Ripe" proglottid that is nearly filled by an expanded uterus containing eggs. (b) Scanning electron micrograph of a tapeworm scolex. (Magnification × 105)

(b)

Parasitic relationships are products of long evolutionary processes and very frequently involve a fairly stable balance between host and parasite. Parasites can withstand the host's defenses effectively enough to become established and reproduce successfully. But a parasite is usually not so successful and aggressive that the host is killed. Thus, a "good" parasite lives at the expense of its host, but does not destroy its own "livelihood" by killing the host. Clearly, any significant shift in the balance between host and parasite would have important evolutionary consequences because it would threaten the survival of either the parasite or host species.

While individual parasites usually do limited damage to a host, heavy parasite loads involving many parasites of one type or simultaneous parasitism by several types of parasites frequently do cause serious illness and death. Sublethal parasite infections interfere with a host's ability to live normally.

Parasitic diseases are important impediments to progress in many of the world's underdeveloped areas, and they produce a tremendous burden of human misery.

Some Evolutionary Relationships of Animals

Before discussion of more complex animals begins, the evolutionary roots of the diversity seen in living animals should be examined.

Animal Origins and "Lower" Invertebrates

Virtually all biologists agree that animals evolved from single-celled, protistan ancestors, but they can only speculate about specific details of early animal evolution. Many biologists think that primitive flagellated protists that formed colonies were the ancestors of the early animals. An important step in this early evolution was the specialization of some cells that function as gametes and others that function as ordinary body cells (**somatic cells**). Such specialization does exist in certain present-day colonial flagellates, so there is at least indirect support for this hypothesis.

It is thought that such colonies of flagellated cells then gave rise to all of the more complex animals. Some biologists think that sponges represent a very early branch in the family tree of animals or that sponges arose separately from the rest of animals because they are so different from all other animal groups (figure 37.21).

Some biologists propose that an organism resembling a present-day planula larva may have been the ancestor of all other animals. Although this hypothesis is reasonable, there is little fossil evidence to support it.

Animal origins are shrouded in a rather misty evolutionary past. There are numerous fossils of members of all of the major invertebrate groups dating from the Cambrian period, which began 600 million years ago, but pre-Cambrian invertebrate fossils are not nearly so numerous. Thus, there really is not a clear fossil record of the early evolution of animals.

Digestive Tracts and Body Cavities

Coelenterates and platyhelminthes have **incomplete digestive tracts** because a single opening must serve as both entrance (**mouth**) and exit (**anus**) of the digestive cavity. Most other animal groups have more complex body plans—the most obvious difference being **complete digestive tracts.** In these animals, food is taken in through a mouth, digestion and absorption occur as food moves along through the digestive tract (**gut**), and undigested materials exit the body through an anus.

In the bodies of platyhelminthes, mesoderm, the middle body layer, is solid tissue filling the space between ectoderm and endoderm. The bodies of more complex animals have a different general pattern of body organization: a body cavity separates internal organs from the outer body wall.

Body cavities are organized in two ways. Some animals, such as nematodes (roundworms), have a **pseudocoelom.** In a pseudocoelom, mesoderm lies beneath the ectoderm and is separated from the endoderm of the digestive tract. Other animals have true **coeloms** and are called **coelomate** animals. A true coelom is a body cavity that is completely lined with mesoderm. That is, there is mesodermal tissue lining the inside of the ectoderm, and mesodermal tissue covering the outer surfaces of the digestive tract. The coelom, therefore, is a space between mesodermal lining and covering layers (figure 37.22).

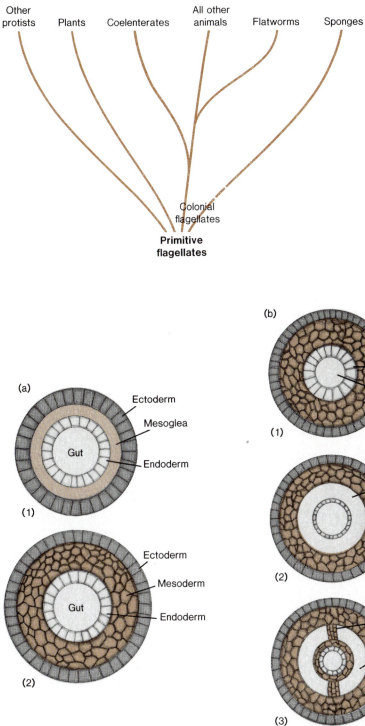

Figure 37.21 A family tree showing proposed evolutionary origins and relationships of all organisms. The broken line indicates that there are questions about sponges' evolutionary relationships to other animals.

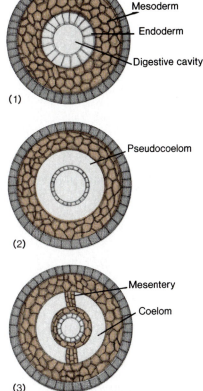

Figure 37.22 Body organization. (a) In coelenterates (1), the ectoderm and mesoderm are separated by mesoglea, which contains scattered cells. In flatworms (2), a solidly packed mesoderm layer lies between the ectoderm and the endoderm. (b) Comparison of mesoderm organization. Flatworms (1) are acoelomate; that is, their mesoderm is packed solidly.

Pseudocoelomate animals (2) have mesodermal tissues such as muscle inside their ectoderm, but not adjacent to their gut endoderm. Coelomate animals (3) also have mesodermal tissue covering their guts. True coeloms have body cavities completely lined by mesodermal tissue. **Mesenteries** hold organs in place within body cavities.

The evolution of coeloms added an important dimension to animal body organization. Muscles that develop from the mesoderm in the body wall and muscles around the gut tube can contract independently. This independent contraction permits food to be moved through the digestive tract without the need for whole body movements.

Coeloms contain fluid that can aid in transport of nutrients, wastes, and gases from place to place in the body. And, in the majority of coelomate animals, a true circulatory system with a pumping heart (or hearts) develops in the space provided by the coelom. Coeloms isolate heart muscle contraction from body wall muscle contraction. As a result, circulation is continued without general body movement and, conversely, body movements do not apply so much pressure that they interfere with heart function. Circulatory systems are essential for all but small, thin-bodied animals. Thus, the evolution of coeloms and the subsequent evolution of efficient circulatory systems were particularly important stages in the evolution of animals with larger bodies.

Protostomes and Deuterostomes

All of the coelomate animals appear to fall into two separate groups, called protostomes and deuterostomes. Much of the evidence for this division is based on differences in the way in which animals in the two groups develop (figure 37.23).

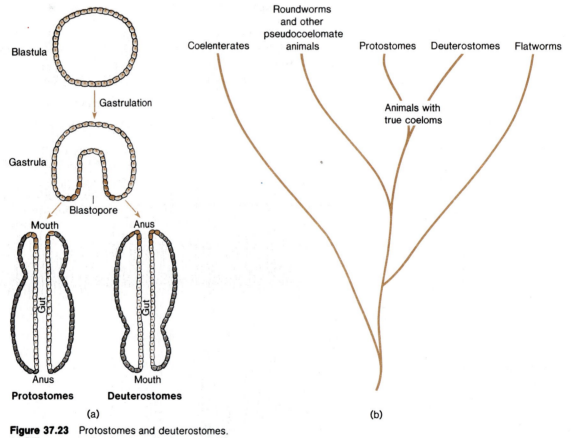

Figure 37.23 Protostomes and deuterostomes. (a) Differences in the developmental pattern of two major groups of animal phyla. In protostome embryos, the blastopore becomes the mouth. In deuterostome embryos, it becomes the anus. (b) "Family tree" of animals showing proposed evolutionary relationships.

During development of an animal zygote (see chapter 23 for more details), a series of mitotic cell divisions and subsequent cell rearrangements produce a hollow sphere of small cells, the blastula. Another set of rearrangements, called gastrulation, converts the blastula into the general form of an animal body. In the first step of gastrulation, one side of the blastula sinks in and forms a pit. This indented part produces the endoderm (gut lining) of the body as it extends further into the interior of the blastula.

An opening called the **blastopore** is left at the surface by this process. In the embryos of some animals, the blastopore becomes the mouth, and a new opening forms the anus. These animals are called **protostomes** (proto = first, stome = mouth). Molluscs, annelids, and arthropods are protostomes. In the embryos of other animals, the blastopore becomes the anus, and a new opening forms the mouth. These animals are called **deuterostomes** (deutero = second, stome = mouth). The most familiar of the deuterostome phyla are the echinoderms and chordates.

Phylum Nematoda (Roundworms)

The **nematodes** (**roundworms**) are a group of elongate, pseudocoelomate worms. They all have remarkably similar body plans, and thus it is not surprising that biologists have difficulty in agreeing on how many nematode species there actually are. Most biologists estimate that there are about 10,000 species, but some estimates range as high as several hundred thousand species.

Nematodes are among the most numerous of animals, and there are nematodes everywhere. They are very common in soil and in the muddy bottoms of freshwater lakes and ocean shores, but they also thrive in some rather exotic environments produced by human activities. For example, some nematodes live in beer-soaked mats in German taverns and in vats of vinegar.

A handful of garden soil can contain hundreds or even thousands of tiny roundworms. Many of these are harmless, but some soil nematodes do serious damage to cultivated plants and cause significant agricultural loss.

Virtually all animals have characteristic parasitic nematodes. For example, nematodes live in earthworms' excretory systems, and a grasshopper's digestive tract may contain coiled nematodes that are several times as long as the host's body. However, nematodes that parasitize vertebrates are better known. The *Ascaris* worm, an intestinal parasite of vertebrates, for instance, is a relatively large roundworm and commonly is studied as a representative nematode (figure 37.24).

Ascaris and other intestinal parasites have tough cuticles that protect them from damage by digestive enzymes. An *Ascaris* infection starts when a vertebrate animal eats soil containing *Ascaris* eggs. The eggs hatch in the digestive tract, the small worms penetrate the intestinal wall, and enter blood vessels. They remain in the circulatory system for a time but eventually crawl into the lungs. They then work their way up the trachea and pass "over the top" into the esophagus. When they reach the gut, they mature and reproduce.

A female *Ascaris* can produce as many as 200,000 eggs per day. Considering this huge reproductive rate and the fact the *Ascaris* eggs remain viable in the soil for several years, it is easy to understand how soil contaminated with feces can be so infective. A hoglot where pigs have lived for several years can contain millions of viable *Ascaris* eggs per square meter.

Although human *Ascaris* infections are not common in North America, hundreds of millions of people around the world are infected by *Ascaris*. There also are several other widespread human nematode parasites. For example, intestinal pinworms infect almost all people (at least 90 percent) in every part of the world sometime during their lifetimes. **Pinworm** infections usually cause more irritation than real damage. Other human nematode parasites, however, can cause more serious problems.

Figure 37.24 *Ascaris,* a parasitic roundworm. (a) Female and male *Ascaris* worms. Mature females are larger than mature males. Males have a curved posterior tip. (b) Internal anatomy of a female *Ascaris.* (c) Cross section of the body of a female *Ascaris.* Because it is extensively coiled, the reproductive tract is cut across several times. Note the pseudocoelomate arrangement. There is mesodermally derived muscle tissue inside the epidermis but no covering mesodermal tissue on the outside surface of the intestine.

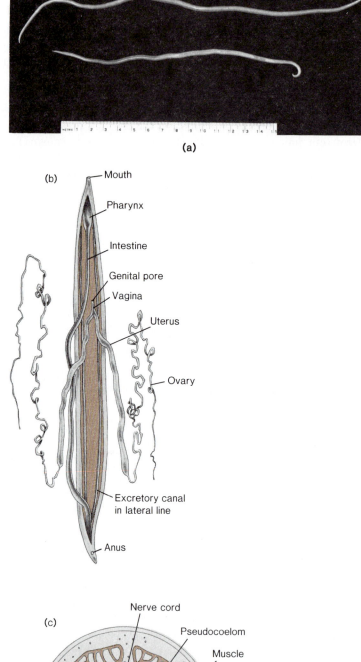

(a)

(b)

- Mouth
- Pharynx
- Intestine
- Genital pore
- Vagina
- Uterus
- Ovary
- Excretory canal in lateral line
- Anus

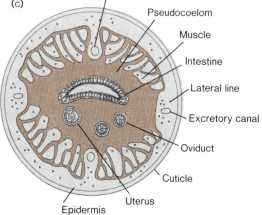

(c)

- Nerve cord
- Pseudocoelom
- Muscle
- Intestine
- Lateral line
- Excretory canal
- Oviduct
- Cuticle
- Uterus
- Epidermis

Trichinosis, an infection resulting from eating poorly cooked meat that contains worm cysts, can cause serious illness and permanent muscle damage (figure 37.25). **Filaria worms** are transmitted by mosquitoes in tropical areas. The worms enter the lymphatic system and cause large fluid accumulations and swellings in various body areas when they block lymphatic vessels (figure 37.26). **Guinea worms** live as larvae inside water fleas *(Cyclops).* They infect people who swallow the *Cyclops* with drinking water. The adults migrate through blood vessels, and the females, which may be a meter long, produce ulcerated cysts in the leg skin. When the leg is in water, hatched larval worms escape from the wound into the water, and the cycle continues if the larvae are eaten by a *Cyclops.*

Rotifers and Gastrotrichs

Some biologists group rotifers and gastrotrichs along with nematodes as classes of a single phylum, the phylum **Aschelminthes,** but they are treated here as separate phyla.

Members of the **phylum Rotifera** are tiny, active animals that commonly are found in fresh water from almost any source. Most **rotifers** are very small, often smaller than large Protozoa. They are sometimes called "wheel animals" because a ring of cilia around a rotifer's mouth looks like a turning wheel when the cilia beat. This beating moves food particles into the rotifer's mouth (figure 37.27).

Rotifers have several unusual features. For example, adult rotifers have constant cell numbers; that is, all cell division stops at some point during development, and the animals complete their growth exclusively by increase in cell size. Thus, rotifers cannot replace lost cells. They have very poor healing powers and no ability to regenerate lost parts.

Reproductive patterns of many rotifer species also are unusual. Tough, resistant eggs hatch in the spring, and female rotifers develop. These females produce female offspring by **parthenogenesis.** This pattern of exclusively female populations continues for several to many generations, but in the autumn a change occurs. Some females produce smaller eggs that develop into small male rotifers. These males mate with females, and the

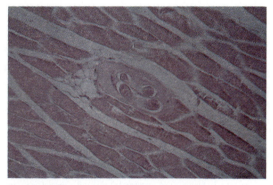

Figure 37.25 Larvae of *Trichinella spiralis,* a roundworm that causes trichinosis, encysted inside muscle. These worms could infect an animal that ate this meat raw or poorly cooked.

Figure 37.26 This man is suffering from **elephantiasis.** The condition results when filaria worms block lymphatic vessels. Fluid accumulates and causes extreme swellings of some body regions.

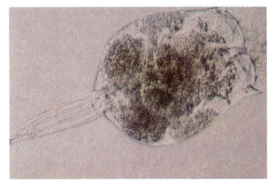

Figure 37.27 A rotifer.

resulting zygotes form the resistant overwintering zygotes. Thus, males appear only at certain times of the year.

Members of the **phylum Gastrotricha** are also tiny aquatic organisms. They glide about propelled by cilia on their body surfaces and feed on debris and small organisms in the water (figure 37.28).

For many years, biologists believed that some of the most common **gastrotrichs** reproduced entirely by parthenogenesis and that sexual reproduction was completely absent. But in 1979, sperm were identified in the bodies of some of these gastrotrichs, and only further research will clarify their reproductive patterns.

Phylum Mollusca

The **phylum Mollusca** includes a number of familiar animals, such as **snails, clams, oysters, squids,** and **octopuses.** There are at least 80,000 species of **molluscs,** and many of those species contain huge numbers of individuals. Thus, the phylum is one of the largest in the animal kingdom.

Molluscs are bilaterally symmetrical, and they have true coeloms, although the coeloms of many modern molluscs are very much reduced. All molluscs have a ventral muscular extension called the **foot** that functions in locomotion or is specialized in some other way—for example, as tentacles in squids. They have a soft **visceral mass** that contains digestive, excretory, and reproductive organs. The phylum name comes from the Latin word *mollis,* meaning soft, and refers to this visceral mass. The visceral mass is enclosed by a heavy fold of tissue, the **mantle,** that also secretes a shell in the many species that have one. Most molluscs have **gills** located in a **mantle cavity** that lies under the mantle. Molluscan gills are gas exchange organs, but in many molluscs, they also filter food out of the water (figure 37.29).

Molluscs have hearts that pump blood through vessels that empty into open spaces, **sinuses,** around body organs. Blood then enters vessels again and returns to the heart. This arrangement is called an **open circulatory system** because blood is not continuously enclosed within blood vessels. Only the cephalopods, among the molluscs, have **closed circulatory systems** with blood continuously enclosed in vessels.

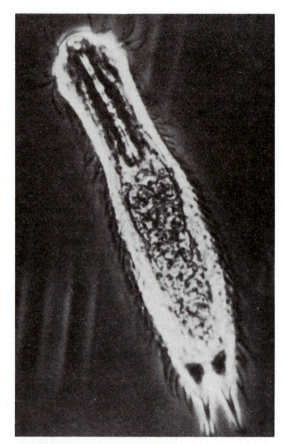

Figure 37.28 A gastrotrich. These small aquatic animals have spines over their body surface. They glide through the water propelled by cilia. (Magnification X 950)

Class Amphineura

The **chitons** are flattened marine molluscs that have a dorsal shell arranged in a series of plates. Chitons use their flat, muscular foot to crawl over surfaces where they feed on algae. A hard, rasping, toothed strap, called a **radula,** can be protruded from the mouth to scrape algae loose from rocks and other surfaces (figure 37.30a).

Class Pelecypoda

The class Pelecypoda includes clams, oysters, and scallops—the two-shelled (**bivalve**) molluscs (figure 37.30b). Pelecypoda means "hatchet foot" and refers to the shape of the foot that is extended and used for digging by many bivalves.

Bivalves are sedentary and feed by moving a stream of water through their gills and filtering out food particles.

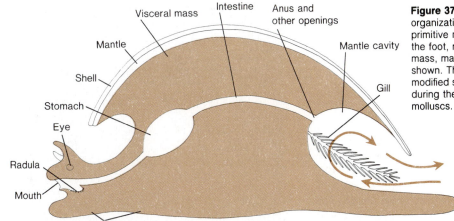

Visceral mass
Intestine
Anus and
other openings
Mantle
Mantle cavity
Shell
Gill
Stomach
Eye
Radula
Mouth
Head-foot

Figure 37.29 The body organization of a hypothetical primitive mollusc. Relationships of the foot, mantle, shell, visceral mass, mantle cavity, and gills are shown. This basic plan has been modified several different ways during the evolution of modern molluscs.

(a)

Figure 37.30 (a) Class Amphineura. A chiton. (b) Class Pelecypoda. The giant clam *Tridacna*. *Tridacna* shells can reach lengths of well over 1 m. Note the open siphons and the coral colonies growing on the clam's shells.

(b)

The two shells of these molluscs are hinged and are closed by powerful muscles. When a bivalve is disturbed, it pulls in and closes its shell firmly. Only a few predators are able to open tightly closed bivalve shells.

Class Gastropoda

Gastropoda means "belly foot" and describes the ventral, flattened foot of the snails, conchs, and other gastropod molluscs (figure 37.31). Most gastropods have coiled shells, and their visceral masses spiral up the inside of these shells. The majority of gastropods move about by waves of muscular contractions passing along their feet. Others, such as the common American coastal mud snail *Nassarius,* lay down a slime track and are moved by cilia on their feet.

Most gastropods browse by rasping off algae with their radulas, but some are predatory. The "oyster drill," *Urosalpinx,* perches on top of an oyster, bores through the shell, and eats the visceral mass.

Helix, the common garden snail, and other terrestrial snails have modified mantle cavities that lack gills. Such a snail mantle cavity is richly supplied with blood vessels and functions as a lung. These **pulmonate** (lunged) snails move air in and out through respiratory pores.

Some gastropods, such as common terrestrial slugs, lack shells. Some other gastropods without shells are the **nudibranchs,** which are called "sea slugs" and "sea hares." Certain of the nudibranchs are among the most colorful and beautiful of all animals.

Class Cephalopoda

The **cephalopod** ("head foot") molluscs are the most advanced and specialized group in the phylum. This class includes octopuses, squids, and nautiluses (figure 37.32). Most of the cephalopods are fast-moving predatory animals with well-developed sense organs, including focusing camera-type eyes that are very similar to those of vertebrates. Cephalopods in general and octopuses in particular have well-developed brains and display complex behavior, including impressive learning ability.

(a)

(b)

(c)

Figure 37.31 Class Gastropoda. (a) A snail. (b) Periwinkles *(Littorina littorea)* on a seashore boulder exposed at low tide. (c) A slug *(Arion).*

(a)

(b)

(c)

Figure 37.32 Class Cephalopoda. (a) An octopus *(Octopus)* moving over the surface of coral in the Pacific Ocean. (b) A squid *(Loligo pealeii).* (c) A nautilus shell cut open to show its chambers. The animal lives in the last chamber of its shell, but it carries the outgrown smaller chambers that it lived in earlier.

Nautiluses are enclosed in shells, but the shells of squids are reduced and internal. Octopuses lack shells entirely. This allows these cephalopods to squeeze their mantle cavities so that water is forced out, thus propelling them rapidly backwards by a sort of jet propulsion. Cephalopods also possess ink sacs from which they can squirt out a cloud of brown or black ink. This action often leaves a would-be predator completely confused.

Cephalopods use the ring of long **tentacles (arms)** around their mouth to capture prey, which they then tear up with a sharp **beak.** Experienced divers fear the beak of an octopus much more than its tentacles.

Although giant octopuses are a standard part of science fiction, octopuses seldom grow to be very large. Some giant squids, however, are enormous; they are by far the largest invertebrate animals. There are accurate records of a captured squid that was 18 m long from tentacle tip to tail and weighed at least two tons. Fairly reliable evidence also indicates that even larger squids may exist in the ocean depths, and there are reports that giant squids actually fight back when attacked by sperm whales.

Evolutionary Relationships of Molluscs

Biologists place the molluscs in a larger grouping of animals called the Protostomia, along with the annelids and arthropods. But most adult molluscs do not obviously resemble the other protostomes. For example, annelids (earthworms and their relatives), who are thought to be molluscs' nearest relatives in evolutionary terms, are **segmented;** that is, annelid bodies have repeating units called segments. This annelid segmentation usually is displayed externally as a series of rings or bulges.

In 1952, ten living specimens of a mollusc species previously presumed extinct for about 500 million years were dredged up from a depth of more than 3,500 m in the ocean near Puerto Rico. When H. Lemche studied the anatomy of this mollusc, named *Neopilina galatheae,* he found that it has five pairs of gills arranged in a repeating unit fashion. The internal anatomy of the animal displays further evidence of segmentation. It contains six pairs of nephridia (excretory organs) and five paired sets of muscles. This shows clearly that *Neopilina* is a segmented animal.

Some biologists believe that *Neopilina* is similar to an early segmented ancestor of the molluscs and that this is evidence that both modern molluscs and modern annelids descended from primitive segmented ancestors (figure 37.33a). They propose that segmentation was lost during evolution of modern molluscs, while it became even more pronounced during evolution of modern annelids.

Not all biologists accept these ideas about *Neopilina* and annelid–mollusc ancestry, but evidence derived from developmental patterns is more widely accepted. For example, annelids and molluscs have larval stages that are very similar. The zygotes of many annelids and many molluscs develop into top-shaped larvae called **trochophores** (figure 37.33b).

Phylum Annelida

The name **annelid** (from the Latin word *annelus,* meaning "little ring") is derived from the most obvious characteristic of this group of animals. Their bodies consist of a series of similar, but not identical, units, or segments. This segmentation is apparent externally as a series of rings. Earthworms probably are the most familiar annelids.

Annelids have true coeloms that are broken up into segmental compartments by partitions called **septa** (singular: **septum**). The digestive tract passes through the septa and runs the length of the body, but many body segments have their own pair of excretory tubules, the **nephridia.** Each segment also receives its own branches of major blood vessels, and the nervous system has an enlarged nerve center called a **ganglion** (plural: **ganglia**) in each segment.

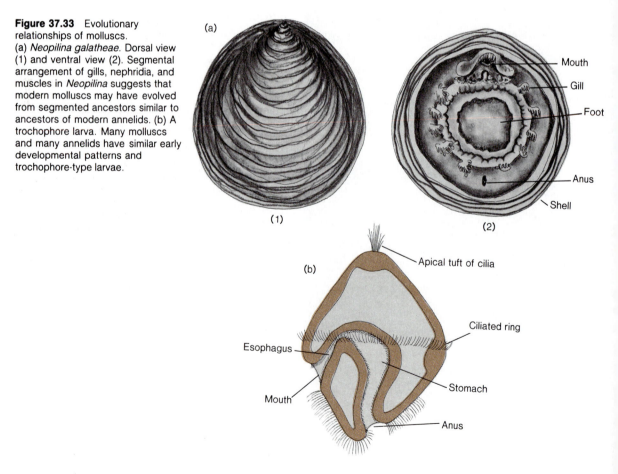

Figure 37.33 Evolutionary relationships of molluscs. (a) *Neopilina galatheae.* Dorsal view (1) and ventral view (2). Segmental arrangement of gills, nephridia, and muscles in *Neopilina* suggests that modern molluscs may have evolved from segmented ancestors similar to ancestors of modern annelids. (b) A trochophore larva. Many molluscs and many annelids have similar early developmental patterns and trochophore-type larvae.

In contrast with the molluscs' open circulatory arrangement in which blood leaves vessels and flows into spaces around body organs, annelids have **closed circulatory systems.** That is, an annelid's blood remains inside blood vessels continuously, and exchanges between the blood and body tissues must be made through the walls of blood vessels. Annelid circulatory systems also carry blood to gills (or to the body surface of annelids that lack gills), where gas exchange occurs.

The annelid coelom is filled with fluid and is important for locomotion. **Circular muscles** in the body wall contract against this fluid and alter the shapes of the segments, causing them to narrow and lengthen. This squeezing action stretches the body out lengthwise. **Longitudinal muscles,** on the other hand, can contract to shorten and thicken the body as they force the fluid-filled segments to expand outward (figure 37.34).

Precisely coordinated use of this dual movement capability is possible because the coelom is partitioned by septa and because muscles in each segment can be contracted independently. Annelids, in general, display well-coordinated movement and strong burrowing capabilities.

Muscle contraction that applies pressure against small, flexible packets of fluid (the fluid within the body segments) is very different from the muscle contraction in animals whose muscles pull against rigid skeletons when they contract. Annelids are said to have **hydraulic (or hydrostatic) skeletons.** Because of their hydraulic skeletons, annelid bodies are much more flexible than the bodies of most other actively moving animals, as body shape changes are severely limited by rigid skeletons.

Although common soil-dwelling earthworms probably are the best-known annelids, a majority of the annelids are aquatic, and a great many of those species are marine.

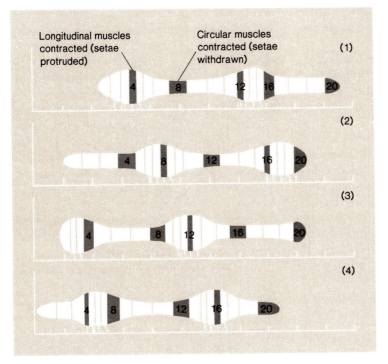

Longitudinal muscles contracted (setae protruded)

Circular muscles contracted (setae withdrawn)

(1)
(2)
(3)
(4)

Figure 37.34 Locomotion in a hypothetical, twenty-segment earthworm as an example of annelid movement. The worm's progress is shown at successive times in sketches (1) through (4). Contractions of circular or longitudinal muscles apply pressure against partitioned, fluid-filled body segments, altering their shapes in specific ways. Setae are bristles that anchor portions of the body that are expanded in circumference. The pattern of alternate anchoring and extending permits the worm to move forward. Such movement is especially effective, for example, as an earthworm burrows through the soil.

Class Polychaeta

The **polychaetes** are marine annelids that have a pair of fleshy, lateral appendages called **parapodia** (singular: **parapodium**) on each of their body segments. Each parapodium has a number of stiff bristles, the **setae,** and the class name comes from this characteristic (polychaetae = "many setae"). Polychaetes have well-developed heads and anterior sense organs.

One group of polychaetes consists of active, free-swimming, crawling, or burrowing animals. Most of the polychaetes in this group are predators.

The other group of polychaetes consists of sedentary, filter-feeding worms that dwell permanently in tubes they have constructed (figure 37.35). Tube-dwelling polychaetes use modified parapodia or other specialized water-moving devices to keep a stream of water flowing through their tubes. They filter food out of this stream and exchange gases with the water as it flows over their gills.

All species of polychaetes have separate sexes, and gametes are released into the water, where fertilization occurs. The gametes of most polychaete species are released from the gonads into the coelom and leave the body through nephridial tubes. Some polychaete bodies fill with gametes that are released when the body bursts open.

Special precise timing mechanisms that assure simultaneous gamete shedding by many individuals have evolved in many polychaete species. A rather spectacular example of this precision is provided by the **Palolo worm,** which lives in coral reefs in the South Pacific. About 99 percent of the worms shed their gametes as they swarm near the surface of the water during a single two-hour period on the day of the last quarter of the moon in November. A high percentage of fertilization is assured since the gametes of all animals are shed in the same place at the same time.

Most polychaete zygotes develop into trochophore-type larvae (see page 1048), which eventually develop into worms.

Polychaetes are considered to be most similar to primitive ancestors of modern annelids, while the other classes of annelids about to be discussed are considered to be more divergent from the ancestral types.

Class Oligochaeta

The most familiar **oligochaetes** are the earthworms that live in moist soils throughout the world, but many oligochaetes also live in the muddy bottoms of freshwater streams, lakes, and ponds.

Oligochaetes differ from the polychaetes in that they lack parapodia. Oligochaetes have fewer setae (oligochaetae = "few setae"), and their setae protrude in clusters directly from the surfaces of their cylindrical bodies. Oligochaetes lack well-developed heads.

Earthworms are often studied as representative oligochaete annelids. Several aspects of earthworm biology, including digestion (figure 10.12), gas exchange (figure 11.6), and circulation (figure 12.3), are considered elsewhere. Earthworms are terrestrial, soil-dwelling animals, but they are terrestrial only in a limited sense because they are restricted to soils containing adequate moisture. Earthworms must maintain a moist body surface if gas exchange is to occur. When they encounter dry conditions, this surface water begins to evaporate, and they quickly dry out. Thus, earthworms are only marginally adapted to terrestrial life because they normally cannot remain exposed to the air above the ground surface for any length of time.

Excess water, however, also poses a threat to earthworms. When rainwater fills their underground burrows and the air spaces among soil particles around their burrows, gas diffusion is not rapid enough to meet their gas exchange requirements. Thus, when air in the soil is displaced by water, as it is during a rainstorm, earthworms move up to the ground surface and remain out until the water percolates down and away from their burrows and the surrounding soil.

Earthworms feed by ingesting quantities of soil containing living and decaying organic matter. They digest and absorb some of this material, but much of what they eat passes on through their bodies. They carry a good deal of this dirt up out of the soil where they leave small piles of dirt ("worm casts") on the soil surface.

This soil-carrying activity of earthworms is important for soil turnover, and their burrowing activity is a factor in soil aeration. Farmers and gardeners are well aware of the beneficial effects of earthworm activity on the soil, and biologists have long been studying earthworms' impressive ability to alter soil features. For example, Charles

(a)

(b)

(c)

(d)

Figure 37.35   Polychaete annelids. (a) The sandworm *(Nereis)*, is an example of an active polychaete that moves around seeking food. Note the characteristic fleshy parapodia on each of the body segments. (b) The tube of a tubeworm *(Chaetopterus)*, which is an example of a sedentary polychaete that feeds on material extracted from the stream of water that it moves through its tube. (c) A specimen of *Chaetopterus* removed from its tube. Its modified, fanlike parapodia are used to move water through the tube. (d) The fanworm *(Sabella)* is another sedentary polychaete. Its segmented body is enclosed in a tube. The worm feeds by sweeping food out of the water using its extended tentacles ("filaments").

Darwin's last book, published in 1881, was a study of earthworms and their activities in the soil. That book was one of the first published studies in quantitative ecology, and in it Darwin presented his calculations concerning earthworms' ability to carry soil to the ground surface. He estimated that earthworms could carry and deposit as much as 7 to 16 English tons of soil per acre (17.5 to 40 metric tons per hectare) on the ground surface annually.

Earthworms and other oligochaetes are hermaphroditic; that is, each individual contains reproductive organs of both sexes. But self-fertilization does not occur; earthworms exchange sperm in copulation. Two worms line up

facing opposite directions. Then the collarlike **clitellums** of the two worms secrete mucus that protects sperm during reciprocal transfers.

Several days after copulation, a worm's clitellum secretes a second mucous sheath, this one having a hard outer covering. The worm then backs out of the sheath. As the sheath passes the openings of ducts from the ovaries and **seminal receptacles,** eggs and sperm (those sperm received earlier during copulation) are released into the sheath, where fertilization occurs. Once the worm slips completely out of the mucous sheath, the sheath closes into a **cocoon.** Development of the zygotes proceeds inside the cocoon, and young worms eventually hatch out of it (figure 37.36).

(a)

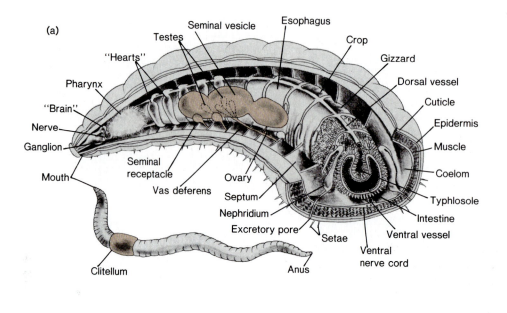

Seminal vesicle

Esophagus

Testes

Crop

"Hearts"

Gizzard

Pharynx

Dorsal vessel

"Brain"

Cuticle

Nerve

Epidermis

Ganglion

Muscle

Mouth

Coelom

Seminal
receptacle

Ovary

Typhlosole

Vas deferens

Septum

Intestine

Nephridium

Ventral vessel

Excretory pore

Setae

Ventral
nerve cord

Clitellum

Anus

(b)

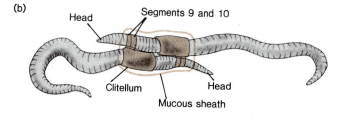

Head

Segments 9 and 10

Clitellum

Head

Mucous sheath

Figure 37.36 Reproduction of the earthworm. (a) Sketch of the internal anatomy of the anterior part of an earthworm's body with reproductive structures shown in color. Earthworms are hermaphroditic. Each worm has both male and female reproductive structures. The small sketch shows the location of the clitellum. (b) Copulation of earthworms. Worms face in opposite directions, and their clitellums secrete a mucous sheath that protects sperm as they are being exchanged. Sperm leave the vas deferens of one worm and enter seminal receptacles of the other worm (see *a*), where they are stored until fertilization. (c) Earthworm cocoons. A mucous sheath, secreted by the clitellum, slips forward as a worm backs up. Eggs from the ovaries and sperm from the seminal receptacles enter the sheath, where fertilization occurs. Where the worm slips out of the sheath, it forms a cocoon enclosing the developing zygotes.

(c)

Class Hirudinea (Leeches)

This relatively small class of annelids has only about 300 species. Most leeches are aquatic, but a few species live in moist terrestrial environments. Members of this class have flattened bodies that are tapered at the ends, and the first and last body segments are modified as suckers that are used for attachment and movement. Although leeches have external segmentation, they do not have internal body partitioning like that of other annelids, such as the earthworms (figure 37.37).

Most leeches are **ectoparasites** that attach to the surface of other animals and feed by sucking their blood. Some leeches feed on snails or other invertebrates, but the majority of leech species prefer vertebrate animals as their hosts.

Leeches are very well adapted for this form of feeding. They attach to their hosts by means of suckers and penetrate the skin using either very sharp jaws or enzymes that erode the skin. Many leeches secrete an anesthetic substance that helps to prevent their host from becoming aware of this activity.

As a leech is sucking blood, it secretes an anticoagulant, called **hirudin,** that prevents blood clotting during feeding. Leech guts and body walls can stretch considerably, and thus a leech can suck enough blood so that it swells to several times its prefeeding size. This adaptation permits leeches to survive for long periods between meals, and some leeches are known to get along nicely without feeding for months.

Leeches played an intriguing role in medical practice until about 100 years ago. Leeches were used to bleed people who were diagnosed as suffering from "bad blood" or an excess of one or another of various "humors." While the idea of bleeding an already weakened patient may now seem a rather bizarre treatment, the choice of leeches as an effective "instrument" for the bleeding process itself was excellent.

Peripatus: The Walking Worm

The "walking worm," *Peripatus,* belongs to the small phylum **Onychophora,** which includes only about seventy living species. These animals live mainly in humid habitats in tropical regions but also are found in some temperate regions of the Southern Hemisphere.

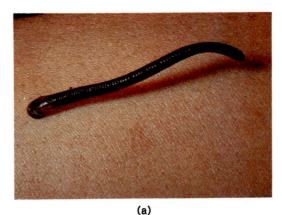

(a)

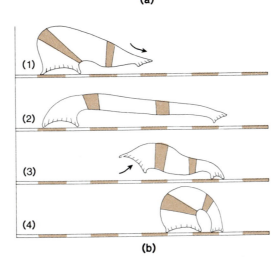

(b)

Figure 37.37 Leeches. (a) A blood-sucking leech on a human arm. (b) Locomotion in a leech. Colored regions are the same body areas in all diagrams. A leech moves by alternately attaching anterior and posterior suckers. Contraction of circular muscles makes the body stretch out and become long and thin. Contraction of longitudinal muscles shortens and thickens the body. Muscles can change body shape because the body is a fluid-filled sac of unchanging volume that acts as a "hydrostatic skeleton."

Onychophorans possess an interesting mixture of characteristics of two other phyla, the annelids and the arthropods (crabs, insects, spiders, and so on). Onychophorans have segmented bodies that closely resemble those of annelids. They have a thin, flexible cuticle that resembles the annelid body covering and a pair of excretory structures that resemble the nephridia of annelids in each body segment.

Onychophorans also have some arthropod-like characteristics. While onychophorans' legs are not jointed as arthropod legs are, they do have claws that resemble those of arthropods. Their respiratory system is a network of tracheal tubes that resemble those of terrestrial arthropods, such as insects. Onychophorans have an open circulatory system in which blood leaves the vessels and enters sinuses around body organs as it does in arthropods (figure 37.38).

As neat and clear-cut as taxonomic groupings of living things may appear, there always are organisms that seem to occupy intermediate positions astride the arbitrary boundary lines, thus frustrating biologists' attempts at categorical divisions that separate organisms into comfortably discrete groups. The mixture of characteristics of the onychophorans illustrates the sometimes confusing complexity of the living world.

The evolutionary position of onychophorans is interesting because they appear to preserve some of the characteristics of the ancient wormlike animals from which modern arthropods arose. Actually, living onychophorans very closely resemble fossil specimens from as long ago as the Cambrian period. Apparently, these "walking worms" have occupied a particular niche and remained relatively unchanged for millions of years, while ancient relatives of theirs were diversifying as the huge and evolutionarily successful group to be discussed next: the phylum Arthropoda. See figure 37.39.

Phylum Arthropoda

The **phylum Arthropoda** includes many common and familiar animals: spiders and scorpions, crabs and lobsters, and insects, as well as many other less familiar creatures. Arthropod literally means "jointed foot," and possession of such jointed appendages is an important characteristic of phylum members.

The phylum Arthropoda is by far the largest animal phylum: it contains about a million described and named species, as many species as there are in all other animal phyla *and* plant divisions combined. In addition, there probably are at least another several hundred thousand, as yet undescribed, species of arthropods.

The evolutionary success of the arthropod phylum is a story of diversity. Arthropods live in a variety of habitats on land, in fresh water, and in the oceans. Life in these many habitats requires a wide range of different specializations. Yet some common features unite the arthropods as a group.

Figure 37.38 The "walking worm" *Peripatus* (phylum Onychophora), showing its body segmentation, short, unjointed legs, and tentacles.

The Arthropod Exoskeleton

In addition to their jointed legs, all arthropods have **exoskeletons** (exo = "outside"), also called **cuticles,** that are composed primarily of **chitin.** Chitin is a polymer made up of nitrogen-containing units that are synthesized using glucose as a starting point. The exoskeleton is secreted by skin cells and remains attached to the skin, forming a tough, protective body armor that covers and protects the body surface.

The exoskeleton functions in movement because body muscles attached to it cause movements by pulling against it when they contract. Such movements are possible because an arthropod's exoskeleton is not a single hard shell, but rather a number of rigid plates connected by softer, flexible hinges at the joints. These joints permit the complex range of movements made possible by the diverse arrays of muscle bands running in many directions in an arthropod's body (figure 37.40).

Hard, chitinous exoskeletons provide general protection for soft body parts, and they represent a crunchy deterrent to attacking predators. But the exoskeleton also provides another, quite different kind of protection: it is impermeable to water and thus is a factor in water balance maintenance. This characteristic impermeability of the exoskeleton is significant for all arthropods, whatever habitat they occupy, but it is crucially important for terrestrial arthropods because it helps to retard evaporative water loss and prevent desiccation, the most urgent physiological problem faced by land animals.

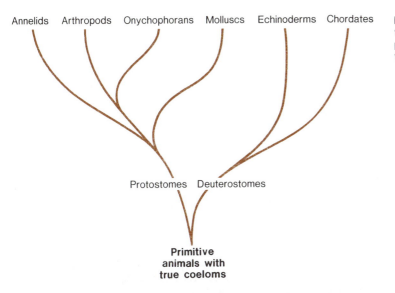

Figure 37.39 Simplified "family tree" of the more complex animal phyla. This sketch is an extension of the sketch in figure 37.21.

Annelids   Arthropods   Onychophorans   Molluscs   Echinoderms   Chordates

Protostomes   Deuterostomes

Primitive animals with true coeloms

(a)

Figure 37.40 The hinged exoskeleton of the arthropods. Hard, protective plates are connected by flexible hinges that permit movement. (a) One of the arthropods, a millipede, walking. (b) A millipede rolled up in its normal behavioral response to danger. Flexible hinges between exoskeleton plates permit this coiling.

(b)

Arthropods' exoskeletons are hard and non-expandable. Thus, growth poses some special problems for arthropods that are solved by **molting,** a periodic shedding of the exoskeleton (figure 37.41). Before molting, an arthropod's skin loosens from the exoskeleton and secretes a new, larger exoskeleton. This new exoskeleton develops in a soft, wrinkled form underneath the old exoskeleton. Then the skin secretes enzymes that partially dissolve the old exoskeleton and greatly weaken it. This allows the animal to break open the exoskeleton and wriggle out of it.

The exoskeleton shedding process is fairly difficult because the exoskeleton fits very snugly along all body contours. In fact, chitinous exoskeleton even lines some of the tracheal tubes that penetrate deep into insects' bodies and function in gas exchange. Even these tracheal tube linings must be shed and replaced at each molt.

Once an arthropod frees itself from its old exoskeleton, the new exoskeleton expands and hardens in a short time, but it takes some time for the animal to grow into its new exoskeleton. Eventually, when it does fill the exoskeleton, the molting process must be repeated for further growth to be possible.

Some arthropods, such as crabs and lobsters, continue to grow and go through molt cycles throughout their lives. In fact, during the times of the year when many lobsters are molting, knowledgeable New Englanders stop buying lobsters for a while. Because lobsters' new exoskeletons expand due to considerable fluid uptake, live weights of newly molted lobsters give a very misleading estimate of actual meat content. Thus, the natives prefer to leave newly molted lobsters for the tourists.

Other arthropods, such as insects and spiders, go through a specific, determined number of molts, and their molts are associated with a regular developmental pattern, as will be discussed later in this chapter.

Other Arthropod Characteristics

Arthropods probably evolved from annelidlike ancestors, and this ancestry is reflected in arthropods' segmented body arrangement. Ancient arthropods had segmented bodies with a pair of jointed appendages on each body segment. For example, the **trilobites,** a group of marine arthropods known only from fossil forms, had such

Figure 37.41 Molting in arthropods. A male Jonah crab *(Cancer borealis)* standing over a recently molted female. Her shed exoskeleton is to the left. The new exoskeleton will increase in size and harden in a few days. Arthropods must molt as they grow because the arthropod exoskeleton is hard and nonexpandable.

a body plan (figure 37.42a). Although the trilobites have been extinct for about 225 million years, they are so well represented in the fossil record that about 3,900 species have been recognized, and even the developmental stages of some species have been studied.

Modern arthropods have fewer body segments than their more primitive ancestors had, and in many modern arthropods, body segments tend to be combined into distinctive body regions. A familiar example of this union of body segments is the **head, thorax,** and **abdomen** arrangement in insect bodies (figure 37.42b). Along with the numerical reduction and grouping of body segments in modern arthropods, there has been an evolutionary trend toward great specialization of the appendages on some body segments and the elimination of appendages on other segments.

Like annelids, arthropods have a solid, double nerve cord that runs along the ventral side of their body. In various body segments, the cord is enlarged by the presence of clusters of nerve cells, the ganglia (figure 37.43). Ganglia function as important reflex centers that control many activities in arthropods' bodies.

(a)

Figure 37.42 Arthropod body plans. (a) "Primitive" segmented arthropod body plan with a pair of jointed appendages on each body segment. Fossil trilobites show this body arrangement. Two longitudinal folds divide the trilobite body into a median lobe and two lateral lobes. This three-part body arrangement suggested the name "trilobite." (b) Many modern arthropods have distinct combined body regions rather than numerous similar body segments. This ventral view of a carpenter ant shows the head-thorax-abdomen arrangement characteristic of the insect. Note that all three pairs of legs are attached to the thorax and that the abdomen has no appendages. This ant has a pair of large antennae on its head.

(b)

Figure 37.43 An arthropod nervous system. This crayfish nervous system is an example of the ventral arthropod nerve cord with ganglia. Larger anterior ganglia function as a "brain." Internally, the ventral nerve cord actually is double, a pair of cords running side by side. Thus, with ganglia, the arthropod nervous system is often described as "ladderlike."

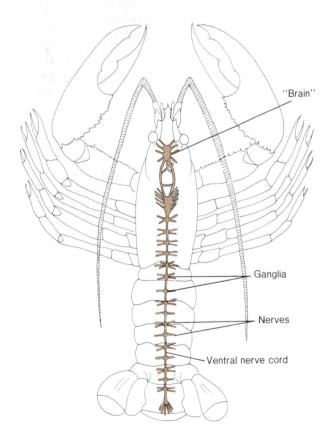

"Brain"

Ganglia

Nerves

Ventral nerve cord

In modern arthropods, there is obvious cephalization, that is, the development of a definite head with specialized sense organs, and these arthropod heads contain large, prominent ganglia that could be called brains. The term "brain," however, should be used only with reservation since the term often is understood with reference to brains of vertebrate animals, where the brain is clearly the dominant functional center of the nervous system. Nervous system activity in arthropods is much less "brain-centered" than nervous system activity in vertebrate animals.

Behavioral patterns of arthropods also differ greatly from patterns observed in familiar vertebrate animals. Arthropod behavioral patterns tend to be very stereotyped and predictable, and to show much less adjustment to varying conditions than vertebrate behavior patterns. Extremely complex systems of communication and social interaction have evolved among such arthropods as ants, termites, and bees. But complex as they may be, the behavior patterns of arthropods are built largely on a foundation of automatic and stereotyped ("prewired") nervous system responses with little or no possibility for modification of behavior through experience and learning.

Arthropods have complete digestive tracts and bilateral symmetry as do most other large, complex animals. Instead of well-developed coeloms, arthropods have a network of spaces around body organs that is known collectively as the **hemocoel** (hemo = blood; coel = cavity).

The blood of arthropods, which is also known as **hemolymph,** circulates through the spaces of the hemocoel as part of the open circulatory system that characterizes the arthropod circulatory arrangement. Hemolymph enters arthropods' hearts through valve-regulated openings, is pumped through short vessels, and then flows out into the spaces of the hemocoel. There it bathes body organs directly. Thus, exchanges between body cells and the hemolymph are not made through capillary walls as they are in animals such as annelids that have closed circulatory systems.

The phylum Arthropoda has two major subgroups (subphyla) that are distinguishable from one another on the basis of several fairly obvious characteristics of their general body organization and their anterior appendages. These subgroups are the subphyla **Chelicerata** and **Mandibulata.**

Chelicerate Arthropods

Chelicerates' bodies are divided into   ajor regions, a **cephalothorax** and an **abdon   s** opposed to the three-part, head-thorax- domen, body arrangement seen in many familiar mandibulates (for example, insects). The characteristic that gives the chelicerates their name is the specialization of their first (most anterior pair) appendages as mouthparts called **chelicerae.** Chelicerae may be either pincerlike or, as they are in spiders, fanglike. They are used for piercing prey animals and in many chelicerate arthropods are associated with poison glands. Besides the chelicerae, there are five more pairs of appendages on the cephalothorax. The second appendages are specialized as a pair of pedipalps, sensitive feeding devices. The remaining four pairs of appendages are walking legs. Chelicerate arthropods do not have antennae.

There are several classes of chelicerate arthropods, but only two of them—class Xiphosura and class Arachnida—are discussed here.

The **class Xiphosura** includes only four genera. One of them, the **horseshoe crab** *Limulus polyphemus,* is common along the east coast of North America. *Limulus* can be found in large numbers on muddy shores, where they plow along through the mud, feeding on debris and organisms that they dredge up (figure 37.44).

Limulus is a member of an ancient group, most of whose members became extinct millions of years ago. *Limulus,* however, seems to be a hardy, well-adapted animal. It is protected by a hard **carapace** that covers its body, and no predators regularly attack it. Adult *Limulus* can survive, out of water and without food, for days or even weeks if only their gills are kept moist. *Limulus* larvae bear a striking superficial resemblance to ancient fossil trilobites. In fact, the *Limulus* larva is called a "trilobite larva." In evolutionary terms, *Limulus* has been around a long time and probably will be for a long time to come.

Figure 37.44   The horseshoe crab *(Limulus polyphemus)*. (a) Dorsal view of male (left) and female (right) crabs during the breeding season. Note the hard, protective carapace. (b) A ventral view of a *Limulus* showing appendages. Note the chelicerae. The pedipalps of *Limulus* are very similar to the walking legs. Pedipalps of many other cheliceral arthropods are much more specialized. The operculum is a plate that covers and protects gills and reproductive structures.

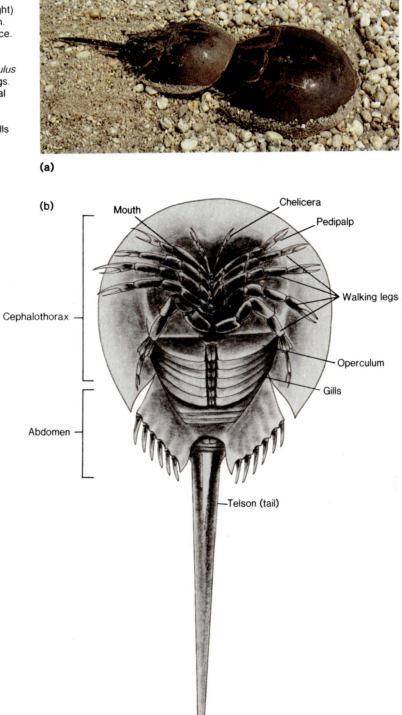

(a)

(b)

The **class Arachnida** includes a number of familiar animals, such as spiders, scorpions, ticks, mites, and daddy longlegs (figure 37.45). In addition to chelicerae, pedipalps, and four pairs of walking legs on their abdomens, some terrestrial arachnids have several pairs of abdominal appendages that are specialized as multi-layered gas exchange organs called **book lungs.**

Arachnids have **simple eyes,** that is, eyes that have a single focusing system, and arachnids, like other chelicerate arthropods, do not have antennae.

Spiders are the best-known arachnids because they are common terrestrial animals. About 32,000 species of spiders have been described so far. Spiders can be very numerous in terrestrial habitats. For example, one study of meadowland in Britain concluded that there were about 2¼ million spiders per acre (about 5½ million per hectare) in some meadows.

Spiders range in size from almost microscopic creatures that are less than 0.5 mm in length to formidable hairy monsters with bodies that are 9 cm long and total leg spans several times that large (figure 37.46).

All spiders are carnivores that live by piercing their prey and sucking out their body fluids; most spiders secrete poison that immobilizes prey animals.

Spiders have **spinnerets,** spinning organs on their abdomen, that secrete a liquid containing an elastic protein that hardens in the air to form a very fine silk thread. Many spiders spin elaborate **webs** of this silk, instinctively tracing out the same intricate geometric patterns that members of their species have spun for hundreds of generations.

(a)

(b)

Figure 37.45   Scorpions, members of the class Arachnida. (a) Dorsal view of a giant scorpion from Kenya. Note the large pedipalps that end in pincers. The scorpion's small chelicerae are not visible in this picture. The last body segment has a sharp stinging barb that injects venom with a stabbing motion. (b) A female scorpion carrying her young on her back. Scorpions brood their eggs inside the female reproductive tract so that young scorpions hatch inside their mother's bodies. At birth, they crawl onto the mother's back, where they remain for about a week until their first molt occurs. Then they leave and become independent.

Figure 37.46   A spider and its web.

Spider webs are important prey capture devices for web-building spiders, but other spiders put the silky threads produced by their spinnerets to other uses. Some larger spiders are active hunters that seek out and overpower their prey. The various hunting spiders use their silk in a number of ways. For example, jumping spiders attach a dragline thread to the ground before leaping at their prey. Other spiders use the silk to line their burrows or to hinge the trapdoors that they pop open when they leap out to catch passing prey. Sometimes, victims are detected by a waiting spider that senses any disturbance of silken lines that radiate away from the burrow openings. Young members of some spider species spin glistening sails that allow them to be wafted along on the breeze for great distances over land and water.

Two groups of tiny arachnids, the **ticks** and **mites**, are specialized for life on a variety of plant and animal hosts, including humans (figure 37.47).

Class Crustacea

Because they are edible, some members of the **class Crustacea**, such as **lobsters, shrimps, crabs,** and **crayfish**, are quite familiar to most people. But **pill bugs**, which are terrestrial, **water fleas, brine shrimp, barnacles**, as well as a host of tiny animals that live as part of the plankton of the world's oceans, also are crustaceans.

Crustacea, along with the insects, millipedes, and centipedes, belong to the second subphylum of the phylum Arthropoda: subphylum Mandibulata. The mandibulate arthropods differ from the chelicerates (horseshoe crabs and arachnids) in several ways.

(a)

(b)

Figure 37.47 Mites and ticks. (a) Scanning electron micrograph of a mite. Note the body hairs. Many mites live as scavengers on the surfaces of plant and animal hosts, but others, such as chiggers, suck blood from their hosts. Some mites are carriers (vectors) of disease-causing microorganisms, which they transmit from one host to another. (Magnification $\times$ 42) (b) The Rocky Mountain wood tick *(Dermacentor andersoni)*. The male is above; the female is below. This tick carries the rickettsia (small bacterium) that causes Rocky Mountain spotted fever in humans. All ticks are parasites. Some of the ticks cause little direct damage themselves, but a number of ticks are vectors of serious animal and human diseases.

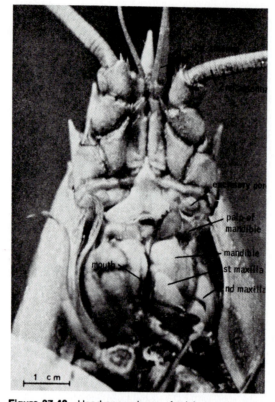

Figure 37.48   Head appendages of a lobster, an example of a mandibulate arthropod. Mandibulates have mandibles as their first mouthparts rather than chelicerae.

Mandibulate arthropods have one or two pairs of antennae, and they have **mandibles** rather than chelicerae. Mandibles usually are modified for biting or chewing, and they never are pincer-like as chelicerae often are. Most mandibulate arthropods have two additional pairs of mouthparts called **maxillae** (figure 37.48).

All crustaceans have two pairs of antennae and the three pairs of characteristic mandibulate arthropod mouthparts. Crustaceans also have appendages on their abdomens. But beyond these characteristics, it is hard to generalize about the class Crustacea because it is a very diverse group of animals (figure 37.49).

The large, familiar, bottom-dwelling crustaceans, such as crabs, lobsters, and crayfish, all belong to a group (order) known as **decapod** ("ten-foot") Crustacea. They are called decapods because they have five pairs of walking legs. Decapod crustaceans have rigid exoskeletons that are heavily impregnated with calcium carbonate. But not all crustaceans have such hard coverings. Many are tiny, soft, delicate creatures (figure 37.50).

Despite a constant demand for the tasty decapods, some of the small, obscure crustaceans actually are more economically important than the ones that are food delicacies. These small crustaceans are extremely important links in aquatic food chains. For example, small shrimp-like crustaceans known as "**krill**" provide food for many oceanic animals. Krill occur in great swarms that can cover areas of the ocean several city blocks large in layers that are 5 m thick. Within such a swarm, there can be more than 60,000 individuals per cubic meter. Such huge masses can provide food sources for even very large animals, despite the small size of the individual crustaceans. For example, blue whales swim into the swarms and filter huge quantities of krill out of the water. A blue whale consumes up to a ton of krill at a single feeding, and it may feed four times a day. Krill is now being considered as a possible direct source of human food, and Russian and Japanese fisheries are experimenting with methods for harvesting krill.

(a)

(b)

(c)

(d)

(e)

(f)

Figure 37.49 Representative members of the class Crustacea. (a) The water flea *Daphnia*, a small freshwater crustacean, photographed by dark-field microscopy. The average *Daphnia* is 1 to 2 mm long. (b) The Japanese spider crab *Macrocheira*, a marine crustacean. (c) *Homarus americanus*, the commercially important lobster that occurs along the northeast coast of the United States. This one was photographed underwater off the coast of Maine. (d) The King crab *Paralithodes*. This tasty crab occurs in the north Pacific and is sometimes called the Alaskan King crab. (e) The hermit crab *Pagarus*. This crab has a very soft abdomen, and it lives inside an empty mollusc shell, which it carries with it. When the hermit crab outgrows its shell, it finds a new shell and abandons the old one. (f) Pill bugs (also called sow bugs or wood lice) are terrestrial crustaceans that live in moist places.

Figure 37.50 Plankton animals, including several kinds of small Crustacea (copepods). Small planktonic crustaceans provide food for many larger animals. The picture also shows an arrow-worm near the center. A coelenterate medusa is near the end of the arrow-worm at the bottom of the picture, and a developing fish embryo is near the other end of the arrow-worm. Diatoms, which appear as fine chains in the picture, provide food for all of these small plankton animals.

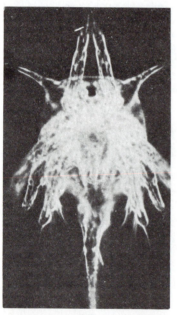

(a)

(b)

Figure 37.51 Barnacles. (a) Larva of the acorn barnacle *Balanus*. Barnacle larvae resemble free-swimming crustaceans. (b) Adult acorn barnacle *Balanus*. Adult barnacles live in shells attached to substrates. Note the extended cirri. This barnacle is feeding. (c) Body organization of an adult barnacle. Cirri sweep food toward the mouth during feeding. Note the presence of a penis and consider the consequences for sexual reproduction of the permanently attached condition of adult barnacles. A barnacle that cannot "reach" another barnacle from its attached position cannot mate. (d) One of a barnacle's cirri. Note that it is divided into two parts that are covered with stiff bristles (setae). The cirri sweep the water very efficiently.

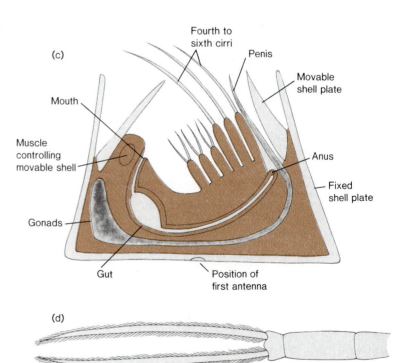

(c)

Fourth to sixth cirri

Penis

Movable shell plate

Mouth

Muscle controlling movable shell

Anus

Fixed shell plate

Gonads

Gut

Position of first antenna

(d)

Most crustaceans are free-swimming throughout their lives, but adults of one group, the **barnacles,** live attached to the substrate. Barnacle larvae are free-swimming, and they closely resemble several other small crustaceans. But, at a certain stage of maturation, a young barnacle settles headfirst on a surface, attaches, and secretes a calcium-containing shell around its body. Some of the barnacle's legs develop into delicate, feathery sweepers (**cirri**) that strain food out of the surrounding water and kick it into the animal's mouth (figure 37.51).

Barnacles attach to rocks, other organisms (for example, snails, crabs, and even whales), wharf pilings, buoys, and ship bottoms. Ships with barnacle-fouled bottoms are slowed considerably because the barnacles interfere with the smooth flow of water over their hulls. Thus, ocean-going ships' bottoms must be scraped periodically to remove barnacles.

Centipedes and Millipedes

The **centipedes** ("hundred legs") and **millipedes** ("thousand legs") are two groups of animals that look alike superficially. Both are terrestrial, and both have an elongate, wormlike body with a head and a trunk consisting of many segments. Both also use tracheal tubes in gas exchange and malpighian tubules in excretion. Yet centipedes and millipedes are different enough to be placed in two separate arthropod classes.

Centipedes (**class Chilopoda**) are active, fast-moving, carnivorous animals that use their first pair of legs, which are modified as poison claws, to kill their prey. Centipedes usually eat small insects and other small invertebrates, but larger centipedes have been known to feed on snakes, mice, and frogs.

Centipedes have antennae, mandibles, and two pairs of maxillae on their head. The centipede body is flattened, and each segment bears one pair of walking legs. Actually, the great majority of centipedes have far less than one hundred legs. Centipedes' reproductive ducts open at the posterior end of their body (figure 37.52).

Millipedes (**class Diplopoda**) do not have nearly a thousand legs, but each body segment behind the first four or five does have two pairs of legs (see figure 37.40). These legs appear to move in waves as a millipede slowly moves along.

Figure 37.52 A centipede (class Chilopoda). A giant South American centipede attacking a frog. Centipedes have paired antennae, poison claws, and flattened bodies with a single pair of walking legs attached to each trunk segment. Compare with the pictures of a millipede in figure 37.40.

Millipedes are herbivores or scavengers that feed on decaying material. They do not have poison claws, and instead of two pairs of maxillae, they have fused mouthparts that function essentially as a lower lip. Millipedes are secretive animals that usually live beneath leaves, stones, or logs, and they avoid trouble by rolling up and feigning death (see figure 37.40).

Several other characteristics further distinguish millipedes from centipedes. Millipedes have rounded, rather than flattened, bodies. They have two pairs of spiracles leading to tracheal tubes in each body segment, while centipedes have only one pair of spiracles per segment. And a millipede's reproductive ducts open near the anterior end of its body.

Class Insecta

The **class Insecta** includes more species than any other class of organisms, and insects are regarded as one of the most successful (possibly *the* most successful) groups of organisms that have ever lived. Ninety percent of the million or so species of arthropods are insects.

Insects occupy almost every kind of freshwater and terrestrial habitat in the world, and they are variously specialized to utilize a tremendous variety of food sources. Some are parasites on animal bodies; some are scavengers that eat dead, decaying organisms; and insects are specialized in countless ways to eat specific parts of plants. For example, the remarkably specialized mouthparts of aphids can penetrate plants

Figure 37.53 Insect structure.
(a) A grasshopper. Note that the three thorax segments—prothorax, mesothorax, and metathorax—each have a pair of walking legs.
(b) Reproductive structures (in color) of a female grasshopper. Sperm are received in the seminal receptacle during mating. The female uses her ovipositor to insert eggs in the ground, where embryos diapause during winter months. (c) Insect (grasshopper) mouthparts in place (left) and spread out (right). Labrum, the ''upper lip,'' does not correspond to mouthparts of arthropods in other classes. The insect labium is a fused structure corresponding to the second pair of maxillae found in other arthropods. (d) Scanning electron micrograph of the compound eye of the fruitfly *Drosophila.* (e) Structure of a compound eye (above) with many ommatidia, and structure of a single ommatidium (below). The corneal lens and the cone focus light on the rhabdom, which is a cylinder consisting of light-sensitive microvilli of receptor cells. Pigment cells form a dark cylinder around the receptor cells that prevents light from passing from one ommatidium to others around it. Each ommatidium receives a separate image, and the insect's nervous system must process information received in multiple images.

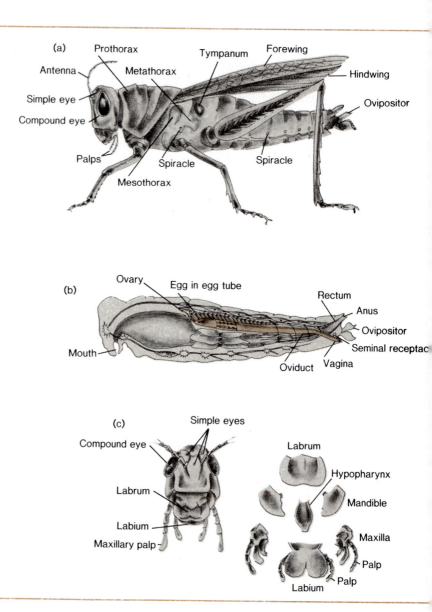

and pierce individual phloem cells without causing an injury response in the normally sensitive cells. Thus, an aphid can feed on a continuing supply of phloem sap (see figure 9.14).

Some insects are smaller than large protozoans and are less than 0.25 mm long, while other living insects have wingspans up to 30 cm. Fossils indicate that some extinct insects were several times that large. It is not clear why none of those gigantic insect species remain today, but biologists speculate that the larger insects of the past competed much more directly with the insects' chief competitors in the terrestrial environment, the vertebrate animals. Smaller insects, on the other hand, generally are able to occupy habitat

niches that do not have enough resources to support most vertebrates. Thus, there may have been selective pressure favoring evolution of many species of small insects.

The Insect Body

A typical adult insect body is divided into a **head,** a **thorax,** and an **abdomen**. The head bears well-developed sense organs, including one pair of antennae, **compound eyes** (eyes with many separate focusing units), and sometimes simple eyes as well. Insect mouthparts include a pair of **mandibles,** a pair of **maxillae,** and a **labium,** which is a sort of lower lip that has evolved through the fusion of a second pair of maxillae (figure 37.53).

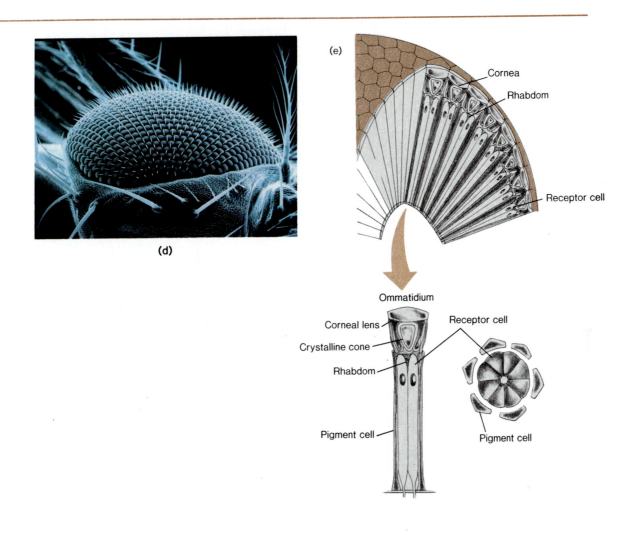

(d)

(e)

Cornea

Rhabdom

Receptor cell

Ommatidium

Corneal lens

Crystalline cone

Rhabdom

Receptor cell

Pigment cell

Pigment cell

Among insects, there are literally thousands of kinds of mouthpart specializations associated with the different feeding habits of members of the class.

The three segments of the thorax each bear a pair of walking legs. These six legs are the only walking legs of adult insects, and possession of six legs is probably the most reliable criterion for identifying a given arthropod as an insect. Many insects also have wings on the second and third thoracic segments and are excellent fliers (figure 37.54).

There are no appendages on the insect abdomen. Many insect abdomens have prominent **spiracles**, the openings leading to the tracheal tubes that function in gas exchange. The abdomen also contains **Malpighian tubules**, the excretory devices that are attached to the posterior part of the digestive tract (see p. 366).

Like other arthropods, the insects have open circulatory systems.

Insects' specialized reproductive strategies are almost as varied as their feeding habits. For example, some insects lay eggs on plants that provide a ready food supply for their developing young. Some insects even use the bodies of other species of insects as "nurseries" for their young.

Figure 37.54 Insect flight. A bee flying among wild aster flowers. The bee's wings appear blurred even in this high-speed, "stop-action" photo because they are beating so rapidly. Bee wings beat at frequencies up to 250 beats per second.

They lay their eggs in or on the bodies of these unfortunate hosts, who eventually serve as a food source for the developing young after they hatch (figure 37.55).

Because it has a nonexpandable exoskeleton, an insect's development must involve a series of molts. But insect development involves much more than simple growth in size; there is also a marked change in body form that converts an immature individual into an adult. This change is called **metamorphosis**.

There are two basic patterns of molting and metamorphosis in insects. In one type of development, the young insect hatches in a form that generally resembles its parents, though it may lack some adult features. This type of young insect is called a **nymph** (figure 37.56a). With each molt, the insect comes to resemble its parents more closely, and it emerges from a final molt as a fully developed adult. This developmental pattern is called **gradual metamorphosis**, and it characterizes the development of insects such as grasshoppers.

The other major category of insect development involves much more drastic and abrupt changes in form. The familiar developmental stages of butterflies and moths—**egg, larva, pupa,** and **adult** (also called **imago**)—illustrate this **complete metamorphosis** (figure 37.56b). A wormlike larva hatches from the egg, grows, molts several times (the actual number of molts depends on the species), and then **pupates**. That is, it produces a hard case around itself.

In many species, the **pupa** may enter a period of suspended activity called **diapause**, in which it has lowered metabolism and increased resistance to environmental stresses, such as the rigors of temperate zone winters.

Dramatic developmental changes occur within the pupal case. Parts of the larval body are broken down and used as raw material for construction of new adult structures. The new structures are produced by the growth of small clusters of cells called **imaginal discs** (so named because they contribute to the body of the imago, the adult insect). Imaginal discs have been inactive since early development, but now they become active to produce antennae, wings, legs, reproductive organs, and other characteristically adult structures. Finally, the imago emerges as a reproductively mature individual.

These complex developmental changes are coordinated by hormonal regulation (see p. 391).

Insect Diversity

Insects are so numerous and so diverse that the study of this one class is a major speciality in biology, and this science, called **entomology**, occupies whole departments in many universities. It takes massive volumes to survey the insects thoroughly, and such detail is far beyond the scope of beginning biology courses. But table 37.1 and figure 37.57 provide a brief introduction to insect diversity by presenting some of the major orders of insects.

Figure 37.55 A blowfly pupal case opened to show developing larvae of the parasitic wasp *Mormoniella*. The female wasp inserts her eggs into the pupa. After the wasp larvae hatch, they feed on the tissues of their host.

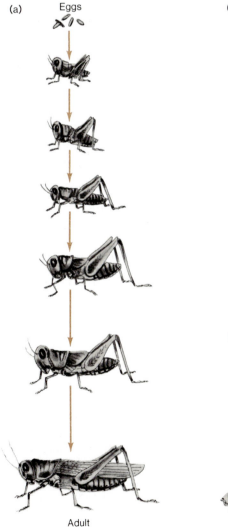

(a) Eggs

Adult

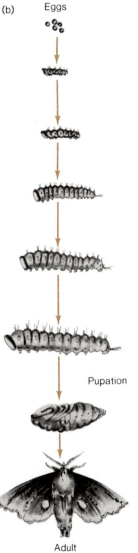

(b) Eggs

Pupation

Adult

Figure 37.56 Insect development. (a) Grasshopper development, an example of gradual metamorphosis. The nymph that hatches generally resembles the adult. Following each molt, the emerging nymph more closely resembles the adult. (b) Moth development, an example of complete metamorphosis. Life history includes several wormlike larval stages, a pupa, and an imago (adult).

Table 37.1
Some Major Orders of the Class Insecta.

Order Thysanura (∼700 species)	Silverfish and bristletails. Fast-running, primitive, wingless insects; chewing mouthparts; long antennae; simple eyes; two or three long, taillike appendages on rear of abdomen. Common in moist environments; some species in houses, particularly bathrooms and basements. Gradual metamorphosis.
Order Odonata (∼5,000 species)	Dragonflies and damselflies. Rapid-flying, predaceous insects; two pairs of long, narrow, net-veined wings; large, highly developed compound eyes; chewing mouthparts. Gradual metamorphosis with immature stages (nymphs) in fresh water.
Order Orthoptera (∼23,000 species)	Grasshoppers, locusts, crickets, mantids, walking-sticks, roaches. Large-headed insects with chewing mouthparts, compound eyes, two or three simple eyes; usually two pairs of wings; membranous hindwings folded beneath narrower, leathery forewings at rest. Gradual metamorphosis.
Order Isoptera (∼1,800 species)	Termites. Soft-bodied, highly social insects with winged and wingless individuals composing the colony; membranous forewings and hindwings of equal size; simple or compound eyes; chewing mouthparts. Live underground or in wood. Gradual metamorphosis.
Order Hemiptera (∼40,000 species)	True bugs. Highly variable body; usually two pairs of wings; forewings thick and leathery at base, membranous at tip; hindwings membranous; piercing/sucking mouthparts; herbivorous or predaceous. Terrestrial or aquatic. Gradual metamorphosis.
Order Homoptera (∼20,000 species)	Cicadas, leaf hoppers, and aphids. Closely related to the Hemiptera. Wings, if present, membranous and held in a tentlike position over the abdomen; mouthparts as a sucking beak. Gradual metamorphosis.
Order Anoplura (∼200 species)	Sucking lice. Ectoparasites of birds and mammals. Flattened bodies; wingless; eyes reduced or absent; piercing/sucking mouthparts; legs and claws used for clinging to host. Gradual metamorphosis.
Order Coleoptera (∼300,000 species)	Beetles and weevils. Largest order of insects. Hard bodies; chewing mouthparts; forewings form hard protective covering for membranous hindwings. Herbivorous or predaceous. Complete metamorphosis.
Order Lepidoptera (∼110,000 species)	Butterflies and moths. Long, soft bodies; two pairs of large wings covered with pigmented scales; compound eyes and antennae well developed. Larvae (caterpillars) have chewing mouthparts and eat plants. Adult mouthparts modified as a coiled proboscis used for sucking flower nectar. Complete metamorphosis.
Order Diptera (∼85,000 species)	True flies: mosquitoes, gnats, midges, horseflies, houseflies. One pair of functional, membranous front wings; reduced, knoblike hind wings (halteres) act as balancing organs; well-developed eyes; piercing, sucking, or sponging mouthparts. Complete metamorphosis.
Order Siphonaptera (∼1,100 species)	Fleas. Intermittent ectoparasites. Very small wingless insects with laterally flattened bodies; long legs well adapted for jumping; piercing and sucking mouthparts used for feeding on the blood of birds and mammals. Complete metamorphosis.
Order Hymenoptera (∼100,000 species)	Wasps, bees, ants, and sawflies. Chewing or chewing/lapping mouthparts; wings membranous when present; some adult females with stinger or piercing ovipositor. Complete metamorphosis.

Figure 37.57 Some representatives of the orders of insects described in table 37.1. Insects are drawn to different scales. (a) Silver fish (order Thysanura). (b) Dragonfly (order Odonata). (c) Camel cricket (order Orthoptera). (See also the grasshopper in figure 37.53a.) (d) Termite soldier (order Isoptera). (e) Cinch bug (order Hemiptera). (f) Human body louse (order Anoplura). (g) Beetle (order Coleoptera). (See also dung beetles in figure 37.59.) (h) Royal walnut butterfly and its caterpillar (order Lepidoptera) (i) Gall gnat (order Diptera). (j) Buffalo treehopper (order Homoptera). (See also the cicadas in figure 37.58.) (k) Flea (order Siphonaptera). (l) Wasp (order Hymenoptera).

Insect Success

The insect exoskeleton is a major factor in the evolutionary success of insects, especially in terrestrial environments. The exoskeleton is virtually impermeable and thus provides a barrier to water loss, which is a continuing problem of terrestrial life. Because the exoskeleton is a solid support against gravity, it permits upright posture, yet it is hinged so that muscles attached to it can cause efficient movements.

But possibly the greatest factor in the overall evolutionary success of insects is the host of marvelous adaptations of insects to almost every imaginable habitat niche. Some insects are large, flying hunters and conspicuous plant eaters; others are microscopic species that live virtually unnoticed in single hair follicles of vertebrate animals, including humans.

Insect adaptations to environment go far beyond immediately obvious factors, such as mouthpart specializations. Insects' entire life cycles are adapted to environmental conditions. Often, larvae and adults use different food sources and thus take advantage of seasonal changes in vegetation. Precisely timed periods of diapause permit insects to avoid the most adverse climatic conditions. Some species diapause as an embryo within an eggshell (grasshoppers); some diapause as pupae (moths); and some diapause as adults (houseflies). Whatever the diapause form, onset is timed to precede adverse conditions, and breaking of diapause coincides precisely with return of favorable conditions for normal activities.

Instead of the common annual cycle, some insect reproductive processes show remarkable timing adaptations that lead to cycles of many years duration. Some species of **cicadas** live underground for seventeen years as larvae. Then they pupate and emerge as vast swarms of adults that complete their mating and egg laying within a few days, leaving the next generation under the ground, not to be seen again for seventeen years. Although many cicada adults are eaten by predators when they emerge, the mass emergence of so many individuals in such a short time "saturates" predation and assures adequate reproductive success to start the new generation (figure 37.58).

Insects also show remarkable behavioral adaptations. There are complex individual behavioral responses, such as the evasive maneuvers by which flying moths avoid hunting bats (p. 847). But possibly behavior adaptations reach their zenith in the social systems of bees, ants, termites, and other colonial insects. Castes of individuals are structurally and behaviorally specialized to perform specific functions in the colony. Both developmental and behavioral differences are based on chemical communication systems in which substances (pheromones) produced by some individuals affect other individuals in the colony. Behavioral forms of communication also coordinate activities in the colony. For example, the **bee dance** permits one bee to communicate distance and direction of a food source to other bees (see p. 839).

Insects and Humans

Some insects play a vital role in the environment as pollinators of plants that depend entirely on insect pollination. Other insects are active in the important natural processes that break down dead plant and animal bodies.

Even dung is processed by insects. For example, in the African plains herds of large grazing animals deposit enormous quantities of dung, and yet there is no obvious accumulation of dung as the herds pass. Busy little dung beetles do a spectacular job of attacking even football-size balls of elephant dung, which can be reduced to a 2 cm-thick mat overnight. Thousands of beetles congregate, cut out small balls of dung, and roll them away (at velocities up to 14 m per minute) to underground burrows, where adult beetles eat them or leave them as food for developing young (figure 37.59). All of this activity hurries the eventual return of vital resources contained in dung to forms that are usable by other organisms.

But even as dung beetles are doing their valuable work in African grazing areas, other insects are destroying more than half of all the crops in the fields of the same African countries.

And so it goes around the world. Insects everywhere eat humans' food and fiber-producing plants in the field and attack agricultural products in storage. Other insects transmit diseases to humans and to animals and plants upon which humans depend. This agricultural destruction and disease transmission has put human beings and insects in a direct conflict situation.

Figure 37.58 An adult seventeen-year cicada *(Magicicada septendecim).* In each population of seventeen-year cicadas, swarms of adults emerge and live for only a few days out of each seventeen-year cycle.

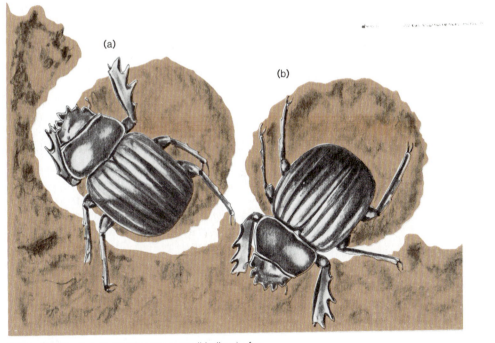

(a)

(b)

Figure 37.59 A dung beetle cuts a small ball out of a mass of dung (a) and quickly rolls the ball away (b) toward an underground burrow.

Humans' battle with insects has featured massive attempts to poison insects selectively (with **insecticides**). Unfortunately, some of these insecticides are not so selective, and poisons intended for harmful insects also kill beneficial insects, and accumulate in the environment, enter food chains, and build up to toxic levels in other animals considered beneficial to humans. These poisons even threaten human health directly in some cases. Insects with resistance to low levels of these insecticides appear very quickly once insecticides are put into use. Humans' response to this resistance typically has been to apply even stronger doses of pesticide, thus further increasing environmental pollution.

In the future humans may be able to fight the battle against insects through **biological control.** Biological controls depend on specific biological attacks on an individual insect species' development or metabolism without use of chemical insecticides (table 37.2). But each strategy for effective biological control requires long and costly research efforts, and the easy way to continue the battle often seems to be application of still greater quantities of insecticides. The long war between humans and some insect species seems likely to continue far into the future.

Table 37.2
Some Strategies for Biological Control of Insects.

Control Involving Other Organisms
1. Introduce predators that selectively attack a particular harmful species.
2. Discover and disseminate viral or bacterial diseases of the insect in question.
3. Breed plants that produce insect-repelling substances without changing output of agricultural products.

Control Involving the Insect's Own Biology
1. Use sex attractants (pheromones) to lure large quantities of adults to traps.
2. For species in which females mate only once, release huge quantities of sterilized males that compete with normal wild males.
3. Apply hormones (for example, juvenile hormone) that disrupt development and prevent attainment of diapause stage by a critical time of year.

Phylum Echinodermata

The **phylum Echinodermata** consists of about 6,000 species, including **sea stars (starfish), brittle stars, sea urchins, sea cucumbers,** and **sea lilies. Echinoderms** are radially symmetrical animals, most of whom have a five-part body organization. They are bottom dwellers that range from shoreline intertidal zones to very deep parts of the sea.

Echinoderms are the closest invertebrate relatives of the vertebrate animals and other members of the phylum to which they belong, the phylum Chordata. Superficially, these radially symmetrical animals would not seem likely to have such a close evolutionary relationship with chordates.

Most of the evidence linking echinoderms and chordates is developmental. Cleavage occurs in the same way in both groups, and other aspects of early development are very similar. The digestive tract forms the same way in both phyla; members of both phyla are **deuterostomes,** that is, animals in which the anus develops from the embryo's blastopore (see p. 1040). Also, mesoderm and coelom development is similar in echinoderms and chordates.

It seems likely that both echinoderms and chordates have descended from common or very similar bilaterally symmetrical ancestors. Echinoderm larvae still are bilaterally symmetrical.

Members of both phyla have an **internal skeleton (endoskeleton)** that develops from mesoderm and is covered by a separate skin. Echinoderm skeletons consist of **calcareous (CaCO$_3$ containing)** plates, and in some echinoderms, especially sea urchins, these plates bear spines that stick out through the delicate skin that covers the skeleton. In fact, the name Echinodermata comes from Greek words meaning "spiny skin."

Echinoderms have no excretory system and a poorly developed circulatory system. An extensive coelom is filled with fluid that bathes body organs. Materials move around the body through the coelomic fluid, and wandering **amoeboid cells** clean up particulate wastes around the body.

Gas exchange takes place through small **gills** that are tiny fingerlike extensions of the skin (see figure 11.7). The coelomic cavity extends into each gill; thus, oxygen can pass through coelomic fluid to body organs, and carbon dioxide can return by the same route.

The echinoderm nervous system has a central ring of nerves and nerve branches extending out into various body divisions, but echinoderms have no brain. Despite the absence of a central control center, echinoderms are capable of coordinated, but slow, responses and body movements.

A unique characteristic of the echinoderms is their **water vascular system.** This system consists of a network of canals that contain watery fluid and extend throughout the body. Numerous tube feet that are attached to these canals function in body movement.

Sexes are separate in echinoderms, and they shed gametes into the water, where fertilization occurs. An echinoderm zygote develops into a free-swimming, bilaterally symmetrical larva that swims by means of cilia and feeds on plankton. These larvae eventually settle to the bottom as they undergo metamorphosis to become radially symmetrical adults (figure 37.60).

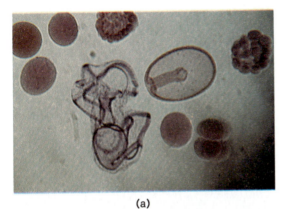

(a)

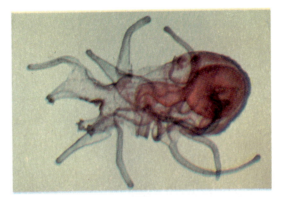

(b)

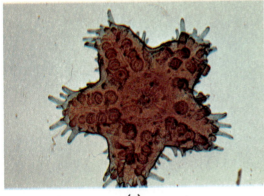

(c)

(d)

Figure 37.60 Starfish development as an example of echinoderm metamorphosis. (a) Stages in embryonic development of the starfish *Asterias*. Zygote and first cleavage, lower right. Advanced cleavage stages, top and upper right. Gastrulation, center right. Larva, center left. (b) Advanced larval stage. (c) A young starfish during metamorphosis. (d) A young starfish shortly after metamorphosis.

Class Asteroidea

The **sea stars** (**starfish**) are the most familiar of the echinoderms. The starfish body is flattened, and it has a **central disc** with five sturdy **arms** (**rays**) extending outward from it. Each arm has rows of tube feet running along a groove on its ventral side.

Tube feet can be extended or shortened, and they can attach firmly to surfaces. Thus, a starfish can move by extending, attaching, and then contracting groups of tube feet (figure 37.61).

Starfish commonly feed on clams, oysters, and other bivalve molluscs (figure 37.62). A starfish's mouth is on the ventral side of its central disc, and its anus is on the dorsal side. To feed, a starfish positions itself over its intended victim, a clam, for example. It attaches some of its tube feet to the clam's shells, and the tube feet begin to contract, pulling on the shells. The clam resists this pulling by forcefully contracting its shell-closing muscles. The starfish usually can win this tug-of-war, however, because it uses its tube feet in relays, by attaching and contracting some

Figure 37.61 Starfish anatomy. (a) Starfish with different arms dissected to show details of different systems. Note that canals of the water vascular system connect to a ring canal in the central disc. Water from outside the body can enter or leave the system via the madreporite. (b) A starfish with one of its arms lifted showing extended tube feet. Contractions of ampullae extend tube feet. The feet then attach and contract, pulling the starfish along.

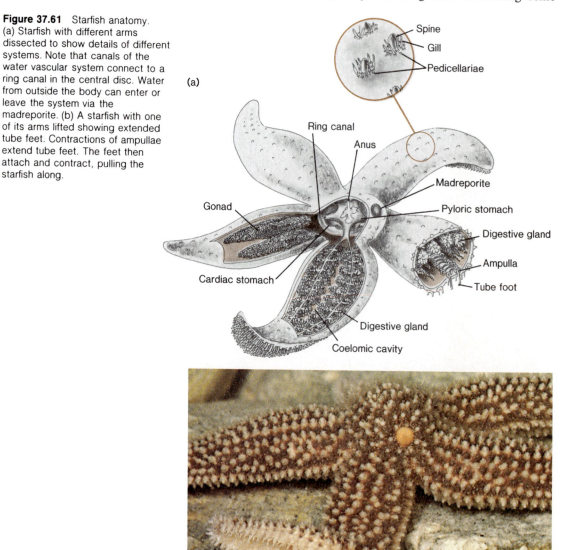

(a)

Spine
Gill
Pedicellariae

Ring canal

Anus

Gonad

Madreporite

Pyloric stomach

Digestive gland

Ampulla

Tube foot

Cardiac stomach

Digestive gland

Coelomic cavity

(b)

while others rest. Eventually, the clam's muscles tire and weaken, and the shell begins to open ever so slightly.

A very small crack between the shells is enough for the starfish, which then pushes the large, lower part of its stomach out of its mouth and through the crack between the clam's shells. Because this everted stomach is inside out, exposed digestive surfaces can be spread over the soft body mass of the clam. The stomach secretes enzymes, and the digestion of the clam's body begins even while the clam is still working to close its shells. Later, partly digested food is taken into the starfish's body, where digestion continues in the upper part of the stomach and out in the digestive glands that extend in pairs into each arm.

Because of their ability to open shells, starfish are very efficient predators on molluscs, and starfish are considered a great threat to shellfish production. In an attempt to rid oyster beds of starfish, oyster fishermen used to chop up captured starfish and throw them back into the sea. Unfortunately, this strategy backfired because

(a)

Figure 37.62 Starfish feeding. A starfish feeds on a bivalve mollusc by pulling with its tube feet until the mollusc can no longer keep its shells closed. (a) A starfish moves into position over a clam. (b) A starfish wrapped around a clam (center) in feeding position.

(b)

starfish have tremendous regenerative capacities, and an entire starfish can regenerate from even a single arm.

Other Classes of Echinoderms

Superficially, some of the other echinoderms look quite different from starfish, but all other echinoderms are also radially symmetrical, and most have five-part body organization.

The **class Ophiuroidea** includes **brittle stars** and **basket stars** (figure 37.63). They have relatively small central discs and relatively long, slender, flexible arms. Brittle stars can move around much more rapidly than starfish because they bend their long arms to push themselves along quickly and are not as dependent on the slow action of tube feet. When brittle stars' long arms are injured, they are discarded, and new ones grow in their place.

Sea urchins and **sand dollars** belong to the **class Echinoidea** (figure 37.64). Sand dollars have skeletal plates that are fused into a single, flattened unit. Sea urchins also have fused skeletal plates, but they have more rounded bodies. Sea urchins have long, movable spines that protrude from their skin. These spines have complex

muscle sets at their bases so that they can be moved in any direction. Coordinated spine movement pushes urchins along and supplements the action of tube feet in movement.

Sea urchins have a complex chewing apparatus called **Aristotle's lantern** that is used to grind up food. Some urchins eat algae scraped from rocks, some eat detritus from the bottom, and others feed on large kelps.

Sea urchins themselves provide food for many other organisms, ranging from starfish to mammals. For example, sea otters float on their back and use stones to crack open urchins held on their chest. Sea urchin eggs are even used for human food, especially in Japan where large quantities are imported each year. Yearly harvests, for export, of as much as a million kilograms of the urchin *Strongylocentrotus* have been made along the west coast of the United States and Canada. Residents in Baja California also consider sea urchin gonads a delicacy. Finally, sea urchin gametes and embryos are favorite research and teaching subjects for biologists around the world. Much current knowledge of fertilization and early development in animals was obtained by studying sea urchins.

(a)

(b)

Figure 37.63 Members of the class Ophiuroidea. (a) Brittle stars. (b) A basket star. The five arms of a basket star branch and rebranch to produce a mass of coils.

(a)

(b)

(c)

(d)

Figure 37.64 Members of the class Echinoidea, sand dollars and sea urchins. (a) Skeleton of the arrowhead sand dollar *Encope*. (b) Skeleton of the sea urchin *Arbacia*. Note the definite five-part organization. (c) Living specimen of *Arbacia* showing spines with extended tube feet among them. (d) Some urchins, such as this *Diadema*, have long, sharp, hollow spines containing an irritant. These spines can inflict painful puncture wounds if the urchin is handled or stepped on. (e) Chewing plates of Aristotle's lantern protruding from the mouth of a sea urchin.

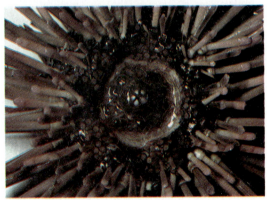

(e)

The **sea cucumbers, class Holothuroidea,** are very different from other echinoderms because they have leathery bodies and very reduced skeletons (figure 37.65a). A sea cucumber lies on its side and traps food particles in mucus on the surface of tentacles that are set in a ring around its mouth. Then it puts one tentacle at a time into its mouth and scrapes off the food.

Sea cucumbers have an unusual arrangement of internal tubules called **respiratory trees** that function in gas exchange. Two main "trunks" lead to branches that end in hollow sacs. The animal pumps water in and out of this system by way of a posterior opening, the **cloaca,** which also serves as the exit from the digestive system (figure 37.65b).

An interesting symbiotic relationship exists between sea cucumbers and a small tropical fish, the pearlfish, which lives in the cloaca and the trunks of sea cucumbers' respiratory trees. A pearlfish swims out at night to feed, then returns to the sea cucumber and backs through the cloaca into the safety of the respiratory tree.

Sea cucumbers also have great regenerative ability, which they sometimes put to unusual use. If stressed or bothered by a predator, a sea cucumber can eviscerate, that is, cast out large parts of its internal organs either by the mouth or cloaca, depending on the species. Often, this gives the sea cucumber an opportunity to move away and begin regeneration of the lost organs, while the startled predator is left to contemplate a writhing mass of body organs.

The final class of echinoderms are the oldest and most primitive in evolutionary terms. The **sea lilies,** members of the **class Crinoidea,** are sessile; that is, they live attached to a substrate by a stalk (figure 37.65c). Unlike all other echinoderms, their mouths are directed upward. Most sea lilies are suspension feeders, and they have branched arms with small appendages on them to sweep food out of the water.

(a)

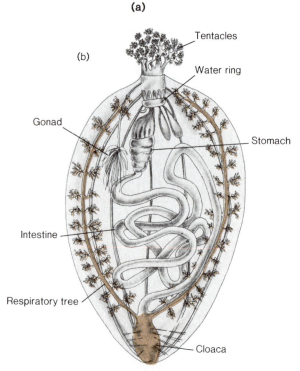

(b)

Tentacles

Water ring

Gonad

Stomach

Intestine

Respiratory tree

Cloaca

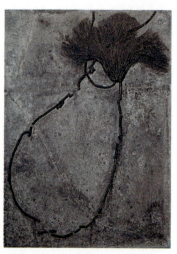

(c)

Figure 37.65 A sea cucumber and a sea lily. (a) A sea cucumber, *Cucumaria* (class Ophiuroidea). (b) Internal structure of the sea cucumber *Thyone,* showing the arrangement of the respiratory tree, which functions in gas exchange. (c) A fossil sea lily (class Crinoidea). Some modern crinoids are quite similar to these very ancient echinoderms.

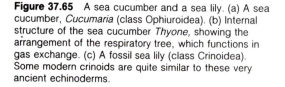

Phylum Chordata

Three primary diagnostic characteristics distinguish animals in the **phylum Chordata** from all other animals.

First, every chordate, at least sometime during its life, has a rodlike **notochord,** a supporting structure that runs along its body just ventral to its nerve cord. This is the structural feature that gives the phylum its name, chordata. Some nonvertebrate chordates have a notochord throughout life. However, in vertebrates, a notochord temporarily provides support for the developing embryo but is replaced during development by a **vertebral column.** The vertebral column is a chain of bones that provides support for the body axis and protection for the nerve tube.

Second, every chordate has a **dorsal hollow nerve tube,** an arrangement that contrasts sharply with the solid, usually ventral, nerve cords seen in the other animal phyla surveyed.

The third of the exclusively chordate characteristics is the possession at some time of **pharyngeal pouches** (figure 37.66). These are lateral expansions of the pharynx portion of the digestive tract. In many chordates, the pouches break through to the outside, and in some of these animals, they produce **gill slits.** But the pouches are only temporary embryonic structures in many vertebrates, such as humans, and they do not break through to the outside.

Chordates also commonly have several or all of the following nonexclusive characteristics: (1) **segmentation,** especially of body muscles; (2) a **tail,** a portion of the body located posterior to the anus; and (3) an **endoskeleton.** While these features further describe chordates, they do not distinguish them from all other animals because members of some other phyla also have one or several of these characteristics.

Echinoderms, Hemichordates, and Chordates: Evolutionary Relatives

As discussed earlier, echinoderms and chordates appear to be related because of some common developmental characteristics, especially the fact that both groups are deuterostomes. But the story of this relationship is not complete without the introduction of another group of animals, the **hemichordates.** The most common hemichordates are the **acorn worms,** marine animals that live in burrows in coastal sand or mud (figure 37.67).

At the anterior end of an acorn worm's body is a muscular **proboscis.** Just behind that proboscis is an enlarged collar region. The mouth opens at the edge of the collar and leads to a **pharynx** with many gill slits in its wall. As water is drawn in through the mouth and forced out through the gill slits, food is trapped in mucus in the pharynx and passed on through the digestive tract. Oxygen and carbon dioxide are exchanged between the moving water and blood in vessels adjacent to the gill slits.

Acorn worms clearly have one of the chordate characteristics, pharyngeal pouches (which develop into their gill slits), but what about the other chordate characteristics? Acorn worms have a diffuse nervous system that takes the form of a hollow cord only in the collar region. They also have a notochordlike structure, but it is found only in the proboscis. Because they somewhat resemble chordates, the hemichordates used to be classified as a subphylum of the phylum Chordata. But this no longer seems justified, and they are classified as the separate **phylum Hemichordata,** a phylum that is distinct from, but closely related to, the phylum Chordata.

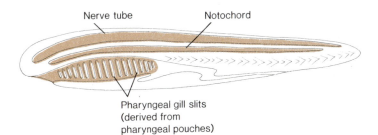

Nerve tube　　　　Notochord

Pharyngeal gill slits
(derived from
pharyngeal pouches)

Figure 37.66 Diagnostic characteristics of chordates. Pharyngeal pouches do not open as gill slits in all chordates. Note that this chordate has segmental muscles and a postanal tail.

Figure 37.67 Hemichordates and
relatives. (a) Model of a
hemichordate *(Dolichoglossus).*
(b) Larva of a hemichordate
(Glossobalanus). Compare with (c)
and with figure 37.60a (starfish
larva). Evolutionary linkage of
hemichordates and echinoderms
depends on developmental patterns
and similarities of larvae. (c) Larva
of a sea cucumber *(Labidoplax).*

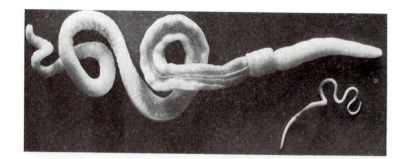

(a)

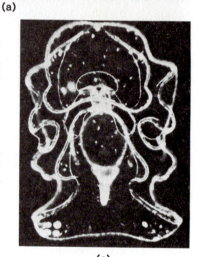

(b)

(c)

Hemichordates and echinoderms also appear quite closely related because they have larvae that are virtually identical (figure 37.67c).
This may indicate that echinoderms and hemichordates descended from a common ancestor,
probably the same group of primitive animals
that gave rise to the chordates as well.

Nonvertebrate Chordates

By now, it should be clear that there are chordates that are not vertebrates. Nonvertebrate
chordates belong to two groups, the **subphylum
Urochordata** and the **subphylum Cephalochordata**.

The **sea squirts (tunicates)** are the most common urochordates. Adult tunicates are sessile
marine animals that, at first glance, might well
be mistaken for sponges, or possibly coelenterates. A tunicate is surrounded by a tough outer
covering, the **tunic**, that is peculiar because it
contains cellulose, a substance very rarely found
in animal bodies.

Tunicates filter food from water pumped
through hundreds of pharyngeal gill slits. But
beyond having pharyngeal gill slits, adult tunicates do not seem to resemble other chordates.
In fact, tunicates are classified as chordates
mainly because their tadpolelike larvae possess
the fundamental chordate characteristics (figure
37.68).

The common names for members of the subphylum Cephalochordata are **amphioxus** or
lancelet. These are small, fishlike, marine animals that can swim freely but often remain stationary, partly buried in the mud. Amphioxus is
a filter feeder with a pharyngeal gill apparatus,
and it has a well-developed hollow nerve cord and
notochord (figure 37.69). Thus, the adult amphioxus displays the fundamental chordate characteristics in a clear and unambiguous fashion.
In fact, it is commonly thought to be quite similar
to the ancient segmented animals from which
vertebrates evolved.

(a)

Figure 37.68 Tunicates. (a) Living specimens of *Ciona intestinalis*, a solitary tunicate (sea squirt). (b) Generalized sketch of the body of a tunicate. Water passes through gill slits into a chamber called the atrium and then out of the body through the excurrent siphon. The anus and reproductive ducts empty into the atrium near the excurrent siphon. (c) Sketch of a tunicate larva ("tadpole") showing that it clearly possesses the basic chordate characteristics. These larvae swim actively until they settle to the bottom and attach as they undergo metamorphosis. A larva loses its tail, notochord, and most of its nervous system during metamorphosis.

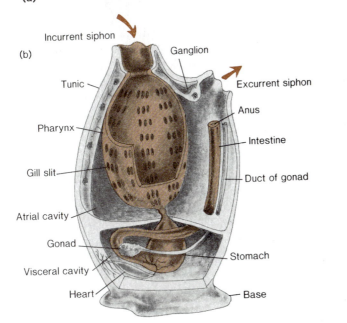

(b)

Incurrent siphon
Ganglion
Tunic
Excurrent siphon
Anus
Pharynx
Intestine
Gill slit
Duct of gonad
Atrial cavity
Gonad
Stomach
Visceral cavity
Heart
Base

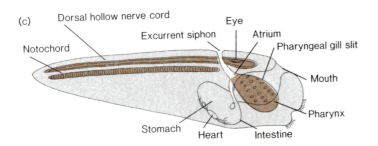

(c)

Dorsal hollow nerve cord
Eye
Excurrent siphon
Atrium
Notochord
Pharyngeal gill slit
Mouth
Pharynx
Stomach
Heart
Intestine

(a)

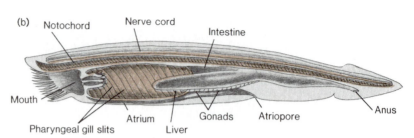

(b) Notochord   Nerve cord   Intestine

Mouth

Pharyngeal gill slits   Atrium   Liver   Gonads   Atriopore   Anus

Figure 37.69 Amphioxus (lancelet), a member of the subphylum Cephalochordata. (a) External view of several individuals, showing segmentally arranged muscles. Sometimes amphioxus swims freely. At other times it continues its filter feeding while partially buried in the bottom. (b) Longitudinal section showing internal anatomy. Water enters mouth, passes through pharyngeal gill slits into atrium, and then leaves body through atriopore.

Vertebrates

The **subphylum Vertebrata** needs little introduction. Humans are vertebrates, common domesticated animals are vertebrates, and the large dominant wild animals in the terrestrial habitat are vertebrates. Many aspects of vertebrate function and structure have been discussed elsewhere in this text.

The most obvious characteristic of vertebrates is that they have a bony or cartilaginous endoskeleton that includes a **vertebral column** ("backbone"), made up of a series of **vertebrae**. Vertebrates have a brain, which is an obvious anterior enlargement of the dorsal, hollow nerve tube. The vertebrate brain is protected by a brain case, the **cranium**. Vertebrates have a high degree of **cephalization** (head development), and most of them have several pairs of sense organs on their heads. Most vertebrates have a tail. A vertebrate digestive system characteristically includes a liver and a pancreas.

There are seven living classes of vertebrates and one class that is extinct and known only from the fossil record. Three living classes and the exclusively fossil class are fishes. These fish classes are the **Agnatha** (lampreys and other jawless fishes), **Placodermi** (extinct, primitive fishes with jaws), **Chondrichthyes** (sharks, rays, and other cartilaginous fishes), and **Osteichthyes** (bony fishes). The four-limbed (**tetrapod**) vertebrates make up the other four vertebrate classes: **Amphibia**, **Reptilia**, **Aves** (birds), and **Mammalia**.

Class Agnatha

Members of the class Agnatha ("without jaws") also are known collectively as the **cyclostomes** ("round mouths") because they are fish that have round, sucking mouths and lack hinged jaws. **Lampreys** and **hagfish** are living members of the class. They have skeletons constructed entirely of cartilage, but this condition may be a result of degenerative evolutionary changes since the earliest fossil jawless fish, the **Ostracoderms**, had bony armor plates that covered their bodies.

Ostracoderms are the oldest vertebrate fossils that have been found, and they date from the Ordovician period, which began about 500 million years ago. The vertebrates are the only major animal group not represented among the fossils of the Cambrian period, which preceded the Ordovician and ended about 500 million years ago. Some primitive vertebrates may have been living during the Cambrian period, but as yet no fossil record of them has been found.

Ostracoderms were small, jawless fishes with bony skeletons, and they probably lived as filter feeders. Modern cyclostomes are cartilaginous fishes that either are parasitic (lampreys) on other fish or scavengers (hagfish) that usually eat dead fish.

A lamprey feeds by attaching its round, sucking mouth to the body of another fish (figure 37.70). It uses its horny tongue to rasp a hole in the host's skin and then proceeds to suck out blood and other body fluid from the host. While a single lamprey attack often is not fatal and some fish have been caught bearing several lamprey scars, repeated lamprey attacks weaken and eventually kill even large fish.

Lampreys enter freshwater streams to breed, and their eggs develop into small filter-feeding larvae that closely resemble amphioxus. After two or three years, the larvae undergo metamorphosis to become adult lampreys and go to sea.

It used to be thought that lampreys had to follow this pattern and that adult lampreys were exclusively marine. But during this century, adult lampreys have become established in the Great Lakes of North America, completing their entire life cycles in fresh water. Lampreys in the Great Lakes have had a devastating effect on fish populations. The Great Lakes lake trout fishing industry has only partly recovered despite a concerted effort to destroy lampreys by poisoning lamprey larvae in all of their breeding streams around the Great Lakes.

Figure 37.70 The round, sucking mouth of a lamprey. Lampreys attach to other fish and feed on their blood and tissue fluids.

Class Placodermi

Although they have been extinct for millions of years, the **placoderms** represent an important milestone in vertebrate evolution because they were the first vertebrate animals with hinged jaws (figure 37.71). Jaws probably evolved from bars that supported gills in the anterior portion of the pharynx of their jawless ancestors.

Placoderms, like the ancient Agnatha, had bony armor plates, and they had five to seven pairs of fins. These characteristics set placoderms apart from modern vertebrates, but all living vertebrates except the cyclostomes are considered to be descendants of such primitive jawed fishes that inhabited fresh water.

One group of their descendants, the class Chondrichthyes, have lost all bony structures and have a completely cartilaginous skeleton. Practically all the chondrichthyes are marine organisms.

Figure 37.71 Painting of an extinct placoderm. Placoderms were the first vertebrates to have hinged jaws.

The other major group of descendants from the primitive jawed fishes were members of the class Osteichthyes (bony fishes). They became a large, successful class occupying all manner of aquatic habitats, both freshwater and marine. Furthermore, some of the primitive Osteichthyes probably were ancestors of all of the four-limbed vertebrates.

Class Chondrichthyes

The class Chondrichthyes ("cartilage fishes") includes **sharks, skates,** and **rays** (figure 37.72). The chondrichthyes are descendants of ancient bony fishes, but all of them have exclusively cartilaginous skeletons. Their bodies are covered with small, toothlike scales called **denticles**. Thus, a shark's skin feels like sandpaper. The menacing teeth of sharks and their relatives are simply larger, specialized versions of these scales.

Rays and skates swim along the bottom and feed on animals, mostly invertebrates, that they dredge up. But most sharks are fast-swimming predators with beautifully streamlined bodies that slip easily through the water.

Sharks hold a terrifying fascination for many people, and they have a generally well-deserved reputation as ferocious and somewhat indiscriminate feeders. Tiger sharks, for example, have been known to bite off and swallow chunks of boats. Great white sharks seem to bite first and then decide whether or not to swallow. This may explain why a number of human victims of great white shark attacks seem to have been "spit out."

However, some of the largest sharks, basking sharks and whale sharks, are not active predators, but instead are filter feeders. The whale shark, which reaches a length of 16 m and thus is the largest known fish, filters huge amounts of

(a)

(b)

(c)

(d)

Figure 37.72 The class Chondrichthyes (cartilaginous fishes). (a) A reef white-tip shark. Note the streamlined body and the separate openings of all the gill slits. (b) Piece of shark skin showing scales (denticles). Note their toothlike shape. (c) A leopard shark. (d) A skate.

water to obtain the great quantities of plankton that it eats.

Sharks have internal fertilization, and the eggs of some species develop inside the female reproductive tract. In those species, little sharks are born fully developed and swim away from their mothers immediately. An immediate escape is advantageous because some sharks do eat their own young just as they would eat any other small fish.

Class Osteichthyes

The class Osteichthyes includes most of the familiar fishes. It is a large, diverse class that may have as many as thirty or forty thousand species, although only about seventeen thousand species have been described so far (figure 37.73).

The earliest Osteichthyes probably lived in freshwater habitats that occasionally became stagnant and oxygen-deficient. In addition to gills, they possessed simple, saclike lungs connected to the anterior end of their digestive tract. These lungs probably functioned as supplementary gas exchange surfaces, allowing them to breathe air when conditions in the water were poor.

Ancient members of the class Osteichthyes fell into two major groups. One group was the **ray-fin fishes**, whose fins were supported only by spinelike rays. The other group included **lungfishes** and **lobe-fin fishes**. Lobe-fin fishes had fins that were set in fleshy lobes that had bony, skeletal supports in them.

Ray-fin fishes were the ancestors of most of the modern Osteichthyes. The ancient lung has converted into a **swim bladder**, a gas-filled sac located in the dorsal part of the body cavity. By secreting gas into the swim bladder or absorbing gas from it, a bony fish can change its buoyancy.

(a)

(b)

Figure 37.73 The class Osteichthyes (bony fishes). (a) A sea perch. Note the flattened scales and the operculum that covers the gill chamber. (b) An angler fish. This deep-sea fish has a "lure" near its mouth that attracts potential prey. (c) A spotted moray eel swimming among sea urchins in the Red Sea. (d) A seahorse.

(c)   (d)

This allows the fish to hover effortlessly at different depths in the water. Modern bony fish have broad, flattened **scales** covering some or all of their body surfaces. Their gills are located in an enclosed **gill chamber** that is covered and protected by a hard bony flap, the **operculum**. Thus, they have only one gill opening on each side instead of separate openings for each gill slit, as sharks have.

Modern ray-fin fishes live in every kind of aquatic habitat, ranging from seawater to brackish water to fresh water, and from frigid Arctic waters to perpetually warm tropical seas. They can be bright and beautifully colored fish that dart about in coral reefs or jet black fish that lurk in the ocean depths. Their range of diversity includes snakelike eels with no fins, comical seahorses that hover in the water, angler fish that dangle bait to tempt other fish within reach, and a host of other specializations.

Only a few descendants of the other major group of ancient Osteichthyes, which included the lobe-fin fishes and lungfishes, still survive. African, Australian, and South American lungfishes are rather obscure animals that live as their ancestors did, in freshwater habitats that often become stagnant or even dry up entirely. The lungfishes regularly breath air to supplement gas exchange in their gills (see box 11.1 on page 303).

From an evolutionary viewpoint, the ancient lobe-fin fishes are more interesting than the lungfishes because some of the lobe-fin fishes were the ancestors of all the land-dwelling vertebrates. The lobe-fin fishes could breathe air while they crawled clumsily on their stumpy fin lobes from pond to pond. This ability, marginal as it was, to survive on land opened a whole new realm of possibilities for lobe-fin fishes and their descendants (figure 37.74).

(a)

(b)

Figure 37.74 Lobe-fin fishes and early amphibians. (a) A lobe-fin fish (*Eusthenopteron*) that probably could move out of the water onto land. (b) An early amphibian (*Diplovertebron*). Notice that its body is very low to the ground and its legs stick out to the sides. It must have been an awkward walker, but its legs were an improvement over lobe-fins for locomotion on land.

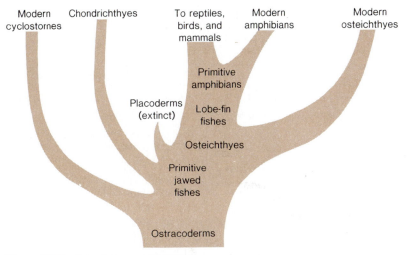

Figure 37.75 A family tree of the fishes.

Figure 37.76 Sketch of a coelacanth (*Latimeria*). This is the only living lobe-fin fish.

Land plants evolved well before land animals so there was an excellent food supply available on land for any animal that could remain ashore long enough to utilize it. Under those conditions, selective pressure obviously favored further development of land-dwelling abilities. Those animals able to spend more time on land could avoid more vigorous competition in the water. It is probable that the first amphibians evolved from some of the lobe-fin fishes. See figure 37.75.

From the discovery of their fossil remains until 1939, all of the lobe-fin fishes were believed to have been extinct for about 75 million years. But in 1939, one was caught off the coast of South Africa. The fishermen who caught it were unaware of its scientific importance, and the first specimen was not in good condition when it reached scientists. Nevertheless, it was identified as a **coelacanth** and given the genus name *Latimeria* (figure 37.76). Since that time, other specimens have been caught and carefully examined. These "living fossils" probably are descendants of relatives of the lobe-fin fishes that gave rise to amphibians and, eventually, to other tetrapods.

Thus, while the lungfishes have remained in the same freshwater environments occupied by their ancient ancestors, the surviving lobe-fin fishes long ago entered the marine environment and "slipped off the continental shelf" into the depths of the ocean, where the coelacanth is found today.

Class Amphibia

The three groups (orders) of modern amphibians are the **salamanders** and other amphibians with tails (order **Urodela**); the **frogs** and **toads** (**order Anura**); and a small group of limbless, burrowing amphibians, known as **caecilians,** that occur only in tropical environments (**order Apoda**) (figure 37.77).

Amphibians probably descended from air-breathing lobe-fin fishes that were able to leave the water for short times (see figure 37.74). The first amphibians were very similar to the lobe-fin fishes but were better adapted for life on land for several reasons. Early amphibians had legs that ended in toes (usually five or less), and they had **pectoral** and **pelvic girdles,** bony structures that supported forelimbs and hindlimbs, respectively, and provided attachment points for muscles that moved the limbs.

Improved locomotion on land and other adaptations for terrestrial life led to a great amphibian diversification during the Carboniferous period that produced many species and large numbers of individuals. Amphibians became the dominant animals of the period (figure 37.78).

But by the end of the Permian period (about 230 million years ago), this "age of amphibians" was ending, and reptiles replaced amphibians as the dominant land animals. During this period of change, many kinds of amphibians became extinct and only a few lines, including those leading to the modern amphibians, continued.

Some factors that limited amphibians' success as terrestrial animals are inherent in their basic biology. Amphibians are closely tied to water or, at least, very moist terrestrial habitats for several reasons. Amphibians have rather simple lungs (p. 305), and most of them depend on cutaneous (skin) exchange as a supplementary gas exchange mechanism. For skin to function efficiently as a gas exchange surface, it must be quite thin and must be kept moist. This combination creates a water loss risk, and amphibians can easily become desiccated and die in dry environments.

(a)

(b)

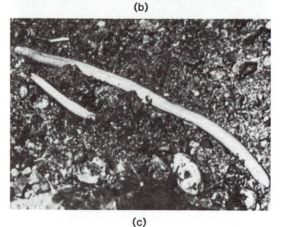

(c)

Figure 37.77 Amphibians. (a) A salamander (class Urodela). (b) A poison arrow frog, a member of the class Anura ("without a tail"). (c) Caecilians, members of the class Apoda ("without limbs").

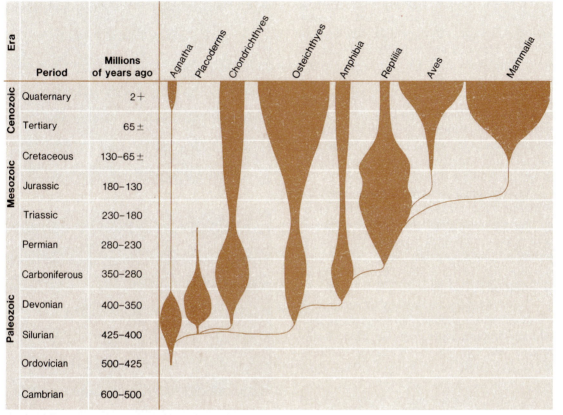

Era	Period	Millions of years ago
Cenozoic	Quaternary	2+
Cenozoic	Tertiary	65 ±
Mesozoic	Cretaceous	130–65 ±
Mesozoic	Jurassic	180–130
Mesozoic	Triassic	230–180
Paleozoic	Permian	280–230
Paleozoic	Carboniferous	350–280
Paleozoic	Devonian	400–350
Paleozoic	Silurian	425–400
Paleozoic	Ordovician	500–425
Paleozoic	Cambrian	600–500

Figure 37.78   Geologic time periods and the evolution of vertebrate classes. The line for each vertebrate class shows evidence from the fossil record regarding the time of origin of the class and changes in abundance of the species in the class. Thicknesses of lines for the different classes are not proportional to one another, however.

However, the most important factor binding amphibians to water is their method of reproduction. Amphibians have external fertilization, with both eggs and sperm being shed into the water. Amphibian eggs, like fish eggs, are enclosed by a jelly coat that provides no protection against desiccation if the eggs are exposed to the air. Young amphibians hatch from their jelly as aquatic larvae (**tadpoles**) with gills. Tadpoles feed and grow in the water. Only after **metamorphosis** do amphibians emerge from the water as air-breathing adults (figure 37.79).

A few amphibians have special reproductive adaptations that partially circumvent this problem. For example, some frogs carry their developing eggs in fluid-filled pouches. Their young proceed to metamorphosis very quickly and emerge as tiny adults ready to live and grow in the terrestrial environment.

But, in the main, amphibians' expansion into various terrestrial habitats has always been limited by their dependence on water. Modern amphibians still live on the borderline between aquatic and terrestrial environments. Their class name, amphibia, means "double life" and accurately describes their life history, which is divided between two different habitats.

The reptiles arose from amphibian ancestors, but they were much better suited to life on land (figure 37.80).

Class Reptilia

Reptiles are truly terrestrial animals because they can complete their entire life cycle in the terrestrial environment. They have internal fertilization, and they lay eggs that are protected by leathery **shells.** Enclosed within such a **cleidoic** ("boxlike") egg, a reptile embryo develops in a sheltered environment, where it is supplied with the water and nutrients required for its development (figure 37.81). **Extraembryonic** ("outside the embryo") **membranes** that develop around the embryo provide for nutrition, for exchange of gases, which diffuse through the shell, and for storage of metabolic wastes that accumulate during development. One of the membranes, the **amnion,** is a sac that fills with fluid and provides a "private pond" within which the embryo develops.

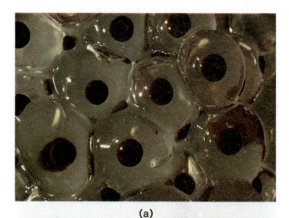

(a)

(b)

(c)

Figure 37.79   The amphibian life history. (a) Eggs of the grass frog (*Rana temporaria*) surrounded by their jelly coats. (b) *Rana temporaria* tadpoles. Tadpoles are herbivores that feed on aquatic plants. (c) Young *Rana temporaria* just after metamorphosis.

Figure 37.80 *Seymouria*, a fossil animal considered to be a possible evolutionary link between amphibians and reptiles.

(a)

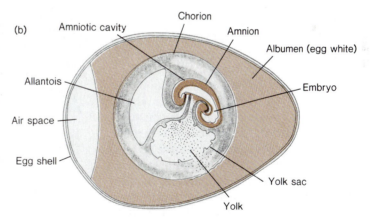

(b)

Figure 37.81 The cleidoic eggs of reptiles, an adaptation for reproduction in a terrestrial environment. (a) Baby kingsnake hatching out of its shell. Note that shells surrounding reptile eggs are leathery and flexible, not brittle like birds' eggs. (b) The arrangement of embryo and extraembryonic membranes in a cleidoic egg. The yolk is drawn smaller than its normal proportions. The chorion encloses the whole system; the yolk sac absorbs nutrients; the allantois stores wastes and blood vessels in its wall function in gas exchange; the amnion encloses amniotic fluid, which surrounds and protects the developing embryo.

Reptiles have more efficient lungs than amphibians and thus do not need extensive cutaneous gas exchange nor thin, moist skin. Reptiles have a thick, scaly skin that contains **keratin**, which helps to make the skin impermeable to water. This is a great help in water conservation, an urgent problem for terrestrial animals.

The four orders of modern reptiles are **turtles** (order **Chelonia**), **crocodiles** and **alligators** (order **Crocodilia**), **lizards** and **snakes** (order **Squamata**), and the **tuatara** (order **Rhynchocephalia**) (figure 37.82). The tuatara is the only surviving member of its order and is found only on islands in New Zealand.

Modern reptiles, however, are only part of the story of reptiles, a story that is complete only in the context of reptiles' history and evolutionary relationships.

The earliest reptiles, called **stem reptiles**, gave rise to several lines of descent (figure 37.83). Some of those lines produced the modern reptiles; other lines led to the mammals and birds, which are evolutionary descendants of reptiles.

Possibly the most fascinating aspect of the history of reptiles, however, was the evolution of the great reptiles that dominated the earth for millions of years during the Mesozoic era and then became extinct at its end (figure 37.84). There were flying reptiles; large, swimming reptiles; and the **dinosaurs**, the reptiles that dominated the terrestrial environment.

The dinosaurs are well represented in the fossil record but still remain shrouded in mystery. One of the great mysteries of the dinosaurs concerns their extinction. One hypothesis is that climatic change brought about their demise because they could not stand cooler temperatures or because their food sources were reduced. Another hypothesis proposes that mammals evolved to a point where they preyed on dinosaur eggs to such an extent that the dinosaurs were no longer able to reproduce themselves. There is even a somewhat more radical suggestion that the dinosaurs died out, not gradually over millions of years, but quickly within a period of hundreds of years or even much less, following some catastrophic event (for example, an asteroid collision) that abruptly altered conditions worldwide.

In an evolutionary sense, the length of time that the dinosaurs dominated the earth is even more impressive than the fact of their extinction.

(a)

(b)

(c)

(d)

Figure 37.82   The orders of reptiles. (a) A turtle (*Gopherus*) of the order Chelonia laying eggs in the Okefenokee swamp. (b) Alligators (*Alligator*) of the order Crocodilia. (c) Brazilian boa constrictor (*Vila*) of the order Squamata. This order also includes lizards. (d) Tuatara (*Sphenedon*) of the order Rhynchocephalia.

Figure 37.83 A family tree of the reptiles, including their relationships to other vertebrate classes.

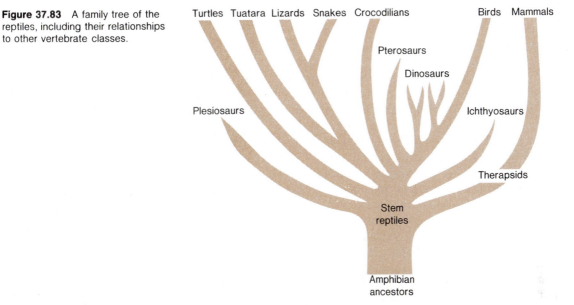

Figure 37.84 Fossil reptiles from the Mesozoic era. (a) A pterosaur. Pterosaurs were flying reptiles that used a broad, flat, skin flap for flying. Pterosaurs were not ancestors of birds, since birds descended from a separate group of reptiles. (b) Plesiosaurs (left) and Ichthyosaurs, swimming reptiles. (c) *Brontosaurus,* a giant herbivorous dinosaur that probably spent much of its time standing partially submerged in water, where its enormous weight (as much as 30 tons) was partially supported. (d) *Triceratops* (left), an armored herbivore, and *Tyrannosaurus* (right), a huge carnivore that was as much as 15 m long and 6 m tall when it stood upright.

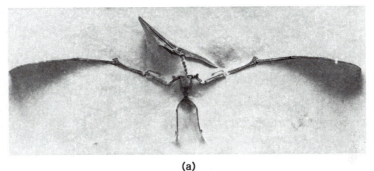

(a)

(b)

(c)

(d)

Their reign lasted about 100 million years. Mammals existed during the dinosaurs' time, but they were small, relatively insignificant animals living where they could in the dinosaurs' world (figure 37.85). Mammals have dominated the terrestrial environment only for the last 65 million years (since the extinction of the dinosaurs). The life-span of the human species is measured only in thousands of years, a moment in time compared to the length of the age of dinosaurs.

Another mystery concerns the basic nature of dinosaurs. They have long been described as stupid, sluggish, slow-moving animals that could function efficiently only in a uniformly warm environment that kept their body temperature high enough to permit a reasonable range of activity. But in the last few years, some biologists have challenged the view that the dinosaurs all were heterotherms, animals that were unable to regulate their body temperatures metabolically. Dinosaurs lived successfully in northern latitudes, where very short winter days would have made it very difficult for them to heat up their giant bodies by solar radiation, even though climates generally were warmer. Present-day reptiles, which all are heterotherms, have very sparsely vascularized bones. But studies of dinosaur bones reveal rich blood vessel networks such as those found in the bones of mammals, which are homeotherms ("warm-blooded animals").

It now appears that some reptiles, who were ancestors of the birds but were not fliers themselves, had feathers or featherlike structures. Birds' feathers are involved in flying, but they also are critically important in body insulation. Thus, these nonflying, feathered reptiles may have been supplied with insulation that aided in maintaining a constant high body temperature. This discovery of feathered, possibly homeothermic, dinosaurs has led some biologists to propose that the dinosaurs are not extinct after all, but that birds are just a highly specialized group of surviving dinosaurs.

Class Aves

Birds are the only modern animals that have **feathers**. Birds are descended from reptiles, and feathers have evolved as modified scales. Scales on the legs and feet of modern birds are reminders of their reptilian ancestry, as are the claws at the ends of their toes (figure 37.86).

Figure 37.85 A therapsid reptile. Mammals evolved from these primitive reptiles early in the Mesozoic era, but mammals remained small animals, living in the shadows of the great reptiles until the end of the era.

Figure 37.86 A representation of *Archeopteryx*, the oldest known fossil bird. *Archeopteryx* had several reptilian characteristics, including jaws with teeth and a long, jointed, reptilian tail. It was about the size of a crow and had rather feeble wings for its size, but it was well-feathered.

Birds, however, are not descended from the group of flying reptiles, the pterosaurs, that thrived during the Mesozoic era. Pterosaurs used a broad flap of skin as the flat surface required for flight. Birds, on the other hand, achieve the same effect with specialized **contour feathers** that are attached to their wings in such a way that they overlap to produce a broad flat surface used in flight. This arrangement has an important advantage over broad skin flaps because damage to a few feathers does not put a bird's wing out of flying commission, as damage to a single broad skin flap might well do.

Feathers also function in body temperature regulation. **Down feathers** and the downy bases of other feathers provide excellent insulation against body heat loss because down traps air in tiny spaces (figure 37.87). This insulating effect

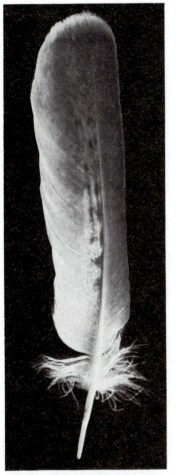

(a)

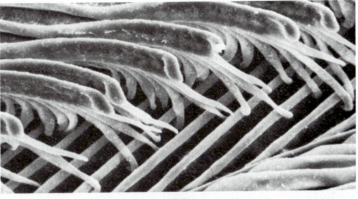

(b)

(c)

(d)

Figure 37.87 Feathers. (a) A contour feather from a red-tailed hawk showing the many **barbs** that branch from the shaft of the feather. Note the downy branches near the base of the feather. (b) Scanning electron micrograph of barbs of a goose flight feather showing the little hooks that interlock the barbs so that each feather is firm. Wing feathers overlap to form a flat surface for flight. (Magnification $\times$ 420) (c) An embryonic down feather. Each of about a dozen barbs has many barbules branching from it. Thousands of down feathers together trap air and provide effective insulation. (Magnification $\times$ 5) (d) Scanning electron micrograph of part of an embryonic down feather showing part of one barb and several barbules. Each barbule is a chain of cells arranged end to end. (Magnification $\times$ 290)

is important because birds are **homeothermic** animals; that is, they maintain a constant, relatively high body temperature by internal regulation. This permits birds to be continuously active, even when there are rather drastic fluctuations in environmental temperature.

Birds are not a very diverse class of animals. This probably is true because a set of adaptations for flight dominates birds' body organization, and there may be limits to the potential range of functional and structural designs for flying vertebrate animals. Some birds cannot fly, and some of those nonflying birds look quite different from flying birds (figure 37.88). Yet, they share some of the fundamental features that are recognized as flight adaptations.

One flight adaptation is that birds' bones are hollow, and thus very light. The bones contain internal air spaces that are connected to the network of air sacs and lungs that makes up the avian gas exchange system (p. 308). These hollow bones are surprisingly strong, however, because small internal support braces span the air spaces at key points where extra strength is needed.

Weight reduction is achieved in other ways as well. A female bird has only one functional ovary and oviduct instead of the paired structures found in other vertebrates. The testes of males remain small except during the breeding season, when they enlarge dramatically.

The general shape of the average bird's body also is a flight adaptation. A bird's head is relatively small and light because all birds have light horny **beaks** instead of jaws with teeth. A slender neck connects the head to a rounded, compact body. The weight of the digestive system, with a **crop** used in food storage and a thick muscular **gizzard** that grinds food, is well centered.

Birds have internal fertilization, and they lay cleidoic (boxlike) eggs that are enclosed in a brittle, calcium-containing shell. Generally, young birds hatch from their shells in a much more helpless condition than do young reptiles. Many newly hatched birds are poor thermoregulators and require brooding by their parents. Nestling birds require feeding by one or both parents and are able to fly away and seek food for themselves only after a period of intensive parental care. Complex hormonal regulation and behavioral responses are involved in nesting behavior and parental care.

(a)

(b)

(c)

Figure 37.88 Birds. (a) A bald eagle. (b) Adelie penguins. Penguins are flightless birds. Their wings are modified as flippers used for swimming. (c) A brown kiwi from New Zealand. The kiwi is a flightless bird whose tiny wings are hidden under its plumage. The kiwi looks shaggy because all of its feathers resemble the juvenile down feathers of other birds. The kiwi uses its long bill to probe for earthworms.

Flight requires well-developed sense organs. Birds have particularly acute vision and excellent visual reflexes. These permit birds to land precisely on small landing targets, such as small tree branches. They also allow predatory birds to swoop from great heights to capture small animals that they have spotted while cruising high above their unsuspecting prey. Other birds even dive into the water to capture fish that they have seen as they fly high above the water's surface (figure 37.89).

Possibly the most remarkable aspect of the biology of birds, however, is the annual **migration** that so many of them accomplish. Some birds undertake migrations of thousands of kilometers over land and ocean, navigating successfully by day and by night, through sunshine and cloudy weather, to reach very specific distant goals. Mysteries still remain regarding bird navigation, but birds almost certainly use celestial navigation and even variations in the earth's magnetic field intensity to find their way on these impressive journeys (p. 844).

Class Mammalia

Mammals are homeothermic vertebrate animals that have **mammary glands**, which produce milk, and body **hair**, which provides insulation against heat loss and thus aids in maintenance of a constant body temperature.

Homeothermic animals must obtain more food than heterothermic animals of comparable size because homeotherms expend considerable energy for heat production. But this disadvantage is outweighed by the fact that mammals, as homeotherms, are able to continue their activities in environments with variable temperatures.

Mammary glands permit female mammals to feed (nurse) their young regularly without having to find food every time that the young are hungry, as nesting birds must do. Also, the nursing relationship cements a behavioral bond between the young mammal and its mother that assures parental care and attention during the most vulnerable period of the young mammal's life. Thus, the relationship increases the reproductive efficiency of mammals because parental care and protection helps to assure that a greater percentage of young mammals survive to reach adulthood than generally is the case in other animals.

(a)

(b)

Figure 37.89 Predatory birds. (a) Owls with prey. (b) Hunting birds in action demonstrate the acute visual reflexes of birds. Here brown pelicans are diving for fish off the Florida Keys.

Another factor that enhances mammalian reproductive efficiency is that the great majority of mammals are **viviparous.** Their embryos develop inside a specialized portion of the female reproductive tract, the **uterus.** A circulatory exchange organ called the **placenta** permits exchange of materials between the bloodstreams of the mother and the developing fetus. This viviparous arrangement shelters developing young from environmental changes and frees adults from having to tend a nest, as birds must do.

Thus, mammals generally produce fewer eggs and, eventually, fewer newborn offspring than other vertebrates. But a larger percentage of the offspring grow to adulthood because of viviparity and parental care.

Mammals' limbs are oriented and connected in ways that permit them to run faster than other vertebrates can, and there are very fast-moving animals among both herbivorous and carnivorous mammals. Associated with their active life styles is the tendency for mammals to have well-developed sense organs and extensive development of brain centers, especially in the cerebral cortex, that are involved in flexible behavior patterns that are modifiable by experience and learning.

Mammals evolved sometime early in the Mesozoic era, most likely as descendants of therapsid reptiles (figure 37.85). Some of the therapsids may even have had hair and been homeothermic. Mammals remained relatively small animals during the Mesozoic era. Following the end of the age of reptiles, however, mammals radiated into many of the habitats previously occupied by the large reptiles. Large herbivorous (grazing) mammals evolved, as did the large carnivorous mammals that prey on herbivorous mammals. Small- and medium-sized mammals also spread into a variety of niches previously occupied by reptiles.

Living mammals are grouped into three subclasses. Two smaller subclasses consist of nonplacental mammals; the third, and largest, subclass contains the placental mammals and includes about 95 percent of living mammals.

Subclass Prototheria: Egg-laying Mammals
Only two kinds of egg-laying (**oviparous**) mammals, or **monotremes,** survive: the **duck-billed platypus** of Australia and the **echidna,** or **spiny anteater,** which lives in Australia and New Guinea (figure 37.90). These animals lay their eggs, which resemble reptile eggs, in burrows in the ground, where they incubate the eggs and brood the young, much as birds do. Although the monotremes are oviparous, they have mammary glands and feed their young milk after they hatch. This curious combination puts the monotremes on the borderline between reptiles and mammals.

Subclass Metatheria: Pouched Mammals The young of pouched, or **marsupial,** animals begin their development inside the female's body, but they are born in a very immature condition. Newborn young enter a pouch, the **marsupium,** on their mother's abdomen. Inside the pouch, they attach to nipples of mammary glands that are located there. They continue their development in the sheltered environment and emerge later when they are much more mature.

Marsupial mammals are found mainly in Australia; only a few marsupials, such as the American opossum, are found outside Australia (figure 37.91). Apparently, ancestral marsupials became isolated as virtually the only mammals in Australia, while marsupial animals elsewhere in the world had to compete with evolving placental mammals. Placental mammals have become the dominant mammals elsewhere, but in the absence of competition, Australian marsupials diversified to fill all of the specialized roles played by placental mammals on other continents.

Some of the marsupial mammals in Australia include the Koala bears, which are tree-climbing marsupial herbivores, and kangaroos, which fill the niche occupied by hoofed grazing mammals (**ungulates**) on other continents. The extinct Tasmanian "wolf" or "tiger" was a carnivorous marsupial mammal about the size of a collie dog. There are even marsupial moles in Australia.

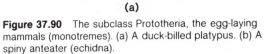

(a)

(b)

Figure 37.90 The subclass Prototheria, the egg-laying mammals (monotremes). (a) A duck-billed platypus. (b) A spiny anteater (echidna).

(a)

(c)

(b)

(d)

Figure 37.91 Subclass Metatheria, the pouched mammals (marsupials). (a) American opossums. Young opossums cling to their mother's back for a time after their period of development inside her pouch has been completed. (b) Koala bears. The koala is a tree-climbing, herbivorous marsupial. (c) A red kangaroo, with its young (a ''joey'') in its pouch. A young kangaroo remains in the pouch even after it is no longer attached to a nipple. It ventures out occasionally and then returns to the pouch. Kangaroos are marsupials that occupy a grazing niche, which is occupied by ungulates on other continents. (d) The Tasmanian ''wolf,'' a marsupial carnivore that became extinct in this century.

Table 37.3
Some Major Orders of Placental Mammals.

Insectivora	Primitive, insect-eating mammals. Moles, shrews, and hedgehogs.
Chiroptera	Flying mammals with a broad skin flap extending from elongated fingers to the body and legs. Bats.
Primates	Omnivorous mammals, opposable thumb and fingers, eyes directed forward, well-developed cerebral cortex. Lemurs, monkeys, apes, and humans.
Edentata	Mammals with few or no teeth. Sloths, anteaters, and armadillos.
Rodentia	Mammals with sharp chisellike incisor teeth that grow continuously. Squirrels, beavers, rats, mice, voles, porcupines, hamsters, chinchillas, and guinea pigs.
Lagomorpha	Mammals with chewing teeth, tails reduced or absent, hindlimbs longer than forelimbs. Hares, rabbits, and pikas.
Proboscidea	Long muscular trunk (proboscis); thick, loose skin; incisors elongated as tusks. Elephants.
Cetacea	Marine mammals with fish-shaped bodies, finlike forelimbs, no hindlimbs, body insulated with a thick layer of fat (blubber). Whales, dolphins, and porpoises.
Perissodactyla	Herbivorous, odd-toed hoofed mammals. Horses, zebras, tapirs, and rhinoceroses.
Artiodactyla	Herbivorous, even-toed hoofed mammals. Cattle, sheep, pigs, giraffes, deer, antelopes, gazelles, hippopotamuses, camels, bison, and llamas.
Carnivora	Carnivorous mammals; sharp, pointed canine teeth and shearing molars. Cats, dogs, foxes, wolves, hyenas, bears, otters, minks, weasels, skunks, badgers, seals, walruses, and sea lions.

Although kangaroos are numerous enough to be a threat to agriculture, western civilization and the introduction of many placental animals have had unfortunate negative effects on many of the other interesting marsupial animals of Australia.

Subclass Eutheria: Placental Mammals During the development of a placental mammal, there is a long period of exchange, which takes place in the placenta, between the mother's bloodstream and the bloodstream of the fetus. During this exchange, the fetus is supplied with nutrients and oxygen and is rid of wastes and carbon dioxide. The pregnancy (**gestation**) periods of some placental mammals are long, and the young are rather mature at birth. Some ungulates, such as horses, can stand and walk within minutes after they are born. Other placental mammals are not nearly so mature at birth. For example, newborn kittens are blind and helpless, and the human infant is totally unable to move around independently or meet any of its own basic needs for a long time after it is born.

Placental mammals are a very diverse group of animals. Some major orders of mammals are listed in table 37.3. Most of them are animals that roam the terrestrial environment, but there are obvious exceptions (figure 37.92).

Bats (**order Chiroptera**) are flying mammals that feed at night, thereby avoiding direct competition with birds, which generally are active in the daytime. Predatory bats are remarkably adapted for catching flying insects. Bats emit sounds and detect the echoes that bounce back from their prey (page 842).

Whales and their relatives (**order Cetacea**) are marine animals. Like all mammals, they have mammary glands and feed their young milk, but most whales have very little body hair. Their bodies are insulated by thick layers of fat under their skin. In recent years many whales have been hunted almost to extinction mainly for the oils extracted from their body fat. Great blue whales are the largest animals that have ever lived, even larger than the largest dinosaurs. They are so large that early whalers could not process them and, therefore, did not attempt to capture them.

(a)

(b)

(c)

(d)

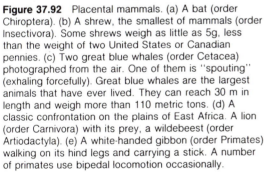

Figure 37.92   Placental mammals. (a) A bat (order Chiroptera). (b) A shrew, the smallest of mammals (order Insectivora). Some shrews weigh as little as 5g, less than the weight of two United States or Canadian pennies. (c) Two great blue whales (order Cetacea) photographed from the air. One of them is "spouting" (exhaling forcefully). Great blue whales are the largest animals that have ever lived. They can reach 30 m in length and weigh more than 110 metric tons. (d) A classic confrontation on the plains of East Africa. A lion (order Carnivora) with its prey, a wildebeest (order Artiodactyla). (e) A white-handed gibbon (order Primates) walking on its hind legs and carrying a stick. A number of primates use bipedal locomotion occasionally.

(e)

But now, whaling "factory ships," especially from Russia and Japan, are seeking out and killing these giants of the deep, and they are now endangered along with the smaller whales.

As they migrate through the oceans, whales sing complex songs that apparently communicate messages great distances through the water to others of their species. The communication and behavior of whales and their relatives, such as dolphins, indicate a high level of intelligence, and people have long dreamed of learning to communicate with these fascinating marine mammals.

In the terrestrial environment, the largest placental mammals are herbivores such as elephants (**order Proboscidea**) and ungulates (hoofed mammals). The ungulates fall into two groups, those with odd numbers of toes (**order Perissodactyla**), such as horses and rhinoceroses, and those with even numbers of toes (**order Artiodactyla**), such as cattle, bison, antelopes, and deer.

In natural ecosystems, the ungulates are prey to the largest members of the **order Carnivora**, which includes the great cats—lions, tigers, cheetahs, and leopards. In terms of physical characteristics, these large, fast, ferocious hunters may represent the pinnacle of evolution of placental mammals, but none of them is the single dominant species on earth today.

That distinction is reserved for the human species. Humans are members of the **order Primates**. They are not the fastest or strongest of animals. They do not have sense organs that are superior to those of all other animals. But they are dominant despite their physical limitations because human evolution has brought one key characteristic to the forefront: their brains are superior to those of other animals. They alone are able to contemplate their existence and the processes that have shaped the living world. Unfortunately, the power generated by their superior brains also has led to alterations of the natural world, many of which threaten their welfare and that of many other species. It now seems that, for better or worse, the quality of future life on earth will be determined by human activities in the coming years.

Summary

This chapter examines and compares the major animal phyla and the evolutionary relationships among the groups.

Simply organized animals are restricted to relatively stable, aquatic environments or sheltered environments inside other organisms (internal parasites).

Terrestrial animals have more complex body organizations, including impermeable body surfaces that separate a stable internal body environment from the changeable external environment. Exchanges between inside and outside take place only in restricted body areas.

All animals are heterotrophs, but animals display a variety of nutritional specializations. Aquatic animals range from sessile filter feeders to active predators. Terrestrial animals include herbivores and the carnivores that feed on the herbivores. All of these animal nutritional strategies are associated with specific structural and functional specializations of the animals' bodies.

Reproductive patterns also are adapted to environmental conditions. This is particularly true of terrestrial animals. Well-adapted terrestrial animals have internal fertilization, which protects gametes from desiccation, and they either lay eggs enclosed by protective shells (for example, insects, arachnids, reptiles, birds) or have reproductive adaptations that permit early development of their young to occur inside the female body (placental mammals).

Questions

1. Many animals that live attached to a substrate or move about very slowly are radially symmetrical. Animals that move about very actively are bilaterally symmetrical. How would you explain this difference?

2. What is the adaptive significance of the very large reproductive potentials of virtually all parasitic animals whose life histories involve several different host animals?

3. Explain why, in the United States, the danger of tapeworm infections is significant while the danger of fluke infections is very small.

4. Describe some advantages and disadvantages of the arthropod exoskeleton as a body surface covering for terrestrial animals.

5. List and explain the three primary diagnostic characteristics of the chordates.

6. What characteristics of amphibians limit their capacity to occupy large portions of the terrestrial environment?

7. What reproductive adaptations of reptiles, birds, and mammals contribute to their success as terrestrial animals?

Suggested Readings

Books

Barnes, R. D. 1980. *Invertebrate zoology*. 4th ed. Philadelphia: Saunders.

Borror, D. J.; DeLong, D. M.; and Triplehorn, C. A. 1976. *An introduction to the study of insects*. 4th ed. New York: Holt, Rinehart & Winston.

Cochran, D. M. 1961. *Living amphibians of the world*. Garden City, N.Y.: Doubleday.

Cockrum, E. L. 1962. *Introduction to mammalogy*. New York: Ronald Press.

Herald, E. S. 1961. *Living fishes of the world*. Garden City, N.Y.: Doubleday.

Romer, A. S., and Parsons, R. S. 1977. *The vertebrate body*. 5th ed. Philadelphia: Saunders.

Russell-Hunter, W. D. 1979. *A life of invertebrates*. New York: Macmillan.

Schmidt, K. P., and Inger, R. F. 1957. *Living reptiles of the world*. Garden City, N.Y.: Doubleday.

Welty, J. C. 1982. *The life of birds*. 3d ed. Philadelphia: Saunders.

Articles

Alvarez, L. W.; Alvarez, W.; Asaro, F.; and Michel, H. V. 1980. Extraterrestrial cause for the Cretaceous-Tertiary extinction. *Science* 208: 1095.

Calder, W. A., III. July 1978. The kiwi. *Scientific American* (offprint 1396).

Caldwell, R. L., and Dingle, H. January 1976. Stomatopods. *Scientific American*.

Goreau, T. F.; Goreau, N. I.; and Goreau, T. J. August 1979. Corals and coral reefs. *Scientific American* (offprint 1434).

Hedgpeth, J. W. 1974. *Animal diversity: Organisms*. Biocore Unit XIV. New York: McGraw-Hill.

Langston, W., Jr. February 1981. Pterosaurs. *Scientific American* (offprint 1492).

Marx, J. L. 1978. Warm-blooded dinosaurs: Evidence pro and con. *Science* 199: 1424.

McWhinnie, M. A., and Denys, C. J. 1980. The high importance of the lowly krill. *Natural History* 89 (3):66.

Nichols, D. 1975. *The uniqueness of the echinoderms*. Carolina Biology Readers no. 53. Burlington, N.C.: Carolina Biological Supply Co.

Roper, C. F. E., and Boss, K. J. April 1982. The giant squid. *Scientific American*.

Würsig, B. March 1979. Dolphins. *Scientific American* (offprint 1424).

Appendix
Classification Summary

This appendix summarizes the major taxonomic groups discussed in the survey of organisms in chapters 34–37 and lists some examples of members of many of the groups. The classification system used here is only one of several systems currently in use by biologists and should not be considered the only possible classification of living things. There is, for example, considerable disagreement among biologists concerning the status of the slime molds and the assignment of various groups of algae to the kingdoms Protista and Plantae.

Botanists use the term "division" for major groups of plants, while zoologists use the term "phylum" for the major groups of animals. This convention has been observed here and has been extended to include the groups of protists that formerly were included in the plant kingdom under the old two-kingdom classification system.

Kingdom Monera
Prokaryotes: Bacteria and cyanobacteria (blue-green algae)

Kingdom Protista
Eukaryotes, unicellular or colonies without tissue differentiation

[Protozoa-heterotrophic, unicellular protists]

Phylum Mastigophora. Flagellated protozoa
Trichonympha, Trypanosoma

Phylum Sarcodina. Pseudopodial protozoa
Amoeba, Entamoeba, Arcella, foraminiferans, radiolarians

Phylum Sporozoa. Spore-forming protozoa
Toxoplasma, Plasmodium

Phylum Ciliophora. Ciliated protozoa
Euplotes, Paramecium, Stentor, Vorticella, Didinium, Tokoprya

[Unicellular algae]

Division Euglenophyta. Euglenoids
Euglena

Division Pyrrophyta. Dinoflagellates
Ceratium, Gonyaulax, Peridinium, Gymnodinium

Division Chrysophyta. Yellow-green and golden-brown algae
Diatoms

[Funguslike protists]

Division Chytridiomycota. Chytrids
Allomyces, Blastocladiella

Division Oomycota. Water molds, late blights, downy mildews
Saprolegnia, Phytophthora, Plasmopara

[Slime molds]

Division Gymnomycota. Plasmodial and cellular slime molds

Kingdom Fungi
Eukaryotic heterotrophs (absorptive nutrition), mycelial organization

Division Zygomycota. Zygospore-forming fungi
Rhizopus, many common fruit and bread molds

Division Ascomycota. Sac-fungi
Saccharomyces and other yeasts, morels, truffles, apple scabs, powdery mildews, Dutch elm disease, ergot disease of rye

Division Basidiomycota. Club fungi
Mushrooms, puffballs, shelf or bracket fungi, rusts, smuts

Division Deuteromycota. "Imperfect" fungi
Penicillium, Aspergillus, Candida

Kingdom Plantae
Multicellular eukaryotes with walled cells, photosynthetic

Division Chlorophyta. Green algae
  Chlamydomonas, Volvox, Ulothrix, Spirogyra, Oedogonium, Ulva

Division Phaeophyta. Brown algae
  Fucus, Macrocystis, Nereocystis, Laminaria, Sargassum

Division Rhodophyta. Red algae

Division Bryophyta. Mosses and liverworts
  Marchantia

Division Tracheophyta. Vascular plants
  Subdivision Psilopsida. "Whisk ferns"
    Psilotum
  Subdivision Lycopsida. Club mosses
    Lycopodium
  Subdivision Sphenopsida. Horsetails
    Equisetum
  Subdivision Pteropsida. Plants with complex conducting
                          systems and large complex
                          leaves
  Class Filicineae. Ferns
    Pteridium
  Class Gymnospermae. "Naked seed" plants
    Cycads, *Ginkgo, Welwitschia, Pinus, Sequoia*
  Class Angiospermae. "Enclosed seed" plants
    Subclass Monocotyledoneae. Monocots
      Lilies, palms, orchids, grasses
    Subclass Dicotyledoneae. Dicots
      Buttercups, maples, carnations, roses

Kingdom Animalia
Multicellular, eukaryotic heterotrophs

Phylum Porifera. Sponges

Phylum Coelenterata.
  Class Hydrozoa. Hydrozoans
    Hydras, *Obelia, Physalia*
  Class Scyphozoa. "True jellyfish"
    Aurelia
  Class Anthozoa. Sea anemones and corals
    Metridium

Phylum Platyhelminthes. Flatworms
  Class Turbellaria. Planarians
    Dugesia
  Class Trematoda. Flukes
    Opisthorchis, Schistosoma
  Class Cestoda. Tapeworms
    Taenia
                    [Protostomes]

Phylum Nematoda. Roundworms
  Ascaris, pinworms, *Trichinella,* filaria worms, guinea worms
Phylum Rotifera. ''Wheel animals''
Phylum Gastrotricha. Gastrotrichs
Phylum Mollusca. Molluscs
    Class Amphineura. Chitons
    Class Pelecypoda. Bivalves
      Clams, oysters, scallops
    Class Gastropoda. ''Belly-foot'' molluscs
      Helix, Nassarius, Urosalpinx, slugs, nudibranchs
    Class Cephalopoda. ''Head-foot'' molluscs
      Octopuses, squids, nautiluses
Phylum Annelida. Segmented worms
    Class Polychaeta. Marine annelids with parapodia
      Nereis, Chaetopterus, Palolo worm
    Class Oligochaeta. Terrestrial and freshwater annelids
    without parapodia
      Earthworms
    Class Hirudinea. Leeches
Phylum Onychophora.
  Peripatus
Phylum Arthropoda. ''Joint-footed'' animals
  Subphylum Chelicerata. First mouthparts are chelicerae;
  antennae are absent
    Class Xiphosura.
      Limulus
    Class Arachnida.
      Spiders, scorpions, ticks, mites, daddy longlegs
  Subphylum Mandibulata. Possess mandibles and not
  chelicerae, antennae are present
    Class Crustacea.
      Crabs, lobsters, crayfish, shrimp, copepods,
      barnacles
    Class Chilopoda.
      Centipedes
    Class Diplopoda.
      Millipedes
    Class Insecta.
      Silverfish, grasshoppers, termites, bugs, beetles,
      butterflies and moths, flies, bees, ants (see p. 1070
      for details on orders of insects)
            [Deuterostomes]
Phylum Echinodermata.
    Class Asteroidea.
      Sea stars
    Class Ophiuroidea.
      Brittle stars and basket stars
    Class Echinoidea.
      Sea urchins and sand dollars
    Class Holothuroidea.
      Sea cucumbers

Phylum Hemichordata. Acorn worms
Phylum Chordata. Chordates
    Subphylum Urochordata.
      Sea squirts
    Subphylum Cephalochordata.
      Amphioxus
    Subphylum Vertebrata.
      Class Agnatha. Cyclostomes
        Lampreys, hagfishes, ostracoderms
      Class Placodermi. Extinct, armored, jawed fish
      Class Chondrichthyes. "Cartilage fishes"
        Sharks, skates, rays
      Class Osteichthyes. "Bony fishes"
        Lobe-fin fishes, lungfishes, ray-fin fishes
      Class Amphibia. Amphibians
          Order Urodela. Tailed amphibians
            Salamanders
          Order Anura. Tail-less amphibians
            Frogs, toads
          Order Apoda. Limbless amphibians
      Class Reptilia. Reptiles
          Order Chelonia.
            Turtles
          Order Crocodilia.
            Crocodiles and alligators
          Order Squamata.
            Snakes and lizards
          Order Rhynchocephalia.
            Tuatara
      Class Aves. Birds
      Class Mammalia. Mammals
        Subclass Prototheria. Egg-laying mammals
          Platypus, echidna
        Subclass Metatheria. Pouched (marsupial) mammals
          Kangaroos, koala bears, opossums
        Subclass Eutheria. Placental mammals
          Shrews, bats, monkeys, apes, humans, rats,
          rabbits, elephants, whales, horses, antelopes, cats,
          dogs, bears, walruses (see page 1102 for details
          on orders of mammals)

Glossary

a- [Gr.]
Not, without, lacking.

ab- [La.]
Off, away, away from.

abdomen [La.]
(1) In vertebrate animals, the posterior portion of the trunk containing visceral organs other than heart and lungs. (2) In invertebrates, especially arthropods, the posterior portion of the body.

abiotic [Gr. *bios:* life]
Nonliving.

acid [La. (1) A substance that can contribute a proton (hydrogen ion, H^+) in a reaction. (2) A substance that dissociates in solution to yield hydrogen ions (H^+), but not hydroxide ions (OH^+). (3) Sometimes used to describe a solution with a pH of less than 7.0.

actin
A protein that along with myosin accomplishes the physical shortening of muscle fibers. Actin also is found in many other kinds of cells that can shorten or contract.

action potential
An abrupt, localized change in the electrical potential difference across a cell membrane. A propagated series of action potentials result in transmission of a nerve impulse along a nerve cell membrane. Also occurs in muscle cells as part of the stimulus to contract.

active site
See catalytic site.

active transport
Movement of a substance across a membrane by a carrier molecule in a process that requires energy expenditure by the cell.

ad- [La.]
To, toward, next to, at.

adaptation
(1) An inherited characteristic that increases an organism's fitness for life in its environment.
(2) Evolutionary changes that increase organisms' fitness to live and reproduce in their environment.

adenosine diphosphate (ADP)
The diphosphate of the nucleoside adenosine. ADP is a product of ATP hydrolysis. ADP is phosphorylated to produce ATP.

adenosine triphosphate (ATP)
The triphosphate of the nucleoside adenosine. Called the "cellular energy currency" because hydrolysis of ATP to ADP and phosphate makes energy available for energy-requiring processes in cells. Phosphorylation of ADP to produce ATP is a key process in cellular energy conversions such as photosynthesis and cell respiration.

ADP
See adenosine diphospate.

adrenal
[La. *ad:* next to; *renes:* kidney]
An endocrine gland of vertebrate animals located just anterior to the kidney.

adrenalin
See epinephrine.

aerobic [Gr. *aer:* air]
(1) In the presence of molecular oxygen (O_2). (2) Biological processes that require oxygen or that can occur in the presence of oxygen.

afferent [La. *ad; ferre:* to carry]
Something that leads or carries toward a given point, as an afferent blood vessel. Opposite of efferent.

alga, pl. **algae** [La., seaweed]
Any one of a diverse group of mainly aquatic photosynthesizing organisms that either are unicellular or are multicellular with little structural and functional differentiation. Various algae are classified in two kingdoms and a number of divisions.

alkaline
(1) Having a pH greater than 7.0.
(2) Basic.

allantois
[Gr. *allantoeides:* sausage-shaped]
A saclike extraembryonic membrane, connected to the embryonic hindgut, that functions in nitrogenous waste shortage in some vertebrates. Allantoic blood vessels connect the embryo with important extraembryonic exchange areas.

alleles
Two or more alternative forms of a gene that can occur at a particular chromosomal locus.

allopatric
[Gr. *allos:* different; La. *patria:* homeland]
Describes populations that occupy physically separated geographic ranges.

allosteric
[Gr. *allos:* different; *stereos:* solid]
Describes a molecule in which the conformation and function of one binding or combining site is affected by the condition of a second binding site.

altruism
Behavior that apparently demonstrates unselfish interest in the welfare of others.

amino acid
An organic compound with an amino group (NH_2) and a carboxyl group (COOH) bonded to the same carbon atom. Amino acids are called the "building blocks" of proteins because proteins are linear polymers of amino acids.

ammonia
NH_3, but usually present as an ammonium ion (NH_4^+) at the pH levels found inside animal bodies. Ammonia is a nitrogenous waste product of deamination reactions in animals.

amnion   [Gr., lamb]
An extraembryonic membrane that forms as a fluid-filled sac surrounding and enclosing the embryo during development in reptiles, birds, and mammals.

amoeboid   [Gr. *amoibē:* change]
Amoebalike, movement of a cell by cytoplasmic flow that produces protrusions called pseudopodia and results in cell shape changes.

amylase   [La. *amylum:* starch]
Enzyme that catalyzes the hydrolysis of starch.

an-   [Gr.]
Without, not.

anaerobic
[Gr. *an:* without; *aer:* air]
(1) In the absence of molecular oxygen (O_2). (2) Biological processes that can occur without oxygen.

analagous
Structures of different organisms that are similar in function and general appearance, but differ in evolutionary origins and fundamental structural plans.

anatomy
(1) The gross structure of an organism. (2) The study of gross structure.

angio-, -angium   [Gr.]
Vessel, container, case.

annual   [La. *annus:* year]
Describes a plant whose entire life, from germination to seed production, is completed in one year. Annual plants die after reproduction or at the end of the growing season.

anterior
[La. *ante:* before, in front of]
The front end of an organism, or toward the front end.

antheridium
A sperm-producing structure in an alga or a plant.

anti-   [Gr.]
Against, opposite.

antibiotic
A chemical substance produced by a microorganism that can kill or inhibit growth of other microorganisms.

antibody
A protein produced by the immune system in response to the presence of an antigen, to which the antibody binds specifically. This binding inactivates the antigen or leads to its destruction.

anticoagulant
A substance that inhibits blood clotting.

antigen
A foreign substance, usually a protein or polysaccharide, that stimulates production of specific antibody molecules that bind to the antigen. Cell surface antigens are particularly important in cellular recognition processes.

anus   [La., ring]
The posterior opening of the digestive tract, through which feces are expelled.

aorta
(1) A major artery in a circulatory system. (2) The main systemic artery.

apical
[La. *apic:* apex, summit, tip]
At or near the tip (apex), as of a plant shoot or root tip.

apo-   [Gr.]
From, off, away from, different.

aquatic   [La. *aqua:* water]
Water, of water, living in water.

aqueous   [La. *aqua:* water]
Watery, dissolved in water, pertaining to water.

archegonium
[Gr. *archegonos:* first offspring or generation]
A multicellular, egg-producing organ in a plant.

archenteron
[Gr. *arche:* first; *enteron:* gut]
An early-developing cavity in an animal embryo that becomes the digestive tract.

arteriole
A small branch of an artery that supplies a capillary bed in a body tissue.

artery
A vessel that carries blood away from the heart toward body tissues.

artifact
(1) A product of human activity or intervention. (2) In science, not an inherent property or part of the process or thing being observed.

-ase
Suffix added to the root name of a compound to identify an enzyme that catalyzes reactions involving the compound, *e.g.,* sucrase, which catalyzes reactions of sucrose.

asexual
A reproductive process that does not involve the union of gametes.

atom
[Gr. *atomos:* cannot be cut, indivisible]
The smallest unit that has the characteristic properties of a chemical element.

atomic number
The number of protons in a particular atomic nucleus.

atomic weight
The average weight of the atoms of an element relative to the weight of atoms of ^{12}carbon (^{12}C), which are arbitrarily assigned the atomic weight of 12.0.

ATP
See adenosine triphosphate.

auto- [Gr.]
Self, same.

autonomic nervous system
A division of the vertebrate nervous system including motor nerves and ganglia that innervate internal organs. Autonomic functions normally are not under voluntary control.

autosome [Gr. *soma:* body]
Any chromosome other than the sex chromosomes.

autotrophic
[Gr. *trophe:* food, nutrition]
Capable of synthesizing all required organic compounds from inorganic substances using an external energy source, most commonly sunlight. Contrast with heterotrophic.

auxin
[Gr. *auxein:* to grow or increase]
One of a class of plant hormones that promote cell elongation and have a variety of other growth-regulating effects.

axon [Gr. *axon:* axis, axle]
Usually the fiber of a nerve cell that carries impulses away from the cell body. Many axons can release a neurotransmitter that carries a message to another nerve cell.

bacillus, pl. bacilli
[La., small rod]
A rod-shaped bacterium.

bacteriophage
[Gr. *phagein:* to eat]
A virus that infects a bacterial cell.

basal [La., base, foundation]
At or near the base or point of attachment, as of a plant shoot or an animal's limb.

base
(1) A substance that can accept a proton (hydrogen ion, H⁺) in a reaction. (2) A substance that dissociates in solution to yield hydroxide ions (OH⁻), but not hydrogen ions (H⁺). (3) Sometimes used to describe a solution with a pH of more than 7.0.

benthic
[Gr. *benthos:* depths of the sea]
Pertaining to or living at the bottom of a body of water.

bi- [La.]
Two, twice, double.

biennial [La. *annus:* year]
A plant that lives through two growing seasons. Characteristically, biennial plants show only vegetative growth during the first season, then flower and set seed during the second. Biennials die after reproduction or at the end of the second growing season.

bilateral symmetry
Describes a body that has two essentially equal, mirror-image sides. A bilaterally symmetrical body can be cut along one (and only one) plane into similar right and left halves.

bile
The complex secretion of the vertebrate liver that is stored in the gall bladder and carried to the duodenum via the common bile duct. Bile contains sodium bicarbonate, bile salts, and bile pigments.

bio- [Gr. *bios:* life]
Life, living.

biological rhythm
Regular cyclic fluctuation in a physiological process or behavioral activity.

biomass
The total weight of all the living things, or of a selected group of living things, in a particular habitat.

biome
A major terrestrial ecosystem characterized by the climax community type that it supports.

biotic
(1) Living, pertaining to life. (2) In a complex system, the living components.

bivalve
A mollusc that has two shells (valves) hinged together; a member of the class Pelecypoda.

blasto- [Gr. *blastos:* bud, sprout]
Embryo, embryonic, part of an embryo.

blastopore
The opening from the interior of the archenteron to the outside of an animal embryo.

blastula
An early stage in development of an animal embryo reached at the end of the period of most active cleavage divisions. In many embryos, the blastula consists of cells organized in a hollow sphere that surrounds a fluid-filled cavity, the blastocoel.

brackish
Describes water that is intermediate in saltiness between fresh water and seawater, such as water in estuaries.

buffer
A chemical system that resists pH change in a solution by binding H⁺ ions in response to increases in the H⁺ ion concentration, or releasing H⁺ ions in response to decreases in the H⁺ ion concentration.

caecum [La. *caecus:* blind]
A saclike pouch off a digestive tract; a blind diverticulum.

calcareous
Composed of calcium carbonate or containing quantities of calcium carbonate.

callus
A mass of undifferentiated tissue growing in response to a wound or in a tissue culture.

calorie [La. *calor:* heat]
The energy unit traditionally used by biologists. The amount of energy, in the form of heat, required to raise the temperature of one gram of water 1°C. Also called a gram calorie or small calorie. Often, the kilocalorie (kcal or Calorie, capitalized) is used in nutrition and metabolism studies. A kilocalorie equals 1,000 gram calories. The modern International System of Units recommends replacement of the calorie by the joule as a unit for energy measurement. *See* joule.

cambium [La. *cambialis:* change]
A layer of meristematic tissue in stems and roots of many vascular plants that gives rise to secondary xylem and phloem.

cAMP
See cyclic adenosine monophosphate.

capillaries   [La. *capillus:* hair]
The smallest of blood vessels, capillaries form diffuse networks in body tissues. Exchanges between blood and tissue cells occur through capillary walls, which are only one cell thick. Capillaries receive blood from arteries and carry it to veins.

carbohydrate
An organic compound consisting of a chain of carbon atoms with hydrogen and oxygen attached, and having a carbon: hydrogen: oxygen ratio of about 1:2:1. Examples include sugars, starch, cellulose, and glycogen.

carcinogenic
Cancer-causing or cancer-promoting.

cardiac   [Gr. *cardia:* heart]
Pertaining to the heart; near or part of the heart.

carnivore
[La. *carni:* flesh; *vorare:* to eat, devour]
An animal that feeds on animals. Contrast with herbivore and omnivore.

cartilage
A vertebrate skeletal tissue that provides support with a degree of flexibility because its intercellular material (matrix) is rubbery, as opposed to the matrix of bone, which is rigid.

catalyst
[Gr. *katalysis:* a dissolving]
A substance that accelerates the rate at which a chemical reaction proceeds toward equilibrium, but itself is not permanently altered by the reaction. Enzymes are catalysts.

catalytic site
The specific portion of an enzyme molecule that binds reactants (substrates) during an enzyme-catalyzed reaction.

caudal   [La. *cauda:* tail]
Pertaining to the tail or posterior end of the body; at, near, or toward the posterior end.

cell cycle
The series of events during an entire cycle from the formation of a new cell until that cell has undergone mitosis.

cellulose
A polysaccharide that makes up fibers that form a major part of the rigid cell walls around plant cells. Cellulose is a polymer of β-glucose units.

central nervous system
(1) In vertebrates, the brain and spinal cord. (2) In general, a concentration of nerve cells, especially interneurons, that exerts a measure of control over the rest of the nervous system.

centri   [La. *centrum:* center]
(1) The center. (2) A point.

centriole
A pair of cylindrical, microtubule-containing structures located together near the nucleus of animal cells and cells of some plants. Centrioles are duplicated before cell division, and a pair of centrioles moves to each of the poles of the developing spindle apparatus. Centrioles appear to be identical with the basal bodies of cilia and flagella.

centromere   [Gr. *meros:* a part]
A small platelike structure located on a constricted area of each chromosome. Spindle microtubules attach to centromeres during mitosis and meiosis. Also called kinetochore.

cephalo-   [Gr. *kephale:* head]
Pertaining to the head, or brain.

cephalothorax
One of the two major body regions (along with the abdomen) in arthropods that do not have the familiar head-thorax-abdomen arrangement seen in insects.

chemoautotroph
An organism that synthesizes organic compounds by chemosynthesis; therefore, an autotrophic organism that is not photosynthetic.

chemoreceptor
A specialized receptor cell that detects and responds to changes in the abundance of a particular chemical substance in its immediate environment.

chemosynthesis
Synthesis of organic compounds from inorganic precursors using energy obtained through oxidation of reduced substances from the environment.

chitin   [Gr. *chiton:* a tunic]
A polymer of nitrogen-containing units that are synthesized using glucose as a precursor. Chitin is a major component of the hard exoskeleton of arthropods, such as insects, and many other invertebrates.

chloro-   [Gr.]
Green.

chlorophyll   [Gr. *phyllon:* leaf]
Any of several similar green pigments that absorb light energy in photosynthesis.

chloroplast
A membrane-bound organelle, found in some eukaryotic cells, that contains chlorophyll and is the site of photosynthesis.

chorion   [Gr.]
A covering layer that encloses an embryo. In reptiles, birds, and mammals, the outer membrane that encloses the embryo and the other extraembryonic membranes; also contributes to placenta formation in mammals.

chrom-, -chrome   [Gr.]
Color, colored, pigment.

chromatid
One of the two strands of a duplicated chromosome that are joined in the regions of their centromeres.

chromosome
A structure in the cell nucleus (nucleoid in prokaryotic cells) on which gene loci are located.

chyme
The thick, soupy mixture of food and gastric secretions that passes from the stomach into the duodenum.

cilium, *pl.* cilia
[La., eyelash, small hair]
A hairlike locomotory structure on the surface of a cell. Characteristically, nine pairs of microtubules are arranged around two central microtubules in each cilium.

class
A taxonomic category between phylum or division and order. Usually, a phylum or division includes several classes, and a class includes several orders.

cleavage
(1) The successive cell divisions that convert a zygote into a multicellular embryo. (2) The actual process of cytoplasmic division (cytokinesis) by furrowing and pinching of the plasma membrane during animal cell division.

cleidoic [Gr., boxlike]
An egg of a terrestrial organism, such as a bird or reptile, with a protective shell that encloses all nutrients and water required by the developing embryo as well as a provision for waste storage so that only gas exchange with the outside environment is required.

climax community
A final, relatively stable stage reached in an ecological succession.

cloaca [La., sewer]
A common exit chamber through which materials from digestive, excretory, and reproductive systems leave the body.

clone [Gr. *klon:* branch or twig]
(1) A group of genetically identical organisms descended by asexual reproduction from a single ancestor. (2) A line of cells descended from a single cell by mitotic cell division.

closed circulatory system
An animal circulatory arrangement in which blood is continuously enclosed within vessels and exchange of materials takes place through capillary walls.

co- [La.]
With, together.

coccus, pl. **cocci** [Gr., a berry]
A spherical bacterium.

codon
The basic unit of genetic coding, consisting of a three-nucleotide sequence in mRNA that specifies a particular amino acid.

coel-, -coel [Gr. *koilos:* hollow]
(1) Hollow. (2) A cavity or chamber.

coelom
A body cavity that is completely lined with mesoderm.

coenocytic
See syncytial.

coenzyme
A nonprotein organic molecule that plays a necessary accessory role in an enzyme's catalytic action, often in the transfer of electrons, atoms, or molecules as part of the enzyme-catalyzed reaction.

collagen
A fibrous protein that is very abundant in animal bodies. It forms extracellular fibers that contribute to the flexibility and tensile strength of many tissues.

colon
The large intestine portion of the vertebrate digestive tract.

com-, con- [La.]
With, together.

common bile duct
The duct that delivers bile from the liver and gallbladder to the duodenum in vertebrate animals.

community An ecological unit composed of all the populations of organisms living and interacting in a given area at a particular time.

complete digestive system
A tubular, flow-through digestive tract with an anterior mouth and a posterior exit (the anus) for digestive wastes.

condensation reaction
A reaction that joins two compounds while removing the components of a water molecule. Thus, each condensation reaction produces one water molecule.

conjugation
[La. *conjugatio:* a blending, joining]
A sexual process that occurs in some microorganisms in which genetic material passes from one cell to another through a tubelike cytoplasmic bridge between cells.

connective tissue
A connecting, supporting, or enclosing tissue in animals in which cells are distributed in a relatively extensive intercellular matrix. The matrix material may be rigid as in bone, flexible as in cartilage, or fluid as in blood.

consumer
An organism that obtains energy and nutrients by ingesting producer organisms or other consumer organisms in the ecosystem; that is, animals that eat plants or other animals.

contractile vacuole
An organelle that accumulates dilute fluid and then expels it through a tiny pore in the cell surface, thus ridding the cell of excess water gained from a hypotonic environment.

copulation [La. *copulare:* to join]
Physical joining of two animals during which sperm are transferred from male to female.

cortex [La., bark]
(1) The outer layer, such as the adrenal cortex or cerebral cortex. (2) In plants, the root or stem tissue between the epidermis and the central vascular tissues.

cotyledon [Gr., cup]
A "seed leaf," a leaflike portion of a plant embryo, which in some plants enlarges and functions as a storage site for nutrients that support early growth after seed germination.

countercurrent exchange
An exchange between two streams of fluid, flowing in opposite directions past each other, that allows for very efficient transfer from one to the other, as in gas exchange in fish gills, heat exchange in many vertebrate limb vessels, and salt exchange in vertebrate kidney tubules.

covalent bond
A chemical bond in which a pair of electrons are shared between two atoms.

crop
An anterior enlargement of a digestive tract that functions as a food storage organ.

crossing over
Exchange of segments between chromatids of synapsed chromosomes during meiosis.

crypt- [Gr. *kryptos:* hidden]
Hidden, concealed.

cutaneous [La. *cutis:* skin]
Pertaining to the skin.

cuticle [La. *cutis:* skin]
(1) A waxy layer on the outer surfaces of plant parts. (2) A relatively impermeable covering layer on the bodies of many invertebrate animals.

cyanobacterium [Gr. *cyano:* blue]
Also called blue-green algae. Very widespread, diverse, and abundant group of organisms that are prokaryotic, but resemble algae in general appearance and photosynthetic mechanisms.

cyclic adenosine monophosphate (cyclic AMP or cAMP)
Compound synthesized from ATP in a reaction catalyzed by adenylate cyclase that functions as a second messenger inside target cells responding to certain hormones.

cyst
[Gr. *kystis:* hollow place, bag]
(1) A capsule that encloses an organism during an inactive period of its life, usually providing protection against adverse environmental conditions. (2) A saclike abnormal growth.

-cyte, cyto-
[Gr. *kytos:* vessel, container]
Cell.

cytokinesis [Gr. *kinesis:* motion]
Division of the cytoplasm of a cell.

cytoplasm
All of the material in a cell except the nucleus.

de- [La.]
Away, from, off, down, out.

deamination
Removal of an amino group ($^-NH_2$) from an organic compound, such as an amino acid.

decarboxylation
Removal of a carboxyl group (^-COOH) from an organic compound. In cellular metabolism, results in release of carbon dioxide.

deciduous
[La. *decidere,* to fall off]
Plants that shed their leaves each year.

dendr-, dendro-
[Gr. *dendron:* tree]
Tree, treelike, branch, branching.

dendrite
Slender, branched extension that functions as a major receptor area of a nerve cell.

deoxyribonucleic acid (DNA)
A molecule that encodes genetic information in cells. DNA consists of two polynucleotide strands arranged in a double helix. The pentose sugar found in DNA nucleotides is deoxyribose.

derm [Gr., skin]
Skin, covering, layer.

dermis
The deeper layer of the skin of vertebrate animals, under the epidermis.

detritus
Dead plant and animal material; decaying organic matter.

deuterostome
One of a group of animals in which the embryonic blastopore produces the anus. Echinoderms and chordates are deuterostomes. Contrast with protostome.

di- [Gr.]
Two, double, separate, apart.

diapause
[Gr. *dia:* through; *pausis:* a stopping]
An extended period of reduced metabolic activity that is a normal part of many animal life cycles. Diapause occurs in many insects.

diaphragm
The dome-shaped muscular floor of the thoracic cavity of mammals that contracts and flattens during inhalation.

diatom [Gr. *diatomos:* cut in half]
A member of the division Chrysophyta, an extremely abundant group of aquatic algae that are the most important producers in marine ecosystems. A diatom is enclosed by a boxlike pair of silica-containing valves.

dicot
A member of one of the two subclasses of angiosperms, the flowering plants. Dicots possess two cotyledons ("seed leaves") and several other distinguishing features.

differentiation
The developmental process by which unspecialized cells and tissues become structurally and functionally specialized.

diffusion
The tendency for a net movement of particles from an area of higher initial concentration to an area of lower initial concentration due to spontaneous random movements of individual particles resulting from thermal agitation.

digestion
(1) Physical and chemical processes that convert ingested food materials into absorbable forms. (2) Enzyme-catalyzed hydrolysis of complex organic compounds that yields absorbable constituent molecules.

diploid (2N)
The condition in which cells have two of each type of chromosome; twice the number of chromosomes found in gametes that have the haploid (N) number of chromosomes.

disaccharide
A double sugar that yields two monosaccharides upon hydrolysis. Sucrose, maltose, and lactose are common disaccharides.

distal [La. *distare:* to stand apart]
Located away from a point of origin or attachment; farther from a central or reference point. Opposite of proximal.

diverticulum
[La. *devertere:* to turn aside]
A sac with only one opening that branches off another structure such as a cavity, duct, or canal. A blind pouch.

division
A major taxonomic grouping applied to plants, autotrophic protists, and fungi that is equivalent to the term "phylum" applied to animals and heterotrophic protists. A division includes several to many classes that share certain basic features and are assumed to have common ancestry.

DNA
See deoxyribonucleic acid.

dormancy [La. *dormire:* to sleep]
A period of suspended activity and growth, usually associated with lowered metabolic rate and increased resistance to environmental stresses.

dorsal [La. *dorsum:* back]
Pertaining to the back, or located at or near the back. Opposite of ventral.

duodenum
[La. *duodecim:* twelve (related to its length, about 12 fingers' breadth)] The first part of the vertebrate small intestine. Ducts carrying bile and pancreatic juice empty into it.

ecdysis
See molting.

ecosystem
[Gr. *oikos:* household or habitation] An ecological unit composed of a community of organisms and the physical features of their environment.

ecto- [Gr.]
Outside, out, outer, external.

ectoderm
(1) The outer tissue layer of an animal embryo. (2) The outer primary body layer (germ layer) that gives rise to the epidermis and the nervous system in vertebrate animals.

-ectomy
[Gr. *ek:* out of; *tomein:* to cut] Removal or excision of a structure, such as adrenalectomy (surgical removal of the adrenal gland).

ectoplasm
The outer, more gellike portion of the cytoplasm adjacent to the plasma membrane. Contrast with endoplasm.

effector
A part of an organism capable of responding to a stimulus, especially a tissue or organ such as a muscle or gland capable of responding to a stimulus delivered by the nervous system.

efferent [La. *ex; ferre,* to carry]
Something that leads or carries away from a given point, such as an efferent blood vessel. Opposite of afferent.

egg
(1) A female gamete, usually nonmotile and larger than a male gamete. Eggs usually contain stored nutrients. (2) Sometimes used as a collective name for the egg proper together with accessory structures that surround the egg and the developing embryo, such as the albumen, shell membranes, and shell of birds' eggs.

electron
A negatively charged subatomic particle.

embryo [Gr.]
(1) A plant or animal at an early stage of development. An embryo develops from a zygote. (2) In human development, the name applied during the first eight weeks after fertilization.

end-, endo- [Gr.]
Within, inner, inside.

endergonic [Gr. *ergon:* work]
Energy-requiring or energy-absorbing, as in an endergonic chemical reaction.

endocrine
[Gr. *krinein:* to separate] Secreting internally. Applied to ductless glands that produce hormones and secrete them directly into the blood.

endocytosis
A process in which extracellular material is enclosed in a vesicle and taken into a cell. Phagocytosis and pinocytosis are forms of endocytosis.

endoderm
(1) The innermost layer of an animal embryo. (2) The inner primary body layer (germ layer) that gives rise to the linings of the digestive tract and its outgrowths, including the respiratory system, in vertebrate animals.

endodermis
A plant tissue that is the innermost layer of the cortex. In roots, endodermis cells' walls fit together tightly to form a barrier, the Casparian strip, between the cortex and the central vascular tissues.

endogenous
Originating inside a cell or body. Opposite of exogenous.

endoplasm
The inner, more fluid portion of the cytoplasm. Contrast with ectoplasm.

endoplasmic reticulum
[La. *reticulum:* network] A system of branched, membranous tubules and sacs in the cytoplasm of eukaryotic cells. Sometimes has a rough surface due to a coating of ribosomes.

endoskeleton
An internal skeleton.

endosperm [Gr. *sperma:* seed]
A tissue found in seeds that functions in nutrient storage.

entropy
A measure of the disorder or randomness of a system.

enzyme [Gr. *zyme:* leaven]
A protein molecule that acts as a catalyst in a chemical reaction.

epi- [Gr.]
Upon, over, outer, outside.

epidermis
The outermost layer or layers of cells of a plant or animal body.

epiglottis
The flap of tissue that during swallowing covers the glottis (the opening from the pharynx into the larynx at the top of the trachea) and prevents food from entering the respiratory passage.

epinephrine
A hormone produced by the adrenal medulla that stimulates physiological changes that prepare the organism to respond to emergency and stressful situations. Also called adrenalin.

epithelium
A type of animal tissue that forms flattened covering and lining layers of external body surfaces and internal cavities.

erythrocyte [Gr. *erythros:* red]
A hemoglobin-containing red blood cell.

estivation [La. *estival:* summer]
A resting state with lowered metabolic rate in which some animals can remain inactive for a long period. Estivation usually occurs when the environment is too dry to support normal life activities. Sometimes called "summer sleep."

estuary
An area where fresh water meets the sea; thus, an area with salinity intermediate between fresh water and seawater.

etiolation
The result of growing a plant in darkness. An etiolated plant is pale in color and has a long, spindly stem, abnormal vascular tissue, and few or no leaves except cotyledons.

eu-   [Gr.]
True, most typical, good, well.

eukaryotic cell
[Gr. *karyon:* nut or kernel]
A cell possessing a membrane-bounded nucleus and various membrane-bounded organelles in its cytoplasm, and having its DNA complexed with histones. None of these characteristics pertain to the prokaryotic cells of bacteria and cyanobacteria (blue-green algae).

evagination   [La. *vagina:* sheath]
(1) Something that is folded or projected outward. (2) The process of becoming folded or projected outward.

eversible
[La. *evertere:* to turn out]
Can be turned inside out; usually something that is protruded at the same time.

evolution   [La. *evolutio:* unrolling]
(1) Changes in the gene pool of a population with time. (2) Descent with modification.

ex-   [La.]
Out, off, from, beyond.

excretion
The set of processes involved in removal of metabolic wastes, excess materials, and toxic substances.

exergonic   [Gr. *ergon:* work]
Energy-releasing, as in an exergonic chemical reaction.

exhalation
Forcing air out of the lungs; breathing out.

exo-   [Gr.]
Out, out of, outside, without.

exocytosis
A process in which an intracellular vesicle fuses with the plasma membrane so that the vesicle's contents are released to the cell's exterior.

exogenous
Originating or coming from a source outside a cell or body. Opposite of endogenous.

exoskeleton
An external skeleton that both supports and covers the body.

extra-   [La.]
Outside, beyond, more, besides.

extracellular digestion
Digestion in which digestive enzymes are secreted into an extracellular space or a specialized digestive cavity where nutrients are hydrolyzed into absorbable forms.

extraembryonic
Pertains to structures outside the body of an embryo that enclose, protect, and supply the embryo.

extrinsic   [La.]
From outside, external to, not a basic part of.

family
A taxonomic category between order and genus. Usually order includes several families, and a family includes several genera.

fauna
All of the animals present in a given area or during a particular period.

feces   [La., dregs]
Undigested materials, various wastes, and intestinal bacteria discharged from the digestive tract.

feedback control
An arrangement in which a control mechanism is itself regulated by the function that it controls.

fertilization
(1) The fusion of two haploid gamete nuclei to form a diploid zygote nucleus. (2) More broadly, the complex set of interactions and responses of gametes, including nuclear fusion.

fetus   [La., pregnant, fruitful]
(1) A developing vertebrate animal after it has completed embryonic development of major organs and systems and before birth or hatching. (2) A developing human from the ninth week after fertilization until birth.

filter feeder
An animal that feeds on small organisms or particles strained out of the water around it or out of a stream of water that it moves through some part of its body.

fixation
Conversion of a substance into a biologically usable form or incorporation of a substance into a biologically usable compound, as in fixation of atmospheric nitrogen by microorganisms or fixation of carbon dioxide during photosynthesis.

flagellum   [La., whip]
A hairlike locomotory structure on the surface of a cell. Longer than a cilium, but possesses the same microtubule arrangement with nine pairs around two central tubules.

flora
All of the plants present in a given area or during a particular period.

food chain
A sequence of organisms through which energy and materials move in an ecosystem. Includes producers and consumers.

free energy
Energy in a system that is available to do work.

fruiting body
A specialized structure produced by a fungus or slime mold on which spores are produced.

gamete   [Gr., wife]
(1) A specialized haploid reproductive cell that fuses with another gamete in fertilization to produce a zygote. (2) In many kinds of organisms, an egg or sperm.

gametophyte   [Gr. *phyton:* plant]
The haploid, gamete-producing phase (generation) in the typical plant life cycle with alternation of haploid and diploid phases (generations).

ganglion   [Gr., knot, swelling]
(1) A mass or cluster of nerve cell bodies. (2) In vertebrate animals, a cluster of nerve cell bodies outside the central nervous system.

gastr-, gastro-
[Gr. *gaster:* stomach, belly]
(1) Pertaining to the stomach.
(2) Ventral part or surface.

gastrovascular cavity
A blind, branched digestive cavity that also serves a circulatory (transport) function in animals that lack a circulatory system.

gastrulation
A set of processes that convert a single-layered embryo, the blastula, into a several-layered body with specialized areas.

-gen, -geny
[Gr., produce, bear, birth]
Producing, production.

gene
A hereditary unit located at a particular site (locus) on a chromosome. In Mendelian genetics, a gene determines the nature of a phenotypic character. Biochemically, a gene is a segment of DNA that encodes information for synthesis of a polypeptide or RNA molecule.

gene pool
The sum of all of the alleles of all genes present in a population.

genotype
The genetic constitution of an organism or a cell, as opposed to the phenotype, which is a set of observable characteristics.

genus, pl. **genera** [La., race]
A taxonomic category between family and species; a group of closely related species.

geo- [Gr., the earth]
Pertaining to the earth, especially to the earth's gravity.

geotaxis
A specific animal orientation with respect to gravity.

geotropism
A plant growth response oriented with respect to gravity.

germ cells
Gametes or the cells that will eventually give rise to gametes.

gestation period
[La. gest: carried]
The time from fertilization to birth; the length of pregnancy.

gill
Projections of the body surface (or digestive tract) of aquatic animals that are specialized for gas exchange.

gizzard
A digestive organ with thick muscular walls that rub together and grind food.

glia
Nonneuron cells that are packed among neurons in the central nervous system. Glia are actually more numerous than neurons in the nervous system.

glottis
The opening from the pharynx into the larynx, which lies at the top of the trachea (windpipe).

glucose [Gr. glykys: sweet]
One of the six-carbon sugars ($C_6H_{12}O_6$) and a key substrate in cellular energy conversion processes.

glycogen
A polysaccharide that is the principal carbohydrate storage form in animals. Glycogen is a polymer made up of glucose units combined by condensation reactions.

Golgi apparatus
A membranous organelle composed of stacked, saclike cisternae that is involved in packaging and secretion of cell products.

gonad [Gr. gone: seed]
(1) A gamete-producing organ in an animal. (2) An ovary or testis.

granum, pl. **grana** [La., grain]
A stack of membranous bags or discs in a chloroplast.

guard cells
Specialized leaf epidermal cells that occur in pairs that surround each stoma and regulate its opening and closing.

habitat [La., to live in]
The kind of surroundings in which individuals of a given animal or plant species normally live.

haploid (N)
The condition in which cells have only one of each type of chromosome; one-half the diploid (2N) number of chromosomes.

helminth [Gr. helminthos: worm]
Worm, pertaining to worms, wormlike.

hem-, hema-, hemat-, hemo-
[Gr. haima: blood]
Blood, related to blood, filled with blood.

hemoglobin [La. globus: ball]
An iron-containing red pigment in blood that functions in oxygen transport.

hemorrhage
A large discharge of blood from a blood vessel, massive bleeding.

hepatic [Gr.]
Pertaining to the liver.

hept-, hepta- [Gr.]
Seven.

herbaceous
[La. herba: grass, herb]
A nonwoody plant.

herbivore
[La. vorare: to eat, devour]
An animal that feeds on plants. Contrast with carnivore and omnivore.

hermaphroditic
[Gr. Hermes: a god; Aphrodite: a goddess]
Describes the condition in which both female and male reproductive organs are present in the same individual.

hetero- [Gr. heteros: different]
Different, other.

heterogamy
[Gr. gamos: union, marriage]
Reproduction involving fusion of two gametes that differ in size and structure.

heterothermic
Having an environmentally influenced, variable body core temperature, not capable of precise metabolic temperature regulation.

heterotrophic
[Gr. trophe: food, nutrition]
Not capable of synthesizing required organic compounds from inorganic substances; requires organic nutrients produced by other organisms. Contrast with autotrophic.

heterozygous [Gr. zygos: yoked]
Having different alleles at the corresponding loci of homologous chromosomes. Opposite of homozygous.

hex-, hexa- [Gr.]
Six, as in hexose (a six-carbon sugar).

hist- [Gr. hostos: web, sheet]
Pertaining to tissue.

homeo-, homo-
[Gr. homos: same]
Same, like, similar.

homeostasis
[Gr. *stasis*: standing, posture]
The dynamic steady state
maintained within living things.

homeothermic   [Gr. therme: heat]
The ability of animals to maintain a
stable body core temperature
despite fluctuations in the
temperature of their immediate
environment.

hominid
A member of the family Hominidae,
the family in which modern humans,
fossil humans, and very closely
related fossil primates are placed.

homologous
(1) Pairs of chromosomes that bear
the same gene loci and synapse
during prophase of the first meiotic
division. (2) In evolution,
fundamentally similar structures
inherited from a common ancestor.

homozygous   [Gr. *zygos*: yoked]
Having identical alleles at the
corresponding loci of homologous
chromosomes. Opposite of
heterozygous.

hormone
[Gr. *hormaein*: to set in motion, to
excite]
A chemical regulator substance that
is produced in an organ or tissue
and diffuses or is transported to
target organs or tissues on which it
has specific effects.

hybrid   [La., mongrel, hybrid]
(1) Offspring of parents that differ in
one or more heritable traits.
(2) Offspring of parents belonging to
two different species or varieties.

hydr-, hydro-   [Gr. *hydro*: water]
(1) Water, fluid. (2) Hydrogen.

hydrolysis   [Gr. *lysis*: loosening]
Splitting of a compound into parts in
a reaction that involves addition of
water, with the H^+ ion being
incorporated in one fragment and
the OH^- ion in the other.

hyper-   [Gr.]
Over, above, excessive, more.

hypertonic   [Gr. *tonos*: tension]
Describes a solution that tends to
gain water osmotically from another
solution that is separated from it by
a selectively permeable membrane.
Tonicity terms (hypertonic,
hypotonic, isotonic) usually are used
specifically with reference to the
effects of solutions surrounding
living cells.

hypha   [Gr. *hyphe*: web]
One of the filaments that makes up
a fungus mycelium.

hypo-   [Gr.]
Under, below, deficient, less.

hypophysis
See pituitary.

hypothalamus
[Gr. *thalamos*: inner chamber or
room]
A region in the floor of the forebrain
that contains reflex centers that
regulate the secretion of pituitary
hormones and various activities of
the autonomic nervous system. It is
involved in regulation of body
temperature and metabolic rate,
water balance, circulation, hunger,
thirst, sleep, and other functions and
activities.

hypothesis
A supposition that is established by
reasoning after consideration of
available evidence and that can be
tested by obtaining more data, often
by experimentation.

hypotonic   [Gr. *tonos*: tension]
Describes a solution that tends to
lose water osmotically to another
solution that is separated from it by
a selectively permeable membrane.
(*See* note under *hypertonic*
regarding usage of tonicity terms.)

ichythy-   [Gr. *ichthyos*: fish]
Pertaining to fish; sometimes fishlike.

immunity
Result of a complex set of
responses through which an animal
exposed to a foreign antigen,
particularly an antigen associated
with a pathogenic microorganism,
develops capabilities to make
effective responses to that antigen
and thus is resistant to damage or
disease that might be caused by
subsequent exposure to it.

immunoglobulin
One of a class of globular proteins
present in blood, tissue fluids, and
various body secretions. Antibodies
are immunoglobulins.

implantation
The entry of a mammalian embryo
into the wall of the uterus.

incomplete digestive system
A gut that is a blind sac and that
has only one opening that must
serve as both entrance for food and
exit for undigested residues.

inf-, infra-   [La.]
Below, under.

inflammation
A tissue and blood vessel response
to infection or injury in which fluid,
plasma proteins, and phagocytic
cells move out of blood vessels into
interstitial spaces, making the
inflamed tissue appear swollen.

ingestion
Taking in food from the outside
environment; eating.

inhalation
Drawing air into the lungs; breathing
in.

inorganic
Not organic; not a carbon
compound (other than carbon
dioxide).

in situ   [La., in place]
In its original, natural, or proper
place.

integument
[La. *integere*: to cover]
A covering, skin, outer surface.

inter-   [La.]
Between, among.

interferon
A substance produced and released
by cells infected by viruses that
protects other cells from virus
infection.

interneuron
A neuron that connects two other
neurons, such as a sensory neuron
and a motor neuron or two other
interneurons. Complex brains contain
huge numbers of interneurons
arranged in very complex structural
and functional networks.

internode   [La. *nodus*: knot]
A length of stem between nodes
(the attachment points of leaves or
buds).

interstitial fluid
The extracellular fluid in the spaces
around body cells.

intra-   [La.]
Within, inside.

intracellular digestion
Digestion in which nutrients are
taken into a cell by phagocytosis or
pinocytosis and enclosed in a
vesicle with which an enzyme-
containing lysosome fuses. Digestion
products are absorbed through the
vesicle membrane.

intrinsic   [La.]
Contained within, inherent in, a basic part of.

invagination   [La. *vagina:* sheath]
(1) Something that is folded or projected inward. (2) The process of becoming folded or projected inward.

invertebrate
[La. *in:* without; *vertebra:* joint, vertebra]
Any animal not possessing a vertebral column. *See* vertebrate.

in vitro   [La., in glass]
In a laboratory culture, in a laboratory vessel, not in place in an intact organism.

in vivo   [La., in the living]
In the living thing, in an intact organism.

ion
An atom or molecule that has an electrical charge because of the loss or gain of one or more electrons.

irritability
Responsiveness to environmental stimuli. Irritability is especially well developed in nerve cells and specialized sensory receptor cells.

iso-   [Gr.]
Equal.

isoenzyme
One of several different molecular forms of an enzyme, all of which catalyze the same reaction. Also called isozymes.

isogamy
[Gr. *gamos:* union, marriage]
Reproduction involving fusion of two gametes that are alike in size and structure.

isotonic   [Gr. *tonos:* tension]
Describes a solution that tends neither to gain nor to lose water osmotically to another solution that is separated from it by a selectively permeable membrane. (*See* note under *hypertonic* regarding usage of tonicity terms.)

isotope   [Gr. *topos:* place]
Atom of an element that differs from another atom of that element in the number of neutrons in its nucleus. Different isotopes have the same atomic number but different mass numbers. Some isotopes are unstable (radioactive) and tend to decay by emission of radiation.

joule
The standard basic unit of energy measurement. 4.186 joules $=$ 1 calorie, and 1 joule $=$ 10^7 ergs.

juxta-   [La.]
Near to, next to.

keratin   [Gr. *keratos:* horn]
A tough, horny, water-insoluble protein produced by epidermal tissues of vertebrate animals. Especially abundant in nails, claws, hair, feathers, beaks, and hooves.

kilo-   [Gr.]
A thousand.

lamella   [La.]
A thin, platelike structure, thin sheet, layer.

larva   [La., ghost]
An immature form of an animal that is very different in appearance from the mature adult.

lateral   [La. *later:* the side]
Located away from a midline. Opposite of medial.

leuk-, leuko-   [Gr. *leukos:* white]
White or clear.

leukocyte
A white blood cell.

lichen   [Gr. *leichen:* a tree moss]
An intimate symbiotic association between an alga and a fungus.

ligament
A piece of tough connective tissue that links one bone to another in a joint.

lignin   [La. *lignum:* wood]
A complex organic polymer produced by some plant cells that adds to the hardness of their walls. Wood is heavily lignified xylem tissue.

linkage
The tendency of certain genes to assort together because their loci are close together on the same chromosome. This results in linked inheritance of phenotypic traits determined by those genes.

lip-, lipo-   [Gr. *lipos:* fat]
Fat, fatlike, fat-containing.

lipase
An enzyme that catalyzes the hydrolysis of fat molecules to fatty acids and glycerol.

lipid
One of a group of organic compounds that are insoluble in water but soluble in hydrophobic solvents. Includes fats, steroids, phospholipids, oils, and waxes.

locus, pl. loci   [La., place]
A specific part of a chromosome that is the site of a particular gene. Any one of the alleles of that gene may occupy its locus.

-logy   [Gr.: *logos*]
Study, study of, discourse.

longitudinal
Along the length, lengthwise, as in a longitudinal section of a body or structure.

lumen   [La., light, opening]
The cavity of space within a tube or sac.

-lysis, lyso-   [Gr. *lysis:* loosening]
Loosening, disintegration.

lysosome   [Gr. *soma:* a body]
A membrane-bound organelle containing hydrolytic enzymes.

macro-   [Gr.]
Large, long.

macrophage
A large phagocytic cell in a body tissue outside the circulatory system. Macrophages develop from circulating monocytes that leave blood vessels and remain in a tissue.

marine   [La. *mari:* sea]
Living in or pertaining to the sea or ocean.

marsupial
[Gr. *marsypion:* little bag]
A pouched mammal; member of the subclass Metatheria.

matrix   [La. *mater:* mother]
The material in which something is embedded or enclosed, or the homogeneous material in spaces among formed elements, as the extracellular matrix of bone or cartilage, or the matrix among membranous elements in a mitochondrion.

medial   [La. *medi:* the middle]
Located at or toward a midline. Opposite of *lateral*.

medulla   [La., marrow, pith]
The inner portion of a structure or organ, such as the adrenal medulla.

mega- [Gr.]
Large, great.

megaspore
A plant spore that will develop into a female gametophyte (megametophyte).

meiosis [Gr., diminution]
A kind of nuclear division, usually as part of two successive cell divisions, that produces nuclei with half the number of chromosomes in the original nucleus.

membrane potential
An electrical charge difference between the inside and outside of a cell; thus, an electrical potential across the plasma membrane.

meristem [Gr. *meristos:* divisible]
A plant tissue made up of undifferentiated cells that function in production of new cells by mitosis.

meso- [Gr.]
Middle.

mesoderm
(1) The middle tissue layer of an animal embryo lying between ectoderm and endoderm. (2) The middle primary body layer (germ layer) that gives rise to skeletal, muscular, and circulatory systems as well as the lining and covering layers in the body cavities of vertebrate animals.

messenger RNA (mRNA)
Single-stranded RNA molecule that has a nucleotide sequence complementary to that of a segment of DNA and specifies amino acid sequence in polypeptides being assembled during translation. Thus, mRNA is a ''messenger'' from the genome to the cytoplasm.

meta- [Gr.]
(1) After, posterior. (2) Change.

metabolism [Gr. *metabol:* change]
(1) The sum of chemical reactions within a cell, or the sum of all cellular activities in an organism. (2) Sometimes used in collective names for particular sets of biochemical reactions, as in amino acid metabolism, lipid metabolism, and nitrogen metabolism.

metamorphosis
[Gr. *morpho:* form, shape]
A change in body organization that transforms an immature animal (usually a larva) into an adult.

micro- [La.]
(1) Small. (2) In units of measurement, one millionth.

microbe [Gr. *bios:* life]
A microscopic organism, especially a bacterium.

microspore
A plant spore that will develop into a male gametophyte. In gymnosperms and angiosperms, microspores develop into pollen grains.

milli- [La.]
One thousandth.

mimicry
An organism presenting a showy appearance that resembles another organism, often one that is not attractive to predators.

mineral [La. *minera:* mine]
An inorganic material.

mitochondrion, pl. mitochondria
[Gr. *mitos:* thread; *chondros:* grain (or cartilage)]
A double-membraned organelle in which aerobic respiration occurs in eukaryotic cells. The site of Krebs cycle reactions and the electron transport system.

mitosis [Gr. *mitos:* thread]
Nuclear division that results in distribution of replicated chromosomes so that each of two daughter nuclei receives a set of chromosomes identical to that of the original nucleus.

molecule
The smallest characteristic unit of a compound. Composed of two or more atoms.

molting
Shedding and replacement of an outer body covering. Used especially in connection with the periodic shedding of the exoskeleton that is necessary for growth of arthropods.

mono- [Gr.]
One, single.

monocot
A member of one of the two subclasses of the angiosperms, the flowering plants. Monocots possess one cotyledon (''seed leaf'') and several other distinguishing features.

monomer [Gr. *meris:* part]
A relatively small molecule that is linked to other molecules to form a polymer.

-morph, morpho- [Gr.]
Form, shape, structure.

morphogenesis
[Gr. *genesis:* origin]
The development of shape and form of an organism.

morphology
[Gr. *logos:* a word, discourse, study of]
(1) The structure of organisms or their parts. (2) The study of structure and form.

motor neuron
A neuron that transmits nerve impulses from the central nervous system toward an effector, such as muscle.

muscle
[La. *musculus:* a little mouse, muscle]
A contractile tissue of animals that contains cells capable of exerting force by shortening when they are appropriately stimulated.

mutation [La. *muta:* change]
A stable, heritable change in the genetic material.

mutualism
A symbiotic relationship in which both organisms benefit from the association.

mycelium [Gr. *mykes:* fungus]
The mass of tubular filaments (hyphae) that constitute a fungus.

myo- [Gr. *mys:* muscle, mouse]
Muscle, having to do with muscle.

NAD
See nicotinamide adenine dinucleotide.

NADP
See nicotinamide adenine dinucleotide phosphate.

nano- [La. *nanus:* dwarf]
One thousand-millionth part (10^{-9}).

nasal [La., nose]
Pertaining to the nose or air passages through the nose.

necrotic [Gr. *necros:* death]
Having dead and degenerating cells.

-nema, -neme, nemato-
[Gr. *nema:* thread]
Threadlike, filamentous.

neo- [Gr.]
New, recent.

neonate [La. *nata:* birth, be born]
A newborn (or newly hatched)
animal, especially a newborn
mammal.

nephr-, nephro-
[Gr. *nephros:* kidney]
Kidney, pertaining to a kidney.

nephron
An individual functional unit of a
vertebrate kidney. Each kidney
contains many nephrons.

nerve [La. *nervus:* nerve, tendon]
A bundle of neuron fibers outside
the central nervous system.

nerve impulse
A chain reaction series of action
potentials conducted along the
membrane of a neuron.

neur-, neuro- [Gr., nerve]
Having do do with nerve cells,
nerves, or the nervous system.

neuron
A nerve cell.

neurosecretion
(1) A substance produced by a
specialized nerve cell
(neurosecretory cell) that is released
in the blood and acts on target cells
elsewhere in the body. (2) The
process of releasing a
neurosecretion.

neurotransmitter
A substance that is released by one
neuron, crosses a synaptic cleft,
binds with a receptor molecule, and
causes a specific ion conductance
change in the membrane of a
second neuron.

neutron
A subatomic particle, found in the
atomic nucleus, that is electrically
uncharged and has approximately
the same mass as a proton.

niche
An ecological description of
structural and functional role of a
particular species in its natural
community.

**nicotinamide adenine dinucleotide
(NAD)**
An organic compound that functions
as an electron acceptor or donor in
various cellular oxidation-reduction
reactions.

**nicotinamide adenine dinucleotide
phosphate (NADP)**
Similar in function to nicotinamide
adenine dinucleotide, but is
structurally different because of
presence of an additional phosphate
group.

nitrogen fixation
The incorporation of nitrogen from
the air into forms that can be used
by organisms.

node [La. *nodus:* knot]
The point of attachment of a leaf or
a bud on a plant's stem.

-nomy [Gr.]
The science of.

nucleic acid
One of several types of large
organic molecules that are polymers
of nucleotides and function in
heredity and genetic expression in
cells. The principal nucleic acids are
deoxyribonucleic (DNA) and
ribonucleic acid (RNA).

nucleoid
A region of a prokaryotic cell where
the chromosome is located. Not
membrane-bounded as in eukaryotic
cells.

nucleolus
[La., a little nut, a kernel]
A dense body, observable within the
nucleus of a eukaryotic cell when it
is not dividing, that is involved in
synthesis of rRNA.

nucleosome
A basic structural unit of
chromosomes in eukaryotic cells,
consisting of an aggregate of eight
histone molecules with a specific
length of DNA wrapped in a helical
coil around it.

nucleotide
An organic compound consisting of
a nitrogen-containing base (either a
purine or a pyrimidine), a five-carbon
sugar, and phosphoric acid. Nucleic
acids are polymers of nucleotides.

nucleus [La., a little nut, a kernel]
(1) A membrane-bounded spherical
body that contains the
chromosomes of a eukaryotic cell.
(2) The central part of an atom.
(3) A cluster of nerve cell bodies
within the central nervous system.

nymph
[Gr. *nympha:* a bride, nymph]
A young insect that generally
resembles its parents at hatching
but lacks some adult characteristics,
which it gains as it undergoes
gradual metamorphosis through a
series of molts.

oct-, octi-, octo- [La., Gr.]
Eight.

olfactory [La. *olfactere:* to smell]
Pertaining to the sense of smell.

olig-, oligo- [Gr.]
Few, small.

omnivore
[La. *omnis:* all; *vorare:* to eat,
devour]
An animal that feeds on both plants
and animals. Contrast with carnivore
and herbivore.

ontogeny
[Gr. *onto:* being, existence; *genesis:*
origin]
The developmental history of an
individual organism.

oo- [Gr. *oion:* egg]
Egg or pertaining to an egg or its
development.

oogamy
[Gr. *ganos:* union, marriage]
A type of sexual reproduction in
which one of the gametes is a large,
nonmotile egg. Oogamy is a form of
heterogamy.

oogonium
An egg-producing cell in an alga.

open circulatory system
An animal circulatory arrangement in
which blood leaves the confines of
vessels and circulates in spaces
among body organs and tissues
before returning to vessels leading
to the heart.

operculum [La., cover, lid]
A platelike structure that covers or
encloses other structures, as the
operculum covering the gills of a
bony fish or the operculum covering
the capsule of a moss sporophyte.

oral [La. *oris:* mouth]
Pertaining to the mouth.

order
A taxonomic category between class and family. Usually, a class includes several orders, and an order includes several families.

organ [Gr. *organon:* tool]
A body part specialized for a particular function or set of functions, usually consisting of several types of tissues.

organelle [La. *-ell:* small]
A distinctive intracellular structure specialized for a particular function.

organic
(1) A chemical compound that contains carbon. (2) Something produced by or derived from living things. (3) Pertaining generally to living things.

organism
A single cell or a multicellular aggregate that constitutes an individual living thing.

osmosis [Gr. *osmos:* thrust, push]
The net movement of a solvent, such as water, through a selectively permeable membrane.

osmotic pressure
The pressure that must be exerted to halt net water movement across a truly selectively permeable membrane that separates solutions having different solute concentrations. Usually measured with distilled water on one side of the membrane. Thus, a measure of the tendency of water to move osmotically.

osteo- [Gr. *osteon:* bone]
Pertaining to bone.

ov-, ovi- [La. *ovum:* egg]
Egg, or pertaining to an egg.

ovary
An egg-producing organ. Eggs are released from animal ovaries, but zygotes begin their development within plant ovaries that are converted into fruits in many flowering plants.

oviduct
A tube that transports eggs away from the ovary.

ovulation
Release of an egg from the ovary.

ovum, pl. ova
An egg cell, a female gamete.

oxidation
A chemical reaction that results in removal of one or more electrons from an atom, ion, or compound. Oxidation of one substance occurs simultaneously with reduction of another.

pale-, paleo- [Gr.]
Ancient, old.

paleontology
The study of fossils and the fossil record of life in past geologic times.

papilla [La. nipple]
A small protuberance; a nipplelike projection.

para- [Gr.]
Beside, near, beyond.

parasit- [Gr.]
(1) Near food. (2) To eat with another or at another's table. (3) A parasite.

parasitism
A symbiotic relationship in which one member benefits at the expense of the other.

parasympathetic nervous system
One of the two divisions of the autonomic (visceral motor) nervous system. Acts antagonistically to the sympathetic division in control of various effectors.

parenchyma
A plant tissue made up of loosely packed, relatively unspecialized cells.

parthenogenesis
[Gr. *parthenos:* virgin]
Development of an egg without fertilization.

pathogenic [Gr. *pathos:* suffering]
Disease-causing.

pelagic [Gr. *pelagos:* ocean]
Oceanic, pertaining to the open sea.

pent-, penta- [Gr.]
Five, as in pentaploid (possessing five sets of chromosomes).

peptide bond
A bond formed between two amino acids in a condensation reaction between the amino group of one and the carboxyl group of the other. Peptide bonds link amino acids in polypeptides.

perennial
[La. *per:* through; *annus:* year]
Describes a plant that lives throughout the year and grows during several to many growing seasons.

peri- [Gr.]
Around, surrounding.

peristalsis
[Gr. *stalsis:* constriction, contraction]
Rhythmic waves of contraction and relaxation that pass along the walls of hollow, tubular structures and function to move their contents along.

peritoneum
A very smooth membrane, of mesodermal origin, lining a wall of the coelomic cavity or covering the surface of an organ.

permeable
[La. *permeare:* to pass through]
Permitting a substance or substances to pass through. Usually pertains to properties of membranes in biology.

pH
Symbol for the negative logarithm of the hydrogen ion [H^+] concentration. pH values are measures of acidity in solutions and range from 0 to 14. pH 7 is neutral, less than 7 is acidic, and more than 7 is basic.

phage
See bacteriophage.

phagocytosis
[Gr. *phagein:* to eat]
Engulfment and intake of solid particles by cells.

pharyngeal
Pertaining to the pharynx.

pharynx
Portion of a digestive tract between the mouth cavity and the esophagus. Also associated with gas movement and exchange functions in vertebrate animals.

phenotype
[Gr. *phainein:* to show]
A set of observable characteristics of an organism or a cell, as opposed to the genotype, which is its genetic constitution.

pheromone
[Gr. *pherein:* to carry, bear]
A substance secreted and released into the environment by one organism that evokes behavioral, reproductive, or developmental responses in other individuals of the same species.

-phil, phili-, philo-
[Gr., love, loving]
Attraction, positive response.

phloem [Gr. *phlois:* bark]
A vascular tissue in plants that transports (translocates) dissolved material from place to place. Phloem may move material either up or down stems and roots.

-phob, -phobia [Gr., fear, dread]
Avoidance, negative response.

-phore [Gr. *pherein:* to carry]
Carrier.

phospholipid
One of a number of phosphate-containing lipids that are important constituents of cell membranes.

phosphorylation
Addition of one or more phosphate groups to a molecule.

photochemical reaction
A chemical reaction that involves absorption or release of light energy.

photoperiodism [Gr. *photos:* light]
Physiological responses of organisms to the lengths of light and dark periods in the twenty-four-hour daily cycle, especially to changes in relative lengths of the two.

photosynthesis
Synthesis of organic compounds from inorganic compounds (commonly CO_2 and water) using light energy absorbed by chlorophyll.

phototropism
[Gr. *tropos:* a turning, change]
A movement or turning in response to a directional light source.

-phyll [Gr. *phyllon:* leaf]
Leaf, of a leaf, relating to leaves.

phylogeny [Gr. *phylon:* tribe]
The evolutionary history of a group of organisms.

phylum [Gr. *phylon:* tribe]
A major taxonomic grouping of organisms, usually including several to many classes of organisms that share certain basic characteristics and are assumed to have a common ancestry.

physiology [Gr. *physis:* nature]
(1) Life functions and processes of cells, tissues, organs, and organisms. (2) The study of the functioning of one or more of these units.

-phyte, phyto- [Gr. *phyton:* plant]
Plant, pertaining to a plant or plants.

phytochrome
A plant pigment that exists in two different forms and is converted from one to the other in reversible reactions involving absorption of red or far-red light energy.

phytoplankton
[Gr. *planktos:* wandering]
Microscopic, free-floating, photosynthetic organisms that function as major producers in freshwater and marine ecosystems.

pigment [La. *pigmentum:* paint]
A colored substance that absorbs light energy of a particular set of wavelengths.

pinocytosis [Gr. *pinein:* to drink]
Engulfment and intake of small quantities of fluid by cells.

pistil [La. *pistillus:* pestle]
The portion of a flower that contains ovules, which are the sites of megaspore production and subsequent female gametophyte development. The pistil consists of stigma, style, and ovary.

pituitary
[La. *pituitarius:* secreting phlegm]
A small gland situated below, and attached to, the hypothalamus. The anterior lobe secretes several hormones that regulate functions of other endocrine glands. Other anterior lobe hormones and the hormones released in the posterior lobe regulate nonendocrine target cells. A functional intermediate lobe is present in some vertebrates.

placenta
[Gr. *plax:* flat object or surface]
An organ formed during mammalian development, made up of tissues of both mother and infant, within which materials are exchanged through a tissue barrier that separates elements of the two circulations.

plankton
[Gr. *planktos:* wandering]
Free-floating organisms, most of which are microscopic, found in freshwater and marine ecosystems. Includes both autotrophic (photosynthetic phytoplankton) and heterotrophic (zooplankton) organisms.

plasm-, plasmo-, -plasm
[Gr. *plasma:* something molded, modeled, or shaped]
Plasma; cytoplasm or part of cytoplasm; formed material.

plasma
The clear, fluid portion of vertebrate blood; that is, blood minus formed elements (cells and platelets).

plasma membrane
The outer membrane of a cell that forms the boundary between the cell and its environment.

plasmid
A relatively small, circular DNA molecule in a bacterial cell that replicates independently of the cell's chromosome and that also can be transferred to another bacterial cell.

plasmodesma,
pl. **plasmodesmata**
[Gr. *desma:* band, bond]
A minute, delicate cytoplasmic connection between adjacent plant cells.

plastid
A plant cell organelle that functions in synthesis (chloroplast) or nutrient storage (leucoplast).

pleura [Gr., the side, a rib]
The smooth membranes that line the thoracic cavity and cover lung surfaces.

pneumonia
A condition, caused by an infection, in which fluid and dead white blood cells accumulate in the alveoli and interfere with gas exchange.

-pod, -podium
[Gr. *pod:* foot; *podion:* little foot]
A foot, leg, extension.

poikilothermic
Having an environmentally influenced, variable body core temperature; not capable of precise metabolic temperature regulation.

polarity
Having parts or ends that have contrasting properties. In molecules, having negative charge at one end, positive at the other. In organisms, having head and tail or base and apex.

pollen grains   [La., fine flour, dust]
Small male gametophytes, developed from microspores, that are released and carried by wind or animal pollinators.

pollination
Transfer of pollen to the stigma of a receptive flower. Pollination is not synonymous with fertilization, which occurs only after further development of the male gametophyte.

poly-   [Gr.]
Many, much.

polymer   [Gr. meris: part]
A large molecular chain of smaller molecular subunits (monomers) linked together.

polypeptide
A molecule composed of many amino acids linked together by peptide bonds.

polyploid
Having more than two complete sets of chromosomes per nucleus.

polysaccharide
A large carbohydrate molecule; that is, a polymer of single sugar monomers linked together by means of condensation reactions. Starch, glycogen, and cellulose are polysaccharides, being polymers of glucose.

polytene chromosome
A chromosome in which there has been repeated replication without mitosis, thus forming giant chromosomes.

population   A group of individuals of the same species that interbreed and occupy a given area at the same time.

portal system
[La. porta: gate, door]
A portion of the circulatory system that begins in a capillary bed and leads not to vessels that carry blood directly toward the heart, but rather to a second capillary bed.

posterior
[La. post: hinder, posterior]
The rear end of an organism, or toward the rear end.

precursor
Something that comes before or precedes. Chemically, a reactant that will be altered or incorporated into another form or compound.

predator
[La. praedatio: plundering]
A free-living organism that feeds on other organisms.

primate
A member of the order Primates, the order of mammals that includes monkeys, apes, and humans.

primitive   [La. primus: first]
(1) Unspecialized, at an early stage of development or evolution. (2) Old, ancient, resembling an ancestral condition.

pro-   [Gr.]
Before, in front of, forward.

proboscis   [Gr. boskein: to feed]
A tubular extension of the head or snout of an animal, usually used in feeding, sometimes as a sucking tube.

procoagulant
A substance that promotes blood clotting.

producer
An organism that synthesizes organic compounds using only inorganic materials and energy from an external source, usually sunlight used in photosynthesis.

prohormone
A molecule that can be converted into an active hormone form.

prokaryotic cell
[Gr. karyon: nut or kernel]
A cell that lacks a membrane-enclosed nucleus and membrane-bounded organelles in its cytoplasm. Bacteria and cyanobacteria (blue-green algae) are prokaryotic cells.

protease
A general name for a proteolytic enzyme, an enzyme that catalyzes hydrolysis of peptide bonds in proteins. Thus, a protein-digesting enzyme.

protein
A large, complex organic molecule composed of one or several long polypeptides.

proteolytic
Protein-hydrolyzing, as in proteolytic enzymes.

proto-   [Gr.]
First, original, primary.

proton
A positively charged subatomic particle found in the atomic nucleus.

protonema   [Gr. nema: a thread]
A simple, end-to-end chain of cells in an alga, or in the early development of a moss or fern gametophyte.

protostome
One of a group of animals in which the mouth develops from the blastopore of the embryo. Molluscs, annelids, and arthropods are protostomes. Compare with deuterostome.

proximal   [La. proxim: nearest]
Located near a point of origin or attachment, nearer a central or reference point. Opposite of distal.

pseudo-   [Gr., false]
False, substituting for, temporary.

pseudocoelom
A body cavity between mesoderm and endoderm; therefore, not completely lined with mesoderm as is a true coelom.

pseudopod, pseudopodium
A temporary membrane extension and cytoplasmic protrusion of an amoeboid cell, which functions in locomotion and phagocytosis.

pulmonary   [La. pulmonis: lung]
Relating to the gas exchange organs, especially to lungs, as in pulmonary arteries and veins.

pupa, pl. **pupae**   [La., doll]
A developmental stage in insects between larval and adult stages. Pupae are immobile (sometimes enclosed in cases) and undergo extensive body reorganization.

purine
A double-ringed nitrogenous base, such as adenine or guanine, that is a component of nucleotides and nucleic acids.

pyrimidine
A single-ringed nitrogenous base, such as cytosine, thymine, or uracil, that is a component of nucleotides and nucleic acids.

pyruvic acid
A three-carbon compound that is the product of the Embden-Meyerhof pathway of reactions.

quantum
A unit of electromagnetic energy. The energy of a quantum is inversely proportional to the wavelength of the radiation.

radi- [La.]
A spoke, ray, radius.

radial symmetry
Describes a body that is arranged around a central axis so that cutting it lengthwise along any plane that runs along that axis results in separation of two similar halves. Contrast with bilateral symmetry.

radiation
Energy transmitted as electromagnetic waves.

radioactive isotope
An unstable isotope that tends to decay by emitting radiation.

reduction
A chemical reaction that results in addition of one or more electrons to an atom, ion, or compound. Reduction of one substance occurs simultaneously with oxidation of another.

reflex [La. *reflexus:* bent back]
An automatic activation of an effector in response to a stimulus detected by a receptor. A neural reflex depends on a functional unit, including sensory and motor elements, and usually one or more interneurons that connect them.

refractory period
A period of time after a response or action during which a cell, tissue, organ, or organism does not show its characteristic response to a stimulus or cannot carry out a normal function.

regeneration
Replacement of a lost structure or body part.

renal [La. *renes:* kidneys]
Pertaining to the kidney.

replication
Of DNA, when the two strands of DNA molecule separate and a complementary strand is assembled for each original strand so that two DNA molecules identical to the original are produced.

reticulum [La.]
A fine network.

rhizoid [Gr. *rhiza:* root]
A colorless, hairlike extension of a plant or fungus that functions in nutrient absorption.

rhizome
[Gr. *rhizoma:* mass of roots]
An underground stem in some plants that gives rise to aboveground leaves. Some rhizomes serve as storage organs or in vegetative reproduction of the plants.

ribonucleic acid (RNA)
A class of nucleic acids that contain the pentose sugar ribose (not deoxyribose as in DNA) and the pyrimidine uracil (not found in DNA), and are involved in the several steps of cellular genetic expression.

ribosome
A minute granule assembled from two subunits composed of RNA and proteins that functions as a site of translation during protein synthesis.

RNA
See ribonucleic acid.

rudiment
[La. *rudis:* unformed, rough]
The first visible evidence of development of a structure or organ in an embryo.

rudimentary
[La. *rudis:* unformed, rough]
Incompletely developed; having a primitive form.

rumen
An enlarged, saclike area of the stomach of cattle and other ruminant mammals that houses symbiotic microorganisms that produce cellulase, an enzyme that catalyzes hydrolysis of cellulose in plants eaten by the ruminants.

salinity
[La. *salin:* salt pit, salt, salty]
Saltiness; measure of concentration of dissolved salts.

saprophyte
A heterotrophic bacterium or a fungus that absorbs nutrients directly from dead and decaying organic material.

sarcomere
[Gr. *sarx:* flesh; *meris:* part]
A basic contractile unit in a skeletal muscle fiber; the portion of the fiber between two Z lines.

savanna
A grassland with occasional trees or clumps of trees.

secretion
(1) Release of a cell product being "exported" from a cell.
(2) Simultaneous secretion by many cells, as in glandular secretion.

section
(1) To cut apart or across. (2) A slice or a representation of a slice cut from a body or structure, such as a cross (transverse) section, which is a section cut at a right angle to the long axis of a body or structure.

sedentary
Relatively inactive, tending to stay or sit in one place.

seed
A plant embryo, together with a food reserve, enclosed in tough, protective seed coats that are derived from part of the ovule.

segmentation
Arrangement of an organism in a series of similar units or segments.

sensory neuron
A neuron that either responds directly to a stimulus or to a change in a specialized receptor cell and carries impulses toward the central nervous system.

septum, pl. **septa** [La., fence]
A dividing wall or membrane, a partition.

serum
[La., whey, watery part of a fluid]
Blood plasma minus the proteins involved in the clotting process.

sessile [La.]
Attached to a surface, sedentary, not free to move around.

sex chromosome
One of the chromosome pair that differs between females and males in animals that have a chromosomal sex determination mechanism. *See* X chromosome and Y chromosome.

sex-linked
Genes that are located on a sex chromosome, mostly on the X chromosome, but a very few on the Y chromosome.

sexual dimorphism
Describes the condition in which males and females of a species show clear and consistent differences in size or other general body characteristics.

shoot
Aboveground portions of a vascular plant; a stem with its branches and leaves.

sieve tube
A linear array of specialized cells (elements) running vertically through the phloem that functions in translocation of solutes.

sinus
[La., a fold, a hollow; curve, bend] (1) A hollow space within a structure. (2) A large passage or channel in the circulatory system.

siphon
A tubular structure through which fluids are drawn in or expelled.

solute
A substance that is dissolved (uniformly dispersed) in another substance (the solvent.)

solution
A homogeneous mixture in which one or more solutes are dissolved in a solvent. Water is the solvent in the great majority of biological solutions.

solvent
A medium in which a solute or several solutes are dissolved.

-soma, somat-, -some
[Gr. *soma:* body]
Body, entity, unit.

somatic cells
All of the cells of the body except the germ cells.

sorus, pl. sori   [Gr., a heap]
A cluster of sporangia on a leaf, such as a fern leaf.

species
A population of morphologically similar organisms that can reproduce sexually among themselves but are reproductively isolated from other organisms. Among microorganisms, a group of organisms that share an extensive set of common characteristics.

sperm   [Gr. *sperma:* seed]
A male gamete, usually motile and smaller than a female gamete.

sphincter   [Gr., band]
A circular muscle that forms a ring around a tubular structure and can close it by contracting.

spindle
The spindle, or spindle apparatus, is a set of microtubules involved in chromosome movements during mitosis or meiosis.

spirillum, pl. spirilla
[La., little coil]
A helical or spiral-shaped bacterium.

sporangium
A structure within which spores are produced.

spore   [Gr. *sporos:* seed]
(1) A reproductive cell that can develop directly into an organism. (2) A specialized bacterial structure that has a low metabolic rate, is protected by a thick capsule, can survive adverse conditions, and germinate under favorable conditions to produce a new cell.

sporophyll   [Gr. *phyllon:* leaf]
A sporangium-bearing leaf. Flowers contain highly specialized and modified sporophylls.

sporophyte
The diploid phase (generation), which produces haploid spores by meiosis, in the typical plant life cycle with alternation of haploid and diploid phases (generations).

stamen   [La., thread, fiber]
A portion of a flower consisting of a filament topped by an anther containing microsporangia within which are produced microspores that develop into pollen grains.

standard metabolism
The resting metabolic rate; the energy spent on ordinary life maintenance functions.

starch
A class of large, polymeric carbohydrate molecules that contain many glucose monomers and function as food-storage substances in plants.

stele
The vascular cylinder at the central core of a root or stem. Includes all tissues inside the endodermis.

stimulus
A change in the environment that is detected by an irritable cell, such as a sensory receptor, or a group of irritable cells.

stom-, -stome, stomo-   [Gr.]
Mouth.

stoma, pl. stomata
A small opening in the epidermis of a leaf or other plant part that is bounded by guard cells that regulate its opening and closing.

stroma
[Gr., anything spread out, a mattress, bed]
The nonmembranous matrix or ground substance within chloroplasts that is the site of several important reactions.

structural gene
A gene that contains coded information for synthesis of a polypeptide chain with a particular amino acid sequence.

sub-   [La.]
Under, below.

substrate
(1) Reactant in an enzyme-catalyzed reaction. (2) The base surface on which an organism lives.

succession
In ecology, a series of progressive changes in the plants and animals making up a community in a particular area.

sucrose
A common disaccharide (double sugar) consisting of a glucose molecule and a fructose molecule linked together. Sucrose is the major transport sugar in plants.

super-, supra-   [La.]
Above, over.

sym-, syn-   [Gr.]
With, together.

symbiosis [Gr. *bios:* life]
A long-term association between two organisms of different species living together in an intimate relationship.

sympathetic nervous system
One of the two divisions of the autonomic (visceral motor) nervous system. Acts antagonistically to the parasympathetic division in control of various effectors.

sympatric [La. *patria:* homeland]
Describes populations that occupy the same geographic range.

synapse
[Gr. *synapsis:* falling together, union]
A specialized junction between nerve cells at which communication of excitation occurs.

synapsis [Gr.]
The pairing of homologous chromosomes during prophase I of meiosis, a process that has no counterpart in mitosis.

syncytial
A multinucleate structure; several to many nuclei scattered in a mass of cytoplasm inside a single plasma membrane.

synergistic [Gr. *ergon:* work]
A factor, substance, or process acting together with and enhancing the effect of another.

syngamy
[Gr. *gamos:* union, marriage]
The union of gametes in sexual reproduction. Fertilization.

systemic circulation
The parts of the circulatory system supplying the remainder of the body apart from the circulation supplying the gas-exchange surfaces (the pulmonary circulation).

tadpole
(1) A tailed amphibian larva.
(2) Sometimes applied to tailed larvae of some members of the subphylum Urochordata.

taiga [Russ.]
A worldwide belt of northern coniferous forest.

target cell
A cell that has appropriate receptors for a hormone and responds physiologically to the hormone.

taxis [Gr., arrangement, order]
A movement in which an organism assumes a specific orientation to a stimulus source, such as a phototaxis (a movement orientated to a light source).

taxonomy [Gr. *taxis:* arrangement]
The science concerned with naming organisms and arranging them in a system of classification based on their similarities and probable evolutionary relationships.

telo- [Gr.]
An end, complete, final.

tendon [La.]
A piece of tough connective tissue that attaches muscle to bone.

tentacle
[La. *tentare:* to handle, touch]
A slender, elongate, flexible process, extending from an animal body, that serves tactile and grasping functions. Tentacles usually occur around or near the mouth.

terrestrial [La.]
On land, land-dwelling, pertaining to the dry land environment.

testis, pl. **testes**
The sperm-producing organ. Also the source of male sex hormone in vertebrate animals.

tetra- [Gr. *tetrart:* the fourth]
Four, as in tetraploid (possessing four sets of chromosomes).

thalamos
[Gr., chamber, inner room]
A major relay center derived from part of the embryonic forebrain that functions as an intermediary between the cerebrum and other portions of the nervous system.

thallus
[Gr. *thallos:* young shoot, twig]
A flattened, sheetlike plant body with relatively little tissue specialization and no differentiation of roots, stems, or leaves.

theory
A well-established scientific concept or set of concepts supported by a large body of data. *Not* a vague speculation or unsupported guess.

thorax [Gr., breastplate]
(1) In insects, the body region between the head and abdomen, to which walking legs and wings are attached. (2) In vertebrates, the anterior portion of the trunk that contains lungs and heart.

thymus
A lymphoid organ that is involved in development and differentiation of immunologic capabilities in vertebrates.

thyroid [Gr. *thyreo:* shield]
An endocrine gland located at the base of the neck in vertebrate animals that produces several hormones, including thyroxin, an iodine-containing hormone involved in metabolic regulation.

tissue [La. *texere:* to weave]
A group of similar specialized cells, along with intercellular material that binds them together, that are organized as a structural and functional unit.

toxin [La.]
A substance, produced by an organism, that is very poisonous to another organism and will cause damage or death.

trachea, pl. **tracheae**
A tubular passage that conducts air inside the body, as the vertebrate windpipe or the tracheal tubes of insects.

tracheid
An elongate, spindle-shaped conducting cell in the xylem of vascular plants. Tracheids overlap at their tapered ends, which have pitted walls that allow fluid to pass from one cell to another.

trans- [La.]
Across, through, beyond.

transcription
[La. *scribere:* to write]
The process that results in the production of a strand of messenger RNA that is complementary to a segment of DNA.

transduction [La. *ducere:* to lead]
Transfer of genetic material from one bacterium to another by a virus.

transfer RNA (tRNA)
One of a group of small RNA molecules, each of which binds a specific amino acid and has a segment that is complementary to a messenger RNA codon. Thus, tRNA functions during polypeptide synthesis (translation) to position appropriate amino acids in the specified sequence.

transformation
(1) A genetic change in cells of a bacterial strain that results when the cells receive genetic information in the form of DNA fragments from another strain, including fragments released from dead cells. (2) The process by which normal body cells are converted into neoplastic (abnormally growing) cells.

translation
The complex interaction of mRNA, ribosomes, and tRNA that results in synthesis of a polypeptide having an amino acid sequence specified by the sequence of codons in mRNA.

translocation
[La. *locare:* to put or place]
(1) The transport of solutes from one part of the plant to another through the phloem. (2) In genetics, the breaking of a chromosome segment that becomes attached to a nonhomologous chromosome.

transpiration
[La. *spirare:* to breathe]
Evaporation of water from plants to the surrounding atmosphere, mainly through stomata.

transverse
[La. *transversere:* to cross]
On a plane that crosses the body from side to side and separates anterior and posterior parts, as a transverse (cross) section.

tri- [La.]
Three, as in triploid (possessing three sets of chromosomes).

trilobite
One of a group of extinct marine arthropods that had two longitudinal folds that appeared to divide the body into three lobes. Trilobites were very abundant during the Paleozoic era.

tropism [Gr. *tropos:* turn]
A turning response by a nonmotile organism to an external stimulus, primarily turning of a plant by a differential growth response.

trypsin
A general proteolytic enzyme that catalyzes hydrolysis of peptide bonds in many different kinds of proteins.

tumor [La. *tumere:* to swell]
An abnormal growth, especially one that produces a definite mass or nodule of tissue.

tundra [Russ.]
A treeless area characterized by short growing seasons, cold temperatures, and little rainfall, but, in the Arctic, wet conditions because of low evaporation, poor drainage, and the presence of permafrost.

turgor pressure
[La. *turgere:* to swell]
The pressure exerted by plant cells against their cell walls because the fluid in their environment usually is hypotonic.

ungulates [La. *ungula:* hoof]
Four-legged mammals with variously fused digits that are protected at their ends by a horny covering, a hoof.

-ura, uro- [Gr.]
Tail; pertaining to the posterior tip of the body.

urea
A water-soluble nitrogenous waste product of mammals and some other animals that functions in ammonia excretion, which is necessary as a result of amino acid deamination.

ureter
The duct that carries urine from the kidney to the urinary bladder (or cloaca) in vertebrate animals.

urethra
The duct that carries urine from the urinary bladder to the exterior in mammals.

uric acid
An insoluble nitrogenous waste product produced by reptiles, birds, and terrestrial arthropods.

uterus [La., womb]
A portion of a female reproductive tract enlarged for egg storage and, in some animals, embryo development. In mammals, the muscular chamber within which embryonic and fetal development take place.

vacuole [La. *vacuus:* empty]
A membrane-bounded, fluid-filled cellular organelle.

valve
(1) A cuplike flap or set of flaps that prevents backward flow of blood in a heart or blood vessel. (2) A hard shell or covering.

vas-, vasa-, vaso-
[La., vessel, duct]
Blood vessel or pertaining to blood vessels.

vascular tissue
[La. *vasculum:* small vessel]
Tissue made up of tubular vessels that function in fluid transport.

vegetative
(1) Plant parts not specialized for sexual reproduction. (2) Asexual, when applied to reproduction.

vein [La. *vena:* vein]
A vessel that carries blood away from body tissues toward the heart.

venous
Pertaining to veins.

ventilation
The process of renewing the gas-carrying medium over the surfaces where gas exchange occurs.

ventral [La. *venter:* belly]
Pertaining to the underside or belly, or located at or near the underside. Opposite of dorsal.

ventricle
[La. *venter:* stomach, belly]
(1) A cavity in an organ, such as one of the four ventricles of the vertebrate brain. (2) The large, muscular pumping chamber that pumps blood out from the heart.

venule
A small vessel that collects blood from a capillary bed and carries it toward a vein.

vertebrate
[La. *vertebra:* joint, vertebra]
A member of the subphylum of the phylum Chordata. Includes animals possessing a vertebral column, a series of skeletal elements that enclose and protect the posterior portion of the nervous system. Fish, amphibians, reptiles, birds, and mammals are vertebrates.

vessel element
An individual vessel cell in the xylem of a vascular plant. During differentiation, a vessel element loses its nucleus and cytoplasm. End walls become perforated or disappear, leaving pipelike end-to-end chains of empty cell walls.

villus, pl. **villi**   [La., shaggy hair]
A minute fingerlike projection that contains blood vessels and increases surface available for absorption or exchange of materials.

virion
A free virus particle. The reproductively inactive but infective form in which a virus exists outside a host cell.

viroid
An infectious agent consisting of only a short strand of nucleic acid, much smaller than the smallest virus.

virus   [La., slimy liquid, poison]
An infectious submicroscopic particle composed of a nucleic acid core and a protein coat. Viruses reproduce only inside living cells.

viscera   [La.]
The internal body organs of an animal.

vitamin   [La. *vita:* life]
One of a set of relatively simple organic molecules that are required in small quantities for various biological processes and must be in an organism's diet because they cannot be synthesized in the organism's body.

viviparous
[La. *vivus:* alive; *parere:* to bring forth, produce]
Bearing young born at a relatively advanced stage after an extended period of development inside a sheltered environment in the female body (the uterus in mammals), where nutrient and other metabolic requirements are met through exchange of materials with the mother, as in the mammalian placenta.

woody
Plant tissue with lignin-containing (lignified) cell walls. The xylem of trees and shrubs is woody tissue.

X chromosome
One of the sex chromosomes. A pair of X chromosomes is present in the cells of one sex (*e.g.,* the female in *Drosophila* and in mammals, the male in birds). One X chromosome and one Y chromosome commonly are present in the other sex.

xylem   [Gr. *xylon:* wood]
A vascular tissue that transports water and mineral solutes upward through the plant body.

Y chromosome
One of the sex chromosomes. One Y chromosome and one X chromosome are present in the cells of one sex (*e.g.,* the male in *Drosophila* and in mammals, the female in birds). Two X chromosomes are present in the other sex. Y chromosomes characteristically bear very few functional gene loci.

yolk
Stored nutrient material in an egg.

zoo-   [Gr.]
(1) Animal, pertaining to animals.
(2) Animallike, motile.

zooplankton
A collective name for the small animals present in the plankton.

zoospore   [Gr. *sporos:* seed]
A flagellated, motile plant spore.

zygote
[Gr. *zygotos:* yoked together]
The diploid (2N) cell formed by the union of two gametes, the product of fertilization.

zymogen
[Gr. *zyme:* ferment, leaven, yeast]
An enzymatically inactive precursor of a digestive enzyme molecule.

Visual Credits

Illustrations

Introduction

Fine Line Illustrations, Inc.: p. 15.

Chapter 1

Fine Line Illustrations, Inc.: 1.1, 1.2, 1.3, 1.4, 1.5(a), 1.6, 1.7, 1.8, 1.9, 1.10, 1.11(a-c), 1.12(a), 1.13, 1.14, 1.15, 1.16, 1.17, 1.18.

Chapter 2

Fine Line Illustrations, Inc.: 2.1, 2.2, 2.3, 2.4(a), 2.5(a), 2.6, 2.7, 2.8, 2.11, 2.12, 2.13(b), 2.14(a), 2.15, 2.16, 2.18(b), 2.19, 2.20, 2.21, 2.23, 2.24(a), 2.26, 2.27. From Albert L. Lehninger, BIOCHEMISTRY, 2d edition, Worth Publishers, New York, 1975, p. 264: 2.4(b). Reprinted with permission of Longman Group Limited from INTRODUCTION TO MOLECULAR BIOLOGY by G. H. Haggis et al., 1964: 2.9(a and b). From BIOCHEMISTRY, 2d ed., by Lubert Stryer. W. H. Freeman and Company. Copyright © 1981: 2.10(a). Reprinted with the permission of the Academic Press from THE PROTEINS by R. E. Dickerson and H. Neurath, eds., 1964: 2.10(c). The quaternary structure of hemoglobin. Adapted by permission from R. E. Dickerson and I. Geis, THE STRUCTURE AND ACTION OF PROTEINS. Benjamin/Cummings, Menlo Park, Calif., 1969. Copyright 1969 by Dickerson and Geis: 2.14(b). Reprinted with permission of Academic Press from article by Pauling and Corey in *Archives of Biochemistry and Biophysics* 65(1956):164:2.18(a). Reprinted from "The Fluid Mosaic Model of the Structure of Cell Membranes" in *Science* 175 (18 February 1972): 720–31. Copyright 1972 by the American Association for the Advancement of Science: 2.25.

Chapter 3

From C. E. Dobell, *Antony van Leeuwenhoek and his "Little Animals."* © 1932. Reprinted, New York: Russell & Russell, 1958: 3.2. Reprinted with permission of Macmillan Publishing Co., Inc. From THE CELL IN DEVELOPMENT AND HEREDITY, 3d ed., by E. B. Wilson. Copyright 1925 by Macmillan Publishing Co., Inc., renewed 1953 by Anna M. K. Wilson: 3.3. Fine Line Illustrations, Inc.: 3.4, 3.5, 3.8, 3.9, 3.10, 3.13, 3.22(b), 3.26. Thomas D. Brock, BIOLOGY OF MICROORGANISMS, 3d ed., © 1979, pp. 774, 775. Reprinted by permission of Prentice-Hall, Inc., Englewood Cliffs, N.J.: 3.6(b). From Mader, Sylvia S., INQUIRY INTO LIFE, 3d ed. © 1976, 1979, 1982, Wm. C. Brown Company Publishers, Dubuque, Iowa. Reprinted by permission: 3.14(a and b), 3.15. John Walters and Associates: 3.16, 3.18, 3.19, 3.20, 3.21, 3.27, 3.28, 3.29, 3.30. From "How Cilia Move" by Peter Satir. Copyright © 1974 by Scientific American, Inc. All rights reserved: 3.24.

Chapter 4

Fine Line Illustrations, Inc.: 4.1, 4.2, 4.3, 4.5, 4.6, 4.7, 4.8, 4.9, 4.10, 4.11, 4.13, 4.14, 4.17, 4.18, 4.19. From BOTANY, Fifth Edition, by Carl L. Wilson, Walter E. Loomis, and Taylor A. Steeves. Copyright 1952 © 1957, 1962, 1967, 1971 by Holt, Rinehart and Winston, Inc. Reprinted by permission of Holt, Rinehart and Winston: 4.16(b and d).

Chapter 5

Fine Line Illustrations, Inc.: 5.1, 5.2, 5.3, 5.5, 5.6, 5.7, 5.8, 5.9, 5.10, 5.11, 5.12, 5.13, 5.14, 5.15, 5.16, 5.17, 5.18, 5.19. Redrawn from *Physical Chemistry*, 3d ed., by Daniels and Alberty, John Wiley and Sons, Inc., 1966. Used by permission: 5.4. Reprinted from R. E. Dickerson and I. Geis, THE STRUCTURE AND ACTION OF PROTEINS. Benjamin/Cummings, Menlo Park, Calif., 1969. Copyright 1969 by Dickerson and Geis: box figure 5.1A.

Chapter 6

Fine Line Illustrations, Inc.: 6.1, 6.2, 6.3, 6.8, 6.9, 6.10, 6.11, 6.12, 6.13, 6.14, 6.15, 6.17, 6.18, 6.19, 6.20(a), 6.21, 6.22, 6.25, 6.27, 6.28, box figure 6.1B. John Walters and Associates: 6.4, 6.5(a), 6.7, 6.26. From PLANT PHYSIOLOGY, Second Edition, by Frank B. Salisbury and Cleon W. Ross. © 1978 by Wadsworth Publishing Company, Inc. Reprinted by permission of Wadsworth Publishing Company, Belmont, California 94002: 6.5(c). Reprinted with permission of Elsevier North-Holland Biomedical Press from an article by P. Jóliot, A. Jóliot, and B. Kok in *Biochemica Et Biophysica Acta* 153 (1968):635–52:6.16. From Albert L. Lehninger, BIOCHEMISTRY, 2d edition, Worth Publishers, New York, 1975, pp. 612–13: 6.20(b). Redrawn from Figure 1, Photorespiration, published and copyrighted by Carolina Biological Supply Company, Burlington, NC: box figure 6.1A.

Chapter 7

Fine Line Illustrations, Inc.: 7.1, 7.2, 7.3, 7.4, 7.5, 7.6, 7.7, 7.8, 7.9, 7.10, 7.11, 7.12, 7.13, 7.14, 7.15, 7.16, 7.17, 7.18, 7.19, 7.20, 7.21, 7.22, 7.23, 7.24, 7.25, 7.26, box figure 7.1A. Knut Schmidt-Nielsen, ANIMAL PHYSIOLOGY, 3d ed., © 1970, pp. 42, 51. Adapted by permission of Prentice-Hall, Inc., Englewood Cliffs, N.J.: 7.27, 7.28.

Chapter 8

Fine Line Illustrations, Inc.: 8.1, 8.2, 8.3, 8.14, 8.17. John Walters and Associates: 8.4(a), 8.5, 8.7, 8.8, 8.9, 8.10, 8.11, 8.12, 8.13. Reprinted with permission of Oxford University Press from *The Physiology of Plants*, edited and translated by A. J. Ewart (1904): 8.16. Modified from PLANT

with permission of Edward Arnold (Publishers) Ltd. from ENDOGENOUS PLANT GROWTH SUBSTANCES, Studies in Biology no. 40, by Hill: 18.10. Reprinted with permission of Academic Press, Inc. from AN INTRODUCTION TO BIOLOGICAL RHYTHMS by John D. Palmer: 18.29. Adapted from BIOLOGICAL SCIENCE, by William T. Keeton, Third Edition, Illustrated by Paula Di Santo Bensadoun, by permission of W. W. Norton and Company, Inc. Copyright © 1980, 1979, 1978, 1972, 1967, by W. W. Norton & Company, Inc.: box figure 18.1A. Reprinted with permission of the University of Chicago Press from K. C. Hamner and J. Bonner, "Photoperiodism in relation to hormones as factors in floral initiation and development," *Botanical Gazette* 100(1938): 388–412. © 1938: box figure 18.1B.

Chapter 19

Fine Line Illustrations, Inc.: 19.1, 19.10, 19.16, 19.20, 19.25, box figure 19.1A. John Walters and Associates: 19.2, 19.3, 19.7.

Chapter 20

Fine Line Illustrations, Inc.: 20.2, 20.3, 20.4, 20.5, 20.6, 20.7, 20.8, 20.9, 20.10, 20.11, 20.12, 20.13, 20.14, 20.15, 20.16, 20.17, 20.20, 20.21. From *Genetics Notes,* 7th Ed. 1976, James F. Crow, Burgess Publishing Company, Minneapolis, Minnesota. Reprinted by permission of the publisher: 20.19.

Chapter 21

Fine Line Illustrations, Inc.: 21.1(b), 21.2, 21.7, 21.8, 21.9, 21.11, 21.12, 21.13, 21.14, 21.16, 21.17, 21.18, 21.19, 21.20. From Avers, *Cell Biology,* 2d ed., copyright 1981 by Willard Grant Press, Boston, MA. Used by permission: 21.3. From PRINCIPLES OF GENETICS, 6th ed., by Gardner/Snustad, Copyright © 1981. Reprinted by permission of John Wiley & Sons, Inc.: 21.21.

Chapter 22

From CELL HEREDITY by R. Sage and F. J. Ryan, Copyright © 1961. Reprinted by permission of John Wiley & Sons, Inc.: 22.1. From *Biology Today* by David Kirk, Copyright © 1979 by David Kirk. Reprinted by permission of CRM books, a Division of Random House, Inc.: 22.3. Fine Line Illustrations, Inc.: 22.4, 22.5, 22.6, 22.7, 22.8, 22.9, 22.10, 22.11, 22.12, 22.13, 22.14, 22.16, 22.17, 22.18, 22.19, 22.20, 22.23, 22.24, 22.25(a and b), box figure 22.1A. Reprinted with permission of Macmillan Publishing Co., Inc. from GENETICS, 2d Edition, by Monroe W. Strickberger. Copyright © 1976, Monroe W. Strickberger: 22.15.

Chapter 23

John Walters and Associates: 23.1, 23.3, 23.5, 23.6, 23.7, 23.8(b), 23.14, 23.18(b), 23.19(c and d), 23.20, 23.21, 23.22, 23.23, 23.24, 23.25. From "The Program of Fertilization" by David Epel, Copyright © 1977 by Scientific American, Inc. All rights reserved: 23.10(a–h). Reprinted with permission of Dr. Paul Gross. From an article by P. R. Gross, C. I. Malkin, and W. A. Mayer in P.N.A.S. 51(1964): 407–14: 23.11. From Johnson, Leland G. and E. Peter Volpe, PATTERNS AND EXPERIMENTS IN DEVELOPMENTAL BIOLOGY. © 1973 Wm. C. Brown Company Publishers, Dubuque, Iowa. Reprinted by permission: 23.17(a and b), 23.18(a and c). From AN INTRODUCTION TO EMBRYOLOGY, Fifth Edition, by B. I. Balinsky, and assisted by B. C. Fabian. Copyright © 1981 by CBS College Publishing Company. Copyright 1954, 1959, 1965, 1970, and 1975 by W. B. Saunders Company. Reprinted by permission of Holt, Rinehart and Winston: 23.18(d).

Chapter 24

John Walters and Associates: 24.1(a), 24.4, 24.5, 24.7, 24.8, 24.9, 24.13, 24.16, 24.17(a–c), 24.23. From Arey, L. B.: *Developmental Anatomy.* 7th edition. Philadelphia, W. B. Saunders Company, 1974: 24.1(b and c). Fine Line Illustrations, Inc.: 24.2, 24.3, 24.6, 24.10, 24.15, 24.18. Drawing by Edith Tagrin reproduced from HUMAN DESIGN by William S. Beck, © 1971 by Harcourt Brace Jovanovich, Inc., by permission of the publisher: 24.17(d).

Chapter 25

John Walters and Associates: 25.1, 25.4, 25.5, 25.6, 25.7, 25.9(a), 25.11, 25.12(a and b), 25.13, 25.14, 25.16, 25.17(c and d), 25.18(b), 25.22(a), 25.23(a), 25.28(b). Fine Line Illustrations, Inc.: 25.2, 25.10, 25.20, 25.24, 25.27. From Volpe, E. P., MAN, NATURE AND SOCIETY, 2d ed., © 1975, 1979 Wm. C. Brown Company Publishers, Dubuque, Iowa. Reprinted by permission: 25.3. Modified from figure appearing in article by F. C. Steward et al. in *Science* 143 (1964): 20–27: 25.8. From AN INTRODUCTION TO EMBRYOLOGY, Fifth Edition, by B. I. Balinsky, and assisted by B. C. Fabian. Copyright © 1981 by CBS College Publishing Company. Copyright 1954, 1959, 1965, 1970, and 1975 by W. B. Saunders Company. Reprinted by permission of Holt, Rinehart and Winston: 25.15, 25.17(a). Reprinted by permission from *Nature* 244 (1973): 492. Copyright © 1973 Macmillan Journals Limited: 25.18(c). From J. W. Saunders, *Animal Morphogenesis,* p. 47 (1968). © J. W. Saunders, 1968: 25.19. After Becker, *Chromosoma* 10 (1959): 654, by permission of Springer-Verlag: 25.23(b).

Clement L. Markert, Heinrich Ursprung, DEVELOPMENTAL GENETICS, © 1971, pp. 43, 44. Adapted by permission of Prentice-Hall, Inc., Englewood Cliffs, N.J.: 25.25(a and b). Used with permission from article by William Zinkham, M.D. in *Pediatrics* 37(1): 120–31, Fig. 5, p. 126, January 1966. Copyright American Academy of Pediatrics 1966: 25.25(c). From Townes and Holtfreter. *Journal of Experimental Zoology* 128: 53–120, 1955. Copyright 1955 by the Wistar Institute Press, Philadelphia. Reprinted with permission: 25.26.

Chapter 26

John Walters and Associates: 26.1, 26.2, 26.4, 26.5, 26.8, 26.9, 26.10, 26.15, 26.17, 26.18(b), 26.19, 26.22, 26.23, 26.27. Fine Line Illustrations, Inc.: 26.3, 26.6(a), 26.7, 26.11, 26.12, 26.13, 26.14, 26.26, 26.29. Reprinted with permission of the Stony Brook Foundation, Inc. from an article by V. V. Brunst in *Quarterly Review of Biology* 25(1950): 1: 26.20. From Borgens, R. B., Vanable, J. W., and Jaffe, L. F. "Bioelectricity and Regeneration," *Bioscience* 29(8) 470–71. Copyright © 1979. American Institute of Biological Sciences. Used with permission: 26.21. From CANCER: SCIENCE AND SOCIETY by John Cairns. W. H. Freeman and Company. Copyright © 1978: 26.25. Reproduced from *CANCER* 16: 379–86, by permission of the Editor and of the Author: 26.30. Reprinted by permission of the American Cancer Society, Inc.: 26.31. Reprinted with permission of Blackie and Son Limited from *Biology of Ageing* by M. J. Lamb: 26.32. From R. Kohn, *Principles of Mammalian Aging,* 2d ed. Englewood Cliffs, N.J.: Prentice-Hall. Copyright © 1971; based on data in H. Brody, *J. Comp. Neurol.* 102:511, 1955: 26.33.

Chapter 27

Fine Line Illustrations, Inc.: 27.2, 27.7, box figure 27.1A. John Walters and Associates: 27.4(a and b), 27.12. Reprinted with permission of Cambridge University Press from DARWIN'S FINCHES by David Lack. Copyright © 1947: 27.5. Adapted from W. M. Levi, THE PIGEON (Sumter, S. C.: Levi Publishing Co., 1957), Figs. 51, 63, 77, 105, 173, 241, and 286. Used by permission of Levi Publishing Co.: 27.6. From Volpe, E. Peter, UNDERSTANDING EVOLUTION, 4th ed., © 1967, 1970, 1977, 1981 Wm. C. Brown Company Publishers, Dubuque, Iowa. Reprinted by permission: 27.9, 27.11. From BIOLOGICAL PRINCIPLES AND PROCESSES, Second Edition, by Claude A. Villee and Vincent G. Dethier. Copyright © 1976 by W. B. Saunders Company. Copyright 1971 by W. B. Saunders Company. Reprinted by permission of Holt, Rinehart & Winston: 27.10(a).

Chapter 28

Fine Line Illustrations, Inc.: 28.1, 28.2, 28.7, 28.9, 28.13. From Volpe, E. Peter, UNDERSTANDING EVOLUTION, 4th ed., © 1967, 1970, 1977, 1981 Wm. C. Brown Company Publishers, Dubuque, Iowa. Reprinted by permission: 28.3, 28.5.

Chapter 29

Fine Line Illustrations, Inc.: 29.3, 29.13. John Walters and Associates: 29.7, 29.8, 29.9, 29.10, 29.15. Reprinted by permission of SCIENCE 81 Magazine, copyright the American Association for the Advancement of Science: 29.11.

Chapter 30

Fine Line Illustrations, Inc.: 30.3, 30.13. From H. S. Jennings, *Behavior of the Lower Organisms* (New York: Columbia University Press, Macmillan Co., 1906): 30.4. From "The Behavior of Lovebirds" by William C. Dilger. Copyright © 1962 by Scientific American, Inc. All rights reserved: 30.6. John Walters and Associates: 30.7, 30.15, 30.16, 30.17, 30.19, 30.23, 30.24. From "How an Instinct Is Learned" by Jack P. Hailman. Copyright © 1969 by Scientific American, Inc. All rights reserved: 30.8. Reprinted from "Ontogeny of Bird Song" by F. Nottebohn in *Science* 167 (13 February 1970): 950–56. Copyright 1970 by the American Association for the Advancement of Science: 30.10. From "The Reproduction Behavior of Ring Doves" by Daniel S. Lehrman. Copyright © 1964 by Scientific American, Inc. All rights reserved: 30.11. Reprinted with permission of Williams & Wilkins Company from "Radioimmunoassay of plasma estradiol during the breeding cycle of the ring dove" by Korenbort, Schomberg, & Erickson in *Endocrinology* 94 (1964): 1126–32. © 1964 The Endocrine Society: 30.12. From *The Study of Instinct* by N. Tinbergen, published by Oxford University Press, 1951: 30.20, 30.21, 30.22.

Chapter 31

Fine Line Illustrations, Inc.: 31.2, 31.3, 31.4, 31.7, 31.16. After C. Kabat and D. R. Thompson, 1963, Wisconsin Quail 1834–1962: Population dynamics and habitat management, Tech. Bull. No. 30, Wisconsin Conservation Department, Madison, Wisc. Used with permission: 31.5. Reprinted from "The Rise and Fall of a Reindeer Herd" by V. C. Scheffer in *Scientific Monthly* 73 (December 1951): 356–62: 31.6. Reprinted from *Population Growth and Land Use* by Colin Clark. © 1967, 1977 by Colin Clark. Reprinted by permission of St. Martin's Press and Macmillan London and Basingstoke: 31.8.

Figures 8 (p. 15), 2 (p. 51), 3 (p. 52), and 4 (p. 52) from THE ECOLOGY OF MAN, 2d edition, by Robert Leo Smith. Copyright 1972, 1976 by Robert Leo Smith. Reprinted by permission of Harper & Row, Publishers, Inc.: 31.9. Reprinted with permission of Hafner Press from *The Struggle for Existence* by G. F. Gause. Copyright © 1934: 31.11. From "Physiological Ecology of Three Codominant Successional Annuals" by N. K. Wieland and F. A. Bazzaz, *Ecology* 56, 681–88. Copyright 1975, the Ecological Society of America. Reprinted by permission: 31.12. Reprinted from "Population Ecology of Some Warblers in Northeastern Coniferous Forests" by R. H. MacArthur, *Ecology* 39, 599–619: 31.13. Reprinted from "The Influence of Interspecific Competition and Other Factors on the Distribution of the Barnacle *Chthamalus Stellatus*" by Joseph H. Connell, *Ecology* 42: 710–23: 31.14. Used by permission of the Division of Agricultural Sciences, University of California: 31.17. Reprinted from "Eye Marks in Vertebrates: Aids to Vision" by R. W. Ficken et al. in *Science* 173 (3 September 1971): 936–39. Copyright 1971 by the American Association for the Advancement of Science: 31.18.

Chapter 32

Figure 10 (p. 16) from THE ECOLOGY OF MAN, 2d edition, by Robert Leo Smith. Copyright © 1972, 1976 by Robert Leo Smith. Reprinted by permission of Harper & Row, Publishers, Inc.: 32.1. Fine Line Illustrations, Inc.: 32.2, 32.10, 32.11, 32.12, 32.14. Reprinted from "Reproductive Strategies and the Co-Occurrence of Climax Tree Species" by L. K. Forcier in *Science* 189 (5 September 1975): 808–10. Copyright 1975 by the American Association for the Advancement of Science: 32.5. From "Ecosystem Persistence and Heterotrophic Regulation" by R. V. O'Neill, *Ecology* 57: 1244–53. Copyright 1976, the Ecological Society of America. Reprinted by permission: 32.9. Reprinted with permission of Macmillan Publishing Co., Inc. from COMMUNITIES AND ECOSYSTEMS by Robert H. Whittaker. Copyright © 1975 by Robert H. Whittaker: 32.13. Figure 2 (p. 177) from THE ECOLOGY OF MAN by Robert Leo Smith. Copyright © 1972 by Robert Leo Smith. Reprinted by permission of Harper & Row, Publishers, Inc.: 32.15.

Chapter 33

Fine Line Illustrations, Inc.: 33.1, 33.4, 33.12, box figure 33.1A. Reprinted from *BioScience*, Vol. no. 28, Page no. 649, 1978, Copyright © 1978 by the American Institute of Biological Sciences. Reprinted with permission of the copyright holder: 33.8.

Chapter 34

Fine Line Illustrations, Inc.: 34.1, 34.8. Reprinted from "New Concepts of Kingdoms of Organisms" by R. H. Whittaker in *Science* 163 (10 January 1969): 150–60. Copyright 1969 by the American Association for the Advancement of Science: 34.2. Reprinted with permission of W. B. Saunders Company from BURROWS TEXTBOOK OF MICROBIOLOGY, 21st ed., by Freeman (1979): 34.3. John Walters and Associates: 34.5, 34.7, 34.11, 34.14, 34.15. From MICROBIOLOGY: MOLECULES, MICROBES AND MAN, Second Edition by Eugene W. Nester, C. Evans Roberts, Brian J. McCarthy, and Nancy N. Pearsall. Copyright © 1978 by Holt, Rinehart and Winston. Copyright © 1973 by Holt, Rinehart and Winston, Inc. Reprinted by permission of Holt, Rinehart and Winston: 34.6. Reprinted with permission from G. Cohan-Bazire and J. London, *J. Bacteriol.* 94:458, American Society for Microbiology: 34.12(b). From Stern, Kingsley R., INTRODUCTORY PLANT BIOLOGY. © 1979 Wm. C. Brown Company Publishers, Dubuque, Iowa. Reprinted by permission: 34.18.

Chapter 35

From Lane, Theodore R. (coordinating editor): *Life: The Individual, the Species,* St. Louis, 1976, The C. V. Mosby Co.: 35.1. Reprinted by permission of *American Scientist,* journal of Sigma Xi, the Scientific Research Society: 35.2. Reprinted with permission of K. Vickerman and F. E. G. Cox (1967) *The Protozoa.* Houghton Mifflin. pp. 15, 25, 46: 35.3(b), 35.9, 35.11, 35.13. Reprinted with permission of Macmillan Publishing Co., Inc. from MICROBIOLOGY: AN INTRODUCTION TO PROTISTA by James Stone Poindexter. Copyright © 1971 by James Stone Poindexter: 35.4(b and c). John Walters and Associates: 35.5, 35.14, 35.17, 35.25. Copyright © 1970 by Macmillan Publishing Co., Inc.: 35.6, 35.12(a and b). Reprinted with permission of Macmillan Publishing Co., Inc. from THE INVERTEBRATES: FUNCTION AND FORM by Irwin W. Sherman and Vilia G. Sherman. Copyright © 1970 by Macmillan Publishing Co., Inc.: 35.10. Reprinted with permission of Macmillan Publishing Co., Inc. from ALGAE AND FUNGI by Constantine J. Alexopoulos and Harold C. Bold. Copyright © 1967, Macmillan Publishing Co., Inc.: 35.19, 35.22. From Stern, Kingsley R., INTRODUCTORY PLANT BIOLOGY, © 1979 Wm. C. Brown Company Publishers, Dubuque, Iowa. Reprinted by permission: 35.21, 35.25.

Chapter 36

Chapter 37

Photos

Part Openings

Part 1: © R. Hamilton Smith. Part 2: © R. C. Patel/Taurus Photos. Part 3: © R. Hamilton Smith. Part 4: © William Bochm/West Stock, Inc. Part 5: © Tom McHugh/Photo Researchers. Part 6: © Michael Quinton. Part 7: Carolina Biological Supply.

Introduction

European Art: p. 2. Courtesy, Museum of the American Indian: p. 3 (top). The Oriental Institute Museum: p. 3 (bottom). Historical Pictures Service, Inc.: p. 5 (top and bottom). Courtesy, National Library of Medicine: pp. 7 (top and middle), 8, 9, 10 (top and bottom). Leonard Lee Rue III/Tom Stack & Assoc.: p. 7 (bottom). Columbia University, Low Memorial Library: p. 11. UPI Photo: p. 12. Robin Moyer/Black Star: p. 14 (top). George Wuerthner/Tom Stack & Associates: p. 14 (bottom). Ed Pacheco: p. 16.

Chapter 1

Manfred Kage/Peter Arnold: 1.5(b). © George Holton/Photo Researchers: 1.11(d). © John Bova/Photo Researchers: 1.12(b).

Chapter 2

S. E. Frederick, Courtesy of E. H. Newcomb, University of Wisconsin: 2.4(c). Robert J. Waaland/Biological Photo Service: 2.5(b). Bob & Marian Francis/Tom Stack & Associates: 2.5(c) (right). L. G. Johnson: 2.5(c)(right). Dave Spier/Tom Stack & Associates: 2.5(c)(left). From BIOCHEMISTRY, 2/e by Lubert Stryer, W. H. Freeman & Company © 1981: 2.10(b). Dr. Gerome Gross: 2.13(a). Wide World Photos: 2.14(c). Prof. M. H. F. Wilkins: 2.17. Courtesy, The Ealing Corporation: 2.18(c). Dr. G. F. Leedale/Biophoto Associates: 2.22. Dr. Vincent Marchesi: 2.24(b).

Chapter 3

Historical Pictures Service, Inc.: 3.1. Courtesy, Carl Zeiss: 3.6(a). Carolina Biological Supply: 3.11. Dr. Paul Heidger: 3.12, 3.14(a and b). Dr. Gordon F. Leedale/Biophoto Associates: 3.15, 3.27(a). Dr. James Burbach: 3.17. Charles Havel/CPH Biomedical Photography: 3.19(a). Courtesy, Herbert W. Israel, Cornell University: 3.20(a). Courtesy, Elias Lazarides, California Institute of Technology: 3.22(a). L. E. Roth, University of Tennessee; Y. Shigenaka, University of Hiroshima; D. J. Pihlaja, Howe Laboratory of Ophthalnology, Boston/Biological Photo Service: 3.23(a and b). Jan Löfberg, Institute of Zoology, Uppsala University, Sweden: 3.25(a). J. Andre' and E. Favret-Fremiet: 3.25(b). Dr. Donald F. Lungdren: 3.30(a).

Chapter 4

Courtesy, S. J. Singer: 4.4. From *The Cell*, D. W. Fawcett 2/e p. 107 figure 52. © W. B. Saunders Company, 1966: 4.12. J. Robert Waaland/Biological Photo Service: 4.15(a). Manfred Kage/Peter Arnold: 4.15(b). Walter H. Hodge/Peter Arnold: 4.16(a). Tom Stack: 4.16(c). Nancy Marcus, Woods Hole Oceanographic Institution: box figure 4.1A.

Chapter 5

Thomas Eisner: box figure 5.2A.

Chapter 6

John Troughton and Leslie Donaldson: 6.4(b), 6.5(b). From *Plant Physiology*, 2/e, by Frank Salisbury and Cleon W. Ross. © 1978 by Wadsworth Publishing Company, Inc. Reprinted by permission of Wadsworth Publishing Company, Belmont, California: 6.5(d). TRIARCH PRODUCTS, Ripon, Wisconsin: 6.6. Dr. Gordon F. Leedale, Biophoto Associates: 6.7. Courtesy, Dr. Melvin Calvin: 6.23. Courtesy, Nancy Vander Sluis: 6.26.

Chapter 7

Charles Havel/CPH Biomedical Photography: 7.14(a). Dr. Donald F. Parsons, New York State Department of Health: 7.15(a). Tom McHugh/Tom Stack and Associates: 7.29. © Lynn Rogers: 7.30.

Chapter 8

John Troughton and Leslie Donaldson: 8.4(b). John W. Kimball, BIOLOGY, 2/e © 1968 Addison-Wesley, Reading, Massachusetts: 8.7(a). Dr. Gordon F. Leedale, Biophoto Associates: 8.15(a), 8.6(b, c, and d). Carolina Biological Supply: 8.15(b). Donald H. Marx, Institute for Mycorrhizal Research and Development: 8.15(c). Prof. E. J. Hewitt: 8.18(a, b, and c).

Chapter 9

Charles Havel/CPH Biomedical Photography: 9.3(a). Carolina Biological Supply: 9.4. Al Bussewitz: 9.5(a). Thomas Eisner: 9.7(c). John Troughton & Leslie Donaldson: 9.10. From *Plant Physiology*, 2/e by Frank Salisbury & Cleon Ross. © 1978 by Wadsworth Publishing Company, Inc. Reprinted by permission of Wadsworth Publishing Co. Belmont, CA: 9.11. Martin H. Zimmerman: 9.14(a and b).

Chapter 10

Lester V. Bergmann Associates, Inc.: 10.4, 10.5(a). Courtesy, Dr. Niilo Hallman: 10.5(b). Gregory Antipa: 10.7(a and b). Tom Stack: 10.11(b). E. S. Ross: 10.14(a). Charles Andrea/Tom Stack & Associates: 10.14(b). © Ira Block/Woodfin Camp: 10.14(c). Wallace Kirkland/Tom Stack & Associates: 10.14(d). Carolina Biological Supply: 10.16. TISSUES & ORGANS: A TEXT ATLAS OF SCANNING ELECTRON MICROSCOPY by Richard G. Kessel & Randy H. Kardon. Published by W. H. Freeman & Co.: 10.18(b), 10.20(b), 10.25(b). Carroll Weiss, Camera M.D. Studios: 10.27. Courtesy, Prof. Avery Gallup: box figure 10.1A. Courtesy, Dr. David Pramer, Rutgers: box figure 10.1B.

Chapter 11

Thomas Eisner: 11.12(c), 11.13(b). Robert Ross: 11.13(a). Prof. Hans-Rainer Düncker: 11.20. TISSUES & ORGANS: A TEXT ATLAS OF SCANNING ELECTRON MICROSCOPY by Richard G. Kessel & Randy H. Kardon. Published by W. H. Freeman & Co.: 11.22(a and b). Martin Rotker/TAURUS PHOTOS: 11.29(a, b, and c).

Chapter 12

From TISSUES AND ORGANS: A TEXT ATLAS OF SCANNING ELECTRON MICROSCOPY by Richard G. Kessel and Randy H. Kardon published by W. H. Freeman: 12.5(b and c), 12.26. Fulton/British Heart Journal 18(1956):12.9. From *A Textbook of Histology* by D. W. Fawcett published by W. B. Saunders: 12.15. Don Fawcett: 12.16(a and b). Carolina Biological Supply: 12.18. Jean-Paul Revel: 12.21. © Leo J. Lebeau/Biological Photo Service: 12.22(a). Abbott Laboratories: 12.23. Eila Kairinen and Emil Bernstein, Gilette Research Institute: 12.24. From *Essentials of Roentgen Interpretation*, 4th edition by Paul and Juhl: 12.27(a). American Heart Association: 12.27(c).

Chapter 13

Carl May/Biological Photo Service: 13.6(a). D. P. Wilson: 13.6(b). W. M. Stevens/Tom Stack and Associates: 13.6(c). Dr. K. H. Hausmann: 13.8. Dr. R. B. Wilson: 13.18(b). Dr. John Crowe: box figures 13.1A, 13.1B. Tom McHugh/Photo Researchers: box figure 13.2A.

Chapter 14

The Bettmann Archives: 14.8. Dr. Joseph T. Bagnara: 14.9. Carolina Biological Supply: 14.12, 14.16. © E. S. Ross: 14.13. Charles Havel/CPH Biomedical Photography: 14.15. From GENERAL ENDOCRINOLOGY, Sixth Edition, by C. Connell Turner and Joseph T. Bagnara.

Chapter 15

Carolina Biological Supply: 15.2. Carl May/Biological Photo Service: 15.11. J. David Robertson: 15.12. Dr. John Heuser: 15.17(a).

Chapter 16

Manfred Kage/Peter Arnold: 16.5.

Chapter 17

Russ Kinne/Photo Researchers: 17.1(a). Edwin Reshke: 17.3. From TISSUES AND ORGANS: A TEXT ATLAS OF SCANNING ELECTRON MICROSCOPY by Richard G. Kessel and Randy H. Kardon. W. H. Freeman Co. © 1979: 17.4. From ANATOMY AND PHYSIOLOGY LAB TEXTBOOK, 2/E Benson & Gunstream, © Wm. C. Brown Company Publishers: 17.7. Dr. H. E. Huxley: 17.15. UPI Photo: box figure 17.1A.

Chapter 18

Keith Gunnar/Tom Stack and Associates: 18.1. Frank B. Salisbury: 18.2(a). L. G. Johnson: 18.9(a). Al Bussowitz: 18.9(b), 18.22. *American Journal of Botany,* Vol. 37, No. 3 March 1950 from an article by J. P. Nitsch entitled "Growth and Morphogenesis of the Strawberry as Related to Auxin": 18.11. Folke Skoog: 18.14. Proceedings of the National Academy of Sciences, 1956. B. Phinney: 18.16. Dr. Sylvan Wittwer: 18.17. From *Phytochrome and Plant Growth* by R. E. Kendricks & B. Frankland. Published by Edward Arnold, Ltd.: 18.20. From PLANT PHYSIOLOGY, second edition, by Frank H. Salisbury and Cleon W. Ross © 1978 by Wadsworth Publishing Company: 18.27. Carolina Biological Supply: 18.32(b).

Chapter 19

Dr. Marta Walters: 19.4, 19.5, 19.8, 19.11, 19.12(a and b). E. J. Dupraw: 19.6. Sparrow, A. H., *et al.,* 1965 Radiat Biol. 5 (Supp): 101–132: 19.9. Prof. Daniel Mazia, Hopkins Marine Station: 19.13. W. P. Wergin, Biological Photo Service: 19.14. Dr. G. Lefevre: 19.15. M. M. Rhoades: 19.17, 19.23, 19.24. Christine Fastnaught: 19.18. D. von Wettstein: 19.19. N. V. Rothwell: 19.21. J. G. Gall: 19.22.

Chapter 20

American Museum of Natural History: 20.1. Courtesy, the Library of Congress Collection: 20.18.

Chapter 21

Ada L. Olins and Donald E. Olins, The University of Tennessee, Oak Ridge National Laboratories: 21.1(a). From *Sex Chromosomes* by Dr. Ursula Mittwoch, Academic Press © 1967 page 104, figure 8.1a: 21.4. Courtesy, the Upjohn Company: 21.5. Dr. Digamber S. Borgoankar, Ph.D.: 21.6. S. D. Sigamoni, Christian Medical College of Vellore, from *The American Journal of Human Genetics,* Vol. 16, No. 4: 21.10. Bob Coyle: 21.15. Courtesy of Dr. Murray L. Barr: box figure 21.2A.

Chapter 22

Courtesy, Dr. Maclyn McCarty, the Rockefeller University: 22.2. Huntington Potter/David Dressler, *Life* Magazine © 1980 Time, Inc.: 22.21. Courtesy, Eli Lily, Company: 22.22. Courtesy, Professor P. E. Polani, Guy's Hospital Medical School, London: 22.25.

Chapter 23

Charles Havel/CPH Biomedical Photography: 23.2(a). Robert A. Ross: 23.2(b), 23.8(a). © E. S. Ross: 23.4(a). Bob Coyle: 23.4(b). Courtesy, Dr. William Byrd, Louisiana State University: 23.9(a). Dr. Gerald Schatten, Florida State University: 23.9(b). Courtesy, Dr. Mia Tegner, Scripps Institution of Oceanography: 23.9(c and d). Courtesy, Dr. Victor D. Vacquier: 23.12, 23.13. Carolina Biological Supply: 23.15. From *Scanning Electron Microscopy in Biology* by Richard G. Kessel, and Ching Y. Shih. © Spring-Verlag, Berlin: 23.16, 23.17(c and d), 23.19(a). Courtesy, Tony Hunt: 23.19(b).

Chapter 24

Landrum B. Shettles, M.D.: 24.11, 24.12(d), 24.14(d), 24.19. American Museum of Natural History: 24.12(a). Carnegie Institution of Washington, Department of Embryology, Davis Division: 24.12(b), 24.14(a, b, and c), 24.20. H. M. Seitz, et al., From "Cleavage of Human Ova in Vitro" *Fertility and Sterility* (1971) 22, p. 255: 24.12(c). Cleveland Health Museum: 24.21. Peter Karas: 24.22. Planned Parenthood of New York City: 24.24. Martin Rotker/TAURUS PHOTOS: 24.25.

Chapter 25

Courtesy, Dr. Kenneth B. Raper: 25.9(b and c). Courtesy, Dr. David Francis: 25.9(d). © David Scharf/Peter Arnold, Inc.: 25.9(e). Courtesy, Johnson and Volpe, *Patterns and Experiments in Developmental Biology* © 1973, Wm. C. Brown Company Publishers: 25.12(c), 25.16(a). Courtesy, K. W. Tosney as printed in N. K. Wessells, *Tissue Interactions and Development* © 1977 by Benjamin Cummings, Menlo Park, California: 25.17(b). Courtesy, Dr. J. C. Smith, Sidney Farber Cancer Institute: 25.17(c). Courtesy, Kathryn W. Tosney: 25.18(a). Courtesy, J. H. Lewis, D. Summerbell and *The Journal of Embryology & Experimental Morphology:* 25.18(b). Paul A. Roberts/Biological Photo Service: 25.21. Courtesy, Dr. Bo Lambert: 25.22(b). Courtesy, Gilula, Reeves, and Steinbach, *Nature,* No. 235: 262–265: 25.28(a). Courtesy, Dr. Clement L. Markert, Yale University: box figure 25.1A.

Chapter 26

Courtesy, Dr. R. Dourmashkin: 26.6(b). Courtesy, Professor L. Brent: 26.16. Charles Havel/CPA Biomedical Photography: 26.18(a). Dr. Richard J. Goss: 26.20(a). Courtesy, Dr. Beth Johnson: 26.24. Courtesy, R. W. Briggs: 26.28. Courtesy, Dr. Gerome Gross: 26.34.

Chapter 27

Courtesy, The National Library of Medicine: 27.1. © George Kleiman/Uniphoto: 27.3(a). © A. Nelson/Tom Stack and Associates: 27.3(b). Leonard Lee Rue/Tom Stack and Associates: 27.3(c). © Rudolph Freund: 27.4(c). © E. S. Ross: 27.8. Courtesy, The American Museum of Natural History: 27.10.

Chapter 28

Charles Havel/CPH Biomedical Photography: 28.4. Wide World Photos: 28.6. Dr. H. B. D. Kettlewell: 28.8. © E. S. Ross: 28.10(a). © Bud Higdon: 28.10(b). Carolina Biological Supply: 28.11(a). © Thomas Eisner: 28.11(b). © Hans Pfletschinger/Peter Arnold: 28.12.

Chapter 29

Hubbard Scientific: 29.1. © W. H. Müller/Peter Arnold: 29.2(a). Alan Nelson/Root Resources: 29.2(b). © Tom McHugh/Photo Researchers: 29.4(a). © Douglas Faulkner: 29.4(b). © Annie Griffiths: 29.4(c). © Russ Kinne/Photo Researchers: 29.5. © E. S. Ross: 29.6, 29.12(b). © National Geographic Society: 29.12(a). David Brill © National Geographic Society: 29.12(c). American Museum of Natural History: 29.14, 29.16. Peabody Museum: 29.17.

Index